-ia, state or condition: insomnia
idi-, *idios,* one's own: idiopathic
ile-, *ileum:* ileocolic sphincter
ili-, ilio-, *ilium*: iliac
in-, in, within, or denoting negative effect: inactivate
infra-, *infra,* beneath: infraorbital
inter-, *inter,* between: interventricular
intra-, *intra,* within: intracapsular
ipsi-, *ipse,* itself: ipsilateral
iso-, *isos,* equal: isotonic
-itis, *-itis,* inflammation: dermatitis
karyo-, *karyon,* body: megakaryocyte
kerato-, *keros,* horn: keratin
kino-, -kinin, *kinein,* to move: bradykinin
lact-, lacto-, -lactin, *lac,* milk: prolactin
lapar-, *lapara,* flank or loins: laparoscopy
-lemma, *lemma,* husk: sarcolemma
leuko-, *leukos,* white: leukocyte
liga-, *ligare,* to bind together: ligase
lip-, lipo-, *lipos,* fat: lipoid
lith-, *lithos,* stone: otolith
lyso-, -lysis, -lyze, *lysis,* dissolution: hydrolysis
macr-, *makros,* large: macrophage
mal-, *mal,* abnormal: malabsorption
mamilla-, *mamilla,* little breast: mamillary
mast-, masto-, *mastos,* breast: mastoid
mega-, *megas,* big: megakaryocyte
melan-, *melas,* black: melanocyte
men-, *men,* month: menstrual
mero-, *meros,* part: merocrine
meso-, *mesos,* middle: mesoderm
meta-, *meta,* after, beyond: metaphase
micr-, *mikros,* small: microscope
mono-, *monos,* single: monocyte
morpho-, *morphe,* form: morphology
multi-, *multus,* many: multicelllular
-mural, *murus,* wall: intramural
myelo-, *myelos,* marrow: myeloblast
myo-, *mys,* muscle: myofilament
narc-, *narkoun,* to numb or deaden: narcotics
nas-, *nasus,* nose: nasolacrimal duct
natri-, *natrium,* sodium: natriuretic
necr-, *nekros,* corpse: necrosis
nephr-, *nephros,* kidney: nephron
neur-, neuro-, *neuron,* nerve: neuromuscular
oculo-, *oculus,* eye: oculomotor
odont-, *odontos,* tooth: odontoid process
-oid, *eidos,* form: odontoid process
oligo-, *oligos,* little, few: oligodendrite
-ology, *logos,* the study of: physiology
-oma, *-oma,* swelling: carcinoma
onco-, *onkos,* mass, tumor: oncology
oo-, *oon,* egg: oocyte
ophthalm-, *ophthalmos,* eye: ophthalmic nerve
-opia, *ops,* eye: optic
orb-, *orbita,* a circle: orbicularis oris
orchi-, *orchis,* testis: orchiectomy
orth-, *orthos,* correct, straight: orthopedist
-osis, *-osis,* state, condition: neurosis
osteon, osteo-, *os,* bone: osteocyte
oto-, *otikos,* ear: otolith
para-, *para,* beyond: paraplegia
patho-, -path, -pathy, *pathos,* disease: pathology
pedia-, *paidos,* child: pediatrician
per-, *per,* through, throughout: percutaneous
peri-, *peri,* around: perineurium
phag-, *phagein,* to eat: phagocyte
-phasia, *phasis,* speech: aphasia
-phil, -philia, *philus,* love: hydrophilic
phleb-, *phleps,* a vein: phlebitis
-phobe, -phobia, *phobos,* fear: hydrophobic
phot-, *phos,* light: photoreceptor
-phylaxis, *phylax,* a guard: prophylaxis
physio-, *physis,* nature: physiology
-plasia, *plasis,* formation: dysplasia
platy-, *platys,* flat: platysma
-plegia, *plege,* a blow, paralysis: paraplegia
-plexy, *plessein,* to strike: apoplexy
pneum-, *pneuma,* air: pneumonia
podo, *podon,* foot: podocyte
-poiesis, *poiesis,* making: hemopoiesis
poly-, *polys,* many: polysaccharide
post-, *post,* after: postanal
pre-, *prae,* before: precapillary sphincter
presby-, *presbys,* old: presbyopia
pro-, *pro,* before: prophase
proct-, *proktos,* anus: proctology
pterygo-, *pteryx,* wing: pterygoid
pulmo-, *pulmo,* lung: pulmonary
pulp-, *pulpa,* flesh: pulpitis
pyel-, *pyelos,* trough or pelvis: pyelitis
quadr-, *quadrans,* one quarter: quadriplegia
re-, *back,* again: reinfection
retro-, *retro,* backward: retroperitoneal
rhin-, *rhis,* nose: rhinitis
-rrhage, *rhegnymi,* to burst forth: hemorrhage
-rrhea, *rhein,* flow, discharge: amenorrhea
sarco-, *sarkos,* flesh: sarcomere
scler-, sclero-, *skleros,* hard: sclera
-scope, *skopeo,* to view: microscope
-sect, *sectio,* to cut: transect
semi-, *semis,* half: semilunar valve
-septic, *septikos,* putrid: antiseptic
-sis, state or condition: metastasis
som-, -some, *soma,* body: somatic
spino-, *spina,* spine, vertebral column: spinodeltoid
-stalsis, *staltikos,* contractile: peristalsis
sten-, *stenos,* a narrowing: stenosis
-stomy, *stoma,* mouth, opening: colostomy
stylo-, *stylus,* stake, pole: styloid
sub-, *sub,* below; subcutaneous
super-, *super,* above or beyond: superficial
supra-, *supra,* on the upper side: supraspinous fossa
syn-, *syn,* together: synthesis
tachy-, *tachys,* swift: tachycardia
telo-, *telos,* end: telophase
tetra-, four: tetralogy of Fallot,
therm-, thermo-, *therme,* heat: thermoregulation
thorac-, thorax, chest: thoracentesis
thromb-, *thrombos,* clot: thrombocyte
-tomy, *temnein,* to cut: appendectomy
tox-, *toxikon,* poison: toxemia
trans-, *trans,* through: transudate
tri-, three: trimester
-trophic, -trophin, -trophy, *trophikos,* nourishing: adrenocorticotrophic
tropho-, *trophe,* nutrition: trophoblast
tropo-, *tropikos,* turning: troponin
uni-, one: unicellular
uro-, -uria, *ouron,* urine: glycosuria
vas-, *vas,* vessel: vascular
zyg-, *zygotos,* yoked: zygote

Essentials of Anatomy & Physiology

Frederic H. Martini, Ph.D.
Edwin F. Bartholomew, M.S.

with

William C. Ober, M.D.
Art coordinator and illustrator

Claire W. Garrison, R.N.
Illustrator

Kathleen Welch, M.D.
Clinical consultant

Ralph T. Hutchings
Biomedical Photographer

Prentice Hall, Upper Saddle River, New Jersey 07458

Library of Congress Cataloging-in-Publication Data

Martini, Frederic, 1947–
Essentials of anatomy and physiology/Frederic H. Martini; with William C. Ober, Claire W. Garrison, illustrator, Kathleen Welch, clinical consultant, Ralph T. Hutchings, biomedical photographer.
p. cm.
Includes index.
ISBN 0-13-400144-3 (alk. paper)
1. Human Physiology. 2. Human Anatomy. I. Bartholomew, Edwin
II. Title
QP36.M42 1997 96-21435
612—dc20

Dedication

To Kitty, P.K., Ivy, and Kate: We couldn't have done this without you. Thank you for your encouragement, patience, and understanding.

Executive Editor: ***David Kendric Brake***
Editorial Director: ***Tim Bozik***
Editor in Chief Development: ***Ray Mullaney***
Editor in Chief: ***Paul Corey***
Senior Development Editor: ***Laura J. Edwards***
Production Editors: ***Joanne E. Jimenez, Karen M. Malley***
Assistant Vice President of Production and Manufacturing: ***David W. Riccardi***
Executive Managing Editor: ***Kathleen Schiaparelli***
Assistant Managing Editor: ***Shari Toron***
Creative Director: ***Paula Maylahn***
Director of Marketing: ***Kelly McDonald***
Cover Design: ***Tamara Newnam-Cavallo***
Interior Design: ***Lorraine Mullaney***
Cover Illustration: ***Abraham Echevarria, Patrice Van Acker***
Manufacturing Manager: ***Trudy Pisciotti***
Copy Editor: ***Margo Quinto***
Illustrators: ***William C. Ober, M.D., Claire W. Garrison, R.N.***
Editorial Assistants: ***Grace Anspake, Mary Hastings***
Art Director: ***Heather Scott***
Page Layout: ***Karen Noferi, Richard Foster, Jeff Henn***
Art Manager: ***Gus Vibal***
Art Coordinator: ***Warren Ramezzana***
Photo Research: ***Stuart Kenter***
Photo Editor: ***Carolyn Gauntt***

Simon & Schuster/A Viacom Company
Upper Saddle River, New Jersey 07458

Printed in the United States of America
10 9 8 7 6 5 4 3 2 1

ISBN 0-13-400144-3

Prentice-Hall International (UK) Limited, *London*
Prentice-Hall of Australia Pty. Limited, *Sydney*
Prentice-Hall Canada Inc., *Toronto*
Prentice-Hall Hispanoamericana, S.A., *Mexico*
Prentice-Hall of India Private Limited, *New Delhi*
Prentice-Hall of Japan, Inc., *Tokyo*
Simon & Schuster Asia Pte. Ltd., *Singapore*
Editora Prentice-Hall do Brasil, Ltda., *Rio de Janeiro*

Text and Illustration Team

Frederic H. Martini received his Ph.D. from Cornell University in Comparative and Functional Anatomy. He has broad interests in vertebrate biology, with special expertise in anatomy, physiology, histology, and embryology. Dr. Martini's publications include journal articles, technical reports, magazine articles, and a book for naturalists on the biology and geology of tropical islands. Working with Professor Michael J. Timmons, he has coauthored an undergraduate textbook, *Human Anatomy* (Prentice Hall, 1995).

Dr. Martini has been involved in teaching undergraduate courses in anatomy and physiology (comparative and/or human) since 1970. During the 1980s he spent his winters teaching courses, including human anatomy and physiology, at Maui Community College, and his summers teaching an upper-level field course in vertebrate biology and evolution for Cornell University, at the Shoals Marine Laboratory (SML). Dr. Martini now teaches part-time in the winters, and devotes most of his attention to developing new approaches to A&P education. His primary interest is in the use of appropriate technologies in creating an integrated learning system. Dr. Martini is a member of the Human Anatomy and Physiology Society, the National Association of Biology Teachers, the American Society of Zoologists, the Society for College Science Teachers, the Western Society of Naturalists, and the National Association of Underwater Instructors.

Edwin F. Bartholomew received his undergraduate degree from Bowling Green State University in Ohio and his M.S. from the University of Hawaii. His interests range widely, from human anatomy and physiology to the marine environment and the "backyard" aquaculture of escargots and ornamental fish. During the last two and a half decades he has taught human anatomy and physiology at both the secondary and undergraduate levels. In addition, he has taught a wide variety of other science courses (from botany to zoology) at Maui Community College. He is presently teaching at historic Lahainaluna High School, the oldest school west of the Rockies. Mr. Bartholomew has written journal articles, a weekly newspaper column, and many magazine articles. He is a member of the Human Anatomy and Physiology Society, the National Association of Biology Teachers, and the American Association for the Advancement of Science.

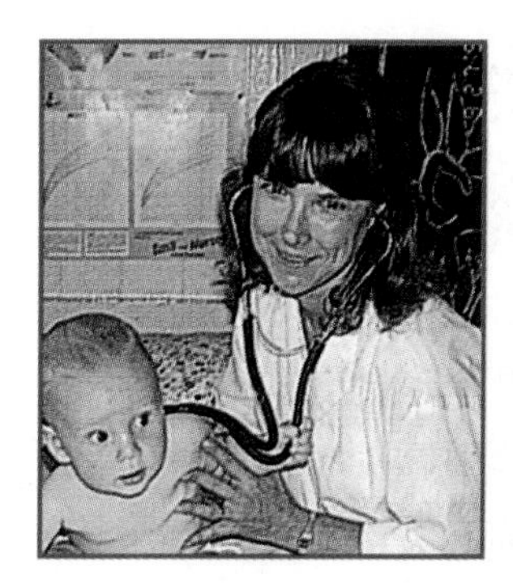

Dr. Kathleen Welch (clinical consultant) received her M.D. from the University of Washington in Seattle, and did her residency at the University of North Carolina in Chapel Hill. For two years she served as Director of Maternal and Child Health at the LBJ Tropical Medical Center in American Samoa, and subsequently was a member of the Department of Family Practice at the Kaiser Permanente Clinic in Lahaina, Hawaii. She has been in private practice since 1987. Dr. Welch is a Fellow of the American Academy of Family Practice. She is also a member of the Hawaii Medical Association and the Human Anatomy and Physiology Society.

Dr. William C. Ober (art coordinator and illustrator) received his undergraduate degree from Washington and Lee University and his M.D. from the University of Virginia in Charlottesville. While in medical school he also studied in the Department of Art as Applied to Medicine at Johns Hopkins University. After graduation Dr. Ober completed a residency in family practice, and is currently on the faculty of the University of Virginia as a Clinical Assistant Professor in the Department of Family Medicine. He is also part of the Core Faculty at Shoals Marine Laboratory, where he teaches biological illustration in the summer program. Dr. Ober now devotes his full attention to medical and scientific illustration.

Claire W. Garrison, R.N. (illustrator) practiced pediatric and obstetric nursing for nearly 20 years before turning to medical illustration as a full-time career. Following a five-year apprenticeship, she has worked as Dr. Ober's associate since 1986. Ms. Garrison is also a Core Faculty member at Shoals.

Texts illustrated by Dr. Ober and Ms. Garrison have received national recognition and awards from the Association of Medical Illustrators (Award of Excellence), American Institute of Graphics Arts (Certificate of Excellence), Chicago Book Clinic (Award for Art and Design), Printing Industries of America (Award of Excellence), and Bookbuilders West. They are also recipients of the Art Directors Award.

Contents in Brief

Contents

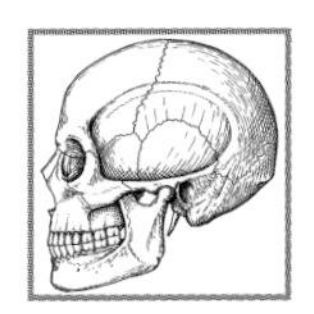

Chapter 6

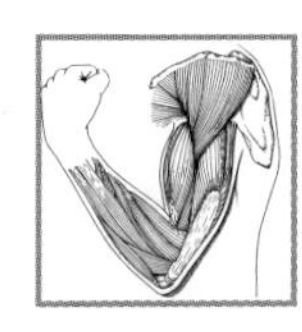

Chapter 7

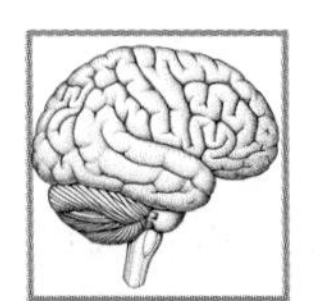

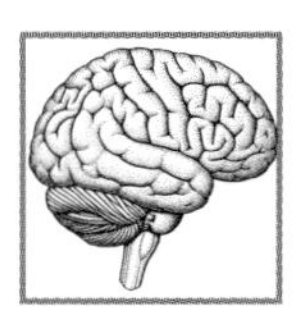

Chapter 9
The Peripheral Nervous System and Integrated Neural Functions 229

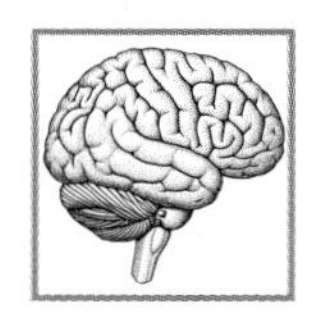

Chapter 10
Sensory Function 249

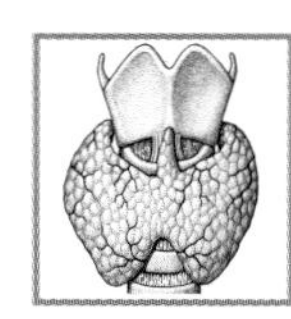

Chapter 11
The Endocrine System 281

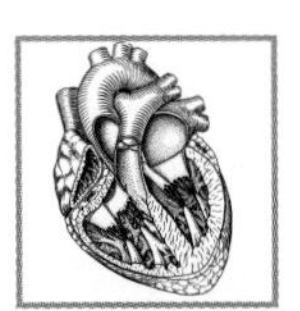

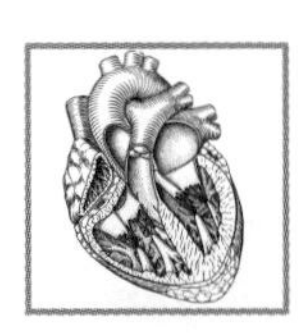

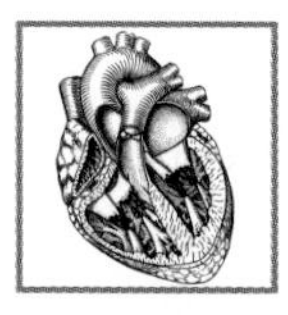

Chapter 14
Blood Vessels and Circulation 347

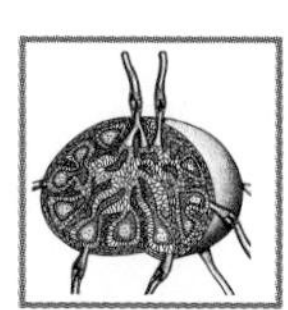

Chapter 15
The Lymphatic System and Immunity 382

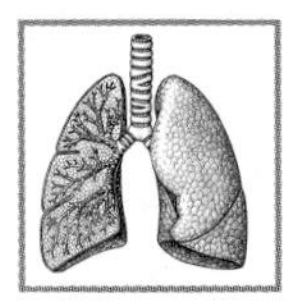

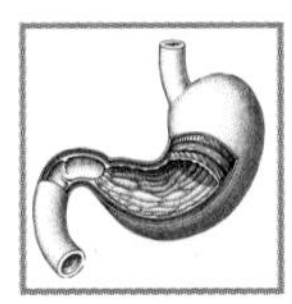

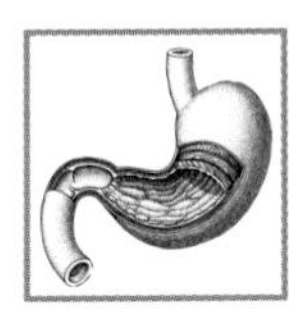

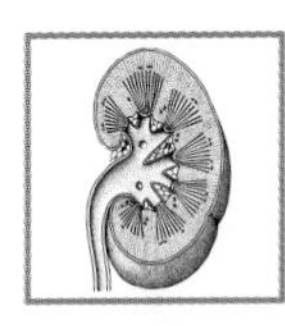

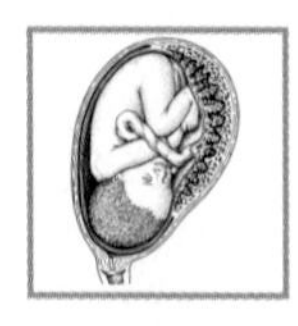

Chapter 20
The Reproductive System 515

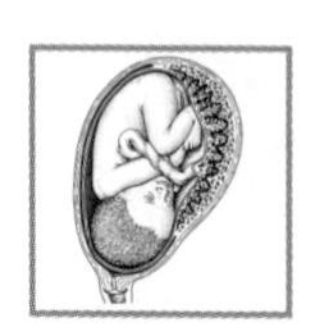

Chapter 21
Development and Inheritance 543

Preface

Pick up a newspaper, turn on the television, or view the latest movie thriller and you will find that the human body is in the spotlight. Wonder drugs and deadly new diseases; sports records and sports injuries; crash diets and artificial sweeteners; cholesterol reduction and calcium supplements; brain structure and human behavior; a vanishing ozone layer and rising health care costs—these are just a few of the topics that are concerned with some aspect of human anatomy or physiology.

The increase in available, useful information concerning the structure and function of the body has the potential to affect every aspect of our lives. There have never been greater opportunities for employment in applied health-related fields, from nursing to sports training, from dietetics to occupational safety. At the same time, however, this information explosion has the potential to overwhelm and frustrate new students of anatomy and physiology. A common student response to this avalanche of information is, "There is no way I can understand this because I don't have the time to learn all those terms!"

No textbook can give a student more time. A carefully crafted text can, however, help students make better use of their time. To succeed in this course, students must do more than develop a large technical vocabulary and retain a large volume of detailed information. They must relate what they learn in various chapters and understand how a small, localized change in structure or function can affect the entire body. In other words, anatomy and physiology students must develop their capacity for critical thinking, abstraction, and concept integration. These skills are important for everyone, not only for those pursuing careers in the health sciences.

Recent educational research has demonstrated that memorization of information does not by itself improve critical thinking skills unless the individual learns the material in a logical framework that stresses concept organization. Unfortunately, introductory students are often unprepared for this type of course, primarily because secondary school courses often stress rote memorization and recital of discrete blocks of information. This text has evolved to help students make the transition from one learning style to another, while making the process as easy and enjoyable as possible.

This textbook introduces the basic concepts and principles important to an understanding of the human body. It has two primary goals:

1. Building a foundation of essential knowledge (What structure is that? How does it work? What happens when it doesn't work?) that will support further courses dealing with specific topics in human anatomy or physiology.
2. Providing a framework for applying, interpreting, and applying related information obtained outside of the classroom.

Our aim has been to present information simply and clearly, with suitable emphasis on the concrete, applied aspects of each topic. Those pursuing careers in the medical or allied health sciences will acquire the background needed to organize and integrate additional information. For those seeking careers outside the biomedical fields, the perception that anatomical and physiological processes are understandable, relevant, and logical should remain intact and valuable long after the origin and insertion of the *latissimus dorsi* muscle have been forgotten.

THEMES

The cornerstone of any introductory anatomy and physiology course is the concept that the human body functions as an integrated unit. That integration exists at all structural levels, from cell to cell, tissue to tissue, organ to organ, and system to system. Homeostasis is maintained through interactions that occur on each of these levels, as well as between levels. When homeostasis breaks down, symptoms of disease appear. The basic concepts of structure and function and homeostasis are introduced early in the text and reinforced in all subsequent chapters.

GENERAL FEATURES OF *ESSENTIALS OF ANATOMY & PHYSIOLOGY*

Essentials of Anatomy & Physiology is intended to inform, develop, and stimulate a continuum of student populations; from those considering or planning careers in health, to those interested in understanding their own body and the basics of health.

Given that not all students comprehend and internalize information in the same way, the pedagogical structure of each chapter now offers more help to the diverse types of students who take this course. Every element from the chapter-opening vignette to the Three Level Review System that ties together the end-of-chapter material, as well as several key ancillaries to the text, are indicative of a sensitivity to the ways students learn and retain information

Emphasis on Concepts

- **Chapter Opening:** Each of the twenty-one chapters in *Essentials of Anatomy & Physiology* opens with a photograph and a narrative that briefly highlights the relationship between the image and the material covered in that particular chapter. This feature is designed to stimulate interest, generate questions, and establish lines of thought that will propel the reader into the chapter with anticipation instead of anxiety.

- **Learning Objectives:** The chapter opener also contains a manageable list of learning objectives. These objectives focus attention on the key concepts presented in the chapter text. Mastering these objectives will provide students with a foundation to build upon, and presenting them at the outset provides a preview of the key concepts that will be developed. To help students find pertinent information when reviewing chapter material, the learning objectives have been integrated with the chapter outline.
- **The Use of Analogies:** Whenever possible, basic physiological principles are related to familiar physical principles or events in everyday life. This helps to create a mental picture that enhances comprehension.
- **An Emphasis on Applied Topics:** The running text often makes reference to concrete, real-life examples that drive home the impact of relatively abstract material. Our text also contains boxed discussions dealing with clinical and applied topics.
- **Boxed Material:** The text includes several different types of boxed discussions. These boxes, located after or immediately adjacent to the relevant narrative, provide useful insights into the relevance or application of important concepts. Because the material is both boxed and categorized by topic, it can be read, if assigned, or ignored without disturbing the flow of essential information in the running text. Although few instructors assign all of the material presented, these discussions, together with the supplemental material in the Applications Manual, address the major clinical conditions and problems affecting each system. This information represents a useful reference for information about personal or family health concerns. With this in mind, care has been taken to ensure that the information contained is current, concise, and accurate.
- **A Discussion of the Effects of Aging on Each System:** These discussions place the physiological changes associated with aging into the context of normal anatomy and physiology. An understanding of the aging process is becoming increasingly important because the proportion of the population over age sixty-five is increasing dramatically.
- **Concept Checkpoints:** Two to four questions are placed near the end of each major section in a chapter. These questions are intended to provide a quick means of checking reading comprehension and improving the ability to integrate the information contained in blocks of text material. For easy reference, the answers are located in the Appendix.
- **Cross-referencing:** A concept link icon (∞) and page reference will be found wherever the development of a new concept builds on material presented earlier in the text.

Outstanding Illustrations

- **Integrated and Comprehensive Illustrations:** The art program and the text evolved together, and the layout helps the reader correlate the information provided by the text and the illustrations.
- **Use of Figures Showing the Relationships between Macroscopic and Microscopic Structure:** Introductory students are most familiar (and most comfortable) with the higher levels of organization, those of the individual or organ system. They are much less familiar with, and considerably more apprehensive of, events at the molecular or cellular level. *Essentials of Anatomy & Physiology* includes keystone figures that bridge the gap between the familiar, macroscopic world and the unfamiliar world of cells and tissues.
- **Use of Concept Maps and Flow Charts:** The illustration program for *Essentials of Anatomy & Physiology* provides visual summaries of organizational and functional relationships within and between vital systems.
- **Systems Integrators:** These figures, found near the ends of chapters dealing with specific systems, reinforce the concept that the body functions as an integrated unit rather than as a set of relatively isolated, independent systems.
- **Figure Dots:** Figure callouts in the text are followed by a red dot (●) that refers the reader to the red dots that precede the Figure captions. The dots in the text provide convenient placemarks for the reader, facilitating the return to the narrative after referencing the appropriate figure.

These features have been specifically designed to address the problems students encounter with this material. The extensive use of concept maps and flow charts provides a visual overview that should assist visual learners in mastering difficult material and relationships. The macro- to micro- figures bring the microscopic world into perspective and make structural and functional relationships easier to understand. Cross-referencing of text and figures, coupled with special system-integration figures, helps students develop an integrated perspective on the functioning of the body.

Vocabulary Development Aids

- **Vocabulary Development and Key Words:** An alphabetical list of relevant prefixes, suffixes, combining forms, and word roots is provided near the beginning of each chapter. Within the chapter text, the related key terms are presented in boldface, along with pronunciation guides and the associated word roots.
- **Additional Vocabulary Development.** Several chapters contain tables that summarize informa-

tion concerning the proper use of terms dealing with anatomical orientation, directional terms, and descriptive terms used when dealing with the skeletal or muscular systems.

Appropriate Clinical Coverage

- **Clinical Discussions:** Boxed Clinical Discussions (CDs) contain clinical material relevant to the preceding text. Each CD focuses attention on one or more disorders that demonstrate the consequences of homeostatic imbalances. The discussion is directly tied to the normal physiological or anatomical material in the adjacent text. The number of CDs varies from chapter to chapter.
- **Clinical Notes:** These notes are isolated from the running text. They contain extended discussions of major clinical disorders, diagnostic procedures, medical topics related to athletic activities, and issues that make headlines or stimulate controversy.
- **Focus Boxes:** These boxes integrate text and art to provide a visual summary of important information.
- ***Applications Manual* References:** The text contains references to (1) extended discussions of applied topics and (2) supplementary photographs that are contained in the *Applications Manual* for *Essentials of Anatomy & Physiology.* The linkage is indicated by an icon (AM) with the title of the relevant discussion shown in blue.

Extensive Chapter Review Material

- **Chapter Review:** Each chapter ends with an extensive Chapter Review comprising features that work together to help students study, review, apply, and integrate new material into the general framework of the course. Each module contains the following elements:
- **Key Terms Review:** The most important key terms in the chapter are listed in this section, along with the relevent page numbers in case a quick review is needed.
- **Study Outline:** The Study Outline at the end of each chapter reviews the major concepts and topics in summary fashion. Relevant page numbers are indicated for major headings, and related key terms are boldfaced. For ease of reference, the related figure and table numbers are indicated as appropriate.
- **Review Questions:** The review questions are organized around a **Three Level Review System** that affords each student the opportunity to review material in increasing levels of difficulty.

  Level One tests the understanding and recall of basic concepts and related terminology.

  Level Two encourages students to combine, integrate, and relate the basic concepts mastered in level one.

  Level Three promotes critical thinking skills at a point where such skills are most effectively developed-after mastering level one and level two material.

THE APPLICATIONS MANUAL

The *Applications Manual* for *Essentials of Anatomy & Physiology* is organized into units, each dealing with a different series of applied topics. Sections within those units are cross-referenced to relevant passages in the text, with page numbers indicated. The units address the following topics:

- *A Foundation for Anatomy and Physiology* provides hints on how to best learn the special language of anatomy and physiology, and introduces the methods by which we understand the human body.
- *An Introduction to Diagnostics* introduces the basic principles involved in the clinical detection and identification of disease states.
- *Applied ResearchTopics* considers principles of chemistry and molecular biology that are important to understanding, diagnosing, or treating homeostatic disorders.
- The *Body Systems* unit contains sections dealing with each of the eleven physiological systems. Each section ends with a series of critical thinking questions, and clinical problems follow groups of systems, such as muscular/skeletal or nervous/endocrine.
- The *Surface Anatomy Cadaver Atlas* contains twenty-four full-color pages of Ralph Hutchings photographs that supplement the images and line art in the text.

OTHER USEFUL FEATURES

- **Endpapers:** The front endpapers contain the most important foreign word roots, prefixes, suffixes, and combining forms encountered in the text. The back endpapers contain a list of common abbreviations and an overview of the contents of the Applications Manual.
- **Appendices:** The appendices contain material that most students and instructors will use at some time in the course.

  Appendix I contains the answers to the concept checkpoints organized by chapter and page number. These answers let students monitor their progress and their abilities to deal with the various types of questions encountered in each chapter.

  Appendix II provides a review of the important systems of weights and measures used in the text. Students are usually advised to review this material while completing the introductory chapter, to avoid confusion and distress later in the text.

Appendix III contains reference tables that report normal physiological values for body fluids.

- **Glossary:** The glossary provides pronunciations and definitions of important terms.

SUPPLEMENTS

Like the textbook itself, the ancillary package has been carefully crafted to meet the needs of the instructor and the student.

For the Instructor

Our new **CD-ROM Image Bank** for anatomy and physiology is a multimedia tool that enables you to easily create customized presentations in the classroom. The Image Bank contains images from the text as well as additional photographs, animations, and video. Included on the Image Bank is **Multimedia Presentation Manager 2.0**, a navigation software package that allows you to perform keyword searches, read and customize notes for each image, organize items in any order, incorporate outside lecture resources, and print a list of all resources chosen with accompanying notes.

A set of 276 full-color **Transparency Acetates,** which includes key illustrations from the main text, will also be available to adopters.

The **Instructor's Resource Guide** is a unique supplement designed to be useful for both the new instructor, as well as one who has taught Anatomy & Physiology for several years. It's your source for lively, unique analogies and teaching tips as well as suggestions from other instructors across the country.

For those wanting more exploration on their own, the **Anatomy & Physiology Home Page** is the place to go. This Web site features an interactive Study Guide and links to "hot" A&P sites on the Internet for students; and for instructors, there will be tips on how to teach A&P using the Web.

The **Test Item File** offers over 3000 questions and parallels the three-level learning system of the text. In addition, 200 pieces of unlabeled text art is included to create labeling exercises for exams.

The **Prentice Hall Custom Test** is an easily accessible software version of the above Test Item File and is designed to operate on all platforms—DOS, Windows, and Macintosh. It offers full mouse support, complete question editing capabilities, random test generation, graphics and printing capabilities.

The **Prentice Hall Laserdisc for Anatomy and Physiology** and its **Bar Code Manual** encourage instructors to bring a wealth of images into the classroom and the lab. The disc features text art, quiz frames (text figures with labels removed), histology slides, 300 cadaver images, pictures of common lab models and equipment and a complete set of cat dissection photographs. The video portion of the disc has high-quality, three-dimensional physiology animations and footage demonstrating lab experiments that are inaccessible to many faculty members (i.e. those utilizing computer analysis, animal specimens, and human body fluids). The disc can also be accessed using the Prentice Hall Multimedia Presentation Manager for Anatomy and Physiology, software that allows instructors to organize visual presentations and create custom overlays for images that accompany lectures. The Presentation Manager has its own User's Guide.

For the Student

An Applications Manual comes packaged with each new copy of the textbook. This unique supplement provides students with access to interesting and relevant clinical and diagnostic information. Critical thinking questions and clinical problems offer students the opportunity to think analytically. Historical background and information on recent research give them a frame of reference for the study of anatomy and physiology. Two atlases, one of color cadaver photographs and one of surface anatomy, are included as additional image resources.

Life on the Internet: Biology—A Student's Guide is an exciting supplement that brings students up to speed on what the Internet is and how to navigate it. This dynamic supplement offers many helpful hints and suggestions for exploring biology resources available on the Internet. This handy guide can be used as an introduction to the Internet for students who have no previous experience on the World Wide Web or as a source of reference for students with more experience.

The Study Guide is an excellent companion to the text and includes a number of labeling exercises and Concept Maps.

The Laboratory Text and Study Guide (by Michael Wood) combines student-referenced laboratory activities, study questions, and a comprehensive art program to create a unique one-stop student study aid.

For students seeking the opportunity to see the dynamics of the most difficult physiological processes, the **Anatomy and Physiology Video Tutor** has it all—even if students don't have access to a computer. This 75-minute video focuses on the concepts that instructors across the country have consistently identified as the most challenging. Physiological processes are demonstrated through the use of top-quality, three-dimensional animations and video footage. On-camera narration and the accompanying frame-referenced study booklet promote repeated concept review.

The *New York Times* **"Themes of the Times"** Program is sponsored jointly by Prentice Hall and the *New York Times* and is designed to enhance student access to current, relevant information. Articles are selected by the text authors and compiled into a free supplement that helps students make the connection between the classroom and the outside world

THE DEVELOPMENT PROCESS

We undertook this project to respond to the concerns and suggestions of anatomy and physiology instructors seeking a text suitable for a one-semester course.

Essentials of Anatomy & Physiology and the associated supplements addresses important educational problems, such as the wide diversity of students' learning types, backgrounds, degrees of preparation, and intellectual abilities, as well as limited instructional time and resources. As a result, it differs in many ways from other essentials of anatomy and physiology texts. The design and implementation of the features unique to our text reflect the combined efforts of the authors, aided by the collective wisdom of many other instructors who were kind enough to assist us.

ACKNOWLEDGMENTS

This textbook represents a group effort, rather than being the product of any single individual. Foremost on the list stand the faculty and reviewers whose advice, comments, and collective wisdom helped to shape the text into its final form. Their interest in the subject, their concern for the accuracy and method of presentation, and their experience with students of widely varying abilities and backgrounds made the review process an educational experience. To these individuals, who carefully recorded their comments, opinions, and sources, we would like to express our sincere thanks and best wishes.

The following individuals devoted large amounts of time reviewing drafts of *Essentials of Anatomy & Physiology:*

Reviewers

Bert Atsma
Union County College

Patricia K. Blaney
Brevard Community College

Douglas Carmichael
Pittsburg State University

Charles Daniels
Kapiolani Community College

Darrell Davies
Kalamazoo Valley Community College

Connie S. Dempsey
Stark State College of Technology

Marty Hitchcock
Gwinnetta Technical Institute

Paul Holdaway
Harper College

Drusilla B. Jolly
Forsyth Technical Community College

Anne L. Lilly
Santa Rosa Junior College

Daniel Mark
Penn Valley Community College

Christine Martin
Stark State College of Technology

Roxine McQuitty
Milwaukee Area Technical College

Sherry Medlar
Southwestern College

Margaret Merkely
Delaware Technical & Community College

Lewis M. Milner
North Central Technical College

Red Nelson
Westark Community College

William F. Nicholson
University of Arkansas-Monticello

Michael Postula
Parkland College

Dell Redding
Evergreen Valley College

Cecil Sahadath
Centennial College

Brian Shmaefsky
Kingwood College

Barbara M. Stout
Stark State College of Technology

Dave Thomas
Fanshawe College

Caryl Tickner
Stark State College of Technology

Frank V. Veselovsky
South Puget Sound Community College

Connie Vinton-Schoepske
Hawkeye Community College

Focus Group Participants

Bert Atsma
Union County College

Connie S. Dempsey
Stark State College of Technology

Anne L. Lilly
Santa Rosa Junior College

Roxine McQuitty
Milwaukee Area Technical College

Barbara M. Stout
Stark State College of Technology

Dave Thomas
Fanshawe College

Technical Reviewers

Brent DeMars
Lakeland Community College

Kathleen Flickinger
Iowa State University

Our gratitude is also extended to the many faculty and students at campuses across the country (and out of the country) whose suggestions and comments stimulated the decision to develop *Essentials of Anatomy and Physiology*.

A text has two components: narrative and visual. In preparing the narrative, we were ably assisted by two development editors. Susan Zorn reviewed an early draft, while Laura Edwards, Development Editor at Prentice Hall, handled the later drafts and assisted us throughout the production process. Laura played a vital role in shaping this text by helping us keep the text organization, general tone, and level of presentation consistent throughout. The accuracy, currency, and clarity of the clinical material in the text and in the *Applications Manual* reflect the detailed clinical reviews performed by Kathleen Welch, M.D.

Virtually without exception, reviewers stressed the importance of accurate, integrated, and visually attractive illustrations in aiding the students to understand essential material. The art program was primarily directed by Bill Ober, M.D. and Claire Garrison, R.N. Many of these figures include color photographs or micrographs collected from a variety of sources. Much of the work in tracking down these materials was performed by Stuart Kenter, whose efforts are greatly appreciated. Many of the light micrographs prepared by the senior author used commercially available slides obtained with the assistance of Carolina Biological Supply and Wards Scientific. The cadaver images and organ photos were provided in large part by Ralph Hutchings, whose artistic abilities and fine eye for detail are both envied and appreciated.

The authors wish to express their appreciation to the editors and support staff at Prentice Hall who made the entire project possible and who kept the text, art, and production programs on schedule and in relative harmony. Special thanks are due to Ray Mullaney, Editor in Chief, College Book Editorial Development, who gave Laura Edwards extra support and latitude; to Tim Bozik, Editorial Director for Engineering, Science, and Mathematics, for his support of the project; to David K. Brake, Executive Editor for Biology, for being the driving force and project coordinator; and to Joanne Jimenez and Karen Malley, Production Editors, for somehow managing to keep people, text, and art moving in the proper directions at the appropriate times. We would also like to thank David Riccardi, Assistant Vice President of Production and Manufacturing, and the pagemaking wizards led by John J. Jordan.

Any errors or oversights within this text are strictly those of the authors, and not the reviewers, artists, or editors. In an effort to improve future editions, we would ask that readers with pertinent information, suggestions, or comments concerning the organization or content of this textbook send their remarks to us care of David Brake, Executive Editor for Biology, Prentice Hall, Inc., 1208 East Broadway, Suite 200, Tempe, AZ 85282. You may also reach us directly, via email, through martini@maui.net. Any and all comments and suggestions will be deeply appreciated and carefully considered in the preparation of the second edition.

TO THE STUDENT

This text was designed to help you master the terminology and basic concepts of human anatomy and physiology. It should also make it possible for you to begin applying what you learn to every-day problems and situations. These aims have helped to shape every aspect of the book.

We have no doubt that you will find the study of anatomy and physiology to be one of the most interesting, challenging, and satisfying of all your educational experiences. Because the subject is so broad, extra care has been taken to give you every possible assistance in organizing new information. Many learning aids are built into the format of the text to make your study of this material easier and more rewarding. This book is meant to be a "machine for learning," one that can help you to focus your efforts and get the most from the time and energy you invest. We encourage you to examine the following overview carefully and to consult your instructor if you have further questions about how to use this textbook.

Best wishes!

Frederic H. Martini
Haiku, Hawaii

Edwin F. Bartholomew
Makawao, Hawaii

Emphasizing Concepts

16 THE RESPIRATORY SYSTEM

This is one way to go nowhere fast—pedaling on a treadmill. All the effort, however, is hardly wasted. The rather awkward-looking array of hoses and wires attached to this rider is designed to monitor respiratory and cardiovascular performance during exercise. Similar equipment can be used in a clinical setting to assess respiratory function in resting or active individuals.

Chapter Outline and Objectives

Chapter Outline and Objectives

Each chapter opens with an outline that gives you an overview of the concepts. Integrated objectives help you structure your reading and keep your attention focused on the key points.

Concept Links

The chain-link icon provides a quick visual signal that new material being presented is related to or builds on earlier discussions.

Vocabulary Development

alveolus, a hollow cavity; *alveolus, alveolar duct*
ateles, imperfect; *atelectasis*
cricoid, ring-shaped; *cricoid cartilage*
ektasis, expansion; *atelectasis*
-ia, condition; *pneumonia*
kentesis, puncture; *thoracentesis*
oris, mouth; *oropharynx*
pneuma, air; *pneumothorax*
pneumon, lung; *pneumonia*
stoma, mouth; *tracheostomy*
thorac-, chest; *thoracentesis*
thyroid, shield-shaped; *thyroid cartilage*

Living cells need energy for maintenance, growth, defense, and replication. Our cells obtain that energy through aerobic respiration, a process that requires oxygen and produces carbon dioxide. ∞ *[p. 414]* Therefore, the cells in the body must have a way to obtain oxygen and eliminate carbon dioxide. Our respiratory exchange surfaces are inside the lungs, where diffusion occurs between the air and the blood. The exchange surfaces are relatively delicate because they must be very thin to encourage rapid diffusion. The cardiovascular system provides the link between the interstitial fluids of the body and the exchange surfaces of the lungs. The circulating blood carries oxygen from the lungs to peripheral tissues; it also accepts and transports the carbon dioxide generated by those tissues, delivering it to the lungs.

Our discussion of the respiratory system begins by following the air as it travels from the exterior toward the alveoli of the lungs. We will then consider the mechanics of breathing, or *pulmonary ventilation* (the physical movement of air into and out of the lungs), and the physiology of respiration, which includes the processes of breathing and gas transport and exchange between the air, blood, and tissues.

FUNCTIONS OF THE RESPIRATORY SYSTEM

The respiratory system performs the following range of functions. It (1) moves air to and from the gas-exchange surfaces where diffusion can occur between air and circulating blood, (2) provides nonspecific defenses against pathogenic invasion, (3) permits vocal communication, and (4) helps control body fluid pH.

ORGANIZATION OF THE RESPIRATORY SYSTEM

The **respiratory system** consists of the nose, nasal cavity, and sinuses; the *pharynx* (throat); the *larynx* (voice box); the *trachea* (windpipe); and the *bronchi* and *bronchioles* (conducting passageways) and *alveoli* (exchange surfaces) of the lungs (Figure 16-1●).

The Respiratory Tract

The **respiratory tract** consists of the airways that carry air to and from the exchange surfaces of the lungs. The respiratory tract can be divided into a *conducting portion* and a *respiratory portion*. The conducting portion begins at the entrance to the nasal cavity and continues through the pharynx, larynx, trachea, bronchi, and the larger bronchioles. The respiratory portion includes the smallest and most delicate bronchioles and the alveoli that are the site of gas exchange.

In addition to delivering air to the lungs, the conducting passageways filter, warm, and humidify the air, thereby protecting the alveoli from debris, pathogens, and environmental extremes. By the time the air reaches the alveoli, most foreign particles and pathogens have been removed, and the humidity and temperature are within acceptable limits.

The Nose

Air normally enters the respiratory system via the paired **external nares** (nostrils) that communicate with the **nasal cavity.** The **vestibule** (VES-ti-būl) is the portion of the nasal cavity contained within the flexible tissues of the external nose. Here coarse hairs guard the nasal cavity from large airborne particles such as sand, dust, and insects.

The maxillary, nasal, frontal, ethmoid, and sphenoid bones form the lateral and superior walls of the nasal cavity. The *nasal septum* divides the nasal cavity into left and right sides. The bony posterior septum includes portions of the vomer and the ethmoid bone. A bony **hard palate,** formed by the palatine and maxillary bones, separates the oral and nasal cavities. A fleshy **soft palate** extends behind the hard palate, marking the boundary line between the superior **nasopharynx** (nā-zō-FĀR-inks) and the rest of the pharynx. The nasal cavity opens into the nasopharynx at the **internal nares** (NĀ-rēz).

Superior, middle, and *inferior nasal conchae* project toward the nasal septum from the lateral walls of the nasal cavity (Figure 16-2●). To pass from the vestibule to the internal nares, air tends to flow between adjacent conchae. As the air eddies and swirls, like water flowing over rapids, small airborne particles come in contact with the mucus that coats the lining of the nasal cavity. In addition to promoting fil-

Vocabulary Aids:

Vocabulary Development

This box follows the chapter opener and lists the important word parts that form the basis of the vocabulary in the chapter.

Key Terms

The most important new terms are highlighted in bold type and often include the **pronunciation.** All key terms are also listed at the end of the chapter for easy review.

Visualizing Structure & Function

Outstanding Anatomy Art and Photos

To understand physiology, you must be able to visualize structures in the human body. Macro-to-micro drawings, coupled with histology or electron micrographs, make the details of anatomy easy to understand.

Figure Reference Locators

Red dots serve as place markers, making it easy to return to your spot in the narrative after you've studied an illustration.

Conceptual Diagrams Show Physiology

Physiological processes are easier to understand with flowcharts and diagrams that link structure and function.

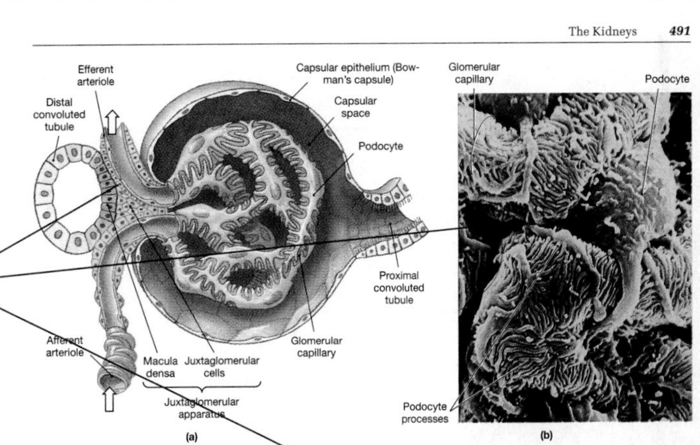

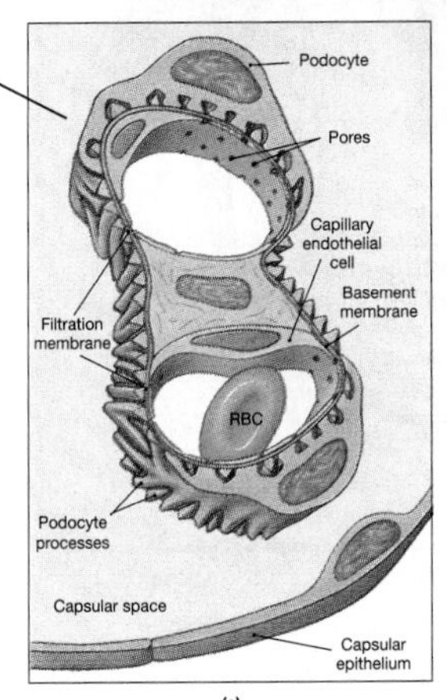

•**Figure 19-5**
The Renal Corpuscle
(a) The renal corpuscle, showing important structural features. **(b)** Electron micrograph of the glomerular surface, showing individual podocytes and their processes. (SEM × 27,248) **(c)** A diagrammatic view of a section from a glomerulus, showing the composition of the filtration membrane.

renal sinus. Figure 19-4• summarizes the functions of the different regions of the nephron and collecting system.

The Renal Corpuscle

The **renal corpuscle** (Figure 19-5•) consists of (1) the capillary knot of the glomerulus and (2) the expanded initial segment of the renal tubule, a region known as **Bowman's capsule**. The glomerulus projects into the Bowman's capsule much as the heart projects into the pericardial cavity (Figure 19-5a•). A *capsular epithelium* lines the wall of the capsule and a *glomerular epithelium* covers the glomerular capillaries. The two are separated by the **capsular space**, which receives the filtrate and empties into the renal tubule. The glomerular epithelium consists of cells called **podocytes** (PŌ-do-sīts, *podon*, foot). Podocytes have long cellular processes that wrap around individual capillaries. A thick basement membrane separates the endothelial cells of the capillaries from the podocytes. The glomerular capillaries are called *fenestrated* (FEN-e-strā-ted; *fenestra*, a window) because their endothelial cells contain pores (Figure 19-5c). To enter the capsular space, a solute must be small enough to pass through (1) the pores of the endothelial cells, (2) the fibers of the basement membrane, and (3) the slits between the slender processes of the podocytes (Figure 19-5b,c•). The fenestrated capil-

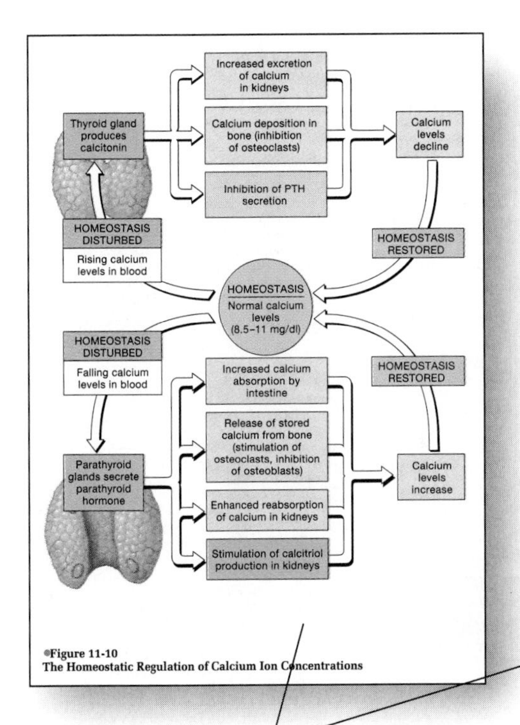

•**Figure 11-10**
The Homeostatic Regulation of Calcium Ion Concentrations

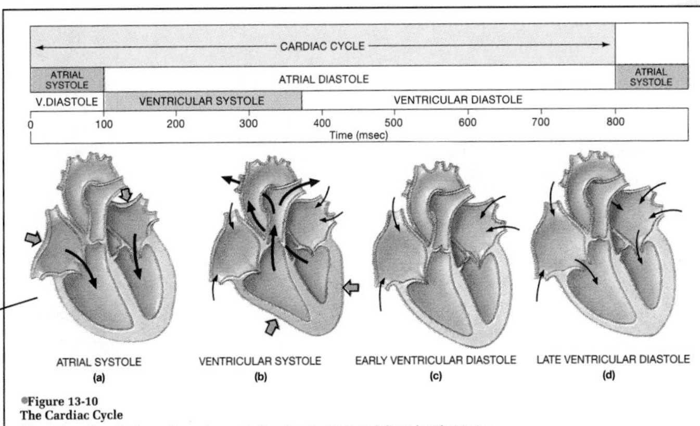

•**Figure 13-10**
The Cardiac Cycle
The atria and ventricles go through repeated cycles of systole and diastole. The timing of systole and diastole differs between the atria and ventricles. A cardiac cycle consists of one period of systole and diastole; we will consider a cardiac cycle as determined by the state of the atria. **(a)** Atrial systole: During this period the atria contract and the ventricles become filled with blood. **(b)** Ventricular systole: Blood is ejected into the pulmonary and aortic trunks. **(c)** Ventricular diastole: Passive filling of the ventricles occurs for the duration of this cycle and through the period of atrial systole in the next cardiac cycle. (d) Condition of the heart at the end of a cardiac cycle, with both the atria and ventricles in diastole.

Integrating Concepts

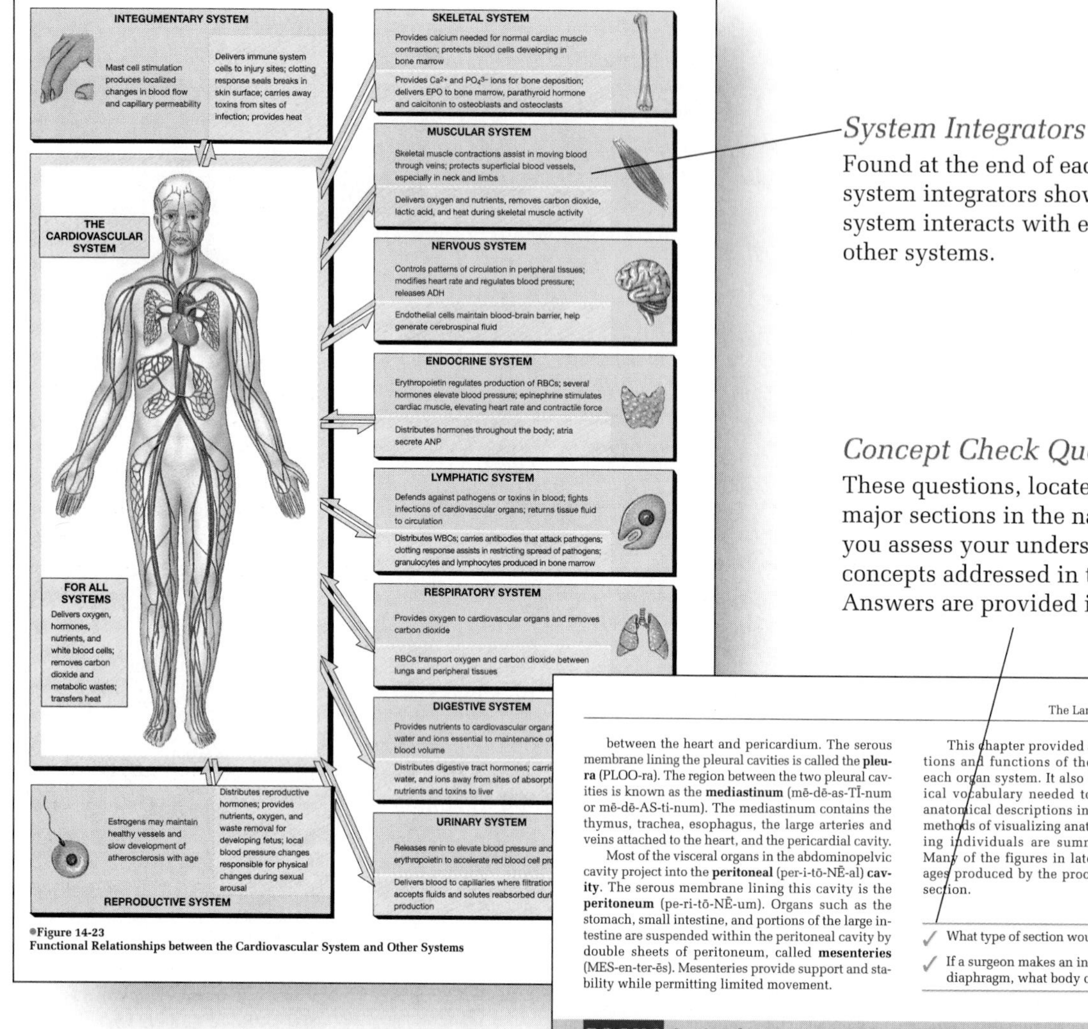

•Figure 14-23
Functional Relationships between the Cardiovascular System and Other Systems

System Integrators

Found at the end of each systems chapter, system integrators show how the particular system interacts with each of the body's other systems.

Concept Check Questions

These questions, located at the ends of major sections in the narrative, will help you assess your understanding of the basic concepts addressed in the previous pages. Answers are provided in Appendix 1.

The Language of Anatomy **21**

between the heart and pericardium. The serous membrane lining the pleural cavities is called the **pleura** (PLOO-ra). The region between the two pleural cavities is known as the **mediastinum** (mē-dē-as-TĪ-num or mē-dē-AS-ti-num). The mediastinum contains the thymus, trachea, esophagus, the large arteries and veins attached to the heart, and the pericardial cavity.

Most of the visceral organs in the abdominopelvic cavity project into the **peritoneal** (per-i-tō-NĒ-al) **cavity**. The serous membrane lining this cavity is the **peritoneum** (pe-ri-tō-NĒ-um). Organs such as the stomach, small intestine, and portions of the large intestine are suspended within the peritoneal cavity by double sheets of peritoneum, called **mesenteries** (MES-en-ter-ēs). Mesenteries provide support and stability while permitting limited movement.

This chapter provided an overview of the locations and functions of the major components of each organ system. It also introduced the anatomical vocabulary needed to follow more detailed anatomical descriptions in later chapters. Modern methods of visualizing anatomical structures in living individuals are summarized on pp. 00–00. Many of the figures in later chapters contain images produced by the procedures outlined in that section.

✓ What type of section would separate the two eyes?

✓ If a surgeon makes an incision just inferior to the diaphragm, what body cavity will be opened?

Focus Boxes

These essays provide illustrated summaries of important processes or clinical conditions. Topics in the text include visual accomodation problems (p. 264), urine formation (p. 498), and bone fractures (p. 117).

FOCUS Sectional Anatomy and Clinical Technology

The term **radiological procedures** includes not only those scanning techniques that involve radioisotopes but also methods that employ radiation sources outside the body. Physicians who specialize in the performance and analysis of these procedures are called **radiologists**. Radiological procedures can provide detailed information about internal systems. Figures 1-11• and 1-12• compare the views provided by several different techniques. These figures include examples of X-rays, CT scans, MRI scans, and ultrasound images. Other examples of clinical technology will be found in later chapters.

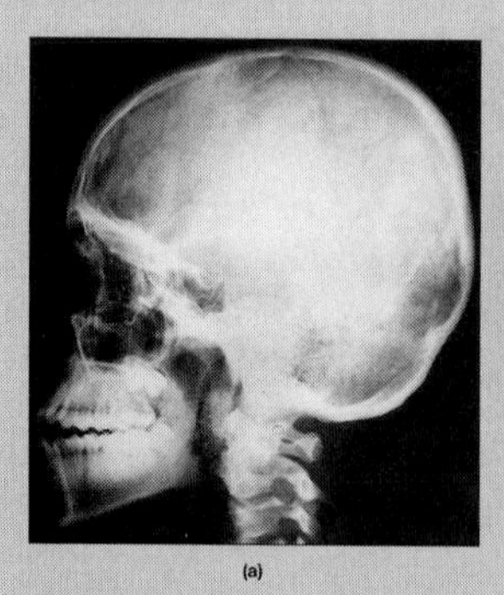

(a)

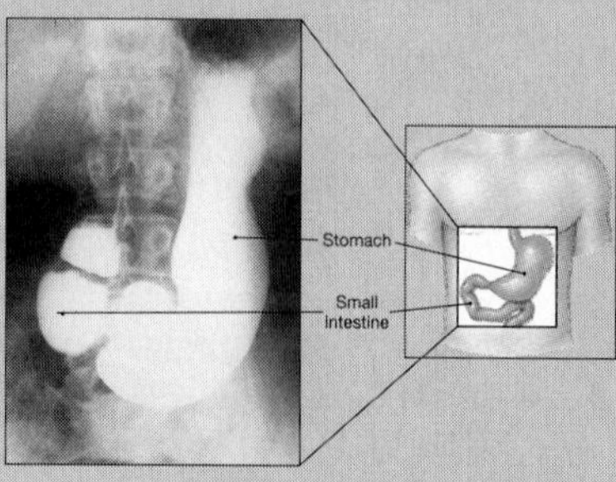

(b)

•Figure 1-11
X-rays.

(a) An X-ray of the skull, taken from the left side. **X-rays** are a form of high-energy radiation that can penetrate living tissues. In the most familiar procedure, a beam of X-rays travels through the body and strikes a photographic plate. All of the projected X-rays do not arrive at the film; some are absorbed or deflected as they pass through the body. The resistance to X-ray penetration is called **radiodensity**. In the human body, the order of increasing radiodensity is as follows: air, fat, liver, blood, muscle, bone. The result is an image with radiodense tissues, such as bone, appearing in white, and less dense tissues in shades of gray to black. The picture is a two-dimensional image of a three-dimensional object; in this image it is difficult to decide whether a particular feature is on the left side (toward the viewer) or on the right side (away from the viewer). **(b)** A barium-contrast X-ray of the upper digestive tract. Barium is very dense, and the contours of the gastric and intestinal lining can be seen outlined against the white of the barium solution.

Relating Clinical Examples

Clinical Note

Checking the Pulse and Blood Pressure

The pulse can be felt within any of the large or medium-sized arteries. The usual procedure involves squeezing an artery with the fingertips against a relatively solid mass, preferably a bone. When the vessel is compressed, the pulse is felt as a pressure against the fingertips.

Figure 14-5a● indicates the locations used to check the pulse. The inside of the wrist is often used because the *radial artery* can easily be pressed against the distal portion of the radius. Other accessible arteries include the *temporal, facial, carotid, brachial, femoral, popliteal,* and *posterior tibial* arteries. Firm pressure exerted at one of these arteries near the base of a limb can reduce or eliminate arterial bleeding distal to the site; the locations are called *pressure points.*

Blood pressure is determined with a *sphygmomanometer* (sfig-mō-ma-NOM-e-ter; *sphygmos*, pulse + *manometer*, device for measuring pressure), as shown in Figure 14-5b●. An inflatable cuff is placed around the arm in such a position that its inflation squeezes the brachial artery. A stethoscope is placed over the artery distal to the cuff, and the cuff is then inflated. A tube connects the cuff to a glass chamber containing liquid mercury, and as the pressure in the cuff rises it pushes the mercury up into a vertical column. A scale along the column permits one to determine the cuff pressure in millimeters of mercury (mm Hg). Inflation continues until cuff pressure is roughly 30 mm Hg above the pressure sufficient to completely collapse the brachial artery, stop the flow of blood, and eliminate the sound of the pulse.

The investigator then slowly lets the air out of the cuff. When the pressure in the cuff falls below systolic pressure, blood can again enter the artery. At first, blood enters only at peak systolic pressures, and the stethoscope picks up the sound of blood pulsing through the artery. As the pressure falls further, the sound changes because the vessel is remaining open for longer and longer periods. When the cuff pressure falls below diastolic pressure, blood flow becomes continuous and the sound of the pulse becomes muffled or disappears completely. Thus the pressure at which the pulse appears corresponds to the peak systolic pressure; when the pulse fades, the pressure has reached diastolic levels. The distinctive sounds heard during this test are called **sounds of Korotkoff** (sometimes spelled *Korotkov* or *Korotkow*). When the blood pressure is recorded, systolic and diastolic pressures are usually separated by a slashmark, as in "120/80" ("one twenty over eighty") or "110/75." A reading of 120/80 would give a pulse pressure of 40 (mm Hg).

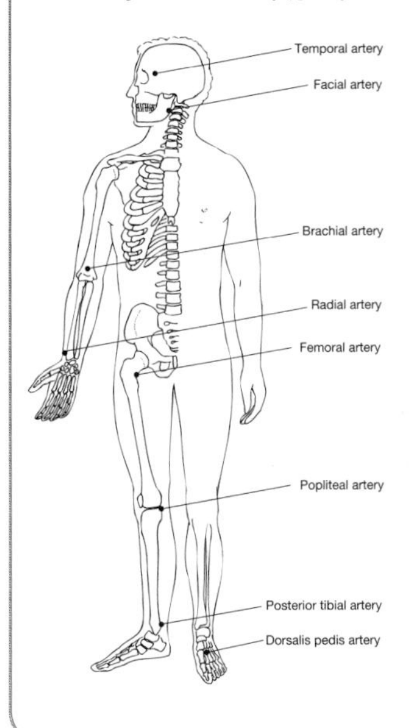

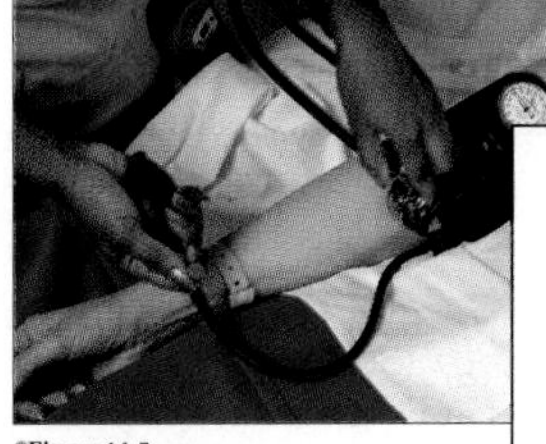

●**Figure 14-5**
Checking the Pulse and Blood Pressure
(a) Pressure points used to check the presence strength of the pulse. **(b)** Use of a sphygmomar to check arterial blood pressure.

Clinical Note

Clinical or health-related topics of particular importance are presented in boxes set off from the main text. These essays cover major diseases, such as lung cancer or AIDS, in addition to other subjects of special interest, such as clinical procedures.

Clinical Discussions

Important clinical topics presented in context are set off by an icon, title, and vertical color bar. These topics have been selected not only for their medical importance, but also to show how an understanding of abnormal conditions can shed light on normal functions, and vice versa.

phocytes are actively dividing. Lymphoid nodules are found beneath the epithelia lining the respiratory, digestive, and urinary tracts. Large nodules in the walls of the pharynx are called **tonsils**. There are usually five tonsils: a single *pharyngeal tonsil*, or *adenoids*, a pair of *palatine tonsils*, and a pair of *lingual tonsils*.

The lymphocytes in a lymphoid nodule are not always able to destroy bacterial or viral invaders, and if pathogens become established in a lymphoid nodule, an infection develops. Two examples are probably familiar to you: *tonsillitis*, an infection of one of the tonsils (usually the pharyngeal tonsil), and *appendicitis*, an infection of lymphoid nodules in the *appendix*, an organ of the digestive tract. AM *Infected Lymphoid Nodules*

Lymphoid Organs

Lymphoid organs have a stable internal structure and are separated from surrounding tissues by a fibrous capsule. Important lymphoid organs include the *lymph nodes*, the *thymus*, and the *spleen*.

Lymph Nodes

Lymph nodes are small, oval lymphoid organs ranging in diameter from 1 to 25 mm. They are covered by a dense, fibrous capsule (Figure 15-5●). One set of lymphatics delivers lymph to a lymph node, and another carries the lymph onward, toward the venous system. The lymph node functions like a kitchen water filter: It filters and purifies the lymph before it reaches the venous system. As lymph flows through a lymph node, at least 99 percent of the antigens present in the arriving lymph will be removed. As the antigens are detected and removed, T cells and B cells are stimulated, and an immune response initiated. Lymph nodes are located in regions where they can detect and eliminate harmful "intruders" before they reach vital organs of the body (Figure 15-1●, p. 384).

SWOLLEN GLANDS

Lymph nodes are often called *lymph glands*, and "swollen glands" usually accompany tissue inflammation or infection. Chronic or excessive enlargement of lymph nodes, a sign called *lymphadenopathy* (lim-fad-e-NOP-a-thē), may occur in response to bacterial or viral infections, endocrine disorders, or cancer.

Since the lymphatic capillaries offer little resistance to the passage of cancer cells, cancer cells often spread along the lymphatics and become trapped in the lymph nodes. Thus an analysis of swollen lymph nodes can provide information on the distribution and nature of the cancer cells, aiding in the selection of appropriate therapies. *Lymphomas*, an important group of lymphatic system cancers, are discussed in the *Applications Manual*. AM Lymphomas

●**Figure 15-5**
Structure of a Lymph Node

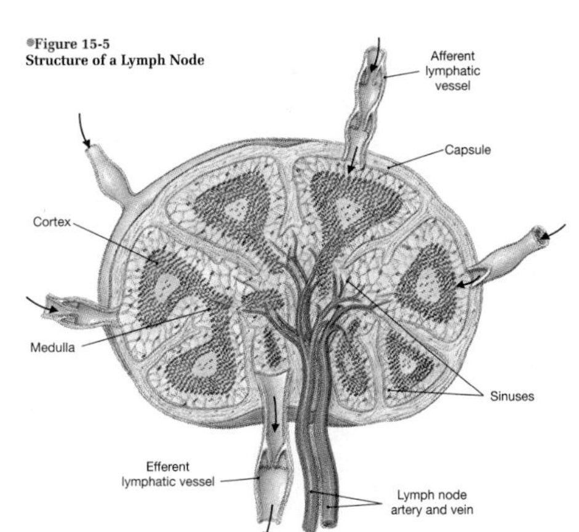

The Thymus

The **thymus**, site of T cell maturation, lies behind the sternum (Figure 15-6●). The thymus reaches its greatest size (relative to body size) in the first year or two after birth and its maximum absolute size during puberty, when it weighs between 30 and 40 g (1.06 to 1.41 oz.). Thereafter the thymus gradually decreases in size.

The thymus has two lobes, each divided into *lobules* by fibrous partitions, or *septae* (*septum*, a wall). Each lobule consists of a densely packed outer *cortex* and a paler central *medulla*. Lymphocytes in the cortex are dividing, and as the T cells mature they migrate into the medulla, eventually entering one of the blood vessels in that region. Other cells within the lobules produce the thymic hormones collectively known as *thymosins*.

Application Manual References

Many clinical topics introduced in the narrative or in Clinical Notes and Discussions are expanded upon in the *Applications Manual* that accompanies each copy of the text. The *Applications Manual* also contains detailed coverage of relevant clinical topics that are not discussed in the text. References to the *Applications Manual* are identified in the text by the AM icon, followed by the title of the relevant discussion in blue.

Reviewing the Concepts

Chapter Review 23

Chapter Review

KEY TERMS

anatomical position, p. 15
anatomy, p. 2
diaphragm, p. 19
frontal plane, p. 17
homeostasis, p. 12
negative feedback, p. 12
peritoneum, p. 21
physiology, p. 2
positive feedback, p. 14
sagittal plane, p. 17
transverse plane, p. 17
viscera, p. 19

SUMMARY OUTLINE

INTRODUCTION p. 2

1. **Biology** is the study of life; one of its goals is to discover the unity and patterns that underlie the diversity of living organisms.
2. All living things from single **cells** to large multicellular organisms, perform the same basic functions: they respond to changes in their environment (i.e., they show adaptability to their environment); they grow and reproduce to create future generations; they are capable of producing movement; and they absorb materials from the environment. Organisms absorb and consume oxygen during respiration, and they discharge waste products during excretion. Digestion occurs in specialized areas of the body to break down complex foods. The circulation forms an internal transportation system between areas of the body.

THE SCIENCES OF ANATOMY AND PHYSIOLOGY p. 2

Anatomical Perspectives p. 3

1. **Anatomy** is the study of internal and external structure and the physical relationships between body parts. **Physiology** is the study of how living organisms perform vital functions. All specific functions are performed by specific structures.
2. The boundaries of **microscopic anatomy** are established by the equipment used. **Cytology** analyzes the internal structure of individual cells. **Histology** examines **tissues** (groups of cells that have specific functional roles). Tissues combine to form **organs,** anatomical units with multiple functions.
3. **Gross (macroscopic) anatomy** considers features visible without a microscope. It includes **surface anatomy** (general form and superficial markings); **regional anatomy** (superficial and internal features in a specific area of the body); and **systemic anatomy** (structure of major organ systems).

Physiology p. 3

4. Human physiology is the study of the functions of the human body. It is based on **cell physiology**, the study of the functions of living cells. **Special physiology** studies the physiology of specific organs. **System physiology** considers all aspects of the function of specific organ systems. **Pathological physiology** studies the effects of diseases on organ or system functions.

LEVELS OF ORGANIZATION p. 4

1. Anatomical structures and physiological mechanisms are arranged in a series of interacting levels of organization. *(Figure 1-1)*

AN INTRODUCTION TO ORGAN SYSTEMS p. 4

1. The major organs of the human body are arranged into 11 organ systems. The organ systems of the human body are the *integumentary, skeletal, muscular, nervous, endocrine, cardiovascular, lymphatic, respiratory, digestive, urinary,* and *reproductive systems. (Figures 1-2, 1-3)*

HOMEOSTASIS AND SYSTEM INTEGRATION p. 12

1. **Homeostasis** is the tendency for physiological systems to stabilize internal conditions; through **homeostatic regulation** these systems adjust to preserve homeostasis.

Homeostatic Regulation p. 12

2. Homeostatic regulation usually involves a **receptor** sensitive to a particular stimulus and an **effector** whose activity affects the same stimulus.
3. **Negative feedback** is a corrective mechanism involving an action that directly opposes a variation from normal limits. *(Figure 1-4)*
4. In **positive feedback** the initial stimulus produces a response that reinforces the stimulus. *(Figure 1-5)*

Homeostasis And Disease p. 14

5. Symptoms of **disease** appear when failure of homeostatic regulation causes organ systems to malfunction.

THE LANGUAGE OF ANATOMY p. 15

Superficial Anatomy p. 15

1. Standard anatomical illustrations show the body in the **anatomical position.** If the figure is shown lying down, it can be either **supine** (face up) or **prone** (face down). *(Figure 1-6; Table 1-1)*
2. **Abdominopelvic quadrants** and **abdominopelvic regions** represent two different approaches to describing anatomical regions of the body. *(Figure 1-7)*
3. The use of special direct... ity when describing anatomical... *Table 1-2)*

Sectional Anatomy p. 16

4. The three **sectional pla... plane, sagittal plane,** and **transv...** tionships between the parts of the... body. *(Figure 1-9; Table 1-3)*
5. **Body cavities** protect de... changes in the size and shape of... **sal body cavity** contains the **cra...** brain) and **spinal cavity** (surroun... **ventral body cavity** surrounds de... diovascular, digestive, urinary, a... *(Figure 1-10a)*
6. During development th... ventral body cavity into the supe... **peritoneal cavities.** By birth, th...

Key Terms

Important terms introduced in the chapter are listed here with a page reference for quick review in context with the relevant material.

Summary Outline

This outline provides a detailed summary of all the sections in the chapter—including page references and all corresponding figure and table numbers.

Chapter Review 45

ATP *(adenosine triphosphate).* When energy is available, cells make ATP by adding a phosphate group to ADP. When energy is needed, ATP is broken down to ADP and phosphate. *(Figure 2-17)*

CHEMICALS AND LIVING CELLS p. 42

14. Biochemical building blocks form functional units called **cells.** *(Figure 2-18; Table 2-7)*

CHAPTER QUESTIONS

LEVEL 1 **Reviewing Facts and Terms**

Match each item in column A with the most closely related item in column B. Use letters for answers in the spaces provided.

Column A	Column B
___ 1. atomic number	a. synthesis
___ 2. covalent bond	b. catalyst
___ 3. ionic bond	c. sharing of electrons
___ 4. catabolism	d. A + B $\rightleftharpoons$ AB
___ 5. anabolism	e. stabilize pH
___ 6. exchange reaction	f. number of protons
___ 7. reversible reaction	g. decomposition
___ 8. acid	h. carbohydrates, lipids, proteins
___ 9. enzyme	i. loss or gain of electrons
___ 10. buffer	j. water, salts
___ 11. organic compounds	k. H^+ donor
___ 12. inorganic compounds	l. AB + CD → AD + CB

13. In atoms, protons and neutrons are found:
(a) only in the nucleus
(b) outside the nucleus
(c) inside and outside the nucleus
(d) in the electron cloud
14. The number and arrangement of electrons in an atom's outer electron shell determines its:
(a) atomic weight (b) atomic number
(c) electrical properties (d) chemical properties
15. The bond between sodium and chlorine in the compound sodium chloride (NaCl) is:
(a) an ionic bond
(b) a single covalent bond
(c) a nonpolar covalent bond
(d) a double covalent bond
16. What is the role of enzymes in chemical reactions?
17. List six elements found in abundance in the body.
18. What four major classes of organic compounds are found in the body?
19. List seven major functions performed by proteins.

LEVEL 2 **Reviewing Concepts**

20. Oxygen has 8 protons, 8 neutrons, and 8 electrons. What is the molecular weight of O_2?
(a) 8 (b) 16
(c) 24 (d) 32
21. Of the following selections, the one that contains only *inorganic* compounds is:
(a) water, electrolytes, oxygen, carbon dioxide
(b) oxygen, carbon dioxide, water, sugars
(c) water, electrolytes, salts, nucleic acids
(d) carbohydrates, lipids, proteins, vitamins
22. Glucose and fructose are examples of:
(a) monosaccharides (simple sugars)
(b) isotopes
(c) lipids
(d) a, b, and c are all correct
23. Explain the differences among (1)nonpolar covalent bonds, (2)polar covalent bonds, and (3)ionic bonds.
24. Why does pure water have a neutral pH?
25. A biologist analyzes a sample that contains an organic molecule and finds the following constituents: carbon, hydrogen, oxygen, nitrogen, and phosphorus. On the basis of this information, is the molecule a carbohydrate, a lipid, a protein, or a nucleic acid?

LEVEL 3 **Critical Thinking and Clinical Applications**

26. The element sulfur has an atomic number of 16 and an atomic weight of 32. How many neutrons are in the nucleus of a sulfur atom? Assuming that sulfur forms covalent bonds with hydrogen, how many hydrogen atoms could bond to one sulfur atom?
27. An important buffer system in the human body involves carbon dioxide (CO_2) and bicarbonate ion (HCO_3^-) as shown below:

$$CO_2 + H_2O \rightleftharpoons H_2CO_3 \rightleftharpoons H^+ + HCO_3^-$$

If a person becomes excited and exhales large amounts of CO_2, how will his body's pH be affected?

Review Questions

Questions are organized in a three-tiered system to help you build your knowledge:

Level 1 questions allow you to test your recall of the chapter's basic information and terminology.

Level 2 questions help you check your grasp of concepts and your ability to integrate ideas presented in different parts of the chapter.

Level 3 questions let you develop your powers of reasoning and analysis by applying chapter material to plausible real-world and clinical situations.

The Applications Manual

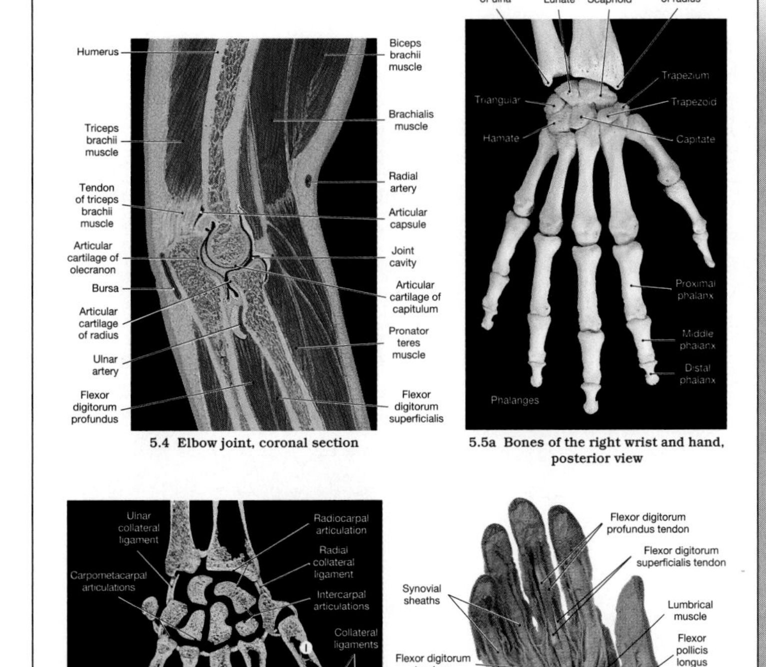

5.4 Elbow joint, coronal section

5.5a Bones of the right wrist and hand, posterior view

5.5b Joints of the right wrist and hand, sectional view

5.5c Muscles, tendons, and ligaments of the right wrist and hand, anterior view

C-12

The unique student *Applications Manual* that accompanies each copy of this text provides a wealth of supplemental material to enrich your students' learning experience. Although all essential principles of anatomy and physiology are covered in the text, the *Applications Manual* allows students to explore many clinical and diagnostic topics in greater depth.

- The introductory sections of the *Applications Manual* include an explanation of the scientific methods and discussions of the applications of chemistry and cell biology to the diagnosis and treatment of disease.

- The Body Systems section parallels the text system by system. Included are more detailed discussions of many clinical topics introduced in the text, as well as discussions of additional diseases, syndromes, and diagnostic techniques not covered in the text.

Body Systems: Clinical and Applied Topics

...atic injuries, such as fractures or dislo... ...nd infections also affect the cartilages, ...and ligaments associated with the bones ...A somewhat different array of conditions ...soft tissues of the bone marrow. Areas of ...marrow contain the stem cells for red ...s, white blood cells, and platelets. The ...row becomes abnormal in diseases of the ...t are characterized by blood cell overpro... ...*leukemia, polycythemia,* p. 96, 91) or ...duction (several *anemias,* p. 92, 95).

...YMPTOMS OF BONE AND ...DISORDERS

...h symptom of a skeletal system disorder is ...e pain and joint pain are common symptoms associated with many bone disorders. As a result, the presence of pain does not provide much help in identifying a specific bone or joint disorder. Chronic, aching bone or joint pain may be tolerated, and a person often will not seek medical assistance until more definitive symptoms appear. This may not occur until the condition is relatively advanced. For example, a symptom that may require immediate attention is a *pathologic fracture.* Pathologic fractures are the result of weakening of the skeleton by disease processes, such as *osteosarcoma* (a bone cancer). These fractures may be caused by physical stresses easily tolerated by normal bones.

EXAMINATION OF THE SKELETAL SYSTEM

The bones of the skeleton cannot be seen without relatively sophisticated equipment. However, there are a number of physical signs that can assist in the diagnosis of a bone or joint disorder. Important factors noted in the physical examination include

1. *Limitation of movement or stiffness:* Many joint disorders, such as the various forms of arthritis, will restrict movement or produce stiffness at one or more joints.
2. *The distribution of joint involvement and inflammation:* In a *monoarthritic* condition, only one joint is affected. In a *polyarthritic* condition, several joints are affected simultaneously.
3. *Sounds associated with joint movement: Bony crepitus* (KREP-i-tus) is a crackling or grating sound generated during movement of an abnormal joint. The sound may result from the movement and collision of bone fragments following an articular fracture or from friction and abrasion at an arthritic joint.
4. *The presence of abnormal bone deposits:* Thickened, raised areas of bone develop around fracture sites during the repair process. Abnormal bone deposits may also develop around the joints in the fingers. These deposits are called *nodules* or *nodes.* When palpated, nodules are solid and painless. Nodules, which can restrict movement, often form at the interphalangeal joints of the fingers in arthritis.
5. *Abnormal posture:* Bone disorders that affect the spinal column can result in abnormal posture. This is most apparent when the condition alters the normal spinal curvature. Examples include *kyphosis, lordosis,* and *scoliosis* (p. 51). A condition involving an intervertebral joint, such as a herniated disc, will also produce abnormal posture and movement.

Inherited Abnormalities in Skeletal Development EAP *p. 114*

There are several inherited conditions that result in abnormal bone formation. Three examples are *osteogenesis imperfecta, Marfan's syndrome,* and *achondroplasia.*

Osteogenesis imperfecta (im-per-FEK-ta) is an inherited condition, appearing in 1 individual in about 20,000, that affects the organization of collagen fibers. Osteoblast function is impaired, growth is abnormal, and the bones are very fragile, leading to progressive skeletal deformation and repeated fractures. Fibroblast activity is also affected, and the ligaments and tendons are very "loose," permitting excessive movement at the joints.

Marfan's syndrome is also linked to defective connective tissue structure. Extremely long and slender limbs, the most obvious physical indication of this disorder, result from excessive cartilage formation at the epiphyseal plates. (Marfan's syndrome is discussed further on p. 103.)

Achondroplasia (a-kon-drō-PLĀ-sē-a) is another condition resulting from abnormal epiphyseal activity. In this case the epiphyseal plates grow unusually slowly, and the individual develops short, stocky limbs. Although there are other skeletal abnormalities, the trunk is normal in size, and sexual and mental development remain unaffected. The adult will be an *achondroplastic dwarf.*

In **osteomalacia** (os-tē-ō-ma-LĀ-shē-ah; *malakia,* softness) the size of the skeletal elements remains the same, but their mineral content decreases, softening the bones. In this condition the osteoblasts are working hard, but the matrix isn't accumulating enough calcium salts. This can occur in adults or children whose diet contains inadequate levels of calcium or vitamin D_3.

Hyperostosis and Acromegaly EAP *p. 114*

The excessive formation of bone is termed **hyperostosis** (hi-per-os-TŌ-sis). In **osteopetrosis** (os-tē-ō-pe-TRŌ-sis; *petros,* stone) the total mass of the skeleton gradually increases because of a decrease in osteoclast activity. Remodeling stops, and the shapes of the bones gradually change. Osteopetrosis in children produces a variety of skeletal deformities. The primary cause for this relatively rare condition is unknown.

- Cross-referencing integrates the text and *Applications Manual,* making it easy to move back and forth between related material in the two books.

- A full-color Cadaver Atlas of dissection photographs allows students to visualize the internal structure of all major body regions and organs. A selection of surface anatomy photographs of live models is included for comparison with the dissection views.

- Critical Thinking Questions at the end of each system help students sharpen their ability to think analytically.

1 An Introduction to Anatomy and Physiology

Every day we perform a balancing act. An amazingly complex performance, it goes on continuously with every part of our physical being involved, from individual atoms to the largest organs and organ systems. Injury, disease, and other stresses, whether they occur inside us or outside, threaten this delicate balance. Adjusting to these changes is crucial, for if the body loses its balance, it may plunge out of control into dysfunction, illness, or even death.

Chapter Outline and Objectives

Vocabulary Development

bios, life; *biology*
cardium, heart; *pericardium*
homeo-, unchanging; *homeostasis*
-logy, study of; *biology*
medianus, situated in the middle; *median*
pathos, disease; *pathology*
peri-, around; *perimeter*
pronus, inclined forward; *prone*
supinus, lying on the back; *supine*
-stasis, standing; *homeostasis*

The world around us contains an enormous diversity of living organisms that vary widely in appearance and lifestyle. Despite their obvious differences, however, all living things perform the same basic functions:

- **Responsiveness**. Organisms respond to changes in their immediate environment; this property is also called irritability. You move your hand away from a hot stove, your dog barks at approaching strangers, fish are scared by loud noises, and tiny amoebas glide toward potential prey. Organisms also make longer-term changes as they adjust to their environments. For example, an animal may grow a heavier coat of fur as winter approaches, or it may migrate to a warmer climate. The capacity to make such adjustments is termed *adaptability*.
- **Growth**. Over a lifetime, organisms grow larger, increasing in size through an increase in the size or number of **cells**, the simplest units of life. Familiar organisms, such as dogs and cats, are composed of billions of cells. In such multicellular organisms, the individual cells become specialized to perform particular functions. This specialization is called *differentiation*.
- **Reproduction**. Organisms reproduce, creating subsequent generations of similar organisms.
- **Movement**. Organisms are capable of producing movement, which may be internal (transporting food, blood, or other materials inside the body) or external (moving through the environment).
- **Metabolism**. Organisms rely on complex chemical reactions to provide the energy for responsiveness, growth, reproduction, and movement. They must also synthesize complex chemicals, such as proteins. The term *metabolism* refers to all of the chemical operations under way in the body. Normal metabolic operations require the absorption of materials from the environment. To generate energy efficiently, most cells require various nutrients, as well as oxygen, a gas. The term *respiration* refers to the absorption, transport, and use of oxygen by cells. Metabolic operations often generate unneeded or potentially harmful waste products that must be eliminated through the process of *excretion*.

Several additional functions can be distinguished when you consider animals as complex as fish, cats, or human beings. For very small organisms, absorption, respiration, and excretion involve the movement of materials across exposed surfaces. But creatures larger than a few millimeters thick seldom absorb nutrients directly from their environment. For example, human beings cannot absorb steaks, apples, or ice cream without processing them first. That processing, called *digestion*, occurs in specialized areas where complex foods are broken down into simpler components that can be absorbed easily. Respiration and excretion are also more complicated for large organisms. Humans have specialized structures responsible for gas exchange (lungs) and waste elimination (kidneys). Finally, because absorption, respiration, and excretion are performed in different portions of the body there must be an internal transportation system, or *circulation*.

Biology, the study of life, includes a number of subspecialties. This text considers two biological subjects, *anatomy* and *physiology*. In the course of these 21 chapters you will become familiar with the basic anatomy and physiology of an unusually large and interesting type of organism, the human being.

THE SCIENCES OF ANATOMY AND PHYSIOLOGY

The word *anatomy* has its origins in Greece, as do many other anatomical terms and phrases. A literal translation would be "to cut open." **Anatomy** is the study of internal and external structure and the physical relationships between body parts. **Physiology**, another adopted Greek phrase, is the study of how living organisms perform the various functions of life. The two subjects are closely interconnected. Anatomical information provides clues about probable functions, and physiological mechanisms can be explained only in terms of the underlying anatomy.

The link between structure and function is always present but not always understood. For example, the anatomy of the heart was clearly described in the fifteenth century, but almost 200 years passed before the pumping action of the heart was demonstrated. On the other hand, many important aspects of cell physiology were recognized decades before the electron microscope revealed the anatomical basis for those functions.

This text will provide a familiarity with anatomical structures and an appreciation of the physiological processes that make human life possible. This information should enable you to understand many kinds of disease processes. In addition, the *Applications Manual* will provide a perspective that should prove useful for making intelligent decisions about your personal health. AM *An Introduction to Diagnostics*

Anatomical Perspectives

Anatomy can be divided into microscopic anatomy and macroscopic (gross) anatomy on the basis of the degree of structural detail under consideration. Other anatomical specialties focus on specific processes or medical applications.

Microscopic Anatomy

Microscopic anatomy considers structures that cannot be seen without magnification. The boundaries of microscopic anatomy are established by the limits of the equipment used. A light microscope reveals basic details about cell structure, whereas an electron microscope can visualize individual molecules only a few nanometers (nm) across. As we proceed through the text, we will be considering details at all levels, from macroscopic to microscopic. (If you are unfamiliar with the terms used to describe measurements and weights over this size range, consult the reference tables in Appendix II.)

Microscopic anatomy can be subdivided into specialties that consider features within a characteristic range of sizes. **Cytology** (sī-TOL-o-jē) analyzes the internal structure of individual cells. The trillions of living cells in our bodies are composed of chemical substances in various combinations, and our lives depend on the chemical processes occurring in those cells. For this reason we will consider basic chemistry (Chapter 2) before examining cell structure (Chapter 3).

Histology (his-TOL-o-jē) takes a broader perspective and examines **tissues**, groups of specialized cells and cell products that work together to perform a particular function. The trillions of cells in the human body can be assigned to four major tissue types, and these tissues are the focus of Chapter 4. **Organs** are anatomical units containing two or more tissue types and capable of performing multiple functions. Examples of organs include the heart, kidney, liver, and brain. Many organs are easily examined without a microscope, and at the organ level we cross the boundary into gross anatomy.

Gross Anatomy

Gross anatomy considers features visible with the unaided eye. There are many ways to approach gross anatomy. **Surface anatomy** refers to the study of general form and superficial markings. **Regional anatomy** considers all of the superficial and internal features in a specific region of the body, such as the head, neck, or trunk. **Systemic anatomy** considers the structure of major *organ systems*, which are groups of organs that function together to produce coordinated effects. For example, the heart, blood, and blood vessels form the *cardiovascular system*, which distributes oxygen and nutrients throughout the body.

Physiology

Physiology examines the function of anatomical structures; it considers the physical and chemical processes responsible for the characteristics of life, or vital functions. Because these functions are complex and much more difficult to examine than most anatomical structures, there are even more specialties in physiology than in anatomy.

The cornerstone of human physiology is **cell physiology**, the study of the functions of living cells. Cell physiology includes events at the chemical and molecular levels: both chemical processes within cells and chemical interactions between cells. **Special physiology** is the study of the physiology of specific organs. Examples include *renal physiology* (kidney function) and *cardiac physiology* (heart function). **System physiology** considers all aspects of the function of specific organ systems. *Respiratory physiology* and *reproductive physiology* are examples. **Pathological physiology** studies the effects of diseases on organ or system functions. (The Greek word *pathos* refers to disease.) Modern medicine depends on an understanding of both normal and pathological physiology; the practitioner must know not only what is wrong but how to correct it.

There are also special topics in physiology that address specific functions of the human body as a whole. These specialties focus on physiological interactions between multiple organ systems. For example, *exercise physiology* studies the physiological adjustments to exercise. Many other such applied topics are discussed in boxes in later chapters.

✓ How are vital functions such as growth, responsiveness, reproduction, and movement dependent upon metabolism?

✓ Would a histologist more likely be considered a specialist in microscopic or gross anatomy? Why?

Figure 1-1
Levels of Organization

Interacting atoms form molecules that combine to form cells, such as heart muscle cells. Groups of cells combine to form tissues with specific functions, such as heart muscle. Two or more tissues combine to form an organ, such as the heart. The heart is one component of the cardiovascular system, which also includes the blood and blood vessels. All of the organ systems combine to create an organism, a living human being.

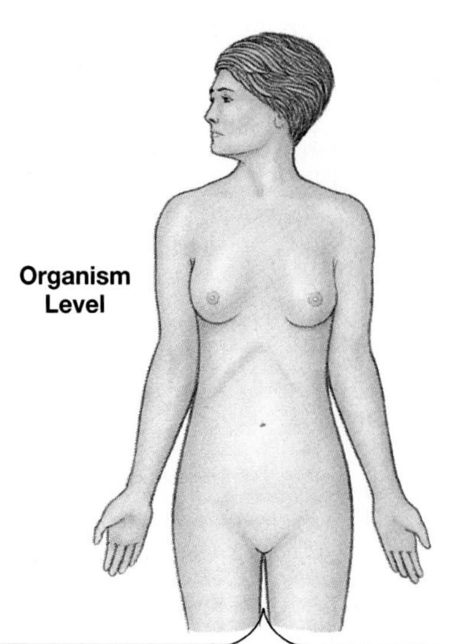

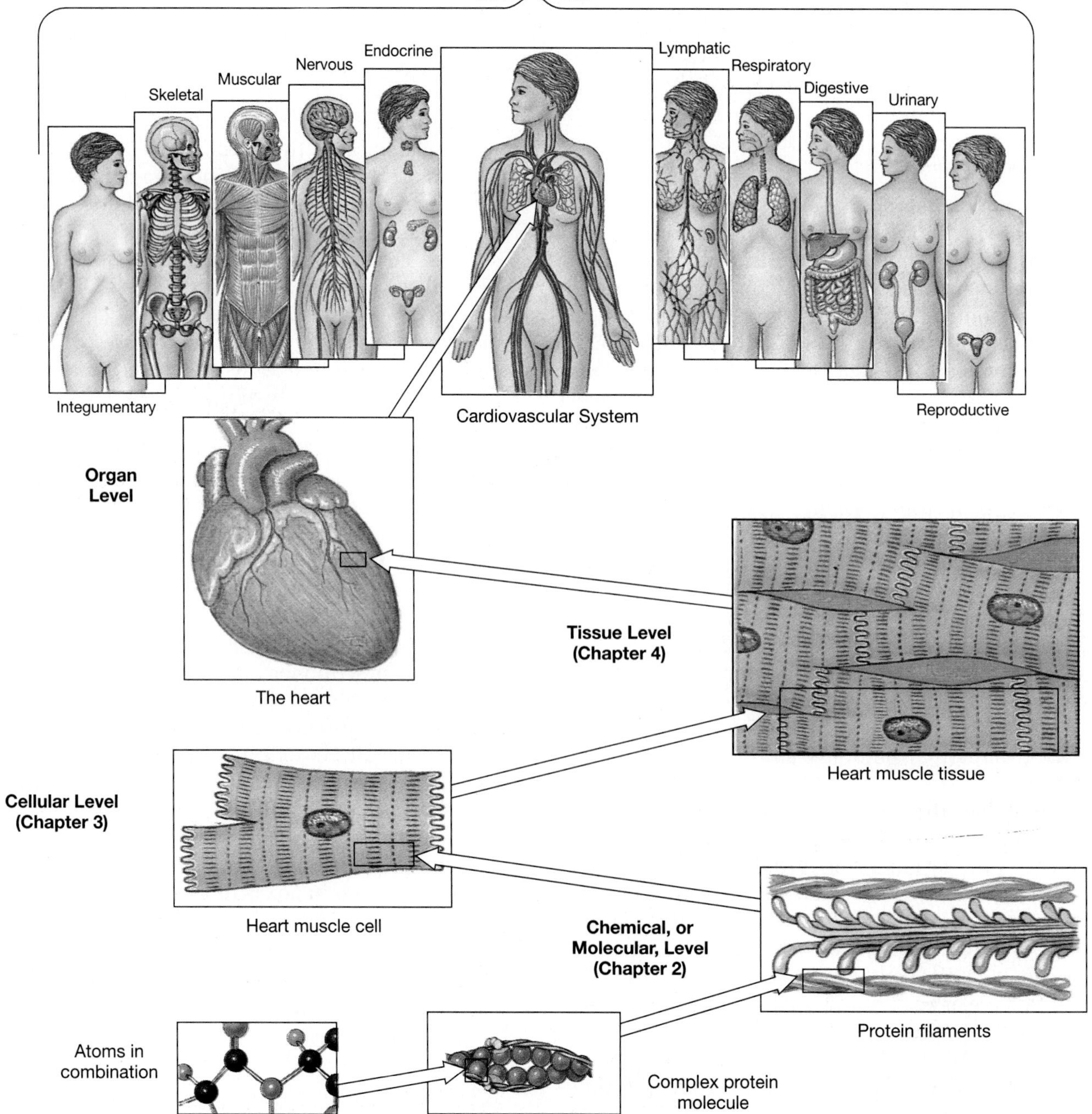

LEVELS OF ORGANIZATION

When considering events from the microscopic to macroscopic scales, we are examining several *levels of organization*. Figure 1-1● presents the relationships among the various levels of organization using the cardiovascular system as an example.

- **Chemical, or Molecular, Level.** *Atoms*, the smallest stable units of matter, combine to form *molecules* with complex shapes. This is the *chemical*, or *molecular*, level of organization.
- **Cellular Level.** Molecules interact to form *organelles*, such as complex protein filaments. Organelles are structural and functional parts of cells. Cells are the smallest living units in the body and they represent the *cellular* level of organization. Heart muscle cells contain large numbers of protein filaments that give them the ability to contract.
- **Tissue Level.** Heart muscle cells are found in one form of *muscle tissue*. Muscle tissue, one of the four tissue types, is an example of the *tissue* level of organization.
- **Organ Level.** Layers of muscle and other tissues form the wall of the heart, a hollow, three-dimensional organ. We are now at the *organ* level of organization.
- **Organ System Level.** Each time it contracts, the heart pushes blood into a network of blood vessels. Together, the heart, blood, and blood vessels form the *cardiovascular system*, an example of the *organ system* level of organization.
- **Organism Level.** All of the organ systems of the body work together to maintain life and health. This brings us to the highest level of organization, that of the *organism*, in this case a human being.

Each level is totally dependent on the others, and damage at the cellular, tissue, or organ level may affect the entire system. A chemical change in heart muscle cells can cause abnormal contractions or even stop the heartbeat. Physical damage to the muscle tissue, as in a chest wound, can make the heart ineffective even when most of the heart muscle cells are intact and uninjured. An inherited abnormality in heart structure can make it an ineffective pump, although the muscle cells and muscle tissue are perfectly normal.

Finally, it should be noted that something that affects the *system* will ultimately affect all of its components. For example, the heart cannot pump blood effectively after a massive blood loss if there is not enough blood to fill the circulatory system. If the heart cannot pump and blood cannot flow, oxygen and nutrients cannot be distributed. In a very short time, the tissue begins to break down as heart muscle cells die from oxygen and nutrient starvation. Of course these changes will not be restricted to the cardiovascular system; all of the cells, tissues, and organs in the body will be damaged.

AN INTRODUCTION TO ORGAN SYSTEMS

Figure 1-2● introduces the 11 organ systems in the human body and indicates their major functions. Figure 1-3● provides an overview of these individual organ systems and their major components.

Organ System	Major Functions
 Integumentary system	Protection from environmental hazards, temperature control
 Skeletal system	Support, protection of soft tissues, mineral storage, blood formation
 Muscular system	Locomotion, support, heat production
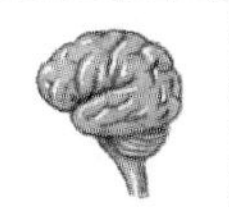 Nervous system	Directing immediate responses to stimuli, usually by coordinating the activities of other organ systems
 Endocrine system	Directing long-term changes in the activities of other organ systems
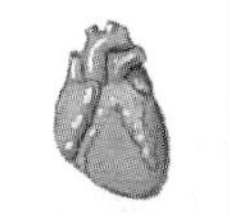 Cardiovascular system	Internal transport of cells and dissolved materials, including nutrients, wastes, and gases
 Lymphatic system	Defense against infection and disease
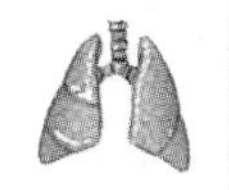 Respiratory system	Delivery of air to sites where gas exchange can occur between the air and circulating blood
 Digestive system	Processing of food and absorption of nutrients, minerals, vitamins, and water
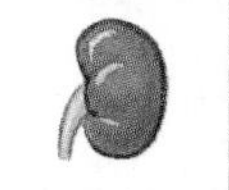 Urinary system	Elimination of excess water, salts, and waste products
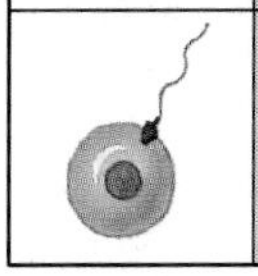 Reproductive system	Production of sex cells and hormones

●**Figure 1-2**
An Introduction to Organ Systems

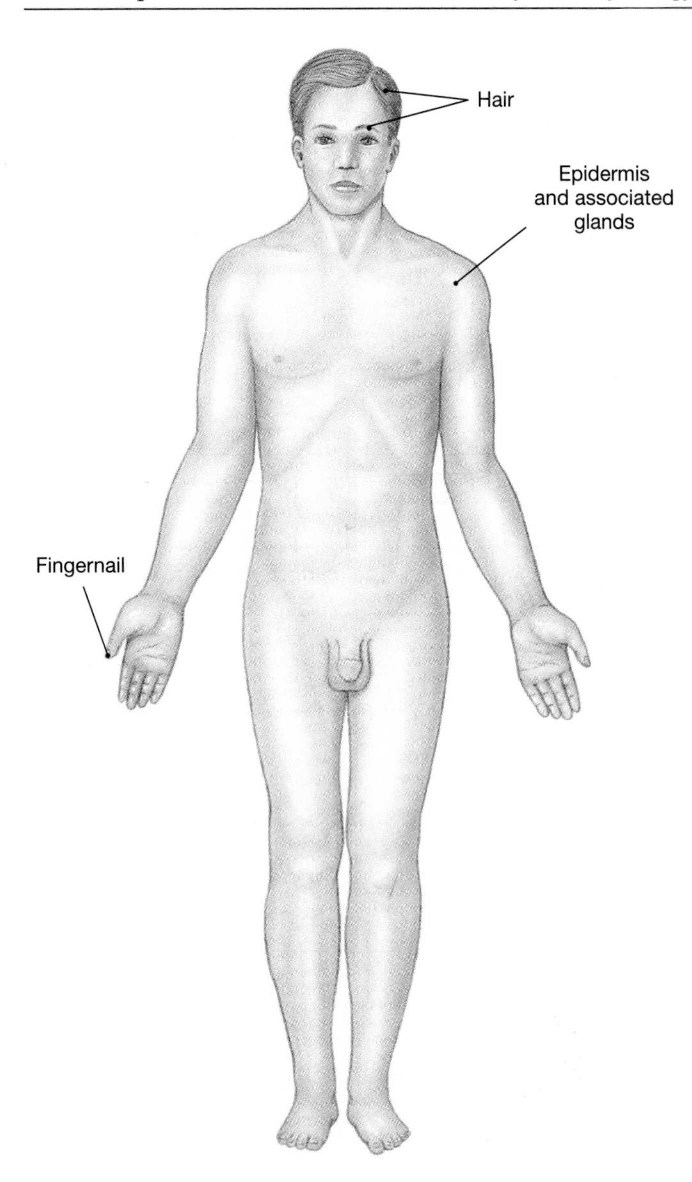

(a) The Integumentary System

AXIAL SKELETON
APPENDICULAR SKELETON
Skull
Supporting bones (scapula and collarbone)
Sternum
Ribs
Upper limb bones
Vertebrae
Sacrum
Supporting bones (hip)
Lower limb bones

(b) The Skeletal System

Organ	*Primary Functions*
EPIDERMIS	Covers surface, protects underlying tissues
DERMIS	Nourishes epidermis, provides strength, contains glands
HAIR FOLLICLES	Produce hair
Hairs	Provide sensation, provide some protection for head
Sebaceous glands	Secrete oil that lubricates hair
SWEAT GLANDS	Produce perspiration for evaporative cooling
NAILS	Protect and stiffen tips of fingers and toes
SENSORY RECEPTORS	Provide sensations of touch, pressure, temperature, pain

Organ	*Primary Functions*
BONES (206), CARTILAGES, AND LIGAMENTS	Support, protect soft tissues; store minerals
Axial skeleton (Skull, vertebrae, sacrum, ribs, sternum)	Protects brain, spinal cord, sense organs, and soft tissues of chest cavity; supports the body weight over the legs
Appendicular skeleton (Limbs and supporting bones)	Provides internal support and positioning of arms and legs; supports and moves axial skeleton
BONE MARROW	Primary site of blood cell production

●FIGURE 1-3
The Organ Systems of the Human Body

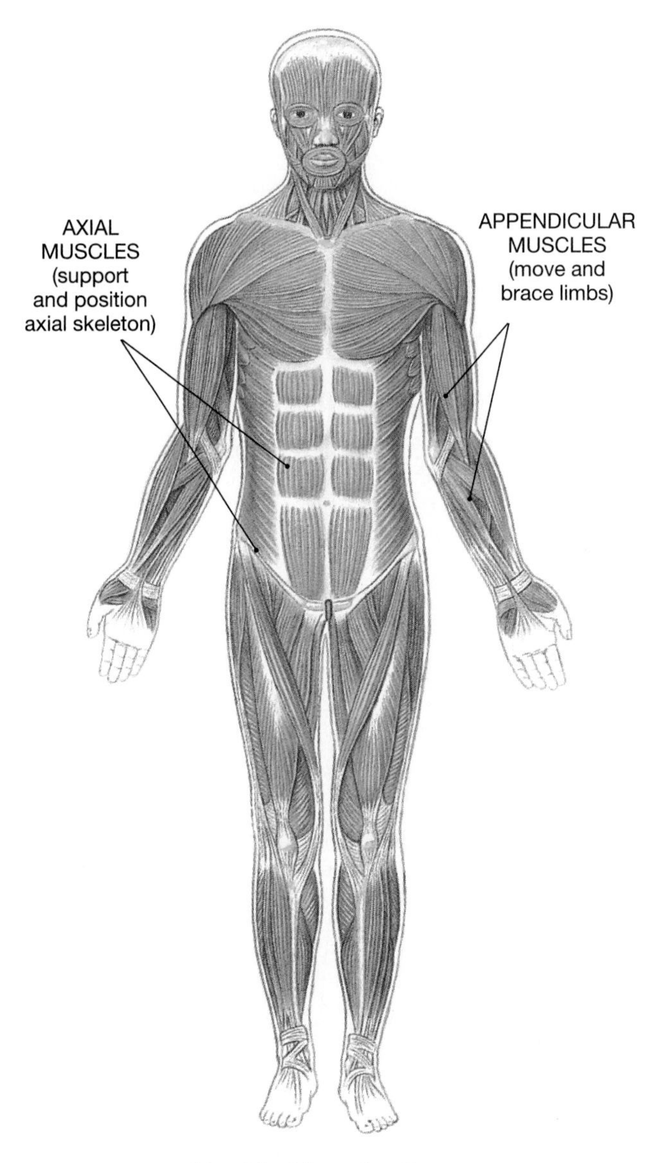

(c) **The Muscular System**

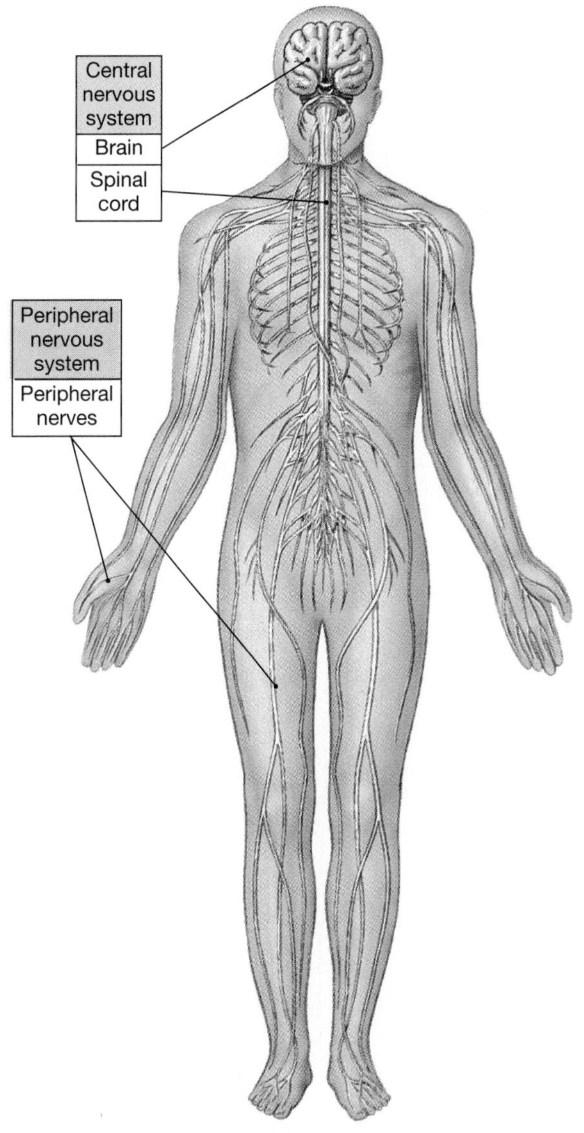

(d) **The Nervous System**

Organ	*Primary Functions*
SKELETAL MUSCLES (700)	Provide skeletal movement, control entrances and exits of digestive tract, produce heat, support skeletal position, protect soft tissues

Organ	*Primary Functions*
CENTRAL NERVOUS SYSTEM (CNS)	Control center for nervous system: processes information, provides short-term control over activities of other systems
Brain	Performs complex integrative functions, controls voluntary activities
Spinal Cord	Relays information to and from the brain, performs less complex integrative functions; directs many simple involuntary activities
PERIPHERAL NERVOUS SYSTEM (PNS)	Links CNS with other systems and with sense organs

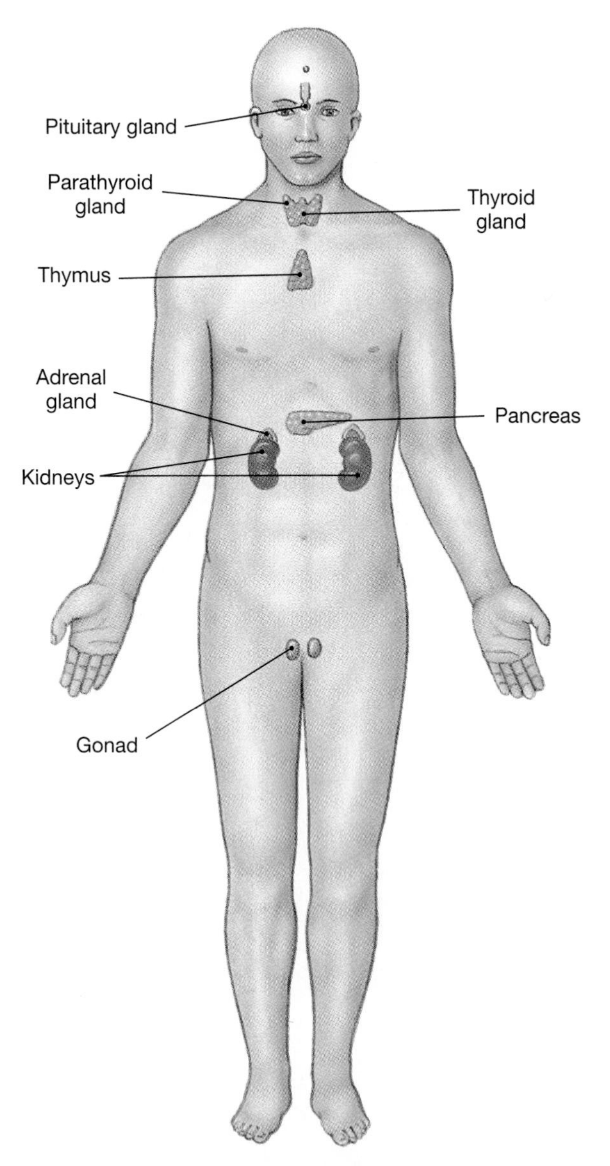

(e) **The Endocrine System**

Organ	*Primary Functions*
PITUITARY GLAND	Controls other glands, regulates growth and fluid balance
THYROID GLAND	Controls tissue metabolic rate and regulates calcium levels
PARATHYROID GLAND	Regulates calcium levels (with thyroid)
THYMUS	Controls white blood cell maturation
ADRENAL GLANDS	Adjust water balance, tissue metabolism, cardiovascular and respiratory activity
KIDNEYS	Control red blood cell production and elevate blood pressure
PANCREAS	Regulates blood glucose levels
GONADS	
Testes	Support male sexual characteristics and reproductive functions *(see Figure 1-3k)*
Ovaries	Support female sexual characteristics and reproductive functions *(see Figure 1-3k)*

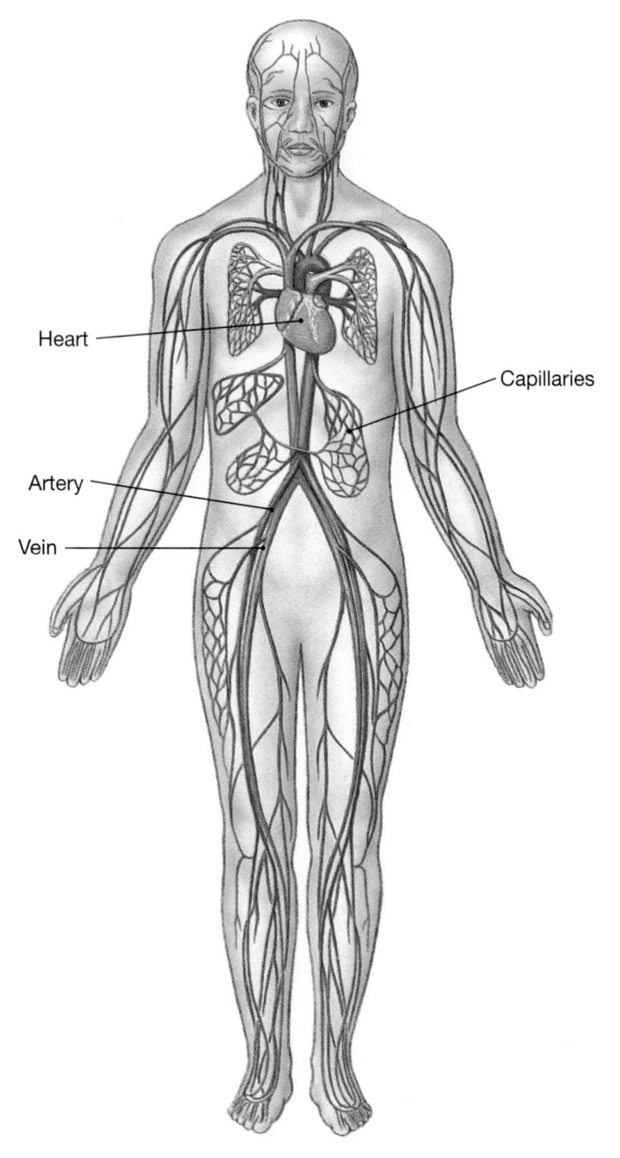

(f) **The Cardiovascular System**

Organ	*Primary Functions*
HEART	Propels blood, maintains blood pressure
BLOOD VESSELS	Distribute blood around the body
Arteries	Carry blood from heart to capillaries
Capillaries	Site of exchange between blood and interstitial fluids
Veins	Return blood from capillaries to heart
BLOOD	Transports oxygen and carbon dioxide, delivers nutrients, removes waste products, assists in defense against disease

FIGURE 1-3 continued

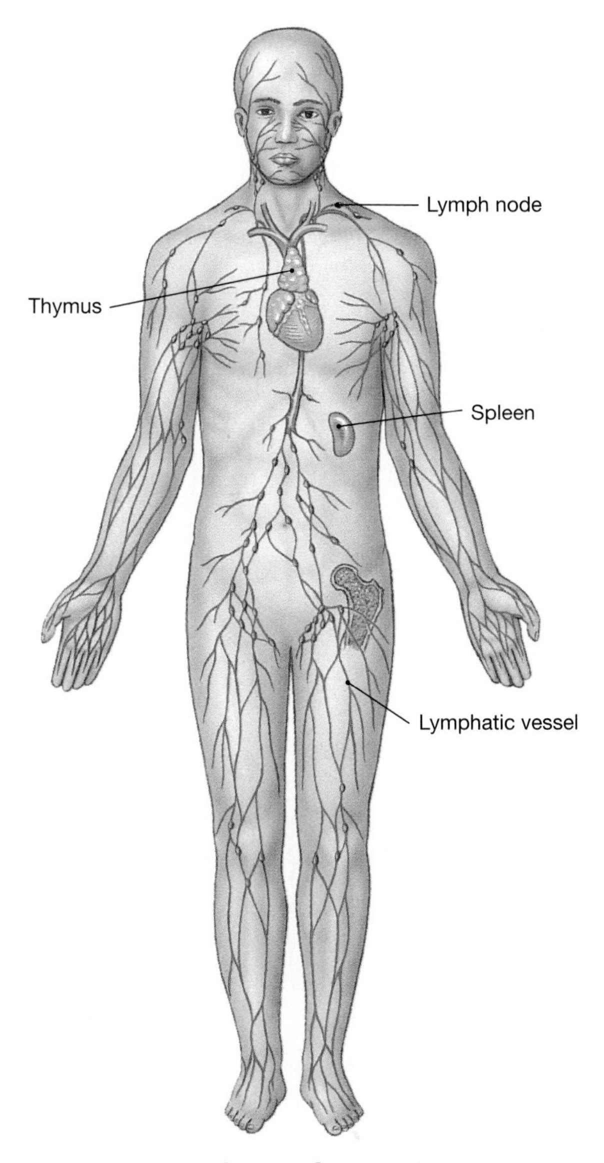

(g) The Lymphatic System

Organ	*Primary Functions*
LYMPHATIC VESSELS	Carry lymph (water and proteins) from peripheral tissues to the veins of the cardiovascular system
LYMPH NODES	Monitor the composition of lymph, stimulate immune response
SPLEEN	Monitors circulating blood, stimulates immune response
THYMUS	Controls development and maintenance of one class of white blood cells (T cells)

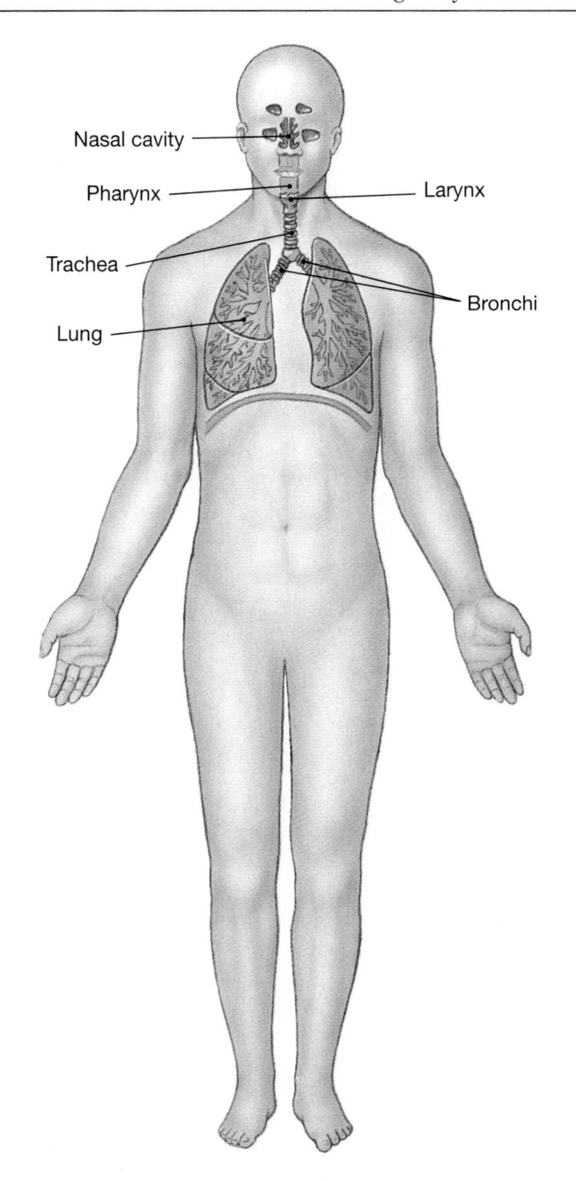

(h) The Respiratory System

Organ	*Primary Functions*
NASAL CAVITIES	Filter, warm, humidify air; detect smells
PHARYNX	Chamber shared with digestive tract; conducts air to larynx
LARYNX	Protects opening to trachea and contains vocal cords
TRACHEA	Filters air, traps particles in mucus; cartilages keep airway open
BRONCHI	Same as trachea
LUNGS	Include airways and alveoli; volume changes responsible for air movement
ALVEOLI	Sites of gas exchange between air and blood

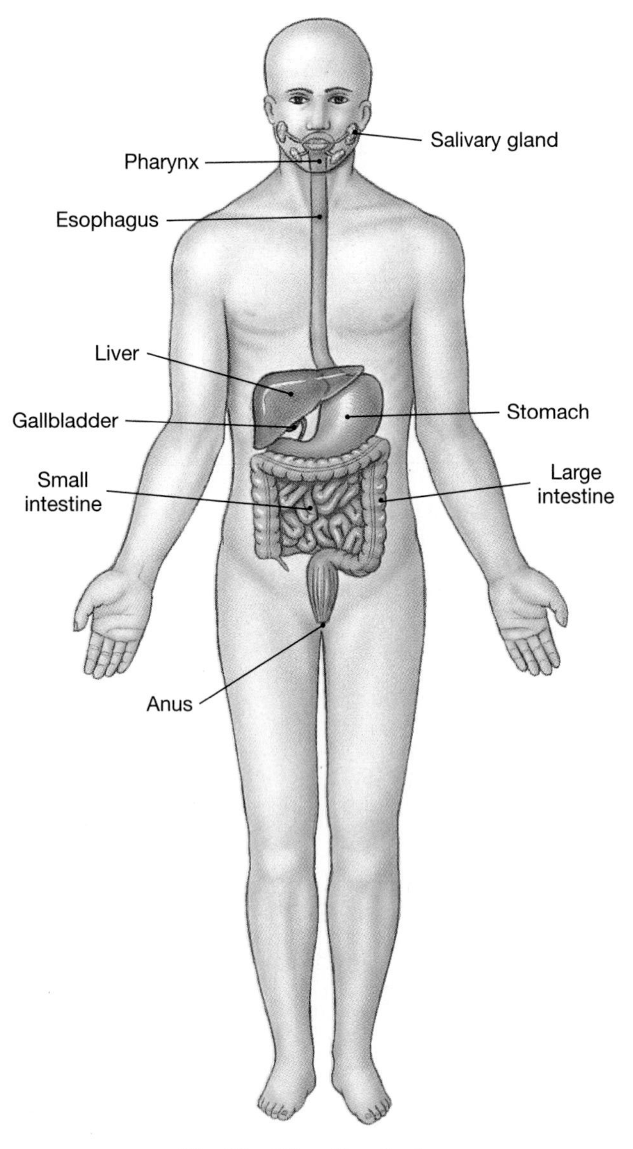

(i) **The Digestive System**

Organ	*Primary Functions*
SALIVARY GLANDS	Provide lubrication, produce buffers and the enzymes that begin digestion
PHARYNX	Passageway connected to esophagus
ESOPHAGUS	Delivers food to stomach
STOMACH	Secretes acids and enzymes
SMALL INTESTINE	Secretes digestive enzymes, absorbs nutrients
LIVER	Secretes bile, regulates blood composition of nutrients
GALLBLADDER	Stores bile for release into small intestine
PANCREAS	Secretes digestive enzymes and buffers; contains endocrine cells *(see Figure 1-3e)*
LARGE INTESTINE	Removes water from fecal material, stores wastes

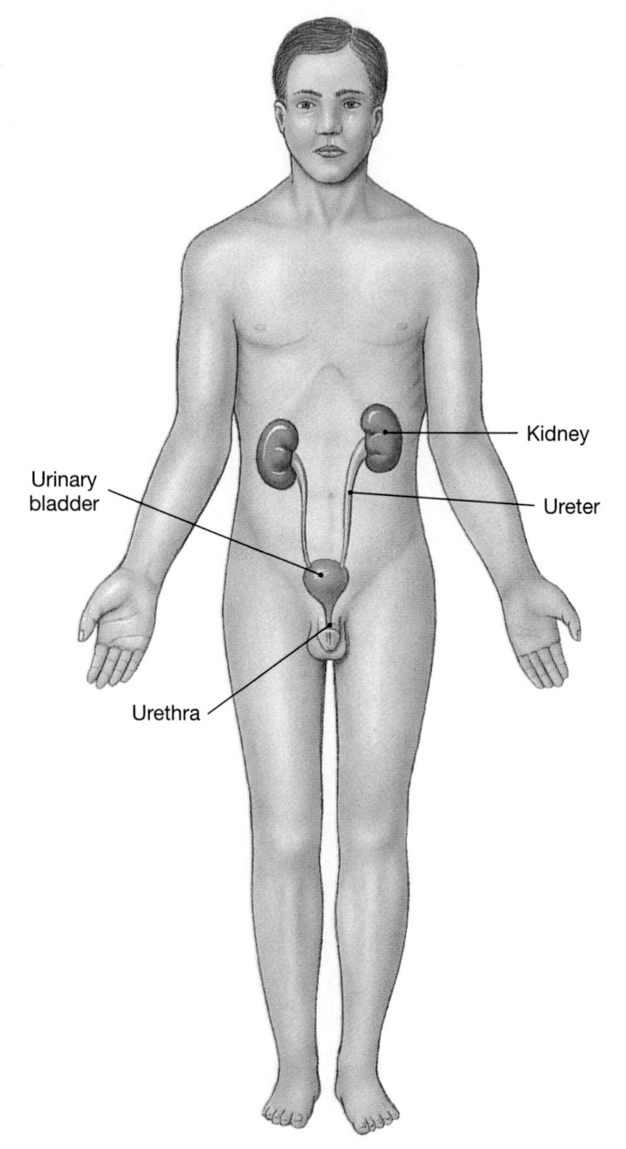

(j) **The Urinary System**

Organ	*Primary Functions*
KIDNEYS	Form and concentrate urine, regulate chemical composition of the blood
URETERS	Conduct urine from kidneys to urinary bladder
URINARY BLADDER	Stores urine for eventual elimination
URETHRA	Conducts urine to exterior

•FIGURE 1-3 continued

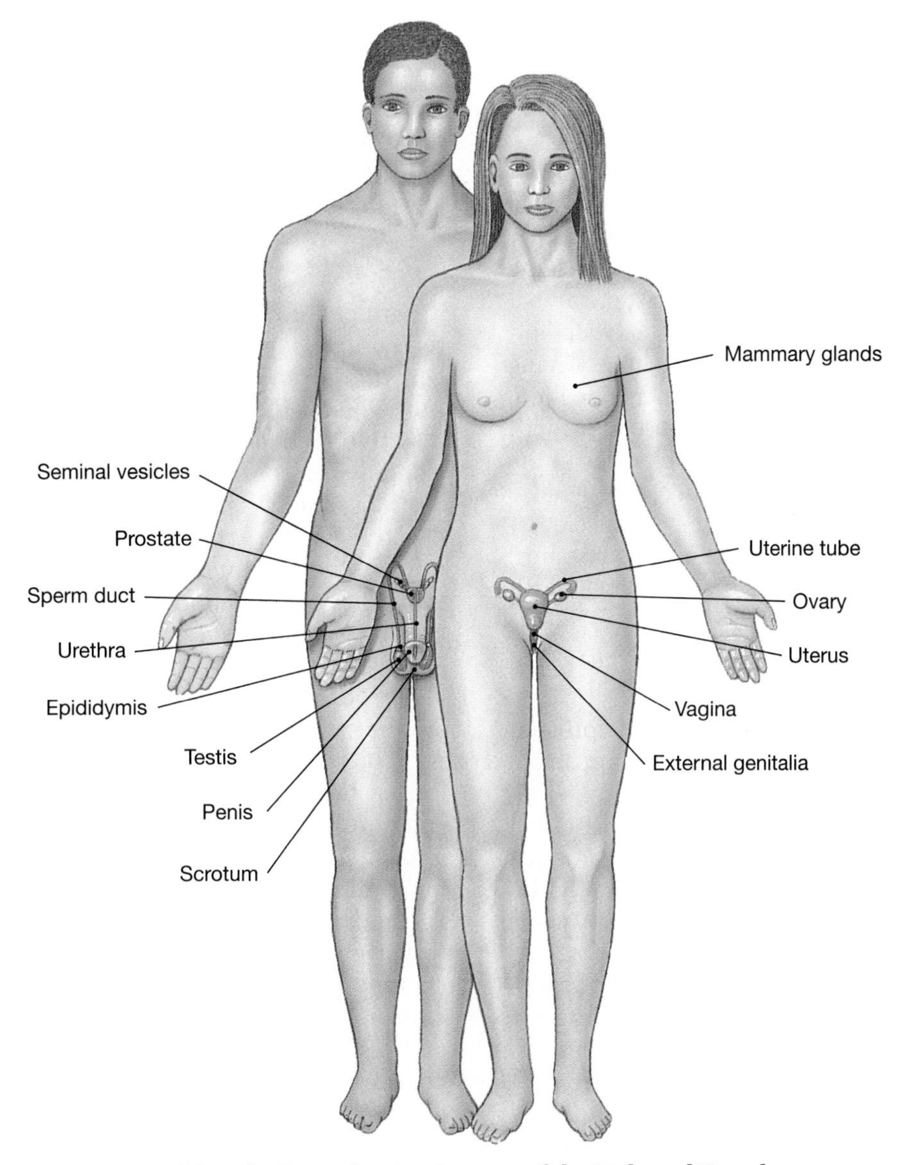

(k) The Reproductive Systems of the Male and Female

Organ	*Primary Functions*
TESTES	Produce sperm and hormones *(see Figure 1-3e)*
ACCESSORY ORGANS	
Epididymis	Site of sperm maturation
Ductus deferens (sperm duct)	Conducts sperm between epididymis and prostate
Seminal vesicles	Secrete fluid that makes up much of the volume of semen
Prostate	Secretes buffers and fluid
Urethra	Conducts semen to exterior
EXTERNAL GENITALIA	
Penis	Erectile organ used to deposit sperm in the vagina of a female; produces pleasurable sensations during sexual act
Scrotum	Surrounds and positions the testes

Organ	*Primary Functions*
OVARIES	Produce ova (eggs) and hormones *(see Figure 1-3e)*
UTERINE TUBES	Deliver ova or embryo to uterus; normal site of fertilization
UTERUS	Site of embryonic development and diffusion between maternal and embryonic bloodstreams
VAGINA	Site of sperm deposition; birth canal at delivery; provides passage of fluids during menstruation
EXTERNAL GENITALIA	
Clitoris	Erectile organ, produces pleasurable sensations during sexual act
Labia	Contain glands that lubricate entrance to vagina
MAMMARY GLANDS	Produce milk that nourishes newborn infant

HOMEOSTASIS AND SYSTEM INTEGRATION

Organ systems are interdependent, interconnected, and packaged together in a relatively small space. The cells, tissues, organs, and systems of the body live together in a shared environment, like the inhabitants of a large city. Just as city dwellers breathe the city air and eat food from local restaurants, cells absorb oxygen and nutrients from the fluids that surround them. All living cells are in contact with blood or some other body fluid, and any change in the composition of these fluids will affect them in some way. For example, changes in the temperature or salt content of the blood could cause anything from a minor adjustment (heart muscle tissue contracts more often, and the heart rate goes up) to a total disaster (the heart stops beating altogether).

Homeostatic Regulation

A variety of physiological mechanisms act to prevent potentially disruptive changes in the environment inside the body. **Homeostasis** (*homeo*, unchanging + *stasis*, standing) refers to the existence of a stable internal environment. To survive, every living organism must maintain homeostasis. The term **homeostatic regulation** refers to the adjustments in physiological systems that preserve homeostasis.

Homeostatic regulation usually involves a **receptor** sensitive to a particular stimulus and an **effector** whose activity has an effect upon the same stimulus. A **control center**, or *integration center*, is placed between the receptor and the effector. You are probably already familiar with several examples of homeostatic regulation, although not in those terms. As an example, consider the operation of the thermostat in a house or apartment (Figure 1-4●).

The thermostat is a control center that monitors room temperature. The gauge on the thermostat establishes the set point, the optimal level for the value under discussion, in this case room temperature. In our example, the set point is at 22°C (around 72°F). The function of the thermostat is to keep room temperature within acceptable limits, usually within a degree or so of the set point. This thermostat receives information from a receptor, a thermometer exposed to air in the room, and it controls two effectors: a heater and an air conditioner. The principle is simple: The heater turns on if the room becomes too cold, and the air conditioner turns on if the room becomes too warm.

When the temperature at the thermometer (a receptor) increases outside of the normal range, the thermostat (a control center) turns on the air conditioner (an effector). The air conditioner then cools the room. When temperature at the thermometer approaches the set point, the thermostat turns off the air conditioner (Figure 1-4a●).

A comparable pattern of events occurs if the temperature drops below normal levels: Temperature falls, the thermostat turns on the heater, the temperature rises, and the thermostat turns off the heater.

Negative Feedback

Regardless of whether the temperature at the receptor rises or falls, *a variation outside of normal limits triggers an automatic response that corrects the situation.* This method of homeostatic regulation is called **negative feedback** because the effector that is activated by the control center opposes the stimulus.

Most homeostatic mechanisms in the body involve negative feedback. For example, consider the control of body temperature, a process called *thermoregulation*. Thermoregulation involves altering the relationship between heat loss, which occurs primarily at the body surface, and heat production, which occurs in all active tissues. In the human body, skeletal muscles are the most important generators of body heat.

The cells of the thermoregulatory control center are located in the brain. Temperature receptors are located in the skin, and the cells in the control center are sensitive to local body temperature. The thermoregulatory center has a set point near 37°C (98.6° F) (Figure 1-4b●). If temperature at the thermoregulatory center rises above 37.2°C, activity in the control center targets two different effectors: (1) smooth muscles in the walls of blood vessels supplying the skin and (2) sweat glands. The blood vessels dilate, increasing blood flow at the body surface, and the sweat glands accelerate their secretion. The skin then acts like a radiator, losing heat to the environment, and the evaporation of sweat speeds the process. When body temperature returns to normal, the control center becomes inactive, and superficial blood flow and sweat gland activity decrease to previous levels.

If temperature at the control center falls below 36.7°C, the control center targets the same two effectors, but this time blood flow to the skin declines, and sweat gland activity decreases. This combination reduces the rate of heat loss to the environment. Because heat production continues, body temperature gradually rises, and once an acceptable temperature has been reached the thermoregulatory center turns itself "off" and both blood flow and sweat gland activity in the skin increase to normal levels.

Homeostatic mechanisms using negative feedback normally ignore minor variations, and they maintain a normal range rather than a fixed value. In the example above, body temperature oscillates around the "ideal" set-point temperature. Thus any measured value, such as body temperature, can vary from moment to moment or day to day for any single individual. The variability between individuals is even

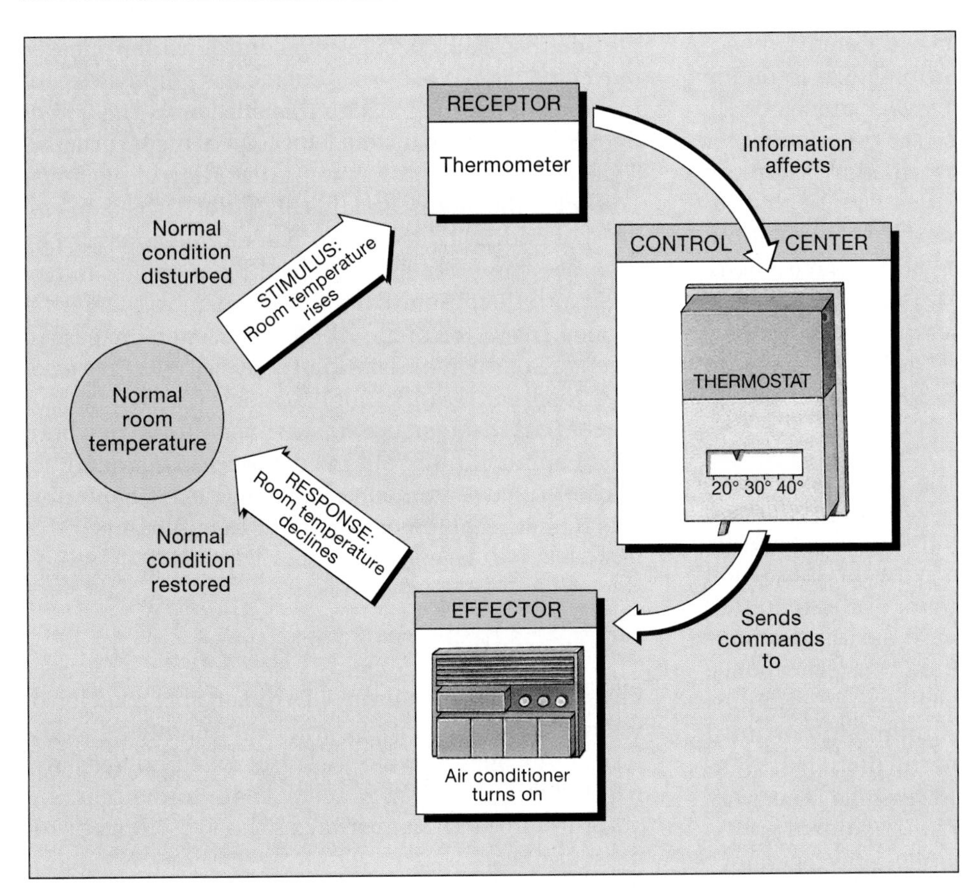

(a)

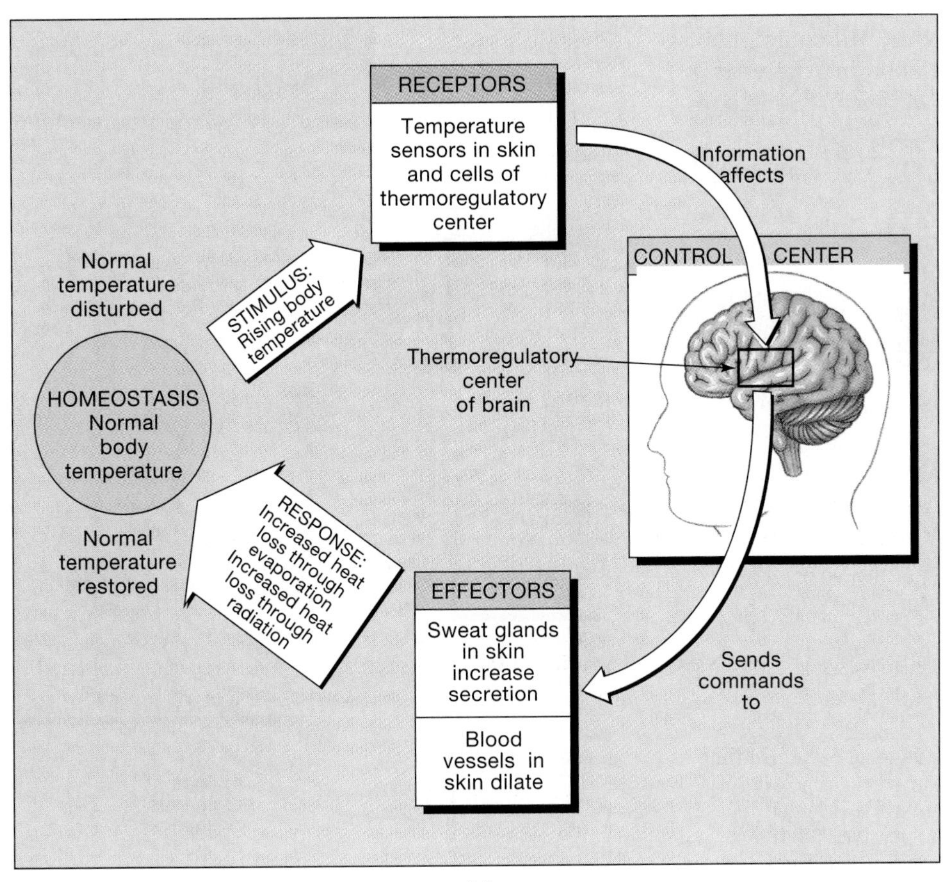

(b)

Figure 1-4
Negative Feedback

In negative feedback, a stimulus produces a response that opposes the original stimulus. **(a)** A thermostat controls heating and cooling systems to keep temperatures within acceptable limits. When the room temperature rises or falls, the thermostat (a control center) triggers an effector response that restores normal temperature. **(b)** Body temperature is regulated by a control center in the brain that functions as a thermostat with a set point of 37°C. If body temperature climbs above 37.2°C, heat loss is increased through enhanced blood flow to the skin and increased sweating.

greater, for each person has slightly different homeostatic set points. It is therefore impractical to define "normal" homeostatic conditions very precisely. By convention, physiological values are reported either as averages, the average value obtained by sampling a large number of individuals, or as a range that includes 95 percent or more of the sample population. For instance, 5 percent of normal adults have a body temperature outside the "normal" range (below 36.7° C or above 37.2°C). But these temperatures are perfectly normal for them, and the variations have no clinical significance.

Positive Feedback

In **positive feedback** *the initial stimulus produces a response that reinforces the stimulus.* For example, suppose that the thermostat was wired so that when the temperature rose it would turn on the heater rather than the air conditioner. In that case the initial stimulus (rising room temperature) would cause a response (heater turns on) that would strengthen the stimulus. The room temperature would continue to rise until some external factor switched off the thermostat, unplugged the heater, or intervened in some other way before the house caught fire and burned down.

Negative feedback provides long-term regulatory control that results in relatively unchanging internal conditions. Positive feedback (Figure 1-5●) is important in controlling necessary changes in the individual that, once initiated, are driven rapidly toward completion. The primary stimulus initiating labor and delivery is distortion of the uterus by the growing fetus. Stretch receptors in the uterine wall are monitored by an endocrine control center in the brain. When sufficient uterine distortion occurs, the control center releases *oxytocin* (oks-i-TŌ-sin), a hormone that stimulates contractions of uterine muscles (the effectors). These contractions begin moving the fetus toward the birth canal. This movement causes extreme distortion of the lower portion of the uterus, and this distortion triggers the release of additional oxytocin. The uterine contractions then become more forceful, leading to greater movement and distortion. Each time the control center responds, the action of the effectors causes an increase in receptor stimulation. This kind of cycle, a *positive feedback loop*, can be broken only by some external force or process; in this instance, the delivery of the newborn infant eliminates the uterine distortion. Labor and delivery will be examined more carefully in Chapter 21. Blood clotting, another important example of positive feedback, will be considered in Chapter 12.

Homeostasis and Disease

Physiological mechanisms do a remarkably good job of maintaining a constant internal environment, regardless of our ongoing activities. But when homeostatic regulation fails, organ systems begin to malfunction and the individual experiences the symptoms of illness, or **disease**.

✓ Why is homeostatic regulation important to human beings?

✓ How is positive feedback helpful in producing necessary changes in individuals?

✓ What happens to the body when homeostasis breaks down?

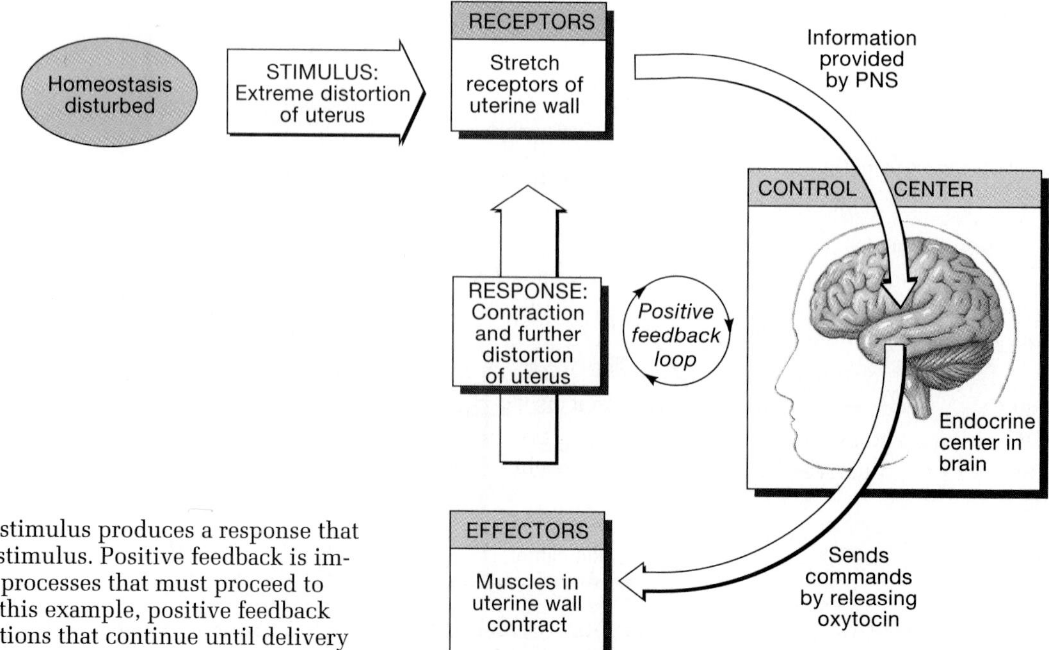

●**Figure 1-5**
Positive Feedback

In positive feedback, a stimulus produces a response that reinforces the original stimulus. Positive feedback is important in accelerating processes that must proceed to completion rapidly. In this example, positive feedback enhances labor contractions that continue until delivery has been completed.

THE LANGUAGE OF ANATOMY

Early anatomists faced serious communication problems. For example, stating that a bump is "on the back" does not give very precise information about its location. So anatomists created maps of the human body, using prominent anatomical structures as landmarks and reporting distances in centimeters or inches. In effect, anatomy uses a special language that must be learned almost at the start.

A familiarity with Latin and Greek roots and patterns makes anatomical terms more understandable. As new terms are introduced, notes on pronunciation and relevant word roots will be provided. Additional information on foreign word roots, prefixes, suffixes, and combining forms can be found on the front endpapers.

Latin and Greek terms are not the only foreign terms imported into the anatomical vocabulary over the centuries, and the vocabulary continues to expand. Many anatomical structures and clinical conditions were initially named after either the discoverer or, in the case of diseases, the most famous victim. Although most commemorative names, or *eponyms*, have been replaced by more precise terms, many are still in use. AM *Learning the Language of Anatomy and Physiology*

Superficial Anatomy

With the exception of the skin and its accessory organs, none of the organ systems can be seen from the body surface. Therefore, you must create your own mental maps and extract information from the terms given in Figures 1-6● and 1-7●. Learning these terms now will help you make sense of many discussions in this book.

Anatomical Landmarks

Standard anatomical illustrations show the human form in the **anatomical position**, with the hands at the sides and the palms facing forward (Figure 1-6●). A person lying down in the anatomical position is said to be **supine** (SŪ-pīne) when lying face up and **prone** when lying face down.

Important anatomical landmarks are also presented in Figure 1-6●. The anatomical terms are given in boldface, the common names in plain type, and the anatomical adjectives in parentheses. You should become familiar with all three terms. For example, the term **brachium** refers to the arm, and later chapters will discuss the brachial artery, brachial nerve, and so forth. You might remember this term more easily if you know that the Latin word *brachium* is also the source of old English and French words meaning "to embrace." Understanding the terms and their origins will help you to remember the location of a particular structure, as well as its name.

Anatomical Regions

Major regions of the body are listed in Table 1-1 and Figure 1-7●. Anatomists and clinicians often need to use regional terms as well as specific landmarks to describe a general area of interest or injury. Two approaches have developed, both concerned with mapping the surface of the abdominopelvic region. Clinicians refer to the **abdominopelvic quadrants:** four segments divided by imaginary lines that intersect at the *umbilicus* (navel). This simple method, shown in Figure 1-7*a*●, is useful for describing aches, pains, and injuries. The location can help the doctor decide the possible cause; for example, tenderness in the right lower quadrant (RLQ) is a symptom of appendicitis, whereas tenderness in the right upper quadrant (RUQ) may indicate gallbladder or liver problems.

Anatomists like to use more precise regional distinctions to describe the location and orientation of internal organs. They recognize nine **abdominopelvic regions** (Figure 1-7b●). Figure 1-7c● shows the relationship between quadrants, regions, and internal organs.

TABLE 1-1 Regions of the Human Body (see Figure 1-6)

Structure	*Area*
Cephalon (head)	Cephalic region
Cervicis (neck)	Cervical region
Thoracis (chest)	Thoracic region
Abdomen	Abdominal region
Pelvis	Pelvic region
Loin (lower back)	Lumbar region
Buttock	Gluteal region
Pubis (anterior pelvis)	Pubic region
Groin	Inguinal region
Axilla (armpit)	Axillary region
Brachium (arm)	Brachial region
Antebrachium (forearm)	Antebrachial region
Manus (hand)	Manual region
Thigh	Femoral region
Leg (anterior)	Crural region
Calf	Sural region
Pes (foot)	Pedal region

Anatomical Directions

Table 1-2 and Figure 1-8● show the principal directional terms and examples of their use. There are many different directional terms, and some can be used interchangeably. For example, *anterior* refers to the front of the body, when viewed in the anatomical position; in human beings, this term is equivalent to *ventral*, which actually refers to the belly. Likewise, *posterior* and *dorsal* are terms that refer to the back of the human body. Although your instructor may have additional recommendations, these are the terms that appear frequently in later chapters. Remember that "left" and "right" always refer to the left and right sides of the subject, not of the observer.

Sectional Anatomy

A presentation in sectional view is sometimes the only way to illustrate the relationships between the parts of a three-dimensional object. An understanding of sectional views has become increasingly important since the development of procedures that enable us to

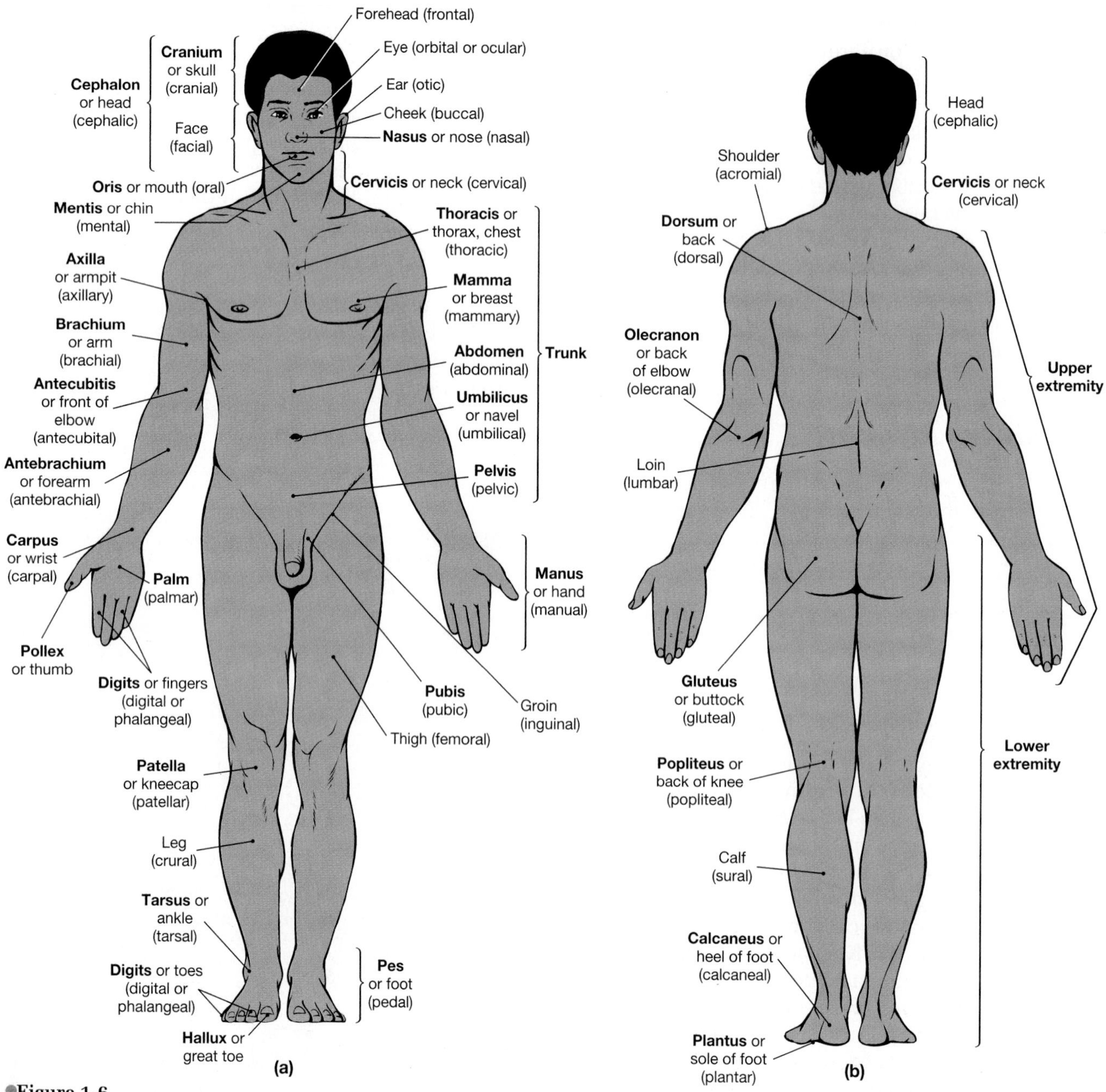

●**Figure 1-6**
Anatomical Landmarks

The anatomical terms are shown in boldface type, the common names are in plain type, and the anatomical adjectives are in parentheses.

see inside the living body without resorting to surgery (see Figure 1-12● for representative views).

Planes and Sections

Any slice through a three-dimensional object can be described with reference to three **sectional planes**, indicated in Table 1-3 and Figure 1-9●.

1. **Tranverse plane.** The transverse plane lies at right angles to the long (head-foot) axis of the body, dividing it into **superior** and **inferior** sections. A cut in this plane is called a **transverse section**, a **horizontal section**, or a *cross-section*.
2. **Frontal Plane.** The **frontal plane**, or **coronal plane**, parallels the long axis of the body. The frontal plane extends from side to side, dividing the body into **anterior** and **posterior** sections.
3. **Sagittal Plane.** The **sagittal plane** also parallels the long axis of the body, but it extends from front to back. A sagittal plane divides the body into *left* and *right* sections. A cut that passes along the midline and divides the body into left and right halves is a **midsagittal section.**

Frontal and sagittal sections are often called *longitudinal sections.*

Body Cavities

Viewed in sections, the human body is not a solid object, like a rock, in which all of the parts are fused together. Many vital organs are suspended in internal chambers called *body cavities*. These cavities have two essential functions: (1) They protect delicate organs, such as the brain and spinal cord, from accidental shocks and cushion them from the thumps and bumps that occur during walking, jumping, and running; and (2) they permit significant changes in the size and shape of visceral organs. For example, because they are situated within body cavities, the lungs, heart, stomach, intestines, urinary bladder, and many other organs can expand and contract without distorting surrounding tissues and disrupting the activities of nearby organs.

Two body cavities form during embryonic development. A **dorsal body cavity** surrounds the brain and spinal cord, and a much larger **ventral body cavity** surrounds developing organs of the respiratory, cardiovascular, digestive, urinary, and reproductive systems.

Dorsal Body Cavities The dorsal body cavity (Figure 1-10a●) is a fluid-filled space whose limits are established by the **cranium**, the bones of the skull that surround the brain, and the spinal vertebrae. The dorsal body cavity is subdivided into the **cranial cavity**, which encloses the brain, and the **spinal cavity**, which surrounds the spinal cord.

●**Figure 1-7**
Abdominopelvic Quadrants and Regions

(a) Abdominopelvic quadrants divide the area into four sections. These terms, or their abbreviations, are most often used in clinical discussions.
(b) More precise regional descriptions are provided by reference to the appropriate abdominopelvic region.
(c) Quadrants or regions are useful because there is a known relationship between superficial anatomical landmarks and underlying organs.

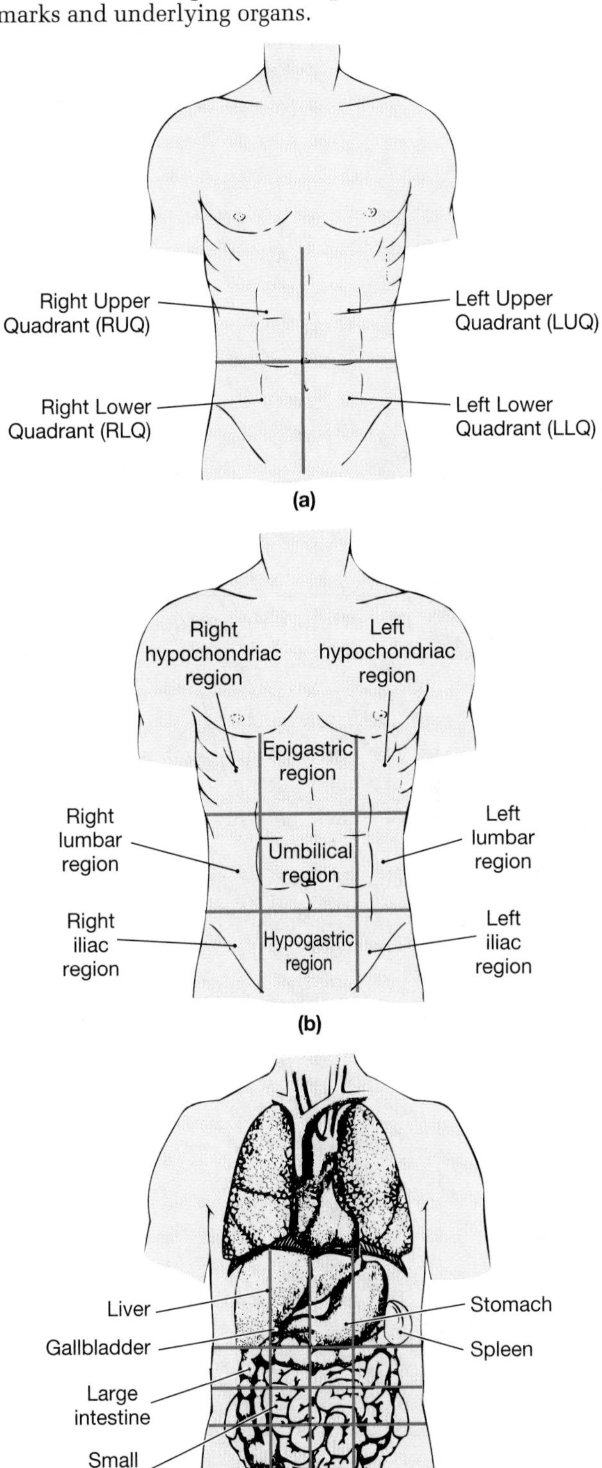

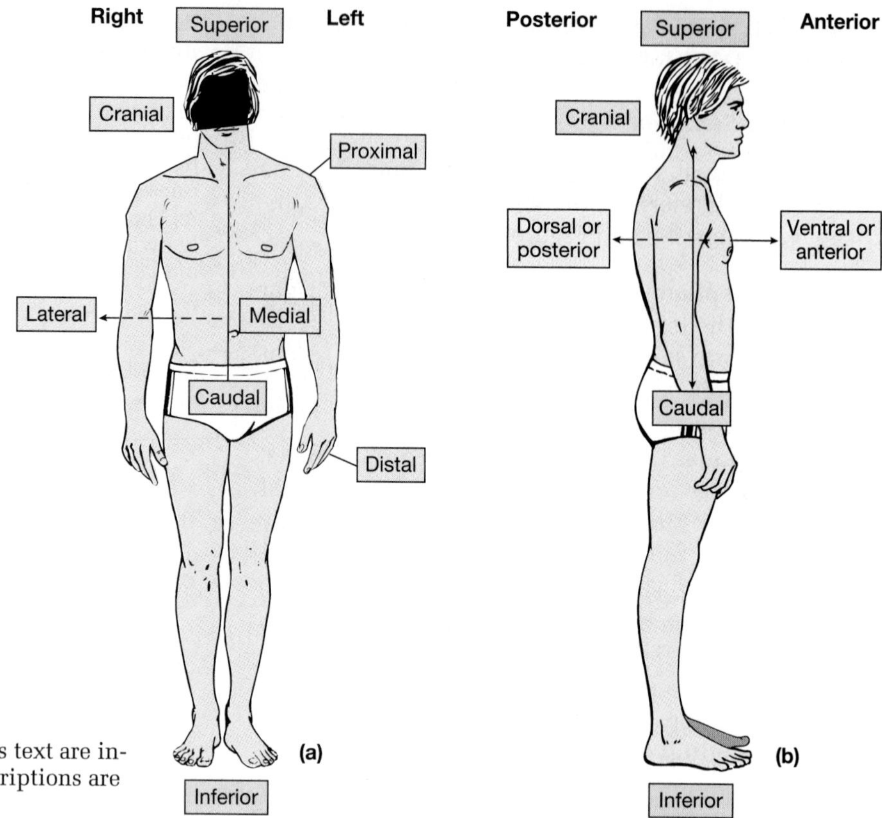

•Figure 1-8
Directional References

Important directional terms used in this text are indicated by arrows; definitions and descriptions are included in Table 1-2.

TABLE 1-2 Directional Terms (see Figure 1-8)

Term	*Region or Reference*	*Example*
Anterior	The front; before	The navel is on the *anterior (ventral)* surface of the trunk.
Ventral	The belly side (equivalent to anterior when referring to human body)	
Posterior	The back; behind	The shoulder blade is located *posterior (dorsal)* to the rib cage.
Dorsal	The back (equivalent to posterior when referring to human body)	The *dorsal* body cavity encloses the brain and spinal cord.
Cranial or cephalic	The head	The *cranial*, or *cephalic*, border of the pelvis is *superior* to the thigh.
Superior	Above; at a higher level (in human body, toward the head)	
Caudal	The tail (coccyx in humans)	The hips are *caudal* to the waist.
Inferior	Below; at a lower level	The knees are *inferior* to the hips.
Medial	Toward the body's longitudinal axis	The *medial* surfaces of the thighs may be in contact; moving medially from the arm across the chest surface brings you to the sternum.
Lateral	Away from the body's longitudinal axis	The thigh articulates with the *lateral* surface of the pelvis; moving laterally from the nose brings you to the eyes.
Proximal	Toward an attached base	The thigh is *proximal* to the foot; moving proximally from the wrist brings you to the elbow.
Distal	Away from an attached base	The fingers are *distal* to the wrist; moving distally from the elbow brings you to the wrist.
Superficial	At, near, or relatively close to the body surface	The skin is *superficial* to underlying structures.
Deep	Farther from the body surface	The bone of the thigh is *deep* to the surrounding skeletal muscles.

Ventral Body Cavities As development proceeds, internal organs grow and change their relative positions. These changes lead to the subdivision of the ventral body cavity. The **diaphragm** (DĪ-a-fram), a flat muscular sheet, divides the ventral body cavity into a superior **thoracic cavity**, enclosed by the chest wall, and an inferior **abdominopelvic cavity**, enclosed by the abdomen and pelvic girdle. The abdominopelvic cavity has two subdivisions. The **abdominal cavity** extends from the inferior surface of the diaphragm to an imaginary line drawn from the inferior surface of the lowest spinal vertebra to the anterior and superior margin of the pelvic girdle. The portion of the ventral body cavity inferior to this imaginary line is the **pelvic cavity**.

The thoracic and abdominopelvic cavities contain spaces lined by a shiny, slippery, and delicate *serous membrane*. The *parietal* portion of a serous membrane forms the outer wall of the body cavity. The *visceral* portion covers the surfaces of internal organs, or **viscera** (VIS-e-ra), where they project into the body cavity. Many visceral organs undergo periodic changes in size and shape; the lungs inflate and deflate with each breath, and the volume of the heart changes during each heartbeat. A covering of serous membrane prevents friction between adjacent viscera and between the visceral organs and the body wall. The spaces between opposing membranes are very small, but they are separated by a thin layer of fluid. To understand the relationship between a visceral organ and the serous membrane, consider the example in Figure 1-10b●. The heart projects into a space known as the **pericardial cavity**. The relationship resembles that of a fist pushing into a balloon. The wrist corresponds to the base of the heart, and the balloon corresponds to the serous membrane lining the pericardial cavity. The serous membrane is called the **pericardium** (*peri-*, around + *cardium*, heart). The layer covering the heart is the **visceral pericardium**, and the opposing surface is the **parietal pericardium**.

The thoracic cavity contains two **pleural cavities**, each surrounding a lung. The spatial relationships between the lungs and the pleural cavities resemble that

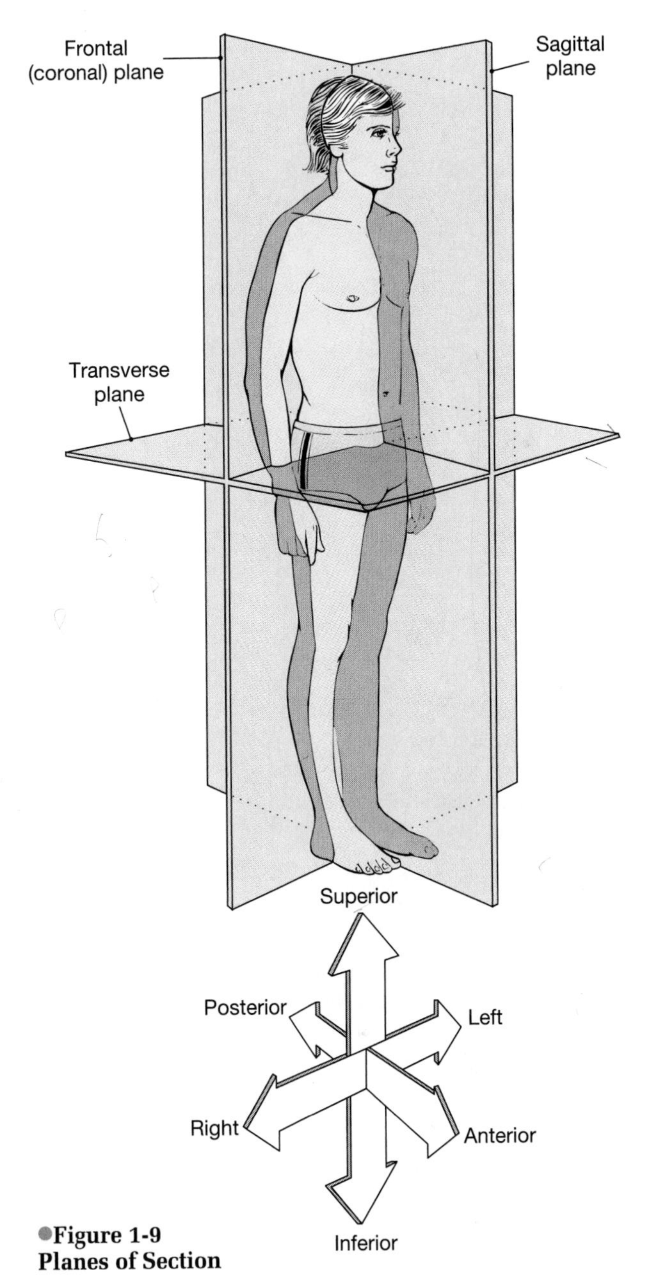

●**Figure 1-9**
Planes of Section
The three primary planes of section are indicated here. Table 1-3 defines and describes them.

TABLE 1-3 Terms That Indicate Planes of Sections (see Figure 1-9)

Orientation of Plane	*Adjective*	*Directional Reference*	*Description*
Parallel to long axis	Sagittal	Sagittally	A *sagittal section* separates right and left portions. You examine a sagittal section, but you section sagittally.
	Midsagittal		In a *midsagittal section* the plane passes through the midline, dividing the body in half and separating right and left sides.
	Frontal or coronal	Frontally or coronally	A *frontal*, or *coronal, section* separates anterior and posterior portions of the body; *coronal* usually refers to sections passing through the skull.
Perpendicular to long axis	Transverse or horizontal	Transversely or horizontally	A *transverse*, or *horizontal, section* separates superior and inferior portions of the body.

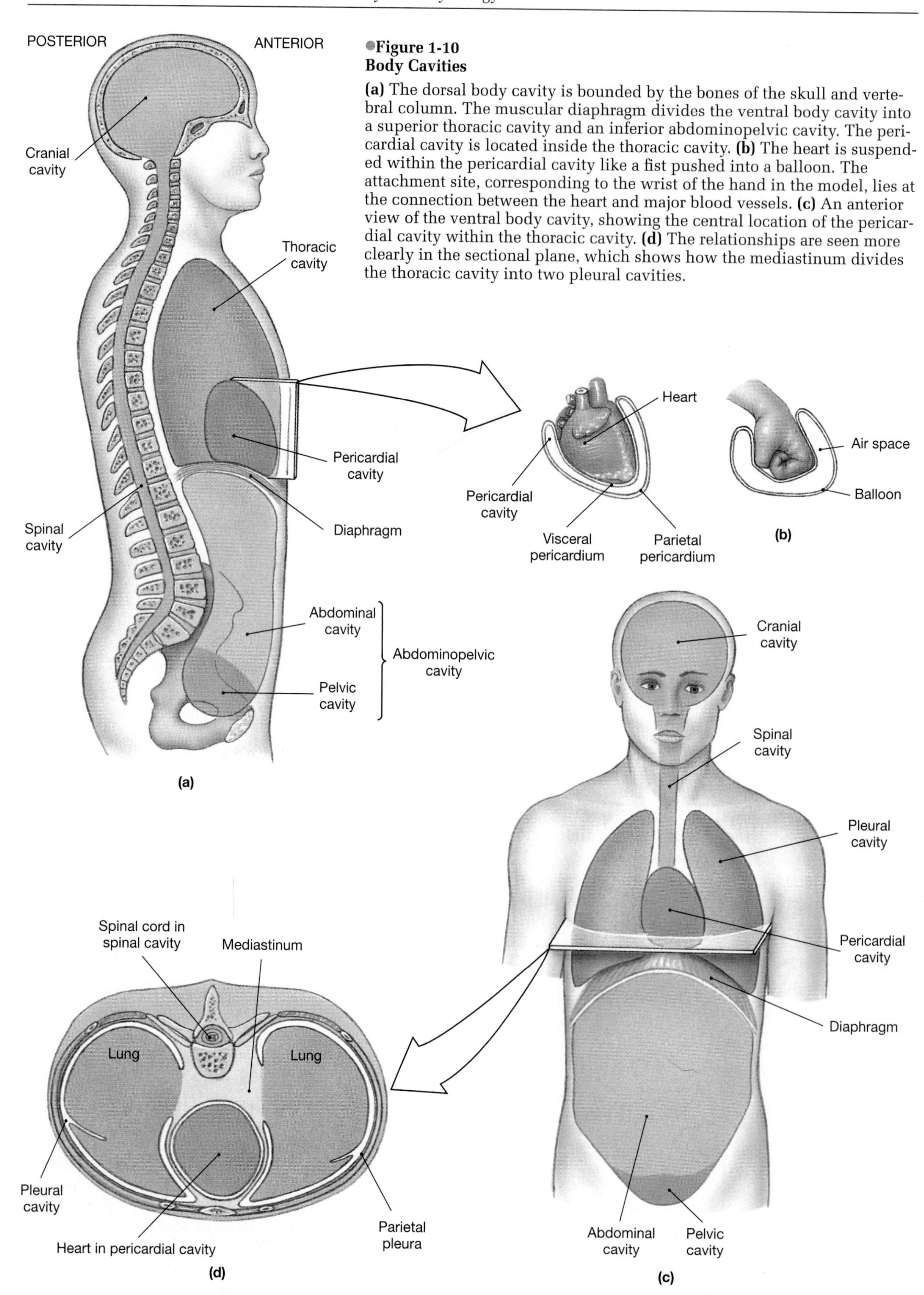

Figure 1-10
Body Cavities

(a) The dorsal body cavity is bounded by the bones of the skull and vertebral column. The muscular diaphragm divides the ventral body cavity into a superior thoracic cavity and an inferior abdominopelvic cavity. The pericardial cavity is located inside the thoracic cavity. **(b)** The heart is suspended within the pericardial cavity like a fist pushed into a balloon. The attachment site, corresponding to the wrist of the hand in the model, lies at the connection between the heart and major blood vessels. **(c)** An anterior view of the ventral body cavity, showing the central location of the pericardial cavity within the thoracic cavity. **(d)** The relationships are seen more clearly in the sectional plane, which shows how the mediastinum divides the thoracic cavity into two pleural cavities.

between the heart and pericardium. The serous membrane lining the pleural cavities is called the **pleura** (PLOO-ra). The region between the two pleural cavities is known as the **mediastinum** (mē-dē-as-TĪ-num or mē-dē-AS-ti-num). The mediastinum contains the thymus, trachea, esophagus, the large arteries and veins attached to the heart, and the pericardial cavity.

Most of the visceral organs in the abdominopelvic cavity project into the **peritoneal** (per-i-tō-NĒ-al) **cavity**. The serous membrane lining this cavity is the **peritoneum** (pe-ri-tō-NĒ-um). Organs such as the stomach, small intestine, and portions of the large intestine are suspended within the peritoneal cavity by double sheets of peritoneum, called **mesenteries** (MES-en-ter-ēs). Mesenteries provide support and stability while permitting limited movement.

This chapter provided an overview of the locations and functions of the major components of each organ system. It also introduced the anatomical vocabulary needed to follow more detailed anatomical descriptions in later chapters. Modern methods of visualizing anatomical structures in living individuals are summarized on pp. 00–00. Many of the figures in later chapters contain images produced by the procedures outlined in that section.

✓ What type of section would separate the two eyes?

✓ If a surgeon makes an incision just inferior to the diaphragm, what body cavity will be opened?

FOCUS Sectional Anatomy and Clinical Technology

The term **radiological procedures** includes not only those scanning techniques that involve radioisotopes but also methods that employ radiation sources outside the body. Physicians who specialize in the performance and analysis of these procedures are called **radiologists**. Radiological procedures can provide detailed information about internal systems. Figures 1-11● and 1-12● compare the views provided by several different techniques. These figures include examples of X-rays, CT scans, MRI scans, and ultrasound images. Other examples of clinical technology will be found in later chapters.

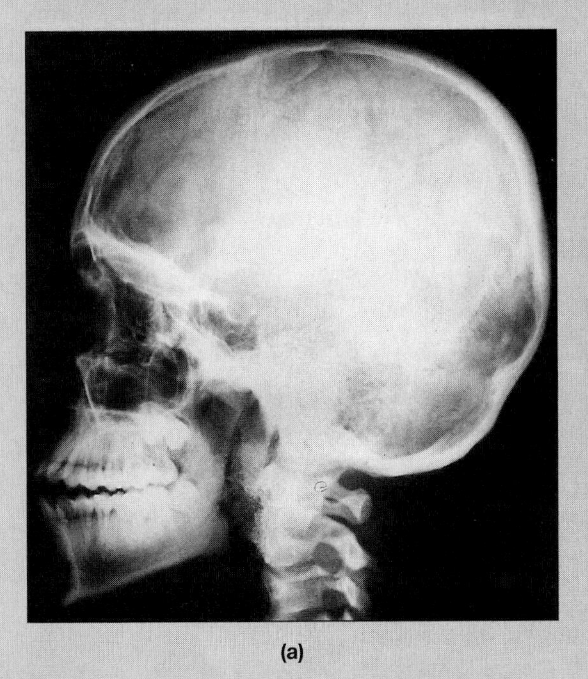

(a)

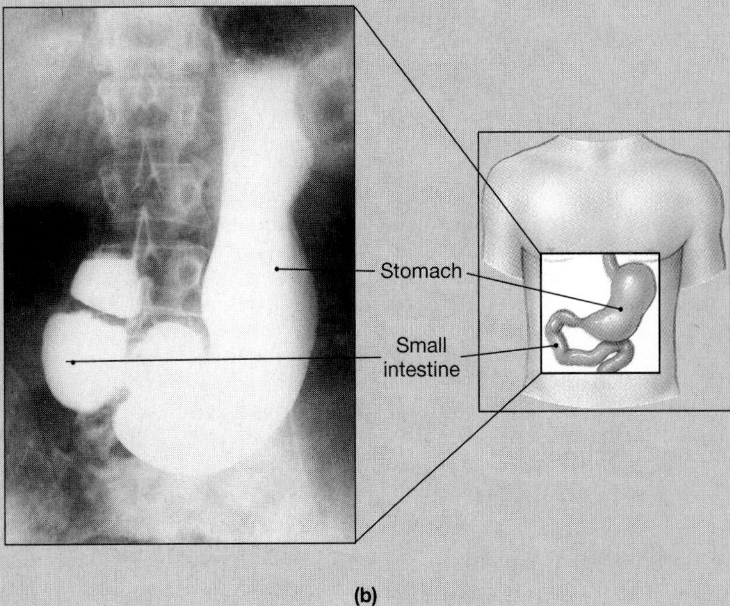

(b)

●**Figure 1-11**
X-rays

(a) An X-ray of the skull, taken from the left side. **X-rays** are a form of high-energy radiation that can penetrate living tissues. In the most familiar procedure, a beam of X-rays travels through the body and strikes a photographic plate. All of the projected X-rays do not arrive at the film; some are absorbed or deflected as they pass through the body. The resistance to X-ray penetration is called **radiodensity**. In the human body, the order of increasing radiodensity is as follows: air, fat, liver, blood, muscle, bone. The result is an image with radiodense tissues, such as bone, appearing in white, and less dense tissues in shades of gray to black. The picture is a two-dimensional image of a three-dimensional object; in this image it is difficult to decide whether a particular feature is on the left side (toward the viewer) or on the right side (away from the viewer). **(b)** A barium-contrast X-ray of the upper digestive tract. Barium is very dense, and the contours of the gastric and intestinal lining can be seen outlined against the white of the barium solution.

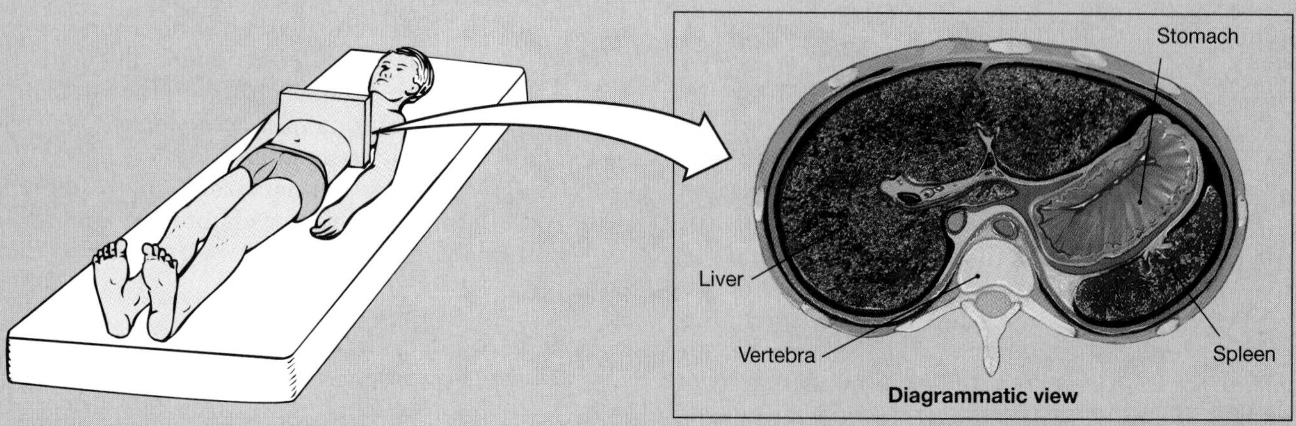

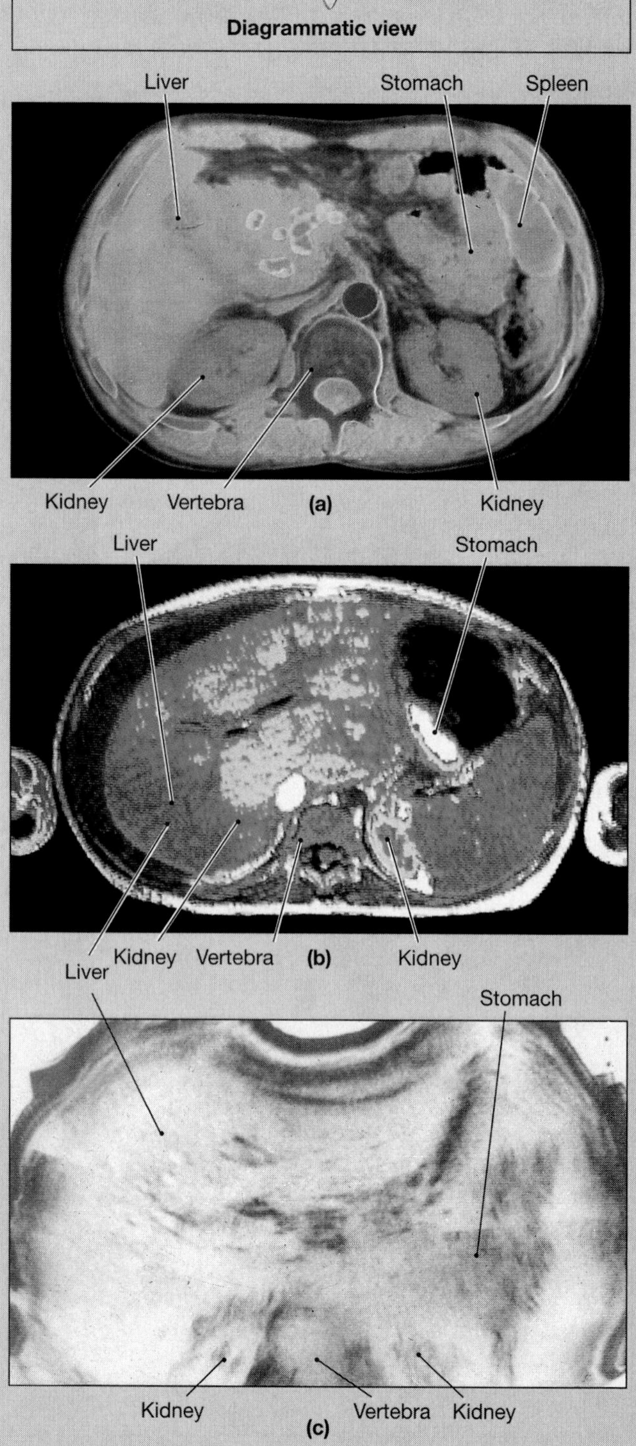

Figure 1-12
Scanning Techniques

(a) A color-enhanced CT scan of the abdomen. **CT** (**C**omputerized **T**omography), formerly called **CAT** (**C**omputerized **A**xial **T**omography), uses computers to reconstruct sectional views. A single X-ray source rotates around the body, and the X-ray beam strikes a sensor monitored by the computer. The source completes one revolution around the body every few seconds; it then moves a short distance and repeats the process. The result is usually displayed as a sectional view in black and white, but it can be colorized for visual effect. CT scans show three-dimensional relationships and soft tissue structure more clearly than standard X-rays. **(b)** A color-enhanced **MRI** scan of the same region. **M**agnetic **R**esonance **I**maging surrounds part or all of the body with a magnetic field about 3000 times as strong as that of the earth. Pulses of radio waves then cause tissues to release energy used to create an image. Details of soft tissue structure are usually much more clearly detailed than in CT scans. Note the differences in detail between this image, the CT scan, and the ultrasound image. **(c)** An **ultrasound** scan of the abdomen. In ultrasound procedures, a small transmitter contacting the skin broadcasts a brief, narrow burst of high-frequency sound and then picks up the echoes. The sound waves are reflected by internal structures, and a picture, or **echogram**, can be assembled from the pattern of echoes. These images lack the clarity of other procedures, but no adverse affects have been reported, and fetal development can be monitored without a significant risk of birth defects. Special methods of transmission and processing permit analysis of the beating heart, without the complications that can accompany dye injections.

Chapter Review

KEY TERMS

anatomical position, p. 15
anatomy, p. 2
diaphragm, p. 19
frontal plane, p. 17
homeostasis, p. 12
negative feedback, p. 12
peritoneum, p. 21
physiology, p. 2
positive feedback, p. 14
sagittal plane, p. 17
transverse plane, p. 17
viscera, p. 19

SUMMARY OUTLINE

INTRODUCTION p. 2

1. **Biology** is the study of life; one of its goals is to discover the unity and patterns that underlie the diversity of living organisms.

2. All living things from single **cells** to large multicellular organisms, perform the same basic functions: they respond to changes in their environment (i.e., they show adaptability to their environment); they grow and reproduce to create future generations; they are capable of producing movement; and they absorb materials from the environment. Organisms absorb and consume oxygen during respiration, and they discharge waste products during excretion. Digestion occurs in specialized areas of the body to break down complex foods. The circulation forms an internal transportation system between areas of the body.

THE SCIENCES OF ANATOMY AND PHYSIOLOGY p. 2

Anatomical Perspectives p. 3

1. **Anatomy** is the study of internal and external structure and the physical relationships between body parts. **Physiology** is the study of how living organisms perform vital functions. All specific functions are performed by specific structures.

2. The boundaries of **microscopic anatomy** are established by the equipment used. **Cytology** analyzes the internal structure of individual cells. **Histology** examines **tissues** (groups of cells that have specific functional roles). Tissues combine to form **organs**, anatomical units with multiple functions.

3. **Gross (macroscopic) anatomy** considers features visible without a microscope. It includes **surface anatomy** (general form and superficial markings); **regional anatomy** (superficial and internal features in a specific area of the body); and **systemic anatomy** (structure of major organ systems).

Physiology p. 3

4. Human physiology is the study of the functions of the human body. It is based on **cell physiology**, the study of the functions of living cells. **Special physiology** studies the physiology of specific organs. **System physiology** considers all aspects of the function of specific organ systems. **Pathological physiology** studies the effects of diseases on organ or system functions.

LEVELS OF ORGANIZATION p. 5

1. Anatomical structures and physiological mechanisms are arranged in a series of interacting levels of organization. *(Figure 1-1)*

AN INTRODUCTION TO ORGAN SYSTEMS p. 5

1. The major organs of the human body are arranged into 11 organ systems. The organ systems of the human body are the *integumentary, skeletal, muscular, nervous, endocrine, cardiovascular, lymphatic, respiratory, digestive, urinary*, and *reproductive systems. (Figures 1-2, 1-3)*

HOMEOSTASIS AND SYSTEM INTEGRATION p. 12

1. **Homeostasis** is the tendency for physiological systems to stabilize internal conditions; through **homeostatic regulation** these systems adjust to preserve homeostasis.

Homeostatic Regulation p. 12

2. Homeostatic regulation usually involves a **receptor** sensitive to a particular stimulus and an **effector** whose activity affects the same stimulus.

3. **Negative feedback** is a corrective mechanism involving an action that directly opposes a variation from normal limits. *(Figure 1-4)*

4. In **positive feedback** the initial stimulus produces a response that reinforces the stimulus. *(Figure 1-5)*

Homeostasis and Disease p. 14

5. Symptoms of **disease** appear when failure of homeostatic regulation causes organ systems to malfunction.

THE LANGUAGE OF ANATOMY p. 15

Superficial Anatomy p. 15

1. Standard anatomical illustrations show the body in the **anatomical position**. If the figure is shown lying down, it can be either **supine** (face up) or **prone** (face down). *(Figure 1-6; Table 1-1)*

2. **Abdominopelvic quadrants** and **abdominopelvic regions** represent two different approaches to describing anatomical regions of the body. *(Figure 1-7)*

3. The use of special directional terms provides clarity when describing anatomical structures. *(Figure 1-8; Table 1-2)*

Sectional Anatomy p. 16

4. The three **sectional planes (frontal** or **coronal plane, sagittal plane**, and **transverse plane)** describe relationships between the parts of the three-dimensional human body. *(Figure 1-9; Table 1-3)*

5. **Body cavities** protect delicate organs and permit changes in the size and shape of visceral organs. The **dorsal body cavity** contains the **cranial cavity** (enclosing the brain) and **spinal cavity** (surrounding the spinal cord). The **ventral body cavity** surrounds developing respiratory, cardiovascular, digestive, urinary, and reproductive organs. *(Figure 1-10a)*

6. During development the **diaphragm** divides the ventral body cavity into the superior **thoracic** and inferior **peritoneal cavities**. By birth, the thoracic cavity contains

two **pleural cavities** (each containing a lung) and a **pericardial cavity** (which surrounds the heart). The peritoneal, or **abdominopelvic, cavity** consists of the **abdominal cavity** and the **pelvic cavity**. *(Figure 1-10b,c,d)*

7. Important **radiological procedures** (which can provide detailed information about internal systems) include **X-rays, CT scans, MRI**, and **ultrasound**. Each technique has its advantages and disadvantages. *(Figures 1-11, 1-12)*

CHAPTER QUESTIONS

LEVEL 1 Reviewing Facts and Terms

Match each item in column A with the most closely related item in column B. Use letters for answers in the spaces provided.

Column A	Column B
___ 1. cytology	a. study of tissues
___ 2. physiology	b. constant internal environment
___ 3. histology	c. face up
___ 4. metabolism	d. study of functions
___ 5. homeostasis	e. positive feedback
___ 6. muscle	f. system
___ 7. heart	g. study of cells
___ 8. endocrine	h. negative feedback
___ 9. temperature regulation	i. brain and spinal cord
___ 10. labor and delivery	j. all chemical activity in body
___ 11. supine	k. thoracic and abdominopelvic
___ 12. prone	l. tissue
___ 13. ventral body cavity	m. serous membrane
___ 14. dorsal body cavity	n. organ
___ 15. pericardium	o. face down

16. The process by which an organism increases the size and/or number of cells is called:
(a) reproduction (b) adaptation
(c) growth (d) metabolism
17. The terms that apply to the front of the body when in anatomical position are:
(a) posterior, dorsal (b) back, front
(c) medial, lateral (d) anterior, ventral
18. A cut through the body that passes perpendicular to the long axis of the body and divides the body into a superior and inferior section is known as a:
(a) sagittal section (b) transverse section
(c) coronal section (d) frontal section
19. The cranial and spinal cavity are found in the:
(a) ventral body cavity (b) thoracic cavity
(c) dorsal body cavity (d) abdominopelvic cavity
20. The diaphragm, a flat muscular sheet, divides the ventral body cavity into a superior __________ cavity and an inferior __________ cavity.
(a) pleural, pericardial
(b) abdominal, pelvic
(c) thoracic, abdominopelvic
(d) cranial, thoracic
21. The mediastinum is the region between the:
(a) lungs and heart (b) two pleural cavities
(c) thorax and abdomen (d) heart and pericardium

LEVEL 2 Reviewing Concepts

22. What basic functions are performed by all living things?
23. Beginning with the molecular level, list in correct sequence the levels of organization from the simplest level to the most complex level.
24. What is homeostatic regulation and what is its physiological importance?
25. How does negative feedback differ from positive feedback?
26. Describe the position of the body when it is in the anatomical position.
27. As a surgeon you perform an invasive procedure that necessitates cutting through the peritoneum. Are you more likely to be operating on the heart or on the stomach?
28. In which body cavity would each of the following organs or systems be found?
(a) brain and spinal cord
(b) cardiovascular, digestive, and urinary systems
(c) heart, lungs
(d) stomach, intestines

LEVEL 3 Critical Thinking and Clinical Applications

29. A hormone called *calcitonin* from the thyroid gland is released in response to increased levels of calcium ions in the blood. If this hormone exerts negative feedback, what effect will its release have on blood calcium levels?
30. An anatomist wishes to make detailed comparisons of medial surfaces of the left and right sides of the brain. This work requires sections that will show the entire medial surface. What kind of sections should be ordered from the lab for this investigation?

2 THE CHEMICAL LEVEL OF ORGANIZATION

Chemically, water is a simple substance: two atoms of hydrogen joined to one of oxygen. Yet when these three atoms are linked by chemical bonds, they produce a substance with many unusual properties. In this chapter you will learn how atoms are bound together to form molecules, the building blocks of living cells. You will also meet the larger molecules that form the structural framework of the body is cells and tissues, and enable our cells to grow, divide, and perform the many other functions that make life possible.

Chapter Outline and Objectives

Vocabulary Development

anabole, a building up; *anabolism*
endo-, inside; *endergonic*
exo-, outside; *exergonic*
hydro-, water
+ **lysis**, breakdown; *hydrolysis*
katabole, a throwing down; *catabolism*
katalysis, dissolution; *catalysis*
lipos, fat; *lipids*
metabole, change; *metabolism*
sakcharon, sugar
+ **mono-**, single; *monosaccharide*
+ **di-**, two; *disaccharide*
+ **poly-**, many; *polysaccharide*

Our study of the human body begins at the most basic level of organization, that of individual atoms and molecules. The characteristics of all living and nonliving things—people, elephants, oranges, oceans, rocks, and air—result from the types of atoms involved and the ways in which those atoms combine and interact. The branch of science that concerns itself with such interactions is **chemistry**. A familiarity with basic chemistry will help us to understand how the properties of atoms can affect the anatomy and physiology of the cells, tissues, organs and organ systems that make up the human body.

ATOMS AND MOLECULES

Matter is anything that occupies space and, on earth, has weight. **Atoms** are the simplest units of matter. Although physicists can split atoms apart, chemical reactions cannot change the basic identity of an atom. Atoms are so small that atomic measurements are most conveniently reported in billionths of a meter. The very largest atoms approach one-half a billionth of a meter (0.5 nanometers) in diameter. One million atoms placed end to end would be no longer than a period on this page.

Atoms contain three major types of *subatomic particles*: **protons**, **neutrons**, and **electrons**. Protons and neutrons are similar in size and mass, but protons have a *positive* electrical charge (p^+), and neutrons are *neutral*—that is, uncharged (n^0). Electrons are much lighter, only 1/1836th as massive as protons, and have a negative electrical charge (e^-). Figure 2-1● is a diagrammatic view of a simple atom of the element *helium*. This atom contains two protons, two neutrons, and two electrons.

Structure of the Atom

All atoms contain protons and electrons, normally in equal numbers. The number of protons in an atom is known as its **atomic number**, and a chemical **element** is a substance that consists entirely of atoms with the same atomic number.

Table 2-1 shows the important elements in the human body. Each element is known worldwide by its own abbreviation, or chemical symbol. Most of the symbols are easily connected with the names of the elements, but a few, such as *Na* for sodium, are abbreviations of their original Latin names, in this case *natrium*.

TABLE 2-1 Principal Elements in the Human Body

Element (% of body weight)	*Significance*
HYDROGEN (H) (9.7)	A component of water and most other compounds in the body.
OXYGEN (O) (65)	A component of water and other compounds; oxygen gas essential for respiration.
CARBON (C) (18.6)	Found in all organic molecules.
NITROGEN (N) (3.2)	Found in proteins, nucleic acids, and other organic compounds.
CALCIUM (Ca) (1.8)	Found in bones and teeth; important for membrane function, nerve impulses, muscle contraction, and blood clotting.
PHOSPHORUS (P) (1)	Found in bones and teeth, nucleic acids, and high-energy compounds.
POTASSIUM (K) (0.4)	Important for proper membrane function, nerve impulses, and muscle contraction.
SODIUM (Na) (0.2)	Important for membrane function, nerve impulses, and muscle contraction.
CHLORINE (Cl) (0.2)	Important for membrane function and water absorption.
MAGNESIUM (Mg) (0.06)	Required for activation of several enzymes.
SULFUR (S) (0.04)	Found in many proteins.
IRON (Fe) (0.007)	Essential for oxygen transport and energy capture.
IODINE (I) (0.0002)	A component of hormones of the thyroid gland.

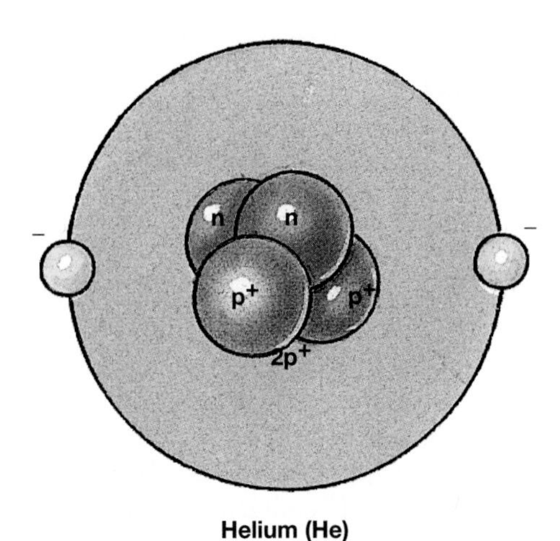

●**Figure 2-1**
Atomic Structure

An atom of helium contains two of each subatomic particle; two protons, two neutrons, and two electrons.

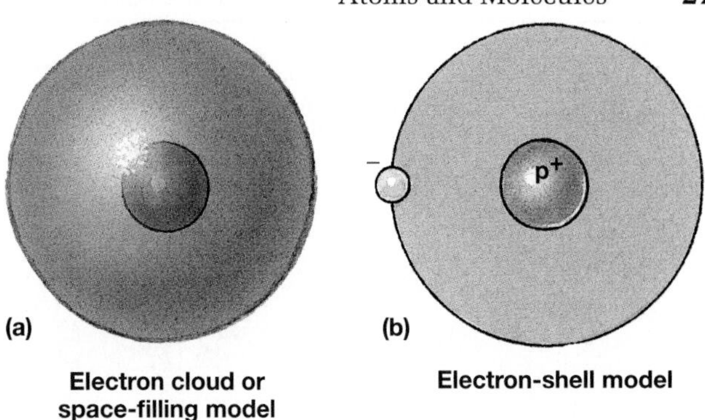

●**Figure 2-2**
Hydrogen Atoms

(a) The electron cloud of a hydrogen atom is formed by the orbiting of an electron around the nucleus. **(b)** A two-dimensional model depicting the electron in an electron shell makes it easier to visualize the components of the atom. A typical hydrogen nucleus contains a single proton and no neutrons.

Hydrogen, the simplest element, has an atomic number of 1 because its atom contains one proton. The proton is located in the center of the atom and forms the **nucleus**. A single electron whirls around the nucleus at high speed, forming an **electron cloud** (Figure 2-2a●). To simplify matters, this cloud is usually represented as a spherical **electron shell** (Figure 2-2b●).

Isotopes

Neutrons are not always present in an atom, but when they are, they are found in the nucleus together with the protons. Unlike protons, neutrons can vary in number, even among atoms of the same element. Atoms of an element whose nuclei contain different numbers of neutrons are called *isotopes*. The presence or absence of neutrons generally has very little effect on the chemical properties of an atom of a particular element. The nuclei of some isotopes may be unstable. Unstable isotopes are *radioactive*; that is, they spontaneously emit subatomic particles. These **radioisotopes** are sometimes used in diagnostic procedures. AM *The Medical Importance of Radioisotopes*

Atomic Weight

The **atomic weight** of an element is roughly equal to the number of protons and neutrons in the most common isotope. For example, hydrogen's atomic weight is approximately 1 (one proton, no neutrons), whereas the atomic weight of carbon is approximately 12 (6 protons, 6 neutrons).

Electrons and Electron Shells

Atoms are electrically neutral because every positively charged proton is balanced by a negatively charged electron. These electrons occupy an orderly series of electron shells, and only the electrons in the outer shell can interact with other atoms. *The number and arrangement of electrons in an atom's outer electron shell determine the chemical properties of that element.*

The stability of the outer electron shell depends on the number of electrons it contains. An atom with a stable outer shell will not interact with other atoms. The first electron shell is filled when it contains two electrons. A hydrogen atom has one electron in this electron shell (Figure 2-2b●), and hydrogen atoms can react with many other atoms. A helium atom has two electrons in this electron shell (Figure 2-1●). Helium is called an *inert gas* because the outer electron shell is full, and therefore stable. Helium atoms will neither react with one another nor combine with atoms of other elements.

The second electron shell can contain up to eight electrons. Carbon, with an atomic number of 6, has six electrons. In a carbon atom the first shell is filled (two electrons) and the second shell contains four electrons (Figure 2-3a●). In a neon atom (atomic number 10), the second shell is filled; neon is another inert gas (Figure 2-3b●).

Chemical Bonds and Chemical Compounds

An atom with a full outer electron shell is very stable and not reactive. Atoms with unfilled outer electron shells can achieve stability by *sharing*, *gaining*, or *losing* electrons through chemical reactions. Many of these chemical reactions produce **molecules**, chemical structures each containing more than one atom. Molecules called **compounds** contain atoms of more than one element. A compound is a new chemical substance with properties that can be quite different from those of its component elements. For example, a mixture of hydrogen and oxygen is a highly flam-

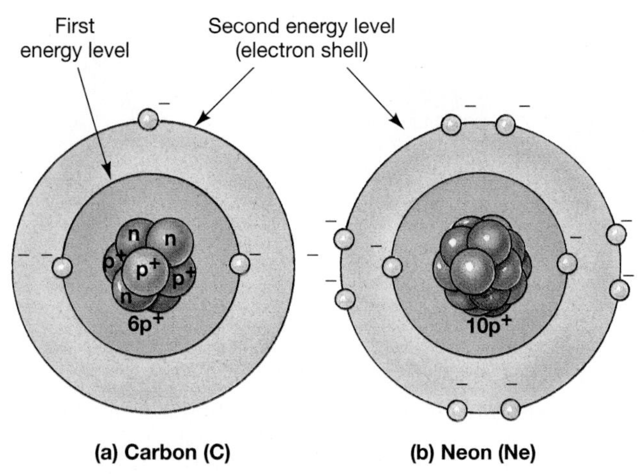

●**Figure 2-3**
Atoms and Electron Shells

(a) The first electron shell can hold only two electrons. In a carbon atom, with six protons and six electrons, the third through sixth electrons occupy the second electron shell. **(b)** The second shell can hold up to eight electrons. A neon atom has 10 protons and 10 electrons; thus both the first and second electron shells are filled. Note that helium, carbon, and neon contain neutrons as well as protons in their nuclei.

mable gas, but combining hydrogen and oxygen atoms produces a compound, water, that can put out fires.

Ionic Bonds

Atoms are electrically neutral because the number of protons (each with a +1 charge) is equal to the number of electrons (each with a –1 charge). If an atom loses an electron, it will exhibit a charge of +1 because there will be one proton without a corresponding electron; losing a second electron would leave the atom with a charge of +2. Similarly, adding one or two extra electrons to the atom will give it a respective charge of –1 or –2. Atoms or molecules that have a (+) or (–) charge are called **ions**. Ions with a positive charge are **cations** (KAT-ī-ons); those with a negative charge are **anions** (AN-ī-ons). Table 2-2 lists several important ions in body fluids.

In an **ionic** (ī-ON-ik) **bond**, anions and cations are held together by the attraction between positive and negative charges. An example of a substance held together by ionic bonds is ordinary table salt (Figure 2-4b●).

The steps in the formation of an ionic bond are illustrated in Figure 2-4a●. In this process, a sodium atom donates an electron to a chlorine atom. This loss of an electron creates a *sodium ion* with a +1 charge and a *chloride ion* with a –1 charge. The two ions do not move apart after the electron transfer because the positively charged sodium ion is attracted to the negatively charged chloride ion. The combination of oppositely charged ions forms the ionic compound *sodium chloride*, the chemical name for table salt.

Covalent Bonds

Another way atoms can complete their outer electron shells is by sharing electrons with other atoms. The result is a molecule held together by **covalent** (kō-VĀ-lent) **bonds**.

For example, individual hydrogen atoms, as diagrammed in Figure 2-2●, are not found in nature. Instead, we find hydrogen molecules (Figure 2-5a●). Molecular hydrogen is a gas present in the atmosphere in very small quantities. The two hydrogen atoms share their electrons, with each electron whirling around both nuclei. The sharing of one pair of electrons creates a **single covalent bond**.

Oxygen, with an atomic number of 8, has two electrons in its first energy level and six in the second. Oxygen atoms (Figure 2-5b●) reach stability by pooling their resources and sharing two pairs of electrons, forming a **double covalent bond**. Molecular oxygen is an atmospheric gas that is very important to living organisms; our cells would die without a constant supply of oxygen.

The chemical reactions in our bodies that consume oxygen also produce a waste product, **carbon dioxide**. The oxygen atoms in a carbon dioxide molecule form double covalent bonds with the carbon atom, as shown in Figure 2-5c●.

Covalent bonds are very strong because the electrons tie the atoms together. In most covalent bonds the atoms remain electrically neutral because the electrons are shared equally. Such equal sharing between carbon atoms creates the stable framework of the large molecules that make up most of the structural components of the human body.

Elements differ in how strongly they hold shared electrons. An unequal sharing creates a **polar covalent bond**. For example, in a molecule of water (see Figure 2-7a●, p. 33), an oxygen atom forms covalent bonds with two hydrogen atoms. The oxygen atom has a much stronger attraction for the shared electrons than do the hydrogen atoms, so their electrons spend most of their time with the oxygen atom. Because of the two extra electrons, the oxygen atom develops a slight negative charge. At the same time, the hydrogen atoms develop a slight positive charge, because their electrons are away part of the time.

TABLE 2-2 The Most Common Ions in Body Fluids

Cations	*Anions*
Na^{+} (sodium)	Cl^{-} (chloride)
K^{+} (potassium)	HCO_3^{-} (bicarbonate)
Ca^{2+} (calcium)	HPO_4^{2-} (biphosphate)
Mg^{2+} (magnesium)	SO_4^{2-} (sulfate)

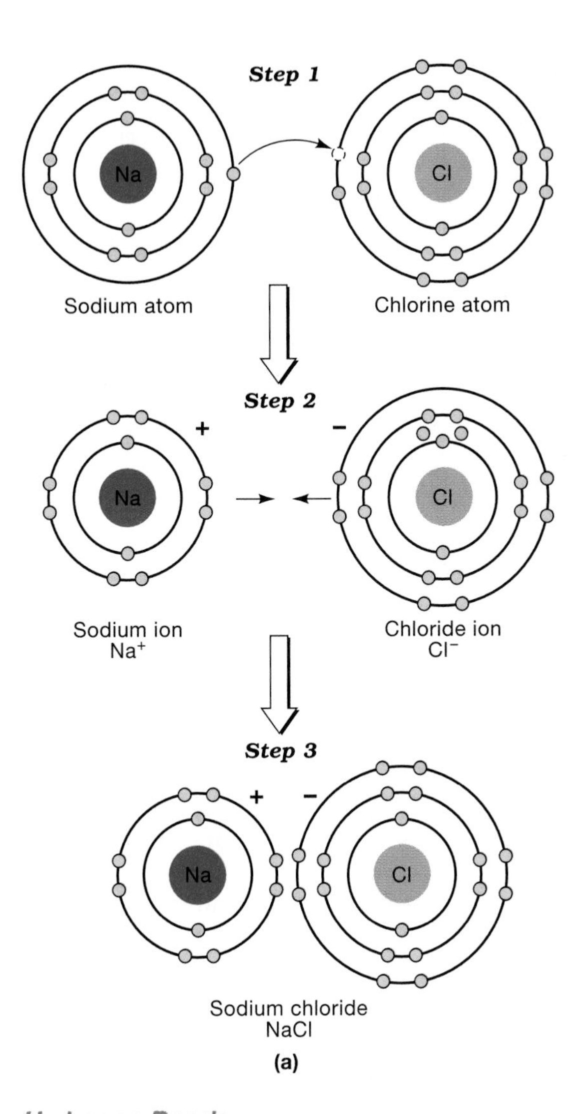

•Figure 2-4
Ionic Bonding

(a) Step 1: A sodium atom loses an electron, which is accepted by a chlorine atom. Step 2: Because the sodium (Na^+) and chloride (Cl^-) ions have opposite charges, they are attracted to one another. Step 3: The association of sodium and chloride ions forms the ionic compound sodium chloride. **(b)** Large numbers of sodium and chloride ions form a crystal of sodium chloride.

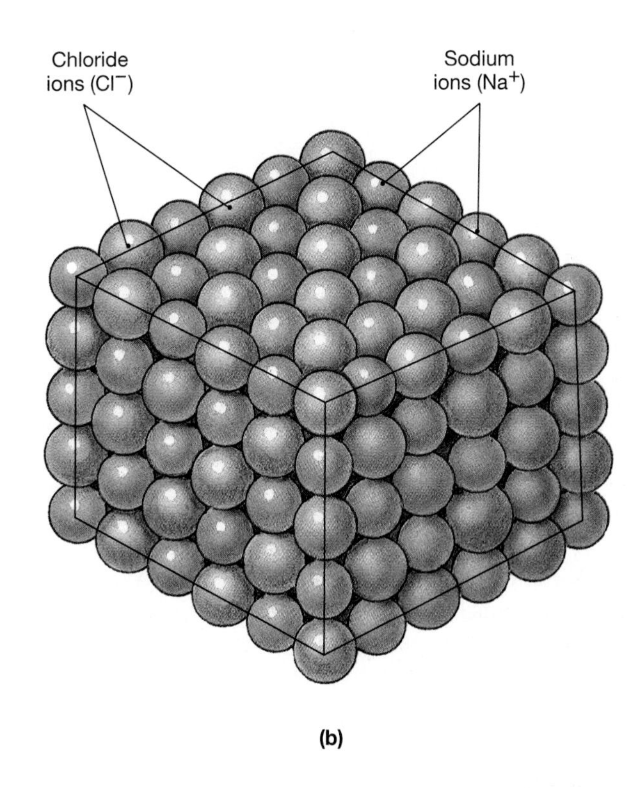

Hydrogen Bonds

In addition to ionic and covalent bonds there are weaker attractive forces that act between atoms within different parts of a large molecule as well as between adjacent molecules. *Hydrogen bonds* are the most important of these attractive forces.

Hydrogen atoms often form polar covalent bonds with the atoms of elements such as oxygen or nitrogen. In the process the hydrogen atom develops a slight positive charge and its partner develops a weak negative charge. A hydrogen bond is the attraction between such a hydrogen atom and a negatively charged atom in another molecule or at another site within the same molecule. Hydrogen bonds do not create molecules, but they can alter molecular shapes or pull different molecules together.

✓ Oxygen and neon are both gases at room temperature. Oxygen combines readily with other elements, but neon does not. Why?

✓ How is it possible for two samples of hydrogen to contain the same number of atoms but have different weights? *(cont.)*

✓ What kind of bond holds atoms in a water molecule together?

CHEMICAL NOTATION

Complex chemical compounds and reactions are most easily described with a simple form of "chemical shorthand" known as chemical notation. The rules of chemical notation are summarized in Table 2-3.

CHEMICAL REACTIONS

Living cells remain alive by controlling internal *chemical reactions.* In every **chemical reaction**, bonds between atoms are broken, and atoms are rearranged into new combinations. In effect, each cell is a chemical factory, carrying out the complex chemical reactions needed to sustain life. The term **metabolism** (meh-TAB-ō-lizm; *metabole*, change) refers to all of the chemical reactions in the body. These reactions release, store, and use energy to maintain *homeostasis* and perform essential functions.

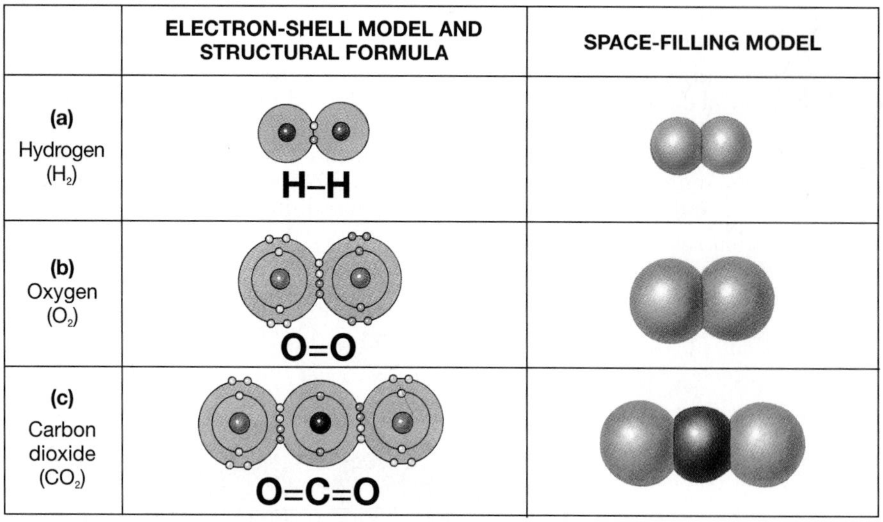

	ELECTRON-SHELL MODEL AND STRUCTURAL FORMULA	SPACE-FILLING MODEL
(a) Hydrogen (H_2)	H–H	
(b) Oxygen (O_2)	O=O	
(c) Carbon dioxide (CO_2)	O=C=O	

•Figure 2-5
Covalent Bonds

(a) In a molecule of hydrogen, two hydrogen atoms share their electrons so that each has a filled outer electron shell. This sharing creates a single covalent bond. **(b)** A molecule of oxygen consists of two oxygen atoms that share two pairs of electrons. The result is a double covalent bond. **(c)** In a molecule of carbon dioxide, a central carbon atom forms double covalent bonds with a pair of oxygen atoms.

Basic Energy Concepts

Most people are familiar with the terms *work, energy* and *heat.* **Work** is movement or a change in physical structure. **Energy** is the capacity to perform work; movement or physical change will not occur unless energy is provided. There are two major types of energy: *kinetic energy* and *potential energy.*

Kinetic energy is the energy of motion. When a car hits a tree, it is kinetic energy that does the damage. **Potential energy** is stored energy. It may result from the position of an object (as when a book sits on a high shelf) or its physical structure (as when a spring is compressed or stretched). Kinetic energy had to be used to lift the book and stretch or compress the spring. The potential energy is converted back into kinetic energy when the book falls or the spring returns to its resting length; the energy released can be used to perform work.

A conversion between potential energy and kinetic energy is not 100 percent efficient. Each time an energy exchange occurs, some of the energy produces **heat**, an increase in random molecular motion. The temperature of an object is directly related to its heat content.

Living cells perform work in many forms. For example, the contraction of a skeletal muscle is a process that requires energy and generates large amounts of heat. The energy comes from the breaking of covalent bonds, utilizing the potential energy contained in the food we eat.

Classes of Reactions

Three classes of chemical reactions are important to the study of physiology: *decomposition reactions, synthesis reactions,* and *exchange reactions.*

Decomposition

A **decomposition reaction** breaks a molecule into smaller fragments. Such reactions occur during digestion when food molecules are broken into smaller pieces. You could diagram a typical decomposition reaction as:

$$AB \rightarrow A + B$$

Catabolism (kah-TAB-o-lizm; *katabole,* a throwing down) refers to the breakdown of complex molecules within cells. A chemical bond contains potential energy that is released when that bond is broken. Our cells can capture some of that energy and use it to power essential functions such as growth, repair, movement, and reproduction.

Synthesis

Synthesis (SIN-the-sis) is the opposite of decomposition. A synthesis reaction assembles larger molecules from smaller components. These relatively simple reactions could be diagrammed as:

$$A + B \rightarrow AB$$

A and B could be individual atoms that combine to form a molecule, or they could be individual molecules combining to form larger, more complex structures. Synthesis always involves the formation of new chemical bonds, whether the reactants are atoms or molecules.

Anabolism (a-NAB-o-lizm; *anabole,* a building up) is the synthesis of new compounds in the body. Because it takes energy to create a chemical bond, anabolism usually represents an uphill struggle. Living cells are constantly balancing their chemical activities, with catabolism providing the energy needed to support synthesis as well as other vital functions.

Exchange Reactions

In an **exchange reaction** parts of the reacting molecules are shuffled around, as in:

$$AB + CD \rightarrow AD + CB$$

TABLE 2-3 Rules of Chemical Notation

1. The abbreviation of an element indicates one atom of that element:

H = an atom of hydrogen; O = an atom of oxygen

2. A number preceding the abbreviation of an element indicates more than one atom:

2 H = two individual atoms of hydrogen
2 O = two individual atoms of oxygen

3. A subscript following the abbreviation of an element indicates a molecule with that number of atoms:

H_2 = a hydrogen molecule composed of two hydrogen atoms
O_2 = an oxygen molecule composed of two oxygen atoms

4. In a description of a chemical reaction, the interacting participants are called *reactants*, and the reaction generates one or more *products*. An arrow indicates the direction of the reaction, from reactants (usually on the left) to products (usually on the right). In the reaction below, two atoms of hydrogen combine with one atom of oxygen to produce a single molecule of water.

$$2\,H + O \rightarrow H_2O$$

5. A superscript plus or minus sign following the abbreviation for an element indicates an ion. A single plus sign indicates an ion with a charge of +1 (loss of one electron). A single minus sign indicates an ion with a charge of –1 (gain of one electron). If more than one electron has been lost or gained, the charge on the ion is indicated by a number preceding the plus or minus.

Na^+ = one sodium ion (has lost 1 electron)
Cl^- = one chloride ion (has gained 1 electron)
Ca^{2+} = one calcium ion (has lost 2 electrons)

6. Chemical reactions neither create nor destroy atoms—they merely rearrange them into new combinations. Therefore, the numbers of atoms of each element must always be the same on both sides of the equation. When this is the case, the equation is *balanced*.

Unbalanced: $H_2 + O_2 \rightarrow H_2O$
Balanced: $2\,H_2 + O_2 \rightarrow 2\,H_2O$

You will notice that there are two products and two reactants. Although the reactants and products contain the same components (A, B, C, and D), they are present in different combinations. In an exchange reaction, the reactant molecules AB and CD break apart (a decomposition) before they interact with one another to form AD and CB (a synthesis). An example of such a reaction is the exchange of the components of sodium hydroxide (NaOH) and hydrochloric acid (HCl) to make table salt and water:

$$NaOH + HCl \rightarrow NaCl + H_2O$$

If breaking the old bonds releases more energy than it takes to create the new ones, the exchange reaction will release energy, usually in the form of heat. Such reactions are said to be **exergonic** (*exo-*, outside). If the energy required for synthesis exceeds the amount released by the associated decomposition reaction, additional energy (usually heat) must be provided. Such reactions are called **endergonic** (*endo-*, inside) because they absorb heat.

Reversible Reactions

Many important biological reactions are freely reversible. Such reactions can be diagrammed as:

$$A + B \leftrightarrow A\,B$$

This equation reminds you that there are really two reactions occurring simultaneously, one a synthesis (A + B → AB) and the other a decomposition (A + B A ← B). At **equilibrium** (ē-kwi-LIB-rē-um) the two rates are in balance. As fast as a molecule of AB forms, another degrades into A + B. As a result, the number of A, B, and AB molecules present at any given moment does not change. Altering the concentrations of one or more of these molecules will temporarily upset the equilibrium. For example, adding additional molecules of A and B will accelerate the synthesis reaction (A + B → AB). As the concentration of AB rises, however, so does the rate of the decomposition reaction (AB → A + B), until a new equilibrium is established.

✓ In living cells, glucose, a six-carbon molecule, is converted into two three-carbon molecules by a reaction that yields energy. How would you classify this reaction?

✓ If the product of a reversible reaction is continuously removed, what do you think the effect will be on the equilibrium?

Acids and Bases

An **acid** is a solute that dissociates to *release* hydrogen ions. (Because a hydrogen ion consists solely of a naked proton, hydrogen ions are often referred to simply as protons, and acids as "proton donors.") Hydrochloric acid (HCl) is an excellent example:

$$HCl \rightarrow H^+ + Cl^-$$

The stomach produces this powerful acid to assist in the breakdown of food.

A **base** is a solute that *removes* hydrogen ions from a solution. Many common bases are compounds that dissociate in solution to liberate a hydroxide ion (OH^-). Hydroxide ions have a strong affinity for hydrogen ions and quickly react with them, tying them up as water molecules, thereby removing them from solution. For example, sodium hydroxide (NaOH) dissociates in solution as:

$$NaOH \rightarrow Na^+ + OH^-$$

Strong bases have a variety of industrial and household uses; drain openers and lye are two familiar examples. The human body contains weak bases that are important in counteracting acids produced by cellular metabolism.

pH

The concentration of hydrogen ions in blood or other body fluids is important because hydrogen ions are extremely reactive. In excessive numbers they can disrupt cell and tissue function by breaking chemical bonds and changing the shapes of complex molecules. The concentration of hydrogen ions must therefore be regulated within relatively narrow limits.

The concentration of hydrogen ions is usually reported in terms of the **pH** of the solution. The pH value is a number between 0 (very acidic) and 14 (very basic). Pure water has a pH of 7. A solution with a pH of 7 is called *neutral* because it contains equal numbers of hydrogen and hydroxide ions. A solution with a pH below 7 is called *acidic* (a-SI-dik), because there are more hydrogen ions than hydroxide ions. A pH above 7 is called *basic*, or *alkaline* (AL-kah-lin), because hydroxide ions outumber hydrogen ions.

The pH values of some common liquids are indicated in Figure 2-6●. The pH of the blood normally ranges from 7.35 to 7.45. Variations in pH outside of this range can damage cells and disrupt normal cellular functions. For example, a blood pH below 7 can produce coma, and a blood pH higher than 7.8 usually causes uncontrollable, sustained muscular contractions.

Buffers and pH

Buffers are compounds that stabilize pH by removing or replacing hydrogen ions. Antacids such as Alka-Seltzer®, Rolaids®, and Tums® are buffers that tie up excess hydrogen ions in the stomach. The normal pH of most body fluids ranges from 7.35 to 7.45. A variety of buffers, including *sodium bicarbonate*, are responsible for regulating pH. We will consider the role of buffers and pH control in Chapter 19.

✓ What is the difference between an acid and a base?

✓ Why would an extreme change in pH of body fluids be undesirable?

✓ How does an antacid decrease stomach discomfort?

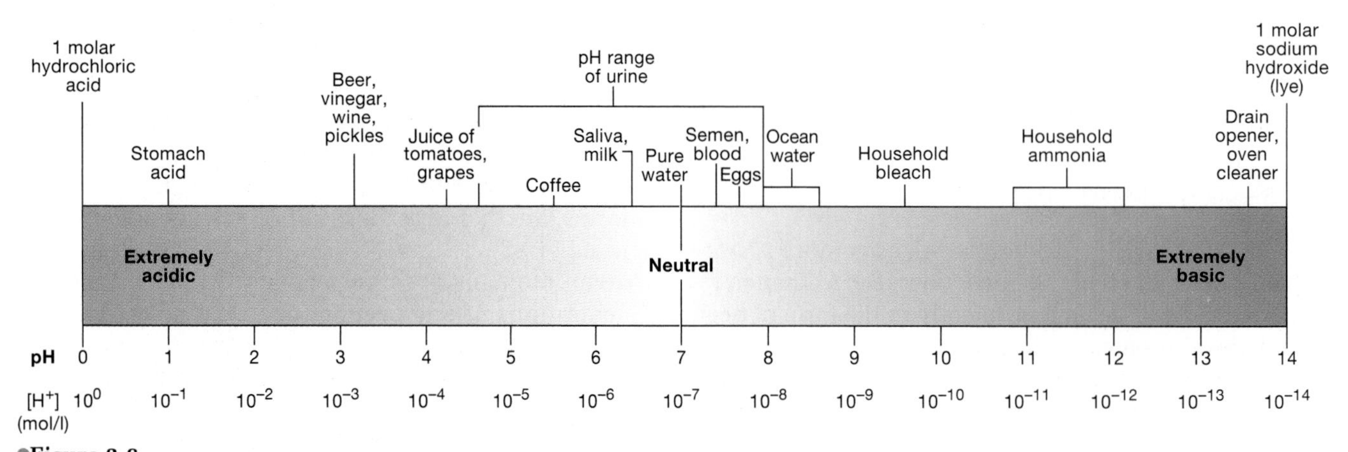

●**Figure 2-6**
pH and Hydrogen Ion Concentration
Note that an increase or decrease of one unit corresponds to a tenfold change in H^+ concentration.

The rest of this chapter focuses on *nutrients* and *metabolites* (me-TAB-o-līts). Nutrients are the essential elements and molecules absorbed from food. Metabolites include all of the molecules synthesized or broken down by chemical reactions inside our bodies. Like all chemical substances, nutrients and metabolites can be broadly categorized as *inorganic* or *organic*. Generally speaking, inorganic compounds are small molecules that do not contain carbon atoms. (The only exception is carbon dioxide, which has long been considered an inorganic compound.) Organic compounds are primarily composed of carbon atoms, and they can be much larger and more complex than inorganic compounds.

INORGANIC COMPOUNDS

The most important inorganic substances in the human body are carbon dioxide, oxygen, water, inorganic acids and bases, and salts.

Carbon Dioxide and Oxygen

Carbon dioxide (CO_2) is produced by our cells in the course of normal metabolic activities. It is transported in the blood and released into the air in the lungs. Oxygen (O_2), an atmospheric gas, is absorbed at the lungs, transported in the blood, and consumed by our cells. The chemical structures of these compounds were introduced earlier in the chapter. 1 [p. 28]

Water and Its Properties

Water, H_2O, is the single most important constituent of the body, accounting for almost two-thirds of its total weight. A drastic change in body water content can have fatal consequences because virtually all physiological systems will be affected.

Two general properties of water are particularly important to our discussion of the human body:

- Water will dissolve a remarkable number of inorganic and organic molecules, creating a *solution*. As they dissolve, the molecules break apart, releasing ions or molecules that become uniformly dispersed throughout the solution. The chemical reactions within living cells occur in solution, and the watery component, or *plasma*, of blood carries dissolved nutrients and waste products throughout the body. Most chemical reactions in the body take place in solution.
- Water has a very high *heat capacity*; it takes a lot of energy to make water boil, and a large amount of energy must be removed before water will freeze. As a result, the water in our cells remains a liquid over a wide range of environmental temperatures. Furthermore, it circulates within the body as the blood transports and redistributes heat. For example, heat absorbed as the blood flows through active muscles will be released when the blood reaches vessels in the relatively cool body surface.

Solutions

A solution consists of a fluid *solvent* and dissolved *solutes*. In biological solutions, the solvent is usually water and the solutes may be inorganic or organic. Inorganic compounds held together by ionic bonds undergo **ionization** (ī-on-i-ZĀ-shun), or *dissociation* (di-sō-sē-Ā-shun), in solution. As shown in Figure 2-7●, ionic bonds are broken apart as individual ions form hydrogen bonds with water molecules. This process produces a mixture of cations and anions so surrounded by water molecules that they are unable to re-form their original bonds.

There are many different methods used to report the concentration of inorganic and organic substances within body fluids. For an overview of these methods, see the *Applications Manual*. AM Solutions and Concentrations

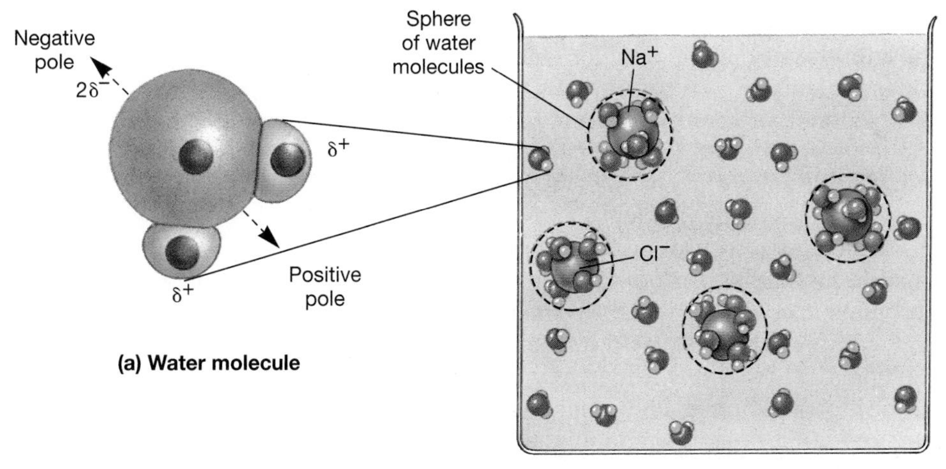

●**Figure 2-7**
Water Molecules and Solutions
(a) In a water molecule, oxygen forms polar covalent bonds with two hydrogen atoms. Because the hydrogen atoms are positioned toward one end of the molecule, the molecule has an uneven distribution of charges that creates positive and negative poles. **(b)** Ionic compounds dissociate in water as the polar water molecules disrupt the ionic bonds. The ions remain in solution because the surrounding water molecules prevent the ionic bonds from re-forming.

Inorganic Acids and Bases

The discussion of acids and bases on p. 32 introduced an inorganic acid, hydrochloric acid (HCl), and an inorganic base, sodium hydroxide (NaOH). There are several other inorganic acids found in body fluids. Examples include *carbonic acid*, *sulfuric acid*, and *phosphoric acid*. These acids, produced during normal metabolism, will be considered further in Chapters 16 and 19.

Salts

A **salt** is an inorganic compound that consists of a cation that is not a hydrogen ion and an anion that is not a hydroxyl ion. Salts are held together by ionic bonds, and in water they dissociate, releasing cations and anions. For example, table salt (NaCl) in solution dissociates into Na^+ and Cl^- ions; these are the most abundant ions in body fluids.

Salts are examples of **electrolytes** (ē-LEK-trō-līts), compounds whose ions will conduct an electrical current in solution. For example, sodium ions (Na^+), potassium ions (K^+), calcium ions (Ca^{2+}), and chloride ions (Cl^-) are released by the dissociation of electrolytes in blood and other body fluids. Alterations in the body fluid concentrations of these ions will disturb almost every vital function. For example, declining potassium levels will lead to a general muscular paralysis, and rising concentrations will cause weak and irregular heartbeats.

ORGANIC COMPOUNDS

Organic compounds contain the elements carbon and hydrogen, and usually oxygen as well. Organic molecules often contain long chains of carbon atoms linked by covalent bonds. These carbon atoms often form additional covalent bonds with hydrogen or oxygen atoms, and less often, with nitrogen, phosphorus, sulfur, iron, or other elements to produce the complex organic molecules characteristic of living organisms.

Although inorganic acids and bases were used as examples earlier in the chapter, there are also important organic acids and bases. For example, *lactic acid* is an organic acid generated by active muscle tissues.

This section focuses on four major classes of large organic molecules: *carbohydrates*, *lipids*, *proteins*, and *nucleic acids*. We will also consider the high-energy compounds that play a crucial role in many of the chemical reactions under way within our cells. In addition, the human body contains small quantities of many other organic compounds whose structures and functions will be considered in later chapters.

Carbohydrates

A **carbohydrate** (kar-bō-HĪ-drāt) molecule consists of carbon, hydrogen, and oxygen in a ratio near 1:2:1. Familiar carbohydrates include the sugars and starches that make up roughly half of the typical American diet. Our tissues can break down most carbohydrates, and, although they sometimes have other functions, carbohydrates are most important as sources of energy. Despite their importance as an energy source, however, carbohydrates account for less than 3 percent of the total body weight. There are three major types of carbohydrates: *monosaccharides*, *disaccharides*, and *polysaccharides*.

Monosaccharides

A **simple sugar**, or **monosaccharide** (mon-ō-SAK-ah-rīd; *mono-*, single + *sakcharon*, sugar), is a carbohydrate containing from three to seven carbon atoms. Included within this group is **glucose** (GLOO-kōs), $C_6H_{12}O_6$ (Figure 2-8●), the most important metabolic "fuel" in the body.

Disaccharides and Polysaccharides

Carbohydrates other than simple sugars are complex molecules composed of monosaccharide building blocks. Through a **dehydration synthesis** reaction that removes water, two simple sugars join together to form a **disaccharide** (dī-SAK-ah-rīd; *di-*, two) (Figure 2-9a●). Disaccharides such as **sucrose** (table sugar) have a sugary taste, and they are also quite soluble. Many foods contain disaccharides, but they must be disassembled before they can be broken down to provide useful energy. The breakdown of a disaccharide into its component simple sugars is an example of **hydrolysis** (hī-DROL-i-sis; *hydro-*, water + *lysis*; breakdown) (Figure 2-9b●). Most popular "junk foods," such as candies and sodas, abound in simple sugars (often fructose) and disaccharides such as sucrose.

Larger carbohydrate molecules are called **polysaccharides** (pol-ē-SAK-ah-rīdz; *poly-*, many). *Starches* are glucose-based polysaccharides important in our diets.

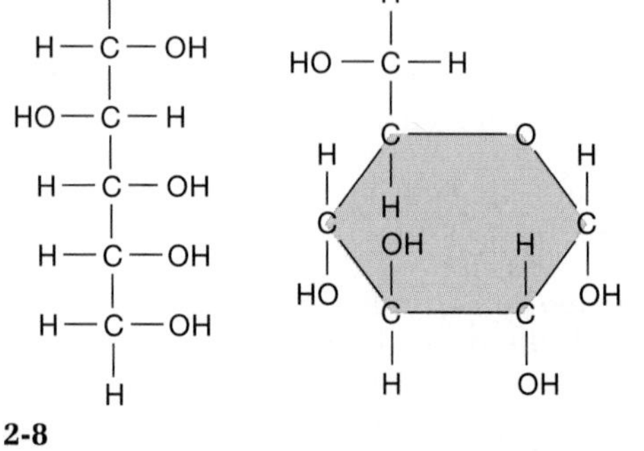

●Figure 2-8
Glucose

(a) The straight-chain structural formula. **(b)** The ring form that is most common in nature.

(a) During dehydration synthesis two molecules are joined by the removal of a water molecule.

(b) Hydrolysis reverses the steps of dehydration synthesis; a complex molecule is broken down by the addition of a water molecule.

●Figure 2-9
The Formation and Breakdown of Complex Sugars and Glycogen

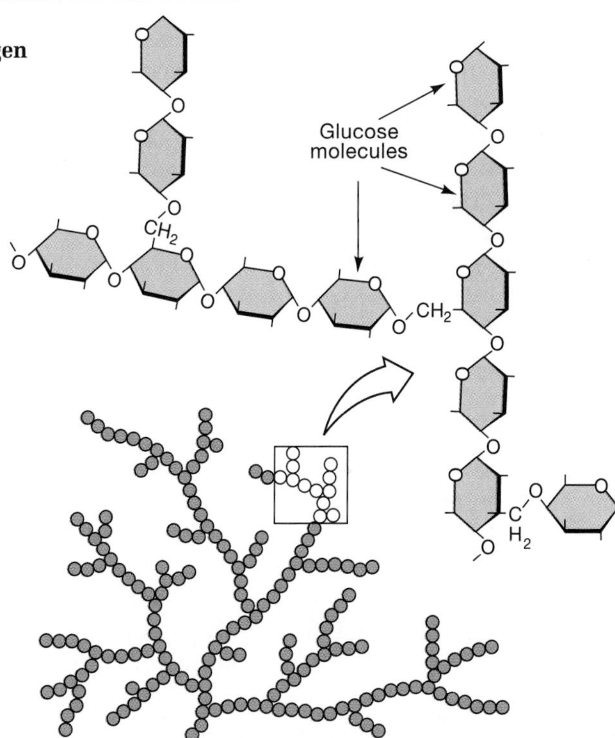

(c) Glycogen, a branching chain of glucose molecules, is stored in muscle cells and liver cells.

Most of the starches in our diet are synthesized by plants. The human digestive tract can break these molecules into simple sugars, and starches such as those found in potatoes and grains are important energy sources.

Glycogen (GLĪ-ko-jen), or *animal starch*, is a branched polysaccharide composed of interconnected glucose molecules (Figure 2-9c●). Like most other large polysaccharides, glycogen will not dissolve in water or other body fluids. Liver and muscle tissues manufacture and store significant amounts of glycogen. When these tissues have a high demand for energy, glycogen molecules are broken down into glucose; when demands are low, the tissues absorb or synthesize glucose and rebuild glycogen reserves.

Table 2-4 summarizes information concerning the carbohydrates.

Some people cannot tolerate sugar for medical reasons; others avoid it because they do not want to gain weight (excess sugars are stored as fat). Many of these people use *artificial sweeteners* in their foods and beverages. These compounds have a very sweet

TABLE 2-4 Carbohydrates in the Body

Structure	*Examples*	*Primary Functions*	*Remarks*
MONOSACCHARIDES (SIMPLE SUGARS)	Glucose, fructose	Energy source	Manufactured in the body and obtained from food; found in body fluids.
DISACCHARIDES	Sucrose, lactose, maltose	Energy source	Sucrose is table sugar, lactose is present in milk; all must be broken down to monosaccharides before absorption.
POLYSACCHARIDES	Glycogen	Storage of glucose molecules	Glycogen is found in animal cells, other starches and cellulose in plant cells.

taste, but they either cannot be broken down in the body or they are used in such small amounts that their breakdown does not contribute to the overall energy balance of the body. AM *Artificial Sweeteners*

Lipids

Lipids (*lipos*, fat) also contain carbon, hydrogen, and oxygen, but because they have relatively less oxygen then carbohydrates, the ratios do not approximate 1:2:1. In addition, lipids may contain small quantities of other elements, such as phosphorus, nitrogen, or sulfur. Familiar lipids include fats, oils, and waxes. Most lipids are insoluble in water, but special transport mechanisms carry them in the circulating blood.

Lipids form essential structural components of all cells. Lipid deposits also serve as energy sources and reserves. On average, lipids provide roughly twice as much energy as carbohydrates, gram for gram, when broken down in the body. For this reason there has been great interest in developing **fat substitutes** that have the taste and texture of lipids, but without the calories. AM *Fat Substitutes*

All together, lipids normally account for roughly 12 percent of our total body weight. There are many different kinds of lipids in the body. The major types, presented in Table 2-5, are *fatty acids*, *glycerides*, *steroids*, and *phospholipids*.

Fatty Acids

Fatty acids are long carbon chains with hydrogen atoms attached. When placed in solution, only the carboxylic acid end (–COOH) of a fatty acid associates with water molecules. The rest of the carbon chain is relatively insoluble.

In a **saturated** fatty acid each carbon atom has four single covalent bonds, allowing it to bond to a maximum number of hydrogen atoms. If any of the carbon-to-carbon bonds are double covalent bonds, then fewer hydrogen atoms are present and the fatty acid is **unsaturated**. Saturated and unsaturated fatty acids are shown in Figure 2-10●. A **polyunsaturated** fatty acid contains multiple unsaturated bonds.

Both saturated and unsaturated fatty acids can be broken down for energy, but a diet containing large amounts of saturated fatty acids increases the risk of

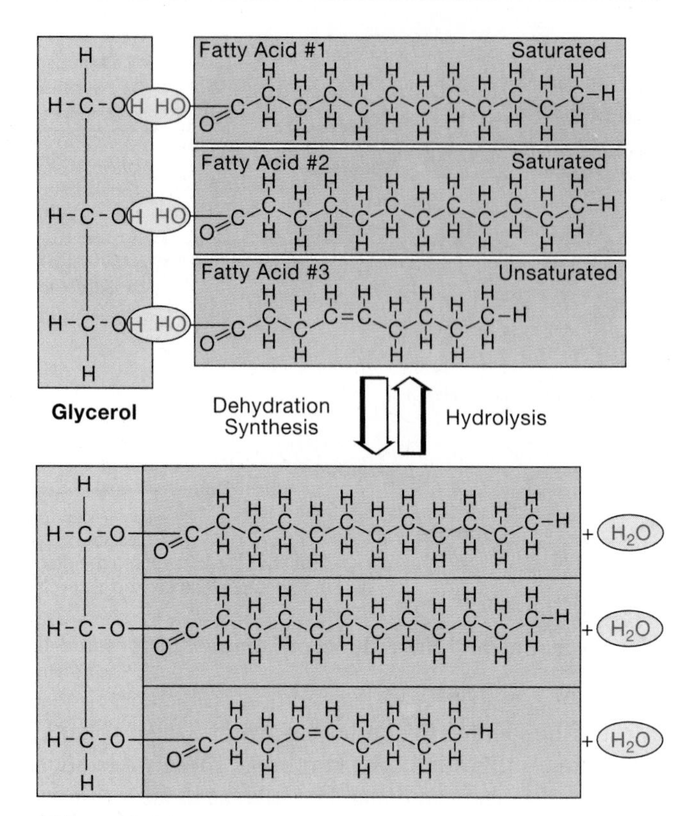

●**Figure 2-10**
Triglyceride Formation
The formation of a triglyceride involves the attachment of three fatty acids to the carbons of a glycerol molecule. This example shows the attachment of one unsaturated and two saturated fatty acids to a glycerol molecule.

TABLE 2-5 Representative Lipids and Their Functions in the Body

Lipid Type	*Examples*	*Primary Functions*	*Remarks*
FATTY ACIDS	Lauric acid	Energy sources	Absorbed from food or synthesized in cells; transported in the blood for use in many tissues.
GLYCERIDES	Monoglycerides, diglycerides, triglycerides	Energy source, energy storage, insulation, and physical protection	Stored in fat deposits; must be broken down to fatty acids and glycerol before they can be used as an energy source.
STEROIDS	Cholesterol	Structural component of cell membranes, hormones, digestive secretions in bile.	All have the same carbon-ring framework.
PHOSPHOLIPIDS		Structural components of cell membranes	

heart disease and other circulatory problems. Butter, fatty meat, and ice cream are popular dietary sources of saturated fatty acids. Vegetable oils such as olive oil or corn oil contain a mixture of unsaturated fatty acids.

Glycerides

Individual fatty acids cannot be strung together in a chain by dehydration synthesis, as simple sugars can. But they can be attached to another compound, **glycerol** (GLI-se-rol), through a similar reaction. In a **triglyceride** (trī-GLI-se-rīd), a glycerol molecule is attached to three fatty acids (see Figure 2-10●). Triglycerides, otherwise known as **neutral fats**, are the most common fats in the body. In addition to serving as an energy reserve, fat deposits under the skin serve as insulation, and a mass of fat around a delicate organ, such as a kidney, provides a protective cushion. *Saturated fats*, triglycerides containing saturated fatty acids, are usually solid at room temperature. *Unsaturated fats*, triglycerides containing unsaturated fatty acids, are usually liquid at room temperature.

Steroids

Steroids are large lipid molecules composed of connected rings of carbon atoms, quite unlike the linear carbon chains of fatty acids. *Cholesterol* (Figure 2-11●) is probably the best-known steroid. All of our cells are surrounded by *cell membranes* that contain cholesterol, and some chemical messengers, or *hormones*, are derived from cholesterol. Examples include the sex hormones *testosterone* and *estrogen.*

The cholesterol needed to maintain cell membranes and manufacture steroid hormones comes from two sources. One source is the diet; meat, cream, and egg yolks are especially rich in cholesterol. The second is the body itself, for the liver can synthesize large amounts of cholesterol. The ability of the body to synthesize this steroid can make it difficult to control blood cholesterol levels by dietary restriction alone. This difficulty can have serious repercussions because a strong link exists between high blood cholesterol concentrations and heart disease. Current nutritional advice suggests reducing cholesterol intake to under 300 mg per day; this amount represents a 40 percent reduction for the average American adult. The connection between blood cholesterol levels and heart disease will be examined more closely in Chapters 12 and 13.

Phospholipids

Phospholipids (FOS-fō-lip-ids) consist of a diglyceride attached to a molecule containing a phosphate group (PO_4^{3-}). The nonlipid portion is soluble in water, whereas the fatty acid portion is relatively insoluble. Phospholipids are the most abundant lipid components of cell membranes.

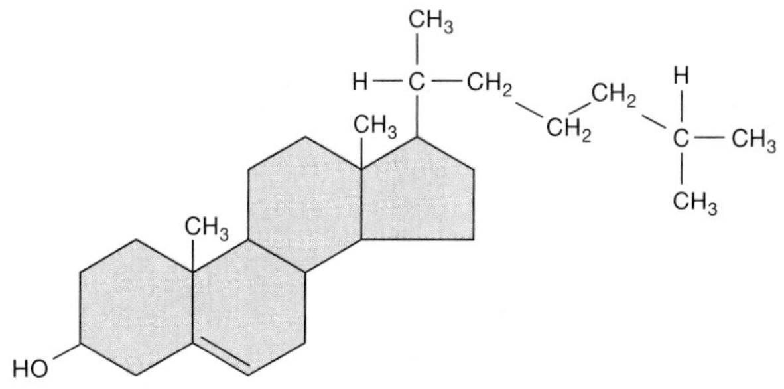

●**Figure 2-11**
Cholesterol Molecule

Cholesterol, like all steroids, contains a complex four-ring structure.

Proteins

Proteins are the most abundant and diverse organic components of the human body. There are roughly 100,000 different kinds of proteins, and they account for about 20 percent of the total body weight. All proteins contain carbon, hydrogen, oxygen, and nitrogen; smaller quantities of sulfur may also be present.

Protein Function

Proteins perform a variety of essential functions. These fall into seven major categories:

1. *Support.* **Structural proteins** create a three-dimensional supporting framework for the body, providing strength, organization, and support for cells, tissues, and organs.
2. *Movement.* **Contractile proteins** are responsible for muscular contraction; related proteins are responsible for the movement of individual cells.
3. *Transport.* Insoluble lipids, respiratory gases, special minerals, such as iron, and several hormones are carried in the blood attached to special **transport proteins.** Other specialized proteins transport materials from one part of a cell to another.
4. *Buffering.* Proteins provide a considerable buffering action, helping to restrict alterations in pH.
5. *Metabolic regulation.* **Enzymes** accelerate chemical reactions in living cells. The sensitivity of enzymes to environmental factors is extremely important in controlling the pace and direction of metabolic operations.
6. *Coordination, communication, and control.* Protein hormones can influence the metabolic activities of every cell in the body or affect the function of specific organs or organ systems.
7. *Defense.* The tough, waterproof proteins of the skin, hair, and nails protect the body from environmental hazards. In addition, proteins known as **antibodies** protect us from disease, and special

clotting proteins restrict bleeding following an injury to the circulatory system.

Protein Structure

Proteins are chains of small organic molecules called **amino acids**. The human body contains significant quantities of 20 different amino acids (Figure 2-12a•). A typical protein contains 1000 amino acids, but the largest protein complexes may have 100,000 or more. The individual amino acids are strung together like beads on a string, with the carboxylic acid group of one amino acid attached to the amino group of another. This connection is called a **peptide bond**. If a molecule consists of two amino acids, it is called a *dipeptide* (Figure 2-12b•). The chain can be lengthened by the addition of more amino acids. **Polypeptides** are long chains of amino acids. Proteins are polypeptide chains containing over 100 amino acids.

The shape of a short peptide chain primarily depends on interactions between amino acids at different sites along the peptide chain (Figure 2-13a•). Large proteins can have complex three-dimensional shapes. In a *globular protein*, the peptide chain folds back on itself, creating a rounded mass such as *myoglobin* (Figure 2-13b•). Myoglobin is a protein found in muscle cells. Complex proteins may consist of several protein subunits. Examples include *hemoglobin*, a globular protein found inside red blood cells (Figure 2-13c•), and *keratin* (Figure 2-13d•), the tough, water-resistant protein found in skin, nails, and hair. Keratin is an example of a *fibrous protein*. In fibrous proteins the polypeptide strands are wound together as in a rope. Fibrous proteins are flexible but very strong.

The shape of a protein determines its functional properties, and the primary determinant of shape is the sequence of amino acids. Combining the 20 amino acids in various combinations creates an almost limitless variety of proteins. Small differences can have large effects; changing one amino acid in a protein containing 10,000 or more may make it incapable of performing its normal function.

The shape of a protein—and thus its function—can be altered by small changes in the ionic composition, temperature, or pH of its surroundings. For example, very high body temperatures (over 43°C, or 110°F) cause death because at these temperatures proteins undergo **denaturation**, a change in their three-dimensional shape. Denatured proteins are nonfunctional, and the loss of structural proteins and enzymes causes irreparable damage to organs and organ systems. You see denaturation in progress each time you fry an egg, for the clear egg white contains abundant dissolved proteins. As the temperature rises, the protein structure changes and eventually the egg proteins form an insoluble white mass.

Enzymes and Chemical Reactions

Most chemical reactions do not occur spontaneously, because they require energy to *activate* the reactant molecules before a reaction can begin. **Activation energy** is the amount of energy required to start a reaction. Figure 2-14a• diagrams the activation energy needed for a typical reaction. Although many reactions can be activated by changes in temperature or pH, such changes are deadly to cells. For example, to break down a complex sugar in the laboratory you must boil it in an acid solution. Cells, however, avoid such harsh requirements by using special proteins called *enzymes* to speed up the reactions that support life. Enzymes belong to a class of substances called

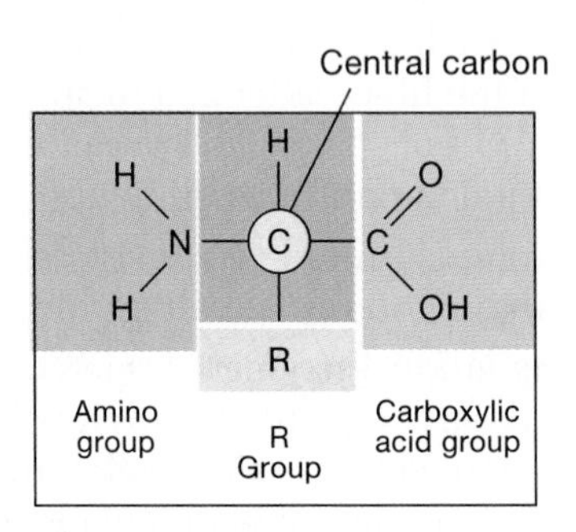

(a) Structure of an amino acid

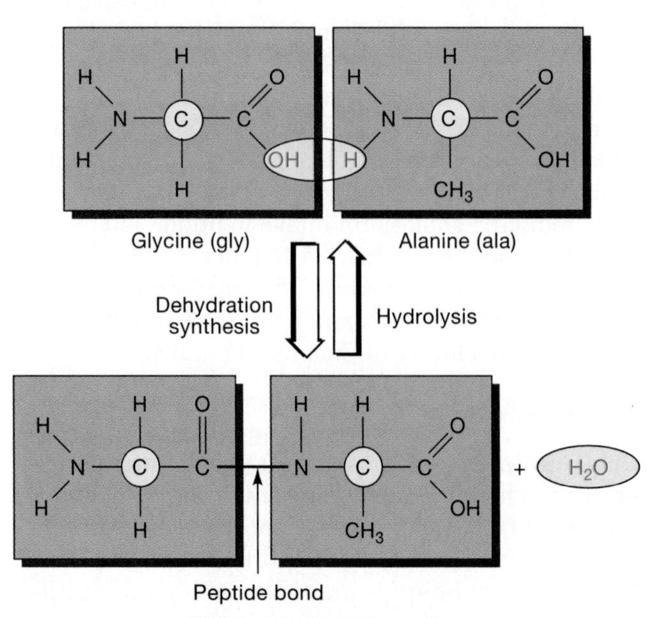

(b) Peptide bond formation

•**Figure 2-12**
Amino Acids and Peptide Bonds

(a) Each amino acid consists of a central carbon atom to which four different groups are attached: a hydrogen atom, an amino group ($-NH_2$), a carboxylic acid group ($-COOH$), and a variable group generally designated as R. **(b)** Peptides form as a dehydration synthesis creates a peptide bond between the carboxyl group of one amino acid and the amino group of another. In this example glycine and alanine are linked to form a dipeptide.

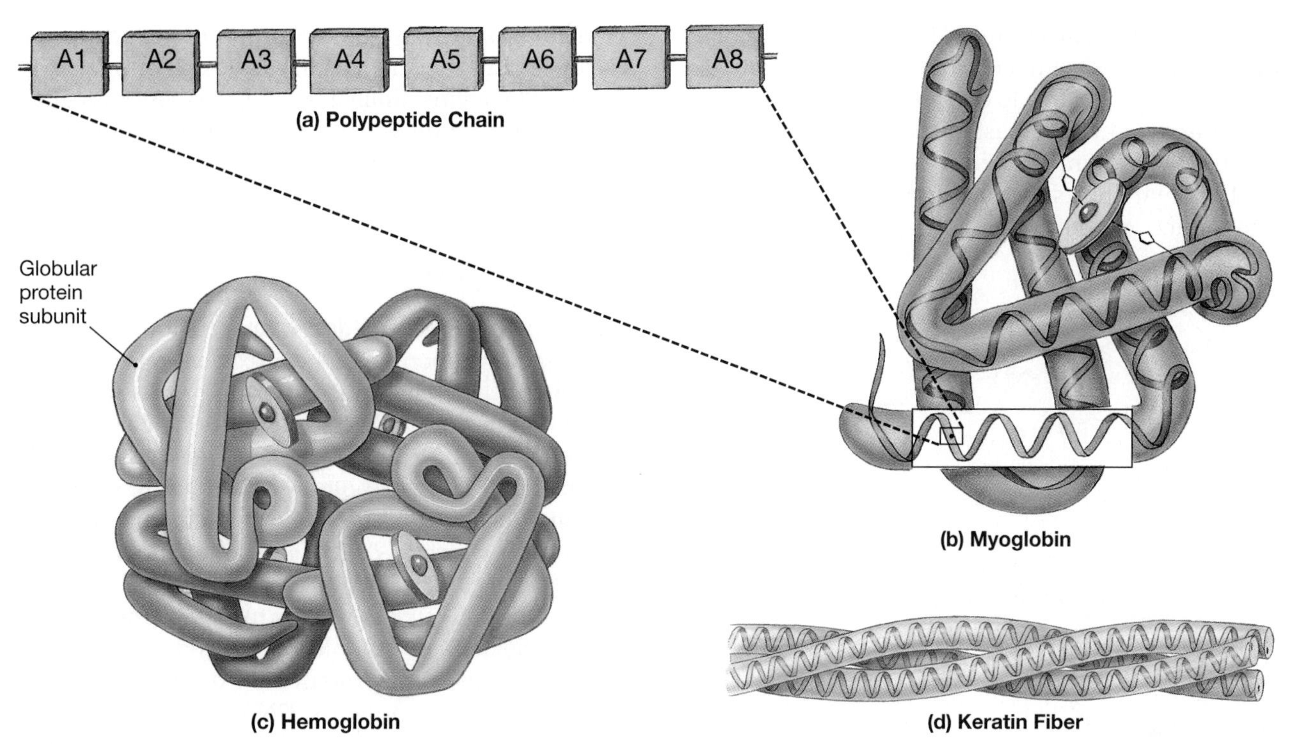

Figure 2-13
Protein Structure

(a) The shape of a polypeptide is determined by its sequence of amino acids. **(b)** Attraction between R groups plays a large role in forming globular proteins. This is myoglobin, a globular protein involved in the storage of oxygen in muscle tissue. **(c)** A single hemoglobin molecule contains four globular subunits, each structurally similar to myoglobin. Hemoglobin transports oxygen in the blood. **(d)** In keratin, three fibrous subunits intertwine like the strands of a rope.

catalysts (KAT-ah-lists; *katalysis*, dissolution): compounds that accelerate chemical reactions without themselves being permanently changed. Living cells create specific enzymes to promote vital reactions.

Figure 2-14b● diagrams the effect of an enzyme on the activation energy of a typical reaction. Lowering the activation energy affects only the *rate* of a reaction, not the direction of the reaction or the products that will be formed. An enzyme cannot bring about a reaction that would otherwise be impossible.

Figure 2-15● shows a simple model of enzyme function. The reactants in an enzymatic reaction, called **substrates**, interact to form a specific **product**. Substrate molecules bind to the enzyme at a particular location called the **active site**. This binding depends on the complementary shapes of the two molecules, much as a key fits in a lock. The shape of the active site is determined

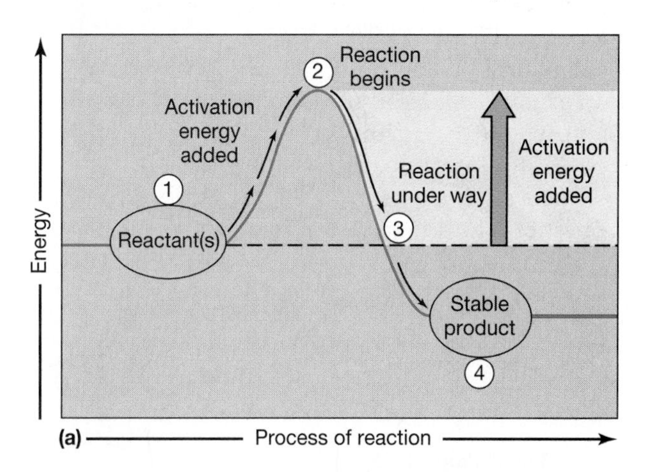

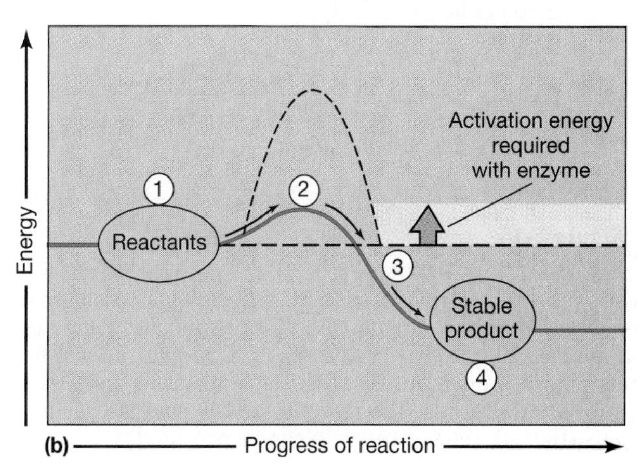

Figure 2-14
Activation Energy and Enzyme Function

(a) Before a reaction can begin, considerable activation energy must be provided. In this diagram the activation energy represents the energy required to proceed from point 1 to point 2. **(b)** The activation energy requirement of the reaction is much lower in the presence of an appropriate enzyme. The lower energy requirement allows the reaction to take place much more rapidly, without the need for extreme conditions that would harm cells.

by the three-dimensional shape of the enzyme molecule. Once the reaction is completed and the products are released, the enzyme is free to catalyze another reaction.

Each enzyme works best at an optimal temperature and pH. As temperatures rise or pH shifts outside of normal limits, proteins change shape, and enzyme function deteriorates.

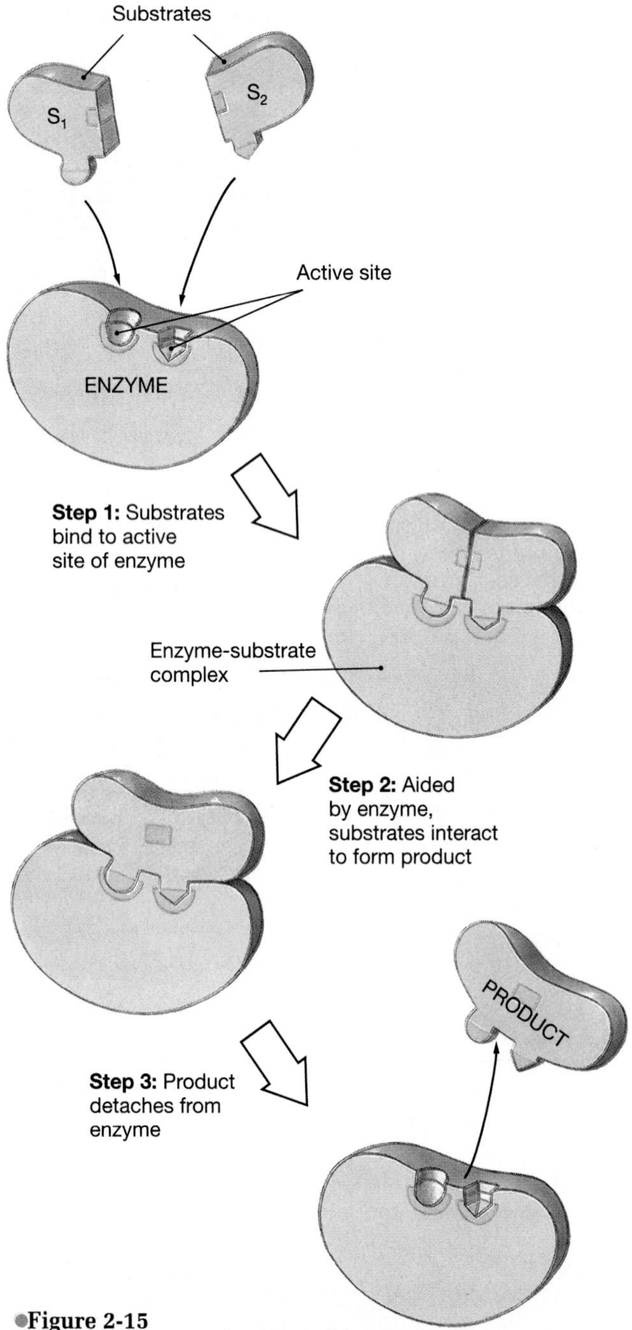

•**Figure 2-15**
Enzyme Structure and Function
Each enzyme contains a specific active site somwehere on its exposed surface. In step 1 a pair of substrate molecules (S_1 and S_2) bind to the active site. In step 2 the substrates interact, forming a product that detaches from the active site (step 3). Because the structure of the enzyme has not been affected, the entire cycle can be repeated.

The complex reactions that support life proceed in a series of interlocking steps, each step controlled by a different enzyme. Such a reaction sequence is called a *metabolic pathway.* We will consider important metabolic pathways in later chapters. AM *Anomalies in Amino Acid Metabolism*

Nucleic Acids

Nucleic (noo-KLĀ-ik) **acids** are large organic molecules composed of carbon, hydrogen, oxygen, nitrogen, and phosphorus. Nucleic acids store and process information at the molecular level inside living cells. There are two classes of nucleic acid molecules, **deoxyribonucleic** (dē-ok-se-rī-bō-noo-KLĀ-ik) **acid**, or **DNA**, and **ribonucleic** (rī-bō-noo-KLĀ-ik) **acid**, or **RNA**.

The DNA in our cells determines our inherited characteristics, such as eye color, hair color, blood type, and so on. It affects all aspects of body structure and function because DNA molecules encode the information needed to build proteins. By directing the synthesis of structural proteins, DNA controls the shape and physical characteristics of our bodies. By controlling the manufacture of enzymes, DNA regulates not only protein synthesis but all aspects of cellular metabolism, including the creation and destruction of lipids, carbohydrates, and other vital molecules.

Several forms of RNA cooperate to manufacture specific proteins using the information provided by DNA. The functional relationships between DNA and RNA will be detailed in Chapter 3.

Structure of Nucleic Acids

A nucleic acid consists of a series of **nucleotides**. A single nucleotide has three basic components: a sugar, a **phosphate group** (PO_4^{3-}), and a **nitrogenous base** (Figure 2-16a•). The sugar is always a five-carbon sugar, either **ribose** (in RNA) or **deoxyribose** (in DNA). There are five nitrogenous bases: **adenine** (A), **guanine** (G), **cytosine** (C), **thymine** (T), and **uracil** (U). Both RNA and DNA contain adenine, guanine, and cytosine. Uracil is found only in RNA, and thymine only in DNA.

Important structural differences between RNA and DNA are listed in Table 2-6. A molecule of RNA consists of a single chain of nucleotides (Figure 2-16a•). A DNA molecule consists of a *pair* of nucleotide chains (Figure 2-16b•) held together by weak bonds between the opposing nitrogen bases. Because of their shapes, adenine can bond only with thymine, and cytosine only with guanine. As a result, adenine-thymine and cytosine-guanine are known as **complementary base pairs**.

The two strands of DNA twist around one another in a **double helix** that resembles a spiral staircase, with the stair steps corresponding to the nitrogen base pairs. Figure 2-16c• presents this three-dimensional view of a DNA molecule.

Figure 2-16 Nucleic Acids: RNA and DNA

Nucleic acids are long chains of nucleotides. Each molecule starts at the sugar-nitrogenous base of the first nucleotide and ends at the phosphate group of the last member of the chain. An RNA molecule **(a)** consists of a single nucleotide chain. Its shape is determined by the sequence of nucleotides and the interactions between them. A DNA molecule **(b)** consists of a pair of nucleotide chains linked by hydrogen bonding between complementary base pairs. **(c)** A three-dimensional model of a DNA molecule shows the double helix formed by the two DNA strands.

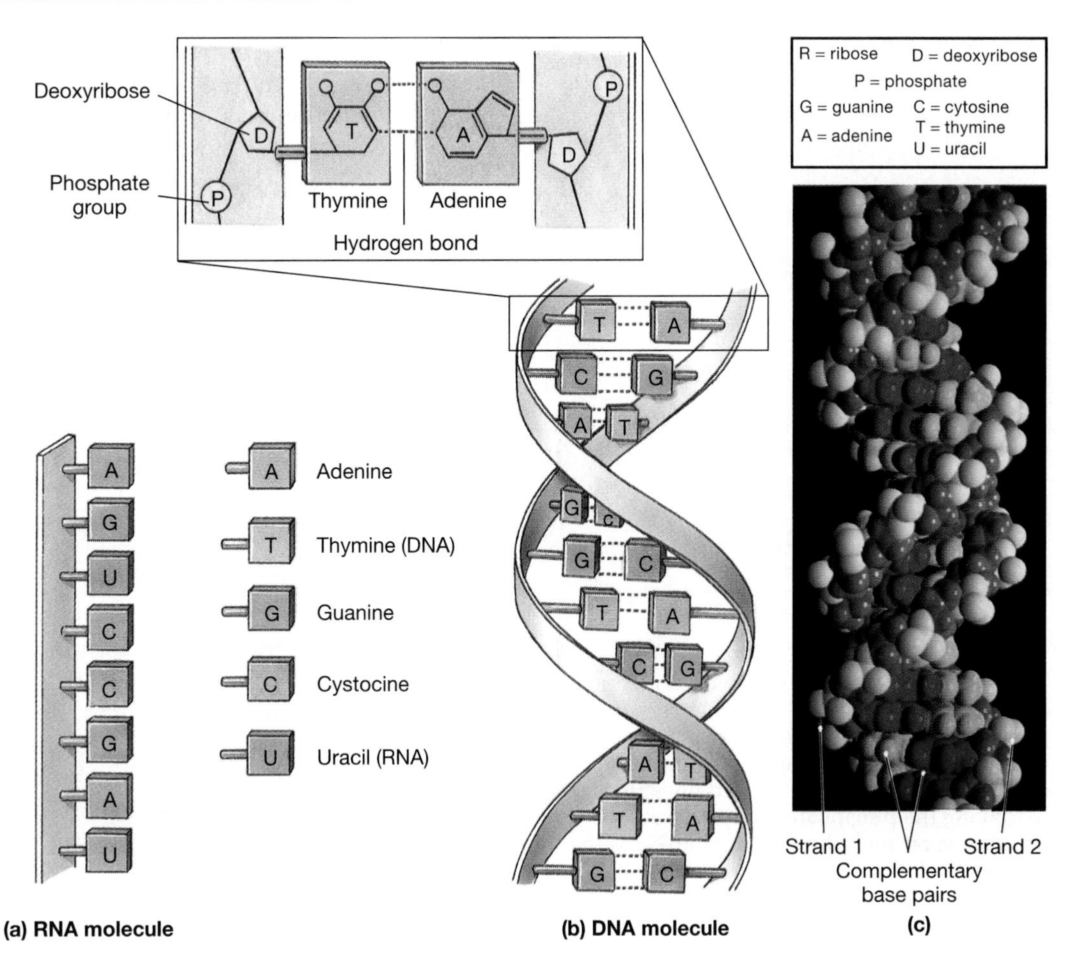

High-Energy Compounds

Catabolism releases energy, and living cells can use that energy in constructive ways. Part of the energy released by catabolic reactions is captured in the creation of **high-energy bonds**. A high-energy bond is a covalent bond that stores an unusually large amount of energy (as would a tightly wound rubber band attached to the propeller of a model airplane). When that bond is later broken, perhaps in a distant portion of the cell, the energy will be released under controlled conditions (as when the propeller is turned by the unwinding rubber band). In our cells, a high-energy bond usually connects a phosphate group (PO_4^{3-}) to an organic molecule, resulting in a **high-energy compound**. One of the most important high-energy compounds in the body is *adenosine triphosphate*, or **ATP**.

TABLE 2-6 Comparison of RNA and DNA

Characteristic	*RNA*	*DNA*
SUGAR	Ribose	Deoxyribose
NITROGENOUS BASES	Adenine Guanine Cytosine Uracil	Adenine Guanine Cytosine Thymine
NUMBER OF NUCLEOTIDES IN TYPICAL MOLECULE	Varies from under 100 nucleotides to around 50,000	Always over 45 million nucleotides
SHAPE OF MOLECULE	Varies, depending on hydrogen bonding along the length of the strand; 3 main types (mRNA, rRNA, tRNA)	Paired strands coiled in a double helix
FUNCTION	Performs protein synthesis as directed by DNA	Stores genetic information that controls protein synthesis by RNA

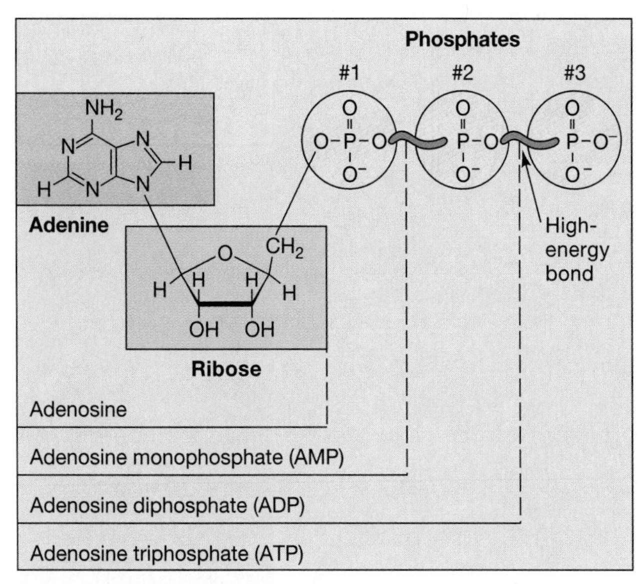

•Figure 2-17
The Structure of ATP

The ATP molecule of an adenosine (adenine and sugar) molecule to which three phosphate groups have been joined. Both the second and third phosphates are bound to the molecule by high-energy bonds. Cells most often store energy by attaching a third phosphate group to ADP. Removing the phosphate group releases the energy for cellular work, including the synthesis of other molecules.

Figure 2-17• details the structure of ATP. Within our cells the conversion of ADP to ATP represents the primary method of energy storage, and the reverse reaction provides a mechanism for controlled energy release. The arrangement can be summarized as:

$$\text{ATP} \leftrightarrow \text{ADP} + \text{phosphate group} + \text{energy}$$

When energy sources are available, our cells make ATP from ADP; when energy is required, the reverse reaction occurs.

CHEMICALS AND LIVING CELLS

Figure 2-18• and Table 2-7 review the major chemical components we have discussed in this chapter. But the human body is more than a collection of chemicals. Biochemical building blocks form functional units called **cells**. Each cell behaves like a miniature organism, responding to internal and external stimuli. A lipid membrane separates the cell from its environment, and internal membranes create compartments with specific functions. Proteins form an internal supporting framework and act as enzymes to accelerate and control the chemical reactions that maintain homeostasis. Nucleic acids direct the synthesis of all cellular proteins, including the enzymes that enable the cell to synthesize a wide variety of other substances. Carbohydrates provide energy for vital activities and form part of specialized compounds, in combination with proteins or lipids. The next chapter considers the combination of these compounds within a living, functional cell.

✓ A food contains organic molecules with the elements C, H, and O in a ratio of 1:2:1. What type of compound is this?

✓ Why does boiling a protein affect its structure and functional properties?

✓ How are DNA and RNA similar?

Chapter Review

KEY TERMS

atom, p. 26
buffer, p. 32
covalent bond, p. 28
decomposition reaction, p. 30
electrolytes, p. 34
electron, p. 26
enzyme, p. 37
ion, p. 28
metabolism, p. 29
molecule, p. 27
neutron, p. 26
proton, p. 26

SUMMARY OUTLINE

INTRODUCTION p. 26

ATOMS AND MOLECULES p. 26

1. **Atoms** are the smallest units of matter; they consist of **protons**, **neutrons**, and **electrons**. *(Figure 2-1)*

Structure of the Atom p. 26

2. An **element** consists entirely of atoms with the same number of protons (**atomic number**). Within an atom, an **electron cloud** surrounds the **nucleus**. *(Figure 2-2; Table 2-1)*

3. The **atomic weight** of an atom is about equal to the total number of protons and neutrons in its nucleus. **Isotopes** are atoms of the same element whose nuclei contain different numbers of neutrons.

4. Electrons occupy a series of **electron shells** around the nucleus. The number of electrons in the outermost electron shell determine an atom's chemical properties. *(Figure 2-3)*

Chemical Bonds and Chemical Compounds p. 27

5. An **ionic bond** results from the attraction between **ions**: atoms that have gained or lost electrons. **Cations** are positively charged, and **anions** are negatively charged. *(Figure 2-4; Table 2-2)*

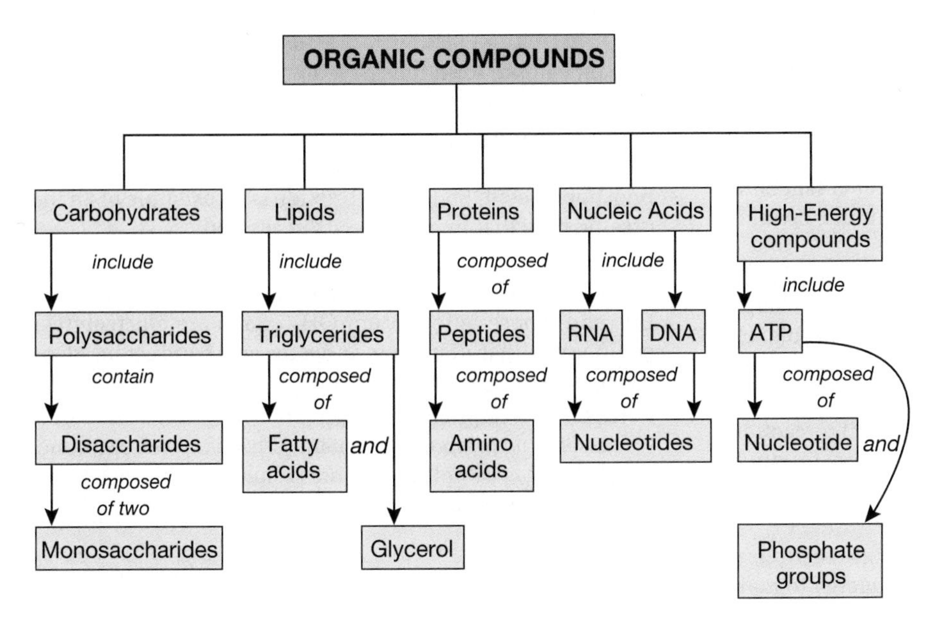

Figure 2-18
Structural Overview of Organic Compounds in the Body

Each of the classes of organic compounds is composed of simple structural subunits. Specific compounds within each class are listed above the basic subunits. Fatty acids are the main subunits of all lipids except steroids, such as cholesterol; only one type of lipid, the triglyceride, is represented here.

TABLE 2-7 Structure and Function of Biologically Important Compounds

Class	*Building Blocks*	*Sources*	*Functions*
INORGANIC			
Water	Hydrogen and oxygen atoms	Absorbed as liquid water or generated via metabolism	Solvent; transport medium for dissolved materials and heat; cooling through evaporation; medium for chemical reactions; reactant in hydrolysis
Acids, bases, salts	H^+, OH^-, various anions and cations	Obtained from the diet or generated via metabolism	Structural components; buffers; sources of ions
Dissolved gases	Oxygen, carbon, nitrogen, and other atoms	Atmosphere	O_2 required for normal cellular metabolism CO_2 generated by cells as a waste product
ORGANIC			
Carbohydrates	C, H, O, sometimes N; CHO in a 1:2:1 ratio	Obtained in diet or manufactured in the body	Energy source; some structural role when attached to lipids or proteins; energy storage
Lipids	C, H, O, sometimes N or P; CHO not in 1:2:1 ratio	Obtained in diet or manufactured in the body	Energy source; energy storage; insulation; structural components; chemical messengers; physical protection
Proteins	C, H, O, N, often S	20 common amino acids; roughly half can be manufactured in the body, others must be obtained in the diet	Catalysts for metabolic reactions; structural components; movement; transport; buffers, defense; control and coordination of activities
Nucleic acids	C, H, O, N, and P; nucleotides composed of phosphates, sugars, and nitrogenous bases	Obtained in diet or manufactured	Storage and processing of genetic information
High-energy compounds	Nucleotides joined to phosphates by high-energy bonds	Synthesized by all cells	Storage or transfer of energy

6. Atoms can combine to form a **molecule**; combinations of atoms of different elements form a **compound**. Some atoms share electrons to form a molecule held together by **covalent bonds**.

7. Sharing one pair of electrons creates a **single covalent bond**; sharing two pairs forms a **double covalent bond**. An unequal sharing of electrons creates a **polar covalent bond**. *(Figure 2-5)*

8. A *hydrogen bond* is the attraction between a hydrogen atom with a slight positive charge and a negatively charged atom in another molecule or within the same molecule. Hydrogen bonds can affect the shapes and properties of molecules.

CHEMICAL NOTATION p. 29

1. *Chemical notation* allows us to describe reactions between reactants that generate one or more products. *(Table 2-3)*

CHEMICAL REACTIONS p. 29

1. **Metabolism** refers to all the chemical reactions in the body. Our cells capture, store, and use energy to maintain homeostasis and support essential functions.

Basic Energy Concepts p. 30

2. **Work** involves movement of an object or a change in its physical structure, and **energy** is the capacity to perform work. There are two major types of energy: kinetic and potential.

3. **Kinetic energy** is the energy of motion. **Potential energy** is stored energy that results from the position or structure of an object. Conversions from potential to kinetic energy are not 100 percent efficient; every energy exchange produces **heat**.

Classes of Reactions p. 30

4. A **chemical reaction** may be classified as a **decomposition**, **synthesis**, or **exchange reaction**. **Exergonic** reactions release heat; **endergonic** reactions absorb heat.

5. Cells gain energy to power their functions by breaking down organic molecules, a process called **catabolism**. Much of this energy supports **anabolism**, the synthesis of new organic molecules.

Reversible Reactions p. 31

6. Reversible reactions consist of simultaneous synthesis and decomposition reactions. At **equilibrium** the rates of these two opposing reactions are in balance.

Acids and Bases p. 32

7. An **acid** releases hydrogen ions, and a **base** removes hydrogen ions from a solution.

pH p. 32

8. The **pH** of a solution indicates the concentration of hydrogen ions it contains. Solutions can be classified as *neutral* ($pH = 7$), *acidic* ($pH < 7$), or *basic (alkaline)* ($pH > 7$) on the basis of pH. *(Figure 2-6)*

9. **Buffers** maintain pH within normal limits (7.35–7.45 in most body fluids) by releasing or absorbing hydrogen ions.

10. *Nutrients* and *metabolites* can be broadly classified as *organic* (carbon-based) or *inorganic*.

INORGANIC COMPOUNDS p. 33

Carbon Dioxide and Oxygen p. 33

1. Living cells in the body generate carbon dioxide and consume oxygen.

Water and Its Properties p. 33

2. Water is the most important inorganic component of the body.

3. Many inorganic compounds will undergo **dissociation** in water to form ions. *(Figure 2-7)*

Inorganic Acids and Bases p. 34

4. Examples of inorganic acids found in the body include hydrochloric acid, carbonic acid, sulfuric acid, and phosphoric acid. Sodium hydroxide is an example of an inorganic base that may form within the body.

Salts p. 34

5. A **salt** is an inorganic compound whose cation is not H^+, and whose anion is not OH^-. Salts are **electrolytes**, compounds that dissociate in water and conduct an electrical current.

ORGANIC COMPOUNDS p. 34

1. *Organic compounds* contain carbon and hydrogen, and usually oxygen as well. Large and complex organic molecules include carbohydrates, lipids, proteins, and nucleic acids.

Carbohydrates p. 34

2. **Carbohydrates** are most important as an energy source for metabolic processes. The three major types are **monosaccharides (simple sugars)**, **disaccharides**, and **polysaccharides**. *(Figures 2-8, 2-9; Table 2-4)*

Lipids p. 36

3. **Lipids** are water-insoluble molecules that include fats, oils, and waxes. There are four important classes of lipids: **fatty acids**, **glycerides**, **steroids**, and **phospholipids**. *(Table 2-5)*

4. **Triglycerides (neutral fats)** consist of three fatty acid molecules attached to a molecule of **glycerol**. *(Figure 2-10)*

5. *Cholesterol* is a precursor of steroid hormones and is an important component of cell membranes. *(Figure 2-11)*

Proteins p. 37

6. Proteins perform a great variety of functions in the body. Important types of proteins include **structural proteins**, **contractile proteins**, **transport proteins**, **enzymes**, and **antibodies**.

7. Proteins are chains of **amino acids** linked by **peptide bonds**. The sequence of amino acids and the interactions of their R groups influence the final shape of the protein molecule. *(Figures 2-12, 2-13)*

8. The shape of a protein determines its functional characteristics. Each protein works best at an optimal combination of temperature and pH.

9. **Activation energy** is the amount of energy required to start a reaction. Proteins called enzymes control many chemical reactions within our bodies. Enzymes are organic *catalysts*—substances that accelerate chemical reactions without themselves being permanently changed. *(Figure 2-14)*

10. The reactants in an enzymatic reaction, called **substrates**, interact to form a **product** by binding to the enzyme at the **active site**. *(Figure 2-15)*

Nucleic Acids p. 40

11. **Nucleic acids** store and process information at the molecular level. There are two kinds of nucleic acids: **deoxyribonucleic acid (DNA)** and **ribonucleic acid (RNA)**. *(Figure 2-16*; Table 2-6)

12. Nucleic acids are chains of **nucleotides**. Each nucleotide contains a sugar, a **phosphate group**, and a **nitrogenous base**. The sugar is always **ribose** or **deoxyribose**. The nitrogenous bases found in DNA are **adenine**, **guanine**, **cytosine**, and **thymine**. In RNA, **uracil** replaces thymine.

High-Energy Compounds p. 41

13. Cells store energy in **high-energy compounds** for later use. The most important high-energy compound is

ATP (*adenosine triphosphate*). When energy is available, cells make ATP by adding a phosphate group to ADP. When energy is needed, ATP is broken down to ADP and phosphate. (*Figure 2-17*)

CHEMICALS AND LIVING CELLS p. 42

14. Biochemical building blocks form functional units called **cells.** (*Figure 2-18; Table 2-7*)

CHAPTER QUESTIONS

LEVEL 1 Reviewing Facts and Terms

Match each item in column A with the most closely related item in column B. Use letters for answers in the spaces provided.

Column A	Column B
___ 1. atomic number	a. synthesis
___ 2. covalent bond	b. catalyst
___ 3. ionic bond	c. sharing of electrons
___ 4. catabolism	d. A + B AB
___ 5. anabolism	e. stabilize pH
___ 6. exchange reaction	f. number of protons
___ 7. reversible reaction	g. decomposition
___ 8. acid	h. carbohydrates, lipids, proteins
___ 9. enzyme	i. loss or gain of electrons
___ 10. buffer	j. water, salts
___ 11. organic compounds	k. H^+ donor
___ 12. inorganic compounds	l. AB + CD → AD + CB

13. In atoms, protons and neutrons are found:
(a) only in the nucleus
(b) outside the nucleus
(c) inside and outside the nucleus
(d) in the electron cloud

14. The number and arrangement of electrons in an atom's outer electron shell determines its:
(a) atomic weight (b) atomic number
(c) electrical properties (d) chemical properties

15. The bond between sodium and chlorine in the compound sodium chloride (NaCl) is:
(a) an ionic bond (b) a single covalent bond
(c) a nonpolar covalent bond (d) a double covalent bond

16. What is the role of enzymes in chemical reactions?

17. List six elements found in abundance in the body.

18. What four major classes of organic compounds are found in the body?

19. List seven major functions performed by proteins.

LEVEL 2 Reviewing Concepts

20. Oxygen has 8 protons, 8 neutrons, and 8 electrons. What is the molecular weight of O_2?
(a) 8 (b) 16
(c) 24 (d) 32

21. Of the following selections, the one that contains only *inorganic* compounds is:
(a) water, electrolytes, oxygen, carbon dioxide
(b) oxygen, carbon dioxide, water, sugars
(c) water, electrolytes, salts, nucleic acids
(d) carbohydrates, lipids, proteins, vitamins

22. Glucose and fructose are examples of:
(a) monosaccharides (simple sugars)
(b) isotopes
(c) lipids
(d) a, b, and c are all correct

23. Explain the differences among (1)nonpolar covalent bonds, (2)polar covalent bonds, and (3)ionic bonds.

24. Why does pure water have a neutral pH?

25. A biologist analyzes a sample that contains an organic molecule and finds the following constituents: carbon, hydrogen, oxygen, nitrogen, and phosphorus. On the basis of this information, is the molecule a carbohydrate, a lipid, a protein, or a nucleic acid?

LEVEL 3 Critical Thinking and Clinical Applications

26. The element sulfur has an atomic number of 16 and an atomic weight of 32. How many neutrons are in the nucleus of a sulfur atom? Assuming that sulfur forms covalent bonds with hydrogen, how many hydrogen atoms could bond to one sulfur atom?

27. An important buffer system in the human body involves carbon dioxide (CO_2) and bicarbonate ion (HCO_3^-) as shown below:

$$CO_2 + H_2O \quad H_2CO_3 \quad H^+ + HCO_3^-$$

If a person becomes excited and exhales large amounts of CO_2, how will his body's pH be affected?

3 CELL STRUCTURE AND FUNCTION

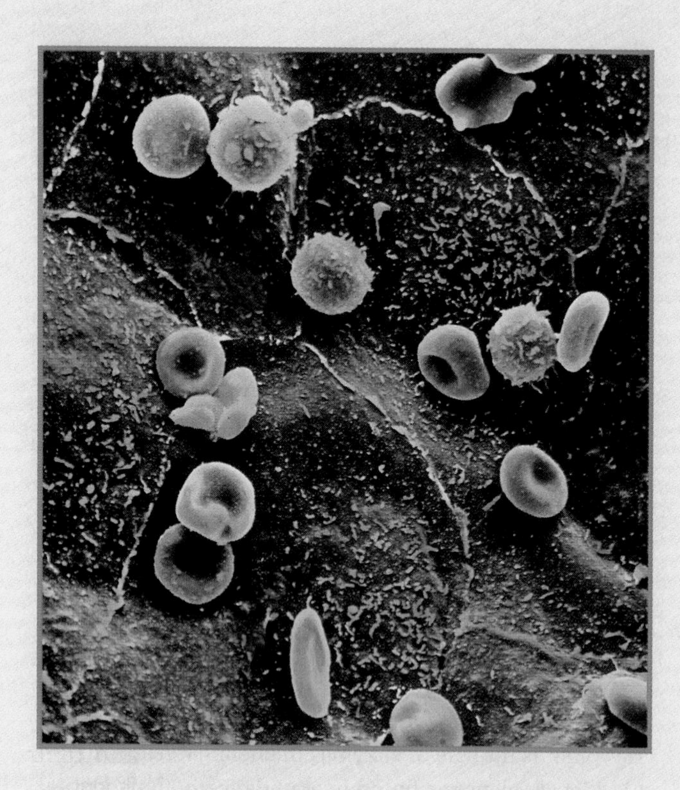

Many familiar structures, from pyramids to patchwork quilts, are made up of numerous small, similar components. The human body is built on the same principle, for it is made up of several trillion tiny units called cells. Cells are the smallest entities that can perform all basic life functions. In addition, each cell in the body performs specialized functions and plays a role in the maintenance of homeostasis. In this chapter we will explore the functional organization of a representative cell.

Chapter Outline and Objectives

Vocabulary Development

aero-, air; *aerobic*
ana-, apart; *anaphase*
chondros, cartilage; *mitochondria*
chroma, color; *chromosome*
cyto-, cell; *cytoplasm*
endo-, inside; *endocytosis*
exo-, outside; *exocytosis*
hemo-, blood; *hemolysis*
hyper-, above; *hypertonic*
hypo-, below; *hypotonic*
inter-, between; *interphase*
interstitium, something standing between; *interstitial fluid*
iso-, equal; *isotonic*
kinesis, motion; *cytokinesis*
meta-, after; *metaphase*
micro-, small; *microtubules*
mitos, thread; *mitochondria*
osmos, thrust; *osmosis*
phagein, to eat; *phagocyte*
podon, foot; *pseudopod*
pro-, before; *prophase*
pseudo-, false; *pseudopod*
reticulum, network; *endoplasmic reticulum*
soma, body; *lysosome*
telos, end; *telophase*
tonos, tension; *isotonic*

As atoms are the building blocks of molecules, cells are the building blocks of the human body. Over the years, biologists have developed the cell theory, which includes the following four basic concepts:

1. Cells are the basic structural units of all plants and animals.
2. Cells are the smallest functioning units of life.
3. Cells are produced only by the division of preexisting cells.
4. Each cell maintains homeostasis.

An individual organism maintains homeostasis only through the combined and coordinated actions of many different types of cells. Figure 3-1● gives some examples of the range of cell sizes and shapes found in the human body.

Numbering in the trillions, the cells of the human body form and maintain anatomical structures and perform physiological functions as different as running and thinking. An understanding of how the human body functions thus requires a familiarity with the nature of cells. AM *The Nature of Pathogens*

STUDYING CELLS

The study of the structure and function of cells is called **cytology** (sī-TOL-ō-jē; *cyto-*, cell + *-logy*, the study of). What we have learned since the 1950s has given us new insights into the physiology of cells and their means of homeostatic control. Acquiring this knowledge depended upon developing better ways of view-

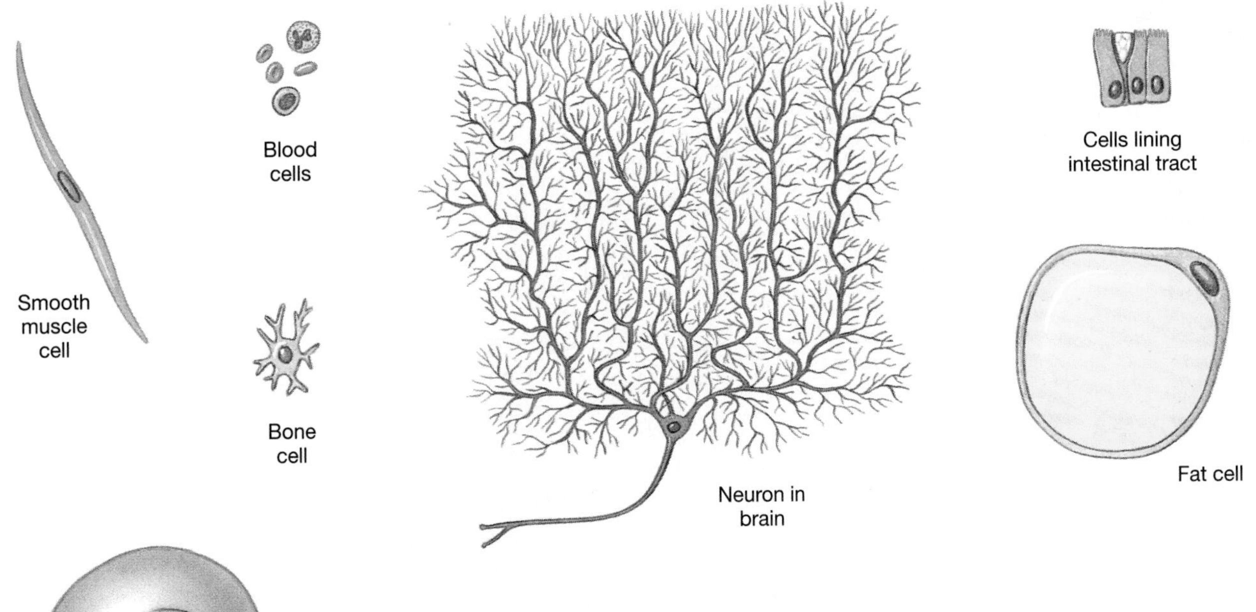

●**Figure 3-1**
The Diversity of Cells in the Human Body
The cells of the body have many different shapes and a variety of special functions. These examples give an indication of the range of forms; all of the cells are shown with the dimensions they would have if magnified approximately 500 times.

ing cells and applying new experimental techniques not only from biology but also chemistry and physics.

The two most common methods used to study cell and tissue structure are *light microscopy* and *electron microscopy*. Before the 1950s most information was obtained through the use of light microscopy. Light microscopy can magnify cellular structures about 1000 times using a series of glass lenses. Many fine details of intracellular structure, however, remained a mystery until investigators began using electron microscopy, a technique that replaced light with a focused beam of electrons. *Transmission electron micrographs* (TEMs) reveal the fine structure of cell membranes and intracellular structures. *Scanning electron micrographs* (SEMs) provide less magnification, but reveal the three-dimensional nature of cell structures. You will see examples of both kinds of electron micrographs, as well as light micrographs, throughout this text.

An Overview of Cellular Anatomy

The "typical" cell is like the "average" person, so any description masks enormous individual variations. Our model cell will share features with most cells of the body without being identical to any. Figure 3-2● shows such a composite cell, and Table 3-1 summarizes the structures and functions of its parts.

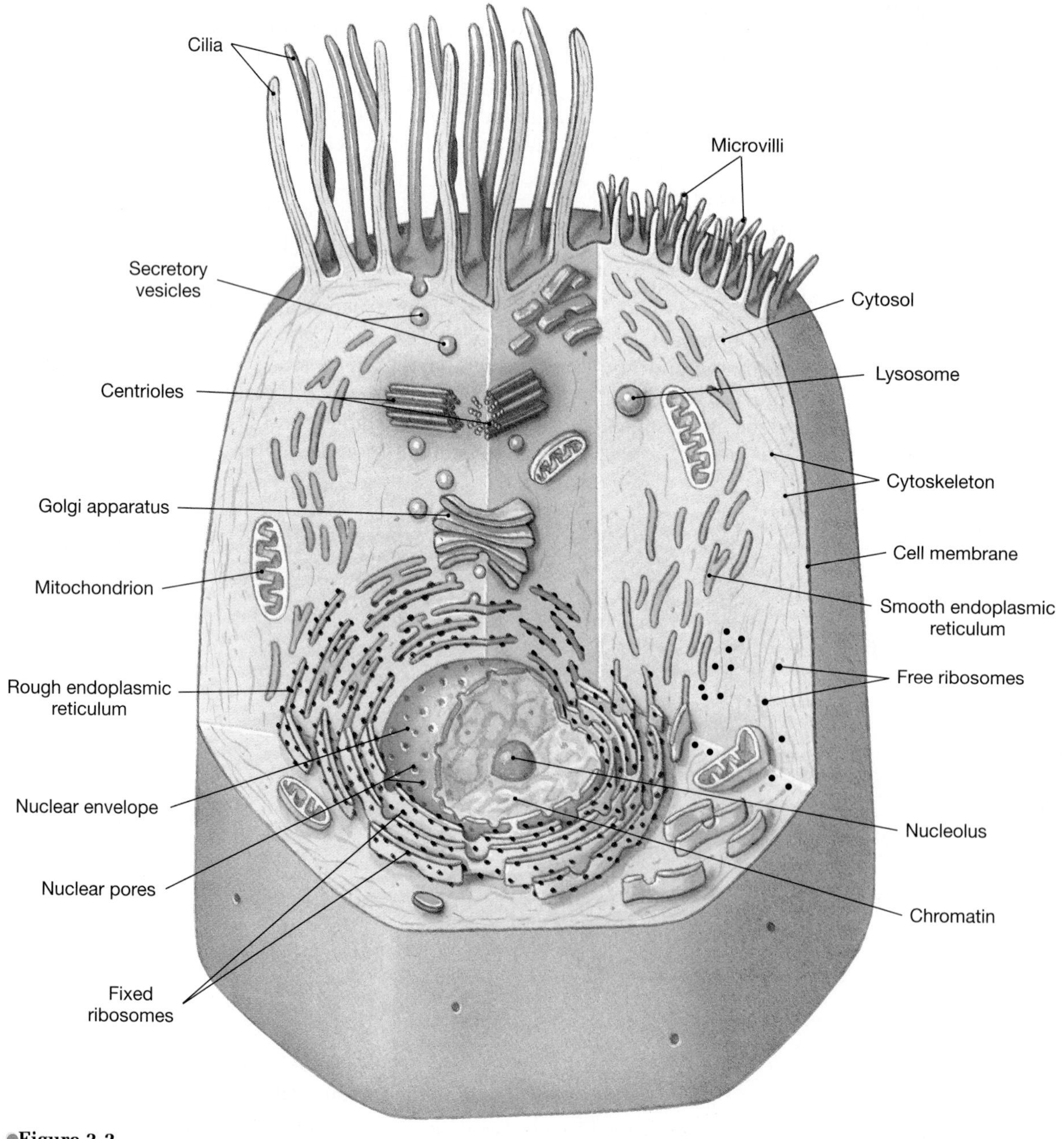

●**Figure 3-2**
Anatomy of a Composite Cell
See Table 3-1 for a summary of the functions associated with the various cell structures.

TABLE 3-1 Organelles of a Representative Cell

Appearance	*Structure*	*Composition*	*Function*
	CELL MEMBRANE	Lipid bilayer, containing phospholipids, steroids, and proteins	Isolation, protection, sensitivity, support; controls entrance/exit of materials
	CYTOSOL	Fluid component of cytoplasm	Distributes materials by diffusion
	NONMEMBRANOUS ORGANELLES		
	Cytoskeleton: Microtubule, Microfilament	Proteins organized in fine filaments or slender tubes	Strength, movement of cellular structures and materials
	Microvilli	Membrane extensions containing microfilaments	Increase surface area to facilitate absorption of extracellular materials
	Cilia	Membrane extensions containing 9 microtubule doublets + a central pair	Movement of materials over surface
	Centrioles	Two centrioles, at right angles; each composed of 9 microtubule triplets	Essential for movement of chromosomes during cell division
	Ribosomes	RNA + proteins; fixed ribosomes bound to endoplasmic reticulum, free ribosomes scattered in cytoplasm	Protein synthesis
	MEMBRANOUS ORGANELLES		
	Endoplasmic reticulum (ER)	Network of membranous channels extending throughout the cytoplasm	Synthesis of secretory products; intracellular storage and transport
	Rough ER	Has ribosomes attached to membranes	Secretory protein synthesis
	Smooth ER	Lacks attached ribosomes	Lipid and carbohydrate synthesis
	Golgi apparatus	Stacks of flattened membranes (saccules) containing chambers (cisternae)	Storage, alteration, and packaging of secretory products and lysosomes
	Lysosomes	Vesicles containing powerful digestive enzymes	Intracellular removal of damaged organelles or of pathogens
	Mitochondria	Double membrane, with inner folds (cristae) enclosing important metabolic enzymes	Produce 95% of the ATP required by the cell
	NUCLEUS	Nucleoplasm containing nucleotides, enzymes, and nucleoproteins; surrounded by double membrane (nuclear envelope)	Control of metabolism; storage and processing of genetic information; control of protein synthesis
	Nucleolus	Dense region in nucleoplasm containing DNA and RNA	Site of rRNA synthesis and assembly of ribosomal subunits

A *cell membrane* separates the cell contents, or *cytoplasm*, from the watery medium that surrounds it, the *extracellular fluid*. The cytoplasm can be further subdivided into a fluid, the *cytosol*, intracellular structures collectively known as *organelles* (or-gan-ELS; "little organs"), and *inclusions*, insoluble materials that may take the form of solid granules or lipid droplets.

THE CELL MEMBRANE

The outer boundary of the cell is formed by a **cell membrane**, or *plasma membrane*. Its four general functions include:

1. **Physical isolation.** The cell membrane is a physical barrier that separates the inside of the cell from the surrounding extracellular fluid.
2. **Regulation of exchange with the environment.** The cell membrane controls the entry of ions and nutrients, the elimination of wastes, and the release of secretory products.
3. **Sensitivity.** The cell membrane is the first part of the cell affected by changes in the extracellular fluid. It also contains a variety of receptors that allow the cell to recognize and respond to specific molecules in its environment. Any alteration in the cell membrane may affect all cellular activities.
4. **Structural support.** Specialized connections between cell membranes or between membranes and extracellular materials give tissues a stable structure.

Membrane Structure

The cell membrane is extremely thin and delicate. Its major components are lipids, proteins, and carbohydrates.

Membrane Lipids

Phospholipids are a major component of cell membranes. In a phospholipid, a phosphate group (PO_4^{3-}) serves as a link between a diglyceride (a glycerol backbone bonded to two fatty acid "tails") and a nonlipid "head". 1 [p. 37] The phospholipids in a cell membrane lie in two distinct layers, with the heads on the outside and the tails on the inside. For this reason, the cell membrane is often called the **phospholipid bilayer** (see Figure 3-3●). Mixed in with the fatty acid tails are cholesterol molecules and small quantities of other lipids.

The lipid tails will not associate with water molecules, and this characteristic allows the cell membrane to act as a selective physical barrier. Ions and water-soluble compounds cannot cross the lipid portion of a cell membrane. Consequently, the cell membrane is very effective in isolating the cytoplasm from the surrounding extracellular fluid.

Membrane Proteins

Several types of proteins are associated with the cell membrane. They may be partially or totally embedded in the phospholipid bilayer or loosely bound to its inner or outer surface. Membrane proteins may function as *receptors, channels, carriers, enzymes, anchors,* or *identifiers*. Table 3-2 provides a functional description and example of each class of membrane protein.

Membrane structure is not rigid, and embedded proteins drift from place to place across the surface of the membrane like ice cubes in a punch bowl. In addition, the composition of the cell membrane can change over time, as membrane is added or removed through processes described later in the chapter.

Membrane Carbohydrates

Carbohydrates and lipids on the outer surface of the membrane (1) are important as cell lubricants and adhesives, (2) act as receptors for extracellular compounds, and (3) are part of a recognition system that keeps the immune system from attacking its own tissues.

Membrane Transport

Precisely which substances enter or leave the cytoplasm is determined by the *permeability* of the cell membrane. If nothing can cross the cell membrane, it is described as *impermeable*. If any substance at all can cross without difficulty, the membrane is *freely permeable*. Cell membranes are **selectively permeable**, permitting the free passage of some materials and restricting the passage of others. Passage selection occurs on the basis of size, electrical charge, molecular shape, lipid solubility, or some combination of factors.

Movement across the membrane may be passive or active. *Active processes,* discussed later in this chapter, require that the cell expend energy, usually in the form of ATP. *Passive processes* move ions or molecules across the cell membrane without any energy expenditure by the cell. Passive processes include *diffusion, osmosis, filtration*, and *facilitated diffusion*.

Diffusion

Ions and molecules are in constant motion, colliding and bouncing off one another and off any obstacles in their paths. **Diffusion** is the net movement of molecules from an area of relatively high concentration (or large number of collisions) to an area of relatively low concentration (or small number of collisions). The difference between the high and low concentrations represents a **concentration gradient,** and diffusion is often described as proceeding "down a concentration gradient" or "downhill." As a result of diffusion, solutes eventually become uniformly distributed, and concentration gradients are eliminated.

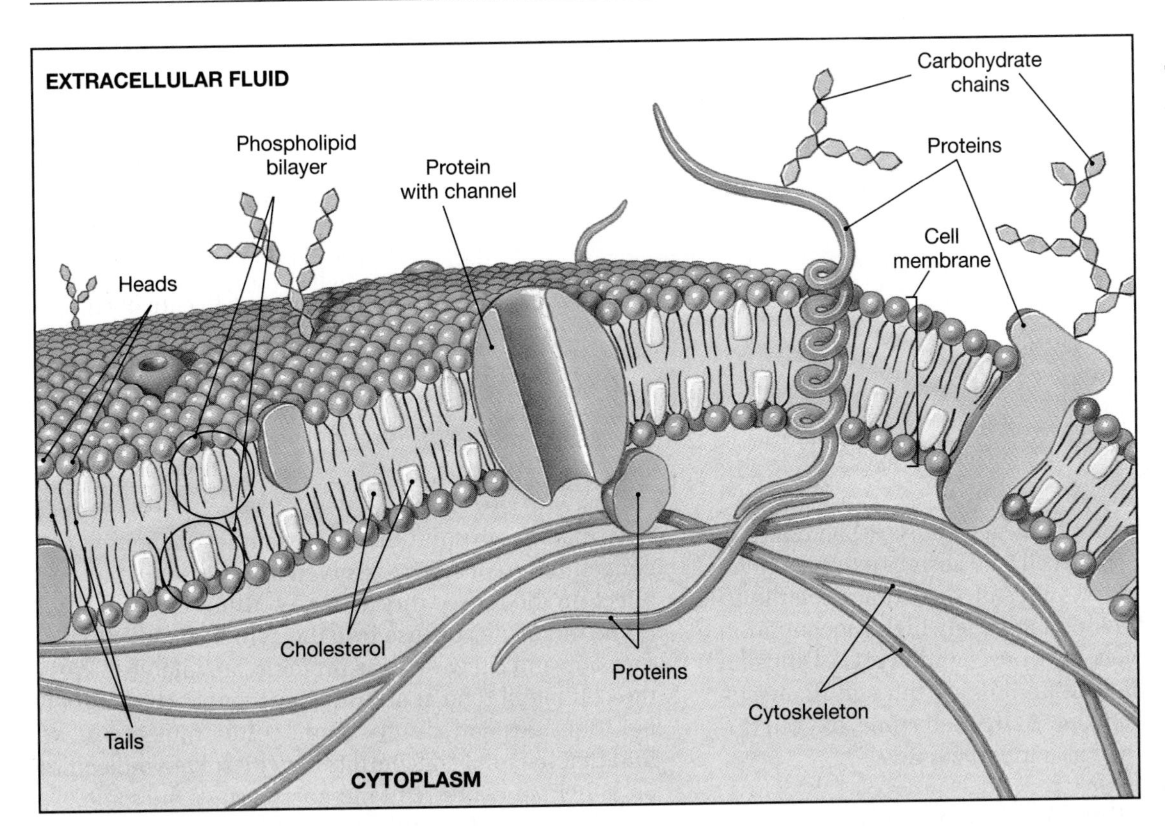

●**Figure 3-3 The Cell Membrane**

Diffusion occurs in air as well as water. The smell of fresh flowers in a vase can sweeten the air in a large room, just as a drop of ink spreads to color an entire glass of water. In each case you begin with an extremely high concentration of molecules in a very localized area. Consider the ink dropped in the water glass, shown in step 1 of Figure 3-4●. Placing that drop in a large volume of clear water establishes a sharp concentration gradient for the ink: the ink concentration is high at the drop and negligible everywhere else. As diffusion proceeds, the ink molecules spread through the solution (step 2) until they are distributed evenly (step 3).

Diffusion is important in body fluids because it tends to eliminate local concentration gradients. For example,

TABLE 3-2 Membrane Proteins

Class	*Function*	*Example*
Receptor proteins	Sensitive to specific extracellular materials that bind to them and trigger a change in a cell's activity.	Binding of the hormone insulin to membrane receptors increases the rate of glucose absorption by the cell.
Channel proteins	Central pore, or channel, permits water and solutes to bypass lipid portion of cell membrane.	Calcium ion movement through channels is involved in muscle contraction and the conduction of nerve impulses.
Carrier proteins	Bind and transport solutes across the cell membrane. This process may or may not require energy.	Carrier proteins bring glucose into the cytoplasm and also transport sodium, potassium, and calcium ions.
Enzymes	Catalyze reactions in the extracellular fluid or within the cell.	Dipeptides are broken down into amino acids by enzymes on the membranes of cells lining the intestinal tract.
Anchor proteins	Attach the cell membrane to other structures and stabilize its position.	Inside the cell, bound to the network of supporting filaments (the cytoskeleton); outside, attach the cell to extracellular protein fibers or to another cell.
Identifier proteins	Identify a cell as self or nonself, normal or abnormal, to the immune system.	One group of such recognition proteins is the *major histocompatibility complex* (MHC) discussed in Chapter 15.

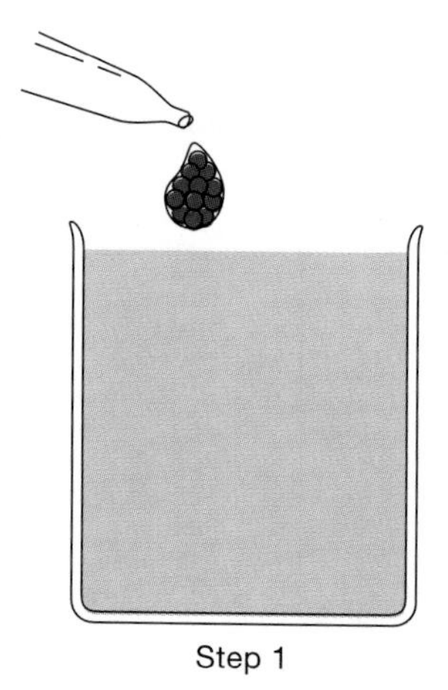

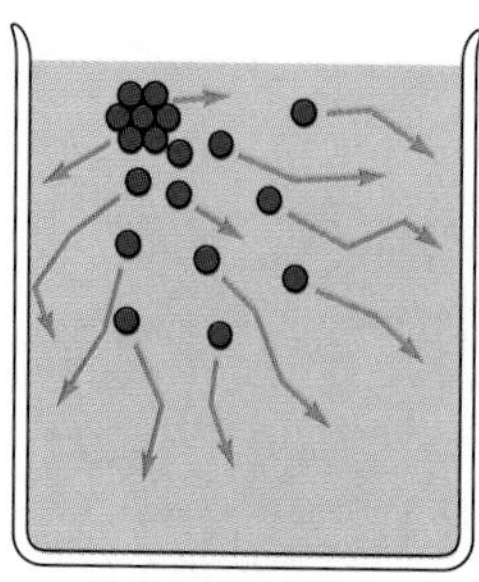

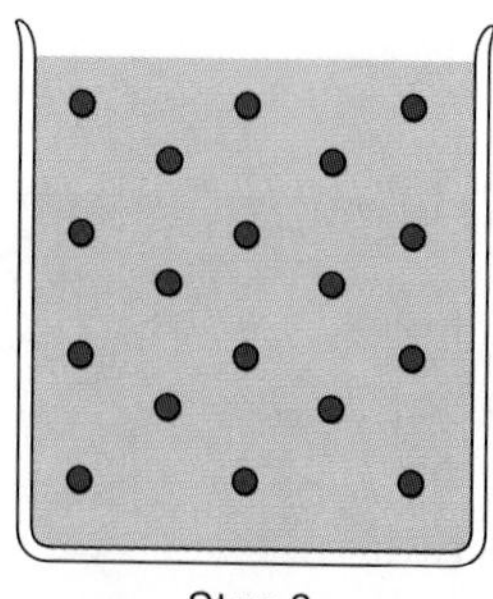

Figure 3-4
Diffusion

Step 1: Placing an ink drop in a glass of water establishes a strong concentration gradient because there are many ink molecules in one location and none elsewhere. **Step 2:** Diffusion occurs, and the ink molecules spread through the solution. **Step 3:** Eventually diffusion eliminates the concentration gradient, and the ink molecules are distributed evenly. Molecular motion continues, but there is no directional movement.

each active cell in your body generates carbon dioxide, which diffuses out of the cell, and absorbs oxygen, which diffuses into the cell. As a result, the extracellular fluid around the cell develops a relatively high concentration of CO_2 and a relatively low concentration of O_2. Diffusion then distributes the carbon dioxide through the tissue and into the bloodstream. At the same time, oxygen diffuses out of the blood and into the tissue.

Diffusion across Cell Membranes Water and dissolved solutes diffuse freely through the extracellular fluids of the body. The cell membrane, however, acts as a barrier that selectively restricts diffusion. Some substances can pass through easily, whereas others cannot penetrate the membrane at all. There are only two ways for an ion or molecule to independently diffuse across a cell membrane: (1) pass through one of the membrane channels or (2) move across the lipid portion of the membrane. Not surprisingly then, the major factors determining whether a substance can diffuse across a cell membrane are its lipid solubility and its size relative to the diameter of the membrane channels (Figure 3-5●).

Lipid solubility Alcohol, fatty acids, and steroids can enter cells easily because they can diffuse through the lipid portions of the membrane. Dissolved gases such as oxygen and carbon dioxide also enter and leave our cells by diffusion through the lipid bilayer.

Size Water-soluble compounds must diffuse through channels in the membrane. These channels are very small, averaging about 0.8 nm in diameter. Water molecules can enter or exit freely, but even a small organic molecule, such as glucose, is too big to fit through the channels.

Osmosis: A Special Type of Diffusion The diffusion of water across a membrane is so important that it is given a special name, **osmosis** (oz-MŌ-sis; *osmos*, thrust). Both intracellular and extracellular fluids are solutions that contain a variety of dissolved materials, or solutes. Each solute tends to diffuse as if it were the only material in solution. For example, changes in the concentration of other dissolved materials will have no effect on the rate or direction of sodium ion diffusion. Some molecules diffuse into the cytoplasm, others diffuse out, and a few, such as proteins, are unable to cross the cell membrane at all. But if we ignore the individual identities and simply count solute molecules, we find that the total concentration of dissolved molecules on either side of the cell membrane stays the same.

This state of equilibrium persists because *the cell membrane is freely permeable to water.* Whenever a concentration gradient exists, water molecules will diffuse rapidly across the cell membrane until the gradient is eliminated. This movement, which eliminates differences in solute concentrations, occurs in response to a concentration gradient for water molecules.

Dissolved solute molecules occupy space that would otherwise be taken up by water molecules. Thus the higher the solute concentration, the lower the water concentration. As a result, *water molecules will tend to diffuse across a membrane toward the solution containing a higher solute concentration.*

Three characteristics of osmosis should be remembered:

1. Osmosis is the diffusion of water molecules across a membrane.
2. Osmosis occurs across a selectively permeable membrane that is freely permeable to water but not to solutes.
3. In osmosis, water will flow across a membrane *toward the solution that has the highest concentration of solutes,* because that is where the concentration of water is lowest.

Osmosis and Osmotic Pressure Figure 3-6● diagrams the process of osmosis. Step 1 shows two solutions (A and B) with differing solute concentrations separated by a selectively permeable membrane. As osmosis occurs, water molecules cross the membrane until the solute concentrations in the two solutions are identical (step 2a). Thus the volume of solution B increases at the ex-

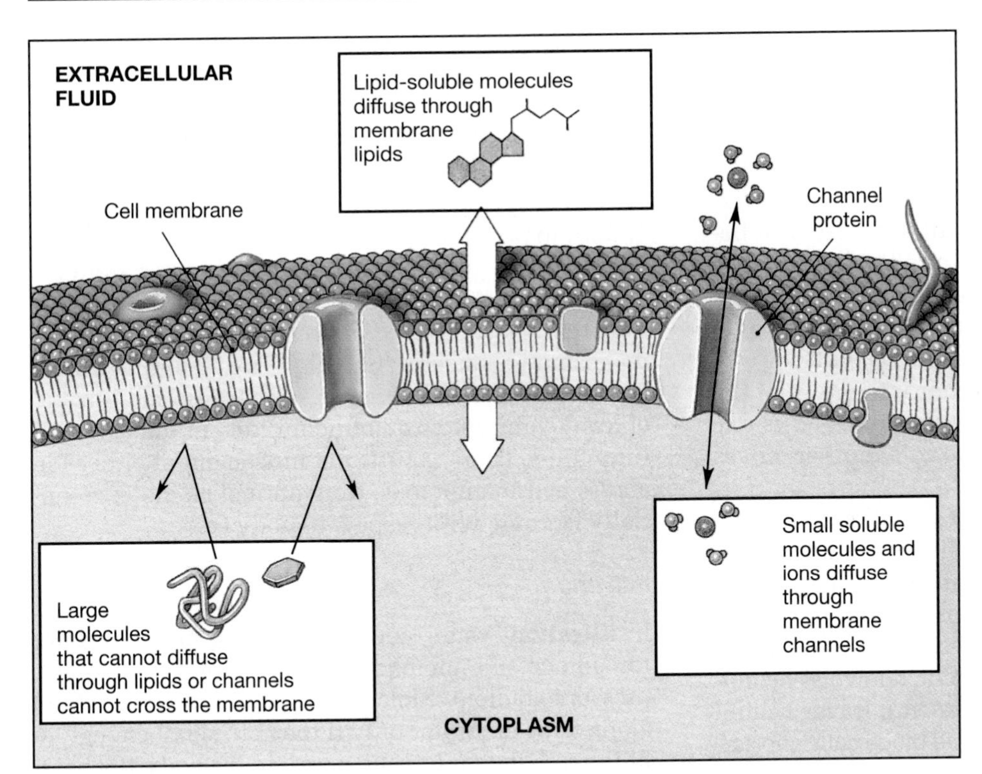

Figure 3-5
Diffusion across Cell Membranes

Movement of a substance across the membrane depends on its lipid solubility and its size.

pense of solution A. The greater the initial difference in solute concentrations, the stronger the osmotic flow. The **osmotic pressure** of a solution is an indication of the force of water movement *into that solution* as a result of its solute concentration. As the solute concentration of a solution increases, so does its osmotic pressure. Osmotic pressure is defined as the amount of pressure needed to stop osmosis. A solution's osmotic pressure can be measured in several ways. For example, a strong enough opposing pressure can prevent the entry of water molecules. Pushing against a fluid generates *hydrostatic pressure.* In step 2b, hydrostatic pressure in solution

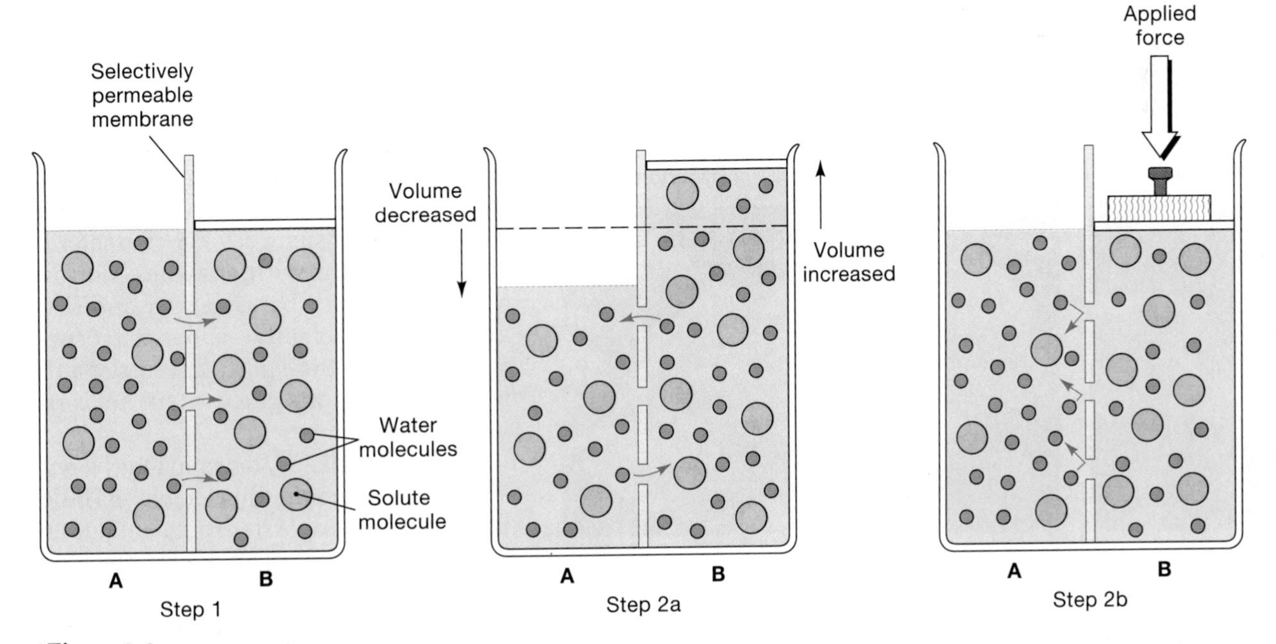

Figure 3-6
Osmosis

Step 1: Two solutions containing different solute concentrations are separated by a selectively permeable membrane. Water molecules (small dots) begin to cross the membrane toward the solution with the higher concentration of solutes (larger circles) (solution B). **Step 2a:** At equilibrium the solute concentrations on the two sides of the membrane are equal. The volume of solution B has increased at the expense of solution A. **Step 2b:** Osmosis can be prevented by resisting the volume change. The osmotic pressure of solution B is equal to the amount of hydrostatic pressure required to stop the osmotic flow.

B, created by the applied force, balances the osmotic pressure, and no net osmotic flow occurs.

Solutions of varying solute concentrations are described as *isotonic, hypotonic,* or *hypertonic* with regard to their effects on the shape or tension of the membrane of living cells. Although the effects of various osmotic solutions are difficult to see in most tissues, they are readily observed in red blood cells.

Figure 3-7a● shows the appearance of a red blood cell immersed in an isotonic solution. An **isotonic** (*iso-*, equal + *tonos*, tension) solution is one that will not cause a net movement of water into or out of the cell. In other words, an equilibrium exists, and as one water molecule moves out of the cell another moves in to replace it.

When a cell is placed in contact with a **hypotonic** (*hypo-*, below) solution, water will flow into the cytoplasm by osmosis. If the difference is substantial, the cell will swell up like a balloon, as shown in Figure 3-7b●. Ultimately the membrane may rupture, or *lyse*. In the case of red blood cells, this event, known as **hemolysis** (*hemo-*, blood + *lysis*, breakdown), leaves behind empty cell membranes known as red blood cell "ghosts."

When a cell is placed in contact with a **hypertonic** (*hyper-*, above) solution, water will flow out of the cell and into the surrounding medium by osmosis, and the cell will shrivel and dehydrate. In the case of red blood cells, this shrinking is called **crenation** (Figure 3-7c●).

It is often necessary to administer large volumes of fluid to people who have had a severe blood loss or who are dehydrated. One fluid often administered is a 0.9 percent (0.9 g/dl) solution of sodium chloride (NaCl). This solution, which approximates the normal osmotic concentration of the extracellular fluids, is called **normal saline**. It is used because sodium and chloride are the most abundant ions in the extracellular fluid. There is little net movement of either ion across cell membranes; thus normal saline is essentially isotonic with respect to body cells.

Filtration

In **filtration**, water and small solute molecules are forced across a membrane because of a hydrostatic pressure gradient. Molecules of solute will be carried along with the water only if they are small enough to fit through the membrane pores. In the body, the heart pushes blood through the circulatory system and generates hydrostatic pressure, or *blood pressure*. Filtration occurs across the walls of small blood vessels, pushing water and dissolved nutrients into the tissues of the body. Filtration across specialized blood vessels in the kidneys is an essential step in the production of urine.

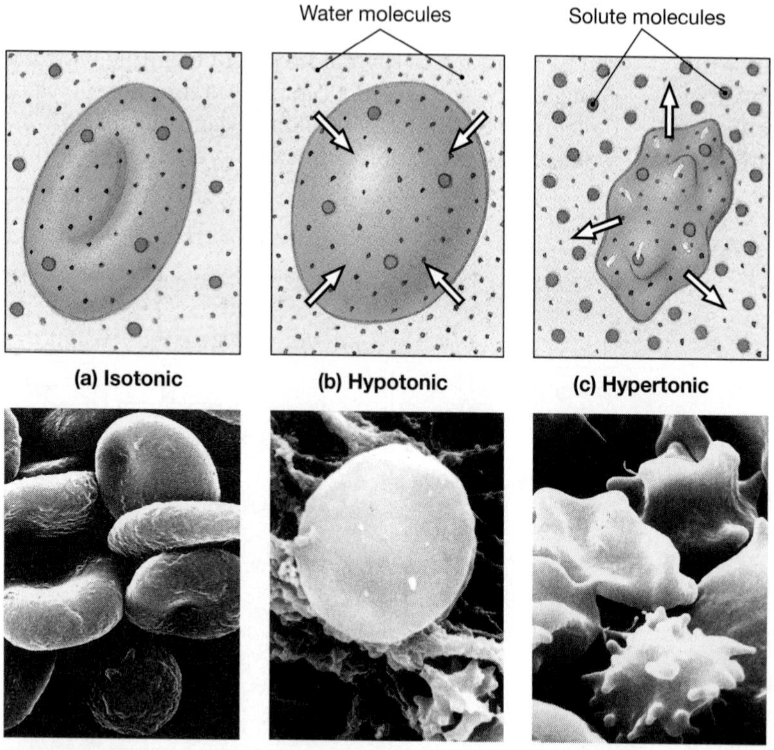

●**Figure 3-7**
Osmotic Flow across Cell Membranes

White arrows indicate the direction of osmotic water movement. **(a)** Because these red blood cells are immersed in an isotonic saline solution, no osmotic flow occurs and the cells have their normal appearance. **(b)** Immersion in a hypotonic saline solution results in the osmotic flow of water into the cells. The swelling may continue until the cell membrane ruptures. **(c)** Exposure to a hypertonic solution results in the movement of water out of the cells. The red blood cells shrivel and become crenated. (SEM × 833)

Carrier-Mediated Transport

Carrier-mediated transport involves the activity of membrane proteins that bind specific ions or organic substrates and move them across the cell membrane. These proteins share several characteristics in common with enzymes. For example, they may be used over and over and are very selective about what they will bind and transport; the carrier protein that transports glucose will not carry other simple sugars.

Carrier-mediated transport may be *passive* (no ATP required) or *active* (ATP-dependent). In passive transport, solutes are typically carried from an area of high concentration to an area of low concentration. Active transport mechanisms may follow or oppose an existing concentration gradient.

Many carrier proteins transport one ion or molecule at a time, but some deal with two solutes simultaneously. In *cotransport*, the carrier transports the two substances in the same direction, either into or out of the cell. In *countertransport*, one substance moves into the cell while the other moves out.

Two major examples of carrier-mediated transport—facilitated diffusion and active transport—are discussed below.

Facilitated Diffusion Many essential nutrients, such as glucose or amino acids, are insoluble in lipids and too large to fit through membrane channels. However, these compounds can be passively transported across the membrane by carrier proteins in a process called **facilitated diffusion.** The molecule to be transported first binds to a **receptor site** on the protein. It is then moved to the inside of the cell membrane and released into the cytoplasm. The process is diagrammed in Figure 3-8●.

As in the case of simple diffusion, no ATP is expended in facilitated diffusion, and the molecules move from an area of higher concentration to one of lower concentration. Facilitated diffusion differs from ordinary diffusion, however, because the rate of transport cannot increase indefinitely; only a limited number of carrier proteins are available in the membrane. Once all of them are operating, any further increase in the concentration of the solute will have no effect on the rate of movement into the cell.

Active Transport In **active transport** the high-energy bond in ATP provides the energy needed to move ions or molecules across the membrane. The process is complex, and specific enzymes must be present in addition to the carrier molecule. The advantage of active transport is that the cell can import or export specific materials *regardless of their intracellular or extracellular concentrations.*

All living cells contain carrier proteins called **ion pumps** that actively transport the cations sodium (Na^+), potassium (K^+), calcium (Ca^{2+}), and magnesium (Mg^{2+}) across their cell membranes. Specialized cells can transport additional ions such as iodide (I^-), chloride (Cl^-), and iron (Fe^{2+}). Many of these carrier proteins move a specific cation or anion in one direction only, either into or out of the cell. In a few cases, one carrier protein will move more than one ion at a time. If one ion moves in one direction and the other moves in the opposite direction, the carrier is called an **exchange pump.**

Sodium and potassium ions are the principal cations in body fluids. Sodium ion concentrations are high in the extracellular fluids, whereas sodium concentrations in the cytoplasm are relatively low. The distribution of potassium in the body is just the opposite—low in the extracellular fluids and high in the cytoplasm. As a result, sodium ions slowly diffuse into the cell, and potassium ions leak out.

Homeostasis within the cell depends on ejecting sodium ions and recapturing lost potassium ions. This exchange is accomplished through the activity of the **sodium-potassium exchange pump.** This ion pump exchanges intracellular sodium for extracellular potassium (Figure 3-9●).

Vesicular Transport

In **vesicular transport,** materials move into or out of the cell through the formation of vesicles, small membranous sacs. There are two major categories of vesicular transport; *endocytosis* and *exocytosis.*

Endocytosis **Endocytosis** (EN-dō-sī-TŌ-sis; *endo-*, inside + *cyte*, cell) is the packaging of extracellular materials in a vesicle at the cell surface for importation *into* the cell. This process may involve relatively large

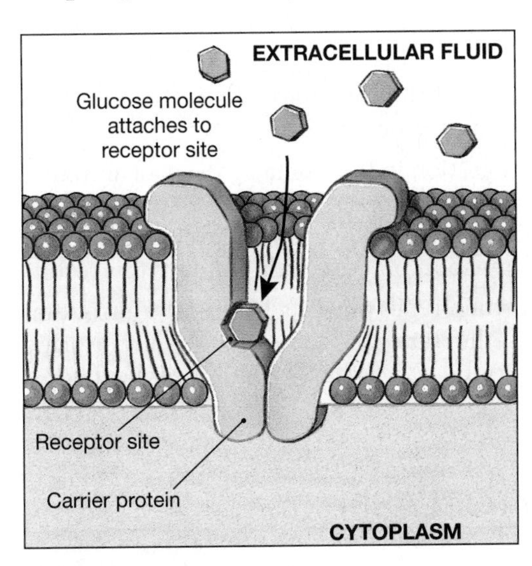

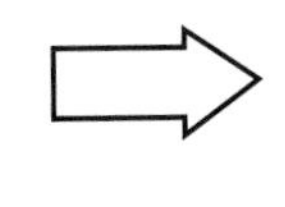

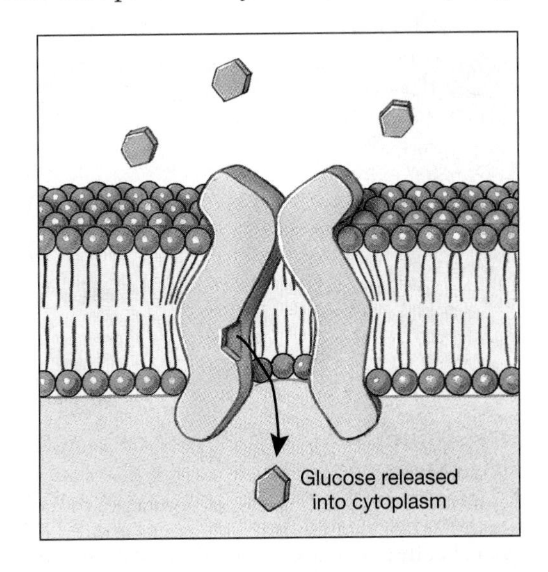

●**Figure 3-8**
Facilitated Diffusion
In this process, an extracellular molecule, such as glucose, binds to a receptor site on a carrier protein. The binding alters the shape of the protein, which then releases the molecule to diffuse into the cytoplasm.

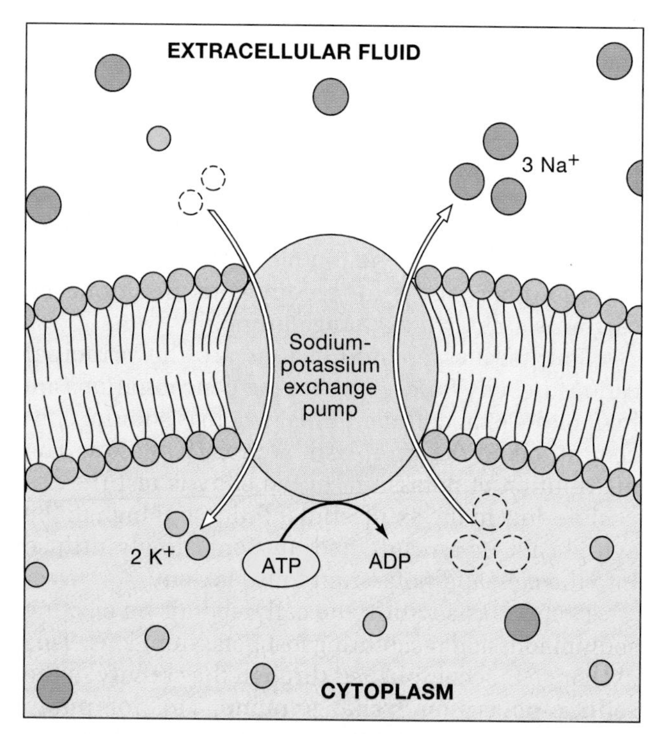

•**Figure 3-9**
The Sodium-Potassium Exchange Pump

The operation of the sodium-potassium exchange pump is an example of active transport because its operation requires the conversion of ATP to ADP.

volumes of extracellular material. There are two major types of endocytosis: *pinocytosis* and *phagocytosis.* Both are active processes that require ATP or other sources of energy.

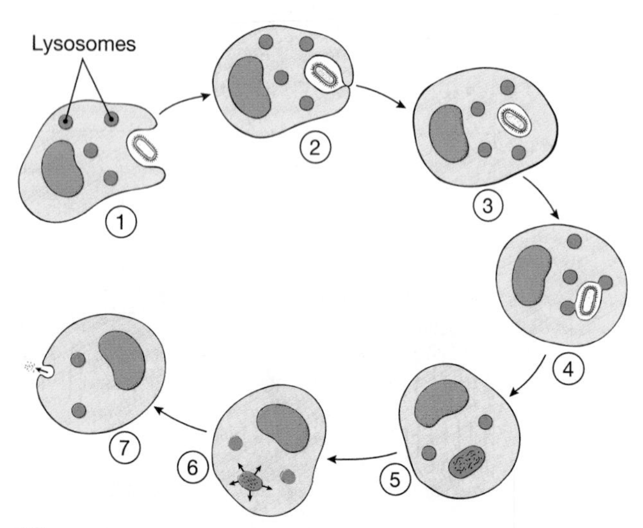

•**Figure 3-10**
Phagocytosis

A phagocytic cell first comes in contact with the foreign object and sends cytoplasmic extensions around it (1). The extensions approach one another (2) and then fuse to trap the material within an endocytic vesicle (3). Lysosomes fuse with this vesicle, activating digestive enzymes that gradually break down the structure of the phagocytized material (4–6). Undissolved residue is then ejected by the cell (7).

- **Pinocytosis** (pi-nō-si-TŌ-sis), or "cell drinking," is the formation of small vesicles filled with extracellular fluid. In this process, common to all cells, a deep groove or pocket forms in the cell membrane and then pinches off.
- **Phagocytosis** (fa-gō-si-TŌ-sis; *phagein*, to eat), or "cell eating," produces vesicles containing *solid objects* (Figure 3-10•). Cytoplasmic extensions called **pseudopodia** (soo-dō-PŌ-dē-a; *pseudo-*, false + *podon*, foot) surround the object, and their membranes fuse to form a vesicle. The vesicle may then fuse with a lysosome, a membranous sac of digestive enzymes, whereupon its contents are broken down.

Most cells display pinocytosis, but phagocytosis, especially the entrapment of living or dead cells, is performed only by specialized cells of the immune system. Phagocytic cells will be considered in chapters dealing with blood cells (Chapter 12) and the immune response (Chapter 15).

Exocytosis **Exocytosis** (EK-sō-sī-TŌ-sis; *exo-*, outside) is the functional reverse of endocytosis. In this process a vesicle created inside the cell fuses with the cell membrane and discharges its contents into the extracellular environment. The ejected material may be a secretory product, such as a hormone (a compound that circulates in the blood and affects cells in other parts of the body), mucus, or waste products remaining from the recycling of damaged organelles.

Many of the transport mechanisms discussed above can be moving materials in and out of the cell at any given moment. These mechanisms are summarized in Table 3-3.

✓ What is the difference between active and passive transport processes?

✓ During digestion in the stomach, the concentration of hydrogen (H^+) ions rises to many times the concentration found in the cells of the stomach. What type of transport process could produce this result?

✓ When certain types of white blood cells encounter bacteria, they are able to engulf them and bring them into the cell. What is this process called?

THE CYTOPLASM

Cytoplasm is a general term for the material inside the cell between the cell membrane and the nucleus. This material can be divided into the cytosol, organelles, and inclusions.

TABLE 3-3 Summary of Mechanisms Involved in Movement across Cell Membranes

Mechanism	*Process*	*Factors Affecting Rate*	*Substances Involved*
DIFFUSION	Molecular movement of solutes; direction determined by relative concentrations	Size of gradient, molecular size, charge, lipid solubility	Small inorganic ions, lipid-soluble materials (all cells)
Osmosis	Movement of water molecules toward solution containing relatively higher solute concentration; requires membrane	Concentration gradient, opposing osmotic or hydrostatic pressure	Water only (all cells)
FILTRATION	Movement of water, usually with solute, by hydrostatic pressure; requires filtration membrane	Amount of pressure, size of pores in filter	Water and small ions (blood vessels)
CARRIER-MEDIATED TRANSPORT			
Facilitated diffusion	Carrier molecules passively transport solutes down a concentration gradient	As above, plus availability of carrier protein	Glucose and amino acids (all cells)
Active transport	Carrier molecules actively transport solutes regardless of any concentration gradients	Availability of carrier, substrate, and ATP	Na^+, K^+, Ca^{2+}, Mg^{2+} (all cells); other solutes by specialized cells
VESICULAR TRANSPORT			
Endocytosis	Creation of vesicles containing fluid or solid material	Stimulus and mechanics incompletely understood; requires ATP	Fluids, nutrients (all cells); debris, pathogens (specialized cells)
Exocytosis	Fusion of vesicles containing fluids and/or solids with the cell membrane	Stimulus and mechanics incompletely understood; requires ATP	Fluids, debris (all cells)

The Cytosol

The **cytosol** is the intracellular fluid, which contains dissolved nutrients, ions, soluble and insoluble proteins, and waste products. It differs in composition from the **interstitial** (in-ter-STISH-al; *interstitium*, something standing between) **fluid**, the extracellular fluid that surrounds most of the cells in the body.

- The cytosol contains a high concentration of potassium, whereas extracellular fluid contains a high concentration of sodium.
- The cytosol contains a relatively high concentration of dissolved proteins, many of them enzymes that regulate metabolic operations. These proteins give the cytosol a consistency that varies between that of thin maple syrup and almost-set gelatin.
- The cytosol contains relatively small quantities of carbohydrates and large reserves of amino acids and lipids. The carbohydrates are broken down to provide energy, and the amino acids are used to manufacture proteins. The lipids stored in the cell are primarily used as an energy source when carbohydrates are unavailable.

The cytosol may also contain insoluble **inclusions**, such as stored nutrients. Examples include glycogen granules in muscle and liver cells and lipid droplets in fat cells.

Organelles

Organelles are structures that perform specific functions essential to normal cell structure, maintenance, and metabolism (see Table 3-1). Membrane-bound organelles include the *nucleus, mitochondria, endoplasmic reticulum, Golgi apparatus,* and *lysosomes.* The membrane isolates the organelle so it can manufacture or store secretions, enzymes, or toxins that might otherwise damage the cell.

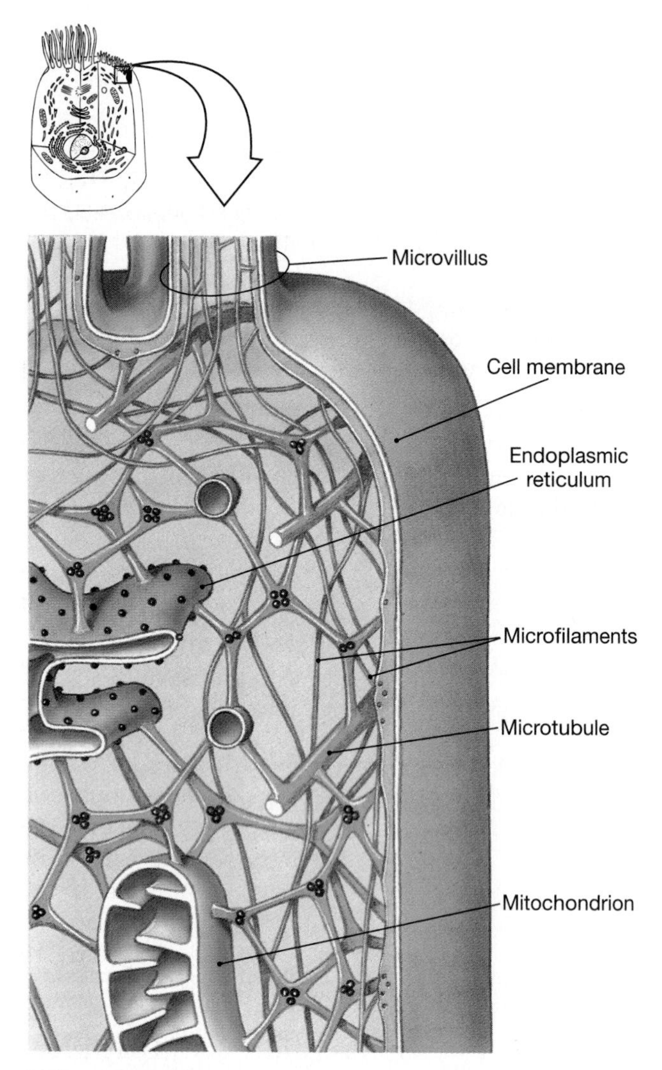

Figure 3-11
The Cytoskeleton

The cytoskeleton provides strength and structural support for the cell and its organelles. Interactions between cytoskeletal components are also important in moving organelles and changing the shape of the cell.

The *cytoskeleton, microvilli, centrioles, cilia, flagella,* and *ribosomes* are organelles without membranes.

The Cytoskeleton

The **cytoskeleton** (Figure 3-11) is an internal protein framework of various threadlike filaments and hollow tubules that gives the cytoplasm strength and flexibility. In most cells, the most important cytoskeletal elements are microfilaments and microtubules.

Microfilaments **Microfilaments** are the thinnest strands, usually composed of the protein **actin**. In most cells they form a dense layer under the cell membrane.

Microfilaments attach the cell membrane to the underlying cytoplasm by forming connections with proteins of the cell membrane. In addition, actin microfilaments can interact with thicker filaments composed of another protein, **myosin**, to produce active movement of a portion of a cell or to change the shape of the entire cell.

Microtubules **Microtubules**, found in all our cells, are hollow tubes built from the globular protein *tubulin*. Microtubules form the primary components of the cytoskeleton, giving the cell strength and rigidity and anchoring the position of major organelles.

During cell division, microtubules form the *spindle apparatus* that distributes the duplicated chromosomes to opposite ends of the dividing cell. This process will be considered in a later section (see p. 66).

Microvilli

Microvilli are small, finger-shaped projections of the cell membrane supported by microfilaments (see Figures 3-2, p. 48, and 3-11). Because they increase the surface area of the membrane, they are common features of cells actively engaged in absorbing materials from the extracellular fluid, such as the cells of the digestive tract and kidneys.

Centrioles, Cilia, and Flagella

In addition to functioning individually in the cytoskeleton, microtubules also interact to form more complex structures known as *centrioles, cilia,* and *flagella.*

Centrioles A **centriole** is a short cylindrical structure composed of microtubules (Figure 3-2, p. 48). All animal cells that are capable of dividing contain a pair of centrioles arranged perpendicular to each other. The centrioles create the spindle fibers that move DNA strands during cell division. Cells that do not divide, such as mature red blood cells and neurons of the brain, lack centrioles.

Cilia **Cilia** (singular *cilium*) are relatively long finger-shaped extensions of the cell membrane (Figure 3-2, p. 48). They are supported internally by a cylindrical array of microtubules. Cilia undergo active movements that require ATP energy. Their movements are coordinated so that their combined efforts move fluids or secretions across the cell surface. For example, cilia lining the respiratory passageways beat in a synchronized manner to move sticky mucus and trapped dust particles toward the throat and away from delicate respiratory surfaces. If the cilia are damaged or immobilized by heavy smoking or some metabolic problem, the cleansing action is lost, and the irritants will no longer be removed. As a result, chronic respiratory infections develop.

Flagella **Flagella** (fla-JEL-ah; singular *flagellum*, whip) resemble cilia but are much longer. Flagella propel a cell through the surrounding fluid, rather than moving the fluid past a stationary cell. Flagella on *pathogenic* (disease-causing) organisms allow them to move through our tissues and body fluids. The sperm cell is the only human cell that has a flagellum. If the flagella are paralyzed or otherwise abnormal, the individual will be sterile, because immobile sperm cannot reach and fertilize an egg.

Ribosomes

Ribosomes are small organelles that manufacture proteins using information provided by the DNA of the nucleus. (This process will be discussed on p. 000.) Each ribosome consists of ribosomal RNA and protein. Ribosomes are found in all cells, but their number varies depending on the type of cell and its activities. For example, liver cells, which manufacture blood proteins, have much greater numbers of ribosomes than do fat cells, which synthesize triglycerides.

There are two major types of ribosomes: *free ribosomes* and *fixed ribosomes.* **Free ribosomes** are scattered throughout the cytoplasm, and the proteins they manufacture enter the cytosol. **Fixed ribosomes** are attached to the endoplasmic reticulum (ER), a membranous organelle discussed herein. Proteins manufactured by fixed ribosomes enter the ER, where they are modified and packaged for export.

✓ Cells lining the small intestine have numerous fingerlike projections on their free surface. What are these structures, and what is their function?

✓ How would the absence of centrioles affect a cell?

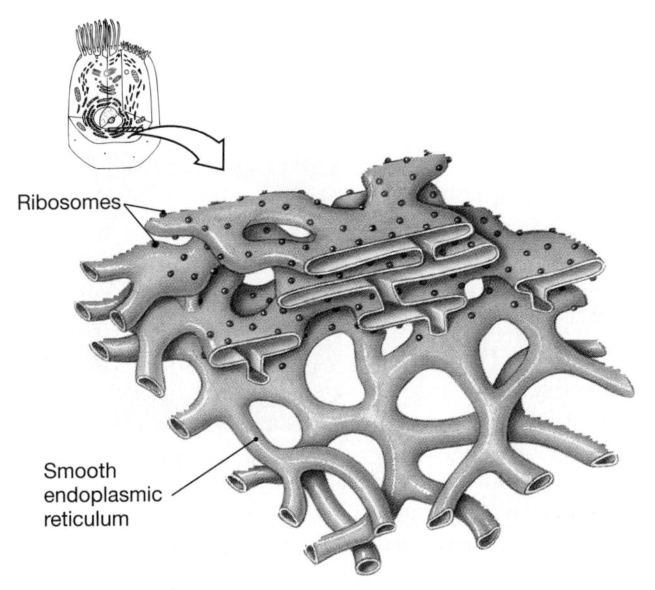

●**Figure 3-12**
The Endoplasmic Reticulum

This diagrammatic sketch indicates the three-dimensional relationships between the rough and smooth endoplasmic reticula.

The Endoplasmic Reticulum

The **endoplasmic reticulum** (en-dō-plaz-mik re-TIK-ū-lum; *reticulum*, network), or **ER**, is a network of intracellular membranes that is connected to the membranous *nuclear envelope* surrounding the nucleus (see Figure 3-2●, p. 48). The endoplasmic reticulum has three major functions:

1. ***Synthesis***. The membrane of the ER manufactures proteins, carbohydrates, and lipids.
2. ***Storage***. The ER can hold or isolate synthesized molecules or materials absorbed from the cytosol that might otherwise affect other cellular operations.
3. ***Transport***. Materials can travel from place to place in the ER.

There are two distinct types of endoplasmic reticulum, **smooth endoplasmic reticulum (SER)** and **rough endoplasmic reticulum (RER)** (Figure 3-12●). The SER, which lacks ribosomes, is the site where lipids and carbohydrates are produced. The membranes of the RER are studded with ribosomes indicating that they participate in protein synthesis. The amount of endoplasmic reticulum and the proportion of RER to SER vary depending on the type of cell and its ongoing activities. For example, pancreatic cells that manufacture digestive enzymes contain an extensive RER, and the SER is relatively small. The proportion is just the reverse in the cells that synthesize steroid hormones in the reproductive system.

The SER has a variety of functions that center around the synthesis of lipids and carbohydrates. These functions include: (1) Synthesis of the phospholipids and cholesterol needed for maintenance and growth of the cell membrane, ER, nuclear membrane, and Golgi apparatus in all cells; (2) Synthesis of steroid hormones, such as *testosterone* and *estrogen* (sex hormones) in cells of the reproductive organs; and (3) Synthesis and storage of glycogen in skeletal muscle and liver cells.

The lipids and carbohydrates produced by the SER and the proteins produced by the ribosomes of the RER may become incorporated into membranes or enter the inner chambers of their respective endoplasmic reticulum. These molecules are then packaged into small membrane sacs that pinch off from the tips of the ER. The sacs, called *transport vesicles,* deliver them to the Golgi apparatus, another membranous organelle, where they are processed further.

The Golgi Apparatus

The **Golgi** (GOL-jē) **apparatus** consists of a set of five to six flattened membrane discs. A single cell may contain several sets, each resembling a stack of dinner plates (see Figure 3-2●, p. 48). The major functions of the Golgi apparatus are: (1) the synthesis and packaging of secretions, such as mucus or enzymes, (2) the renewal or modification of the cell membrane; and (3) the packaging of special enzymes for use in the cytosol.

The Golgi apparatus receives substances from the ER and also sends substances to the cell surface through the formation, movement, and fusion of vesicles. Figure 3-13● illustrates this process. Transport vesicles formed at the ER empty their contents into the Golgi apparatus as they fuse with its membranes. Enzymes in the Golgi apparatus modify the newly arrived molecules as they move through succeeding membranes toward the cell surface.

Ultimately the modified proteins, lipids, or carbohydrates within the Golgi apparatus are packaged within vesicles. Three types of vesicles leave the Golgi apparatus. One class of vesicles contains secretions that will be discharged from the cell. These are called **secretory vesicles**, and secretion occurs through exocytosis at the cell surface (Figure 3-13b●). A second class of vesicles does not contain secretions, but the vesicle membrane will be incorporated into the cell membrane. Because the Golgi apparatus continually adds new membrane to the cell surface in this way, the properties of the cell membrane can change. For example, receptors can be added or removed, making the cell more or less sensitive to a particular stimulus. A third class of vesicles will remain in the cytoplasm. These vesicles, called *lysosomes*, contain digestive enzymes.

Lysosomes

Lysosomes (LĪ-so-sōmz; *lyso-*, breakdown + *soma*, body) are vesicles filled with digestive enzymes. Lysosomes perform the cleanup and recycling functions within the cell. Their enzymes are activated when they fuse with the membranes of damaged organelles, such as mitochondria or fragments of the endoplasmic reticulum. Then the enzymes break down the lysosomal contents. Nutrients reenter the cytosol through passive or active transport processes, and the remaining material is eliminated by exocytosis.

Lysosomes also function in the defense against disease. Through endocytosis, cells may engulf bacteria, fluids, and organic debris from their surroundings into vesicles formed at the cell surface. Lysosomes fuse with vesicles created in this way, and the digestive enzymes then break down the contents and release usable substances such as sugars or amino acids.

Lysosomes perform essential recycling functions by removing damaged or dead cells. Within such cells, lysosome membranes disintegrate, releasing active enzymes into the cytosol. These enzymes rapidly destroy the proteins and organelles of the cell, a process called **autolysis** (aw-TAH-li-sis; *auto-*, self). Because the breakdown of lysosomal membranes can destroy a cell, lysosomes have been called cellular "suicide packets." We do not know how to control lysosomal activities or why the enclosed enzymes do not digest the lysosomal walls unless the cell is damaged. AM *Lysosomal Storage Diseases*

Mitochondria

Mitochondria (mī-tō-KON-drē-ah; singular *mitochondrion; mitos,* thread + *chondros*, cartilage) are small organelles that contain enzymes that regulate the reactions that provide energy for the cell. Mitochondria have an unusual double membrane: an outer membrane surrounding the entire organelle and an inner membrane containing numerous folds, called *cristae*. The energy-producing enzymes occur on the cristae, which provide a large surface area for reactions to take place, and within the fluid they enclose, the *matrix* (Figure 3-14●). The

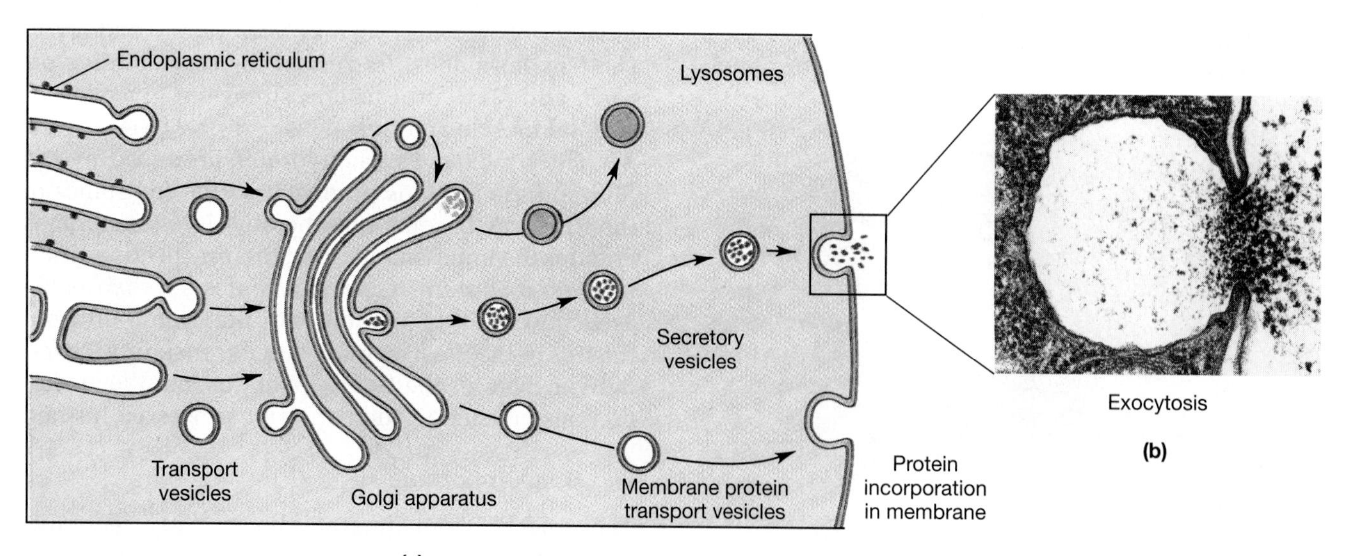

●**Figure 3-13**
The Golgi Apparatus
(a) Function of the Golgi apparatus. **(b)** Exocytosis of secretions at the cell surface.

number of mitochondria in a particular cell varies depending on its energy demands. For example, red blood cells have none, but mitochondria may account for 20 percent of the volume of an active liver cell.

Although most of the chemical reactions that release energy occur in the mitochondria, most of the cellular activities that require energy occur in the surrounding cytoplasm. Cells must therefore store energy in a form that can be moved from place to place. Energy is stored and transferred in the *high-energy bond* of *adenosine triphosphate,* or *ATP,* as discussed in Chapter 2. 1 [p. 41] Living cells break the high-energy phosphate bond under controlled conditions, reconverting ATP to ADP and releasing energy for their use.

Mitochondrial Energy Production Most cells generate ATP and other high-energy compounds through the breakdown of carbohydrates, especially glucose. Although most of the actual energy production occurs inside mitochondria, the first steps take place in the cytosol. In this reaction sequence, called *glycolysis,* six-carbon glucose molecules are broken down into three-carbon molecules of pyruvic acid. These molecules are then absorbed by the mitochondria. If glucose or other carbohydrates are not available, mitochondria can absorb and utilize small carbon chains produced by the breakdown of proteins or lipids.

Because the key reactions involved in mitochondrial activity consume oxygen, the process of mitochondrial energy production is known as **aerobic metabolism** (*aero-*, air + *bios*, life). Aerobic metabolism in mitochondria produces about 95 percent of the energy needed to keep a cell alive. Aerobic metabolism is discussed in more detail in Chapters 7 and 18.

There are several inheritable disorders that result from abnormal mitochondrial activity. The mitochondria involved have defective enzymes that reduce their ability to generate ATP. Cells throughout the body may be affected, but symptoms involving muscle cells, nerve cells, and the light receptor cells in the eye are most commonly seen because these cells have especially high energy demands. AM *Mitochondrial DNA, Disease, and Evolution*

✓ Cells in the ovaries and testes contain large amounts of smooth endoplasmic reticulum (SER). Why?

✓ Microscopic examination of a cell reveals that it contains many mitochondria. What does this observation imply about the cell's energy requirements?

THE NUCLEUS

The **nucleus** is the control center for cellular operations, for here is where the genetic information (DNA) is stored. Most cells contain a single nucleus, but there are exceptions. For example, skeletal muscle cells have many nuclei, and mature red blood cells have none. Figure 3-15● details the structure of a typical nucleus. A **nuclear envelope** consisting of a double membrane surrounds the nucleus and its fluid contents, the **nucleoplasm,** from the cytosol. The nucleoplasm contains ions, enzymes, RNA and DNA nucleotides, proteins, small amounts of RNA, and DNA.

Chemical communication between the nucleus and the cytosol occurs through **nuclear pores.** These pores, which cover about 10 percent of the surface of the nucleus, are large enough to permit the movement of ions and small molecules but too small for the passage of proteins or DNA.

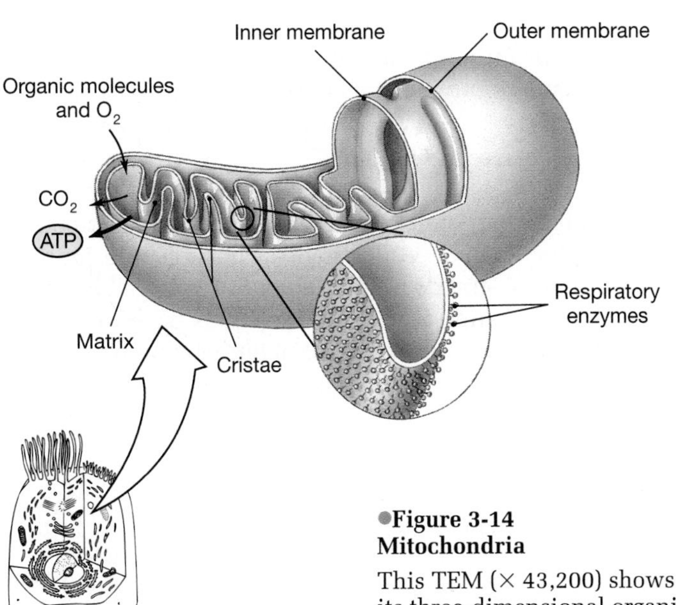

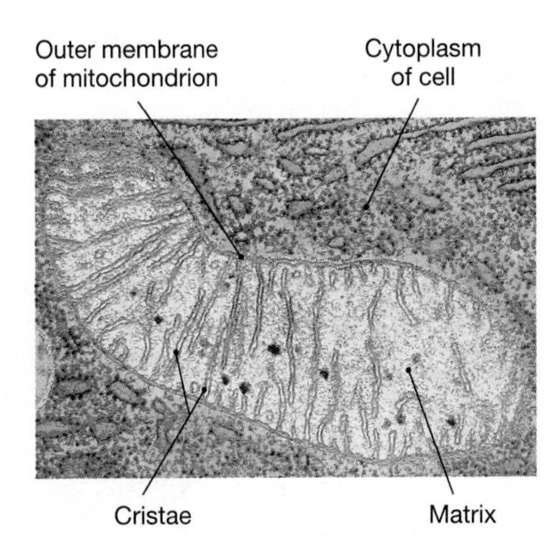

●**Figure 3-14**
Mitochondria

This TEM (× 43,200) shows a typical mitochondrion in section, and the sketch details its three-dimensional organization. Mitochondria absorb short carbon chains, ADP, and oxygen and generate carbon dioxide, water, and ATP.

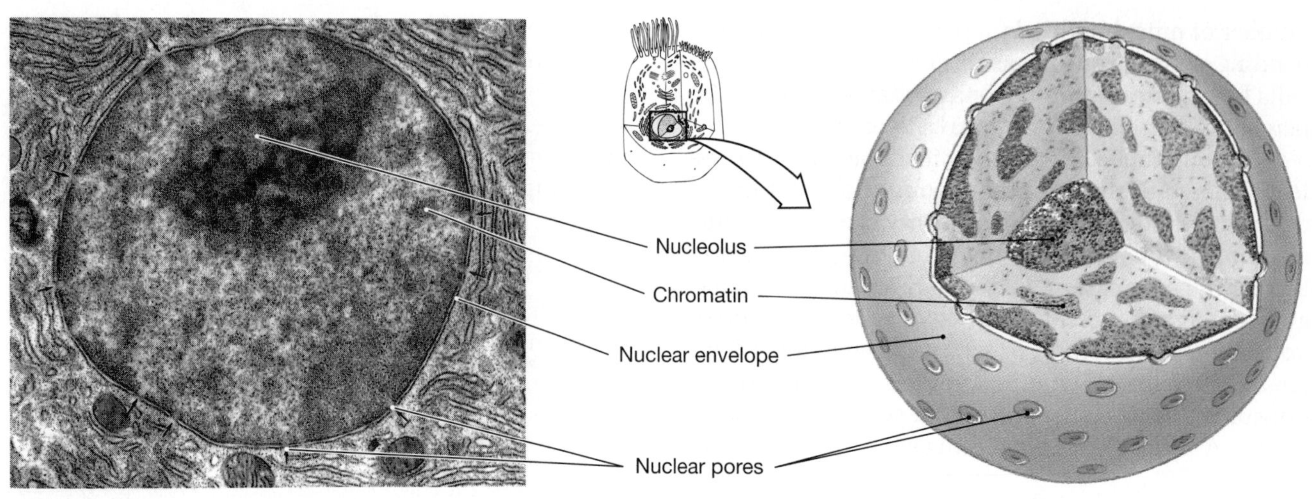

Figure 3-15
The Nucleus
TEM showing important nuclear structures. (TEM × 4828)

Chromosome Structure

The DNA within the cell nucleus interacts with special proteins to form complex structures known as **chromosomes** (*chroma,* color). Each nucleus contains 23 pairs of chromosomes; one member of each pair is derived from our mother and one from our father. The structure of a typical chromosome is diagrammed in Figure 3-16.

Each chromosome contains DNA strands bound to special proteins called *histones.* At intervals, the DNA strands wind around the histones, coiling up the DNA. The degree of coiling determines whether the chromosome is long and thin or short and fat. Chromosomes in a dividing cell are very tightly coiled, and they can be seen clearly as separate structures in light or electron micrographs. In cells that are not dividing, the DNA is loosely coiled, forming a tangle of fine filaments known as **chromatin.**

All vital cellular activities involve proteins, and proteins make up 15–30 percent of the weight of each cell. The nucleus controls cellular operations through its regulation of protein synthesis; the DNA strands of our chromosomes contain the information needed to synthesize at least 100,000 different proteins. (The details of this process are discussed in the next section.) Most nuclei also contain one to four dark-staining regions called **nucleoli** (noo-KLĒ-ō-lī; singular *nucleolus*). Nucleoli are organelles that synthesize the components of ribosomes. For this reason, they are most prominent in cells that manufacture large amounts of proteins, such as muscle or liver cells.

The Genetic Code

The basic structure of nucleic acids was described in Chapter 2. [p. 40] A single DNA molecule consists of a pair of strands held together by hydrogen bonding between complementary nitrogenous bases. Information is stored in the sequence of nitrogenous bases (adenine, A; thymine, T; cytosine, C; and guanine, G) along

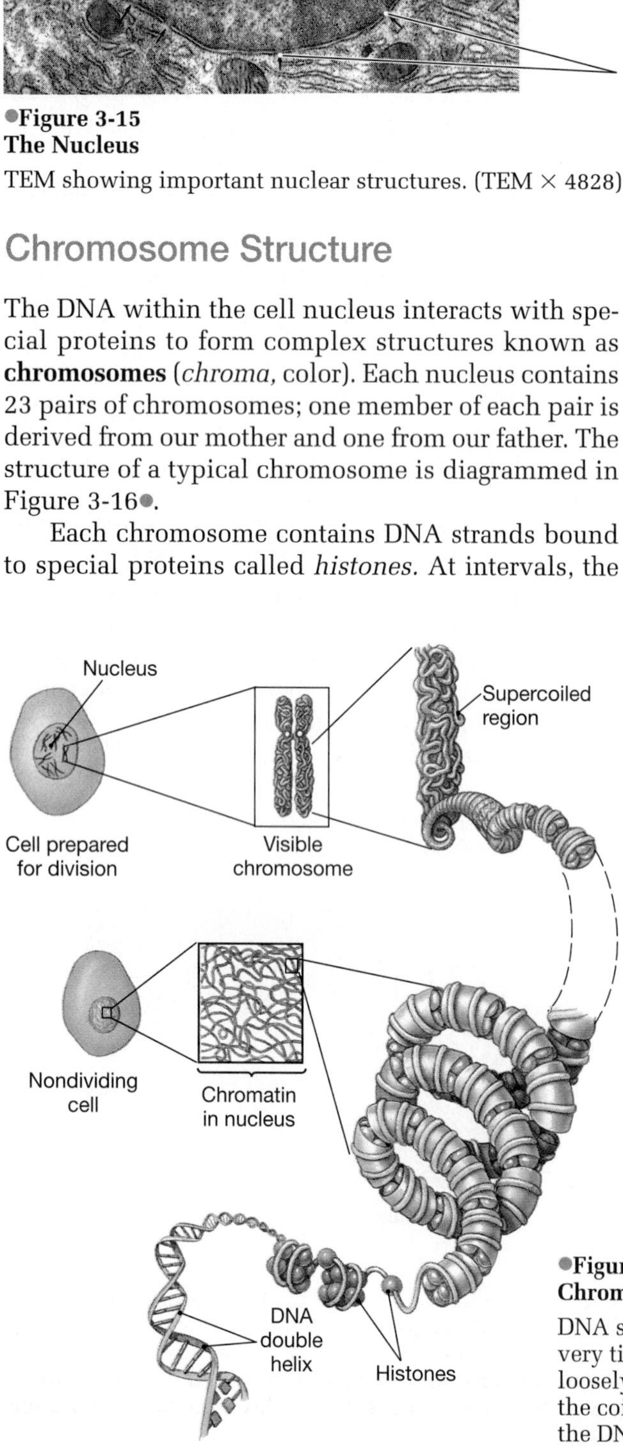

Figure 3-16
Chromosome Structure
DNA strands wound around histone proteins form coils that may be very tight or rather loose. In cells that are not dividing, the DNA is loosely coiled, forming a tangled network known as chromatin. When the coiling becomes tighter, as it does in preparation for cell division, the DNA becomes visible as distinct structures called chromosomes.

the length of DNA strands. This information-storage system is known as the **genetic code**.

The genetic code is called a *triplet code* because a sequence of three nitrogenous bases can specify the identity of a single amino acid. Each **gene** consists of all the triplets needed to produce a specific peptide chain. Because one triplet controls a single amino acid, the number of triplets (and the size of the gene) varies depending on the length of the peptide that will be produced. Each gene also contains special segments responsible for regulating its own activity. In effect these triplets say, "Do (or do not) read this message," "Message starts here," and "Message ends here."

DNA FINGERPRINTING

All of the nucleated cells in the body carry identical copies of the 46 chromosomes present in the fertilized egg at the time of conception. But the DNA nucleotide sequences do vary from individual to individual, and the chances of any two individuals, other than identical twins, having the same pattern is less than one in 9 billion. In other words, it is extremely unlikely that you will ever encounter someone else who has the same pattern of repeating nucleotide sequences present in your DNA.

An individual's identification can therefore be made on the basis of a pattern of DNA analysis, just as it can on the basis of a fingerprint. Skin scrapings, blood, semen, hair, or other tissues can be used as a sample source. Information from DNA fingerprinting has already been used to convict persons committing violent crimes, such as rape or murder.

Transcription

Ribosomes, the organelles of protein synthesis, are found in the cytoplasm, whereas the genes remain in the nucleus. This separation between the manufacturing site and the DNA's protein blueprint is overcome by the movement of a molecular messenger, a single strand of RNA known as **messenger RNA (mRNA)**. The process of mRNA formation is called **transcription** because the newly formed mRNA is copying the information contained in the gene. Figure 3-17● details this process.

Each DNA strand contains thousands of individual genes. When a gene is activated, an enzyme, *RNA polymerase,* binds to the initial segments of the gene. This enzyme promotes the synthesis of a mRNA, using nucleotides complementary to those in the

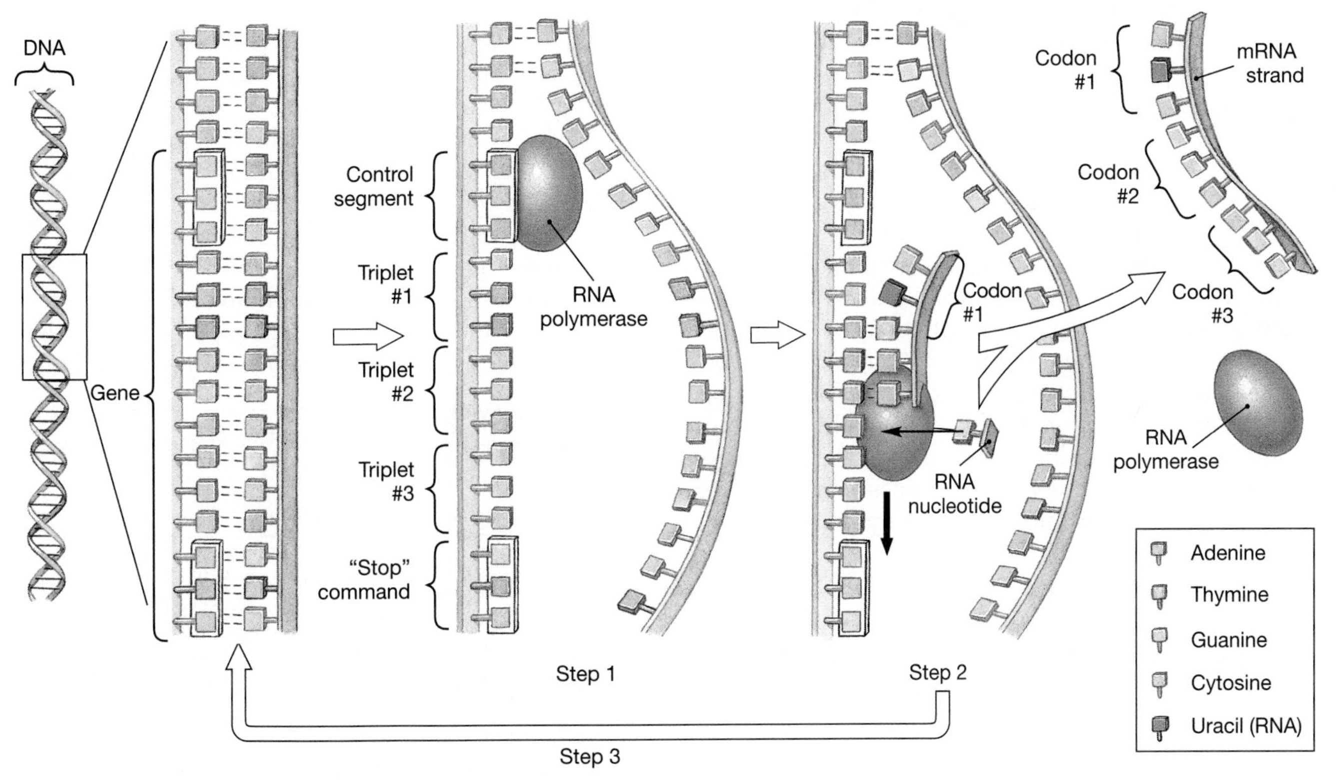

●**Figure 3-17**
Transcription

This diagram shows a small portion of a single DNA molecule, containing a single gene available for transcription. **Step 1:** The two DNA strands separate, and RNA polymerase binds to the control segment of the gene. **Step 2:** The RNA polymerase moves from one triplet to another along the length of the gene. At each site, complementary RNA nucleotides form hydrogen bonds with the DNA nucleotides of the gene. The RNA polymerase then strings the arriving nucleotides together into a strand of mRNA. **Step 3:** Upon reaching the stop signal at the end of the gene, the RNA polymerase and the mRNA strand detach, and the two DNA strands reassociate.

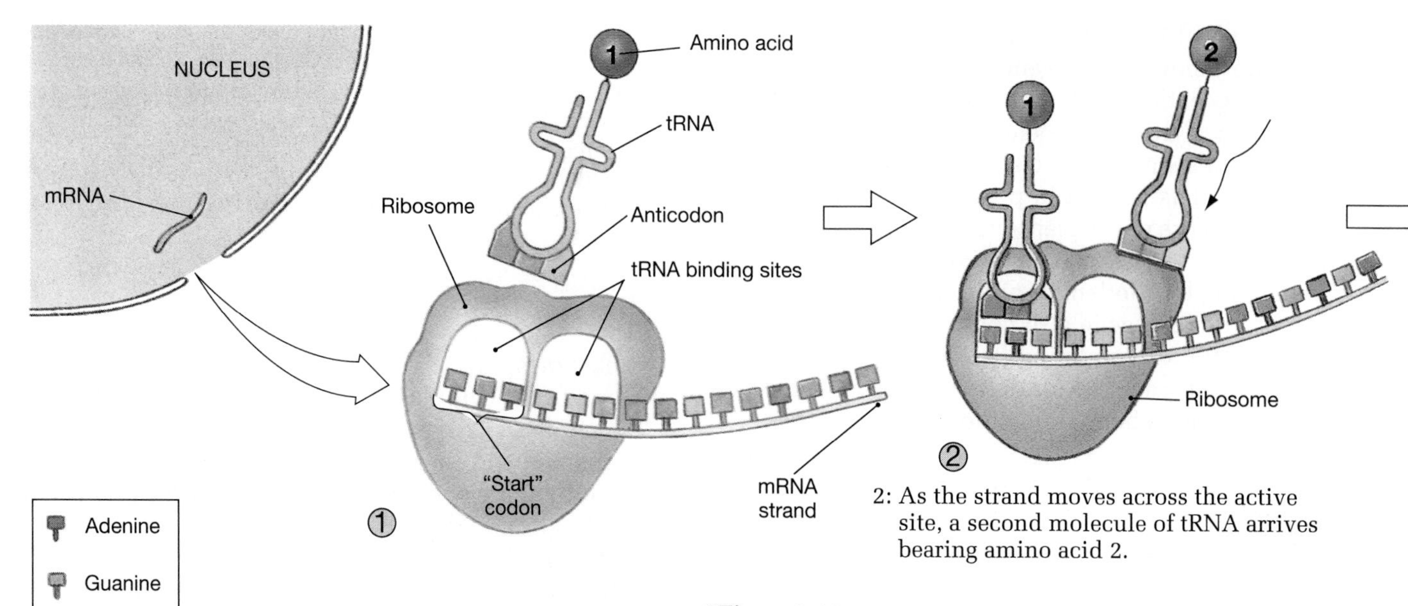

1: A molecule of tRNA brings amino acid 1 to the active binding site; the anticodon of the tRNA must be complementary to the first codon on the mRNA strand.

2: As the strand moves across the active site, a second molecule of tRNA arrives bearing amino acid 2.

Figure 3-18
Translation

Once transcription has been completed, the mRNA moves into the cytoplasm and interacts with the ribosome as described in steps 1–4.

gene. The nucleotides involved are those characteristic of RNA, not DNA; RNA polymerase may attach adenine, guanine, cytosine, or uracil (U), but never thymine. Thus wherever an A occurs in the DNA strand, RNA polymerase will attach a U rather than a T. The mRNA strand thus contains a sequence of nitrogenous bases that are complementary to those of the gene. A sequence of three nitrogen bases along the new mRNA strand represents a **codon** (KŌ-don) that is complementary to the corresponding triplet along the gene. At the DNA "stop here" command, the enzyme and the mRNA strand detach, and the complementary DNA strands reassociate.

The mRNA formed in this way may be altered before it leaves the nucleus. For example, some triplets may be removed, creating a shorter strand of mRNA. After modifications have been performed, the completed, functional mRNA strand passes through one of the nuclear pores and enters the cytoplasm.

Translation

Translation is the synthesis of a polypeptide using the information provided by the sequence of codons along the mRNA strand. Every amino acid has at least one unique and specific codon; Table 3-4 includes several examples. During translation the sequence of codons will determine the sequence of amino acids in the polypeptide.

Translation is initiated when the newly synthesized mRNA binds with a ribosome (Figure 3-18). Molecules of **transfer RNA (tRNA)** then deliver amino acids that will be used by the ribosome to assemble a peptide chain. There are more than 20 different types of transfer RNA, at least one for each amino acid used in protein synthesis. Each tRNA molecule contains a complementary trio of nitrogenous bases, known as an **anticodon**, that will bind to a specific codon on the mRNA.

Translation begins at the start codon of the mRNA strand, with the arrival and binding of the first tRNA. That tRNA carries a specific amino acid (1). A second tRNA then arrives, carrying a different amino acid (2), and binds to the second codon. Ribosomal enzymes now remove amino acid 1 from the first tRNA and attach it to amino acid 2 with a peptide bond. The first tRNA then detaches from the ribosome and reenters

TABLE 3-4 Examples of the Triplet Code

DNA Triplet	*mRNA Codon*	*tRNA Anticodon*	*Amino Acid or Instruction*
AAA	UUU	AAA	Phenylalanine
AAT	UUA	AAU	Leucine
ACA	UGU	ACA	Cysteine
CAA	GUU	CAA	Valine
GGG	CCC	GGG	Proline
CGA	GCU	CGA	Alanine
CGG	GCC	CGG	Alanine
CGC	GCG	CGC	Alanine
TAC	AUG	UAC	Initiator codon
ATT	UAA	[none]	Terminator codon
ATC	UAG	[none]	Terminator codon
ACT	UGA	[none]	Terminator codon

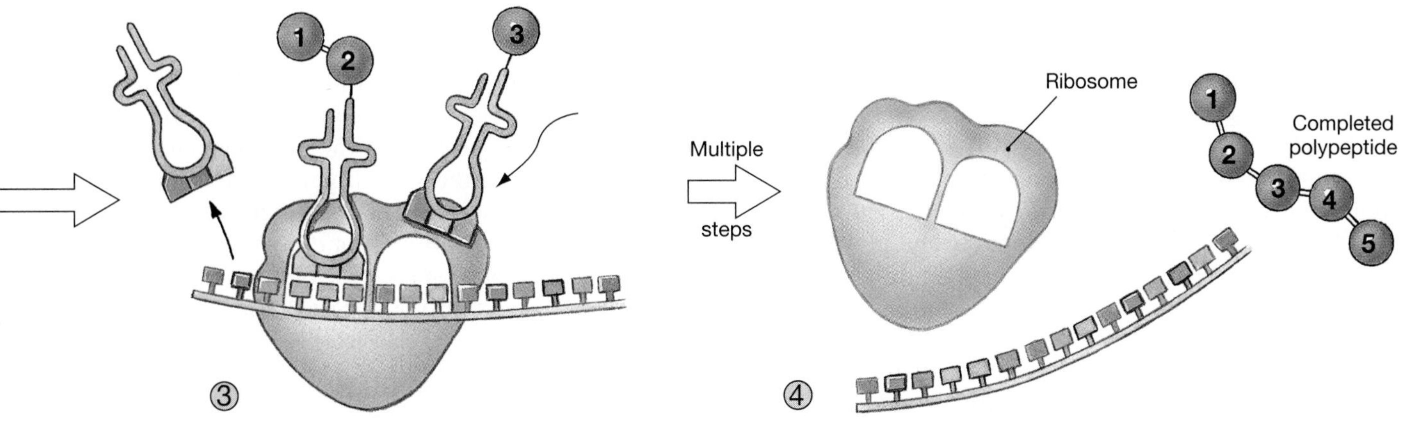

3: Enzymes of the ribosome break the linkage between tRNA-1 and amino acid 1 and join amino acids 1 and 2 with a peptide bond. The ribosome shifts one codon to the right, tRNA 1 departs, and a third tRNA arrives.

4: This process continues until the ribosome reaches the stop codon. The ribosome then breaks the connection between the last tRNA molecule and the peptide chain. The ribosome disengages, leaving the mRNA strand intact.

the cytosol, where it can pick up another amino acid molecule and repeat the process. The ribosome now moves one codon farther along the length of the mRNA strand, and a third tRNA arrives, bearing amino acid 3. The ribosomal enzymes remove the dipeptide (1-2) from the second tRNA and attach it to amino acid 3. The second tRNA is then released, and the ribosome moves one codon farther along the mRNA strand. Amino acids will continue to be added to the growing peptide chain in this way until the ribosome reaches the stop codon. The ribosome then detaches, leaving an intact strand of mRNA and a completed polypeptide.

Translation proceeds swiftly, producing a typical protein (about 1000 amino acids) in around 20 seconds. The polypeptide begins as a simple linear strand, but a more complex structure develops as it grows longer.

MUTATIONS

Mutations are permanent alterations in a cell's DNA that affect the nucleotide sequence of one or more genes. The simplest is a point mutation, a change in a single nucleotide that affects one codon. With roughly 3 billion nucleotides in the DNA of a human cell, a single mistake might seem relatively unimportant. But over 100 inherited disorders have been traced to abnormalities in enzyme or protein structure that reflect alterations in the nucleotide sequence of specific genes. A change in the amino acid sequence of a single structural protein or enzyme can prove fatal. For example, several cancers and two potentially lethal blood disorders, *thalassemia* and *sickle cell anemia*, result from variations in amino acid sequence caused by changing one nucleotide within the associated gene. More elaborate mutations can affect chromosomal structure and even break a chromosome apart.

Mutations are most likely to occur in cells undergoing cell division, with the result that a single cell or group of daughter cells may be changed. If the mutations occur early in development, every cell in the body may be affected. Our increasing understanding of genetic structure is opening the possibility for diagnosing and correcting some of these problems. For a discussion of the principles and technologies involved, see the *Applications Manual.* AM *Genetic Engineering and Gene Therapy*

✓ How does the nucleus control the activities of the cell?

✓ What process would be affected by the lack of the enzyme RNA polymerase?

✓ During the process of transcription, a nucleotide was deleted from a mRNA sequence that coded for a polypeptide. What effect would this deletion have on the amino acid sequence of the polypeptide?

CELL GROWTH AND DIVISION

Between fertilization and physical maturity a human being goes from a single cell to roughly 75 trillion cells. This amazing increase in numbers occurs through a form of cellular reproduction called **cell division**. Even when development has been completed, cell division continues to be essential to survival as it replaces old and damaged cells.

Central to cell reproduction is the accurate duplication of the cell's genetic material and its distribution to the two new daughter cells formed by division. This process is called **mitosis** (mī-TŌ-sis). Mitosis occurs during the division of the **somatic** (*soma*, body), or nonreproductive, cells of the body. The **reproductive cells** give rise to sperm or ova (eggs) through a distinct process, **meiosis** (mī-Ō-sis), that will be described in Chapter 20.

Most cells spend only a small part of their life cycle engaged in cell division, or mitosis. For most of their lives, cells are in **interphase**, an interval of time between cell divisions when they perform normal functions.

Interphase

A cell in interphase is not necessarily preparing for mitosis. Some mature cells, such as skeletal muscle cells and many nerve cells, never undergo mitosis or cell division. A cell that is going to divide must first manufacture enough organelles and cytosol to make two functional cells. These preparations may take hours, days, or weeks to complete, depending on the type of cell and the situation. For example, there are cells in the lining of the digestive tract that divide every few days throughout life, whereas there are cells in other tissues that divide only under special circumstances, such as following an injury.

Once these preparations have been completed, the cell replicates the DNA in the nucleus, a process that takes 6–8 hours. The goal of **DNA replication** is to copy the genetic information in the nucleus so that one set of chromosomes can be given to each of the two cells produced. This process, diagrammed in Figure 3-19●, starts when the complementary strands begin to separate and unwind. Molecules of the enzyme *DNA polymerase* then bind to the exposed nitrogen bases. As a result, complementary nucleotides within the nucleoplasm attach to the exposed nitrogen bases of the DNA strand and form a pair of identical DNA molecules. Shortly after DNA replication has been completed, the process of mitosis begins.

Mitosis

Mitosis is a process that separates and encloses the duplicated chromosomes of the original cell into two identical nuclei. Separation of the cytoplasm to form two separate and distinct cells involves a separate but related process known as *cytokinesis* (sī-tō-ki-NĒ-sis; *cyto-*, cell + *kinesis*, motion). For convenience, mitosis is divided into four stages: *prophase, metaphase, anaphase,* and *telophase* (Figure 3-20●).

Stage 1: Prophase

Prophase (PRŌ-fāz; *pro-*, before) begins when the chromosomes coil so tightly that they become visible as individual structures. As a result of DNA replication there are now two copies of each chromosome. Each copy, called a **chromatid** (KRŌ-ma-tid), is connected to its duplicate at a single point, the **centromere** (SEN-trō-mēr).

As the chromosomes appear, the two pairs of centrioles move toward opposite poles of the nucleus. An array of microtubules, called **spindle fibers**, extend between the centriole pairs. Prophase ends with the disappearance of the nuclear envelope.

Stage 2: Metaphase

Metaphase (MET-a-fāz; *meta-*, after) begins after the disintegration of the nuclear envelope. The spindle fibers now enter the nuclear region and the chromatids

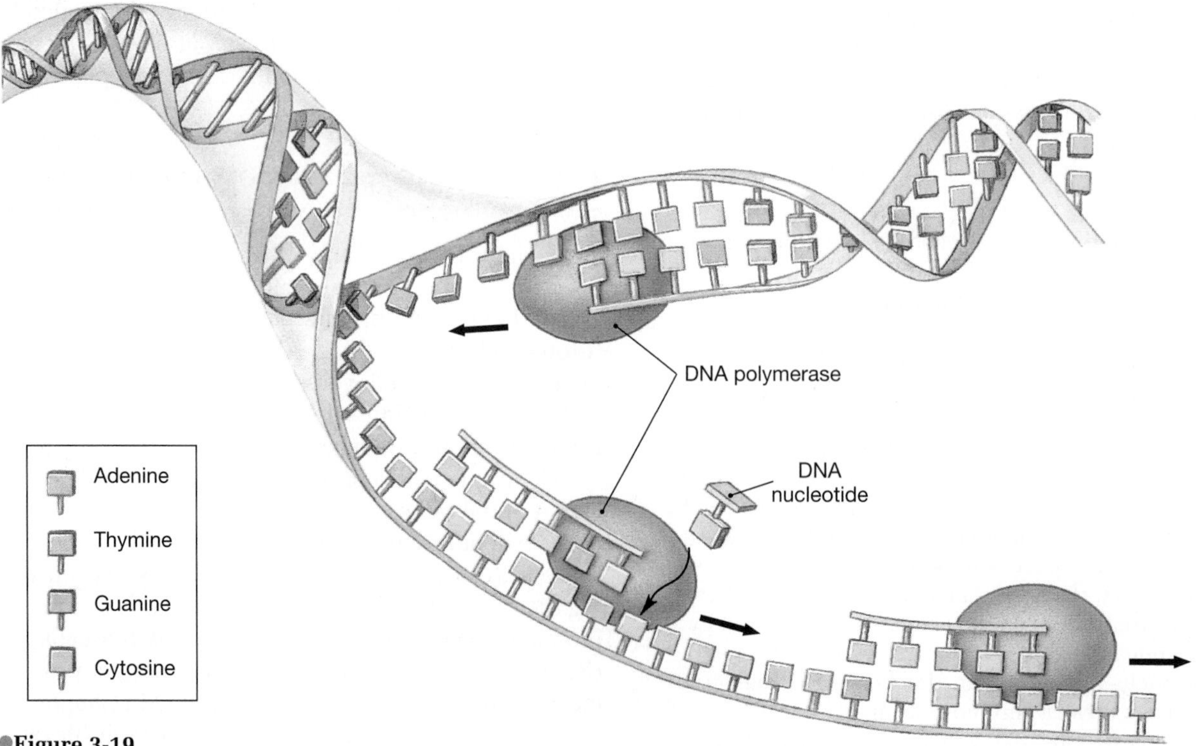

●**Figure 3-19**
DNA Replication
In replication, the DNA strands unwind and DNA polymerase begins attaching complementary DNA nucleotides along each strand. This process produces two identical copies of the original DNA molecule.

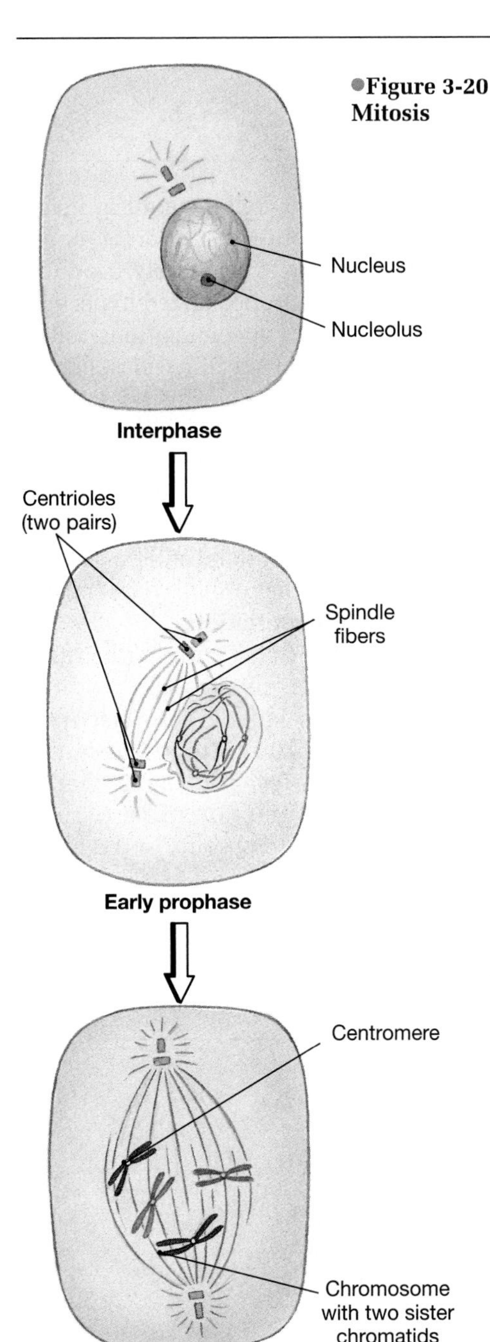

●**Figure 3-20**
Mitosis

become attached to them. Once attachment has been completed, the chromatids move to a narrow central zone called the **metaphase plate.** Metaphase ends when all of the chromatids are aligned in the plane of the metaphase plate.

Stage 3: Anaphase

Anaphase (AN-uh-fāz; *ana-*, apart) begins when the centromere of each chromatid pair splits, and the chromatids separate. The two **daughter chromosomes** are now pulled toward opposite ends of the cell. Anaphase ends when the daughter chromosomes arrive near the centrioles at opposite ends of the cell.

Stage 4: Telophase

During **telophase** (TEL-o-fāz; *telos*, end), the cell prepares to return to the interphase state. The nuclear membranes form, the nuclei enlarge, and the chromosomes gradually uncoil. Once the chromosomes have relaxed and the fine filaments of chromatin become visible, nucleoli reappear and the nuclei resemble those of interphase cells.

Cytokinesis

Telophase marks the end of mitosis proper, but the daughter cells have yet to complete their physical separation. This separation process, called **cytokinesis,** usually begins in late anaphase. As the daughter chromosomes near the ends of the spindle apparatus, the cytoplasm constricts along the plane of the metaphase plate. This process continues throughout telophase and is usually completed sometime after the nuclear membrane has re-formed. The completion of cytokinesis marks the end of the process of cell division.

Cell Division and Cancer

Within normal tissues, the rate of cell division is balanced with the rate of cell loss. If this balance breaks down, abnormal cell growth and cell division will enlarge the tissue and form a **tumor,** or **neoplasm.** In a **benign tumor** the abnormal cells remain consolidated and seldom threaten an individual's life. Surgery can usually re-

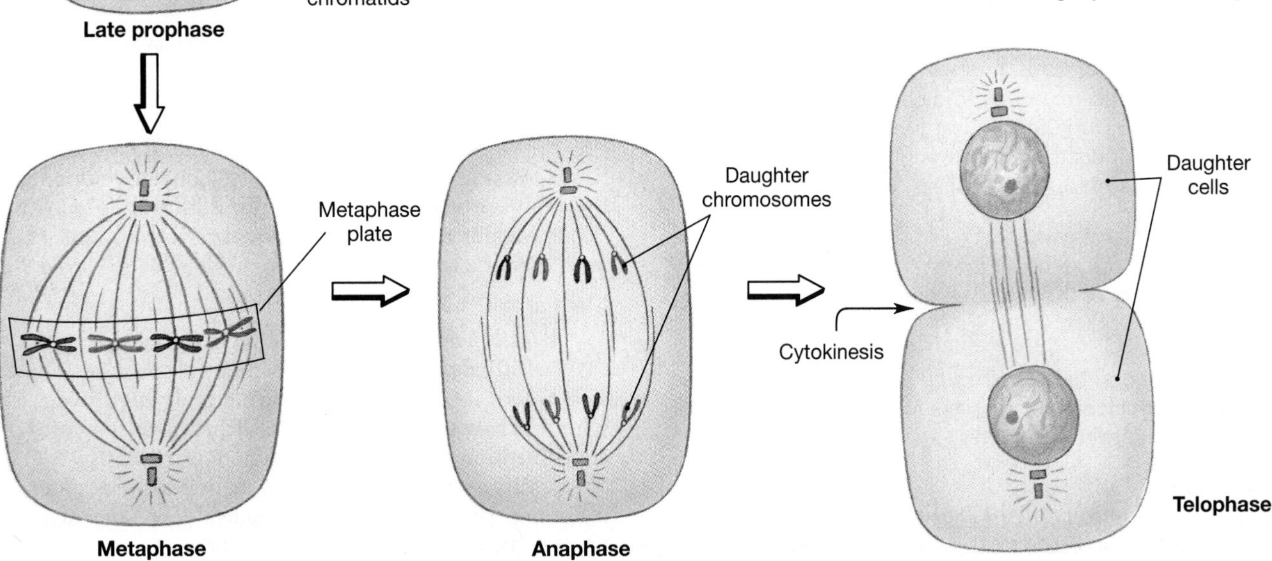

move the tumor if it begins to disturb the functions of the surrounding tissue.

Cells in a **malignant tumor**, however, no longer respond to normal control mechanisms. These cells spread not only into nearby tissue but also to other tissues and organs. This spread is called **metastasis** (me-TAS-ta-sis), a process quite difficult to control. Once in a new location, the metastatic cells produce secondary tumors.

The term **cancer** refers to an illness characterized by malignant cells. Such cancer cells lose their resemblance to normal cells and cause organs to dysfunction as their numbers increase. Cancer cells use energy less efficiently than normal cells, and they grow and multiply at the expense of normal tissues. The cancer cells steal nutrients from normal cells, and this accounts for the starved appearance of many patients in the late stages of cancer. For a detailed discussion of cancer causes, treatments, and statistics, see the *Applications Manual.* AM Cancer

✓ What major event occurs during interphase of cells preparing to undergo mitosis?

✓ List the four stages of mitosis.

✓ What would happen if spindle fibers failed to form in a cell during mitosis?

CELL DIVERSITY AND DIFFERENTIATION

Fertilization produces a single cell with all of its genetic potential intact. There follows a period of repeated cell divisions, a process that ultimately produces trillions of cells. These cells are specialized to perform particular functions, and their specializations reflect the activation or deactivation of specific genes. For example, liver cells, fat cells, and neurons contain the same chromosomes and genes, but each cell type has a different set of genes available for transcription. The other genes in the nucleus have been deactivated or "turned off." When a gene is deactivated, the cell loses the ability to create a particular protein, and thus to perform any functions involving that protein. The specialization process is called **differentiation.**

Differentiation begins early in embryonic development, as the number of cells increases. It produces specialized cells with limited capabilities. These cells form organized collections known as *tissues,* each with discrete functional roles. The next chapter examines the structure and function of tissues and considers the role of tissue interactions in the maintenance of homeostasis.

Chapter Review

KEY TERMS

active transport, p. 55
chromosomes, p. 62
cytoplasm, p. 56
diffusion, p. 50
endocytosis, p. 55
exocytosis, p. 56
gene, p. 63
interstitial fluid, p. 57
mitochondria, p. 60
mitosis, p. 65
nucleus, p. 61
osmosis, p. 52

SUMMARY OUTLINE

INTRODUCTION p. 47

1. Contemporary cell theory incorporates several basic concepts: (1) Cells are the building blocks of all plants and animals; (2) cells are the smallest functioning units of life; (3) cells are produced by the division of preexisting cells; (4) each cell maintains homeostasis. (*Figure 3-1*)

STUDYING CELLS p. 47

1. Electron microscopes are important tools used in **cytology,** the study of the structure and function of cells.

An Overview of Cellular Anatomy p. 48

2. A cell floats in the *extracellular fluid.* The cell's outer boundary, the **cell membrane**, separates the *cytoplasm* from the extracellular fluid. (*Figure 3-2; Table 3-1*)

THE CELL MEMBRANE p. 50

1. The functions of the cell membrane include: (1) physical isolation; (2) control of the exchange of materials with the cell's surroundings; (3) sensitivity; (4) structural support.

Membrane Structure p. 50

2. The cell membrane, or *plasma membrane,* contains lipids, proteins, and carbohydrates. Its major components, lipid molecules, form a **phospholipid bilayer.** (*Figure 3-3*)

3. Membrane proteins may function as receptors, channels, carriers, enzymes, anchors, or identifiers. (*Table 3-2*)

Membrane Transport p. 50

4. Cell membranes are **selectively permeable.**

5. **Diffusion** is the net movement of material from an area where its concentration is relatively high to an area where its concentration is lower. Diffusion occurs until the **concentration gradient** is eliminated. (*Figures 3-4, 3-5*)

6. Diffusion of water across a membrane in response to differences in concentration is **osmosis**. The force of movement is **osmotic pressure.** (*Figures 3-6, 3-7*)

7. In **filtration,** hydrostatic pressure forces water across a membrane; if membrane pores are large enough, molecules of solute will be carried along.

8. **Facilitated diffusion** is a type of **carrier-mediated transport** and requires the presence of membrane carrier proteins. (*Figure 3-8*)

9. **Active transport** mechanisms consume ATP but are independent of concentration gradients. Some **ion pumps** are **exchange pumps**. (*Figure 3-9*)

10. In **vesicular transport**, material moves into or out of a cell in membranous sacs. Movement into the cell occurs through **endocytosis**, an active process that includes **pinocytosis** ("cell-drinking") and **phagocytosis** ("cell-eating"). Movement out of the cell occurs through **exocytosis.** (*Figure 3-10; Table 3-3*)

THE CYTOPLASM p. 56

1. The cytoplasm surrounds the nucleus and contains a fluid **cytosol**, intracellular structures called **organelles,** and **inclusions.**

The Cytosol p. 57

2. The cytosol differs in composition from the **interstitial fluid** that surrounds most cells of the body.

Organelles p. 57

3. **Membranous organelles** are surrounded by lipid membranes that isolate them from the cytosol. They include the endoplasmic reticulum, the nucleus, the Golgi apparatus, lysosomes, and mitochondria. (*Table 3-1*)

4. **Nonmembranous organelles** are always in contact with the cytosol. They include the cytoskeleton, microvilli, centrioles, cilia, flagella, and ribosomes. (*Table 3-1*)

5. The **cytoskeleton** gives the cytoplasm strength and flexibility. Its two main components are **microfilaments** and **microtubules.** (*Figure 3-11*)

6. **Microvilli** are small projections of the cell membrane that increase the surface area exposed to the extracellular environment. (*Figure 3-11*)

7. **Centrioles** direct the movement of chromosomes during cell division.

8. **Cilia** beat rhythmically to move fluids or secretions across the cell surface.

9. **Flagella** move a cell through surrounding fluid, rather than moving fluid past a stationary cell.

10. **Ribosomes** are intracellular factories that manufacture proteins. There are **free ribosomes** in the cytoplasm and **fixed ribosomes** attached to the endoplasmic reticulum.

11. The **endoplasmic reticulum (ER)** is a network of intracellular membranes. There are two types: rough and smooth. **Rough endoplasmic reticulum (RER)** contains ribosomes and is involved in protein synthesis. **Smooth endoplasmic reticulum (SER)** does not; it is involved in lipid and carbohydrate synthesis. (*Figure 3-12*)

12. The **Golgi apparatus** packages lysosomes and **secretory vesicles.** Secretions are discharged from the cell via exocytosis. (*Figure 3-13*)

13. **Lysosomes** are vesicles filled with digestive enzymes. Their functions include ridding the cell of bacteria and debris.

14. **Mitochondria** are double-membraned organelles responsible for 95 percent of the ATP production within a typical cell. High-energy bonds within adenosine triphosphate, or ATP, provide energy for cellular activities. The production of ATP in mitochondria involves **aerobic metabolism.** (*Figure 3-14*)

THE NUCLEUS p. 61

1. The **nucleus** is the control center for cellular operations. It is surrounded by a **nuclear envelope,** through which it communicates with the cytosol through **nuclear pores.** (*Figure 3-15*)

Chromosome Structure p. 62

2. The nucleus controls the cell by directing the synthesis of specific proteins using information stored in the DNA of **chromosomes.** (*Figure 3-16*)

The Genetic Code p. 62

3. The cell's information storage system, the **genetic code,** is called a *triplet code* because a sequence of three nitrogenous bases identifies a single amino acid. Each **gene** consists of all the triplets needed to produce a specific peptide chain. (*Table 3-4*)

Transcription p. 63

4. **Transcription** is the process of forming a strand of **messenger RNA (mRNA),** which carries instructions from the nucleus to the cytoplasm. (*Figure 3-17*)

Translation p. 64

5. During **translation** a functional polypeptide is constructed using the information from an mRNA strand. Each trio of nitrogen bases along the mRNA strand is a **codon**; the sequence of codons determines the sequence of amino acids in the polypeptide. (*Figure 3-18*)

6. Molecules of **transfer RNA (tRNA)** bring amino acids to the ribosomes involved in translation. (*Figure 3-18*)

CELL GROWTH AND DIVISION p. 65

1. **Mitosis** refers to the nuclear division of somatic cells. Sex cells (sperm and eggs) are produced by **meiosis**.

Interphase p. 66

2. Most somatic cells spend most of their time in **interphase.** Cells preparing for mitosis undergo **DNA replication** in this phase. (*Figure 3-19*)

Mitosis p. 66

3. Mitosis proceeds in four stages: **prophase, metaphase, anaphase,** and **telophase**. (*Figure 3-20*)

Cytokinesis p. 67

4. During the process of **cytokinesis,** the cytoplasm is divided, producing two identical daughter cells.

Cell Division and Cancer p. 67

5. Abnormal cell growth and division forms *benign* or *malignant tumors* within a tissue. **Cancer** is a disease characterized by the presence of malignant tumors; over time, the cancer cells tend to spread to new areas of the body.

CELL DIVERSITY AND DIFFERENTIATION p. 68

1. **Differentiation** is the specialization that produces cells with limited capabilities. These specialized cells form organized collections called tissues, each of which has specific functional roles.

CHAPTER QUESTIONS

LEVEL 1 Reviewing Facts and Terms

Match each item in column A with the most closely related item in column B. Use letters for answers in the spaces provided.

Column A	Column B
___ 1. filtration	a. water out of cell
___ 2. osmosis	b. passive carrier-mediated transport
___ 3. hypotonic solution	c. endocytosis, exocytosis
___ 4. hypertonic solution	d. movement of water
___ 5. isotonic solution	e. hydrostatic pressure
___ 6. facilitated diffusion	f. normal saline
___ 7. carrier proteins	g. ion pump
___ 8. vesicular transport	h. water into cell
___ 9. cytosol	i. manufacture proteins
___ 10. cytoskeleton	j. digestive enzymes
___ 11. microvilli	k. internal protein framework
___ 12. ribosomes	l. control center for cellular operations
___ 13. mitochondria	m. intracellular fluid
___ 14. lysosomes	n. DNA strands
___ 15. nucleus	o. cristae
___ 16. chromosomes	p. synthesize components of ribosomes
___ 17. nucleoli	q. increase cell surface area

18. The study of the structure and function of cells is called:
    (a) histology
    (b) cytology
    (c) physiology
    (d) biology
19. The proteins in the cell membranes may function as:
    (a) receptors and channels
    (b) carriers and enzymes
    (c) anchors and identifiers
    (d) a, b, and c are correct
20. All of the following membrane transport mechanisms are passive processes *except*:
    (a) diffusion
    (b) facilitated diffusion
    (c) vesicular transport
    (d) filtration
21. ______________ ion concentrations are high in the extracellular fluids, and ______________ ion concentrations are high in the cytoplasm.
    (a) calcium, magnesium
    (b) chloride, sodium
    (c) potassium, sodium
    (d) sodium, potassium
22. Structures that perform specific functions within the cell are:
    (a) organs (b) organisms
    (c) organelles (d) cytosomes
23. The construction of a functional polypeptide using the information provided by an mRNA strand is:
    (a) translation (b) transcription
    (c) replication (d) gene activation
24. The term *differentiation* refers to:
    (a) the loss of genes from cells
    (b) the acquisition of new functional capabilities by cells
    (c) the production of functionally specialized cells
    (d) the division of genes among different types of cells
25. What are the four general functions of the cell membrane?
26. By what four major transport mechanisms do things get into and out of cells?
27. What are the three major functions of the endoplasmic reticulum?
28. List the four stages of mitosis in their correct sequence.

LEVEL 2 Reviewing Concepts

29. Diffusion is important in body fluids because it tends to:
    (a) increase local concentration gradients
    (b) eliminate local concentration gradients
    (c) move substances against concentration gradients
    (d) create concentration gradients
30. When a cell is placed in a ______________ solution it will lose water through osmosis. The process results in the ______________ of red blood cells.
    (a) hypotonic, crenation (b) hypertonic, crenation
    (c) isotonic, hemolysis (d) hypotonic, hemolysis
31. Suppose that a DNA segment has the following nucleotide sequence: CTC ATA CGA TTC AAG TTA. Which of the following nucleotide sequences would be found in a complementary mRNA strand?
    (a) GAG UAU GAU AAC UUG AAU
    (b) GAG TAT GCT AAG TTC AAT
    (c) GAG UAU GCU AAG UUC AAU
    (d) GUG UAU GGA UUG AAC GGU

32. How many amino acids are coded in the DNA segment in the previous question?
    (a) 18
    (b) 9
    (c) 6
    (d) 3
33. What are the similarities between facilitated diffusion and active transport? What are the differences?
34. How does the cytosol differ in composition from the interstitial fluid?
35. Differentiate between transcription and translation.
36. List the stages of mitosis, and briefly describe the events that occur in each.
37. What is cytokinesis, and what role does it play in the cell cycle?

LEVEL 3 Critical Thinking and Clinical Applications

38. Experimental evidence shows that the transport of a certain molecule exhibits the following characteristics: (1) The molecule moves along its concentration gradient; (2) at concentrations above a given level there is no increase in the rate of transport; and (3) cellular energy is not required for transport to occur. What type of transport process is at work?
39. Two solutions, A and B, are separated by a semipermeable barrier. Over a period of time, the level of fluid on side A increases. Which solution initially had the higher concentration of solute?

4 THE TISSUE LEVEL OF ORGANIZATION

Exotic creatures on the deep sea floor? Well, no. The "creatures" are much closer to home. These are cells that line the airway leading to your lungs. The "tentacles" are cilia that help to remove dirt and harmful microbes from inhaled air. Notice that the surface is formed by more than one type of cell. Groups of cells specialized to perform a particular set of functions are found in tissues. Each of the body's is several different tissues has its own distinctive structure and its own role to play in maintaining homeostasis. We will meet them all in this chapter.

Chapter Outline and Objectives

Vocabulary Development

a-, without; *avascular*
apo-, from; *apocrine*
cardium, heart; *pericardium*
chondros, cartilage; *perichondrium*
dendron, a tree; *dendrites*
desmos, ligament; *desmosome*
glia, glue; *neuroglia*
histos, tissue; *histology*
holos, entire; *holocrine*
hyalos, glass; *hyaline cartilage*
inter-, between; *interstitial*
krinein, to secrete; *exocrine*
lacus, pool; *lacunae*
meros, part; *merocrine*
neuro, nerve; *neuron*
os, bone; *osseous tissue*
peri-, around; *perichondrium*
phagein, to eat; *macrophage*
pleura, rib; *pleural membrane*
sistere, to set; *interstitial*
soma, body; *desmosome*
squama, plate or scale; *squamous*
syn-, together; *synapse*
vas, vessel; *vascular*

No single body cell is able to perform the many functions of the human body. Instead, through the process of differentiation each cell specializes to perform a relatively restricted range of functions. Although there are trillions of cells in the human body, there are only about 200 different types of cells. These cell types combine to form **tissues**, collections of specialized cells and cell products that perform a limited range of functions. There are four **primary tissue types:** *epithelia, connective tissue, muscle tissue*, and *neural tissue* (Figure 4-1●).

Epithelia are layers of cells that cover exposed surfaces and line internal passageways and body cavities. The surface of the skin is an example of an epithelium. *Connective tissues* fill internal spaces, provide structural support, a framework for communication within the body, and store energy. *Muscle tissue* has the ability to contract and produce active movement. *Neural tissue* analyzes and conducts information from one part of the body to another in the form of electrical impulses. These tissues, in varying combinations, form the organs of the body.

The study of tissues, called **histology** (*histos*, tissue), provides beautiful examples of the interplay of form and function. This chapter will examine the characteristics of each major tissue type and the relationship between their highly diverse cells and tissue function. Later chapters will consider the patterns of tissue interaction in various organs and systems in greater detail.

EPITHELIA

Epithelia (e-pi-THĒ-lē-a) are layers of cells that form barriers with specific properties. Important characteristics of epithelia include the following:

- An epithelium always has a *free surface* exposed to the environment or to some internal chamber or passageway.
- An epithelium is attached to underlying connective tissue at a *basement membrane.*
- An epithelium does not contain blood vessels. Because of this **avascular** (ā-VAS-kū-lar; *a-*, without + *vas*, vessel) condition, epithelial cells must obtain nutrients from deeper tissues or from their exposed surfaces.

Epithelia cover both external and internal body surfaces. In addition to covering the skin, epithelia line internal passageways that communicate with the outside world, such as the digestive, respiratory, reproductive, and urinary tracts. These epithelia form selective barriers that separate the deep tissues of the body from the external environment.

Epithelia also line internal cavities and passageways, such as the chest cavity, fluid-filled chambers in the brain, eye, and inner ear, and the inner surfaces of blood vessels and the heart. These epithelia prevent friction, regulate the fluid composition of internal cavities, and restrict communication between the blood and tissue fluids.

Functions of Epithelia

Epithelia perform four essential functions which can be summarized as follows:

1. **Providing physical protection.** Epithelia protect exposed and internal surfaces from abrasion, dehydration, and destruction by chemical or biological agents. For example, as long as it remains intact, the epithelium of your skin resists impacts and scrapes, restricts water loss, and prevents invasion of internal tissues by bacteria.
2. **Controlling permeability.** Any substance that enters or leaves the body must cross an epithelium. Some epithelia are relatively impermeable, whereas others are easily crossed by compounds as large as proteins.
3. **Providing sensations.** Specialized epithelial cells can detect changes in the environment and relay information about such changes to the nervous system. For example, touch receptors in the deepest layers of the epithelium of the skin respond by stimulating neighboring sensory nerves.

4. **Producing specialized secretions.** Epithelial cells that produce secretions are called **gland cells.** In a **glandular epithelium** most or all of the cells actively produce secretions. These secretions are classified according to their discharge location:

   - **Exocrine** (*exo-*, outside + *krinein*, to secrete) secretions are discharged onto the surface of the skin or other epithelial surface. Enzymes entering the digestive tract, perspiration on the skin, and milk produced by mammary glands are examples.
   - **Endocrine** (*endo-*, inside) secretions are released into the surrounding tissues and blood. These secretions, called **hormones**, regulate or coordinate the activities of other tissues, organs, and organ systems. (Hormones are discussed further in Chapter 11.) Endocrine secretions are produced in organs such as the pancreas, thyroid, and pituitary gland.

Intercellular Connections

To be effective in protecting other tissues, epithelial cells must remain attached to one another. If an epithelium is damaged or the connections are broken, it is no longer an effective barrier. For example, when the epithelium of the skin is damaged by a burn or abrasion, disease-causing bacteria can enter underlying tissues and cause a dangerous infection. Undamaged epithelia form effective barriers because the epithelial cells are held together by an *intercellular cement* (composed of a protein-polysaccharide mixture) and a variety of cell connections, or *junctions.* Three such junctions are *gap junctions, tight junctions*, and *desmosomes.*

At a **gap junction** (Figure 4-2a●), two cells are held together by interlocked membrane proteins. Because these are channel proteins, the result is a narrow passageway that lets small solutes, such as ions, pass from cell to cell. Gap junctions interconnect cells in some epithelia, but they are most abundant in cardiac muscle and smooth muscle tissue, where they link adjacent muscle cells.

At a **tight junction** (Figure 4-2b●),the outermost lipid layers of the opposing cell membranes have fused. Tight junctions prevent the passage of water and solutes between the cells. These junctions are common between epithelial cells exposed to harsh chemicals or powerful enzymes. For example, tight junctions between epithelial cells lining the digestive tract keep digestive enzymes, stomach acids, or waste products from damaging underlying tissues.

At **desmosomes** (DEZ-mō-sōmz; *desmos*, ligament + *soma*, body), two cell membranes are locked together by intercellular cement and a network of fine protein filaments (Figure 4-2c●). Desmosomes are very strong, and the connection can resist stretching and twisting. In the skin, these links are so strong that dead cells are usually shed in thick sheets, rather than individually.

Cells lining the digestive tract and other passageways are held together near their surfaces by tight junctions and desmosomes. In addition, the opposing cell membranes are interlocked, and the combination of junctions, intercellular cement, and interlocking holds epithelial cells together and makes such an epithelium particularly effective as a physical barrier.

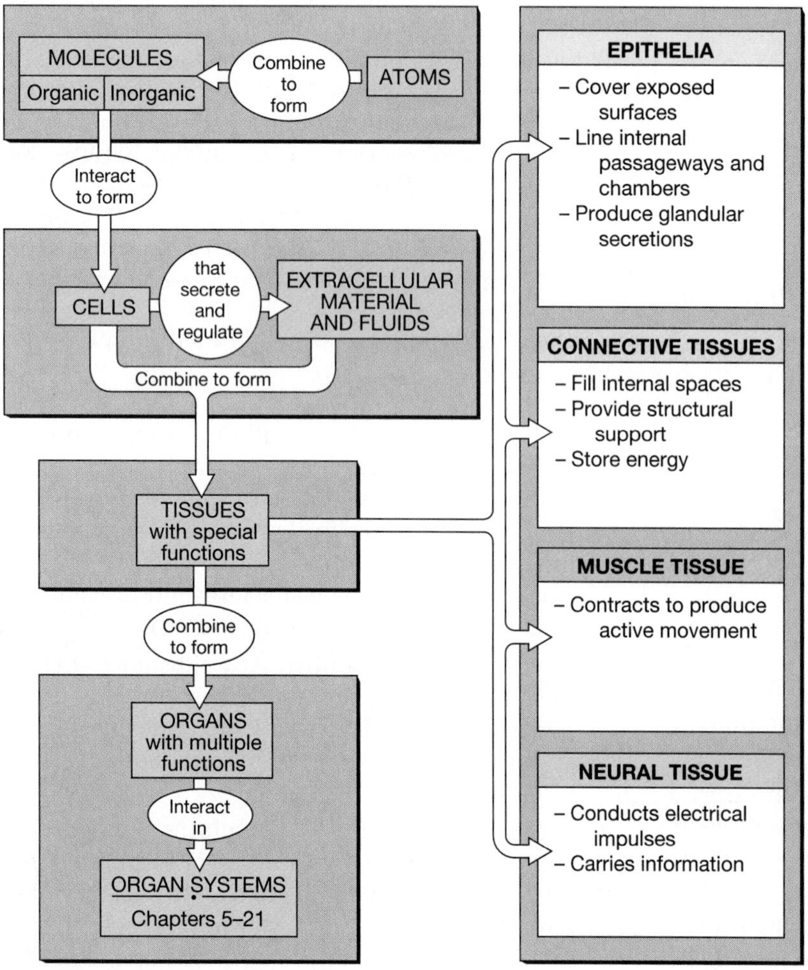

The Epithelial Surface

Many epithelia have *microvilli* on their exposed surfaces. 1 [p. 58] They may vary in number from just a few, to so many that they carpet the entire surface (Figure 4-3●). Microvilli are espe-

●**Figure 4-1**
An Orientation to the Tissues of the Body

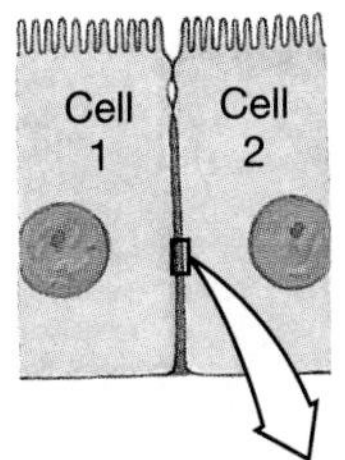

●**Figure 4-2**
Cell Attachments

(a) At a gap junction, binding of membrane proteins creates a cytoplasmic connection between two cells. **(b)** A tight junction is formed by fusion of the outer layers of two cell membranes. **(c)** A desmosome has a highly organized network of protein filaments.

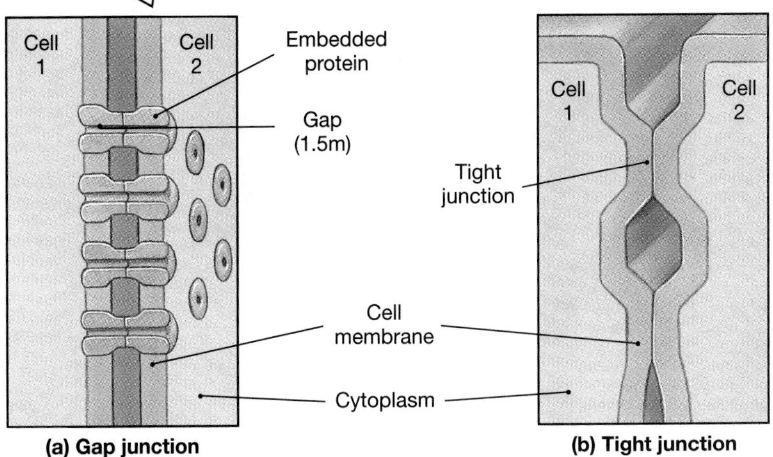

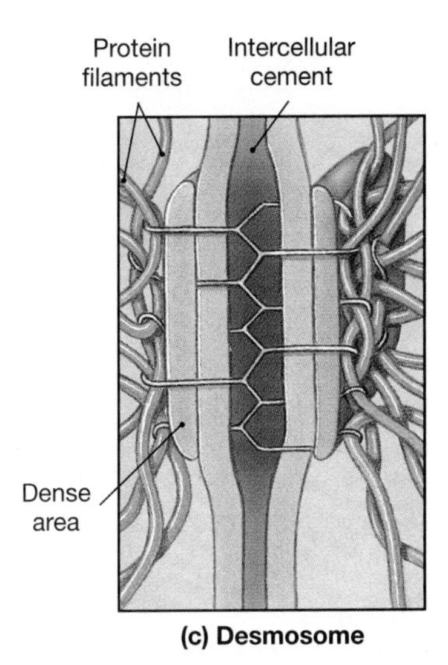

cially abundant on epithelial surfaces where absorption and secretion take place, such as along portions of the digestive and urinary tracts. These epithelial cells specialize in the active and passive transport of materials across their cell membranes. A cell with microvilli has at least 20 times the surface area of a cell without them; the greater the surface area of the cell membrane, the more transport proteins will be exposed to the extracellular environment. The diagram in Figure 4-3● also shows elongated microvilli, called *stereocilia*, that are found only within portions of the male reproductive system and in receptors of the inner ear.

Some epithelia contain *cilia* on their exposed surfaces. ∞ *[p. 58]* A typical cell within a *ciliated epithelium* has roughly 250 cilia that beat in a coordinated fashion to move materials across the epithelial surface. For example, the ciliated epithelium that lines the respiratory tract moves mucus-trapped irritants away from the lungs and toward the throat.

The Basement Membrane

Epithelial cells must not only hold onto one another but must remain firmly connected to the rest of the body. This function is performed by the **basement membrane** that lies between the epithelium and underlying connective tissues (Figure 4-3●). There are no cells within the basement membrane, which consists of a network of protein fibers. The epithelial cells adjacent to the basement membrane are firmly attached to its protein fibers. In addition to providing strength and resisting distortion, the basement membrane also provides a barrier that restricts the movement of proteins and other large molecules from the underlying connective tissue into the epithelium.

Epithelial Renewal and Repair

To maintain its structure, the epithelium must continually repair and renew itself by replacing exposed cells. Epithelial cells may survive for just a day or two, for they are lost or destroyed by exposure to disruptive enzymes, toxic chemicals, pathogenic bacteria, or mechanical abrasion. The only way the epithelium can survive is by replacing itself over time through the continual division of unspecialized cells known as **stem cells**, or **germinative cells**. These are found in the deepest layers of the epithelium, near the basement membrane.

Classification of Epithelia

Epithelia are classified according to the number of cell layers and the shape of the exposed cells. This classification scheme recognizes two types of layering (*simple* and *stratified*) and three cell shapes (*squamous, cuboidal*, and *columnar*).

Cell Layers

If there is only a single layer of cells covering the basement membrane, the epithelium is a **simple epithelium**. Simple epithelia are relatively thin, and the nuclei of the individual cells form a rough line above the basement membrane. Since a single layer of cells cannot provide much mechanical protection, simple

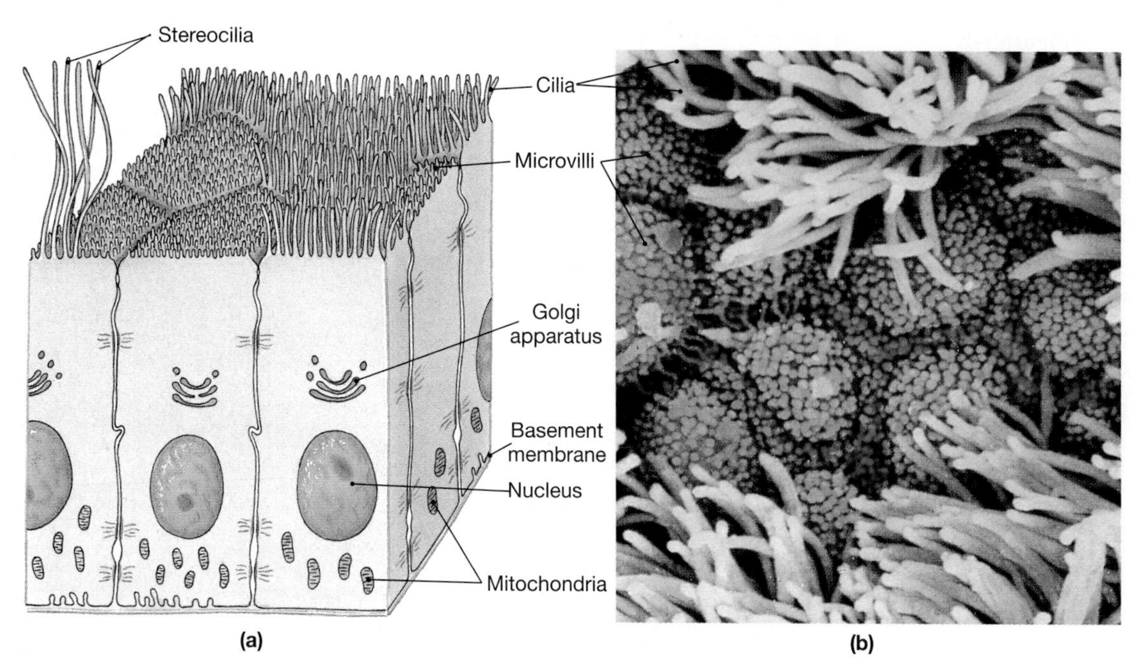

Figure 4-3
The Surfaces of Epithelial Cells
(a) The inner and outer surfaces of most epithelia are specialized for specific functions. The free surface frequently bears microvilli; less often, this surface may have cilia or (very rarely) stereocilia. (All three would not normally be found on the same group of cells but are depicted here for purposes of illustration.) Mitochondria are typically concentrated near the base of the cell, probably to provide energy for the cell's transport activities. **(b)** An SEM showing the surface of a ciliated epithelium that lines most of the respiratory tract. The small, bristly areas are microvilli found on the exposed surfaces of mucus-producing cells that are scattered among the ciliated epithelial cells. (SEM × 13,469)

epithelia are found only in protected areas inside the body. They line internal compartments and passageways, including the body cavities and the interior of the heart and blood vessels.

Simple epithelia are characteristic of regions where secretion or absorption occurs, such as the lining of the digestive and urinary tracts and the gas-exchange surfaces of the lungs. In such places, thinness is an advantage, for it reduces the diffusion time for materials crossing the epithelial barrier.

A **stratified epithelium** provides a greater degree of protection because it has several layers of cells above the basement membrane. Stratified epithelia are usually found in areas subject to mechanical or chemical stresses, such as the surface of the skin and the linings of the mouth and anus.

Cell Shape

In sectional view, the cells at the surface of the epithelium usually have one of three basic shapes.

1. **Squamous**. In a **squamous epithelium** (SKWĀ-mus; *squama*, a plate or scale), the cells are thin and flat and the nucleus occupies the thickest portion of each cell. Viewed from the surface, the cells look like fried eggs laid side by side.
2. **Cuboidal**. The cells of a **cuboidal epithelium** resemble little hexagonal boxes; in typical sectional view, however, they appear square, and the nuclei near the center of each cell form a neat row.
3. **Columnar**. In a **columnar epithelium** the cells are also hexagonal, but taller and more slender. The nuclei are crowded into a narrow band close to the basement membrane, and the height of the epithelium is several times the distance between two nuclei.

The two basic epithelial layouts (simple and stratified) and the three possible cell shapes (squamous, cuboidal, and columnar) enable one to describe almost every epithelium in the body. We will focus here on only a few major types of epithelia; additional examples will be encountered in later chapters.

Simple Squamous Epithelia

Simple epithelia are shown in Figure 4-4. A delicate, **simple squamous epithelium** (Figure 4-4a) is found in protected regions where absorption takes place or where a slick, slippery surface reduces friction. Examples include (1) portions of the kidney tubules, (2) the exchange surfaces of the lungs, (3) the lining of body cavities, and (4) the lining of blood vessels and the heart.

Simple Cuboidal Epithelia

A **simple cuboidal epithelium** (Figure 4-4b) provides limited protection and occurs in regions where se-

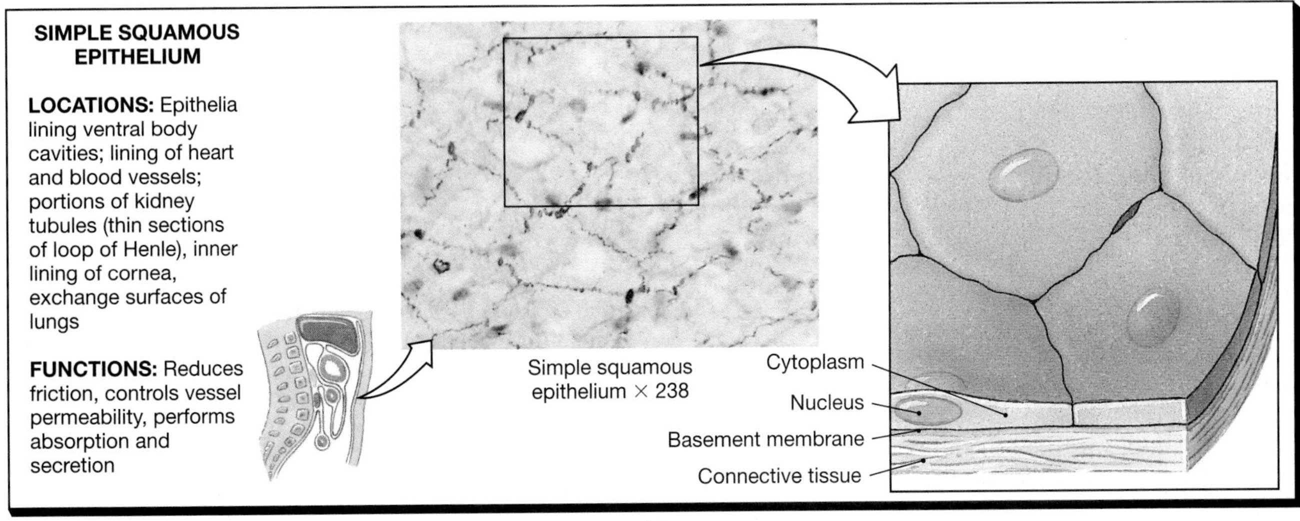
SIMPLE SQUAMOUS EPITHELIUM
LOCATIONS: Epithelia lining ventral body cavities; lining of heart and blood vessels; portions of kidney tubules (thin sections of loop of Henle), inner lining of cornea, exchange surfaces of lungs
FUNCTIONS: Reduces friction, controls vessel permeability, performs absorption and secretion
Simple squamous epithelium × 238
Cytoplasm
Nucleus
Basement membrane
Connective tissue

(a)

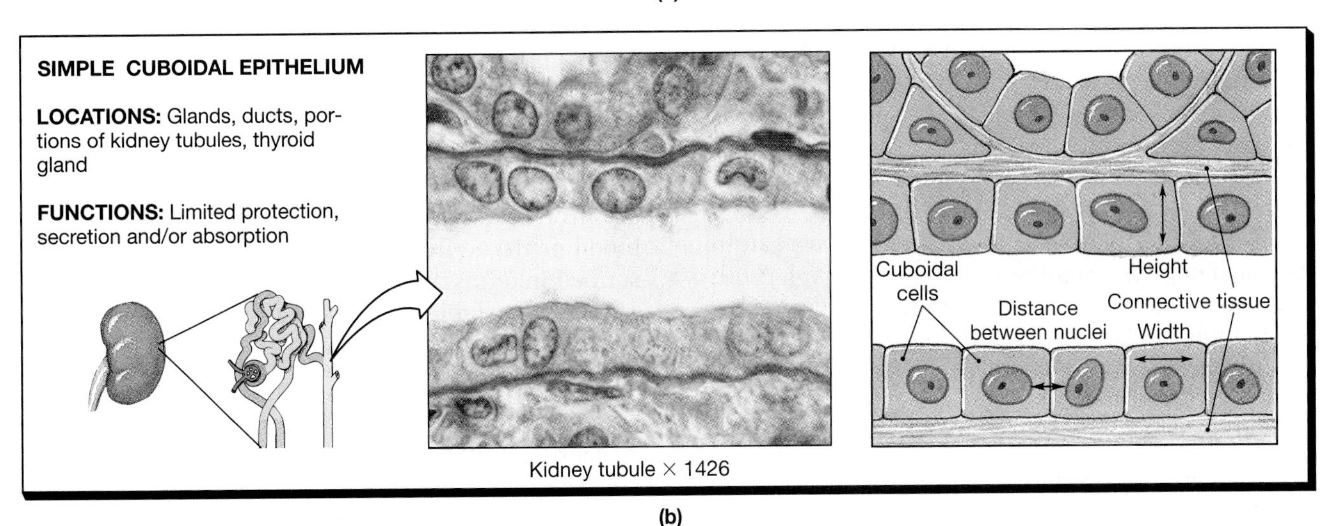
SIMPLE CUBOIDAL EPITHELIUM
LOCATIONS: Glands, ducts, portions of kidney tubules, thyroid gland
FUNCTIONS: Limited protection, secretion and/or absorption
Kidney tubule × 1426
Cuboidal cells
Distance between nuclei
Height
Connective tissue
Width

(b)

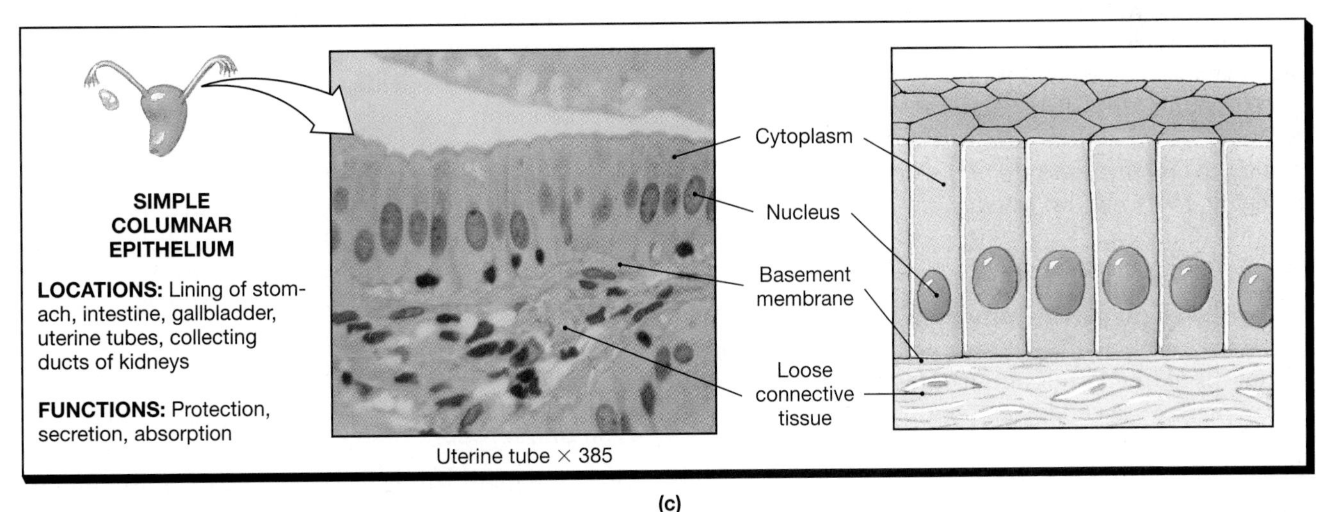
SIMPLE COLUMNAR EPITHELIUM
LOCATIONS: Lining of stomach, intestine, gallbladder, uterine tubes, collecting ducts of kidneys
FUNCTIONS: Protection, secretion, absorption
Uterine tube × 385
Cytoplasm
Nucleus
Basement membrane
Loose connective tissue

(c)

Figure 4-4
Simple Epithelia

(a) A superficial view of the simple squamous epithelium that lines the peritoneal cavity. The three-dimensional drawing shows the epithelium in superficial and sectional view. **(b)** A section through the cuboidal epithelial cells of a kidney tubule. The diagrammatic view emphasizes structural details that permit the classification of an epithelium as cuboidal. **(c)** Micrograph showing the characteristics of a simple columnar epithelium. In the diagrammatic sketch, note the relationships between the height and width of each cell; the relative size, shape, and location of nuclei; and the distance between adjacent nuclei. Compare with Figure 4-4b.

cretion or absorption takes place. These functions are enhanced by larger cells that can better accommodate the necessary organelles. Simple cuboidal epithelia secrete enzymes and buffers in the pancreas and salivary glands and line the ducts that discharge these secretions. Simple cuboidal epithelia also line portions of the kidney tubules involved in the production of urine.

Simple Columnar Epithelia

A **simple columnar epithelium** (Figure 4-4c•) provides some protection and may also be encountered in areas where absorption or secretion occurs. This type of epithelium lines the stomach, the intestinal tract, and many excretory ducts.

Pseudostratified Epithelia

Portions of the respiratory tract contain a columnar epithelium that includes a mixture of cell types. Because the nuclei are situated at varying distances from the surface, the epithelium has a layered appearance. Yet it is not a stratified epithelium because all of the cells contact the basement membrane. Because it looks stratified but isn't, it is known as a **pseudostratified columnar epithelium** (Figure 4-5a•). A ciliated pseudostratified columnar epithelium lines most of the nasal cavity, the trachea (windpipe) and bronchi, and portions of the male reproductive tract.

Transitional Epithelia

A **transitional epithelium** (Figure 4-5b•) lines the ureters and urinary bladder, where significant changes in volume occur. In an empty urinary bladder, the epithelium seems to have many layers, and the outermost cells appear rounded or cuboidal. The layered appearance results from overcrowding; the actual structure of the epithelium can be seen in the full bladder, when the pressure of the urine has stretched the lining to its natural thickness.

Stratified Squamous Epithelia

A **stratified squamous epithelium** (Figure 4-5c•) is found where mechanical stresses are severe. The surface of the skin and the lining of the mouth, esophagus, and anus are good examples.

EXFOLIATIVE CYTOLOGY

Exfoliative cytology (eks-FŌ-li-a-tive; *ex-*, from + *folium*, leaf) is the study of cells shed or collected from epithelial surfaces. The cells may be examined for a variety of reasons, including checking for cellular changes that indicate cancer formation and identifying the pathogens involved in an infection. The cells are collected either by sampling the fluids that cover the epithelia lining the respiratory, digestive, urinary, or reproductive tract or by removing fluid from one of the ventral body cavities. The sampling procedure is often called a *Pap test*, named after Dr. George Papanicolaou. Probably the most familiar Pap test is the test for cervical cancer, which involves scraping a small number of cells from the tip of the *cervix*, a portion of the uterus that projects into the vagina. Exfoliative cytology can also be used in criminal investigations, when skin cells are recovered at a crime scene.

Amniocentesis is another important test that relies on exfoliative cytology. In this procedure, exfoliated epithelial cells are collected from a sample of amniotic fluid, the fluid that surrounds and protects a developing fetus. Examination of these cells can determine if the fetus has a genetic abnormality, such as *Down syndrome*, that affects chromosomal number or structure.

Glandular Epithelia

Many epithelia contain gland cells that produce exocrine or endocrine secretions. Exocrine secretions are produced by exocrine glands that discharge their products through a duct, or tube, onto some external or internal surface. Endocrine secretions (*hormones*) are produced by ductless glands and released into blood or tissue fluids. Exocrine glands are often described in terms of their *mode of secretion* or the *type of secretion*. Table 4-1• summarizes this information and provides specific examples.

Mode of Secretion

A glandular epithelial cell may use one of three methods to release its secretions: *merocrine secretion, apocrine secretion*, or *holocrine secretion*.

In **merocrine secretion** (MER-o-krin; *meros*, part + *krinein*, to secrete) the product is released through exocytosis. ∞ [p. 56] This method (Figure 4-6a•) is the most common mode of secretion. **Apocrine secretion** (AP-ō-krin; *apo-*, off) involves the loss of both cytoplasm and the secretory product (Figure 4-6b•). The outermost portion of the cytoplasm becomes packed with secretory vesicles before it is shed. Whereas merocrine and apocrine secretions leave the cell intact and able to continue secreting, **holocrine secretion** (HOL-ō-krin; *holos*, entire) does not (Figure 4-6c•). Instead, the entire cell becomes packed with secretions and then bursts apart.

Type of Secretion

There are many kinds of exocrine secretions, all performing a variety of functions. Examples include enzymes entering the digestive tract, perspiration on the skin, and the milk produced by mammary glands.

Exocrine glands may be categorized by the type or types of secretions produced. For example, *serous glands* secrete a watery solution containing enzymes, and *mucous glands* secrete a thick, slippery mucus. *Mixed glands* contain more than one type of gland

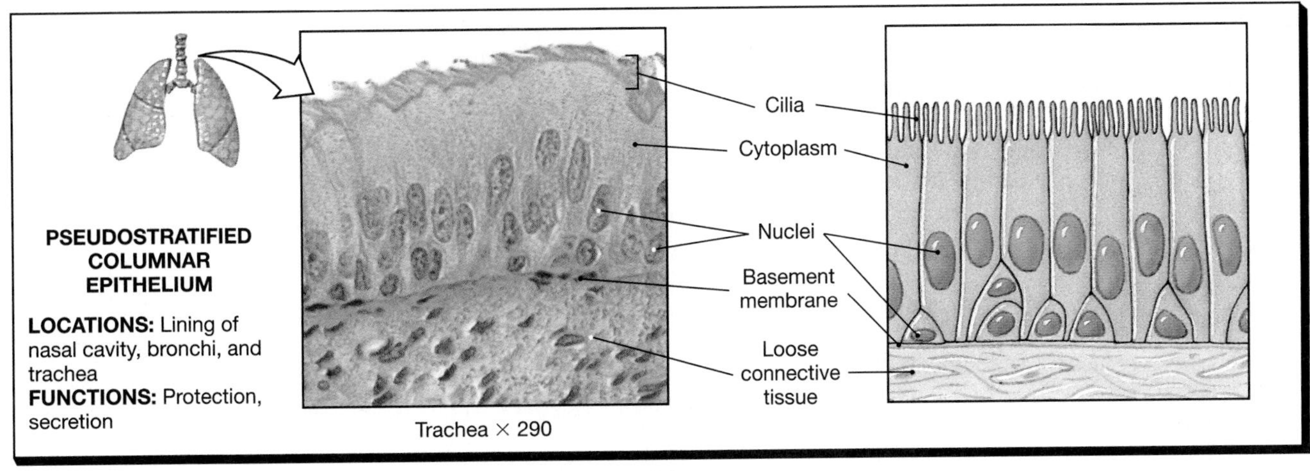

(a)

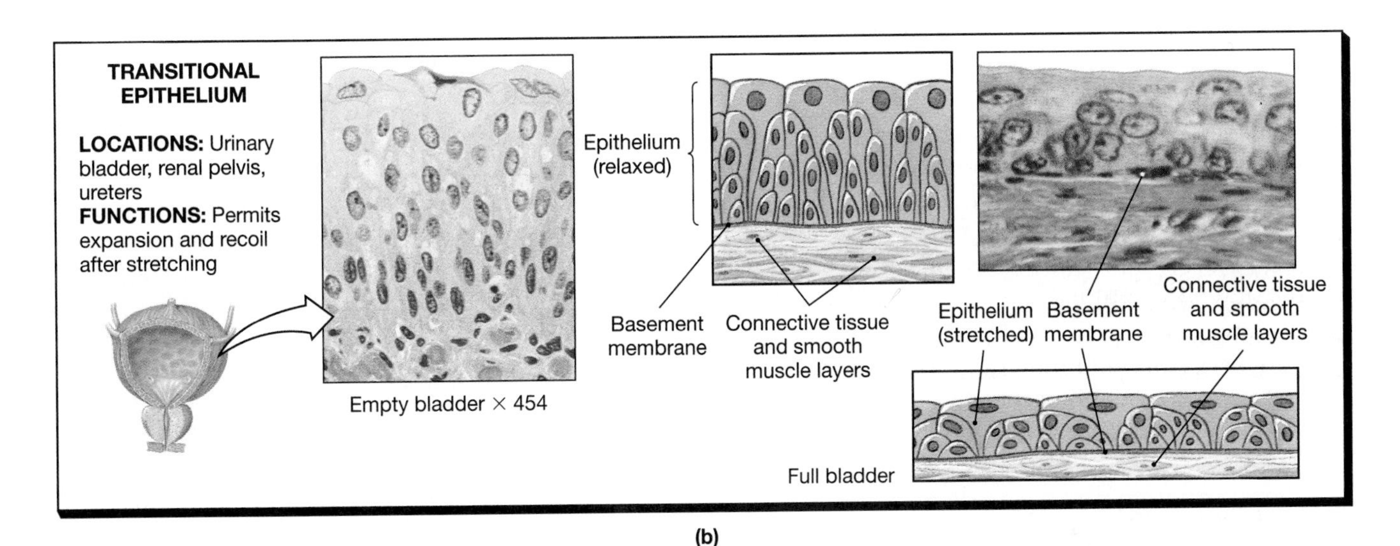

(b)

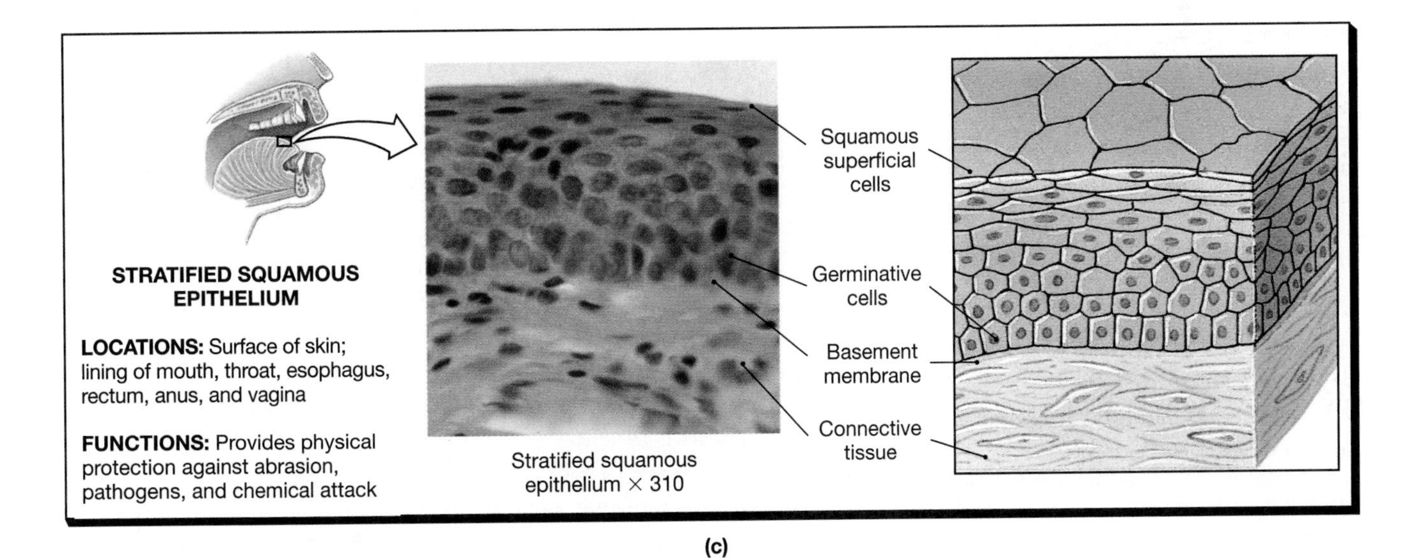

(c)

•Figure 4-5
Stratified Epithelia

(a) The pseudostratified, ciliated, columnar epithelium of the respiratory tract. Note the uneven layering of the nuclei. **(b)** At left, the lining of the empty urinary bladder, showing transitional epithelium in the relaxed state. At right, the lining of the full bladder, showing the effects of stretching on the arrangement of cells in the epithelium. **(c)** A sectional view of the stratified squamous epithelium that covers the tongue.

TABLE 4-1 A Classification of Exocrine Glands

Feature	*Description*	*Example*
MODE OF SECRETION		
Merocrine	Secretion occurs via exocytosis.	Mucus in digestive and respiratory tracts
Apocrine	Secretion occurs via loss of cytoplasm containing secretory product.	Milk in breasts; viscous underarm perspiration
Holocrine	Secretion occurs via loss of entire cell containing secretory product.	Skin oils and waxy coating of hair (produced by sebaceous glands of the skin)
TYPE OF SECRETION		
Serous	Watery solution containing enzymes.	Parotid salivary gland
Mucous	Thick, slippery mucus.	Sublingual salivary gland
Mixed	Produces more than one type of secretion.	Submandibular salivary gland (serous and mucous)

cell and may produce two different exocrine secretions or both exocrine and endocrine secretions.

✓ You look at a tissue under a microscope and see a simple squamous epithelium. Can it be a sample of the skin surface?

✓ Secretory cells associated with hair follicles fill with secretions and then rupture, releasing their contents. What kind of secretion is this?

✓ What physiological functions are enhanced by epithelial cells bearing microvilli and cilia?

CONNECTIVE TISSUES

Connective tissues are deep tissues that are never exposed to the environment outside the body. They have many important functions, including:

- **Providing support and protection.** The minerals and fibers produced by connective tissue cells establish a bony structural framework for the body, protect delicate organs, and surround and interconnect other tissue types.

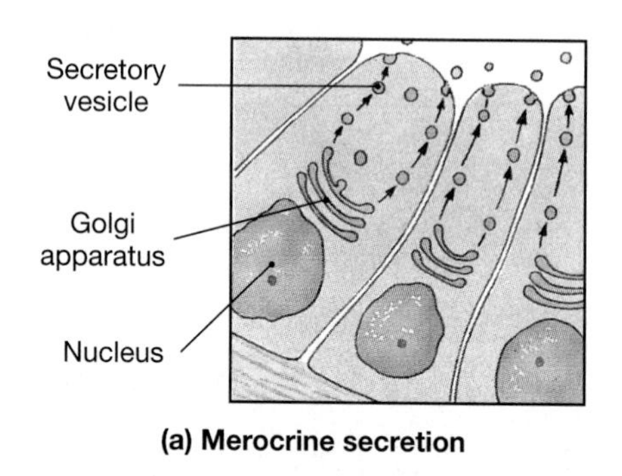

(a) Merocrine secretion

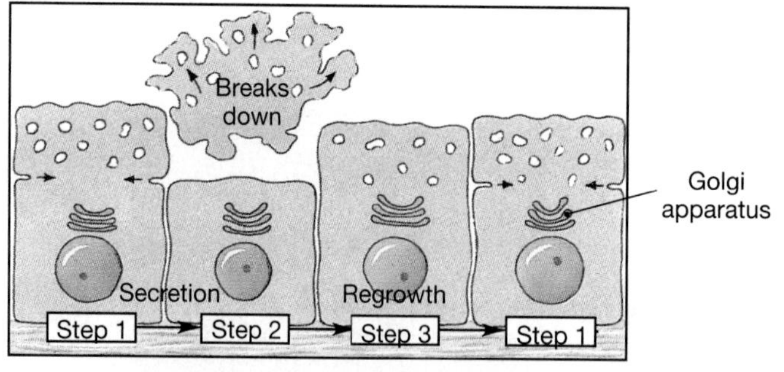

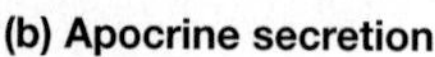

(b) Apocrine secretion

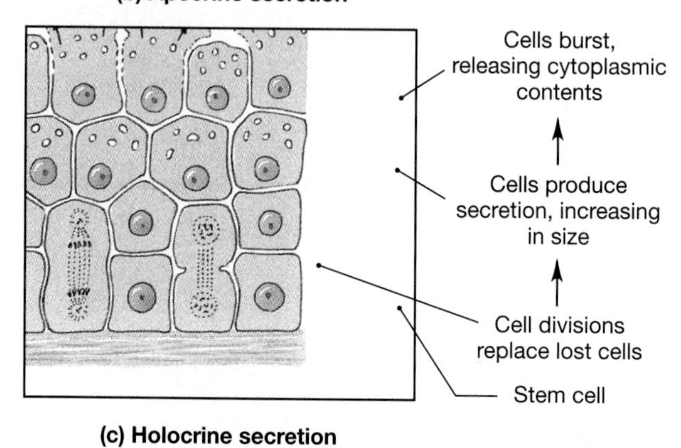

(c) Holocrine secretion

•Figure 4-6
Mechanisms of Glandular Secretion

(a) In merocrine secretion, secretory vesicles are discharged at the surface of the gland cell through exocytosis. **(b)** Apocrine secretion involves the loss of cytoplasm. Inclusions, secretory vesicles, and other cytoplasmic components are shed in the process. The gland cell then undergoes a period of growth and repair before releasing additional secretions. **(c)** Holocrine secretion occurs as superficial gland cells break apart. Continued secretion involves the replacement of these cells through the mitotic divisions of underlying stem cells.

- **Transporting materials.** Fluid connective tissue provides an efficient means to move dissolved materials from one region of the body to another.
- **Storing energy reserves.** Fats are stored in connective tissue cells called *adipose cells* until needed.
- **Defending the body.** Specialized connective tissue cells respond to invasions by microorganisms through cell-to-cell interactions and the production of *antibodies*.

Connective tissues are the most diverse tissues of the body. Bone, blood, and fat are familiar connective tissues that have very different functions and properties. All connective tissues have three basic components: (1) specialized cells, (2) protein fibers, and (3) a **ground substance**, a fluid that varies in consistency. The extracellular fibers and ground substance constitute the **matrix** that surrounds the cells. Whereas epithelial tissue consists almost entirely of cells, the extracellular matrix accounts for most of the volume of connective tissues.

Classification of Connective Tissues

Several classes of connective tissue are recognized on the basis of the physical properties of their ground substance (Figure 4-7•).

- **Connective tissue proper** refers to connective tissues with many types of cells and fibers surrounded by a syrupy ground substance. Examples include the tissue that underlies the skin, fatty tissue, and *tendons* and *ligaments*.
- **Fluid connective tissues** have a distinctive population of cells suspended in a watery ground substance that contains dissolved proteins. There are two fluid connective tissues, *blood* and *lymph*.
- **Supporting connective tissues** are of two types, *cartilage* and *bone*. These tissues have a less diverse cell population than connective tissue proper and a matrix of dense ground substance and closely packed fibers. The fibrous matrix of bone is said to be **calcified** because it contains mineral deposits, primarily calcium salts, which give the bone strength and rigidity.

Connective Tissue Proper

Connective tissue proper (Figure 4-8•) contains fibers, a syrupy ground substance, and a varied cell population. That population includes the following cell types:

- **Fibroblasts** (FĪ-brō-blasts) are the most abundant cells in connective tissue proper. They are responsible for the production and maintenance of the connective tissue fibers and the ground substance.
- **Macrophages** (MAC-rō-fā-jez; *phagein*, to eat) are scattered among the fibers. These cells phagocytize damaged cells or pathogens that enter the tissue and release chemicals that mobilize the immune system. When an infection occurs, additional macrophages are drawn to the affected area.
- **Fat cells** are known as *adipose cells*, or simply **adipocytes** (AD-i-pō-sīts). A typical adipocyte contains such a large droplet of lipid that the nucleus and other organelles are squeezed to one side of the cell. The number of fat cells varies from one connective tissue to another, from one region of the body to another, and from individual to individual.

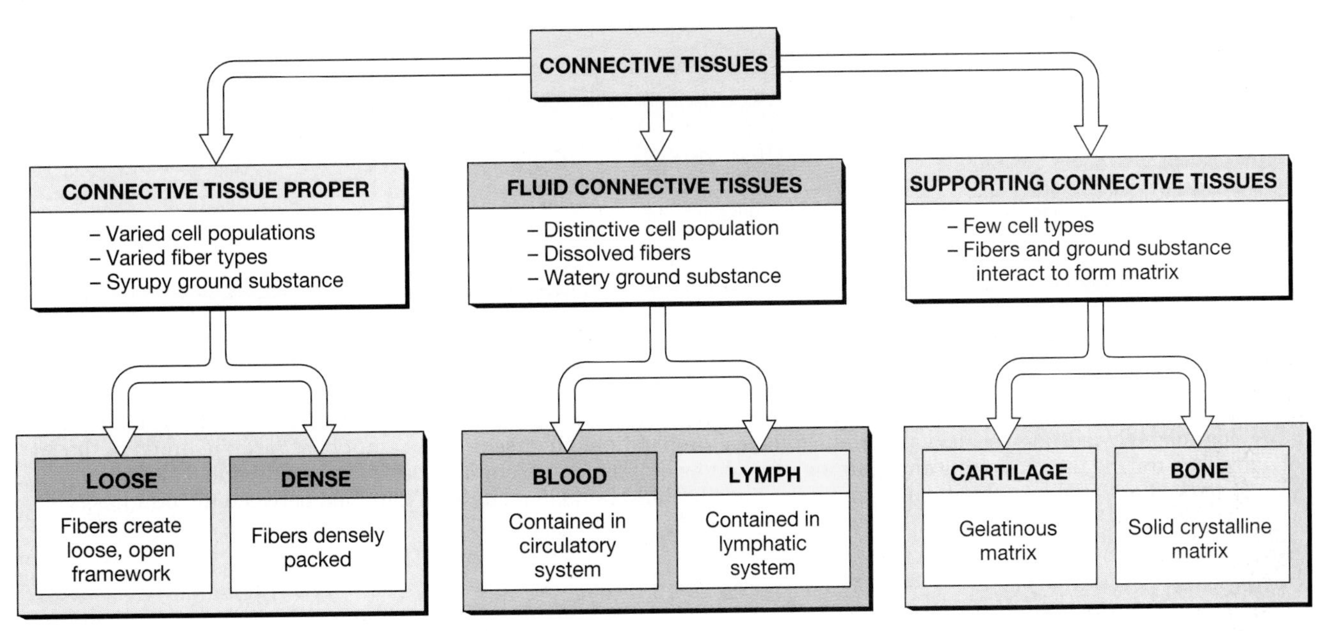

•**Figure 4-7**
Major Types of Connective Tissue

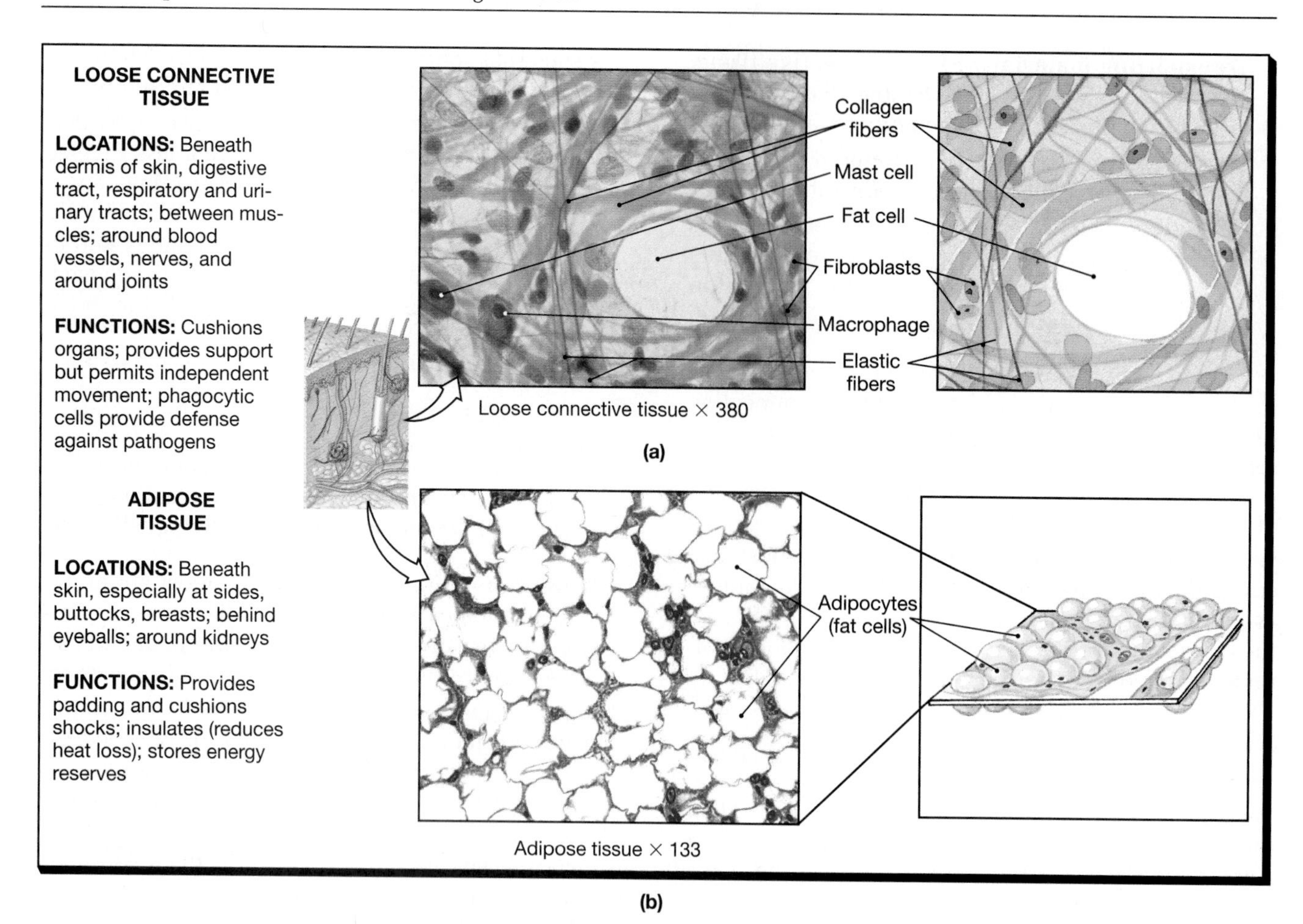
LOOSE CONNECTIVE TISSUE
LOCATIONS: Beneath dermis of skin, digestive tract, respiratory and urinary tracts; between muscles; around blood vessels, nerves, and around joints
FUNCTIONS: Cushions organs; provides support but permits independent movement; phagocytic cells provide defense against pathogens
Collagen fibers
Mast cell
Fat cell
Fibroblasts
Macrophage
Elastic fibers
Loose connective tissue × 380
(a)
ADIPOSE TISSUE
LOCATIONS: Beneath skin, especially at sides, buttocks, breasts; behind eyeballs; around kidneys
FUNCTIONS: Provides padding and cushions shocks; insulates (reduces heat loss); stores energy reserves
Adipocytes (fat cells)
Adipose tissue × 133

(b)

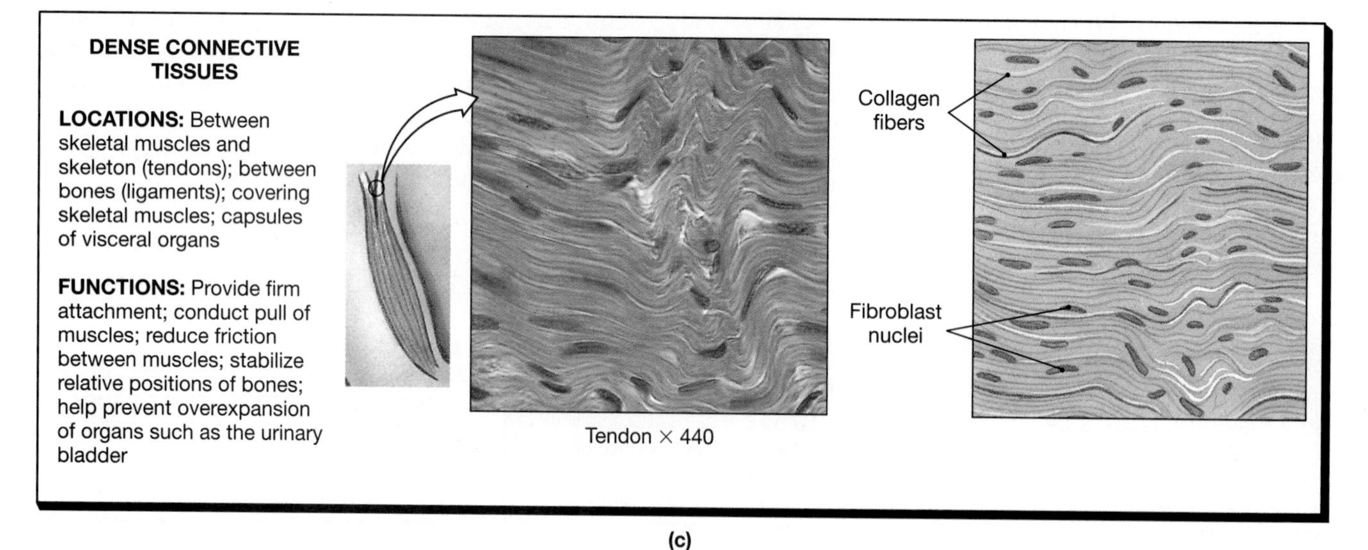
DENSE CONNECTIVE TISSUES
LOCATIONS: Between skeletal muscles and skeleton (tendons); between bones (ligaments); covering skeletal muscles; capsules of visceral organs
FUNCTIONS: Provide firm attachment; conduct pull of muscles; reduce friction between muscles; stabilize relative positions of bones; help prevent overexpansion of organs such as the urinary bladder
Collagen fibers
Fibroblast nuclei
Tendon × 440

(c)

•Figure 4-8
Connective Tissue Proper: Loose and Dense Connective Tissue

(a) All of the cells of connective tissue proper are found in loose connective tissue. **(b)** Adipose tissue is loose connective tissue dominated by adipocytes. In standard histological preparations the tissue looks empty because the lipids in the fat cells dissolve during the sectioning and staining procedures. **(c)** The dense connective tissue in a tendon. Notice the densely packed, parallel bundles of collagen fibers. The fibroblast nuclei can be seen flattened between the bundles.

- **Mast cells** are small, mobile connective tissue cells often found near blood vessels. The cytoplasm of a mast cell is packed with vesicles filled with chemicals that are released to begin the body's defensive activities after an injury or infection, as discussed later in the chapter.

In addition to mast cells and free macrophages, both phagocytic and antibody-producing white blood cells may move through the connective tissue. Their numbers increase markedly if the tissue is damaged, as does the production of **antibodies**, proteins that destroy invading microorganisms or foreign substances. ∞ [p. 37]

Connective Tissue Fibers

There are three basic types of fibers: *collagen, elastic,* and *reticular.* All three types are formed from protein subunits.

- **Collagen fibers** are long, straight, and unbranched. Collagen fibers, the most common fibers in connective tissue proper, are strong but flexible.
- **Elastic fibers** contain the protein *elastin.* They are branched and wavy, and after stretching will return to their original length.
- **Reticular fibers** (*reticulum,* network), the least common of the three, are thinner than collagen fibers and commonly form a branching, interwoven framework in various organs.

Ground Substance

Ground substance fills all the spaces between cells and surrounds all the connective tissue fibers. In normal connective tissue proper it is clear, colorless, and similar in consistency to maple syrup.

MARFAN'S SYNDROME

Marfan's syndrome is an inherited condition caused by the production of an abnormal form of *fibrillin,* a carbohydrate-protein complex important to normal connective tissue strength and elasticity. Because connective tissues are found in most organs, the effects of this defect are widespread. The most visible sign of Marfan's syndrome involves the skeleton; individuals with Marfan's syndrome are usually tall, with abnormally long arms, legs, and fingers. But the most serious consequences involve the cardiovascular system. Roughly 90 percent of the people with Marfan's syndrome have abnormal cardiovascular systems. The most dangerous result is that the weakened connective tissues in the walls of major arteries, such as the aorta, may burst, causing a sudden, fatal loss of blood.

Connective tissue proper can be divided into *loose connective tissues* and *dense connective tissues* on the basis of the relative proportions of cells, fibers, and ground substance. *Loose connective tissues* are the packing material of the body. These tissues fill spaces between organs, provide cushioning, and support epithelia. They also anchor blood vessels and nerves, store lipids, and provide a route for the diffusion of materials. *Dense connective tissues* are tough, strong, and durable. They resist tension and distortion, and interconnect bones and muscles. Dense connective tissue also forms a thick layer, called a *capsule,* that surrounds visceral organs, such as the liver, kidneys, and spleen, and that also encloses joint cavities.

Loose Connective Tissue

Loose connective tissue, or *areolar tissue* (*areola,* little space), is the least specialized connective tissue in the adult body (Figure 4-8a●). It contains all of the cells and fibers found in any connective tissue proper in addition to an extensive circulatory supply.

Loose connective tissue forms a layer that separates the skin from underlying muscles, providing both padding and a considerable amount of independent movement. For example, pinching the skin of the arm does not distort the underlying muscle.

The ample blood supply in this tissue carries wandering cells to and from the tissue and provides for the metabolic needs of nearby epithelial tissue.

Adipose Tissue

Adipose tissue, or fat, is a loose connective tissue containing large numbers of fat cells, or *adipocytes* (Figure 4-8b●). The difference between loose connective tissue and adipose tissue is one of degree—a loose connective tissue is called adipose tissue when it becomes dominated by fat cells. Adipose tissue provides another source of padding and shock absorption for the body. It also acts as an insulating blanket that slows heat loss through the skin and functions in energy storage.

Adipose tissue is common under the skin of the sides, buttocks, and breasts. It fills the bony sockets behind the eyes, surrounds the kidneys, and dominates extensive areas of loose connective tissue in the pericardial and peritoneal (abdominal) cavities.

ADIPOSE TISSUE AND WEIGHT LOSS

Adipocytes are metabolically active cells—their lipids are continually being broken down and replaced. When nutrients are scarce, adipocytes deflate like collapsing balloons. This is what occurs during a weight-loss program. Because the cells are not killed, merely reduced in size, the lost weight can easily be regained in the same areas of the body.

Although adipocytes are incapable of dividing, an excess of nutrients can cause the division of connective tissue stem cells, which then differentiate into additional fat cells. As a result, areas of loose connective tissue can become adipose tissue in times of nutritional plenty. In the procedure known as *liposuction*, unwanted adipose tissue is surgically removed. Because adipose tissue can regenerate through differentiation of stem cells, liposuction provides only a temporary solution to the problem.

Dense Connective Tissues

Dense connective tissues consist mostly of collagen fibers; they may also be called *fibrous tissues*. **Tendons** (Figure 4-8c●) are cords of dense connective tissue that attach skeletal muscles to bones. The collagen fibers run along the length of the tendon and transfer the pull of the contracting muscle to the bone. **Ligaments** (LIG-a-ments) are bundles of fibers that connect one bone to another. Ligaments often contain elastic fibers as well as collagen fibers and thus can tolerate a modest amount of stretching.

Fluid Connective Tissues

Blood and *lymph* are connective tissues that contain distinctive collections of cells in a fluid matrix. Under normal conditions, the proteins dissolved in this watery ground substance do not form large insoluble fibers.

A single cell type, the *red blood cell*, accounts for almost half the volume of blood. Red blood cells transport oxygen in the blood. The watery ground substance, called **plasma**, also contains small numbers of *white blood cells*, which are important components of the immune system, and *platelets*, which function in blood clotting (Figure 4-9a●).

The extracellular fluid of the body is composed of *plasma, interstitial fluid*, and *lymph*. Plasma, confined to the vessels of the circulatory system, is kept in constant motion by contractions of the heart. A network of **arteries** carries blood away from the heart and toward fine, thin-walled vessels called **capillaries. Veins** collect and return blood to the heart, completing the circuit. In the tissues, filtration moves water and small solutes out of the capillaries and into the **interstitial fluid** (in-ter-STISH-al; *inter*, between + *sistere*, to set) that surrounds the cells of other tissues.

The fluid connective tissue called lymph forms as interstitial fluid enters small passageways, or *lymphatics*, that return it to the circulatory system. Figure 4-9● diagrams this flow pattern.

Supporting Connective Tissues

Cartilage and bone are called supporting connective tissues because they provide a strong framework that supports the rest of the body. In these connective tis-

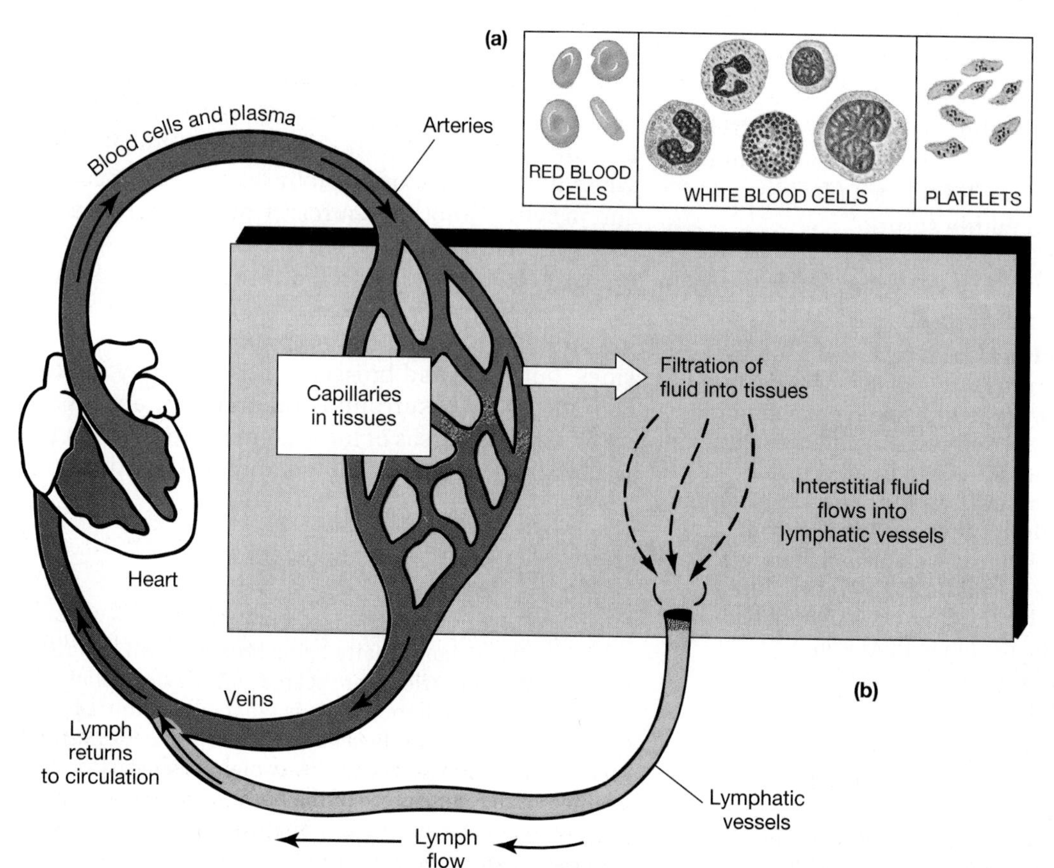

●**Figure 4-9 Fluid Connective Tissues**

(a) Cells of the fluid connective tissues. **(b)** Blood travels through the circulatory system, pushed by the contractions of the heart. In capillaries, hydrostatic (blood) pressure forces fluid and dissolved solutes out of the circulatory system. This fluid mixes with the interstitial fluid already in the tissue. Interstitial fluid slowly enters lymphatic vessels; now called lymph, it travels along the lymphatics and reenters the circulatory system at one of the veins that return blood to the heart.

sues the matrix contains numerous fibers and, in some cases, deposits of insoluble calcium salts.

Cartilage

The matrix of **cartilage** consists of a firm gel containing embedded fibers. **Chondrocytes** (KON-drō-sīts), the only cells found within the matrix, live in small pockets known as lacunae (la- KOO-nē; *lacus*, pool). Because cartilage lacks blood vessels, chondrocytes must obtain nutrients and eliminate waste products by diffusion through the matrix. Structures of cartilage are set apart from surrounding tissues by a **perichondrium** (per-i-KON-drē-um; *peri-*, around + *chondros*, cartilage) which does contain blood vessels. AM *"Jaws" and the Fight against Cancer*

Types of Cartilage There are three major types of cartilage (Figure 4-10 ●): **hyaline cartilage, elastic cartilage**, and **fibrocartilage**.

- Hyaline cartilage (HĪ-a-lin; *hyalos*, glass) is the most common type of cartilage (Figure 4-10a●). Tough and somewhat flexible, this type of cartilage connects the ribs to the sternum (breastbone), supports the conducting passageways of the respiratory tract, and covers the surfaces of bones within joints.
- Elastic cartilage (Figure 4-10b●) contains numerous elastic fibers that make it extremely resilient and flexible. Elastic cartilage supports the external flap (*pinna*) of the outer ear, the epiglottis, and the tip of the nose.
- Fibrocartilage has little ground substance, and the matrix is dominated by collagen fibers (Figure 4-10c●). These fibers are densely interwoven, making this tissue extremely durable and tough. Fibrocartilaginous pads lie between the vertebrae of the spinal column, between the bones of the pelvis, and around or within a few joints and tendons. In these positions they resist compression, absorb shocks, and prevent damaging bone-to-bone contact. Cartilages in general heal poorly, and damaged fibrocartilages in joints such as the knee can interfere with normal movements.

CARTILAGES AND KNEE INJURIES

The knee is an extremely complex joint that contains both hyaline cartilage and fibrocartilage. The hyaline cartilage covers bony surfaces, and pads of fibrocartilage within the joint prevent bone contact when movements are under way. Many sports injuries involve tearing of the cartilage pads. This loss of cushioning places more strain on the cartilages within joints and leads to further joint damage. Because cartilages are avascular, they heal poorly, and joint cartilages heal even more slowly than other cartilages. Surgery usually produces only a temporary or incomplete repair.

Bone

Because the detailed histology of bone, or **osseous tissue** (OS-ē-us; *os*, bone), will be considered in Chapter 6, this discussion will focus on significant differences between cartilage and bone. The matrix of bone consists of hard calcium compounds and flexible collagen fibers. This combination gives bone truly remarkable properties, making it both strong and resistant to shattering. In its overall properties, bone can compete with the best steel-reinforced concrete.

The general organization of bone can be seen in Figure 4-11●. Lacunae within the matrix contain bone cells, or **osteocytes** (OS-tē-ō-sīts). The lacunae surround the blood vessels that branch through the bony matrix. Diffusion cannot occur through the bony matrix, but osteocytes obtain nutrients via cytoplasmic extensions that reach blood vessels and other osteocytes. The passageways through which these cytoplasmic processes run are called **canaliculi** (kan-a-LIK-ū-lē; little canals) because they form a branching network within the bony matrix.

Except within joint cavities, where opposing surfaces are coated with cartilage, each bone is surrounded by a fibrous **periosteum** (per-ē-OS-tē-um). Unlike cartilage, bone is constantly being changed, or remodeled, to such an extent that complete repairs can be made even after severe damage has occurred. Table 4-2● summarizes the similarities and differences between cartilage and bone.

✓ Lack of vitamin C in the diet interferes with the ability of fibroblasts to produce collagen. What effect might this interference have on connective tissue?

✓ Chemical analysis of a connective tissue reveals that the tissue contains primarily triglycerides. What tye of connective tissue is this?

✓ Why does cartilage heal so slowly?

MEMBRANES

Some anatomical terms have more than one meaning, depending on the context. One such term is *membrane*. For example, at the cellular level, membranes are lipid bilayers that restrict the passage of ions and other solutes. *[p. 50]* At the tissue level, membranes again form a barrier or interface, such as the basement membranes that separate epithelia from connective tissues. At still another level, epithelia and connective tissues combine to form membranes that cover and protect other structures and tissues in the body. There are four such types of membranes: *mucous membranes, serous membranes, the cutaneous membrane*, and *synovial membranes* (Figure 4-12●).

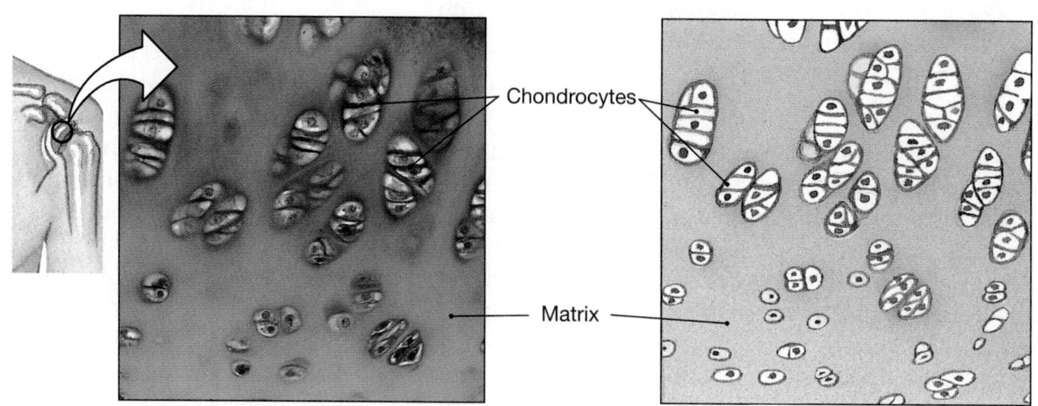
HYALINE CARTILAGE
LOCATIONS: Between tips of ribs and bones of sternum; covering bone surfaces at synovial joints; supporting larynx (voicebox), trachea, and bronchi; forming part of nasal septum
FUNCTIONS: Provides stiff but somewhat flexible support; reduces friction between bony surfaces
Chondrocytes
Matrix
Hyaline cartilage × 500

(a)

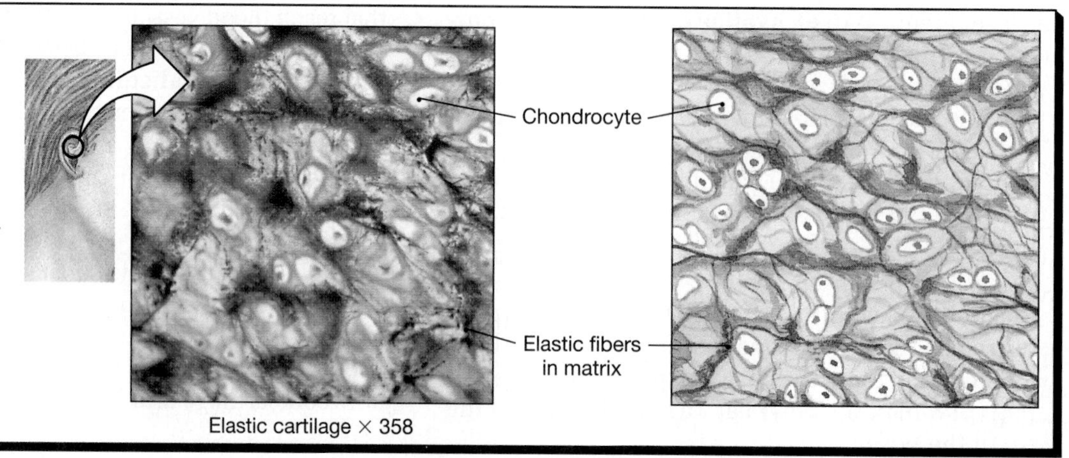
ELASTIC CARTILAGE
LOCATIONS: Pinna of external ear; tip of nose; epiglottis
FUNCTIONS: Provides support, but tolerates distortion without damage and returns to original shape
Chondrocyte
Elastic fibers in matrix
Elastic cartilage × 358

(b)

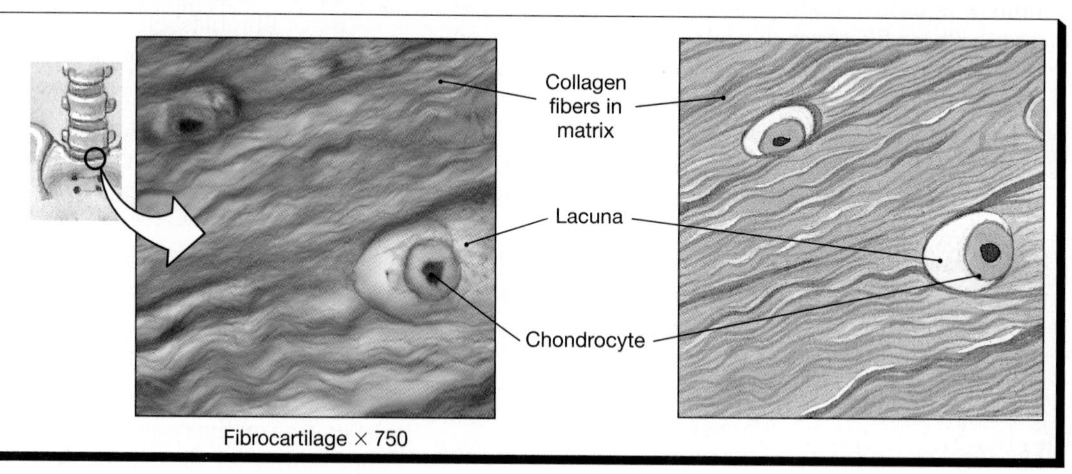
FIBROCARTILAGE
LOCATIONS: Intervertebral discs separating vertebrae along spinal column; pads within knee joint; between pubic bones of pelvis
FUNCTIONS: Resists compression; prevents bone-to-bone contact; limits relative movement
Collagen fibers in matrix
Lacuna
Chondrocyte
Fibrocartilage × 750

(c)

•Figure 4-10
Types of Cartilage

(a) Hyaline cartilage. Note the translucent matrix and the absence of prominent fibers. **(b)** Elastic cartilage. The closely packed elastic fibers are visible between the chondrocytes. **(c)** Fibrocartilage. The collagen fibers are extremely dense, and the chondrocytes are relatively far apart.

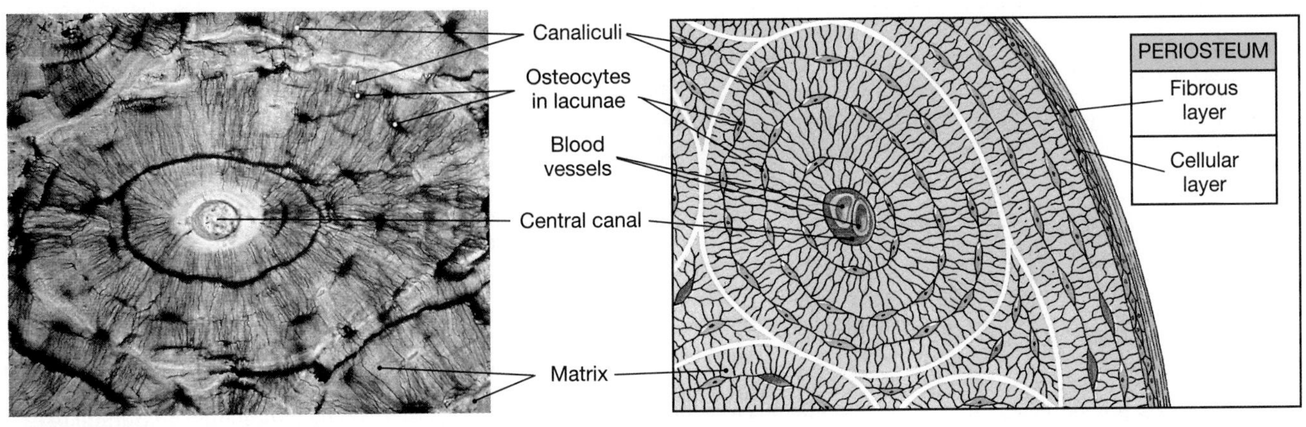

●Figure 4-11
Bone
The osteocytes in bone are usually organized in groups around a central space that contains blood vessels. For the photomicrograph, a sample of bone was ground thin enough to become transparent. Bone dust filled the lacunae and the central canal, making them appear dark.

Mucous Membranes

Mucous membranes (Figure 4-12a●) line cavities that communicate with the exterior, including the digestive, respiratory, reproductive, and urinary tracts. The epithelial surfaces are kept moist at all times, typically by mucous secretions or by exposure to fluids such as urine or semen.

Many mucous membranes are lined by simple epithelia that perform absorptive or secretory functions, such as the simple columnar epithelium of the digestive tract. However, other types of epithelia may be involved. For example, a stratified squamous epithelium covers the mucous membrane of the mouth, and the mucous membrane along most of the urinary tract contains a transitional epithelium.

Serous Membranes

Serous membranes line the sealed, internal cavities of the body. There are three serous membranes, each consisting of a simple epithelium supported by loose connective tissue (Figure 4-12b●). The **pleura** (PLOO-ra; *pleura*, rib) lines the pleural cavities and covers the lungs. The **peritoneum** (pe-ri-tō-NĒ-um) lines the peritoneal cavity and covers the surfaces of enclosed organs such as the liver and stomach. The **pericardium** (pe-ri-KAR-dē-um) lines the pericardial cavity and covers the heart.

A serous membrane has **parietal** and **visceral** portions. ∞ *[p. 19]* The parietal portion lines the outer wall of the internal chamber, and the visceral portion covers organs within the body cavity. For example,

TABLE 4-2 A Comparison of Cartilage and Bone

Characteristic	*Cartilage*	*Bone*
STRUCTURAL FEATURES		
Cells	Chondrocytes in lacunae	Osteocytes in lacunae
Ground substance	Protein-polysaccharide gel	Insoluble salts (calcium phosphate and calcium carbonate)
Fibers	Collagen, elastic, reticular fibers (proportions vary)	Collagen fibers predominate
Vascularity	None	Extensive
Covering	Perichondrium	Periosteum
Strength	Limited: bends easily but hard to break	Strong: resists distortion until breaking point is reached
METABOLIC FEATURES		
Oxygen demands	Relatively low	Relatively high
Nutrient delivery	By diffusion through matrix	By diffusion through cytoplasm and fluid in canaliculi
Repair capabilities	Limited ability	Extensive ability

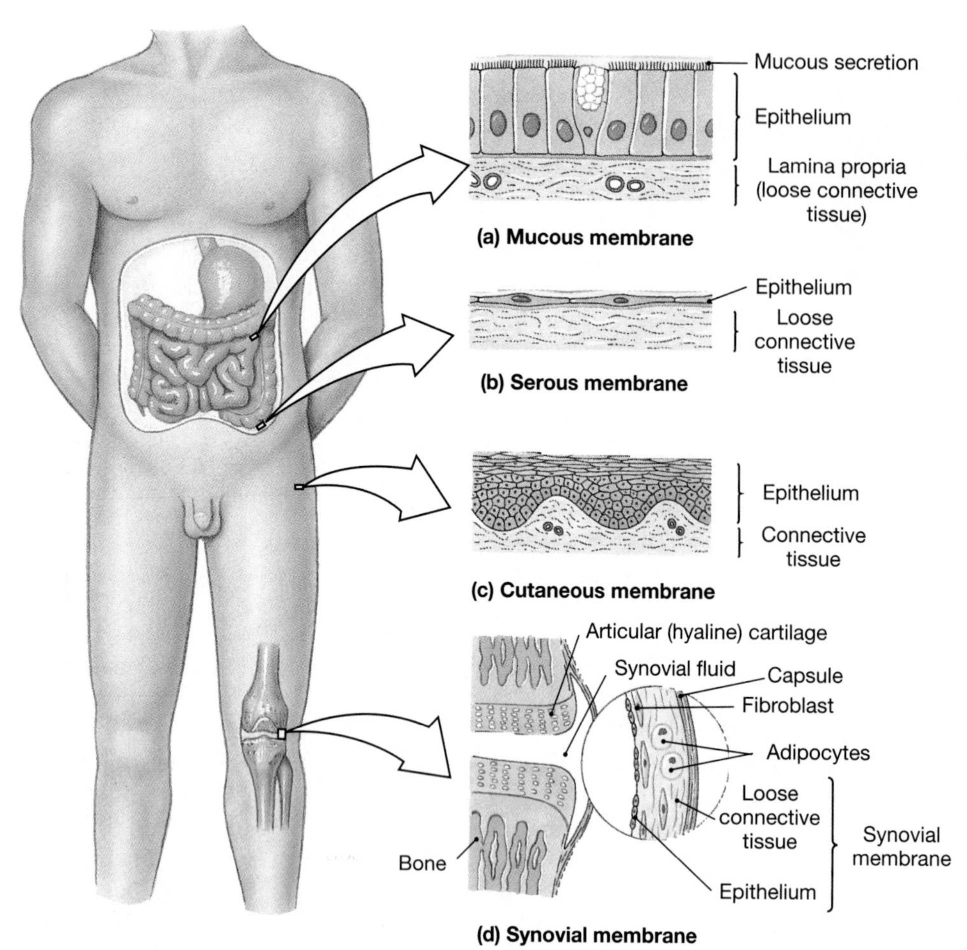

Figure 4-12 Membranes

(a) Mucous membranes are coated with the secretions of mucous glands. Mucous membranes with a simple columnar epithelium line most of the digestive and respiratory tracts and portions of the reproductive tract. **(b)** Serous membranes line the ventral body cavities (the peritoneal, pleural, and pericardial cavities). **(c)** The cutaneous membrane of the skin covers the outer surface of the body. **(d)** Synovial membranes line joint cavities and produce the fluid within the joint.

the visceral pericardium covers the heart, and the parietal pericardium lines the inner surfaces of the pericardial sac that surrounds the pericardial cavity. *Serous fluid* covering the surfaces of the visceral and parietal membranes minimizes the friction between these opposing surfaces.

Cutaneous Membrane

The **cutaneous membrane** of the skin (Figure 4-12c●) covers the surface of the body. It consists of a stratified squamous epithelium and the underlying connective tissues. In contrast to serous or mucous membranes, the cutaneous membrane is thick, relatively waterproof, and usually dry. The skin is discussed in detail in Chapter 5.

Synovial Membranes

Bones of the skeleton contact one another at joints, also called **articulations** (ar-tik-ū-LĀ-shuns). The type of connective tissue at a joint may restrict or enhance its movement. If the joint is mobile, the bony surfaces do not come into direct contact with one another. If they did, abrasion and impacts would damage the opposing surfaces, and smooth movement would be almost impossible. Instead, the ends of the bones are covered with hyaline cartilage and separated by a viscous synovial fluid produced by the **synovial membrane** (sin-Ō-vē-al) (Figure 4-12d●), which lines the joint cavity. Unlike the other three membranes, the synovial membrane consists primarily of loose connective tissue, and the epithelial layer is incomplete.

- ✓ How does a cell membrane differ from a tissue-level membrane?
- ✓ Serous membranes produce fluids. What is their function?
- ✓ Why do you find the same epithelial organization in the mucous membranes of the pharynx, esophagus, anus, and vagina?

MUSCLE TISSUE

Muscle tissue is specialized for contraction. A large skeletal muscle cell may be 100 micrometers (μm; 1 μm = 1/25,000 in.) in diameter and 25 cm (10 in.)

long. Because skeletal muscle cells are relatively long and slender, they are usually called *muscle fibers.*

Muscle cell contraction involves interaction between filaments of *myosin* and *actin*, proteins found in the cytoskeletons of many cells. ∞ [p. 58] In muscle cells, however, the filaments are more numerous and arranged so that their interaction produces a contraction of the entire cell.

There are three types of muscle tissue: *skeletal, cardiac,* and *smooth muscle tissue.* The contraction mechanism is the same in all of them, but the organization of their actin and myosin filaments differs. Because each type will be examined in later chapters, notably Chapter 7, this discussion will focus on general characteristics rather than specific details.

Skeletal Muscle Tissue

Skeletal muscle tissue contains very large, multinucleated fibers (cells) tied together by loose connective tissue. The collagen and elastic fibers surrounding each cell and group of cells blend into those of a tendon that conducts the force of contraction, usually to a bone of the skeleton. Contractions of muscle tissue cause the bones to move.

Because the actin and myosin filaments are arranged in organized groups, skeletal muscle fibers appear to be marked by a series of bands known as *striations* (Figure 4-13a●). Skeletal muscle fibers will not usually contract unless stimulated by nerves. Since the nervous system provides voluntary control over its activities, skeletal muscle is described as *striated voluntary muscle.*

Cardiac Muscle Tissue

Cardiac muscle tissue (Figure 4-13b●) is found only in the heart. Cardiac muscle cells are much smaller than skeletal muscle fibers, and each cardiac muscle cell usually has a single nucleus. Cardiac muscle cells are interconnected at **intercalated discs**, specialized attachment sites containing gap junctions and desmosomes. The muscle cells branch, forming a network that efficiently conducts the force and stimulus for contraction from one area of the heart to another.

Unlike skeletal muscle, cardiac muscle does not rely on nerve activity to start a contraction. Instead, specialized cells, called pacemaker cells, establish a regular rate of contraction. Although the nervous system can alter the rate of pacemaker activity, it does not provide voluntary control over individual cardiac muscle cells. In short, cardiac muscle can be considered as *striated involuntary muscle.*

Smooth Muscle Tissue

Smooth muscle tissue (Figure 4-13c●) can be found in the walls of blood vessels; around hollow organs such as the urinary bladder; and in layers around the respiratory, circulatory, digestive, and reproductive tracts.

A smooth muscle cell is small and slender, tapering to a point at each end. There is one nucleus in each smooth muscle cell. Unlike skeletal and cardiac muscle, the actin and myosin filaments in smooth muscle cells are scattered throughout the cytoplasm, and there are no striations.

Smooth muscle cells may contract on their own, or their contractions may be triggered by neural activity. The nervous system usually does not provide voluntary control over smooth muscle contractions, and smooth muscle is therefore categorized as *nonstriated involuntary muscle.*

NEURAL TISSUE

Neural tissue is specialized for the conduction of electrical impulses that convey information or instructions from one region of the body to another. Most of the neural tissue (98 percent) is concentrated in the brain and spinal cord, the control centers for the nervous system.

Neural tissue contains two basic types of cells: **neurons** (NOO-rons; *neuro-*, nerve) and several different kinds of supporting cells, or **neuroglia** (noo-RŌG-lē-a; *glia*, glue). Neurons transmit the actual signals as electrical events affecting their cell membranes. The neuroglia provide physical support for neural tissue, maintain the chemical composition of the tissue fluids, and defend the tissue from infection.

A typical neuron has a cell body, or **soma** (SŌ-ma; *soma*, body), that contains the nucleus (Figure 4-14●). The stimulus that results in the production of an electrical impulse usually affects the cell membrane of one of the **dendrites** (DEN-drīts; *dendron*, a tree). Stimulation alters the permeability of the cell membrane, eventually producing an electrical impulse that is conducted along the length of the axon. **Axons,** which may reach a meter in length, are often called *nerve fibers.* Each axon ends at a specialized intercellular junction called a **synapse** (SIN-aps; *syn-*, together). Chapter 8, which considers the properties of neural tissue, provides more detail on the structure and function of synapses.

✓ What type of muscle tissue has small, spindle-shaped cells with single nuclei and no obvious banding pattern?

✓ Our voluntary control is restricted to which type of muscle tissue?

✓ Why are skeletal muscle cells and axons also called fibers?

SKELETAL MUSCLE TISSUE

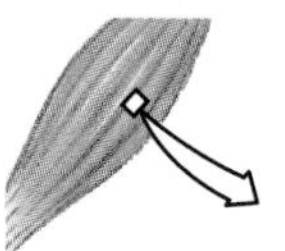

LOCATIONS: Combined with connective tissues and nervous tissue in skeletal muscles, organs such as the skeletal muscles of the limbs

FUNCTIONS: Moves or stabilizes the position of the skeleton; guards entrances and exits to the digestive, respiratory, and urinary tracts; generates heat; protects internal organs

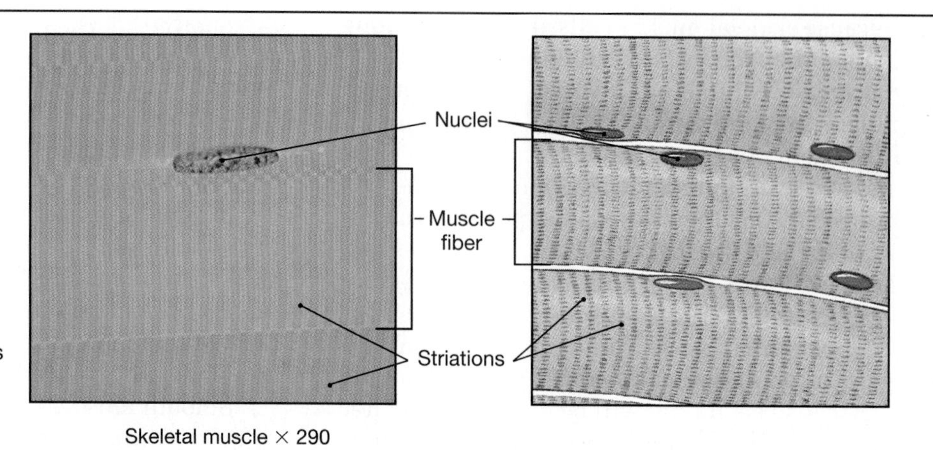

Skeletal muscle × 290

(a)

CARDIAC MUSCLE TISSUE

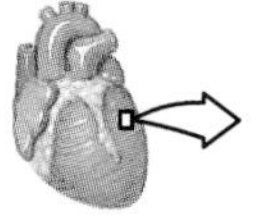

LOCATION: Heart

FUNCTIONS: Circulates blood; maintains blood (hydrostatic) pressure

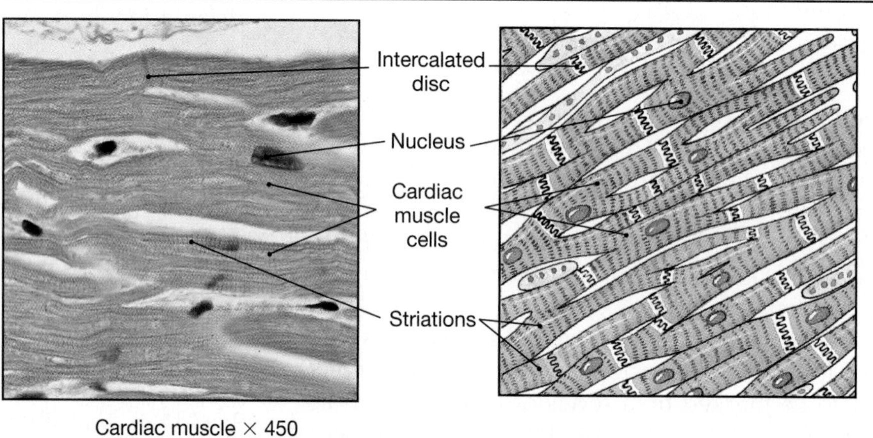

Cardiac muscle × 450

(b)

SMOOTH MUSCLE TISSUE

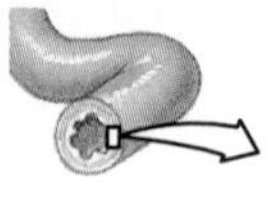

LOCATIONS: Encircles blood vessels; found in the walls of digestive, respiratory, urinary, and reproductive organs

FUNCTIONS: Moves food, urine, and reproductive tract secretions; controls diameters of respiratory passageways; regulates diameter of blood vessels; and contributes to regulation of tissue blood flow

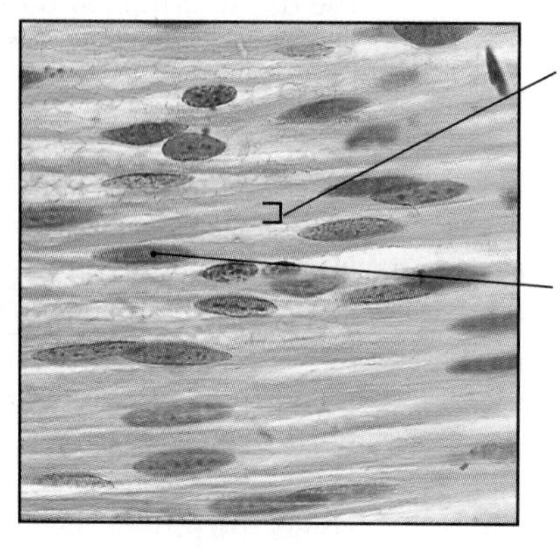

Smooth muscle cell

Nucleus

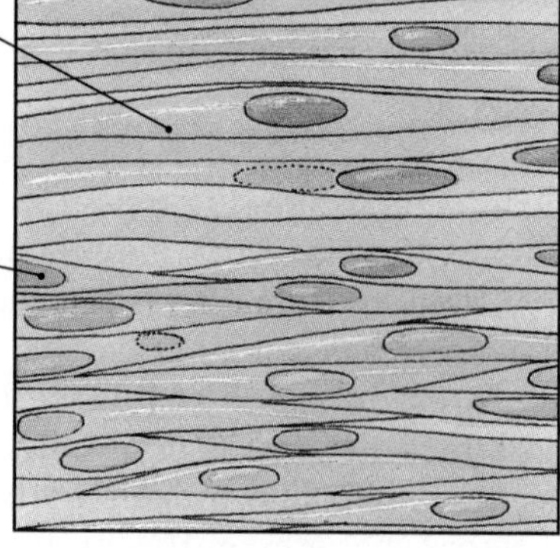

Smooth muscle × 300

(c)

•Figure 4-13
Muscle Tissue

(a) Skeletal muscle fibers. Note the large fiber size, prominent banding pattern, multiple nuclei, and unbranched arrangement. **(b)** Cardiac muscle cells. Cardiac muscle cells differ from skeletal muscle fibers in three major ways: size (cardiac muscle cells are smaller), organization (cardiac muscle cells branch), and number of nuclei (a typical cardiac muscle cell has one centrally placed nucleus). Both contain actin and myosin filaments in an organized array that produces striations. **(c)** Smooth muscle cells. Smooth muscle cells are small and spindle-shaped, with a central nucleus. They do not branch, and there are no striations.

TISSUE INJURIES AND REPAIRS

Tissues in the body are not independent of each other; they combine to form organs with diverse functions. Any injury to the body affects several tissue types simultaneously, and these tissues must respond in a coordinated manner to preserve homeostasis.

The restoration of homeostasis following a tissue injury involves two related processes. First, the area is isolated from neighboring healthy tissue while damaged cells, tissue components, and any dangerous microorganisms are cleaned up. This phase, which coordinates the activities of several different tissues, is called **inflammation**, or the *inflammatory response*. Inflammation begins immediately after an injury and produces several familiar sensations, including swelling, warmth, redness, and pain. An infection is an inflammation resulting from the presence of pathogens, such as bacteria.

Second, the damaged tissues are replaced or repaired to restore normal function. This repair process is called **regeneration**. Inflammation and regeneration are controlled at the tissue level. The two phases overlap; isolation of the area of damaged tissue establishes a framework that guides the cells responsible for reconstruction, and repairs are under way well before cleanup operations have ended. Later chapters, especially Chapter 15, will examine inflammation and regeneration in more detail. AM *Tissue Structure and Disease*

TISSUES AND AGING

Tissues change with age, and there is a decrease in the speed and effectiveness of tissue repairs. In general, repair and maintenance activities throughout the body slow down, and a combination of hormonal changes and alterations in lifestyle affect the structure and chemical composition of many tissues. Epithelia get thinner and connective tissues more fragile. Individuals bruise easily and bones become brittle; joint pains and broken bones are common complaints. Cardiac muscle fibers and neurons cannot be replaced, and cumulative losses from relatively minor damage can contribute to major health problems such as cardiovascular disease or deterioration in mental function.

In future chapters we will consider the effects of aging on specific organs and systems. Some of these changes are genetically programmed. For example, as people age their chondrocytes produce a slightly different form of the gelatinous compound making up the cartilage matrix. This difference in composition probably accounts for the observed increase in thickness and stiffness of cartilages in older people. Other age-related changes in tissue structure have multiple causes. The age-related reduction in bone strength in women, a condition called *osteoporosis*, is often caused by a combination of inactivity, low dietary calcium levels, and a reduction in circulating *estrogens* (sex hormones). A program of exercise, calcium supplements, and hormonal replacement therapies can usually maintain normal bone structure for many years.

Aging and Cancer Incidence

Cancer rates increase with age, and roughly 25 percent of all Americans develop cancer at some point in their lives. It has been estimated that 70–80 percent of

●Figure 4-14 Neural Tissue

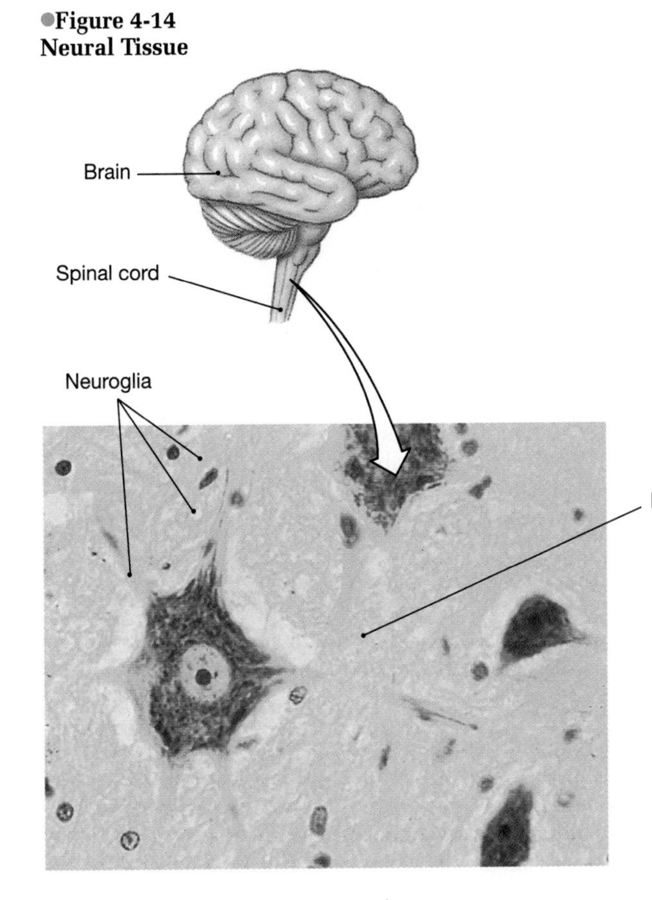

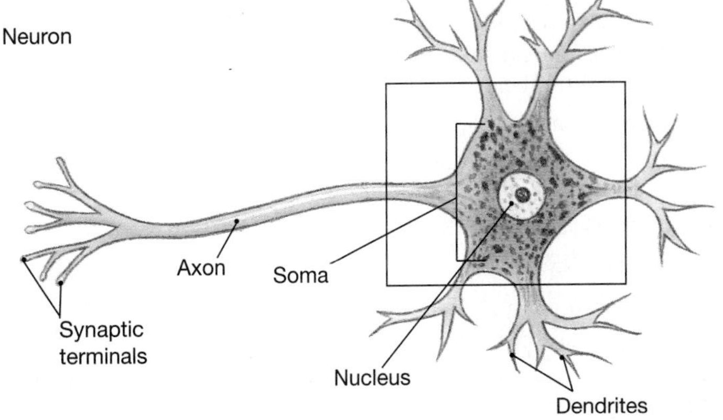

cancer cases result from chemical exposure, environmental factors, or some combination of the two, and 40 percent of these cancers are caused by cigarette smoke. Each year over 500,000 Americans are killed by cancer, making this Public Health Enemy #2, second only to heart disease. The *Applications Manual* contains a detailed discussion of cancer development, growth, and treatment. AM *Cancer*

Chapter Review

KEY TERMS

connective tissue, p. 81
epithelium, p. 73
fibroblasts, p. 81
gland cells, p. 74
inflammation, p. 91
macrophage, p. 81
mucous membrane, p. 87
muscle tissue, p. 88
neural tissue, p. 89
neuron, p. 89
serous membrane, p. 87
tissue, p. 73

SUMMARY OUTLINE

INTRODUCTION p. 73

1. **Tissues** are collections of specialized cells and cell products that are organized to perform a relatively limited number of functions. There are four **primary tissue types:** epithelia, connective tissues, muscle tissue, and neural tissue. **Histology** is the study of tissues. (*Figure 4-1*)

EPITHELIA p. 73

1. An **epithelium** is an **avascular** layer of cells that forms a barrier that has certain properties.

Functions of Epithelia p. 73

2. Epithelia provide physical protection, control permeability, provide sensations, and produce specialized secretions. Gland cells are epithelial cells that produce secretions.

3. **Exocrine** secretions are released onto body surfaces; **endocrine** secretions, known as **hormones,** are released by gland cells into the surrounding tissues.

Intercellular Connections p. 74

4. The individual cells that make up tissues attach to each other or to extracellular protein fibers in three ways: gap junctions, tight junctions, and desmosomes. (*Figure 4-2*)

5. In a **gap junction,** two cells are held together by interlocked membrane proteins, forming a narrow passageway.

6. At a **tight junction,** there is a partial fusion of the two cell membranes; these are the strongest intercellular connections.

7. A **desmosome** has a very thin layer of intercellular cement between the cell membranes, reinforced by a network of protein fibers.

The Epithelial Surface p. 74

8. Epithelial cells may have cilia or microvilli. The coordinated beating of the cilia on a ciliated epithelium moves materials across the epithelial surface. (*Figure 4-3*)

The Basement Membrane p. 75

9. The inner surface of each epithelium is connected to a noncellular **basement membrane.**

Epithelial Renewal and Repair p. 75

10. Divisions by **stem cells,** or **germinative cells,** continually replace the short-lived epithelial cells.

Classification of Epithelia p. 75

11. Epithelia are classified on the basis of the number of cell layers and the shape of the exposed cells.

12. A **simple epithelium** has a single layer of cells covering the basement membrane; a **stratified epithelium** has several layers. In a **squamous epithelium** the cells are thin and flat. Cells in a **cuboidal epithelium** resemble little hexagonal boxes; those in a **columnar epithelium** are taller and more slender. (*Figures 4-4, 4-5*)

Glandular Epithelia p. 78

13. A glandular epithelial cell may release its secretions through merocrine, apocrine, or holocrine mechanisms. (*Figure 4-6*)

14. In **merocrine secretion,** the most common method of secretion, the product is released through exocytosis. **Apocrine secretion** involves the loss of both secretory product and cytoplasm. Unlike the first two methods, **holocrine secretion** destroys the cell, which it becomes packed with secretions and finally bursts.

15. Exocrine secretions may be *serous* (watery, usually containing enzymes), *mucous* (thick and slippery), or *mixed* (containing enzymes and lubricants). (*Table 4-1*)

CONNECTIVE TISSUES p. 80

1. Connective tissues are internal tissues with many important functions: establishing a structural framework; transporting fluids and dissolved materials; protecting delicate organs; supporting, surrounding, and interconnecting tissues; storing energy reserves; and defending the body from microorganisms.

2. All connective tissues have specialized cells, extracellular protein fibers, and a **ground substance.** The protein fibers and ground substance constitute the **matrix.**

Classification of Connective Tissues p. 81

3. **Connective tissue proper** refers to connective tissues that contain varied cell populations and fiber types surrounded by a syrupy ground substance. (*Figure 4-7*)

4. **Fluid connective tissues** have a distinctive population of cells suspended in a watery ground substance con-

taining dissolved proteins. The two types are blood and lymph. (*Figure 4-7*)

5. **Supporting connective tissues** have a less diverse cell population than connective tissue proper and a dense matrix that contains closely packed fibers. The two types of supporting connective tissues are cartilage and bone. (*Figure 4-7*)

Connective Tissue Proper p. 81

6. Connective tissue proper contains fibers, a viscous ground substance, and a varied cell population.

7. There are three types of fiber in connective tissue: **collagen fibers, reticular fibers**, and **elastic fibers**.

8. Connective tissue proper is classified as **loose** or **dense connective tissues**. Loose connective tissues include loose connective tissue, or areolar tissue, and **adipose tissue**. (*Figure 4-8a,b*)

9. Most of the volume in dense connective tissue consists of fibers. Dense connective tissues form **tendons** and **ligaments**. (*Figure 4-8c*)

Fluid Connective Tissues p. 84

10. *Blood* and *lymph* are connective tissues that contain distinctive collections of cells in a fluid matrix. (*Figure 4-9*)

11. Blood contains red blood cells, white blood cells, and platelets. Its watery ground substance is called **plasma**.

12. **Arteries** carry blood from the heart and toward **capillaries**, where water and small solutes move into the **interstitial fluid** of surrounding tissues. **Veins** return blood to the heart.

13. Lymph forms as interstitial fluid enters the **lymphatics**, which return lymph to the circulatory system.

Supporting Connective Tissues p. 84

14. Cartilage and bone are called supporting connective tissues because they support the rest of the body.

15. The matrix of **cartilage** consists of a firm gel and cells called **chondrocytes**. A fibrous **perichondrium** separates cartilage from surrounding tissues. There are three types of cartilage: **hyaline cartilage, elastic cartilage**, and **fibrocartilage**. (*Figure 4-10*)

16. Chondrocytes rely upon diffusion through the avascular matrix to obtain nutrients.

17. Bone, or **osseous tissue**, has a matrix consisting of collagen fibers and calcium salts, which give it unique properties. (*Figure 4-11; Table 4-2*)

18. **Osteocytes** depend on diffusion through **canaliculi** for nutrient intake.

19. Each bone is surrounded by a **periosteum**.

MEMBRANES p. 85

1. Membranes form a barrier or interface. Epithelia and connective tissues combine to form membranes that cover and protect other structures and tissues. There are four types of membranes: *mucous, serous, cutaneous*, and *synovial*. (*Figure 4-12*)

Mucous Membranes p. 87

2. **Mucous membranes** line cavities that communicate with the exterior. Their surfaces are normally moistened by mucous secretions.

Serous Membranes p. 87

3. **Serous membranes** line internal cavities and are delicate, moist, and very permeable.

Cutaneous Membrane p. 88

4. The **cutaneous membrane** covers the body surface. Unlike serous and mucous membranes, it is relatively thick, waterproof, and usually dry.

Synovial Membranes p. 88

5. **Synovial membranes**, located at joints, or articulations, produce synovial fluid in joint cavities. Synovial fluid helps lubricate the joint and promotes smooth movement.

MUSCLE TISSUE p. 89

1. Muscle tissue, consisting of muscle fibers, is specialized for contraction. There are three types of muscle tissue: **skeletal, cardiac,** and **smooth muscle tissues.** (Figure 4-13)

Skeletal Muscle Tissue p. 89

2. Skeletal muscle tissue contains very large fibers tied together by collagen and elastic fibers. Skeletal muscle fibers have a striped appearance because of the organization of contractile proteins. The stripes are called *striations*. Because we can control the contraction of skeletal muscle fibers through the nervous system, skeletal muscle can be considered *striated voluntary muscle*.

Cardiac Muscle Tissue p. 89

3. Cardiac muscle tissue is found only in the heart. The nervous system does not provide voluntary control over cardiac muscle cells. Thus, cardiac muscle is *striated involuntary muscle*.

Smooth Muscle Tissue p. 89

4. Smooth muscle tissue is found in the walls of blood vessels, around hollow organs, and in layers around various tracts. It is classified as *nonstriated involuntary muscle*.

NEURAL TISSUE p. 89

1. **Neural tissue** is specialized to conduct electrical impulses that convey information from one area of the body to another.

2. Cells in neural tissue are either neurons or neuroglia. **Neurons** transmit information as electrical impulses in their cell membranes. Several kinds of **neuroglia** serve both supporting and defense functions. (*Figure 4-14*)

3. A typical neuron has a **soma, dendrites**, and an **axon** that ends at a **synapse**.

TISSUE INJURIES AND REPAIRS p. 91

1. Any injury affects several tissue types simultaneously, and they respond in a coordinated manner. Homeostasis is restored in two processes: inflammation and regeneration.

2. **Inflammation**, or the *inflammatory response*, isolates the injured area while damaged cells, tissue components, and any dangerous microorganisms are cleaned up.

3. **Regeneration** is the repair process that restores normal function.

TISSUES AND AGING p. 91

1 Tissues change with age. Repair and maintenance grow less efficient, and the structure and chemical composition of many tissues are altered.

Aging and Cancer Incidence p. 91

2. Cancer incidence increases with age, with roughly three-quarters of all cases caused by exposure to chemicals or environmental factors.

CHAPTER QUESTIONS

LEVEL 1 Reviewing Facts and Terms

Match each item in column A with the most closely related item in column B. Use letters for answers in the spaces provided.

Column A	Column B
___ 1. histology	a. repair and renewal
___ 2. microvilli	b. ligament
___ 3. gap junction	c. endocrine secretion
___ 4. tight junction	d. absorption and secretion
___ 5. germinative cells	e. fat cells
___ 6. destroys gland cell	f. holocrine secretion
___ 7. hormones	g. study of tissues
___ 8. adipocytes	h. tendon
___ 9. bone-to-bone attachment	i. intercellular connection
___ 10. muscle-to-bone attachment	j. interlocking of membrane proteins
___ 11. skeletal muscle	k. intercalated discs
___ 12. cardiac muscle	l. striated, voluntary

13. The four basic tissue types found in the body are:
(a) epithelia, connective, muscle, neural
(b) simple, cuboidal, squamous, stratified
(c) fibroblasts, adipocytes, melanocytes, mesenchymal
(d) lymphocytes, macrophages, microphages, adipocytes

14. Long microvilli incapable of movement are called:
(a) cilia (b) flagella
(c) stereocilia (d) a, b, and c are correct

15. The most abundant connections between cells in the superficial layers of the skin are:
(a) intermediate junctions
(b) gap junctions
(c) desmosomes
(d) tight junctions

16. The three cell shapes making up epithelial tissue are:
(a) simple, stratified, transitional
(b) simple, stratified, pseudostratified
(c) hexagonal, cuboidal, spherical
(d) cuboidal, squamous, and columnar

17. The type of tissue that contains a fluid known as the ground substance is
(a) epithelial (b) neural
(c) muscle (d) connective

18. The three major types of cartilage found in the body are:
(a) collagen, reticular, elastic
(b) areolar, adipose, reticular
(c) hyaline, elastic, fibrocartilage
(d) keratin, reticular, elastic

19. The primary function of serous membranes in the body is:
(a) to minimize friction between opposing surfaces
(b) to line cavities that communicate with the exterior
(c) to perform absorptive and secretory functions
(d) to cover the surface of the body

20. Large muscle fibers that are multinucleated, striated, and voluntary are found in:
(a) cardiac muscle tissue (b) skeletal muscle tissue
(c) smooth muscle tissue (d) a, b, and c are correct

21. Intercalated discs and pacemaker cells are characteristic of:
(a) smooth muscle tissue (b) cardiac muscle tissue
(c) skeletal muscle tissue (d) a, b, and c are correct

22. Axons, dendrites, and a soma are characteristics of cells found in:
(a) neural tissue (b) muscle tissue
(c) connective tissue (d) epithelial tissue

23. What are the four essential functions of epithelial tissue?

24. What three types of layering make epithelial tissue recognizable?

25. What three basic components are found in connective tissues?

26. What fluid connective tissues and supporting connective tissues are found in the human body?

27. What four kinds of membranes composed of epithelial and connective tissue cover and protect other structures and tissues in the body?

28. What two cell populations make up neural tissue? What is the function of each?

LEVEL 2 Reviewing Concepts

29. In surfaces of the body where mechanical stresses are severe, the dominant epithelium is:
    (a) stratified squamous epithelium
    (b) simple cuboidal epithelium
    (c) simple columnar epithelium
    (d) stratified cuboidal epithelium
30. Why does holocrine secretion require continuous cell division?
31. What is the difference between an exocrine and an endocrine secretion?
32. A significant structural feature in the digestive system is the presence of tight junctions located near the exposed surfaces of cells lining the digestive tract. Why are these junctions so important?
33. Why are infections always a serious threat after a severe burn or abrasion?
34. What characteristics make the cutaneous membranes different from the serous and mucous membranes?

LEVEL 3 Critical Thinking and Clinical Applications

35. A biology student accidentally loses the labels of two prepared slides she is studying. One is a slide of animal intestine and the other of animal esophagus. You volunteer to help her sort them out. How would you decide which slide is which?
36. You are asked to develop a scheme that can be used to identify the three different types of muscle tissue in two steps. What would the two steps be?

5 THE INTEGUMENTARY SYSTEM

How many of the sun lovers who crowd the beaches on a warm day ever stop to consider the contributions and sacrifices made by their skin? This remarkable structure absorbs ultraviolet radiation, prevents dehydration, preserves normal body temperature, and tolerates the chafing and abrasion of the sand. Although few people think of it in these terms, the skin is actually an organ—the largest organ of the body. In this chapter we will examine its varied functions.

Chapter Outline and Objectives

1. Describe the general functions of the integumentary system.
2. Describe the main structural features of the epidermis and explain their functional significance.
3. Explain what accounts for individual and racial differences in skin, such as skin color.
4. Describe how the integumentary system helps regulate body temperature.
5. Discuss the effects of ultraviolet radiation on the skin and the role played by melanocytes.
6. Discuss the functions of the skin's accessory structures.
7. Describe the mechanisms that produce hair and that determine hair texture and color.

8. Explain how the skin responds to injury and repairs itself.

9. Summarize the effects of the aging process on the skin.

Vocabulary Development

cornu, horn; *stratum corneum*
epi-, above or over; *epidermis*
facere, to make; *cornified*
germinare, to start growing; *stratum germinativum*
keros, horn; *keratin*
kyanos, blue; *cyanosis*
luna, moon; *lunula*
onyx, nail; *eponychium*
papilla, a nipple-shaped mound; *dermal papillae*

The integumentary system, consisting of the skin, hair, nails, and various glands, is the most visible organ system of the body. Because of its visibility, we devote a lot of time to its upkeep. Washing the face and hands, brushing or trimming hair, clipping nails, taking showers, and applying deodorant are activities that modify the appearance or properties of the skin. And when something goes wrong with the skin, the effects are immediately apparent. Even a relatively minor skin condition, such as mild acne, will be noticed at once, whereas more serious problems in other systems are often ignored. The skin, however, may also provide visible signs of major systemic disorders through changes in its color, flexibility, or sensitivity.

This chapter focuses on the important structural and functional relationships in the skin. In the process, it demonstrates patterns that apply to tissue and organ interactions in other systems.

INTEGUMENTARY STRUCTURE AND FUNCTION

The **integumentary system**, or simply the *integument*, has two major components: the *cutaneous membrane* and the *accessory structures*.

The **cutaneous membrane**, or *skin*, is an organ composed of the superficial epithelium, or **epidermis** (*epi-*, above), and the underlying connective tissues of the **dermis**. The **accessory structures** include hair, nails, and a variety of multicellular exocrine glands.

The general structure of the integument is shown in Figure 5-1●. Beneath the dermis, the loose connective tissue of the **subcutaneous layer** (or *hypodermis*) attaches the integument to deeper structures, such as muscles or bones. Although often not considered to be part of the integumentary system, this layer will be considered here because of its extensive interconnections with the dermis.

As we will see, the five major functions of the various parts of the integument are these:

1. ***Protection.*** The skin covers and protects underlying tissues and organs from impacts, chemicals, and infections, and prevents the loss of body fluids.
2. ***Temperature maintenance.*** The skin maintains normal body temperature by regulating heat gain or loss to the environment.
3. ***Storage of nutrients.*** The deeper portions of the dermis typically contain a large reserve of lipids, in the form of adipose tissue.
4. ***Sensory reception.*** Receptors in the integument detect touch, pressure, pain, and temperature stimuli and relay that information to the nervous system.
5. ***Excretion and secretion.*** The integument excretes salts, water, and organic wastes, and produces milk, a specialized exocrine secretion.

Each of these functions will be explored more fully as we discuss the individual components of the system.

The Epidermis

The epidermis consists of a *stratified squamous epithelium.* ∞ [p. 78] Several different cell layers are present, but the precise boundaries between them are often difficult to see in a light micrograph. The majority of the body is covered by **thin skin.** In a sample of thin skin the epidermis is a mere 0.08 mm thick. In **thick skin**, found on the palms of the hands and soles of the feet, the epidermis may be as much as six times thicker. The terms *thin* and *thick* refer only to the relative thickness of the epidermis, not to the integument as a whole.

Layers of the Epidermis

Refer to Figure 5-2● as we describe the layers in a section of thick skin. Beginning at the basement membrane and traveling toward the free surface, we find the *stratum germinativum*, the *stratum spinosum*, the *stratum granulosum*, the *stratum lucidum*, and the *stratum corneum*.

Stratum Germinativum The deepest epidermal layer is called the **stratum germinativum** (STRA-tum jer-mi-nā-TĒ-vum; *germinare*, to start growing), or *stratum basale*. This layer is firmly attached to the basement membrane that separates the epidermis from the loose connective tissue of the adjacent dermis. The stratum germinativum forms **epidermal ridges** that extend into the dermis, increasing the area of contact between the two regions. Dermal projections called **dermal papillae** (singular *papilla*, a nipple-shaped mound) extend between adjacent ridges (Figures 5-1●, 5-2●). There are no blood vessels in the epidermis, and the epidermal cells must obtain nutrients delivered by dermal blood vessels. The combination of ridges and papillae increases the surface area for diffusion between the dermis and epidermis.

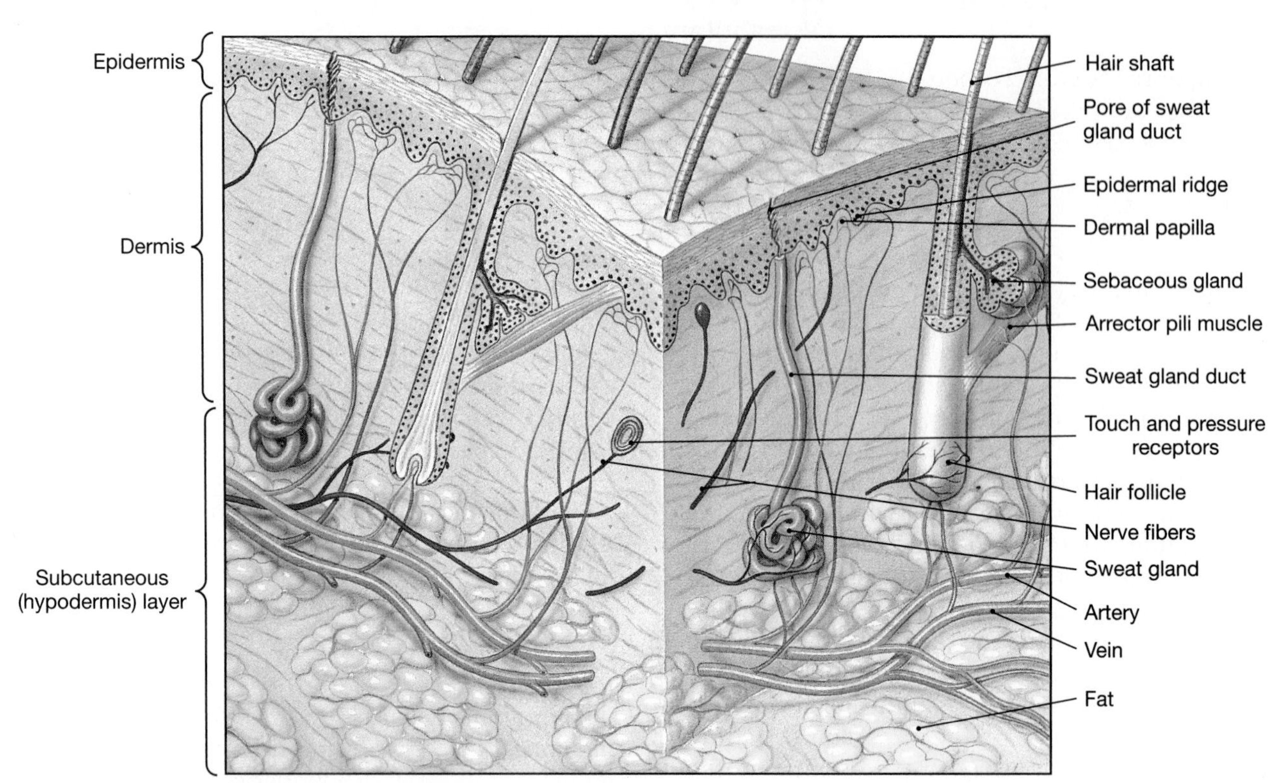

Figure 5-1
Components of the Integumentary System
Relationships among the major components of the integumentary system (with the exception of nails, shown in Figure 5-7).

The contours of the skin surface follow the ridge patterns. This pattern is most easily observed on the thick skin of the palms and soles, where the ridges form complex whorls. The superficial ridges on the palms and soles increase the surface area of the skin and increase friction, ensuring a secure grip. Ridge shapes are genetically determined: Those of each person are unique and do not change in the course of a lifetime. Fingerprints are ridge patterns on the tips of the fingers that can be used to identify individuals; they have been used in criminal investigations for over a century.

Large stem cells dominate the stratum germinativum, making it the layer where new cells are generated and begin to grow. The divisions of these cells replace more superficial cells that are lost or shed at the epithelial surface. The stratum germinativum also contains *melanocytes*, whose cytoplasmic processes extend between epithelial cells in this layer, and receptors that provide information about objects touching the skin. Melanocytes synthesize *melanin*, a yellow-brown to black pigment that colors the epidermis.

Stratum Spinosum Each time a stem cell divides, one of the daughter cells enters the next layer, the **stratum spinosum** (spiny layer), where it may continue to divide. The stratum spinosum is several cells thick, and the cells are bound together by desmosomes. The name of this layer is based on the fact that the desmosomes and other cytoskeletal structures make the cells appear shrunken and spiny when prepared for viewing under a microscope.

Stratum Granulosum The **stratum granulosum** consists of cells displaced from the spinosum layer. By the time epithelial cells reach this layer, most have stopped dividing, and they begin manufacturing large quantities of proteins and enzymes.

Stratum Lucidum In the thick skin of the palms and soles, a glassy **stratum lucidum** (clear layer) covers the stratum granulosum. The cells in this layer are flattened and densely packed. The cytoplasm contains enzymes involved in the production of the fibrous protein **keratin** (KER-a-tin; *keros*, horn), introduced in Chapter 2. ∞ [p. 38]

Stratum Corneum The most superficial layer of the epidermis, the **stratum corneum** (KOR-nē-um; *cornu*, horn), consists of flattened and dead epithelial cells that have accumulated large amounts of keratin.

Keratin is extremely strong, light, flexible, durable, and water-resistant. In the human body, keratin not only coats the surface of the skin, but forms the basic structure of hair, calluses, and nails. In other animals, it is even more versatile, making up cow horns and hooves, bird feathers, reptile scales, porcupine quills, the armor of armadillos, and baleen plates in the mouths of whales.

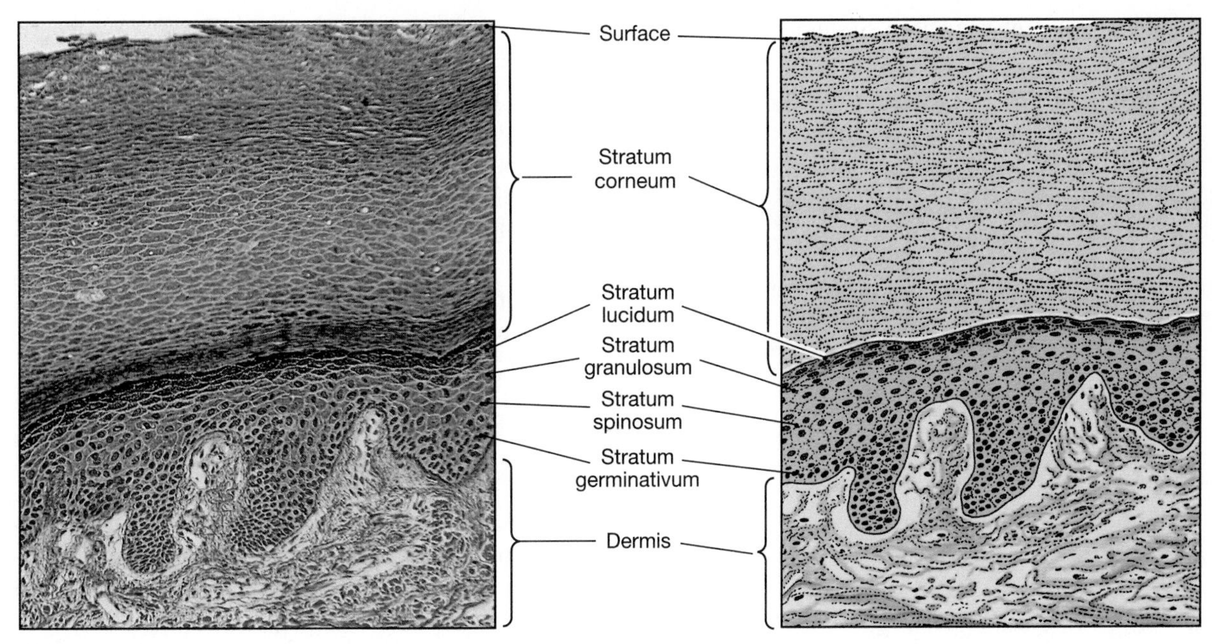

●**Figure 5-2**
Layers of the Epidermis (LM × 90)

It takes 2–4 weeks for a cell to move from the stratum germinativum to the stratum corneum. During this time, the cell is displaced from its oxygen and nutrient supply, becomes packed with keratin, and finally dies. The dead cells usually remain in the stratum corneum for an additional 2 weeks before they are shed or washed away. As superficial layers are lost, new layers arrive from the underlying strata. Thus the deeper layers of the epithelium and underlying tissues remain protected by superficial layers of dead, durable, and expendable cells. AM *Disorders of Keratin Production*

Permeability of the Epidermis

An epithelium containing large amounts of keratin is said to be **keratinized** (ker-A-tin-īzed), or **cornified** (KŌR-ni-fīd; *cornu*, horn + *facere*, to make). A cornified epithelium is only around 1/100,000th as permeable to water and electrolytes as are other epithelia. Normally the stratum corneum is relatively dry; only around 10 percent of the cell weight is water. This dryness makes it unattractive to many bacteria, and thus a good protective barrier. AM *Synthetic Skin*

DRUG ADMINISTRATION THROUGH THE SKIN

Drugs in oils or other lipid-soluble carriers can penetrate the epidermis. The movement is slow, particularly through the layers of cell membranes in the stratum corneum, but once a drug reaches the underlying tissues it will be absorbed into the circulation. Placing a drug in a solvent that is lipid-soluble can assist its movement through the lipid barriers. Drugs can also be administered by packaging them in *liposomes*, artificially produced fat droplets. Liposomes containing DNA fragments have been used experimentally to introduce normal genes into abnormal human cells. For example, genes carried by liposomes have been used to alter the cell membranes of skin cancer cells so that they will be attacked by the immune system. Another experimental procedure involves creating a temporary change in skin permeability by administering a brief pulse of electricity. The electrical pulse temporarily changes the positions of the cells in the stratum corneum, creating channels that allow drug penetration.

A useful technique for long-term drug administration involves placing a sticky patch containing a drug over an area of thin skin. Because the rate of diffusion is relatively slow, the patch must contain an extremely high concentration of the drug. This procedure, called *transdermal administration*, has the advantage that a single patch may work for several days, making daily pills unnecessary. AM *Transdermal Medications*

Skin Color

The color of the epidermis is caused by the interaction between (1) pigment composition and concentration and (2) the dermal blood supply.

Pigmentation The epidermis contains variable quantities of two pigments, *carotene* and *melanin*. **Carotene** (KAR-ō-tēn) is an orange-yellow pigment that normally accumulates inside epidermal cells. Carotene pigments are found in a variety of orange-colored vegetables, such as carrots and squashes. **Melanin** is a brown, yellow-brown, or black pigment produced by **melanocytes**. Melanocytes (Figure 5-3●) manufacture and store melanin and inject that pigment into the epithelial cells of the stratum germinativum and stratum spinosum. This transfer of pigmentation colors the entire epidermis. Melanocyte activity slowly increases in response to sunlight exposure, peaking around 10 days after the initial exposure.

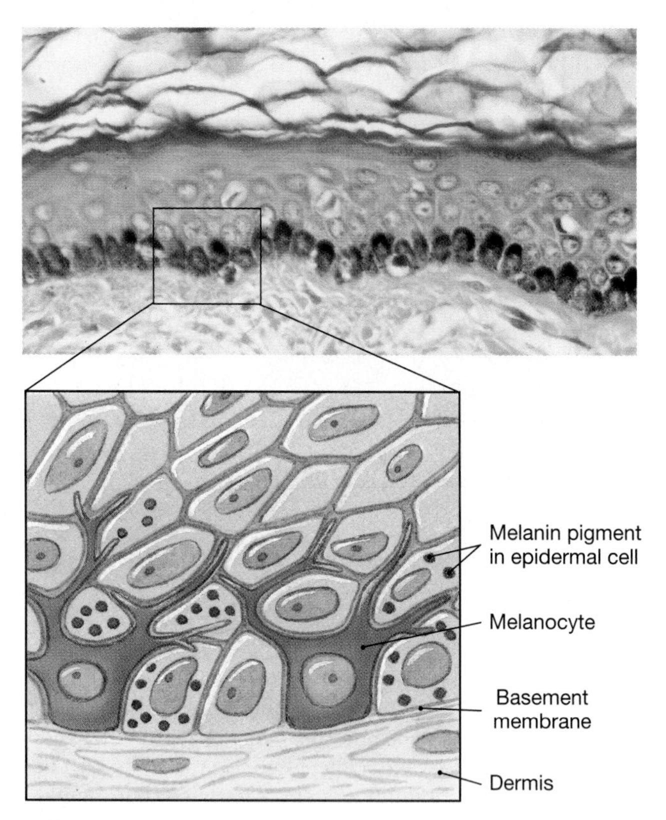

Figure 5-3
Melanocytes

The micrograph and accompanying drawing indicate the location and orientation of melanocytes in the deeper layers of the epidermis of a black person.

Sunlight contains significant amounts of **ultraviolet** (UV) **radiation.** A small amount of UV radiation is beneficial, for it stimulates the synthesis of *vitamin* D_3 in the epidermis; this process is discussed in a later section. Too much ultraviolet radiation, however, produces immediate effects of mild or even serious burns. Melanin helps prevent skin damage by absorbing ultraviolet radiation before it reaches the deep layers of the epidermis and dermis. Within the epidermal cells, melanin concentrates around the nuclear envelope and absorbs the UV before it can damage the nuclear DNA.

Despite the presence of melanin, long-term damage can result from repeated exposure, even in darkly pigmented individuals. For example, alterations in the underlying connective tissues lead to premature wrinkling, and skin cancers can result from chromosomal damage in germinative cells or melanocytes. One of the major consequences of the global depletion of the ozone layer in the upper atmosphere will be a sharp increase in the rate of skin cancers. AM *Skin Cancers and the Ozone Hole*

The ratio between melanocytes and germinative (basal) cells ranges between 1:4 and 1:20, depending on the region of the body surveyed. The observed differences in skin color between individuals and even races do not reflect different *numbers* of melanocytes, but merely different levels of melanin production. Even the melanocytes of **albino** individuals are distributed normally, although they are incapable of producing melanin.

Dermal Circulation Blood with abundant oxygen is bright red, and blood vessels in the dermis normally give the skin a reddish tint that is most easily seen in lightly pigmented individuals. When those vessels are dilated, as during inflammation, the red tones become much more pronounced. When the vessels are temporarily constricted, as when one is frightened, the skin becomes relatively pale. During a sustained reduction in circulatory supply, the blood in the skin loses oxygen and changes color to a much deeper red tone. The skin then takes on a bluish coloration that is called **cyanosis** (sī-a-NŌ-sis; *kyanos*, blue). In individuals of any skin color, cyanosis is most apparent in areas of thin skin, such as the lips, ears, or beneath the nails. It can be a response to extreme cold or a result of circulatory or respiratory disorders, such as heart failure or severe asthma.

In general, pigment content can overshadow other factors; for example, circulatory changes have less visible effect on skin color in dark-skinned individuals. But, although their skin pigments may obscure localized inflammation or cyanosis, color changes can usually be seen through the nails, where the epidermis lacks dark pigments. AM *Abnormal Skin Pigmentation*

The Epidermis and Vitamin D_3

Although strong sunlight can damage epithelial cells and deeper tissues, limited exposure to sunlight is very beneficial. When exposed to ultraviolet radiation, epidermal cells in the stratum spinosum and stratum germinativum convert a steroid related to cholesterol into **vitamin D_3**. This product is absorbed, modified, and released by the liver and then converted by the kidneys into *calcitriol*, a hormone essential for the absorption of calcium and phosphorus by the small intestine. An inadequate supply of vitamin D_3 leads to impaired bone maintenance and growth.

✓ Excessive shedding of cells from the outer layer of skin in the scalp causes dandruff. What is the name of this layer of skin?

✓ As you pick up a piece of lumber, a splinter pierces the palm of your hand and lodges in the third layer of the epidermis. Identify this layer.

✓ Why does exposure to sunlight or tanning lamps cause the skin to become darker?

The Dermis

The **dermis** lies beneath the epidermis. It has two major components, a superficial *papillary layer* and a deeper *reticular layer*.

Layers of the Dermis

The **papillary layer**, named after the dermal papillae, consists of loose connective tissue that supports and nourishes the epidermis. This region contains the capillaries and nerves supplying the surface of the skin.

The deeper **reticular layer** consists of an interwoven meshwork of dense, irregular connective tissue. Bundles of collagen fibers leave the reticular layer to blend into those of the papillary layer above, so the boundary line between these layers is indistinct. Collagen fibers of the reticular layer also extend into the subcutaneous layer (*hypodermis*) below. This layer provides support and attachment for the dermis while also allowing flexibility and independent movement. AM *Dermatitis*

Other Dermal Components

In addition to protein fibers, the dermis contains a mixed cell population that includes all of the cells of connective tissue proper (Chapter 4). *[p. 81]* Accessory organs of epidermal origin, such as hair follicles and sweat glands, extend into the dermis (see Figure 5-1●, p. 98).

Other systems communicate with the skin via the dermis. For example, the reticular and papillary layers contain a network of blood vessels (cardiovascular system), lymphatics (lymphatic system), and nerve fibers (nervous system).

Blood vessels provide nutrients and oxygen and remove carbon dioxide and waste products. Both the blood vessels and the lymphatics help local tissues defend and repair themselves after an injury or infection. The nerve fibers control blood flow, adjust gland secretion rates, and monitor sensory receptors in the dermis and the deeper layers of the epidermis. These receptors, which provide sensations of touch, pain, pressure, and temperature, will be detailed in Chapter 10.

The Subcutaneous Layer

An extensive network of connective tissue fibers attaches the dermis to the subcutaneous layer. The boundary between these two layers is indistinct, and, although the subcutaneous layer is not actually a part of the integument, it is important in stabilizing the position of the skin in relation to underlying tissues and organs.

The subcutaneous layer (or **hypodermis**) consists of loose connective tissue with many fat cells. These adipose cells provide infants and small children with a layer of "body fat" over the entire body that serves to reduce heat loss, provide an energy reserve, and absorb shocks from inevitable tumbles.

As maturation proceeds, the distribution of subcutaneous fat changes, according to the sex of the individual. Men tend to accumulate such fat at the neck, upper arms, along the lower back, and over the buttocks, and women in the breasts, buttocks, hips, and thighs. Both sexes, however, may accumulate distressing amounts in the abdominal hypodermis, producing a prominent "pot belly."

The hypodermis is quite elastic. Below its superficial region with its large blood vessels, the hypodermis contains no vital organs and few capillaries. The lack of vital organs makes **subcutaneous injection** a useful method for administering drugs (thus the familiar term *hypodermic needle*).

Accessory Structures

Accessory structures include hair follicles, sebaceous glands, sweat glands, and nails.

Hair Follicles

Hairs project above the surface of the skin almost everywhere except over the sides and soles of the feet, the palms of the hands, the sides of the fingers and toes, the lips, and portions of the external genital organs. Hairs originate in complex organs called **hair follicles**.

Structure of Hair Follicles Hair follicles project deep into the dermis and often extend into the underlying subcutaneous layer (Figure 5-4●). The walls of each follicle contain all the cell layers found in the epidermis. Hair is formed at the deepest portion of the follicle, where basal cells divide. As keratinization occurs, the follicle produces the cylindrical shaft of a hair. The **shaft** of the hair that projects above the skin surface consists entirely of dead, keratinized cells.

Differences in the appearance of hair among individuals result from the size of the follicles, the activity of follicular cells, and the shapes of the hairs. For example, straight hairs are round in cross section, whereas curly ones are rather flattened.

Functions of Hair The 5 million hairs on the human body have important functions. The roughly 100,000 hairs on the head protect the scalp from ultraviolet light, cushion light blows to the head, and provide insulating benefits for the skull. The hairs guarding the entrances to the nostrils and external ear canals help prevent the entry of foreign particles and insects, and eyelashes perform a similar function for the surface of the eye. A sensory nerve fiber is associated with the base of each hair follicle. As a result, the movement of the shaft of even a single hair can be felt at the conscious level.

This sensitivity provides an early-warning system that may help to prevent injury. For example, you may be able to swat a mosquito before it reaches the skin surface.

Ribbons of smooth muscle, called **arrector pili** (a-REK-tōr PĪ-lī) muscles (Figure 5-4●), extend from the papillary dermis to a connective tissue sheath that surrounds each hair follicle. When stimulated, the arrector pili pull on the follicles and elevate the hairs. Contraction may be caused by emotional states, such as fear or rage. Contractions also occur as a response to cold, producing the characteristic "goose bumps" associated with shivering. In a furry mammal, this action increases the thickness of the insulating coat, rather like putting on an extra sweater. Although we do not receive any comparable insulating benefits, the reflex persists.

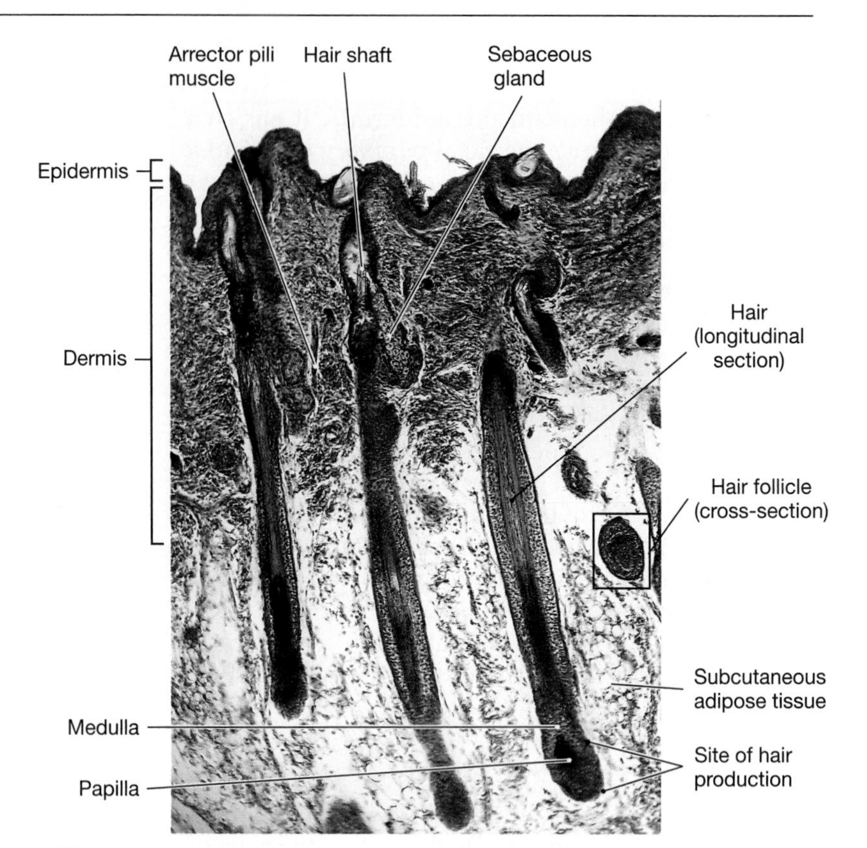

●**Figure 5-4**
Hair Follicles

A light micrograph showing the sectional appearance of the skin of the scalp. Note the many hair follicles and the way they extend into the dermis. (LM × 73)

Color of Hair Hair color reflects differences in the type and amount of pigment produced by melanocytes at the papilla. Although these characteristics are genetically determined, the condition of your hair may also be influenced by hormonal or environmental factors. As pigment production decreases with age, the hair color lightens toward gray. White hair results from the presence of air bubbles within the hair shaft. Because the hair itself is dead and inert, changes in coloration are gradual. Unless bleach is used, it is not possible for hair to "turn white overnight," as some horror stories would have us believe.

HAIR LOSS

Many people are subject to anxiety attacks when they find hairs clinging to their hairbrush instead of to their heads. Such hair loss is not a sign of approaching baldness, but merely a reflection that hair growth occurs in cycles. For example, a hair in the scalp grows for 2–5 years (at a rate of around 0.33 mm/day) and then may rest for a comparable period of time. When another growth cycle begins, the follicle produces a new hair, and the old hair gets pushed toward the surface to be shed.

On the average, about 50 hairs are lost each day, but several factors may affect this rate. Sustained losses of over 100 hairs per day usually indicate that something is wrong. Temporary increases in hair loss can result from drugs, dietary factors, radiation, high fever, stress, and hormonal factors related to pregnancy.

In males, changes in the level of circulating sex hormones can affect the scalp, causing a shift in production from normal hair to fine "peach fuzz" hairs. This alteration is called *male pattern baldness.* AM *Baldness and Hirsutism*

✓ What condition is produced by the contraction of the arrector pili muscles?

✓ A person suffers a burn on the forearm that destroys the epidermis and the deep dermis. When the injury heals, would you expect to find hair growing again in the area of the injury?

✓ Describe the functions of the subcutaneous layer.

Sebaceous Glands

The integument contains two types of exocrine glands, *sebaceous glands* and *sweat glands.* **Sebaceous** (sē-BĀ-shus) **glands** (Figure 5-5●) are holocrine glands that discharge a waxy, oily secretion into hair follicles. Several sebaceous glands may communicate with a single follicle by means of short ducts. The gland cells manufacture large quantities of lipids as they mature, and their eventual death releases the lipids into the open passageway of the gland. Contraction of the arrector pili muscle that elevates the hair squeezes the sebaceous gland, forcing the waxy secretions onto the surface of the skin. This secretion, called *sebum* (SĒ-bum), lubricates the hair shaft to prevent its drying and breaking, and its low pH inhibits the growth of some types of bacteria.

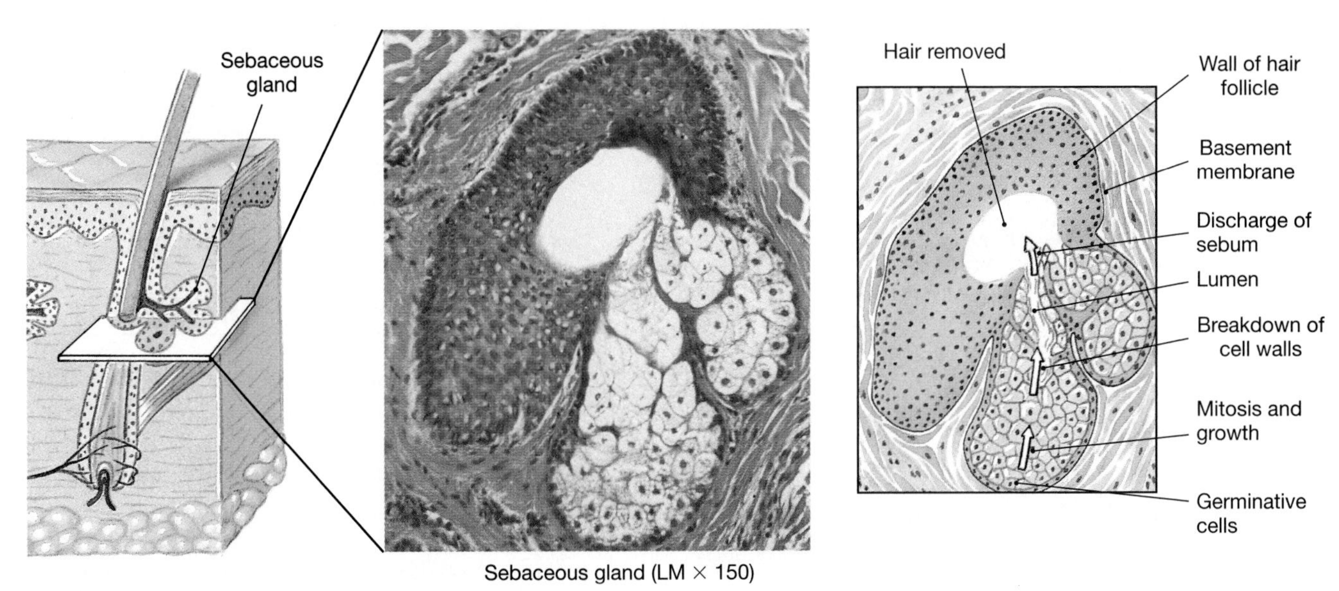

●**Figure 5-5**
Sebaceous Glands and Hair Follicles
The structure of a sebaceous gland and accompanying hair follicle. (LM × 150)

Sebaceous glands are very sensitive to changes in the concentrations of sex hormones, and their secretory activities accelerate at puberty. For this reason an individual with large sebaceous glands may be especially prone to develop **acne** during adolescence. In this condition, sebaceous ducts become blocked and secretions accumulate, causing inflammation and providing a fertile environment for bacterial infection. AM *Acne*

Sweat Glands

The integument contains two different populations of sweat glands, *apocrine sweat glands* and *merocrine sweat glands*. The differences between apocrine and merocrine secretion were discussed in Chapter 4 (see Table 4-1), and both gland types are shown in Figures 5-6●. *[p. 80]*

Apocrine Sweat Glands In the armpits, around the nipples, and in the groin, **apocrine sweat glands** communicate with hair follicles. At puberty, these coiled tubular glands begin discharging a sticky, cloudy secretion that becomes odorous when broken down by bacteria. In other mammals, this odor is an important form of communication; in our civilization, whatever function it has is masked by products such as deodorants. Other products, such as antiperspirants, contain astringent compounds that contract the skin and its sweat gland openings, thus decreasing the quantity of both apocrine and merocrine secretions.

Merocrine Sweat Glands **Merocrine sweat glands**, or *eccrine* (EK-rin) *sweat glands*, are far more numerous and widely distributed than apocrine glands. The adult integument contains around 3 million eccrine glands. Palms and soles have the highest numbers; it has been estimated that the palm of the hand has about 500 glands per square centimeter (3000 per square inch).

Merocrine sweat glands are coiled tubular glands that discharge their secretions directly onto the surface of the skin. The primary functions of their secretions, or perspiration, are to cool the surface of the skin and to reduce body temperature. When a person sweats in the hot sun, all the merocrine glands are working together. The blood vessels beneath the epidermis are flushed with blood, and the skin assumes a reddish coloration. The skin surface is warm and wet, and as the moisture evaporates the skin cools. If

●**Figure 5-6**
Sweat Glands

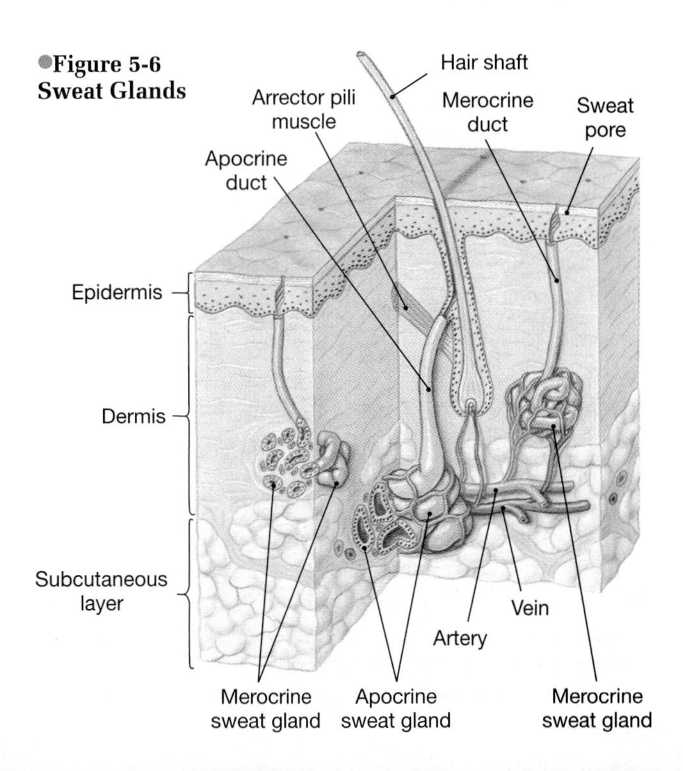

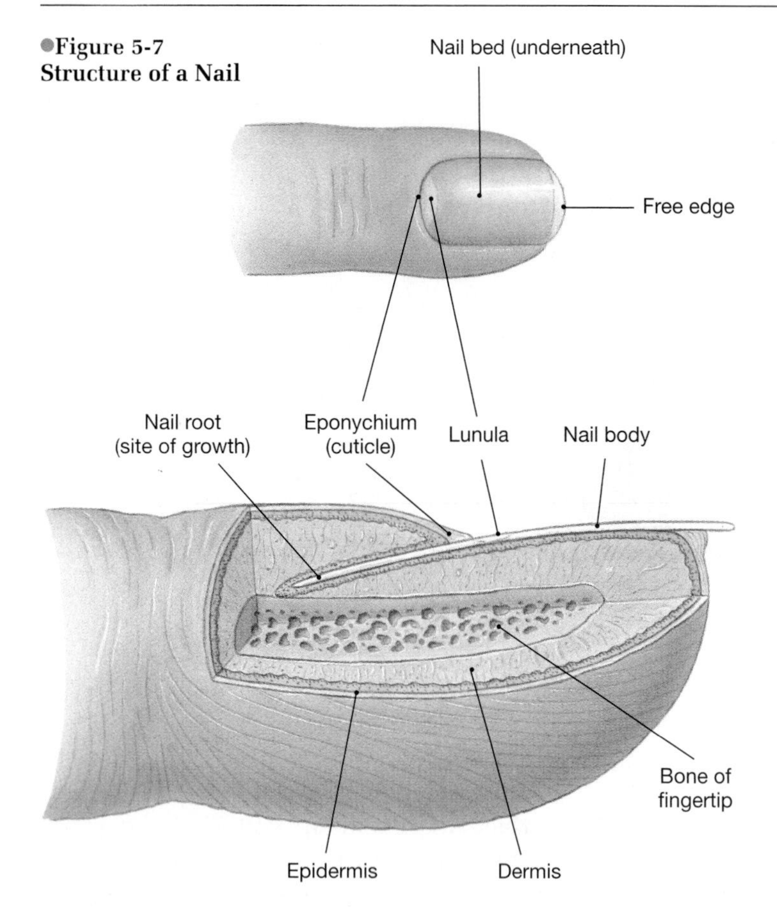

●**Figure 5-7**
Structure of a Nail

body temperature falls below normal, perspiration ceases, blood flow to the skin declines, and the cool, dry surfaces release little heat into the environment. The role of the skin in the regulation of body temperature was considered in Chapter 1 (see Figure 1-4●, p. 13); Chapter 18 will examine this process in greater detail.

The perspiration produced by merocrine glands is a clear secretion that is more than 99 percent water, but it does contain a mixture of electrolytes, metabolites, and waste products such as urea. The electrolytes give sweat its salty taste. When all of the merocrine sweat glands are working at maximum, the rate of perspiration may exceed a gallon (about 4 liters) per hour, and dangerous fluid and electrolyte losses can occur. For this reason marathon runners and other athletes in endurance sports must drink fluids at regular intervals.

Nails

Nails form over the tips of the fingers and toes, where they protect the exposed tips and help limit their distortion when they are subjected to mechanical stress—as when running or grasping objects. The visible **nail body** consists of a dense mass of dead, keratinized cells. The structure of a nail can be seen in Figure 5-7●. The body of the nail covers the **nail bed**, but nail production occurs at the **nail root**, an epithelial fold not visible from the surface. A portion of the stratum corneum of the fold extends over the exposed nail nearest the root, forming the **cuticle**, or *eponychium* (ep-ō-NIK-ē-um; *epi-*, over + *onyx*, nail). Underlying blood vessels give the nail its pink color, but near the root these vessels may be obscured, leaving a pale crescent known as the **lunula** (LOO-nu-la; *luna*, moon). The **nail body** is recessed beneath the level of the surrounding epithelium.

INJURY AND REPAIR

The skin can regenerate effectively even after considerable damage has occurred. Skin regeneration can occur because stem cells that persist in both its epithelial and connective tissue components divide to replace lost epidermal and dermal cells. This process can be slow, and when large surface areas are involved, problems of infection and fluid loss complicate the situation. The relative speed and effectiveness of skin repair vary depending on the type of wound involved. A slender, straight cut, or *incision*, may heal relatively quickly compared with a deep scrape, or *abrasion*, which involves a much greater area. For a discussion of how the skin repairs itself after injury, see the *Applications Manual.* AM *Injury and Repair* and *A Classification of Wounds*

Burns are relatively common injuries that result from exposure of the skin to heat, radiation, electrical shock, or strong chemical agents. The severity of the burn reflects the depth of penetration and the total area affected. The most common descriptive terms refer to the depth of penetration, and these are detailed in Table 5-1. The larger the area affected, the greater the impact on integumentary function. For a comprehensive discussion of burns and their clinical effects, see the *Applications Manual.* AM *Burns and Grafts*

AGING AND THE INTEGUMENTARY SYSTEM

Aging affects all the components of the integumentary system. The major changes including the following:

- *Skin injuries and infections become more common.* Such problems are more likely because the epidermis thins as germinative cell activity declines.
- *The sensitivity of the immune system is reduced.* The number of macrophages and other cells of the immune system decreases to around 50 percent of levels seen at maturity. This loss further encourages skin damage and infection.
- *Muscles become weaker, and bone strength decreases.* Such changes are caused by a decline in vitamin D_3 production of around 75 percent.

TABLE 5-1 A Classification of Burns

Classification	*Damage Report*	*Appearance and Sensation*
FIRST-DEGREE BURN	*Killed*: superficial cells of epidermis *Injured*: deeper layers of epidermis, papillary dermis	Inflamed, tender
SECOND-DEGREE BURN	*Killed*: superficial and deeper cells of epidermis; dermis may be affected *Injured*: damage may extend into reticular layer of the dermis, but many accessory structures unaffected	Blisters, pain
THIRD-DEGREE BURN	*Killed*: all epidermal and dermal cells *Injured*: hypodermal and deeper tissues and organs	Charred, no sensation at all

- *Sensitivity to sun exposure increases.* Lesser amounts of melanin are produced because melanocyte activity declines. The skin of Caucasians becomes very pale.
- *The skin becomes dry and often scaly.* Glandular activity declines, reducing sebum production and perspiration (see below).
- *Hair thins and changes color.* Follicles stop functioning or produce thinner, finer hairs. With decreased melanocyte activity, these hairs are gray or white.
- *The integument weakens, and sagging and wrinkling occur.* The dermis becomes thinner, and the elastic fiber network decreases in size. The integument therefore becomes weaker and less resilient. These effects are most pronounced in areas exposed to the sun.
- *Ability to lose heat lessens.* The blood supply to the dermis is reduced at the same time that sweat glands become less active. This combination makes the elderly less able to lose body heat, and overexertion or overexposure to warm temperatures can cause dangerously high body temperatures.
- *Skin repairs proceed relatively slowly.* For example, it takes 3–4 weeks to complete the repairs to a blister site in a person age 18–25. The same repairs at age 65–75 take 6–8 weeks. Because repairs are slow, recurrent infections may occur.
- *Secondary sexual characteristics in hair and body fat distribution begin to fade.* Because of lowered levels of sex hormones, people age 90–100 of both sexes and all races look very much alike.

✓ What will happen if the duct of an infected sebaceous gland becomes blocked?

✓ Deodorants are used to mask the effects of secretions from what type of skin gland?

✓ Older people do not tolerate the summer heat as well as they did when they were young, and they are more prone to heat-related illness. What accounts for this change?

INTEGRATION WITH OTHER SYSTEMS

Although the integumentary system can function independently, many of its activities are integrated with those of other systems. Figure 5-8● diagrams the major functional relationships. The role of the skin in temperature control, through interactions with the nervous and cardiovascular systems, was detailed in Chapter 1 (see Figure 1-4●, p. 13).

Chapter Review

KEY TERMS

cutaneous membrane, p. 97
dermis, p. 97
epidermis, p. 97
hair, p. 101
hair follicle, p. 101
integument, p. 97
keratin, p. 98
melanin, p. 99
nail, p. 104
sebaceous glands, p. 102
stratum, p. 97
subcutaneous layer, p. 101

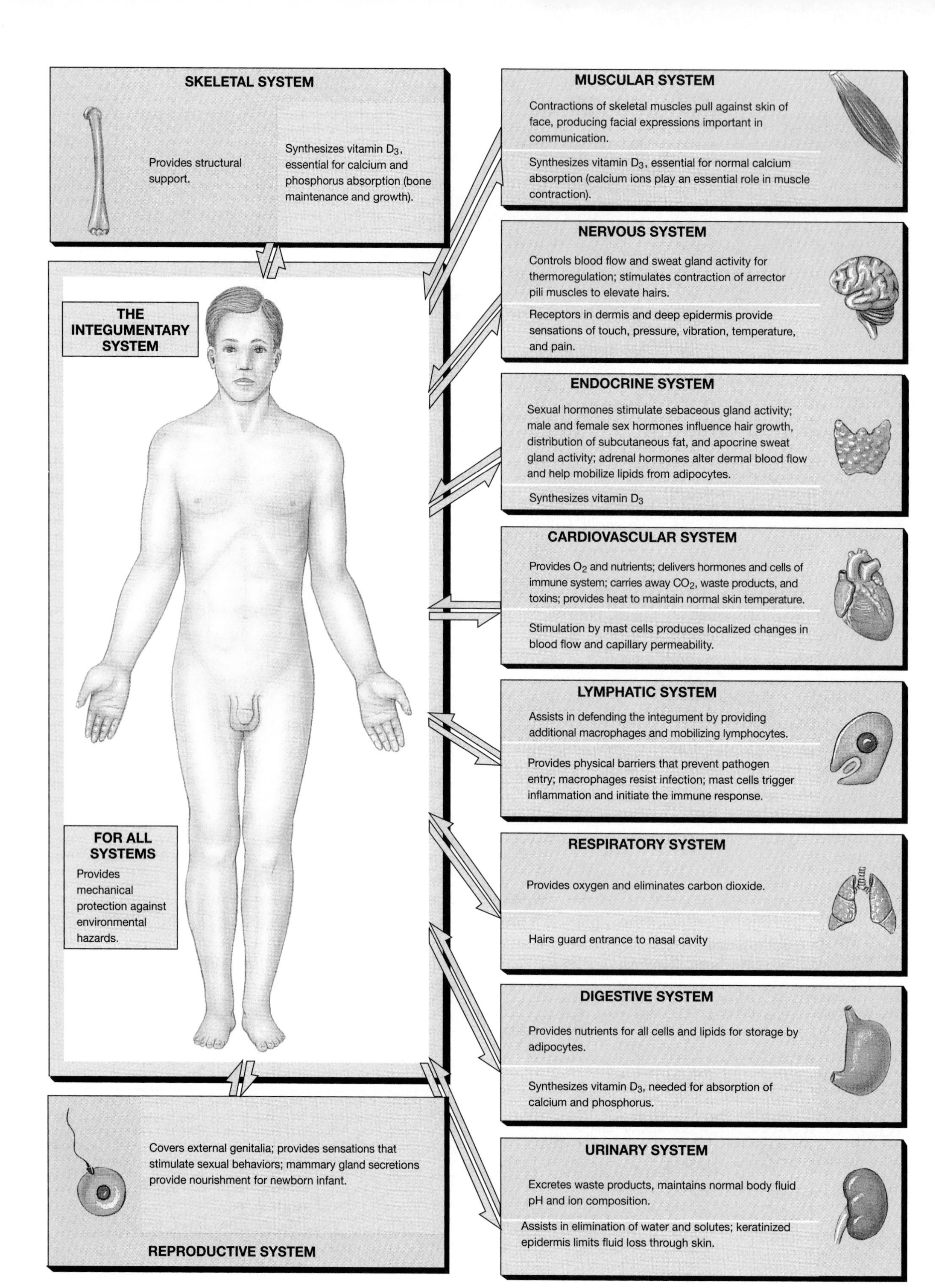

SKELETAL SYSTEM
Provides structural support.
Synthesizes vitamin D_3, essential for calcium and phosphorus absorption (bone maintenance and growth).
MUSCULAR SYSTEM
Contractions of skeletal muscles pull against skin of face, producing facial expressions important in communication.
Synthesizes vitamin D_3, essential for normal calcium absorption (calcium ions play an essential role in muscle contraction).
NERVOUS SYSTEM
Controls blood flow and sweat gland activity for thermoregulation; stimulates contraction of arrector pili muscles to elevate hairs.
Receptors in dermis and deep epidermis provide sensations of touch, pressure, vibration, temperature, and pain.
THE INTEGUMENTARY SYSTEM
ENDOCRINE SYSTEM
Sexual hormones stimulate sebaceous gland activity; male and female sex hormones influence hair growth, distribution of subcutaneous fat, and apocrine sweat gland activity; adrenal hormones alter dermal blood flow and help mobilize lipids from adipocytes.
Synthesizes vitamin D_3
CARDIOVASCULAR SYSTEM
Provides O_2 and nutrients; delivers hormones and cells of immune system; carries away CO_2, waste products, and toxins; provides heat to maintain normal skin temperature.
Stimulation by mast cells produces localized changes in blood flow and capillary permeability.
LYMPHATIC SYSTEM
Assists in defending the integument by providing additional macrophages and mobilizing lymphocytes.
Provides physical barriers that prevent pathogen entry; macrophages resist infection; mast cells trigger inflammation and initiate the immune response.
FOR ALL SYSTEMS
Provides mechanical protection against environmental hazards.
RESPIRATORY SYSTEM
Provides oxygen and eliminates carbon dioxide.
Hairs guard entrance to nasal cavity
DIGESTIVE SYSTEM
Provides nutrients for all cells and lipids for storage by adipocytes.
Synthesizes vitamin D_3, needed for absorption of calcium and phosphorus.
Covers external genitalia; provides sensations that stimulate sexual behaviors; mammary gland secretions provide nourishment for newborn infant.
REPRODUCTIVE SYSTEM
URINARY SYSTEM
Excretes waste products, maintains normal body fluid pH and ion composition.
Assists in elimination of water and solutes; keratinized epidermis limits fluid loss through skin.

Figure 5-8
Functional Relationships between the Integumentary System and Other Systems

SUMMARY OUTLINE

INTRODUCTION p. 97

INTEGUMENTARY STRUCTURE AND FUNCTION p. 97

1. The **integumentary system** consists of the **cutaneous membrane**, which includes the **epidermis** and **dermis**, and the **accessory structures**. Underneath lies the **subcutaneous layer**. (*Figure 5-1*)

The Epidermis p. 97

2. **Thin skin** covers most of the body; heavily abraded body surfaces may be covered by **thick skin**.

3. Cell divisions in the **stratum germinativum** replace more superficial cells.

4. As epidermal cells age they pass through the **stratum spinosum**, the **stratum granulosum**, the **stratum lucidum** (in thick skin), and the **stratum corneum**. In the process they accumulate large amounts of **keratin**. Ultimately the cells are shed or lost. (*Figure 5-2*)

5. **Epidermal ridges**, such as those on the palms and soles, improve our gripping ability and increase the skin's sensitivity.

6. The color of the epidermis depends on two factors: blood supply and pigment composition and concentration. **Melanocytes** protect us from **ultraviolet radiation**. (*Figure 5-3*)

7. Epidermal cells synthesize vitamin D_3 when exposed to sunlight.

The Dermis p. 101

8. The **dermis** consists of the **papillary layer** and the deeper **reticular layer**.

9. The papillary layer of the dermis contains blood vessels, lymphatics, and sensory nerves. This layer supports and nourishes the overlying epidermis. The reticular layer consists of a meshwork of collagen and elastic fibers oriented to resist tension in the skin.

10. Components of other systems (cardiovascular, lymphatic, and nervous) that communicate with the skin are present in the dermis.

The Subcutaneous Layer p. 101

11. The subcutaneous layer, or **hypodermis**, stabilizes the skin's position against underlying organs and tissues.

Accessory Structures p. 101

12. Hairs originate in complex organs called **hair follicles**. Each hair has a **shaft** composed of dead keratinized cells. (*Figure 5-4*)

13. The **arrector pili** muscles can elevate the hairs.

14. Our hairs grow and are shed according to a cyclic pattern. A single hair grows for 2–5 years and is subsequently shed.

15. **Sebaceous glands** discharge the waxy **sebum** into hair follicles. (*Figure 5-5*)

16. **Apocrine sweat glands** produce an odorous secretion; the more numerous **merocrine sweat glands** produce a watery secretion. (*Figure 5-6*)

17. Nails are sheets of dense, keratinized cells. They provide support for the tips of the fingers and toes. Nail production occurs at the **nail root**. (*Figure 5-7*)

INJURY AND REPAIR p. 104

1. The skin can regenerate effectively even after considerable damage.

2. Burns are relatively common injuries characterized by damage to layers of the epidermis and perhaps the dermis. (*Table 5-1*)

AGING AND THE INTEGUMENTARY SYSTEM p. 104

1. Aging affects all the components of the integumentary system. (*Figure 5-8*)

INTEGRATION WITH OTHER SYSTEMS p. 105

1. Many of the integumentary system's functions are integrated with those of other systems.

CHAPTER QUESTIONS

LEVEL 1 Reviewing Facts and Terms

Match each item in column A with the most closely related item in column B. Use letters for answers in the spaces provided.

Column A	Column B
___ 1. cutaneous membrane	a. arrector pili
___ 2. carotene	b. cyanosis
___ 3. melanocytes	c. perspiration
___ 4. epidermal layer containing stem cells	d. sebum
___ 5. smooth muscle	e. stratum corneum
___ 6. epidermal layer of flattened and dead cells	f. skin
___ 7. bluish coloration of the skin	g. orange-yellow pigment
___ 8. sebaceous glands	h. bone growth
___ 9. merocrine (eccrine) glands	i. pigment cells
___ 10. vitamin D_3	j. stratum germinativum

11. The two major components of the integumentary system are:
    (a) the cutaneous membrane and the accessory structures
    (b) the epidermis and the hypodermis
    (c) the hair and the nails
    (d) the dermis and the subcutaneous layer
12. The fibrous protein that forms the basic structural component of hair and nails is:
    (a) collagen (b) melanin
    (c) elastin (d) keratin
13. The two types of exocrine glands in the skin are:
    (a) merocrine and sweat glands
    (b) sebaceous and sweat glands
    (c) apocrine and sweat glands
    (d) eccrine and sweat glands
14. The following are all accessory structures *except*:
    (a) nails (b) hair
    (c) dermal papillae (d) sweat glands
15. Sweat glands that communicate with hair follicles in the armpits and produce an odorous secretion are:
    (a) apocrine glands
    (b) merocrine glands
    (c) sebaceous glands
    (d) a, b, and c are correct
16. The reason older persons are more sensitive to sun exposure and more likely to experience sunburn is that with age:
    (a) melanocyte activity declines
    (b) vitamin D_3 production declines
    (c) glandular activity declines
    (d) skin thickness decreases
17. What two skin pigments are found in the epidermis?
18. What two major layers constitute the dermis, and what components are found in each layer?
19. What two different groups of sweat glands are contained in the integument?

LEVEL 2 Reviewing Concepts

20. During transdermal administration of drugs, why are fat-soluble drugs more desirable than those that are water-soluble?
21. In our society, a tan body is associated with good health. However, medical research constantly warns about the dangers of excessive exposure to the sun. What are the benefits of a tan?
22. Why is a subcutaneous injection with a hypodermic needle a useful method for administering drugs?
23. Why does skin sag and wrinkle as a person ages?

LEVEL 3 Critical Thinking and Clinical Applications

24. A new mother notices that her 6-month-old child has a yellow-orange complexion. Fearful that the child may have jaundice (a condition caused by a toxic yellow-orange pigment in the blood), she takes him to her pediatrician. After examining the child, the pediatrician declares him perfectly healthy and advises the mother to watch the child's diet. Why?
25. Vanessa remarks that her 80-year-old grandmother keeps her thermostat set at 80°F and wears a sweater on balmy spring days. When she asks her grandmother why, her grandmother tells her that she is cold. Vanessa can't understand this and asks you for an explanation. What would you tell Vanessa?

6 The Skeletal System

The Tin Man wore his skeleton on the outside. We have internal skeletons made of bone, a remarkable tissue that is strong, immune to rust (unlike the Tin Man is), and capable of repairing itself even after serious injury.

Chapter Outline and Objectives

Vocabulary Development

ab-, from; *abduction*
acetabulum, a vinegar cup; *acetabulum* of the hip joint
ad-, toward, to; *adduction*
amphi-, on both sides; *amphiarthrosis*
arthros, joint; *synarthrosis*
blast, precursor; *osteoblast*
circum-, around; *circumduction*
clast, break; *osteoclast*
clavius, clavicle; *clavicle*
concha, shell; middle *concha*
corona, crown; *coronoid fossa*
cranio-, skull; *cranium*
cribrum, sieve; *cribriform plate*
dens, tooth; *dens*
dia-, through; *diarthrosis*
duco, to lead; *adduction*
e-, out; *eversion*
gennan, to produce; *osteogenesis*
gomphosis, a bolting together; *gomphosis*
in-, into; *inversion*
infra-, beneath; *infraspinous fossa*
lacrimae, tears; *lacrimal* bones
lamella, thin plate; *lamellae* of bone
malleolus, little hammer; medial *malleolus*
meniscus, crescent; *menisci*
osteon, bone; *osteocytes*
penia, lacking; *osteopenia*
planta, sole; *plantar*
porosus, porous; *osteoporosis*
septum, wall; *nasal septum*
stylos, pillar; *styloid process*
supra-, above; *supraspinous fossa*
sutura, a sewing together; *suture*
teres, cylindrical; ligamentum *teres*
trabecula, wall; *trabeculae* in spongy bone
trochlea, pulley; *trochlea*
vertere, to turn; *inversion*

The skeleton has many functions, but the most obvious is supporting the weight of the body. This support is provided by bones, structures as strong or stronger than reinforced concrete but considerably lighter. Unlike concrete, bones can be remodeled and reshaped to meet changing metabolic demands and patterns of activity. Bones work together with muscles to maintain body position and to produce controlled, precise movements. With the skeleton to pull against, contracting muscles can make us sit, stand, walk, or run.

The skeletal system includes the bones of the skeleton and the cartilages, ligaments, and other connective tissues that stabilize or connect them. This system performs the following functions:

1. **Support.** The skeletal system provides structural support for the entire body. Individual bones or groups of bones provide a framework for the attachment of soft tissues and organs.
2. **Leverage.** The bones of the skeleton function as *levers* that can change the magnitude and direction of the forces generated by skeletal muscles. The movements produced range from the delicate motion of a fingertip to powerful changes in the position of the entire body.
3. **Protection.** Delicate tissues and organs are often surrounded by skeletal elements. The ribs protect the heart and lungs, the skull encloses the brain, the vertebrae shield the spinal cord, and the pelvis cradles delicate digestive and reproductive organs.
4. **Storage.** The calcium salts of bone represent a valuable mineral reserve that maintains normal concentrations of calcium and phosphate ions in body fluids. In addition, fat cells in areas of *yellow marrow* store lipids as an energy reserve.
5. **Blood cell production.** Red blood cells and other blood elements are produced within the *red marrow* that fills the internal cavities of many bones. The role of the bone marrow in blood cell formation will be discussed in later chapters dealing with the cardiovascular and lymphatic systems (Chapters 12–15).

Structure of Bone

Bone tissue, or **osseous tissue,** is one of the supporting connective tissues, and it contains specialized cells, extracellular fibers, and a ground substance. In supporting connective tissues, the fibers and ground substance interact to form a *matrix.* The distinctive solid, stony character of bone results from the deposition of calcium salts within the matrix. Calcium phosphate, $Ca_3(PO_4)_2$, accounts for almost two-thirds of the weight of bone. The remaining third is dominated by collagen fibers, with osteocytes and other cell types providing only around 2 percent of the mass of a bone.

Macroscopic Features of Bone

The bones of the human skeleton have four general shapes: *long, short, flat,* and *irregular* (Figure 6-1●). **Long bones** are longer than they are wide, whereas **short bones** are of roughly equal dimensions. Examples of long bones include bones of the limbs such as the arm (*humerus*) and thigh (*femur*). **Short bones** include the bones of the wrist (*carpals*) and ankles (*tarsals*). **Flat bones** are thin and relatively broad, such as the *parietal bones* of the skull, the ribs, and the shoulder blades (*scapulae*). **Irregular bones** have complex shapes that do not fit easily into any other category. An example would be one of the *vertebrae* of the spinal column.

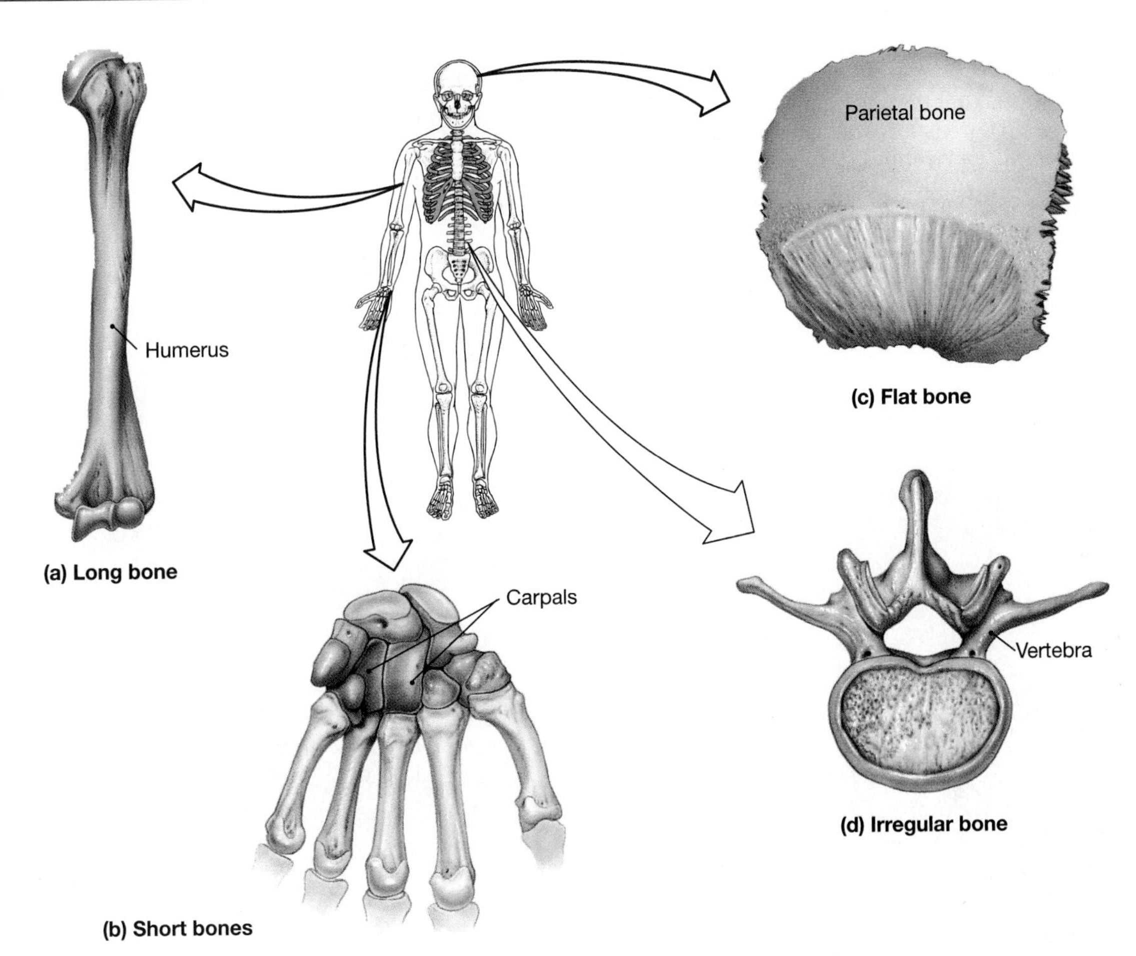

●**Figure 6-1**
Shapes of Bones

The typical features of a long bone such as the humerus are shown intact in Figure 6-1a● and in longitudinal section in Figure 6-2●. A long bone has a central shaft, or **diaphysis** (dī-A-fi-sis), and expanded ends, or **epiphyses** (ē-PIF-i-sēz). The diaphysis surrounds a central *marrow cavity.* The epiphyses of adjacent bones articulate with each other and are covered by *articular cartilages.* As will be discussed below, growth in the length of an immature long bone occurs at the junctions between the epiphyses and the diaphysis.

The two types of bone tissue are visible in Figure 6-2●: *compact (dense) bone* and *spongy (cancellous) bone.* **Compact bone,** or *dense bone,* is relatively solid, whereas **spongy bone,** or *cancellous* (KAN-sel-us) *bone,* resembles a network of bony rods or struts separated by spaces that are normally filled with bone marrow. Both compact and spongy bone are present in the humerus; compact bone forms the diaphysis, and spongy bone fills the epiphyses.

The outer surface of a bone is covered by a **periosteum** that consists of a fibrous outer layer and a cellular inner layer (see Figure 6-3●, p. 112). The periosteum isolates the bone from surrounding tissues, provides a route for circulatory and nervous supply, and actively participates in bone growth and repair. The fibers of *tendons* and *ligaments* intermingle with those of the periosteum, attaching skeletal muscles to bones and one bone to another.

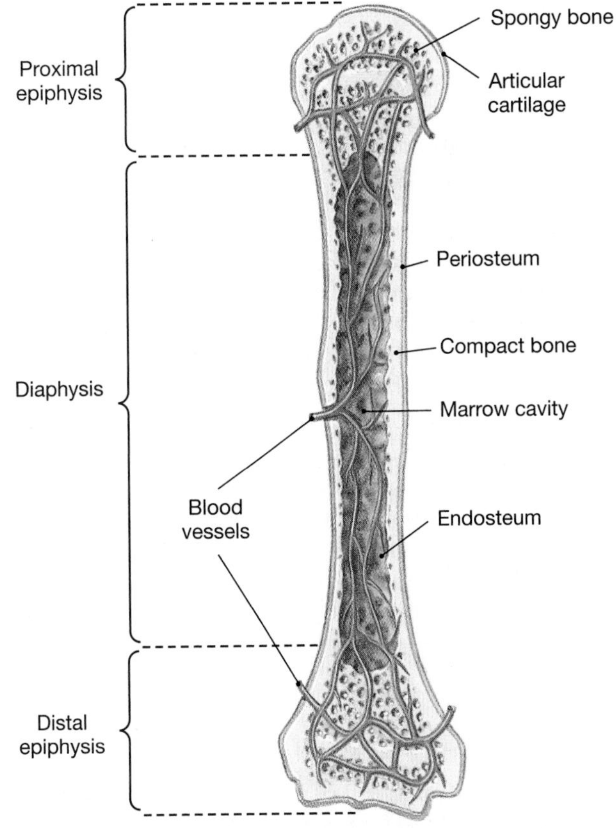

●**Figure 6-2**
Structure of a Long Bone

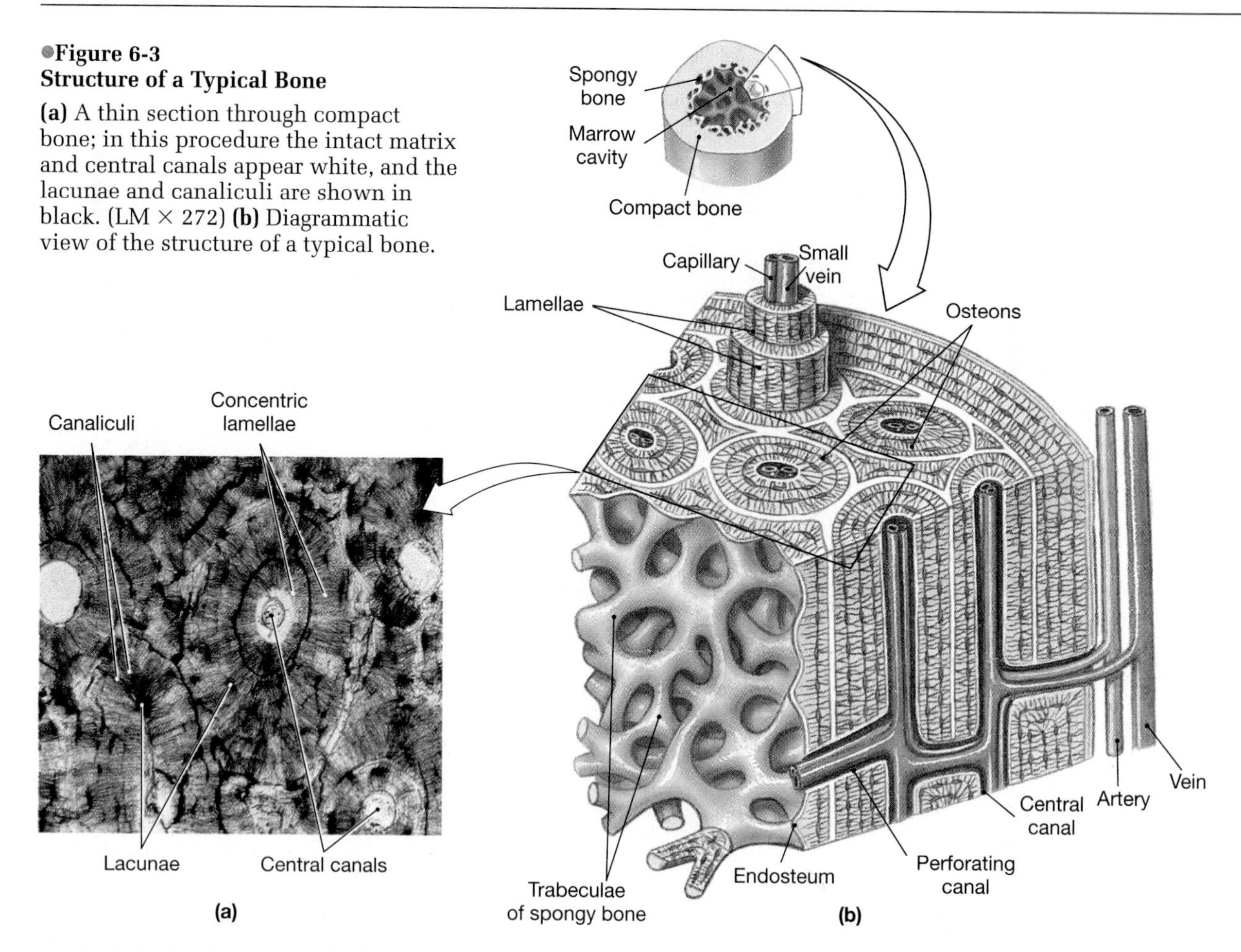

●Figure 6-3
Structure of a Typical Bone
(a) A thin section through compact bone; in this procedure the intact matrix and central canals appear white, and the lacunae and canaliculi are shown in black. (LM × 272) **(b)** Diagrammatic view of the structure of a typical bone.

Inside the bone, a cellular **endosteum** lines the marrow cavity and other inner surfaces. The endosteum is active during bone growth and whenever repair or remodeling is under way.

Microscopic Features of Bone

The general histology of bone was introduced in Chapter 4 (see Figure 4-11●, p. 87). Both compact and spongy bone contain bone cells, or **osteocytes** (OS-tē-ō-sīts; *osteon*, bone), in small pockets called **lacunae** (la-KOO-nē). Lacunae are found between narrow sheets of calcified matrix that are known as **lamellae** (lah-MEL-lē; *lamella*, thin plate). Small channels, called **canaliculi** (ka-na-LIK-ū-lē), radiate through the matrix, interconnecting lacunae and linking them to nearby blood vessels. The canaliculi contain cytoplasmic extensions of the osteocytes. Nutrients from the blood and waste products from the osteocytes diffuse through the fluid that surrounds these cells and their extensions.

Compact and Spongy Bone

The basic functional unit of compact bone, the **osteon** (OS-tē-on), or *Haversian system*, is shown in Figure 6-3●. Within an osteon, the osteocytes are arranged in concentric layers around a **central canal**, or *Haversian canal*, that contains one or more blood vessels. The lamellae are cylindrical, oriented parallel to the long axis of the central canal. **Perforating canals**, or *canals of Volkmann*, provide passageways for linking the blood vessels of the central canals with those of the periosteum or the marrow cavity.

Spongy bone has a quite different lamellar arrangement and no osteons. Instead, the lamellae form rods or plates called **trabeculae** (tra-BEK- ū-lē; *trabecula*, a wall). Frequent branchings of the thin trabeculae create an open network. Canaliculi radiating from the lacunae of spongy bone end at the exposed surfaces of the trabeculae where nutrients and wastes diffuse between the marrow and osteocytes.

A layer of compact bone covers bone surfaces everywhere except inside *joint capsules*, where articular cartilages protect opposing surfaces. Compact bone is usually found where stresses come from a limited range of directions. The limb bones, for example, are built to withstand forces applied at either end. Because the osteons are parallel to the long axis of the shaft, a limb bone does not bend, even when a large force is applied to either end. However, a much smaller force applied to the side of the shaft can break the bone.

Bones do bend in disorders that reduce the amount of calcium salts in the skeleton. As the proportion of calcium to collagen decreases, the bones become very flexible, and they become less able to resist compression and tension. An example of such a condition is *rickets*, a childhood disorder detailed in a later section, which results from a deficiency of vitamin D_3. Affected individuals develop a bowlegged appearance as the leg bones bend under the weight of the body.

In contrast, spongy bone is found where bones are not heavily stressed or where stresses arrive from many directions. For example, spongy bone is present at the epiphyses of long bones, where stresses are transferred across joints. In addition, spongy bone is much lighter than compact bone. This reduces the weight of the skeleton and makes it easier for muscles to move the bones. Finally, the trabecular framework protects the cells of the bone marrow, and areas of spongy bone, such as the epiphyses of the femur, are important sites of blood cell formation.

Cells in Bone

Although osteocytes are the most abundant cells in bone, other cell types are also present. These cells, called *osteoclasts* and *osteoblasts*, are associated with the endosteum that lines the inner cavities of both compact and spongy bone and the cellular layer of the periosteum. Thus, there are three primary cell types in bone,

1. **Osteocytes** are mature bone cells. Osteocytes maintain normal bone structure by recycling the calcium salts in the bony matrix around themselves and assisting in repairs.
2. **Osteoclasts** (OS-tē-ō-klasts; *clast*, break) are giant cells with 50 or more nuclei. Acids secreted by osteoclasts dissolve the bony matrix through *osteolysis* and release the stored minerals. This process helps regulate calcium and phosphate concentrations in body fluids.
3. **Osteoblasts** (OS-tē-ō-blasts; *blast*, precursor) are cells that are responsible for the production of new bone, a process called *osteogenesis* (os-tē-ō-JEN-e-sis; *gennan*, to produce). At any given moment, osteoclasts are removing matrix and osteoblasts are adding to it. When an osteoblast becomes completely surrounded by calcified matrix, it differentiates into an osteocyte.

✓ How would the strength of a bone be affected if the ratio of collagen to calcium increased?

✓ A sample of bone shows concentric layers surrounding a central canal. Is it from the shaft or the end of a long bone?

✓ If the activity of osteoclasts exceeds the activity of osteoblasts in a bone, how will the mass of the bone be affected?

BONE DEVELOPMENT AND GROWTH

The growth of the skeleton determines the size and proportions of the body. Skeletal growth begins about 6 weeks after fertilization, when the embryo is approximately 12 mm long. (Before this time all supporting elements are cartilaginous.) Bone growth continues through adolescence, and portions of the skeleton usually do not stop growing until ages 18 to 25. This section considers the physical process of *osteogenesis* (bone formation) and growth. The next section examines the maintenance and turnover of mineral reserves in the adult skeleton.

During development, cartilage or connective tissue stem cells are replaced by bone. This process of replacing other tissues with bone is called **ossification**. There are two major forms of ossification. In *intramembranous ossification*, bone develops within sheets or membranes of connective tissue. In *endochondral ossification* bone replaces an existing cartilage model.

Intramembranous Ossification

Intramembranous (in-tra-MEM-bra-nus) **ossification** begins when stem cells within connective tissue, normally the deeper layers of the dermis, differentiate into osteoblasts. The osteoblasts cluster together and secrete the organic components of the matrix. This mixture then becomes mineralized through the crystallization of calcium salts. The place where ossification first occurs is called an **ossification center**. As ossification proceeds, some osteoblasts become trapped inside bony pockets and change into osteocytes.

Bone growth is an active process, and osteoblasts require oxygen and a reliable supply of nutrients. Blood vessels that branch between the advancing bone growth meet these demands. Although initially the intramembranous bone resembles spongy bone, further remodeling around the trapped blood vessels can produce compact bone. Several flat bones of the skull, the lower jaw, and the collarbones (*clavicles*) are formed in this way.

Endochondral Ossification

Most of the bones of the skeleton are formed through the **endochondral ossification** (en-dō-KON-dral; *endo*, inside + *chondros*, cartilage) of cartilaginous models. By the time an embryo is 6 weeks old, the rapidly growing cartilaginous models of the limb bones begin to be replaced by bone. Steps in their growth and ossification are diagrammed in Figure 6-4●. Bone formation begins near the middle of the shaft, at a *primary center of ossification*, and proceeds toward either end. Eventually, *sec-*

ondary ossification centers develop in the epiphyses. At this stage the bone of the shaft and the bone of each epiphysis are separated by areas of cartilage known as **epiphyseal plates**. On the shaft side of the epiphyseal plate, osteoblasts are continually invading the cartilage and converting it to bone. But on the epiphyseal side of the plate, new cartilage is being produced at the same rate. As a result, the shaft grows longer, but the epiphyseal plate remains. This process could be compared to a jogger following someone on a bicycle. The jogger keeps advancing, but the bike stays a few feet ahead.

When sex hormone production increases at puberty, bone growth accelerates dramatically, and osteoblasts begin to produce bone faster than epiphyseal cartilage expands. In effect, the jogger speeds up, moving closer to the bicycle ahead. Over time, the epiphyseal plates narrow and finally ossify, or "close." This period of sudden growth ends as the individual reaches sexual and physical maturity. The location of the plate can still be detected in X-rays as a distinct *epiphyseal line* that remains after epiphyseal growth has ended.

While the bone elongates, it also grows larger in diameter. The diameter enlarges at its outer surface through a process called *appositional growth*. This enlargement occurs as periosteal cells develop into osteoblasts and produce additional bony matrix. As new bone is deposited on the outer surface of the shaft, the inner surface is eroded by osteoclasts, and the marrow cavity gradually enlarges.

Bone Growth and Body Proportions

The timing of epiphyseal closure varies from bone to bone and individual to individual. The toes may complete their ossification by age 11, whereas portions of the pelvis or the wrist may continue to enlarge until age 25. The epiphyseal plates in the arms and legs usually close by age 18 (women) or 20 (men). Differences in sex hormones account for variations in body size and proportions between men and women. AM *Hyperostosis and Acromegaly*

Requirements for Normal Bone Growth

Normal osteogenesis cannot occur without a reliable source of minerals, especially calcium salts. During prenatal development these minerals are absorbed

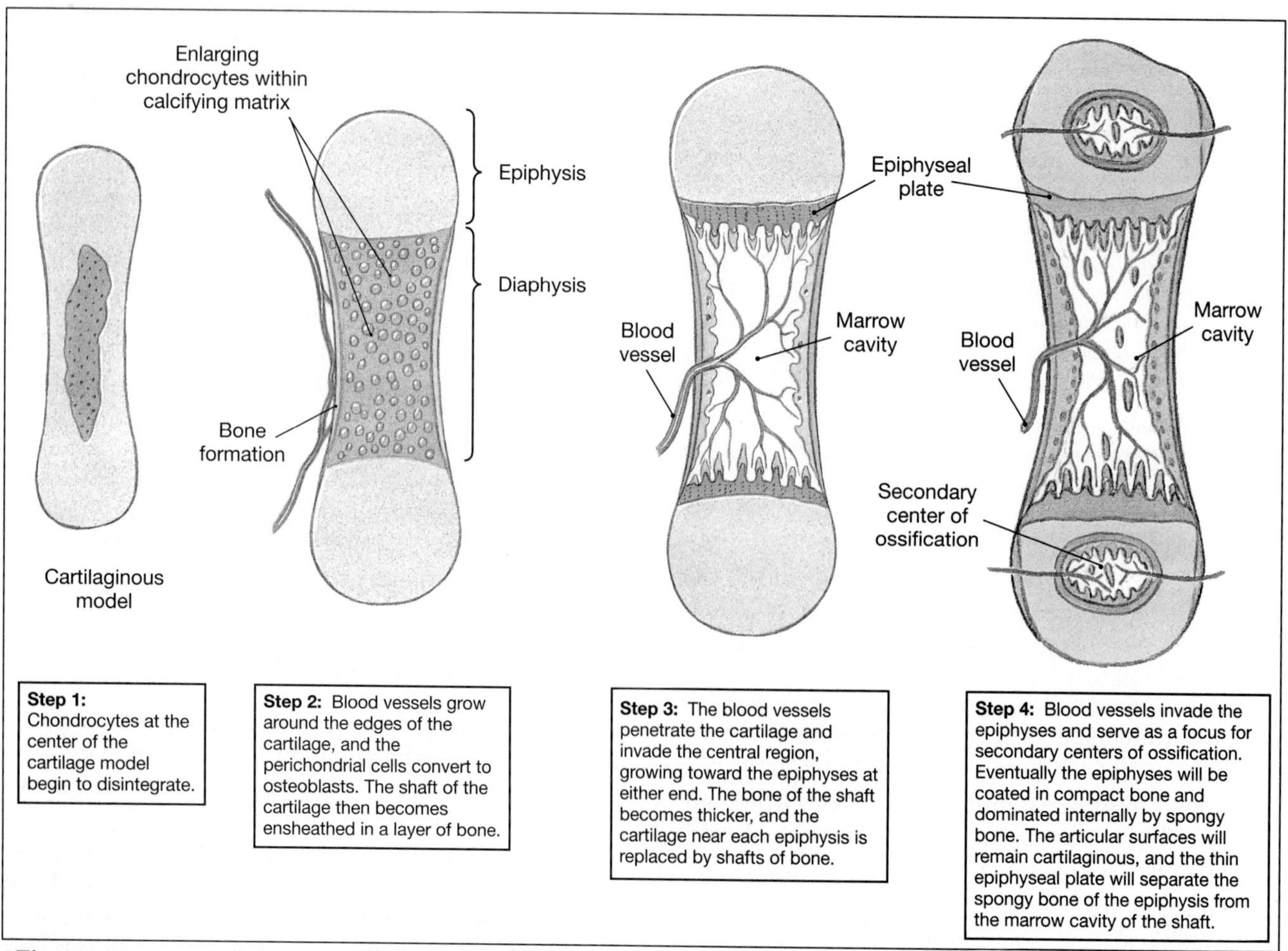

Enlarging chondrocytes within calcifying matrix
Epiphysis
Diaphysis
Bone formation
Cartilaginous model
Epiphyseal plate
Blood vessel
Marrow cavity
Blood vessel
Marrow cavity
Secondary center of ossification
Step 1: Chondrocytes at the center of the cartilage model begin to disintegrate.
Step 2: Blood vessels grow around the edges of the cartilage, and the perichondrial cells convert to osteoblasts. The shaft of the cartilage then becomes ensheathed in a layer of bone.
Step 3: The blood vessels penetrate the cartilage and invade the central region, growing toward the epiphyses at either end. The bone of the shaft becomes thicker, and the cartilage near each epiphysis is replaced by shafts of bone.
Step 4: Blood vessels invade the epiphyses and serve as a focus for secondary centers of ossification. Eventually the epiphyses will be coated in compact bone and dominated internally by spongy bone. The articular surfaces will remain cartilaginous, and the thin epiphyseal plate will separate the spongy bone of the epiphysis from the marrow cavity of the shaft.

Figure 6-4
Endochondral Ossification

from the mother's bloodstream. The demands are so great that the maternal skeleton often loses bone mass during pregnancy. From infancy to adulthood, the diet must provide adequate amounts of calcium and phosphate, and the individual must be able to absorb and transport these minerals to sites of bone formation.

Vitamin D_3 plays an important role in normal calcium metabolism by stimulating the absorption and transport of calcium and phosphate ions. This vitamin can be obtained from dietary supplements or manufactured by epidermal cells exposed to ultraviolet radiation. ∞ *[p. 100]* After vitamin D_3 has been processed in the liver, the kidneys convert a derivative of this vitamin into **calcitriol**, a hormone that stimulates the absorption of calcium and phosphate ions across the intestinal lining. *Rickets,* mentioned earlier in this chapter, is a condition marked by a softening and bending of bones that occurs in growing children, usually as a result of vitamin D_3 deficiency.

Vitamins A and **C** are also essential for normal bone growth and remodeling. For example, a deficiency of vitamin C will result in *scurvy.* One of the primary symptoms of this condition is a reduction in osteoblast activity that leads to weak and brittle bones. In addition to vitamins, hormones such as growth hormone, thyroid hormones, sex hormones, and those involved with calcium metabolism are essential to normal skeletal growth and development.

✓ How could X-rays of the femur be used to determine whether a person had reached full height?

✓ In the Middle Ages, choirboys were sometimes castrated (had their testes removed) to prevent their voices from changing. How would castration have affected their height?

✓ Why are pregnant women given calcium supplements and encouraged to drink milk even though their skeletons are fully formed?

REMODELING AND HOMEOSTATIC MECHANISMS

Of the five major functions of the skeleton discussed earlier in this chapter, support and storage depend on the dynamic nature of bone. In the adult, osteocytes in lacunae maintain the surrounding matrix, continually removing and replacing the surrounding calcium salts. But osteoclasts and osteoblasts also remain active, even after the epiphyseal plates have closed. Normally their activities are balanced: As one osteon forms through the activity of osteoblasts another is destroyed by osteoclasts. The turnover rate for bone is quite high, and in adults roughly 18 percent of the protein and mineral components are removed and replaced each year. Every part of every bone may not be affected, as there are regional and even local differences in the rate of turnover. For example, the spongy bone in the head of the femur may be replaced two or three times each year, whereas the compact bone along the shaft remains largely untouched.

Remodeling and Support

Regular mineral turnover gives each bone the ability to adapt to new stresses. Heavily stressed bones become thicker, stronger, and develop more pronounced surface ridges, whereas bones not subjected to ordinary stresses become thin and brittle. Regular exercise is therefore an important stimulus that maintains normal bone structure.

Degenerative changes in the skeleton occur after relatively brief periods of inactivity. For example, using a crutch while wearing a cast takes the loading off the injured leg. After a few weeks, the unstressed leg will lose up to about a third of its bone mass. The bones rebuild just as quickly when they again carry their normal load.

Homeostasis and Mineral Storage

The bones of the skeleton are more than just racks to hang muscles on. They are important mineral reservoirs, and calcium is the most abundant mineral in the human body. A typical human body contains 1–2 kg (2.2–4.4 lb) of calcium, with 99 percent of it deposited in the skeleton.

Calcium ion concentrations must be closely controlled to prevent damage to essential physiological systems. Small variations from normal concentrations will have some effect on cellular operations, and larger changes can cause a clinical crisis. Neurons and muscle cells are particularly sensitive to changes in the concentration of calcium ions. If the calcium concentration in body fluids increases by 30 percent, neurons and muscle cells become relatively unresponsive. If calcium levels decrease by 35 percent, they become so excitable that convulsions may occur. A 50 percent reduction in calcium concentrations usually causes death. Such gross disturbances in calcium metabolism are relatively rare, for the calcium ion concentrations are so closely regulated that daily fluctuations of more than 10 percent are very unusual.

Parathyroid hormone (PTH) and *calcitriol* are hormones that work together to elevate calcium levels in body fluids. Their actions are opposed by *calcitonin,* which depresses calcium levels in body fluids. These hormones and their regulation are detailed in Chapter 11.

By providing a calcium reserve, the skeleton helps maintain calcium homeostasis in body fluids. This function can have a direct effect on the shape and strength of the bones in the skeleton. When large numbers of calcium ions are mobilized, the bones become weaker; when calcium salts are deposited, the bones become more massive.

Injury and Repair

Despite its mineral strength, bone cracks or even breaks if subjected to extreme loads, sudden impacts, or stresses from unusual directions. All such cracks and breaks in bones constitute a **fracture**. Fractures are classified according to their external appearance, the site of the fracture, and the nature of the break in the bone. Important fracture types are indicated in the Focus box on p. 117.

Bones will usually heal even after they have been severely damaged, as long as the circulatory supply and the cellular components of the endosteum and periosteum survive. The stages in this process, which may take from 4 months to well over a year, are diagrammed in Figure 6-5●. When the remodeling is complete, the fragments of dead bone and the trabecular bone of the calluses will be gone, and only living compact bone will remain. The repair may be "good as new," with no sign that a fracture ever occurred, but often the bone will be slightly thicker than normal at the fracture site. *Stimulation of Bone Growth and Repair*

AGING AND THE SKELETAL SYSTEM

The bones of the skeleton become thinner and relatively weaker as a normal part of the aging process. Inadequate ossification is called **osteopenia** (os-tē-ō-Pē-nē-a; *penia*, lacking), and all people become slightly osteopenic as they age. The reduction in bone mass occurs because between the ages of 30 and 40 osteoblast activity begins to decline while osteoclast activity continues at normal levels. Once the reduction begins, women lose roughly 8 percent of their skeletal mass every decade, whereas men's skeletons deteriorate at the slower rate of about 3 percent per decade. All parts of the skeleton are not equally affected. Epiphyses, vertebrae, and the jaws lose more than their fair share, resulting in fragile limbs, a reduction in height, and the loss of teeth.

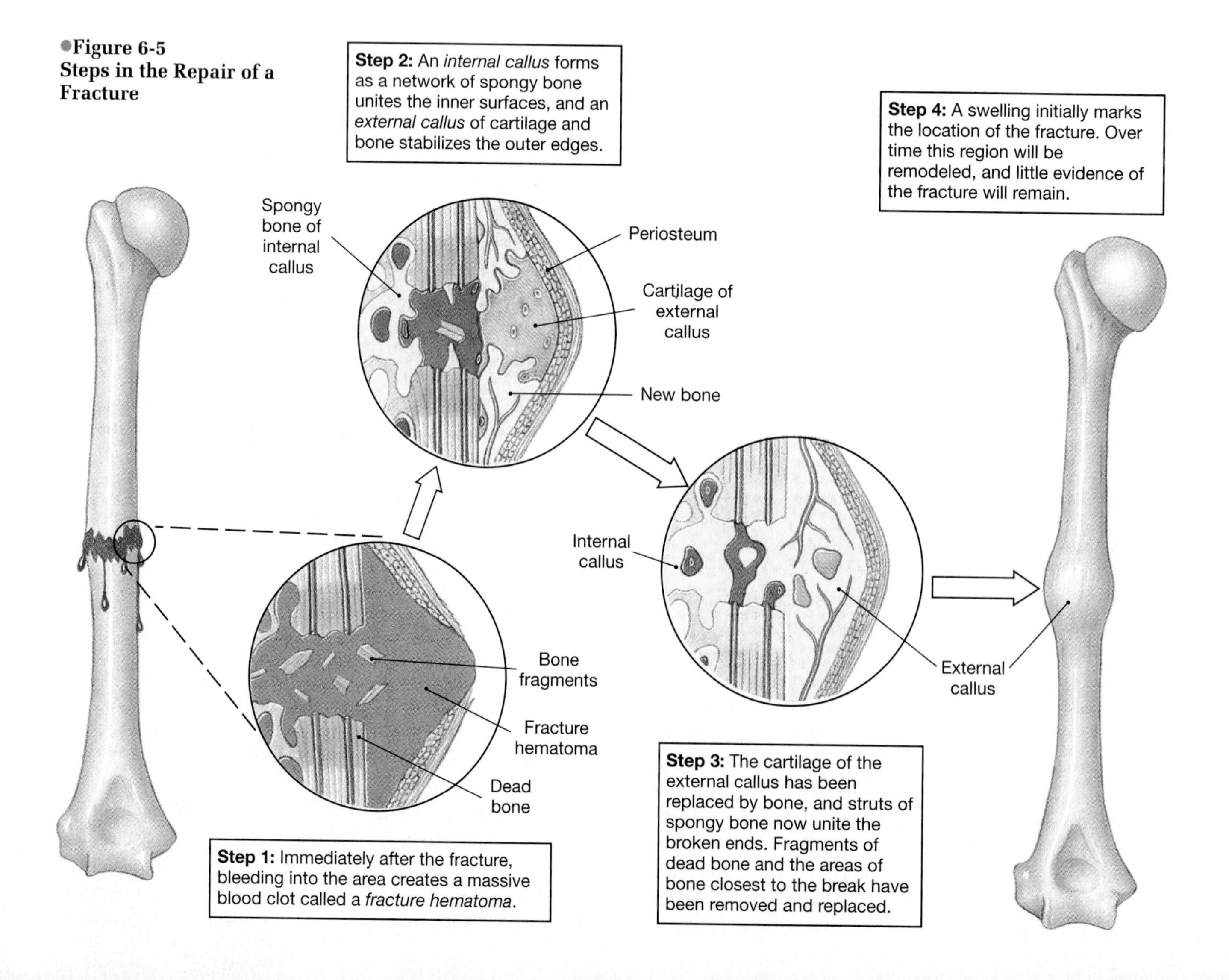

Step 2: An internal callus forms as a network of spongy bone unites the inner surfaces, and an external callus of cartilage and bone stabilizes the outer edges.
Step 4: A swelling initially marks the location of the fracture. Over time this region will be remodeled, and little evidence of the fracture will remain.
Spongy bone of internal callus
Periosteum
Cartilage of external callus
New bone
Internal callus
External callus
Bone fragments
Fracture hematoma
Dead bone
Step 3: The cartilage of the external callus has been replaced by bone, and struts of spongy bone now unite the broken ends. Fragments of dead bone and the areas of bone closest to the break have been removed and replaced.
Step 1: Immediately after the fracture, bleeding into the area creates a massive blood clot called a fracture hematoma.

●Figure 6-5 Steps in the Repair of a Fracture

FOCUS A Classification of Fractures

Fractures are classified according to their external appearance, the site of the fracture, and the nature of the crack or break in the bone. Important fracture types are indicated below, with representative X-rays. Many fractures fall into more than one category. For example, a Colles' fracture is a transverse fracture, but depending on the injury, it may also be a comminuted fracture that can be either open or closed. Closed, or simple, fractures are completely internal; they do not involve a break in the skin. Open, or compound, fractures project through the skin; they are more dangerous because of the possibility of infection or uncontrolled bleeding.

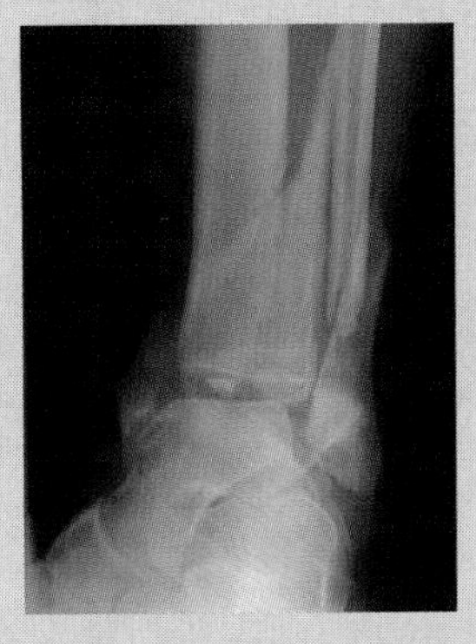

A **Pott's fracture** occurs at the ankle and affects both bones of the leg.

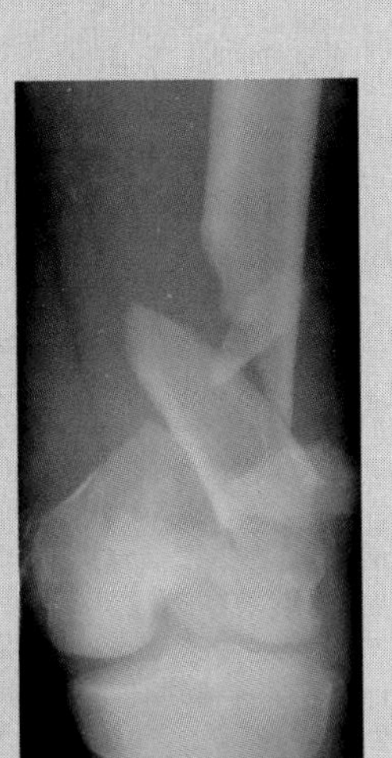

Comminuted fracture of distal femur

Comminuted fractures shatter the affected area into a multitude of bony fragments.

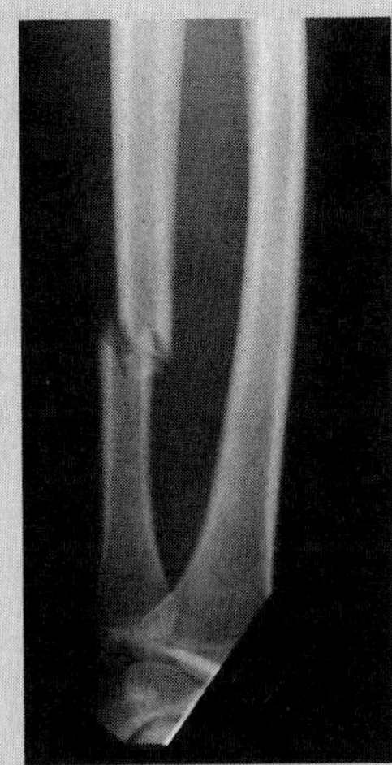

Transverse fractures break a shaft of a bone across its long axis.

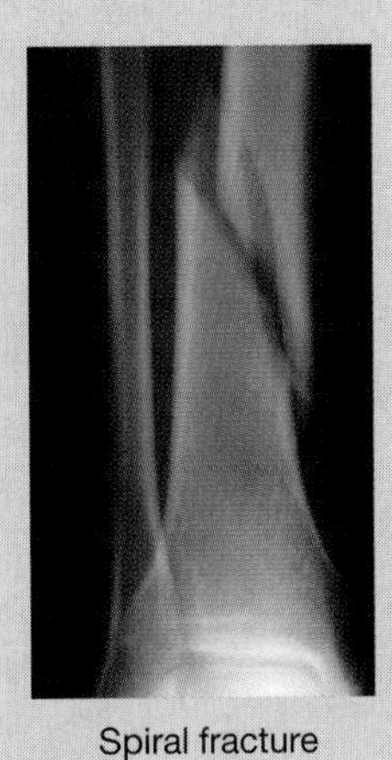

Spiral fracture of tibia

Spiral fractures, produced by twisting stresses, spread along the length of the bone.

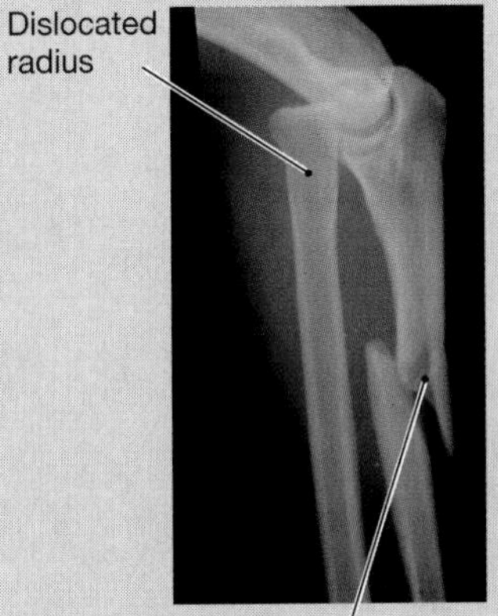

Displaced fractures produce new and abnormal arrangements of bony elements.

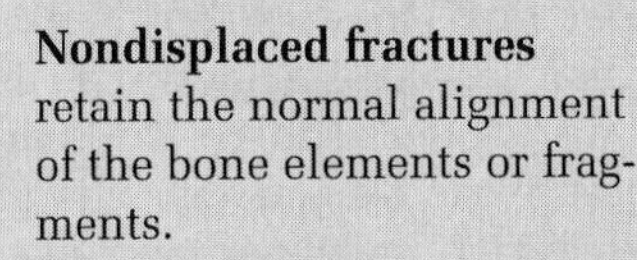

Nondisplaced fractures retain the normal alignment of the bone elements or fragments.

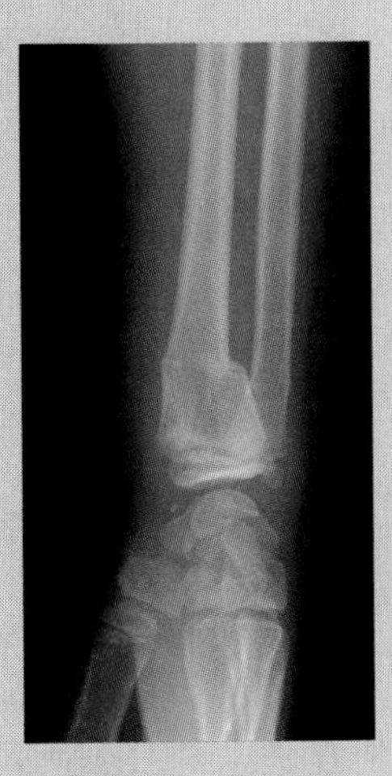

A **Colles' fracture** is a break in the distal portion of the radius, the slender bone of the forearm; it is often the result of reaching out to cushion a fall.

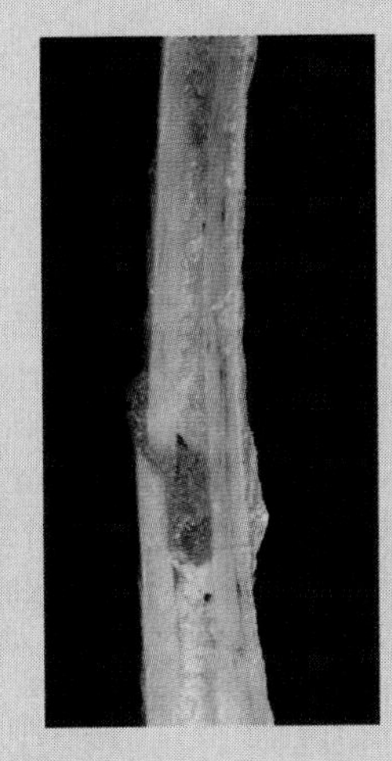

In a **greenstick fracture,** only one side of the shaft is broken, and the other is bent; this type usually occurs in children, whose long bones have yet to fully ossify.

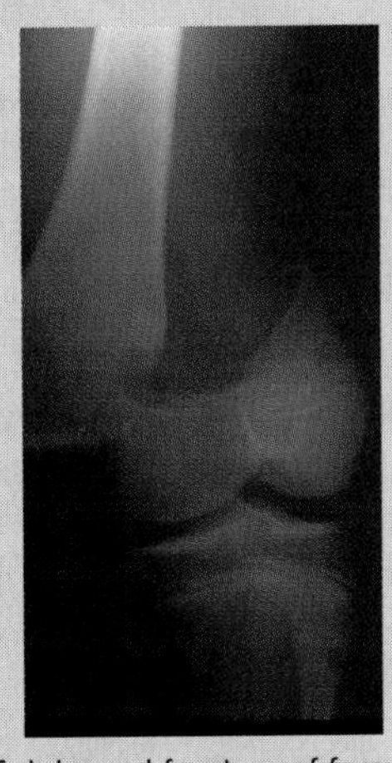

Epiphyseal fracture of femur

Epiphyseal fractures usually occur where the matrix is undergoing calcification and chondrocytes are dying. A clean transverse fracture along this line usually heals well. Fractures between the epiphysis and the epiphyseal plate can permanently halt further longitudinal growth unless carefully treated; often surgery is required.

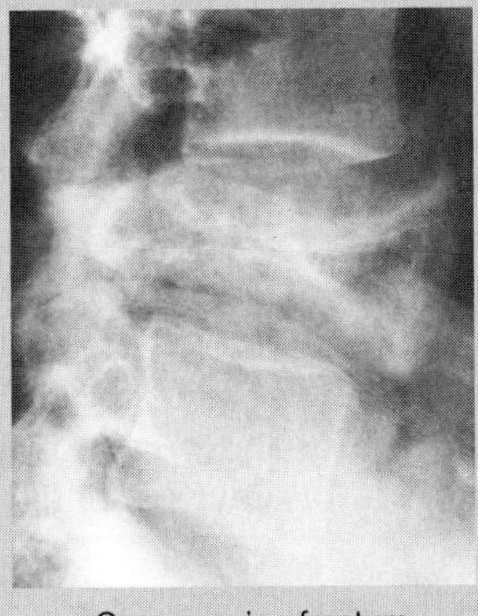

Compression fracture of vertebra

Compression fractures occur in vertebrae subjected to extreme stresses, as when landing on your seat after a fall.

TABLE 6-1 An Introduction to Skeletal Terminology

General Description	*Anatomical Term*	*Definition*
Elevations and projections (general)	Process	Any projection or bump
	Ramus	An extension of a bone making an angle to the rest of the structure
Processes formed where tendons or ligaments attach	Trochanter	A large, rough projection
	Tuberosity	A smaller, rough projection
	Tubercle	A small, rounded projection
	Crest	A prominent ridge
	Line	A low ridge
Processes formed for articulation with adjacent bones	Head	The expanded articular end of an epiphysis, separated from the shaft by a narrower neck
	Condyle	A smooth, rounded articular process
	Trochlea	A smooth, grooved articular process shaped like a pulley
	Facet	A small, flat articular surface
	Spine	A pointed process
Depressions	Fossa	A shallow depression
	Sulcus	A narrow groove
Openings	Foramen	A rounded passageway for blood vessels and/or nerves
	Fissure	An elongated cleft
	Meatus	A canal leading through the substance of a bone
	Sinus or antrum	A chamber within a bone, normally filled with air

OSTEOPOROSIS

Osteoporosis (os-tē-ō-por-ō-sis; *porosus*, porous) is a condition that produces a reduction in bone mass great enough to compromise normal function. The difference between the "normal" osteopenia of aging and the clinical condition of osteoporosis is a matter of degree. It is estimated that 29 percent of women between the ages of 45 and 79 can be considered osteoporotic. The increase in incidence after menopause has been linked to decreases in the production of estrogens (female sex hormones). The incidence of osteoporosis in men of the same age is estimated at 18 percent.

Because bones are more fragile, they break easily and do not repair well. Vertebrae may collapse, distorting the vertebral articulations and putting pressure on spinal nerves. Therapies that boost estrogen levels, dietary changes to elevate calcium levels in the blood, and exercise that stresses bones and stimulates osteoblast activity appear to slow but not completely prevent the development of osteoporosis.

AM *Osteoporosis and Age-Related Skeletal Abnormalities*

✓ Why would you expect the arm bones of a weight lifter to be thicker and heavier than those of a jogger?

✓ Why is the condition known as osteoporosis more common in older women than older men?

AN OVERVIEW OF THE SKELETON

Skeletal Terminology

Each of the bones in the human skeleton not only has a distinctive shape but also has characteristic external features. For example, elevations or projections form where tendons and ligaments attach and where adjacent bones articulate. Depressions and openings indicate sites where blood vessels and nerves lie alongside or penetrate the bone. These external landmarks are called **bone markings**, or *surface features*. The most common terms used to describe bone markings are detailed in Table 6-1.

Skeletal Divisions

The skeletal system (Figure 6-6●) consists of 206 separate bones and a number of associated cartilages. It is divided into *axial* and *appendicular divisions*. The **axial skeleton** forms the longitudinal axis of the body. This division's 80 bones can be subdivided into the **skull** (22 bones), plus the **auditory ossicles** (6) and **hyoid bone** (1) associated with the skull, the **vertebral column** (26), and the **thoracic** (*rib*) **cage** composed of the **ribs** (24) and the **sternum** (1).

The **appendicular skeleton** includes the bones of the arms and legs and those of the **pectoral** and **pelvic girdles**, which attach the limbs to the trunk. All together there are 126 appendicular bones; 32 are associated with each upper limb, and 31 with each lower limb.

The sections that follow discuss the structure of a typical skeleton. When viewed in detail, however, no two skeletons are exactly alike, owing to differences in body size, weight, sex, race, medical history, and other factors. A detailed analysis of a skeleton can actually provide important information about an individual; for this reason, skeletal analysis plays an important

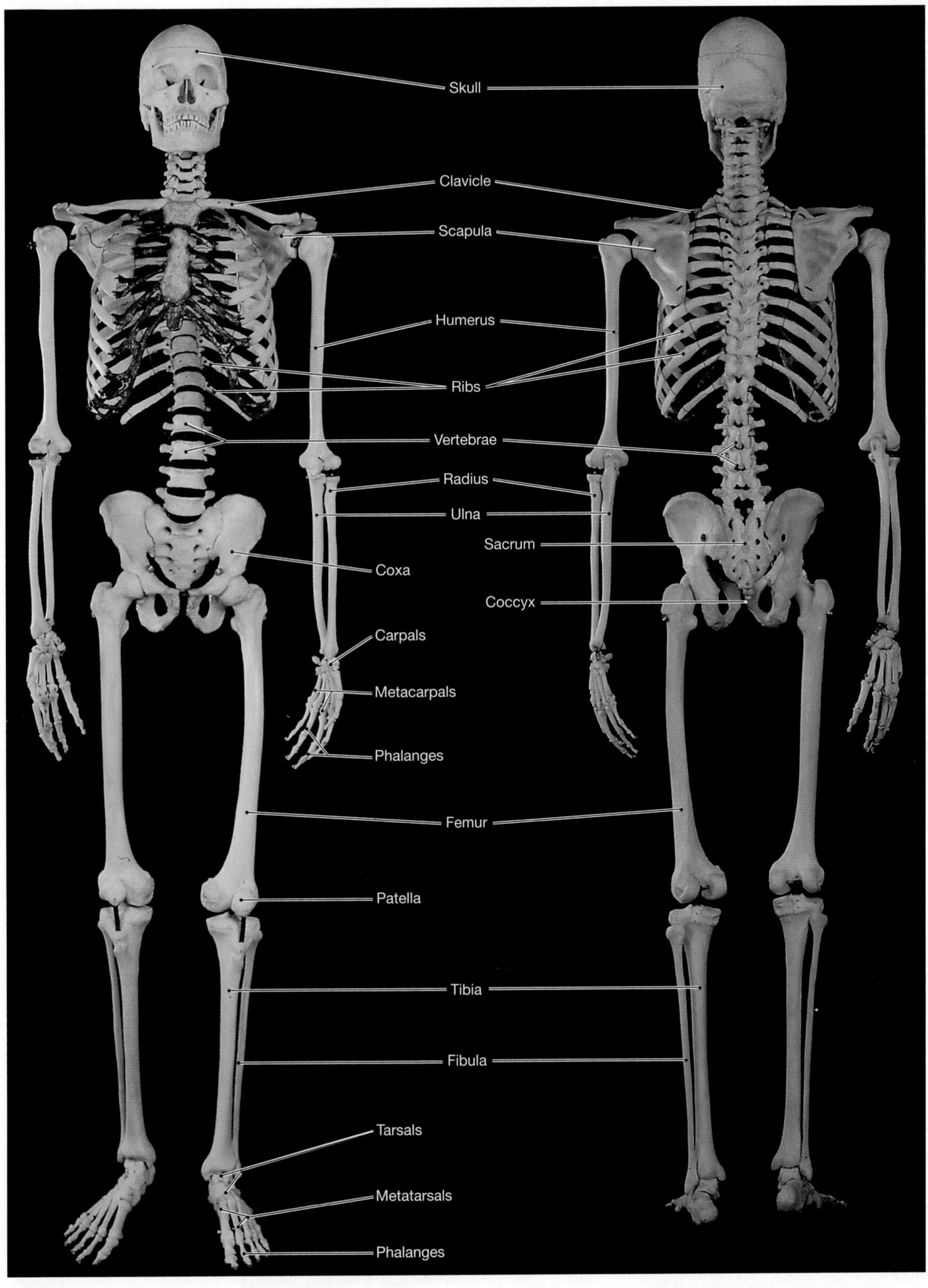

Figure 6-6 The Skeleton

role in anthropology, as well as in criminal investigations. AM *Individual Variations in the Skeleton*

THE AXIAL DIVISION

The axial skeleton creates a framework that supports and protects organ systems in the dorsal and ventral body cavities. In addition, it provides an extensive surface area for the attachment of muscles that (1) adjust the positions of the head, neck, and trunk; (2) perform respiratory movements; and (3) stabilize or position elements of the appendicular skeleton.

The Skull

The bones of the skull protect the brain and support delicate sense organs involved with vision, hearing, balance, olfaction (smell), and gustation (taste). The skull is made up of 22 bones: 8 form the *cranium*, and 14 are associated with the *face*. Seven additional bones are associated with the skull: 6 *auditory ossicles*, tiny bones involved in sound detection, are encased by the *temporal bones* of the skull, and the *hyoid bone* has ligamentous connections to the inferior surface of the skull.

The cranium encloses the cranial cavity, a fluid-filled chamber that cushions and supports the brain. The outer surface of the cranium provides an extensive area for the attachment of muscles that move the eyes, jaws, and head.

Refer to Figures 6-7, 6-8, and 6-9• during the following discussion of the skull. Figures 6-7 and 6-8• illustrate various external views, and Figure 6-9• reveals important internal features through horizontal (transverse) and sagittal sections.

The Bones of the Cranium

The Frontal Bone The **frontal bone** of the cranium forms the forehead and the superior surface of each eye socket, or **orbit** (Figures 6-7 and 6-8•). A **supraorbital foramen** pierces the bony ridge above each orbit, forming a passageway for blood vessels and nerves passing to or from the forehead. (Sometimes the ridge has a deep groove, called a *supraorbital notch*, rather than a foramen, but the function is the same.) Above the orbit, the frontal bone contains air-filled internal chambers that communicate with the nasal cavity. These **frontal sinuses** (Figure 6-9b•) make the bone lighter and produce mucus that cleans and moistens the nasal cavities.

The Parietal Bones On either side of the skull, a **parietal** (pa-RĪ-e-tal) **bone** (Figures 6-8, 6-9•) is found posterior to the frontal bone. Together the parietal bones form the roof and the superior walls of the cranium. The parietal bones interlock along the **sagittal**

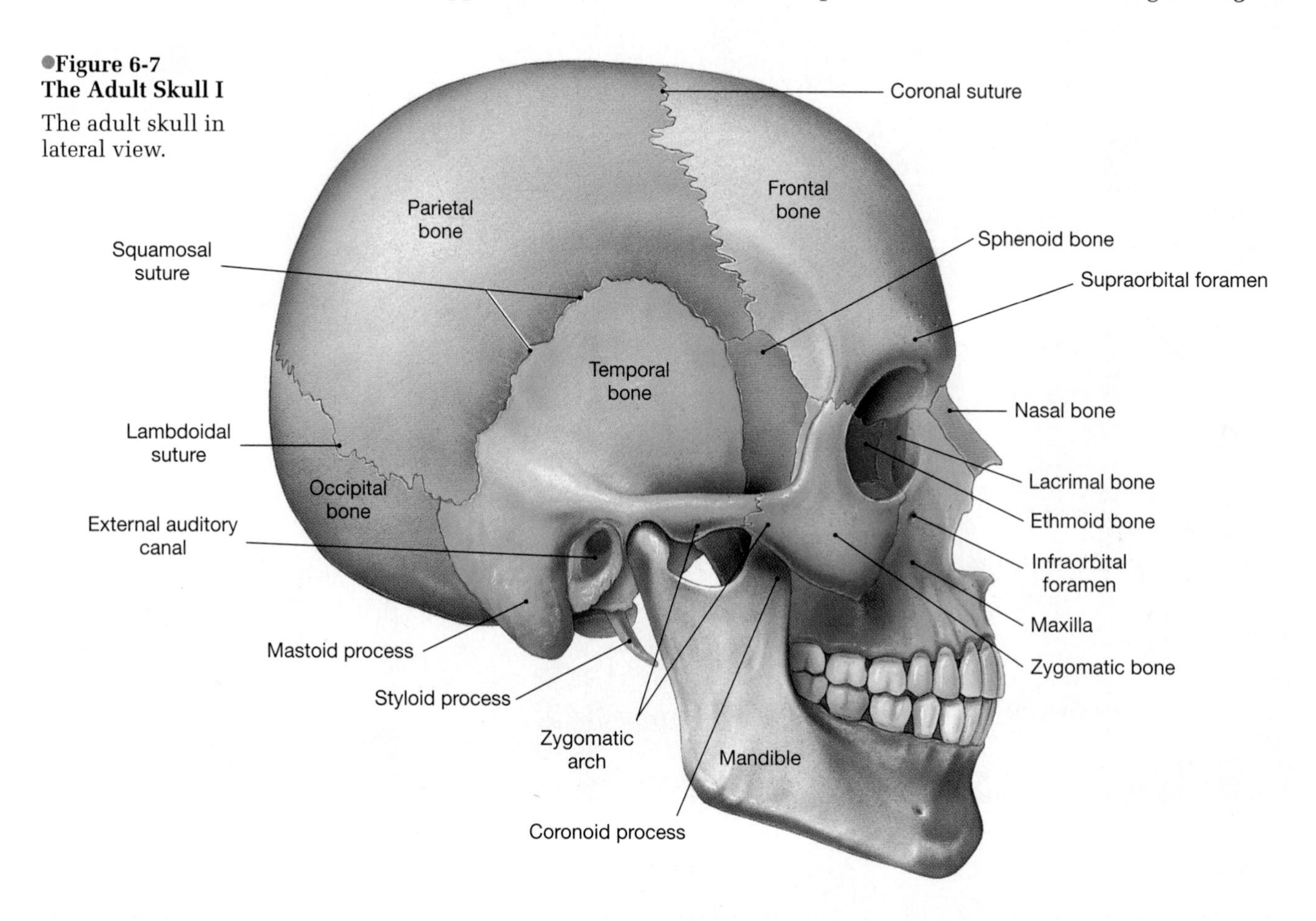

•**Figure 6-7 The Adult Skull I**
The adult skull in lateral view.

Figure 6-8
The Adult Skull II

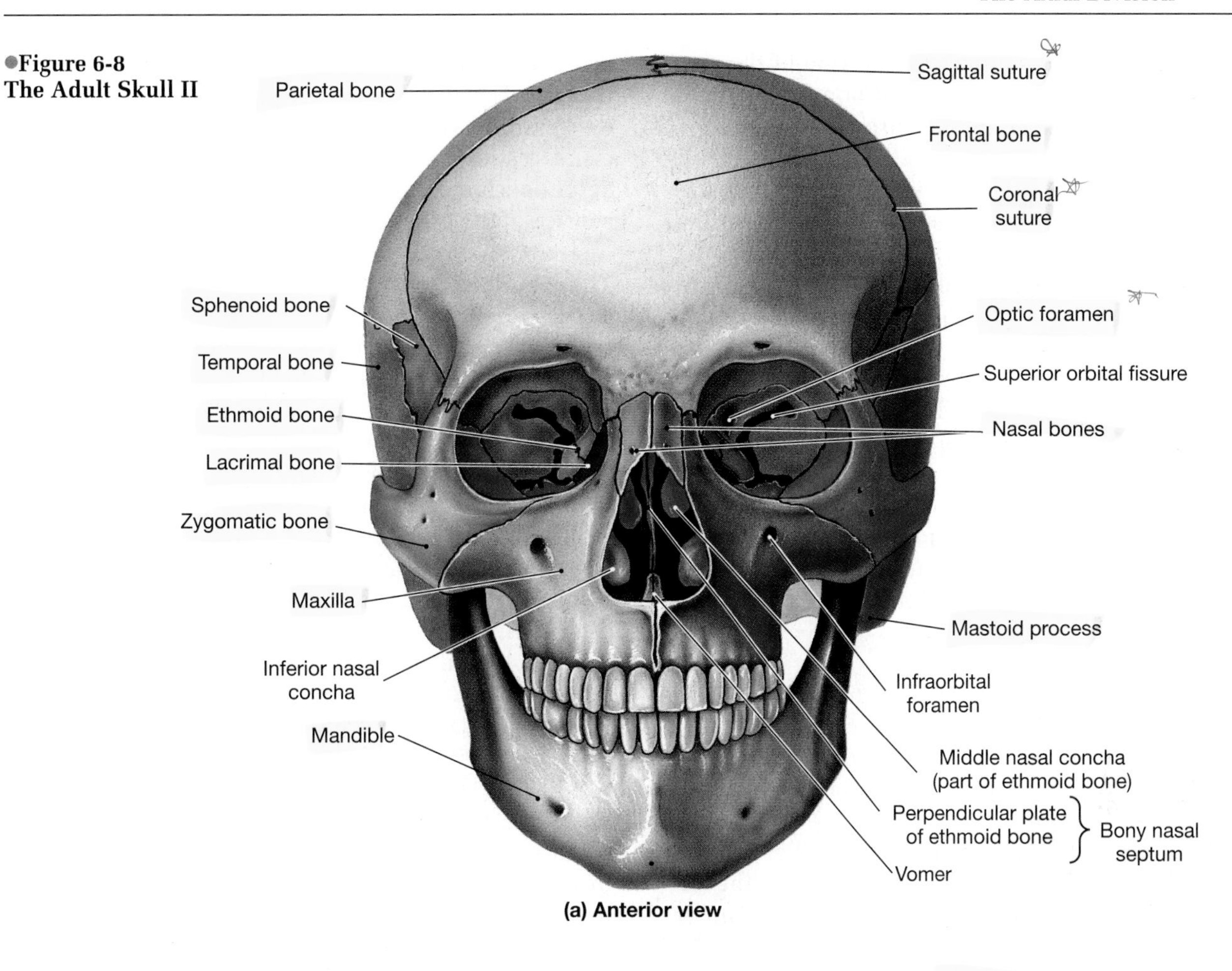
Parietal bone
Sagittal suture
Frontal bone
Coronal suture
Sphenoid bone
Temporal bone
Ethmoid bone
Lacrimal bone
Zygomatic bone
Maxilla
Inferior nasal concha
Mandible
Optic foramen
Superior orbital fissure
Nasal bones
Mastoid process
Infraorbital foramen
Middle nasal concha (part of ethmoid bone)
Perpendicular plate of ethmoid bone
Vomer
Bony nasal septum

(a) Anterior view

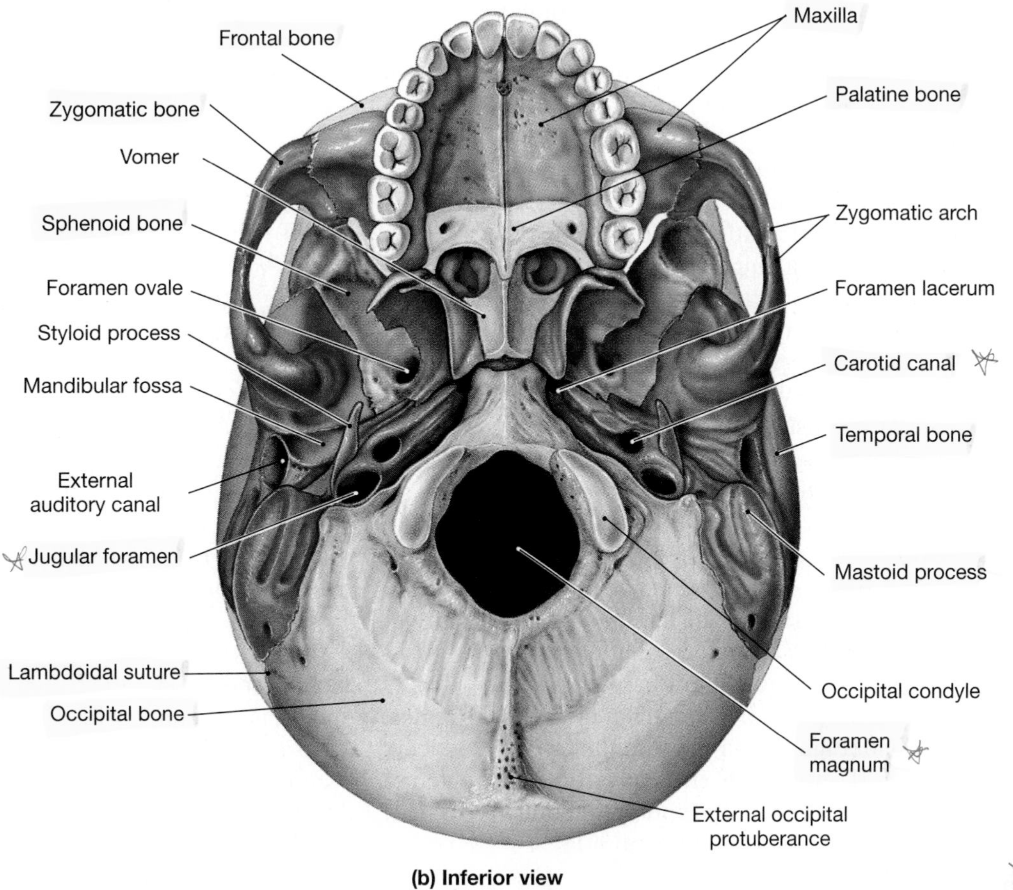
Frontal bone
Zygomatic bone
Vomer
Sphenoid bone
Foramen ovale
Styloid process
Mandibular fossa
External auditory canal
Jugular foramen
Lambdoidal suture
Occipital bone
Maxilla
Palatine bone
Zygomatic arch
Foramen lacerum
Carotid canal
Temporal bone
Mastoid process
Occipital condyle
Foramen magnum
External occipital protuberance

(b) Inferior view

✱ 4 foramena

suture, which extends along the midline of the cranium. Anteriorly, the two parietal bones articulate with the frontal bone along the **coronal suture**.

The Occipital Bone The occipital bone forms the posterior and inferior portions of the cranium. Along its superior margin, the occipital bone contacts the two parietal bones at the **lambdoidal** (lam-DOYD-al) **suture** (Figure 6-7•). Figure 6-8b• presents an inferior view of the skull, showing the orientation of the occipital bone and its relationships with other bones in the floor of the cranium. The occipital bone surrounds the **foramen magnum**, the opening that connects the cranial cavity with the spinal cavity. The spinal cord passes through the foramen magnum to connect with the inferior portion of the brain. On either side of the foramen magnum are the **occipital condyles**, the sites of articulation between the skull and the vertebral column.

The Temporal Bones Lying below the parietal bones and contributing to the sides and base of the cranium are the **temporal bones**. As can be seen in Figure 6-7•, the temporal bones contact the parietal bones along the **squamosal suture**.

The temporal bones display a number of distinctive anatomical landmarks. One of them, the **external auditory canal** leads to the *eardrum*, or **tympanic membrane**. The eardrum separates the external auditory canal from the *middle ear cavity* that contains the *auditory ossicles*, or *ear bones*. The structure and function of the tympanic membrane, middle ear cavity, and auditory ossicles will be considered in Chapter 10.

Anterior to the external auditory canal is a transverse depression, the **mandibular fossa** (Fig. 6-8b•), which marks the point of articulation with the lower jaw (mandible). The prominent bulge just posterior and inferior to the entrance to the external auditory canal is the **mastoid process**. Adjacent to the base of the mastoid process is the long, sharp **styloid process** (STĪ-loyd; *stylos*, pillar). The mastoid process provides a site for the attachment of muscles that rotate or extend the head. The styloid process anchors muscles and ligaments associated with the tongue and hyoid bone.

The Sphenoid Bone The **sphenoid** (SFĒ-noid) **bone** forms part of the floor of the cranium. It also acts as a bridge uniting the cranial and facial bones, and it braces the sides of the skull. The general shape of the sphenoid has been compared to a giant bat, with its wings extended. The wings can be seen most clearly on the superior surface (Figure 6-9a•). From the front (Figure 6-8a•) or side (Figure 6-7•), it is covered by other bones. Like the frontal bone, the sphenoid bone also contains a pair of sinuses, called **sphenoidal sinuses** (Figure 6-9b•).

The lateral "wings" of the sphenoid extend to either side from a central depression (Figure 6-9a•). This depression is called the **sella turcica** (TUR-si-ka) (Turk's saddle). In life, it encloses the pituitary gland, an endocrine organ that is connected to the inferior surface of the brain by a narrow stalk of neural tissue.

The Ethmoid Bone The **ethmoid bone** is anterior to the sphenoid bone. The ethmoid bone consists of two honeycombed masses of bone. It forms part of the cranial floor, contributes to the medial surfaces of the orbit of each eye (Figure 6-8a•), and forms the roof and sides of the nasal cavity. A prominent ridge, the **crista galli** (Figure 6-9a•), or "cock's comb," projects above the superior surface of the ethmoid. Holes in the **cribriform plate** (*cribrum*, sieve) on either side permit passage of sensory nerves traveling from olfactory (smell) receptors in the nasal cavity to the brain.

The lateral portions of the ethmoid contain the **ethmoidal sinuses** that drain into the nasal cavity. Projections called the **superior** and **middle conchae** (KONG-kē; *concha*, shell) extend into the nasal cavity toward the *nasal septum* (*septum*, wall) that divides the nasal cavity into left and right portions (Figures 6-8a, 6-9b,c•). The conchae slow the movement of air through the nasal cavity, allowing time for the air to become warm, moist, and clean before it reaches the delicate portions of the respiratory tract. The **perpendicular plate** of the ethmoid extends inferiorly from the crista galli, passing between the conchae to contribute to the nasal septum (Figure 6-8a•).

✓ The mastoid and styloid processes are found on which of the skull bones?

✓ What bone contains the depression called the sella turcica? What is located in the depression?

✓ Which bone of the cranium articulates directly with the vertebral column?

The Bones of the Face

The facial bones protect and support the entrances to the digestive and respiratory tracts. They also provide areas for the attachment of muscles that control our facial expressions and help us manipulate food. Of the 14 facial bones, only the lower jaw, or *mandible*, is movable.

The Maxillary Bones The **maxillary** (MAK-si-ler-ē) **bones**, or **maxillae**, articulate with all other facial bones except the mandible. The maxillary bones form (1) the floor and medial portion of the rim of the orbit (Figure 6-8a•), (2) the walls of the nasal cavity, and (3) the anterior roof of the mouth, or *hard palate* (Figure 6-9b•). The maxillary bones contain large **maxillary sinuses**

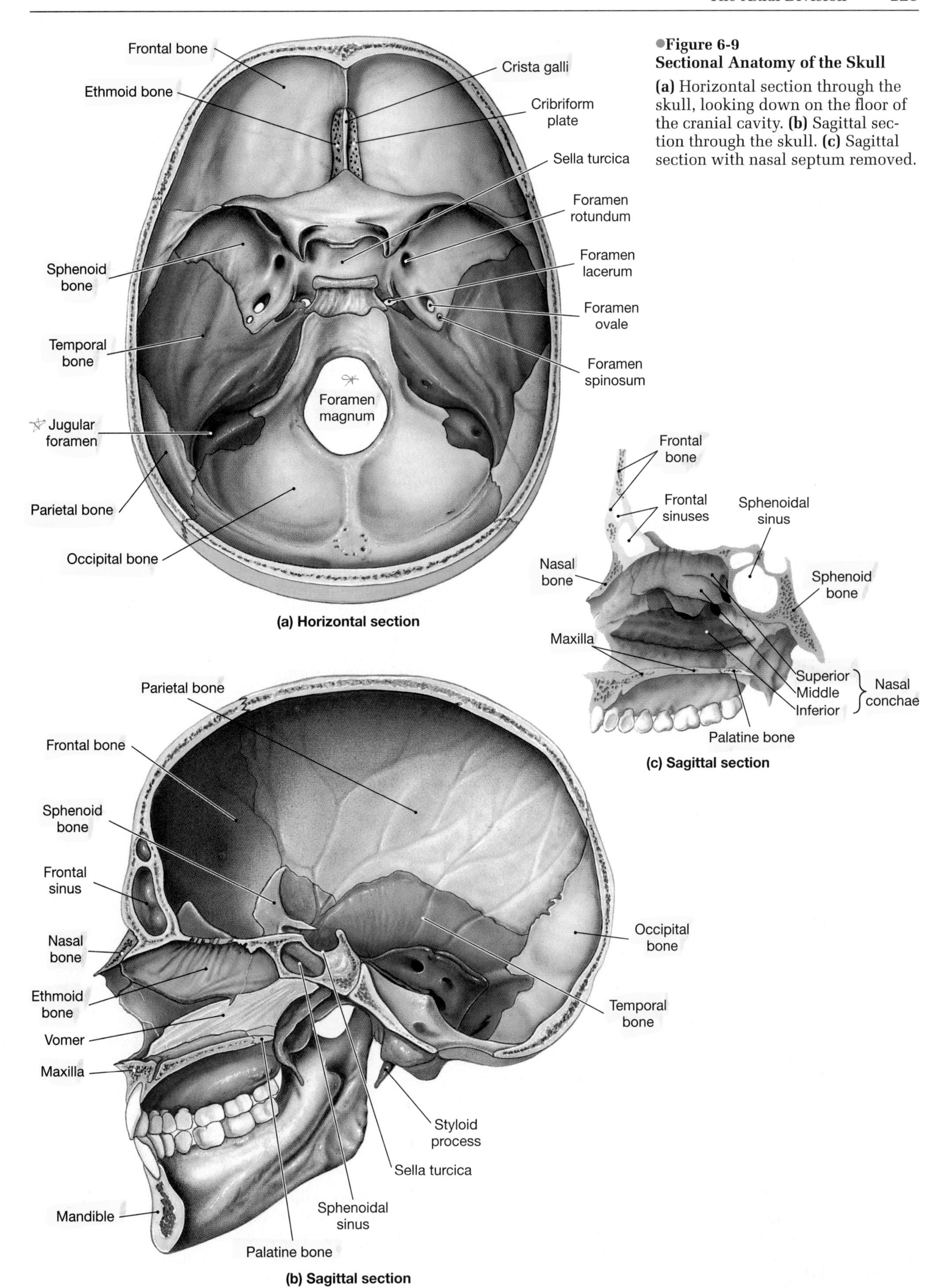

Figure 6-9
Sectional Anatomy of the Skull

(a) Horizontal section through the skull, looking down on the floor of the cranial cavity. **(b)** Sagittal section through the skull. **(c)** Sagittal section with nasal septum removed.

that lighten the portion of the maxillae above the embedded teeth. Infections of the gums or teeth can sometimes spread into the maxillary sinuses, increasing pain and making treatment more complicated.

The Palatine Bones The paired **palatine bones** (Figures 6-8b, 6-9b●) form the posterior surface of the bony, hard palate, or "roof of the mouth." The superior surfaces of the palatines form the floor of the nasal cavity.

The Vomer The inferior margin of the **vomer** articulates with the paired palatine bones. The vomer supports a prominent partition that forms part of the nasal septum, along with the ethmoid bone (Figures 6-8a, 6-9b●).

The Zygomatic Bones On each side of the skull a **zygomatic** (zī-go-MA-tik) **bone** articulates with the frontal bone and the maxilla to complete the lateral wall of the orbit (Figures 6-7, 6-8a●). Along its lateral margin each zygomatic bone gives rise to a slender bony extension that curves laterally and posteriorly to meet a process from the temporal bone. Together these processes form the **zygomatic arch**, or *cheekbone.*

The Nasal Bones Forming the bridge of the nose midway between the orbits, the **nasal bones** articulate with the superior frontal bone and the maxillary bones (Figures 6-7, 6-8a●).

The Lacrimal Bones The **lacrimal** (*lacrimae*, tears) **bones** are found within the orbit on its medial surface. They articulate with the frontal, ethmoid, and maxillary bones (Figures 6-7, 6-8a●).

The Inferior Nasal Conchae The paired **inferior nasal conchae** (Figure 6-8a●) project from the lateral walls of the nasal cavity. Their shape helps slow airflow and deflects arriving air toward the olfactory (smell) receptors located near the upper portions of the nasal cavity.

The Nasal Complex The **nasal complex** includes the bones that form the superior and lateral walls of the nasal cavities and the sinuses that drain into them. The ethmoid and vomer form the **nasal septum** that separates the left and right portions of the nasal cavity (Figure 6-8a●). The frontal, sphenoid, ethmoid, and maxillary bones contain air-filled chambers collectively known as the **paranasal sinuses.** Sinuses make the skull bones lighter and provide an extensive area of mucous epithelium. The mucous secretions are released into the nasal cavities, and the ciliated epithelium passes the mucus back toward the throat, where it is eventually swallowed. Incoming air is humidified and warmed as it flows across this carpet of mucus, and foreign particles, such as dust or bacteria, become trapped in the sticky mucus and swallowed. This mechanism helps protect more delicate portions of the respiratory tract. AM *Septal Defects and Sinus Problems*

The Mandible The broad **mandible** is the bone of the lower jaw. It forms a broad, horizontal curve that on either side extends into two vertical processes. The

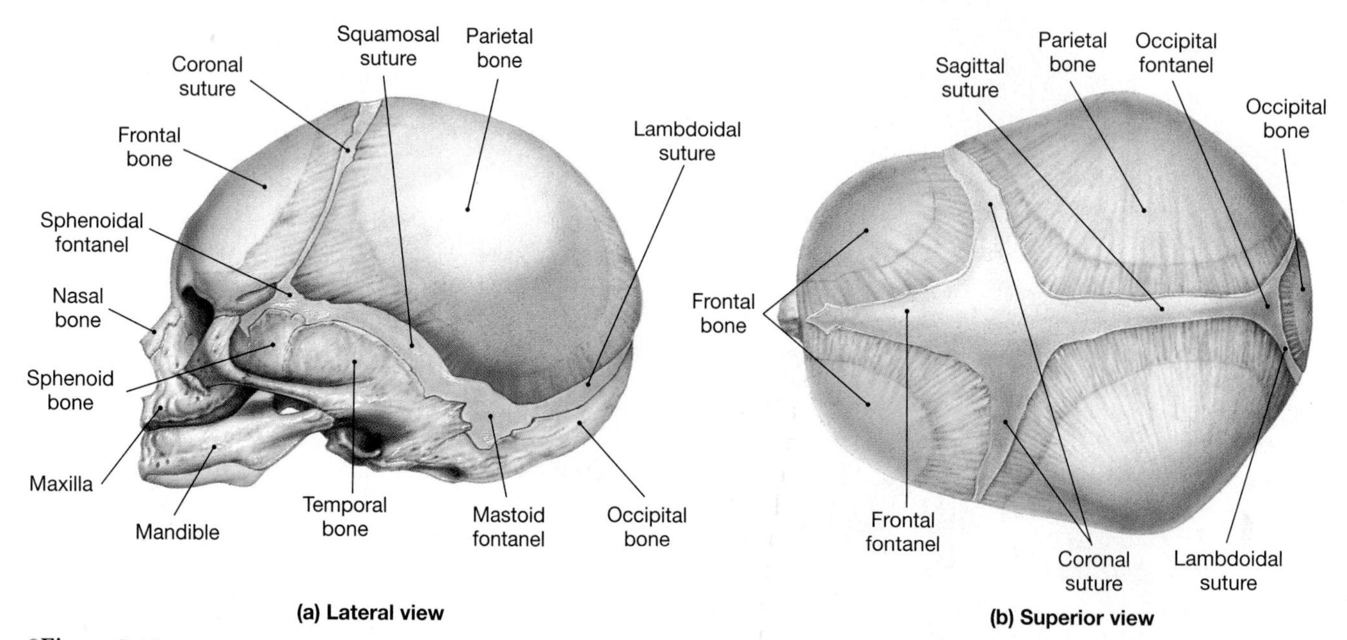

●**Figure 6-10**
The Skull of an Infant

(a) Lateral view. An infant's skull contains more individual bones than an adult's skull. Many of these bones eventually fuse to create the adult skull. The flat bones of the skull are separated by areas of fibrous connective tissue called fontanels that allow for cranial expansion and distortion during birth. By about age 4 these areas will disappear, and skull growth will be completed. **(b)** Superior view.

more posterior process articulates with the mandibular fossa of the temporal bone on that side. This articulation is quite mobile, and the disadvantage of such mobility is that the jaw can easily be dislocated. AM *TMJ Syndrome*

The *temporalis muscle* of the skull originates along the lateral surface of the temporal bone and inserts on the **coronoid** (kō-RŌ-noid) **processes** of the mandible. The temporalis is the most powerful muscle involved in closing the mouth.

The Hyoid Bone

The small, U-shaped hyoid bone hangs below the skull, suspended by ligaments from the styloid processes of the temporal bones. The hyoid (1) serves as a base for muscles associated with the tongue and *larynx* (voicebox) and (2) supports and stabilizes the position of the larynx.

- ✓ During baseball practice, a ball hits Casey in the eye, fracturing the bones directly above and below the orbit. Which bones were broken?
- ✓ What are the functions of the paranasal sinuses?
- ✓ Why would a fracture of the coronoid process of the mandible make it difficult to close the mouth?
- ✓ What symptoms would you expect to see in a person suffering from a fractured hyoid bone?

The Skulls of Infants and Children

Many different centers of ossification are involved in the formation of the skull, but as the fetus develops, the individual centers begin to fuse. This fusion produces a smaller number of composite bones. For example, the sphenoid begins as 14 separate ossification centers but ends as just one. At birth the fusion process has yet to be completed, and there are two frontal bones, four occipital bones, and several sphenoid and temporal elements.

The skull organizes around the developing brain, and as the time of birth approaches the brain enlarges rapidly. Although the bones of the skull are also growing, they fail to keep pace, and at birth the cranial bones are connected by areas of fibrous connective tissue known as **fontanels** (fon-tah-NELS). These connections are quite flexible, and the skull can be distorted without damage. Such distortion normally occurs during delivery and eases the passage of the infant along the birth canal. Figure 6-10● indicates the prominent fontanels and the appearance of the skull at birth.

The Neck and Trunk

The rest of the axial skeleton is subdivided on the basis of vertebral structure, as indicated in Figure 6-11●. The 7 **cervical vertebrae** of the neck (abbreviated as C_1 to C_7) extend inferiorly as far as the trunk. Each of the 12 **thoracic vertebrae** (T_1 to T_{12}) articulates with one or more pairs of **ribs**. The 5 **lumbar vertebrae** (L_1 to L_5) continue caudally, the fifth articulating with the **sacrum**, a single bone formed by the fusion of 5 vertebrae. The small **coccyx** (KOK-siks) also consists of fused vertebrae. The total length of the vertebral column of an adult averages 71 cm (28 in.).

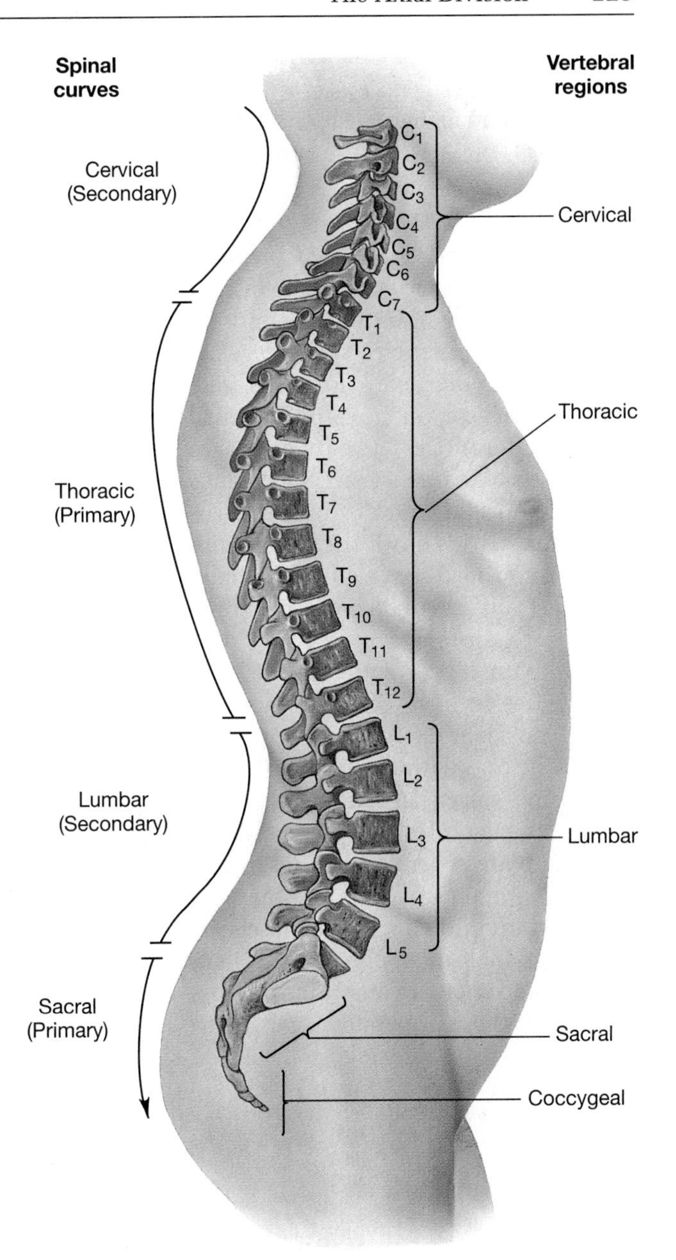

●**Figure 6-11**
The Vertebral Column
The major divisions of the vertebral column, showing the four spinal curves.

Spinal Curvature

The vertebrae do not form a straight and rigid structure. A side view of the spinal column reveals four **spinal curves**, shown in Figure 6-11●. The *thoracic*

•Figure 6-12
Typical Vertebrae of the Cervical, Thoracic, and Lumbar Regions
Each vertebra is shown in superior view.

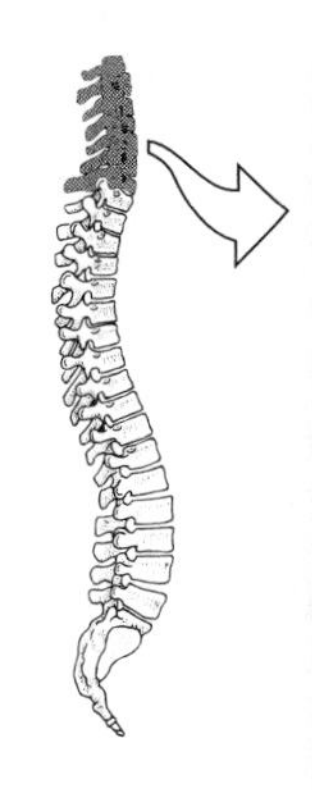

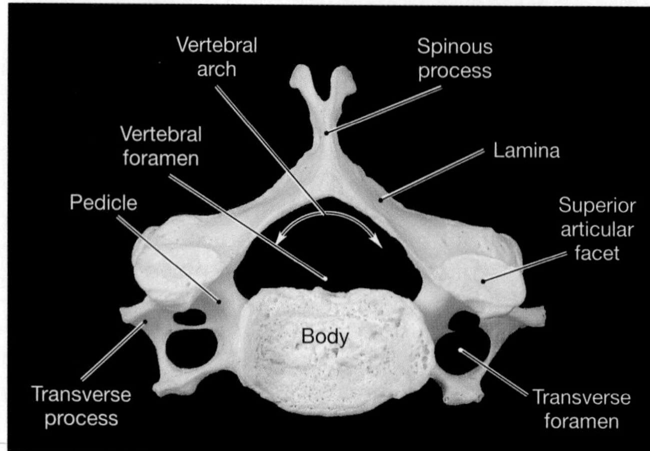

(a) Typical cervical vertebra

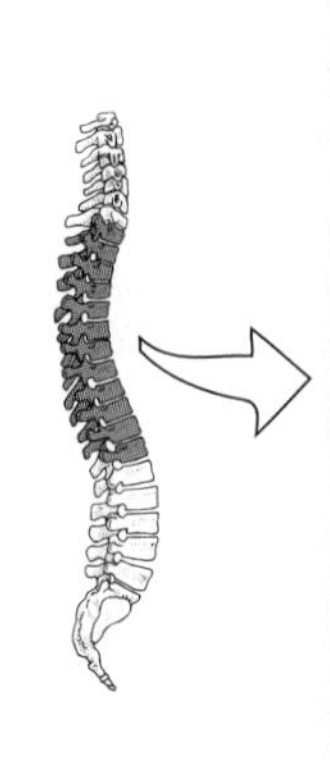

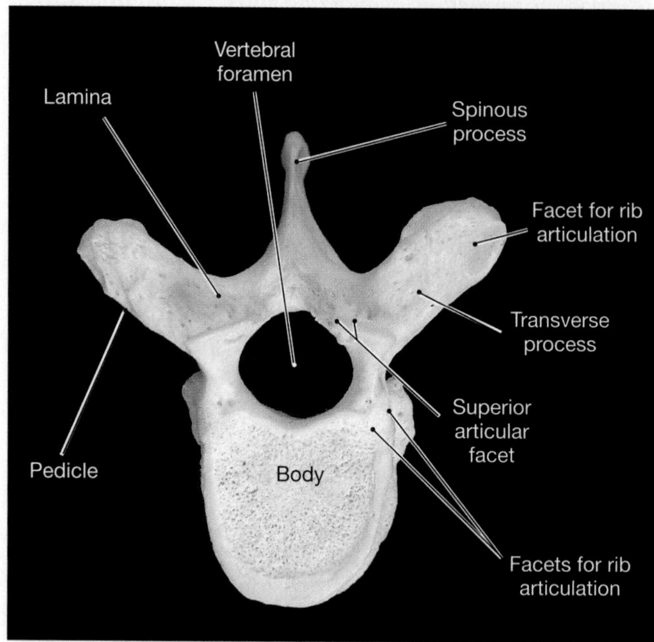

(b) Typical thoracic vertebra

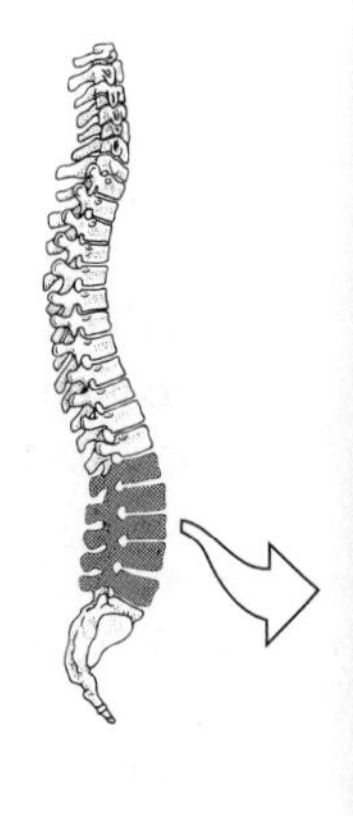

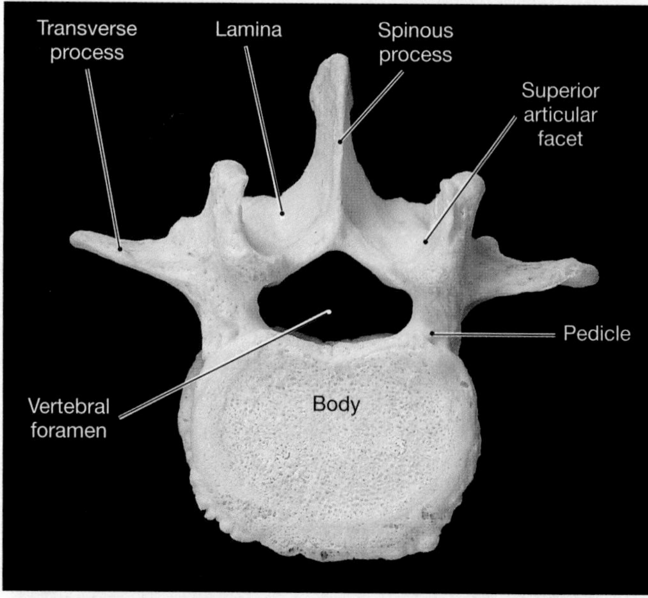

(c) Typical lumber vertebra

and *sacral curves* are called **primary curves** because they appear late in fetal development, as the thoracic and abdominal viscera enlarge. The *cervical* and *lumbar curves,* known as **secondary curves,** do not appear until months after birth. The cervical curve develops as the infant learns to balance the head upright, and the lumbar curve develops with the ability to stand. When standing, the weight of the body must be transmitted through the spinal column to the pelvic girdle and ultimately to the legs. Yet most of the body weight lies in front of the spinal column. The secondary curves bring that weight in line with the body axis. All four spinal curves are fully developed by the time a child is 10 years old.

There are several abnormal distortions of spinal curvature that may appear during childhood and adolescence. Examples include *kyphosis* (kī-FŌ-sis; exaggerated thoracic curvature), *lordosis* (lor-DŌ-sis; exaggerated lumbar curvature), and *scoliosis* (skō-lē-Ō-sis; an abnormal lateral curvature). AM *Kyphosis, Lordosis, Scoliosis*

Vertebral Anatomy

Figure 6-12• shows representative vertebrae from different regions of the verteberal column. A comparison of these vertebrae reveals a large number of similar features. The more massive, weight-bearing portion of a vertebra is called the **body**. Extending posteriorly from the sides of vertebral body are the **pedicles** (PE-di-kls). **Transverse processes** projecting laterally or dorsolaterally from the pedicles serve as sites for muscle attachment. The pedicle on each side supports a **lamina,** which unites with the lamina of the opposite side to form the **spinous process**, or *spinal process.* The spinous processes form the bumps that can be felt along the midline of your back. The **articular processes** of successive vertebrae contact one another at the **articular facets**.

The pedicles and laminae create the **vertebral arch** that forms the lateral and posterior walls of the **vertebral foramen**. The vertebral foramen encloses a portion of the **vertebral canal** that in life contains the spinal cord. **Intervertebral foramina** between successive vertebrae permit the passage of nerves running to or from the enclosed spinal cord.

Adjacent vertebrae are connected by longitudinal ligaments. The bony faces of the vertebral bodies usually do not contact one another because an **intervertebral disc** of fibrocartilage lies between them. Intervertebral discs are not found in the sacrum and coccyx, where the vertebrae have fused, or between the first and second cervical vertebrae.

An intervertebral disc consists of an extensive region of fibrocartilage that surrounds a soft, gelatinous mass. The intervertebral discs act as shock absorbers, compressing and distorting when stressed. This change prevents bone-to-bone contact that might damage the vertebrae or jolt the spinal cord and brain. These discs make a significant contribution to an individual's height; they account for roughly one-quarter of the length of the spinal column above the sacrum. Part of the loss in height that accompanies aging results from the decreasing size and resiliency of the intervertebral discs.

Although all vertebrae have many similar characteristics, some regional differences reflect differences in function. The structural differences among the vertebrae are discussed next.

The Cervical Vertebrae

The seven cervical vertebrae extend from the head to the thorax. A typical cervical vertebra is illustrated in Figure 6-12a●. Notice that the body of the vertebra is relatively small compared with the size of the vertebral foramen. At this level the spinal cord still contains most of the axons that connect the brain to the rest of the body. As you continue inferiorly along the vertebral canal, the diameter of the spinal cord decreases, and so does the size of the vertebral foramen. At the same time, the vertebral bodies gradually enlarge, because they must bear more weight.

Distinctive features of a typical cervical vertebra include: (1) an oval, concave vertebral body; (2) a relatively large vertebral foramen; (3) a stumpy spinous process, usually with a notched tip; and (4) round **transverse foramina** within the transverse processes. These foramina protect important blood vessels supplying the brain.

The first two vertebrae have unique characteristics that allow for specialized movements. The **atlas** (C_1) holds up the head, articulating with the occipital condyles of the skull. It is named after Atlas, a figure in Greek mythology who held up the world. The articulation between the occipital condyles and the atlas permits nodding (as when indicating "yes") but prevents twisting. The atlas in turn forms a pivot joint with the **axis** (C_2) through a projection on the axis

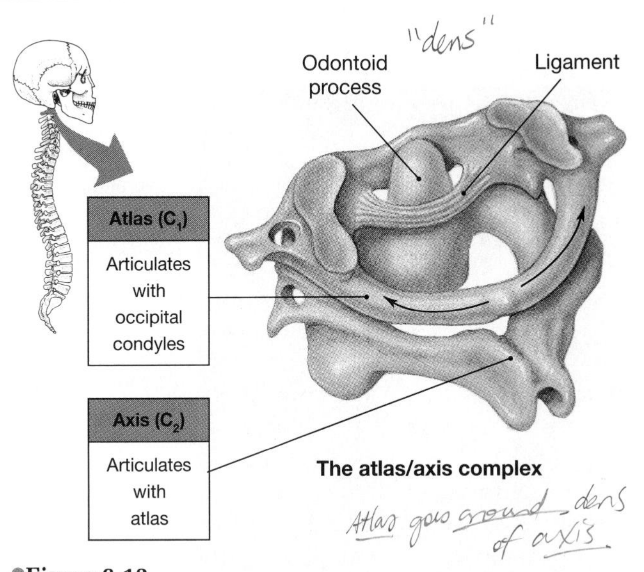

●**Figure 6-13**
The Atlas and Axis
The articulation between the atlas (C_1) and the axis (C_2).

called the **odontoid process**, or *dens* (denz; tooth). This articulation, which permits rotation (as when shaking the head to indicate "no"), is shown in Figure 6-13●.

The Thoracic Vertebrae

There are 12 thoracic vertebrae. Distinctive features of a thoracic vertebra (Figure 6-12b●) include: (1) a distinctive heart-shaped body that is more massive than that of a cervical vertebra; (2) a large, slender spinous process that points inferiorly; and (3) articular surfaces on the body and, in most cases, on the transverse processes for articulation with one or more pairs of ribs.

The Lumbar Vertebrae

The distinctive features of lumbar vertebrae (Figure 6-12c●) include: (1) a vertebral body that is thicker and more oval than that of a thoracic vertebra; (2) a relatively massive, stumpy spinous process that projects posteriorly, providing surface area for the attachment of the lower back muscles; and (3) bladelike transverse processes that lack articulations for ribs.

The lumbar vertebrae are the most massive and least mobile, for they support most of the body weight. As you increase the loading on the vertebrae, the intervertebral discs become increasingly important as shock absorbers. The lumbar discs, which are subjected to the most pressure, are the thickest of all. The lumbar articulations restrict the stresses on the discs by limiting vertebral motion.

Shortly after physical maturity is reached, the gelatinous mass within each disc begins to degenerate, and the "cushion" becomes less effective. Over the same period, the outer fibrocartilage loses its elasticity. If the stresses are sufficient, the inner mass

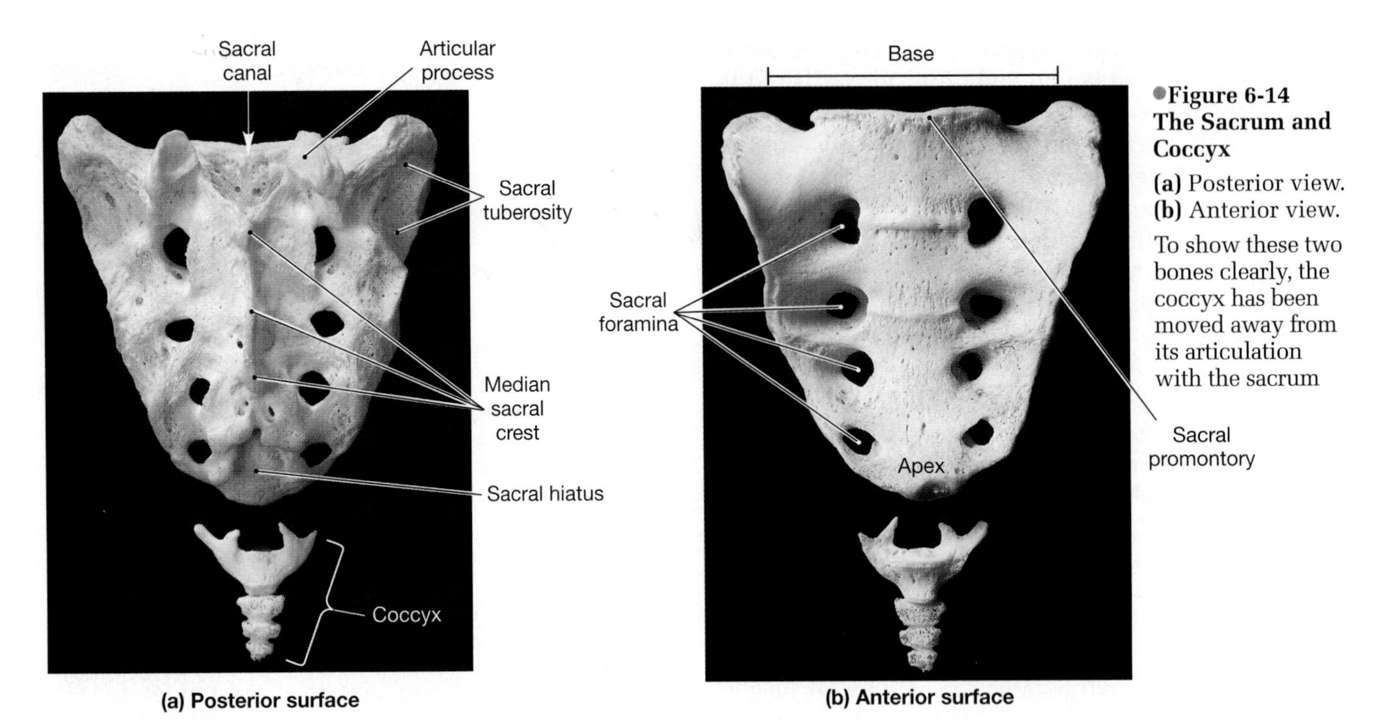

Figure 6-14 The Sacrum and Coccyx
(a) Posterior view. **(b)** Anterior view. To show these two bones clearly, the coccyx has been moved away from its articulation with the sacrum

may break through the surrounding fibrocartilage and protrude beyond the intervertebral space. This condition, called a *herniated disc*, further reduces disc function. The term *slipped disc* is often used to describe this problem, although the disc does not actually slip. AM *Problems with Intervertebral Discs*

The Sacrum and Coccyx

The *sacrum* consists of the fused elements of five sacral vertebrae. This structure protects the reproductive, digestive, and excretory organs and attaches the axial skeleton to the appendicular skeleton by articulation with the pelvic girdle. The broad surface area of the sacrum provides an extensive area for the attachment of muscles, especially those responsible for leg movement. Figure 6-14● shows the posterior and anterior surfaces of the sacrum.

Because the sacrum resembles a triangle, the narrow, caudal portion is called the **apex**, and the broad superior surface is the **base**. The **articular processes** form articulations with the last lumbar vertebra. The **sacral canal** begins between those processes and extends the length of the sacrum. Nerves and membranes that line the vertebral canal in the spinal cord continue into the sacral canal. A prominent bulge at the anterior tip of the base, the **sacral promontory**, is an important landmark during pelvic examinations and during labor and delivery.

Before birth, five vertebrae fuse to form the sacrum, and their spinal processes form a series of elevations along the median sacral crest. The inferior end of the sacral canal is covered by connective tissues. On either side of the sacrum, the **sacral foramina** represent the intervertebral foramina, now enclosed by the fused sacral bones. Along its lateral border, a thickened, flattened area marks the *sacroiliac joint*, the site of articulation with the *coxae* (hip bones). AM *Spina Bifida*

The *coccyx* provides an attachment site for a muscle that closes the anal opening. The 3–5 (most often four) coccygeal vertebrae do not complete their fusion until late in adulthood. In elderly people, the coccyx may also fuse with the sacrum.

The Thorax

The skeleton of the chest, or thorax, consists of the thoracic vertebrae, the ribs, and the sternum. The ribs and the sternum form the *thoracic cage*, or *rib cage*, and establish the contours of the thoracic cavity. The thoracic cage protects the heart, lungs, and other internal organs and serves as a base for muscles involved with respiration.

The Ribs and Sternum Ribs, or *costal bones,* are elongate, flattened bones that originate on or between the thoracic vertebrae and end in the wall of the thoracic cavity. There are 12 pairs of ribs in each sex (Figure 6-15●). The first seven pairs are called **true ribs.** These ribs reach the anterior body wall and are connected to the sternum by separate cartilaginous extensions, the **costal cartilages.** Ribs 8 to 12 are called the **false ribs** because they do not attach directly to the sternum. The costal cartilages of ribs 8 to 10 fuse together. This fused cartilage merges with the costal cartilage of rib 7 before it reaches the sternum. The last two pairs of ribs are called **floating ribs** because they have no connection with the sternum.

The adult sternum has three parts. The broad, triangular **manubrium** (ma-NŪ-brē-um) articulates (1)

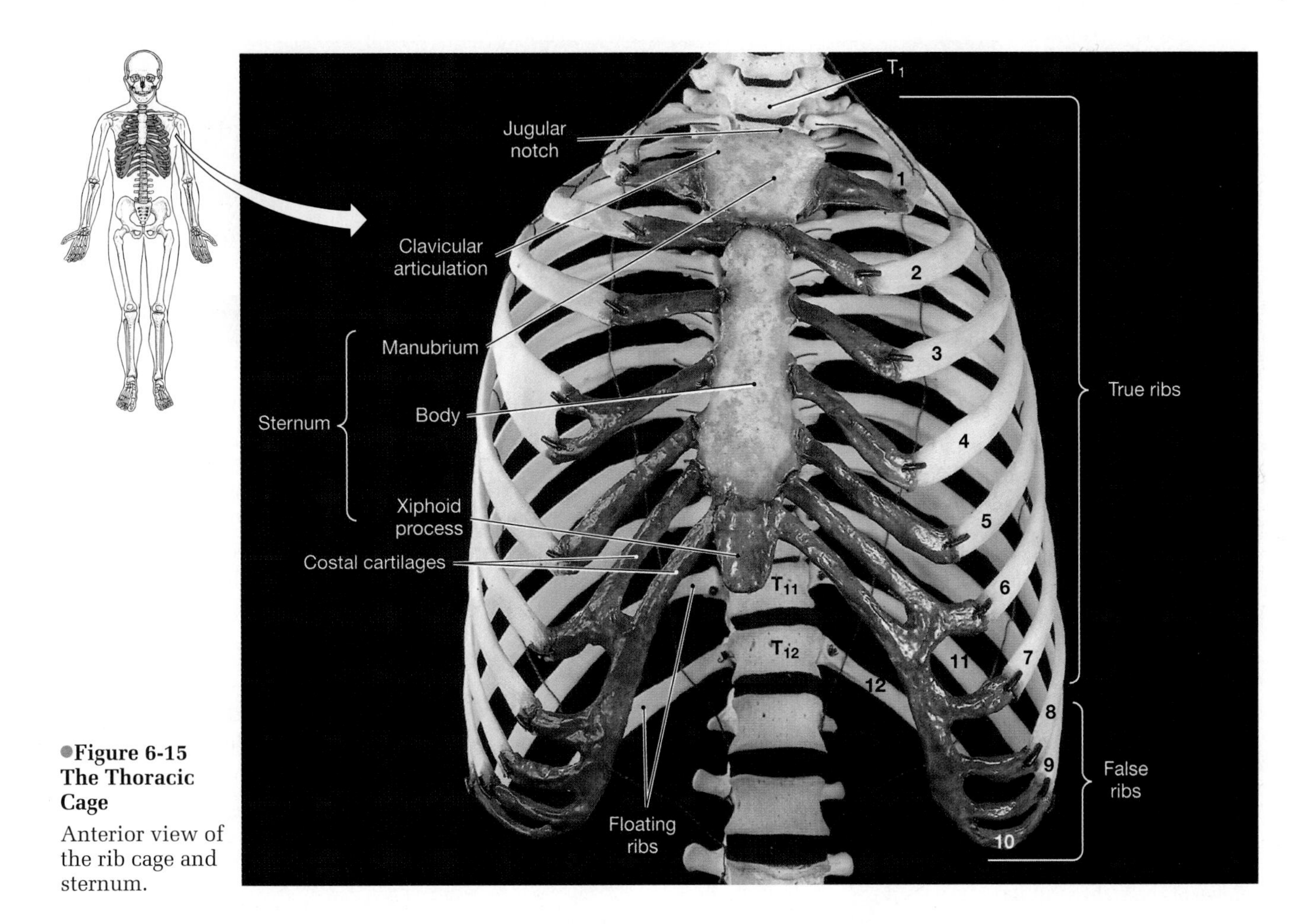

•Figure 6-15 The Thoracic Cage
Anterior view of the rib cage and sternum.

with the clavicles of the appendicular skeleton and (2) with the cartilages of the first pair of ribs. The elongate **body** ends at the slender **xiphoid** (ZĪ-foid) **process**. Ossification of the sternum begins at six to ten different centers, and fusion is not completed until at least age 25. The xiphoid process is usually the last of the sternal components to ossify and fuse. Impact or strong pressure can drive it into the liver, causing severe damage. For this reason, the hand must be properly positioned during cardiopulmonary resuscitation (CPR).

With their complex musculature, dual articulations at the vertebrae, and flexible connection to the sternum, the ribs are quite mobile. Because they are curved, their movements affect both the width and the depth of the thoracic cage, increasing or decreasing its volume.

✓ Joe suffered a hairline fracture at the base of the odontoid process. What bone is fractured and where would you find it?

✓ Improper administration of CPR (cardiopulmonary resuscitation) could result in a fracture of what bone?

✓ In adults, five large vertebrae fuse to form what single structure?

THE APPENDICULAR DIVISION

The appendicular skeleton includes the bones of the upper and lower limbs and the pectoral and pelvic girdles that connect the limbs to the trunk.

The Pectoral Girdle

Each upper limb articulates with the trunk at the **shoulder girdle**, or *pectoral girdle*. The pectoral girdle consists of a broad, flat **scapula** (*shoulder blade*) and the short **clavicle** (*collarbone*). The clavicle articulates with the manubrium of the sternum; this is the *only* direct connection between the pectoral girdle and the axial skeleton. Skeletal muscles support and position the scapula, which has no bony or ligamentous bonds to the thoracic cage.

Movements of the clavicle and scapula position the shoulder joint and provide a base for arm movement. Once the shoulder joint is in position, muscles that originate on the pectoral girdle help to move the arm. The surfaces of the scapula and clavicle are therefore extremely important as sites for muscle attachment.

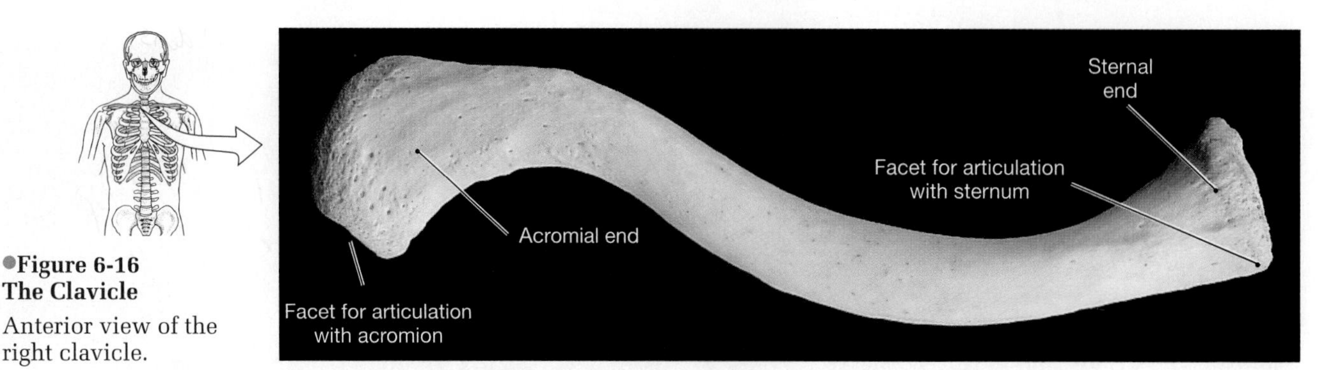

Figure 6-16 The Clavicle
Anterior view of the right clavicle.

The Clavicle

The S-shaped clavicle bone, shown in Figure 6-16, articulates with the manubrium component of the sternum and the *acromion* of the scapula. The smooth superior surface of the clavicle lies just beneath the skin. The rough inferior surface of the acromial end is marked by prominent lines and tubercles that indicate the attachment sites for muscles and ligaments.

The clavicle is the only firm attachment between the axial skeleton and the shoulder girdle and upper limb. It is small and light, and its curved shape makes it relatively fragile. As a result, clavicular fractures are very common injuries.

The clavicle limits the range of motion of the shoulder. People with inherited developmental abnormalities that reduce or eliminate the clavicles have completely mobile scapulae and can swing their shoulders medially almost far enough to meet in front of the sternum.

The Scapula

Figure 6-17 details the anatomy of the right scapula. Its anterior face forms a broad triangle (Figure 6-17a)

Figure 6-17 The Scapula
(a) Anterior, **(b)** lateral, and **(c)** posterior views of the right scapula.

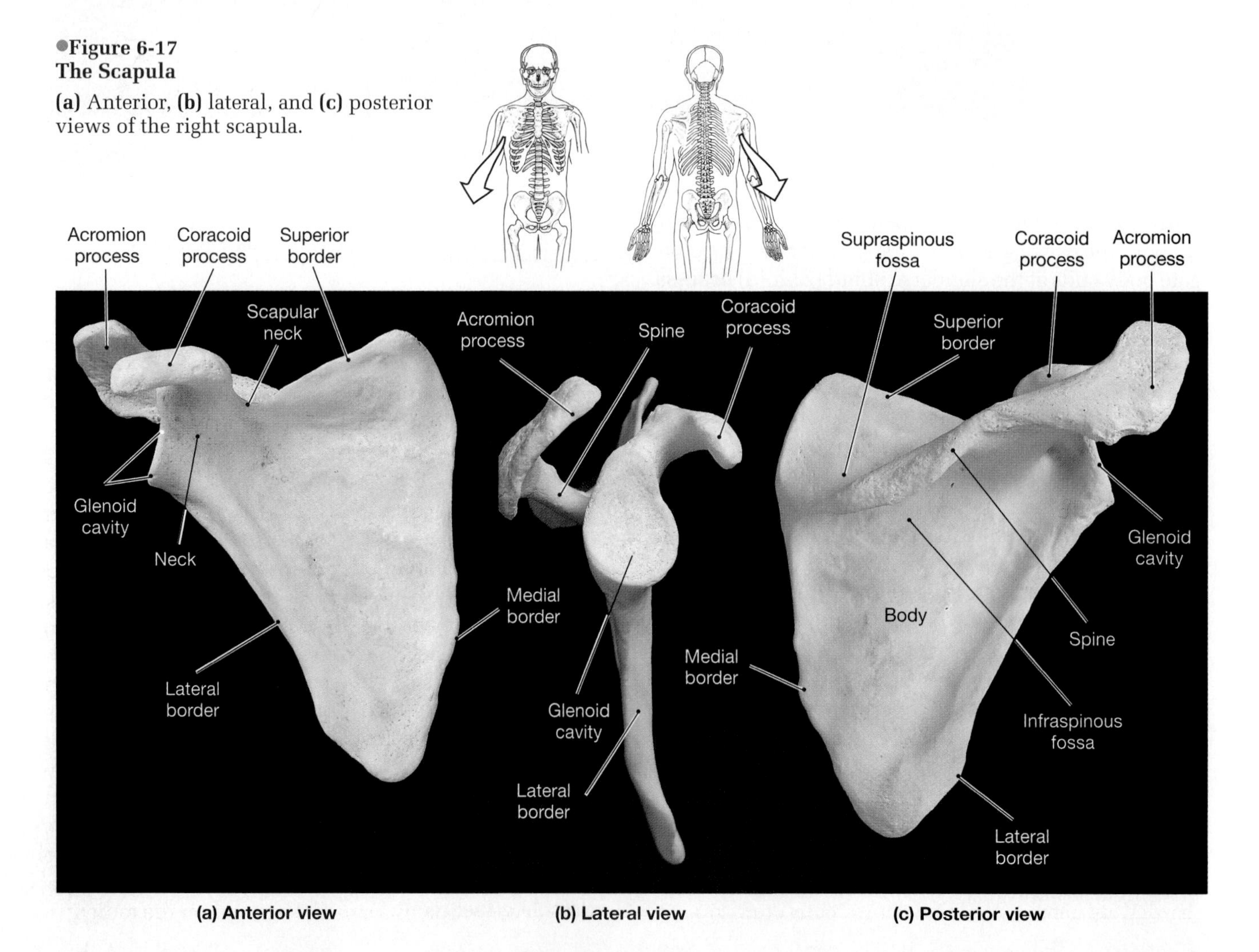

bounded by the **superior, medial**, and **lateral borders**. Muscles that position the scapula attach along these edges. The intersection of the lateral and superior borders thickens into the shallow, cup-shaped **glenoid cavity**, or **glenoid fossa**. At the glenoid cavity, the scapula articulates with the proximal end of the humerus to form the *shoulder joint*. The bone surrounding the glenoid cavity attaches to the body of the scapula at the *scapular neck*.

Figure 6-17b● shows a lateral view of the scapula and the two large processes that extend over the glenoid fossa. The smaller, anterior projection is the **coracoid** (kō-RA-koid) **process**. The **acromion** (a-KRŌ-mē-on) is the larger posterior process. If you run your fingers along the superior surface of the shoulder joint, you will feel this process. The acromion articulates with the distal end of the clavicle.

The **scapular spine** (Figure 6-17c●) divides the posterior surface of the scapula into two regions. The area superior to the spine is the **supraspinous fossa** (*supra-*, above); the *supraspinatus muscle* attaches here. The region below the spine is the **infraspinous fossa** (*infra-*, beneath), home of the *infraspinatus muscle*. Both muscles are attached to the humerus, the proximal bone of the upper limb.

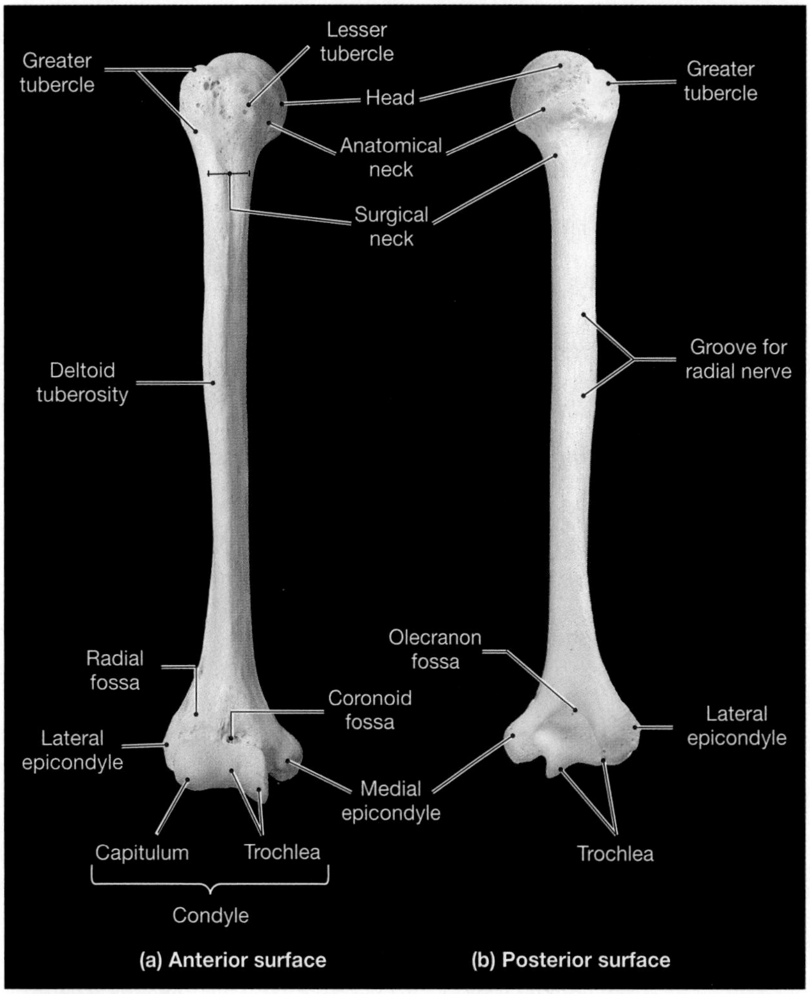

●**Figure 6-18**
The Humerus
Major landmarks on **(a)** the anterior and **(b)** the posterior surface of the right humerus.

The Upper Limb

The upper limb consists of the arm and forearm. The arm contains a single bone, the **humerus**, which extends from the scapula to the elbow. At its proximal end, the round **head** of the humerus articulates with the scapula. At its distal end, it articulates with the bones of the forearm, the *radius* and *ulna*.

The Humerus

Figure 6-18● illustrates the anatomy of the *humerus*. The prominent **greater tubercle**, located near the rounded head, establishes the contour of the shoulder (Figure 6-18a●). The **lesser tubercle** lies more anteriorly, separated from the greater tubercle by a deep groove. Muscles are attached to both tubercles, and a large tendon runs along the groove. The *anatomical neck* lies between the tubercles and below the surface of the head. Distal to the tubercles, the narrow *surgical neck* corresponds to the region of growing bone. It earned its name by being a common fracture site.

The proximal **shaft** of the humerus is round in section. The elevated **deltoid tuberosity** that runs along the lateral border of the shaft is named after the *deltoid muscle* that attaches to it.

Distally, the posterior surface of the shaft flattens and the humerus expands to either side, forming a broad triangle (Figure 6-18b●). **Medial** and **lateral epicondyles** project to either side, providing additional surface area for muscle attachment, and the smooth, articular **condyle** dominates the inferior surface of the humerus.

A low ridge crosses the condyle, dividing it into two distinct regions. The **trochlea** is the large medial portion shaped like a spool or pulley (*trochlea* is the Latin word for a pulley). The trochlea extends from the base of the **coronoid fossa** (KŌR-ō-noyd; *corona*, crown) on the anterior surface to the **olecranon fossa** on the posterior surface. These depressions accept projections from the surface of the ulna as the elbow reaches its limits of motion. The **capitulum** forms the lateral region of the condyle. A shallow **radial fossa** proximal to the capitulum accommodates a small projection on the radius.

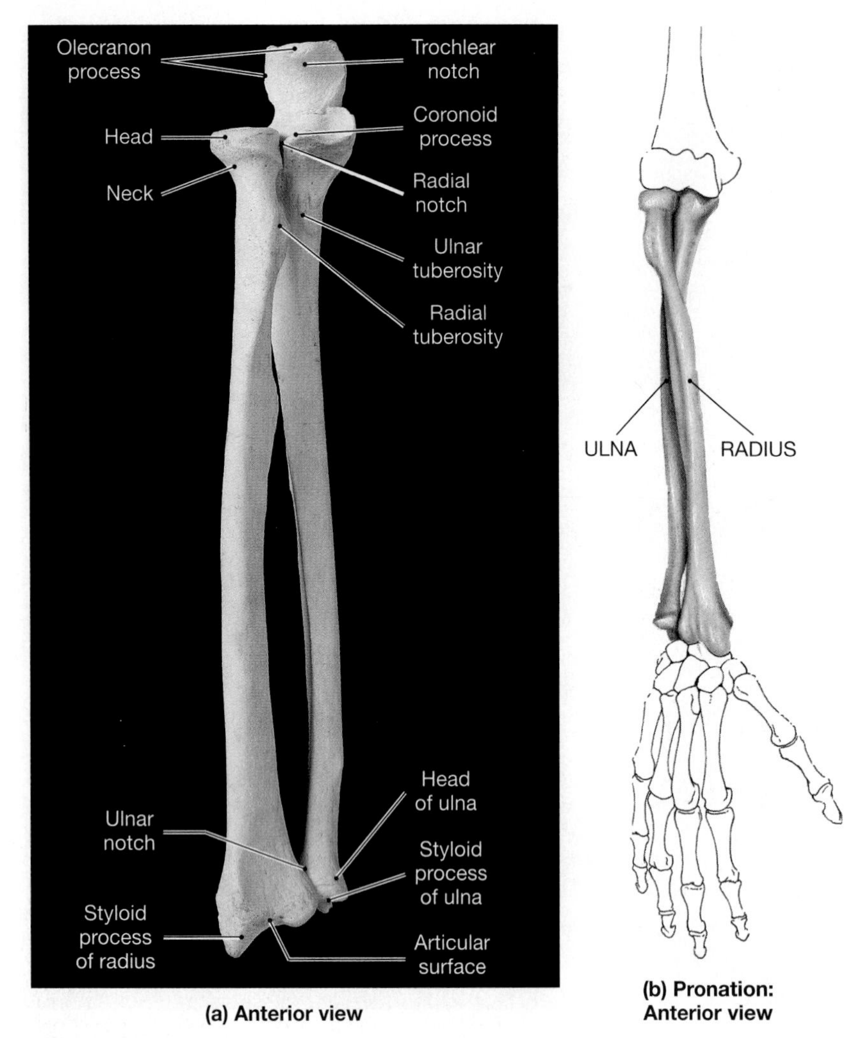

•**Figure 6-19**
The Radius and Ulna

(a) The radius and ulna are shown in anterior view. **(b)** Note the changes that occur during pronation.

The Radius and Ulna

The **radius** and **ulna** are the bones of the forearm. In the anatomical position, the radius lies along the lateral (thumb) side of the forearm. The ulna forms the medial support of the forearm. The structure of these bones can be seen in Figure 6-19a•.

The elbow is formed by the superior projection of the ulna, the **olecranon** (ō-LEK-ra-non) **process**. On its anterior surface, the **trochlear notch** articulates with the trochlea of the humerus to form the elbow joint. The olecranon process forms the superior lip of the notch, and the **coronoid process** provides a prominent inferior margin. When the elbow joint is fully extended, the arm and forearm form a straight line, and the olecranon process swings into the olecranon fossa on the posterior face of the humerus. A muscle that attaches to the ulna at the **ulnar tuberosity** swings the forearm toward the arm, a movement called *flexion*. At the limit of motion, the coronoid process projects into the coronoid fossa on the anterior surface of the humerus.

Lateral to the coronoid process, a smooth **radial notch** accommodates the head of the radius. A fibrous sheet connects the lateral margin of the ulna to the radius along its length. The ulnar shaft ends at a disc-shaped head whose posterior margin supports a short **styloid process**. The distal end of the ulna is separated from the wrist joint by a pad of cartilage, and only the expansive distal portion of the radius participates in the wrist joint. The **styloid process** of the radius assists in the stabilization of the joint by preventing lateral movement of the bones of the wrist (*carpals*).

Near the elbow, a prominent **radial tuberosity** marks the attachment site of another powerful muscle, the *biceps brachii*, that flexes the forearm. A narrow *neck* extends from the tuberosity to the head of the radius. The disc-shaped head articulates with the capitulum of the humerus at the elbow joint and with the ulna at the radial notch. This proximal articulation with the ulna allows the radius to roll across the ulna, rotating the palm in a movement known as *pronation* (Figure 6-19b•).

The Wrist and Hand

There are 27 bones in the hand (Figure 6-20•), supporting the wrist, palm, and fingers. The eight bones of the wrist, or **carpus**, form two rows. There are four proximal **carpals**: (1) *scaphoid*, (2) *lunate*, (3) *triangular* (or *triquetral*), and (4) *pisiform* (PI-si-form). There are also four distal carpals: (1) *trapezium*, (2) *trapezoid*, (3) *capitate*, and (4) *hamate*. A fibrous capsule, reinforced by broad ligaments, surrounds the wrist complex and stabilizes the positions of the individual carpals.

There are 14 phalangeal bones in each hand. Five **metacarpals** (met-a-KAR-pals) articulate with the distal carpals and form the palm of the hand. The metacarpals in turn articulate with the finger bones, or **phalanges** (fa-LAN-jēz). Four of the fingers contain three phalanges (proximal, middle, and distal), but the thumb, or *pollex*, has only two (proximal and distal).

✓ Why would a broken clavicle affect the mobility of the scapula?

✓ The rounded projections on either side of the elbow are parts of what bone?

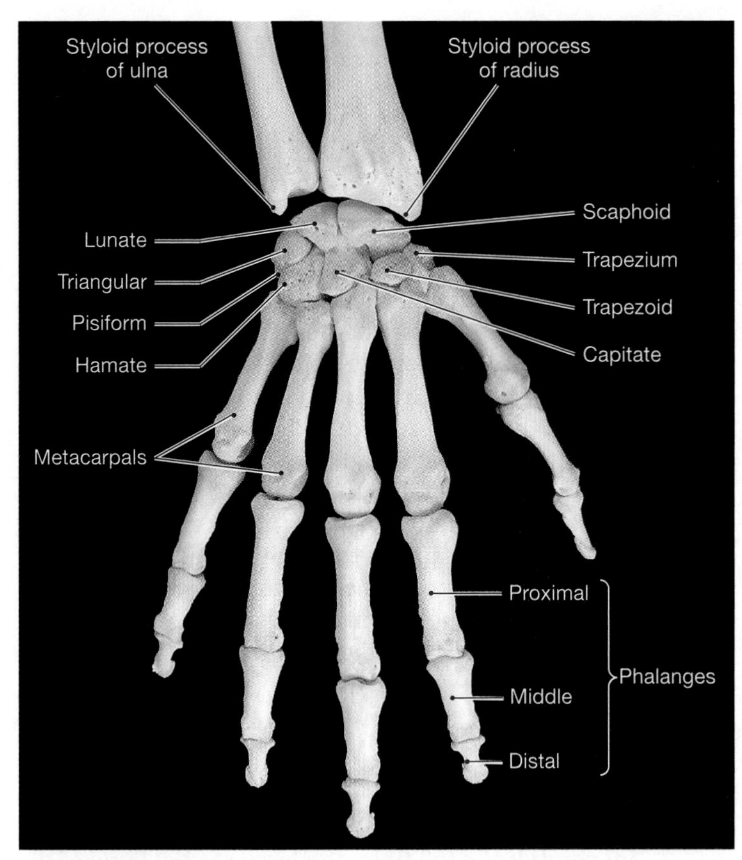

●**Figure 6-20**
Bones of the Wrist and Hand, Posterior View

The Pelvic Girdle

The *pelvic girdle* articulates with the thigh bones. Because of the stresses involved in weight bearing and locomotion, the bones of the pelvic girdle and lower limbs are more massive than those of the pectoral complex. The pelvic girdle is also much more firmly attached to the axial skeleton. Dorsally, the two halves of the pelvic girdle contact the lateral surfaces of the sacrum. Ventrally, the pelvic elements are interconnected by a fibrocartilage pad.

The pelvic girdle consists of two large hip bones, or **coxae** (Figure 6-21a●). Each coxa forms through the fusion of three bones, an **ilium** (IL-ē-um), an **ischium** (IS-kē-um), and a **pubis** (PYŪ-bis). Dorsally the hipbones articulate with the sacrum at the **sacroiliac joint**. Ventrally the coxae are connected at the *pubic symphysis*. At the hip joint on either side, the head of the femur (thighbone) articulates with the curved surface of the **acetabulum** (a-se-TAB-ū-lum; *acetabulum*, a vinegar cup) (Figure 6-21b●).

The Coxa

The *ilium* is the most superior and largest coxal bone. Above the acetabulum, the ilium forms a broad, curved surface that provides an extensive area for the attachment of muscles, tendons, and ligaments. The superior margin of the ilium, the **iliac crest,** marks the sites of attachments of both ligaments and muscles. Near the superior and posterior margin of the acetabulum, the ilium fuses with the *ischium*. The roughened inferior surface of the ischium supports the body's weight when sitting.

The fusion of a narrow branch of the ischium with a branch of the *pubis* completes the encirclement of the **obturator** (OB-tū-rā-tor) **foramen**. This space is closed by a sheet of collagen fibers whose inner and

●**Figure 6-21**
The Pelvis
(a) Components of the pelvis. **(b)** Anterior view of the pelvis.

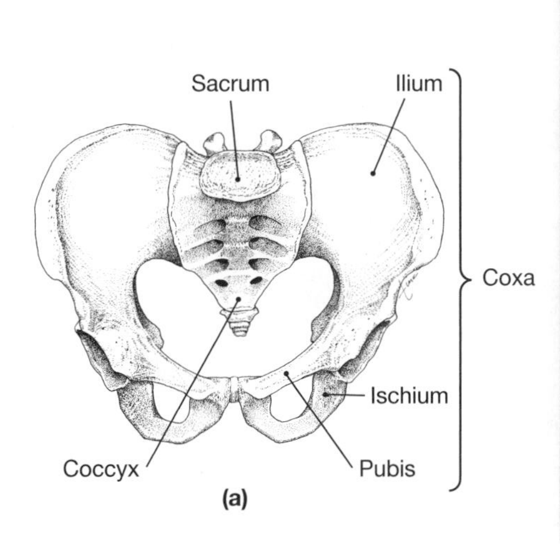

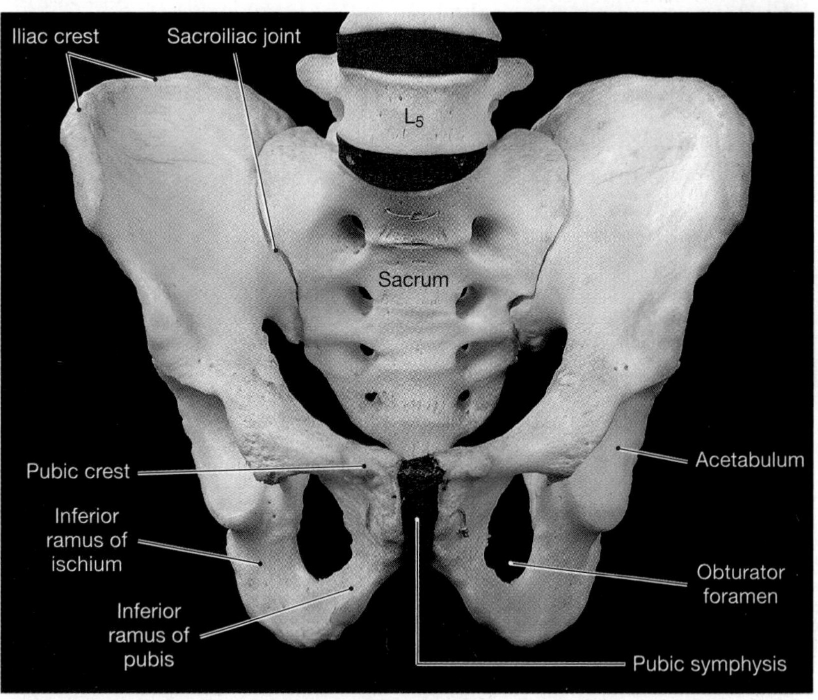

(b) Anterior view

outer surfaces provide a firm base for the attachment of muscles and visceral structures.

The anterior and medial surface of the pubis contains a roughened area that marks the **pubic symphysis**, an articulation with the pubis of the opposite side. The pubic symphysis limits movement between the two pubic bones.

The Pelvis

The **pelvis** consists of the coxae, the sacrum, and the coccyx (see Figure 6-21a●). It is thus a composite structure that includes portions of both the appendicular and axial skeletons. An extensive network of ligaments connects the lateral borders of the sacrum with the iliac crests, the inferior surfaces of the ischia, and the superior border of the pubic bones. Other ligaments tie the ilia to the posterior lumbar vertebrae. These interconnections increase the structural stability of the pelvis.

The shape of the pelvis of a female is somewhat different from that of a male (Figure 6-21c●). Some of the differences are the result of variations in body size and muscle mass. Others are adaptations for childbearing and are necessary (1) to support the weight of the developing fetus and (2) to ease passage of the newborn through the pelvis during delivery.

The Lower Limb

The skeleton of the lower limb includes (1) the *femur*, the bone of the thigh, (2) the *tibia* and *fibula*, the bones of the leg, and (3) the bones of the ankle and foot.

The Femur

The **femur,** or *thighbone,* is the longest, heaviest, and strongest bone in the body (Figure 6-22●). Distally, the femur articulates with the tibia of the leg at the knee joint. The rounded epiphysis, or head, of the femur articulates with the pelvis at the acetabulum. The **greater trochanter** arises lateral to the juncture of the neck and the shaft; the **lesser trochanter** originates along the crest near the medial surface of the femur. Both trochanters develop where large tendons attach to the femoral shaft. On the posterior surface of the femur, a stout ridge, the **linea aspera**, marks the attachment of powerful muscles that pull the shaft of the femur toward the midline, a movement called *adduction* (*ad-*, toward + *duco*, to lead).

The proximal femoral shaft is round in cross-section. Moving distally, the shaft becomes more flattened and ends in two large **articular condyles.** A pair of **epicondyles** bulge to either side of the round condylar surfaces. The articular condyles merge anteriorly to produce an articular surface with elevated lateral borders. This is the **patellar surface** over which the **patella** (*kneecap*) glides.

The Tibia and Fibula

Figure 6-23● shows the structure of the **tibia** and **fibula**. The condyles of the femur articulate with the **tibial condyles** of the **tibia**, the large medial bone of the lower leg, also known as the shinbone. A ligament from the patella attaches to the **tibial tuberosity** just below the knee joint.

A projecting **anterior crest** extends almost the entire length of the anterior surface. The tibia broadens at its distal end into a large process, the **medial malleolus** (ma-LĒ-o-lus; *malleolus*, hammer). The inferior surface of the tibia forms a joint with the proximal bone of the ankle; the medial malleolus provides medial support for the ankle.

The slender **fibula** parallels the lateral border of the tibia. The fibular **head** articulates along the lateral margin of the tibia, inferior and slightly posterior to

Figure 6-21 (*continued*)
(c) Anatomical differences between the pelvis of a male and that of a female.

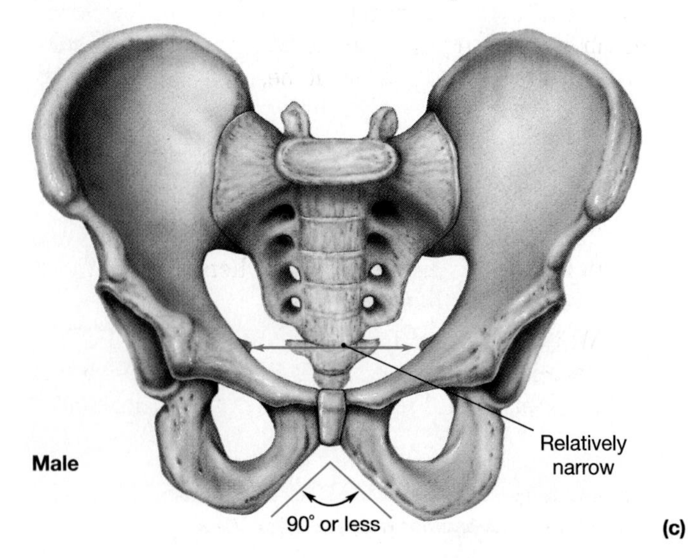

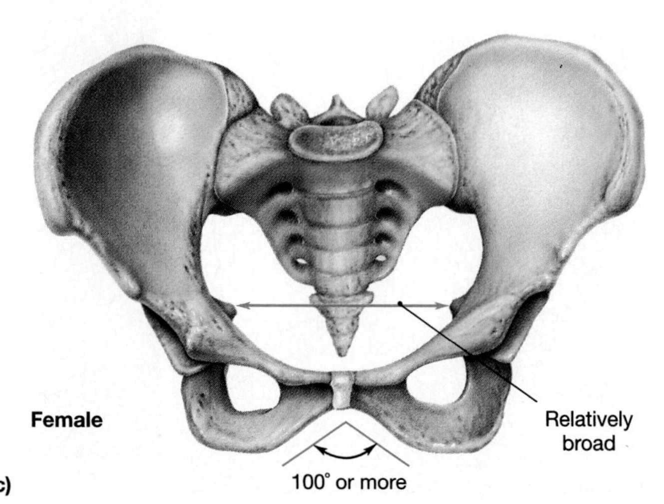

(c)

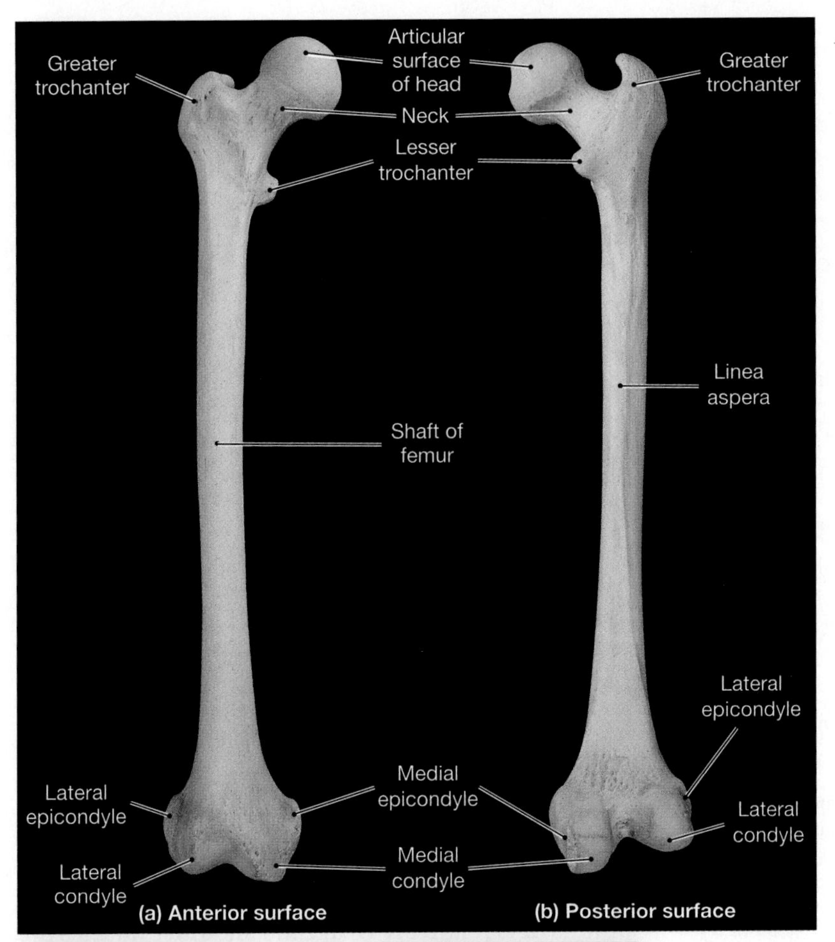

●**Figure 6-22**
The Femur.

Bone markings on the right femur are presented as seen from **(a)** the anterior and **(b)** the posterior surface.

the lateral condyle. The fibula does not participate in the knee joint, and it does not bear weight. However, it is an important surface for muscle attachment, and the distal **lateral malleolus** provides lateral stability to the ankle. A fibrous membrane extending between the two bones helps stabilize their relative positions and provides additional surface area for muscle attachment.

The Ankle and Foot

The ankle, or **tarsus**, includes seven separate **tarsal bones**: the (1) *talus,* (2) *calcaneus,* (3) *navicular,* (4) *cuboid,* (5) *first cuneiform*, (6) *second cuneiform*, and (7) *third cuneiform* (Figure 6-24●). Only the proximal tarsal, the **talus**, articulates with the tibia and fibula. The talus then passes the weight to the ground via other bones of the foot.

When you are standing normally, most of your weight is transmitted to the ground through the talus to the large **calcaneus** (kal-KĀ-nē-us), or *heel bone.* The posterior projection of the calcaneus receives the composite **calcaneal tendon**, or *Achilles tendon,* of the calf muscles that raise the heel and depress the sole (plantar flexion). The rest of the body weight is passed through the cuboid and cuneiforms to the **metatarsals**, which support the sole of the foot.

The basic organizational pattern at the metatarsals and phalanges of the foot resembles that of the hand. The metatarsals are numbered 1 to 5 from medial to lateral, and their distal ends form the ball of the foot. The same number of phalanges present in the thumb (2) and fingers (3 each) also make up the great toe, or *hallux*, and other toes. AM *Problems with the Ankle and Foot*

✓ What three bones make up the coxa?

✓ The fibula does not participate in the knee joint nor does it bear weight, but when it is fractured walking is difficult. Why?

✓ While jumping off the back steps at his house, 10-year-old Joey lands on his right heel and breaks his foot. What foot bone is most likely broken?

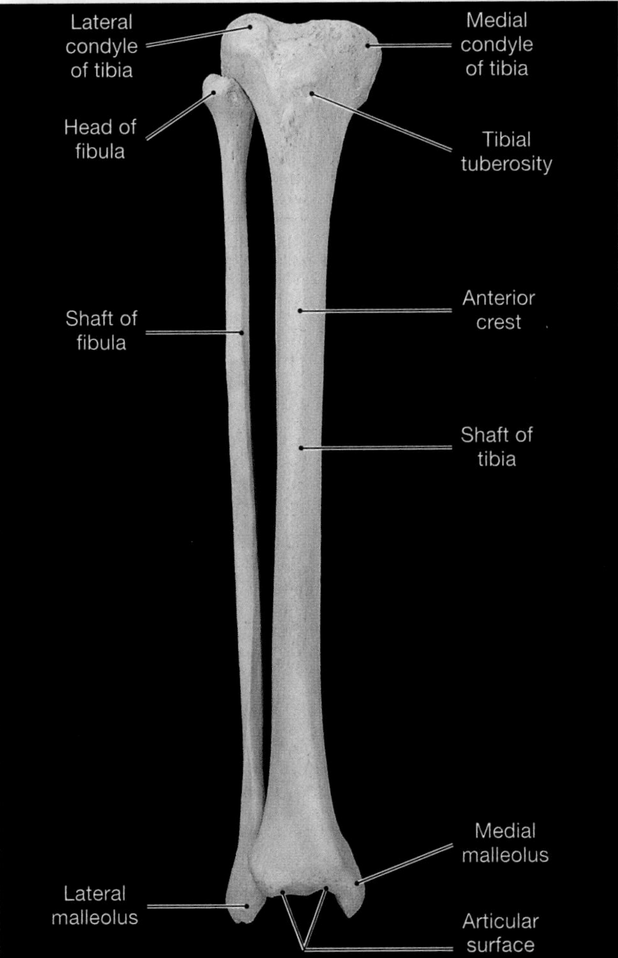

●**Figure 6-23**
The Right Tibia and Fibula, Anterior View

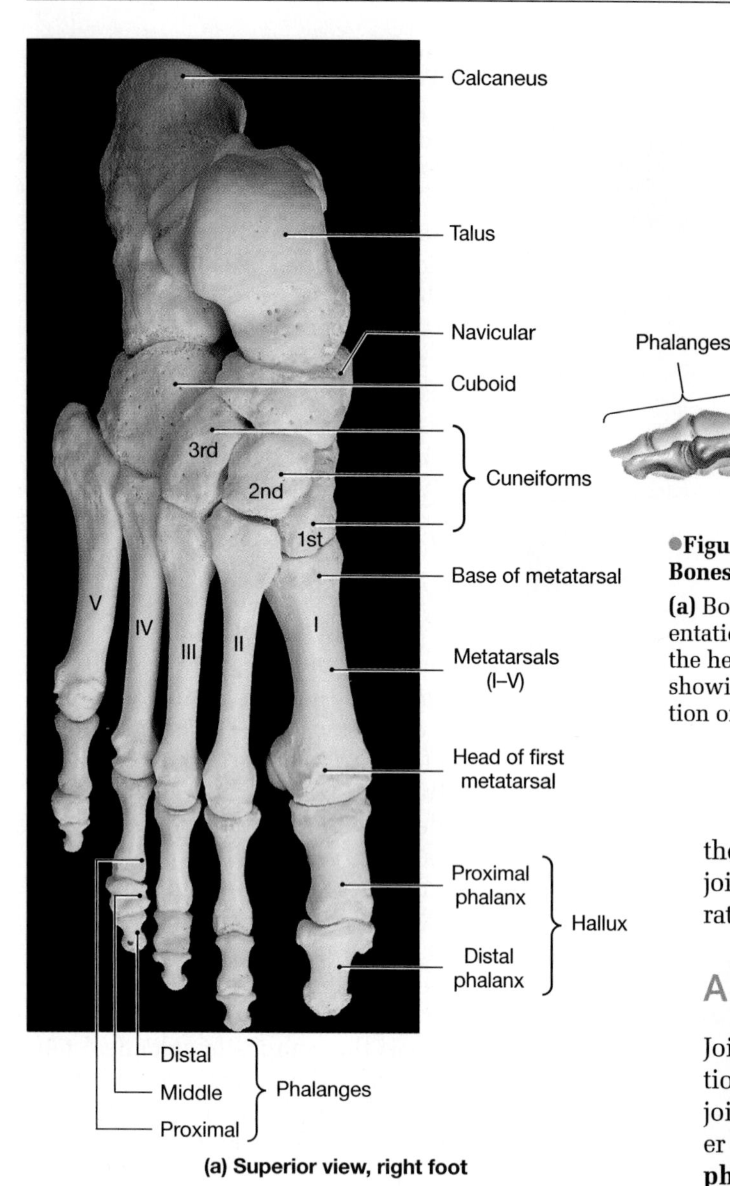

Figure 6-24
Bones of the Ankle and Foot
(a) Bones of the right foot as viewed from above. Note the orientation of the tarsals, which convey the weight of the body to the heel and the plantar surfaces of the foot. **(b)** Medial view, showing the relative positions of the tarsals and the orientation of the foot bones.

ARTICULATIONS

Joints, or **articulations**, exist wherever two bones meet. The function of each joint depends on its anatomy. Each joint reflects a workable compromise between the need for strength and the need for mobility. When movement is not required, or when relative movement could actually be dangerous, joints can be very strong. For example, joints such as the sutures of the skull are so intricate and extensive that they lock the elements together as if they were a single bone. This rigidity is crucial because the sutures weld the cranium into a solid case that encloses and protects the delicate tissues of the brain. At other joints, movement is more important than strength. The interconnections can then be less extensive, and the joint correspondingly weaker. For example, the articulation at the shoulder permits a range of arm movement that is limited more by the surrounding muscles than by joint structure. The joint itself is relatively weak, and shoulder injuries are rather common.

A Classification of Joints

Joints can be classified according to the range of motion they permit (see Table 6-2, p. 138). An immovable joint is a **synarthrosis** (sin-ar-THRŌ-sis; *syn-*, together + *arthros*, joint); a slightly movable joint is an **amphiarthrosis** (am-fē-ar-THRŌ-sis; *amphi-*, on both sides); and a freely movable joint is a **diarthrosis** (dī-ar-THRŌ-sis; *dia-*, through).

Subdivisions are further recognized within each of these three major categories. Synarthrotic or amphiarthrotic joints are classified according to the type of connective tissue binding them together, such as *fibrous* or *cartilaginous*. Diarthrotic joints, however, are subdivided according to the types, or ranges, of movement permitted.

Immovable Joints (Synarthroses)

At a synarthrosis, the bony edges are quite close together and may even interlock. A **suture** (*sutura*, a sewing together) is a synarthrotic joint between the bones of the skull. The edges of the bones are interlocked and bound together by dense connective tissue. In another synarthrosis, called a **gomphosis** (gom-FŌ-sis; *gomphosis*, a bolting together), a ligament binds each tooth in the mouth within a bony socket (*alveolus*).

An epiphyseal plate also represents an articulation between two bones, even though the two are part of the same skeletal element. Such a rigid, cartilaginous connection characterizes a **synchondrosis** (sin-kon-DRŌ-sis; *syn*, together + *chondros*, cartilage).

Slightly Movable Joints (Amphiarthroses)

An amphiarthrosis permits very limited movement, and the bones are usually farther apart than they are at a synarthrosis. The bones may be connected by collagen fibers or cartilage. At a **syndesmosis** (sin-dez-MŌ-sis; *desmos*, a band or ligament) they are connected by a ligament. Examples include the articulations between the two bones of the leg, the tibia and fibula. At a **symphysis** the bones are separated by a broad disc or pad of fibrocartilage. The articulations between the spinal vertebrae and the anterior connection between the two pelvic bones, or coxae, are examples of symphyses.

Freely Movable Joints (Diarthroses)

Diarthroses, or **synovial** (si-NŌ-vē-al) **joints**, permit a wide range of motion. The basic structure of a synovial joint was introduced in Chapter 4 during our discussion of synovial membranes. ∞ *[p. 88]* Figure 6-25a● details the structure of a representative synovial joint.

Synovial joints are typically found at the ends of long bones, such as those of the arms and legs. Under normal conditions the bony surfaces do not contact one another, for they are covered with special **articular cartilages**. The joint is surrounded by a fibrous **joint (articular) capsule**, and the inner surfaces of the joint cavity are lined with a synovial membrane. **Synovial fluid** diffuses across the synovial membrane and provides lubrication that reduces the friction between the moving surfaces in the joint.

In complex joints such as the knee, additional padding lies between the opposing articular surfaces. An example of such shock-absorbing, fibrocartilage pads are the **menisci** (men-IS-kē; *meniscus*, crescent), shown in Figure 6-25b●. Also present in such joints are **fat pads**, which protect the articular cartilages and act as packing material. When the bones move, the fat pads fill in the spaces created as the joint cavity changes shape.

The joint capsule that surrounds the entire joint is continuous with the periostea of the articulating bones. In addition, **ligaments** that join bone to bone may be found on the outside or inside the joint capsule. Where a tendon or ligament rubs against other tissues, **bursae,** small pockets containing synovial fluid, form to reduce friction and act as shock absorbers. Bursae are characteristic of many synovial joints and may also appear around tendon sheaths, covering a bone, or within other connective tissues exposed to friction or pressure.

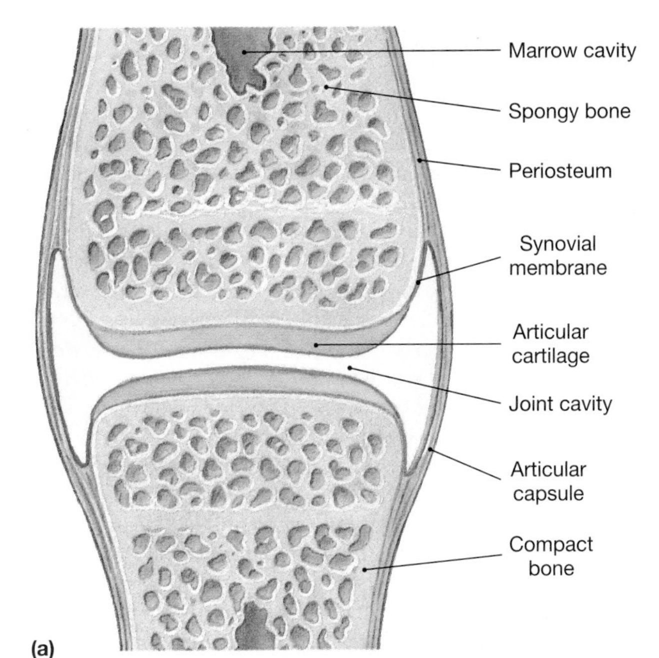

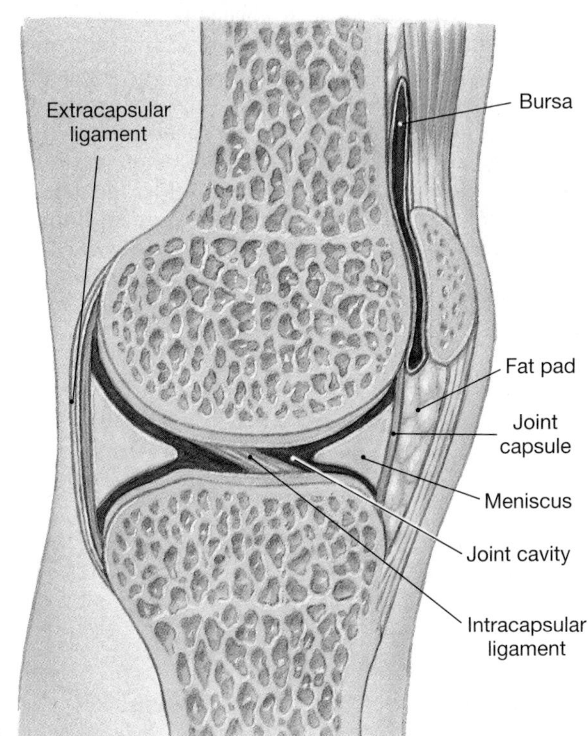

●**Figure 6-25**
The Structure of a Synovial Joint
(a) Diagrammatic view of a simple articulation. **(b)** A sectional view of the knee joint.

Articular Form and Function

In discussions of motion at synovial joints, phrases such as "bend the leg" or "raise the arm" are not sufficiently precise. Anatomists use descriptive terms that have specific meanings. We will consider these movements with regard to the basic categories of movement considered above.

RHEUMATISM AND ARTHRITIS

Rheumatism (ROO-ma-tizm) is a general term that indicates pain and stiffness affecting the skeletal or muscular systems or both. There are several major forms of rheumatism. **Arthritis** (ar-THRĪ-tis) includes all of the rheumatic diseases that affect synovial joints. Arthritis always involves damage to the articular cartilages, but the specific cause may vary. For example, arthritis can result from bacterial or viral infection, injury to the joint, metabolic problems, or severe physical stresses.

Osteoarthritis (os-tē-ō-ar-THRĪ-tis), also known as *degenerative arthritis,* or *degenerative joint disease (DJD),* usually affects individuals age 60 or older. DJD may result from cumulative wear and tear at the joint surfaces or from genetic factors affecting collagen formation. In the U.S. population, 25 percent of women and 15 percent of men over 60 years of age show signs of this disease. *Rheumatoid arthritis* is an inflammatory condition that affects roughly 2.5 percent of the adult population. At least some cases result when the immune response mistakenly attacks the joint tissues. Allergies, bacteria, viruses, and genetic factors have all been proposed as contributing to or triggering the destructive inflammation.

Regular exercise, physical therapy, and drugs that reduce inflammation, such as aspirin, can slow the progress of the disease. Surgical procedures can realign or redesign the affected joint, and in extreme cases involving the hip, knee, elbow, or shoulder the defective joint can be replaced by an artificial one. Additional information concerning the various forms of arthritis can be found in the *Applications Manual.* AM *Rheumatism, Arthritis, and Synovial Function*

Types of Movement

Gliding In **gliding**, two opposing surfaces slide past one another. Gliding occurs between the surfaces of articulating carpals and tarsals, and between the clavicles and the sternum. The movement can occur in almost any direction, but the amount of movement is slight and rotation is usually prevented by the capsule and associated ligaments.

Angular Motion Examples of angular motion include *flexion, extension, adduction,* and *abduction.* The description of each movement is based on reference to an individual in the anatomical position.

Flexion/Extension. **Flexion** (FLEK-shun) can be defined as movement in the anterior-posterior plane that *reduces the angle between the articulating elements.* **Extension** occurs in the same plane, but it *increases the angle between articulating elements* (Figure 6-26a●). When you bring your head toward your chest, you flex the head. When you bend down to touch your toes, you flex the spine. Extension reverses these movements.

Flexion at the shoulder or hip moves the limbs forward, whereas extension moves them back. Flexion of the wrist moves the palm forward, and extension moves it back. In each of these examples, extension can be continued past the anatomical position, in which case **hyperextension** occurs. You can also hyperextend the

TABLE 6-2 A Functional Classification of Articulations

Functional Category	*StructuralÅ Category*	*Description*	*Example*
SYNARTHROSIS (no movement)	**Fibrous**		
	Suture	Fibrous connections plus interdigitation	Between the bones of the skull
	Gomphosis	Fibrous connections plus insertion in alveolus	Between the teeth and jaws
	Cartilaginous		
	Synchondrosis	Interposition of cartilage plate	Epiphyseal plates
AMPHIARTHROSIS (little movement)	**Fibrous**		
	Syndesmosis	Ligamentous connection	Between the tibia and fibula
	Cartilaginous		
	Symphysis	Connection by a fibrocartilage pad	Between right and left halves of pelvis; between adjacent vertebrae of spinal column
DIARTHROSIS (free movement)	**Synovial**	Complex joint bounded by joint capsule and containing synovial fluid	Numerous; subdivided by range of movement

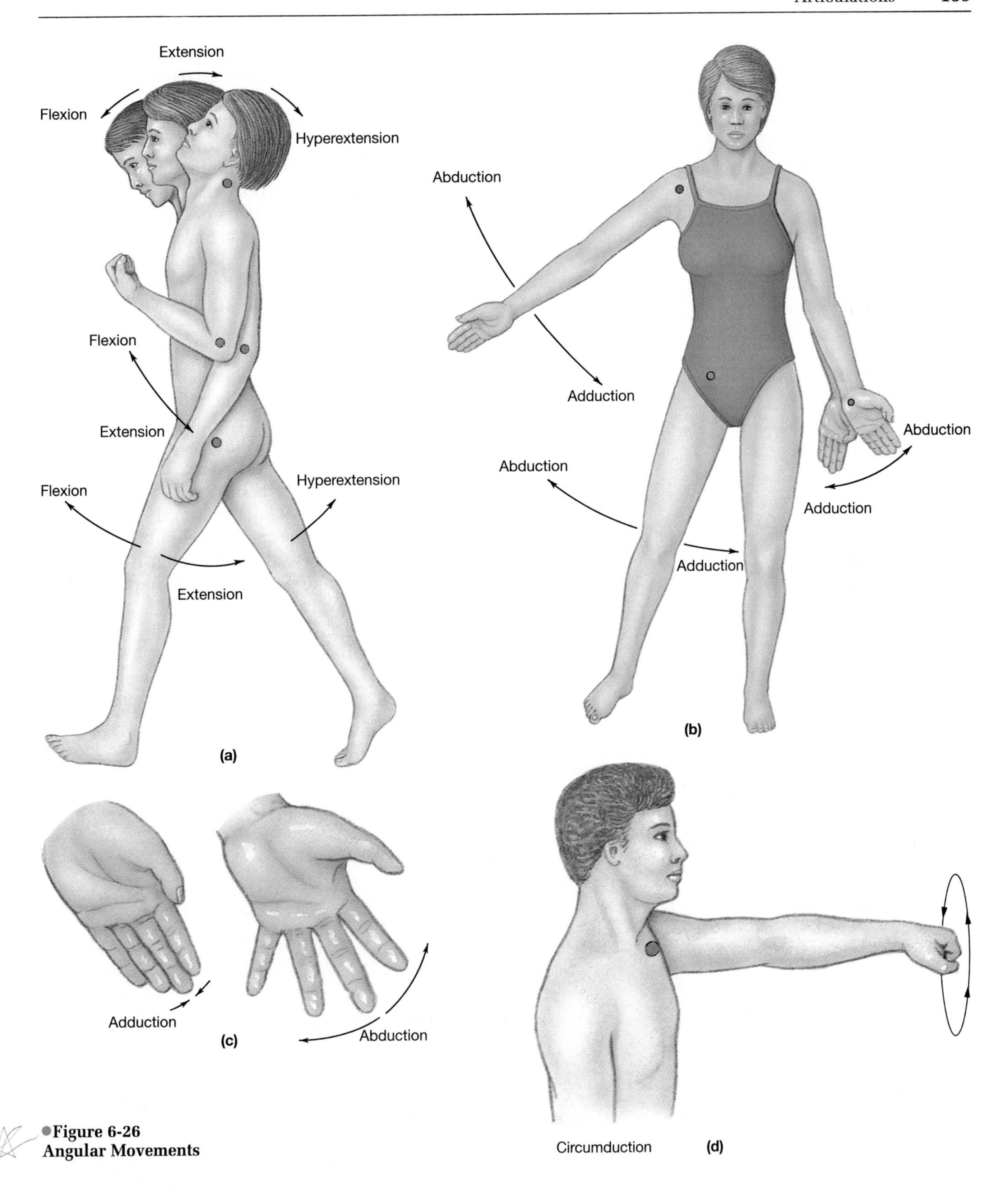

●**Figure 6-26**
Angular Movements

head, a movement that allows you to gaze at the ceiling. Hyperextension of other joints is usually prevented by ligaments, bony processes, or soft tissues.

Abduction/Adduction. **Abduction** (*ab-*, from) is movement *away from the longitudinal axis of the body* in the frontal plane. For example, swinging the upper limb to the side is abduction of the limb (Figure 6-26b●); moving it back constitutes **adduction**. Adduction of the wrist moves the heel of the hand toward the body, whereas abduction moves it farther away. Spreading the fingers or toes apart abducts them, because they move *away* from a central digit (finger or toe), as in Figure 6-26c●. Bringing them together con-

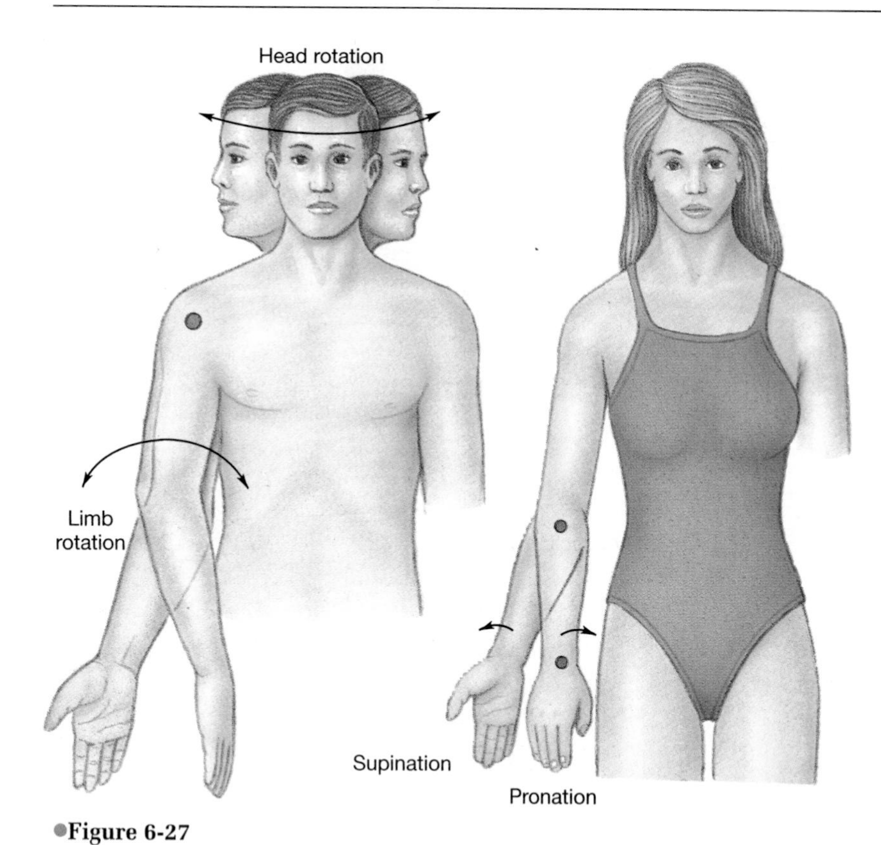

●**Figure 6-27**
Rotational Movements

stitutes adduction. Abduction and adduction always refer to movements of the appendicular skeleton.

Circumduction. A special type of angular motion, **circumduction** (*circum*, around), is shown in Figure 6-26d●. A familiar example of circumduction is moving your arm in a loop, as when drawing a large circle on a chalkboard.

Rotation Rotational movements are also described with reference to a figure in the anatomical position. Rotation involves turning around the longitudinal axis of the body or limb. For example, you may rotate your head to look to one side and rotate your arm to screw in a light bulb. Rotational movements are illustrated in Figure 6-27●.

Pronation/Supination. The articulations between the radius and ulna permit the rotation of the distal end of the radius across the anterior surface of the ulna. This rotation moves the wrist and hand from palm-facing-front to palm-facing-back. This motion is called **pronation** (prō-NĀ-shun). The opposing movement in which the palm is turned forward, is **supination** (su-pi-NĀ-shun).

Special Movements A number of special terms apply to specific articulations or unusual types of movement (Figure 6-28●).

Inversion/Eversion. **Inversion** (*in-*, into + *vertere*, to turn), is a twisting motion of the foot that turns the sole inward. The opposite movement is called **eversion** (ē-VER-shun; *e-*, out).

●**Figure 6-28**
Special Movements

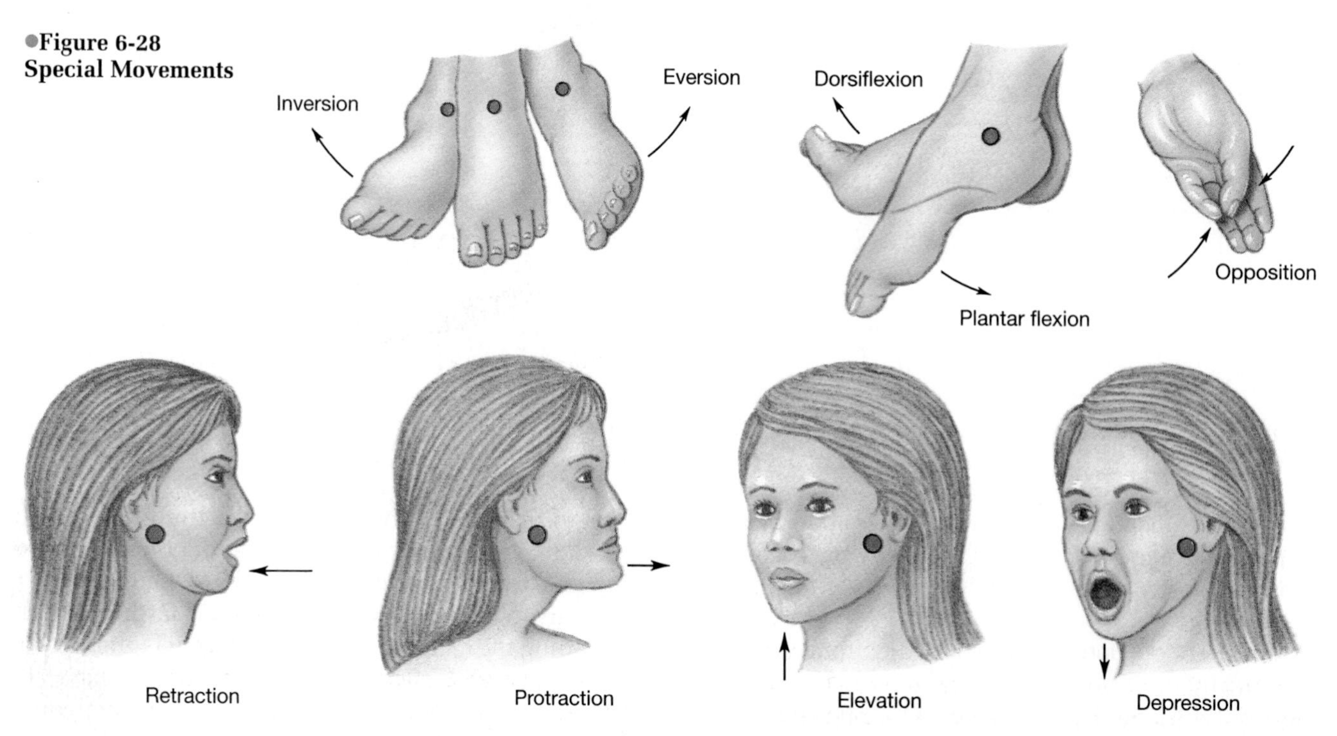

Dorsiflexion/Plantar flexion. These terms also refer to movements of the foot. Dorsiflexion is flexion of the ankle and elevation of the sole, as when "digging in the heels." **Plantar flexion** (*planta*, sole), the opposite movement, extends the ankle and elevates the heel, as when standing on tiptoes.

Opposition. **Opposition** is the special movement of the thumb that enables it to grasp and hold an object.

Protraction/Retraction. **Protraction** entails moving a part of the body anteriorly in the horizontal plane. **Retraction** is the reverse movement. You protract your jaw when you grasp your upper lip with your lower teeth, and you protract your clavicles when you cross your arms.

Elevation/Depression. These occur when a structure moves in a superior or inferior direction. You depress your mandible when you open your mouth, and elevate it as you close it.

A Functional Classification of Synovial Joints

Synovial joints can be described as **gliding, hinge, pivot, ellipsoidal, saddle**, or **ball-and-socket** joints on the basis of the shapes of the articulating surfaces. Each type of joint permits a different type and range of motion.

Gliding Joints Gliding joints (Figure 6-29a●) have flattened or slightly curved faces. The relatively flat articular surfaces slide across one another, but the amount of movement is very slight. Although rotation is theoretically possible at such a joint, ligaments usually prevent or restrict such movement. Gliding joints are found at the ends of the clavicles, between the carpals, between the tarsals, and between the articular facets of adjacent spinal vertebrae.

Hinge Joints Hinge joints (Figure 6-29b●) permit angular movement in a single plane, like the opening and closing of a door. Examples include the joint between the occipital bone and atlas, in the axial skeleton, and the elbow, knee, ankle, and interphalangeal joints of the appendicular skeleton.

Pivot Joints Pivot joints (Figure 6-29c●) permit only rotation. A pivot joint between the atlas and axis allows you to rotate your head to either side, and another between the head of the radius and the proximal shaft of the ulna permit pronation and supination of the palm.

Ellipsoidal Joints In an ellipsoidal joint (Figure 6-29d●), an oval articular face nestles within a depression on the opposing surface. With such an arrangement, angular motion occurs in two planes, along or across the length of the oval. Ellipsoidal joints connect the fingers and toes with the metacarpals and metatarsals.

Saddle Joints Saddle joints (Figure 6-29e●) have articular faces that resemble saddles. Each face is concave on one axis and convex on the other, and the opposing faces nest together. This arrangement permits angular motion, including circumduction, but prevents rotation. The carpometacarpal joint at the base of the thumb is the best example of a saddle joint, and "twiddling your thumbs" will demonstrate the possible movements.

Ball and Socket Joints In a ball-and-socket joint (Figure 6-29f●), the round head of one bone rests within a cup-shaped depression in another. All combinations of movements, including circumduction and rotation, can be performed at ball-and-socket joints. Examples include the shoulder and hip joints.

- ✓ In a newborn infant, the large bones of the skull are joined by fibrous connective tissue. What type of joint is this?
- ✓ These bones later grow, interlock, and form immovable joints. What type of joints are these?
- ✓ Give the proper term for each of the following types of motion: (a) moving your arm away from the midline of the body, (b) turning your palms so that they face forward, (c) bending your elbow.

Representative Articulations

This section considers examples of articulations that demonstrate important functional principles. We will first consider the *intervertebral articulations* of the axial skeleton. We will then proceed to a discussion of the synovial articulations of the appendicular skeleton: the shoulder and elbow of the upper limb and the hip and knee of the lower limb.

Intervertebral Articulations

The vertebrae articulate with one another in two ways: (1) at gliding joints between **superior** and **inferior articular processes** and (2) at symphyseal joints between the vertebral bodies. The articulations between the superior and inferior articular processes of adjacent vertebrae permit small movements associated with flexion and rotation of the vertebral column. Little gliding occurs between adjacent vertebral bodies. Figure 6-30● illustrates the structure of these joints.

As noted earler, the vertebrae are separated and cushioned by pads called *intervertebral discs*. Each intervertebral disc consists of a tough outer layer of fibrocartilage whose collagen fibers attach the discs to adjacent vertebrae. The fibrocartilage surrounds a soft, elastic and gelatinous core, which gives the disc resiliency and enables it to act as a shock absorber. AM *Problems with Intervertebral Discs*

Figure 6-29
A Functional Classification of Synovial Joints

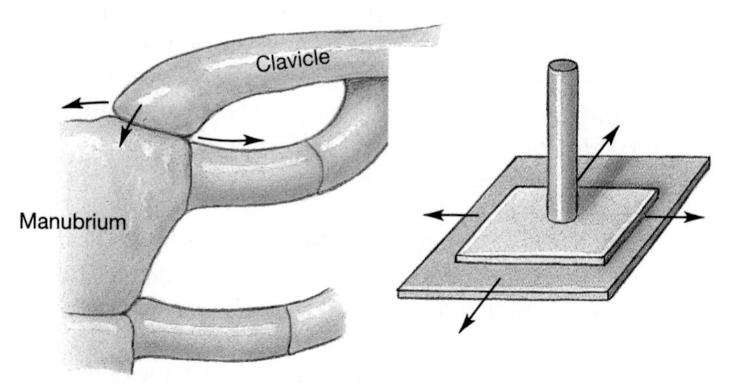

(a) Gliding joint

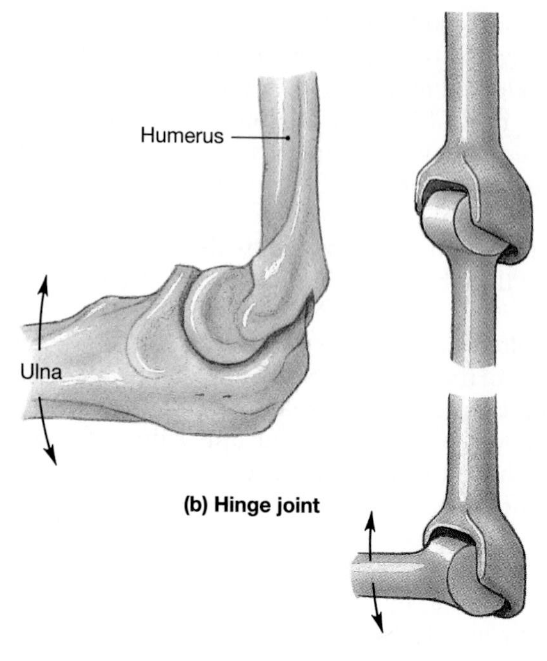

(b) Hinge joint

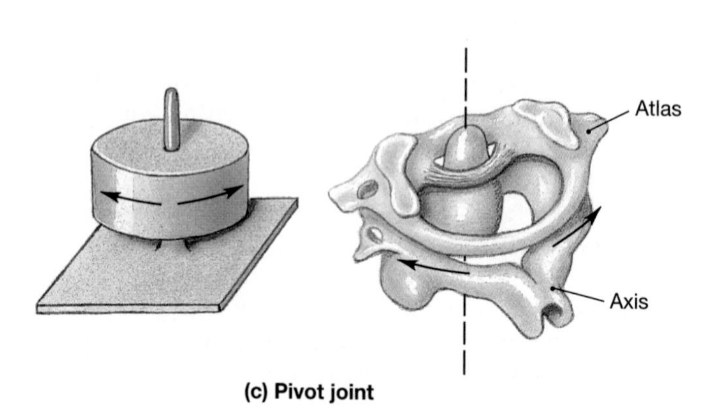

(c) Pivot joint

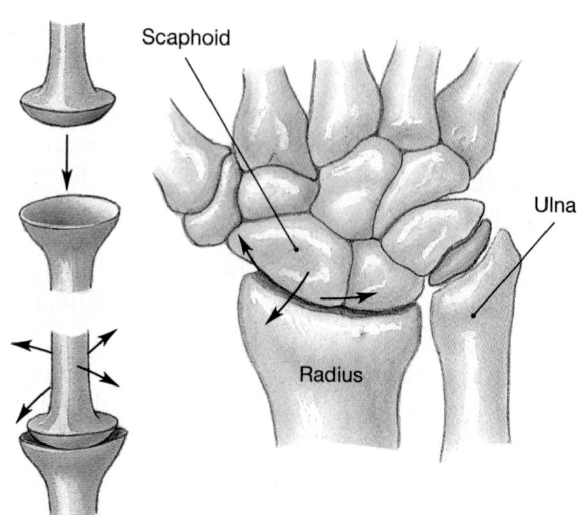

(d) Ellipsoidal joint

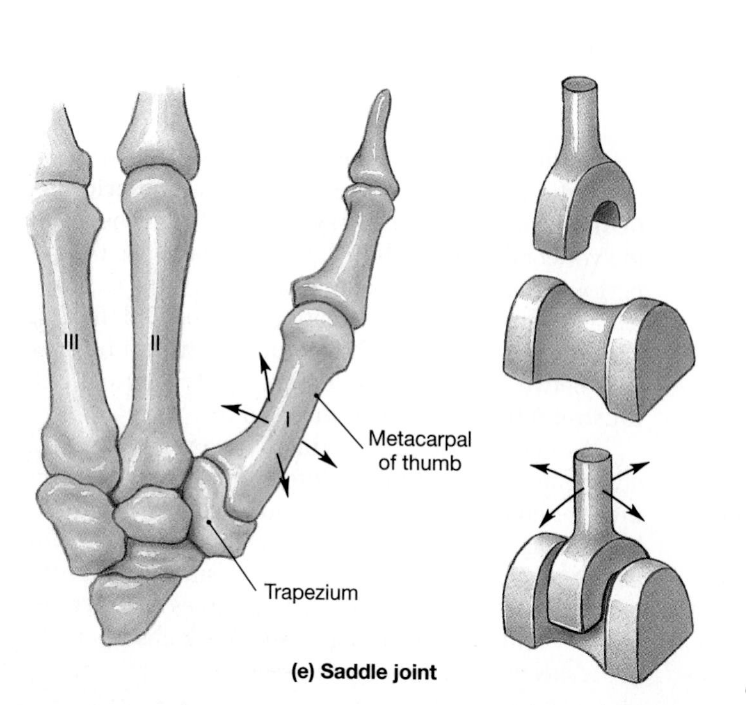

(e) Saddle joint

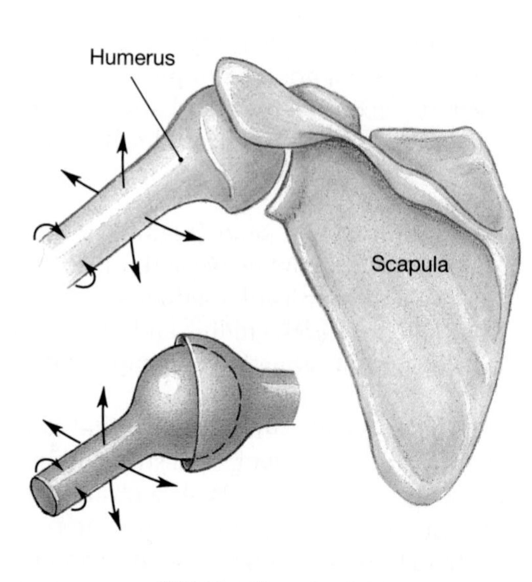

(f) Ball-and-socket joint

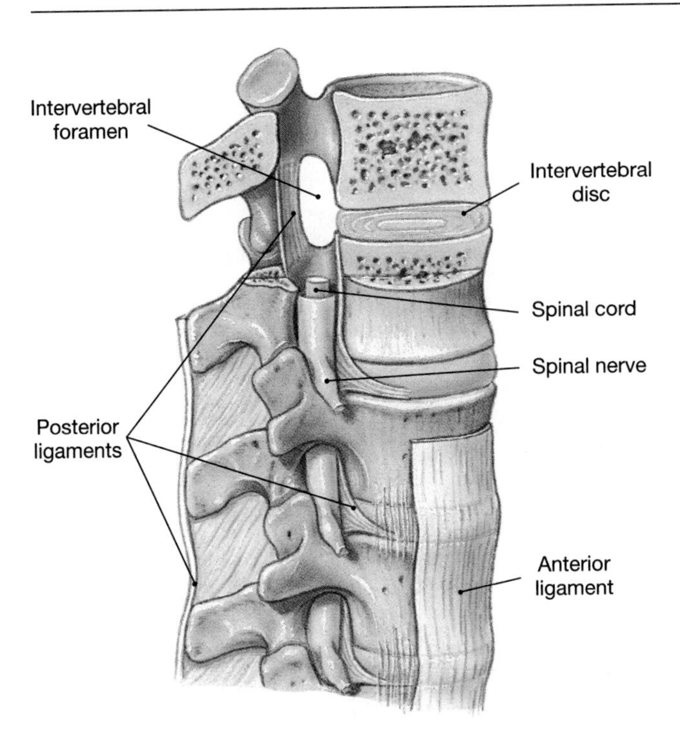

●**Figure 6-30**
Intervertebral Articulations

Articulations of the Upper Limb

The shoulder, elbow, and wrist are responsible for positioning the hand, which performs precise and controlled movements. The shoulder has great mobility, the elbow has great strength, and the wrist makes fine adjustments in the orientation of the palm and fingers.

The Shoulder Joint The shoulder joint permits the greatest range of motion of any joint in the body. Because it is also the most frequently dislocated joint, it provides an excellent demonstration of the principle that strength and stability must be sacrificed to obtain mobility.

Figure 6-31● details the structure of the shoulder joint. The relatively loose articular capsule extends from the scapular neck to the humerus, and this oversized capsule permits an extensive range of motion. As at other joints, bursae at the shoulder reduce friction where large muscles and tendons pass across the joint capsule. The bursae of the shoulder are especially large and numerous. Several bursae are associated with the capsule, the processes of the scapula, and large shoulder muscles. Inflammation of any of these bursae, a condition called *bursitis*, can restrict motion and produce pain.

Muscles that move the humerus actually do more to stabilize the shoulder joint than all its ligaments and capsular fibers combined. Powerful muscles originating on the trunk, shoulder girdle, and humerus cover the anterior, superior, and posterior surfaces of the capsule. These muscles form the *rotator cuff*, a group of muscles that can swing the arm through an impressive range of motion.

The Elbow Joint Figure 6-32● diagrams the structure of the elbow joint, also known as the **olecranal joint**. The ulna is more important at the elbow than the radius. A large muscle that extends the elbow, the *triceps brachii*, attaches to the rough surface of the olecranon process. Arising on the front of the arm, the smaller *brachialis muscle* attaches to the ulnar tuberosity. Contraction of this muscle flexes the elbow.

The elbow joint is extremely stable because (1) the bony surfaces of the humerus and ulna interlock; (2) the articular capsule is very thick; and (3) the capsule is reinforced by stout ligaments. Nevertheless, the joint can be damaged by severe impacts or unusual stresses. When a you fall on your hand with a partially flexed elbow, powerful contractions of the muscles that extend the elbow can break the ulna at the center of the trochlear notch.

✓ Would a tennis player or a jogger be more likely to develop inflammation of the subacromial bursa? Why?

✓ Mary falls on her hands with her elbows slightly flexed. After the fall, she can't move her left arm at the elbow. If a fracture exists, what bone is most likely broken?

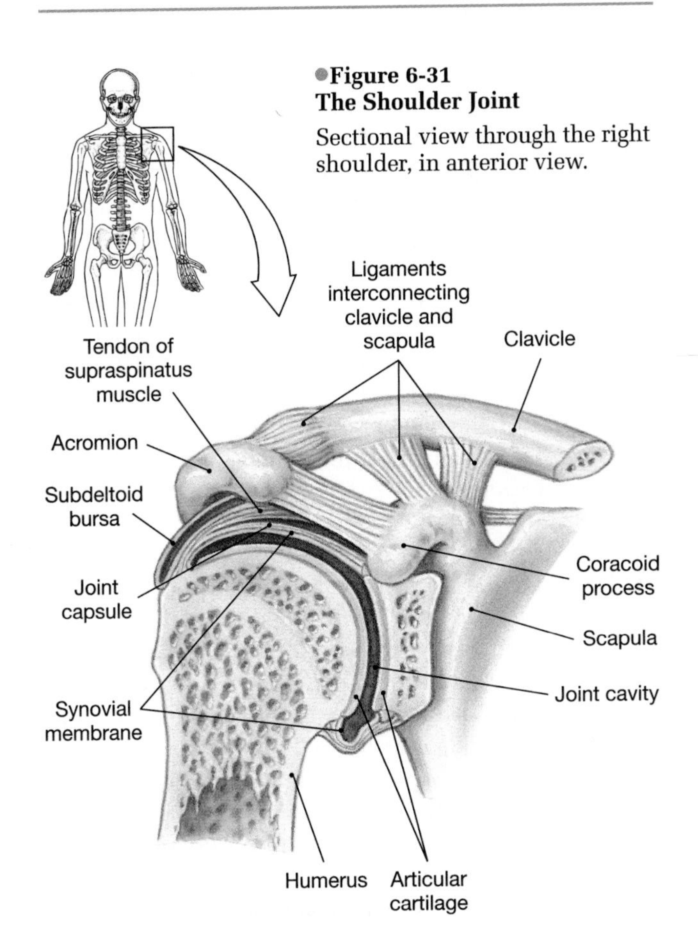

●**Figure 6-31**
The Shoulder Joint
Sectional view through the right shoulder, in anterior view.

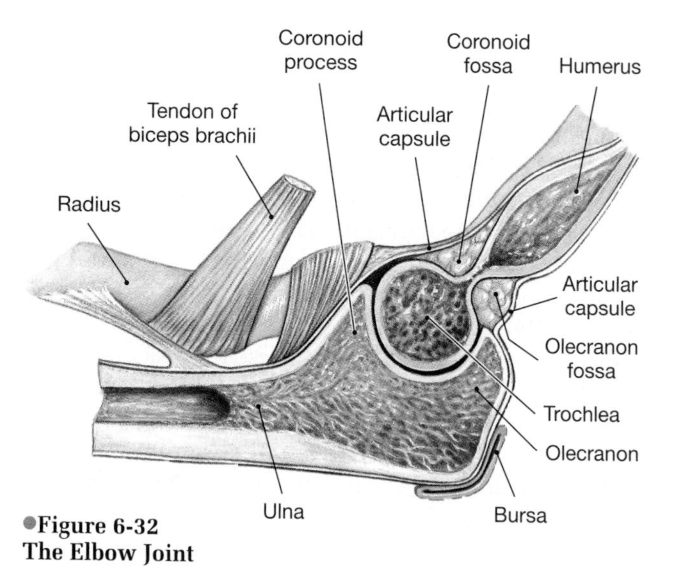

●**Figure 6-32**
The Elbow Joint
Longitudinal section through the right elbow.

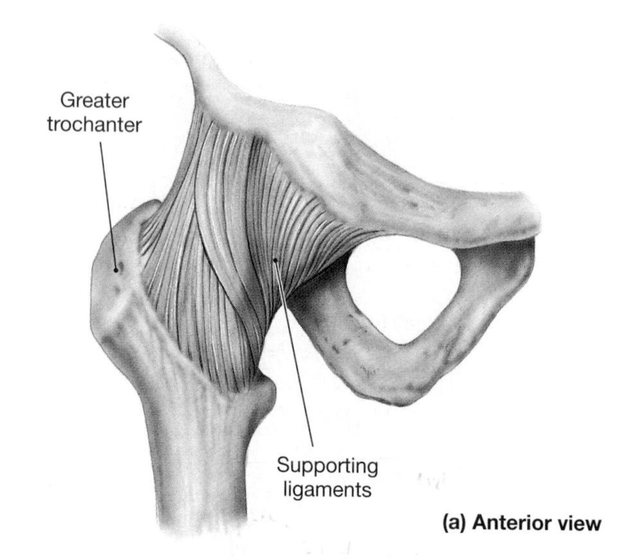

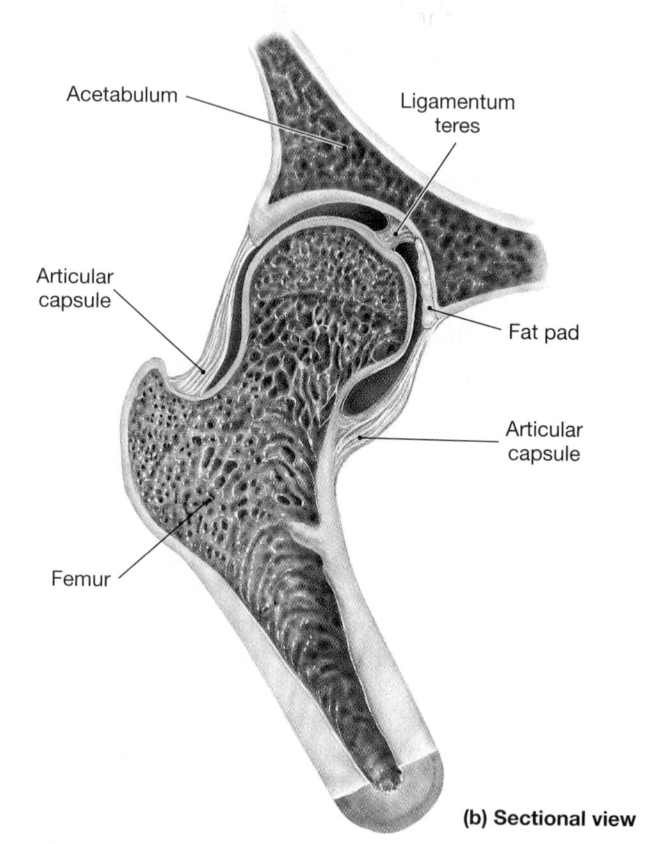

●**Figure 6-33**
The Hip Joint
(a) Anterior view of the right hip joint. This joint is extremely strong and stable, in part because of the massive capsule and surrounding ligaments. **(b)** Sectional view.

Articulations of the Lower Limb

The joints of the hip, ankle, and foot are sturdier than those at corresponding locations in the arms, and their ranges of motion are less. The knee has a range of motion comparable to that of the elbow, but it is subjected to much greater forces and so is less stable.

The Hip Joint Figure 6-33● details the structure of the hip joint. The articulating surface of the acetabulum is covered by a fibrocartilage pad along its edges, a fat pad covered by synovial membrane in its central portion, and a ligament. This combination of coverings and membranes resists compression, absorbs shocks, and stretches and distorts without damage.

Compared with that of the shoulder, the articular capsule of the hip joint, a ball-and-socket diarthrosis, is denser and stronger. It extends from the lateral and inferior surfaces of the pelvic girdle to the femur and encloses both the femoral head and neck. This arrangement helps keep the head from moving away from the acetabulum. Three broad ligaments reinforce the articular capsule, while a fourth, the *ligamentum teres* (*teres*, cylindrical), originates inside the acetabulum and attaches to the center of the femoral head. Additional stabilization comes from the bulk of the surrounding muscles.

The combination of an almost complete bony socket, a strong articular capsule, supporting ligaments, and muscular padding makes this an extremely stable joint. Fractures of the femoral neck or between the trochanters are actually more common than hip dislocations. Although flexion, extension, adduction, abduction, and rotation are permitted, the total range of motion is considerably less than that of the shoulder. Hip flexion is the most important normal movement, and the primary limits are imposed by the surrounding muscles. Other directions of movement are restricted by ligaments and capsular fibers.

HIP FRACTURES

Hip fractures are most often suffered by individuals over 60 years of age, when osteoporosis has weakened the thighbones. These injuries may be accompanied by dislocation of the hip or pelvic fractures. For individuals with osteoporosis, healing proceeds very slowly. In addition, the powerful muscles that surround the joint can easily prevent proper alignment of the bone fragments. Trochanteric fractures usually heal well, if the joint can be stabilized; steel frames, pins, screws, or some combination of those devices may be needed to preserve alignment and permit healing to proceed normally.

Severe hip fractures are most common among those over age 60, but in recent years the frequency of hip fractures has increased dramatically among young, healthy professional athletes. Probably the best known example is the case of Bo Jackson, whose condition is discussed in the *Applications Manual.* AM *Hip Fractures, Aging, and Professional Athletes*

The Knee Joint The hip joint passes weight to the femur, and at the knee joint the femur transfers the weight to the tibia. The shoulder is mobile; the hip, stable; and the knee...? If you had to choose one word, it would probably be "complicated." Although the knee functions as a hinge joint, the articulation is far more complex than that of the elbow or even the ankle. The rounded femoral condyles roll across the top of the tibia, so the points of contact are constantly changing. Important features of the knee joint are indicated in Figure 6-34●.

Structurally, the knee resembles three separate joints, two between the femur and tibia (medial to medial condyle and lateral to lateral condyle) and one between the patella and the femur. There is no single unified capsule, nor is there a common synovial cavity. A pair of fibrocartilage pads, the **medial** and **lateral menisci**, lie between the femoral and tibial surfaces. They act as cushions and conform to the shape of the articulating surfaces as the femur changes position. Prominent **fat pads** provide padding around the margins of the joint and assist the bursae in reducing friction between the patella and other tissues.

Ligaments stabilize the anterior, posterior, medial, and lateral surfaces of this joint, and a complete dislocation of the knee is an extremely rare event. The tendon from the muscles responsible for extending the knee passes over the anterior surface of the joint. The patella is embedded within this tendon, and the **patellar ligament** continues its attachment on the anterior surface of the tibia. This ligament provides support to the front of the knee joint. Posterior ligaments between the femur and the heads of the tibia and fibula reinforce the back of the knee joint. The lateral and medial surfaces of the knee joint are reinforced by another pair of ligaments. These ligaments stabilize the joint at full extension.

Additional ligaments are found inside the joint capsule (Figure 6-34b●). Inside the joint, a pair of ligaments, the *anterior cruciate* and *posterior cruciate*, cross one another as they attach the tibia to

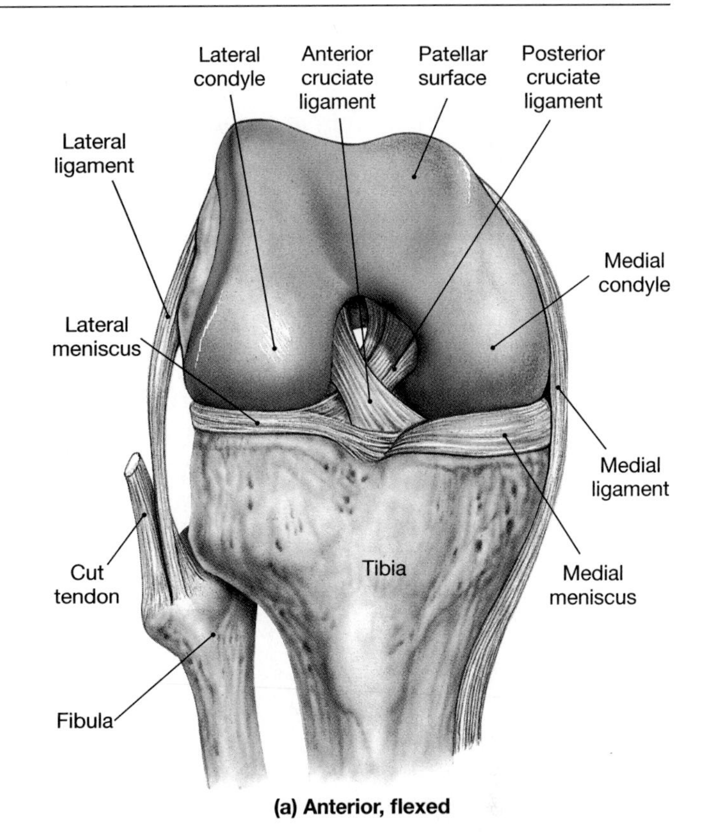

(a) Anterior, flexed

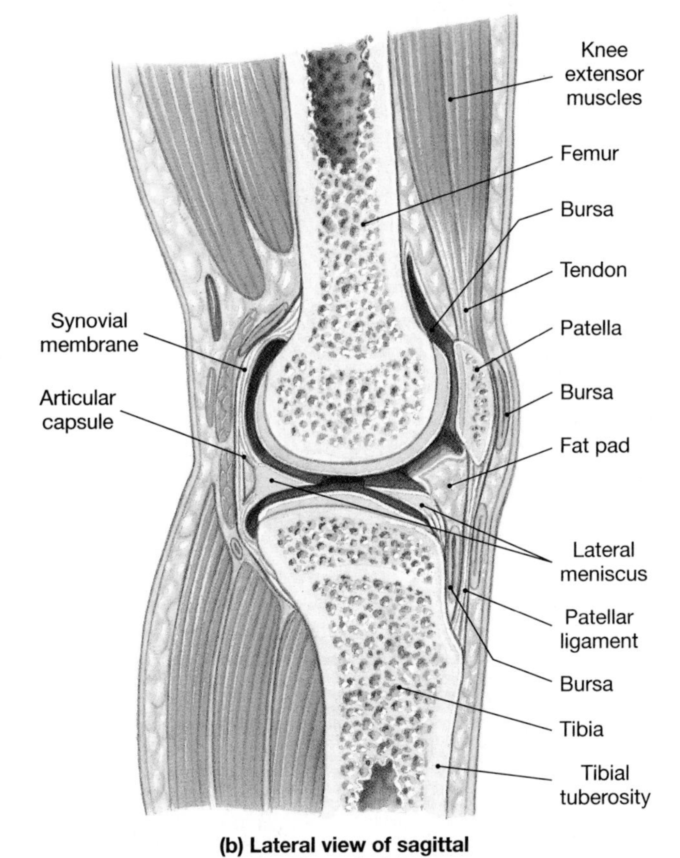

(b) Lateral view of sagittal section through right knee

●**Figure 6-34**
The Knee Joint
(a) The flexed right knee as seen in section, showing major anatomical features. **(b)** The extended knee in sectional view.

INTEGUMENTARY SYSTEM

Synthesizes vitamin D_3, essential for calcium and phosphorus absorption (bone maintenance and growth)

Provides structural support

MUSCULAR SYSTEM

Stabilizes bone positions; tension in tendons stimulates bone growth and maintenance

Stores calcium needed for normal muscle contraction; bones act as levers to produce body movements

NERVOUS SYSTEM

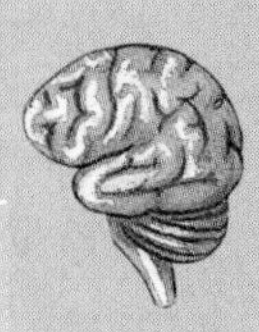

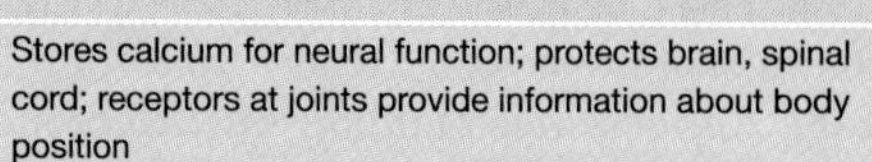

Regulates bone position by controlling muscle contractions

Stores calcium for neural function; protects brain, spinal cord; receptors at joints provide information about body position

ENDOCRINE SYSTEM

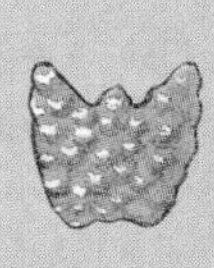

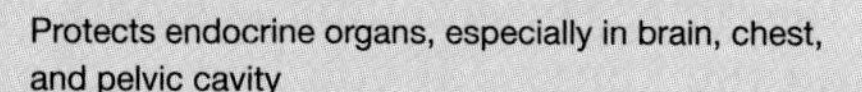

Skeletal growth regulated by growth hormone, thyroid hormones, and sex hormones; calcium mobilization regulated by parathyroid hormone and calcitonin

Protects endocrine organs, especially in brain, chest, and pelvic cavity

THE SKELETAL SYSTEM

FOR ALL SYSTEMS

Provides mechanical support
Stores energy reserves
Stores calcium and phosphate reserves

CARDIOVASCULAR SYSTEM

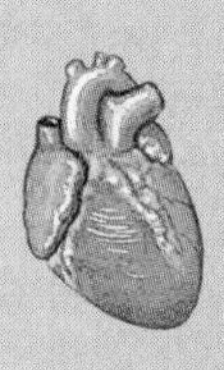

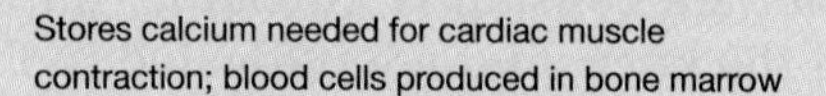

Provides oxygen, nutrients, hormones, blood cells; removes waste products and carbon dioxide

Stores calcium needed for cardiac muscle contraction; blood cells produced in bone marrow

LYMPHATIC SYSTEM

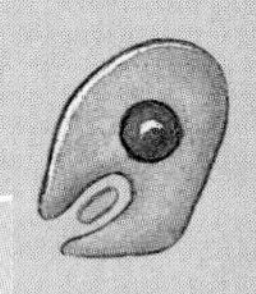

Lymphocytes assist in the defense and repair of bone following injuries

Lymphocytes and other cells of the immune response are produced and stored in bone marrow

RESPIRATORY SYSTEM

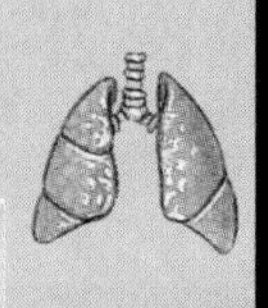

Provides oxygen and eliminates carbon dioxide

Movements of ribs important in breathing; axial skeleton surrounds and protects lungs

DIGESTIVE SYSTEM

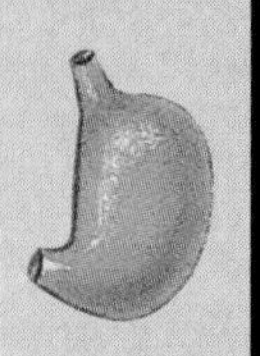

Provides nutrients including calcium and phosphate ions.

Ribs protect portions of liver, stomach, and intestines

URINARY SYSTEM

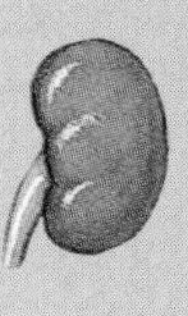

Conserves calcium and phosphate ions needed for bone growth, disposes of waste products

Axial skeleton provides some protection for kidneys and ureters; pelvis protects urinary bladder and proximal urethra

REPRODUCTIVE SYSTEM

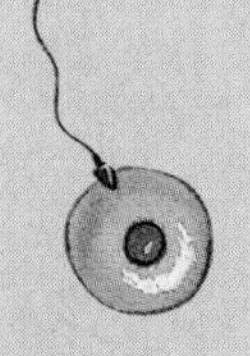

Sexual hormones stimulate growth and maintenance of bones; surge of sex hormones at puberty causes acceleration of growth and closure of epiphyseal plates

Pelvis protects reproductive organs of female, protects portion of ductus deferens and accessory glands in male

Figure 6-35
Functional Relationships between the Skeletal System and Other Systems

the femur. The term *cruciate* is derived from the Latin word *crucialis*, meaning a cross. These ligaments limit the anterior and posterior movement of the femur.

✓ Why is a complete dislocation of the knee joint an infrequent event?

✓ What symptoms would you expect to see in an individual who has damaged the menisci of the knee joint?

INTEGRATION WITH OTHER SYSTEMS

Although the bones may seem inert, you should now realize that they are quite dynamic structures. The entire skeletal system is intimately associated with other systems. For example, bones are attached to the muscular system, extensively connected to the cardiovascular and lymphatic systems, and largely under the physiological control of the endocrine system. These functional relationships are diagrammed in Figure 6-35●.

Chapter Review

KEY TERMS

appendicular skeleton, p. 118
articulation, p. 136
axial skeleton, p. 120
fracture, p. 116
ligament, p. 137
marrow, p. 110
meniscus, p. 137
ossification, p. 113
osteocyte, p. 112
osteon, p. 112
periosteum, p. 111
synovial fluid, p. 137

SUMMARY OUTLINE

INTRODUCTION p. 110

1. The skeletal system includes the bones of the skeleton and the cartilages, ligaments, and other connective tissues that stabilize or interconnect bones. Its functions include structural support, storage, blood cell production, protection, and leverage.

STRUCTURE OF BONE p. 110

1. **Osseous tissue** is a supporting connective tissue with a solid *matrix.*

Macroscopic Features of Bone p. 110

2. General categories of bones are: **long bones, short bones, flat bones,** and **irregular bones.** *(Figure 6-1)*

3. The features of a long bone include a **diaphysis, epiphyses,** and a central marrow cavity. (*Figure 6-2*)

4. There are two types of bone: **compact**, or *dense,* **bone** and **spongy**, or *cancellous,* **bone.**

5. A bone is covered by a **periosteum** and lined with an **endosteum.**

Microscopic Features of Bone p. 112

6. Both types of bone contain **osteocytes** in **lacunae.** Layers of calcified matrix are **lamellae**, interconnected by **canaliculi.** (*Figure 6-3*)

7. The basic functional unit of compact bone is the **osteon**, containing osteocytes arranged around a **central canal.**

8. Spongy bone contains **trabeculae**, often in an open network.

9. Compact bone is found where stresses come from a limited range of directions; spongy bone is located where stresses are few or come from many different directions.

10. Cells other than osteocytes are also present in bone. **Osteoclasts** dissolve the bony matrix through *osteolysis.* **Osteoblasts** synthesize the matrix in the process of *osteogenesis.*

BONE DEVELOPMENT AND GROWTH p. 113

1. **Ossification** is the process of converting other tissues to bone.

Intramembranous Ossification p. 113

2. **Intramembranous ossification** begins when stem cells within connective tissue differentiate into osteoblasts and can produce spongy or compact bone.

Endochondral Ossification p. 113

3. **Endochondral ossification** begins by forming a cartilaginous model that is gradually replaced by bone. (*Figure 6-4*)

Bone Growth and Body Proportions p. 114

4. There are differences between bones and between individuals regarding the timing of epiphyseal closure.

Requirements for Normal Bone Growth p. 114

5. Normal osteogenesis requires a reliable source of minerals, vitamins, and hormones.

REMODELING AND HOMEOSTATIC MECHANISMS p. 115

1. The organic and mineral components of bone are continually recycled and renewed through the process of remodeling.

Remodeling and Support p. 115

2. The shapes and thicknesses of bones reflect the stresses applied to them. Mineral turnover allows bone to adapt to new stresses.

Homeostasis and Mineral Storage p. 115

3. Calcium is the most common mineral in the human body, with roughly 99 percent of it located in the skeleton.

Injury and Repair p. 116

4. A **fracture** is a crack or break in a bone. Repair of a fracture involves the formation of a fracture hematoma, an *external callus*, and an *internal callus*. (*Figure 6-5*) (*Focus: A Classification of Fractures*)

Aging and the Skeletal System p. 116

5. Effects of aging on the skeleton can include **osteopenia** and **osteoporosis.**

AN OVERVIEW OF THE SKELETON p. 118

Skeletal Terminology p. 118

1. **Bone markings** can be used to describe and identify specific bones. (*Table 6-1*)

Skeletal Divisions p. 118

2. The skeletal system consists of the axial skeleton and the appendicular skeleton. The **axial skeleton** can be subdivided into the **skull**, the **auditory ossicles** (ear bones) and **hyoid**, the **thoracic** (rib) **cage** composed of the **ribs** and **sternum**, and the **vertebral column**. (*Figure 6-6*)

3. The **appendicular skeleton** includes the arms, the legs, and the pectoral and pelvic girdles.

THE AXIAL DIVISION p. 120

The Skull, p. 120

1. The **cranium** encloses the **cranial cavity**, a division of the dorsal body cavity that encloses the brain.

2. The **frontal bone** forms the forehead and superior surface of each **orbit.** (*Figures 6-7, 6-8, 6-9*)

3. The **parietal bones** form the upper sides and roof of the cranium. (*Figures 6-7, 6-9*)

4. The **occipital bone** surrounds the foramen magnum and articulates with the sphenoid, temporal, and parietal bones to form the back of the cranium. (*Figures 6-7, 6-8, 6-9*)

5. The **temporal bones** help form the sides and base of the cranium and fuse with the parietal bones along the squamosal suture. (*Figures 6-7, 6-8, 6-9*)

6. The **sphenoid bone** acts as a bridge uniting the cranial and facial bones. (*Figures 6-7, 6-8, 6-9*)

7. The **ethmoid bone** stabilizes the brain and forms the roof and sides of the nasal cavity. Its **cribriform plate** contains perforations for olfactory nerves, and the **perpendicular plate** forms part of the nasal septum. (*Figures 6-7, 6-8, 6-9*)

8. The left and right **maxillae**, or **maxillary bones**, articulate with all the other facial bones except the mandible. (*Figures 6-7, 6-8, 6-9*)

9. The **palatine bones** form the posterior portions of the hard palate. (*Figures 6-8, 6-9*)

10. The **vomer** forms the lower portion of the nasal septum. (*Figures 6-8, 6-9*)

11. The **zygomatic bones** help complete the orbit and together with the temporal bones form the **zygomatic arch** (cheekbone). (*Figures 6-7, 6-8*)

12. The **nasal bones** articulate with the superior frontal bone and maxillary bones. (*Figures 6-7, 6-8, 6-9*)

13. The **lacrimal bones** are within the orbit on its medial surface. (*Figures 6-7, 6-8*)

14. The **inferior nasal conchae** inside the nasal cavity aid the **superior** and **middle conchae** of the ethmoid bone to slow incoming air. (*Figure 6-8a, 6-9c*)

15. The **nasal complex** includes the bones that form the superior and lateral walls of the nasal cavity and the sinuses that drain into them. The **nasal septum** divides the nasal cavities. Together the **frontal, sphenoidal, ethmoidal,** and **maxillary sinuses** make up the **paranasal sinuses**. (*Figures 6-8, 6-9*)

16. The **mandible** is the lower jaw. (*Figures 6-7, 6-8, 6-9*)

17. The hyoid bone is suspended below the skull by ligaments from the styloid processes of the temporal bones.

18. Fibrous connections at **fontanels** permit the skulls of infants and children to continue growing. (*Figure 6-10*)

The Neck and Trunk p. 125

19. There are 7 cervical vertebrae, 12 **thoracic vertebrae** (which articulate with **ribs**), and 5 **lumbar vertebrae** (which articulate with the sacrum). The **sacrum** and **coccyx** consist of fused vertebrae. (*Figure 6-11*)

20. The spinal column has four **spinal curves**, which accommodate the unequal distribution of body weight and keep it in line with the body axis. (*Figure 6-11*)

21. A typical vertebra has a **body** and a **vertebral arch**; it articulates with other vertebrae at the **articular processes.** Adjacent vertebrae are separated by an **intervertebral disc**. (*Figure 6-12*)

22. Cervical vertebrae are distinguished by the shape of the body and **transverse foramina** on either side. (*Figures 6-12, 6-13*)

23. Thoracic vertebrae have distinctive heart-shaped bodies. (*Figure 6-12*)

24. The lumbar vertebrae are the most massive and least mobile; they are subjected to the greatest strains, and a herniated disc can occur. (*Figure 6-12*)

25. The *sacrum* protects reproductive, digestive, and excretory organs. It articulates with the *coccyx* at the **apex** and with the last lumbar vertebra at the **base**. (*Figure 6-14*)

26. The skeleton of the thorax consists of the thoracic vertebrae, the ribs, and the sternum. The ribs and sternum form the *rib cage*. (*Figure 6-15*)

27. Ribs 1 to 7 are **true ribs**. Ribs 8 to 12 lack direct connections to the sternum and are called **false ribs**; they include two pairs of **floating ribs**. The medial end of each rib articulates with a thoracic vertebrae. (*Figure 6-15*)

28. The sternum consists of a **manubrium**, a **body**, and a **xiphoid process**. (*Figure 6-15*)

THE APPENDICULAR DIVISION p. 129

The Pectoral Girdle p. 129

1. Each arm articulates with the trunk at the **shoulder**, or **pectoral, girdle**, which consists of the **scapula** and **clavicle.** (*Figures 6-6, 6-16, 6-17*)

2. The clavicle and scapula position the shoulder joint, help move the arm, and provide a base for arm movement and muscle attachment. (*Figures 6-16, 6-17*)

3. The scapula articulates with the **humerus** at the shoulder joint. Both the **coracoid process** and the **acromion** are attached to ligaments and tendons. The **scapular spine** crosses the posterior surface of the scapular body. (*Figure 6-17*)

The Upper Limb p. 131

4. The **greater tubercle** and **lesser tubercle** of the humerus are important sites for muscle attachment. Other prominent landmarks include the **deltoid tuberosity**, the **medial** and **lateral epicondyles**, and the articular **condyle.** (*Figure 6-18*)

5. Distally the humerus articulates with the radius and ulna. The medial **trochlea** extends from the **coronoid fossa** to the **olecranon fossa**. (*Figure 6-18*)

6. The **radius** and **ulna** are the bones of the forearm. The olecranon fossa accommodates the **olecranon process** during extension of the arm. The coronoid and radial fossae accommodate the **coronoid process** of the ulna. (*Figure 6-19*)

7. The bones of the wrist form two rows of **carpals**. Four of the fingers contain three **phalanges**; the thumb has only two. (*Figure 6-20*)

The Pelvic Girdle p. 133

8. The **pelvic girdle** consists of two **coxae**; each coxa forms through the fusion of an ilium, an ischium, and a pubis. (*Figure 6-21*)

9. The largest coxal bone, the **ilium**, fuses with the **ischium**, which in turn fuses with the **pubis**. The **pubic symphysis** limits movement between the pubic bones. (*Figure 6-21*)

10. The **pelvis** consists of the coxae, the sacrum, and the coccyx. (*Figure 6-21*)

The Lower Limb p. 134

11. The **femur**, or thighbone, is the longest bone in the body. It articulates with the **tibia** at the knee joint. A ligament from the **patella**, the kneecap, attaches at the **tibial tuberosity.** (*Figures 6-22, 6-23*)

12. Other tibial landmarks include the **anterior crest** and the **medial malleolus**. The **fibular head** articulates with the tibia below the knee, and the **lateral malleolus** stabilizes the ankle. (*Figure 6-23*)

13. The ankle includes seven bones, or **tarsals**; only the **talus** articulates with the tibia and fibula. When standing normally, most of our weight is transferred to the **calcaneus**, and the rest is passed on to the **metatarsals**. (*Figure 6-24*)

14. The basic organizational pattern of the metatarsals and phalanges of the foot resembles that of the hand.

ARTICULATIONS p. 136

A Classification of Joints p. 136

1. **Articulations** (joints) exist wherever two bones interact. Immovable joints are **synarthroses**, slightly movable joints are **amphiarthroses**, and those that are freely movable are called **diarthroses**. (*Table 6-2*)

2. Examples of synarthroses include a **suture**, a **gomphosis**, and a **synchondrosis**.

3. Examples of amphiarthroses are a **syndesmosis** and a **symphysis**.

4. The bony surfaces at diarthroses, or **synovial joints**, are covered by **articular cartilages**, lubricated by **synovial fluid**, and enclosed within a **joint capsule**. Other synovial structures can include **menisci, fat pads**, and various ligaments. (*Figure 6-25*)

Articular Form and Function p. 137

5. Important terms that describe dynamic motion are **flexion, extension, hyperextension, rotation, circumduction, abduction,** and **adduction**. (*Figures 6-26, 6-27*)

6. The bones in the forearm permit **pronation** and **supination**. (*Figure 6-27*)

7. The ankle undergoes **dorsiflexion** and **plantar flexion**. Movements of the foot include **inversion** and **eversion**. **Opposition** is the thumb movement that enables us to grasp objects. (*Figure 6-28*)

8. **Protraction** involves moving something forward; **retraction** involves moving it back. **Depression** and **elevation** occur when we move a structure down and up. (*Figure 6-28*)

9. Major types of joints include **gliding joints, hinge joints, pivot joints, ellipsoidal joints, saddle joints**, and **ball-and-socket joints**. (*Figure 6-29*)

Representative Articulations p. 141

10. The articular processes form gliding joints with those of adjacent vertebrae. The bodies form symphyseal joints. They are separated by intervertebral discs. (*Figure 6-30*)

11. The shoulder joint is formed by the glenoid fossa and the head of the humerus. It is extremely mobile and for that reason also unstable and easily dislocated. (*Figure 6-31*)

12. Bursae at the shoulder joint reduce friction from muscles and tendons during movement. (*Figure 6-31*)

13. The elbow joint permits only flexion and extension. It is extremely stable because of extensive ligaments and the shapes of the articulating elements. (*Figure 6-32*)

14. The hip joint is formed by the union of the acetabulum with the head of the femur. This ball-and-socket diarthrosis permits flexion and extension, adduction and abduction, circumduction, and rotation. (*Figure 6-33*)

15. The knee joint is a complicated hinge joint. The patella, or kneecap, is embedded within a tendon that supports the front of the joint. (*Figure 6-34*)

INTEGRATION WITH OTHER SYSTEMS p. 147

1. The skeletal system is dynamically associated with other systems. (*Figure 6-35*)

CHAPTER QUESTIONS

LEVEL 1 Reviewing Facts and Terms

Match each item in column A with the most closely related item in column B. Use letters for answers in the spaces provided.

Column A	Column B
____ 1. osteocytes	a. abduction
____ 2. diaphysis	b. heelbone
____ 3. auditory ossicles	c. ball-and-socket joints
____ 4. cribriform plate	d. bone-dissolving cells
____ 5. osteoblasts	e. hinge joints
____ 6. C_1	f. axis
____ 7. C_2	g. immovable joint
____ 8. hip and shoulder	h. bone shaft
____ 9. patella	i. mature bone cells
____ 10. calcaneus	j. bone-producing cells
____ 11. synarthrosis	k. atlas
____ 12. moving the hand into a palm-front position	l. olfactory nerves
____ 13. osteoclasts	m. ear bones
____ 14. raising the arm laterally	n. supination
____ 15. elbow and knee	o. kneecap

16. The bones of the skeleton store energy reserves as lipids in areas of:
    (a) red marrow
    (b) yellow marrow
    (c) the matrix of bone tissue
    (d) the ground substance

17. The two types of osseous tissue are:
    (a) compact bone and spongy bone
    (b) dense bone and compact bone
    (c) spongy bone and cancellous bone
    (d) a, b, and c are correct

18. The basic functional units of mature compact bone are:
    (a) lacunae (b) osteocytes
    (c) osteons (d) canaliculi

19. The axial skeleton consists of the bones of the:
    (a) pectoral and pelvic girdles
    (b) skull, thorax, and vertebral column
    (c) arm, legs, hand, and feet
    (d) limbs, pectoral girdle, and pelvic girdle

20. The appendicular skeleton consists of the bones of the:
    (a) pectoral and pelvic girdles
    (b) skull, thorax, and vertebral column
    (c) arm, legs, hand, and feet
    (d) limbs, pectoral girdle, and pelvic girdle

21. Which of the following contains *only* bones of the cranium?
    (a) frontal, parietal, occipital, sphenoid
    (b) frontal, occipital, zygomatic, parietal
    (c) occipital, sphenoid, temporal, parietal
    (d) mandible, maxilla, nasal, zygomatic

22. Of the following bones, which one is unpaired?
    (a) vomer (b) maxilla
    (c) palatine (d) nasal

23. At the glenoid cavity, the scapula articulates with the proximal end of the:
    (a) humerus (b) radius
    (c) ulna (d) femur

24. In anatomical position, the ulna lies:
    (a) medial to the radius (b) lateral to the radius
    (c) inferior to the radius (d) superior to the radius

25. Each coxa of the pelvic girdle consists of three fused bones:
    (a) ulna, radius, humerus
    (b) ilium, ischium, pubis
    (c) femur, tibia, fibula
    (d) hamate, capitate, trapezium

26. Joints typically found at the end of long bones are:
    (a) synarthroses (b) amphiarthroses
    (c) diarthroses (d) sutures

27. The function of the synovial fluid is:
    (a) to nourish chondrocytes
    (b) to provide lubrication
    (c) to absorb shock
    (d) a, b, and c are correct

28. Abduction and adduciton always refer to movements of the:
    (a) axial skeleton
    (b) appendicular skeleton
    (c) skull
    (d) vertebral column

29. Standing on tiptoe is an example of a movement called:
    (a) elevation
    (b) dorsiflexion
    (c) plantar flexion
    (d) retraction

30. What are the five primary functions of the skeletal system?

31. What is the primary difference between intramembranous ossification and endochondral ossification?

32. What unique characteristic of the hyoid bone makes it different from all the other bones in the body?

33. What two primary functions are performed by the thoracic cage?

34. What two large scapular processes are associated with the shoulder joint?

LEVEL 2 Reviewing Concepts

35. Why are stresses or impacts to the side of the shaft in a long bone more dangerous than stress applied to the long axis of the shaft?
36. During the growth of a long bone, how is the epiphysis forced farther from the shaft?
37. Why are ruptured intervertebral discs more common in lumbar vertebrae and dislocations and fractures more common in cervical vertebrae?
38. Why are clavicular injuries common?
39. What is the difference between the *pelvic girdle* and the *pelvis*?
40. How do articular cartilages differ from other cartilages in the body?
41. What is the significance of the fact that the pubic symphysis is a slightly movable joint?

LEVEL 3 Critical Thinking and Clinical Applications

42. While playing on her swingset, 10-year-old Sally falls and breaks her right leg. At the emergency room, the doctor tells her parents that the proximal end of the tibia where the epiphysis meets the diaphysis is fractured. The fracture is properly set and eventually heals. During a routine physical when she is 18, Sally learns that her right leg is 2 cm shorter than her left, probably because of her accident. What might account for this difference?
43. Tess is diagnosed with a disease that affects the membranes surrounding the brain. The doctor tells her family that the disease is caused by an airborne virus. Explain how this virus could have entered the cranium.
44. While working at an excavation, an archaeologist finds several small skull bones. She examines the frontal, parietal, and occipital bones and concludes that the skulls are those of children not yet 1 year old. How can she tell their ages from examining the bones?
45. Frank Fireman is fighting a fire in a building when part of the ceiling collapses and a beam strikes him on his left shoulder. He is rescued by his friends, but he has a great deal of pain in his shoulder and cannot move his arm properly, especially in the anterior direction. His clavicle is not broken, and his humerus is intact. What is the probable nature of Frank's injury?
46. Ed "overturns" his ankle while playing tennis. He experiences swelling and pain, but after examination, he is told that there are no torn ligaments and the structure of the ankle is not affected. On the basis of the symptoms and the examination results, what do you think happened to Ed's ankle?

7 The Muscular System

Of the 700 or so skeletal muscles of the body, bodybuilders concentrate on the largest, most prominent muscles responsible for powerful movements of the axial and appendicular skeleton.

Chapter Outline and Objectives

Vocabulary Development

aer, air; *aerobic*
an, not; *anaerobic*
bi, two; *biceps*
caput, head; *biceps*
clavius, clavicle; *clavicle*
di, two; *digastricus*
epi-, on; *epimysium*
ergon, work; *synergist*
fasciculus, a bundle; *fascicle*
galea, a helmet; *galea aponeurotica*
gaster, stomach; *gastrocnemius*
iso-, equal; *isometric*
kneme, knee; *gastrocnemius*
lemma, husk; *sarcolemma*
meros, part; *sarcomere*
metric, measure; *isometric*
mys, muscle; *epimysium*
peri-, around; *perimysium*
platys, flat; *platysma*
sarkos, flesh; *sarcolemma*
syn-, together; *synergist*
tetanos, convulsive tension; *tetanus*
tonos, tension; *isotonic*
trope, a turning; *tropomyosin*

It is hard to imagine what life would be like without muscle tissue. We would be unable to sit, stand, walk, speak, or grasp objects. Blood would not circulate, because there would be no heartbeat to propel it through the vessels. The lungs could not rhythmically empty and fill, nor could food move through the digestive tract. Muscle tissue, one of the four primary tissue types, consists chiefly of elongated muscle cells that are highly specialized for contraction. The three types of muscle tissue, *skeletal muscle, cardiac muscle*, and *smooth muscle*, were introduced in Chapter 4. ∞ [p. 89] These muscle tissues share four basic properties:

1. **Excitability** is the ability to respond to stimulation. For example, skeletal muscles normally respond to stimulation by the nervous system, and some smooth muscles respond to circulating hormones.
2. **Contractility** is the ability to shorten actively and exert a pull, or tension, that can be harnessed by connective tissues.
3. **Extensibility** is the ability to continue to contract over a range of resting lengths. For example, a smooth muscle cell can be stretched to several times its original length and still contract on stimulation.
4. **Elasticity** is the ability of a muscle to rebound toward its original length after a contraction.

This chapter begins with the organization of skeletal muscle tissue. Although most of the muscle tissue in the body is skeletal muscle, this section will also consider cardiac and smooth muscle tissue. We will then proceed to a consideration of the functional organization of the *muscular system*.

FUNCTIONS OF SKELETAL MUSCLE

Skeletal muscle tissue, connective tissues, and neural tissue combine to form **skeletal muscles**, contractile organs that are directly or indirectly attached to bones. The muscular system includes approximately 700 skeletal muscles. These muscles perform the following functions:

1. **Produce movement**. Muscle contractions pull on tendons and move the bones of the skeleton.
2. **Maintain posture and body position**. Without constant muscular tension we could not sit upright without collapsing or stand without toppling over.
3. **Support soft tissues**. The abdominal wall and the floor of the pelvic cavity consist of layers of muscle that support the weight of visceral organs and shield internal tissues from injury.
4. **Guard entrances and exits**. Skeletal muscles guard openings to the digestive and urinary tracts and provide voluntary control over swallowing, defecating, and urination.
5. **Maintain body temperature**. Muscle contractions require energy, and whenever energy is used in the body, some of it is converted to heat. The heat lost by working muscles keeps our body temperature in the normal range.

THE ANATOMY OF SKELETAL MUSCLE

When naming structural features of muscles and their components, anatomists often used the Greek words *sarkos* (flesh) and *mys* (muscle). These word roots will be encountered in the following discussions of the anatomy of skeletal muscle.

Gross Anatomy

Figure 7-1● illustrates the appearance and organization of a typical skeletal muscle. A skeletal muscle contains connective tissues, blood vessels, nerves, and skeletal muscle tissue.

Connective Tissue Organization

Three layers of connective tissue are part of each muscle: an outer *epimysium*, a central *perimysium*, and an inner *endomysium* (Figure 7-1a●). The entire muscle is surrounded by the **epimysium** (ep-i-MĪS-ē-um; *epi-*, on + *mys*, muscle), a layer of collagen fibers that separates the muscle from surrounding tissues and organs.

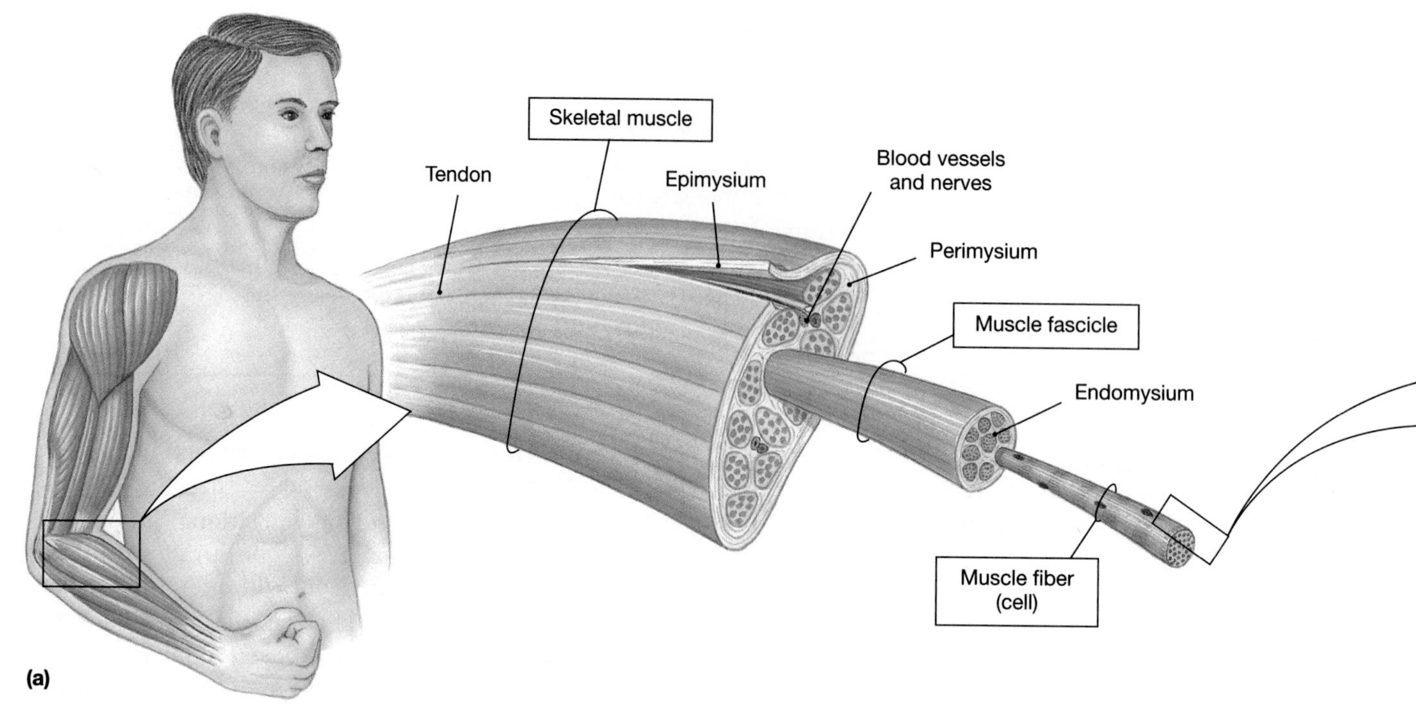

●**Figure 7-1**
Organization of Skeletal Muscles

(a) Gross organization of a skeletal muscle. **(b)** Structure of a skeletal muscle fiber. **(c)** Diagrammatic organization of a sarcomere, part of a single myofibril. **(d)** The sarcomere in part (c), stretched to the point where thick and thin filaments no longer overlap. (This cannot happen in an intact muscle fiber.) **(e)** Structure of a thin filament. **(f)** Structure of a thick filament.

At each end of the muscle, the epimysial fibers come together to form tendons that attach skeletal muscles to bones. [p. 84] The tendon fibers are interwoven into the periosteum of the bone, providing a firm attachment. Any contraction of the muscle will exert a pull on its tendon and in turn on the attached bone.

The connective tissue fibers of the **perimysium** (per-i-MĪS-ē-um; *peri-*, around) divide the skeletal muscle into a series of compartments, each containing a bundle of muscle fibers called a **fascicle** (FA-sik-ul; *fasciculus*, a bundle). In addition to collagen and elastic fibers, the perimysium contains blood vessels and nerves that supply the fascicles.

Within a fascicle, the **endomysium** (en-dō-MĪS-ē-um; *endo-*, inside) surrounds each skeletal muscle fiber and ties adjacent muscle fibers together. Stem cells scattered among the muscle fibers can help repair damaged muscle tissue. AM *Trichinosis, Necrotizing Fasciitis,* and *Fibromyalgia and Chronic Fatigue Syndrome*

Nerves and Blood Vessels

Skeletal muscles are often called *voluntary muscles* because their contractions can occur under voluntary control. Many of these skeletal muscles may also be controlled involuntarily. For example, skeletal muscles involved with breathing, such as the *diaphragm*, usually work under involuntary control. Each skeletal muscle fiber is controlled by a **motor neuron** whose cell body lies within the central nervous system. Axons, or *nerve fibers*, of the motor neurons involved with a particular skeletal muscle penetrate the epimysium, branch through the perimysium, and enter the endomysium to control individual muscle fibers. Since muscle contraction requires a tremendous amount of energy, an extensive network of blood vessels delivers the oxygen and nutrients needed for the production of ATP in active skeletal muscles.

Microanatomy

Skeletal muscle fibers (Figure 7-1a,b●) are quite different from the "typical" cell described in Chapter 3. One obvious difference is their enormous size. For example, a skeletal muscle fiber from a leg muscle could have a diameter of 100 µm and a length equal to that of the entire muscle (30–40 cm, or 10–16 in.). In addition, each skeletal muscle fiber is *multinucleate*, containing hundreds of nuclei just beneath the cell membrane. The genes contained in these nuclei direct the production of enzymes and structural proteins required for normal contraction, and the presence of multiple copies of these genes speeds up the process.

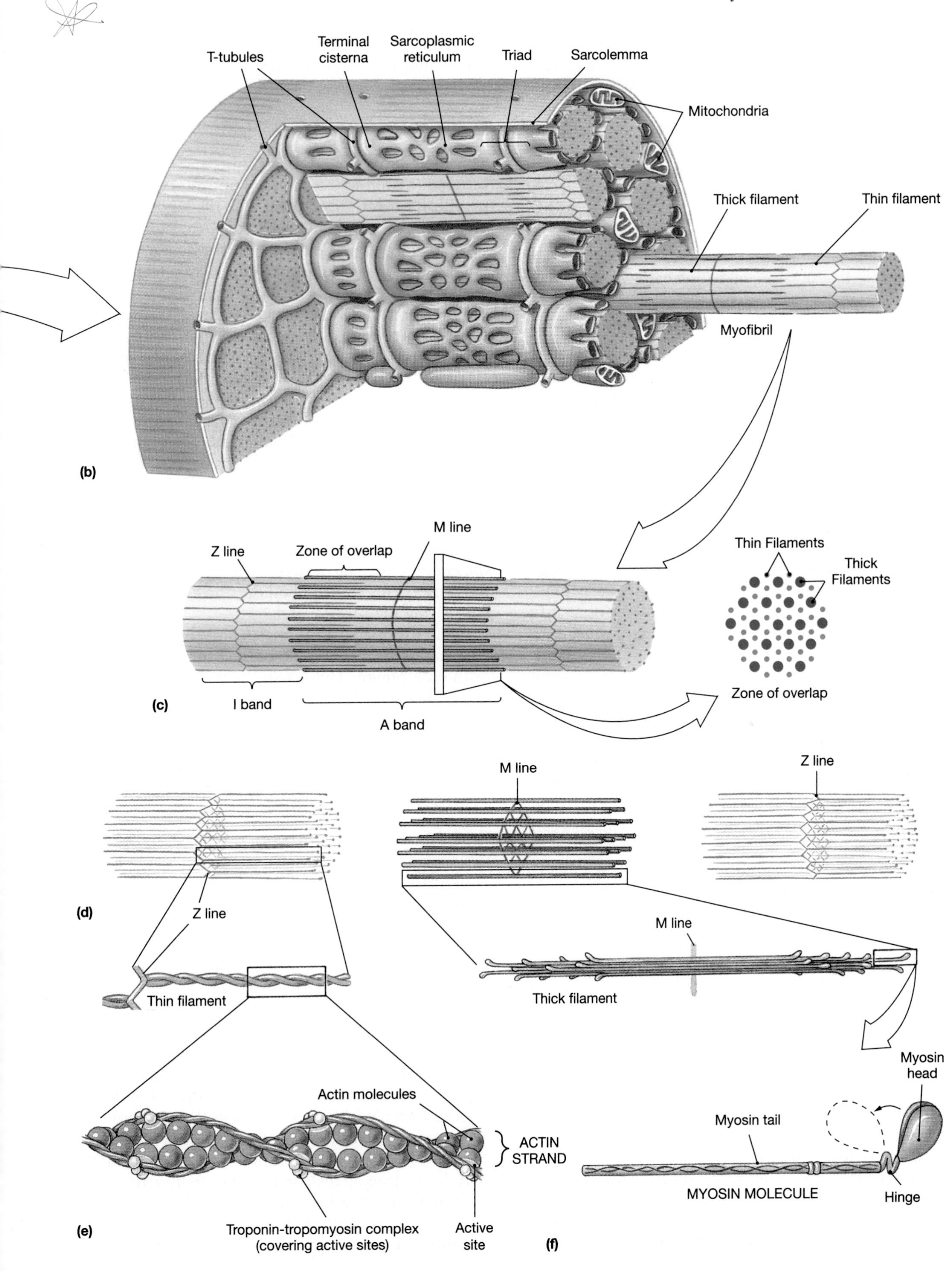
T-tubules
Terminal cisterna
Sarcoplasmic reticulum
Triad
Sarcolemma
Mitochondria
Thick filament
Thin filament
Myofibril
(b)
M line
Z line
Zone of overlap
Thin Filaments
Thick Filaments
Zone of overlap
(c)
I band
A band
M line
Z line
(d)
Z line
M line
Thin filament
Thick filament
Actin molecules
ACTIN STRAND
Myosin tail
Myosin head
MYOSIN MOLECULE
Hinge
(e)
Troponin-tropomyosin complex (covering active sites)
Active site
(f)

The cell membrane, or **sarcolemma** (sar-cō-LEM-a; *sarkos*, flesh + *lemma*, husk) of a muscle fiber surrounds the cytoplasm, or **sarcoplasm** (SAR-kō-plazm). Openings scattered across the surface of the sarcolemma lead into a network of narrow tubules called **transverse tubules**, or **T tubules**. Transverse tubules begin at the sarcolemma and extend into the sarcoplasm at right angles to the membrane surface. They are filled with extracellular fluid, and they form passageways through the muscle fiber, like a series of tunnels through a mountain. The T tubules play a major role in coordinating the contraction of the muscle fiber.

Transverse Tubules and the Sarcoplasmic Reticulum

A muscle fiber contraction occurs through the orderly interaction of both electrical and chemical events. Electrical events at the sarcolemma trigger a contraction by altering the chemical environment everywhere inside the muscle fiber. The electrical "message" is distributed by the transverse tubules that extend deep into the sarcoplasm of the muscle fiber. There they encircle the individual **myofibrils**, cylindrical structures that are responsible for the contraction of the muscle fiber.

As a T tubule encircles a myofibril it makes close contact with expanded chambers of the **sarcoplasmic reticulum**, a specialized form of endoplasmic reticulum. These chambers are called *cisternae*. As it encircles a myofibril, a transverse tubule lies sandwiched between a pair of cisternae (Figure 7-1b●), forming a *triad*.

These cisternae contain high concentrations of calcium ions. The calcium ion concentration in the cytoplasm of all cells is kept very low. Most cells, including skeletal muscle fibers, pump calcium ions across their cell membranes and into the extracellular fluid. Skeletal muscle fibers, however, also actively transport calcium ions into the cisternae of the sarcoplasmic reticulum. A muscle contraction begins when the stored calcium ions are released by the cisternae.

Myofibrils and Myofilaments

The sarcoplasm contains hundreds to thousands of myofibrils. Each myofibril is a cylinder 1–2 μm in diameter and as long as the entire muscle fiber. Myofibrils are responsible for muscle fiber contraction. Because they are attached to the sarcolemma at each end of the cell, their contraction shortens the entire cell. Scattered between the myofibrils are mitochondria and glycogen granules. The breakdown of glycogen and the activity of mitochondria provide the ATP needed to power muscular contractions.

Myofibrils are bundles of **myofilaments**, protein filaments consisting primarily of the proteins *actin* and *myosin*. Actin molecules are found in **thin filaments**, and the myosin molecules are found in **thick filaments**. Myofilaments are organized in repeating functional units called **sarcomeres** (SAR-kō-mērz; *sarkos*, flesh + *meros*, part).

Sarcomere Organization

The arrangement of thick and thin filaments within a sarcomere produces a banded appearance. All of the myofibrils are arranged parallel to the long axis of the cell, with their sarcomeres lying side by side. As a result, the entire muscle fiber has a banded, or striated, appearance corresponding to the bands of the individual sarcomeres (Figure 7-1b●).

Each myofibril consists of a linear series of approximately 10,000 sarcomeres. *The sarcomere is the smallest functional unit of the muscle fiber; interactions between the thick and thin filaments of sarcomeres are responsible for muscle contraction.*

Figure 7-1c● diagrams the external and internal structure of an individual sarcomere. Each sarcomere has a resting length of about 2.6 μm. The thick filaments lie in the center of the sarcomere. Thin filaments at either end of the sarcomere are attached to interconnecting filaments that make up the **Z lines**. From the Z lines, the thin filaments extend toward the center of the sarcomere, passing among the thick filaments. You can understand the relationship by comparing Figure 7-1c and 7-1d●.

The differences in the size and density of thick and thin filaments account for the banded appearance of the sarcomere. The **A band** is the area containing thick filaments. The region between two successive A bands—including the Z line—is the **I band**. (It may help you to remember that the A band appears dArk and the I band lIght in a light micrograph.)

Thin and Thick Filaments

Each thin filament consists of a twisted strand of actin molecules (Figure 7-1e●). Each actin molecule has an **active site** capable of interacting with myosin. In a resting muscle, the active sites along the thin filaments are covered by strands of the protein **tropomyosin** (trō-pō-MĪ-o-sin; *trope*, turning). The tropomyosin strands are held in position by molecules of **troponin** (TRŌ-pō-nin) that are bound to the actin strand.

Thick filaments are composed of myosin molecules, each of an attached *tail* and a free globular *head*. The myosin molecules are oriented away from the center of the sarcomere, with the heads projecting outward (Figure 7-1f●). It is the myosin heads that interact with actin molecules during a contraction. This interaction cannot occur unless the troponin changes position, moving the tropomyosin and exposing the active sites.

Calcium is the key that unlocks the active sites and starts a contraction. When calcium ions bind to troponin, the protein changes shape, swinging the tropomyosin away from the active sites. Cross-bridge binding can then occur, and a contraction begins. The cisternae of the sarcoplasmic reticulum are the source of the calcium that triggers muscle contraction, and the effect on the sarcomere is almost instantaneous.

Sliding Filaments

When a sarcomere contracts, the Z lines move closer together as the thin filaments slide toward the center of the sarcomere, parallel to the thick filaments. Figure 7-2● presents a two-dimensional view of this process. The sliding occurs through interactions between the thick and thin filaments. It begins when the myosin heads of thick filaments bind to active sites on thin filaments, in much the same way that a substrate molecule binds to the active site of an enzyme. When they connect thick filaments and thin filaments, the myosin heads are called **cross-bridges**. When a cross-bridge binds to an active site, it pivots toward the center of the sarcomere, pulling the thin filament in that direction. The cross-bridge then detaches and returns to its original position, ready to repeat a cycle of "attach, pivot, detach, and return" like a person pulling in a rope one-handed.

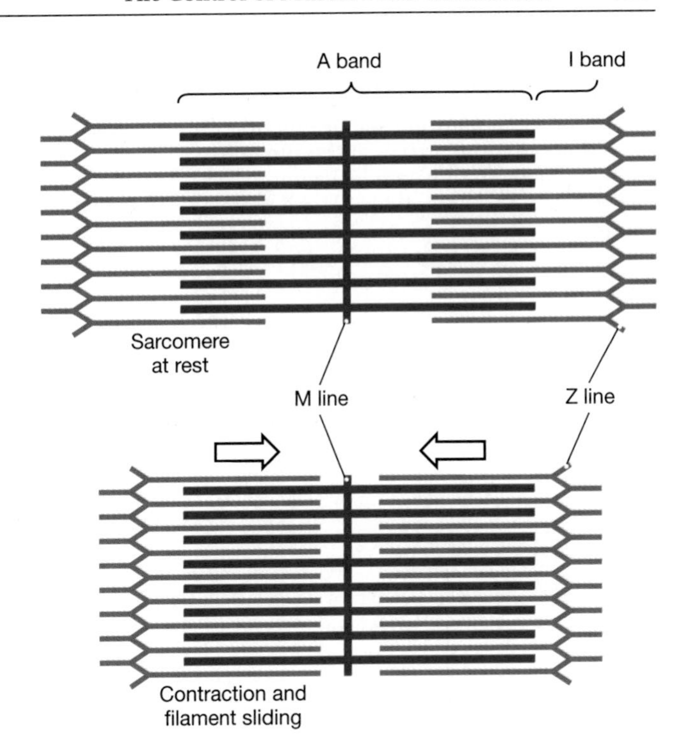

●Figure 7-2
Changes in the Appearance of a Sarcomere during Contraction of a Skeletal Muscle Fiber

During a contraction the A band stays the same width, but the Z lines move closer together and the I band gets smaller.

THE MUSCULAR DYSTROPHIES

Abnormalities in the genes that code for structural and functional proteins in muscle fibers are responsible for a number of inherited diseases collectively known as the **muscular dystrophies** (DIS-trō-fēz). These conditions, which cause a progressive muscular weakness and deterioration, are the result of abnormalities in the sarcolemma or in the structure of internal proteins. The best known example is *Duchenne's muscular dystrophy*, which typically affects male children age 3–7. AM *The Muscular Dystrophies*

✓ How would severing the tendon that was attached to a muscle affect the ability of the muscle to move a body part?

✓ Why does skeletal muscle appear striated when viewed with a microscope?

✓ Where would you expect to find the greatest concentration of calcium ions in resting skeletal muscle?

THE CONTROL OF MUSCLE FIBER CONTRACTION

Chapter 8 examines neural physiology in detail, so our discussion here will focus on neural stimulation of skeletal muscles. Before proceeding, however, you should take a moment to review the details of neuron structure that were introduced in Chapter 4. ∞ *[p. 89]*

Structure of the Neuromuscular Junction

Communication between the nervous system and a skeletal muscle fiber occurs at a specialized intercellular connection known as a **neuromuscular junction**. Each skeletal muscle fiber is controlled by a motor neuron at a single neuromuscular junction midway along its length. Figure 7-3b● summarizes key features of this structure. Each axon or branch of an axon ends at a **synaptic knob**. The cytoplasm of the synaptic knob contains mitochondria and vesicles filled with molecules of **acetylcholine** (as-ē-til-KŌ-lēn), or **ACh**. ACh is an example of a *neurotransmitter*, a chemical released by a neuron to change the activities of other cells. The release of ACh from the synaptic knob results in changes in the sarcolemma that trigger the contraction of the muscle fiber.

A narrow space, the **synaptic cleft**, separates the synaptic knob from the sarcolemma. This portion of the membrane, which contains receptors that bind ACh, is known as the **motor end plate**. The motor end plate has deep creases that increase the membrane surface area and the number of ACh receptors. Both the synaptic cleft and motor end plate contain the enzyme **acetylcholinesterase** (**AChE**, or *cholinesterase*), which breaks down molecules of ACh.

Figure 7-3
Neural Control of Muscle Contraction

(a) An SEM of a neuromuscular junction. **(b)** An action potential arrives at the synaptic knob. **(c)** Steps in the chemical communication between the synaptic knob and motor end plate.

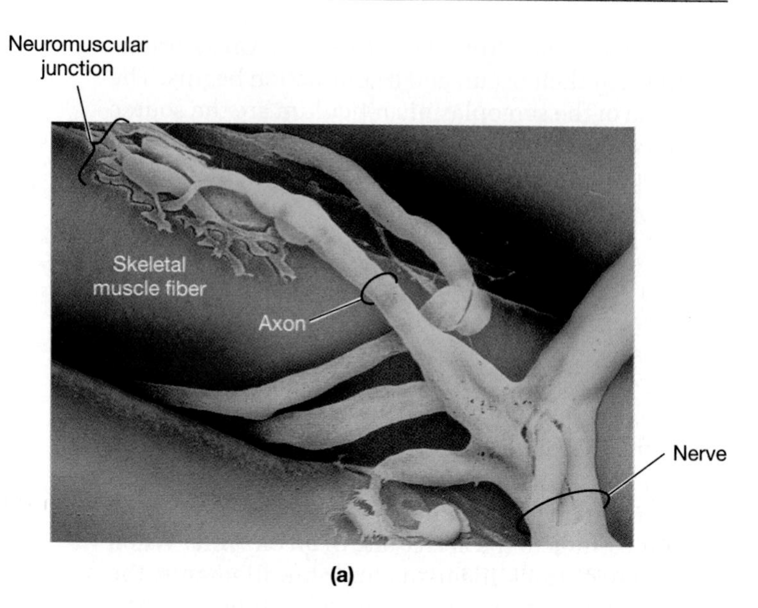

(a)

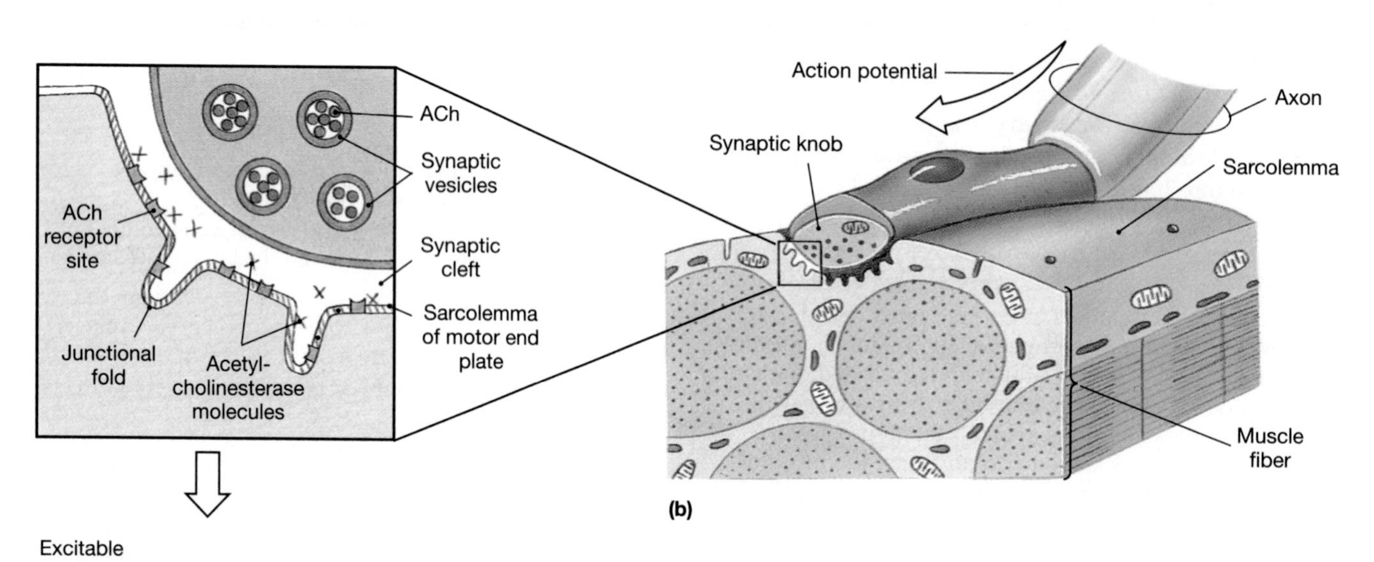

(b)

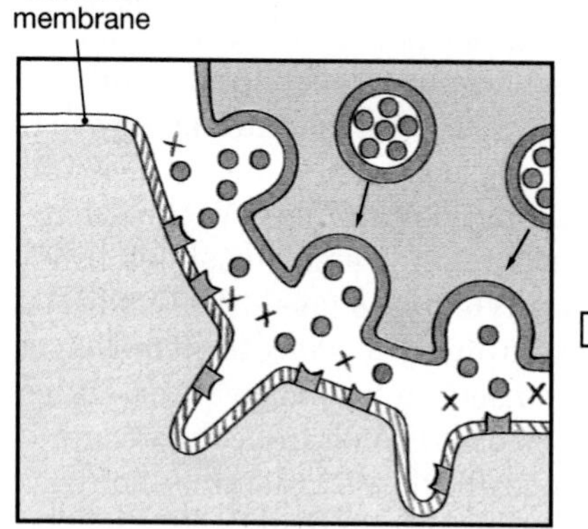

Step 1: Release of Acetylcholine. Vesicles in the synaptic knob release their contents into the synaptic cleft.

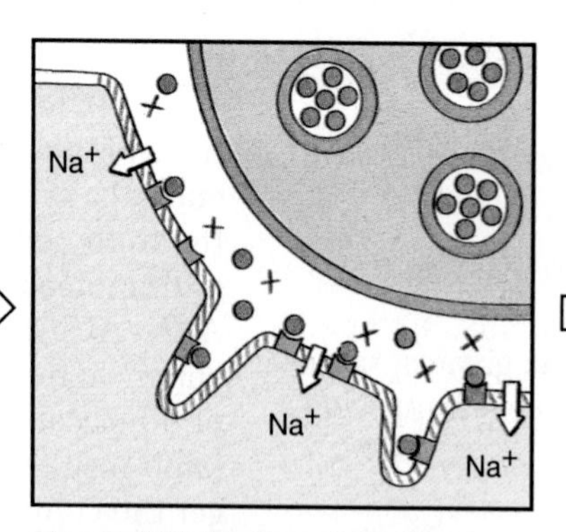

Step 2: ACh Binding at the Motor End Plate. The binding of ACh to the receptors changes the membrane permeability and induces an action potential in the sarcolemma.

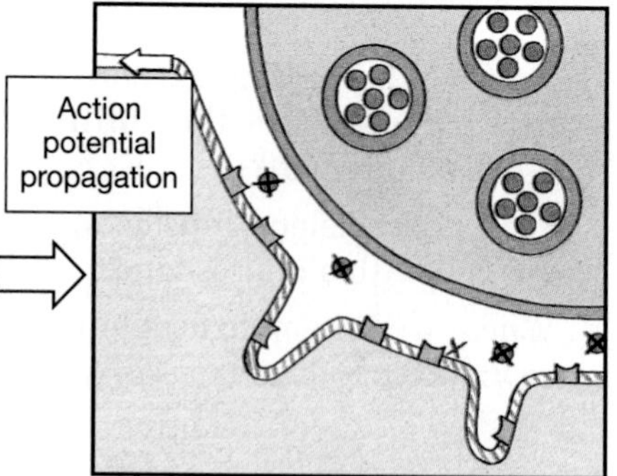

Step 3: Action Potential Conduction by the Sarcolemma. The action potential spreads across the membrane surface and travels down the transverse tubules, triggering the release of calcium ions at the terminal cisternae. While this occurs AChE removes the acetylcholine from the synaptic cleft.

(c)

Function of the Neuromuscular Junction

Neurons control skeletal muscle fibers by stimulating the production of an electrical impulse, or *action potential*, in the sarcolemma. The sequence of events, detailed in Figure 7-3c, can be summarized as follows:

STEP 1: Release of Acetylcholine. An action potential travels along the axon of a motor neuron. When this impulse reaches the synaptic knob, vesicles in the synaptic knob release acetylcholine into the synaptic cleft.

STEP 2: ACh Binding at the Motor End Plate. The ACh molecules diffuse across the synaptic cleft and bind to ACh receptors on the sarcolemma. This event changes the permeability of the membrane to sodium ions. It is this sudden rush of sodium ions into the sarcoplasm that produces an action potential in the sarcolemma. (We will examine the formation and conduction of action potentials more closely in the next chapter.)

STEP 3: Action Potential Conduction by the Sarcolemma. The action potential spreads over the entire sarcolemmal surface. It also travels down all of the transverse tubules toward the cisternae that encircle the sarcomeres of the muscle fiber. The passage of an action potential triggers a sudden, massive release of calcium ions by the cisternae.

As the calcium ion concentration rises, active sites are exposed on the thin filaments, cross-bridge interactions occur, and a contraction begins. Because all of the cisternae in the muscle fiber are affected, this contraction is a combined effort involving every sarcomere on every myofibril. While the contraction process gets under way, the acetylcholine is being broken down by acetylcholinesterase.

The Contraction Cycle

In the resting sarcomere each cross-bridge is bound to a molecule of ADP and phosphate (PO_4^{3-}), the products released by the breakdown of a molecule of ATP. In addition to binding the breakdown products, the cross-bridge stores the energy released by the rupture of the high-energy bond. In effect, the resting cross-bridge is "primed" for a contraction, like a cocked pistol or a set mouse trap.

The contraction process involves five interlocking steps which are diagrammed in Figure 7-4●:

STEP 1: Exposure of the active site following the binding of calcium ions (Ca^{2+}) to troponin.

STEP 2: Attachment of the myosin cross-bridge to the exposed active site on the thin filaments.

STEP 3: Pivoting of the attached myosin head toward the center of the sarcomere and the release of ADP and a phosphate group. This step uses the energy that was stored in the myosin molecule at rest.

STEP 4: Detachment of the cross-bridges when the myosin head binds another ATP molecule.

STEP 5: Reactivation of the detached myosin head, as it splits the ATP and captures the released energy. The entire cycle can now be repeated, beginning with step 2.

This cycle is broken when calcium ion concentrations return to normal resting levels, primarily through active transport into the sarcoplasmic reticulum. If a single action potential sweeps across the sarcolemma, calcium ion removal occurs very rapidly, and the contraction will be very brief. A sustained contraction will occur only if action potentials occur one after another, and calcium loss from the cisternae continues.

RIGOR MORTIS

When death occurs, circulation ceases and the skeletal muscles are deprived of nutrients and oxygen. Within a few hours, the skeletal muscle fibers have run out of ATP, and the sarcoplasmic reticulum becomes unable to remove calcium ions from the sarcoplasm. Calcium ions diffusing into the sarcoplasm from the extracellular fluid or leaking out of the sarcoplasmic reticulum then trigger a sustained contraction. Without ATP, the cross-bridges cannot detach from the active sites, and the muscle locks in the contracted position. All of the body's skeletal muscles are involved, and the individual becomes "stiff as a board." This physical state, called **rigor mortis**, lasts until the lysosomal enzymes released by autolysis break down the myofilaments 15–25 hours later.

Muscle Contraction: A Summary

Table 7-1 provides a visual summary of the contraction process, from ACh release to the end of the contraction.

INTERFERENCE WITH NEURAL CONTROL MECHANISMS

Anything that interferes with either neural function or excitation-contraction coupling will cause muscular paralysis. Two examples are worth noting:

Cases of *botulism* result from the consumption of contaminated canned or smoked food. The toxin, produced by bacteria, prevents the release of ACh at the synaptic terminals, leading to a potentially fatal muscular paralysis. AM *Botulism*

The progressive muscular paralysis seen in *myasthenia gravis* results from the loss of ACh receptors at the motor end plate. The primary cause is a misguided

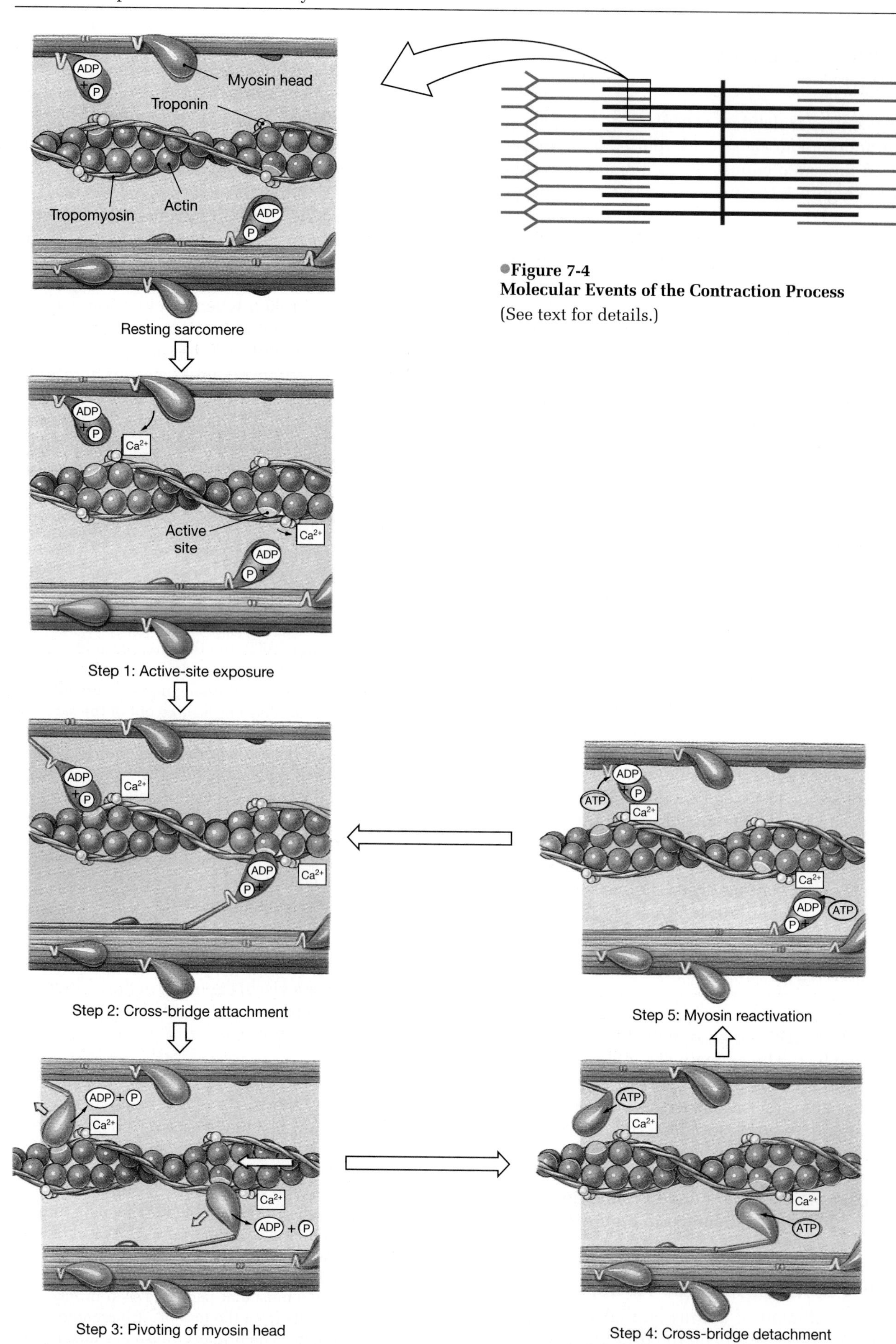

Figure 7-4
Molecular Events of the Contraction Process
(See text for details.)

TABLE 7-1 A Summary of the Steps Involved in Skeletal Muscle Contraction

Key steps in the initiation of a contraction include:

1. At the neuromuscular junction, ACh released by the synaptic knob binds to receptors on the sarcolemma.
2. The resulting change in the membrane potential of the muscle fiber leads to the production of an action potential that spreads across its entire surface and reaches the triads via the transverse tubules.
3. The sarcoplasmic reticulum releases stored calcium ions, increasing the calcium concentration of the sarcoplasm in and around the sarcomeres.
4. Calcium ions bind to troponin, producing a change in the orientation of the troponin-tropomyosin complex that exposes active sites on the thin (actin) filaments.
5. Repeated cycles of cross-bridge binding, pivoting, and detachment occur, powered by the breakdown of ATP. These events produce filament sliding, and the muscle fiber shortens.

This process continues for a brief period, until:

6. Action potential generation ceases as ACh is removed by acetylcholinesterase.
7. The sarcoplasmic reticulum reabsorbs calcium ions, and the concentration of calcium ions in the sarcoplasm declines.
8. When calcium ion concentrations approach normal resting levels, the troponin-tropomyosin complex returns to its normal position. This change covers the active sites and prevents further cross-bridge interaction.
9. Without cross-bridge interactions, further sliding will not take place, and the contraction will end.
10. Muscle relaxation occurs, muscle returns passively toward resting length.

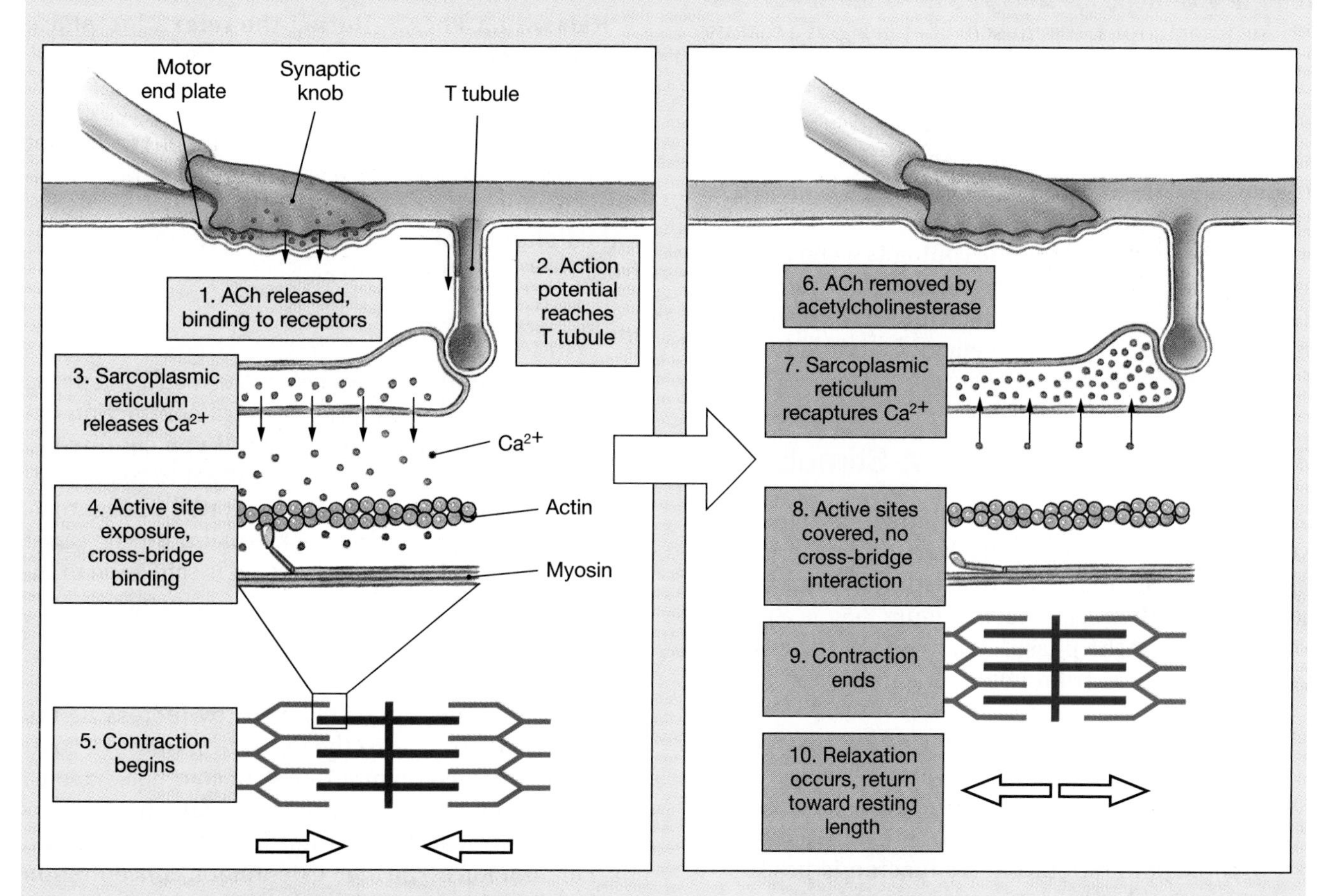

Steps that initiate a contraction.

Steps that end the contraction.

attack on the ACh receptors by the immune system. Genetic factors play a role in predisposing individuals to develop this condition. AM *Myasthenia gravis*

✓ How would a drug that interferes with cross-bridge formation affect muscle contraction?

✓ What would you expect to happen to a resting skeletal muscle if the sarcolemma suddenly became very permeable to calcium ions?

✓ Predict what would happen to a muscle if the motor end plate did not contain acetylcholinesterase.

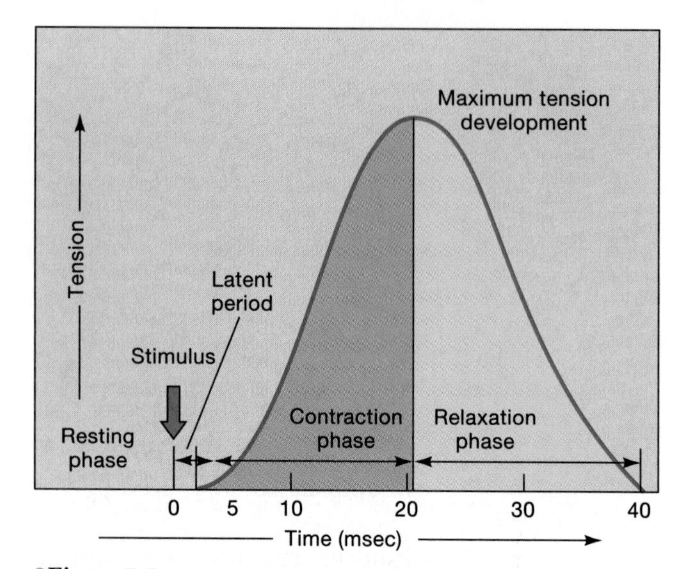

Figure 7-5
The Twitch and Development of Tension
Myogram showing the time course of a single twitch contraction in the gastriocnemius muscle.

MUSCLE MECHANICS

Now that we are familiar with muscle contraction of individual muscle fibers, we can examine the performance of *skeletal muscles*, the organs of the muscular system. In this section we will consider the coordinated contractions of an entire population of muscle fibers.

The amount of tension produced by an individual muscle fiber depends solely on the number of cross-bridge interactions. If a muscle fiber at a given resting length is stimulated to contract, it will always produce the same amount of tension. There is no mechanism to regulate the amount of tension produced in that contraction: The muscle fiber is either "ON" (producing tension) or "OFF" (relaxed). This is known as the **all-or-none principle.**

An entire skeletal muscle contracts when its component muscle fibers are stimulated. The amount of tension produced in the skeletal muscle *as a whole* is determined by (1) the frequency of stimulation and (2) the number of muscle fibers activated.

The Frequency of Muscle Stimulation

A **twitch** is a single stimulus-contraction-relaxation sequence in a muscle fiber. Its duration can be as brief as 7.5 msec, as in an eye muscle fiber, or up to 100 msec, as in calf muscle fibers. Figure 7-5 is a graph, or **myogram**, of the phases of a twitch in the *gastrocnemius muscle,* a prominent calf muscle.

- **Latent Period.** The **latent period** begins at stimulation and typically lasts about 2 msec. Over this period the action potential sweeps across the sarcolemma, and calcium ions are released by the sarcoplasmic reticulum. No tension is produced by the muscle fiber because the contractile mechanism is not yet activated.
- **Contraction Phase.** In the **contraction phase**, tension rises to a peak. Throughout this period the cross-bridges are interacting with the active sites on the actin filaments.
- **Relaxation Phase.** During the **relaxation phase** muscle tension falls to resting levels as the cross-bridges detach.

A single stimulation produces a single twitch, but twitches in a skeletal muscle do not accomplish anything useful. All normal activities involve more sustained muscle contractions. Such contractions result from repeated stimulations.

Incomplete Tetanus

If a second stimulus arrives before the relaxation phase has ended, a second, more powerful contraction occurs, as indicated in Figure 7-6a. If you continue to stimulate the muscle, never allowing it to relax completely, tension will peak as illustrated in Figure 7-6b. A muscle producing peak tension during rapid cycles of contraction and relaxation is said to be in **incomplete tetanus** (*tetanos*, convulsive tension).

Complete Tetanus

Complete tetanus can be obtained by increasing the rate of stimulation until the relaxation phase is completely eliminated (Figure 7-6c). In complete tetanus, the action potentials are arriving so fast that the sarcoplasmic reticulum does not have time to reclaim the calcium ions. The high calcium ion concentration in the cytoplasm prolongs the state of contraction, making it continuous. Virtually all normal muscular contractions involve the complete tetanus of the participating muscle units.

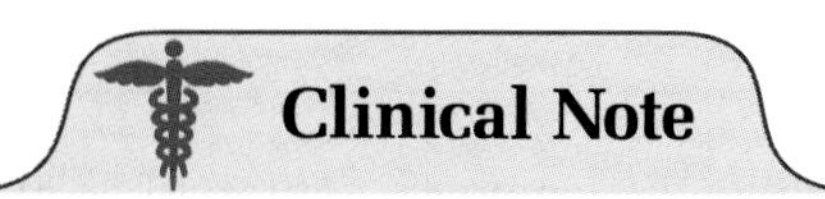

Tetanus

Children are often told to watch out for rusty nails. Parents are not worrying about the rust or the nail, but about infection with a very common bacterium, *Clostridium tetani*. This bacterium can cause **tetanus**, a disease that has no relationship to the normal response to neural stimulation. The *Clostridium* bacteria, although found virtually everywhere, can thrive only in tissues that contain abnormally low amounts of oxygen. For this reason, a deep puncture wound, such as that from a nail, carries a much greater risk than a shallow, open cut that bleeds freely.

When active in body tissues, these bacteria release a powerful toxin that affects the central nervous system. Motor neurons are particularly sensitive to it, and their stimulation produces a sustained, powerful contraction of skeletal muscles throughout the body. This toxin also has a direct effect on skeletal muscle fibers, producing contraction even in the absence of neural commands.

After exposure, the incubation period (the time before symptoms develop) is usually less than 2 weeks. The most common complaints are headache, muscle stiffness, and difficulty in swallowing. Because it soon becomes difficult to open the mouth, this disease is also called *lockjaw*. Widespread muscle spasms usually develop within 2–3 days of the initial symptoms, and they often continue for a week before subsiding. After 2–4 weeks, symptoms in surviving patients disappear with no aftereffects.

Although severe tetanus has a 40–60 percent mortality rate, immunization is effective in preventing the disease. There are approximately 500,000 cases of tetanus worldwide each year, but only about 100 of them occur in the United States, thanks to an effective immunization program. (Most readers will have had "tetanus shots" or "tetanus boosters" sometime within the last 10 years.) Severe symptoms in unimmunized patients can be prevented by early administration of an antitoxin, usually *human tetanus immune globulin*. Such treatment does not reduce symptoms that have already appeared, however, and there is no generally effective treatment for this disease.

The Number of Muscle Fibers Involved

We have a remarkable ability to control the amount of tension exerted by our skeletal muscles. During a normal movement, our muscles contract smoothly, not jerkily, because activated muscle fibers are responding in complete tetanus. The *total force* exerted by the muscle as a whole depends on how many muscle fibers are activated.

A typical skeletal muscle contains thousands of muscle fibers. Although some motor neurons control a single muscle fiber, most control hundreds or thousands of muscle fibers through multiple synaptic knobs. All of the muscle fibers controlled by a single motor neuron constitute a **motor unit**.

The size of a motor unit indicates how fine the control of movement can be. In the muscles of the eye, where precise control is extremely important, a motor neuron may control two or three muscle fibers. We have much less precise control over our leg muscles, where up to 2,000 muscle fibers may respond to the call of a single motor neuron.

When a decision is made to perform a specific movement, specific groups of motor neurons within the central nervous system are stimulated. The stimulated neurons do not respond spontaneously, and over time the number of activated motor units gradually increases. The smooth but steady increase in muscular tension produced by increasing the number of active motor units is called **recruitment**.

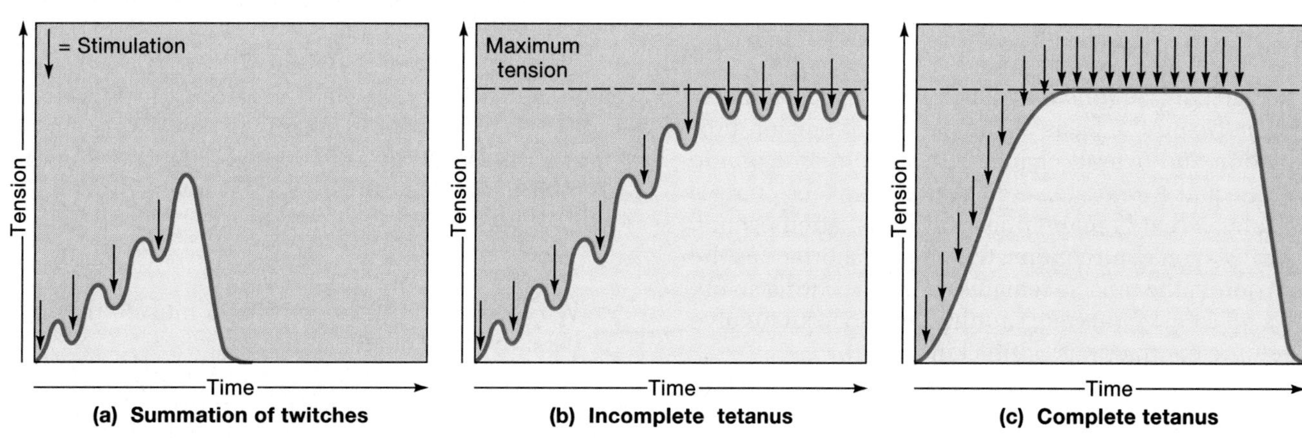

Figure 7-6
Effects of Repeated Stimulations

(a) Tension rises when successive stimuli arrive before relaxation has been completed. **(b)** Incomplete tetanus occurs if the rate of stimulation increases further. Tension production will rise to a peak, and the periods of relaxation will be very brief. **(c)** In complete tetanus, the frequency of stimulation is so high that the relaxation phase has been completely eliminated and tension plateaus at maximal levels.

Peak tension production occurs when all of the motor units in the muscle are contracting in complete tetanus. Such contractions do not last long, however, because the muscle fibers soon use up their available energy supplies. During a sustained **tetanic contraction**, motor units are activated on a rotating basis, so that some are resting while others are contracting.

Muscle Tone

Some of the motor units within any particular muscle are always active, even when the entire muscle is not contracting. Their contractions do not produce enough tension to cause movement, but they do tense and firm the muscle. This resting tension in a skeletal muscle is called **muscle tone**. A muscle with little muscle tone appears limp and flaccid, whereas one with moderate muscle tone is quite firm and solid.

Stimulated muscle fibers can enlarge and become more powerful through a process called *hypertrophy*. At the other extreme, a skeletal muscle that is not stimulated by a motor neuron on a regular basis will **atrophy**: Its muscle fibers will become smaller and weaker. Individuals paralyzed by spinal injuries or other damage to the nervous system gradually lose muscle tone and volume in the areas affected. Even a temporary reduction in muscle use can lead to muscular atrophy, as is easily seen by comparing limb muscles before and after a cast has been worn. Muscle atrophy is initially reversible, but dying muscle fibers are not replaced, and in extreme atrophy the functional losses are permanent. That is why physical therapy is so important in cases where patients are temporarily unable to move normally. AM *Polio*

Isotonic and Isometric Contractions

Muscle contractions may be classified as *isotonic* or *isometric* on the basis of the pattern of tension production and overall change in shape. In an **isotonic** (*iso-*, equal + *tonos*, tension) contraction, tension rises to a level that is maintained until relaxation occurs. During such contractions, the muscle shortens as the tension in the muscle remains constant. Lifting an object off a desk, walking, running, and so forth involve isotonic contractions of this kind.

In an **isometric** (*metric*, measure) contraction tension continues to rise but the muscle as a whole does not change in length. Examples of isometric contractions include pushing against a wall or trying to pick up a large car. These are rather unusual movements, but many of the everyday reflexive muscle contractions that keep our bodies upright when standing or sitting involve the isometric contractions of muscles that oppose gravity.

Normal daily activities involve a combination of isotonic and isometric muscular contractions. As you sit reading this text, isometric contractions of postural muscles stabilize your vertebrae and maintain your upright position. When you next turn a page, the movements of your arm, forearm, hand, and fingers are produced by isotonic contractions.

Muscle Elongation

Upon entering the relaxation phase (see Figure 7-5●, p. 162), a muscle fiber is at its contracted length. How then does it return to its original (uncontracted) length? There is no active mechanism for muscle fiber elongation; contraction is active, but elongation is passive. After a contraction, a muscle fiber usually returns to its original length through a combination of (1) elastic forces and (2) the movements of opposing muscles.

Elastic Forces

Every time a muscle fiber contracts, its organelles are compressed and the extracellular fibers, such as those of the endomysium, are stretched. When the contraction ends the intracellular and extracellular elements rebound to their original dimensions, gradually returning the muscle fiber to its original resting length.

Opposing Muscle Movements

Much more rapid returns to resting length result from the contraction of opposing muscles. For example, contraction of the *biceps brachii* muscle on the anterior part of the arm flexes the elbow; contraction of the *triceps brachii* muscle on the posterior surface of the arm extends the elbow. When the biceps brachii contracts, the triceps brachii is stretched; when the biceps brachii relaxes, contraction of the triceps brachii extends the elbow and stretches the muscle fibers of the biceps brachii to their original length.

✓ Why is it difficult to contract a muscle that has been overstretched?

✓ A motor unit from a skeletal muscle contains 1500 muscle fibers. Would this muscle be involved in fine, delicate movements or powerful, gross movements? Explain.

✓ Is it possible for a muscle to contract without shortening? Explain.

THE ENERGETICS OF MUSCULAR ACTIVITY

Muscle contraction requires large amounts of energy. For example, an active skeletal muscle fiber may require some 600 trillion molecules of ATP each second, not including the energy needed to pump the calcium ions back into the sarcoplasmic reticulum. Although resting skeletal muscle cells contain large energy reserves in the form of ATP and other high-energy compounds, the reserves are only enough to sustain a contraction until additional ATP can be generated. Throughout the rest of the contraction the muscle fiber will generate ATP at roughly the same rate as it is used.

ATP and CP Reserves

The primary function of ATP is the transfer of energy from one location to another rather than the long-term storage of energy. At rest, a skeletal muscle fiber produces more ATP than it needs, and under these conditions ATP transfers energy to another high-energy compound, **creatine phosphate (CP)** (Figure 7-7a●). During a contraction each cross-bridge breaks down ATP, producing ADP and a phosphate group. The energy stored in creatine phosphate is then used to "recharge" the ADP back to ATP.

A resting skeletal muscle fiber contains about six times as much creatine phosphate as ATP. But when a muscle fiber is contracting repeatedly, both of these energy reserves will be exhausted in about 30 seconds. At such times, the muscle fiber must then rely on other mechanisms to convert ADP to ATP.

ATP Generation

As you may recall from Chapter 3, most cells in the body generate ATP through **aerobic metabolism** (*aer*, air) in mitochondria. ∞ *[pp. 60–61]*

Aerobic Metabolism

Aerobic metabolism normally provides 95 percent of the ATP needed by a resting cell. In this process, mitochondria absorb oxygen, ADP, phosphate ions, and organic substrates from the surrounding cytoplasm. The organic substrates are carbon chains produced by disassembling carbohydrates, lipids, or proteins. The absorbed molecules then enter the *TCA cycle*, where they are completely broken down. The carbon atoms and oxygen atoms are released as carbon dioxide (CO_2). The hydrogen atoms are shuttled to *respiratory enzymes* on the mitochondrial cristae, and they ultimately combine with oxygen to form water (H_2O).

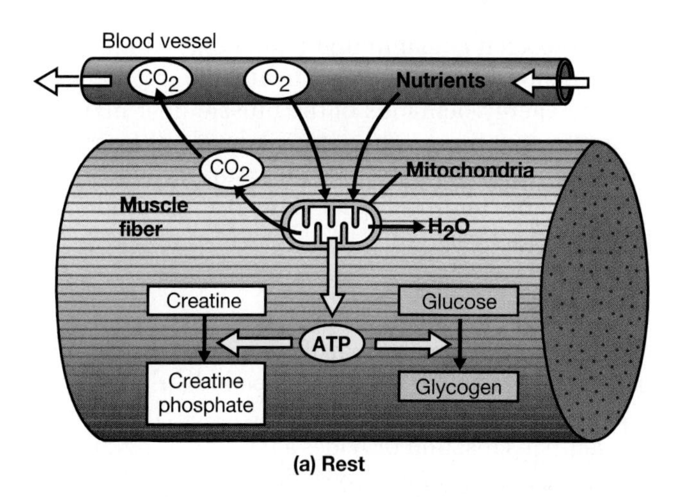

(a) Rest

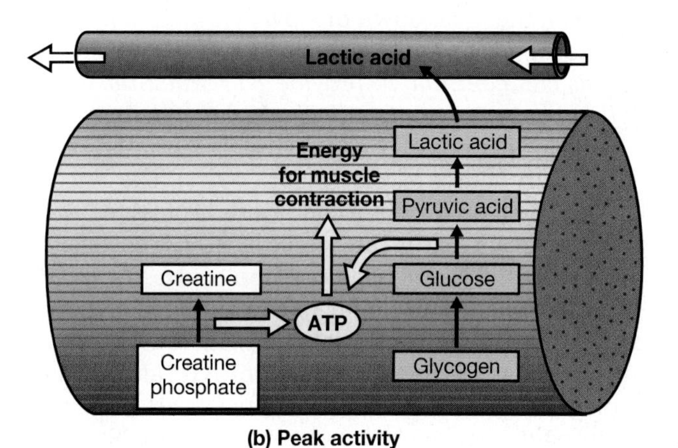

(b) Peak activity

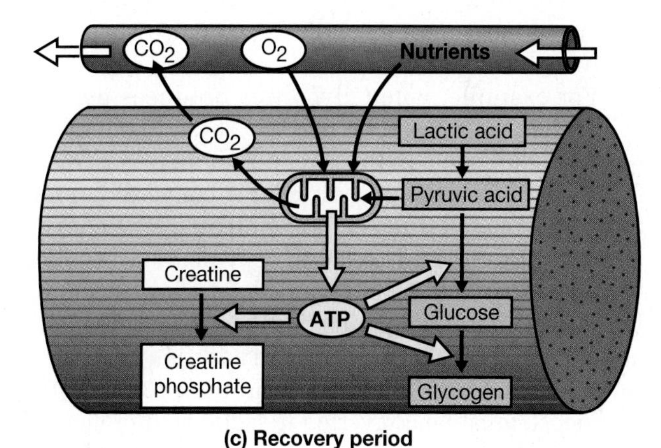

(c) Recovery period

●**Figure 7-7**
Muscle Metabolism

(a) A resting muscle, maintaining its energy reserves. **(b)** A muscle at peak activity, using its energy reserves. **(c)** The recovery period and the rebuilding of energy reserves.

Along the way, large amounts of energy are released and used to make ATP.

The maximum rate of ATP generation within mitochondria is limited by the availability of oxygen. A sufficient supply of oxygen becomes a problem as the energy demands of the muscle fiber increase. Although oxygen consumption and energy production by mitochondria can increase to 40 times resting levels, the energy demands of the muscle fiber may increase by 120 times. Thus at peak levels of exertion, mitochondrial activity provides only around one-third of the required ATP. The rest is produced through *glycolysis*.

Glycolysis

Glycolysis is the breakdown of glucose to pyruvic acid in the cytoplasm of the cell. It is called an *anaerobic* (*an-*, not + *aer-*, air) *process* because it does not require oxygen. This reaction provides a net gain of 2 ATP and generates 2 molecules of **pyruvic acid**. The ATP yield of glycolysis is much lower than that of aerobic metabolism; the breakdown of 2 pyruvic acid molecules in mitochondria would generate 34 ATP. However, *glycolysis can continue to provide ATP when the availability of oxygen limits mitochondrial activity.*

During periods of peak activity, glycolysis becomes the primary source of ATP (Figure 7-7b●). The glucose broken down under these conditions is obtained from glycogen reserves in the sarcoplasm. Glycogen is a polysaccharide chain of glucose molecules. ∞ [p. 35] Typical skeletal muscle fibers contain large glycogen reserves in the form of insoluble granules. When the muscle fiber begins to run short of ATP and CP, enzymes break the glycogen molecules apart, releasing glucose that can be used to generate more ATP.

The anaerobic process of glycolysis enables the cell to continue generating ATP when mitochondrial activity alone cannot meet the demand. However, this pathway has its drawbacks:

For example, when glycolysis produces pyruvic acid faster than it can be used by the mitochondria, pyruvic acid levels in the sarcoplasm increase. Under these conditions the pyruvic acid is converted to **lactic acid**, a related three-carbon molecule. The conversion of pyruvic acid to lactic acid poses a problem because lactic acid is an organic acid whose accumulation can cause dangerous changes in the pH inside and outside of the muscle fiber. In addition, glycolysis is inefficient. Under anaerobic conditions 18 molecules of glucose must be converted to lactic acid molecules to obtain the same amount of energy produced by the aerobic catabolism of a single glucose molecule.

Muscle Fatigue

A skeletal muscle fiber is **fatigued** when it can no longer contract despite continued neural stimulation. Muscle fatigue may be caused by the exhaustion of energy reserves or the buildup of lactic acid.

If the muscle contractions use ATP at or below the maximum rate of mitochondrial ATP generation, the muscle fiber can function aerobically. Under these conditions, fatigue will not occur until glycogen and other reserves such as lipids and amino acids are depleted. This type of fatigue affects the muscles of long-distance athletes, such as marathon runners, after hours of exertion.

When a muscle produces a sudden, intense burst of activity, the ATP is provided by glycolysis. After a relatively short time (seconds to minutes), the rising lactic acid levels lower the tissue pH, and the muscle can no longer function normally. Athletes running sprints, such as the 100-yard dash, suffer from this type of muscle fatigue.

The Recovery Period

When a muscle fiber contracts, the conditions in the sarcoplasm are changed. For example, energy reserves are consumed, heat is released, and lactic acid may be present. In the **recovery period**, conditions inside the muscle are returned to normal, preexertion levels. During the recovery period, the muscle's metabolic activity focuses on the removal of lactic acid and the replacement of intracellular energy reserves, and the body as a whole loses the heat generated during intense muscular contraction.

Lactic Acid Recycling

The reaction that converts pyruvic acid to lactic acid is freely reversible. During the recovery period, when lactic acid concentration is high, the lactic acid is converted back to pyruvic acid. This pyruvic acid can then be used (1) to synthesize glucose and (2) to generate ATP through mitochondrial activity. The ATP produced is used to convert creatine to creatine phosphate and to store the newly synthesized glucose as glycogen (Figure 7-7c●).

During the recovery period, the body's oxygen demand goes up considerably. The extra oxygen is consumed by liver cells, as they produce ATP for the conversion of lactic acid back to glucose, and by muscle cells, as they restore their reserves of ATP, creatine phosphate, and glycogen. The additional oxygen required during the recovery period is often called an *oxygen debt*. While that debt is being repaid, the breathing rate and depth are increased; that is why you continue to breathe heavily for some time even after you stop exercising.

Heat Loss

Muscular activity generates substantial amounts of heat that warms the sarcoplasm, interstitial fluid, and circulating blood. Since muscle makes up a large fraction of the total mass of the body, muscle contractions play an important role in the maintenance of normal body temperature. For example, shivering can help keep you warm in a cold environment. But, when skeletal muscles are contracting at peak levels, body temperature soon begins to climb. In response, blood flow to the skin increases, promoting heat loss through mechanisms described in Chapters 1 and 5. ∞ [pp. 12, 103–4]

Muscle Performance

Muscle performance can be considered in terms of sheer **power**, the maximum amount of tension produced by a particular muscle or muscle group, and **endurance**, the amount of time for which the individual can perform a particular activity. Two major factors determine the capabilities of a particular skeletal muscle: (1) the types of muscle fibers within the muscle and (2) physical conditioning or training.

Types of Skeletal Muscle Fiber

There are two contrasting types of skeletal muscle fibers in the human body: fast fibers and slow fibers.

Fast Fibers

Most of the skeletal muscle fibers in the body are called **fast fibers** because they can contract in 0.01 second or less following stimulation. Fast fibers are large in diameter; they contain densely packed myofibrils, large glycogen reserves, and relatively few mitochondria. The tension produced by a muscle fiber is directly proportional to the number of sarcomeres, so fast-fiber muscles produce powerful contractions. However, because these contractions use ATP in massive amounts, prolonged activity is primarily supported by glycolysis, and fast fibers fatigue rapidly.

Slow Fibers

Slow fibers are only about half the diameter of fast fibers, and they take three times as long to contract after stimulation; however, they can continue contracting for extended periods, long after a fast muscle would have become fatigued. Three specializations make this possible:

1. Slow muscle tissue contains a more extensive network of capillaries than does typical fast muscle tissue, so oxygen supply is dramatically increased.
2. Slow muscle fibers contain the red pigment **myoglobin** (MĪ-ō-glō-bin), a globular protein structurally related to hemoglobin, the oxygen-carrying pigment found in the blood. ∞ [p. 38] Since myoglobin also binds oxygen molecules, resting slow muscle fibers contain oxygen reserves that can be mobilized during a contraction.
3. Slow muscle fibers contain a relatively larger number of mitochondria than do fast muscle fibers.

Distribution of Muscle Fibers and Muscle Performance

The percentage of fast and slow muscle fibers in a particular skeletal muscle can be quite variable. Muscles dominated by fast fibers appear pale, and they are often called **white muscles**. Chicken breasts contain "white meat" because chickens use their wings only for brief intervals, as when fleeing from a predator, and the power for flight comes from fast fibers in their breast muscles. The extensive blood vessels and myoglobin in slow muscle fibers give them a reddish color, and muscles dominated by slow fibers are therefore known as **red muscles**. Chickens walk around all day, and the movements are performed by the slow muscle fibers in the "dark meat" of their legs.

Most human muscles contain a mixture of both fiber types, and so appear pink. However, there are no slow fibers in muscles of the eye and hand, where swift but brief contractions are required. Many back and calf muscles are dominated by slow fibers; these muscles contract almost continuously to maintain an upright posture. The percentage of fast versus slow fibers in each muscle is genetically determined, but the fatigue resistance of fast muscle fibers can be increased through athletic training.

Physical Conditioning

Physical conditioning and training schedules enable athletes to improve both power and endurance. In practice, the training schedule varies depending on whether the activity is primarily supported by aerobic or anaerobic energy production.

Anaerobic endurance is the ability to support sustained, powerful muscle contractions through anaerobic mechanisms. Examples of activities that require anaerobic endurance include a 50-yard dash or swim, a pole vault, or a weight-lifting competition. Such activities are performed by fast muscle fibers. Athletes training to develop anaerobic endurance perform frequent, brief, intensive workouts. The net effect is an enlargement, or **hypertrophy**, of the stimulated muscle as seen in a champion weight lifter or body builder.

Aerobic endurance refers to the length of time for which a muscle can continue to contract while supported by mitochondrial activities. Because mitochondrial activities yield relatively large quantities of ATP, muscle contractions can continue for an extended period. Training to improve aerobic endurance usually involves sustained low levels of muscular activity. Examples include jogging, distance swimming, and other exercises that do not require peak tension production. Because glucose is a preferred energy source, aerobic athletes, such as marathon runners, often "load" or "bulk up" on carbohydrates on the day before an event.

✓ Why would a sprinter experience muscle fatigue before a marathon runner would?

✓ Which activity would be more likely to create an oxygen debt, swimming laps or lifting weights?

✓ What type of muscle fibers would you expect to predominate in the large leg muscles of someone who excels at endurance activities such as cycling or long-distance running?

CARDIAC AND SMOOTH MUSCLE TISSUE

Cardiac muscle tissue and smooth muscle tissue were introduced in Chapter 4. Table 7-2 compares skeletal, cardiac, and smooth muscle tissue in greater detail.

Cardiac Muscle Tissue

Cardiac muscle cells are relatively small and usually have a single, centrally placed nucleus. Cardiac muscle tissue is found only in the heart.

Differences between Cardiac Muscle and Skeletal Muscle

Figure 7-8a● provides an overview of the structure of a cardiac muscle cell. Like skeletal muscle fibers, cardiac muscle cells contain an orderly arrangement of myofibrils and are striated, but significant differences exist in their structure and function. The most obvious structural difference is that cardiac muscle cells are branched, and each cardiac cell contacts several others at specialized sites called **intercalated** (in-TER-ka-lā-ted) **discs** (see Figure 4-13●, p. 90). These cellular connections contain gap junctions that provide a means for the movement of ions and small molecules and the rapid passage of action potentials from cell to cell, resulting in their simultaneous contraction. Because the myofibrils are also attached to the intercalated discs, the cells "pull together" quite efficiently.

There are also several important functional differences between cardiac and skeletal muscle:

- Cardiac muscle tissue contracts without neural stimulation, a property called *automaticity*. The timing of contractions is normally determined by specialized cardiac muscle cells called **pacemaker cells**.
- Cardiac muscle cell contractions last roughly 10 times as long as those of skeletal muscle fibers.
- The properties of cardiac muscle cell membranes differ from those of skeletal muscle fibers. As a result, cardiac muscle tissue cannot undergo tetanus, or sustained contraction. This property is important because a heart in tetany could not pump blood.
- Cardiac muscle cells rely on aerobic metabolism to obtain the energy needed to continue contracting. The sarcoplasm thus contains large numbers of mitochondria and abundant reserves of myoglobin (to store oxygen).

TABLE 7-2 A Comparison of Skeletal, Cardiac, and Smooth Muscle Tissue

Property	*Skeletal Muscle Fiber*	*Cardiac Muscle Cell*	*Smooth Muscle Cell*
Fiber dimensions (diameter × length)	100 μm × up to 30 cm	15 μm × 100 μm	5–10 μm × 30–200 μm
Nuclei	Multiple, near sarcolemma	Usually single, centrally located	Single, centrally located
Filament organization	Sarcomeres along myofibrils	Sarcomeres along myofibrils	Scattered throughout sarcoplasm
Control mechanism	Neural, at single neuromuscular junction	Automaticity (pacemaker cells)	Automaticity (pacesetter cells), neural or hormonal control
Ca^{2+} source	Release from sarcoplasmic reticulum	Across sarcolemma and release from sarcoplasmic reticulum	Across sarcolemma
Contraction	Rapid onset, may be tetanized, rapid fatigue	Slower onset, cannot be tetanized, resistant to fatigue	Slow onset, may be tetanized, resistant to fatigue
Energy source	Aerobic metabolism at moderate levels of activity; glycolysis (anaerobic) during peak activity	Aerobic metabolism, usually lipid or carbohydrate substrates	Primarily aerobic metabolism

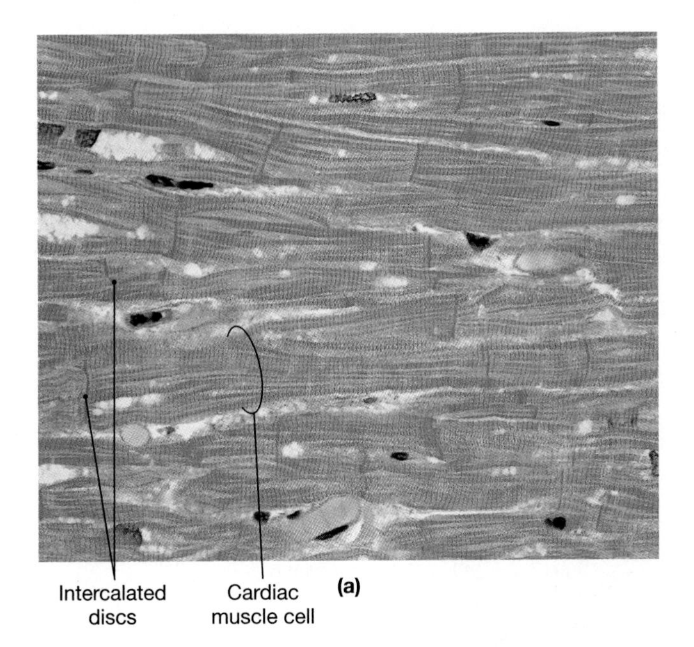

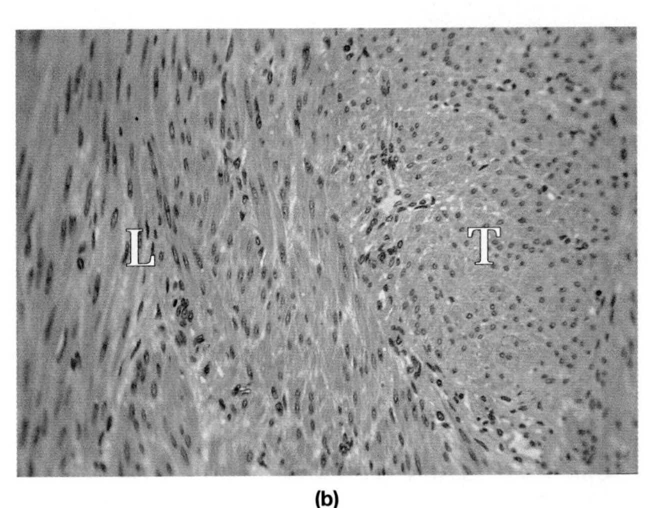

Figure 7-8
Cardiac and Smooth Muscle Tissue

(a) Cardiac muscle tissue as seen with the light microscope. Note the striations and the intercalated discs. **(b)** Many visceral organs contain layers or sheets of smooth muscle fibers oriented in different ways. This view shows smooth muscle cells in longitudinal (L) and transverse (T) sections.

Smooth Muscle Tissue

Smooth muscle cells are similar in size to cardiac muscle cells and also contain a single, centrally located nucleus within each spindle-shaped cell (Figure 7-8b●). Smooth muscle tissue is found within almost every organ, forming sheets, bundles, or sheaths around other tissues. In the skeletal, muscular, nervous, and endocrine systems, smooth muscles around blood vessels regulate blood flow through vital organs. In the digestive and urinary systems, rings of smooth muscles, called sphincters, regulate movement along internal passageways.

Differences between Smooth Muscle and Other Muscle Tissues

Structural Differences Actin and myosin are present in all three muscle types. In skeletal and cardiac muscle cells, these proteins are organized into sarcomeres. The internal organization of a smooth muscle cell is very different from that of skeletal or cardiac muscle cells.

- There are no myofibrils, sarcomeres, or striations in smooth muscle tissue.
- The thin filaments of a smooth muscle cell are anchored within the cytoplasm and to the sarcolemma.
- Adjacent smooth muscle cells are bound together at these anchoring sites, thus transmitting the contractile forces throughout the tissue.

Functional Differences Smooth muscle tissue differs from other muscle types in major ways.

- Calcium ions trigger the contraction through a different mechanism than that found in the other muscle types.
- Smooth muscle cells are able to contract over a greater range of lengths than skeletal or cardiac muscle because the actin and myosin filaments are not rigidly organized. This property is important because layers of smooth muscle are found in the walls of organs such as the urinary bladder and stomach, which undergo large changes in volume.
- Many smooth muscle cells are not innervated by motor neurons, and the muscle cells contract automatically, or in response to environmental or hormonal stimulation. When smooth muscle fibers are innervated by motor neurons, the neurons involved are not under voluntary control.

✓ How do intercalated discs enhance the functioning of cardiac muscle tissue?

✓ Why are cardiac and smooth muscle contractions more affected by changes in extracellular calcium ions than are skeletal muscle contractions?

✓ Smooth muscle can contract over a wider range of resting lengths than skeletal muscle. Why?

THE MUSCULAR SYSTEM

The muscular system (Figure 7-9●) includes all of the skeletal muscles that can be controlled voluntarily. The general appearance of each of the nearly 700 skeletal muscles provides clues to its primary function. Muscles involved with locomotion and posture work across joints, producing skeletal movement. Those that support soft tissue form slings or sheets between rel-

Figure 7-9
An Overview of the Major Skeletal Muscles
(a) Anterior view. **(b)** Posterior view.

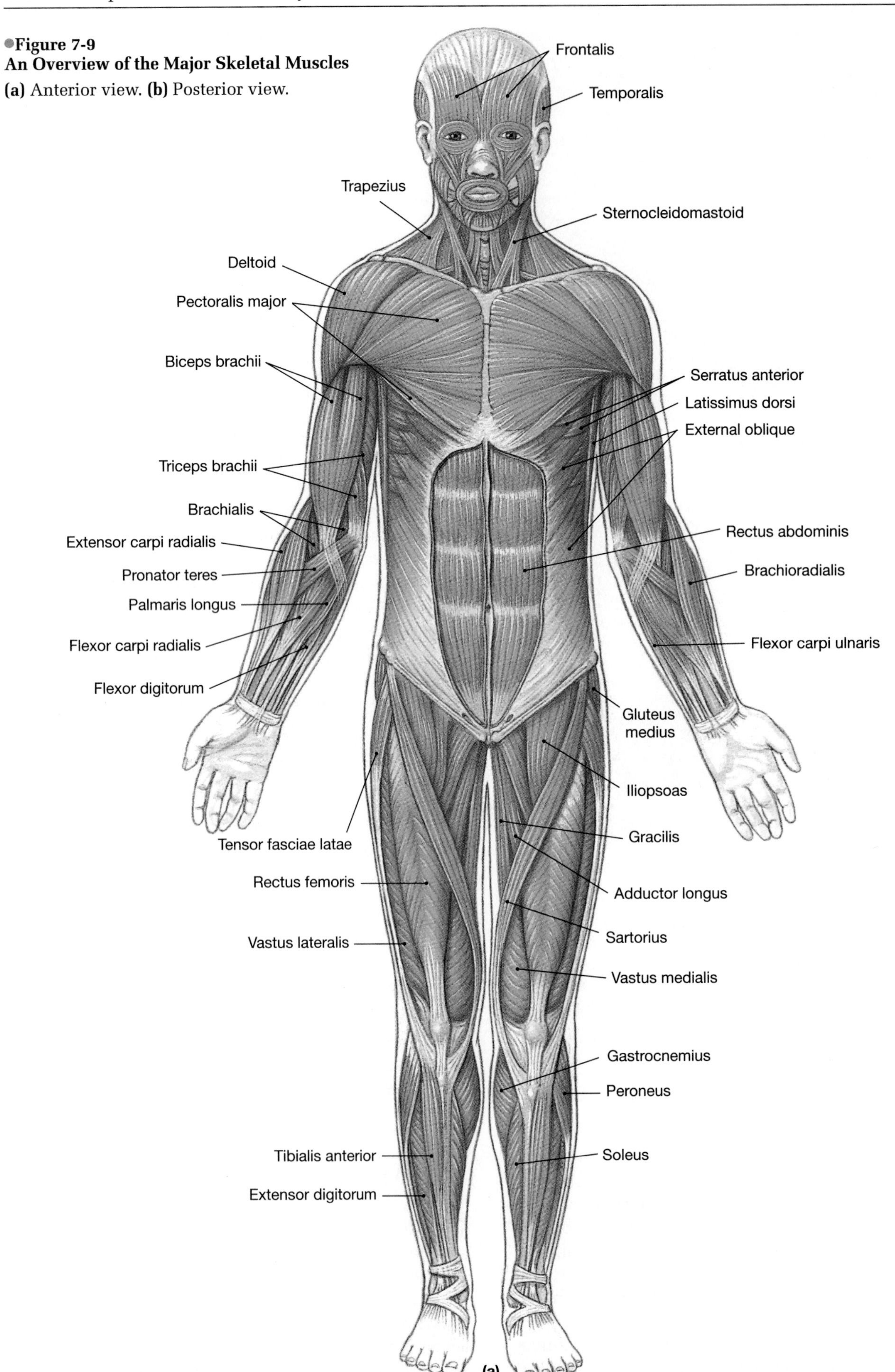

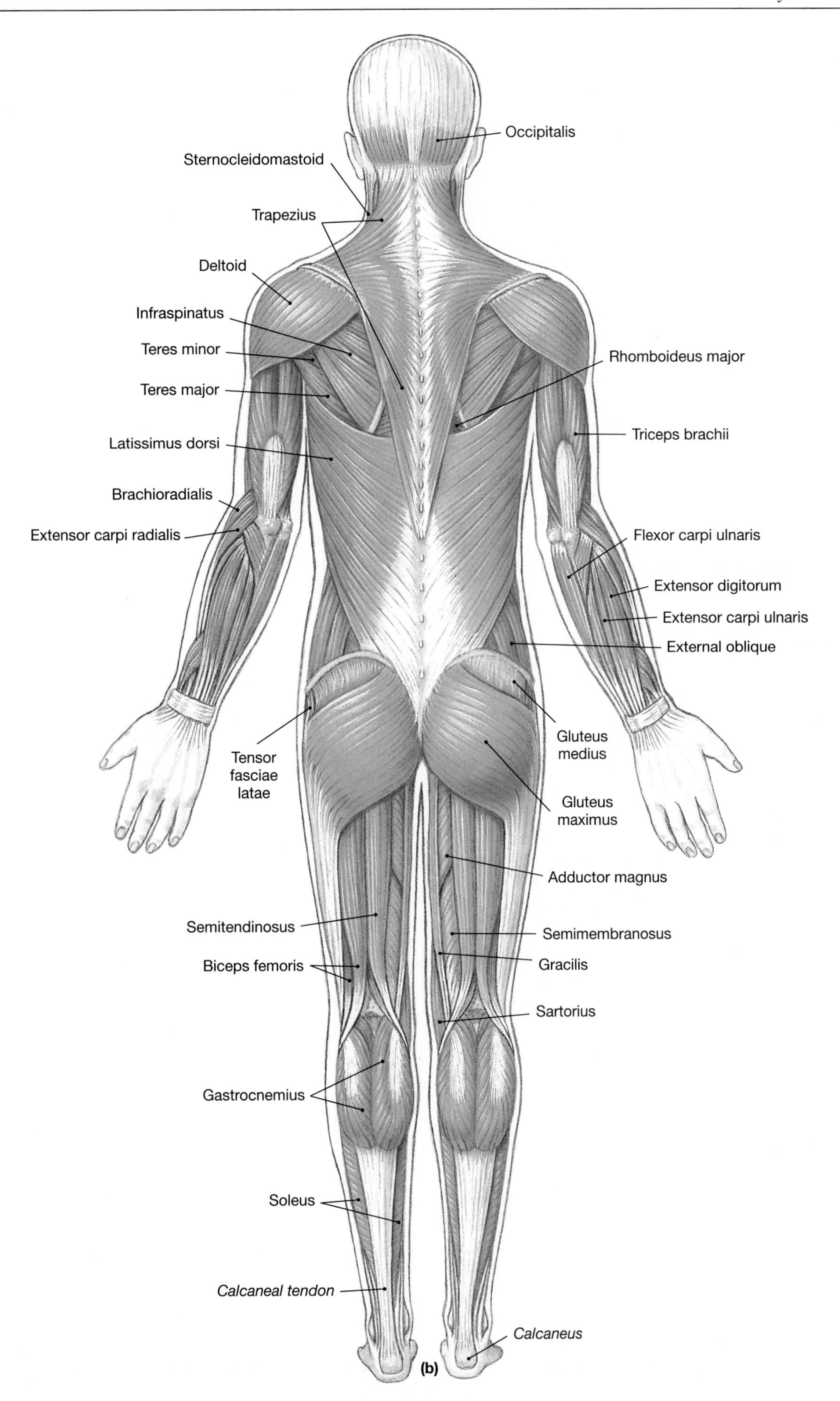
Occipitalis
Sternocleidomastoid
Trapezius
Deltoid
Infraspinatus
Teres minor
Teres major
Rhomboideus major
Triceps brachii
Latissimus dorsi
Brachioradialis
Extensor carpi radialis
Flexor carpi ulnaris
Extensor digitorum
Extensor carpi ulnaris
External oblique
Gluteus medius
Tensor fasciae latae
Gluteus maximus
Adductor magnus
Semitendinosus
Semimembranosus
Biceps femoris
Gracilis
Sartorius
Gastrocnemius
Soleus
Calcaneal tendon
Calcaneus
(b)

atively stable bony elements, whereas those that guard an entrance or exit completely encircle the opening.

Origins, Insertions, and Actions

Each muscle begins at an **origin**, ends at an **insertion**, and contracts to produce a specific **action**. In general, the origin end remains stationary while the insertion moves. For example, the *triceps brachii* inserts on the olecranon process and originates closer to the shoulder. Such determinations are made during normal movement.

Almost all skeletal muscles either originate or insert upon the skeleton. When they move the bone, they may produce *flexion, extension, adduction, abduction, protraction, retraction, elevation, depression, rotation, circumduction, pronation, supination, inversion,* or *eversion.* (You may wish to review Figures 6-26 to 6-28●, pp. 139–40 at this time).

Muscles can be grouped according to their **primary actions**.

- A **prime mover**, or **agonist**, is a muscle whose contraction is chiefly responsible for producing a particular movement. The *biceps brachii* is an example of prime mover that produces flexion of the forearm.
- **Antagonists** are prime movers whose actions oppose that of the agonist under consideration. The *triceps brachii* is a prime mover that extends the forearm. It is therefore an antagonist of the biceps brachii, and the biceps brachii is an antagonist of the triceps brachii. Agonists and antagonists are functional opposites—if one produces flexion, the other will have extension as its primary action.
- When a **synergist** (*syn-*, together + *ergon*, work) contracts, it assists the prime mover in performing that action. Synergists may provide additional pull near the insertion or stabilize the point of origin. In many cases, they are most useful at the start of a movement, when the prime mover is stretched and its power is relatively low. For example, the *deltoid* muscle acts to lift the arm away from the body (abduction). A smaller muscle, the *supraspinatus*, assists the deltoid in starting this movement.

Names of Skeletal Muscles

The names of our muscles were assigned by anatomists of the past whose memories were not any better than ours. Rather than writing the answers on their sleeves, they created terms that helped them remember what they were talking about. Table 7-3, which summarizes muscle terminology, will serve as a useful reference as you go through the rest of this chapter.

Some names, often with Greek or Latin roots, refer to the orientation of the muscle fibers. For example, *rectus* means "straight," and *rectus muscles* are parallel muscles whose fibers generally run along the long axis of the body, as in *rectus abdominis*. In a few cases, a muscle is such a prominent feature that the regional name alone can identify it, such as the *temporalis* of the head. Other muscles are named after structural features; for example, a *biceps muscle*, such as the *biceps brachii*, has two tendons of origin (*bi-*, two + *caput*, head), whereas the *triceps* has three. Table 7-3 also lists names reflecting shape, length, size, and whether a muscle is visible at the body surface (*externus, superficialis*) or lying beneath (*internus, profundus*). Superficial muscles that position or stabilize an organ are called *extrinsic muscles*; those that operate within an organ are called *intrinsic muscles*.

The first part of many names indicates the origin and the second part the insertion of the muscle. The *sternohyoid*, for example, originates at the sternum and inserts on the hyoid bone. Other names may indicate the primary function of the muscle as well. For example, the *extensor carpi radialis* is a muscle found along the radial (lateral) border of the forearm, and its contraction produces extension of the carpus (wrist).

The separation of the skeletal system into axial and appendicular divisions provides a useful guideline for subdividing the muscular system as well.

- The **axial musculature** arises on the axial skeleton. It positions the head and spinal column and also moves the rib cage, assisting in the movements that make breathing possible. It does not play a role in movement or support of the pectoral or pelvic girdles or appendages. This category encompasses roughly 60 percent of the skeletal muscles in the body.
- The **appendicular musculature** stabilizes or moves components of the appendicular skeleton.

✓ What type of muscle would you expect to find guarding the opening between the stomach and the small intestine?

✓ What muscle would be the antagonist of the *biceps brachii*?

✓ What does the name *flexor carpi radialis* tell you about this muscle?

The Axial Musculature

The axial muscles fall into four logical groups based on location, function, or both. The first group includes the *muscles of the head and neck* that are not associated with the spinal column. These muscles include those that are responsible for facial expression, chewing, and swallowing. The second group, the *muscles of the spine*, includes flexors and extensors of the head, neck, and spinal column. The third group, the *oblique and rectus muscles*, form the muscular walls

TABLE 7-3 Muscle Terminology

Terms Indicating Direction Relative to Axes of the Body	*Terms Indicating Specific Regions of the Body**	*Terms Indicating Structural Characteristics of the Muscle*	*Terms Indicating Actions*
Anterior (front)	Abdominis (abdomen)	**Origin**	**General**
Externus (superficial)	Anconeus (elbow)	Biceps (two heads)	Abductor
Extrinsic (outside)	Auricularis (auricle of ear)	Triceps (three heads)	Adductor
Inferioris (inferior)	Brachialis (brachium)	Quadriceps (four heads)	Depressor
Internus (deep, internal)	Capitis (head)		Extensor
Intrinsic (inside)	Carpi (wrist)	**Shape**	Flexor
Lateralis (lateral)	Cervicis (neck)	Deltoid (triangle)	Levator
Medialis/medius (medial, middle)	Cleido/clavius (clavicle)	Orbicularis (circle)	Pronator
Obliquus (oblique)	Coccygeus (coccyx)	Pectinate (comblike)	Rotator
Posterior (back)	Costalis (ribs)	Piriformis (pear-shaped)	Supinator
Profundus (deep)	Cutaneous (skin)	Platys- (flat)	Tensor
Rectus (straight, parallel)	Femoris (femur)	Pyramidal (pyramid)	
Superficialis (superficial)	Genio- (chin)	Rhomboideus (rhomboid)	**Specific**
Superioris (superior)	Glosso/glossal (tongue)	Serratus (serrated)	Buccinator (trumpeter)
Transversus (transverse)	Hallucis (great toe)	Splenius (bandage)	Risorius (laugher)
	Ilio- (ilium)	Teres (long and round)	Sartorius (like a tailor)
	Inguinal (groin)	Trapezius (trapezoid)	
	Lumborum (lumbar region)		
	Nasalis (nose)	**Other Striking Features**	
	Nuchal (back of neck)	Alba (white)	
	Oculo- (eye)	Brevis (short)	
	Oris (mouth)	Gracilis (slender)	
	Palpebrae (eyelid)	Lata (wide)	
	Pollicis (thumb)	Latissimus (widest)	
	Popliteus (behind knee)	Longissimus (longest)	
	Psoas (loin)	Longus (long)	
	Radialis (radius)	Magnus (large)	
	Scapularis (scapula)	Major (larger)	
	Temporalis (temples)	Maximus (largest)	
	Thoracis (thoracic region)	Minimus (smallest)	
	Tibialis (tibia)	Minor (smaller)	
	Ulnaris (ulna)	-tendinosus (tendinous)	
	Uro- (urinary)	Vastus (great)	

**For other regional terms, refer to Figure 1-6, p. 16, which deals with anatomical landmarks.*

of the thoracic and abdominopelvic cavities. The fourth group, the *muscles of the pelvic floor*, extend between the sacrum and pelvic girdle, forming the muscular *perineum* that closes the pelvic outlet.

Muscles of the Head and Neck

The muscles of the head and neck are shown in Figure 7-10● and detailed in Table 7-4. The muscles of the face originate on the surface of the skull and insert into the dermis of the skin. When they contract, the skin moves. The largest group of facial muscles is associated with the mouth. The **orbicularis oris** constricts the opening, and other muscles move the lips or the corners of the mouth. The **buccinator** (BUK-si-nā-tor), one of the muscles associated with the mouth, compresses the cheeks, as when pursing the lips and blowing forcefully. (*Buccinator* translates as "trumpet player." During chewing, contraction and relaxation of the buccinator move food back across the teeth from the space inside the cheeks. The chewing motions are primarily produced by contractions of

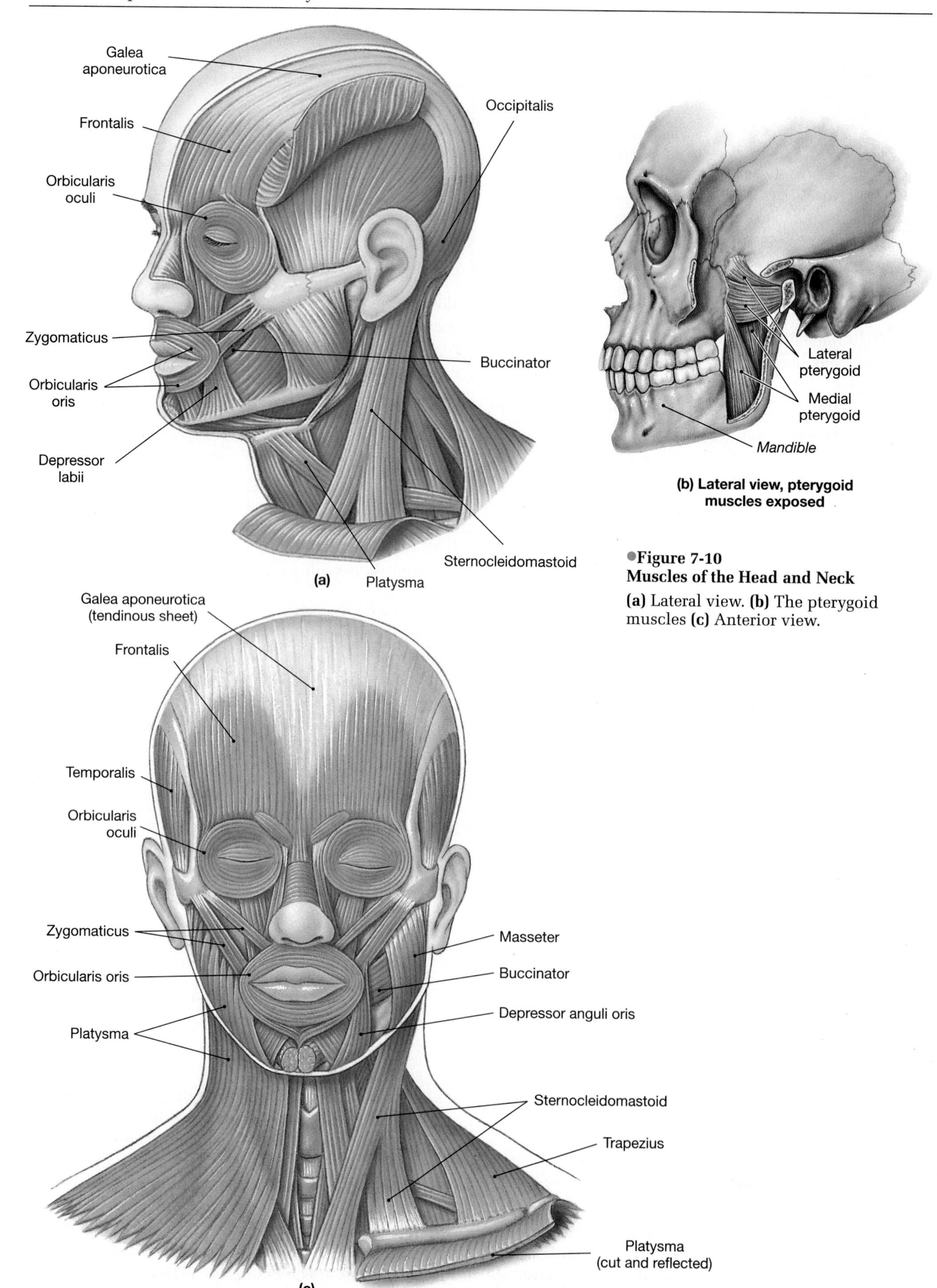

Figure 7-10
Muscles of the Head and Neck
(a) Lateral view. **(b)** The pterygoid muscles **(c)** Anterior view.

TABLE 7-4 Muscles of the Head and Neck

Region/Muscle	*Origin*	*Insertion*	*Action*
MOUTH			
Buccinator	Maxillae and mandible	Blends into fibers of orbicularis oris	Compresses cheeks
Orbicularis oris	Maxillae and mandible	Lips	Compresses, purses lips
Depressor anguli oris	Mandibular body	Skin at angle of mouth	Depresses corner of mouth
Zygomaticus	Zygomatic bone	Angle of mouth	Draws corner of mouth back and up
EYE			
Orbicularis oculi	Medial margin of orbit	Skin around eyelids	Closes eye
SCALP			
Frontalis	Galea aponeurotica	Skin of eyebrow and bridge of nose	Raises eyebrows, wrinkles forehead
Occipitalis	Occipital bone	Galea aponeurotica	Tenses, retracts scalp
LOWER JAW			
Masseter	Zygomatic arch	Lateral surface of mandible	Elevates mandible
Temporalis	Along temporal lines of skull	Coronoid process of mandible	Elevates mandible
Pterygoids	Inferior processes of sphenoid	Median surface of mandible	Elevate, protract, and/or move mandible sideways
NECK			
Platysma	From cartilage of second rib to acromion of scapula	Mandible and skin of cheek	Tenses skin of neck, depresses mandible
Digastricus	Mastoid region of temporal and inferior surface of chin	Hyoid bone	Depresses mandible and/or elevates larynx
Mylohyoid	Medial surface of mandible	Median connective tissue band	Elevates floor of mouth and hyoid, and/or depresses mandible
Sternohyoid	Clavicle and sternum	Hyoid bone	As above
Sternothyroid	Dorsal surface of sternum and 1st rib	Thyroid cartilage of larynx	As above
Stylohyoid	Styloid process of temporal bone	Hyoid bone	Elevates larynx
Sternocleidomastoid	Superior margins of sternum and clavicle	Mastoid region of skull	Together they flex the neck; alone one side bends head toward shoulder and turns face to opposite side

the **masseter**, assisted by the **temporalis** and the **pterygoid** muscles used in various combinations.

Smaller groups of muscles control movements of the eyebrows and eyelids, the scalp, the nose, and the external ear. The **platysma** (pla-TIZ-ma; *platys*, flat) covers the ventral surface of the neck, extending from the base of the neck to the mandible and the corners of the mouth.

The muscles of the neck control the position of the larynx, depress the mandible, tense the floor of the mouth, and provide a stable foundation for muscles of the tongue and pharynx. These muscles include the following:

- The **digastricus**, which has two bellies (*di-*, two + *gaster*, stomach) opens the mouth by depressing the mandible.
- The broad, flat **mylohyoid** provides a muscular floor to the mouth and supports the tongue.

- The **stylohyoid** forms a muscular connection between the hyoid apparatus and the styloid process of the skull.
- The **sternocleidomastoid** (ster-nō-klī-dō-MAS-toid) extends from the clavicles and the sternum to the mastoid region of the skull. It can rotate the head or flex the neck.

✓ If you were contracting and relaxing your masseter muscle, what would you probably be doing?

✓ What facial muscle would you expect to be well developed in a trumpet player?

Muscles of the Spine

The muscles of the spine are covered by more superficial back muscles, such as the trapezius and latissimus dorsi (see Figure 7-9b●) The spinal extensors, or *erector spinae,* act to maintain an erect spinal column and head. Moving laterally from the spine, these muscles can be subdivided into *spinalis, longissimus,* and *iliocostalis* divisions (Figure 7-11● and Table 7-5). In the lower lumbar and sacral regions the border between the longissimus and iliocostalis muscles becomes indistinct, and they are sometimes known as the *sacrospinalis* muscles. When contracting together, these muscles extend the spinal column. When only the muscles on one side contract, the spine is bent laterally.

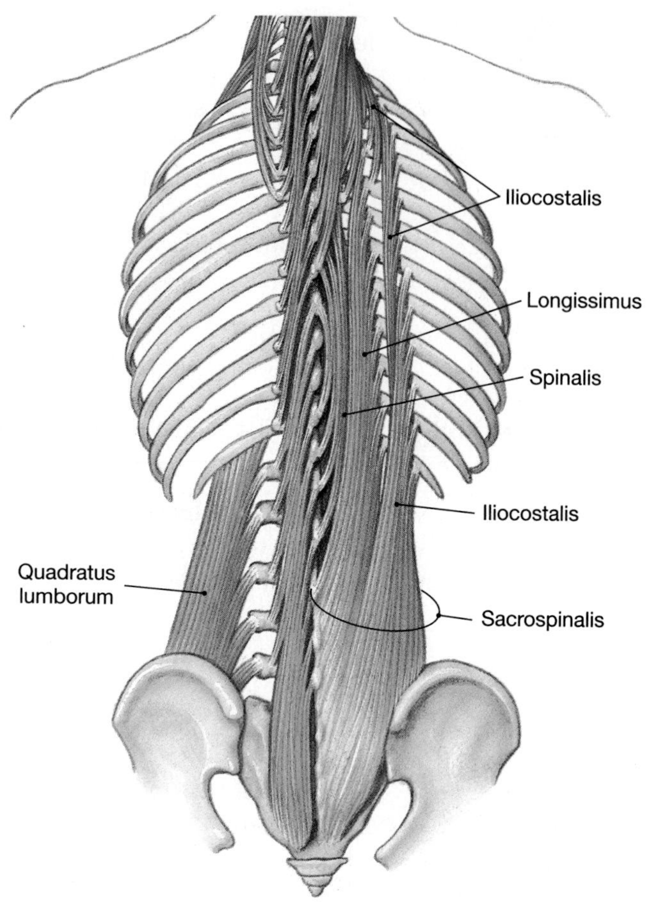

●**Figure 7-11**
Muscles of the Spine

TABLE 7-5 Muscles of the Spine

Region/Muscle	*Origin*	*Insertion*	*Action*
SPINAL EXTENSORS			
Spinalis group	Spinous processes and transverse processes of cervical and thoracic vertebrae	Base of skull and spinous processes of cervical and upper thoracic vertebrae	The two sides act together to extend head or spinal column; either alone extends and tilts head or rotates spinal column to that side
Longissimus group	Processes of lower cervical, thoracic, and upper lumbar vertebrae	Mastoid processes of temporal bone, transverse processes of cervical vertebrae and inferior surfaces of ribs	The two sides act together to extend head or spinal column; either alone rotates and tilts head or spinal column to that side
Iliocostalis group	Superior borders of ribs and iliac crest	Transverse processes of cervical vertebrae and inferior surfaces of ribs	Extends or bends spinal column and moves ribs
SPINAL FLEXOR			
Quadratus lumborum	Iliac crest	Last rib and transverse processes of lumbar vertebrae	Together they depress ribs, flex spine; one side alone flexes laterally

The Axial Muscles of the Trunk

The *oblique* and *rectus muscles* form the muscular walls of the thoracic and abdominopelvic cavities between the first thoracic vertebra and the pelvis. In the thoracic area, these muscles are partitioned by the ribs, but over the abdominal surface they form broad muscular sheets (Figure 7-12● and Table 7-6). The oblique muscles can compress underlying structures or rotate the spinal column, depending on whether one or both sides are contracting. The rectus muscles are important flexors of the spinal column, opposing the erector spinae.

The axial muscles of the trunk include (1) the external and internal **intercostals** and **obliques**, (2) the **transversus abdominis**, (3) the **rectus abdominis**, (4) the muscular **diaphragm** that separates the thoracic and abdominopelvic cavities, and (5) muscles that form the floor of the pelvic cavity.

Muscles of the Pelvic Floor

The floor of the pelvic cavity is called the **perineum** (Figure 7-13● and Table 7-7). It is formed by a broad sheet of muscles that connects the sacrum and coccyx to the ischium and pubis. These muscles support the organs of the pelvic cavity and control the movement of materials through the urethra and anus.

HERNIAS

When the abdominal muscles contract forcefully, pressure in the abdominopelvic cavity can skyrocket, and those pressures are applied to internal organs. If the individual exhales at the same time, the pressure is relieved, because the diaphragm can move upward as the lungs collapse. But during vigorous isometric exercises or when lifting a weight while holding one's breath, pressure in the abdominopelvic cavity can rise high enough to cause a variety of problems, among them the development of a *hernia.*

A **hernia** develops when an organ protrudes through an abnormal opening. The most common hernias are *inguinal hernias* and *diaphragmatic hernias.* Inguinal hernias typically occur in males, at the site where blood vessels, nerves, and reproductive ducts pass through the abdominal wall en route to the testes. Diaphragmatic hernias develop when visceral organs, such as a portion of the stomach, are forced into the left pleural cavity. If herniated structures become trapped or twisted, surgery may be required to prevent serious complications. AM *Hernias*

✓ Damage to the external intercostal muscles would interfere with what important process?

✓ If someone hit you in your rectus abdominis muscle, how would your body position change?

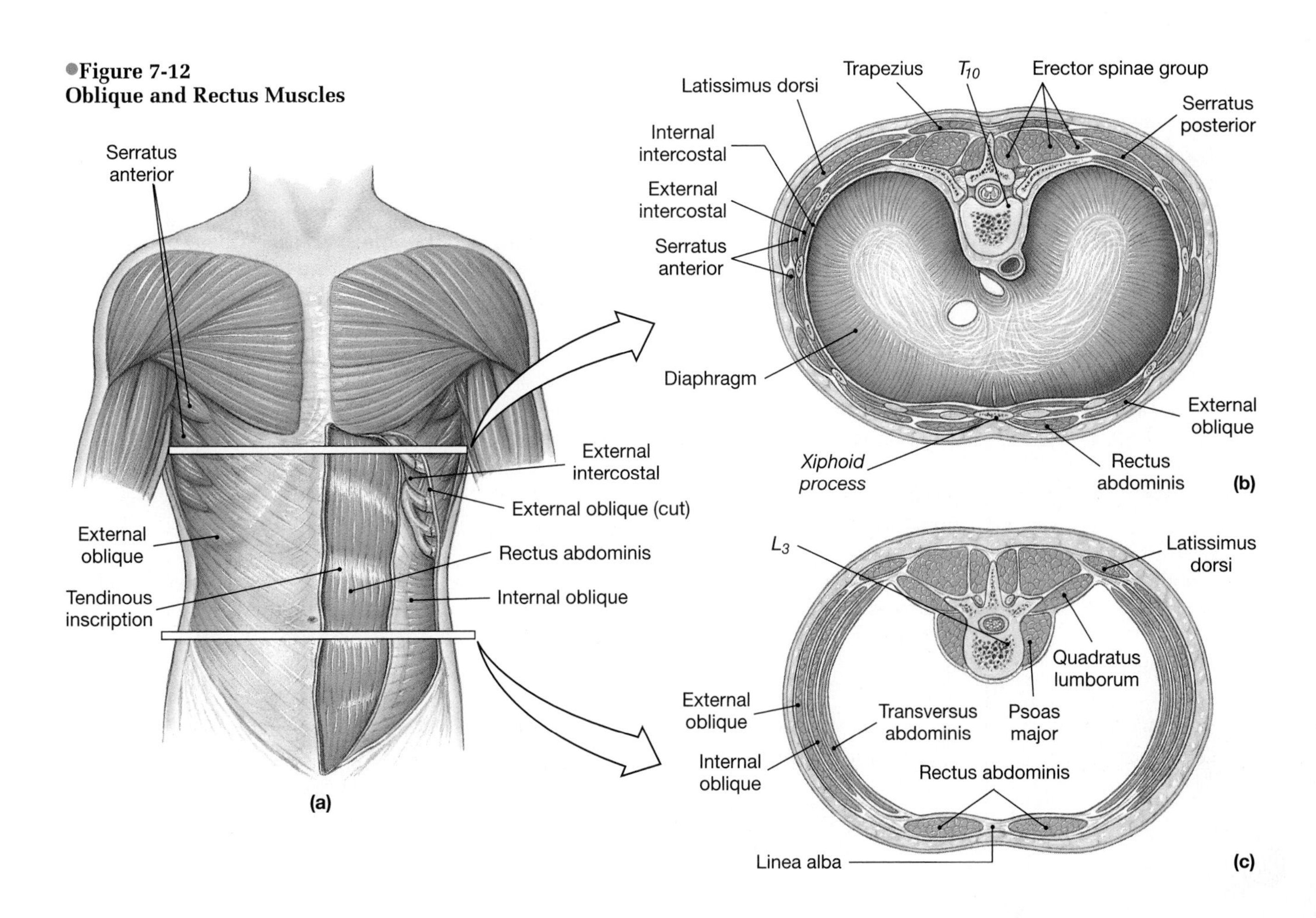

●**Figure 7-12**
Oblique and Rectus Muscles

TABLE 7-6 Axial Muscles of the Trunk

Region/Muscle	*Origin*	*Insertion*	*Action*
THORACIC REGION			
External intercostals	Inferior border of each rib	Superior border of next rib	Elevate ribs
Internal intercostals	Superior border of each rib	Inferior border of the previous rib	Depress ribs
Diaphragm	Xiphoid process, cartilages of ribs 4–10, and anterior surfaces of lumbar vertebrae	Central tendinous sheet	Contraction expands thoracic cavity, compresses abdominopelvic cavity
ABDOMINAL REGION			
External oblique	Lower eight ribs	Linea alba and iliac crest	Compresses abdomen, depresses ribs, flexes or bends spine
Internal oblique	Iliac crest and adjacent connective tissues	Lower ribs, xiphoid of sternum, and linea alba	As above
Transversus abdominis	Cartilages of lower ribs, iliac crest, and adjacent connective tissues	Linea alba and pubis	Compresses abdomen
Rectus abdominis	Superior surface of pubis around symphysis	Inferior surfaces of costal cartilages (ribs 5–7) and xiphoid process of sternum	Depresses ribs, flexes vertebral column

TABLE 7-7 Muscles of the Perineum

Muscle	*Origin*	*Insertion*	*Action*
Bulbocavernosus: **male**	Base of penis; fibers cross over urethra	Midline and central tendon of perineum	Compresses base, stiffens penis, ejects urine or semen
female	Base of clitoris; fibers run on either side of urethral and vaginal openings	Central tendon of perineum	Compresses and stiffens clitoris, narrows vaginal opening
Ischiocavernosus	Inferior medial surface of ischium	Symphysis pubis anterior to base of penis or clitoris	Compresses and stiffens penis or clitoris
Transverse perineus	Inferior, medial surface of ischium	Central tendon of perineum	Stabilizes central tendon of perineum
Urethral sphincter:			
male	Inferior, medial surfaces of ischium and pubis	Midline at base of penis; inner fibers encircle urethra	Closes urethra, compresses prostate and bulbourethral glands
female	As above	Midline; inner fibers encircle urethra	Closes urethra, compresses vagina and greater vestibular glands
Anal sphincter	Via tendon from coccyx	Encircles anal opening	Closes anal opening
Levator ani	Ischial spine and pubis	Coccyx	Tenses floor of pelvis, supports pelvic organs, flexes coccyx, elevates and retracts anus

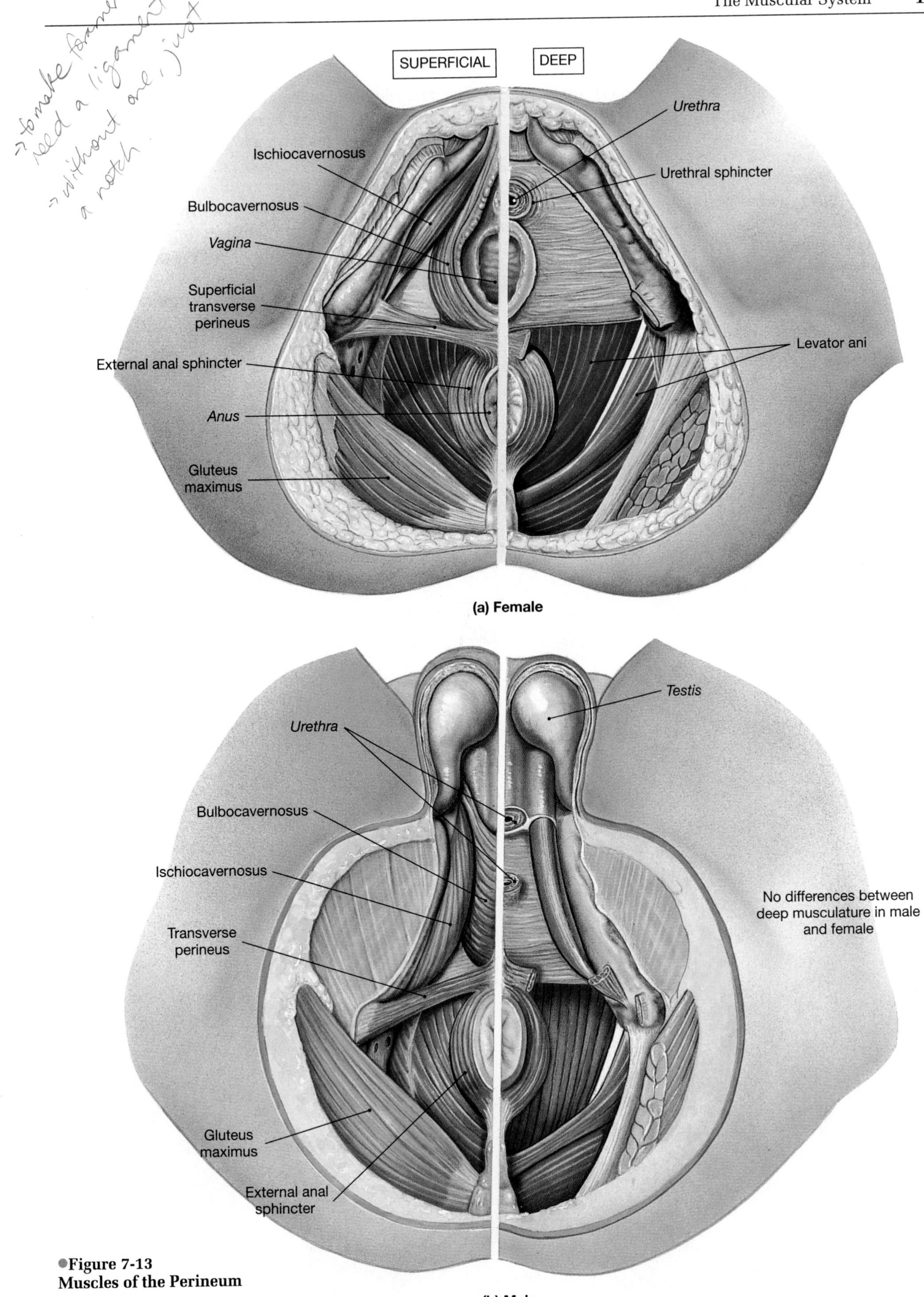

Figure 7-13
Muscles of the Perineum
(a) Female. **(b)** Male.

The Appendicular Musculature

The appendicular musculature includes (1) the muscles of the shoulders and upper limbs and (2) the muscles of the pelvic girdle and lower limbs. There are few similarities between the two groups, because the functions and required ranges of motion are very different. In addition to increasing the mobility of the upper limb, the muscular connections between the pectoral girdle and the axial skeleton must act as shock absorbers. For example, people who are jogging can still perform delicate hand movements because the muscular connections between the axial and appendicular skeleton smooth out the bounces in their stride. In contrast, the pelvic girdle has evolved to transfer weight from the axial to the appendicular skeleton. A muscular connection would reduce the efficiency of the transfer, and the emphasis is on sheer power rather than versatility.

Muscles of the Shoulder and Upper Limb

The large, superficial **trapezius** muscles cover the back and portions of the neck, reaching to the base of the skull. These muscles form a broad diamond (Figure 7-14a● and Table 7-8). Its actions are quite varied because specific regions can be made to contract independently. The **rhomboideus** muscles and the **levator scapulae** are covered by the trapezius. Contraction of the rhomboids adducts the scapula, pulling it toward the center of the back. The levator scapulae elevates the scapula, as when you shrug your shoulders.

On the chest, the **serratus anterior** originates along the anterior surfaces of several ribs and inserts along the vertebral border of the scapula. When the serratus anterior contracts, it pulls the shoulder forward. The **pectoralis minor** attaches to the coracoid process of the scapula. When it contracts, it depresses and protracts the scapula.

●**Figure 7-14**
Muscles of the Shoulder
(a) Posterior view. **(b)** Anterior view.

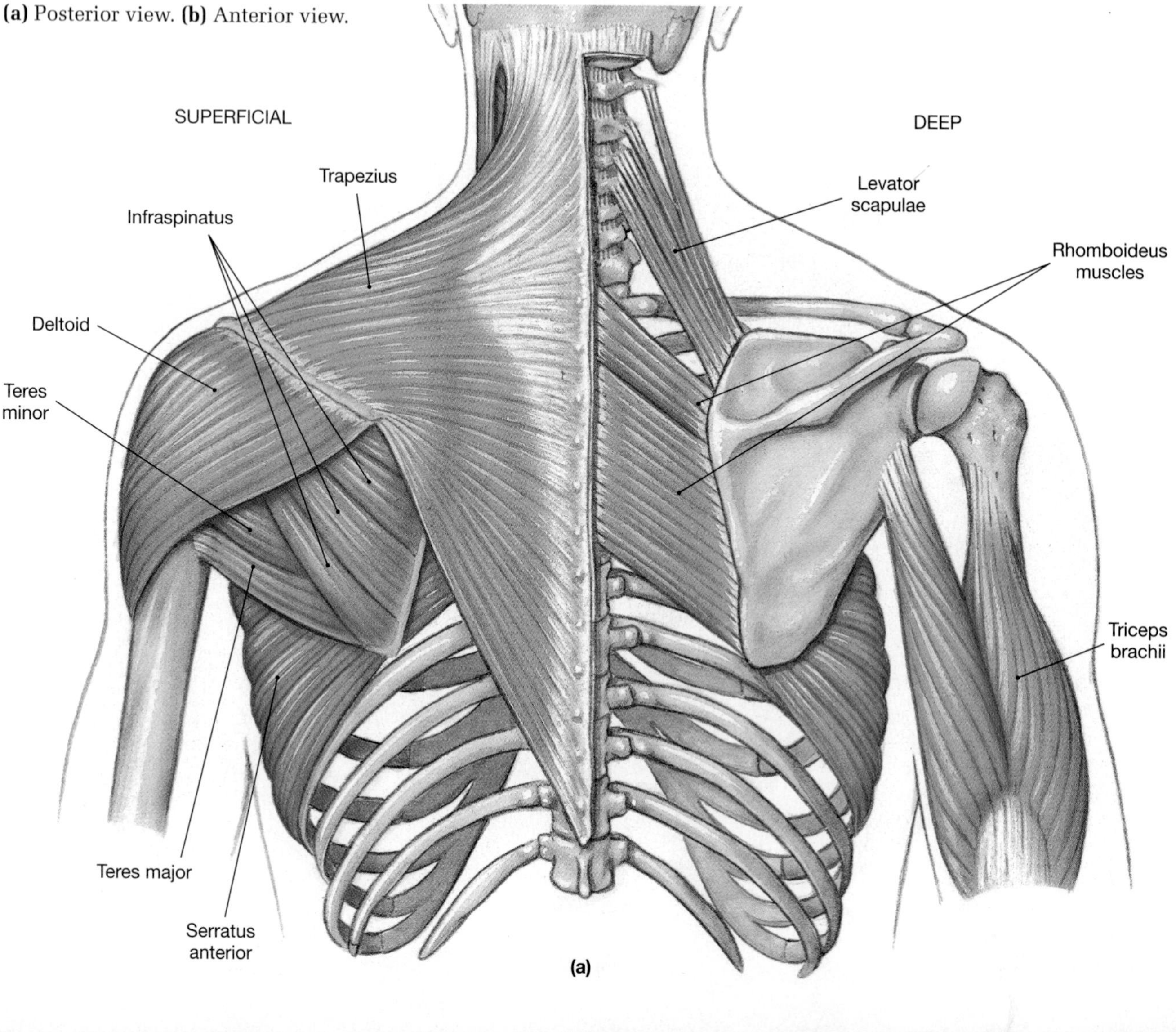

TABLE 7-8 Muscles of the Shoulder

Muscle	*Origin*	*Insertion*	*Action*
Levator scapulae	Posterior surface of first 4 cervical vertebrae	Vertebral border of scapula	Elevates scapula
Pectoralis minor	Anterior surfaces of ribs 3–5	Coracoid process of scapula	Depresses and protracts shoulder; rotates scapula laterally; elevates ribs if scapula is stationary
Rhomboideus muscles	Spinous processes of lower cervical and upper thoracic vertebrae	Vertebral border of scapula	Adducts and rotates scapula laterally
Serratus anterior	Anterior and superior margins of ribs 1–9	Anterior surface of vertebral border of scapula	Protracts shoulder, abducts and medially rotates scapula
Subclavius	First rib	Clavicle	Depresses and protracts shoulder
Trapezius	Occipital bone and spinous processes of thoracic vertebrae	Clavicle and scapula (acromion and scapular spine)	Depends on active region and state of other muscles; may elevate, adduct, depress, or rotate scapula and/or elevate clavicle; can also extend head and neck

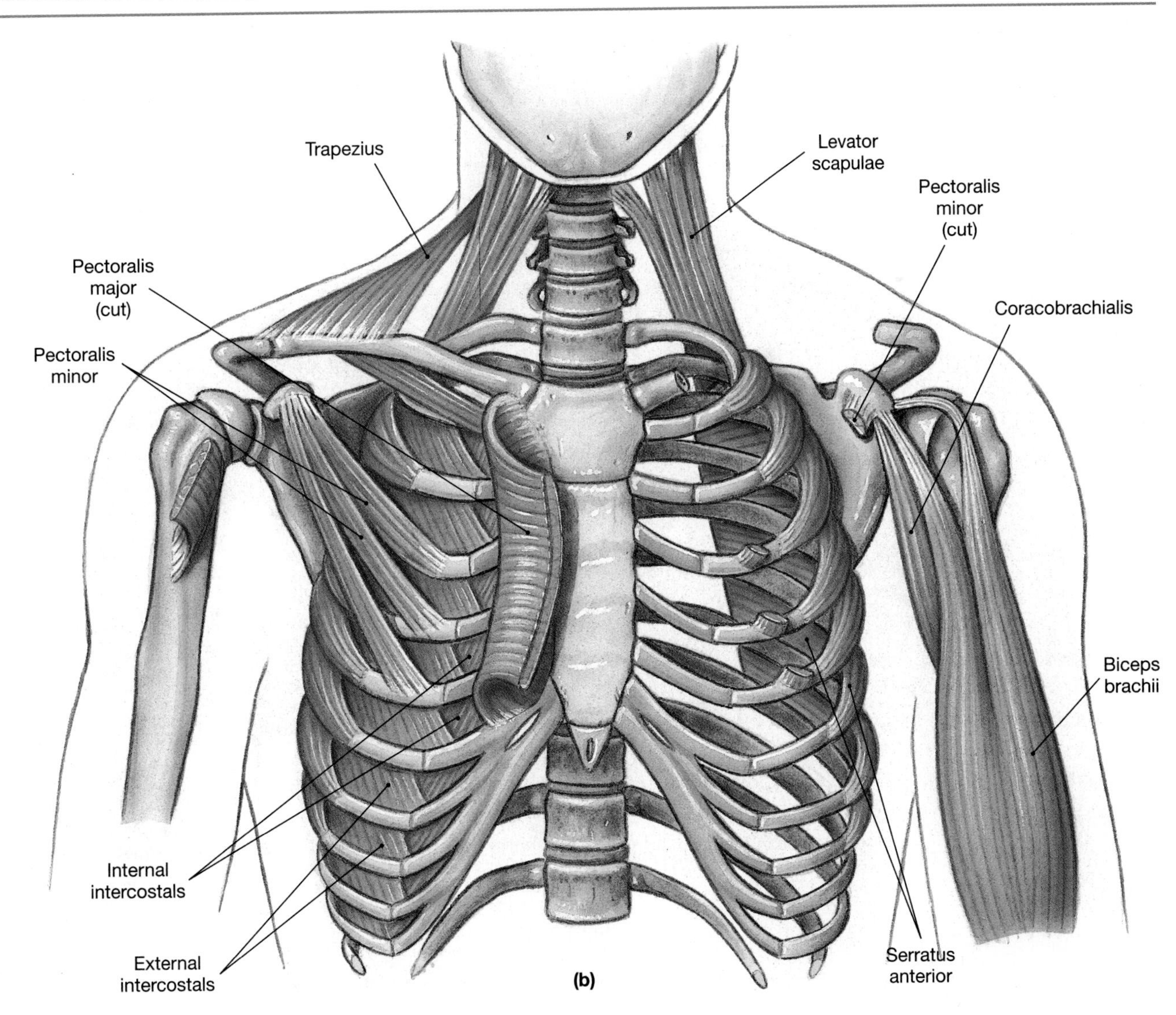

(b)

Muscles That Move the Arm The muscles that move the arm (Figure 7-15● and Table 7-9) are easiest to remember when grouped by primary actions.

- The **deltoid** is the major abductor of the arm, and the **supraspinatus** assists at the start of this movement.
- The **subscapularis, teres major, infraspinatus**, and **teres minor** rotate the arm.
- The **pectoralis major** extends between the chest and the greater tubercle of the humerus; the **latissimus dorsi** extends between the thoracic vertebrae and the lesser tubercle of the humerus. The pectoralis major flexes the arm, and the latissimus dorsi extends it. The two muscles also work together to adduct and rotate the arm.

These muscles provide substantial support for the loosely built shoulder joint. The tendons of the deltoid, supraspinatus, infraspinatus, subscapularis, teres major, and teres minor blend with and support the capsular fibers that enclose the shoulder joint. They are the muscles of the *rotator cuff*, a frequent site of sports injuries. As noted in Chapter 6, these muscles must stabilize the shoulder joint while controlling an extensive range of movement. *[p. 129]* Powerful, repetitive arm movements, such as pitching a fastball at 96 mph for nine innings, can place intolerable strains on the muscles of the rotator cuff, leading to a muscle strain (a tear or break in the muscle), *bursitis*, and other painful injuries.

SPORTS INJURIES

Exercise carries risks because of the stresses placed on muscles, joints, and connective tissues. Many Americans participate in exercise programs and sports on a regular basis; more than 30 million Americans go jogging, and millions more participate in various amateur and professional sports. As a result, *sports injuries* are very common, and sports medicine has become an active area of professional and academic research interest. For information on the incidence and classification of sports injuries, see the *Applications Manual*. AM *Sports Injuries*

Muscles That Move the Forearm and Wrist Although most of the muscles that insert upon the fore-

TABLE 7-9 Muscles That Move the Arm

Muscle	*Origin*	*Insertion*	*Action*
Coracobrachialis	Coracoid process	Medial margin of shaft of humerus	Adducts and flexes humerus
Deltoid	Clavicle and scapula (acromion and adjacent scapular spine)	Deltoid tuberosity of humerus	Abducts arm
Latissimus dorsi	Spinous processess of lower thoracic vertebrae, ribs, and lumbar vertebrae	Lesser tubercle, intertubercular groove of humerus	Extends, adducts, and medially rotates humerus
Pectoralis major	Cartilages of ribs 2–6, body of sternum, and clavicle	Greater tubercle of humerus	Flexes, adducts, and medially rotates humerus
ROTATOR CUFF MUSCLES			
Supraspinatus	Supraspinous fossa of scapula	Greater tubercle of humerus	Abducts arm
Infraspinatus	Infraspinous fossa of scapula	Greater tubercle of humerus	Lateral rotation of humerus
Subscapularis	Subscapular fossa of scapula	Lesser tubercle of humerus	Medial rotation of humerus
Teres minor	Axillary border of scapula	Greater tubercle of humerus	Lateral rotation of humerus
Teres major	Inferior angle of scapula	Intertubercular groove of humerus	Adducts and medially rotates arm

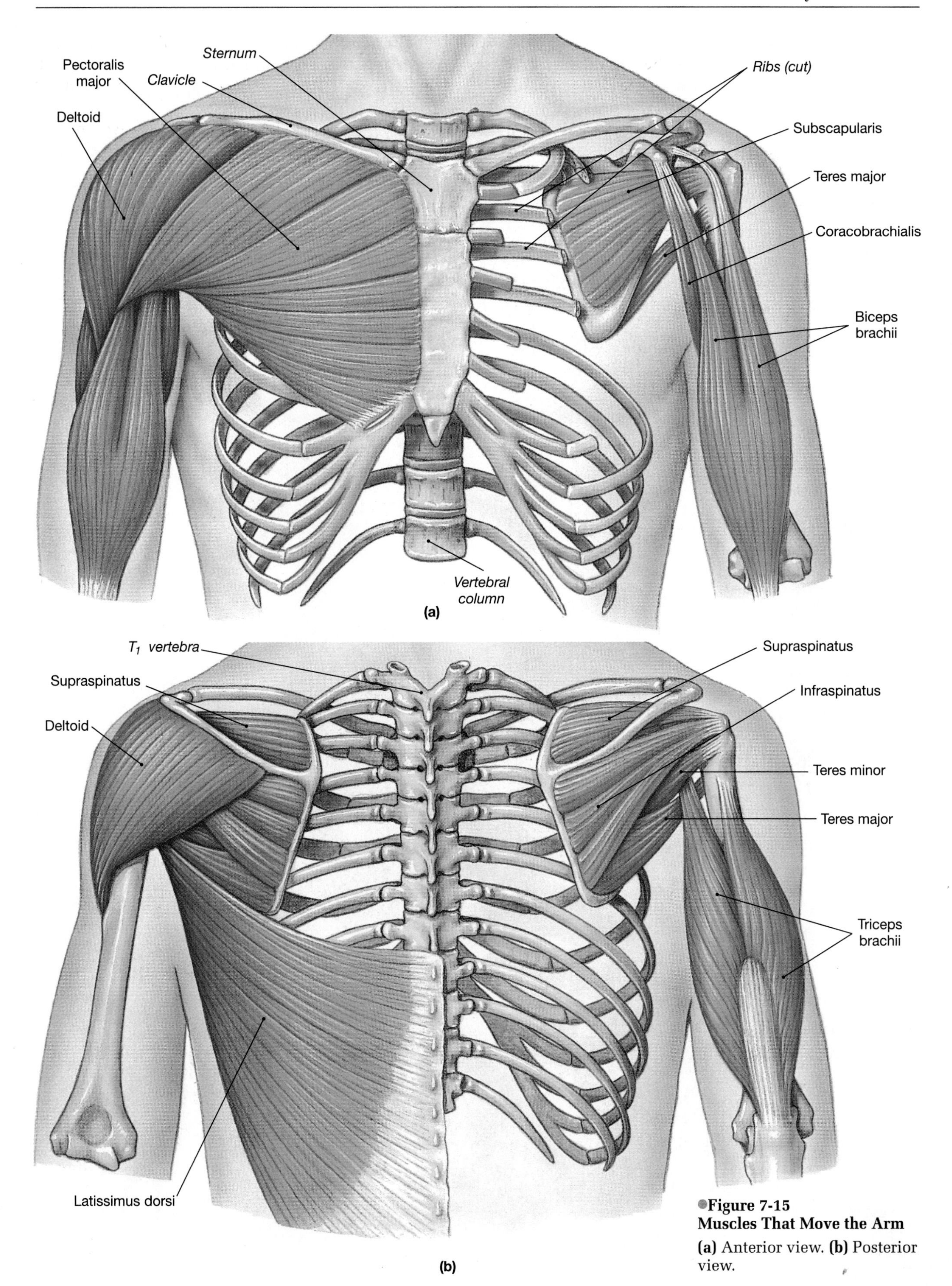

Figure 7-15
Muscles That Move the Arm
(a) Anterior view. **(b)** Posterior view.

arm and wrist (Figure 7-16● and Table 7-10) originate on the humerus, there are two noteworthy exceptions. The **biceps brachii** and **triceps brachii**, which insert upon the bones of the forearm, originate on the scapula. Although their contractions can have a secondary effect on the shoulder, their primary actions are on the elbow. The triceps brachii extends the forearm when, for example, we do push-ups. The biceps brachii both flexes and supinates the forearm. With the forearm pronated (palm facing back), the biceps brachii cannot function effectively. As a result, we are strongest when flexing the supinated forearm; the biceps brachii then makes a prominent bulge.

Other important muscles include the following:

- The **brachialis** and **brachioradialis** also flex the forearm, opposed by the triceps brachii.
- The **flexor carpi ulnaris**, the **flexor carpi radialis**, and the **palmaris longus** are superficial muscles that work together to produce flexion of the wrist. Because they originate on opposite sides of the humerus, the flexor carpi radialis flexes and abducts while the flexor carpi ulnaris flexes and adducts.
- The **extensor carpi radialis** muscles and the **extensor carpi ulnaris** have a similar relationship; the former produces extension and abduction, the latter extension and adduction.
- The **pronators** and the **supinator** rotate the radius; the supinator may be assisted by the biceps brachii.

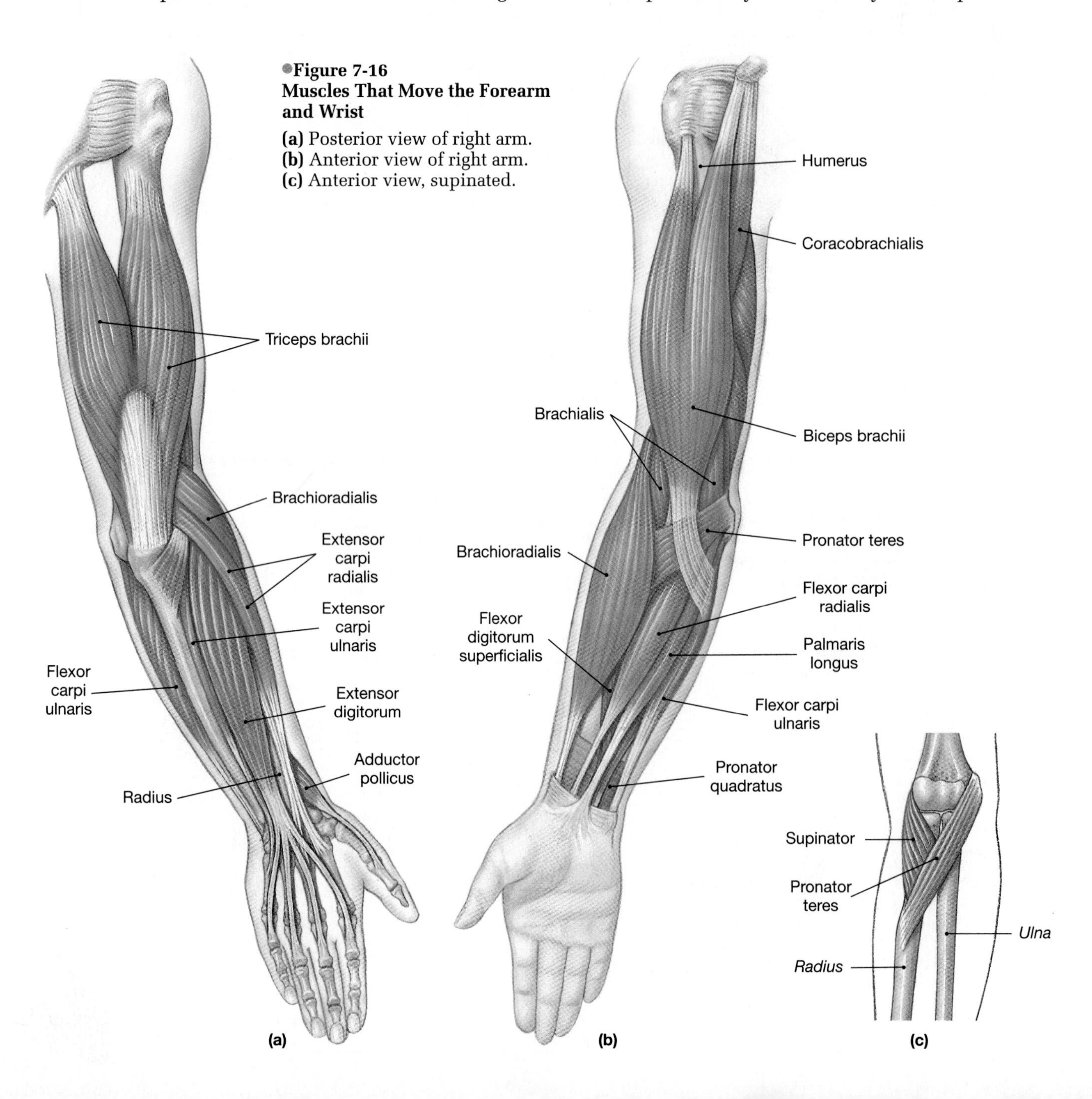

●Figure 7-16
Muscles That Move the Forearm and Wrist
(a) Posterior view of right arm.
(b) Anterior view of right arm.
(c) Anterior view, supinated.

TABLE 7-10 Muscles That Move the Forearm, Wrist, and Hand

Muscle	*Origin*	*Insertion*	*Action*
PRIMARY ACTION AT THE ELBOW			
Flexors			
Biceps brachii	From the coracoid process (short head) and body (long head) of scapula	Tuberosity of radius	Flexes and supinates forearm
Brachialis	Anterior, distal surface of humerus	Tuberosity of ulna	Flexes forearm
Brachioradialis	Lateral epicondyle of humerus	Styloid process of radius	As above
Extensors			
Triceps brachii	Superior, posterior, and lateral margins of humerus, and the scapula	Olecranon process of ulna	Extends forearm
PRONATORS/ SUPINATOR			
Pronator quadratus	Medial surface of distal portion of ulna	Anterior and lateral surface of distal portion of radius	Pronates forearm
Pronator teres	Medial epicondyle of humerus and coronoid process of ulna	Distal lateral surface of radius	As above
Supinator	Lateral epicondyle of humerus and ulna	Anterior and lateral surface of radius distal to the radial tuberosity	Supinates forearm
PRIMARY ACTION AT THE WRIST			
Flexors			
Flexor carpi radialis	Medial epicondyle of humerus	Bases of 2nd and 3rd metacarpals	Flexes and abducts palm
Flexor carpi ulnaris	Medial epicondyle of humerus and adjacent surfaces of ulna	Bases of metacarpals (3rd to 5th), pisiform and hamate, bone 5	Flexes and adducts palm
Palmaris longus	Medial epicondyle of humerus	A tendinous sheet on the palm	Flexes palm
Extensors			
Extensors carpi radialis	Distal lateral surface and lateral epicondyle of humerus	Bases of 2nd and 3rd metacarpals	Extends and abducts palm
Extensor carpi ulnaris	Lateral epicondyle of humerus and adjacent surface of ulna	Base of 5th metacarpal	Extends and adducts palm
ACTION AT THE HAND			
Extensor digitorum	Lateral epicondyle of humerus	Posterior surfaces of the phalanges	Extends fingers and palms
Flexor digitorum	Proximal medial and anterior surface of ulna and radius; medial epicondyle of humerus	Distal phalanges	Flexes fingers

Muscles That Move the Palm and Fingers The muscles of the forearm detailed in Table 7-10 perform flexion and extension of the fingers. These muscles stop before reaching the hand, and only their tendons cross the wrist. These are relatively large muscles, and keeping them clear of the joints ensures maximum mobility at both the wrist and hand. The tendons that cross the dorsal and ventral surfaces of the wrist pass through *tendon sheaths*, elongate bursae that reduce friction. Inflammation of tendon sheaths can restrict movement and irritate the *median nerve*, a nerve that innervates the palm of the hand. Chronic pain, often associated with weakness in the hand muscles, is the result. This condition is known as *carpal tunnel syndrome*. AM *Carpal Tunnel Syndrome*

✓ What muscle are you using when you shrug your shoulders?

✓ Sometimes baseball pitchers will suffer from rotator cuff injuries. What muscles are involved in this type of injury? *(cont.)*

✓ Injury to the flexor carpi ulnaris would impair what two movements?

Muscles of the Lower Limb

The muscles of the lower limb can be divided into three functional groups: (1) muscles that move the thigh, (2) muscles that move the leg, and (3) muscles that move the ankles, feet, and toes.

Muscles That Move the Thigh The muscles that move the thigh are detailed in Figure 7-17● and Table 7-11.

- **Gluteal muscles** (Figure 7-17a●) cover the lateral surfaces of the ilia. The **gluteus maximus** is the largest and most posterior of the gluteal muscles, which extend, rotate, and abduct the thigh.
- The adductors of the thigh (Figure 7-17b●) include the **adductor magnus**, the **adductor brevis**, the **adductor longus**, and the **gracilis** (GRAS-i-lis).

TABLE 7-11 Muscles That Move the Thigh

Group/Muscle	*Origin*	*Insertion*	*Action*
GLUTEAL GROUP			
Gluteus maximus	Iliac crest of ilium, sacrum, and coccyx	Iliotibial tract and gluteal tuberosity of femur	Extends and laterally rotates thigh
Gluteus medius	Anterior iliac crest and lateral surface of ilium	Greater trochanter of femur	Abducts and medially rotates thigh
Gluteus minimus	Lateral surface of ilium	Greater trochanter of femur	Abducts and medially rotates thigh
Tensor fasciae latae	Iliac crest and surface of ilium between anterior iliac spines	Iliotibial tract	Flexes, abducts, and medially rotates thigh; tenses fascia lata, which laterally supports the thigh
ADDUCTOR GROUP			
Adductor brevis	Ramus of pubis	Linea aspera of femur	Adducts thigh
Adductor longus	Ramus of pubis	As above	Adducts, flexes, and medially rotates thigh
Adductor magnus	Ramus of pubis	As above	Adducts thigh; anterior portion flexes thigh, posterior portion extends thigh
Gracilis	Inferior rami of pubis and ischium	Anterior surface of tibia inferior to medial condyle	Flexes leg and adducts thigh
ILIOPSOAS			
Iliacus	Medial surface of ilium	Femur distal to lesser trochanter; tendon fused with that of psoas major	Flexes hip and/or lumbar spine
Psoas major	Anterior surfaces and transverse processes of lumbar vertebrae	Femur distal to lesser trochanter in company with iliacus	As above

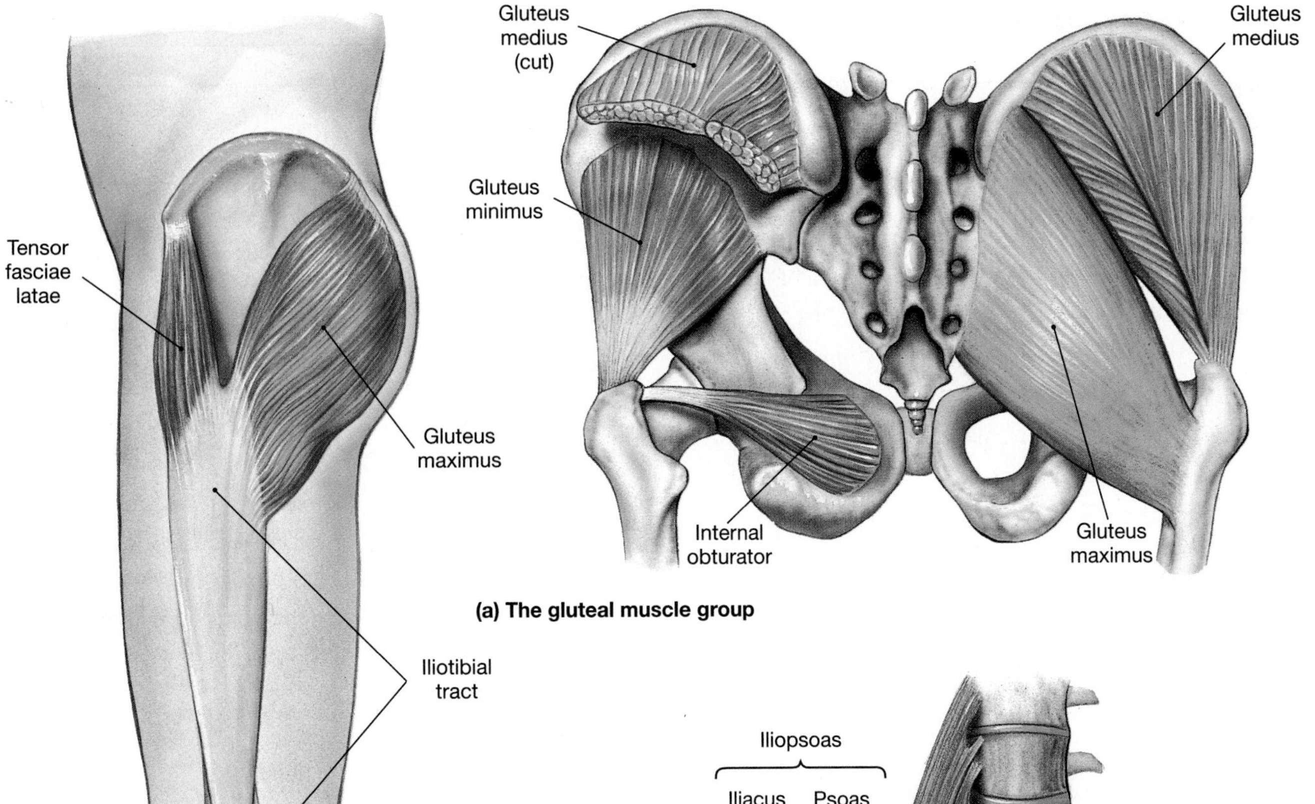

(a) The gluteal muscle group

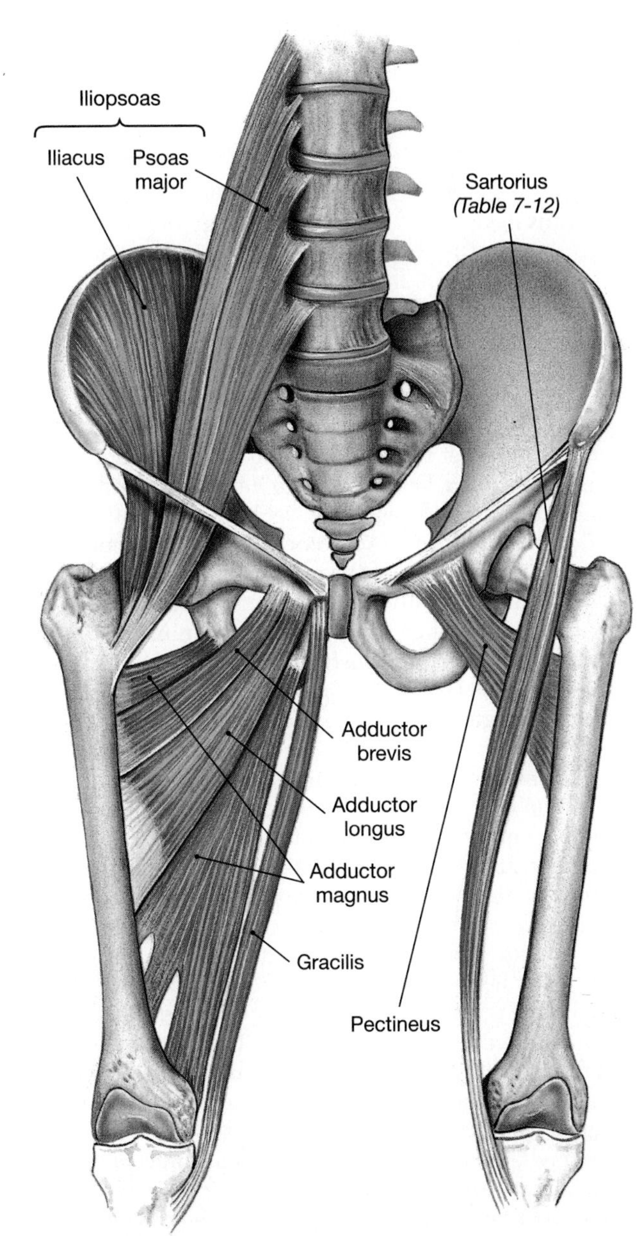

(b) The iliopsoas muscle and the adductor group

●**Figure 7-17**
Muscles That Move the Thigh
(a) The gluteal muscle group. **(b)** The iliopsoas muscle and the adductor group.

When an athlete suffers a *pulled groin*, the problem is a strain in one of these adductor muscles.

- The largest flexor of the thigh is the **iliopsoas** muscle. The iliopsoas is really two muscles, the **psoas major** and the **iliacus**, that share a common insertion at the greater trochanter.

Muscles That Move the Leg The general pattern of muscle distribution in the lower limb is that extensors are found along the anterior and lateral surfaces of the limb, and flexors lie along the posterior and medial surfaces.

- The flexors of the leg (Figure 7-18a●) include three muscles collectively known as the *hamstrings* (the **biceps femoris**, the **semimembranosus**, and the **semitendinosus**) and the **sartorius**.
- Collectively the *knee extensors* are known as the **quadriceps femoris**. The **vastus** muscles and the **rectus femoris** insert upon the patella, which is attached to the tibial tuberosity by the patellar ligament.

The flexors and extensors of the leg are detailed in Table 7-12.

TABLE 7-12 Muscles That Move the Leg

Muscle	*Origin*	*Insertion*	*Action*
FLEXORS			
Biceps femoris	Inferior surface of ischium and linea aspera of femur	Head of fibula, lateral condyle of tibia	Flexes leg, extends and adducts thigh
Semimembranous	Inferior surface of ischium	Posterior surface of medial condyle of tibia	Flexes leg, extends, adducts, and medially rotates thigh
Semitendinosus	Inferior surface of ischium	Proximal, posterior, and medial surface of tibia	As above
Sartorius	Anterior superior spine of ilium	Medial surface of tibia near tibial tuberosity	Flexes leg, flexes and laterally rotates thigh
EXTENSORS			
Rectus femoris	Anterior inferior spine and superior acetabular rim of ilium	Tibial tuberosity via patellar ligament	Extends leg, flexes thigh
Vastus intermedius	Anterior and lateral surface of femur along linea aspera	As above	Extends leg
Vastus lateralis	Anterior and inferior to greater trochanter of femur and along linea aspera	As above	As above
Vastus medialis	Entire length of linea aspera of femur	As above	As above

INTRAMUSCULAR INJECTIONS

Drugs are frequently injected into tissues, rather than directly into the circulation. This method enables the physician to introduce a large amount of a drug at one time, yet have it enter the circulation gradually. In an **intramuscular (IM) injection** the drug is introduced into the mass of a large skeletal muscle. Uptake is usually faster and accompanied by less tissue irritation than when drugs are administered *intradermally* or *subcutaneously* (injected into the dermis or subcutaneous layers). ∞ *[p. 101]* Up to 5 ml of fluid may be injected at one time, and multiple injections are possible.

The most common complications involve accidental injection into a blood vessel or piercing of a nerve. The sudden entry of massive quantities of a drug into the bloodstream can have unpleasant or even fatal consequences; damage to a nerve can cause motor paralysis or sensory loss. As a result, the site of injection must be selected with care. Bulky muscles that contain few large vessels or nerves make ideal targets, and the gluteus medius or the posterior, lateral, superior portion of the gluteus maximus is often selected. The deltoid muscle of the arm, about 2.5 cm (1 in.) distal to the acromion, is another popular site. Probably the most satisfactory from a technical point of view is the vastus lateralis of the thigh, for an injection into this thick muscle will not encounter vessels or nerves. This is the preferred injection site in infants and young children whose gluteal and deltoid muscles are relatively small.

Muscles That Move the Foot and Toes Muscles that move the foot and toes are shown in Figure 7-19● and detailed in Table 7-13. Most of the muscles that move the ankle produce the plantar flexion involved with walking and running movements.

- The large **gastrocnemius** (gas-trok-NĒ-mē-us; *gaster*, stomach + *kneme*, knee) of the calf is assisted by the underlying **soleus** muscle. These muscles share a common tendon, the **calcaneal tendon**, or *Achilles tendon*.
- A pair of deep **peroneus** muscles produce eversion as well as plantar flexion.
- Inversion is caused by contraction of the **tibialis** muscles; the large **tibialis anterior** opposes the gastrocnemius and soleus, and dorsiflexes the foot.

Important digital muscles originate on the surface of the tibia, the fibula, or both. Several smaller muscles originate on the bones of the tarsus and foot, and their contractions move the toes.

✓ You often hear of athletes suffering a "pulled hamstring." To what does this phrase refer?

✓ How would you expect a torn calcaneal tendon to affect movement of the foot?

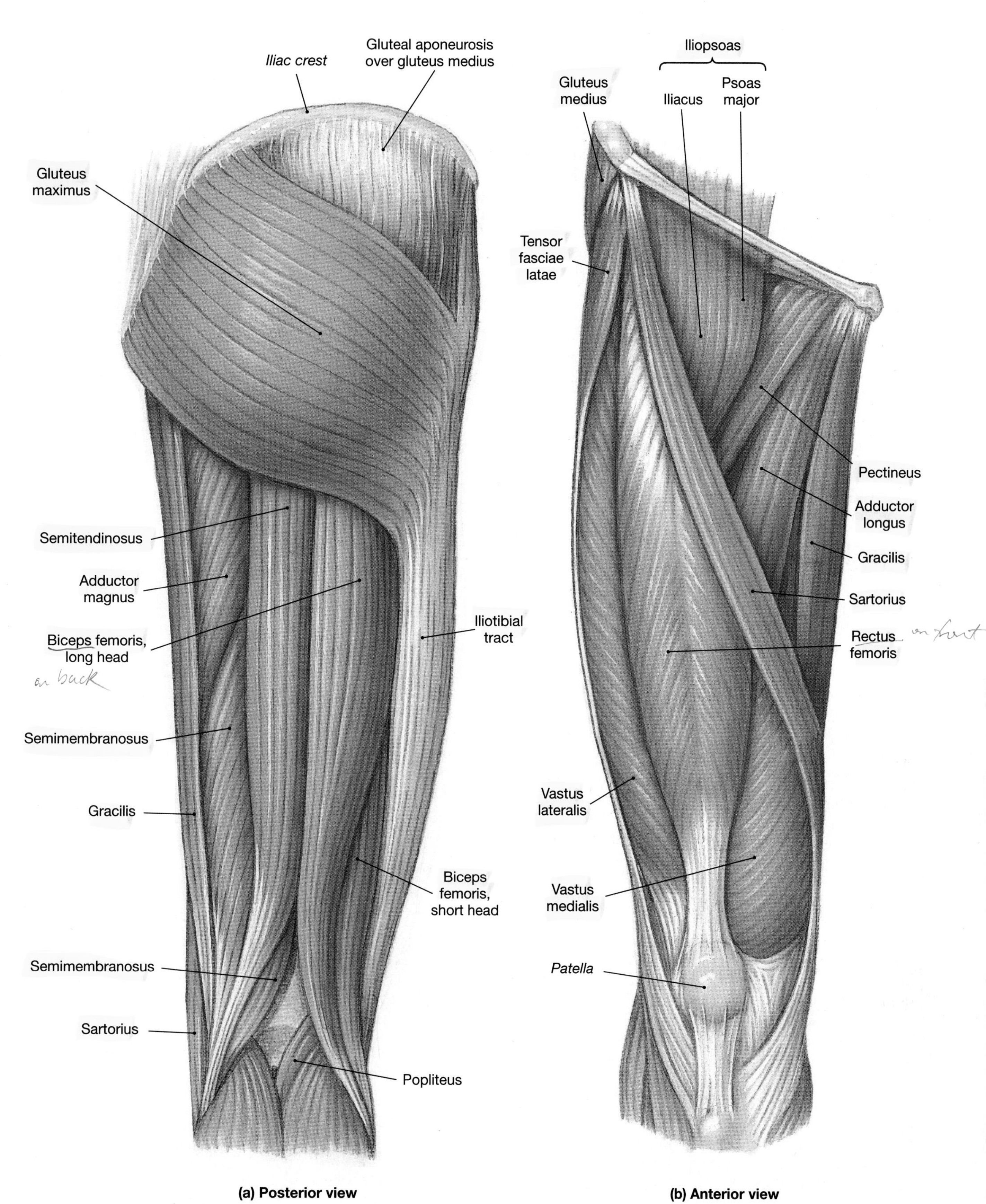

(a) Posterior view

(b) Anterior view

Figure 7-18
Muscles That Move the Leg
(a) Posterior view of right thigh. **(b)** Anterior view of right thigh.

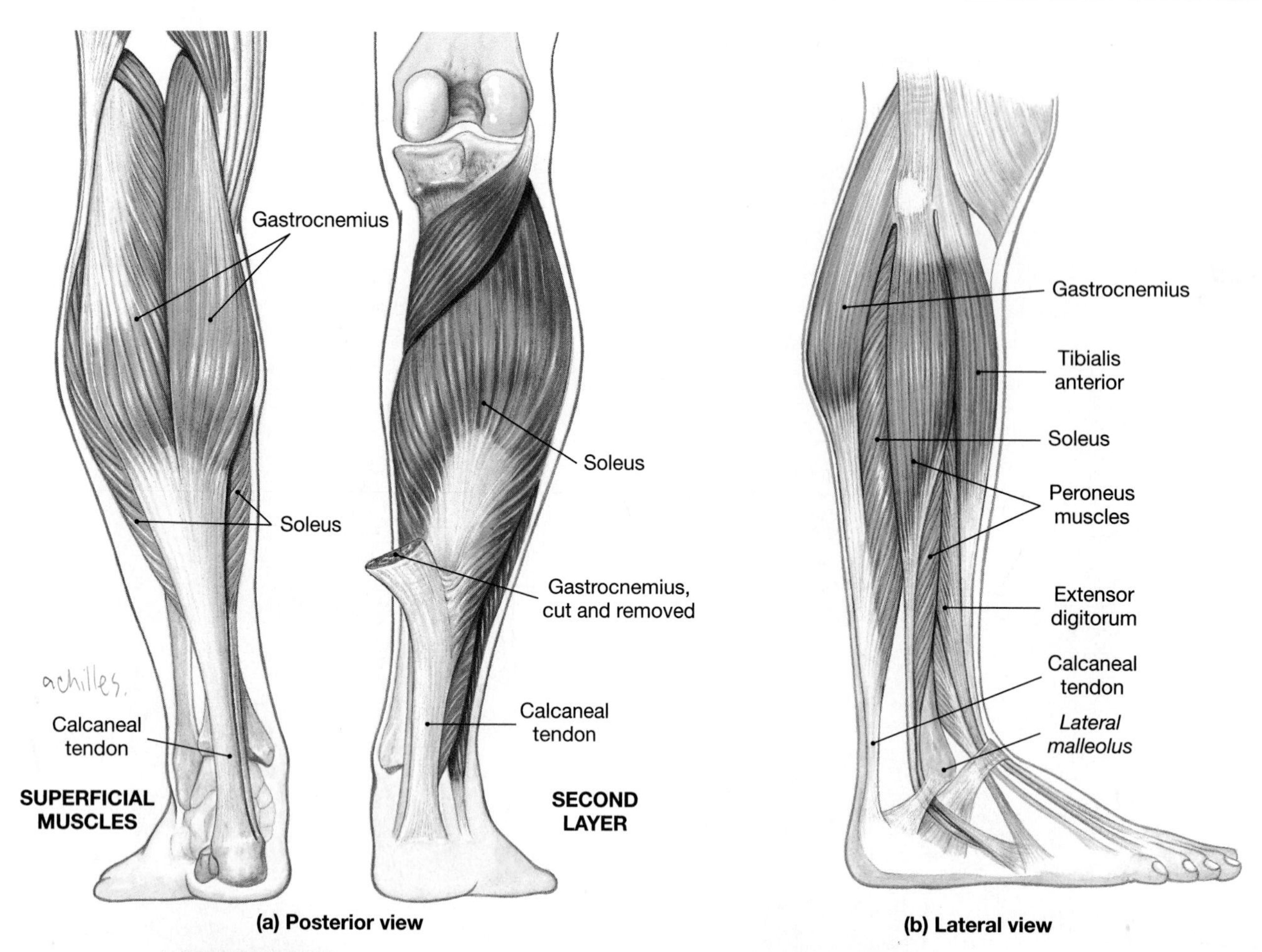

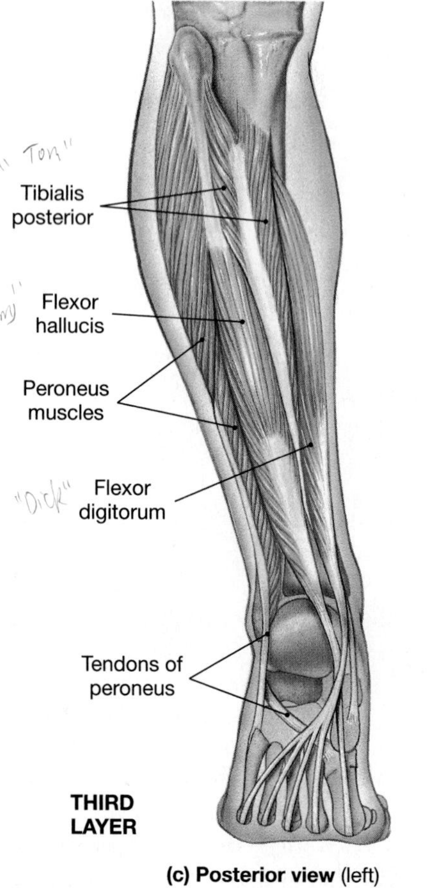

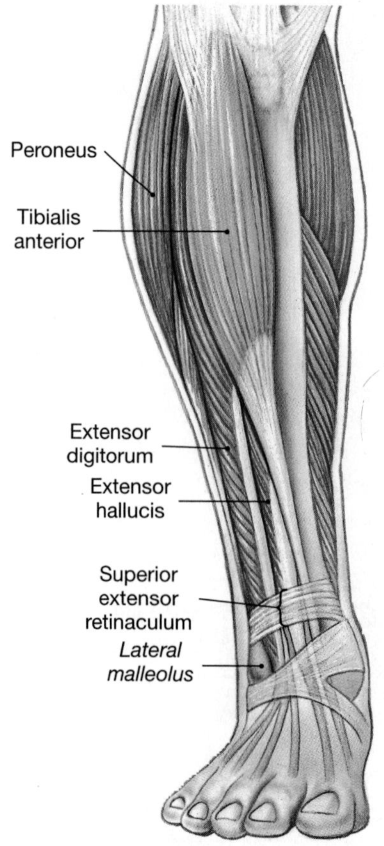

Figure 7-19 Muscles That Move the Foot and Toes

(a) Posterior view, superficial muscles. **(b)** Lateral view. **(c)** Posterior view, deep muscles. **(d)** Anterior view.

TABLE 7-13 Muscles That Move the Foot and Toes

Muscle	*Origin*	*Insertion*	*Action*
DORSIFLEXORS			
Tibialis anterior	Lateral condyle and proximal shaft of tibia	Base of 1st metatarsal	Dorsiflexes foot
PLANTAR FLEXORS			
Gastrocnemius	Above femoral condyles	Calcaneus via calcaneal tendon	Plantar flexes, inverts, and adducts foot; flexes leg
Peroneus	Fibula and lateral condyle of tibia	Bases of 1st and 5th metatarsals	Everts and plantar flexes foot
Soleus	Head and proximal shaft of fibula, and adjacent shaft of tibia	Calcaneus via calcaneal tendon	Plantar flexes, inverts, and adducts foot
Tibialis posterior	Connective tissue membrane and adjacent shafts of tibia and fibula	Tarsals and metatarsals	Adducts and inverts foot
ACTION AT THE TOES			
Flexors			
Flexor digitorum	Posterior and medial surface of tibia	Inferior surface of phalanges, toes 2–5	Flexes toes 2–5
Flexor hallucis	Posterior surface of fibula	Inferior surface, last phalanx of great toe	Flexes great toe
Extensors			
Extensor digitorum	Lateral condyle of tibia, anterior surface of fibula	Superior surfaces of phalanges, toes 2–5	Extends toes 2–5
Extensor hallucis	Anterior surface of fibula	Superior surface, terminal phalanx of great toe	Extends great toe

AGING AND THE MUSCULAR SYSTEM

As the body ages, there is a general reduction in the size and power of all muscle tissues. The effects on the muscular system can be summarized as follows:

1. *Skeletal muscle fibers become smaller in diameter.* The overall effects are a reduction in muscle strength and endurance and a tendency to fatigue rapidly. Because cardiovascular performance also decreases with age, blood flow to active muscles does not increase with exercise as rapidly as it does in younger people.
2. *Skeletal muscles become smaller and less elastic.* Aging skeletal muscles develop increasing amounts of fibrous connective tissue, a process called *fibrosis*. Fibrosis makes the muscle less flexible, and the collagen fibers can restrict movement and circulation.
3. *The tolerance for exercise decreases.* A lower tolerance for exercise results in part from the tendency for rapid fatigue, and in part from the reduction in thermoregulatory ability (described in Chapters 1 and 5), which leads to overheating.
4. *The ability to recover from muscular injuries decreases.* When an injury occurs, repair capabilities are limited, and scar tissue formation is the usual result.

The rate of decline in muscular performance is the same in all individuals, regardless of their exercise patterns or lifestyle. Therefore to be in good shape late in life, one must be in very good shape early in life. Regular exercise helps control body weight, strengthens bones, and generally improves the quality of life at all ages. Extremely demanding exercise is not as important as regular exercise. In fact, extreme exercise in the elderly may lead to problems with tendons, bones, and joints. Although it has obvious effects on the quality of life, there is no clear evidence that exercise prolongs life expectancy.

THE MUSCULAR SYSTEM

FOR ALL SYSTEMS
Generates heat that maintains normal body temperature

INTEGUMENTARY SYSTEM

Removes excess body heat; synthesizes vitamin D_3 for Ca^{2+} and PO_4^{3-} absorption; protects underlying muscles

Skeletal muscles pulling on skin of face produce facial expressions

SKELETAL SYSTEM

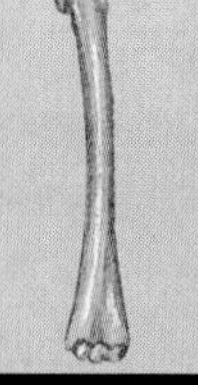

Maintains normal calcium and phosphate levels in body fluids; supports skeletal muscles; provides sites of attachment

Provides movement and support; stresses exerted by tendons maintain bone mass; stabilizes bones and joints

NERVOUS SYSTEM

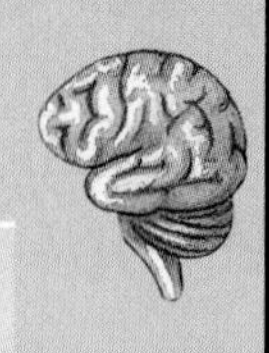

Controls skeletal muscle contractions; adjusts activities of respiratory and cardiovascular systems during periods of muscular activity

Muscle spindles monitor body position; facial muscles express emotions; intrinsic laryngeal muscles permit speech

ENDOCRINE SYSTEM

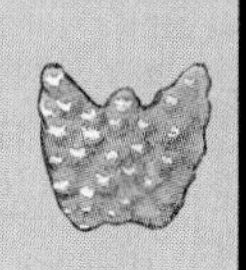

Hormones adjust muscle metabolism and growth; parathyroid hormone and calcitonin regulate calcium and phosphate ion concentrations

Skeletal muscles provide protection for some endocrine organs

CARDIOVASCULAR SYSTEM

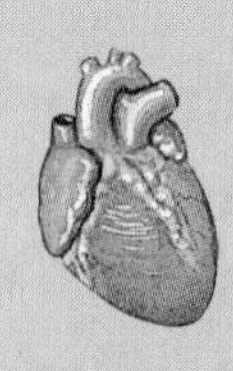

Delivers oxygen and nutrients; removes carbon dioxide, lactic acid, and heat

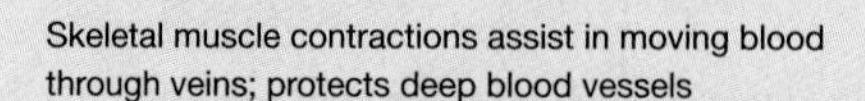

Skeletal muscle contractions assist in moving blood through veins; protects deep blood vessels

LYMPHATIC SYSTEM

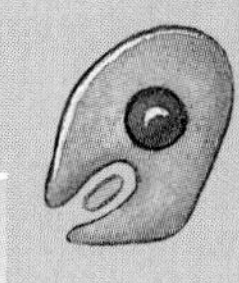

Defends skeletal muscles against infection and assists in tissue repairs after injury

Protects superficial lymph nodes and the lymphatic vessels in the abdominopelvic cavity

RESPIRATORY SYSTEM

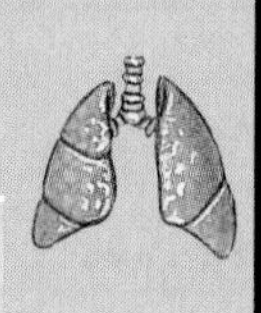

Provides oxygen and eliminates carbon dioxide

Muscles generate CO_2, control entrances to respiratory tract; fill and empty lungs, control airflow through larynx, and produce sounds

DIGESTIVE SYSTEM

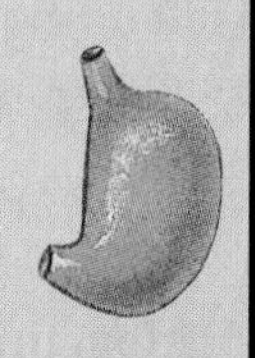

Provides nutrients; liver regulates blood glucose and fatty acid levels and removes lactic acid from circulation

Protects and supports soft tissues in abdominal cavity; controls entrances and exits to digestive tract

URINARY SYSTEM

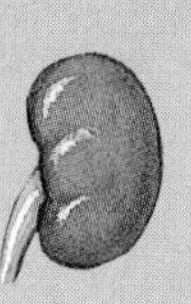

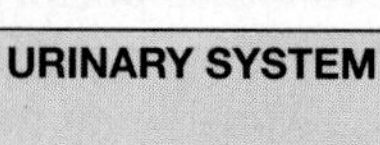

Removes waste products of protein metabolism; assists in regulation of calcium and phosphate concentrations

External sphincter controls urination by constricting urethra

REPRODUCTIVE SYSTEM

Reproductive hormones accelerate skeletal muscle growth

Contractions of skeletal muscles eject semen from male reproductive tract; muscle contractions during sexual act produce pleasurable sensations

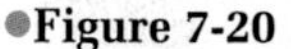

•Figure 7-20
Functional Relationships between the Muscular System and Other Systems

INTEGRATION WITH OTHER SYSTEMS

To operate at maximum efficiency, the muscular system must be supported by many other systems. The changes that occur during exercise provide a good example of such interaction. As noted in earlier sections, active muscles consume oxygen and generate carbon dioxide and heat. Responses of other systems include the following:

1. *Cardiovascular system.* The blood vessels dilate in the active muscles and the skin, and the heart rate increases. These adjustments accelerate oxygen delivery and carbon dioxide removal at the muscle and bring heat to the skin for radiation into the environment.
2. *Respiratory system.* The rate and depth of respiration increases. Air moves in and out of the lungs more quickly, keeping pace with the increased rate of blood flow through the lungs.
3. *Integumentary system.* Blood vessels dilate, and sweat gland secretion increases. This combination helps promote evaporation at the skin surface and removes the excess heat generated by muscular activity.
4. *Nervous and endocrine systems.* These systems direct the responses of other systems by controlling heart rate, respiratory rate, and sweat gland activity.

The muscular system has extensive interactions with other systems even at rest. Figure 7-20• summarizes the range of interactions between the muscular system and other vital systems.

Chapter Review

KEY TERMS

aerobic, p. 165
anaerobic, p. 166
glycolysis, p. 166
insertion, p. 172
lactic acid, p. 166
motor unit, p. 163
myofilaments, p. 156
origin, p. 172
prime mover, p. 172
sarcomere, p. 156
synergist, p. 172
tetanic contraction, p. 162

SUMMARY OUTLINE

INTRODUCTION p. 153

1. There are three types of muscle tissue: skeletal muscle, cardiac muscle, and smooth muscle. The muscular system includes all the skeletal muscle tissue that can be controlled voluntarily.

FUNCTIONS OF SKELETAL MUSCLE p. 153

1. **Skeletal muscles** attach to bones directly or indirectly and perform these functions: (1) produce skeletal movement; (2) maintain posture and body position; (3) support soft tissues; (4) guard entrances and exits; (5) maintain body temperature.

THE ANATOMY OF SKELETAL MUSCLE p. 153

Gross Anatomy p. 153

1. Each muscle fiber is surrounded by an **endomysium.** Bundles of muscle fibers are sheathed by a **perimysium,** and the entire muscle is covered by an **epimysium.** At the end of the muscle is a **tendon.** (*Figure 7-1a*)

Microanatomy p. 154

2. A muscle cell has a cell membrane, or **sarcolemma, sarcoplasm** (cytoplasm), and a **sarcoplasmic reticulum,** similar to the smooth endoplasmic reticulum of other cells. **Transverse tubules (T tubules)** and **myofibrils** aid in contraction. Filaments in a myofibril are organized into repeating functional units called **sarcomeres.** (*Figure 7-1b–d*)

3. Myofilaments consist of **thin filaments** (actin) and **thick filaments** (myosin). (*Figure 7-1e,f*)

4. The relationship between the thick and thin filaments changes as the muscle contracts and shortens. The Z lines move closer together as the thin filaments slide past the thick filaments. (*Figure 7-2*)

5. The contraction process involves **active sites** on thin filaments and **cross-bridges** of the thick filaments. For each cross-bridge, sliding involves repeated cycles of "attach, pivot, detach, and return." At rest the necessary interactions are prevented by **tropomyosin** and **troponin** proteins on the thin filaments.

THE CONTROL OF MUSCLE FIBER CONTRACTION p. 157

1. Neural control of muscle function involves a link between electrical activity in the sarcolemma and the initiation of a contraction.

Structure of the Neuromuscular Junction p. 157

2. Each skeletal muscle fiber is controlled by a neuron at a **neuromuscular junction**; the junction includes the **synaptic knob,** the **synaptic cleft,** and the **motor end plate. Acetylcholine (ACh)** and **acetylcholinesterase (AChE)** play a role in the chemical communication between the synaptic knob and muscle fiber. (*Figure 7-3a,b*)

Function of the Neuromuscular Junction p. 159

3. When an *action potential* arrives at the synaptic knob, Acetylcholine (ACh) is released into the synaptic cleft. Binding of ACh to receptors on the motor end plate leads to the generation of an action potential in the sarcolemma. The passage of an action potential along a transverse tubule triggers the release of calcium ions from the cisternae of the sarcoplasmic reticulum. (*Figure 7-3c*)

The Contraction Cycle p. 159

4. A contraction involves a repeated cycle of "attach, pivot, detach, and return." It begins when calcium ions are released by the sarcoplasmic reticulum. The calcium ions bind to troponin, which changes position and moves tropomyosin away from the active sites of actin. Cross-bridge binding of myosin heads to actin can now occur. After binding, each myosin head pivots at its base, pulling the actin filament toward the center of the sarcomere. (*Figure 7-4*)

Muscle Contraction: A Summary p. 159

5. A summary of the contraction process, from ACh release to the end of the contraction, is shown in *Table 7-1*.

MUSCLE MECHANICS p. 162

1. There is no mechanism to regulate the amount of tension produced in the contraction of an individual muscle fiber: It is either "ON" (producing tension) or "OFF" (relaxed). This is known as the **all-or-none principle**.

2. Both the number of activated muscle fibers and their rate of stimulation control the tension developed by an entire skeletal muscle.

The Frequency of Muscle Stimulation p. 162

3. A muscle fiber **twitch** (a single stimulus-contraction-relaxation sequence) consists of a *latent period*, a *contraction phase*, and a *relaxation phase*. (*Figure 7-5*)

4. Repeated stimulation before the relaxation phase ends may produce **incomplete tetanus** (in which tension will peak because the muscle is never allowed to relax completely), or **complete tetanus** (in which the relaxation phase is completely eliminated). Almost all normal muscular contractions involve the complete tetanus of motor units in the participating muscles. (*Figure 7-6*)

The Number of Muscle Fibers Involved p. 163

5. The number and size of a muscle's **motor units** indicate how precisely controlled its movements are.

6. An increase in muscle tension is produced by increasing the number of motor units through **recruitment**.

7. Resting **muscle tone** stabilizes bones and joints. Inadequate stimulation causes muscles to undergo **atrophy**.

Isotonic and Isometric Contractions p. 164

8. Normal activities usually include both **isotonic** contractions (in which the tension in a muscle remains constant as the muscle shortens) and **isometric** contractions (in which tension rises but the length of the muscle remains constant).

Muscle Elongation p. 164

9. Contraction is active, but elongation of a muscle fiber is a passive process that can occur either through elastic forces or through the contraction of opposing muscles.

THE ENERGETICS OF MUSCULAR ACTIVITY p. 165

1. Muscle contractions require large amounts of ATP energy.

ATP and CP Reserves p. 165

2. ATP is an energy-transfer molecule, not an energy-storage molecule. **Creatine phosphate (CP)** can release stored energy to convert ADP to ATP. A resting muscle cell contains many times more CP than ATP.

ATP Generation p. 165

3. At rest or moderate levels of activity, aerobic metabolism in mitochondria can provide most of the necessary ATP to support muscle contractions. (*Figure 7-7a*)

4. When a muscle fiber runs short of ATP and CP, enzymes can break down glycogen molecules to release glucose that can be broken down via **glycolysis**.

5. At peak levels of activity the cell relies heavily on the anaerobic process of glycolysis to generate ATP, because the mitochondria cannot obtain enough oxygen to meet the existing ATP demands. (*Figure 7-7b*)

Muscle Fatigue p. 166

6. A fatigued muscle can no longer contract, because of changes in pH, a lack of energy, or other problems.

The Recovery Period p. 166

7. The **recovery period** begins immediately after a period of muscle activity, and continues until conditions inside the muscle have returned to preexertion levels. The **oxygen debt** created during exercise is the amount of oxygen used in the recovery period to restore normal conditions. (*Figure 7-7c*)

MUSCLE PERFORMANCE p. 167

1. Muscle performance can be considered in terms of **power** (the maximum amount of tension produced by a particular muscle or muscle group) and **endurance** (the duration of muscular activity).

Types of Skeletal Muscle Fiber p. 167

2. The two types of human skeletal muscle fibers are **fast fibers** and **slow fibers**.

3. Fast fibers are large in diameter, contain densely packed myofibrils, large reserves of glycogen, and few mitochondria. They produce rapid and powerful contractions of relatively short duration.

4. Slow fibers are smaller in diameter and take three times as long to contract after stimulation. Specializations such as an extensive capillary supply, abundant mitochondria, and high concentrations of **myoglobin** enable them to contract for long periods of time.

Physical Conditioning p. 167

5. **Anaerobic endurance** is the ability to support sustained, powerful muscle contractions through anaerobic mechanisms. Training to develop anaerobic endurance can lead to **hypertrophy** (enlargement) of the stimulated muscles.

6. **Aerobic endurance** is the time over which a muscle can continue to contract while supported by mitochondrial activities.

CARDIAC AND SMOOTH MUSCLE TISSUE p. 168

Cardiac Muscle Tissue p. 168

1. Cardiac muscle cells and skeletal muscle fibers differ structurally in terms of (1) size, (2) the number and location of nuclei, (3) their relative dependence on aerobic metabolism when contracting at peak levels, and (4) the presence or absence of **intercalated discs**. (*Figure 7-8a; Table 7-2*)

2. Cardiac muscle cells have automaticity and do not require neural stimulation to contract. Their contractions last longer than those of skeletal muscles, and cardiac muscle cannot be tetanized.

Smooth Muscle Tissue p. 169

3. Smooth muscle is nonstriated, involuntary muscle tissue that can contract over a greater range of lengths than skeletal muscle cells. (*Figure 7-8b; Table 7-2*)

4. Many smooth muscle fibers lack direct connections to motor neurons; those that are innervated are not under voluntary control.

THE MUSCULAR SYSTEM p. 169

1. The muscular system includes approximately 700 skeletal muscles that can be voluntarily controlled. (*Figure 7-9*)

Origins, Insertions, and Actions p. 172

2. Each muscle may be identified by its **origin**, **insertion**, and **primary action**. A muscle may be classified as a **prime mover** or **agonist**, a **synergist**, or an **antagonist**.

Names of Skeletal Muscles p. 172

3. The names of muscles often provide clues to their location, orientation, or function. (*Table 7-3*)

4. The **axial musculature** arises on the axial skeleton; it positions the head and spinal column and moves the rib cage. The **appendicular musculature** stabilizes or moves components of the appendicular skeleton.

The Axial Musculature p. 172

5. The axial muscles fall into four logical groups based on location and/or function. These groups include muscles of (1) the head and neck, (2) the spine, (3) the trunk, and (4) the pelvic floor.

6. The muscles of the head include the **orbicularis oris**, **buccinator**, frontalis and occipitalis, **masseter**, **temporalis**, **pterygoids**, and **platysma**. (*Figure 7-10; Table 7-4*)

7. The muscles of the neck include the **digastricus**, **mylohyoid**, **stylohyoid**, and **sternocleidomastoid**. (*Figure 7-10; Table 7-4*)

8. The extensor muscles of the spine, or *erector spinae*, can be classified into the *spinalis*, *longissimus*, and *iliocostalis* divisions. In the lower lumbar and sacral regions the longissimus and iliocostalis are sometimes called the *sacrospinalis* muscles. (*Figure 7-11; Table 7-5*)

9. The muscles of the trunk include the **oblique** and **rectus** muscles. The thoracic region muscles include the **intercostal** and **transversus** muscles. The **external intercostals** and the **internal intercostals** are important in respiratory movements of the ribs. Also important to respiration is the **diaphragm**. (*Figure 7-12; Table 7-6*)

10. The muscular floor of the pelvic cavity is called the **perineum**. These muscles support the organs of the pelvic cavity and control the movement of materials through the urethra and anus. (*Figure 7-13; Table 7-7*)

The Appendicular Musculature p. 180

11. The **trapezius** and the sternocleidomastoid together affect the position of the shoulder, head, and neck. Other muscles inserting on the scapula include the **rhomboideus**, the **levator scapulae**, the **serratus anterior**, and the **pectoralis minor**. (*Figure 7-14; Table 7-8*)

12. The **deltoid** and the **supraspinatus** are important arm abductors. The **subscapularis**, **teres major**, **infraspinatus**, and **teres minor** rotate the arm. (*Figure 7-15; Table 7-9*)

13. The **pectoralis major** flexes the arm, and the **latissimus dorsi** extends it. Both muscles adduct and rotate the arm. (*Figure 7-15; Table 7-9*)

14. The primary actions of the **biceps brachii** and the **triceps brachii** (long head) affect the elbow. The **brachialis** and **brachioradialis** flex the forearm. The **flexor carpi ulnaris**, the **flexor carpi radialis**, and the **palmaris longus** cooperate to flex the wrist. They are opposed by the **extensor carpi radialis** and the **extensor carpi ulnaris**. The **pronator** muscles pronate the forearm, opposed by the **supinator** and the biceps brachii. (*Figure 7-16; Table 7-10*)

15. **Gluteal muscles** cover the lateral surfaces of the ilia. They extend, abduct, and rotate the thighs. (*Figure 7-17a; Table 7-11*)

16. The four adductors of the thigh are the **adductor magnus**, **adductor longus**, **adductor brevis**, and **gracilis**. (*Figure 7-17b; Table 7-11*)

17. The **psoas major** and the **iliacus** merge to form the **iliopsoas** muscle, a powerful flexor of the thigh. (*Figure 7-17b; Table 7-11*)

18. The flexors of the leg, include the hamstrings (**biceps femoris**, **semimembranosus**, and **semitendinosus**) and **sartorius**. (*Figure 7-18a; Table 7-12*)

19. The knee extensors are known as the **quadriceps femoris**. This group includes the three **vastus** muscles and the **rectus femoris**. (*Figure 7-18b; Table 7-12*)

20. The **gastrocnemius** and **soleus** muscles produce plantar flexion. A pair of **peroneus** muscles produce eversion as well as plantar flexion. The **tibialis anterior** performs dorsiflexion. (*Figure 7-19; Table 7-13*)

21. Control of the phalanges is provided by muscles originating at the tarsals and metatarsals. (*Figure 7-19; Table 7-13*)

AGING AND THE MUSCULAR SYSTEM p. 191

1. The aging process reduces the size, elasticity, and power of all muscle tissues. Exercise tolerance and the ability to recover from muscular injuries both decrease.

INTEGRATION WITH OTHER SYSTEMS p. 193

1. To operate at maximum efficiency, the muscular system must be supported by many other systems. Even at rest, it interacts extensively with other systems. (*Figure 7-20*)

CHAPTER QUESTIONS

LEVEL 1 Reviewing Facts and Terms

Match each item in column A with the most closely related item in column B. Use letters for answers in the spaces provided.

Column A	Column B
____ 1. epimysium	a. resting tension
____ 2. fascicle	b. contractile units
____ 3. endomysium	c. thin filaments
____ 4. motor end plate	d. surrounds muscle fiber
____ 5. transverse tubule	e. enlargement
____ 6. actin	f. surrounds muscle
____ 7. myosin	g. slow fibers
____ 8. extensor of the leg	h. thick filaments
____ 9. sarcomeres	i. muscle bundle
____ 10. tropomyosin	j. hamstring muscles
____ 11. recruitment	k. covers active sites
____ 12. muscle tone	l. conducts action potentials
____ 13. white muscles	m. fast fibers
____ 14. flexor of the leg	n. quadriceps muscles
____ 15. red muscles	o. multiple motor unit summation
____ 16. hypertrophy	p. binds ACh

17. A skeletal muscle contains:
    (a) connective tissues
    (b) blood vessels and nerves
    (c) skeletal muscle tissue
    (d) a, b, and c are correct
18. The type of contraction in which the tension rises but the resistance does not move is called:
    (a) a wave summation (b) a twitch
    (c) an isotonic contraction (d) an isometric contraction
19. What are the five functions of skeletal muscle?
20. What five interlocking steps are involved in the contraction process?
21. What forms of energy reserves are found in resting skeletal muscle cells?
22. What two mechanisms are used to generate ATP from glucose in muscle cells?
23. What is the functional difference between the axial musculature and the appendicular musculature?

LEVEL 2 Reviewing Concepts

24. Areas of the body where no slow fibers would be found include the:
    (a) back and calf muscles (b) eye and hand
    (c) chest and abdomen (d) a, b, and c are correct
25. Describe the basic sequence of events that occur at a neuromuscular junction.
26. Why is the multinucleate condition important in skeletal muscle fibers?
27. The muscles of the spine include many dorsal extensors but few ventral flexors. Why?
28. What specific structural characteristic makes voluntary control of urination and defecation possible?
29. What types of movements are affected when the hamstrings are injured?

LEVEL 3 Critical Thinking and Clinical Applications

30. Many potent insecticides contain toxins called *organophosphates* that interfere with the action of the enzyme acetylcholinesterase. Terry is using an insecticide containing organophosphates and is very careless. He does not use gloves or a dust mask and absorbs some of the chemical through his skin. He inhales a large amount as well. What symptoms would you expect to observe in Terry as a result of the organophosphate poisoning?
31. Mary has just completed a 10-km race. Thirty minutes later she begins to notice soreness and stiffness in her leg muscles. She wonders whether she may have damaged the muscle somehow during the race. She visits her doctor, who orders a blood test. How could the doctor tell from a blood test whether muscle damage had occurred?
32. Jeff is interested in building up his leg muscles, specifically the quadriceps group. What exercises would you recommend to help Jeff accomplish his goal?

8 NEURAL TISSUE AND THE CENTRAL NERVOUS SYSTEM

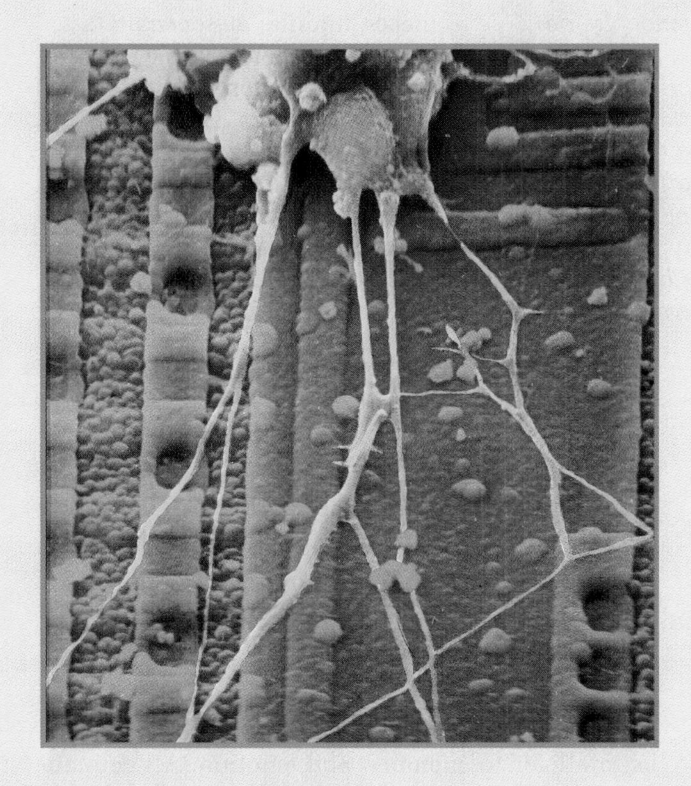

If you think of the nervous system as an organic computer, individual neurons are the "chips" that make it work. This scanning electron micrograph shows a human neuron growing on the surface of a silicon computer chip.

Chapter Outline and Objectives

Vocabulary Development

a-, without; *aphasia*
af, to; *afferent*
amygdale, almond; *amygdaloid bodies*
arachne, spider; *arachnoid membrane*
astro-, star; *astrocyte*
ataxia, a lack of order; *ataxia*
cauda, tail; *cauda equina*
cephalo-, head; *diencephalon*
choroid, a vascular coat; *choroid plexus*
colliculus, a small hill; *superior colliculus*
commissura, a joining together; *commissure*
cortex, rind; *neural cortex*
cyte, cell; *astrocyte*
dia, through; *diencephalon*
dura, hard; *dura mater*
equus, horse; *cauda equina*
ef-, ex-, from; *efferent*
extero-, outside; *exteroceptor*
ferre, to carry; *afferent*
glia, glue; *neuroglia*
hypo-, below; *hypothalamus*
inter-, between; *interneurons*
lexis, diction; *dyslexia*
limbus, a border; *limbic system*
mamilla, a little breast; *mamillary bodies*
mater, mother; *dura mater*
meso-, middle; *mesencephalon*
neuro-, nerve; *neuron*
nigra, black; *substantia nigra*
oligo-, few; *oligodendrocytes*
phasia, speech; *aphasia*
pia, delicate; *pia mater*
plexus, a network; *choroid plexus*
saltare, to leap; *saltatory*
vas, vessel; *vasomotor*

Two organ systems, the *nervous system* and the *endocrine system*, coordinate organ system activities in response to changing environmental conditions. The nervous system controls relatively swift but brief responses to stimuli, whereas the endocrine system usually controls processes that are slower but longer-lasting. For example, the nervous system adjusts body position and moves your eyes across this page, while the endocrine system is regulating the daily rate of energy use by the entire body. We will consider the mechanics of endocrine regulation in Chapter 11.

The nervous system is complex and versatile. As you read these words and think about them, at the involuntary level your nervous system is also monitoring the external environment and internal systems, and issuing appropriate commands as needed to maintain homeostasis. In a few hours, at mealtime or while you are sleeping, the pattern of nervous system activity will be very different. The change from one pattern of activity to another can be almost instantaneous because neural function relies on electrical events that proceed at great speed.

This chapter first examines neurons and other cells of the nervous system and then discusses the basic organization of the *central nervous system*, the brain and spinal cord.

The Nervous System

The **nervous system** includes all the neural tissue in the body. Neural tissue, introduced in Chapter 4, consists of two kinds of cells, *neurons* and *neuroglia*. ∞ *[p. 89]* The nervous system (1) monitors the internal and external environments, (2) integrates sensory information, and (3) coordinates the voluntary and involuntary responses of many other organ systems. All of these functions are performed by neurons, which are supported and protected by surrounding neuroglia.

The two major anatomical subdivisions of the nervous system are detailed in Figure 8-1●. The **central nervous system (CNS)**, consisting of the *brain* and *spinal cord*, is responsible for integrating and coordinating sensory data and motor commands. The CNS is also the seat of higher functions, such as intelligence, memory, and emotion. All communication between the CNS and the rest of the body occurs over the **peripheral nervous system (PNS)**. The peripheral nervous system includes all the neural tissue *outside* the CNS. Its **afferent division** (*af-*, to + *ferre*, to carry) brings sensory information to the CNS, and its **efferent division** (*ef-*, from) carries motor commands to muscles and glands. Within the efferent division, the **somatic nervous system (SNS)** provides voluntary control over skeletal muscles, and the *visceral motor system*, or **autonomic nervous system (ANS)**, provides automatic, involuntary regulation of smooth muscle, cardiac muscle, and glandular activity or secretions.

Cellular Organization in Neural Tissue

Neural tissue includes two distinct cell populations: *neurons* and supporting cells, or *neuroglia*. **Neurons** (*neuro-*, nerve) are the basic units of the nervous system. All neural functions involve neurons' communicating with one another and with other cells. The **neuroglia** (noo-RŌG- lē-a; *glia*, glue), regulate the environment around the neurons, provide a supporting framework for neural tissue, and act as phagocytes. Although they are much smaller cells, neuroglia, also called *glial cells*, far outnumber the neurons.

Neurons

Functional Classification of Neurons

Neurons can be sorted into three functional groups: (1) *sensory neurons*, (2) *motor neurons*, and (3) *interneurons*.

Sensory Neurons **Sensory neurons** of the afferent division convey information from both the external and internal environments to other neurons inside the CNS. Sensory neurons form the afferent division of the PNS, and there are approximately 10 million sensory neurons in the human body.

The processes of sensory neurons extend between a *sensory receptor* and the spinal cord or brain. Stimulation of a receptor usually leads to the stimulation of the sensory neuron. The receptor itself may be a process of a sensory neuron or a specialized cell that communicates with the sensory neuron. Receptors are broadly categorized as *exteroceptors*, *proprioceptors*, and *interoceptors*. **Exteroceptors** (*extero-*, outside) provide information about the external environment in the form of touch, temperature, and pressure sensations and the more complex senses of sight, smell, hearing, and touch. **Proprioceptors** (prō-prē-ō-SEP-torz) monitor the position of skeletal muscles and joints. **Interoceptors** monitor the activities of the digestive, respiratory, cardiovascular, urinary, and reproductive systems and provide sensations of taste, deep pressure, and pain.

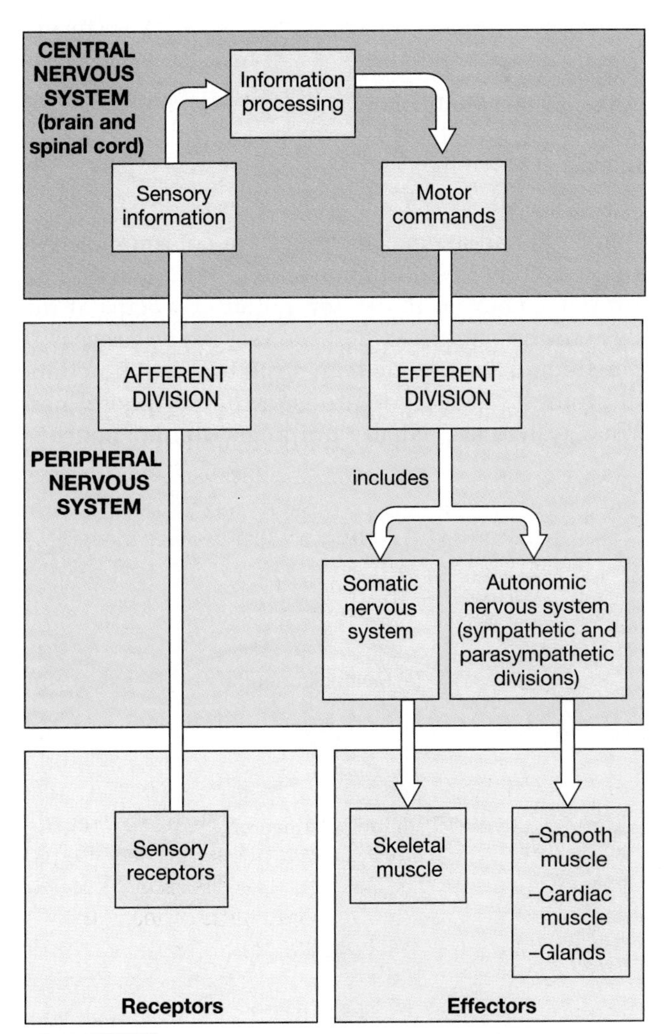

Figure 8-1
Functional Overview of the Nervous System

Motor Neurons The half a million **motor neurons** of the efferent division carry instructions *from* the CNS *to other* tissues, organs, or organ systems. The peripheral targets are called *effectors* because they change their activities in response to the commands issued by the motor neurons. For example, a skeletal muscle is an effector that contracts on neural stimulation. There are two efferent divisions in the PNS, each targeting a separate class of effectors. The **somatic motor neurons** of the *somatic nervous system* innervate skeletal muscles, and the **visceral motor neurons** of the *autonomic nervous system* innervate other peripheral effectors, such as cardiac muscle, smooth muscle, and glands.

Interneurons The 20 billion **interneurons**, or *association neurons*, are located entirely within the brain and spinal cord. Interneurons, as the name implies (*inter-*, between), interconnect other neurons. Interneurons are responsible for the analysis of sensory inputs and the coordination of motor outputs. The more complex the response to a given stimulus, the greater the number of interneurons involved.

The General Structure of Neurons

As we saw in Figure 4-14, (p. 91), a representative neuron has (1) a cell body, or **soma**; (2) several branching, sensitive **dendrites** that receive incoming signals; and (3) an elongate **axon** that carries outgoing signals toward (4) one or more **synaptic knobs**. At each synaptic knob, the neuron communicates with another cell. Neurons can have a variety of shapes; Figure 8-2 shows a *multipolar neuron*, the most common type of neuron in the CNS.

The soma of a typical neuron contains a relatively large, round nucleus with a prominent nucleolus. There are usually no centrioles, organelles that in other cells form the spindle fibers that move chromosomes during cell division. Most neurons lose their centrioles during differentiation and become incapable of undergoing mitosis. As a result, most neurons lost to injury or disease cannot be replaced.

The soma also contains the organelles that provide energy and synthesize organic compounds. The numerous mitochondria, free and fixed ribosomes, and membranes of the rough endoplasmic reticulum (RER) give the cytoplasm a coarse, grainy appearance. The soma contains clusters of rough ER and free ribosomes. These dense clusters, known as **Nissl bodies**, give a gray color to areas containing neuron cell bodies.

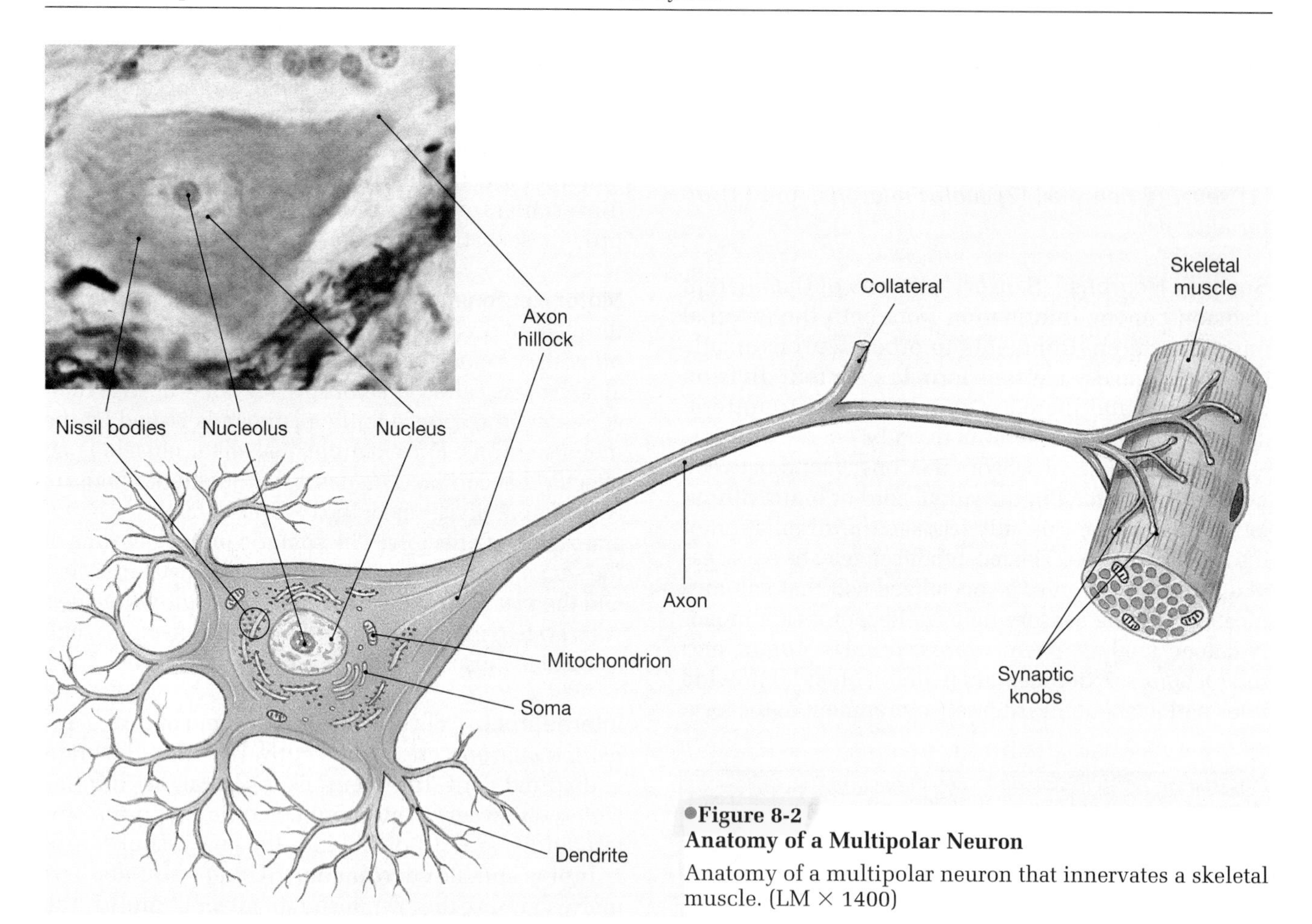

•**Figure 8-2**
Anatomy of a Multipolar Neuron
Anatomy of a multipolar neuron that innervates a skeletal muscle. (LM × 1400)

Projecting from the soma are a variable number of dendrites and a single large axon. The cell membrane of the dendrites and soma is sensitive to chemical, mechanical, or electrical stimulation. In a process described later, such stimulation often leads to the generation of an electrical impulse, or *action potential*, that is conducted along the axon. The base of the axon is attached to the soma at a thickened region known as the *axon hillock*. Action potentials begin at the boundary between the axon hillock and the axon. The axon may branch along its length, producing branches called *collaterals*. Expanded synaptic knobs are found at the tips of each branch. A synaptic knob is part of a **synapse**, a site where a neuron communicates with another cell. How signals are transmitted from cell to cell across the synapse will be the focus of a later section.

Structural Classification of Neurons

The neuron described above is called a **multipolar neuron** because there are multiple processes extending away from the cell body. Multipolar neurons (Figure 8-3a•) are very common within the CNS. For example, all of the motor neurons that control skeletal muscles are multipolar.

Other types of neurons may have fewer extensions from the cell body. In a **unipolar neuron**, the dendritic and axonal processes are continuous, and the cell body lies off to one side (Figure 8-3b•). In a unipolar neuron, the action potential begins at the base of the dendrites, and the rest of the process is considered an axon on both structural and functional grounds. Sensory neurons of the peripheral nervous system are usually unipolar. **Bipolar neurons**

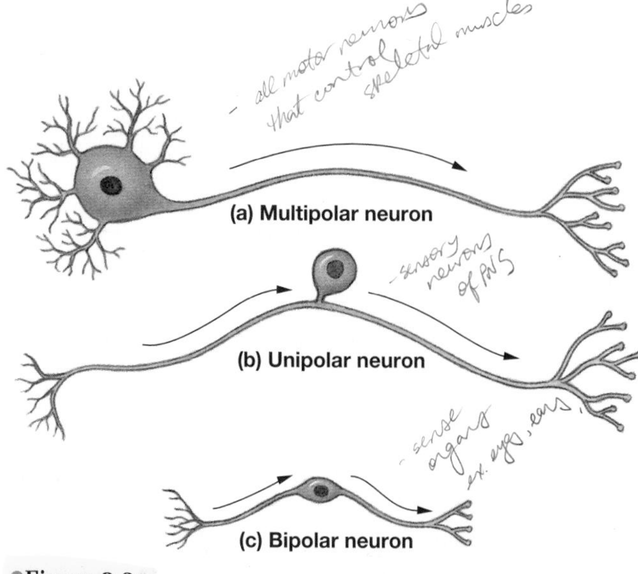

•**Figure 8-3**
A Structural Classification of Neurons

(Figure 8-3c●) have two processes, one dendrite and one axon, with the soma between them. Bipolar neurons are important components of special sense organs such as the eye and ear.

Neuroglia

Neuroglia are found in both the CNS and PNS, but the CNS has the greatest diversity of glial cells. There are four types of glial cells in the central nervous system (Figure 8-4●):

1. **Astrocytes** (AS-trō-sīts; *astro-*, star + *cyte*, cell) are the largest and most numerous neuroglia. Astrocytes secrete chemicals vital to the maintenance of the *blood-brain barrier* that isolates the CNS from chemicals in the general circulation. Astrocytes also create a structural framework for the CNS and perform repairs in damaged neural tissues.

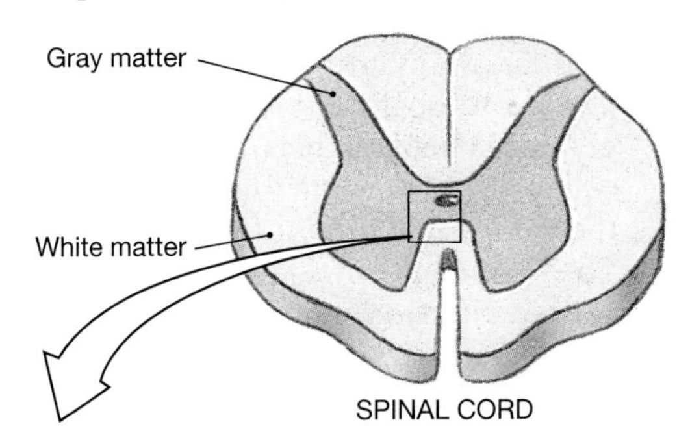

●**Figure 8-4**
Neuroglia in the CNS

A diagrammatic view of neural tissue in the CNS, showing relationships between neuroglia and neurons.

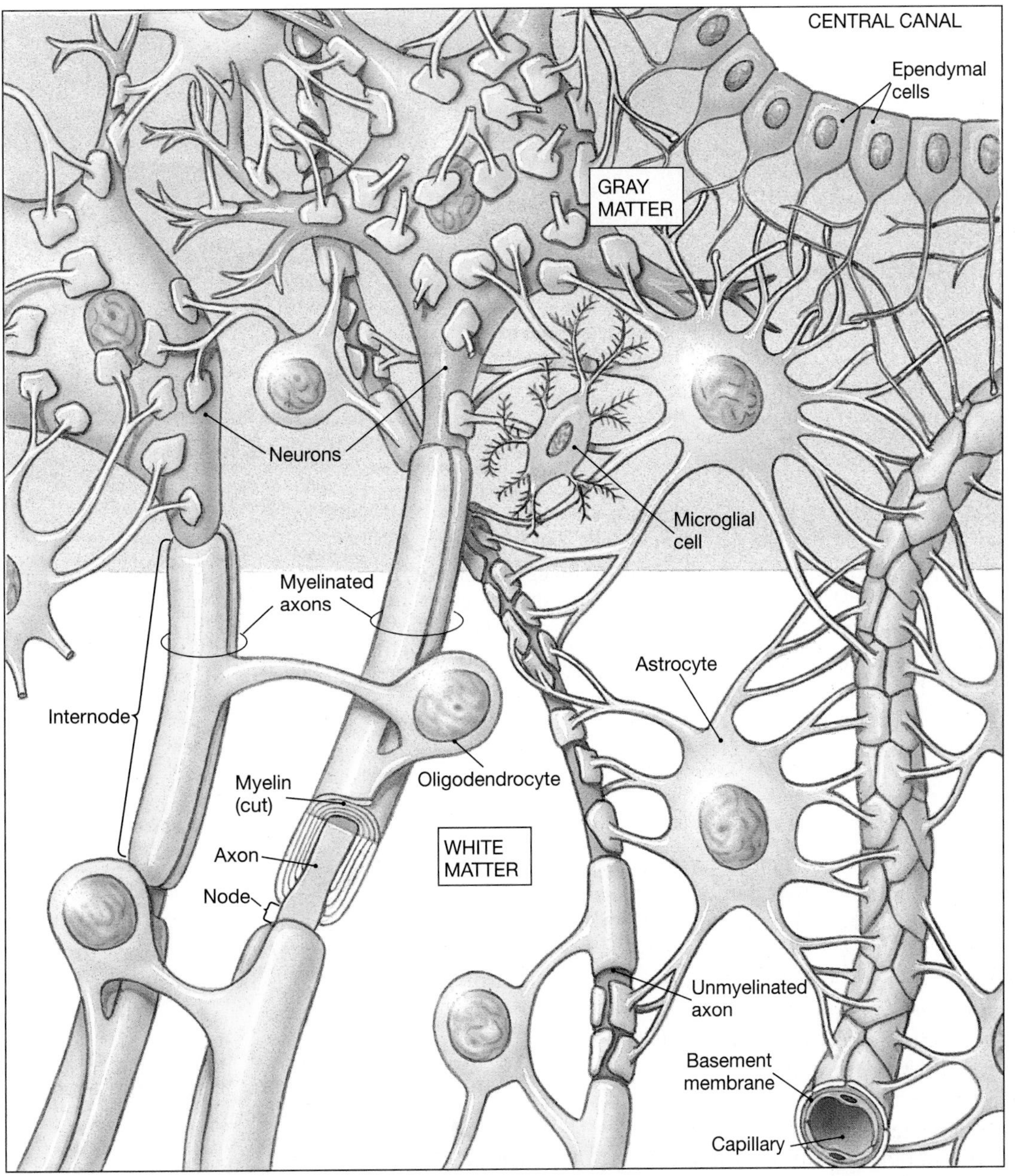

2. **Oligodendrocytes** (o-li-gō-DEN-drō-sīts; *oligo-*, few) have cytoplasmic extensions that can wrap around axons, creating a membranous sheath called **myelin** (Figure 8-4●). Myelin improves the speed of impulse conduction along the axon, by a mechanism detailed later in the chapter. Many oligodendrocytes are needed to coat an entire axon with myelin. The gaps between adjacent cell processes are called *nodes*, or the *nodes of Ranvier* (RAHN-vē-a). The areas covered in myelin are called *internodes.* An axon coated with myelin is said to be **myelinated.** Not every axon in the CNS is myelinated, and those without a myelin coating are said to be **unmyelinated.** Myelin is lipid-rich, and on dissection areas of the CNS containing myelinated axons appear glossy white. These areas constitute the **white matter** of the CNS, whereas areas of **gray matter** are dominated by neuron cell bodies.
3. **Microglia** (mī-KRŌG-lē-a) are the smallest and rarest of the neuroglia in the CNS. Microglia are phagocytic white blood cells that have migrated across capillary walls in the CNS, and they perform protective functions similar to those of white blood cells.
4. **Ependymal cells** line the *central canal* of the spinal cord and the chambers, or *ventricles*, of the brain. This lining is called the **ependyma** (ep-EN-di-mah). In some regions of the brain, the ependyma produces *cerebrospinal fluid (CSF)*, and the cilia on ependymal surfaces in other locations help circulate this fluid within and around the CNS.

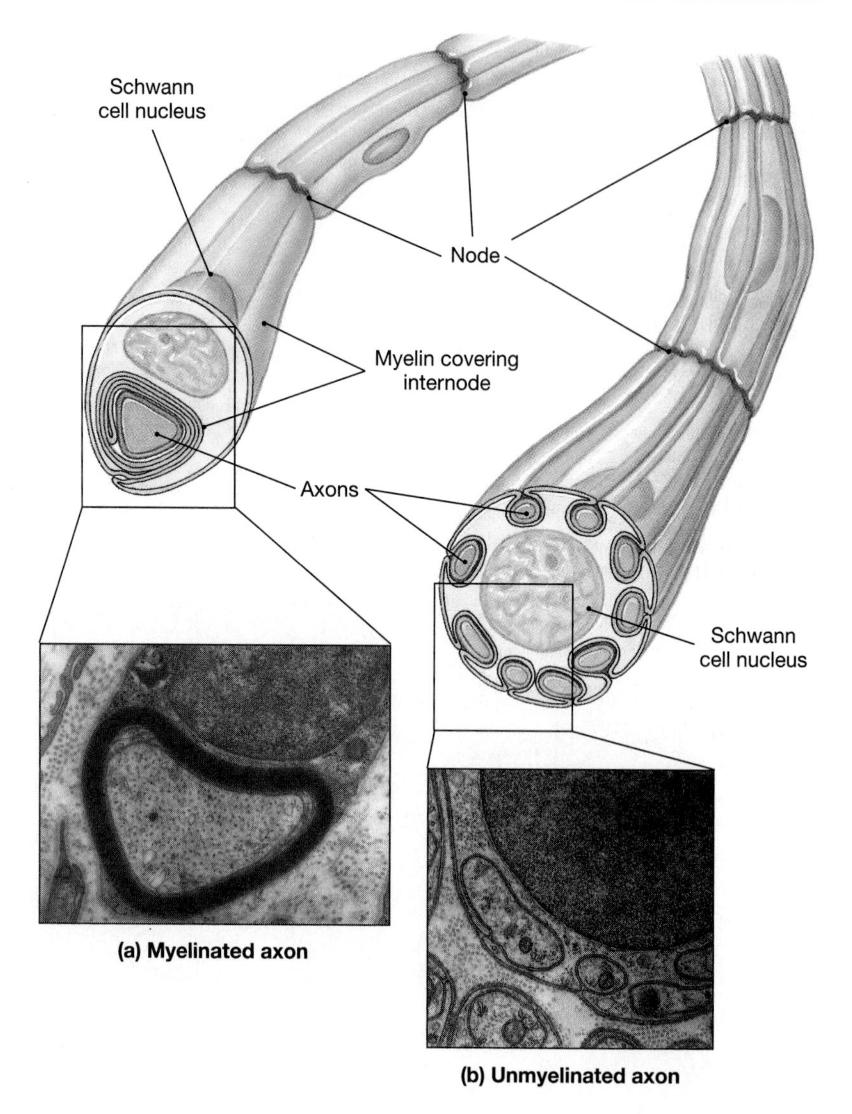

●**Figure 8-5**
Schwann Cells and Peripheral Axons
(a) A single Schwann cell forms the myelin sheath around a portion of a single axon. This arrangement differs from the way myelin forms inside the CNS; compare with *Figure 8-4*. (TEM × 14,048). **(b)** A single Schwann cell can encircle several unmyelinated axons. Unlike the situation inside the CNS (*Figure 8-4*), every axon in the PNS is completely enclosed by glial cells.

Schwann cells are the most important glial cells in the PNS. Schwann cells cover every axon outside of the CNS, whether it is myelinated or unmyelinated. In creating a myelin sheath, a Schwann cell wraps itself around a segment of a single axon (Figure 8-5a●). Although the mechanism of myelination differs between the CNS and PNS, myelination in both divisions creates nodes and internodes, and the presence of myelin—however formed—increases the rate of impulse conduction. A Schwann cell may surround portions of several different unmyelinated axons (Figure 8-5b●).

✓ What would damage to the afferent division of the nervous system affect?

✓ Examination of a tissue sample shows unipolar neurons. Are these more likely to be sensory neurons or motor neurons?

✓ What type of glial cell would you expect to find in large numbers in brain tissue from a person suffering from a CNS infection?

NEUROPHYSIOLOGY

The sensory, integrative, and motor functions of the nervous system are dynamic and ever-changing. All of the important communications between neurons and other cells occur at membrane surfaces, through changes in the membrane potential. These membrane changes are electrical events that proceed at great speed.

The Membrane Potential

The cell membrane of an undisturbed cell has an excess of positive charges on the outside and an excess of negative charges on the inside. Such an uneven distribution of charges is known as a *potential difference*, and the size of the potential difference is measured in terms of *volts*. Because the charges are separated by a cell membrane, the potential difference across the cell membrane of a living cell is called a **membrane potential**, or *transmembrane potential*. The **resting potential**, or membrane potential, of an undisturbed cell is very small. For example, the resting potential of a neuron averages about 0.070 volts, versus around 1.5 volts for a flashlight battery. Because they are so small, membrane potentials are usually reported in *millivolts* (mV, thousandths of volts) rather than volts. The resting potential of a neuron is –70 mV, with the minus sign indicating that the inside of the cell membrane contains an excess of negative charges as compared with the outside.

Factors Responsible for the Membrane Potential

As noted in earlier chapters, the intracellular and extracellular fluids differ markedly in ionic composition. For example, the extracellular fluid contains relatively high concentrations of sodium ions (Na^+) and chloride ions (Cl^-), whereas the intracellular fluid contains high concentrations of potassium ions (K^+) and negatively charged proteins (Pr^-).

The intracellular and extracellular fluids are separated by the cell membrane. The proteins cannot cross the membrane at all, and the ions can enter or leave the cell only by passing through membrane channels. ∞ [p. 52] There are many different types of channels in the membrane; some are always open (*leak channels*), and others open or close under specific circumstances (*gated channels*).

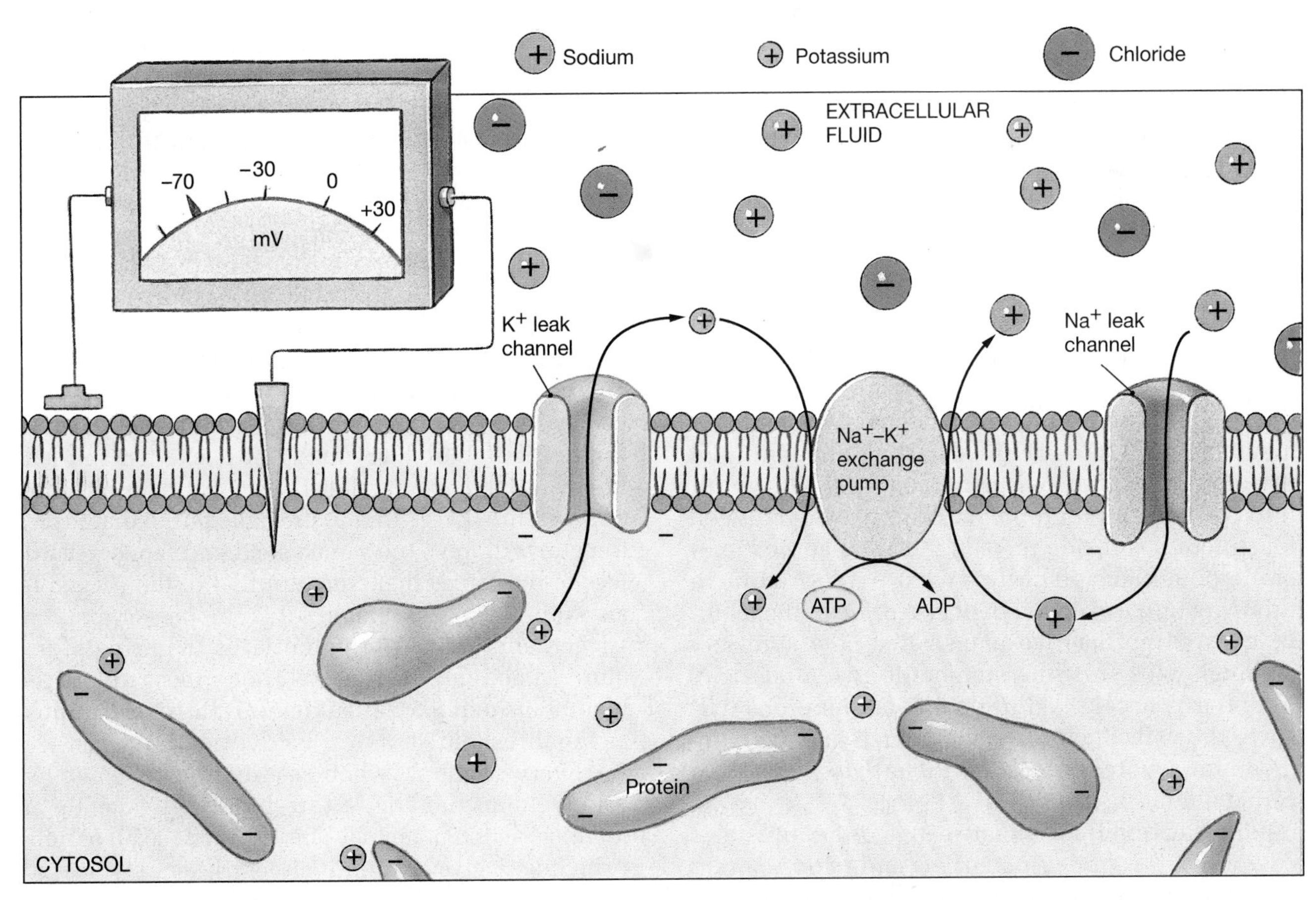

Figure 8-6
The Cell Membrane at the Resting Potential

Because the intracellular concentration of potassium ions is relatively high, potassium ions tend to diffuse out of the cell. This movement is driven by the concentration gradient for potassium ions. Similarly, the concentration gradient for sodium ions tends to promote their movement into the cell. However, the cell membrane is significantly more permeable to potassium ions than to sodium ions. As a result, potassium ions diffuse out of the cell faster than sodium ions enter the cytoplasm. The cell therefore experiences a net loss of positive charges, and as a result the interior of the cell membrane contains an excess of negative charges, primarily from negatively charged proteins.

The resting potential remains stable over time because the cell membrane contains the sodium-potassium exchange pump, an ion pump introduced in Chapter 3. ∞ *[p. 55]* At a membrane potential of –70 mV, the rate of sodium entry to potassium loss can be precisely balanced by the sodium-potassium exchange pump. Figure 8-6● presents a diagrammatic view of the cell membrane at the resting potential.

Changes in the Membrane Potential

Every living cell has a resting potential. Any stimulus that (1) alters membrane permeability to sodium or potassium or (2) alters the activity of the exchange pump will disturb the resting potential of a cell. Examples of stimuli that can affect the membrane potential include exposure to specific chemicals, mechanical pressure, changes in temperature, or shifts in the extracellular ion concentrations. Any change in the resting potential can have an immediate effect on the cell. For example, in a skeletal muscle fiber, permeability changes in the sarcolemma trigger a contraction. ∞ *[p. 159]*

In most cases, a stimulus opens gated ion channels that are closed when the cell membrane is at the normal resting potential. The opening of these channels accelerates the movement of ions across the cell membrane, and this movement changes the membrane potential. For example, the opening of gated sodium channels will accelerate sodium entry into the cell. As the number of positively charged ions increases inside the cell membrane, the membrane potential will shift toward 0 mV. A shift in this direction is called a *depolarization* of the membrane. A stimulus that opens gated potassium ion channels will shift the membrane potential away from 0 mV, because additional potassium ions will leave the cell. Such a change, which may take the membrane potential from –70 mV to –80mV, is called a *hyperpolarization*

Information transfer between neurons and other cells involves *graded potentials* and *action potentials*. **Graded potentials** affect only a limited portion of the cell membrane. For example, if a chemical stimulus applied to the cell membrane of a neuron opens gated sodium ion channels at a single site, the sodium ions entering the cell will depolarize the membrane at that location. The sodium ions will then diffuse along the inner surface of the membrane in all directions. Because the sodium ions are spreading out, the degree of depolarization decreases with distance away from the point of entry.

Graded potentials occur in the membranes of all living cells in response to environmental stimuli. These changes can trigger shifts in cellular function, and motor neurons control other cells by producing graded potentials in their cell membranes. For example, chemicals released by motor neurons produce graded potentials in the membranes of gland cells, and these potential changes can stimulate or inhibit glandular secretion. However, graded potentials affect too small an area to have an effect on the activities of relatively enormous cells, such as skeletal muscle fibers or neurons. In these cells, graded potentials can affect operations in distant portions of the cell only if they lead to the production of an *action potential*, an electrical signal that affects the entire membrane surface.

An **action potential** is a conducted change in the permeability of the cell membrane. Skeletal muscle fibers and axons have membranes that will conduct action potentials. In a skeletal muscle fiber, the action potential begins at the neuromuscular junction and travels across the entire membrane surface, including the T tubules. ∞ *[p. 159]* The resulting ion movements trigger a contraction. In an axon, an action potential usually begins near the axon hillock and travels along the length of the axon toward the synaptic knobs, where its arrival activates the synapses.

Action potentials are created by the opening and closing of sodium and potassium channels in response to a local depolarization. The graded depolarization acts like pressure on the trigger of a gun. A gun fires only after a certain minimum pressure has been applied to the trigger. It does not matter whether the pressure builds gradually or is exerted suddenly—when the pressure reaches a critical point, the gun will fire. Every time it fires, the bullet that leaves the gun will have the same speed and range, regardless of the forces that were applied to the trigger. In an axon, the graded potential is the pressure on the trigger, and the action potential is the firing of the gun. An action potential will not appear unless the membrane depolarizes sufficiently, to a level known as the **threshold**.

Every stimulus, whether minor or extreme, that brings the membrane to threshold will generate an identical action potential. This is called the **all or none principle**: A given stimulus either triggers a typical action potential or does not produce one at all. The all or none principle applies to all excitable membranes.

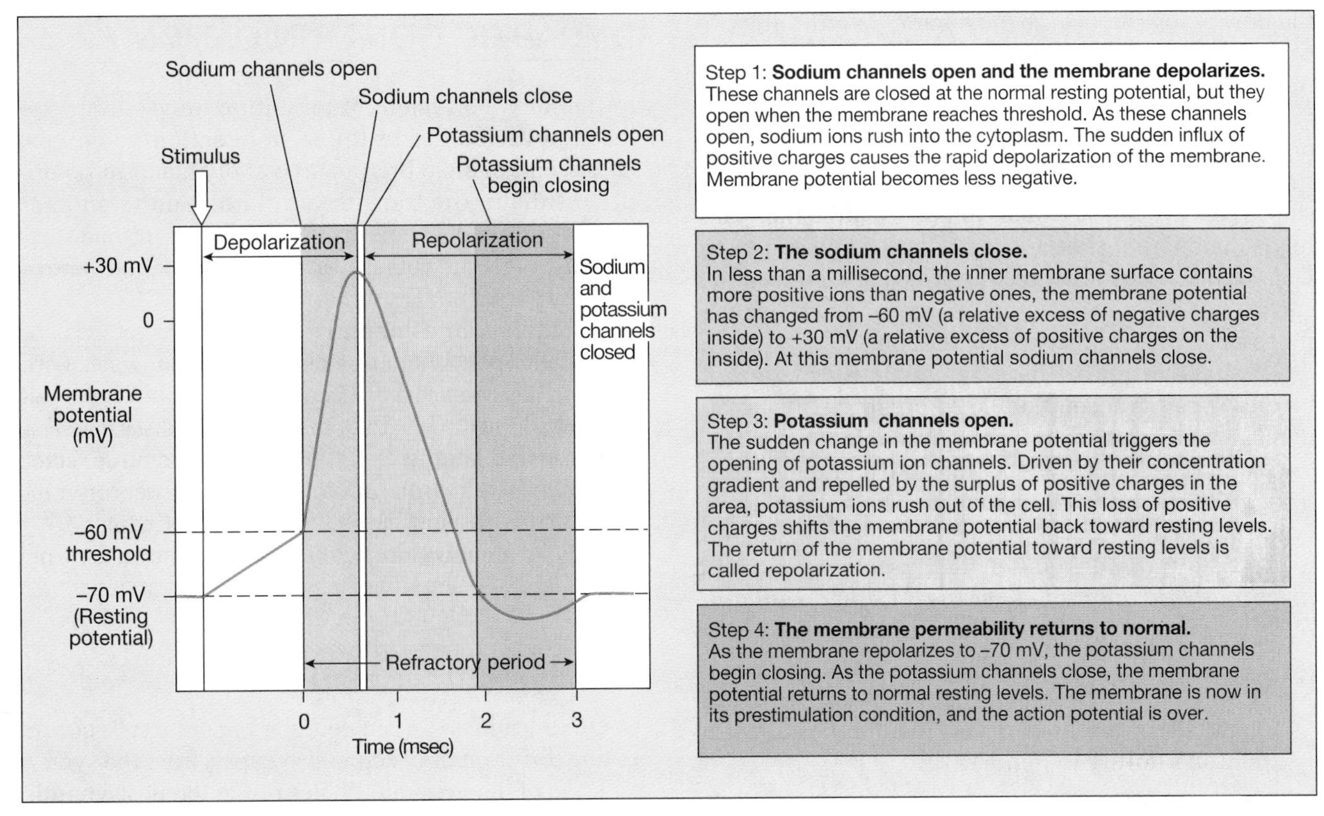

Sodium channels open
Sodium channels close
Potassium channels open
Potassium channels begin closing
Stimulus
Depolarization
Repolarization
Sodium and potassium channels closed
+30 mV
0
Membrane potential (mV)
–60 mV threshold
–70 mV (Resting potential)
Refractory period
0
1
2
3
Time (msec)
Step 1: Sodium channels open and the membrane depolarizes.
These channels are closed at the normal resting potential, but they open when the membrane reaches threshold. As these channels open, sodium ions rush into the cytoplasm. The sudden influx of positive charges causes the rapid depolarization of the membrane. Membrane potential becomes less negative.
Step 2: The sodium channels close.
In less than a millisecond, the inner membrane surface contains more positive ions than negative ones, the membrane potential has changed from –60 mV (a relative excess of negative charges inside) to +30 mV (a relative excess of positive charges on the inside). At this membrane potential sodium channels close.
Step 3: Potassium channels open.
The sudden change in the membrane potential triggers the opening of potassium ion channels. Driven by their concentration gradient and repelled by the surplus of positive charges in the area, potassium ions rush out of the cell. This loss of positive charges shifts the membrane potential back toward resting levels. The return of the membrane potential toward resting levels is called repolarization.
Step 4: The membrane permeability returns to normal.
As the membrane repolarizes to –70 mV, the potassium channels begin closing. As the potassium channels close, the membrane potential returns to normal resting levels. The membrane is now in its prestimulation condition, and the action potential is over.

●**Figure 8-7**
An Action Potential

A sufficiently strong depolarizing stimulus will bring the membrane potential to threshold and trigger an action potential.

The Generation of an Action Potential

An action potential begins when the cell membrane depolarizes to threshold. Figure 8-7● diagrams the steps involved in the generation of an action potential, beginning with a graded depolarization to threshold and ending with a return to the resting potential.

From the moment that the sodium channels open at threshold until repolarization is completed, the membrane cannot respond normally to further stimulation. This is the **refractory period**. The refractory period limits the number of action potentials that can be generated in an excitable membrane. (The maximum rate of action potential generation is 500–1000 per second.)

Potentially deadly forms of human poisoning result from eating seafood containing *neurotoxins*, poisons that primarily affect neurons. Several neurotoxins, such as *tetrodotoxin* (TTX) from puffer fish, either block open sodium channels or prevent their opening. Motor neurons cannot function under these conditions, and death may result from paralysis of the respiratory muscles.

Conduction of an Action Potential

An action potential initially involves a relatively small segment of the total membrane surface. But unlike graded potentials, which diminish rapidly with distance, action potentials affect the entire membrane surface. The basic mechanism of action potential conduction along unmyelinated and myelinated axons is shown in Figure 8-8●. (The areas of the membrane that are in the refractory period are shaded.)

At a given site, for a brief moment at the peak of the action potential, the inside of the cell membrane contains an excess of positive ions. Because opposite charges attract one another, these ions immediately begin spreading along the inner surface of the membrane, drawn to the surrounding negative charges. This *local current* depolarizes adjacent portions of the membrane, and when threshold is reached action potentials occur at these locations (Figure 8-8a●).

The process continues in a chain reaction that soon reaches the most distant portions of the cell membrane. This form of action potential transmission is known as **continuous conduction**. You might compare continuous conduction to a person walking heel-to-toe; progress is made in a series of small steps. Continuous conduction occurs along unmyelinated axons at a speed of around 1 meter per second (2 mph).

In a myelinated fiber, the axon is wrapped in layers of myelin. This wrapping is complete except at the nodes, where adjacent glial cells contact one another. Between the nodes, the lipids of the myelin sheath block the flow of ions across the membrane. As a result, continuous conduction cannot occur. Instead, when an action potential occurs at the base of the axon, the local current skips the internode and depolarizes the closest node to threshold. The action potential jumps from node to node, rather than proceeding in a series of small steps. This process is called **saltatory conduction**, taking its name from *saltare*, the Latin word meaning "to leap." Saltatory conduction, illustrated in Figure 8-8b●, carries nerve impulses along an axon roughly seven times as fast as continuous conduction.

Demyelination is the progressive destruction of myelin sheaths, accompanied by inflammation, axon damage, and scarring of neural tissue. In most cases, the result is a gradual loss of sensation and motor control that leaves affected regions numb and paralyzed. Four important demyelinating disorders considered in the *Applications Manual* are *heavy metal poisoning, diphtheria, multiple sclerosis (MS),* and *Guillain-Barré syndrome.* AM *Demyelination Disorders*

✓ How would a chemical that blocks the sodium channels in the neuron cell membranes affect a neuron's ability to depolarize?

✓ Two axons are tested for propagation velocities. One carries action potentials at 50 meters per second, the other at 1 meter per second. Which axon is myelinated?

Synaptic Communication

In the nervous system, information moves from one location to another in the form of action potentials. At the end of an axon, the arrival of an action potential results in the transfer of information to another neuron or effector cell. The information transfer occurs through the release of chemicals called **neurotransmitters** from the synaptic knob.

Neurons can communicate with other neurons or other cell types. When one neuron communicates with another, the synapse may occur on a dendrite, on the cell body, or along the length of the axon. Synapses between a neuron and another cell type are called **neuroeffector junctions**. At a *neuromuscular junction* the neuron communicates with a muscle cell, as we saw in Chapter 7. [p. 157] At a *neuroglandular junction* a neuron controls or regulates the activity of a secretory cell.

Structure of a Synapse

Communication between neurons and other cells occurs in one direction only across a synapse, from the synaptic knob of the **presynaptic neuron** to the **postsynaptic neuron** (Figure 8-9●) or other cell type. The opposing cell membranes are separated by a narrow **synaptic cleft**.

Each synaptic knob contains mitochondria, synaptic vesicles, and areas of endoplasmic reticulum. Every

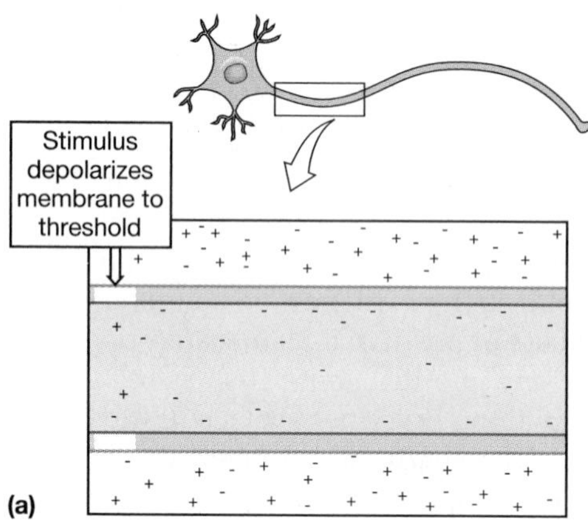

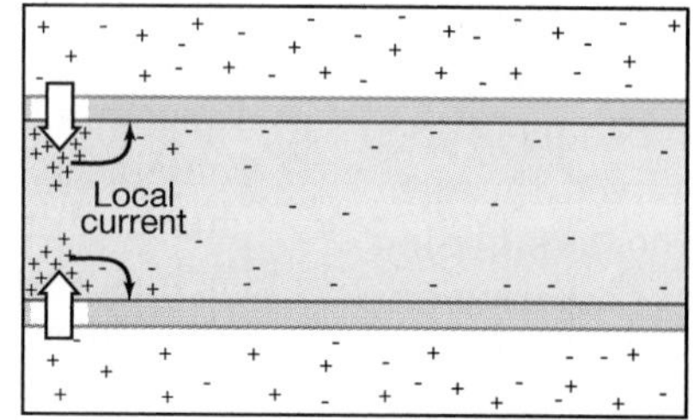

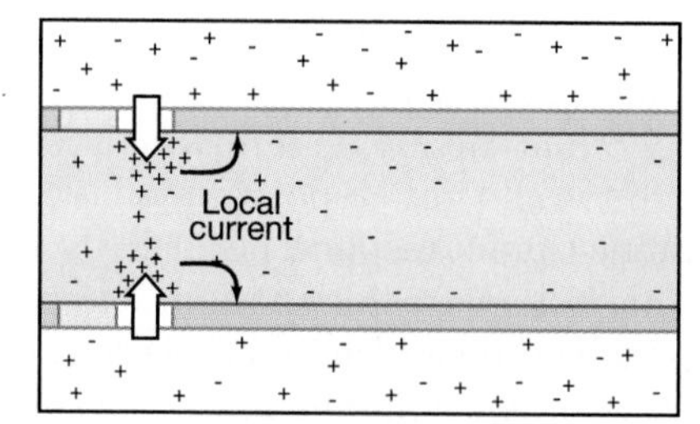

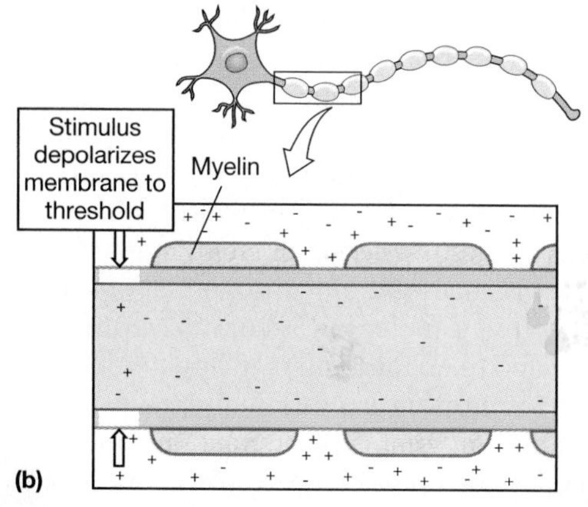

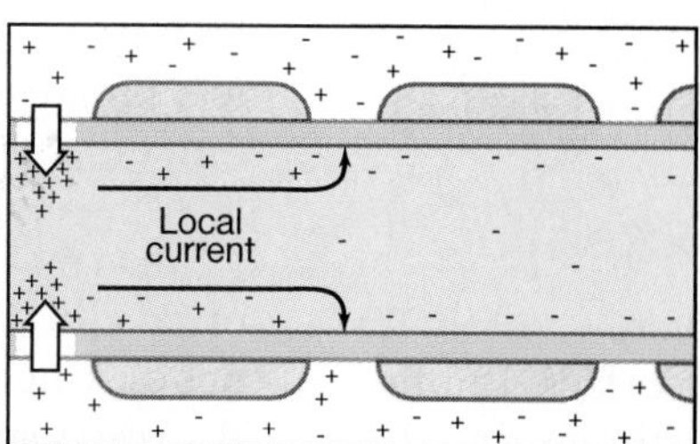

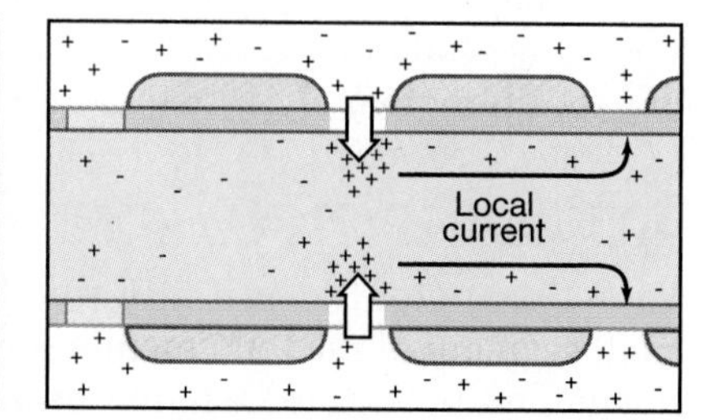

●**Figure 8-8**
Action Potential Conduction over Unmyelinated and Myelinated Axons
(a) Continuous conduction along an unmyelinated axon. **(b)** Saltatory conduction along a myelinated axon. (The pale violet membrane areas are in the refractory period.)

synaptic vesicle contains several thousand molecules of a specific neurotransmitter, and on stimulation many of these vesicles release their contents into the synaptic cleft. The neurotransmitter then diffuses across the synaptic cleft and binds to receptors on the postsynaptic membrane.

Synaptic Events

There are many different neurotransmitters. The neurotransmitter *acetylcholine*, or *ACh*, is released at *cholinergic synapses.* Cholinergic synapses are common inside and outside of the CNS; the neuromuscular junction described in Chapter 7 is one example. *[p. 159]* Figure 8-10● and Table 8-1 summarize the events that occur at a cholinergic synapse following the arrival of an action potential at the presynaptic neuron.

Neurotransmitters

Although acetylcholine is a common neurotransmitter, there are many others. For example, **norepinephrine** (nōr-ep-i-NEF-rin), or **NE**, is important in the brain and in portions of the autonomic nervous system. It is also called *noradrenaline*, and synapses releasing NE are described as **adrenergic**. Like ACh, NE usually has a temporary excitatory effect on the postsynaptic membrane; it is then broken down by enzymes such as *monoamine oxidase.*

Dopamine (DŌ-pah-mēn), **gamma aminobutyric** (GAM-ma a-MĒ-nō-bū-TIR-ik) **acid**, also known as **GABA**, and **serotonin** (ser-ō-TŌ-nin) are CNS neurotransmitters whose effects are usually inhibitory. There are also many other neurotransmitters (at least 50) whose functions are poorly understood. *Drugs and Synaptic Function*

✓ A neurotransmitter causes potassium channels to open but not sodium channels. What effect would this neurotransmitter produce at the postsynaptic membrane?

TABLE 8-1 Sequence of Events at a Typical Cholinergic Synapse

Step 1:
- An arriving action potential depolarizes the synaptic knob and the presynaptic membrane.

Step 2:
- Calcium ions enter the cytoplasm of the synaptic knob.
- ACh release occurs through diffusion and exocytosis of neurotransmitter vesicles.

Step 3:
- ACh diffuses across the synaptic cleft and binds to receptors on the postsynaptic membrane.
- Chemically regulated sodium channels on the postsynaptic surface are activated, producing a graded depolarization.
- ACh release ceases because calcium ions are removed from the cytoplasm of the synaptic knob.

Step 4:
- The depolarization ends as ACh is broken down into acetate and choline by AChE.
- The synaptic knob reabsorbs choline from the synaptic cleft and uses it to resynthesize ACh.

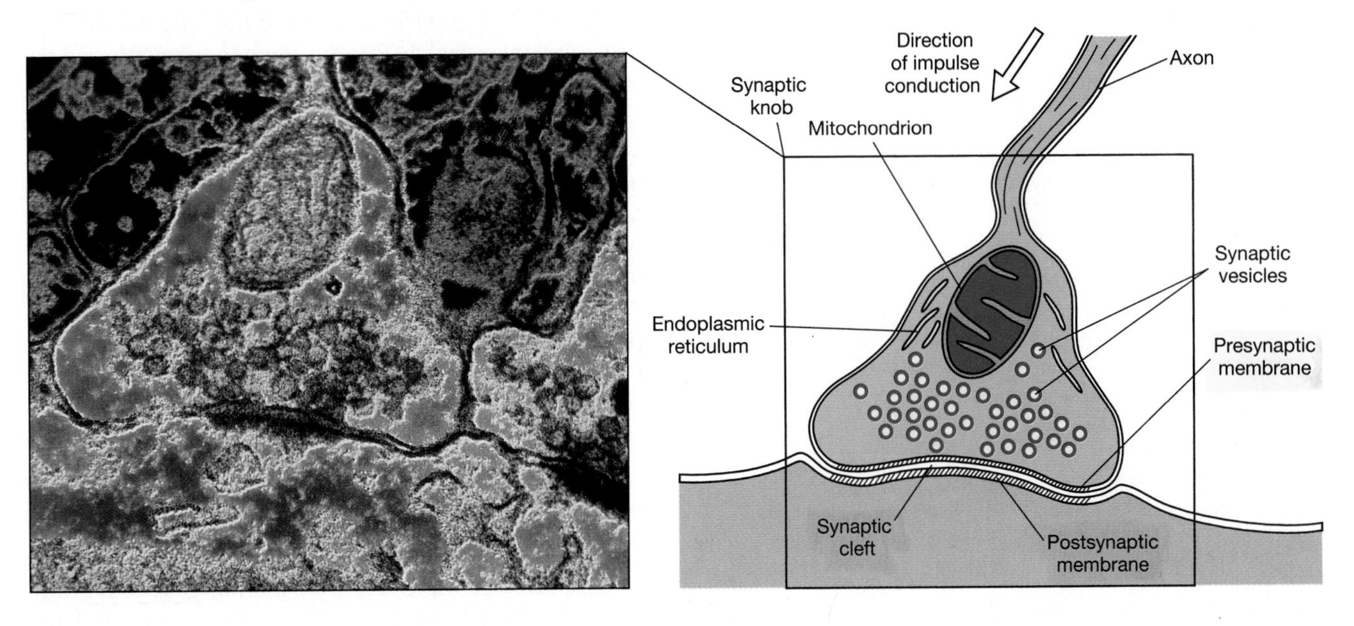

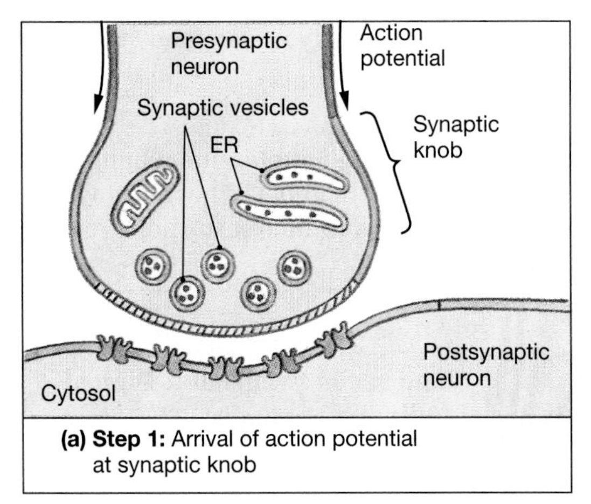

(a) **Step 1:** Arrival of action potential at synaptic knob

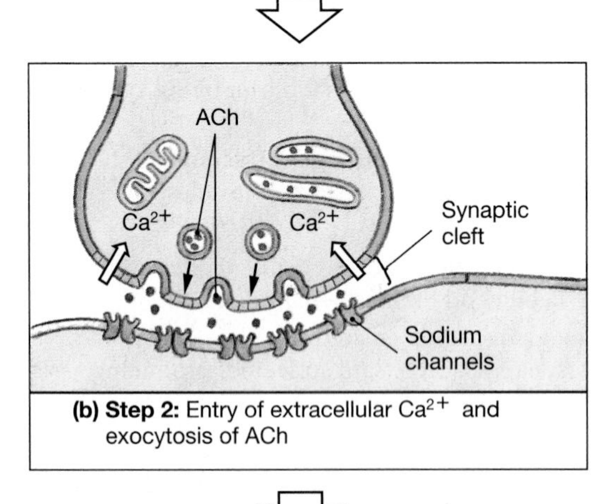

(b) **Step 2:** Entry of extracellular Ca^{2+} and exocytosis of ACh

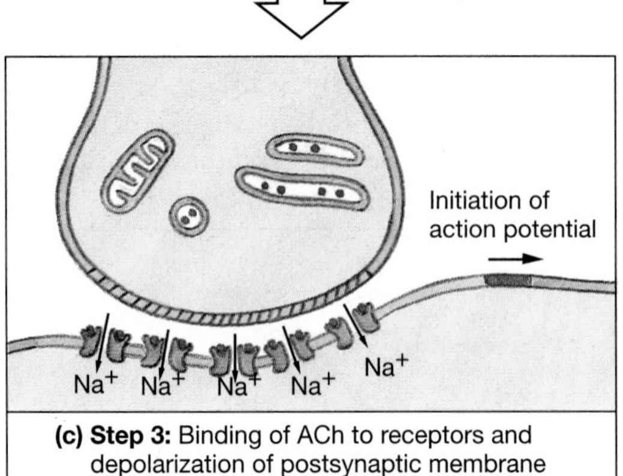

(c) **Step 3:** Binding of ACh to receptors and depolarization of postsynaptic membrane bring adjacent segment to threshold

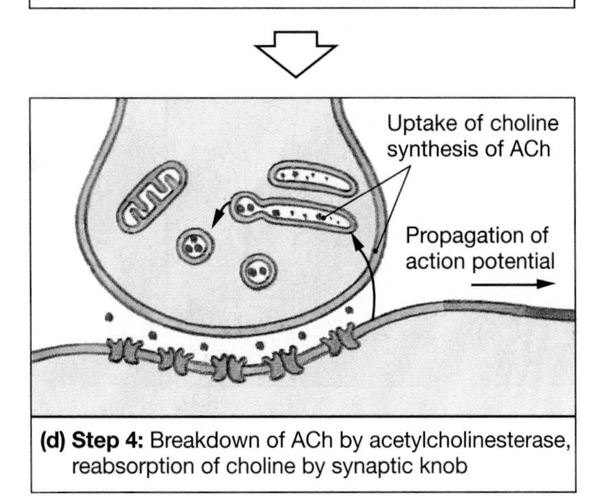

(d) **Step 4:** Breakdown of ACh by acetylcholinesterase, reabsorption of choline by synaptic knob

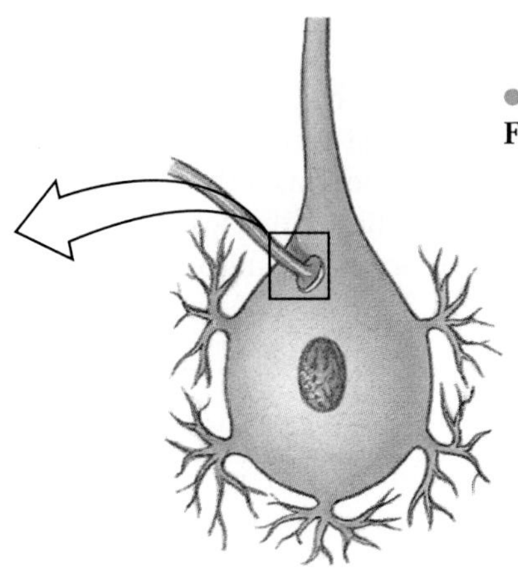

Figure 8-10
Function of a Cholinergic Synapse

NEURONAL POOLS

Interneurons are organized into functional groups, or **neuronal pools**. Each pool has a limited number of input sources and output destinations, and each may contain excitatory and inhibitory neurons. The output of one pool may (1) stimulate or depress the activity of other pools or (2) exert direct control over motor neurons and peripheral effectors.

Neurons communicate with one another in several patterns. In **divergence**, information spreads from one neuron to several neurons (Figure 8-11a). Divergence usually occurs when sensory neurons bring information into the CNS, because the sensory information must be distributed to neuronal pools throughout the spinal cord and brain. For example, when you accidentally touch a hot stove, divergence is what allows you to feel the pain and flex your arm at the same time.

In **convergence** (Figure 8-11b), several neurons synapse on the same postsynaptic neuron. Convergence makes possible both voluntary and involuntary control of some body processes. For example, the movements of your diaphragm and ribs are now being involuntarily controlled by respiratory centers in the brain. These centers activate or inhibit motor neurons in the spinal cord that control the respiratory muscles. But the same movements can be controlled voluntarily, as when you take a deep breath and hold it. Although voluntary commands originate in a different neuronal pool, they influence the same motor neurons.

Parallel processing (Figure 8-11c) occurs when several neuronal pools receive the same information at the same time. Thanks to parallel processing, many different responses occur simultaneously. For example, stepping on a tack stimulates sensory neurons that distribute the information to a number of neuronal pools. As a result, within a heartbeat you can withdraw your foot, shift your weight, move your arms, feel the pain, cry "ouch," and remember the last time you stepped on a tack, all at the same time.

AN INTRODUCTION TO REFLEXES

Conditions inside or outside the body can change rapidly and unexpectedly. **Reflexes** are automatic motor responses, triggered by specific stimuli, that help us preserve homeostasis by making rapid adjustments in the function of our organs or organ

●**Figure 8-11**
Organization of Neuronal Pools

(a) Divergence, a mechanism for spreading stimulation to multiple neurons or neuronal pools in the CNS. **(b)** Convergence, a mechanism providing input to a single neuron from multiple sources. **(c)** Parallel processing, in which neurons or pools process information simultaneously.

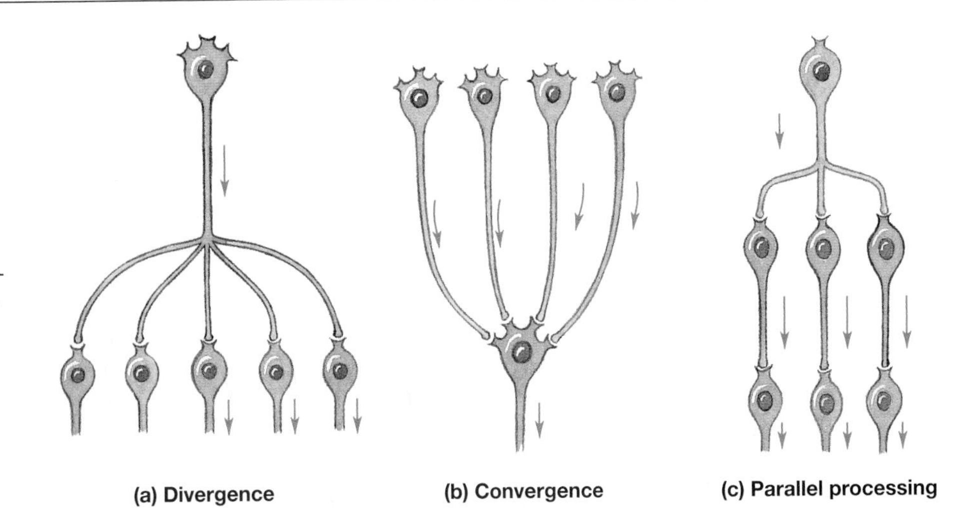

systems. Such involuntary responses include the control of heart rate, blood pressure, swallowing, and sneezing. The response shows little variability—when a particular reflex is activated it always produces the same motor response. *Spinal reflexes* are processed within the spinal cord; reflexes processed in the brain are called *cranial reflexes.* Specific examples of these reflexes will be considered in Chapter 9.

The Reflex Arc

The "wiring" of a single reflex is called a **reflex arc**. A reflex arc begins at a receptor and ends at an effector, such as a muscle or gland. Figure 8-12● diagrams the five steps involved in a neural reflex: (1) *arrival of a stimulus and activation of a receptor,* (2) *activation of a sensory neuron,* (3) *information processing,* (4) *activation of a motor neuron,* and (5) *response by an effector* (muscle or gland).

A reflex response usually removes or opposes the original stimulus; in Figure 8-12● the contracting muscle pulls the hand away from the painful stimulus. This reflex arc is therefore an example of a *negative feedback control mechanism.* ∞ [p. 12] By opposing potentially harmful changes in the internal or external environment, reflexes play an important role in homeostatic maintenance.

✓ How would synapse function be affected by blocking the uptake of calcium at the presynaptic membrane of a cholinergic synapse?

✓ What is the minimum number of neurons needed for a reflex arc?

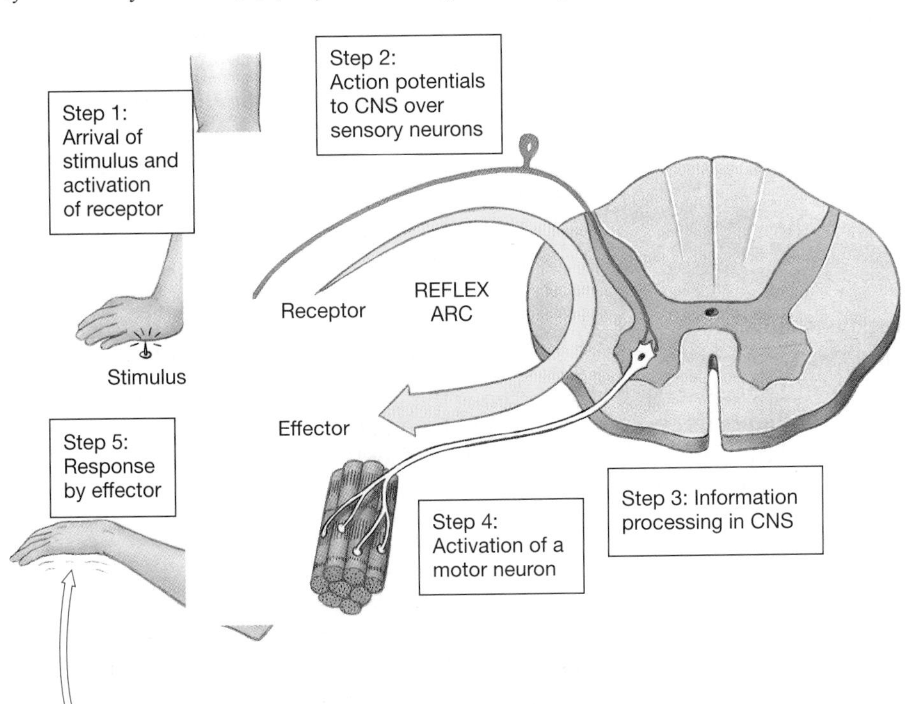

●**Figure 8-12**
Components of a Reflex Arc

AN INTRODUCTION TO THE ANATOMY OF THE NERVOUS SYSTEM

The basic functioning of the nervous system depends upon the action potentials generated at the level of the individual neuron. More complex functions of the nervous system depend on the interactions between neurons in neuronal pools, and the most complex interactions occur in the spinal cord and brain. Neurons and their axons are not randomly scattered in the CNS and PNS. Instead, they form masses or bundles with distinct anatomical boundaries and are identified by specific terms. These terms will be used in all later chapters, and a brief overview at this time will prove helpful.

In the PNS:

- **Ganglia** contain a group of neuron cell bodies.
- **Nerves** consist of bundles of axons, with spinal nerves connected to the spinal cord and cranial nerves connected to the brain.

In the CNS:

- A **center** is a collection of nerve cell bodies that share a particular function. A center with a discrete anatomical boundary is called a **nucleus**. Portions of the brain surface are covered by a thick blanket of gray matter. This layer is called **neural cortex** (*cortex*, rind). The term *higher centers* refers to the most complex integration centers, nuclei, and cortical areas in the brain.
- **Tracts** are bundles of axons inside the CNS that share common origins, destinations, and functions. Tracts in the spinal cord form larger groups, called **columns**.
- **Pathways** link the centers of the brain with the rest of the body. For example, **sensory pathways** distribute information from sensory receptors to processing centers in the brain, and **motor pathways** begin at motor centers in the CNS and end at the skeletal muscles they control. These relationships are diagrammed in Figure 8-13●.

The central nervous system consists of the spinal cord and brain. These masses of neural tissue are extremely delicate and must be protected against shocks, infection, and other dangers. In addition to glial cells within the neural tissue, the CNS is protected by a series of covering layers, the *meninges*, and by the special properties of the *blood-brain barrier*.

The Meninges

Neural tissue has a very high metabolic rate and requires abundant nutrients and a constant supply of oxygen. At the same time, the CNS must be isolated from a variety of compounds in the blood that could interfere with its complex operations. These diverse needs have resulted in a variety of special adaptations for the protection and support of the brain. The delicate neural tissues must also be defended against damaging contacts with the surrounding bones. This defense is provided by a series of specialized membranes, the **meninges** (men-IN-jēz) (Figure 8-14●). Blood vessels branching within these layers also deliver oxygen and nutrients to the CNS. At the foramen magnum of the skull, the meninges covering the brain are continuous with the meninges that surround the spinal cord. There are three meningeal layers: the *dura mater*, the *arachnoid*, and the *pia mater*. The meninges cover the cranial and spinal nerves as they penetrate the skull or pass through the intervertebral foramina, becoming continuous with the connective tissues surrounding the peripheral nerves.

The Dura Mater

The tough, fibrous **dura mater** (DŪ-ra MĀ-ter; *dura*, hard + *mater*, mother) forms the outermost covering of the central nervous system. The dura mater surrounding the brain consists of two fibrous layers, with the outermost fused to the periosteum of the skull. The inner and outer layers are separated by a slender gap that contains tissue fluids and blood vessels, including the large veins known as *dural sinuses*. At several locations, the innermost layer of the dura mater extends deep into the cranial cavity, providing additional stabilization and support to the brain.

Between the dura mater of the spinal cord and the walls of the vertebral canal lies the **epidural space**, which contains loose connective tissue, blood vessels, and adipose tissue. Injecting an anesthetic into the epidural space produces a temporary sensory and motor paralysis known as an *epidural block*. This technique has the advantage of affecting only the spinal nerves in the immediate area of the injection. Epidural blocks in the lower lumbar or sacral regions may be used to control pain during childbirth. AM *Spinal Anesthesia*

EPIDURAL AND SUBDURAL DAMAGE

A head injury that damages cerebral blood vessels may cause bleeding into the epidural or subdural spaces. These are serious conditions because the blood entering these spaces compresses and distorts the relatively soft tissues of the brain. The symptoms vary depending on whether the damaged vessel is an artery or a vein. AM *Epidural and Subdural Hemorrhages*

The Arachnoid

A narrow **subdural space** separates the inner surface of the dura mater from the second meningeal layer, the **arachnoid** (a-RAK-noyd; *arachne*, spider). This intervening space contains a small quantity of lymphatic fluid, which reduces friction between the op-

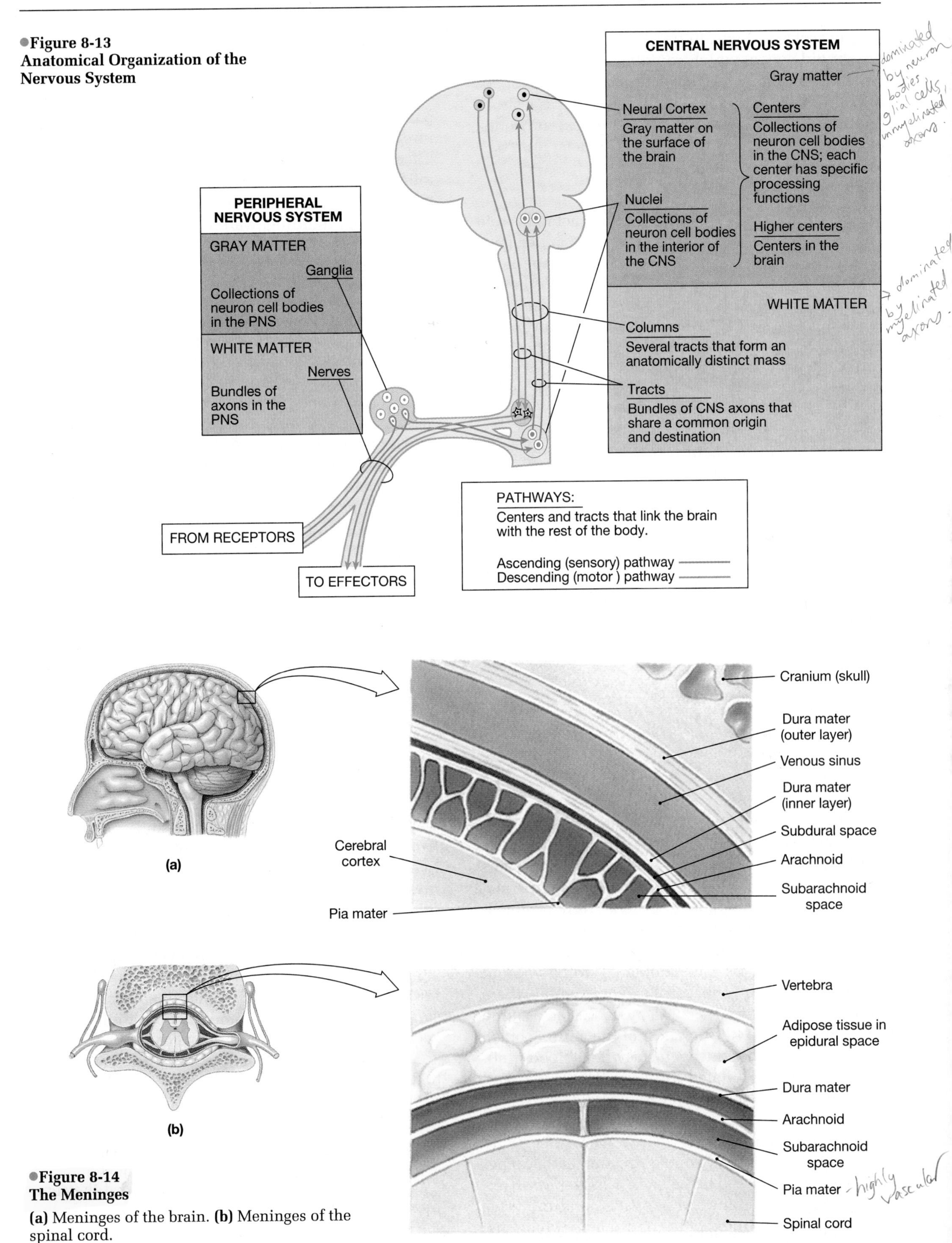
CENTRAL NERVOUS SYSTEM
Gray matter
dominated by neuron bodies, glial cells, unmyelinated axons.
Neural Cortex
Gray matter on the surface of the brain
Centers
Collections of neuron cell bodies in the CNS; each center has specific processing functions
Nuclei
Collections of neuron cell bodies in the interior of the CNS
Higher centers
Centers in the brain
WHITE MATTER
dominated by myelinated axons.
Columns
Several tracts that form an anatomically distinct mass
Tracts
Bundles of CNS axons that share a common origin and destination
PERIPHERAL NERVOUS SYSTEM
GRAY MATTER
Ganglia
Collections of neuron cell bodies in the PNS
WHITE MATTER
Nerves
Bundles of axons in the PNS
FROM RECEPTORS
TO EFFECTORS
PATHWAYS:
Centers and tracts that link the brain with the rest of the body.
Ascending (sensory) pathway
Descending (motor) pathway
Cranium (skull)
Dura mater (outer layer)
Venous sinus
Dura mater (inner layer)
Subdural space
Arachnoid
Subarachnoid space
Cerebral cortex
Pia mater
(a)
Vertebra
Adipose tissue in epidural space
Dura mater
Arachnoid
Subarachnoid space
Pia mater - highly vascular
Spinal cord
(b)

Figure 8-13
Anatomical Organization of the Nervous System

Figure 8-14
The Meninges
(a) Meninges of the brain. **(b)** Meninges of the spinal cord.

posing surfaces. Beneath the arachnoid lies the **subarachnoid space**, which contains a delicate web of collagen and elastic fibers. The subarachnoid space is filled with **cerebrospinal fluid** that acts as a shock absorber and transports dissolved gases, nutrients, chemical messengers, and waste products.

The Pia Mater

The subarachnoid space separates the arachnoid from the innermost meningeal layer, the **pia mater** (*pia*, delicate + *mater*, mother). Unlike more superficial meninges, the pia mater is firmly bound to the underlying neural tissue. The blood vessels servicing the brain and spinal cord are found in this layer. The pia mater of the brain is highly vascular, and large vessels branch over the surface of the brain, supplying the superficial areas of neural cortex. This extensive circulatory supply is extremely important, for the brain has a very high rate of metabolism; at rest, the 1.4 kg (3.1 lb.) brain uses as much oxygen as 28 kg (61.6 lb.) of skeletal muscle.

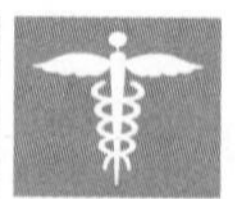

MENINGITIS

Meningitis is the inflammation of the meningeal membranes following bacterial or viral infection. Meningitis is dangerous because it can disrupt the normal circulatory and cerebrospinal fluid supplies, damaging or killing neurons and glial cells in the affected areas. Although the initial diagnosis may specify the meninges of the spinal cord (*spinal meningitis*) or brain (*cerebral meningitis*), in later stages the entire meningeal system is usually affected. AM *Meningitis*

The Blood-Brain Barrier

The neural tissue in the CNS is isolated from the general circulation by the *blood-brain barrier*. This barrier is maintained by astrocytes, whose secretions cause the capillaries of the CNS to become impermeable to many compounds. In general, only lipid-soluble compounds can diffuse into the interstitial fluid of the brain and spinal cord. Water-soluble compounds cannot cross the endothelial lining without the assistance of specific carriers. For example, there are separate transport systems for glucose, large amino acids, and glycine (the smallest of the amino acids). Most of the transport mechanisms of the blood-brain barrier involve facilitated diffusion (Chapter 3) and so occur down concentration gradients. ∞ [p. 55]

We will now take a closer look at the organization of the central nervous system. We will begin by examining the relatively simple structure of the spinal cord and then proceed to the brain. Chapter 9 will examine the peripheral nervous system and the ways the CNS and PNS interact to control body functions.

THE SPINAL CORD

The spinal cord serves as the major highway for the passage of sensory impulses to the brain and motor impulses from the brain. In addition, the spinal cord integrates information on its own and controls spinal reflexes, ranging from withdrawal from pain (Figure 8-12●, p. 209) to complex reflex patterns involved with sitting, standing, walking, and running.

Gross Anatomy

The spinal cord, detailed in Figure 8-15● measures approximately 45 cm (18 in.) in length. The posterior surface of the spinal cord has a shallow groove, the *posterior median sulcus*, and a deeper crease along the anterior surface is the *anterior median fissure*. With two exceptions, the diameter of the cord decreases in size as it extends toward the sacral region.

The two exceptions are regions concerned with the sensory and motor control of the limbs. The *cervical enlargement* supplies nerves to the shoulder girdle and upper limbs, and the *lumbar enlargement* provides innervation to the pelvis and lower limbs. Below the lumbar enlargement, the spinal cord becomes tapered and conical. A slender strand of fibrous tissue extends from the inferior tip of the spinal cord to the coccyx, serving as an anchor that prevents superior movement.

The entire spinal cord consists of 31 segments, each identified by a letter and number designation. Every spinal segment is associated with a pair of **dorsal root ganglia** that contain the cell bodies of sensory neurons. The **dorsal roots**, which contain the axons of these neurons, bring sensory information to the spinal cord. A pair of **ventral roots** contain the axons of CNS motor neurons that control muscles and glands. On either side, the dorsal and ventral roots from each segment leave the vertebral column between adjacent vertebrae at the *intervertebral foramen*. The spinal nerves on either side form outside of the vertebral canal, where the ventral and dorsal roots unite, so that the dorsal root ganglion lies between the pedicles of succeeding vertebra. (You may wish to review the description of vertebral anatomy in Chapter 6.) ∞ [p. 126] Distal to each dorsal root ganglion the sensory and motor roots are bound together into a single **spinal nerve**. All spinal nerves are classified as **mixed nerves**, because they contain both sensory and motor fibers.

The adult spinal cord extends only to the level of the first or second lumbar vertebrae. When seen in gross dissection, the long ventral and dorsal roots inferior to the tip of the spinal cord reminded early anatomists of a horse's tail. With that in mind, they called the complex the *cauda equina* (KAW-da ek-WĪ-na; *cauda*, tail + *equus*, horse).

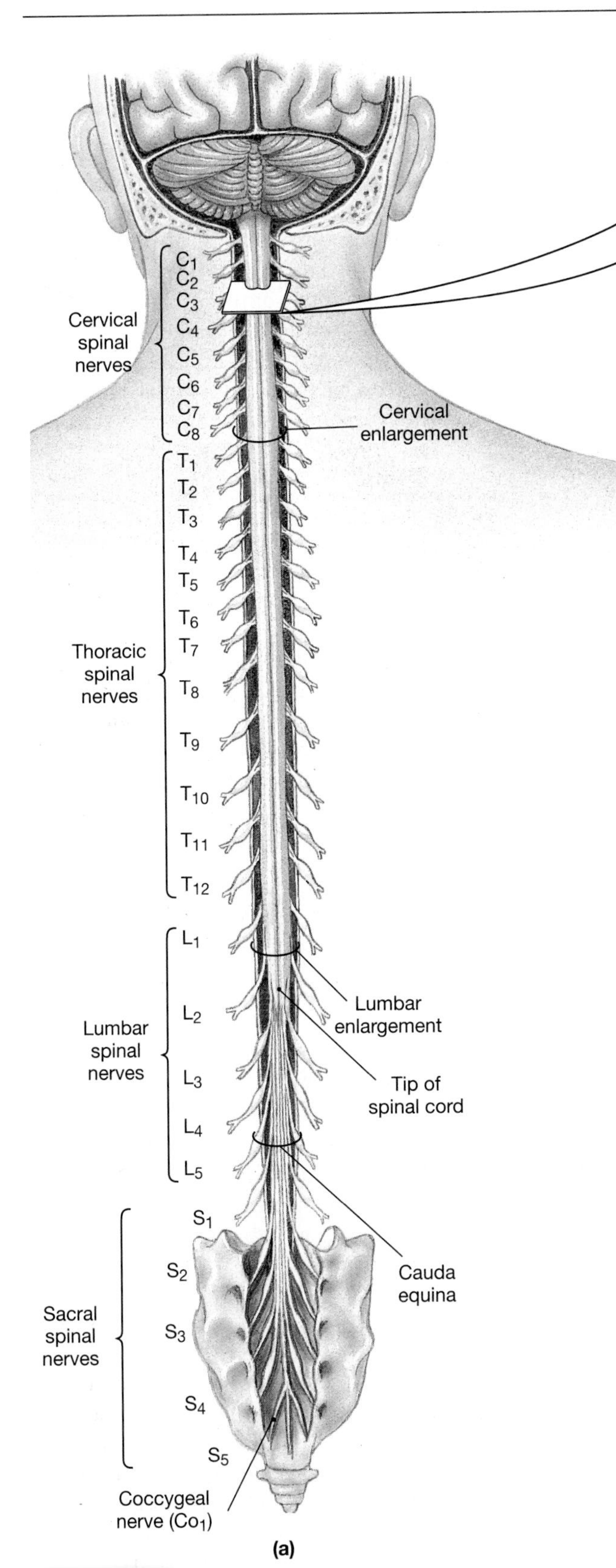

•Figure 8-15
Gross Anatomy of the Spinal Cord

(a) Superficial anatomy and orientation of the adult spinal cord. The numbers to the left identify the spinal nerves. **(b)** Cross-section through the cervical region of the spinal cord, showing the arrangement of gray and white matter.

Sectional Anatomy

The anterior median fissure and the posterior median sulcus mark the division between left and right sides of the spinal cord (Figure 8-16•). The *gray matter* is dominated by the cell bodies of neurons and glial cells. It forms a rough H, or butterfly shape, around the narrow central canal. Projections of gray matter, called *horns*, extend outward into the *white matter*, which contains large numbers of myelinated and unmyelinated axons.

The *posterior gray horns*, seen in Figure 8-16•, contain sensory nuclei, whereas the *anterior gray horns* are concerned with the motor control of skeletal muscles. Nuclei in the *lateral gray horns* contain the visceral motor neurons that control smooth muscle, cardiac muscle, and glands. The *gray commissures* above and below the central canal interconnect the horns on either side of the spinal cord.

The white matter can be divided into a half-dozen regions, or *columns* (Figure 8-16b•). The *posterior white columns* extend between the posterior gray horns and the posterior median sulcus. The *anterior white columns* lie between the anterior gray horns and the anterior median fissure; they are interconnected by the *anterior white commissure*. The white matter between the anterior and posterior columns makes up the *lateral white columns*.

Each column contains tracts whose axons carry either sensory data or motor commands. Small tracts carry sensory or motor signals between segments of the spinal cord, and larger tracts connect the spinal cord with the brain. **Ascending tracts** carry sensory information up toward the brain, and **descending tracts** convey motor commands down into the spinal cord.

Injuries affecting the spinal cord or cauda equina will produce symptoms of sensory loss or motor paralysis that reflect the specific nuclei, tracts, or spinal

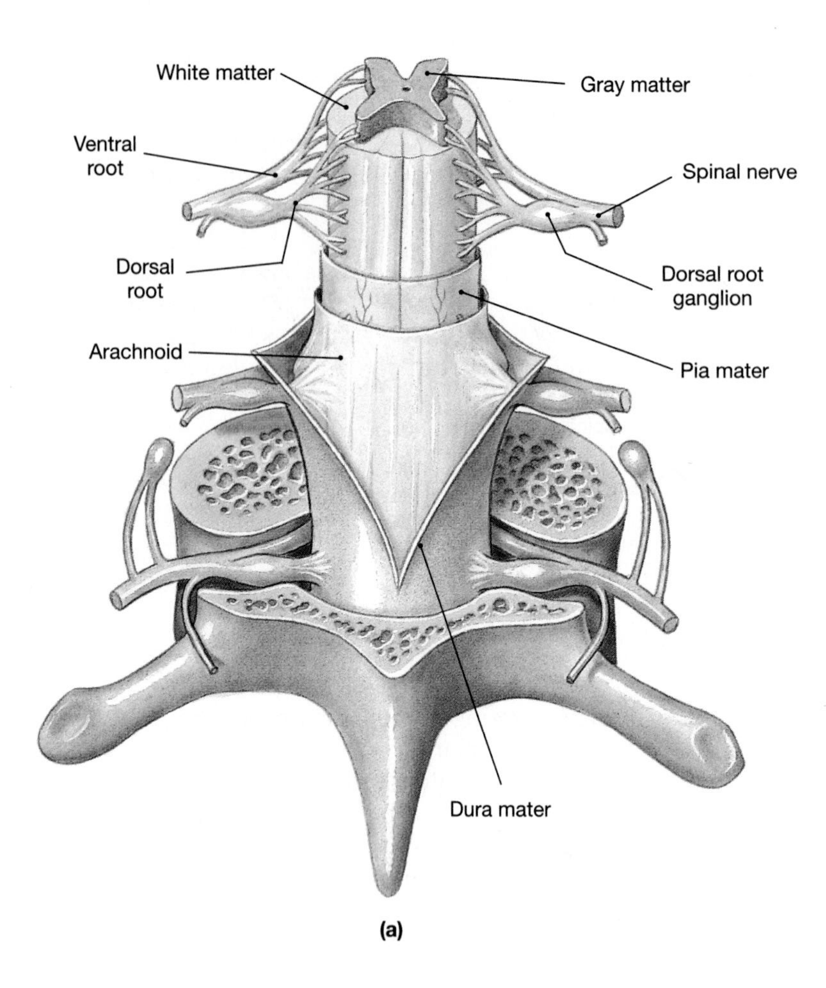

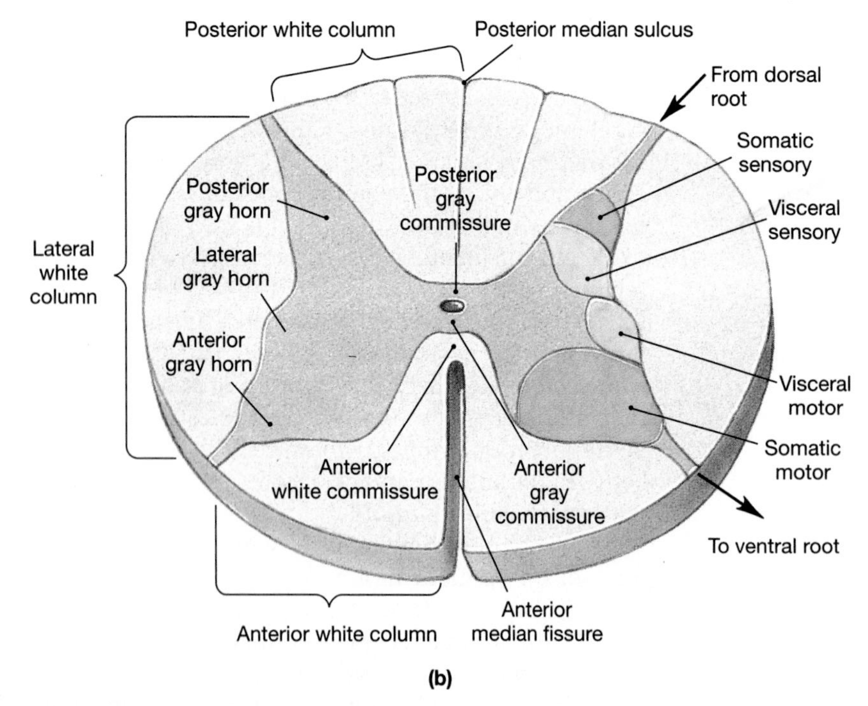

•Figure 8-16
Sectional Anatomy of the Spinal Cord
(a) Posterior view of the spinal cord and spinal nerve roots within the vertebral canal. **(b)** Diagrammatic sectional view through the spinal cord, showing major landmarks.

nerves involved. A general paralysis can result from severe damage to the spinal cord in an auto crash or other accident, and the damaged tracts seldom undergo even partial repairs. Extensive damage at the fourth or fifth cervical vertebra will eliminate sensation and motor control of the upper and lower extremities. The extensive paralysis produced is called *quadriplegia. Paraplegia*, the loss of motor control of the lower limbs, may follow damage to the thoracic vertebrae. AM *Spinal Cord Injuries and Experimental Treatments*

✓ Damage to which root of a spinal nerve would interfere with motor function?

✓ A patient suffering from polio has lost the use of his leg muscles. In what area of the spinal cord would you expect to locate the virally infected motor neurons in this individual?

✓ Why are spinal nerves also called mixed nerves?

✓ Where is the cerebrospinal fluid that surrounds the spinal cord located?

THE BRAIN

The brain is far larger and more complex than the spinal cord, and its responses to stimuli are more versatile. It contains roughly 35 billion neurons organized into hundreds of neuronal pools; it has a complex three-dimensional structure and performs a bewildering array of functions. All of our dreams, passions, plans, and memories are the result of brain activity.

The adult human brain contains almost 98 percent of the neural tissue in the body. A representative adult brain weighs 1.4 kg (3 lb) and has a volume of 1200 cc (71 in.^{3}). There is considerable individual variation, and the brains of males are generally about 10 percent larger than those of females, because of differences in average body size. There is no correlation between brain size and intelligence, and people with the smallest (750 cc) and largest (2100 cc) brains are functionally normal.

Major Divisions of the Brain

There are six major regions in the adult brain: (1) the *cerebrum*, (2) the *diencephalon*, (3) the *midbrain*, (4) the *pons*, (5) the *cerebellum*, and (6) the *medulla oblongata*. Major landmarks are indicated in Figure 8-17●.

Viewed from the superior surface, the **cerebrum** (SER-e-brum) can be divided into large, paired **cerebral hemispheres**. Conscious thought processes, sensations, intellectual functions, memory storage and retrieval, and complex motor patterns originate in the cerebral hemispheres. The hollow **diencephalon** (dī-en-SEF-a-lon; *dia-*, through + *cephalo*, head) is connected to the cerebrum. Its sides form the **thalamus**, which contains relay and processing centers for sensory information. A narrow stalk connects the floor of the diencephalon, or **hypothalamus** (*hypo-*, below),

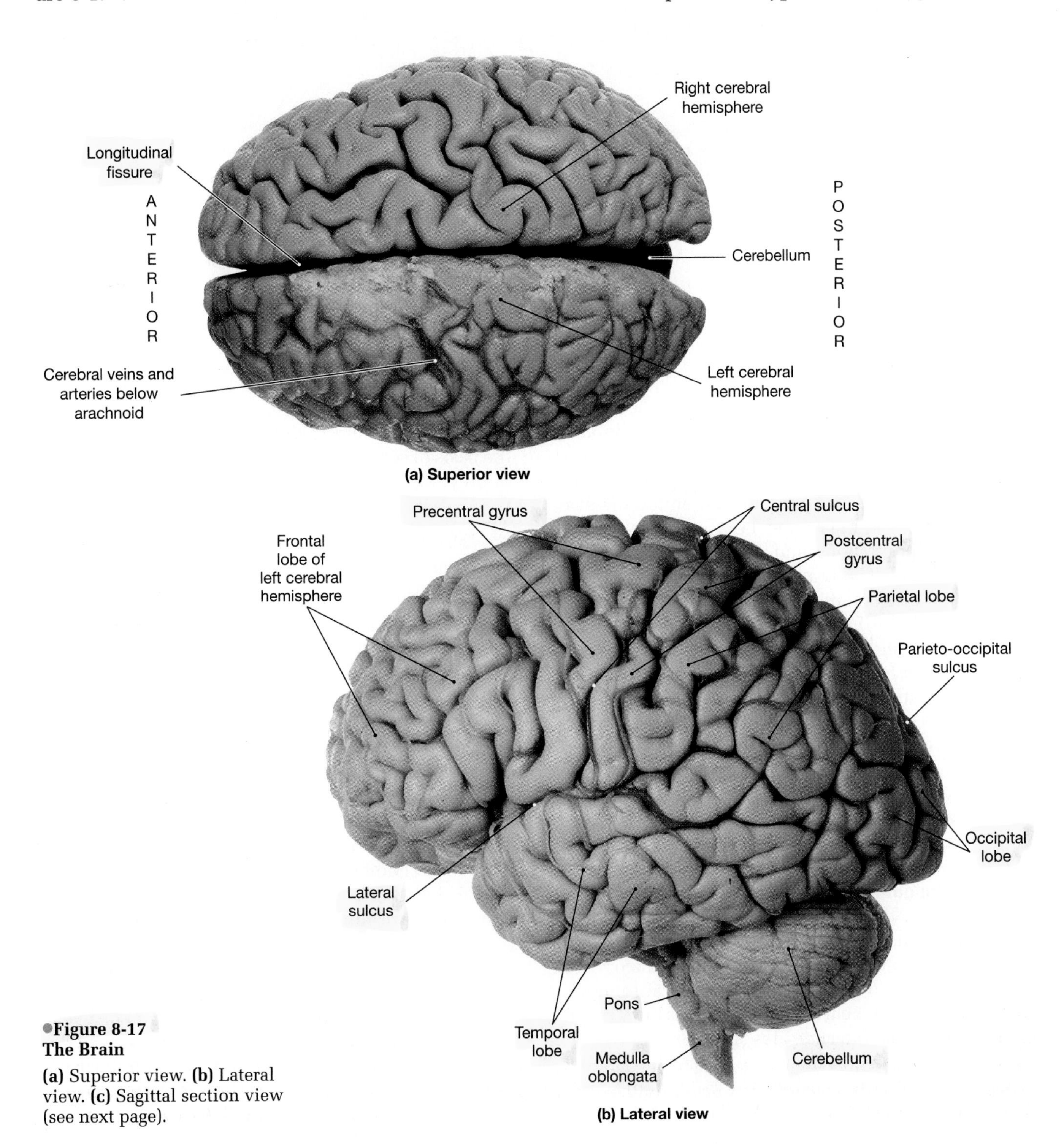

●**Figure 8-17**
The Brain
(a) Superior view. **(b)** Lateral view. **(c)** Sagittal section view (see next page).

FIGURE 8-17 continued

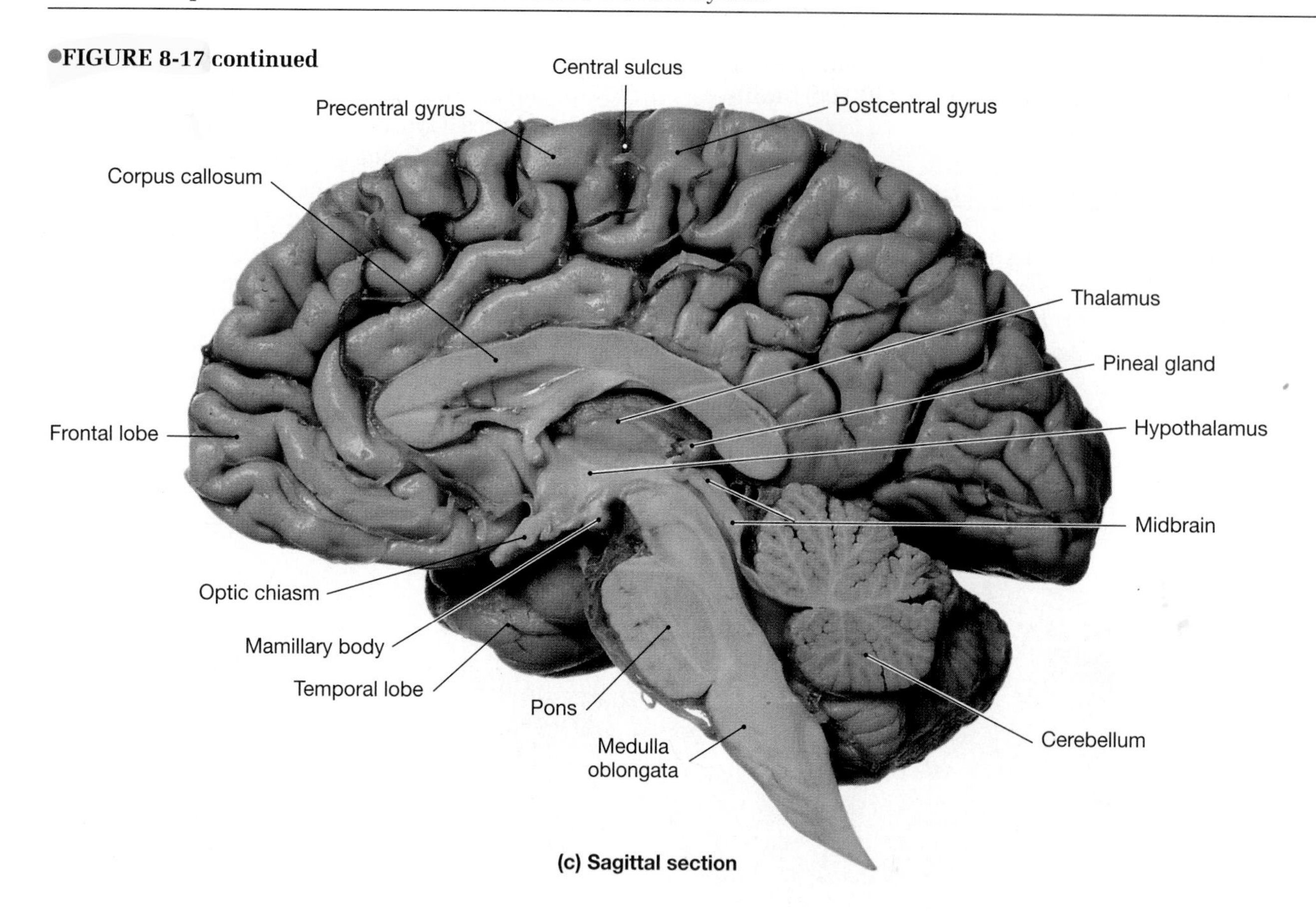

(c) Sagittal section

to the *pituitary gland*. The hypothalamus contains centers involved with emotions, autonomic function, and hormone production. The pituitary gland, the primary link between the nervous and endocrine systems, is discussed in Chapter 11.

The midbrain, pons, and medulla oblongata form the **brain stem**. The brain stem contains important processing centers and relay stations for information headed to or from the cerebrum or cerebellum. Nuclei in the **midbrain**, or *mesencephalon* (mez-en-SEF-a-lon; *meso-*, middle), process visual and auditory information and generates involuntary motor responses. This region also contains centers involved with the maintenance of consciousness. The term **pons** refers to a bridge, and the pons of the brain connects the *cerebellum* to the brain stem. In addition to tracts and relay centers, this region of the brain also contains nuclei involved with somatic and visceral motor control. The pons is also connected to the **medulla oblongata**, the segment of the brain that is attached to the spinal cord. The medulla oblongata relays sensory information to the thalamus and other brain stem centers; it also contains major centers concerned with the regulation of autonomic function, such as heart rate, blood pressure, respiration, and digestive activities.

The large cerebral hemispheres and the smaller hemispheres of the **cerebellum** (ser-e-BEL-um) almost completely cover the brain stem. The cerebellum adjusts voluntary and involuntary motor activities on the basis of sensory information and stored memories of previous movements.

✓ Describe one major function of each of the six regions of the brain.

✓ The pituitary gland links the nervous and endocrine systems. To what portion of the diencephalon is it attached?

The Ventricles of the Brain

The brain and spinal cord are hollow, with internal cavities filled with cerebrospinal fluid. The passageway within the spinal cord is a narrow **central canal**. The brain has a central passageway that expands to form four chambers, called **ventricles** (VEN-tri-kls) (Figure 8-18a●). The largest of these, the two **lateral ventricles**, are found in each cerebral hemisphere (Figure 8-18b●). There is no direct connection between them, but an opening, the *interventricular foramen*, allows each of them to communicate with the **third ventricle** of the diencephalon. Instead of a ventricle, the midbrain has a slender canal known as the *mesencephalic aqueduct (cerebral aqueduct)*, which connects the third ventricle with the **fourth ventricle** of the pons and upper portion of the medulla oblon-

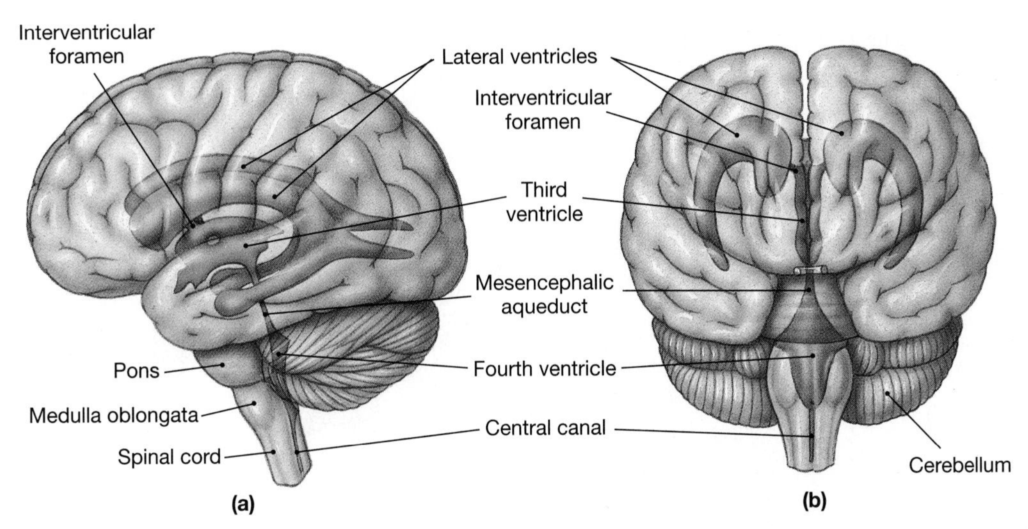

Figure 8-18 Ventricles of the Brain

(a) Orientation and extent of the ventricles as seen through a transparent brain in lateral view. **(b)** Anterior view of the ventricles, showing the relationships between the lateral ventricles and the third ventricle.

gata. Within the medulla oblongata the fourth ventricle narrows and becomes continuous with the central canal of the spinal cord.

Cerebrospinal Fluid

Cerebrospinal fluid, or **CSF**, provides cushioning for delicate neural structures. It also provides support, because the brain essentially floats in the cerebrospinal fluid. A human brain weighs about 1400 g (3.1 lbs.) in air, but only about 50 g (1.76 oz.) when supported by the cerebrospinal fluid. Finally, the CSF transports nutrients, chemical messengers, and waste products. Except at the choroid plexus, the ependymal lining is freely permeable, and the CSF is in constant chemical communication with the interstitial fluid of the CNS.

Cerebrospinal fluid is produced at the **choroid plexus** (*choroid*, a vascular coat + *plexus*, a network), which extends into each of the four ventricles (Figure 8-19a). The capillaries of the choroid plexus are covered by large ependymal cells that secrete cerebrospinal fluid at a rate of about 500 ml/day. The total volume of CSF at any given moment is approximately 150 ml; this means that the entire volume of CSF is replaced roughly every 8 hours. Despite this rapid turnover, the composition of CSF is closely regulated, and the rate of removal normally keeps pace with the rate of production. If it does not, a variety of clinical problems may appear. AM *Hydro cephalus*

Figure 8-19 diagrams the circulation of cerebrospinal fluid. CSF forms at the choroid plexus and circulates between the different ventricles, passes along the central canal, and enters the subarachnoid space. Once inside the subarachnoid space, the CSF circulates around the spinal cord and cauda equina and across the surfaces of the brain. Between the cerebral hemispheres, slender extensions of the arachnoid penetrate the inner layer of the dura mater. These **arachnoid granulations** (Figure 8-19b) project into the *superior sagittal sinus*, a large cerebral vein. Diffusion across the arachnoid granulations returns excess cerebrospinal fluid to the venous circulation.

Because free exchange occurs between the interstitial fluid and CSF, changes in CNS function may produce changes in the composition of the CSF. Samples of the CSF can be obtained through a *lumbar puncture*, or *spinal tap*, providing useful clinical information concerning CNS injury, infection, or disease. AM *Lumbar Puncture and Myelography*

The Cerebrum

The cerebrum, the largest region of the brain, is the site where conscious thought and intellectual functions originate. Much of the cerebrum is involved in receiving somatic sensory information and then exerting voluntary or involuntary control over somatic motor neurons. In general, we are aware of these events. However, most sensory processing and all visceral motor (autonomic) control occur elsewhere in the brain, usually outside our conscious awareness.

The cerebrum includes gray matter and white matter. Gray matter is found in a superficial layer of neural cortex and in deeper *cerebral nuclei*. The *central white matter*, composed of myelinated axons, lies beneath the neural cortex and surrounds the cerebral nuclei.

Structure of the Cerebral Hemispheres

Figure 8-20 presents a diagrammatic superficial view of the cerebrum. A thick blanket of neural cortex known as the **cerebral cortex** covers the paired **cerebral hemispheres** that form the superior surface of the cerebrum. This outer surface forms a series of elevated ridges, or **gyri** (JĪ-rē), separated by shallow depressions, called **sulci** (SUL-kē), or deeper grooves, called **fissures**. Gyri increase the surface area of the cerebral hemispheres and the number of neurons in the cortex.

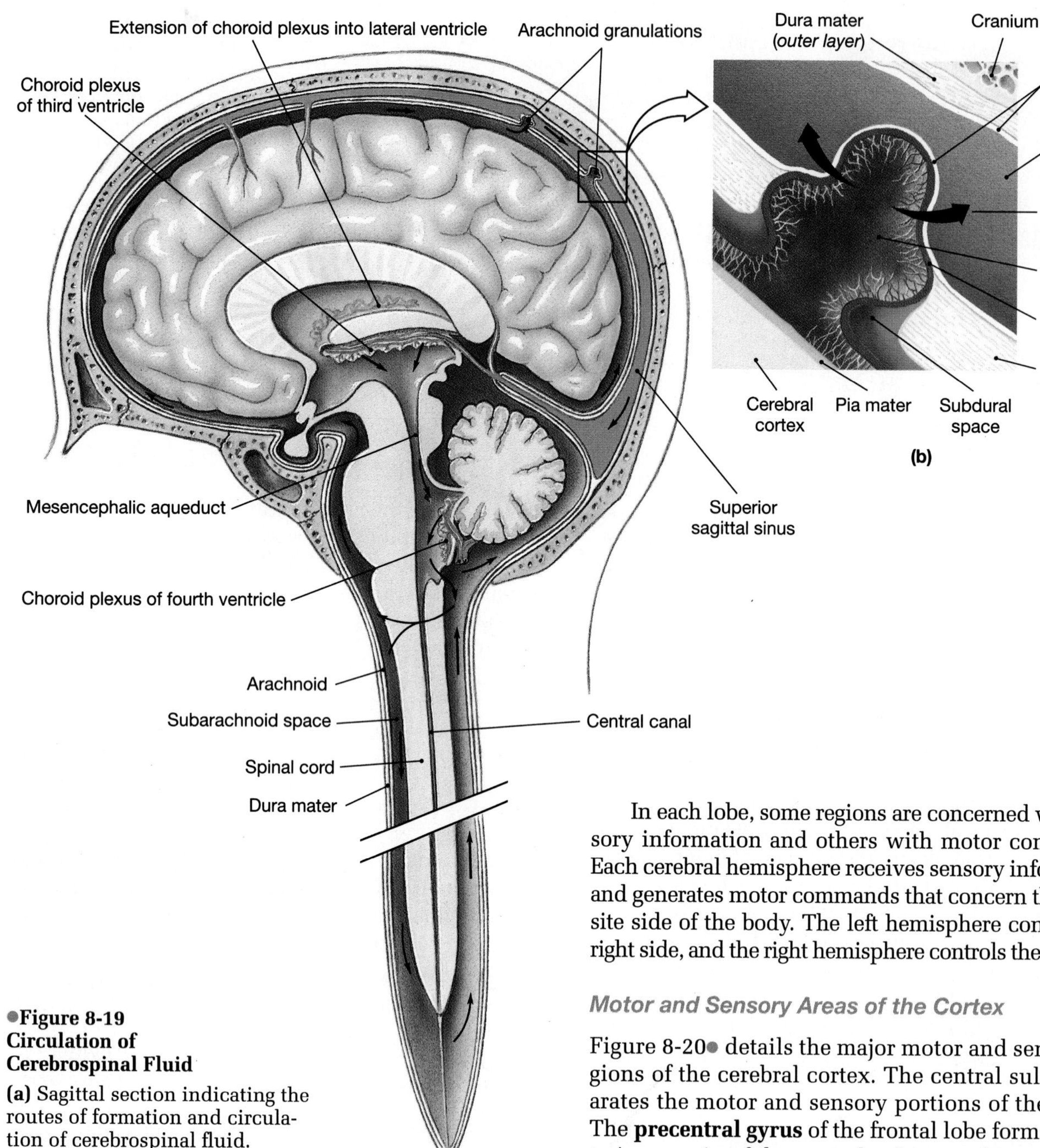

•Figure 8-19 Circulation of Cerebrospinal Fluid
(a) Sagittal section indicating the routes of formation and circulation of cerebrospinal fluid.
(b) Orientation of the arachnoid granulations.

The two cerebral hemispheres are separated by a deep **longitudinal fissure**. Each cerebral hemisphere can be divided into **lobes** named after the overlying bones of the skull (Figures 8-17, p. 215, and 8-20•). Extending laterally from the longitudinal fissure is a deep groove, the **central sulcus**. Anterior to this is the **frontal lobe**, bordered inferiorly by the **lateral sulcus**. The cortex inferior to the lateral sulcus is the **temporal lobe** which, in turn, overlaps the **insula** (IN-su-la), an "island" of cortex that is otherwise hidden. The **parietal lobe** extends between the central sulcus and the **parieto-occipital sulcus**. What remains is the **occipital lobe**.

In each lobe, some regions are concerned with sensory information and others with motor commands. Each cerebral hemisphere receives sensory information and generates motor commands that concern the opposite side of the body. The left hemisphere controls the right side, and the right hemisphere controls the left side.

Motor and Sensory Areas of the Cortex

Figure 8-20• details the major motor and sensory regions of the cerebral cortex. The central sulcus separates the motor and sensory portions of the cortex. The **precentral gyrus** of the frontal lobe forms the anterior margin of the central sulcus, and its surface is the **primary motor cortex**. Neurons of the primary motor cortex direct voluntary movements by controlling somatic motor neurons in the brain stem and spinal cord.

The **postcentral gyrus** of the parietal lobe forms the posterior margin of the central sulcus, and its surface contains the **primary sensory cortex**. Neurons in this region receive somatic sensory information from touch, pressure, pain, taste, and temperature receptors. We are consciously aware of these sensations because the brain stem nuclei relay sensory information to the primary sensory cortex.

Sensory information concerning sensations of sight, sound, and smell arrive at other portions of the cerebral cortex. The **visual cortex** of the occipi-

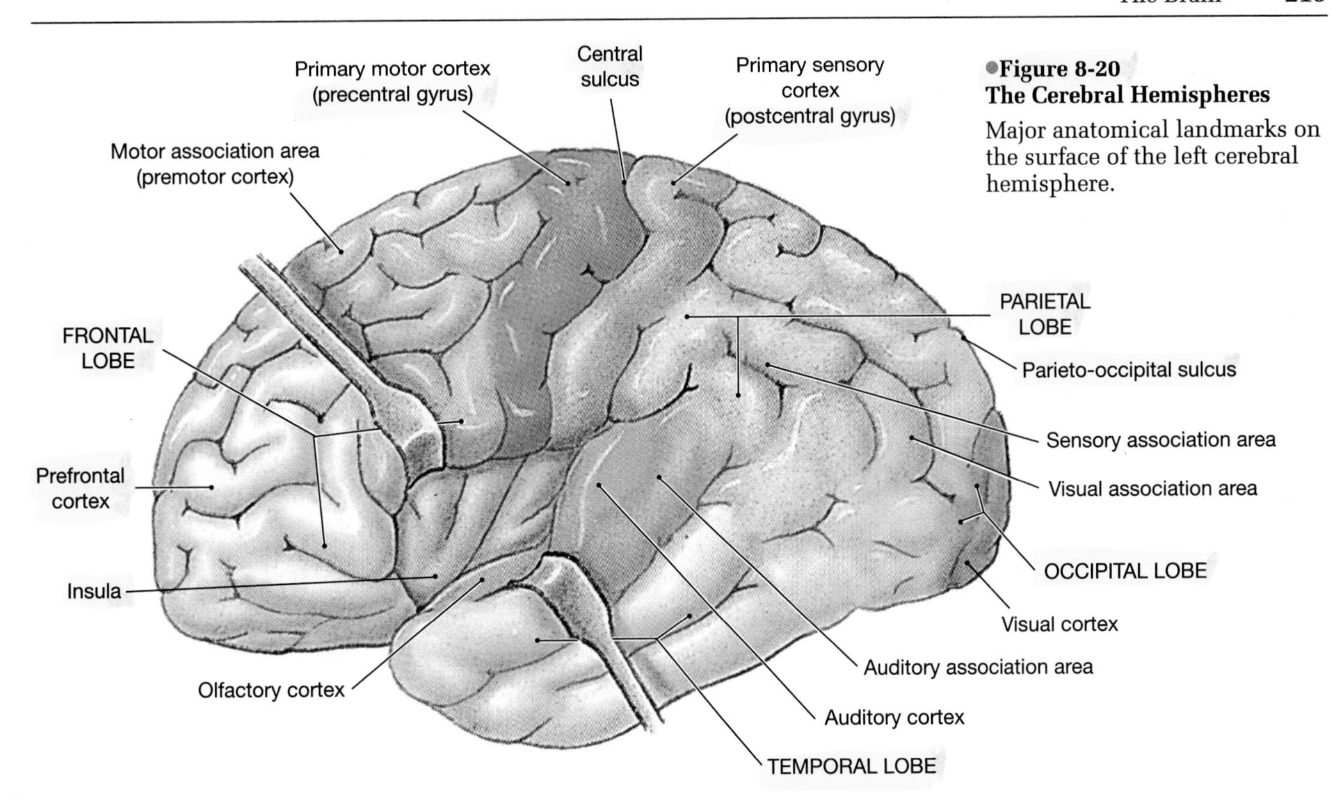

●**Figure 8-20**
The Cerebral Hemispheres
Major anatomical landmarks on the surface of the left cerebral hemisphere.

tal lobe receives visual information, and the **auditory cortex** and **olfactory cortex** of the temporal lobe receive information concerned with hearing and smell respectively.

Association Areas

The sensory and motor regions of the cortex are connected to nearby **association areas** that interpret incoming data or coordinate a motor response. For example, the **premotor cortex**, or *motor association area*, is responsible for coordinating learned movements, such as the eye movements involved in following the lines on this page. The functional distinctions between the motor and sensory association areas are most evident after localized brain damage has occurred. For example, someone with damage to the premotor cortex might understand written letters and words but be unable to read owing to an inability to track along the lines on a printed page. In contrast, someone with a damaged **visual association area** can scan the lines of a printed page but cannot figure out what the letters mean.

Cortical Connections

The various regions of the cerebral cortex are interconnected by the white matter that lies beneath the cerebral cortex. This white matter interconnects areas within a single cerebral hemisphere and links the two hemispheres across the **corpus callosum** (Figure 8-17c●, p. 216). Other bundles of axons link the cerebral cortex with the diencephalon, brain stem, cerebellum, and spinal cord.

Cerebral Processing Centers

There are also "higher-order" integrative centers that receive information through axons from many different association areas. These regions, shown in Figure 8-21●, control extremely complex motor activities and perform complicated analytical functions. They may be found in the lobes and cortical areas of both cerebral hemispheres, or they may be restricted to either the left or the right side.

The General Interpretive Area The *general interpretive area* receives information from all the sensory association areas. This center is present in only one hemisphere, usually the left. Damage to this area affects the ability to interpret what is read or heard, even though the words are understood as individual entities. For example, an individual might understand the meaning of the words "sit" and "here" but be totally bewildered by the instruction "Sit here."

The Speech Center The general interpretive area is connected to the *speech center (Broca's area)*, which lies along the edge of the premotor cortex in the same hemisphere as the general interpretive area. The speech center regulates the patterns of breathing and vocalization needed for normal speech. The corresponding regions on the opposite hemisphere are not inactive, but their functions are less well defined. Damage to the speech center can manifest itself in various ways. Some people have difficulty speaking even when they know exactly what words to use; others talk constantly but use all the wrong words.

The Prefrontal Cortex The *prefrontal cortex* of the frontal lobe coordinates information from the secondary and special association areas of the entire cortex. In doing so, it performs such abstract intellectual functions as predicting the future consequences of events or actions. The prefrontal cortex also has connections with other portions of the brain, such as the limbic system, discussed later. Feelings of frustration, tension, and anxiety are generated at the prefrontal cortex as it interprets ongoing events and predicts future situations or consequences. If the connections between the prefrontal cortex and other brain regions are severed, the tensions, frustrations, and anxieties are removed. Earlier in this century this rather drastic procedure, called a *prefrontal lobotomy*, was used to "cure" a variety of mental illnesses, especially those associated with violent or antisocial behavior.

APHASIA AND DYSLEXIA

Aphasia (*a*-, without + *phasia*, speech) is a disorder affecting the ability to speak or read. Extreme aphasia, or *global aphasia*, results from extensive damage to the general interpretive area or to the associated sensory tracts. Affected individuals are totally unable to speak, to read, or to understand or interpret the speech of others. Global aphasia often accompanies a severe stroke or tumor that affects a large area of cortex including the speech and language areas. Lesser degrees of aphasia often follow minor strokes, with no initial period of global aphasia. Individuals can understand spoken and written words, and many recover completely.

Dyslexia (*lexis*, diction) is a disorder affecting the comprehension and use of words. *Developmental dyslexia* affects children; some estimates show that up to 15 percent of children in the United States suffer from some degree of dyslexia. These children have difficulty reading and writing, even though their other intellectual functions may be normal or above normal. Recent evidence suggests that at least some forms of dyslexia result from problems in processing and sorting visual information.

Memory

What was the topic of the last sentence you read? What do your parents look like? What is your Social Security number? What does a red traffic light mean? What does a hot dog taste like? Answering these questions involves accessing *memories*, stored bits of information gathered through prior experience. **Fact memories** are specific bits of information, such as the color of a stop sign or the smell of perfume. **Skill memories** are learned motor behaviors. For example, you can probably remember how to light a match or open a screw-top jar. With repetition, skill memories become incorporated at the unconscious level. Examples would include the complex motor patterns involved in skiing, playing the violin, and similar activities. Skill memories related to programmed behaviors, such as eating, are stored in appropriate portions of the brain stem. Complex skill memories involve an interplay between the cerebullum and cerebral cortex.

Two classes of memories exist. **Short-term memories**, or *primary memories*, do not last long, but while they persist the information can be recalled immediately. Primary memories contain small bits of information, such as a person's name or a telephone number. Repeating a phone number or other bit of information reinforces the original short-term memory and helps ensure its conversion to a long-term memory. **Long-term memories** remain for much longer periods, in some cases for an entire lifetime. Some long-term memories fade with time and may require considerable effort to recall. Other long-term memories seem to be part of consciousness, such as your name or the contours of your own body.

The conversion from short-term to long-term memory is called *memory consolidation*. Although much remains to be learned about this process, it is clear that anatomical and physiological changes occur at the cellular level. For example, the rates of neurotransmitter production and release in the stimulated neurons. Efficient conversion from short-term to long-term memory storage takes time, usually at least an hour and often longer. Whether or not that conversion occurs depends on several factors, including the nature, intensity, and frequency of the original stimulus. Very strong, repeated, or exceedingly pleasant (or unpleasant) events are excellent candidates for conversion to long-term memories. Drugs that stimulate the CNS, such as caffeine and nicotine, may enhance memory consolidation.

Most long-term memories are stored in the cerebral cortex. Conscious motor and sensory memories are referred to the appropriate association areas. For example, visual memories are stored in the visual association area, and memories of voluntary motor activity are kept in the premotor cortex. Special portions of the occipital and temporal lobes retain the memories of faces, voices, and words. In at least some cases, a specific memory probably reflects the activity of a single neuron. For example, in one portion of the temporal lobe an individual neuron responds to the sound of one word and ignores others.

Amnesia refers to the loss of memory from disease or trauma. The type of memory loss depends upon the specific regions of the brain affected. For example, damage to the auditory association areas may make it difficult to remember sounds. Damage to thalamic and limbic structures, expecially the hippocampus, will affect memory storage and consolidation. AM *Amnesia*

Hemispheric Specialization

Although the two hemispheres are very similar in appearance, there are significant functional differences between them. The assignment of a specific function to a region of the cerebral cortex is not always straightforward. Not only may one region have several different functions, some aspects of cortical function, such as consciousness, cannot easily be assigned to any single region.

Figure 8-21● indicates the major functional differences between the hemispheres. Higher-order centers in the left and right hemispheres have different but complementary functions. In most people, the left hemisphere contains the general interpretive and speech centers, and this is the hemisphere responsible for reading, writing, and speaking. The left hemisphere is also important in performing analytical tasks, such as mathematical calculations and logical decision making. Although the left hemisphere was formerly called the *dominant hemisphere*, a more appropriate term is the *categorical hemisphere*, because the right hemisphere also has many important functions.

The right cerebral hemisphere analyzes sensory information and relates the body to the sensory environment. Interpretive centers in this hemisphere permit one to identify familiar objects by touch, smell, taste, or feel. Because it is concerned with spatial relationships and analyses, the term *representational hemisphere* is used to refer to the right hemisphere. Interestingly, there may be a link between handedness and sensory/spatial abilities. An unusually high percentage of musicians and artists are left-handed; the complex motor activities performed by these individuals are directed by the primary motor cortex and association areas on the right hemisphere.

Hemispheric specialization does not mean that the two hemispheres function independently of each other. As noted above, the white fibers of the corpus callosum link the two hemispheres, including their sensory information and motor commands. The corpus callosum alone contains over 200 million axons, carrying an estimated 4 billion impulses per second!

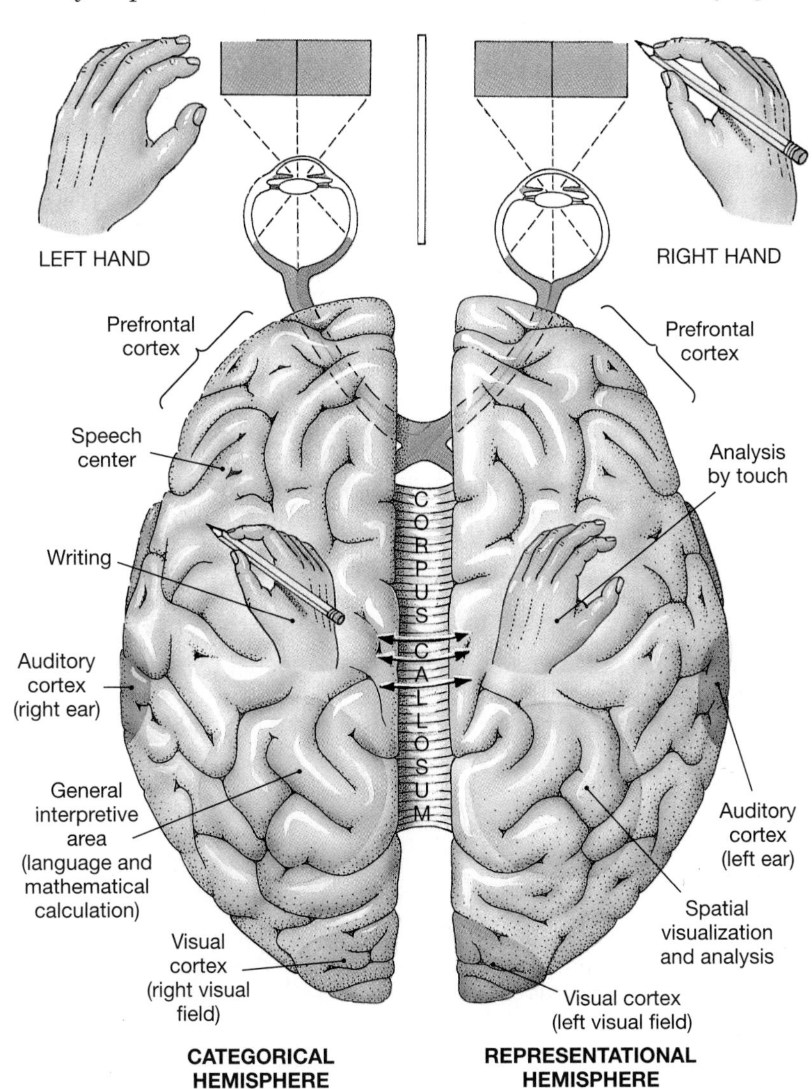

●**Figure 8-21**
Hemispheric Specialization
Functional differences between the left and right cerebral hemispheres.

The Electroencephalogram

The specific areas just described were mapped by direct stimulation in patients undergoing brain surgery. Noninvasive methods are also used to correlate the activity of different cortical regions with function. One of the most common noninvasive methods involves monitoring the electrical activity of the brain.

Neural function depends on electrical events within the cell membrane. The brain contains billions of nerve cells, and their activity generates an electrical field that can be measured by placing electrodes on the brain or the outer surface of the skull. The electrical activity changes constantly as nuclei and cortical areas are stimulated or quiet down. An **electroencephalogram (EEG)** is a printed record of this electrical activity over time. The patterns of **brain waves** that are observed can be correlated with the individual's level of consciousness. Electroencephalograms can also provide useful diagnostic information regarding brain disorders. Four types of brain wave patterns are shown in Figure 8-22●.

The Cerebral Nuclei

The **cerebral nuclei**, also known as *basal ganglia*, lie within the central white matter of each cerebral hemisphere as diagrammed in Figure 8-23●

(a) Alpha waves	Alpha waves are characteristic of normal resting adults
(b) Beta waves	Beta waves typically accompany intense concentration
(c) Theta waves	Theta waves are seen in children and in frustrated adults
(d) Delta waves	Delta waves occur in deep sleep and in certain pathological states 1 sec

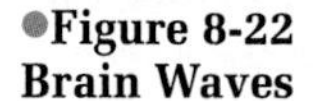

•Figure 8-22
Brain Waves

The cerebral nuclei play an important roll in the involuntary control of skeletal muscle tone and the coordination of learned movement patterns. These nuclei do not start a movement—that decision is a voluntary one—but once a movement is under way the cerebral nuclei provide pattern and rhythm. For example, during a simple walk the cerebral nuclei control the cycles of arm and thigh movements that occur between the time the decision is made to "start walking" and the time the "stop" order is given.

SEIZURES

A **seizure** is a temporary cerebral disorder accompanied by abnormal, involuntary movements, unusual sensations, or inappropriate behavior. Clinical conditions characterized by seizures are known as seizure disorders, or *epilepsies*. Seizures of all kinds are accompanied by a marked change in the pattern of electrical activity monitored in an electroencephalogram. The alteration begins in one portion of the cerebral cortex but may subsequently spread across the entire cortical surface, like a wave on the surface of a pond. The nature of the symptoms produced depends on the region of the cortex involved. If the seizure affects the primary motor cortex, involuntary movements will occur; if it affects the auditory cortex, the individual will hear strange sounds. AM *Seizures and Epilepsy*

The Limbic System

The **limbic system** (LIM-bik; *limbus*, a border), shown in Figure 8-24•, includes several cerebral nuclei, gyri, and tracts along the border between the cerebrum and diencephalon. This system includes cerebral centers concerned with (1) the sense of smell and (2) long-term memory storage. The *amygdaloid bodies* and the *hippocampus*, for example, play a vital role in learning and the storage of long-term memories. Damage to the hippocampus in Alzheimer's disease interferes with memory storage and retrieval. The limbic system also includes hypothalamic centers responsible for (1) emotional states, such as rage, fear, and sexual arousal, and (2) the control of reflex movements that can be consciously activated. For example, the limbic system includes the *mamillary bodies* (MAM-i-lar-ē; *mamilla*, a little breast) of the hypothalamus. These nuclei control reflex movements associated with eating, such as chewing, licking, and swallowing.

- ✓ How would decreased diffusion across the arachnoid granulations affect the volume of cerebrospinal fluid in the ventricles?
- ✓ Mary suffers a head injury that damages her primary motor cortex. Where is this area located?
- ✓ What senses would be affected by damage to the temporal lobes of the cerebrum?

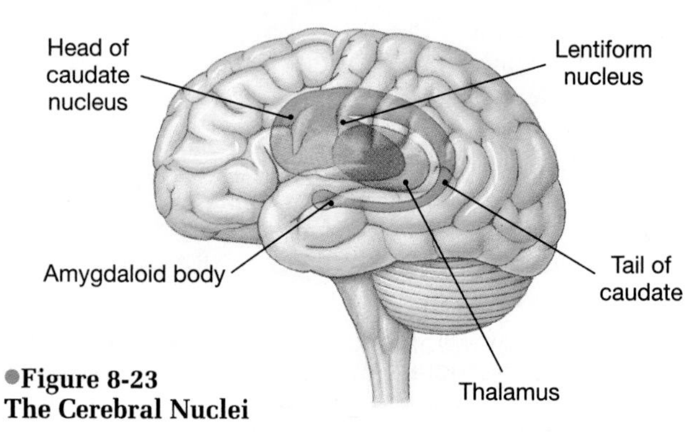

•Figure 8-23
The Cerebral Nuclei

The relative positions of the cerebral nuclei in the intact brain.

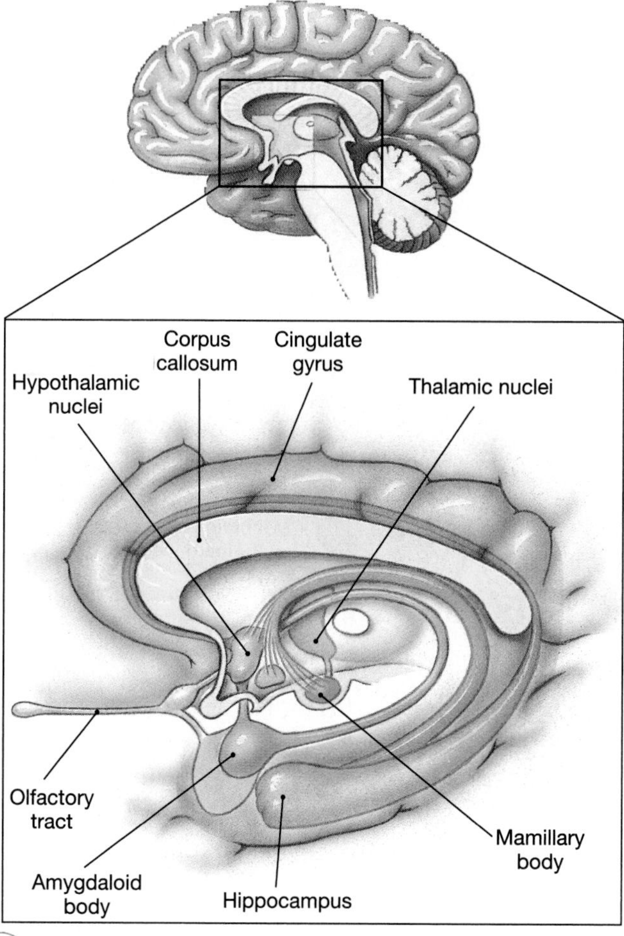

Figure 8-24
The Limbic System
A three-dimensional reconstruction of the limbic system, showing the relationships between the major components.

The Diencephalon

The diencephalon (Figure 8-25●) provides switching and relay centers that integrate the conscious and unconscious sensory and motor pathways. The diencephalon contains a central chamber, the *third ventricle*, that is filled with cerebrospinal fluid. The diencephalic roof contains (1) an extensive area of *choroid plexus* and (2) the *pineal gland*, an endocrine structure that secretes the hormone *melatonin*. Among other functions, melatonin is important in regulating day-night cycles.

The Thalamus

The thalamus (Figure 8-25●) is the final relay point for all ascending sensory information, other than olfactory, that will reach our conscious awareness. It acts as a filter, passing on to the primary sensory cortex only a small portion of the arriving sensory information. The rest is relayed to the cerebral nuclei and centers in the brain stem. The thalamus also plays a role in the coordination of voluntary and involuntary motor commands.

The Hypothalamus

The hypothalamus (1) contains centers associated with the emotions of rage, pleasure, pain, thirst, hunger, and sexual arousal; (2) adjusts and coordinates the activities of autonomic centers in the pons and medulla oblongata; (3) coordinates neural and endocrine activities; (4) produces a variety of hormones, including *antidiuretic hormone (ADH)* and *oxytocin*; (5) coordinates voluntary and autonomic functions; and (6) maintains normal body temperature.

The Midbrain

The midbrain (Figure 8-25●) contains various nuclei and bundles of ascending and descending nerve fibers. It includes two pairs of sensory nuclei, or *colliculi* (kol-IK-ū-lī; singular *colliculus*, a small hill), concerned with the processing of visual and auditory sensations. These midbrain nuclei direct the involuntary motor reflexes to sudden visual and auditory stimuli (such as a blinding flash of light or a loud noise). The midbrain also contains motor nuclei for two of the cranial nerves (N III, IV) concerned with eye movements.

The midbrain is also headquarters to one of the most important brain stem components, the **reticular formation**, which is involved in the regulation of many involuntary functions. The reticular formation is a network of interconnected nuclei that extends the length of the brain stem. The reticular formation of the midbrain contains the *reticular activating system (RAS)*. The output of this system directly affects the activity of the cerebral cortex. When the RAS is inactive, so are we; when the RAS is stimulated, so is our state of attention or wakefulness.

The maintenance of muscle tone and posture is due to midbrain nuclei that integrate information from the cerebrum and cerebellum and issue the appropriate involuntary motor commands. Other midbrain nuclei, such as the *substantia nigra* (NĪ-grah; black), play an important role in regulating the motor output of the cerebral nuclei. If the substantia nigra is damaged, there is a gradual increase in muscle tone and the appearance of symptoms characteristic of *Parkinson's disease*. AM *The Cerebral Nuclei and Parkinson's Disease*

The Pons

The pons (Figure 8-25●) links the cerebellum with the mesencephalon, diencephalon, cerebrum, and spinal cord. One group of nuclei within the pons includes the sensory and motor nuclei for four of the cranial nerves (N V–VIII). Other nuclei are concerned with the involuntary control of the pace and depth of respiration. Tracts passing through the pons link the cerebellum with the brain stem, cerebrum, and spinal cord.

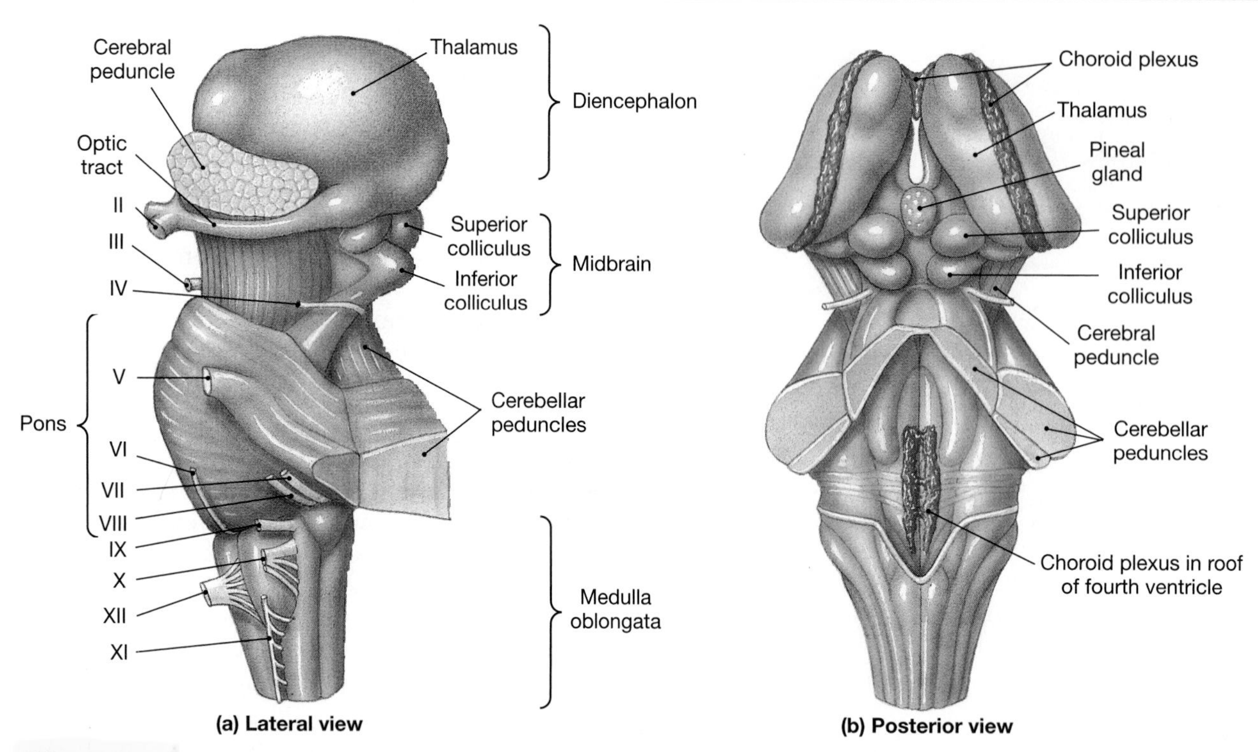

•Figure 8-25
The Diencephalon and Brain Stem
(a) Lateral view, as seen from the left side. Roman numerals indicate the positions of cranial nerves. **(b)** Posterior view.

The Cerebellum

The cerebellum (Figure 8-17•, p. 215) performs two important functions: (1) It makes rapid adjustments in muscle tone and position to maintain balance and equilibrium by modifying the activity of nuclei in the brain stem; and (2) it programs and fine-tunes voluntary and involuntary movements. These functions are performed indirectly, by regulating activity along motor pathways at the cerebrum and brain stem. The tracts that link the cerebellum with these different regions are the **cerebellar peduncles** (Figure 8-25•). Like the cerebral hemispheres, the cerebellar hemispheres are composed of white matter covered by a layer of neural cortex called the *cerebellar cortex.*

The cerebellum may be permanently damaged by trauma or stroke or temporarily affected by drugs such as alcohol. These alterations can produce *ataxia* (a-TAK-sē-a; *ataxia*, a lack of order), a disturbance in balance. AM *Cerebellar Dysfunction*

The Medulla Oblongata

The medulla oblongata (Figure 8-25•) physically connects the brain with the spinal cord, and many of its functions are directly related to this fact. For example, all communication between the brain and spinal cord involves tracts that ascend or descend through the medulla oblongata. These tracts often synapse in the medulla oblongata, in sensory or motor nuclei that act as relay stations and processing centers. In addition to these nuclei, the medulla oblongata contains sensory and motor nuclei associated with five of the cranial nerves (N VIII–XII).

Within the medulla oblongata, a variety of nuclei and centers representing the reticular formation regulate vital autonomic functions. These reflex centers receive inputs from cranial nerves, the cerebral cortex, and the brain stem, and their output controls or adjusts the activities of one or more peripheral systems. The *cardiovascular centers* adjust heart rate, the strength of cardiac contractions, and the flow of blood through peripheral tissues. On functional grounds, the cardiovascular centers may be subdivided into a *cardiac center* regulating the heart rate and a *vasomotor center* controlling peripheral blood flow. The *respiratory rhythmicity center* sets the basic pace for respiratory movements, and its activity is adjusted by the respiratory centers of the pons.

✓ The thalamus acts as a relay point to all but what type of sensory information?

✓ What area of the diencephalon would be stimulated by changes in body temperature?

✓ The medulla oblongata is one of the smallest sections of the brain, yet damage there can cause death, whereas similar damage in the cerebrum might go unnoticed. Why?

Chapter Review

KEY TERMS

action potential, p. 204
cerebrospinal fluid, p. 212
cerebrum, p. 215
limbic system, p. 222
meninges, p. 210
multipolar neuron, p. 200
myelin, p. 200
neuroglia, p. 198
neurotransmitter, p. 206
reflex, p. 208
somatic nervous system, p. 198
synapse, p. 206

SUMMARY OUTLINE

INTRODUCTION p. 198

1. Two organ systems, the nervous and endocrine systems, coordinate organ system activity. The nervous system provides swift but brief responses to stimuli; the endocrine system adjusts metabolic operations and directs long-term changes.

THE NERVOUS SYSTEM p. 198

1. The **nervous system** includes all the neural tissue in the body. Its anatomical divisions include the **central nervous system (CNS)** (the brain and spinal cord) and the **peripheral nervous system (PNS)** (all of the neural tissue outside the CNS).

2. Functionally, it can be divided into an **afferent division**, which brings sensory information to the CNS and an **efferent division**, which carries motor commands to muscles and glands. The efferent division includes the **somatic nervous system (SNS)** (voluntary control over skeletal muscle contractions) and the **autonomic nervous system (ANS)** (automatic, involuntary regulation of smooth muscle, cardiac muscle, and glandular activity). (*Figure 8-1*)

CELLULAR ORGANIZATION IN NEURAL TISSUE p. 198

1. There are two types of cells in neural tissue: **neurons**, which are responsible for information transfer and processing, and **neuroglia**, or *glial cells*, which provide a supporting framework and act as phagocytes.

Neurons p. 199

2. **Sensory neurons** form the afferent division of the PNS and deliver information to the CNS. **Motor neurons** stimulate or modify the activity of a peripheral tissue, organ, or organ system. **Interneurons (association neurons)** may be located between sensory and motor neurons; they analyze sensory inputs and coordinate motor outputs.

3. Sensory recpetors are classified according to their source of stimuli: (1) **exteroceptors** provide information about the external environment; (2) **proprioceptors** monitor the position of the body's skeletal muscles and joints; and, (3) **interoceptors** monitor activities of various organ systems of the body.

4. A typical neuron has a **soma** (cell body), an **axon**, and several branching, sensitive **dendrites**. (*Figure 8-2*)

5. **Synaptic knobs** occur at the end of axons. (*Figure 8-2*)

6. Neurons may be described as **unipolar**, **bipolar**, or **multipolar**. (*Figure 8-3*)

Neuroglia p. 201

7. There are four types of neuroglia in CNS: (1) **astrocytes** (largest and most numerous); (2) **oligodendrocytes**, which are responsible for the **myelination** of CNS axons; (3) **microglia**, phagocytic white blood cells; and (4) **ependymal cells**, with functions related to the *cerebrospinal fluid (CSF)*. (*Figure 8-4*)

8. Nerve cell bodies in the PNS are clustered into *ganglia* (singular *ganglion*). Their axons are covered by myelin wrappings of **Schwann cells**. (*Figure 8-5*)

NEUROPHYSIOLOGY p. 203

The Membrane Potential p. 203

1. The membrane, or **resting, potential** of an undisturbed nerve cell is due to a balance between the rate of sodium ion entry and potassium ion loss and to the sodium-potassium exchange pump. Any stimulus that affects this balance will alter the resting potential of the cell. (*Figure 8-6*)

2. An **action potential** appears when the membrane depolarizes to a level known as the **threshold**. The steps involved include: opening of sodium channels and membrane depolarization; sodium channels closing and potassium channels opening; and return to normal permeability. (*Figure 8-7*)

Conduction of an Action Potential, p. 205

3. In **continuous conduction**, an action potential spreads across the entire excitable membrane surface in a series of small steps. (*Figure 8-8a*)

4. During **saltatory conduction**, the action potential appears to leap from node to node, skipping the intervening membrane surface. (*Figure 8-8b*)

SYNAPTIC COMMUNICATION p. 206

1. A **synapse** is a site where intercellular communication occurs through the release of chemicals called **neurotransmitters**. A synapse where neurons communicate with other cell types is a **neuroeffector junction**.

Structure of a Synapse p. 206

2. Communication moves from the **presynaptic neuron** to the **postsynaptic neuron** over the **synaptic cleft**. (*Figure 8-9*)

Synaptic Events p. 207

3. *Cholinergic synapses* release the neurotransmitter **acetylcholine (ACh)**. (*Figure 8-10; Table 8-1*)

Neurotransmitters p. 207

4. Other neurotransmitters include **norepinephrine (NE)**, **dopamine**, **gamma aminobutyric acid (GABA)**, and **serotonin**. NE usually has an excitatory effect on postsynaptic membranes; while dopamine, GABA, and serotonin are typically inhibitory.

NEURONAL POOLS p. 208

1. The roughly 20 billion interneurons can be classified into **neuronal pools** (groups of interconnected neurons with specific functions).

2. **Divergence** is the spread of information from one neuron to several, or from one pool to several pools. In **convergence**, several neurons synapse on the same postsynaptic neuron. In **parallel processing**, neuronal pools process the same information simultaneously. (*Figure 8-11*)

AN INTRODUCTION TO REFLEXES p. 208

1. A **reflex** is an automatic, involuntary motor response that helps preserve homeostasis by rapidly adjusting the functions of organs or organ systems.

The Reflex Arc p. 209

2. A **reflex arc** is the "wiring" of a single reflex.

3. There are five steps involved in a neural reflex: (1) arrival of a stimulus and activation of a receptor, (2) activation of a sensory neuron, (3) information processing, (4) activation of a motor neuron, (5) response by an effector. (*Figure 8-12*)

AN INTRODUCTION TO THE ANATOMY OF THE NERVOUS SYSTEM p. 210

1. The functions of the nervous system as a whole depend on interactions between neurons in neuronal pools. In the PNS, spinal nerves communicate with the spinal cord, and cranial nerves are connected to the brain.

2. In the CNS, a collection of neuron cell bodies that share a particular function is called a **center**. A center with a discrete anatomical boundary is called a **nucleus**. Portions of the brain surface are covered by a thick layer of gray matter called the **neural cortex**. (*Figure 8-13*)

3. The white matter of the CNS contains bundles of axons, or **tracts**, that share common origins, destinations, and functions. Tracts in the spinal cord form larger groups, called **columns**. (*Figure 8-13*)

4. **Sensory** (ascending) **pathways** carry information from peripheral sensory receptors to the brain; **motor** (descending) **pathways** extend from CNS centers concerned with motor control to the associated skeletal muscles. (*Figure 8-13*)

The Meninges p. 210

5. Special covering membranes, the **meninges**, protect and support the spinal cord and delicate brain. The cranial meninges (the dura mater, arachnoid, and pia mater) are continuous with those of the spinal cord, the spinal meninges. (*Figure 8-14*)

6. The **dura mater** covers the brain and spinal cord. The **epidural space** separates the spinal dura mater from the walls of the vertebral canal.

7. Beneath the inner surface of the dura mater lies the **arachnoid** (the second meningeal layer), and the **subarachnoid space**. The latter contains **cerebrospinal fluid (CSF)**, which acts as a shock absorber and a diffusion medium for dissolved gases, nutrients, chemical messengers, and waste products.

8. The **pia mater**, the innermost meningeal layer, is bound to the underlying neural tissue.

The Blood-Brain Barrier p. 212

9. The **blood-brain barrier** isolates neural tissue from the general circulation.

THE SPINAL CORD p. 212

1. In addition to relaying information to and from the brain, the spinal cord integrates and processes information on its own.

Gross Anatomy p. 212

2. The adult spinal cord includes localized enlargements that provide innervation to the limbs. (*Figure 8-15a*)

3. The spinal cord has 31 segments, each associated with a pair of **dorsal root ganglia** and their **dorsal roots** and a pair of **ventral roots**. (*Figures 8-15b; 8-16a*)

Sectional Anatomy p. 213

4. The white matter contains myelinated and unmyelinated axons; the gray matter contains cell bodies of neurons and glial cells. The projections of gray matter toward the outer surface of the spinal cord are called horns. (*Figure 8-16b*)

THE BRAIN p. 214

Major Divisions of the Brain p. 215

1. There are six regions in the adult brain: cerebrum, diencephalon, midbrain, cerebellum, pons, and medulla oblongata. (*Figure 8-17*)

2. Conscious thought, intellectual functions, memory, and complex involuntary motor patterns originate in the **cerebrum**. The **cerebellum** adjusts voluntary and involuntary motor activities on the basis of sensory data and stored memories. (*Figure 8-17*)

3. The walls of the **diencephalon** form the **thalamus**, which contains relay and processing centers for sensory data. The **hypothalamus** contains centers involved with emotions, autonomic function, and hormone production. (*Figure 8-17c*)

4. Three regions make up the **brain stem**. The **midbrain** processes visual and auditory information and generates involuntary somatic motor responses. The **pons** connects the cerebellum to the brain stem and is involved with somatic and visceral motor control. The spinal cord connects to the brain at the **medulla oblongata**, which relays sensory information and regulates autonomic functions. (*Figure 8-17c*)

The Ventricles of the Brain p. 216

5. The central passageway of the brain expands to form four chambers called **ventricles**. Cerebrospinal fluid continually circulates from the ventricles and central canal of the spinal cord into the subarachnoid space of the meninges that surround the CNS. (*Figure 8-18*)

The Cerebrum p. 217

6. The cortical surface contains **gyri** (elevated ridges) separated by **sulci** (shallow depressions) or deeper grooves (**fissures**). The **longitudinal fissure** separates the two **cerebral hemispheres**. The **central sulcus** marks the boundary between the **frontal lobe** and the **parietal lobe**. Other sulci form the boundaries of the **temporal lobe** and the **occipital lobe**. (*Figures 8-17; 8-20*)

7. Each cerebral hemisphere receives sensory information and generates motor commands that concern the opposite side of the body.

8. The **primary motor cortex** of the **precentral gyrus** directs voluntary movements. The **primary sensory cortex** of the **postcentral gyrus** receives somatic sensory information from touch, pressure, pain, taste, and temperature receptors.

9. **Association areas**, such as the **visual association area** and **premotor cortex** (motor association area), control our ability to understand sensory information and coordinate a motor response.

10. "Higher-order" integrative centers receive information from many different association areas and direct complex motor activities and analytical functions. (*Figure 8-21*)

11. The left hemisphere is usually the *categorical hemisphere*, which contains the general interpretive and speech centers and is responsible for language-based skills. The right hemisphere, or *representational hemisphere*, is concerned with spatial relationships and analyses. (*Figure 8-21*)

The Electroencephalogram p. 221

12. An **electroencephalogram (EEG)** is a printed record of **brain waves. Alpha waves**, found in normal adults under resting conditions, are replaced by **beta waves** during times of stress or tension. **Theta waves** appear in the brains of children and in frustrated adults and may indicate a brain disorder. **Delta waves** appear in the brains of infants and in the brains of adults during deep sleep or in cases of brain damage. (*Figure 8-22*)

The Cerebral Nuclei p. 221

13. The **cerebral nuclei** (basal ganglia) lie within the central white matter and aid in the coordination of learned movement patterns and other somatic motor activities. (*Figure 8-23*)

The Limbic System p. 222

14. The **limbic system** includes the *hippocampus*, which is involved in memory and learning, and the *mamillary bodies*, which control reflex movements associated with eating. The functions of the limbic system involve emotional states and related behavioral drives. (*Figure 8-24*)

The Diencephalon p. 223

15. The diencephalon provides the switching and relay centers necessary to integrate the conscious and unconscious sensory and motor pathways. The diencephalic roof contains the *pineal gland* and a vascular network that produces cerebrospinal fluid. (*Figure 8-25*)

16. The thalamus is the final relay point for ascending sensory information. It acts as a filter, passing on only a small portion of the arriving sensory information to the cerebral cortex, relaying the rest to the cerebral nuclei and centers in the brain stem. (*Figure 8-25*)

17. The hypothalamus contains important control and integrative centers. It can (1) produce emotioins and behavioral drives; (2) control autonomic function; (3) coordinate activities of the nervous and endocrine systems; (4) secrete hormones; (5) coordinate voluntary and autonomic functions; and (6) regulate body temperature.

The Midbrain p. 223

18. The midbrain contains two pairs of sensory nuclei, the *colliculi*, that receive visual and auditory information. It is also the center of the **reticular formation**, nuclei that affect cerebral activity. (*Figure 8-25*)

The Pons p. 223

19. The pons contains (1) sensory and motor nuclei for four cranial nerves, (2) nuclei concerned with involuntary control of respiration, and (3) ascending and descending tracts. (*Figure 8-25*)

The Cerebellum p. 224

20. The cerebellum oversees the body's postural muscles and programs and tunes voluntary and involuntary movements. The **cerebellar peduncles** link the cerebellum with the brain stem, cerebrum, and spinal cord. (*Figures 8-17; 8-25*)

The Medulla Oblongata p. 224

21. The medulla oblongata connects the brain to the spinal cord. Its nuclei relay information from the spinal cord and brain stem to the cerebral cortex. Its reflex centers, including the cardiovascular centers and the respiratory rhythmicity center, control or adjust the activities of peripheral systems. (*Figure 8-25*)

CHAPTER QUESTIONS

LEVEL 1 Reviewing Facts and Terms

Match each item in column A with the most closely related item in column B. Use letters for answers in the spaces provided.

Column A

___ 1. neuroglia
___ 2. microglia
___ 3. sensory neurons
___ 4. motor neurons
___ 5. ganglia
___ 6. oligodendrocytes
___ 7. ascending tracts
___ 8. descending tracts
___ 9. saltatory conduction
___ 10. continuous conduction
___ 11. dura mater
___ 12. pia mater
___ 13. choroid plexus
___ 14. cerebellum
___ 15. Nissl bodies
___ 16. hypothalamus
___ 17. medulla oblongata

Column B

a. coat CNS axons with myelin
b. carry sensory information to the brain
c. occurs along unmyelinated axons
d. outermost covering of brain and spinal cord
e. production of CSF
f. supporting cells
g. phagocytic white blood cells
h. occurs along myelinated axons
i. link between nervous and endocrine systems
j. carry motor commands to spinal cord
k. efferent division of the PNS
l. aggregations of RER and ribosomes
m. masses of neuron cell bodies
n. connects the brain to the spinal cord
o. innermost meningeal layer
p. afferent division of the PNS
q. maintains muscle tone and posture

18. Regulation by the nervous system provides:
    (a) relatively slow but long-lasting responses to stimuli
    (b) swift, long-lasting responses to stimuli
    (c) swift but brief responses to stimuli
    (d) relatively slow, short-lived responses to stimuli
19. All the motor neurons that control skeletal muscles are:
    (a) multipolar neurons
    (b) myelinated bipolar neurons
    (c) unipolar, unmyelinated sensory neurons
    (d) propriocaptors
20. Depolarization of a neuron cell membrane will shift the membrane potential toward:
    (a) 0 mV (b) –70 mV
    (c) –90 mV (d) a, b, and c are correct
21. The primary determinant of the resting membrane potential is the:
    (a) membrane permeability to sodium
    (b) membrane permeability to potassium
    (c) intracellular negatively charged proteins
    (d) negatively charged chloride ions in the extracellular fluid
22. The structural and functional link between the cerebral hemispheres and the components of the brain stem is the:
    (a) neural cortex (b) medulla oblongata
    (c) mesencephalon (d) diencephalon
23. The ventricles in the brain are filled with:
    (a) blood (b) cerebrospinal fluid
    (c) air (d) neural tissue
24. Reading, writing, and speaking are dependent on processing in the:
    (a) right cerebral hemisphere
    (b) left cerebral hemisphere
    (c) prefrontal cortex
    (d) postcentral gyrus
25. Establishment of emotional states and related behavioral drives are functions of the:
    (a) limbic system (b) pineal gland
    (c) mamillary bodies (d) thalamus
26. The final relay point for ascending sensory information that will be projected to the primary sensory cortex is the:
    (a) hypothalamus (b) thalamus
    (c) spinal cord (d) medulla oblongata
27. What two integrated steps are necessary for transfer of nerve impulses from neuron to neuron?
28. State the all-or-none principle regarding action potentials.
29. What are the primary functions of the cerebrum?

LEVEL 2 Reviewing Concepts

30. A graded potential:
    (a) decreases with distance from the point of stimulation
    (b) spreads passively because of local currents
    (c) may involve either depolarization or hyperpolarization
    (d) a, b, and c are correct
31. The loss of positive ions from the interior of a neuron produces:
    (a) depolarization (b) threshold
    (c) hyperpolarization (d) an action potential
32. What purpose do axon collaterals serve in the nervous system?
33. What would happen if the ventral root of a spinal nerve was damaged or transected?
34. Stimulation of what part of the brain would produce sensations of hunger and thirst?
35. What major part of the brain is associated with respiratory and cardiac activity?
36. Multiple sclerosis (MS) is a demyelination disorder. How does this condition produce muscular paralysis and sensory losses?

LEVEL 3 Critical Thinking and Clinical Applications

37. If neurons in the central nervous system lack centrioles and are unable to divide, how can a person develop brain cancer?
38. Myelination of peripheral neurons occurs rapidly through the first year of life. How can this process explain the increased abilities of infants during their first year?
39. A police officer has just stopped Bill on suspicion of driving while intoxicated. The officer asks Bill to walk the yellow line on the road and then asks him to place the tip of his index finger on the tip of his nose. How would these activities indicate Bill's level of sobriety? What part of the brain is being tested by these activities?

9 THE PERIPHERAL NERVOUS SYSTEM AND INTEGRATED NEURAL FUNCTIONS

Our perceptions of the world around us depend on thousands of interactions among neurons within the central nervous system. We seldom realize how complex these processes are unless they go wrong in some way. This child has dyslexia, a condition characterized by difficulties with the recognition and use of words. Although the cause of dyslexia remains a mystery, there is general agreement that it results from problems with the integration and processing of visual or auditory information.

This chapter introduces the sensory and motor pathways involved in such activities as writing, and considers the neural basis for higher-order functions, such as memory and learning.

Chapter Outline and Objectives

Introduction, p. 230

The Peripheral Nervous System, p. 230

The Cranial Nerves, p. 230

The Spinal Nerves, p. 232

Nerve Plexuses, p. 233

1. Identify the cranial nerves and relate each pair of cranial nerves to its principal functions.
2. Relate the distribution pattern of spinal nerves to the regions they innervate.

The CNS and PNS: Integrated Functions, p. 233

Simple Reflexes, p. 234

Complex Reflexes, p. 234

Integration and Control of Spinal Reflexes, p. 236

3. Distinguish between the motor responses produced by simple and complex reflexes.
4. Explain how higher centers control and modify reflex responses.

Sensory and Motor Pathways, p. 236

Sensory Pathways, p. 236

Motor Pathways, p. 237

5. Identify the principal sensory and motor pathways.
6. Explain how we can distinguish among sensations from different areas of the body.

The Autonomic Nervous System, p. 238

The Sympathetic Division, p. 240

The Parasympathetic Division, p. 240

Relationships between the Sympathetic and Parasympathetic Divisions, p. 243

7. Compare the autonomic nervous system with the other divisions of the nervous system.
8. Explain the functions and structures of the sympathetic and parasympathetic divisions.
9. Discuss the relationship between the sympathetic and parasympathetic divisions and explain the implications of dual innervation.

Aging and the Nervous System, p. 243

Age-Related Changes, p. 244

10. Summarize the effects of aging on the nervous system.

Interactions with Other Systems, p. 244

11. Discuss the interrelationships among the nervous system and other organ systems.

Vocabulary Development

chiasm, a crossing; *optic chiasm*
cochlea, snail shell; *cochlear nerve*
glossus, tongue; *glossopharyngeal nerve*
mono-, one; *monosynaptic*
murus, wall; *intramural ganglia*
poly-, many; *polysynaptic*
trochlea, a pulley; *trochlear nerve*
vagus, wandering; *vagus nerve*
vestibulum, a cavity; *vestibular nerve*

The peripheral nervous system (PNS) is the link between the neurons of the central nervous system (CNS) and the rest of the body; all sensory information and motor commands are carried by axons of the PNS. As a result, although the PNS contains less than 2 percent of the neural tissue in the body, it is absolutely vital to the function of the nervous system. The essential communication between the CNS and PNS occurs over tracts and nuclei that relay sensory information and motor commands. *Pathways* consist of the nuclei and tracts that link the brain with the rest of the body. Sensory and motor pathways involve a series of synapses, one after the other. For example, a sensation carried by an ascending sensory pathway may be relayed across synapses in the medulla oblongata and the thalamus before reaching the cerebral cortex and our conscious awareness. At each synapse there are opportunities for divergence and for the distribution of information to neuronal pools operating at an involuntary level. For example, while you are consciously planning a response to a stumble, centers in the brain stem and cerebellum may already be issuing the motor commands necessary to prevent a fall.

This chapter begins with a consideration of the structure of the peripheral nervous system. We will then examine the integrated functioning of the nervous system, beginning with simple spinal reflexes and proceeding to the organization of representative sensory and motor pathways that produce conscious sensations and provide control over skeletal muscles. We will conclude with an overview of the autonomic nervous system, which coordinates cardiovascular, respiratory, digestive, excretory, and reproductive functions, usually without instructions or interference from the conscious mind.

THE PERIPHERAL NERVOUS SYSTEM

The PNS is dominated by axons heading to or from the central nervous system. These axons, bundled together and wrapped in connective tissue, form **peripheral nerves**, or simply *nerves*. The PNS also contains both the cell bodies and the axons of sensory neurons and motor neurons of the autonomic nervous system. The cell bodies are clustered together in masses called **ganglia** (singular *ganglion*). The peripheral nervous system includes *cranial nerves* and *spinal nerves*.

The Cranial Nerves

The cranial nerves are components of the peripheral nervous system that connect to the brain rather than to the spinal cord. The 12 pairs of cranial nerves, shown in Figure 9-1●, are numbered according to their position along the longitudinal axis of the brain. The prefix N designates a cranial nerve and Roman numerals are used to distinguish the individual nerves. For example, N I refers to the first pair of cranial nerves, the *olfactory nerves*.

Distribution and Function

Functionally, each nerve can be classified as primarily sensory, motor, or mixed (sensory and motor). It should be noted, however, that cranial nerves often have secondary functions. For example, several cranial nerves (N III, VII, IX, and X) also distribute autonomic fibers to PNS ganglia, just as spinal nerves deliver them to ganglia along the spinal cord. The distribution and functions of the cranial nerves are described below and summarized in Table 9-1.

The Olfactory Nerves (N I): The first pair of cranial nerves, the **olfactory nerves**, are the only cranial nerves attached to the cerebrum. The rest originate or terminate within nuclei of the brain stem. These nerves carry special sensory information responsible for the sense of smell. The olfactory nerves reach the brain by penetrating the cribriform plate to synapse in the *olfactory bulbs*.

The Optic Nerves (N II): The **optic nerves** (N II) carry visual information from the eyes. After passing through the optic foramina of the orbits, they intersect at the **optic chiasm** ("crossing") (Figure 9-1a●) before they continue to the *lateral geniculate* nuclei of the thalamus.

The Oculomotor Nerves (N III): The motor nuclei controlling the third cranial nerves are in the midbrain. Each **oculomotor nerve** (N III) innervates four of the six muscles that move an eyeball (the *superior, medial,* and *inferior rectus* muscles and the *inferior oblique* muscle). These nerves also carry autonomic (*parasympathetic*) fibers to intrinsic eye muscles that control (1) the amount of light entering the eye and (2) the shape of the lens.

The Trochlear Nerves (N IV): The **trochlear nerves**, (TRŌK-lē-ar; *trochlea*, a pulley), the smallest of the cranial nerves, innervate the *superior oblique* muscles of the eyes. The motor nuclei lie in the midbrain.

The Trigeminal Nerves (N V): The pons contains the nuclei associated with three cranial nerves (N V, VI, and VII) and part of a fourth (N VIII). The **trigeminal** (trī-JEM-i-nal) **nerves** are the largest of the cranial nerves. These nerves provide sensory information from the head and face and motor control over the muscles of mastication. As the name implies, the *tri*geminal has three major branches. The *ophthalmic branch* provides sensory information from the orbit of

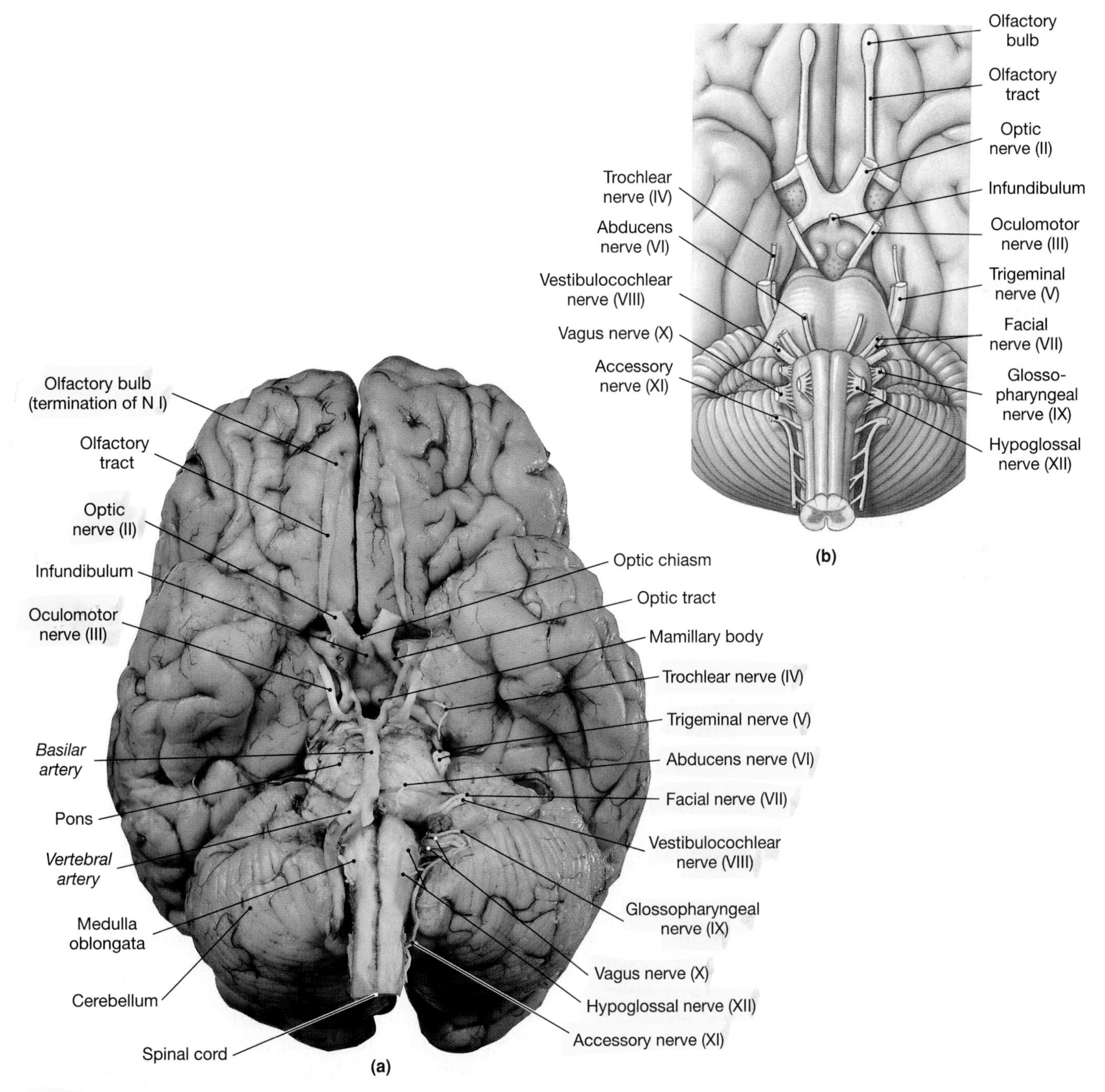

•Figure 9-1
The Cranial Nerves

(a) Inferior view of the brain. **(b)** Diagrammatic view, showing the attachment of the 12 pairs of cranial nerves.

the eye, the nasal cavity and sinuses, and the skin of the forehead, eyebrows, eyelids, and nose. The *maxillary branch* provides sensory information from the lower eyelid, upper lip, cheek, and nose. It also monitors the upper gums and teeth, the palate, and portions of the pharynx. The *mandibular branch*, the largest of the three, provides sensory information from the skin of the temples, the lower gums and teeth, the salivary glands, and the anterior portions of the tongue. It also provides motor control over the chewing muscles (the *temporalis, masseter,* and *pterygoid* muscles). ∞ [p. 175]

The Abducens Nerves (N VI): The **abducens** (ab-Doo-senz) **nerves** innervate only the *lateral rectus*, the sixth of the extrinsic eye muscles. The nerve emerges at the border between the pons and the medulla oblongata and reaches the orbit of the eye in company with the oculomotor and trochlear nerves.

The Facial Nerves (N VII): The **facial nerves** are mixed nerves of the face whose sensory and motor roots emerge from the side of the pons. The motor fibers produce facial expressions by controlling the superficial muscles of the scalp and face and deep muscles near the ear. The sensory fibers monitor proprioceptors in the facial muscles, provide deep pressure sensations over the face, and the taste information from receptors along the anterior two-thirds of the tongue.

The Vestibulocochlear Nerves (N VIII): The **vestibulocochlear nerves** monitor the sensory receptors of the inner ear. Each vestibulocochlear nerve has two components: (1) a **vestibular nerve** (*vestibulum*, a cavity), which originates at the *vestibule*, the portion of the inner ear concerned with balance sensations, conveys information on position, movement, and balance; and (2) the **cochlear nerve** (KOK-lē-ar; *cochlea*, snail shell), which monitors the receptors of the *cochlea*, the portion of the inner ear responsible for the sense of hearing.

The Glossopharyngeal Nerves (N IX): The medulla oblongata contains the sensory and motor nuclei for the ninth, tenth, eleventh, and twelfth cranial nerves. The **glossopharyngeal nerves** (glos-ō-fah-RIN-jē-al; *glossus*, tongue) are mixed nerves. The sensory portion provides taste sensations from the posterior third of the tongue and monitors the blood pressure and dissolved gas concentrations within major blood vessels. The motor portion controls the pharyngeal muscles involved in swallowing.

The Vagus Nerves (N X): The **vagus** (VĀ-gus; *vagus*, wandering) **nerves** provide sensory information from the ear canals, the diaphragm, and taste receptors in the pharynx and from visceral receptors along the esophagus, respiratory tract, and abdominal organs as far away as the last portions of the large intestine. This sensory information is vital to the autonomic control of visceral function, but because it often fails to reach the cerebral cortex, we are not consciously aware of the sensations the vagus nerves provide. The motor components of the vagus nerves control skeletal muscles of the soft palate, pharynx, and esophagus and affect cardiac muscle, smooth muscle, and glands of the esophagus, stomach, intestines, and gallbladder.

The Accessory Nerves (N XI): The **accessory nerves**, sometimes called the *spinal accessory nerves*, differ from other cranial nerves in that some of their motor fibers originate in the lateral gray horns of the first five cervical vertebrae, in addition to the medulla oblongata. The *medullary branch* innervates the voluntary swallowing muscles of the soft palate and pharynx and the laryngeal muscles that control the vocal cords and produce speech. The *spinal branch controls* the *sternocleidomastoid* and *trapezius* muscles associated with the pectoral girdle. ∞ [p. 180]

The Hypoglossal Nerves (N XII): The **hypoglossal** (hī-pō-GLOS-al) **nerves** provide voluntary control over the skeletal muscles of the tongue.

Few people are able to remember the names, numbers, and functions of the cranial nerves without a struggle. Many use phrases in which the first letter of each word represents the cranial nerves, such as *Oh, Once One Takes The Anatomy Final, Very Good Vacations Are Heavenly.*

✓ John is experiencing problems in moving his tongue. His doctor tells him it is due to pressure on a cranial nerve. Which cranial nerve is involved?

✓ What symptoms would you associate with damage to the abducens nerve (N VI)?

The Spinal Nerves

The 31 pairs of spinal nerves can be grouped according to the region of the vertebral column from which they originate (Figure 9-2●). There are 8 pairs of cervical nerves (C_1–C_8), 12 pairs of thoracic nerves (T_1–T_{12}), 5 pairs of lumbar nerves (L_1–L_5), 5 pairs of sacral nerves (S_1–S_5), and 1 pair of coccygeal nerves (Co_1). Each pair of spinal nerves monitors a specific region of the body surface, and damage or infection of a spinal nerve will produce a characteristic loss of sensation in specific parts of the skin. For example, in *shingles*, a virus that infects dorsal root ganglia causes a painful rash whose distribution corresponds to that of the affected sensory nerves. AM *Shingles and Hansen's Disease*

MULTIPLE SCLEROSIS

Multiple sclerosis (MS) produces muscular paralysis and sensory losses through demyelination. [p. 206] The initial symptoms appear as the result of myelin degeneration within the white matter of the lateral and posterior columns of the spinal cord or along tracts within the brain. During subsequent attacks the effects may become more widespread, and the cumulative sensory and motor losses can eventually lead to a generalized muscular paralysis. AM *Multiple Sclerosis*

Nerve Plexuses

During development, skeletal muscles often fuse together, forming larger muscles innervated by nerve trunks containing axons derived from several spinal nerves. These compound nerve trunks originate at networks called **nerve plexuses** The plexuses and major peripheral nerves are shown in Figure 9-2●.

The **cervical plexus** innervates the muscles of the neck and extends into the thoracic cavity to control the diaphragm. The **brachial plexus** innervates the shoulder girdle and upper limb. The **lumbosacral plexus** supplies the pelvic girdle and lower limb. It can be further subdivided into a *lumbar plexus* and a *sacral plexus*. Table 9-2 lists the spinal nerve plexuses and describes some of the major nerves and their distribution.

Peripheral *nerve palsies*, also known as *peripheral neuropathies*, are characterized by regional losses of sensory and motor function as the result of nerve trauma or compression. You have experienced a mild, temporary palsy if your arm or leg has ever "fallen asleep" after leaning or sitting in an uncomfortable position. AM *Palsies*

THE CNS AND PNS: INTEGRATED FUNCTIONS

The central nervous system and peripheral nervous system can be studied separately, but they function together. This section considers the ways the CNS and PNS interact. We begin with simple reflex responses to stimulation.

TABLE 9-1 The Cranial Nerves

Cranial Nerves (Number)	*Primary Function*	*Innervation*
Olfactory (I)	Special sensory	Olfactory epithelium
Optic (II)	Special sensory	Retina of eye
Oculomotor (III)	Motor	Inferior, medial, superior rectus and intrinsic muscles of eye
Trochlear (IV)	Motor	Superior oblique muscle of eye
Trigeminal (V)	Mixed	Areas associated with the jaws: *Sensory* from orbital structures, nasal cavity, skin of forehead, upper eyelid, eyebrows, nose, lips, gums and teeth; cheek, palate, pharynx, and tongue *Motor* to chewing (temporalis, masseter, pterygoids) muscles
Abducens (VI)	Motor	Lateral rectus muscle of eye
Facial (VII)	Mixed	*Sensory* to taste receptors on the anterior 2/3 of tongue *Motor* to muscles of facial expression, lacrimal gland, submandibular gland, sublingual salivary glands
Vestibulocochlear (Acoustic) (VIII)	Special sensory	Cochlea (receptors for hearing) Vestibule (receptors for motion and balance)
Glossopharyngeal (IX)	Mixed	*Sensory* from posterior 1/3 of tongue; pharynx and palate (part); blood pressure and composition *Motor* to pharyngeal muscles, parotid salivary gland
Vagus (X)	Mixed	*Sensory* from pharynx; pinna and external meatus; diaphragm; visceral organs in thoracic and abdominopelvic cavities *Motor* to palatal and pharyngeal muscles and visceral organs in thoracic and abdominopelvic cavities
Accessory (XI)	Motor	Voluntary muscles of palate, pharynx, and larynx; sternocleidomastoid and trapezius muscles
Hypoglossal (XII)	Motor	Tongue muscles

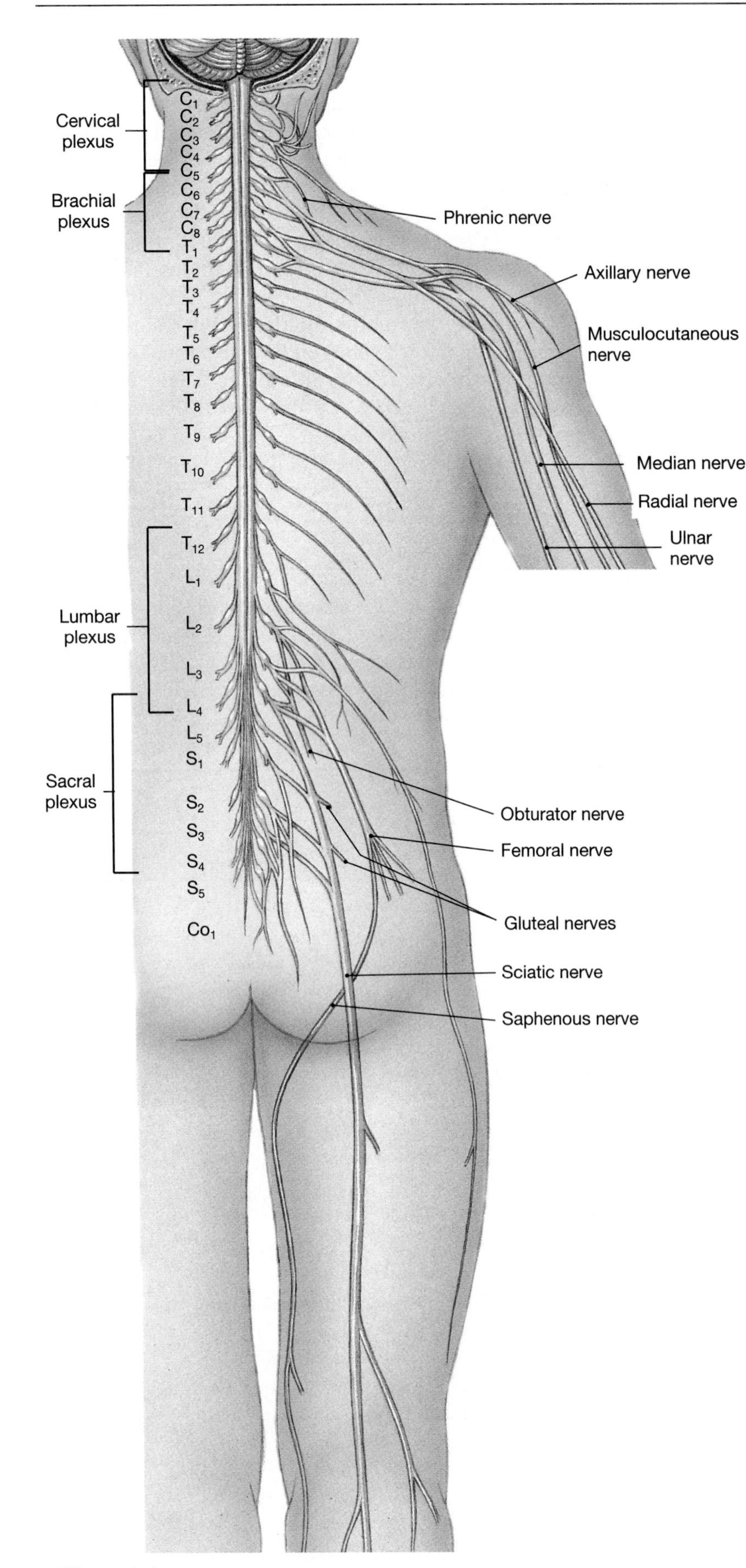

●**Figure 9-2**
Peripheral Nerves and Plexuses

Simple Reflexes

A **reflex arc**, introduced in Chapter 8, includes a receptor, a sensory neuron, a motor neuron, and an effector; interneurons may be found between the sensory neuron and the motor neuron. In the simplest reflex arc, a sensory neuron synapses directly on a motor neuron. ∞ *[p. 209]* Such a reflex is called a **monosynaptic reflex**. Because there is only one synapse, monosynaptic reflexes control the most rapid, stereotyped motor responses of the nervous system. The best known example is the *stretch reflex.*

The Stretch Reflex

The **stretch reflex** provides automatic regulation of skeletal muscle length. The sensory receptors in the stretch reflex are called **muscle spindles**, bundles of small, specialized skeletal muscle fibers scattered throughout the skeletal muscles. The stimulus (increasing muscle length) activates a sensory neuron that triggers an immediate motor response (contraction of the stretched muscle) that counteracts the stimulus.

Stretch reflexes are important in the maintenance of normal posture and balance and in making automatic adjustments in muscle tone. Physicians can use the sensitivity of the stretch reflex to test the general condition of the spinal cord, peripheral nerves, and muscles. For example, in the **knee jerk**, or *patellar reflex*, a sharp rap on the patellar ligament stretches muscle spindles in the quadriceps (thigh) muscles. With so brief a stimulus, the reflexive contraction occurs unopposed and produces a noticeable kick. If this contraction shortens the muscle spindles below their original resting lengths, the sensory nerve endings are compressed, the sensory neuron inhibited, and the leg drops back.

Complex Reflexes

Many spinal reflexes may have at least one interneuron placed between the sensory afferent and the motor efferent. Because there are more synapses, these **polysynaptic reflexes** have a longer delay between a stimulus and

TABLE 9-2 Nerve Plexuses

Plexus	*Major Nerve*	*Distribution*
Cervical Plexus (C_1–C_5)	Phrenic nerve	Diaphragm
	Other branches	Muscles of the neck; skin of upper chest, neck, and ears
Brachial Plexus (C_5–T_1)	Axillary nerve	Deltoid and teres minor muscles; skin of shoulder
	Musculocutaneous nerve	Flexor muscles of the arm and forearm; skin on lateral surface of forearm
	Median nerve	Flexor muscles of forearm and hand; skin over lateral surface of hand
	Radial nerve	Extensor muscles of the arm, forearm, and hand; skin on posterolateral surface of the arm
	Ulnar nerve	Flexor muscles of forearm and small digital muscles; skin of medial surface of hand
Lumbosacral Plexus Lumbar = T_{12}–L_4 Sacral = L_4–S_4	Obtutrator nerve	Adductors of thigh; skin over medial surface of thigh
	Femoral nerve	Adductors of thigh, extensors of leg; skin over medial surfaces of thigh and leg
	Gluteal nerve	Adductors and extensors of thigh
	Sciatic nerve	Flexors of leg and foot, flexors and extensors of toes
	Saphenous nerve	Skin over medial surface of leg

response. But they can produce far more complicated responses because the interneurons can control several different muscle groups simultaneously.

Withdrawal Reflexes

Withdrawal reflexes move affected parts of the body away from a source of stimulation. The strongest withdrawal reflexes are triggered by painful stimuli, but these reflexes are also initiated by the stimulation of touch or pressure receptors. The **flexor reflex** is a withdrawal reflex affecting the muscles of a limb. Stepping on a tack produces a dramatic flexor reflex in the affected limb (Figure 9-3●). When the pain receptors in the foot are stimulated, the sensory neurons activate interneurons in the spinal cord that stimulate motor neurons in the anterior gray horns. The result is a contraction of flexor muscles that yanks the foot off the ground.

When a specific muscle contracts, opposing muscles are stretched. For example, the flexor muscles that bend the knee are opposed by *extensor muscles* that straighten it out. A potential conflict exists here: Contraction of a flexor muscle should theoretically trigger a stretch reflex in the extensors that would

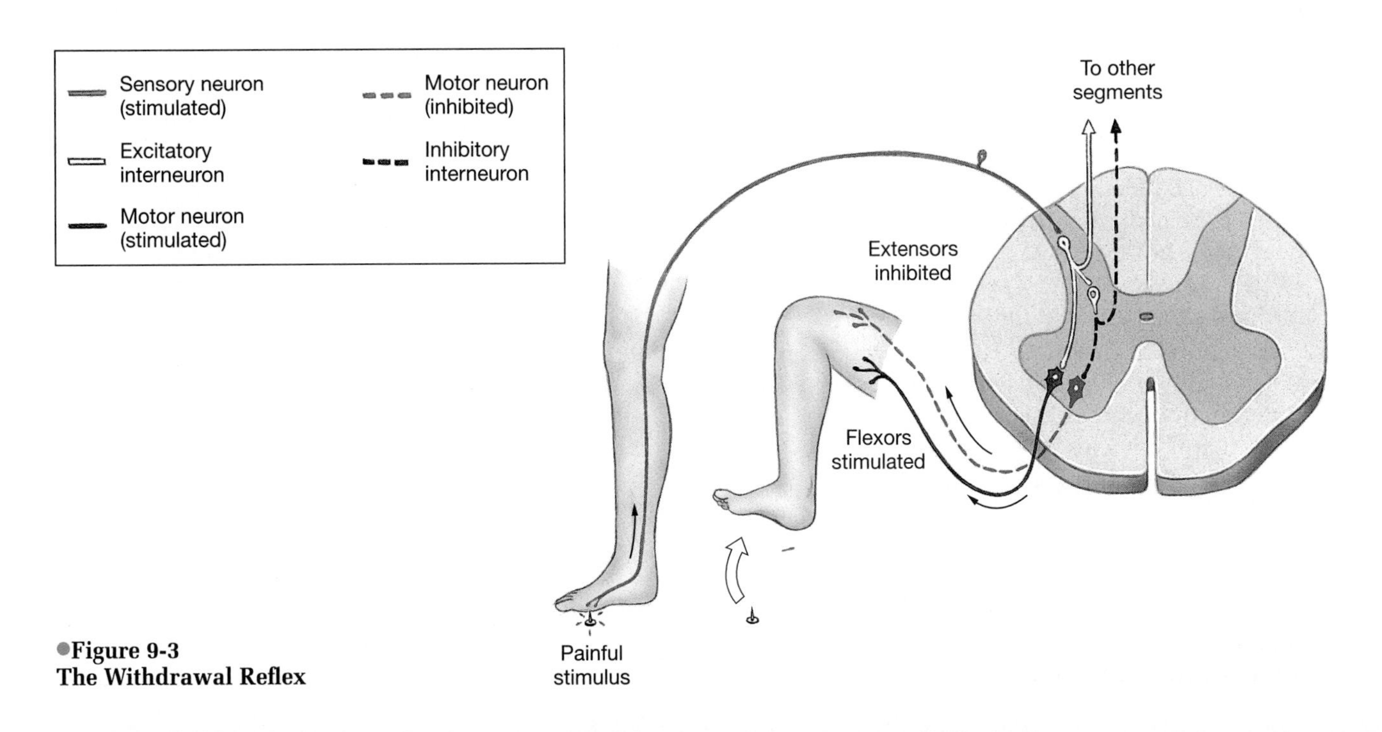

●**Figure 9-3**
The Withdrawal Reflex

cause them to contract, opposing the movement that is under way. Interneurons in the spinal cord prevent such competition through **reciprocal inhibition**. When one set of motor neurons is stimulated,those controlling antagonistic muscles are inhibited.

Integration and Control of Spinal Reflexes

Although reflexes are automatic, higher centers in the brain influence these responses by stimulating or inhibiting the interneurons and motor neurons involved. The sensitivity of a reflex can thus be modified. For example, an effort to pull apart clasped hands elevates the general state of stimulation along the spinal cord, leading to the exaggeration of spinal reflexes.

Other descending fibers have an inhibitory effect on spinal reflexes. Stroking an infant's foot on the side of the sole produces a fanning of the toes known as the **Babinski sign**, or *positive Babinski reflex*. This response disappears as descending inhibitory synapses develop, so that in the adult the same stimulus produces a curling of the toes, called a **plantar reflex**, or *negative Babinski reflex*, after about a 1-second delay. If either the higher centers or the descending tracts are damaged, the Babinski sign will reappear. As a result, this reflex is often tested if CNS injury is suspected.

When higher centers issue motor commands, they can activate the complex motor patterns already programmed into the spinal cord. By making use of preexisting patterns, a relatively small number of descending fibers can control complex motor functions. For example, the motor patterns for walking, running, and jumping are primarily directed by neuronal pools in the spinal cord. The descending pathways facilitate, inhibit, or fine-tune the established patterns.

Many reflexes can be assessed through careful observation and the use of simple tools. The procedures are easy to perform, and the results can provide valuable information about damage to the spinal cord or spinal nerves. By testing a series of spinal and cranial reflexes, a physician can assess the function of sensory pathways and motor centers throughout the spinal cord and brain. AM *Reflexes and Diagnostic Testing* and *Abnormal Reflex Activity*

✓ What common reflex is used by doctors to test the general condition of the spinal cord, peripheral nerves, and muscles?

✓ Polysynaptic reflexes can produce more complicated responses than monosynaptic reflexes. Why?

✓ After suffering an injury to his back, Tom exhibits a positive Babinski reflex. What does this reaction imply about Tom's injury?

SENSORY AND MOTOR PATHWAYS

As already mentioned, the communication between the CNS, the PNS (peripheral nervous system), and organs and systems occurs over pathways, nerve tracts, and nuclei that relay sensory and motor information. The major ascending (sensory) and descending (motor) tracts of the spinal cord are named with regard to the destinations of the axons. If the name of a tract begins with *spino-*, it must *start* in the spinal cord and *end* in the brain, and it therefore carries sensory information. If the name of a tract ends in *-spinal,* its axons must *start* in the higher centers and *end* in the spinal cord, bearing motor commands. The rest of the tract name indicates the associated nucleus or cortical area of the brain.

Table 9-3 lists examples of sensory and motor pathways and their functions.

Sensory Pathways

Sensory receptors monitor conditions in the body or the external environment. This information, called a **sensation**, arrives in the form of action potentials in an

TABLE 9-3 Sensory and Motor Pathways

Pathway	*Function*
Sensory (Ascending)	
Posterior column pathway	Delivers highly localized sensations of fine touch, pressure, vibration, and proprioception to the primary sensory cortex
Spinothalamic pathway	Delivers poorly localized sensations of touch, pressure, pain, and temperature to the primary sensory cortex
Spinocerebellar pathway	Delivers proprioceptive information concerning the positions of muscles, bones, and joints to the cerebellar cortex
Motor (Descending)	
Pyramidal system (voluntary)	Voluntary control of skeletal muscles throughout the body
Extrapyramidal system (involuntary)	Regulation of skeletal muscle tone, controls reflexive skeletal muscle responses to equilibrium sensations and to sudden or strong visual and auditory stimuli

afferent (sensory) fiber. Most of the processing of arriving sensations occurs in centers along the sensory pathways in the spinal cord or brain stem; only about 1 percent of the information provided by afferent fibers reaches the cerebral cortex and our conscious awareness. For example, we usually do not feel the clothes we wear or hear the hum of the engine in our car.

The Posterior Column Pathway

One example of an ascending, sensory pathway is the **posterior column pathway** (Figure 9-4●). It sends highly localized ("fine") touch, pressure, vibration, and proprioceptive (position) sensations along the dorsal roots of spinal nerves. The sensations ascend within this column to the medulla oblongata, where they cross over to the opposite side of the brain stem before continuing on to the thalamus. In the thalamus, sensations arriving over the posterior column pathway are sorted according to the region of the body involved and then are sent to the primary sensory cortex. The sensations arrive organized in such a way that sensory information from the toes arrives at one end of the primary sensory cortex and information from the head arrives at the other.

As a result, the sensory cortex contains a miniature map of the body surface. That map is distorted because the area of sensory cortex devoted to a particular region is proportional not to its size, but to the number of sensory receptors it contains. In other words, it takes many more cortical neurons to process sensory information arriving from the tongue, which has tens of thousands of taste and touch receptors, than it does to analyze sensations originating on the back, where touch receptors are few and far between.

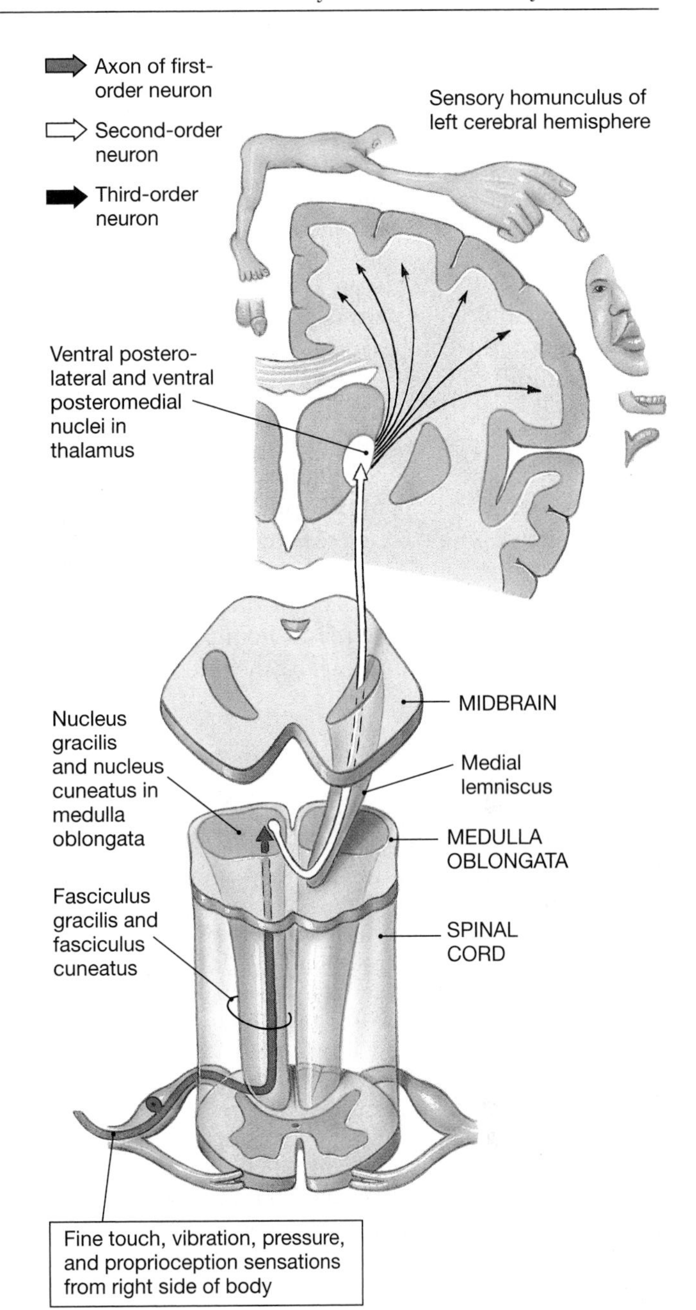

●**Figure 9-4**
The Posterior Column Pathway
The posterior column pathway delivers fine touch, vibration, and proprioception information to the primary sensory cortex of the cerebral hemisphere on the opposite side of the body. (For clarity, this figure shows only the pathway for sensations originating on the right side of the body.)

Motor Pathways

In response to information provided by sensory systems, the CNS issues motor commands that are distributed by the *somatic nervous system (SNS)* and the *autonomic nervous system (ANS)*. The SNS, under voluntary control, issues somatic motor commands that direct the contractions of skeletal muscles. The motor commands of the autonomic nervous system, which are issued outside our conscious awareness, controls the smooth and cardiac muscles, glands, and fat cells.

Two integrated motor pathways, the *pyramidal system* and the *extrapyramidal system*, provide control over skeletal muscles. Refer to Table 9-3 for examples and functions of these motor pathways. We will focus on the pyramidal system, which provides voluntary control.

The Pyramidal System

The *pyramidal system* provides voluntary control of skeletal muscles. Figure 9-5● shows the motor pathway providing voluntary control over the right side of the body. The neurons of the primary motor cortex are organized into a miniature map of the body. It is just as distorted as the sensory map on the primary sensory cortex; the proportions indicate the number of motor units present in that portion of the body. For example, the grossly oversized hands provide an indication of how many different muscles and motor units are involved in writing, grasping, and manipulating objects in our environment.

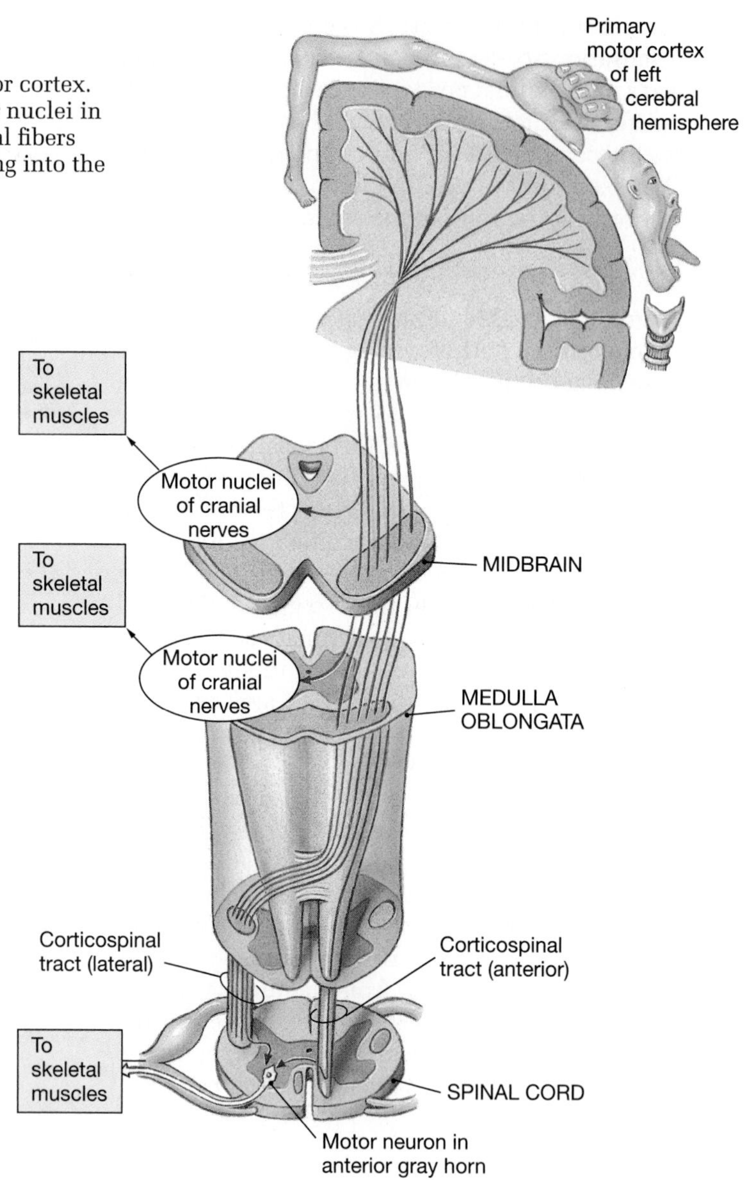

●**Figure 9-5**
The Pyramidal System

The pyramidal system originates at the primary motor cortex. Axons of the pyramidal cells descend to reach motor nuclei in the brain stem and spinal cord. Most of the pyramidal fibers cross over in the medulla oblongata before descending into the spinal cord as the corticospinal tracts.

The pyramidal system begins at the *pyramidal cells* of the cerebral cortex. The axons of the pyramidal cells extend into the brain stem and spinal cord to synapse on somatic motor neurons. For example, the *corticospinal tracts* (Figure 9-5●) synapse with motor neurons in the anterior gray horns of the spinal cord. All of the corticospinal tracts cross over to reach motor neurons on the opposite side of the body. As a result, the left side of the body is controlled by the right cerebral hemisphere, and the right side is controlled by the left cerebral hemisphere.

CEREBRAL PALSY AND HUNTINGTON'S DISEASE

The term *cerebral palsy* refers to a number of disorders affecting voluntary motor performance that appear during infancy or childhood and persist throughout the life of the affected individual. The cause may be trauma associated with premature or unusually stressful birth, maternal exposure to drugs, including alcohol, or a genetic defect that causes improper development of the motor pathways. Problems with labor and delivery result from compression or interruption of placental circulation or oxygen supplies. If the oxygen concentration in the fetal blood declines significantly for as little as 5–10 minutes, CNS function may be permanently impaired. The cerebral cortex, cerebellum, cerebral nuclei, hippocampus, and thalamus are likely targets, producing abnormalities in motor performance, involuntary control of posture and balance, memory, speech, and learning abilities.

Huntington's disease is an inherited condition characterized by the progressive degeneration of neurons in the frontal lobes and cerebal nuclei. The reason for this destruction is unknown. As the condition progresses, the individual has difficulties in performing voluntary and involuntary movements, and there is a gradual decline in intellectual abilities. AM *Huntington's Disease*

✓ As a result of pressure on her spinal cord, Jill cannot feel touch or pressure on her legs. What sensory pathway is being compressed?

✓ What is the anatomical reason for the left side of the brain controlling motor function on the right side of the body? *(cont.)*

✓ An injury to the superior portion of the motor cortex would affect what part of the body?

THE AUTONOMIC NERVOUS SYSTEM

Using the pathways already discussed, we can respond to sensory information and exert voluntary control over the activities of our skeletal muscles. Yet our conscious sensations, plans, and responses represent only a tiny fraction of the activities of the nervous system. In practical terms, conscious activities have little to do with our immediate or long-term survival, and the adjustments made by the autonomic nervous system are much more important. Without the ANS, a simple night's sleep would be a life-threatening event.

Chapter 8 introduced the two efferent divisions of the nervous system, the *somatic nervous system (SNS)* and the *autonomic nervous system*. The SNS controls skeletal muscles over the pyramidal and extrapyramidal motor pathways. The ANS is concerned with the involuntary regulation of smooth muscle, cardiac muscle, and glands. There are clear anatomical differences between the SNS and ANS (Figure 9-6●). In the ANS, there is always a synapse between the central nervous system and the peripheral effector. The motor neurons in the CNS, known as **preganglionic neurons**, send their axons, called *preganglionic fibers*, to *autonomic ganglia* outside the CNS. In these ganglia, the axons of the preganglionic neurons synapse with **ganglionic neurons**. The axons of these neurons leave the ganglia, and these *postganglionic fibers* innervate cardiac muscle, smooth muscles, glands, and adipose tissues.

The ANS consists of two divisions. Preganglionic fibers from the thoracic and lumbar spinal segments synapse in ganglia near the spinal cord. These axons and ganglia are part of the **sympathetic division** of the ANS. The sympathetic division is often called the "fight or flight" system because it usually stimulates tissue metabolism, increases alertness, and generally prepares the body to deal with emergencies.

Preganglionic fibers originating in the brain and the sacral spinal segments synapse on neurons of **intramural ganglia** (*murus*, wall), located inside the tissues of visceral organs. These components are part of the **parasympathetic division** of the ANS, often regarded as the "rest and repose" system because it conserves energy and promotes sedentary activities, such as digestion.

The sympathetic and parasympathetic divisions affect target organs through the controlled release of specific neurotransmitters by the postganglionic fibers. Whether the result is a stimulation or inhibition of activity depends on the response of the membrane

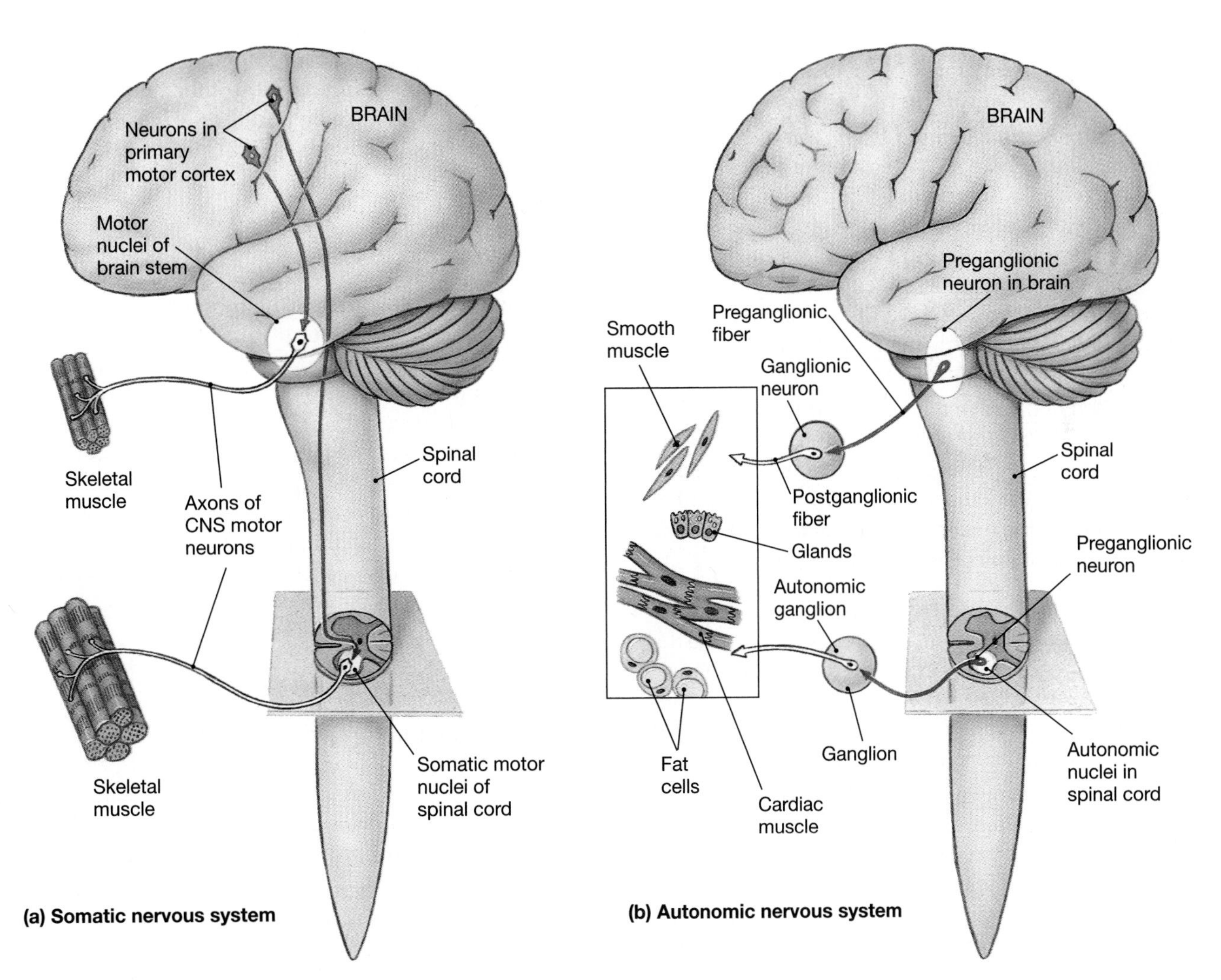

●**Figure 9-6**
The Somatic and Autonomic Nervous Systems
(a) In the SNS, a motor neuron in the central nervous system has direct control over skeletal muscle fibers. **(b)** In the ANS, the preganglionic neurons synapse with ganglionic neurons that innervate effectors, such as smooth muscle, cardiac muscle, glands, and fat cells.

receptor to the presence of the neurotransmitter. Some general patterns are worth noting:

- All preganglionic autonomic fibers are *cholinergic*: They release acetylcholine (ACh) at their synaptic terminals. The effects are always excitatory.
- Postganglionic parasympathetic fibers are also cholinergic, but the effects may be excitatory or inhibitory, depending on the nature of the receptor of the target cell.
- Most postganglionic sympathetic terminals are *adrenergic*: They release norepinephrine (NE). The effects are usually excitatory.

The Sympathetic Division

The sympathetic division, diagrammed in Figure 9-7●, consists of the following:

- *Preganglionic neurons located between segments T_1 and L_2 of the spinal cord.* These neurons are situated in the lateral gray horns, and their short axons enter the ventral roots of these segments.
- *Ganglionic neurons located in ganglia near the vertebral column.* There are two different types of sympathetic ganglia. *Sympathetic chain ganglia* lie on either side of the vertebral column. These ganglia contain neurons that control effectors in the body wall and inside the thoracic cavity. Unpaired *collateral ganglia* lie anterior to the vertebral column. These contain ganglionic neurons that innervate tissues and organs in the abdominopelvic cavity.
- *Specialized neurons in a modified ganglion in the interior of each adrenal gland,* a region known as the *adrenal medulla.* The ganglionic neurons of the medullae have very short axons, and when stimulated they release the neurotransmitters norepinephrine and epinephrine into the general circulation.

The sympathetic division shows extensive divergence, and a single sympathetic motor neuron inside the CNS can produce a complex and coordinated response.

The Sympathetic Chain

From spinal segments T_1 to L_2, sympathetic preganglionic fibers join the ventral root of each spinal nerve. All of these fibers then exit the spinal nerve to enter the sympathetic chain ganglia. For motor commands to the body wall, a synapse occurs at the chain ganglia, and then the postganglionic fibers return to the spinal nerve for distribution. For the thoracic cavity, a synapse also occurs at the chain ganglia, but the postganglionic fibers then form *autonomic nerves* that go directly to their targets (Figure 9-7●).

The sympathetic innervation provided by the spinal nerves stimulates sweat gland activity and arrector pili muscles (producing "goosebumps"), reduces circulation to the skin and body wall, accelerates blood flow to skeletal muscles, releases stored lipids from adipose tissue, and dilates the pupils. The activation of the autonomic nerves accelerates the heart rate, increases the force of cardiac contractions, and dilates the respiratory passageways. All of these changes prepare the individual for sudden, intense physical activity.

The Collateral Ganglia

The abdominopelvic viscera receive sympathetic innervation over preganglionic fibers from lower thoracic and upper lumbar segments that pass through the sympathetic chain without synapsing, and instead synapse within separate *collateral ganglia.* The nerves traveling to the collateral ganglia are known as *splanchnic nerves.* The three collateral ganglia and the organs their postganglionic fibers innervate are diagrammed in Figure 9-7●.

The fibers leaving the collateral ganglia extend throughout the abdominopelvic cavity. In general, they reduce the blood flow and energy use by visceral organs that are not important to short-term survival, such as the digestive tract, and they stimulate the release of stored carbohydrate and lipid reserves.

The Adrenal Medullae

Preganglionic fibers entering each adrenal gland proceed to its center, to the region called the **adrenal medulla**. There they synapse on modified neurons that perform an endocrine function. When stimulated, these cells release the neurotransmitters norepinephrine (NE) and epinephrine (E) into surrounding capillaries, which carry them throughout the body. In general, their effects resemble those produced by the stimulation of sympathetic postganglionic fibers. However, (1) they also affect cells not innervated by sympathetic postganglionic fibers, and (2) their effects last much longer than those produced by direct sympathetic innervation.

The Parasympathetic Division

The parasympathetic division of the ANS includes the following:

- *Preganglionic neurons in the brain stem (midbrain, pons, and medulla oblongata) and in the lateral gray horns of sacral segments S_2 to S_4.*
- *Ganglionic neurons in peripheral ganglia located within or adjacent to the target organs.* The preganglionic fibers of the parasympathetic division do not diverge as extensively as do those of the sympathetic division. As a result, the effects of parasympathetic stimulation are more specific and localized than those of the sympathetic division.

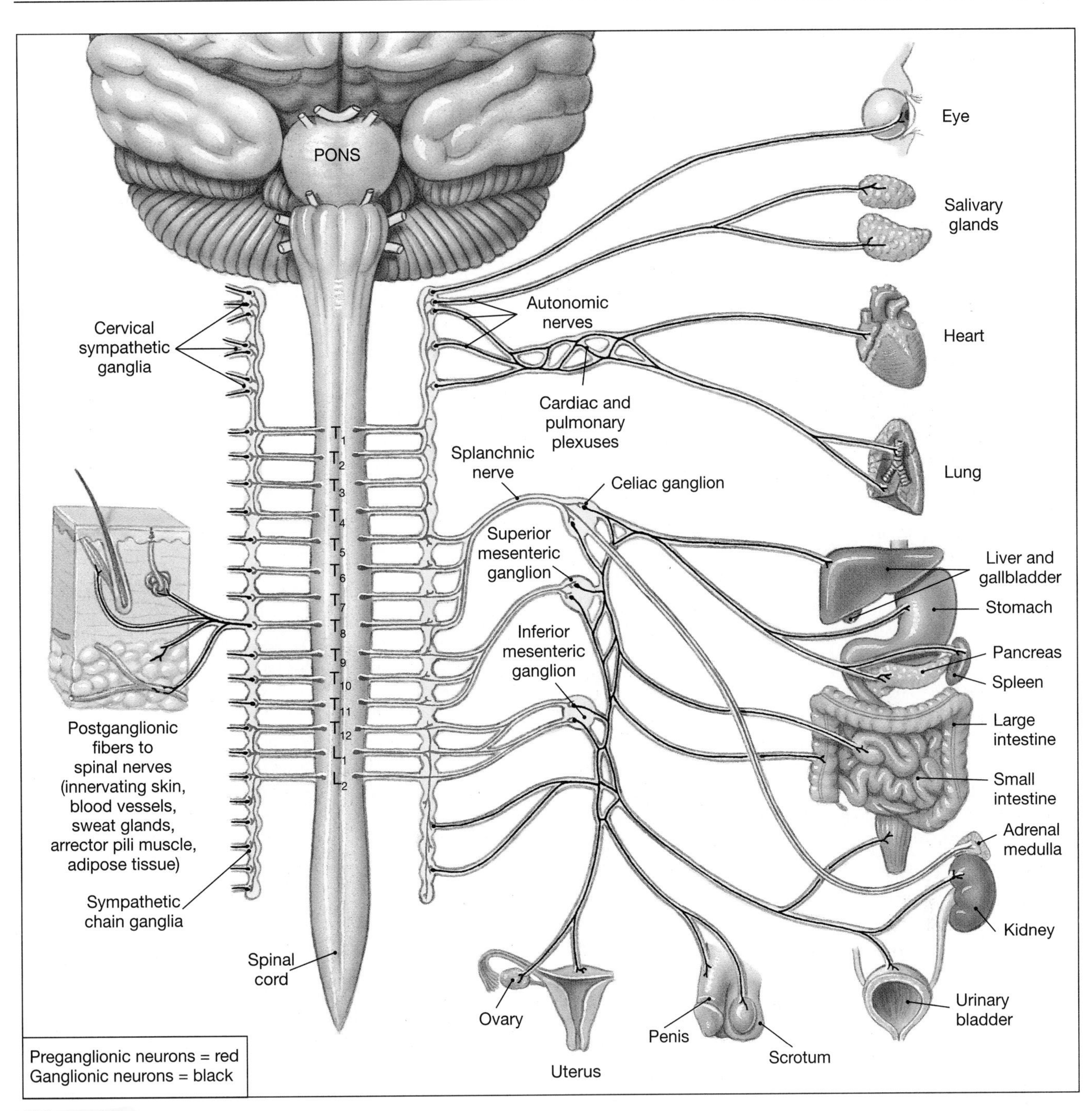

Figure 9-7
The Sympathetic Division
The distribution of sympathetic fibers is the same on both sides of the body. For clarity, the innervation of somatic structures is shown to the left and the innervation of visceral structures to the right.

Organization

Figure 9-8 diagrams the pattern of parasympathetic innervation. Preganglionic fibers leaving the brain travel within cranial nerves III (oculomotor), VII (facial), IX (glossopharyngeal), and X (vagus). These fibers synapse in ganglia located in peripheral tissues, and short postganglionic fibers then continue to their targets. The vagus nerves provide preganglionic parasympathetic innervation to ganglia within organs of the thoracic and abdominopelvic cavity as distant as the last segments of the large intestine. The vagus nerves provide roughly 75 percent of all parasympathetic outflow and innervate most of the organs in the thoracic and abdominopelvic cavities.

The sacral parasympathetic outflow leaves the sacral segments of the spinal cord, and the preganglionic fibers form distinct **pelvic nerves** that innervate intramural ganglia in the kidney and bladder, the last segments of the large intestine, and the sex organs.

General Functions

Among other things, the parasympathetic division constricts the pupils, increases secretions by the digestive glands, increases smooth muscle activity of the digestive tract, stimulates defecation and urination, constricts respiratory passageways, and reduces heart rate and contraction force.

These functions center on relaxation, food processing, and energy absorption. Stimulation of the parasympathetic division leads to a general increase in the nutrient content of the blood. Cells throughout the body respond to this increase by absorbing nutrients and using them to support growth and the storage of energy reserves. The effects of parasympathetic stimulation are usually brief and are restricted to specific organs and sites.

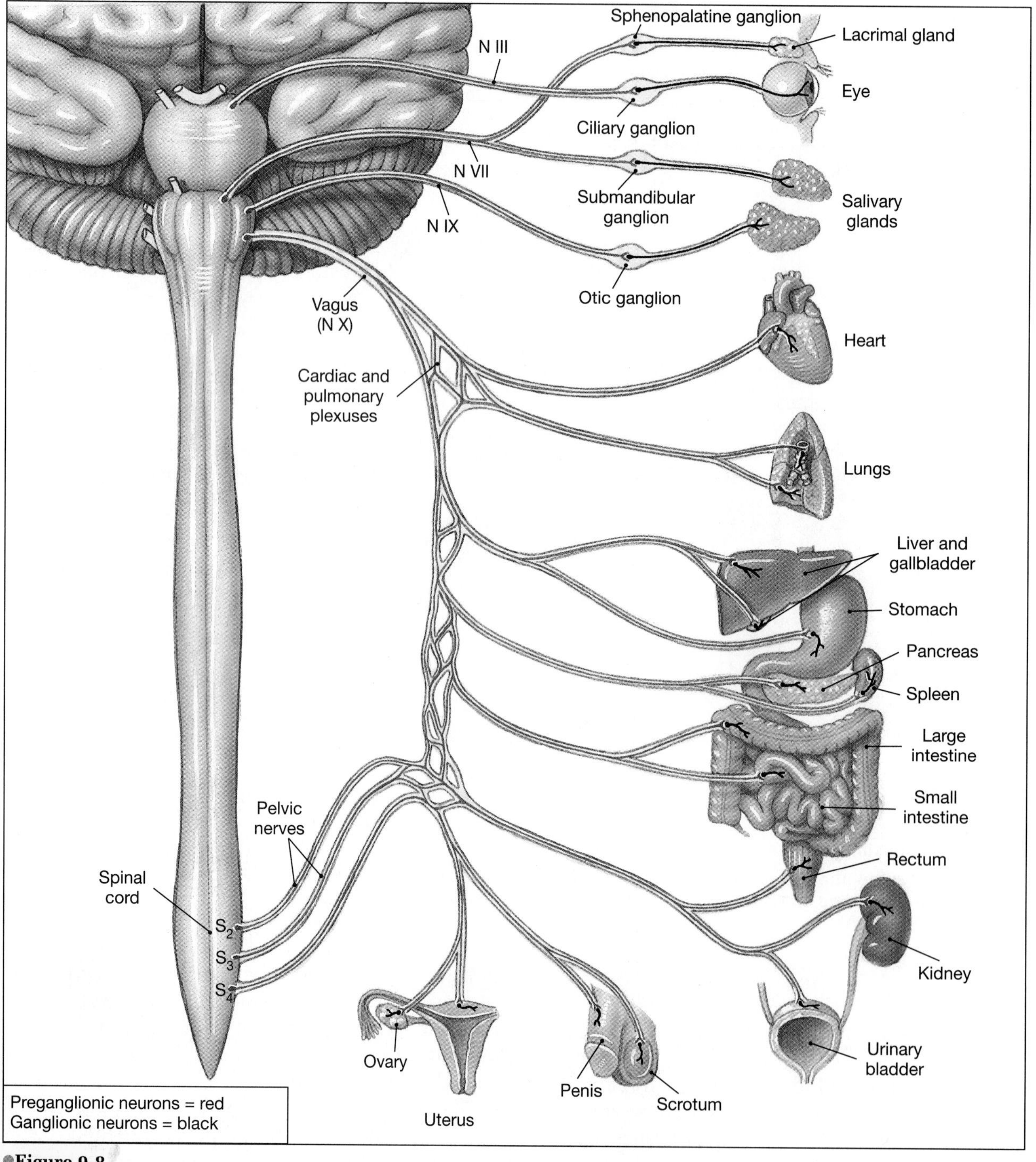

Figure 9-8
The Parasympathetic Division
The distribution of parasympathetic fibers is the same on both sides of the body.

Relationships between the Sympathetic and Parasympathetic Divisions

The sympathetic division has widespread impact, reaching visceral and somatic structures throughout the body, whereas the parasympathetic division innervates only visceral structures serviced by the cranial nerves or lying within the abdominopelvic cavity. Although some organs are innervated by one division or the other, most vital organs receive **dual innervation**—that is, instructions from both autonomic divisions. Where dual innervation exists, the two divisions often have opposing effects. Table 9-4 provides examples of different organs and the effects of either single or dual innervation.

✓ While out for a brisk walk, Jill is suddenly confronted by an angry dog. Which division of the autonomic nervous system is responsible for the physiological changes that occur as Jill turns and runs from the angry animal? *(cont.)*

✓ Why is the parasympathetic division of the ANS sometimes referred to as the anabolic system?

✓ What effect would loss of sympathetic stimulation have on blood flow to a tissue?

✓ What physiological changes would you expect to observe in a patient who is about to have a root canal procedure done and who is quite anxious about the procedure?

AGING AND THE NERVOUS SYSTEM

The aging process affects all bodily systems, and the nervous system is no exception. Anatomical and physiological changes begin shortly after maturity (probably by age 30) and accumulate over time. Although an estimated 85 percent of the elderly (above age 65) lead relatively normal lives, there are noticeable changes in mental performance and CNS functioning.

TABLE 9-4 The Effects of the Sympathetic and Parasympathetic Divisions of the ANS on Various Organs

Structure	*Sympathetic Innervation Effect*	*Parasympathetic Innervation Effect*
EYE	Dilation of pupil Focusing for distance vision	Constriction of pupil Focusing for near vision
SKIN		
Sweat glands	Increases secretion	None (not innervated)
Arrector pili muscles	Contraction, erection of hairs	None (not innervated)
TEAR GLANDS	None (not innervated)	Secretion
CARDIOVASCULAR SYSTEM		
Blood vessels	Vasoconstriction and vasodilation	None (not innervated)
Heart	Increases heart rate, force of contraction, and blood pressure	Decreases heart rate, force of contraction, and blood pressure
ADRENAL GLANDS	Secretion of epinephrine and norepinephrine by adrenal medullae	None (not innervated)
RESPIRATORY SYSTEM		
Airways	Increases diameter	Decreases diameter
Respiratory rate	Increases rate	Decreases rate
DIGESTIVE SYSTEM		
General level of activity	Decreases activity	Increases activity
Liver	Glycogen breakdown, glucose synthesis and release	Glycogen synthesis
SKELETAL MUSCLES	Increases force of contraction, glycogen breakdown	None (not innervated)
URINARY SYSTEM		
Kidneys	Decreases urine production	Increases urine production
Bladder	Constricts sphincter, relaxes urinary bladder	Tenses urinary bladder, relaxes sphincter to eliminate urine

Age-Related Changes

Age-related anatomical changes in the nervous system that are commonly seen include:

- *A reduction in brain size and weight, primarily from a decrease in the volume of the cerebral cortex.* The brains of elderly individuals have narrower gyri and wider sulci than those of young persons, and the subarachnoid space is larger.
- *A reduction in the number of neurons.* Brain shrinkage has been linked to a loss of cortical neurons, although evidence exists that neuronal loss does not occur (at least to the same degree) in brain stem nuclei.
- *A decrease in blood flow to the brain.* With age, fatty deposits gradually accumulate in the walls of blood vessels. Like a kink in a garden hose or a clog in a drain, these deposits reduce the rate of blood flow through arteries. (This process, called *arteriosclerosis*, affects arteries throughout the body; it is discussed further in Chapter 14.) The reduction in blood flow does not cause a cerebral crisis, but it does increase the chances that the individual will suffer a stroke.
- *Changes in synaptic organization of the brain.* The number of dendritic branchings and interconnections appears to decrease. Synaptic connections are lost, and the rate of neurotransmitter production declines.
- *Intracellular and extracellular changes in CNS neurons.* Many neurons in the brain begin accumulating abnormal intracellular deposits, such as pigments or abnormal proteins that have no known function. The significance of these abnormalities remains to be determined. There is evidence that these changes occur in all aging brains (see below), but when present in excess they seem to be associated with clinical abnormalities.

As a result of these anatomical changes, neural function is impaired. For example, memory consolidation, the conversion of short-term memory to long-term memory (such as repeating a telephone number), often becomes more difficult. Other memories, especially those of the recent past, also become harder to access. The sensory systems, notably hearing, balance, vision, smell, and taste, become less acute. Light must be brighter, sounds louder, and smells stronger before they are perceived. Reaction times are slowed, and reflexes—even some withdrawal reflexes—become weaker or even disappear. There is a decrease in the precision of motor control, and it takes longer to perform a given motor pattern than it did 20 years earlier.

For roughly 85 percent of the elderly, these changes do not interfere with their abilities to function in society. But for as yet unknown reasons, many become incapacitated by progressive CNS changes.

ALZHEIMER'S DISEASE

By far the most common such incapacitating condition is **Alzheimer's disease,** a progressive disorder characterized by the loss of higher cerebral functions. This is the most common cause of **senile dementia**, or "senility." The first symptoms usually appear at 50 to 60 years of age, although the disease occasionally affects younger individuals. Alzheimer's disease has widespread impact on the elderly; an estimated 2 million people in the United States, including roughly 15 percent of those over 65, suffer from some form of the condition, and it causes approximately 100,000 deaths each year.

In its characteristic form, Alzheimer's disease produces a gradual deterioration of mental organization. The afflicted individual loses memories, verbal and reading skills, and emotional control. As memory losses continue to accumulate, problems become more severe. The affected person may forget relatives, a home address, or how to use the telephone. The loss of memory affects both intellectual and motor abilities, and a patient with severe Alzheimer's disease has difficulty in performing even the simplest motor tasks. Although by this time victims are relatively unconcerned about their mental state or motor abilities, the condition can have devastating emotional effects on the immediate family. AM *Alzheimer's Disease*

INTERACTIONS WITH OTHER SYSTEMS

The relationships between the nervous system and other organ systems is diagrammed in Figure 9-9●. Many of these interactions will be explored in greater detail in later chapters.

Chapter Review

KEY TERMS

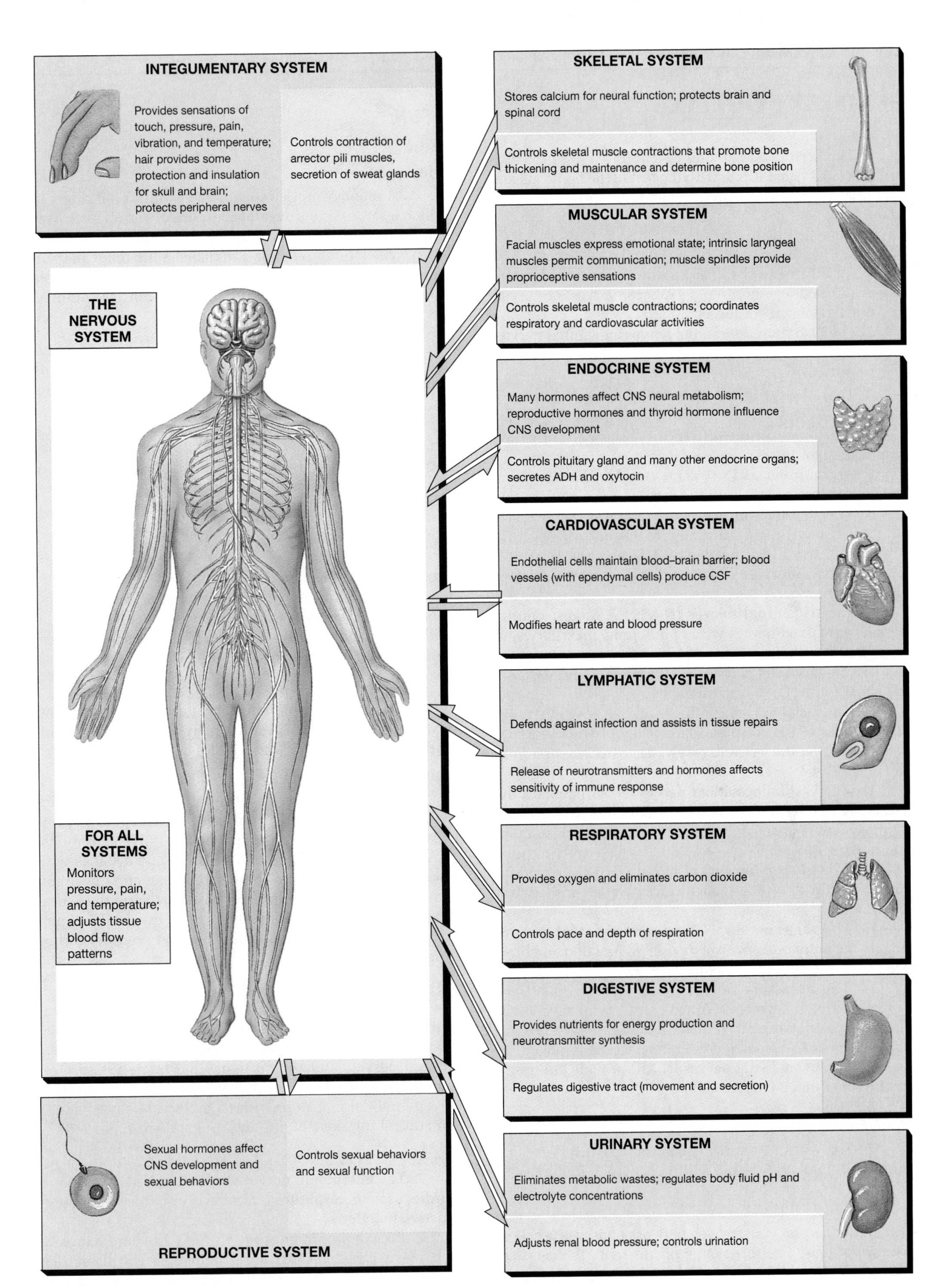

INTEGUMENTARY SYSTEM
Provides sensations of touch, pressure, pain, vibration, and temperature; hair provides some protection and insulation for skull and brain; protects peripheral nerves
Controls contraction of arrector pili muscles, secretion of sweat glands
SKELETAL SYSTEM
Stores calcium for neural function; protects brain and spinal cord
Controls skeletal muscle contractions that promote bone thickening and maintenance and determine bone position
MUSCULAR SYSTEM
Facial muscles express emotional state; intrinsic laryngeal muscles permit communication; muscle spindles provide proprioceptive sensations
Controls skeletal muscle contractions; coordinates respiratory and cardiovascular activities
THE NERVOUS SYSTEM
ENDOCRINE SYSTEM
Many hormones affect CNS neural metabolism; reproductive hormones and thyroid hormone influence CNS development
Controls pituitary gland and many other endocrine organs; secretes ADH and oxytocin
CARDIOVASCULAR SYSTEM
Endothelial cells maintain blood–brain barrier; blood vessels (with ependymal cells) produce CSF
Modifies heart rate and blood pressure
LYMPHATIC SYSTEM
Defends against infection and assists in tissue repairs
Release of neurotransmitters and hormones affects sensitivity of immune response
FOR ALL SYSTEMS
Monitors pressure, pain, and temperature; adjusts tissue blood flow patterns
RESPIRATORY SYSTEM
Provides oxygen and eliminates carbon dioxide
Controls pace and depth of respiration
DIGESTIVE SYSTEM
Provides nutrients for energy production and neurotransmitter synthesis
Regulates digestive tract (movement and secretion)
Sexual hormones affect CNS development and sexual behaviors
Controls sexual behaviors and sexual function
REPRODUCTIVE SYSTEM
URINARY SYSTEM
Eliminates metabolic wastes; regulates body fluid pH and electrolyte concentrations
Adjusts renal blood pressure; controls urination

●**Figure 9-9**
Functional Relationships between the Nervous System and other Systems

SUMMARY OUTLINE

INTRODUCTION p. 230

1. The peripheral nervous system (PNS) links the central nervous system (CNS) with the rest of the body; all sensory information and motor commands are carried by axons of the PNS.

THE PERIPHERAL NERVOUS SYSTEM p. 230

1. Structures of the PNS include sensory and motor axons bundled together into **peripheral nerves**, or nerves, and clusters of cell bodies, or **ganglia.**

2. The PNS includes cranial nerves and spinal nerves.

The Cranial Nerves p. 230

3. There are 12 pairs of cranial nerves. (*Figure 9-1; Table 9-1*)

4. The **olfactory nerves** (N I) carry sensory information responsible for the sense of smell.

5. The **optic nerves** (N II) carry visual information from special sensory receptors in the eyes.

6. The **oculomotor nerves** (N III) are the primary sources of innervation for the muscles that move the eyeball.

7. The **trochlear nerves** (N IV), the smallest cranial nerves, innervate the superior oblique muscles of the eyes.

8. The **trigeminal nerves** (N V), the largest cranial nerves, are mixed nerves with ophthalmic, maxillary, and mandibular branches.

9. The **abducens nerves** (N VI) innervate the sixth extrinsic eye muscle, the lateral rectus.

10. The **facial nerves** (N VII) are mixed nerves that control muscles of the scalp and face. They provide pressure sensations over the face and receive taste information from the tongue.

11. The **vestibulocochlear nerves** (N VIII) contain the vestibular nerves, which monitor sensations of balance, position, and movement, and the cochlear nerves, which monitor hearing receptors.

12. The **glossopharyngeal nerves** (N IX) are mixed nerves that innervate the tongue and pharynx and control swallowing.

13. The **vagus nerves** (N X) are mixed nerves that are vital to the autonomic control of visceral function and have a variety of motor components.

14. The **accessory nerves** (N XI) have a medullary branch, which innervates voluntary swallowing muscles of the soft palate and pharynx, and a spinal branch, which controls muscles associated with the pectoral girdle.

15. The **hypoglossal nerves** (N XII) provide voluntary control over tongue movements.

The Spinal Nerves p. 232

16. There are 31 pairs of spinal nerves: 8 cervical, 12 thoracic, 5 lumbar, 5 sacral, and, 1 coccygeal. (*Figure 9-2*)

Nerve Plexuses p. 233

17. A complex, interwoven network of nerves is called a **nerve plexus**. The three large plexuses are the **cervical plexus**, the **brachial plexus**, and the **lumbosacral plexus**. The latter can be divided into the *lumbar plexus* and the *sacral plexus*. (*Figure 9-2; Table 9-2*)

THE CNS AND PNS: INTEGRATED FUNCTIONS p. 233

Simple Reflexes p. 234

1. A **monosynaptic reflex** is the simplest reflex arc, in which a sensory neuron synapses directly on a motor neuron that acts as the processing center.

2. The **stretch reflex** is a monosynaptic reflex that automatically regulates skeletal muscle length and muscle tone. The sensory receptors involved are **muscle spindles.**

Complex Reflexes p. 234

3. **Polysynaptic reflexes**, which have at least one interneuron placed between the sensory afferent and the motor efferent, have a longer delay between stimulus and response than a monosynaptic synapse.

4. Polysynaptic reflexes can also produce more complicated responses. Examples include the **withdrawal reflexes**, which move affected portions of the body away from a source of stimulation; the **flexor reflex** is a withdrawal reflex affecting the muscles of a limb. (*Figure 9-3*)

Integration and Control of Spinal Reflexes p. 236

5. The brain can facilitate or inhibit reflex motor patterns based in the spinal cord.

6. Motor control involves a series of interacting levels; monosynaptic reflexes form the lowest level, while at the highest level are the centers in the brain that can enhance, inhibit, or build upon programmed reflexes.

SENSORY AND MOTOR PATHWAYS p. 236

1. The essential communication between the CNS and PNS occurs over pathways that relay sensory information and motor commands. (*Table 9-3*)

Sensory Pathways p. 236

2. A **sensation** arrives in the form of an action potential in an afferent fiber. The **posterior column pathway** carries fine touch, pressure, and proprioceptive sensations. The axons ascend within this pathway and synapse with neurons in the medulla. These axons then cross over and travel on to the thalamus. The thalamus sorts the sensations according to the region of the body involved and projects them to specific regions of the primary sensory cortex. (*Figure 9-4; Table 9-3*)

Motor Pathways p. 237

3. The neurons of the primary motor cortex are **pyramidal cells**; the **pyramidal system** provides voluntary skeletal muscle control. The corticobulbar tracts terminate at the cranial nerves, and the corticospinal tracts synapse on motor neurons in the anterior gray horns of the spinal cord. The pyramidal system provides a rapid, direct mechanism for controlling skeletal muscles. (*Figure 9-5; Table 9-3*)

THE AUTONOMIC NERVOUS SYSTEM p. 238

1. The autonomic nervous system (ANS) coordinates cardiovascular, respiratory, digestive, excretory, and reproductive functions.

2. **Preganglionic neurons** in the CNS send axons to synapse on **ganglionic neurons** in **autonomic ganglia** outside the CNS. The axons of the ganglionic neurons (postganglionic fibers) innervate cardiac muscle, smooth muscles, glands, and adipose tissues. (*Figure 9-6*)

3. Preganglionic fibers from the thoracic and lumbar segments form the **sympathetic division** ("fight or flight" system) of the ANS. Preganglionic fibers leaving the brain and sacral segments form the **parasympathetic division** ("rest and repose" system).

The Sympathetic Division p. 240

4. The sympathetic division consists of preganglionic neurons between segments T_1 and L_2, ganglionic neurons in ganglia near the vertebral column, and specialized neurons inside the adrenal gland. There are two types of sympathetic ganglia: paired **sympathetic chain ganglia** and unpaired **collateral ganglia**. (*Figure 9-7*)

5. Some postganglionic fibers enter each spinal nerve to provide innervation to the body wall and limbs. Postganglionic fibers targeting thoracic cavity structures form autonomic nerves that go directly to their visceral destination. Preganglionic fibers running between the sympathetic chain ganglia interconnect them to form an elongated sympathetic chain. (*Figure 9-7*)

6. The abdominopelvic viscera receive sympathetic innervation via preganglionic fibers that synapse within collateral ganglia. The preganglionic fibers that innervate the three collateral ganglia form the *splanchnic nerves*. (*Figure 9-7*)

7. Preganglionic fibers entering the adrenal glands synapse within the **adrenal medullae**. During sympathetic activation these endocrine organs secrete epinephrine and norepinephrine into the bloodstream.

8. In a crisis, the entire division responds, producing increased alertness, a feeling of energy and euphoria, increased cardiovascular and respiratory activity, and general elevation in muscle tone.

The Parasympathetic Division p. 240

9. The parasympathetic division includes preganglionic neurons in the brain stem and sacral segments of the spinal cord and ganglionic neurons in peripheral ganglia located within or next to target organs. (*Figure 9-8*)

10. Preganglionic fibers leaving the sacral segments form **pelvic nerves**. (*Figure 9-8*)

11. The effects produced by the parasympathetic division center on relaxation, food processing, and energy absorption.

12. The effects of stimulation are usually brief and restricted to specific sites.

Relationships between the Sympathetic and Parasympathetic Divisions p. 243

13. The sympathetic division has widespread impact, reaching visceral and somatic structures throughout the body.

14. The parasympathetic division innervates only visceral structures serviced by cranial nerves or lying within the abdominopelvic cavity. Organs with **dual innervation** receive instructions from both divisions. (*Table 9-4*)

AGING AND THE NERVOUS SYSTEM p. 243

Age-Related Changes p. 244

1. Age-related changes in the nervous system include (1) reduction in brain size and weight, (2) reduction in number of neurons, (3) decreased blood flow to the brain, (4) changes in synaptic organization of the brain, and (5) intracellular and extracellular changes in CNS neurons.

INTERACTIONS WITH OTHER SYSTEMS p. 244

1. The nervous system monitors pressure, pain, and temperature and adjusts tissue blood flow for all systems. (*Figure 9-9*)

CHAPTER QUESTIONS

LEVEL 1 Reviewing Facts and Terms

Match each item in column A with the most closely related item in column B. Use letters for answers in the spaces provided.

Column A

___ 1. olfactory nerve
___ 2. optic nerve
___ 3. vestibulocochlear nerve
___ 4. hypoglossal nerve
___ 5. sympathetic division
___ 6. parasympathetic division
___ 7. somatic nervous system
___ 8. autonomic nervous system
___ 9. monosynaptic reflex
___ 10. polysynaptic reflex
___ 11. negative Babinski reflex
___ 12. dual innervation
___ 13. Babinski sign

Column B

a. "fight or flight"
b. controls smooth and cardiac muscle and glands
c. sensory, vision
d. stretch reflex
e. opposing effects
f. motor, tongue movements
g. equilibrium, hearing
h. controls contractions of skeletal muscles
i. sensory, smell
j. "rest and repose"
k. plantar reflex
l. flexor reflex
m. fanning of toes on infant's foot

14. Spinal nerves are called mixed nerves because:
(a) they contain sensory and motor fibers
(b) they exit at intervertebral foramina
(c) they are associated with a pair of dorsal root ganglia
(d) they are associated with a pair of dorsal and ventral roots

15. If a tract name begins with *spino-*, it:
(a) starts in the brain and ends in the spinal cord
(b) carries motor commands from the brain to the spinal cord
(c) carries sensory informatin from the brain to the spinal cord
(d) starts in the spinal cord and ends in the brain

16. The contraction of flexor muscles and the relaxation of extensor muscles illustrates the principle of:
    (a) reverberating circuitry (b) generalized facilitation
    (c) reciprocal inhibition (d) reinforcement
17. There is always a synapse between the CNS and the peripheral effector in the:
    (a) autonomic nervous system
    (b) somatic nervous system
    (c) reflex arc
    (d) a, b, and c are correct
18. All preganglionic autonomic fibers release ________ at their synaptic terminals, and the effects are always ________.
    (a) norepinephrine; inhibitory
    (b) norepinephrine; excitatory
    (c) acetylcholine; excitatory
    (d) acetylcholine; inhibitory
19. Approximately 75 percent of parasympathetic outflow is provided by the:
    (a) vagus nerve
    (b) sciatic nerve
    (c) glossopharyngeal nerves
    (d) pelvic nerves
20. Using the mnemonic device "Oh, Once One Takes The Anatomy Final, Very Good Vacations Are Heavenly," list the 12 pairs of cranial nerves and their functions.
21. What are pyramidal cells, and what is their function?
22. How does the emergence of sympathetic fibers from the spinal cord differ from the emergence of parasympathetic fibers?

LEVEL 2 Reviewing Concepts

23. Dual innervation refers to situations in which:
    (a) vital organs receive instructions from sympathetic and parasympathetic fibers
    (b) the atria and ventricles of the heart receive autonomic stimulation from the same nerves
    (c) sympathetic and parasympathetic fibers have similar effects
    (d) a, b, and c are correct
24. Why is response time in a monosynaptic reflex much faster than response time in a polysynaptic reflex?
25. Compare the general effects of the sympathetic and parasympathetic divisions of the ANS.
26. Why is the adrenal medulla considered to be a modified sympathetic ganglion?
27. You are alone in your home late at night when you hear what sounds like breaking glass. What physiological effects would this experience probably produce, and what would be their cause?

LEVEL 3 Critical Thinking and Clinical Applications

28. In some very severe cases of stomach ulcers, the branches of the vagus nerve (N X) that lead to the stomach are surgically severed. How would this procedure help control the ulcers?
29. Improper use of crutches can produce a condition known as crutch paralysis, which is characterized by a lack of response by the extensor muscles of the arm and a condition known as wrist drop. Which nerve is involved?
30. While playing football, Ramon is tackled hard and suffers an injury to his left leg. As he tries to get up, he finds that he cannot flex his left thigh or extend the leg. What nerve is damaged, and how would this damage affect sensory perception in the left leg?

10 Sensory Function

Our comprehension of the world around us is based on information provided by our senses. For example, when we can see something, it must be there because seeing is believing…or is it? The more you study this painting the more confusing it becomes. This chapter examines the way receptors provide sensory information, as well as the sensory pathways that distribute this information to provide us with our senses of smell, taste, vision, equilibrium, and hearing.

Chapter Outline and Objectives

Vocabulary Development

baro-, pressure; *baroreceptors*
circa, about; *circadian*
circum-, around; *circumvallate papillae*
dies, day; *circadian*
emmetro-, proper measure; *emmetropia*
incus, anvil; *incus* (auditory ossicle)
labyrinthos, a network of canals; *labyrinth*
lithos, a stone; *otolith*
macula, spot; *macula lutea*
malleus, a hammer; *malleus* (auditory ossicle)
myein, to shut; *myopia*
noceo, hurt; *nociceptor*
ops, eye; *myopia*
oto-, ear; *otolith*
presbys, old man; *presbyopia*
stapes, stirrup; *stapes* (auditory ossicle)
tectum, roof; *tectorial membrane*
vallum, wall; *circumvallate papillae*

Our knowledge of the world around us is limited to those characteristics that stimulate our sensory receptors. Although we may not realize it, our picture of the environment is incomplete. Colors invisible to us guide insects to flowers, and sounds and smells we cannot detect are regular information to dogs, cats, and dolphins. Moreover, our senses are sometimes deceptive: In phantom limb pain, a person "feels" pain in a missing limb, and during an epileptic seizure an individual may experience sights, sounds, or smells that have no physical basis.

All sensory information is picked up by *sensory receptors,* specialized cells that monitor internal and external conditions. The simplest receptors are the dendritic processes of sensory neurons. A dendritic array of this kind is called a *free nerve ending.* Free nerve endings are not protected by accessory structures, and they are sensitive to many different stimuli. For example, the same free nerve endings in the skin may provide the sensation of pain in response to crushing, heat, or a cut. Other receptors are especially sensitive to one kind of stimulus. For example, a touch receptor is very sensitive to pressure but relatively insensitive to chemical stimuli; a taste receptor is sensitive to dissolved chemicals but insensitive to pressure. The most complex receptors, such as the visual receptors of the eye, are protected by accessory cells and layers of connective tissue. Not only are these receptor cells specialized to detect light, they are seldom exposed to any stimulus *except* light.

All sensory information arrives at the CNS in the form of action potentials in an afferent (sensory) fiber. In general, the stronger the stimulus, the higher the frequency of action potentials. When sensory information arrives at the CNS, it is routed according to the location and nature of the stimulus. For example, touch, pressure, pain, temperature, and taste sensations arrive at the primary sensory cortex; visual, auditory, and olfactory information reach the visual, auditory, and olfactory regions of the cortex. Because the CNS interprets the nature of sensory information entirely on the basis of the area of the brain stimulated, it cannot tell the difference between a "true" sensation and a "false" one. For example, when rubbing the eyes, one often "sees" flashes of light. Although the stimulus is mechanical rather than visual, any activity along the optic nerve is projected to the visual cortex and experienced as a visual perception.

Adaptation is a reduction in sensitivity in the presence of a constant stimulus. Familiar examples of adaptation include stepping into a hot bath or jumping into a cold lake. Shortly afterward, neither temperature seems as extreme as it did initially. Adaptation reduces the amount of information arriving at the cerebral cortex. Most sensory information is routed to centers along the spinal cord or brain stem, potentially triggering involuntary reflexes, such as the withdrawal reflex. Only about 1 percent of the information provided by afferent fibers reaches the cerebral cortex and our conscious awareness.

Output from higher centers, however, can increase receptor sensitivity or facilitate transmission along a sensory pathway. For example, the *reticular activating system* in the midbrain helps focus attention and thus heightens or reduces awareness of arriving sensations. [p. 223] This adjustment of sensitivity can occur under conscious or unconscious direction. When we "listen carefully," our sensitivity to and awareness of auditory stimuli increase. The reverse occurs when we enter a noisy factory or walk along a crowded city street, as we automatically "tune out" the high level of background noise.

The *general senses* are senses of temperature, pain, touch, pressure, vibration, and proprioception (body position). The receptors for the general senses are scattered throughout the body. The **special senses** are smell (*olfaction*), taste (*gustation*), balance (*equilibrium*), hearing, and vision. The receptors for the five special senses are concentrated within specific structures, the **sense organs.** This chapter explores both the general senses and the special senses. AM *Problems with Sensory Systems*

THE GENERAL SENSES

Receptors for the general senses are scattered throughout the body and are relatively simple in structure. These receptors are classified according to the nature of the stimulus that excites them; important classes include receptors sensitive to pain (*nociceptors*); temperature (*thermoreceptors*); distortion, contact, and pressure (*mechanoreceptors*); and chemical stimuli (*chemoreceptors*).

Pain

Pain receptors, or **nociceptors** (nō-sē-SEP-tōrz; *noceo*, hurt), are especially common in the superficial portions of the skin, in joint capsules, within the periostea of bones, and around the walls of blood vessels. There are few nociceptors in other deep tissues or in most visceral organs.

Once pain receptors in a region are stimulated, two types of axons carry the painful sensations. Myelinated fibers carry very localized sensations of **fast pain,** or *prickling pain,* such as that caused by an injection or deep cut. These sensations reach the CNS quickly, leading to somatic reflexes and stimulation of the primary sensory cortex. Slower, unmyelinated fibers carry sensations of **slow pain,** or *burning and aching pain.* Unlike fast pain sensations, slow pain sensations enable one to determine only the general area involved.

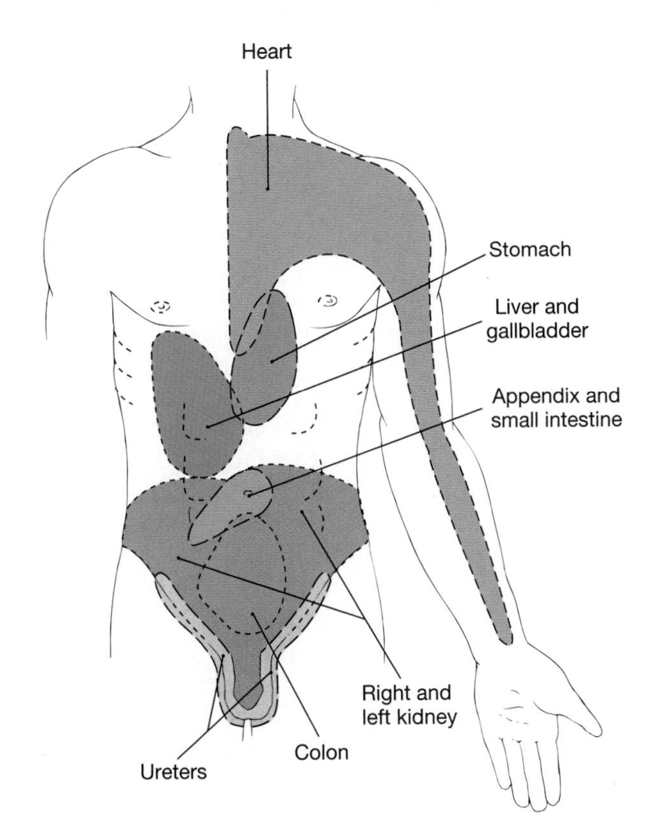

●**Figure 10-1**
Referred Pain

Pain sensations originating in visceral organs are often perceived as involving specific regions of the body surface innervated by the same spinal nerves.

Pain sensations from visceral organs are often perceived as originating in more superficial regions, generally regions innervated by the same spinal nerves. The perception of pain coming from parts of the body that are not actually stimulated is called **referred pain.** The precise mechanism responsible for referred pain remains to be determined, but several clinical examples are shown in Figure 10-1●. Cardiac pain, for example, is often perceived as originating in the upper chest and left arm.

Although pain receptors continue to respond as long as the painful stimulus remains, the *awareness* of the pain can decrease over time because of the inhibition of pain centers in the thalamus, reticular formation, lower brain stem, and spinal cord. AM *The Control of Pain*

Temperature

Temperature receptors, or **thermoreceptors,** are free nerve endings that are scattered immediately beneath the surface of the skin. They are also found in skeletal muscles, in the liver, and in the hypothalamus. Cold receptors are three or four times as numerous as warm receptors. There are no known structural differences between warm and cold thermoreceptors.

Temperature sensations are relayed along the same pathways that carry pain sensations. They are distributed to the reticular formation, the thalamus, and, to a lesser extent, the primary sensory cortex. Thermoreceptors are very active when the temperature is changing, but they quickly adapt to a stable temperature. When we enter an air-conditioned classroom on a hot summer day or a toasty lecture hall on a brisk fall evening, the temperature seems unpleasant at first, but the discomfort fades as adaptation occurs.

Touch, Pressure, and Position

Mechanoreceptors are sensitive to stimuli that distort their cell membranes. They contain *mechanically regulated ion channels,* which open or close in response to stretching, compression, twisting, or other distortions of the membrane. There are three classes of mechanoreceptors: *tactile (touch) receptors, baroreceptors (pressure),* and *proprioceptors (position).*

Tactile Receptors

Tactile receptors provide sensations of touch, pressure, and vibration. The distinctions between these are hazy, for a touch also represents a pressure, and a vibration consists of an oscillating touch/pressure stimulus. **Fine touch** and **pressure receptors** provide detailed information about a source of stimulation, including its exact location, shape, size, texture, and

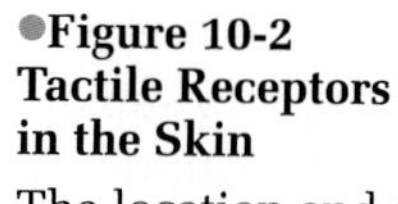

Figure 10-2 Tactile Receptors in the Skin

The location and general appearance of six important tactile receptors.

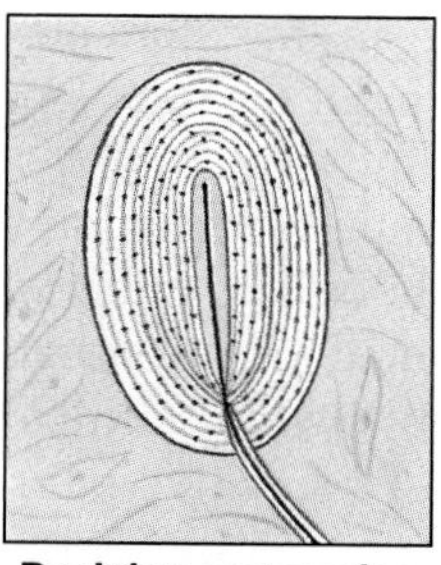

Pacinian corpuscles are large receptors most sensitive to pulsing or vibrating stimuli; they are common in the skin of the fingers, breasts, and external genitalia and in joint capsules, mesenteries, and the wall of the urinary bladder.

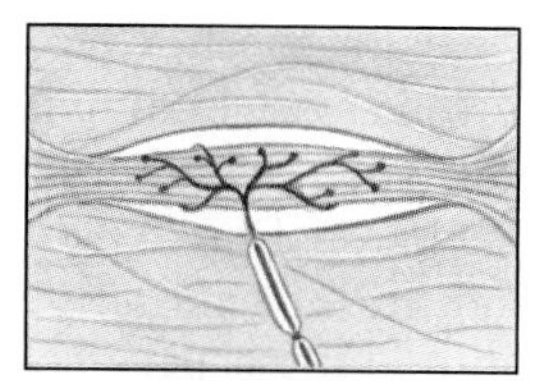

Ruffini corpuscles detect pressure and distortion of the dermis.

Merkel Cells

Merkel's discs are nerve endings that contact specialized epithelial cells *(Merkel cells)* sensitive to fine touch.

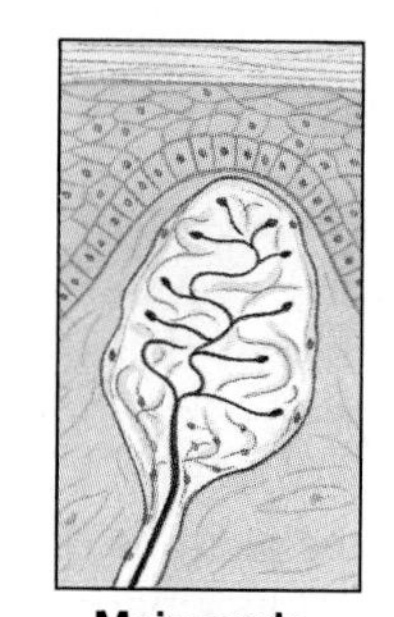

Meissner's corpuscles provide touch sensations from the eyelids, lips, fingertips, nipples, and external genitalia.

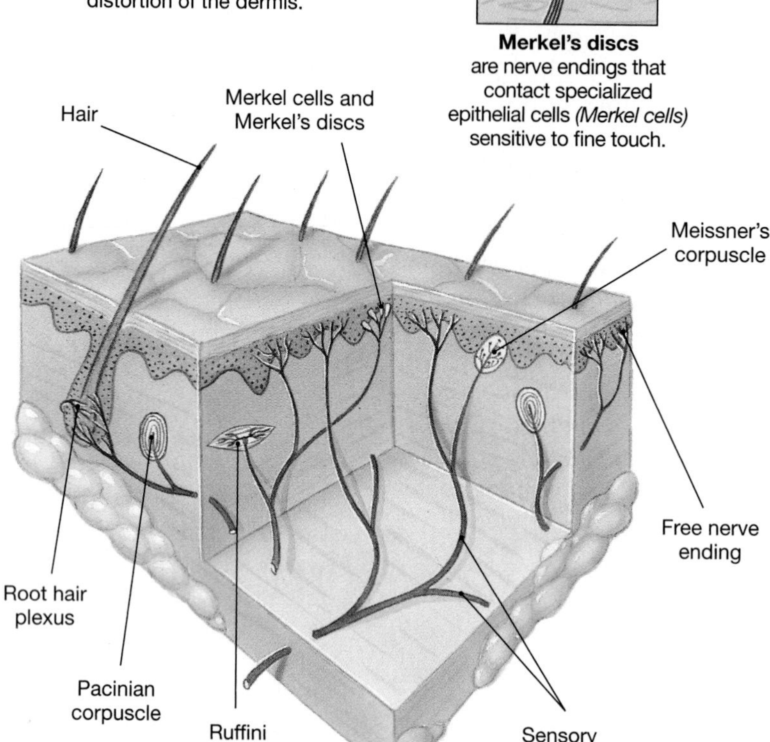

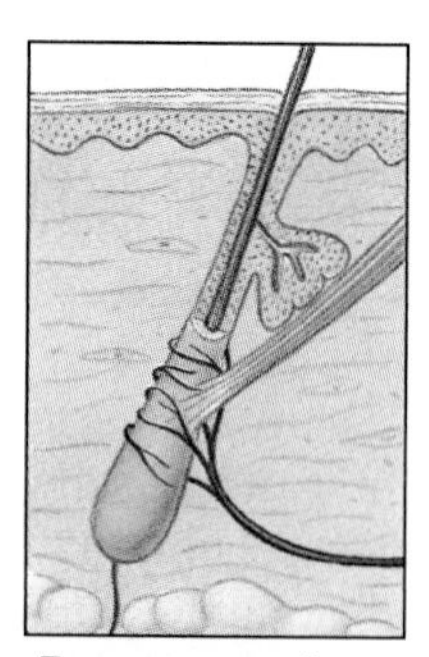

Free nerve endings of the **root hair plexus** are distorted by movement of the hairs.

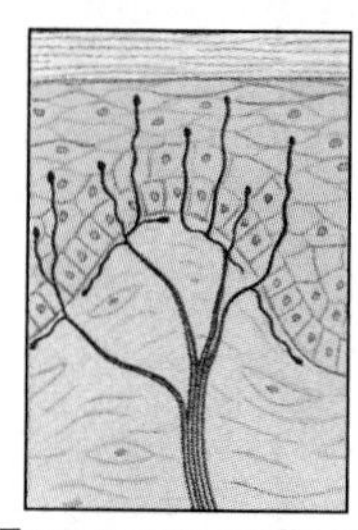

Free nerve endings are found between epidermal cells and are sensitive to a variety of stimuli.

movement. **Crude touch** and **pressure receptors** provide poor localization and little additional information about the stimulus.

Tactile receptors range in complexity from free nerve endings to specialized sensory complexes with accessory cells and supporting structures. Figure 10-2● shows six types of tactile receptors in the skin. Tactile sensations travel through the posterior column and spinothalamic pathways, as discussed in Chapter 9. ∞ *[pp. 236–237]* Tactile sensitivities may be altered by peripheral infection, disease processes, and damage to sensory afferents or central pathways. Several clinical tests may be used to evaluate tactile sensitivity as part of a physical examination. AM *Assessment of Tactile Sensitivities*

Baroreceptors

Baroreceptors (bar-ō-rē-SEP-tōrz; *baro-*, pressure) monitor changes in pressure. The receptor itself consists of free nerve endings that branch within the elastic tissues in the wall of a distensible organ, such as a blood vessel or a portion of the respiratory, digestive, or urinary tract. When the pressure changes, the elastic walls of these tracts stretch or recoil. This movement distorts the dendritic branches and alters the rate of action potential generation. Baroreceptors respond immediately to a change in pressure, but they adapt rapidly, and the output along the afferent fibers gradually returns to "normal."

Figure 10-3● gives examples of important baroreceptor functions. Baroreceptors monitor blood pressure in the walls of major vessels, including the carotid artery (at the *carotid sinus*) and the aorta (at the *aortic sinus*). The information plays a major role in regulating cardiac function and adjusting blood flow to vital tissues. Baroreceptors in the lungs monitor the degree of lung expansion. This information is relayed to the respiratory rhythmicity center, which sets the pace of respiration. Baroreceptors in the digestive and urinary tracts trigger various visceral reflexes, including those of urination and defecation.

Proprioceptors

Proprioceptors monitor the position of joints, the tension in tendons and ligaments, and the state of muscular contraction. In general, proprioceptors do not adapt

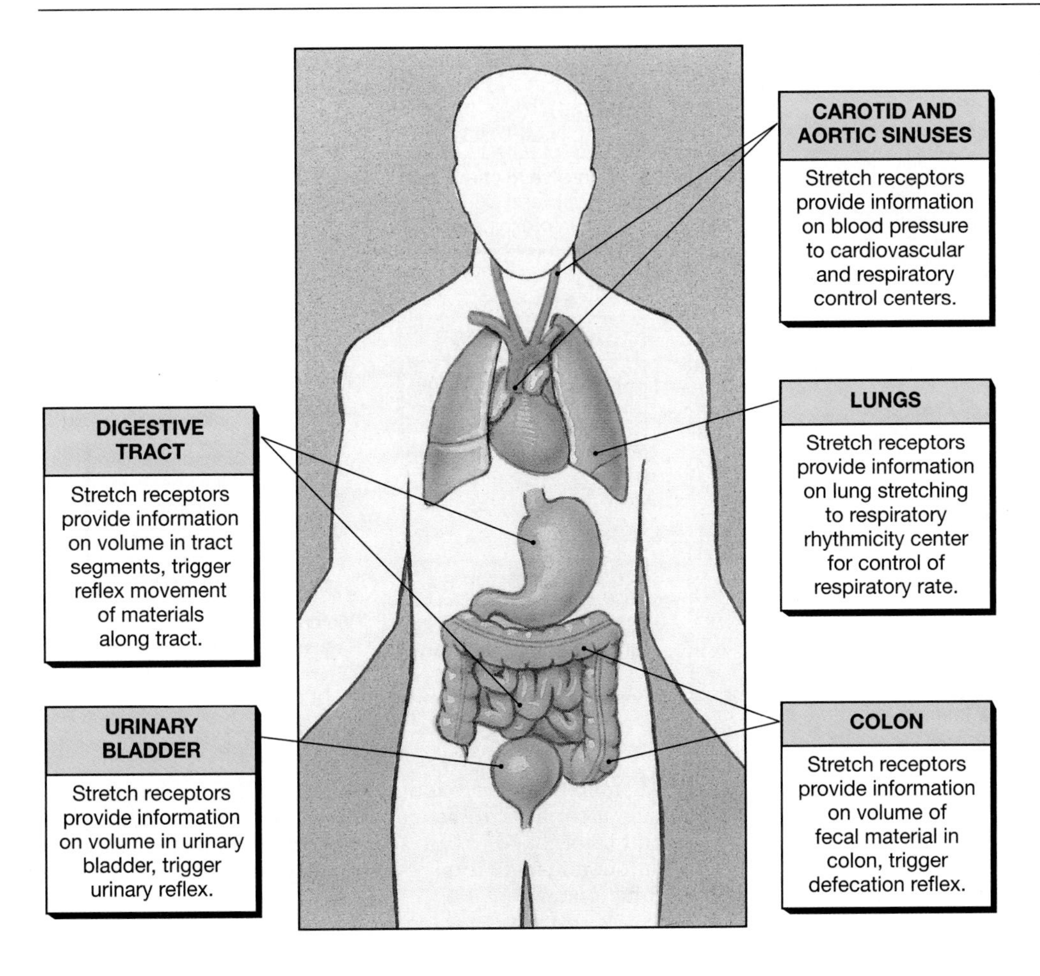

CAROTID AND AORTIC SINUSES
Stretch receptors provide information on blood pressure to cardiovascular and respiratory control centers.
LUNGS
Stretch receptors provide information on lung stretching to respiratory rhythmicity center for control of respiratory rate.
DIGESTIVE TRACT
Stretch receptors provide information on volume in tract segments, trigger reflex movement of materials along tract.
URINARY BLADDER
Stretch receptors provide information on volume in urinary bladder, trigger urinary reflex.
COLON
Stretch receptors provide information on volume of fecal material in colon, trigger defecation reflex.

●Figure 10-3 Baroreceptors and the Regulation of Autonomic Functions

Baroreceptors provide information essential to the regulation of autonomic activities, including respiration, digestion, urination, and defecation.

to constant stimulation. Of all the general sensory receptors, they are the most structurally and functionally complex. Two representative examples are *tendon organs,* which monitor the strain on a tendon, and *muscle spindles,* which monitor the length of a skeletal muscle.

Chemical Detection

In general, **chemoreceptors** respond only to water-soluble and lipid-soluble substances that are dissolved in the surrounding fluid. Adaptation usually occurs over a few seconds following stimulation. Except for the special senses of taste and smell, there are no well-defined chemosensory pathways in the brain or spinal cord. The locations of important chemosensory receptors are shown in Figure 10-4●. Neurons within the respiratory centers of the brain respond to the concentration of hydrogen ions (pH) and carbon dioxide molecules in the cerebrospinal fluid. Other receptors in the periphery monitor the oxygen concentration of arterial blood. These chemoreceptive neurons are found within the **carotid bodies,** near the origin of the internal carotid arteries on each side of the neck, and in the **aortic bodies** between the major branches of the aortic arch. The afferent fibers leaving the carotid and aortic bodies reach the respiratory centers by traveling along the ninth (glossopharyngeal) and tenth (vagus) cranial nerves. These chemoreceptors play an important role in the reflexive control of respiration and cardiovascular function.

SMELL

The sense of smell, or *olfaction,* is provided by paired **olfactory organs.** These organs, shown in Figure 10-5●, are located in the nasal cavity on either side of the nasal septum just inferior to the cribriform plate of the ethmoid bone. Each olfactory organ consists of an **olfactory epithelium,** which contains the **olfactory receptors,** supporting cells, and *basal cells* (stem cells). Beneath the basement membrane, large **olfactory glands** produce a pigmented mucus that covers the epithelium. The olfactory glands produce a continual stream of mucus that passes across the surface of the olfactory organ, preventing the buildup of potentially dangerous or overpowering stimuli and keeping the area moist and free from dust or other debris. Once the compounds have reached the olfactory organs, they must diffuse into the mucus before they can stimulate the olfactory receptors.

The olfactory receptors are highly modified neurons. The exposed tip of each receptor forms a prominent knob that provides a base for elongate cilia. All together there are somewhere between 10 and 20 million ol-

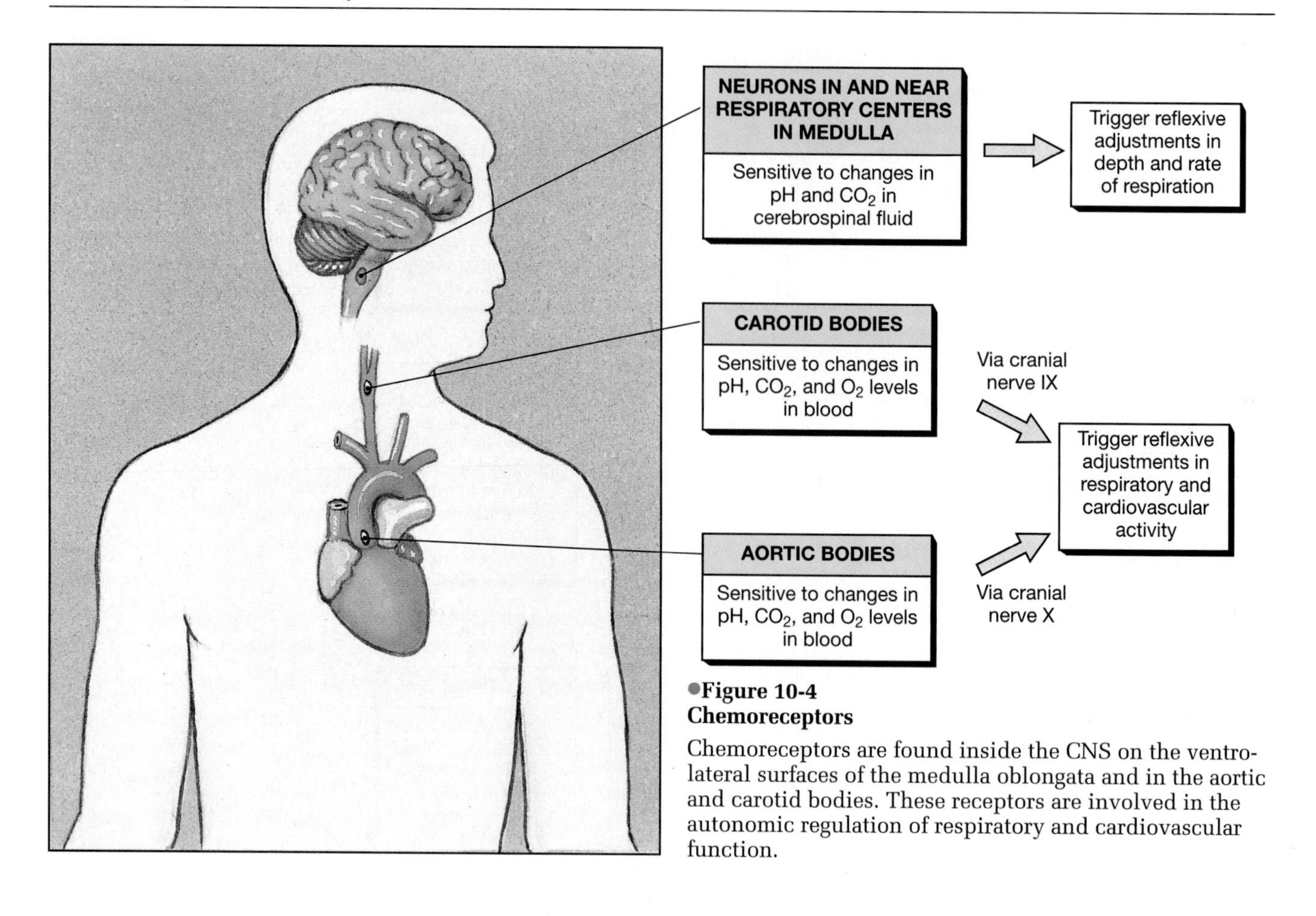

Figure 10-4
Chemoreceptors

Chemoreceptors are found inside the CNS on the ventrolateral surfaces of the medulla oblongata and in the aortic and carotid bodies. These receptors are involved in the autonomic regulation of respiratory and cardiovascular function.

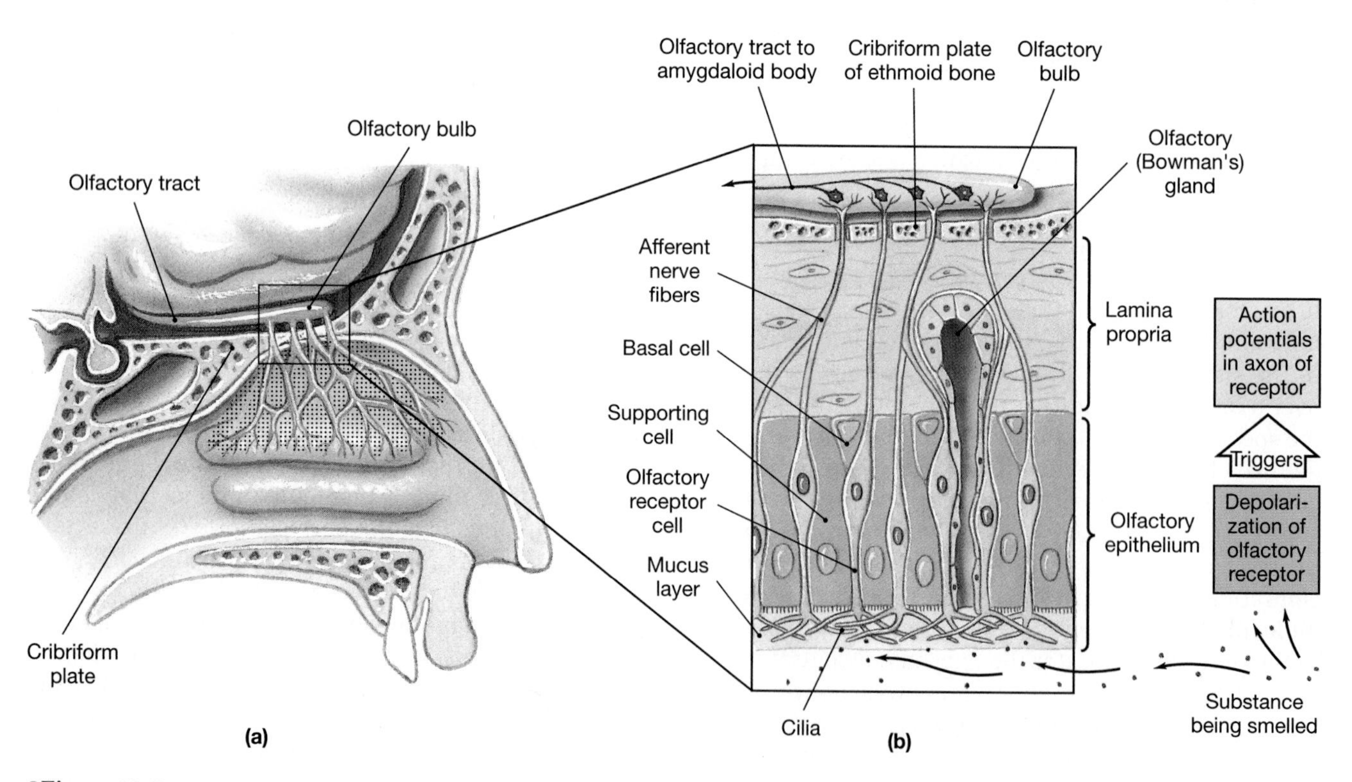

Figure 10-5
The Olfactory Organs

(a) The structure of the olfactory organ on the right side of the nasal septum. **(b)** An olfactory receptor is a modified neuron with multiple cilia extending from its free surface.

factory receptors packed into an area of roughly 5 cm^2. Nevertheless, our olfactory sensitivities cannot compare with those of other vertebrates such as dogs, cats, or fishes. A German shepherd sniffing for smuggled drugs or explosives has an olfactory receptor surface 72 times greater than that of the nearby Customs inspector.

Olfactory reception occurs as dissolved chemicals interact with receptors on the surfaces of the cilia. This interaction changes the permeability of the receptor membrane, producing action potentials. This information is relayed to the central nervous system, which interprets the smell on the basis of the particular pattern of receptor activity.

Olfactory Pathways

The axons leaving the olfactory epithelium collect into 20 or more bundles that penetrate the cribriform plate of the ethmoid to reach the **olfactory bulbs** of the cerebrum. Axons leaving the olfactory bulb travel along the olfactory tract to reach the olfactory cortex, the hypothalamus, and portions of the limbic system.

Olfactory stimuli are the only type of sensory information that reaches the cerebral cortex without first synapsing in the thalamus. The extensive limbic and hypothalamic connections help explain the profound emotional and behavioral responses that can be produced by certain smells. The perfume industry understands the practical implications of these connections, spending billions of dollars to develop odors that trigger sexual responses.

TASTE

The **gustatory** (GUS-ta-tōr-ē), or **taste, receptors** are distributed over the surface of the tongue and adjacent portions of the pharynx and larynx in individual **taste buds.** The taste buds are particularly well protected from the excessive mechanical stress due to chewing, for they lie along the sides of epithelial projections called **papillae** (pa-PIL-lē). The greatest number of taste buds are associated with the large *circumvallate papillae* that form a V that points toward the attached base of the tongue (Figure 10-6●).

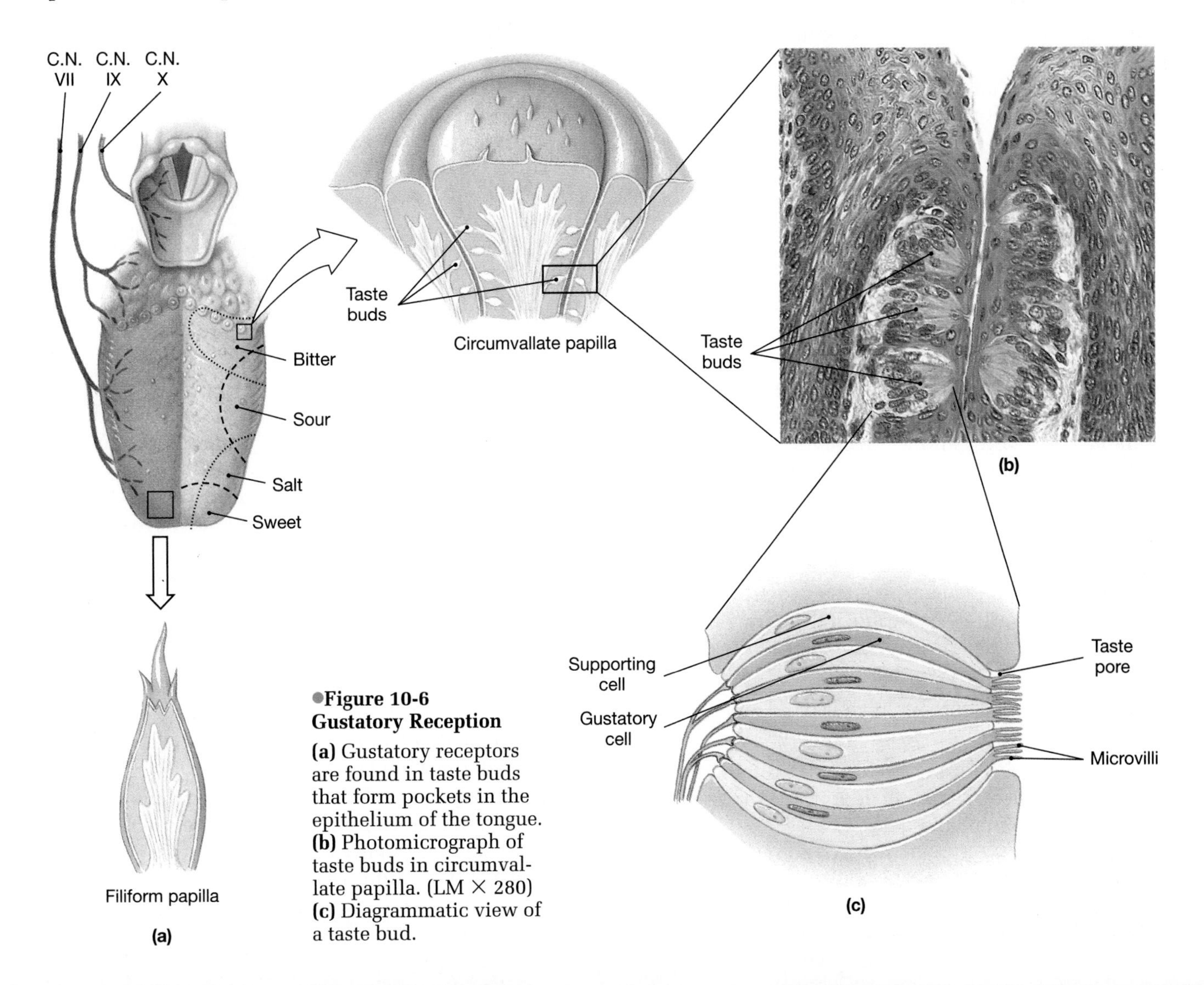

●**Figure 10-6 Gustatory Reception**

(a) Gustatory receptors are found in taste buds that form pockets in the epithelium of the tongue. **(b)** Photomicrograph of taste buds in circumvallate papilla. (LM × 280) **(c)** Diagrammatic view of a taste bud.

Each taste bud contains slender sensory receptors, known as **gustatory cells,** and supporting cells. Each gustatory cell extends slender microvilli, sometimes called *taste hairs,* into the surrounding fluids through a narrow opening, the **taste pore.** The mechanism behind gustatory reception seems to parallel that of olfaction. Dissolved chemicals contacting the taste hairs stimulate a change in the membrane potential of the taste cell, which leads to action potentials in the sensory neuron.

There are four **primary taste sensations**: sweet, salt, sour, and bitter. Each taste bud shows a particular sensitivity to one of these tastes, and a sensory map of the tongue indicates that each is concentrated in a different area. The threshold for receptor stimulation varies for each of the primary taste sensations, and the taste receptors respond most readily to unpleasant rather than pleasant stimuli.

Taste Pathways

Taste buds are monitored by the seventh (facial), ninth (glossopharyngeal), and tenth (vagus) cranial nerves. The sensory afferents synapse within a nucleus in the medulla, the axons of the postsynaptic neurons synapse in the thalamus, and the information is projected to the appropriate portions of the primary sensory cortex.

In assembling a conscious perception of taste, the information received from the taste buds is correlated with other sensory data. Our perception of the general texture of the food, together with the taste-related sensations of "peppery" or "burning," result from the stimulation of general sensory afferents contained in the trigeminal nerve (N V). In addition, information from the olfactory receptors plays an overwhelming role in taste perception. We are several thousand times more sensitive to "tastes" when our olfactory organs are fully functional. If a person has a cold, and airborne molecules cannot reach the olfactory receptors, meals taste dull and unappealing even when the taste buds are responding normally.

✓ When you first enter the A & P laboratory for dissection, you are very aware of the odor, but by the end of the lab period the smell doesn't seem nearly as strong. Why?

✓ When the nociceptors in your hand are stimulated, what sensation do you perceive?

✓ What would happen to an individual if the information from proprioceptors in the legs was blocked from reaching the CNS?

✓ If you completely dry the surface of the tongue, then place salt or sugar crystals on it, they cannot be tasted. Why not?

VISION

Humans rely more on vision than on any other sense. Our visual receptors are contained in elaborate structures, the eyes, which enable us not only to detect light but to create detailed visual images. We will begin our discussion of these complex organs by considering the *accessory structures* of the eye that provide protection, lubrication, and support.

Accessory Structures of the Eye

The **accessory structures** of the eye include the (1) eyelids and associated exocrine glands, (2) the superficial epithelium of the eye, (3) structures associated with the production, secretion, and removal of tears, and (4) the extrinsic eye muscles.

The eyelids can close firmly to protect the delicate surface of the eye and, by their continual blinking movements, keep the surface lubricated and free from dust and debris. The upper and lower eyelids are connected at the **medial canthus** (KAN-thus) and the **lateral canthus** (Figure 10-7●). The eyelashes are very robust hairs that help prevent foreign particles and insects from reaching the surface of the eye.

Several types of glands protect the eye and its accessory structures. At the medial canthus, the **lacrimal caruncle** (KAR-unk-ul) contains glands that produce thick secretions that contribute to the gritty deposits occasionally found after a night's sleep. The accessory glands of the eye sometimes become infected by bacteria. An infection in a sebaceous gland of one of the eyelashes, or in one of the adjacent sweat glands, produces a painful localized swelling known as a *sty.*

The anterior surface of the eye is covered by the **conjunctiva** (kon-junk-TĪ-va), a distinctive epithelium continuous with the inner lining of the eyelids. The conjunctiva extends to the delicate *corneal epithelium* that covers the transparent **cornea** (KOR-nē-a). The conjunctiva contains many free nerve endings and is very sensitive. The painful condition of **conjunctivitis**, or "pink-eye," results from damage to and irritation of the conjunctival surface. The most obvious symptom results from dilation of the blood vessels beneath the conjunctival epithelium. AM *Conjunctivitis*

Above the eyeball, a dozen or more ducts from the **lacrimal gland,** or tear gland, empty into the pocket between the eyelid and the eye. This gland nestles within a depression in the frontal bone, just inside the orbit and superior and lateral to the eyeball. The lacrimal gland normally provides the key ingredients and most of the volume of the tears that bathe the conjunctival surfaces. Its secretions are watery, slightly alkaline, and contain *lysozyme,* an enzyme that attacks bacteria. Tears reduce friction, remove debris, prevent bacterial infection, and provide nutrients and oxygen to the conjunctiva. The blinking of the eye sweeps the tears across the surface of the eye to the medial canthus. Two small pores direct the tears into the **lacrimal canals.** These passageways end at the **lacrimal sac,** and from there the **nasolacrimal duct** carries the tears to the nasal cavity.

The Extrinsic Eye Muscles

Six **extrinsic eye muscles,** or *oculomotor* (ok-ū-lō-MŌ-ter) *muscles,* originate on the surface of the orbit and control the position of the eye. These muscles are the **inferior rectus, lateral rectus, medial rectus, superior rectus, inferior oblique,** and **superior oblique** (Figure 10-8● and Table 10-1).

Anatomy of the Eye

The eyes are extremely specialized visual organs, more versatile and adaptable than the most expensive cameras, yet light, compact, and durable. Each eye is roughly spherical, with a diameter of nearly 2.5 cm (1 in.), and weighs around 8 g (0.28 oz). The eyeball shares space within the orbit with the extrinsic eye muscles, the lacrimal gland, and the various cranial nerves and blood vessels that service the eye and adjacent areas of the orbit and face. A mass of *orbital fat* provides padding and insulation.

The eyeball is hollow, and the interior can be divided into two cavities (Figure 10-9●). The large **posterior cavity** is also called the *vitreous chamber* because it contains the gelatinous *vitreous body.* The smaller **anterior cavity** is subdivided into the *anterior chamber* and the *posterior chamber.* The shape of the eye is stabilized in part by the vitreous body and the clear *aqueous humor* that fills the anterior cavity. The wall of the eye contains three distinct layers, or *tunics:* an outer *fibrous tunic,* an intermediate *vascular tunic,* and an inner *neural tunic.*

The Fibrous Tunic

The **fibrous tunic,** the outermost layer covering the eye, consists of the *sclera* and the *cornea.* The fibrous tunic (1) provides mechanical support and some degree of physical protection, (2) serves as an attachment site for the extrinsic eye muscles, and (3) assists in the focusing process.

On the anterior surface of the eye, the loose connective tissue beneath the conjunctiva covers the **sclera** (SKLER-a), a layer of dense fibrous connective tissue (Figure 10-9●). The network of small capillaries that lies beneath the conjunctiva usually does not carry enough blood to lend an obvious color to the sclera, and the collagen fibers are visible as the "white of the eye." The six extrinsic eye muscles insert upon the sclera.

The transparent cornea is continuous with the sclera, but the collagen fibers of the cornea are organized into a series of layers that do not interfere with the passage of light. Because the cornea has only

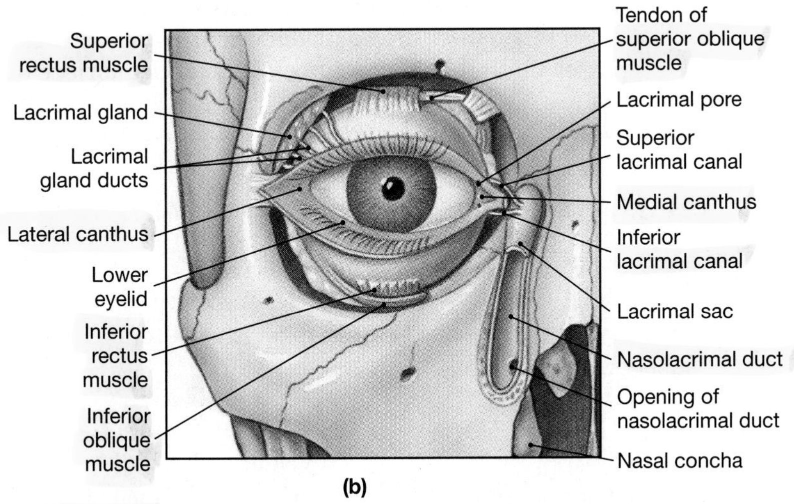

●Figure 10-7
Accessory Structures of the Eye
(a) Gross and superficial anatomy of the accessory structures. **(b)** Details of the lacrimal apparatus.

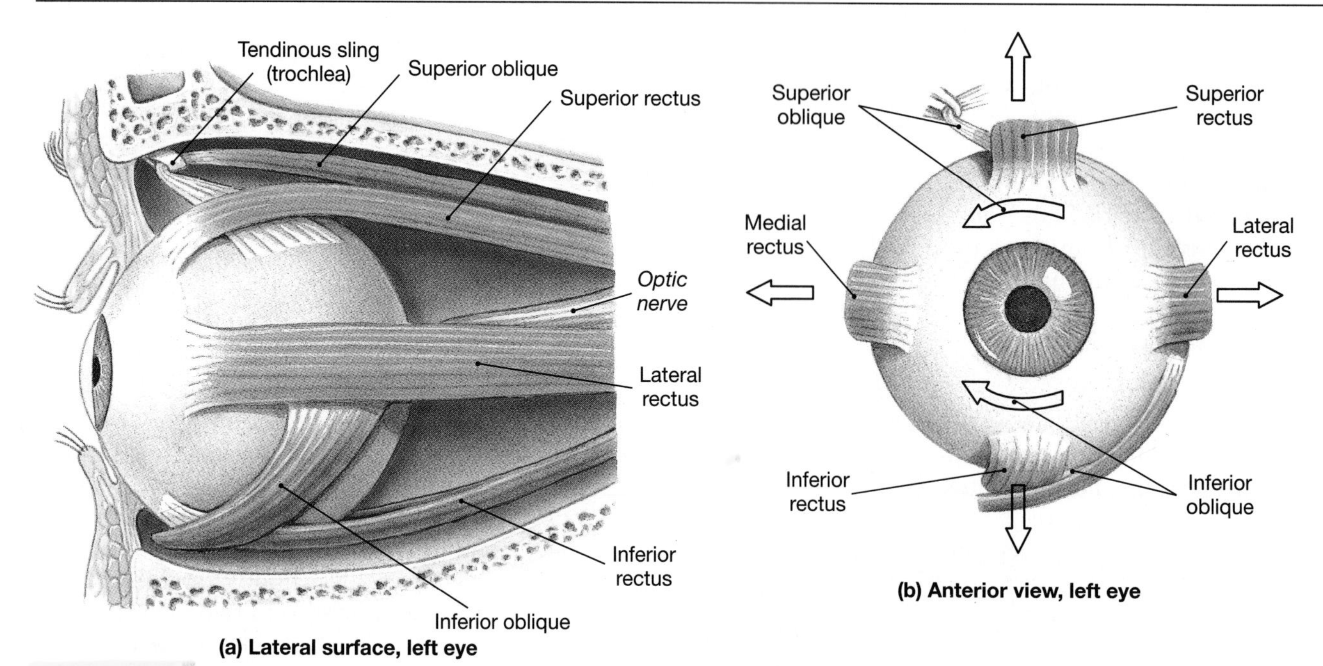

Figure 10-8
The Extrinsic Eye Muscles

a limited ability to repair itself, corneal injuries must be treated immediately to prevent serious vision losses. Restoration of vision after corneal scarring has occurred usually requires replacement of the cornea through a *corneal transplant.* AM *Corneal Transplants*

The Vascular Tunic

The **vascular tunic** contains numerous blood vessels, lymphatics, and all of the *intrinsic eye muscles.* The functions of this layer include (1) providing a route for blood vessels and lymphatics that supply tissues of the eye, (2) regulating the amount of light entering the eye, (3) secreting and reabsorbing the *aqueous humor* that circulates within the eye, and (4) controlling the shape of the lens, an essential part of the focusing process.

The vascular tunic includes the *iris,* the *ciliary body,* and the *choroid* (Figure 10-9●). Visible through the transparent corneal surface, the **iris** contains blood vessels, pigment cells, and two layers of smooth muscle fibers. When these muscles contract, they change the diameter of the central opening, or **pupil,** of the iris. Dilation and constriction are controlled by the autonomic nervous system in response to sudden changes in light intensity. Exposure to bright light produces a rapid reflexive decrease in pupillary diameter, under parasympathetic stimulation. A sudden reduction in light levels produces a much slower pupillary dilation, under the control of the sympathetic division.

The thickness of the iris and the number and distribution of pigment cells determine its apparent color. When there are no pigment cells in the iris, light passes through it and bounces off its inner surface of

TABLE 10-1 Extrinsic Eye Muscles

Muscle	*Origin*	*Insertion*	*Action (Figure 10-8b)*
Inferior rectus	Sphenoid around optic foramen	Inferior, medial surface of eyeball	Eye looks down
Medial rectus	As above	Medial surface of eyeball	Eye rotates medially
Superior rectus	As above	Superior, lateral surface of eyeball	Eye looks up
Inferior oblique	Maxilla at front of orbit	Inferior, lateral surface of eyeball	Eye rolls, looks up and to the side
Superior oblique	Sphenoid around optic foramen	Superior, medial surface of eyeball	Eye rolls, looks down and to the side
Lateral rectus	As above	Lateral surface of eyeball	Eye rotates laterally

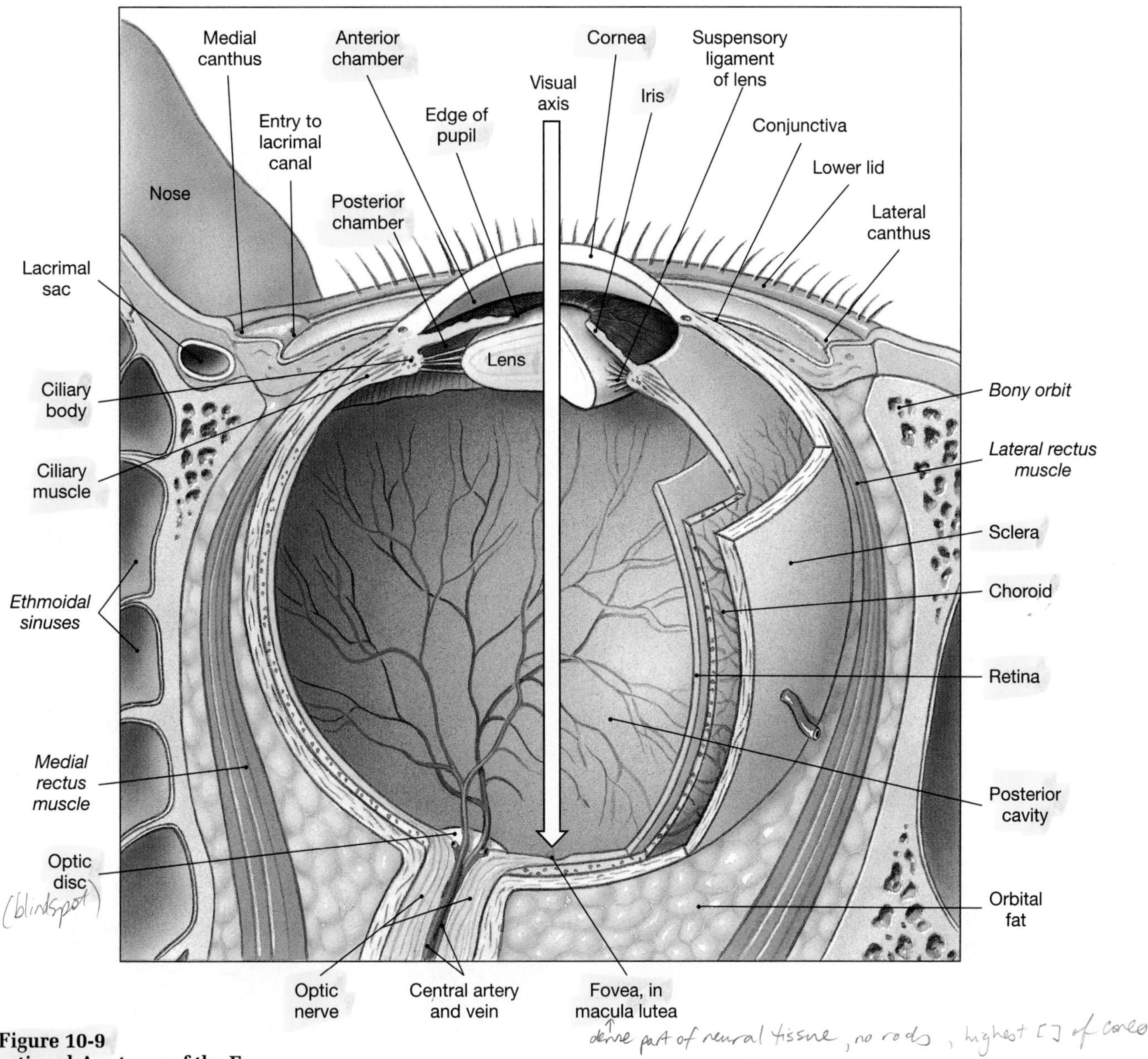

●**Figure 10-9**
Sectional Anatomy of the Eye
Landmarks and features of the eye in a horizontal section through the right eye.

pigmented epithelium. The eye then appears blue. Individuals with gray, brown, and black eyes have increasing numbers, respectively, of pigment cells in the body and surface of the iris.

Along its outer edge, the iris attaches to the anterior portion of the **ciliary body,** the bulk of which consists of the *ciliary muscle,* a muscular ring that projects into the interior of the eye. The ciliary body begins at the junction between the cornea and sclera and extends to the scalloped border that also marks the anterior edge of the *retina.* Posterior to the iris, the surface of the ciliary body is thrown into folds called *ciliary processes.* The **suspensory ligaments** of the lens attach to these processes. These fibers position the lens so that light passing through the pupil passes through the center of the lens.

The **choroid** separates the fibrous and neural tunics posterior to the ciliary body (Figure 10-9●). The choroid contains a capillary network that delivers oxygen and nutrients to the retina.

The Neural Tunic

The **neural tunic,** or **retina,** consists of a thin, outer **pigment layer** and a thick inner layer, the **neural retina.** The neural retina contains the visual receptors and associated neurons. The pigment layer absorbs light after it passes through the receptor layer. The neural retina contains (1) the photoreceptors that respond to light, (2) supporting cells and neurons that perform preliminary processing and integration of visual information, and (3) blood vessels supplying tissues lining the posterior cavity. The neural retina forms a cup that establishes the posterior and lateral boundaries of the posterior cavity (Figure 10-9●).

Retinal Organization The retina contains several layers of cells (Figure 10-10a●). The outermost layer, closest to the pigment layer, contains the photoreceptors. There are two types of photoreceptors, **rods** and **cones.** Rods do not discriminate between different colors of light. They are very light-sensitive and enable us to see in dimly lit rooms, at twilight, or in pale moonlight. Cones provide us with color vision. There are three types of cones, and their stimulation in various combinations provides the perception of different colors. Cones give us sharper, clearer images, but they require more intense light than rods. If you sit outside at sunset (or sunrise) you will probably be able to tell when your visual system shifts from cone-based vi-

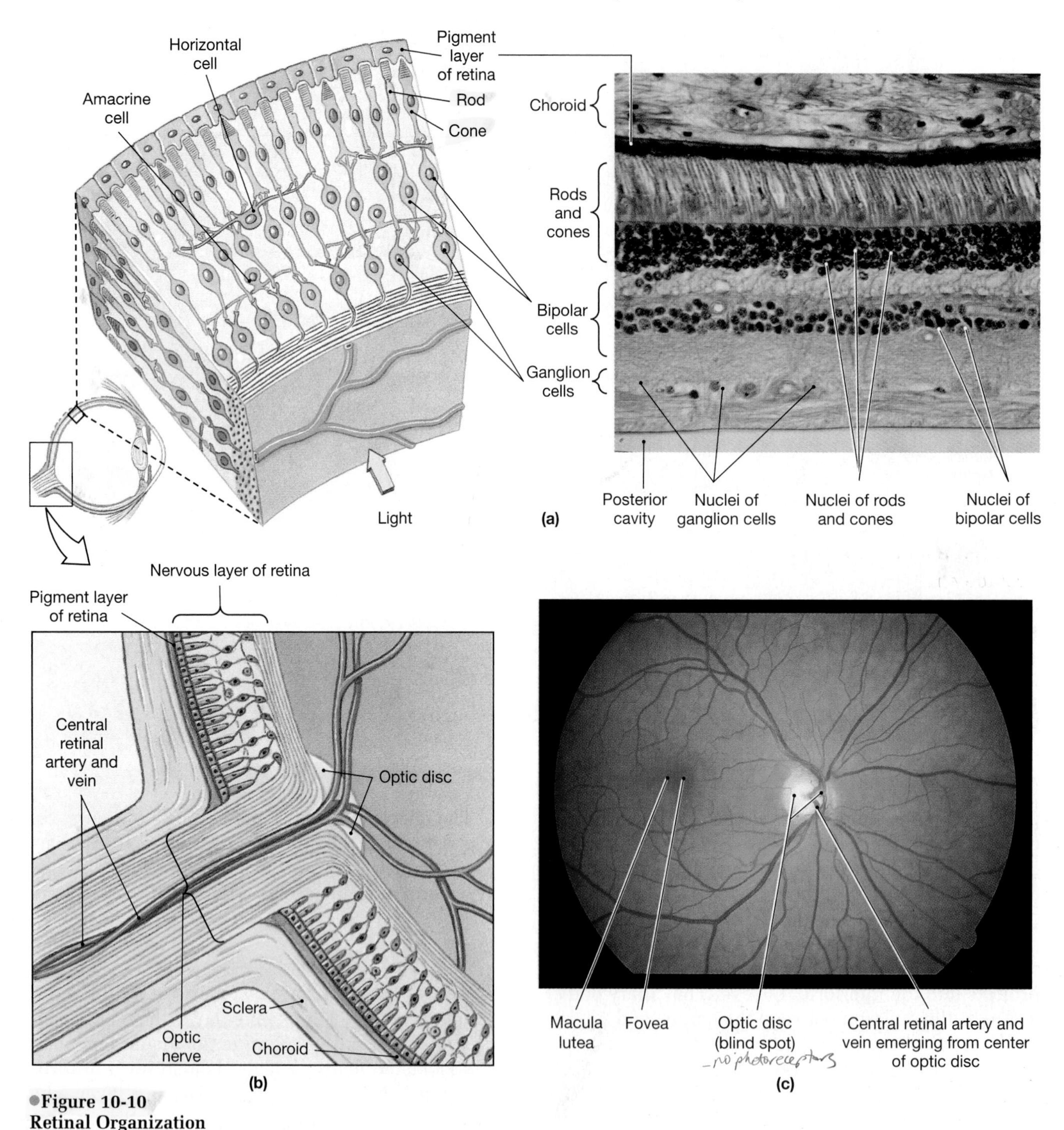

●**Figure 10-10**
Retinal Organization

(a) Cellular organization of the retina. Note that the photoreceptors are located closer to the choroid rather than near the posterior cavity. (LM × 290) **(b)** The optic disc in diagrammatic section. **(c)** A photograph of the retina as seen through the pupil of the eye.

light. Cones provide us with color vision. There are three types of cones, and their stimulation in various combinations provides the perception of different colors. Cones give us sharper, clearer images, but they require more intense light than rods. If you sit outside at sunset (or sunrise) you will probably be able to tell when your visual system shifts from cone-based vision (clear images in full color) to rod-based vision (relatively grainy images in black and white).

Rods and cones are not evenly distributed across the retina. If you consider the retina as a cup, the approximately 125 million rods are found on the sides and the roughly 6 million cones dominate the bottom. There are no rods in the region where the visual image arrives after passing through the cornea and lens. This area is the **macula lutea** (LOO-tē-a; yellow spot). The highest concentration of cones is found in the central portion of the macula lutea, an area called the **fovea** (FŌ-vē-a; shallow depression), or *fovea centralis* (Figure 10-10c●). The fovea is the center of color vision and the site of sharpest vision; when you look directly at an object its image falls upon this portion of the retina.

You are probably already aware of the visual consequences of this distribution. During the day, when there is enough light to stimulate the cones, you see a very good image. In very dim light, cones simply cannot function. For example, when you try to stare at a dim star, you are unable to see it. But if you look a little to one side, rather than directly at the star, you will see it quite clearly. Shifting your gaze moves the image of the star from the fovea, where it does not provide enough light to stimulate the cones, to the edges of the retina, where it stimulates the more sensitive rods.

The rods and cones synapse with roughly 6 million **bipolar cells.** Bipolar cells in turn synapse within the layer of **ganglion cells** that faces the posterior cavity. The axons of the ganglion cells deliver the sensory information to the brain. *Horizontal cells* and *amacrine* (AM-a-krīn) *cells* can regulate communication between photoreceptors and ganglion cells, adjusting the sensitivity of the retina. The effect could be compared to adjusting the contrast setting on a television. Their activities play an important role in the eye's adjustment to dim or brightly lit environments.

The Optic Disc Axons from an estimated 1 million ganglion cells converge on the **optic disc,** a circular region just medial to the fovea. The optic disc is the origin of the optic nerve (N II) (Figure 10-10b●). From this point, the axons turn, penetrate the wall of the eye, and proceed toward the diencephalon. Blood vessels that supply the retina pass through the center of the optic nerve and emerge on the surface of the optic disc (Figure 10-10b,c●). There are no photoreceptors or other retinal structures at the optic disc. Because light striking this area goes unnoticed, it is commonly called the **blind spot.** You do not notice a blank spot in the visual field because involuntary eye movements keep the visual image moving and allow the brain to fill in the missing information. A simple experiment, shown in Figure 10-11●, will demonstrate the presence and location of the blind spot.

The Chambers of the Eye

The ciliary body and lens divide the interior of the eye into a large posterior cavity, also called the *vitreous chamber,* and a smaller anterior cavity. The anterior cavity is further subdivided into the **anterior chamber,** which extends from the cornea to the iris, and a **posterior chamber** between the iris and the ciliary body and lens. The anterior and posterior chambers are filled with a fluid called *aqueous humor.* Aqueous humor circulates within the anterior cavity, passing from the posterior to the anterior chamber via the pupil of the eye (Figure 10-12●). The posterior cavity is filled with a gelatinous substance known as the *vitreous body.* The presence of the vitreous body helps maintain the shape of the eye and holds the retina against the choroid.

Aqueous Humor **Aqueous humor** forms at the ciliary processes through active secretion into the posterior chamber (Figure 10-12●). Pressure exerted by this fluid helps maintain the shape of the eye, and the circulation of aqueous humor transports nutrients and wastes. In the anterior chamber near the edge of the iris, the aqueous humor passes into a passageway, the *canal of Schlemm,* that returns it to the venous system. Interference with the normal circulation and reabsorption of aqueous humor leads to an elevation in the

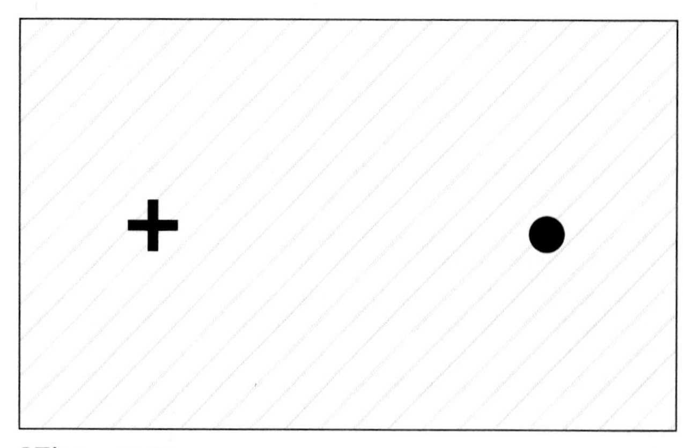

●**Figure 10-11**
The Optic Disc

Close your left eye and stare at the cross with your right eye, keeping it in the center of your field of vision. Begin with the page a few inches away and gradually increase the distance. The dot will disappear when its image falls on the blind spot. To check the blind spot in the left eye, close your right eye and repeat this sequence staring at the dot.

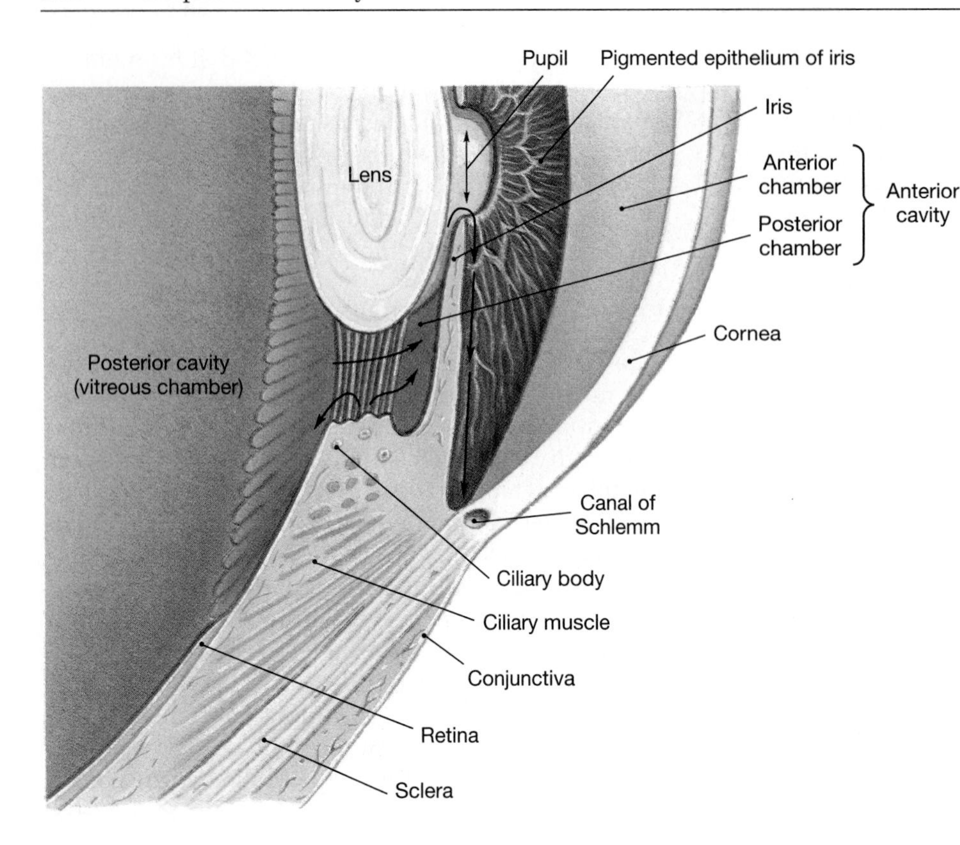

Figure 10-12
Eye Chambers and the Circulation of Aqueous Humor
The lens is suspended between the posterior cavity and the posterior chamber. Its position is maintained by the suspensory ligaments that attach the lens to the ciliary body. Aqueous humor secreted at the ciliary body circulates through the posterior and anterior chambers before being reabsorbed via the canal of Schlemm.

body of the choroid. The primary function of the lens is to focus the visual image on the retinal receptors. It does so by changing its shape.

Structure of the Lens The transparent lens consists of organized, concentric layers of cells wrapped in a dense fibrous capsule. The capsule is elastic, and unless an outside force is applied it will contract and make the lens spherical. However, tension in the suspensory ligaments can overpower the elastic capsule and pull the lens into the shape of a flattened oval.

CATARACTS

The transparency of the lens depends on a precise combination of structural and biochemical characteristics. When that balance is disturbed, the lens loses its transparency, and the abnormal lens is known as a **cataract.** Cataracts may result from drug reactions, injuries, or radiation, but *senile cataracts* are the most common form. As aging proceeds, the lens becomes less elastic, takes on a yellowish hue, and eventually begins to lose its transparency. As the lens becomes opaque, or "cloudy," the person needs brighter and brighter reading lights, and visual clarity begins to fade. When the lens becomes completely opaque, the person is functionally blind, even though the retinal receptors are alive and well. Modern surgical procedures involve removing the lens, either intact or in pieces after shattering it with high-frequency sound. The missing lens can be replaced by an artificial substitute, and vision can then be fine-tuned with glasses or contact lenses.

Accommodation As in a camera, the arriving image must be in focus if it is to provide any useful information. "In focus" means that the rays of light arriving from the object strike the sensitive surface of the film precisely ordered so as to form a miniature image of the original. If they are not perfectly focused, the image will be blurry. Focusing normally occurs in two steps, as light passes through the cornea and lens.

Light is bent, or *refracted,* when it passes between media of differing densities. In the human eye, the greatest amount of refraction occurs when light passes from the air into the cornea. Additional refraction takes place when the light enters the relatively dense lens. The lens provides the additional refraction needed to focus the light rays from an object toward a specific *focal point.* The distance between the center of the lens and the focal point is the *focal distance,* which is determined by (1) *the distance of the object from the lens* (focal distance increases as an object moves closer) and (2) *the shape of the lens* (the rounder the lens, the shorter the focal distance). In the eye, the lens changes shape to keep the focal distance constant, thereby keeping the image focused on the retina. **Accommodation** (Figure 10-13) is the process of focusing an image on the retina by changing the shape of the lens. During accommodation, the lens becomes rounder to focus the image of a nearby object on the retina, and it flattens when focusing on a distant object.

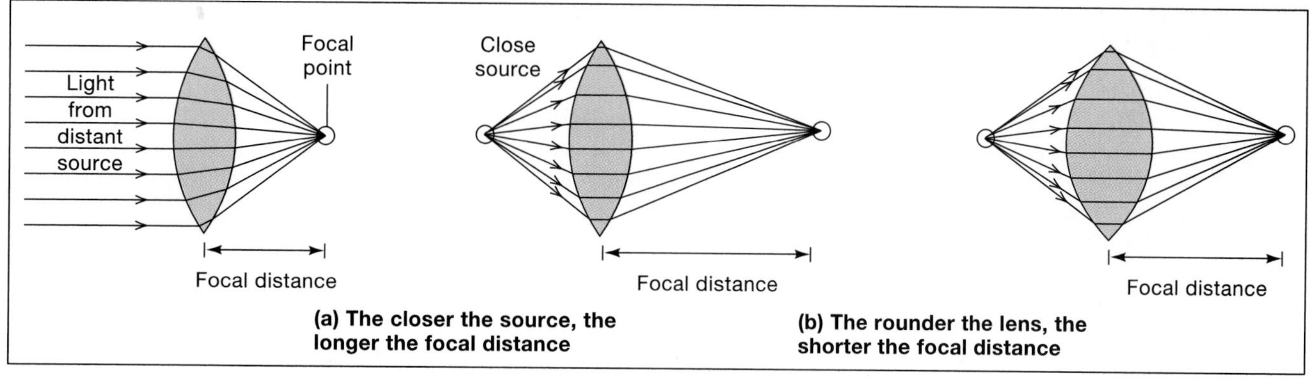

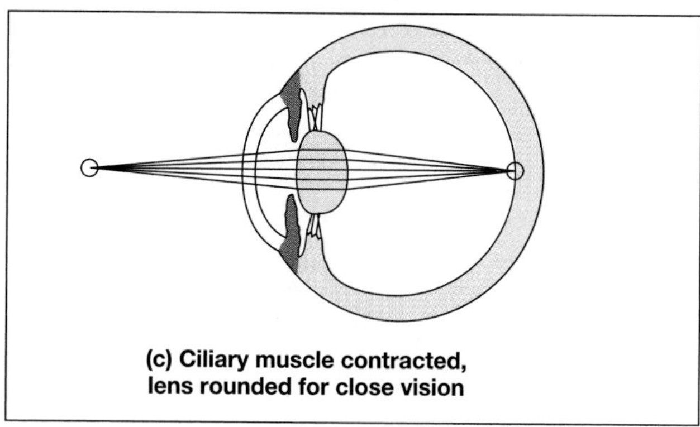

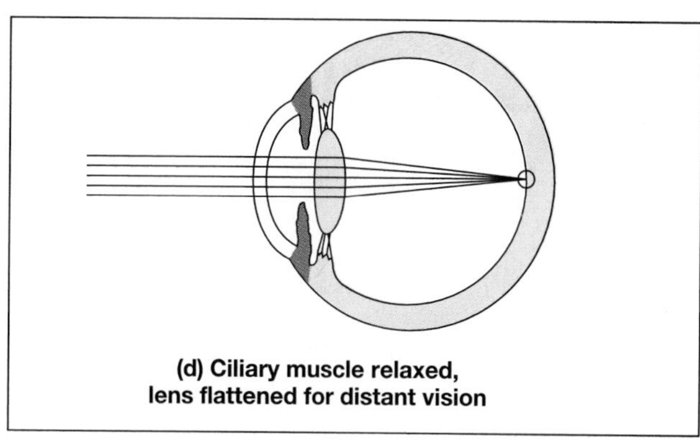

•Figure 10-13
Image Formation and Visual Accommodation
(a) A lens refracts light toward a specific (focal) point. The distance from the center of the lens to that point is the focal distance of the lens. Light from a distant source arrives with all of the light waves traveling parallel to one another. Light from a nearby source, however, will still be diverging or spreading out from its source when it strikes the lens. Note the difference in focal distance after refraction. **(b)** The rounder the lens, the shorter the focal distance. **(c)** When the ciliary muscle contracts, the suspensory ligaments allow the lens to round up. **(d)** For the eye to form a sharp image, the focal distance must equal the distance between the center of the lens and the retina. The lens compensates for variations in the distance between the eye and the object in view by changing its shape. When the ciliary muscle relaxes, the ligaments pull against the margins of the lens and flatten it.

The lens is held in place by the suspensory ligaments that originate at the ciliary body. Smooth muscle fibers in the ciliary body encircle the lens. As you view a nearby object, your ciliary muscles contract and the ciliary body moves toward the lens (Figure 10-13c•). This movement reduces the tension in the suspensory ligaments, and the elastic capsule pulls the lens into a more spherical shape. When you view a distant object, the ciliary muscle relaxes, the suspensory ligaments pull at the circumference of the lens, and the lens becomes relatively flat (Figure 10-13d•).

VISUAL ACUITY

Clarity of vision, or visual acuity, is rated on the basis of the sight of a "normal" person. A person whose vision is rated 20/20 is seeing details at a distance of 20 feet as clearly as a "normal" individual would. Vision noted as 20/15 is better than average, for at 20 feet the person is able to see details that would be clear to a normal eye only at a distance of 15 feet. Conversely, a person with 20/30 vision must be 20 feet from an object to discern details that a person with normal vision could make out at a distance of 30 feet.

When visual acuity falls below 20/200, even with the help of glasses or contact lenses, the individual is considered to be legally blind. There are probably fewer than 400,000 legally blind people in the United States; more than half are over 65 years of age. Common causes of blindness include diabetes mellitus, cataracts, glaucoma, corneal scarring, retinal detachment, accidental injuries, and hereditary factors that are as yet poorly understood.

Visual Physiology

The rods and cones of the retina are called **photoreceptors** because they detect *photons,* basic units of light. Our eyes are sensitive to the spectrum of **visible light.** This spectrum, seen in a rainbow, can be remembered by the acronym "ROY G. BIV" (red, orange, yellow, green, blue, indigo, violet). The property of color depends on the wavelength of the light. The smaller the energy content, the longer the wavelength. Photons of red light carry the least energy and have

FOCUS Accommodation Problems

In the normal eye, when the ciliary muscles are relaxed and the lens is flattened, a distant image will be focused on the retinal surface (Figure 10-14a●), a condition called **emmetropia** (*emmetro-*, proper measure). However, irregularities in the shape of the lens or cornea can affect the clarity of the visual image. This condition, called *astigmatism*, can usually be corrected by glasses or special contact lenses.

Figure 10-14● diagrams two other common problems with the accommodation mechanism. If the eyeball is too deep, the image of a distant object will form in front of the retina, and the retinal picture will be blurry and out of focus (Figure 10-14b●). Vision at close range will be normal, because the lens will be able to round up as needed to focus the image on the retina. As a result, such individuals are said to be "nearsighted." Their condition is more formally termed **myopia** (*myein*, to shut + *ops*, eye). Myopia can be corrected by placing a diverging lens in front of the eye (Figure 10-14c●).

If the eyeball is too shallow, **hyperopia** results (Figure 10-14d●). The ciliary muscles must contract to focus even a distant object on the retina, and at close range the lens cannot provide enough refraction. These individuals are said to be "farsighted" because they can see distant objects most clearly. Older individuals become farsighted as their lenses lose elasticity; this form of hyperopia is called **presbyopia** (*presbys*, old man). Hyperopia can be treated by placing a converging lens in front of the eye (Figure 10-14e●).

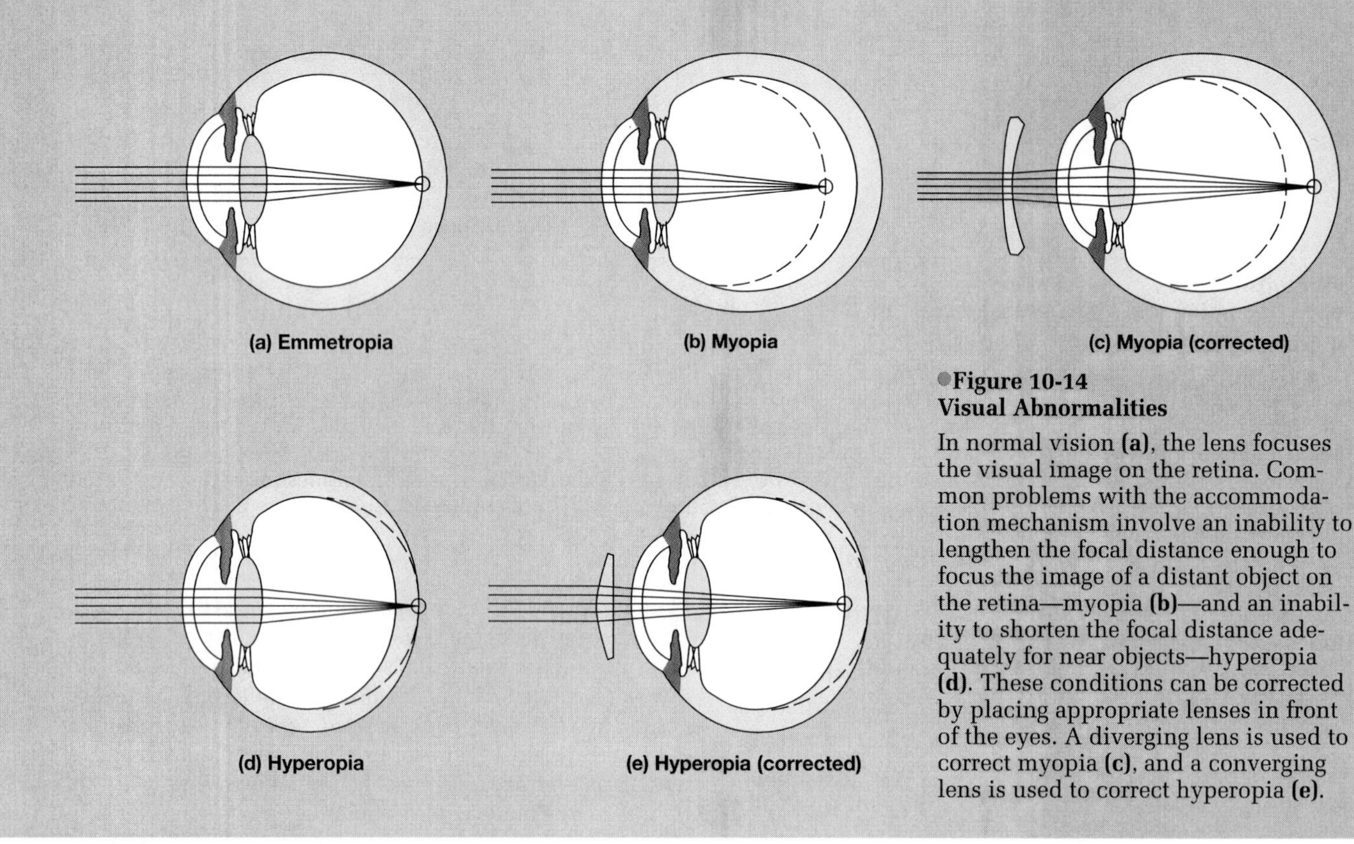

●**Figure 10-14**
Visual Abnormalities

In normal vision **(a)**, the lens focuses the visual image on the retina. Common problems with the accommodation mechanism involve an inability to lengthen the focal distance enough to focus the image of a distant object on the retina—myopia **(b)**—and an inability to shorten the focal distance adequately for near objects—hyperopia **(d)**. These conditions can be corrected by placing appropriate lenses in front of the eyes. A diverging lens is used to correct myopia **(c)**, and a converging lens is used to correct hyperopia **(e)**.

the longest wavelength. Photons from the violet portion of the spectrum carry the most energy and have the shortest wavelength.

Rods and Cones

Rods provide the CNS with information on the presence or absence of photons, without regard to wavelength. As a result, they do not discriminate between different colors of light. They are very sensitive, however, and enable us to see in dimly lit rooms, at twilight, or in pale moonlight.

Cones provide information on the wavelength of photons. Because cones are less sensitive than rods, they function only in relatively bright light. There are three types of cones: blue cones, green cones, and red cones. Each type is sensitive to a different range of wavelengths of light, and their stimulation in various combinations accounts for the perception of colors. Persons unable to distinguish certain colors have a form of *color blindness.* The standard tests for color vision involve picking numbers or letters out of a complex and colorful picture, such as the one in Figure 10-15●. Color blindness occurs because one or more

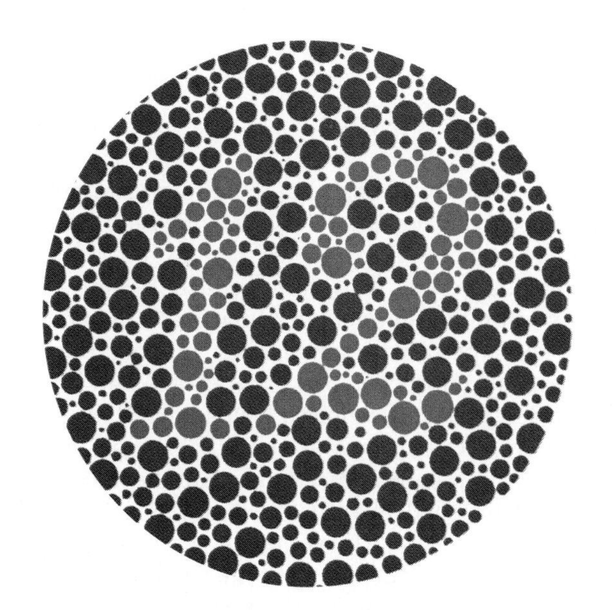

●**Figure 10-15**
Cones and Color Vision
Part of a standard test for color vision. Lack of one or more populations of cones will produce an inability to distinguish the patterned images.

Photoreception begins when a photon strikes a rhodopsin molecule in the outer segment of a photoreceptor. Upon the absorption of light, a change in the shape of the retinal component activates opsin. Activation of opsin starts a chain of enzymatic events that alters the rate of neurotransmitter release. This change is the signal that light has struck a photoreceptor at that particular location on the retina.

Shortly after the conformational change occurs, the rhodopsin molecule begins to break down into

classes of cones are absent or nonfunctional. In the most common condition, the red cones are missing, and the individual cannot distinguish red light from green light. Ten percent of all men show some color blindness, whereas the incidence among women is only around 0.67 percent. Total color blindness is extremely rare; only 1 person in 300,000 has no cone pigments of any kind.

Photoreceptor Function

Figure 10-16● compares the structure of rods and cones. The outer portion of a photoreceptor contains hundreds to thousands of flattened membranous discs. The inner portion of a photoreceptor communicates with a bipolar cell. In the dark, each photoreceptor continually releases neurotransmitters across these synapses. The arrival of a photon initiates a chain of events that alters the membrane potential of the photoreceptor and changes the rate of neurotransmitter release.

Visual Pigments

Light absorption, the first key step in the process of photoreception, requires special organic compounds called **visual pigments.** These compounds, found in the outer segments of all photoreceptors, are derivatives of the compound **rhodopsin** (rō-DOP-sin). Rhodopsin consists of a protein, the enzyme **opsin,** bound to a pigment, **retinal** (RET-i-nal), synthesized from **vitamin A.** The pigment retinal is identical in both rods and cones, but a different form of opsin is found in the rods and each of the three types of cones (red, blue, and green).

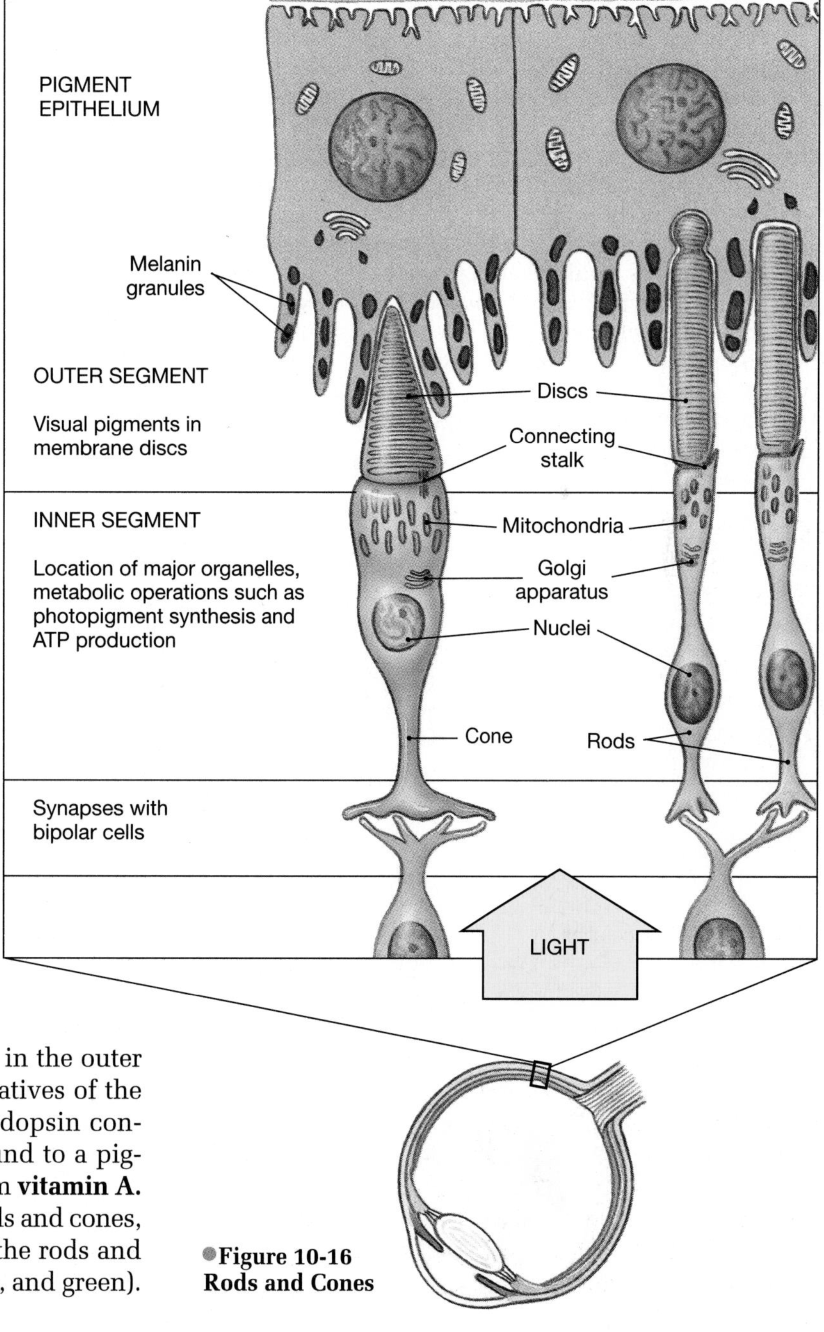

●**Figure 10-16**
Rods and Cones

retinal and opsin, a process known as *bleaching* (Figure 10-17●). The retinal must be converted to its former shape before it can recombine with opsin. This conversion requires energy in the form of ATP, and it takes time. Bleaching contributes to the lingering visual impression after a camera flash goes off. After an intense exposure to light, a photoreceptor cannot respond to further stimulation until its rhodopsin molecules have been regenerated. As a result, a "ghost" image remains on the retina.

NIGHT BLINDNESS

The visual pigments of the photoreceptors are synthesized from vitamin A. The body contains vitamin A reserves sufficient for several months, and a significant amount is stored in the cells of the pigment layer of the neural tunic. If dietary sources are inadequate, these reserves are gradually exhausted, and the amount of visual pigment in the photoreceptors begins to decline. Daylight vision is affected, but during the day the light is usually bright enough to stimulate whatever visual pigments remain within the densely packed cone population. As a result, the problem first becomes apparent at night, when the dim light proves insufficient to activate the rods. This condition, known as night blindness, can be treated by administration of vitamin A. The carotene pigments found in many vegetables can be converted to vitamin A within the body. Carrots are a particularly good source of carotene, which explains the old saying that carrots are good for your eyes.

The Visual Pathway

The visual pathway begins at the photoreceptors and ends at the visual cortex of the cerebral hemispheres. In other sensory pathways we have examined, there is at most one synapse between a receptor and a sensory neuron that delivers information to the CNS. In the visual pathway, the message must cross two synapses (photoreceptor to bipolar cell, and bipolar cell to ganglion cell) before it heads toward the brain. Axons from the entire population of ganglion cells converge on the optic disc, penetrate the wall of the eye, and proceed toward the diencephalon as the optic nerve (N II). The two optic nerves, one from each eye, reach the diencephalon at the optic chiasm (Figure 10-18●). From this point approximately one-half of the fibers proceed toward the lateral geniculate on the same side of the brain, while the other half cross over to reach the *lateral geniculate* nucleus of the thalamus on the opposite side. The lateral geniculates act as switching and processing centers that relay visual information to reflex centers in the brain stem as well as to the cerebral cortex.

The sensation of vision arises from the integration of information arriving at the visual cortex of the cerebrum. The visual cortex of the each occipital lobe contains a sensory map of the entire field of vision. As with the primary sensory cortex, the map does not faithfully duplicate the relative areas within the sensory field. For example, the area assigned to the fovea covers about 35 times the sur-

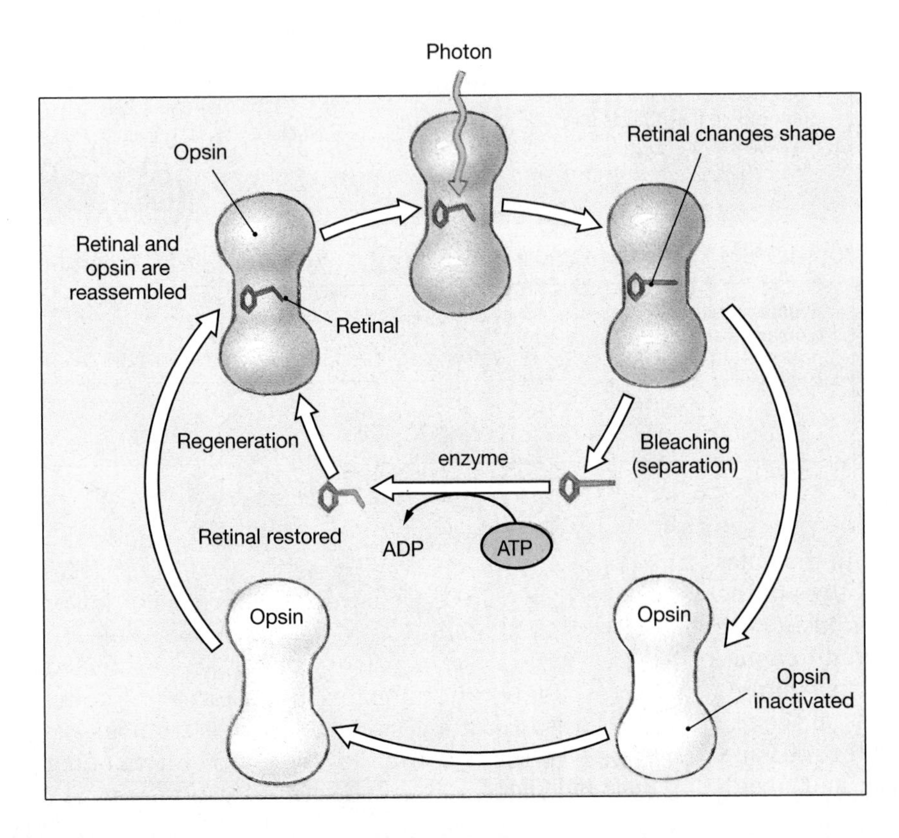

●**Figure 10-17**
Bleaching and Recovery of Visual Pigments

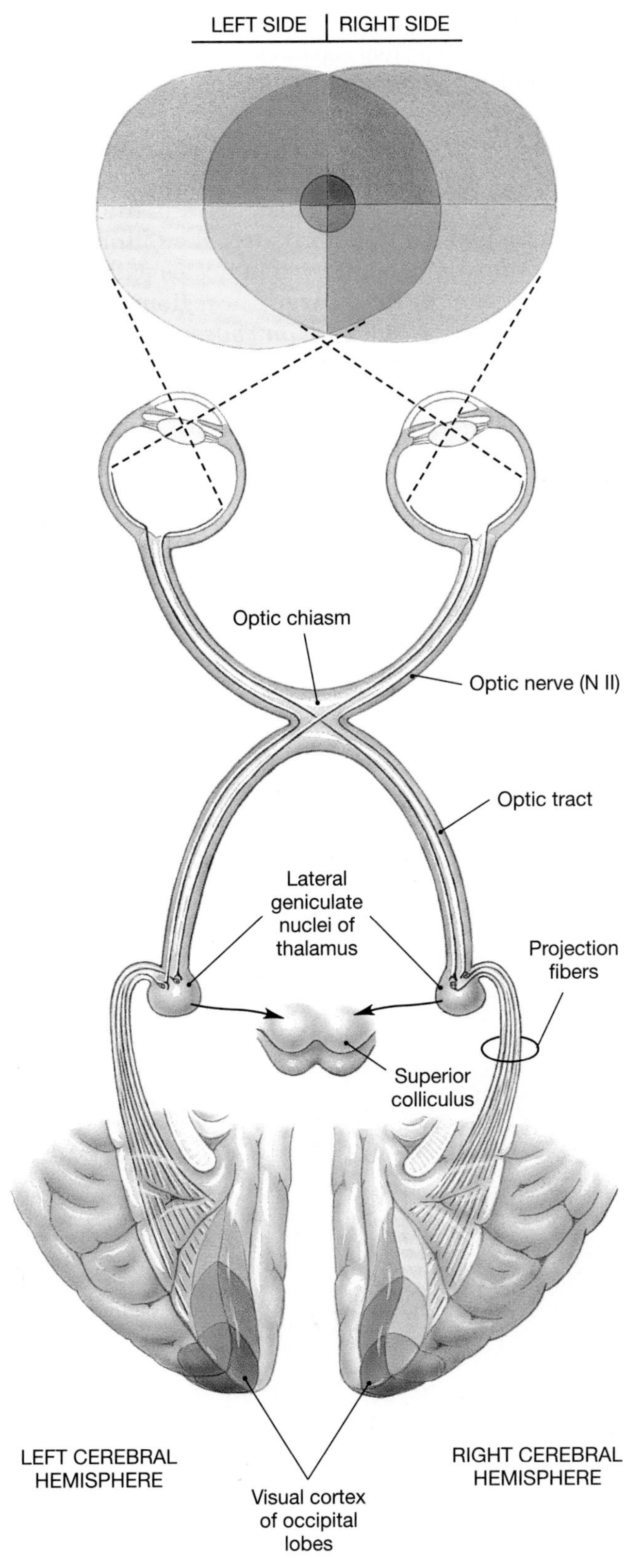

•Figure 10-18
The Visual Pathways

At the optic chiasm, a partial crossover of nerve fibers occurs. As a result, each hemisphere receives visual information from the lateral half of the retina on that side and from the medial half of the retina on the opposite side. Visual association areas integrate this information to develop a composite picture of the entire visual field.

face it would cover if the map were proportionally accurate.

Many centers in the brain stem receive visual information, either from the lateral geniculates or over collaterals from the optic tracts. For example, some collaterals that bypass the lateral geniculates synapse in the hypothalamus. Visual inputs there and at the *pineal gland* establish a daily pattern of activity that is tied to the day-night cycle. This *circadian* (*circa,* about + *dies*, day) *rhythm* affects metabolic rate, endocrine function, blood pressure, digestive activities, the awake-sleep cycle, and other processes. ∞ [p. 223]

✓ What layer of the eye would be the first to be affected by inadequate tear production?

✓ When the lens is very round, are you looking at an object that is close to you or distant from you?

✓ If a person was born without cones in her eyes, would she still be able to see? Explain.

✓ How could a diet that is deficient in vitamin A affect vision?

EQUILIBRIUM AND HEARING

The senses of equilibrium and hearing are provided by the *inner ear,* a receptor complex located in the temporal bone of the skull. ∞ [p. 122] The basic receptor mechanism for these senses is the same. The receptors, or *hair cells,* are simple mechanoreceptors. The complex structure of the inner ear and the different arrangements of accessory structures account for the abilities of the hair cells to respond to different stimuli, and thus to provide the input for two different senses:

- Equilibrium, which informs us of the position of the body in space, by monitoring gravity, linear acceleration, and rotation.
- Hearing, which enables us to detect and interpret sound waves.

Anatomy of the Ear

The ear is divided into three anatomical regions: the *external ear,* the *middle ear,* and the *inner ear* (Figure 10-19•). The external ear is the visible portion of the ear, and it collects and directs sound waves to the eardrum. The middle ear is a chamber located within a thickened portion of the temporal bone. Structures within the middle ear collect and amplify sound waves and transmit them to a portion of the inner ear concerned with hearing. The inner ear also contains the sensory organs responsible for equilibrium sensations.

The External Ear

The **external ear** includes the fleshy **pinna,** or *auricle,* that surrounds the entrance to the **external auditory canal.** The pinna, which is supported by elastic cartilage, protects the opening of the canal and provides directional sensitivity to the ear. Sounds coming from behind the head are partially blocked by the pinna; sounds coming from the side are collected and channeled into the external auditory canal. (When you "cup your ear" with your hand to hear a faint sound more clearly, you are exaggerating this effect.) **Ceruminous** (se-ROO-mi-nus) **glands** along the external auditory canal secrete a waxy material (cerumen) that slows the growth of microorganisms and reduces the chances of infection. In addition, small, outwardly projecting hairs help prevent the entry of foreign objects or insects. The external auditory canal ends at the eardrum, also called the **tympanum,** or *tympanic membrane.* The tympanum is a thin sheet that separates the external ear from the middle ear (Figure 10-19●).

The Middle Ear

The **middle ear,** or *tympanic cavity,* is filled with air. It is separated from the external auditory canal by the tympanum, but it communicates with the superior portion of the pharynx, a region known as the *nasopharynx,* and with *air cells* within the mastoid process of the temporal bone. The connection with the nasopharynx is the **auditory tube,** also called the *pharyngotympanic tube* or the *Eustachian tube* (Figure 10-19●). The auditory tube permits the equalization pressure on either side of the eardrum. Unfortunately, it can also allow microorganisms to travel from the nasopharynx into the tympanic cavity. This can lead to an unpleasant middle ear infection known as *otitis media.* AM *Otitis Media and Mastoiditis*

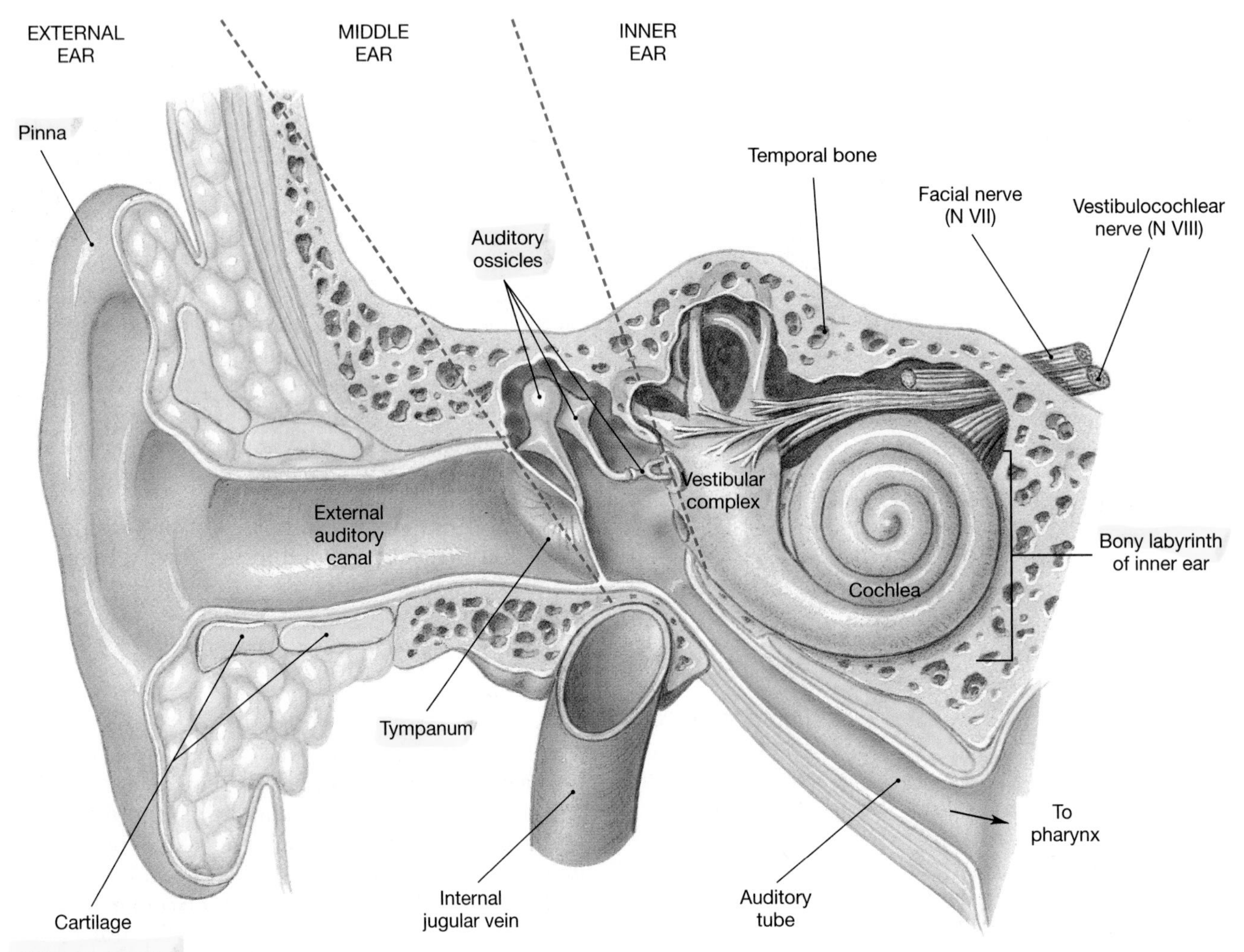

●**Figure 10-19**
Anatomy of the Ear
The orientation of the external, middle, and inner ear.

The Auditory Ossicles The middle ear contains three tiny ear bones, collectively called **auditory ossicles.** The ear bones connect the tympanum with the receptor complex of the inner ear (Figure 10-20●). The three auditory ossicles are the *malleus,* the *incus,* and the *stapes.* The **malleus** (*malleus,* a hammer) attaches at three points to the interior surface of the tympanum. The middle bone, the **incus** (*incus,* an anvil), attaches the malleus to the inner **stapes** (*stapes,* a stirrup). The base of the stapes almost completely fills the *oval window* of the inner ear.

Vibration of the tympanum converts arriving sound energy into mechanical movements of the auditory ossicles. The auditory ossicles act as levers that conduct the in-out vibrations to the fluid-filled inner chamber of the inner ear. Because the tympanum is larger and heavier than the oval window, the amount of movement increases markedly from tympanum to oval window.

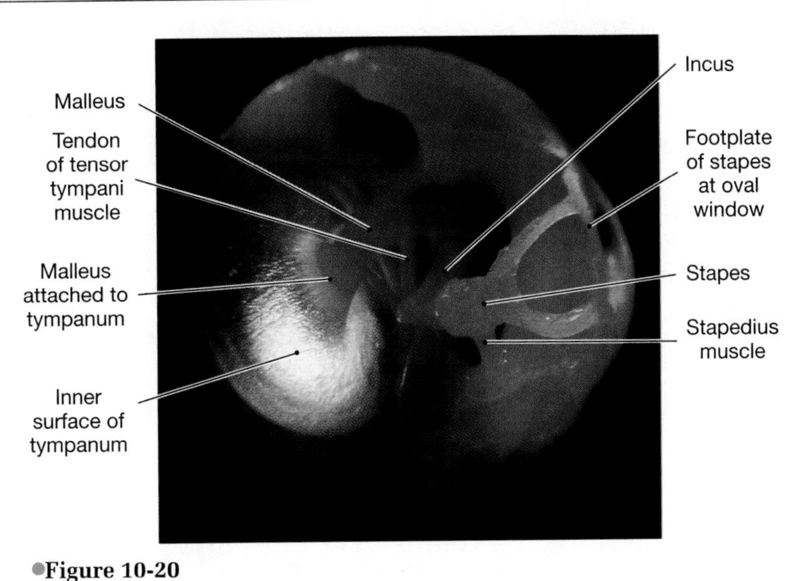

●**Figure 10-20**
The Middle Ear
The tympanum and auditory ossicles.

This magnification in movement is responsible for our ability to hear very faint sounds. It can also be a problem, however, when we are exposed to very loud noises. Within the tympanic cavity, two small muscles serve to protect the eardrum and ossicles from violent movements under noisy conditions. The *tensor tympani* (TEN-sor/tim-PAN-ē) *muscle* increases the tension, or stiffness, of the tympanum and reduces the amount of possible movement. The *stapedius* (stā-PĒ-dē-us) *muscle* pulls on the stapes, thereby reducing its movement at the oval window.

The Inner Ear

The senses of equilibrium and hearing are provided by the receptors of the **inner ear** (Figure 10-21a●). The receptors lie within a collection of fluid-filled tubes and chambers known as the **membranous labyrinth** (*labyrinthos,* a network of canals). The membranous labyrinth contains a fluid, called **endolymph** (EN-dō-limf). The **bony labyrinth** is a shell of dense bone that surrounds and protects the membranous labyrinth. Its inner contours closely follow the contours of the membranous labyrinth (Figure 10-21●), while its outer walls are fused with the surrounding temporal bone (see Figure 10-19●). Between the bony and membranous labyrinths flows another fluid, the **perilymph** (PER-i-limf).

The bony labyrinth can be subdivided into three parts as seen in Figure 10-21●.

1. **Vestibule**. The vestibule (VES-ti-būl), includes a pair of membranous sacs, the **saccule** (SAK-ūl) and the **utricle** (Ū-tre-kl). Receptors in the saccule and utricle provide sensations of gravity and linear acceleration.
2. **Semicircular canals.** The **semicircular canals** enclose slender *semicircular ducts.* Receptors in the semicircular ducts are stimulated by rotation of the head. The combination of vestibule and semicircular canals is called the **vestibular complex,** because the fluid-filled chambers within the vestibule are broadly continuous with those of the semicircular canals.
3. **Cochlea**. The bony **cochlea** (KOK-lē-a; *cochlea,* snail shell) contains the **cochlear duct** of the membranous labyrinth. Receptors within the cochlear duct provide the sense of hearing. The cochlear duct sits sandwiched between a pair of perilymph-filled chambers, and the entire complex is coiled around a central bony hub.

The walls of the bony labyrinth consist of dense bone everywhere except at two small areas near the base of the cochlear spiral. The **round window** is a thin, membranous partition that separates perilymph within the cochlea from the air within the middle ear. The margins of the **oval window** are firmly attached to the base of the stapes. When a sound vibrates the tympanum, the movements are conducted over the malleus and incus to the stapes. Movement of the stapes ultimately leads to the stimulation of receptors within the cochlear duct, and we hear the sound.

Receptor Function in the Inner Ear The basic receptors of the inner ear are called **hair cells** (Figure 10-21b●). Each hair cell communicates with a senso-

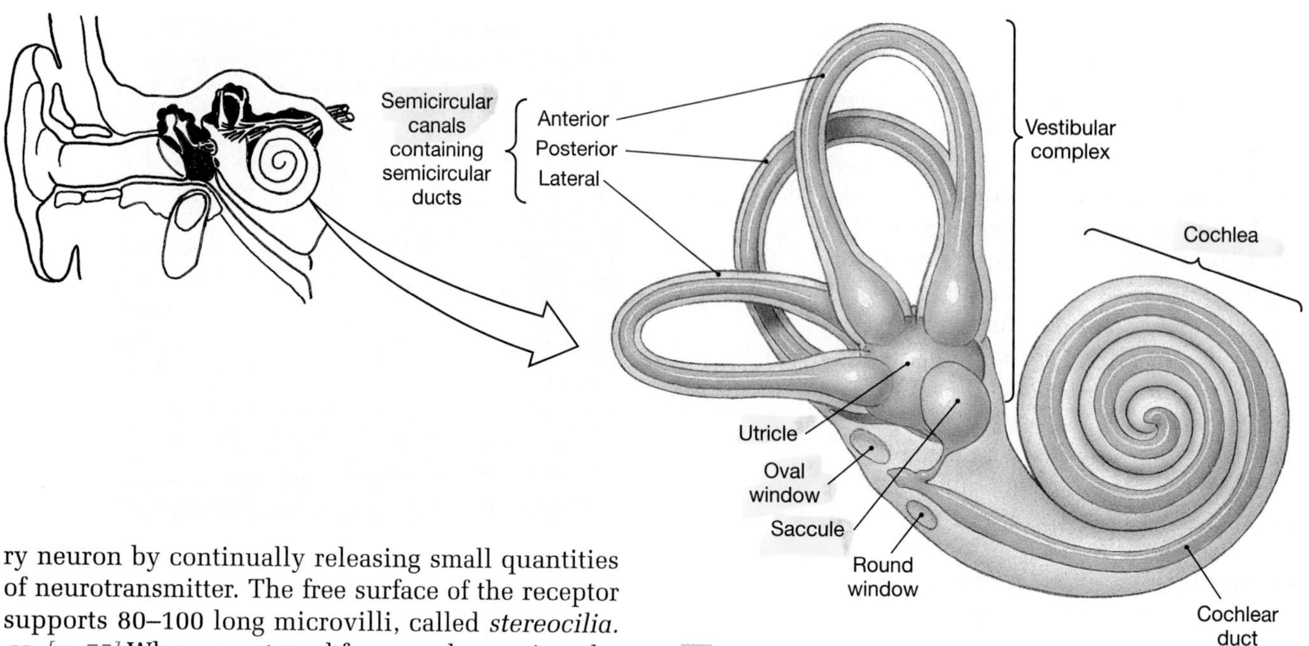

ry neuron by continually releasing small quantities of neurotransmitter. The free surface of the receptor supports 80–100 long microvilli, called *stereocilia.* *[p. 75]* When an external force pushes against the stereocilia, the movement distorts the cell surface and changes the membrane potential of the hair cell and alters its rate of neurotransmitter release.

Equilibrium

Equilibrium sensations are provided by receptors of the vestibular complex. The semicircular ducts provide information concerning rotational movements of the head. For example, when you turn your head to the left, receptors in the semicircular ducts tell you how rapid the movement is, and in what direction. The saccule and the utricle provide information about your position with respect to gravity. If you stand with your head tilted to one side, these receptors will report the angle involved and whether your head tilts forward or backward. These receptors are also stimulated by sudden changes in velocity. For example, when your car accelerates, the saccular and utricular receptors give you the impression of increasing speed.

The Semicircular Ducts: Rotational Motion

Receptors in the semicircular ducts respond to rotational movements. Figures 10-21a• and 10-22a• illustrate the **anterior, posterior,** and **lateral semicircular ducts** and their continuity with the utricle. Each semicircular duct contains a swollen region, the *ampulla,* which contains the sensory receptors. Hair cells are attached to the wall of the ampulla, with their stereocilia embedded in a gelatinous structure that nearly fills the ampulla (Figure 10-22b•). When the head rotates in the plane of the canal, movement of the endolymph pushes against this structure and stimulates the hair cells.

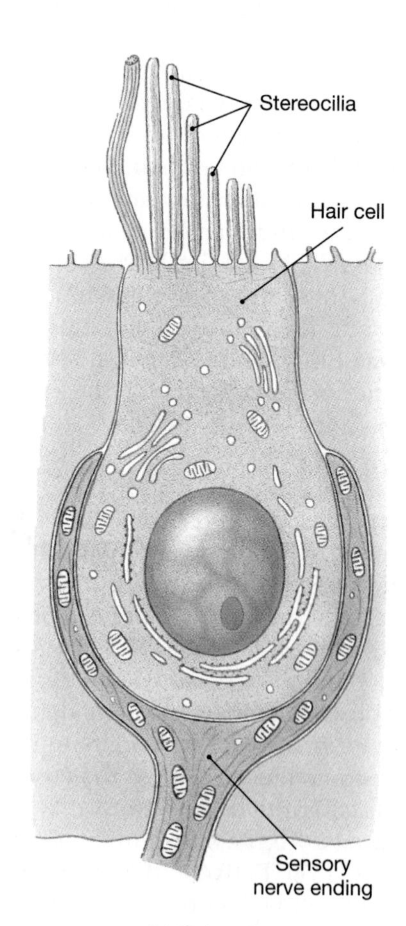

•Figure 10-21
The Inner Ear

(a) Anterior view of the bony labyrinth, showing the outline of the enclosed membranous labyrinth. **(b)** A representative hair cell (receptor) from the vestibular complex.

Each semicircular duct responds to one of three possible rotational movements. To distort the cupula and stimulate the receptors, endolymph must flow along the axis of the duct; that flow will occur only when there is rotation in that plane. A horizontal rotation, as in shaking the head "no," stimulates the hair cells of the lateral semicircular duct. Nodding "yes" excites receptors of the anterior duct, and tilting the head from side to side activates the receptors in the posterior duct. The three planes monitored by the semicircular ducts correspond to the three dimensions in the world around us, and they can provide accurate information about even the most complex movements.

The Vestibule: Gravity and Linear Acceleration

Receptors in the utricle and saccule respond to gravity and linear acceleration. As depicted in Figure 10-22a●, the hair cells of the utricle and saccule are clustered in oval **maculae** (MAK-ū-lē; *macula,* spot). As in the ampullae, the hair cell processes are embedded in a gelatinous mass, but the macular receptors lie under a thin layer containing densely packed mineral crystals. One of these *otoliths* (*oto-*, ear + *lithos*, a stone) can be seen in Figure 10-22d●. When the head is in the normal, upright position the otolith sits atop the macula. The weight presses down on the macular surface, pushing the sensory hairs downward rather than to one side or another. When the head is tilted, the pull of gravity on the otolith shifts the mass to the side. This shift distorts the sensory hairs, and the change in receptor activity tells the CNS that the head is no longer level (Figure 10-22d●).

Otoliths are relatively dense and heavy, and they are connected to the rest of the body only by the sensory processes of the macular cells. So whenever the rest of the body makes a sudden movement, the otoliths lag behind. For example, when an elevator starts downward, we are immediately aware of it because the otoliths no longer push so forcefully against the surfaces of the receptor cells. Once they catch up, and the elevator has reached a constant speed, we are no longer aware of any movement until the elevator brakes to a halt. As the body slows down, the otoliths press harder against the hair cells, and we "feel" the force of gravity increase.

A similar mechanism accounts for our perception of linear acceleration in a car that speeds up suddenly. The otoliths lag behind, distorting the sensory hairs and changing the activity in the sensory neurons. A comparable otolith movement occurs when the chin is raised and gravity pulls the otoliths backward. The brain decides whether the arriving sensations indicate acceleration or a change in head position on the basis of visual information. Flight simulators and some new arcade games take advantage of this mechanism; they provide visual images that will make the brain interpret a tilt as an acceleration. AM *Vertigo*

Central Processing of Vestibular Sensations

Hair cells of the vestibule and semicircular canals are monitored by sensory neurons whose fibers form the **vestibular branch** of the vestibulocochlear nerve, N VIII. These fibers synapse on neurons within the *vestibular nuclei* at the boundary between the pons and medulla. The two vestibular nuclei (1) integrate the sensory information arriving from each side of the head; (2) relay information to the cerebellum; (3) relay information to the cerebral cortex, providing a conscious sense of position and movement; and (4) send commands to motor nuclei in the brain stem and spinal cord. These reflexive motor commands are distributed to the motor nuclei for cranial nerves involved with eye, head, and neck movements (N III, IV, VI, and XI). Descending instructions along the *vestibulospinal tracts* of the spinal cord adjust peripheral muscle tone to complement the reflexive movements of the head or neck.

Hearing

The receptors of the cochlear duct provide us with a sense of hearing that enables us to detect the quietest whisper yet remain functional in a crowded, noisy room. The receptors responsible for auditory sensations are hair cells similar to those of the vestibular complex. However, their placement within the cochlear duct and the organization of the surrounding accessory structures shield them from stimuli other than sound. In conveying vibrations from the tympanum to the oval window, the auditory ossicles convert sound energy (pressure waves) in air to pressure pulses in the perilymph of the cochlea. These pressure pulses stimulate hair cells along the cochlear spiral. The *frequency* of the perceived sound is determined by which part of the cochlear duct is stimulated. The *intensity* (volume) of the perceived sound is determined by *how many* of the hair cells at that location are stimulated.

The Cochlear Duct

In sectional view (Figure 10-23a●, p. 274) the cochlear duct, or *scala media,* lies between a pair of perilymphatic chambers, the **vestibular duct** (*scala vestibuli*) and the **tympanic duct** (*scala tympani*). The vestibular and tympanic ducts are interconnected at the tip of the cochlear spiral. The outer surfaces of these ducts are encased by the bony labyrinth everywhere except at the oval window

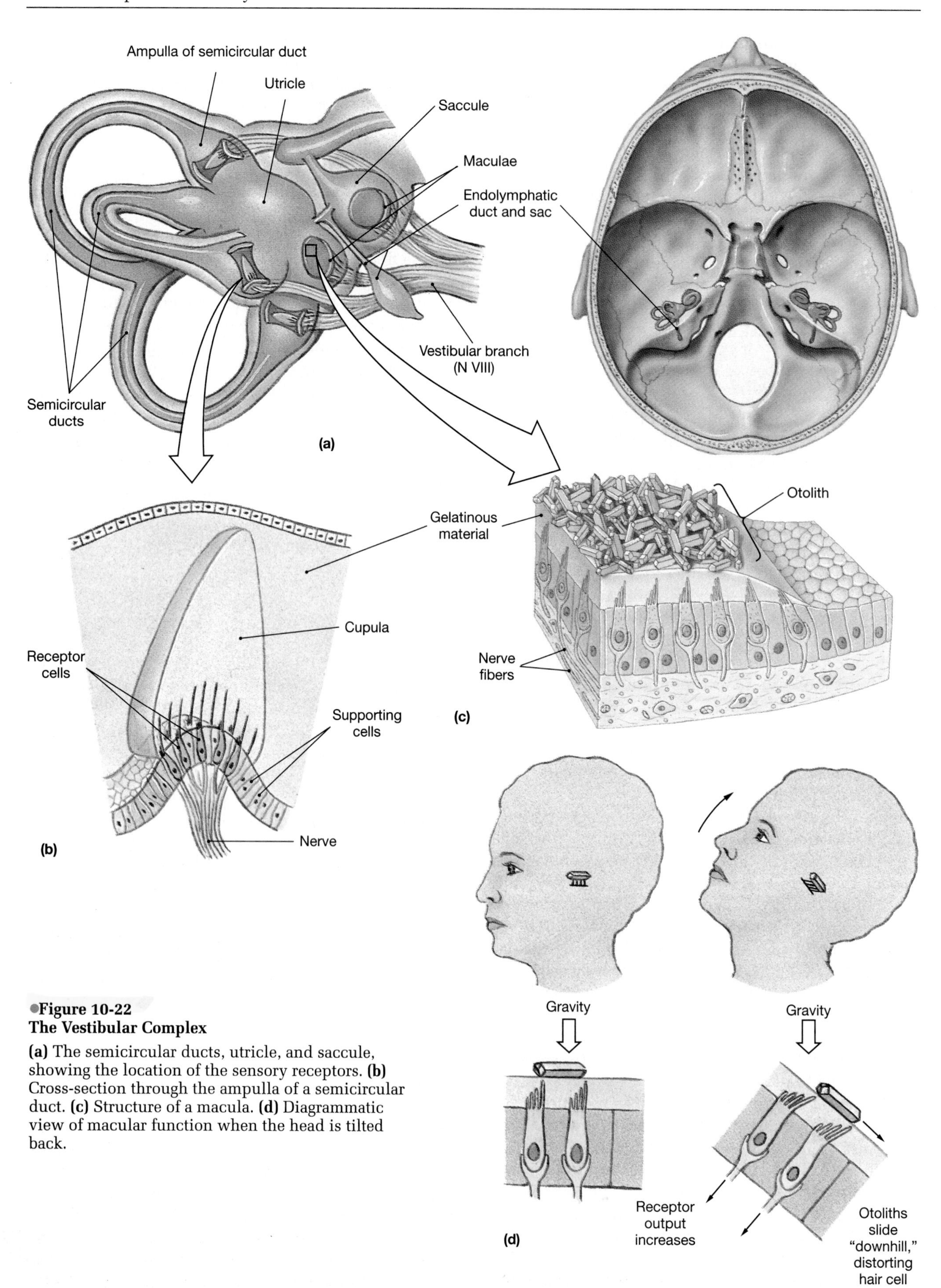

Figure 10-22
The Vestibular Complex

(a) The semicircular ducts, utricle, and saccule, showing the location of the sensory receptors. **(b)** Cross-section through the ampulla of a semicircular duct. **(c)** Structure of a macula. **(d)** Diagrammatic view of macular function when the head is tilted back.

(base of the vestibular duct) and the round window (base of the tympanic duct).

The Organ of Corti The hair cells of the cochlear duct are found in the **organ of Corti**, or *spiral organ* (Figure 10-23b●). This sensory structure sits above the **basilar membrane** that separates the cochlear duct from the tympanic duct. The hair cells are arranged in a series of longitudinal rows, with their stereocilia in contact with the overlying **tectorial membrane** (tek-TŌR-ē-al; *tectum,* roof). This membrane is firmly attached to the inner wall of the cochlear duct. When a portion of the basilar membrane bounces up and down, the stereocilia of the hair cells are distorted as they are pushed up against the tectorial membrane. The basilar membrane moves in response to pressure waves within the perilymph. These waves are produced when sounds arrive at the tympanum; to understand how these waves develop, we must consider the basic properties of sound.

The Hearing Process

Hearing is the detection of sound, which consists of pressure waves conducted through air, water, and solids. Physicists use the term **cycles** rather than waves, and the number of cycles per second (cps), or **hertz (Hz)**, represents the **frequency** of the sound. What we perceive as the **pitch** of a sound (how high or low it is) is our sensory response to its frequency. A *high-frequency* sound (high pitch) might have a frequency of 15,000 Hz or more; a very *low-frequency* sound (low pitch) could have a frequency of 100 Hz or less.

The process of hearing can be divided into six basic steps, diagrammed in Figure 10-23c● and summarized in Table 10-2.

Step 1: *Sound waves arrive at the tympanum.* Sound waves enter the external auditory canal and travel toward the tympanum. Sound waves approaching the side of the head have direct access to the tympanum on that side, whereas sounds arriving from another direction must bend around corners or pass through the pinna or other body tissues.

Step 2: *Vibration of the tympanic membrane causes movement of the auditory ossicles.* The tympanum provides the surface for sound collection, and it vibrates to sound waves with frequencies between approximately 20 and 20,000 Hz (in a young child). When the tympanum vibrates, so does the malleus and, via their articulations, the incus and stapes.

Step 3: *Movement of the stapes at the oval window establishes pressure waves in the perilymph of the vestibular duct.* Movement of the stapes at the oval window applies pressure to the perilymph of the vestibular duct. Because the rest of the cochlea is sheathed in bone, pressure applied at the oval window can be relieved only at the round window. When the stapes moves inward, the round window bulges outward.

Step 4: *The pressure waves distort the basilar membrane on their way to the round window of the tympanic duct.* These pressure waves cause movement in the basilar membrane. The location of maximum stimulation varies depending on the frequency of the sound. High-frequency sounds, which have a very short wavelength, vibrate the basilar membrane near the oval window. The lower the frequency of the sound, the longer the wavelength, and the farther away from the oval window the area of maximum distortion will be. The actual *amount* of movement at a given location will depend on the amount of force applied by the stapes. The louder the sound, the greater the movement of the basilar membrane.

Step 5: *Vibration of the basilar membrane causes vibration of hair cells against the tectorial membrane.* Vibration of the affected region of the basilar membrane moves hair cells against the tectorial membrane. The resulting displacement of the hair cells' stereocilia stimulates sensory neurons. The hair cells are arranged in several rows. A very soft sound may stimulate only a few hair cells in a portion of one row. As the volume of a sound increases, not only do these hair cells become more active, but additional hair cells—at first in the same row, and then in adjacent rows—are stimulated as well. The number of hair cells responding in a given region of the organ of Corti thus provides information on the volume of the sound.

Step 6: *Information concerning the region and intensity of stimulation is relayed to the CNS over the cochlear branch of N VIII.*The cell bodies of the sensory neurons that monitor the cochlear hair cells are found at the center of the bony cochlea (Figure 10-23a●) in the **spiral ganglion.** The information is carried to the cochlear nuclei of the medulla oblongata, for subsequent distribution to other centers in the brain.

Auditory Pathways

Hair cell stimulation activates sensory neurons whose cell bodies are in the adjacent spiral ganglion. Their afferent fibers form the **cochlear branch** (Figure 10-24●) of the vestibulocochlear nerve (N VIII). These axons enter the medulla oblongata and synapse at the *cochlear nucleus.* From here the information crosses to the opposite side of the brain and ascends to the *inferior colliculus* of the midbrain. This processing center coordinates a number of responses to acoustic stimuli, including auditory reflexes involving skele-

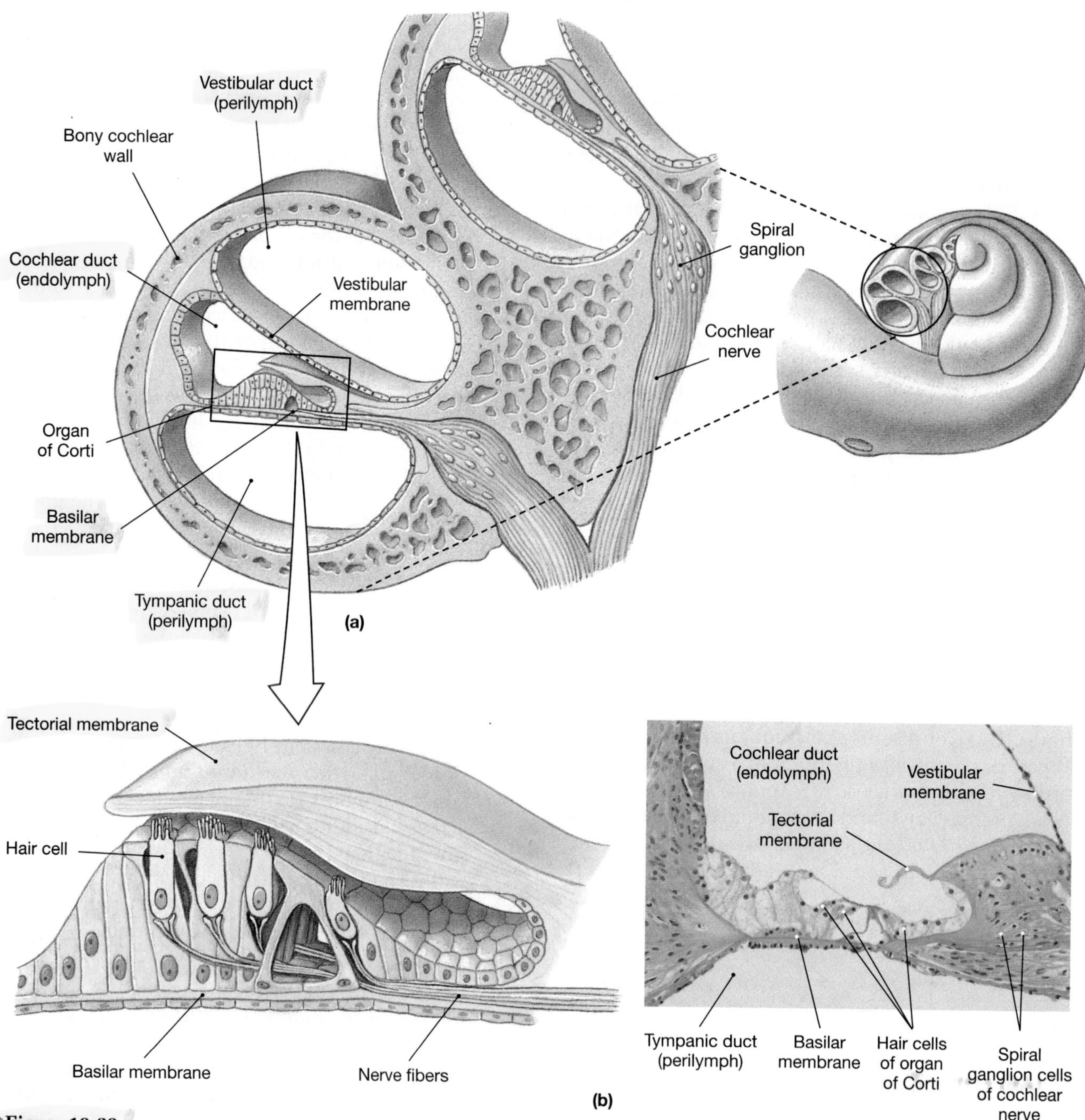

Figure 10-23
The Cochlea and Organ of Corti
(a) Structure of the cochlea as seen in section. **(b)** The three-dimensional structure of the tectorial membrane and hair cell complex of the organ of Corti. **(c)** Steps in the reception of sound and the process of hearing.

tal muscles of the head, face, and trunk. For example, these reflexes automatically change the position of the head in response to a sudden loud noise.

Before reaching the cerebral cortex and our conscious awareness, ascending auditory sensations synapse in the thalamus. Thalamic fibers then deliver the information to the auditory cortex of the temporal lobe. In effect, the auditory cortex contains a map of the organ of Corti. High-frequency sounds activate one portion of the cortex and low-frequency sounds affect another. If the auditory cortex is damaged, the individual will respond to sounds and have normal acoustic reflexes, but sound interpretation and pattern recognition will be difficult or impossible. Damage to the adjacent association area leaves the ability to detect the tones and patterns, but produces an inability to comprehend their meaning.

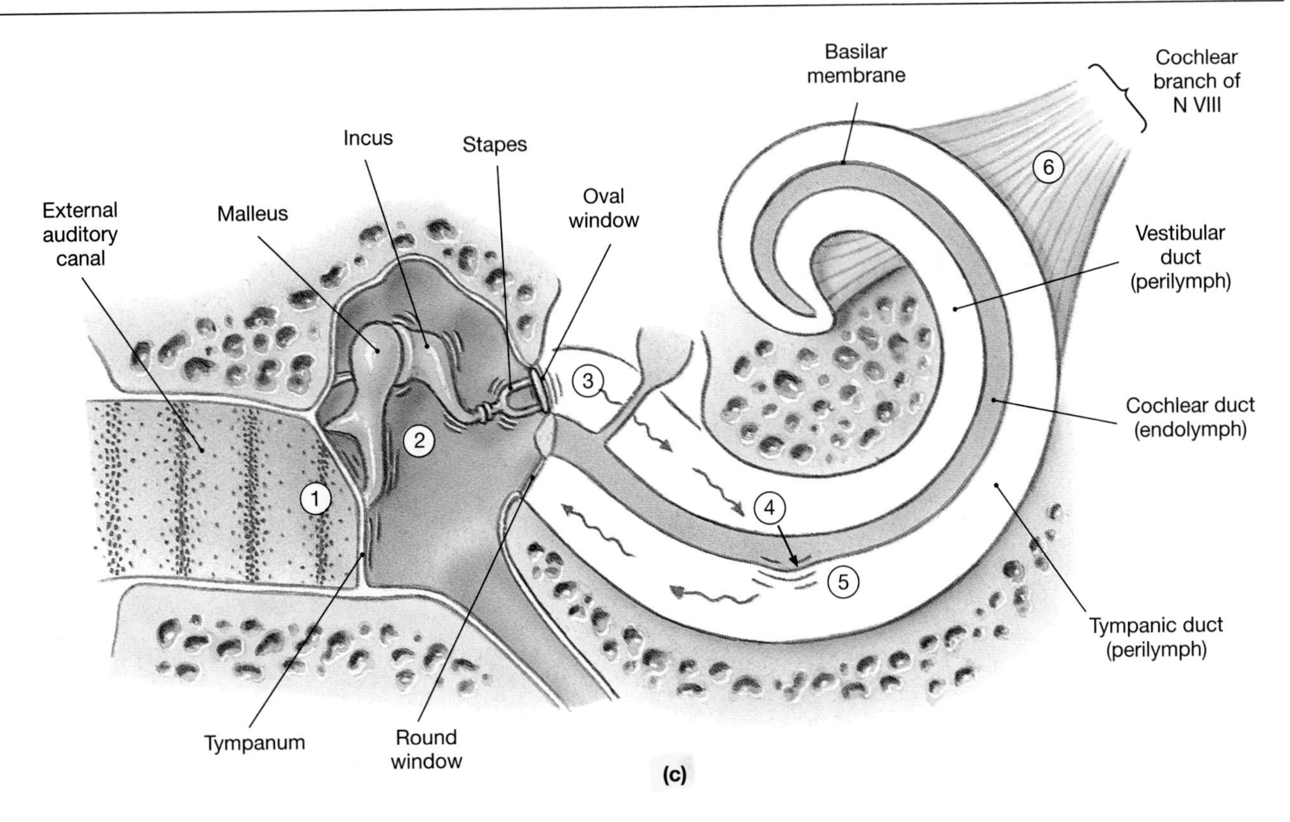

(c)

Auditory Sensitivity

Our hearing abilities are remarkable, though it is difficult to assess the absolute sensitivity of the system. From the softest audible sound to the loudest tolerable blast represents a trillionfold increase in power. Theoretically, if we were to remove the stapes, the receptor mechanism is so sensitive that we could hear the sound of air molecules bouncing off the oval window, responding to displacements as small as one-tenth the diameter of a hydrogen atom. We never utilize the full potential of this system because body movements and our internal organs produce squeaks, groans, thumps, and other sounds that are tuned out by adaptation. When other environmental noises fade away, the level of adaptation drops and the system becomes increasingly sensitive. If we relax in a quiet room, our heartbeat seems to get louder and louder as the auditory system adjusts to the level of background noise.

TABLE 10-2 Steps in the Production of an Auditory Sensation

1. Sound waves arrive at the tympanum.
2. Movement of the tympanum causes displacement of the auditory ossicles.
3. Movement of the stapes at the oval window establishes pressure waves in the perilymph of the vestibular duct.
4. The pressure waves distort the basilar membrane on their way to the round window of the tympanic duct.
5. Vibration of the basilar membrane causes vibration of hair cells against the tectorial membrane.
6. Information concerning the region and intensity of stimulation is relayed to the CNS over the cochlear branch of N VIII.

HEARING DEFICITS

There are probably over 6 million people in the United States alone who have at least a partial hearing deficit. **Conductive deafness** results from conditions in the middle ear that block the normal transfer of vibration from the tympanum to the oval window. Plugging of the external auditory canal by accumulated wax or trapped water may cause a temporary hearing loss. Scarring or perforation of the tympanum and immobilization of one or more of the auditory ossicles are more serious examples of conduction deafness.

In **nerve deafness,** the problem lies within the cochlea or somewhere along the auditory pathway. The vibrations are reaching the oval window and entering the perilymph, but the receptors either cannot respond or their response cannot reach its central destinations. For example, very loud (high-intensity) sounds can produce nerve deafness by breaking stereocilia off the surfaces of the hair cells. (The reflex contraction of the tensor tympani and stapedius muscles in response to a dangerously loud noise occurs in less than 0.1 second, but this may not be fast enough.) Drugs such as the aminoglycoside antibiotics (*neomycin* or *gentamicin*) may diffuse into the endolymph and kill the hair cells. Because hair cells and sensory nerves can also be dam-

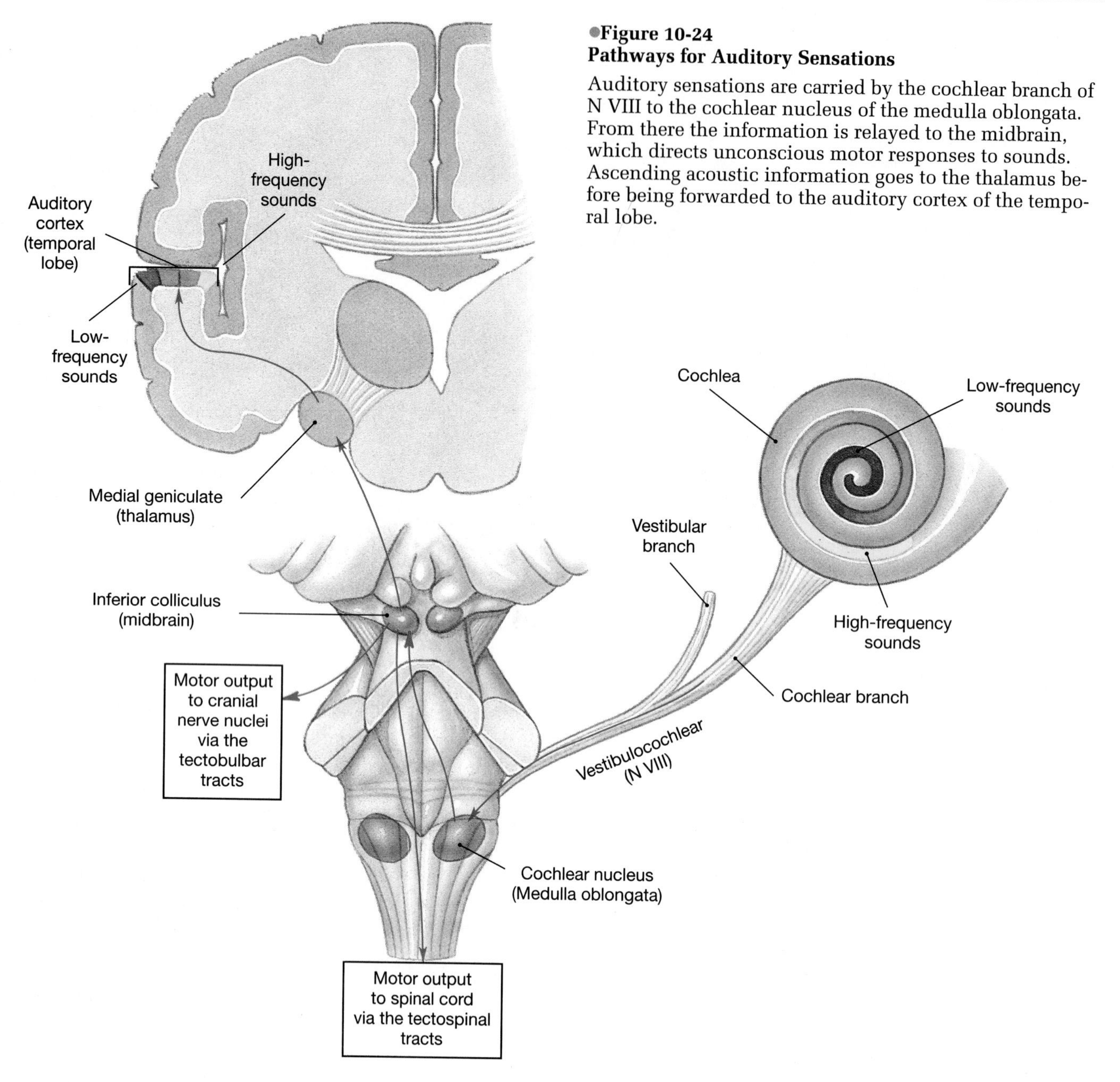

•Figure 10-24
Pathways for Auditory Sensations

Auditory sensations are carried by the cochlear branch of N VIII to the cochlear nucleus of the medulla oblongata. From there the information is relayed to the midbrain, which directs unconscious motor responses to sounds. Ascending acoustic information goes to the thalamus before being forwarded to the auditory cortex of the temporal lobe.

aged by bacterial infection, the potential side effects must be balanced against the severity of infection.

There are many treatment options for conductive deafness; treatment options for nerve deafness are relatively limited. Because many of these problems become progressively worse, early diagnosis improves the chances for successful treatment. AM *Testing and Treating Hearing Deficits*

✓ If the round window were not able to bulge out with increased pressure in the perilymph, how would sound perception be affected?

✓ How would loss of stereocilia from the hair cells of the organ of Corti affect hearing?

Chapter Review

KEY TERMS

cochlea, p. 269
fovea, p. 261
gustation, p. 255
iris, p. 258
macula, p. 271
nociceptors, p. 251
olfaction, p. 253
proprioception, p. 250
pupil, p. 258
retina, p. 259
sclera, p. 257
visual accommodation, p. 262

SUMMARY OUTLINE

INTRODUCTION p. 250

1. The general senses are temperature, pain, touch, pressure, vibration, and proprioception; receptors for these sensations are distributed throughout the body. Receptors for the **special senses** (smell, taste, sight, balance, and hearing) are located in specialized areas or in **sense organs.**

2. A sensory receptor is a specialized cell that, when stimulated, sends a sensation to the CNS. The simplest receptors are **free nerve endings;** the most complex have specialized accessory structures that isolate them from most all but specific types of stimuli.

3. Sensory information is relayed in the form of action potentials in an afferent (sensory) fiber. In general, the larger the stimulus, the greater the frequency of action potentials. The CNS interprets the nature of the arriving sensory information on the basis of the area of the brain stimulated.

4. **Adaptation** (a reduction in sensitivity in the presence of a constant stimulus) may involve changes in receptor sensitivity or inhibition along the sensory pathways.

THE GENERAL SENSES p. 250

Pain p. 251

1. **Nociceptors** respond to a variety of stimuli usually associated with tissue damage. There are two types of these painful sensations: **fast pain,** or prickling pain, and **slow pain,** or burning and aching pain.

2. The perception of pain coming from parts of the body that are not actually stimulated is called **referred pain.** (*Figure 10-1*)

Temperature p. 251

3. **Thermoreceptors** respond to changes in temperature.

Touch, Pressure, and Position p. 251

4. **Mechanoreceptors** respond to physical distortion, contact, or pressure on their cell membranes; **tactile receptors** to touch, pressure, and vibration; **baroreceptors** to pressure changes in the walls of blood vessels, the digestive and urinary tracts, and the lungs; and **proprioceptors** to positions of joints and muscles.

5. **Fine touch** and **pressure receptors** provide detailed information about a source of stimulation; **crude touch** and **pressure receptors** are poorly localized. Important tactile receptors include the *root hair plexus, Merkel's discs, Meissner's corpuscles, Pacinian corpuscles, and Ruffini corpuscles.* (*Figure 10-2*)

6. Baroreceptors monitor changes in pressure; they respond immediately but adapt rapidly. Baroreceptors in the walls of major arteries and veins respond to changes in blood pressure. Receptors along the digestive tract help coordinate reflex activities of digestion. (*Figure 10-3*)

7. Proprioceptors monitor the position of joints, tension in tendons and ligaments, and the state of muscular contraction. Proprioceptors include *tendon organs and muscle spindles.*

Chemical Detection p. 253

8. In general, **chemoreceptors** respond to water-soluble and lipid-soluble substances that are dissolved in the surrounding fluid. They monitor the chemical composition of body fluids. (*Figure 10-4*)

SMELL p. 253

1. The **olfactory organs** contain the **olfactory epithelium** with **olfactory receptors** (neurons sensitive to chemicals dissolved in the overlying mucus), supporting cells, and *basal* (stem) *cells.* Their surfaces are coated with the secretions of the **olfactory glands.** (*Figure 10-5*)

2. The olfactory receptors are modified neurons. Our olfactory sensitivities are much lower than those of many other vertebrates.

Olfactory Pathways p. 255

3. The olfactory system is very sensitive; its extensive limbic and hypothalamic connections help explain the emotional and behavioral responses that can be produced by certain smells.

TASTE p. 255

1. **Gustatory (taste) receptors** are clustered in **taste buds,** each of which contains **gustatory cells,** which extend taste hairs through a narrow **taste pore.** (*Figure 10-6*)

2. Taste buds are associated with epithelial projections (**papillae**) on the superior surface of the tongue. (*Figure 10-6*)

3. The **primary taste sensations** are sweet, salt, sour, and bitter.

Taste Pathways p. 256

4. The taste buds are monitored by cranial nerves that synapse within a nucleus of the medulla oblongata.

VISION p. 256

Accessory Structures of the Eye p. 256

1. The **accessory structures** of the eye include the eyelids, the eyelashes, various exocrine glands, and the extrinsic eye muscles.

2. An epithelium called the **conjunctiva** covers the exposed surface of the eye except over the transparent **cornea.**

3. The secretions of the **lacrimal gland** bathe the conjunctiva; these secretions are slightly alkaline and contain a *lysozyme* (an enzyme that attacks bacteria). Tears reach the nasal cavity after passing through the, the **lacrimal canals,** the **lacrimal sac,** and the **nasolacrimal duct.** (*Figure 10-7*)

4. **Six extrinsic eye muscles** control external eye movements: the **inferior** and **superior rectus, lateral** and **medial rectus,** and **superior** and **inferior obliques.** (*Figure 10-8; Table 10-1*)

Anatomy of the Eye p. 257

5. The eye has three layers: an outer **fibrous tunic,** a vascular tunic, and an inner neural tunic. Most of the ocular surface is covered by the **sclera** (a dense fibrous connective tissue), which is continuous with the **cornea.** (*Figure 10-9*)

6. The **vascular tunic** includes the **iris,** the **ciliary body,** and the **choroid.** The iris forms the boundary between the anterior and posterior chambers. The ciliary body contains the *ciliary muscle* and the *ciliary processes,* which attach to the **suspensory ligaments** of the **lens.** (*Figure 10-9*)

7. The **neural tunic** consists of an outer **pigment layer** and an inner **neural retina;** the latter contains visual receptors and associated neurons. (*Figures 10-9, 10-10*)

8. From the photoreceptors, the information is relayed to **bipolar cells,** then to **ganglion cells,** and to the brain via the optic nerve. Horizontal cells and amacrine cells modify the signals passed between other retinal components. (*Figure 10-10*)

9. The ciliary body and lens divide the interior of the eye into a large **posterior cavity** and a smaller **anterior cavity.** The anterior cavity is subdivided into the **anterior chamber,** which extends from the cornea to the iris, and a **posterior chamber** between the iris and the ciliary body and lens. The posterior chamber contains the *vitreous body,* a gelatinous mass that helps stabilize the shape of the eye and supports the retina. (*Figure 10-12*)

10. **Aqueous humor** circulates within the eye and reenters the circulation after diffusing through the walls of the anterior chamber and into veins of the sclera through the *canal of Schlemm.* (*Figure 10-12*)

11. The lens, held in place by the suspensory ligaments, focuses a visual image on the retinal receptors. Light is *refracted* (bent) when it passes through the cornea and lens. During **accommodation** the shape of the lens changes to focus an image on the retina. (*Figures 10-13, 10-14*)

Visual Physiology p. 263

12. Light is radiated in waves with a characteristic wavelength. A *photon* is a single energy packet of visible light. There are two types of photoreceptors (visual receptors of the retina): **rods** and **cones.** Rods respond to almost any photon, regardless of its energy content; cones have characteristic ranges of sensitivity. Many cones are densely packed within the **fovea** (the central portion of the **macula lutea**), the site of sharpest vision. (*Figures 10-15, 10-16*)

13. Each photoreceptor contains membranous **discs** containing **visual pigments.** Light absorption occurs in the visual pigments, which are derivatives of **rhodopsin** (**opsin** plus a pigment, **retinal,** that is synthesized from **vitamin A**). A photoreceptor responds to light by changing its rate of neurotransmitter release and thereby altering the activity of a bipolar cell. (*Figure 10-17*)

The Visual Pathway p. 266

14. The message is relayed from photoreceptors to bipolar cells to ganglion cells within the retina. The axons of ganglion cells converge at the optic disc and leave the eye as the optic nerve. A partial crossover occurs at the optic chiasm before the information reaches the lateral geniculate of the thalamus on each side of the brain. From the lateral geniculate, visual information is relayed to the visual cortex of the occipital lobe, which contains a sensory map of the field of vision. (*Figure 10-18*)

EQUILIBRIUM AND HEARING p. 267

1. The senses of equilibrium and hearing are provided by the receptors of the **inner ear** (also known as the **membranous labyrinth**). Its chambers and canals contain the fluid **endolymph.** The **bony labyrinth** surrounds and protects the membranous labyrinth, and the space between contains the fluid **perilymph.** The bony labyrinth can be subdivided into the **vestibule,** the **semicircular canals** (receptors in the vestibule and semicircular canals provide the sense of equilibrium) and the **cochlea** (these receptors provide the sense of hearing). The structures and air spaces of the **external ear** and **middle ear** help capture and transmit sound to the cochlea. (*Figures 10-19, 10-21*)

Anatomy of the Ear p. 267

2. The external ear includes the **pinna** that surrounds the entrance to the **external auditory canal** that ends at the **tympanum** (eardrum). (*Figure 10-19*)

3. The middle ear communicates with the nasopharynx via the **auditory tube** (*pharyngotympanic tube or Eustachian tube*). The middle ear encloses and protects the **auditory ossicles,** which connect the tympanum with the receptor complex of the inner ear. (*Figures 10-19, 10-20*)

4. The vestibule includes a pair of membranous sacs, the **saccule** and **utricle,** whose receptors provide sensations of gravity and linear acceleration. The semicircular canals contain the **semicircular ducts,** whose receptors provide sensations of rotation. The cochlea contains the **cochlear duct,** an elongated portion of the membranous labyrinth. (*Figure 10-21a*)

5. The basic receptors of the inner ear are **hair cells** whose surfaces support *stereocilia.* Hair cells provide information about the direction and strength of mechanical stimuli. (*Figure 10-21b*)

Equilibrium p. 270

6. The **anterior, posterior,** and **lateral semicircular ducts** are attached to the utricle. Each semicircular duct contains an *ampulla* with sensory receptors. Here the stereocilia contact a gelatinous mass that is distorted when endolymph flows along the axis of the duct. (*Figures 10-21, 10-22a,b*)

7. In the saccule and utricle, hair cells cluster within **maculae,** where their cilia contact *otoliths* (densely packed mineral crystals). When the head tilts, the mass of otoliths shifts, and the resulting distortion in the sensory hairs signals the CNS. (*Figure 10-22c,d*)

8. The vestibular receptors activate sensory neurons whose axons form the **vestibular branch** of the vestibulocochlear nerve (N VIII), synapsing within the *vestibular nuclei.*

Hearing p. 271

9. Sound waves travel toward the tympanum, which vibrates; the auditory ossicles conduct the vibrations to the inner ear. Movement at the oval window applies pressure to the perilymph of the **vestibular duct.** (*Figure 10-23; Table 10-2*)

10. Pressure waves distort the **basilar membrane** and push the hair cells of the **organ of Corti** against the **tectorial membrane.** The *tensor tympani* and *stapedius muscles* contract to reduce the amount of motion when very loud sounds arrive. (*Figure 10-23*)

11. The sensory neurons are located in the **spiral ganglion** of the cochlea. Afferent fibers of sensory neurons form the **cochlear branch** of the vestibulocochlear nerve (N VIII), synapsing at the *cochlear nucleus.* (*Figure 10-24*)

CHAPTER QUESTIONS

LEVEL 1 Reviewing Facts and Terms

Match each item in column A with the most closely related item in column B. Use letters for answers in the spaces provided.

Column A	Column B
___ 1. myopia	a. pain receptors
___ 2. fibrous tunic	b. free nerve endings
___ 3. nociceptors	c. sclera and cornea
___ 4. proprioceptors	d. rotational movements
___ 5. cones	e. provide information on joint position
___ 6. accommodation	f. color vision
___ 7. tympanum	g. site of sharpest vision
___ 8. thermoreceptors	h. active in dim light
___ 9. rods	i. eardrum
___ 10. olfaction	j. change in lens shape to focus retinal image
___ 11. fovea	k. nearsighted
___ 12. hyperopia	l. farsighted
___ 13. maculae	m. smell
___ 14. semicircular ducts	n. gravity and acceleration receptors

15. Regardless of the nature of a stimulus, sensory information must be sent to the CNS in the form of:
(a) dendritic processes
(b) action potentials
(c) neurotransmitter molecules
(d) generator potentials

16. A reduction in sensitivity in the presence of constant stimulus is called:
(a) transduction (b) sensory coding
(c) line labeling (d) adaptation

17. Mechanoreceptors that detect pressure changes in the walls of blood vessels and in portions of the digestive, reproductive, and urinary tracts are:
(a) tactile receptors (b) baroreceptors
(c) proprioceptors (d) free nerve endings

18. Examples of proprioceptors that monitor the position of joints and the state of muscular contraction are:
(a) Pacinian and Meissner's corpuscles
(b) carotid and aortic sinuses
(c) Merkel's discs and Ruffini corpuscles
(d) tendon organs and muscle spindles

19. When chemicals dissolve in the nasal cavity, they stimulate:
(a) gustatory cells (b) olfactory hairs
(c) rod cells (d) tactile receptors

20. The taste sensation of sweetness is experienced on the:
(a) posterior part of the tongue
(b) anterior part of the tongue
(c) right and left lateral sides of the tongue
(d) the middle part of the tongue

21. The purpose of tears produced by the lacrimal apparatus is to:
(a) keep conjunctival surfaces moist and clean
(b) reduce friction and remove debris from the eye
(c) provide nutrients and oxygen to the conjunctional epithelium
(d) a, b, and c are correct

22. The thickened gel-like fluid that helps support the structure of the eyeball is known as the:
(a) vitreous humor (b) aqueous humor
(c) ora serrata (d) perilymph

23. The retina is considered to be a component of the:
(a) vascular tunic (b) fibrous tunic
(c) neural tunic (d) a, b, and c are correct

24. At sunset or sunrise your visual system adapts to:
(a) fovea vision (b) rod-based vision
(c) macular vision (d) cone-based vision

25. The malleus, incus, and stapes are the tiny ear bones located in the:
(a) outer ear (b) middle ear
(c) inner ear (d) membranous labyrinth

26. Receptors in the saccule and utricle provide sensations of:
(a) balance and equilibrium
(b) hearing
(c) vibration
(d) gravity and linear acceleration

27. The organ of Corti is located within the ________ of the inner ear.
(a) utricle (b) bony labyrinth
(c) vestibule (d) cochlea

28. What three types of mechanoreceptors respond to stretching, compression, twisting, or other distortions of the cell membrane?

29. Identify six types of tactile receptors found in the skin and their sensitivities.

30. (a) What structures make up the fibrous tunic of the eye? (b) What are the functions of the fibrous tunic?

31. What structures are included as parts of the vascular tunic of the eye?

32. What six basic steps are involved in the process of hearing?

LEVEL 2 Reviewing Concepts

33. The CNS interprets sensory information entirely on the basis of the:
    (a) strength of the action potential
    (b) number of generator potentials
    (c) area of brain stimulated
    (d) a, b, and c are correct
34. If the auditory cortex is damaged, the individual will respond to sounds and have normal acoustic reflexes, but:
    (a) the sounds may produce nerve deafness
    (b) the auditory ossicle may be immobilized
    (c) sound interpretation and pattern recognition may be impossible
    (d) normal transfer of vibration to the oval window is inhibited
35. Distinguish between the general senses and the special senses in the human body.
36. In what form does the CNS receive a stimulus detected by a sensory receptor?
37. Why are olfactory sensations long-lasting and an important part of our memories and emotions?
38. Jane makes an appointment with the optometrist for a vision test. Her test results are reported as 20/15. What does this test result mean? Is a rating of 20/20 better or worse?

LEVEL 3 Critical Thinking and Clinical Applications

39. You are at a park watching some deer 35 feet away from you when your friend taps you on the shoulder to ask a question. As you turn to look at your friend who is standing 2 feet away, what changes will occur regarding your eyes?
40. After attending a Fourth of July fireworks extravaganza, Millie finds it difficult to hear normal conversation, and her ears keep "ringing." What is causing her hearing problems?
41. After riding the express elevator from the twentieth floor to the ground floor, for a few seconds you still feel as if you are descending, even though you have obviously come to a stop. Why?

11 THE ENDOCRINE SYSTEM

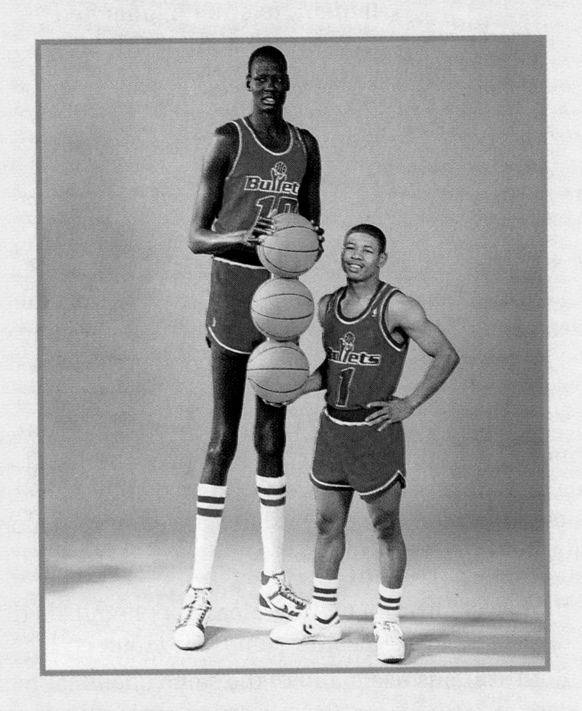

These two basketball players, Manute Bol and Muggsy Bogues, differ in height by more than two feet. Differences in genetically programmed levels of hormones, especially growth hormone, were probably responsible for this difference.

Chapter Outline and Objectives

Vocabulary Development

ad-, to or toward; *adrenal*
andros, man; *androgen*
angeion, vessel; *angiotensin*
corpus, body; *corpus luteum*
diabetes, to pass through; *diabetes*
erythros, red; *erythropoietin*
infundibulum, funnel; *infundibulum*
insipidus, tasteless; *diabetes insipidus*
lac, milk; *prolactin*
mellitum, honey; *diabetes mellitus*
natrium, sodium; *natriuretic*
ouresis, making water; *polyuria*
oxy-, quick; *oxytocin*
poiesis, making; *erythropoietin*
pro-, before; *prolactin*
renes, kidneys; *adrenal*
synairesis, a drawing together; *synergistic*
teinein, to stretch; *angiotensin*
thyreos, an oblong shield; *thyroid*
tokos, childbirth; *oxytocin*
tropos, turning; *gonadotropins*

To function effectively, every cell in the body must communicate with its neighbors and with cells and tissues in distant portions of the body. Most of the communication involves the release and receipt of chemical messages. Each living cell is continually "talking" to its neighbors by releasing chemicals into the extracellular fluid. These chemicals let cells know what their neighbors are doing at any given moment, and the result is the coordination of tissue function at the local level.

The nervous system acts like a telephone company, carrying specific "messages" from one location to another inside the body. The source and the destination are quite specific, and the effects are short-lived. This form of communication is ideal for crisis management; if you are in danger of being hit by a speeding bus, the nervous system can coordinate and direct your leap to safety. Once the crisis is over, and the neural circuit quiets down, things soon return to normal.

In cellular communication, hormones are like addressed letters, and the circulatory system is the postal service. A hormone released into the circulation will be distributed throughout the body. Each hormone has specific *target cells* that will respond to its presence. These cells possess the receptors needed to bind and "read" the hormonal message. Although a cell may be exposed to the mixture of hormones in circulation at any given moment, it will respond only to those hormones it can bind and read. The other hormones will be treated like junk mail, and ignored.

Because the target cells can be anywhere in the body, a single hormone can alter the metabolic activities of multiple tissues and organs simultaneously. These effects may be slow to appear, but they often persist for days. This persistence makes hormones effective in coordinating cell, tissue, and organ activities on a sustained, long-term basis. For example, circulating hormones keep body water content and levels of electrolytes and organic nutrients within normal limits 24 hours a day throughout our entire lives.

While the effects of a single hormone persist, a cell may receive additional instructions from other hormones. The result will be a further modification in cellular operations. Gradual changes in the quantities and identities of circulating hormones can produce complex changes in physical structure and physiological capabilities. Examples include the processes of embryological and fetal development, growth, and puberty.

When viewed from a general perspective, the differences between the nervous and endocrine systems seem relatively clear. In fact, these broad organizational and functional distinctions are the basis for treating them as two separate systems. Yet when considered in detail, the two systems are organized along parallel lines. For example:

- Both systems rely on the release of chemicals that bind to specific receptors on their target cells.
- Both systems use many of the same chemical messengers; for example, norepinephrine and epinephrine are called hormones when released into the circulation, and neurotransmitters when released across synapses.
- Both systems are primarily regulated by negative feedback control mechanisms.
- Both systems share a common goal: to coordinate and regulate the activities of other cells, tissues, organs, and systems and preserve homeostasis.

This chapter introduces the components and functions of the endocrine system and explores the interactions between the nervous and endocrine systems. Subsequent chapters will consider specific endocrine organs, hormones, and functions in greater detail.

AN OVERVIEW OF THE ENDOCRINE SYSTEM

The endocrine system includes all of the endocrine cells and tissues of the body. As noted in Chapter 4, *endocrine cells* are glandular secretory cells that release their secretions internally rather than onto an epithelial surface. This feature distinguishes them from *exocrine cells,* which secrete onto epithelial surfaces. ∞ [p. 74] The chemicals released by endocrine cells may affect only adjacent cells, as in the case of the "local hormones" known as *prostaglandins,* or they may affect cells throughout the body. **Hormones** are chemical messengers that are released in one tissue and transported via the circulation to reach target cells in other tissues.

The components of the endocrine system are introduced in Figure 11-1●. This figure also lists the major hormones produced in each endocrine tissue and organ. Some of these organs, such as the pituitary gland, have endocrine secretion as a primary function; others, such as the pancreas, have many other functions in addition to endocrine secretion.

Structure of Hormones

Hormones can be divided into three different groups on the basis of chemical structure: *amino acid derivatives, peptide hormones,* and *lipid derivatives.*

1. *Amino acid derivatives.* Some hormones are relatively small molecules that are structurally similar to amino acids. (Amino acids, the building blocks of proteins, were introduced in Chapter 2.) ∞ [p. 38] This group includes *epinephrine, norepinephrine,* the *thyroid hormones,* and the pineal hormone *melatonin.*
2. *Peptide hormones.* **Peptide hormones** consist of chains of amino acids. These molecules range from short amino acid chains, such as *ADH* and *oxytocin,* to polypeptides such as *growth hormone* and *prolactin.* This is the largest class of hormones and includes all of the hormones secreted by the hypothalamus, pituitary gland, heart, kidneys, thymus, digestive tract, and pancreas.
3. *Lipid derivatives.* There are two classes of lipid-based hormones: (1) *steroid hormones,* derived from cholesterol, and (2) those derived from *arachidonic acid,* a 20-carbon fatty acid. **Steroid hormones** are lipids structurally similar to cholesterol, a lipid introduced in Chapter 2. ∞ [p. 37] Steroid hormones are released by the reproductive organs and the adrenal glands. The fatty acid–based compounds, which include the **prostaglandins,** coordinate cellular activities and affect enzymatic processes, such as blood clotting, that occur in extracellular fluids.

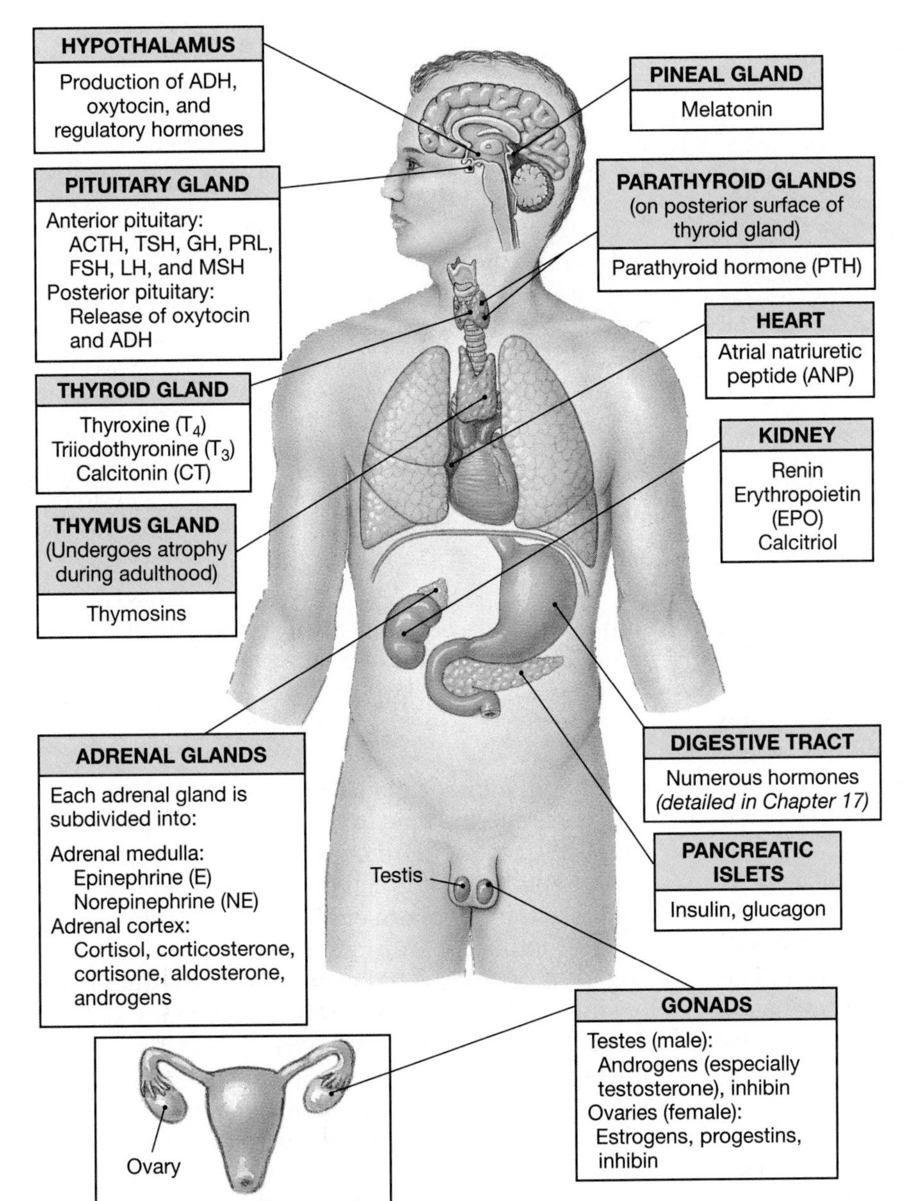

●**Figure 11-1**
The Endocrine System

Mechanisms of Hormonal Action

All cellular structures and functions are determined by proteins. Structural proteins determine the general shape and internal structure of a cell, and enzymes direct its metabolic activities. Hormones alter cellular operations by changing the *identities, activities,* or *quantities* of important enzymes and structural proteins in various **target cells.** The sensitivity of a target cell is determined by the presence or absence of a specific **receptor complex,** either on the cell membrane or in the cytoplasm, with which the hormone interacts (Figure 11-2●).

Hormones and the Cell Membrane

Epinephrine, norepinephrine, and peptide hormones cannot diffuse

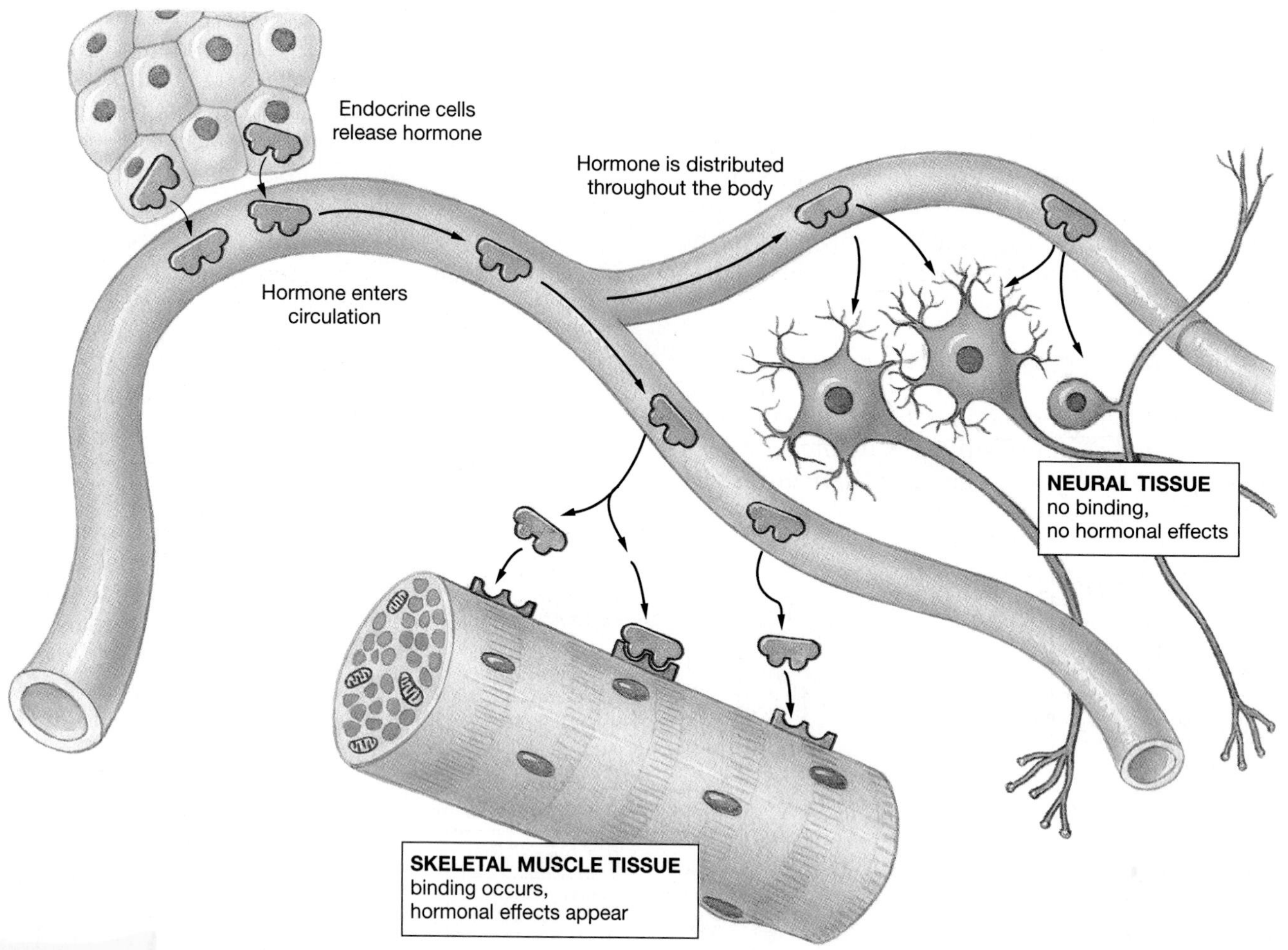

Figure 11-2
Hormones and Target Cells

For a hormone to affect a target cell, that cell must have receptors that can bind the hormone and initiate a change in cellular activity. This hormone affects skeletal muscle tissue but not neural tissue because only the muscle tissue has the appropriate receptors.

through a cell membrane. These hormones, called **first messengers,** target receptors on the cell membrane. When a first messenger binds to an appropriate receptor, it triggers the appearance of a **second messenger** in the cytoplasm. The second messenger may function as an enzyme activator or inhibitor, but the net result will be a change in the cell's metabolic activities.

One of the most important second messengers is **cyclic-AMP (cAMP)**, and the enzyme activated by the receptor is called **adenylate cyclase.** When activated, adenylate cyclase converts ATP to cyclic-AMP. The specific response of the target cell to cAMP depends on the nature of the enzymes already present in the cytoplasm. As a result, a single hormone can have one effect in one target tissue and quite different effects in other target tissues. The effects of cAMP are usually very short-lived, because another enzyme, *phosphodiesterase,* quickly breaks down cAMP.

Cyclic-AMP is one of the most common second messengers, but there are many others. Important examples include calcium ions and the high-energy compound *cyclic-GMP.*

Hormones and Intracellular Receptors

The thyroid hormones and steroid hormones cross the cell membrane and bind to intracellular receptors (Figure 11-3). Steroid hormones diffuse rapidly through the lipid portion of the cell membrane and bind to receptors in the cytoplasm or nucleus. The hormone-receptor complex then binds to DNA segments and triggers the activation or inactivation of specific genes. By altering the rate of mRNA transcription in the nucleus, steroid hormones can change the structure or function of the cell. For example, the hormone *testosterone* stimulates the production of enzymes and proteins in skeletal muscle fibers, increasing muscle size and strength.

Thyroid hormones cross the cell membrane through diffusion or carrier-mediated transport. Once within the cell, these hormones bind to receptors within the nu-

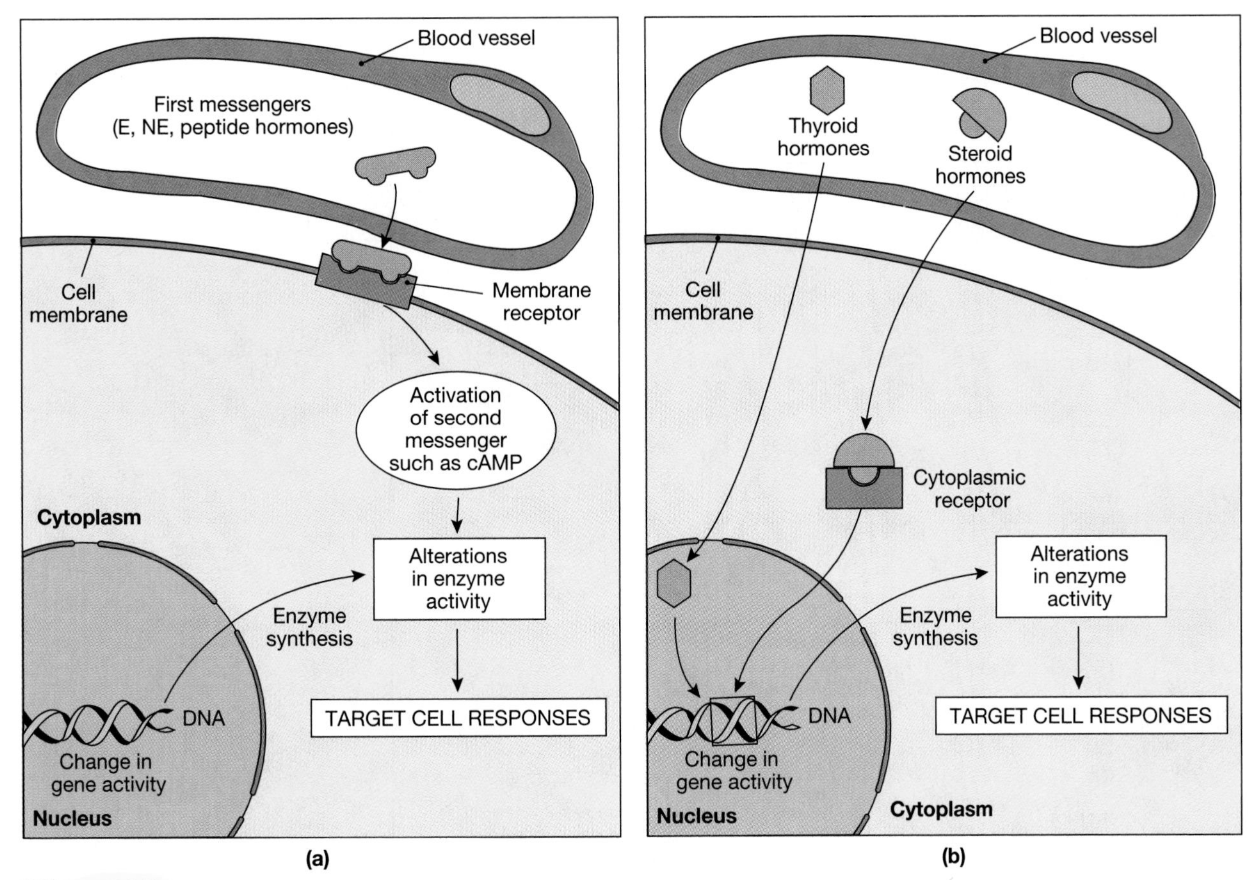

Figure 11-3
Mechanisms of Hormone Action
(a) Epinephrine (E), norepinephrine (NE), and peptide hormones act through second messengers released when the hormones bind to receptors at the membrane surface. **(b)** Steroid hormones bind to receptors in the cytoplasm that then enter the nucleus. Thyroid hormones proceed directly to the nucleus to reach hormonal receptors.

cleus and activate specific genes. The result is an increase in metabolic activity due to changes in the nature or number of enzymes in the cytoplasm.

Control of Endocrine Activity

Endocrine activity may be controlled directly, by a change in the composition of the extracellular fluid, or indirectly, by the hypothalamus. Negative feedback mechanisms provide the basis for the control of endocrine activity. ∞ *[p. 12]*

Direct Negative Feedback Control

In direct negative feedback control, the endocrine cells respond to a change in the composition of the extracellular fluid by releasing their hormone into the circulatory system. The released hormone stimulates target cells to restore homeostasis. For example, consider the control of calcium levels by *parathyroid hormone* and *calcitonin*. When circulating calcium levels decline, parathyroid hormone is released, and the responses of target cells elevate blood calcium levels. When calcium levels rise, calcitonin is released, and responses of target cells lower blood calcium levels.

The Hypothalamus and Endocrine Regulation

Coordinating centers in the hypothalamus regulate the activities of the nervous and endocrine systems in three ways (Figure 11-4):

1. The hypothalamus contains autonomic centers that control the endocrine cells of the adrenal medullae through sympathetic innervation. When the sympathetic division is activated, the adrenal medullae release hormones into the bloodstream.
2. The hypothalamus itself acts as an endocrine organ, releasing the hormones *ADH* and *oxytocin* into the circulation at the posterior pituitary.
3. The hypothalamus secretes **regulatory hormones,** special hormones that regulate the activities of endocrine cells in the anterior pituitary gland. There are two classes of regulatory hormones: (1) *releasing hormones (RH)* stimulate production of one or more hormones in the anterior pituitary, and (2) **inhibiting hormones (IH)** prevent the synthesis and secretion of pituitary hormones.

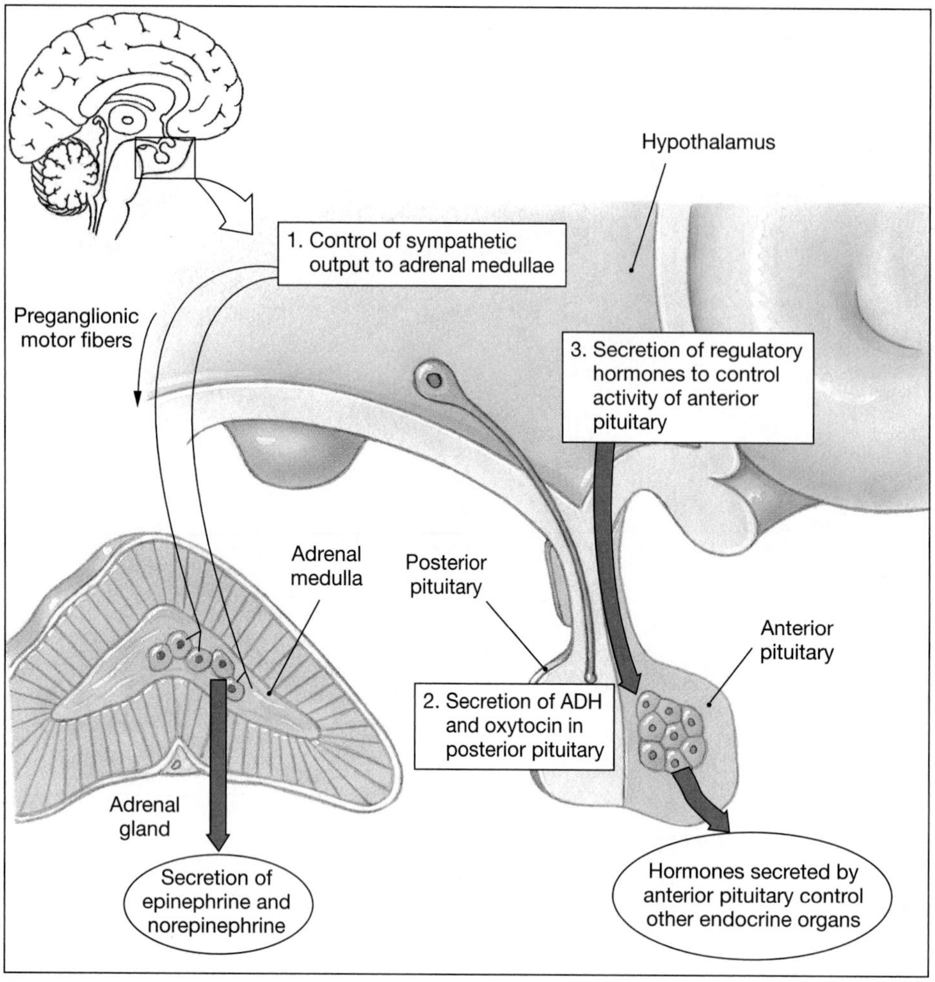

●Figure 11-4
Three Methods of Hypothalamic Control over Endocrine Organs

THE PITUITARY GLAND

The pituitary gland, or **hypophysis** (hī-POF-i-sis), secretes nine different hormones. All are peptide hormones that bind to membrane receptors and use cyclic-AMP as a second messenger. This small, oval gland lies nestled within the *sella turcica,* a depression in the sphenoid bone of the skull (Figure 11-5●). ∞ [p. 122] The pituitary gland hangs beneath the hypothalamus, connected by a slender stalk, the **infundibulum** (in-fun-DIB-ū-lum; funnel). The pituitary gland has a complex structure, with distinct anterior and posterior regions.

The Anterior Pituitary

The **anterior pituitary** contains endocrine cells surrounded by an extensive capillary network. The capillary network, part of the *hypophyseal portal system,* provides an entry into the circulatory system for the endocrine secretions of the anterior pituitary.

The Hypophyseal Portal System

The regulatory hormones produced by the hypothalamus regulate the activities of the anterior pituitary. These hormones, released by hypothalamic neurons near the attachment of the infundibulum, enter a network of unusually permeable capillaries. Before leaving the hypothalamus, this capillary network unites to form a series of larger vessels that descend to the anterior pituitary and form a second capillary network.

This circulatory arrangement, illustrated in Figure 11-6●, is very unusual. A typical artery usually conducts blood from the heart to a capillary network, and a typical vein carries blood from a capillary network back to the heart. The vessels between the hypothalamus and the anterior pituitary, by contrast, carry blood from one capillary network to another. Blood vessels that link two capillary networks are called *portal vessels,* and the entire complex is termed a **portal system.**

Portal systems ensure that all of the blood entering the portal vessels will reach the intended target cells before returning to the general circulation. Portal vessels are named after their destinations, so this particular network of vessels represents the **hypophyseal portal system.**

Hypothalamic Control of the Anterior Pituitary

An endocrine cell in the anterior pituitary may be controlled by releasing hormones, inhibiting hormones, or some combination of the two. The regula-

●**Figure 11-5**
Anatomy and Orientation of the Pituitary Gland
(LM × 62)

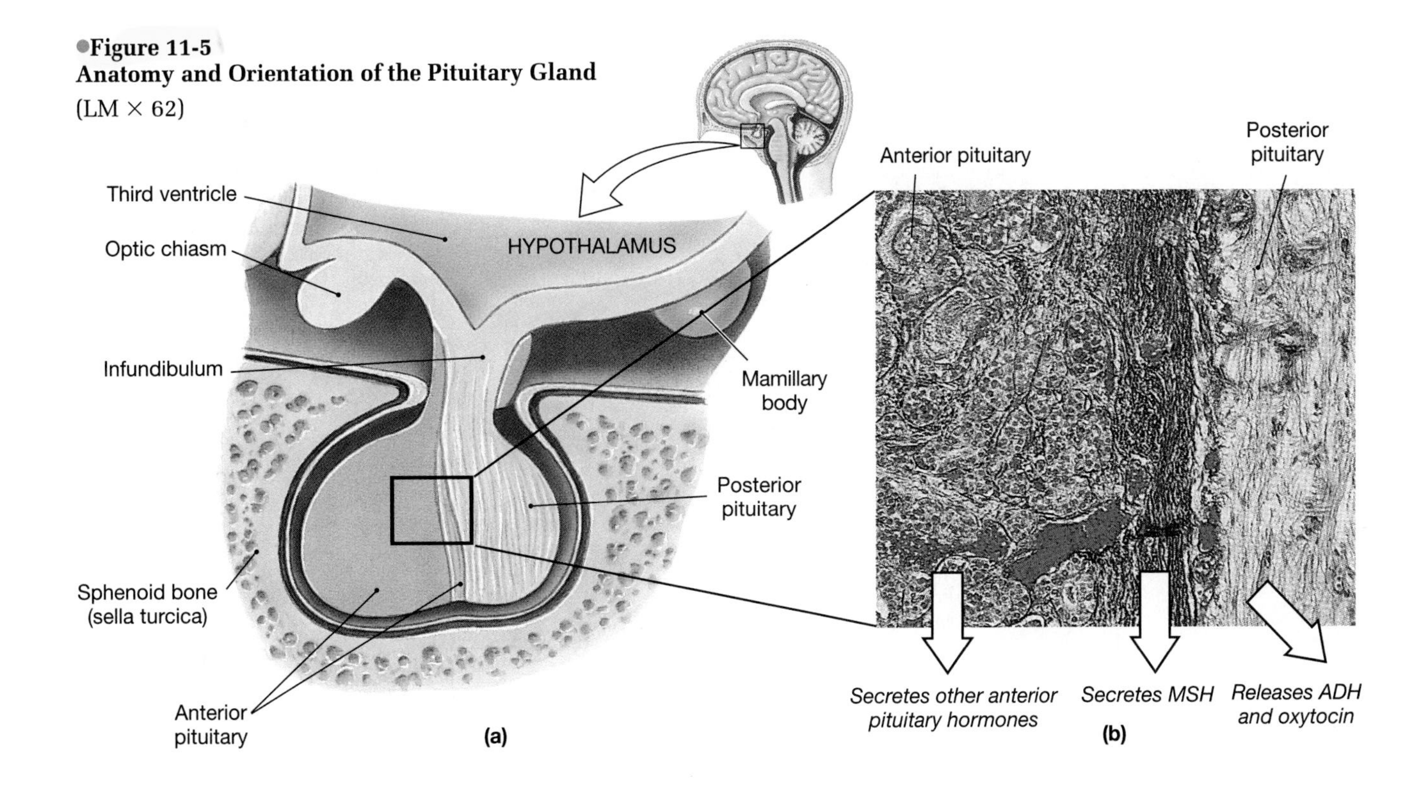

tory hormones released at the hypothalamus are transported directly to the anterior pituitary via the hypophyseal system.

The rate of regulatory hormone secretion by the hypothalamus is regulated through negative feedback mechanisms. The basic regulatory patterns are diagrammed in Figure 11-7●; these will be referenced in the following description of pituitary hormones. Many of these hormones are called *-tropins,* from the Greek word *tropos (*turning), because they turn on (activate) other tissues and organs.

Hormones of the Anterior Pituitary

Seven hormones are produced by the anterior pituitary. Of those, four regulate the production of hormones by other endocrine glands.

1. **Thyroid-stimulating hormone (TSH)** targets the thyroid gland and triggers the release of thyroid hormones. The thyroid hormones released inhibit the pituitary production of TSH and the hypothalamic centers producing the associated releasing hormone. This pattern of regulatory control is detailed in Figure 11-7a●.
2. **Adrenocorticotropic hormone (ACTH)** stimulates the release of steroid hormones by the adrenal glands. ACTH specifically targets cells producing hormones called **glucocorticoids** (gloo-kō-KŌR-ti-koids) that affect glucose metabolism. The feedback control mechanism is comparable to that for TSH (Figure 11-7a●).
3. **Follicle-stimulating hormone (FSH)** promotes egg development in women and stimulates the secretion of *estrogens,* steroid hormones produced by ovarian cells. In men, FSH production supports sperm production in the testes. The feedback control mechanism is comparable to that for TSH (Figure 11-7a●).

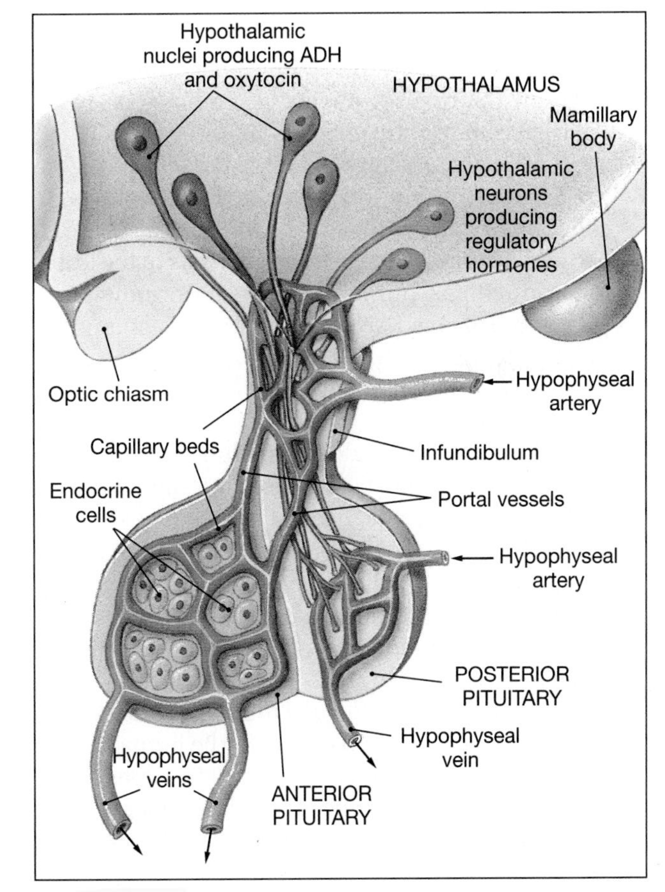

●**Figure 11-6**
The Hypophyseal Portal System

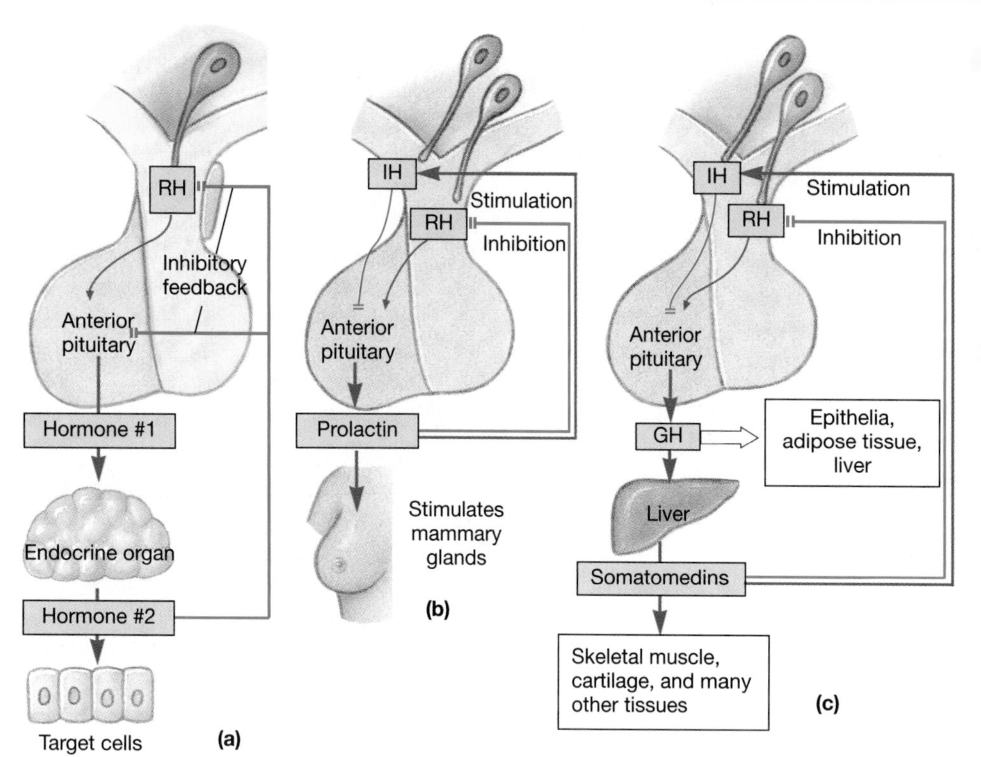

Figure 11-7 Feedback Control of Endocrine Secretion

(a) Typical pattern of regulation when multiple endocrine organs are involved. In these cases, the hypothalamus produces a releasing hormone to stimulate hormone production by other glands, and control is through negative feedback. **(b)** In some cases both a releasing hormone and an inhibiting hormone are produced by the hypothalamus; when one is stimulated, the other is inhibited. This diagram summarizes the control of prolactin production by the anterior pituitary. **(c)** The regulation of growth hormone follows the basic pattern shown in part **(b)**, but intermediate steps are involved.

4. **Luteinizing** (LOO-tē-in-ī-zing) **hormone (LH)** induces ovulation in women and promotes the ovarian secretion of estrogens and the **progestins** (such as *progesterone)* that prepare the body for possible pregnancy. In men, the same hormone was once called *interstitial cell–stimulating hormone (ICSH)* because it stimulates the *interstitial cells* of the testes to produce sex hormones. These hormones are called **androgens** (AN-drō-jenz; *andros,* man); the most important of them is *testosterone.* The feedback control mechanism is comparable to that of TSH (Figure 11-7a).

   FSH and LH are called **gonadotropins** (gō-nad-ō-TRŌ-pinz) because they regulate the activities of the male and female sex organs (gonads).
5. **Prolactin** (prō-LAK-tin; *pro-*, before + *lac,* milk), or **PRL,** stimulates the development of the mammary glands and the production of milk. Although PRL exerts the dominant effect on the glandular cells, normal development of the mammary glands is regulated by the interaction of a number of hormones. Prolactin has no known effects in the human male. The regulation of prolactin release involves interactions between releasing and inhibiting hormones from the hypothalamus. The regulatory pattern is diagrammed in Figure 11-7b.
6. **Growth hormone (GH),** also called *human growth hormone (hGH)* or *somatotropin* (*soma,* body), stimulates cell growth and replication by accelerating the rate of protein synthesis. Although virtually every tissue responds to some degree, skeletal muscle cells and chondrocytes (cartilage cells) are particularly sensitive to levels of growth hormone.

   The stimulation of growth by GH involves two different mechanisms. The primary mechanism, which is indirect, is best understood. Liver cells respond to the presence of growth hormone by synthesizing and releasing **somatomedins,** or *insulin-like growth factors (IGF),* hormones that bind to receptor sites on a variety of cell membranes. Somatomedins increase the rate of amino acid uptake and their incorporation into new proteins. These effects develop almost immediately after GH release occurs, and they are particularly important after a meal, when the blood contains high concentrations of glucose and amino acids. The regulatory mechanism involved in controlling GH production is summarized in Figure 11-7c.

   The direct actions of GH usually do not appear until after blood glucose and amino acid concentrations have returned to normal levels. In epithelia and connective tissues, GH stimulates stem cell divisions and the differentiation of daughter cells. GH also has metabolic effects in adipose tissue and in the liver. In adipose tissue, it stimulates the breakdown of stored fats and the release of fatty acids into the blood. In the liver, GH stimulates the breakdown of glycogen reserves and the release of glucose into the circulation. GH thus plays a role in mobilizing energy reserves. The interactions between growth hormone and other hormones during normal development and maturation will be discussed in a later section.
7. **Melanocyte-stimulating hormone (MSH)** stimulates the melanocytes of the skin, increasing their production of melanin. MSH is important in the

control of skin and hair pigmentation in fishes, amphibians, reptiles, and many mammals. The MSH-producing cells of the pituitary gland in adult humans are virtually nonfunctional, and the circulating blood usually does not contain MSH. However, MSH is secreted by the human pituitary (1) during fetal development, (2) in very young children, (3) in pregnant women, and (4) in some disease states. The functions of MSH under these circumstances is not known. Administration of a synthetic form of MSH causes darkening of the skin, and it has been suggested as a means of obtaining a "sunless tan." A commercial version of this product is now pending approval by the U.S. Food and Drug Administration (FDA).

The Posterior Pituitary

The **posterior pituitary** contains the axons from two different groups of hypothalamic neurons. One group manufactures ADH and the other oxytocin; both groups are located within the hypothalamus. Their products are transported within axons along the infundibulum to the posterior pituitary, as indicated in Figure 11-6●, p. 287.

Antidiuretic hormone (ADH) is released in response to stimuli such as a rise in the concentration of electrolytes in the blood or a fall in blood volume or pressure. The primary function of ADH is to decrease the amount of water lost at the kidneys. With losses minimized, any water absorbed from the digestive tract will be retained, reducing the concentration of electrolytes. ADH also causes the constriction of peripheral blood vessels, which helps to increase blood pressure. The production of ADH is inhibited by alcohol, which explains the increased fluid excretion that follows the consumption of alcoholic beverages.

In women, **oxytocin** (*oxy-*, quick + *tokos*, childbirth) stimulates smooth muscle cells in the uterus and special contractile cells surrounding the secretory cells of the mammary glands. Until the last stages of pregnancy, the uterine muscles are insensitive to oxytocin, but they become more sensitive as the time of delivery approaches. The stimulation of uterine muscles by oxytocin helps maintain and complete normal labor and childbirth (discussed in Chapter 21). After delivery, oxytocin functions in the "milk let-down" reflex. In this reflex, detailed in Chapter 21, oxytocin secreted in response to suckling triggers the release of milk into large collecting chambers.

In the male, oxytocin stimulates the smooth muscle contraction in the walls of the prostate gland. This action may be important in *emission,* the ejection of prostatic secretions, spermatozoa, and the secretions of other glands into the male reproductive tract before ejaculation occurs.

DIABETES INSIPIDUS

There are several forms of **diabetes** (*diabetes,* to pass through), all characterized by excessive urine production (*polyuria*). Although diabetes can be caused by physical damage to the kidneys, most forms are the result of endocrine abnormalities. The two most important forms are *diabetes insipidus* and *diabetes mellitus.* (Diabetes mellitus is described on p. 298.)

Diabetes insipidus (*insipidus,* tasteless) develops when the posterior pituitary no longer releases adequate amounts of ADH. Water conservation at the kidneys is impaired, and excessive amounts of water are lost in the urine. As a result, an individual with diabetes insipidus is constantly thirsty, but the fluids consumed are not retained by the body. Mild cases may not require treatment, as long as fluid and electrolyte intake keep pace with urinary losses. In severe diabetes insipidus, the fluid losses can reach 10 liters per day, and a fatal dehydration will occur unless treatment is provided. One innovative treatment method involves administering a synthetic form of ADH, **desmopressin acetate** (DDAVP) in a nasal spray. The drug enters the bloodstream after diffusing through the nasal epithelium.

Figure 11-8● and Table 11-1 summarize important information concerning the hormonal products of the pituitary gland.

- ✓ Why is cyclic-AMP described as a "second messenger"?
- ✓ If a person were suffering from dehydration, how would this condition affect the level of ADH released by the posterior pituitary?
- ✓ A blood sample shows elevated levels of somatomedins. What pituitary hormone would you expect to be elevated as well?
- ✓ What effect would elevated levels of cortisol, a hormone from the adrenal gland, have on the level of ACTH?

THE THYROID GLAND

The thyroid gland is located just below the **thyroid** ("shield-shaped") **cartilage** that dominates the anterior surface of the larynx. The thyroid gland has a deep red coloration because of the large number of blood vessels servicing the glandular cells.

Thyroid Follicles and Thyroid Hormones

The thyroid gland contains large numbers of spherical **thyroid follicles,** shown in sectional view in Figure 11-9●. The glandular lining of a follicle consists of a simple cuboidal epithelium. A network of capillaries

●Figure 11-8
Pituitary Hormones and Their Targets

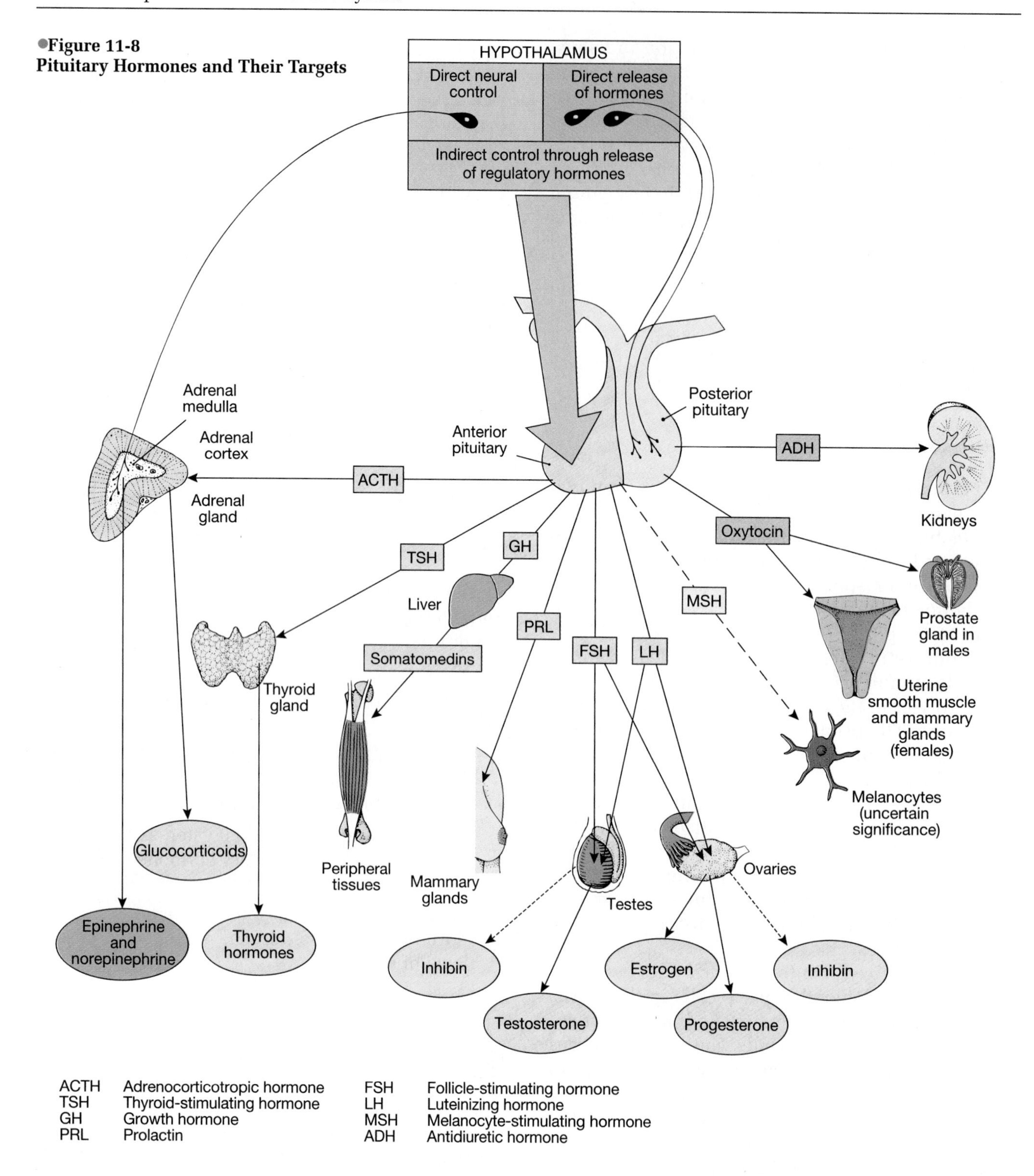

surrounds each follicle, delivering nutrients and regulatory hormones to the glandular cells and accepting their secretory products and metabolic wastes.

Thyroid hormones are stored within the follicles. Under TSH stimulation, the epithelial cells take hormones from the follicles and release them. The thyroid hormones are structural derivatives of the amino acid **tyrosine,** to which three or four iodine atoms have been attached. The hormone **thyroxine** (thī-ROKS-ēn), or **TX**, contains four atoms of iodine; it is also known as *tetraiodothyronine* (tet-ra-ī-ō-dō-THĪ-rō-nēn), or $\mathbf{T_4}$. Thyroxine accounts for roughly 90 percent of all thyroid secretions. **Triiodothyronine,** or $\mathbf{T_3}$, is a related, more potent molecule containing three iodine atoms.

Thyroid hormones affect almost every cell in the body because they diffuse through the cytoplasm to reach receptor sites in the nucleus (see Figure 11-3b●, p. 285). Thyroid hormones activate genes coding for the synthesis of enzymes involved in glycolysis and energy production, resulting in an increase in cellu-

TABLE 11-1 The Pituitary Hormones

Region	*Hormone*	*Target*	*Hormonal Effects*
Anterior pituitary	Thyroid-stimulating hormone (TSH)	Thyroid gland	Secretion of thyroid hormones
	Adrenocorticotropic hormone (ACTH)	Adrenal cortex	Glucocorticoid secretion
	Gonadotropic hormones:		
	Follicle-stimulating hormone (FSH)	Follicle cells of ovaries in female	Estrogen secretion, follicle development
		Sustentacular cells of testes in male	Sperm maturation
	Luteinizing hormone (LH) or interstitial cell-stimulating hormone (ICSH)	Follicle cells after ovulation in female	Formation of corpus luteum and progestin secretion
		Interstitial cells of testes in male	Androgen secretion
	Prolactin (PRL)	Mammary glands	Production of milk
	Growth hormone (GH)	All cells	Growth, protein synthesis, lipid mobilization and catabolism
	Melanocyte-stimulating hormone (MSH)	Melanocytes of skin	Increased melanin synthesis, dispersion in epidermis; not significant in nonpregnant adults
Posterior pituitary	Antidiuretic hormone (ADH)	Kidneys	Reabsorption of water; elevation of blood volume and pressure
	Oxytocin	Uterus, mammary glands in female	Labor contractions, milk ejection
		Prostate gland in male	Smooth muscle contractions, ejection of secretions

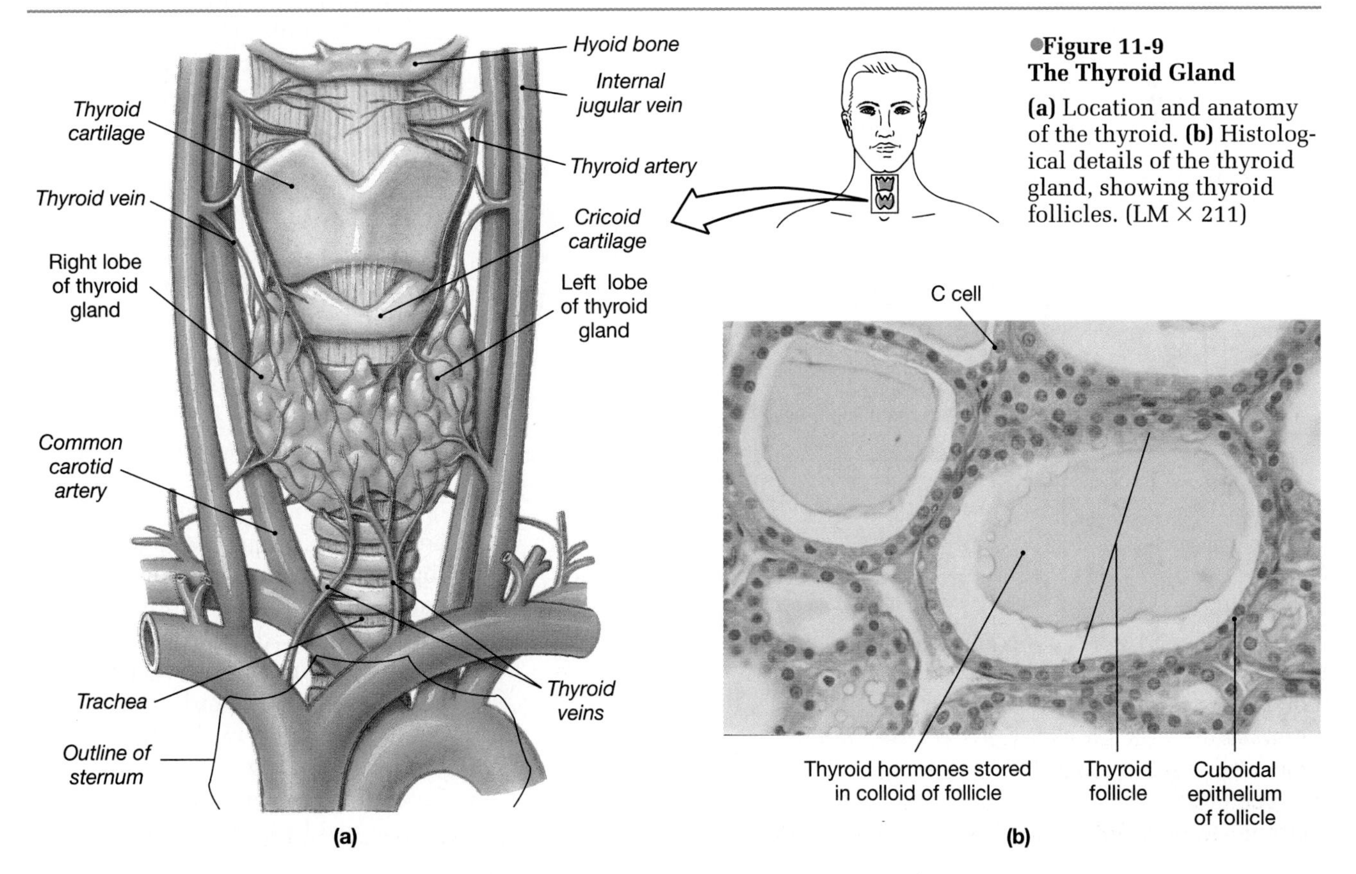

Figure 11-9 The Thyroid Gland

(a) Location and anatomy of the thyroid. **(b)** Histological details of the thyroid gland, showing thyroid follicles. (LM × 211)

lar rates of metabolism and oxygen consumption. Because the cell consumes more energy, and energy use is measured in *calories,* this is called the **calorigenic effect.** When the metabolic rate increases, more heat is generated, and body temperature rises. In growing children, thyroid hormones are also essential to normal development of the skeletal, muscular, and nervous systems.

Normal production of thyroid hormones establishes the background rates of cellular metabolism. These hormones exert their primary effects on active tissues and organs, including skeletal muscles, the liver, the heart, and the kidneys. Overproduction or underproduction of thyroid hormones can therefore cause very serious metabolic problems. In many parts of the world, inadequate dietary iodine leads to an inability to synthesize thyroid hormones. Under these conditions, TSH stimulation continues, and the thyroid follicles become distended with nonfunctional secretions. The result is a swollen and enlarged thyroid gland, or *goiter.* This is seldom a problem in the United States because the typical American diet provides roughly three times the minimum daily requirement of iodine, thanks to the addition of iodine to table salt ("iodized salt"). AM *Thyroid Gland Disorders*

The C Cells of the Thyroid Gland: Calcitonin

A second population of endocrine cells lies sandwiched between the cuboidal follicle cells and their basement membrane. These are the **C cells**, which produce the hormone **calcitonin (CT).** Calcitonin helps regulate calcium ion concentrations in body fluids. As Figure 11-10 illustrates, the C cells release calcitonin when the calcium ion concentration of the blood rises above normal. The target organs are the bones and the kidneys. Calcitonin reduces calcium levels by inhibiting osteoclasts and stimulating calcium excretion at the kidneys. The resulting reduction in the calcium ion concentrations eliminates the stimulus and "turns off" the C cells.

Several chapters have dealt with the importance of calcium ions in controlling muscle and nerve cell activities. Calcium ion concentrations also affect the sodium permeabilities of excitable membranes. At high calcium ion concentrations, sodium permeability decreases and membranes become less responsive. Such problems are prevented by the secretion of calcitonin under appropriate conditions. However, under normal conditions, calcium ion levels seldom rise enough to trigger calcitonin secretion. Most homeostatic adjustments are intended to prevent lower than normal calcium ion concentrations. Low calcium concentrations are dangerous because sodium permeabilities then increase and muscle cells and neurons become extremely excitable. If calcium levels fall too far, convulsions or muscular spasms will occur. The parathyroid glands and their hor-

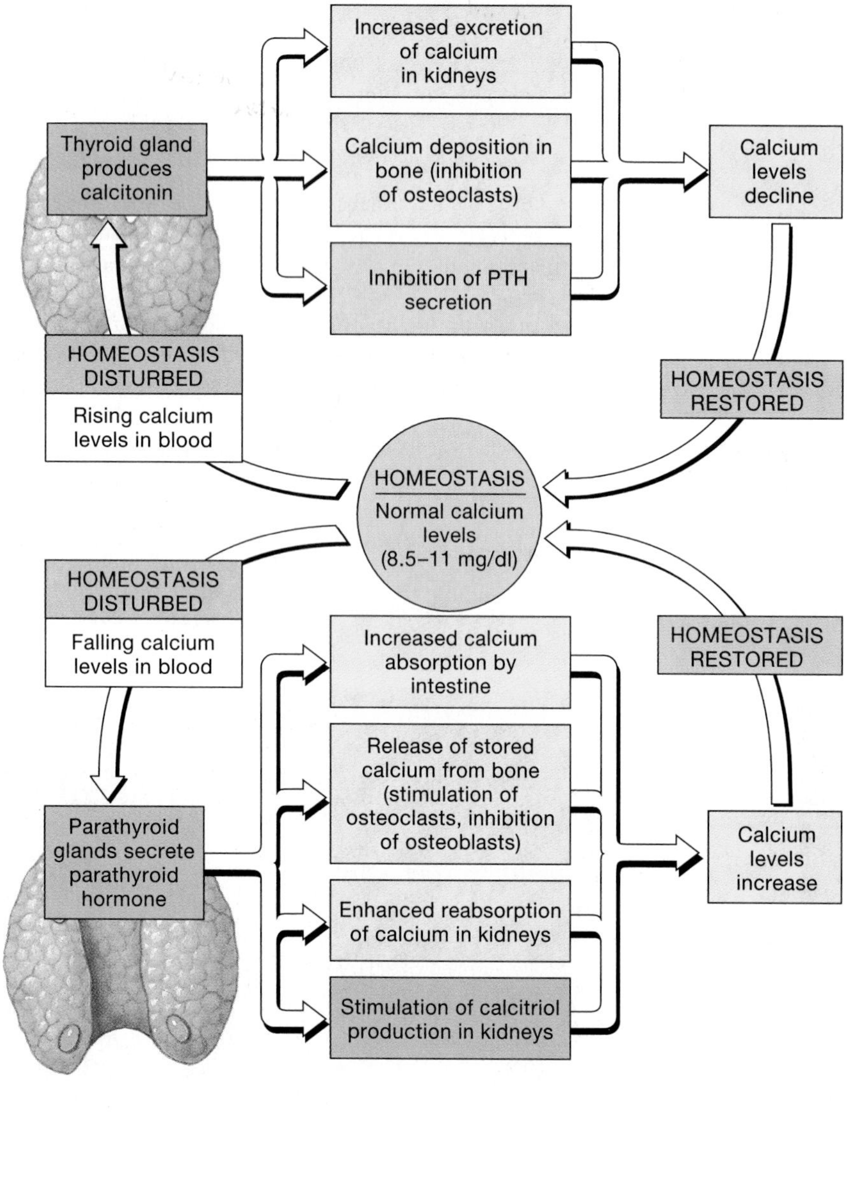

Figure 11-10
The Homeostatic Regulation of Calcium Ion Concentrations

mone secretions are responsible for preventing such disastrous events.

THE PARATHYROID GLANDS

Two tiny pairs of parathyroid glands are embedded in the posterior surfaces of the thyroid gland. The gland cells are separated by the dense capsular fibers of the thyroid. The histological appearance of a parathyroid gland is shown in Figure 11-11●. There are at least two different cell populations found in the parathyroid. The **chief cells** produce parathyroid hormone; the functions of the other cell type are unknown.

Like the C cells of the thyroid, the chief cells monitor the circulating concentration of calcium ions. When the calcium concentration falls below normal, the chief cells secrete **parathyroid hormone (PTH),** or *parathormone* (see Figure 11-10●). Although parathyroid hormone acts on the same target organs as calcitonin, it produces the opposite effects. PTH stimulates osteoclasts, inhibits osteoblasts, promotes absorption of calcium by the intestines, and reduces urinary excretion of calcium ions until blood concentrations return to normal. AM *Disorders of Parathyroid Function*

THE THYMUS

The thymus is embedded in a mass of connective tissue inside the thoracic cavity, usually just behind the sternum. In a newborn infant, the thymus is relatively enormous, often extending from the base of the neck to the superior border of the heart. As the child grows, the thymus continues to enlarge slowly, reaching its maximum size just before puberty, at a weight of around 40 g (1.4 oz.). After puberty, it gradually diminishes in size; by age 50 the thymus may weigh less than 12 g (0.4 oz.).

The thymus produces several hormones, collectively known as the **thymosins** (thī-MŌ-sins). The thymosins play a key role in the development and maintenance of normal immunological defenses. It has been suggested that the gradual decrease in the size and secretory abilities of the thymus may make the elderly more susceptible to disease. The histological organization of the thymus and the functions of the thymosins will be further considered in Chapter 15.

THE ADRENAL GLANDS

A single **adrenal gland** caps the superior border of each kidney (Figure 11-12●). Because of their location, the adrenals are also called the *suprarenal glands* (soo-pra-RĒ-nal; *supra-*, above + *renes,* kidneys). Each adrenal gland can be divided into two parts, a superficial **adrenal cortex** and an inner **adrenal medulla.**

The Adrenal Cortex

The adrenal cortex has a grayish yellow coloration because of the presence of stored lipids, especially cholesterol and various fatty acids. The adrenal cortex produces more than two dozen different steroid hormones, collectively called *adrenocortical steroids,* or simply **corticosteroids.** These hormones are vital; if the adrenal glands are destroyed or removed, corticosteroids must be administered or the individual will not survive. Overproduction or underproduction of any of the corticosteroids will have severe conse-

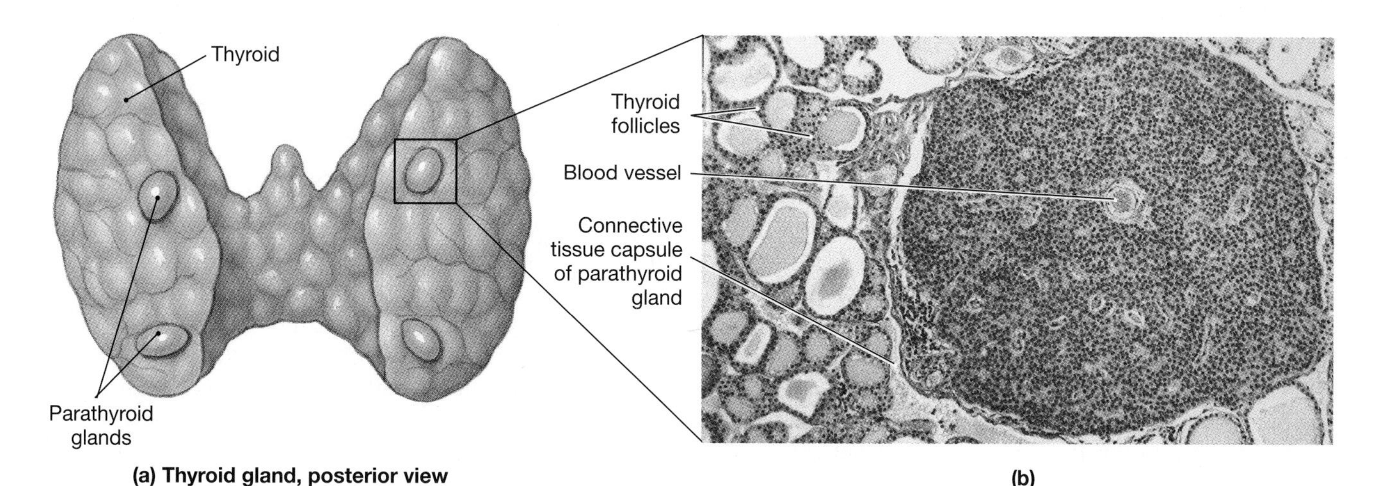

●**Figure 11-11**
The Parathyroid Glands

(a) Location of parathyroids on posterior surface of thyroid lobes. **(b)** Photomicrograph showing both parathyroid and thyroid tissues. (LM × 94)

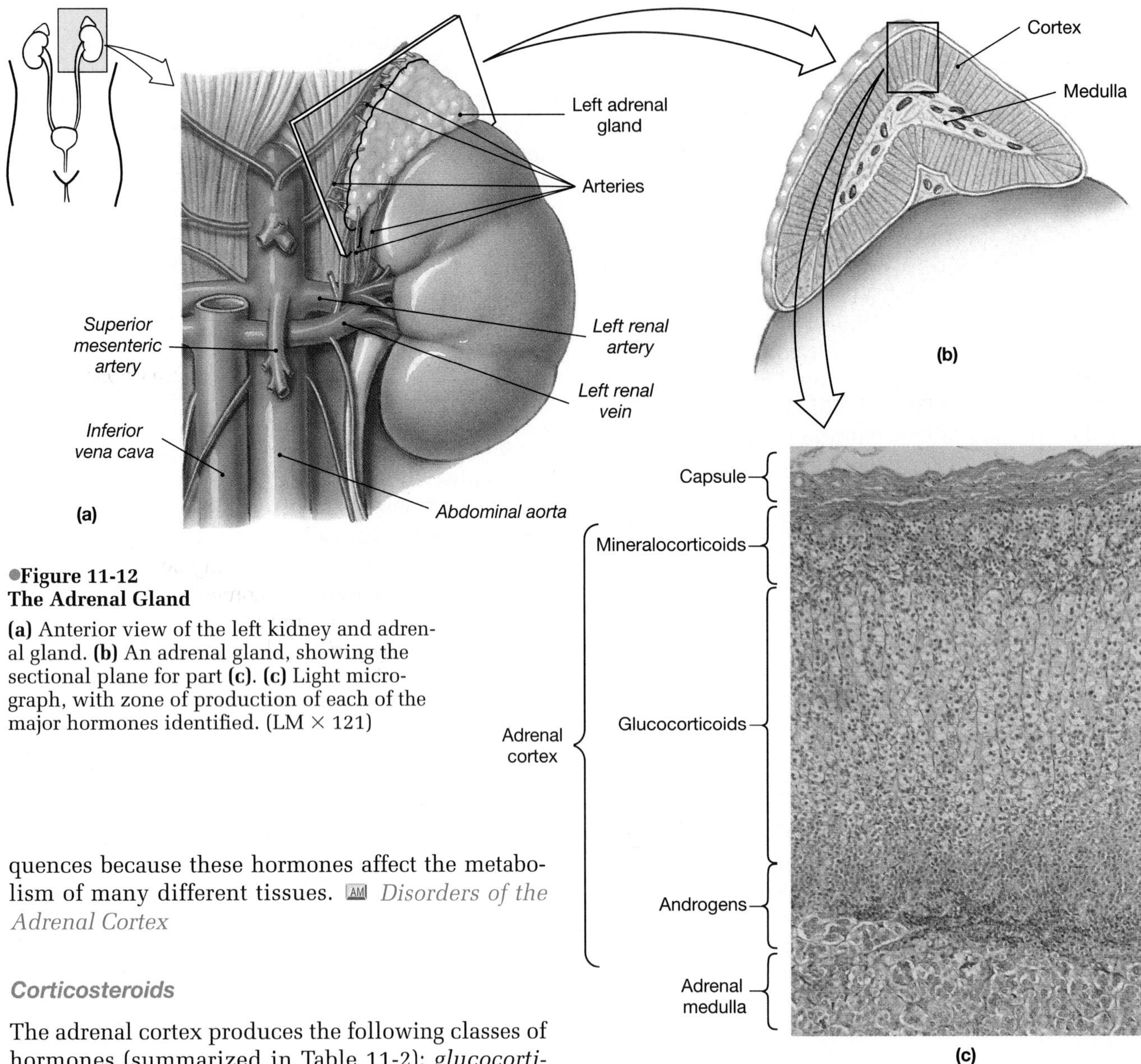

Figure 11-12
The Adrenal Gland

(a) Anterior view of the left kidney and adrenal gland. **(b)** An adrenal gland, showing the sectional plane for part **(c)**. **(c)** Light micrograph, with zone of production of each of the major hormones identified. (LM × 121)

quences because these hormones affect the metabolism of many different tissues. AM *Disorders of the Adrenal Cortex*

Corticosteroids

The adrenal cortex produces the following classes of hormones (summarized in Table 11-2): *glucocorticoids, mineralocorticoids,* and *androgens.*

Glucocorticoids The steroid hormones collectively known as **glucocorticoids (GCs)** affect glucose metabolism. **Cortisol** (KŌR-ti-sol; also called *hydrocortisone*), **corticosterone** (kor-ti-KOS-te-rōn), and **cortisone** are the three most important glucocorticoids. These hormones, secreted under ACTH stimulation, accelerate the rates of glucose synthesis and glycogen formation, especially within the liver. Simultaneously, adipose tissue responds by releasing fatty acids into the blood, and other tissues begin to break down fatty acids instead of glucose. Glucocorticoids also have *anti-inflammatory activity;* they suppress the activities of white blood cells and other components of the immune system. Glucocorticoid creams are often used to control irritating allergic rashes, such as those produced by poison ivy, and injections of glucocorticoids may be used to control more severe allergic reactions. Because they slow wound healing and suppress immune defenses, topical steroids are used to treat superficial rashes but are never applied to open wounds.

Mineralocorticoids Corticosteroids known as **mineralocorticoids (MCs)** affect the electrolyte composition of bodily fluids. **Aldosterone,** the principal mineralocorticoid, targets kidney cells that regulate the ionic composition of the urine. It causes the retention of sodium ions and water, reducing fluid losses in the urine. Aldosterone also reduces sodium and water losses at the sweat glands, salivary glands, and along the digestive tract. The sodium ions recovered are exchanged for potassium ions, so aldosterone also lowers potassium ion concentrations in the extracellular fluid. Aldosterone secretion occurs in response to (1) stimulation by the hormone *angiotensin II*

TABLE 11-2 The Adrenal Hormones

Region	*Hormone*	*Target*	*Effects*
Cortex	Mineralocorticoids (MC), primarily aldosterone	Kidneys	Increases reabsorption of sodium ions and water from the urine; accelerates urinary loss of potassium ions
	Glucocorticoids (GC): cortisol (hydrocortisone), corticosterone, cortisone	Most cells	Releases amino acids from skeletal muscles, lipids from adipose tissues; promotes liver glycogen and glucose formation; promotes peripheral utilization of lipids; anti-inflammatory effects
	Androgens		Uncertain significance under normal conditions
Medulla	Epinephrine (E, adrenaline), norepinephrine (NE, noradrenaline)	Most cells	Increased cardiac activity, blood pressure, glycogen breakdown, blood glucose; release of lipids by adipose tissue (*see* Chapter 17)

(*angeion,* vessel + *teinein,* to stretch) and (2) high extracellular potassium levels. The appearance of angiotensin II in the circulation involves a series of steps that begins with the secretion of an enzyme, *renin,* by kidney cells. The mechanism will be detailed further in Chapters 14 and 19.

Androgens The adrenal cortex in both sexes produces small quantities of sex hormones called *androgens.* Androgens are produced in large quantities by the testes of males, and the importance of the small adrenal production in both sexes remains uncertain.

The Adrenal Medulla

The adrenal medulla has a reddish brown coloration partly because of the many blood vessels in this area. It contains large, rounded cells similar to those found in other sympathetic ganglia, and these cells are innervated by preganglionic sympathetic fibers. (The nature of the innervation of the adrenal medulla and its relationship to the sympathetic division of the ANS was discussed in Chapter 9.) [p. 240]

The adrenal medulla contains two populations of secretory cells, one producing *epinephrine* (E, or adrenaline) and the other *norepinephrine* (NE, or noradrenaline). These hormones are normally released at a low rate, but sympathetic stimulation accelerates the rate of discharge dramatically.

Epinephrine makes up 75–80 percent of the secretions from the medulla; the rest is norepinephrine. These hormones accelerate cellular energy utilization and mobilize energy reserves. Receptors for epinephrine and norepinephrine are found on skeletal muscle fibers, in adipose tissues, and in the liver. Secretion by the medulla triggers a mobilization of glycogen reserves in skeletal muscles and accelerates the breakdown of glucose to provide ATP. This combination increases muscular power and endurance. In adipose tissue, stored fats are broken down to fatty acids, and in the liver, glycogen molecules are converted to glucose. The fatty acids and glucose are then released into the circulation for use by peripheral tissues. The heart also responds to adrenal medulla hormones with an increase in the rate and force of cardiac contractions.

The metabolic changes that follow epinephrine and norepinephrine release are at their peak 30 seconds after adrenal stimulation, and they linger for several minutes thereafter. As a result, the effects produced by stimulation of the adrenal medulla outlast the other signs of sympathetic activation. AM *Disorders of the Adrenal Medullae*

- ✓ What symptoms would you expect to see in an individual whose diet lacks iodine?
- ✓ When a person's thyroid gland is removed, signs of decreased thyroid hormone concentration do not appear until about one week later. Why?
- ✓ Removal of the parathyroid glands would result in a decrease in the blood of what important mineral?
- ✓ What effect would elevated cortisol levels have on the level of glucose in the blood?

THE KIDNEYS

The kidneys are not primarily endocrine organs, but they release three hormones, *calcitriol, erythropoietin,* and *renin.* Calcitriol is important to calcium ion homeostasis. Erythropoietin and renin are involved in the regulation of blood pressure and blood volume.

Calcitriol is a steroid hormone secreted by the kidney in response to the presence of parathyroid hormone (PTH). Its synthesis is dependent on the availability of vitamin D_3, which may be synthesized in the skin or absorbed from the diet. Vitamin D_3 is absorbed by the liver and converted to an intermediary product that is released into the circulation and absorbed by the kidneys. Calcitriol stimulates the absorption of calcium and phosphate ions along the digestive tract. The effects of PTH on calcium ion absorption result at least partially via stimulation of calcitriol release.

Erythropoietin (e-rith-rō-poi-Ē-tin; *erythros,* red + *poiesis,* making), or **EPO,** is a peptide hormone released by the kidney in response to low oxygen levels in kidney tissues. EPO stimulates the production of red blood cells by the bone marrow. The increase in the number of erythrocytes elevates blood volume to some degree, and, because these cells transport oxygen, the increase in their number improves oxygen delivery to peripheral tissues. EPO will be considered in greater detail when we discuss the formation of blood cells in Chapter 12.

Renin is released by kidney cells in response to a decline in blood volume or blood pressure or both. Once in the bloodstream, renin functions as an enzyme that starts an enzymatic chain reaction that leads to the formation of the hormone **angiotensin II.** Angiotensin II has several functions, including the stimulation of aldosterone production by the adrenal cortex. The renin-angiotensin system will be considered further in Chapters 14 and 19.

THE HEART

The endocrine cells in the heart are cardiac muscle cells in the walls of the *right atrium,* the chamber that receives venous blood. If the blood volume becomes too great, these cardiac muscle cells are excessively stretched. Under these conditions, they release *atrial natriuretic peptide (ANP)* (nā-trē-ū-RET-ik; *natrium,* sodium + *ouresis,* making water). This hormone lowers blood volume and reduces the stretching of the cardiac muscle cells in the atrial walls. We will consider the actions of this hormone in greater detail when we discuss the control of blood pressure and volume in Chapter 14. ∞ [p. 361]

ENDOCRINE TISSUES OF THE DIGESTIVE SYSTEM

The linings of the digestive tract, the liver, and the pancreas produce a variety of exocrine secretions that are essential to the normal breakdown and absorption of food. Although the pace of digestive activities can be affected by the autonomic nervous system, most digestive processes are controlled locally. The various components of the digestive tract communicate with one another by means of hormones that will be considered in Chapter 17. One digestive organ, the pancreas, produces two hormones with widespread effects.

The Pancreas

The **pancreas** (Figure 11-13●) lies in the J-shaped loop between the stomach and small intestine. It is a slender, usually pink organ with a nodular (lumpy) consistency, and it contains both exocrine and endocrine cells. The **exocrine pancreas,** discussed further in Chapter 17, produces large quantities of an alkaline, enzyme-rich fluid that is secreted into the digestive tract.

Cells of the **endocrine pancreas** form clusters known as **pancreatic islets,** or the *islets of Langer-*

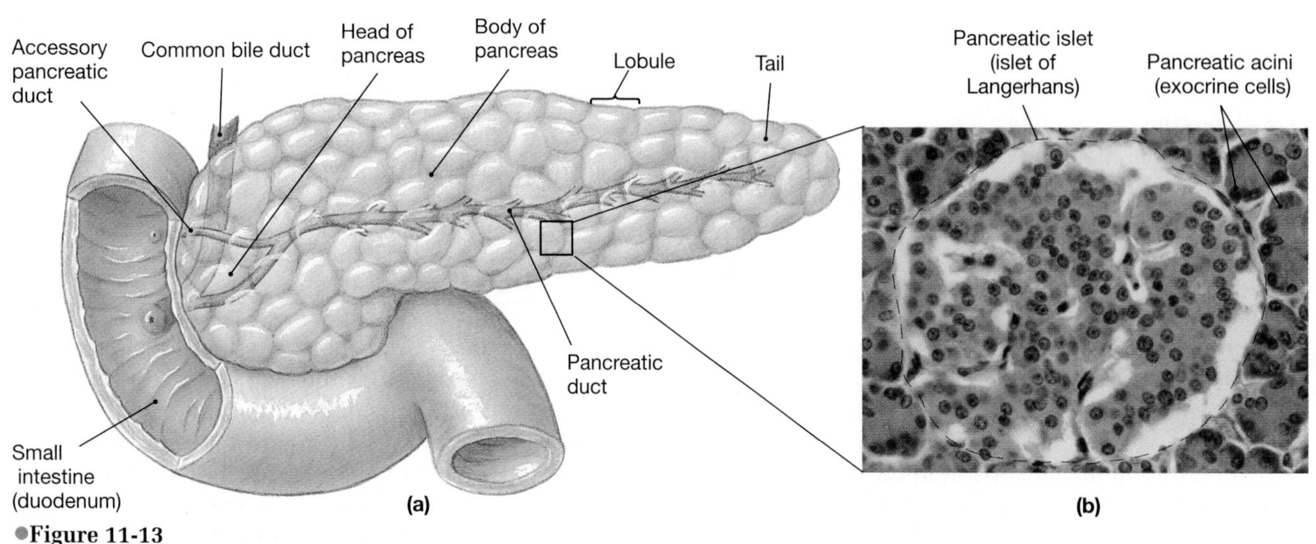

●**Figure 11-13**
The Endocrine Pancreas
(a) Gross anatomy of the pancreas. **(b)** A pancreatic islet surrounded by exocrine-secreting cells. (LM × 276)

hans (LAN-ger-hanz). The islets are scattered among the exocrine cells and account for only about 1 percent of the pancreatic cell population. Each islet contains several different cell types. The two most important are **alpha cells,** which produce the hormone **glucagon** (GLOO-ka-gon), and **beta cells,** which secrete **insulin** (IN-su-lin). Glucagon and insulin regulate blood glucose concentrations in the same way that parathyroid hormone and calcitonin control blood calcium levels.

Regulation of Blood Glucose Concentrations

Figure 11-14● diagrams the mechanism of hormonal regulation of blood glucose levels. Glucose is the preferred energy source for most cells in the body, and under normal conditions the only energy source for neurons. When blood glucose levels rise, beta cells release insulin, and this hormone stimulates glucose transport into its target cells. Almost all cells in the body are affected; the only exceptions are (1) neurons and red blood cells, which cannot metabolize other nutrients, and (2) epithelial cells of the kidney tubules and intestinal lining, where glucose is reabsorbed (kidneys) or obtained from the diet (intestines). When glucose is abundant, all cells use it as an energy source and stop breaking down amino acids and lipids.

The ATP generated by the breakdown of glucose molecules is used to build proteins and to increase energy reserves, and most cells increase their rates of protein synthesis in response to insulin. A secondary effect is an increase in the rate of amino acid transport across cell membranes. Insulin also stimulates fat cells to increase their rates of triglyceride (fat) synthesis and storage. In the liver and in skeletal muscles, insulin also accelerates the formation of glycogen. In summary, when glucose is abundant, insulin secretion stimulates glucose utilization to support growth and to establish glycogen and fat reserves.

When glucose levels decline, insulin secretion is suppressed, and so is glucose transport into its target cells. These cells now shift over to other energy sources, such as fatty acids. At the same time, the alpha cells release glucagon, and energy reserves are mobilized. Skeletal muscles and liver cells break down glycogen, adipose tissue releases fatty acids, and proteins are broken down into their component amino acids. The liver takes in the lipids and amino acids and converts them to glucose that can be released into the circulation. As a result, blood glucose concentrations rise toward normal levels. The interplay between insulin and glucagon both stabilizes blood glucose levels and prevents competition between neural tissue and other tissues for limited glucose supplies.

Pancreatic alpha and beta cells are sensitive to blood glucose concentrations, and their regulatory activities are not under the direct control of other endocrine or nervous components. Yet, because the islet cells are extremely sensitive to variations in blood glucose levels, any hormone that affects blood glucose concentrations will indirectly affect the production of insulin and glucagon.

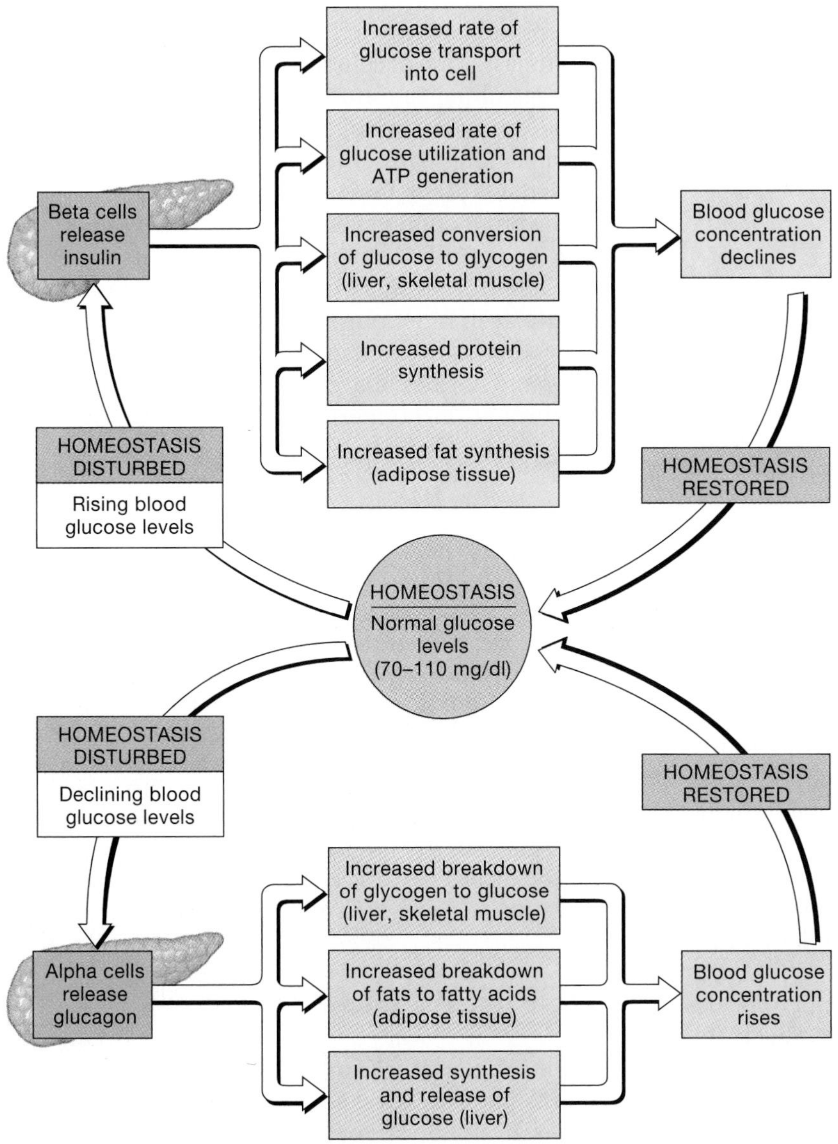

●Figure 11-14
Regulation of Blood Glucose Concentrations

DIABETES MELLITUS

Blood glucose levels are usually very closely regulated by insulin and glucagon. Whether glucose is absorbed across the digestive tract or manufactured and released by the liver, very little leaves the body intact once it has entered the circulation. Glucose does enter the urine, but the cells lining the excretory passageways of the kidneys usually reabsorb virtually all of it.

Diabetes mellitus (me-LĪ-tus; *mellitum*, honey) is characterized by glucose concentrations that are high enough to overwhelm the reabsorption capabilities of the kidneys. Glucose appears in the urine (*glycosuria*), and urine production becomes excessive (*polyuria*). Other metabolic products, such as fatty acids and other lipids, are also present in abnormal concentrations.

Diabetes mellitus may be caused by genetic abnormalities, and some of the genes responsible have been identified. For example, genes that result in inadequate insulin production, the synthesis of abnormal insulin molecules, or the production of defective receptor proteins will produce the same basic symptoms. Diabetes mellitus may also appear as the result of other pathological conditions, injuries, immune disorders, or hormonal imbalances. There are two major types of diabetes mellitus: *insulin-dependent (Type I) diabetes* and *non-insulin-dependent (Type II) diabetes.* AM *Diabetes mellitus*

ENDOCRINE TISSUES OF THE REPRODUCTIVE SYSTEM

The endocrine tissues of the reproductive system are primarily restricted to the male and female reproductive organs, the testes and ovaries. Details of the anatomy of the reproductive organs and the endocrinological control of reproductive function will be considered in Chapter 20.

The Testes

In the male, the **interstitial cells** of the testis produce the steroid hormones known as *androgens.* **Testosterone** (tes-TOS-ter-ōn) is the most important androgen. This hormone promotes the production of functional sperm, maintains the secretory glands of the male reproductive tract, and determines secondary sexual characteristics such as the distribution of facial hair and body fat. Testosterone also affects metabolic operations throughout the body, notably stimulating protein synthesis and muscle growth, and produces aggressive behavioral responses. During embryonic development, the production of testosterone affects the development of male reproductive ducts, external genitalia, and CNS structures, including hypothalamic nuclei, that will later affect sexual behaviors.

Sustentacular cells are directly associated with the formation of functional sperm in the testes. These cells also secrete a hormone called **inhibin.** Inhibin production, which occurs under FSH stimulation, depresses the secretion of FSH by the anterior pituitary. Throughout adult life, these two hormones interact to maintain sperm production at normal levels.

The Ovaries

In the ovaries, female sex cells, or *ova,* develop in specialized structures called **follicles,** under stimulation by FSH. Follicle cells surrounding the ova produce **estrogens** (ES-trō-jenz). These steroid hormones support the maturation of the eggs and stimulate the growth of the uterine lining. Under FSH stimulation, follicular cells secrete inhibin, which suppresses FSH release through a feedback mechanism comparable to that in males. After ovulation has occurred, the follicular cells reorganize into a **corpus luteum**. The luteal cells then begin to release a mixture of estrogens and progestins, especially **progesterone** (prō-JES-ter-ōn). Progesterone accelerates the movement of fertilized eggs along the uterine tubes and prepares the uterus for the arrival of a developing embryo. Progesterone, along with other hormones, also causes an enlargement of the mammary glands.

The production of androgens, estrogens, and progestins is controlled by regulatory hormones released by the anterior pituitary gland. During pregnancy, the placenta itself functions as an endocrine organ, working together with the ovaries and the pituitary gland to promote normal fetal development and delivery.

Information concerning the reproductive hormones can be found in Table 11-3.

ENDOCRINOLOGY AND ATHLETIC PERFORMANCE

Despite being banned by the International Olympic Committee, the United States Olympic Committee, the NCAA, and the NFL, and condemned by the American Medical Association and the American College of Sports Medicine, a significant number of amateur and professional athletes continue to use hormones to improve their performance. Although synthetic forms of testosterone are used most often, young athletes may use any combination of testosterone, growth hormone, EPO, and complementary drugs such as GHB and clenbuterol. AM *Endocrinology and Athletic Performance*

THE PINEAL GLAND

The pineal gland lies in the roof of the thalamus. It contains neurons, glial cells, and secretory cells that synthesize the hormone **melatonin** (mel-a-TŌ-nin). Collaterals (axonal branches of neurons) from the visual pathways enter the pineal gland and affect the

TABLE 11-3 Hormones of the Reproductive System

Structure/Cells	*Hormone*	*Primary Target*	*Effects*
TESTES			
Interstitial cells	Androgens	Most cells	Support functional maturation of sperm, protein synthesis in skeletal muscles, male secondary sexual characteristics, and associated behaviors
Sustenacular cells	Inhibin	Anterior pituitary	Inhibits secretion of FSH
OVARIES			
Follicular cells	Estrogens	Most cells	Support follicle maturation, female secondary sexual characteristics, and associated behaviors
	Inhibin	Anterior pituitary	Inhibits secretion of FSH
Corpus luteum	Progestins	Uterus, mammary glands	Prepare uterus for implantation; prepare mammary glands for secretory functions

rate of melatonin production. Melatonin production is lowest during daylight hours and highest in the dark of night.

Several functions have been suggested for melatonin secretion in humans.

- In a variety of other mammals, melatonin slows the maturation of sperm, eggs, and reproductive organs. The significance of this effect remains uncertain, but there is circumstantial evidence that melatonin may play a role in the timing of human sexual maturation. For example, melatonin levels in the blood decline at puberty, and pineal tumors that eliminate melatonin production will cause premature puberty in young children.
- Melatonin is a very effective *antioxidant* that may protect CNS neurons from *free radicals,* such as nitric oxide (NO) or hydrogen peroxide (H_2O_2), that may be generated in active neural tissue. (Free radicals are highly reactive atoms or molecules that contain unpaired electrons in their outer electron shell.)
- Because of the cyclical nature of its activity, the pineal gland may also be involved with the establishment or maintenance of basic *circadian rhythms,* daily changes in physiological processes that follow a regular pattern. Increased melatonin secretion in darkness has been suggested as a primary cause for *seasonal affective disorder.* This condition, characterized by changes in mood, eating habits, and sleeping patterns, can develop during the winter in high latitudes, where sunshine is meager or lacking altogether. AM *Light and Behavior*

✓ What pancreatic hormone would cause skeletal muscle and liver cells to break down glycogen to glucose?

✓ What effect would increased levels of glucagon have on the amount of glycogen stored in the liver?

✓ Increased amounts of light would inhibit the production of which hormone?

PATTERNS OF HORMONAL INTERACTION

Although hormones are usually studied individually, the extracellular fluids contain a mixture of hormones whose concentrations change daily and even hourly. When a cell receives instructions from two different hormones at the same time, there are several possible results.

1. The two hormones may have opposing, or **antagonistic,** effects, as in the case of parathyroid hormone and calcitonin or insulin and glucagon.
2. The two hormones may have additive, or **synergistic** (sin-er-JIS-tik; *synairesis,* a drawing together), effects. Sometimes the net result is not only greater than the effect that each would produce acting alone, but is actually greater than the *sum* of their individual effects. An example of such a effect is stimulation of mammary gland development by prolactin, estrogens, progestins, and growth hormone.

3. One hormone can have a **permissive** effect on another. In such cases, the first hormone is needed for the second to produce its effect. For example, epinephrine by itself has no apparent effect on energy production. It will exert its effect only if thyroid hormones are present in normal concentrations.
4. Hormones may also produce different but complementary results in specific tissues and organs. These **integrative** effects are important in coordinating the activities of diverse physiological systems.

This section will discuss how hormones interact to control normal growth, reaction to stress, alteration of behavior, and effects of aging. More detailed discussions will be found in chapters dealing with cardiovascular function, metabolism, excretion, and reproduction.

Hormones and Growth

Normal growth requires the cooperation of several endocrine organs. Five hormones are especially important, although many others have secondary effects on growth rates and patterns.

1. **Growth hormone.** Growth hormone helps maintain normal blood glucose concentrations and mobilizes lipid reserves stored in adipose tissues. It is not the primary hormone involved, however, and an adult with a growth hormone deficiency but normal levels of thyroxine, insulin, and glucocorticoids will have no physiological problems. The effects of GH on protein synthesis and cellular growth are most apparent in children, in whom GH supports muscular and skeletal development. Undersecretion or oversecretion of GH can lead to *pituitary dwarfism* or *gigantism.*
2. **Thyroid hormones.** Normal growth also requires appropriate levels of thyroid hormones. If these hormones are absent for the first year after birth, the nervous system fails to develop normally, producing mental retardation. If thyroxine concentrations decline later in life but before puberty, normal skeletal development will not continue.
3. **Insulin**. Growing cells need adequate supplies of energy and nutrients. Without insulin, produced by the pancreas, the passage of glucose and amino acids across cell membranes will be drastically reduced or eliminated.
4. **Parathyroid hormone**. Parathyroid hormone promotes the absorption of calcium salts across the lining of the digestive tract, thereby maintaining normal calcium levels in the circulation, a requirement for normal bone growth.
5. **Gonadal hormones**. The activity of osteoblasts in key locations and the growth of specific cell populations are affected by the presence or absence of sexual hormones. The differential growth induced by these hormones changes skeletal proportions and triggers the development of secondary sexual characteristics.

Hormones and Behavior

As we have seen, many endocrine functions are regulated by the hypothalamus, and hypothalamic neurons monitor the levels of many circulating hormones. Other portions of the central nervous system are also quite sensitive to hormonal stimulation.

The clearest demonstrations of the effects of specific hormones involve individuals whose endocrine glands are oversecreting or undersecreting. Normal changes in circulating hormone levels can also cause behavioral changes. Chapter 6 noted that one of the triggers for closure of the epiphyses is the increase in sexual hormone production at the time of puberty. ∞ *[p. 114]* In *precocious* (premature) *puberty,* sexual hormones are produced at an inappropriate time, perhaps as early as 5 or 6 years of age. The affected children not only begin to develop adult secondary sexual characteristics, but also undergo significant behavioral changes. The "nice little kid" disappears, and the child becomes aggressive and assertive. These behavioral alterations represent the effects of sexual hormones on CNS function. Thus, behaviors that in normal teenagers are usually attributed to external factors, such as peer pressure, actually have some physiological basis as well. In the adult, changes in the mixture of hormones reaching the CNS can have significant effects on intellectual capabilities, memory, learning, and emotional states.

Hormones and Aging

The endocrine system shows relatively few functional changes with age. The most dramatic exception is the decline in the concentration of reproductive hormones. Effects of these hormonal changes on the skeletal system were noted in Chapter 6; further discussion will be found in Chapter 21. ∞ *[p. 116]*

Blood and tissue concentrations of many other hormones, including TSH, thyroid hormones, ADH, PTH, prolactin, and glucocorticoids, remain unchanged. But, while hormone levels may remain within normal limits, some endocrine tissues become less responsive to stimulation. For example, in elderly individuals less GH and insulin are secreted after a carbohydrate-rich meal or in a glucose tolerance test.

Finally, it should be noted that age-related changes in other tissues affect their abilities to *respond* to hormonal stimulation. As a result, peripheral tissues may become less responsive to some hormones. This loss

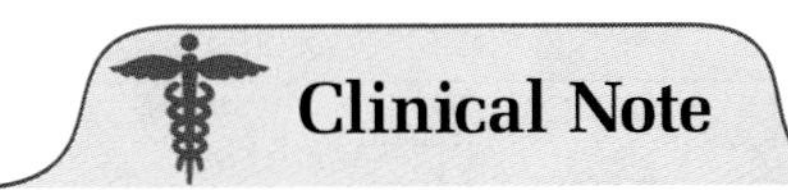

Clinical Note

Hormones and Stress

Any condition within the body that threatens homeostasis is a form of **stress.** Stresses are produced by the action of *stressors* that may be (1) physical, such as illness or injury; (2) emotional, such as depression or anxiety; (3) environmental, such as extreme heat or cold; or (4) metabolic, such as acute starvation. The stresses produced may be opposed by specific homeostatic adjustments. For example, a decline in body temperature will result in responses such as changes in the pattern of circulation, or shivering, in an attempt to restore normal body temperature.

In addition, the body has a general response to stress that can occur while other, more specific, responses are under way. *All stressors produce the same basic pattern of hormonal and physiological adjustments.* These responses are part of the stress response, also known as the **general adaptation syndrome,** or **GAS.** The GAS can be divided into three phases: the *alarm phase,* the *resistance phase,* and the *exhaustion phase* (Figure 11-15●).

The **alarm phase** is an immediate response to the stress, under the direction of the sympathetic division of the autonomic nervous system. During this phase, energy reserves are mobilized, mainly in the form of glucose, and the body prepares for any physical activities needed to eliminate or escape from the source of the stress. Epinephrine is the dominant hormone of the alarm phase, and its secretion accompanies the sympathetic activation that produces the "fight or flight" response discussed in Chapter 9. ∞ *p. 239*

If a stress lasts longer than a few hours, the individual will enter the **resistance phase** of the GAS. Glucocorticoids are the dominant hormones of the resistance phase, but many other hormones are involved. Throughout the resistance phase, energy demands remain higher than normal because of the background production of glucocorticoids, epinephrine, growth hormone, and thyroid hormones.

The endocrine secretions of the resistance phase coordinate three integrated actions to maintain adequate levels of glucose in the blood: (1) the mobilization of lipid and protein reserves, (2) the conservation of glucose for neural tissues, and (3) the synthesis and release of glucose by the liver. When the resistance phase ends, homeostatic regulation breaks down and the **exhaustion phase** begins. Unless corrective actions are taken almost immediately, the failure of one or more organ systems will prove fatal. Although a single cause, such as heart failure, may be listed as the cause of death, the underlying problem is the inability to support the endocrine and metabolic adjustments of the resistance phase.

ALARM PHASE

"Fight or Flight"
Immediate short-term responses to crises

Brain
General sympathetic activation
Sympathetic stimulation
Epinephrine
Norepinephrine
Adrenal medullae

1. Mobilization of glucose reserves
2. Changes in circulation
3. Increases in heart and respiratory rates
4. Increased energy use by all cells

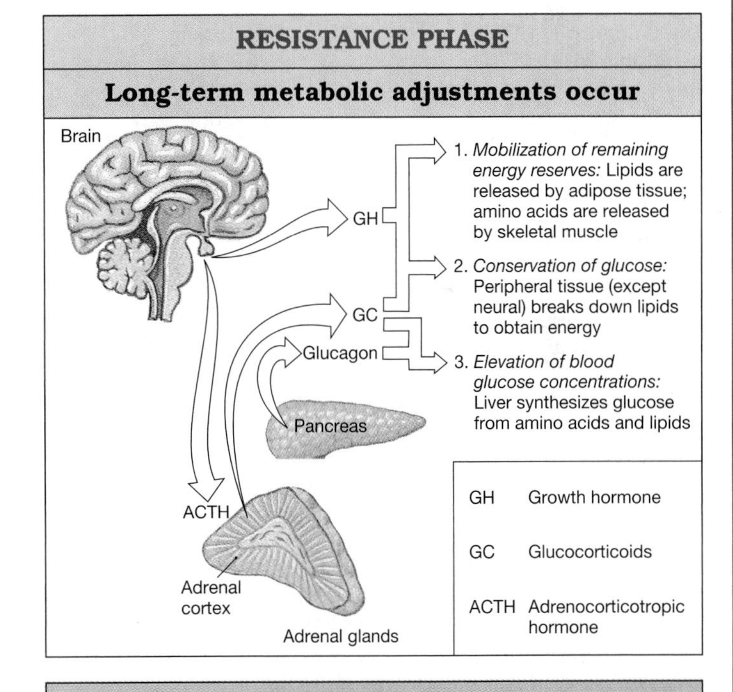

EXHAUSTION PHASE

Collapse of vital systems

Causes may include:
— Exhaustion of lipid reserves
— Inability to produce glucocorticoids
— Failure of electrolyte balance
— Cumulative structural or functional damage to vital organs

●**Figure 11-15**
The General Adaptation Syndrome

of sensitivity has been documented for glucocorticoids and ADH.

ENDOCRINE DISORDERS

The symptoms of endocrine disorders can usually be assigned to one of two basic categories: symptoms of underproduction (inadequate hormonal effects) or symptoms of overproduction (excessive hormonal effects). The observed symptoms may reflect either abnormal hormone production (hyposecretion or hypersecretion) or abnormal cellular sensitivity. These conditions are interesting because they highlight the significance of normally "silent" hormonal contributions. A summary of endocrine disorders in terms of underproduction and overproduction is presented in the *Applications Manual.* AM *Endocrine Disorders*

INTERACTIONS WITH OTHER SYSTEMS

The relationships between the endocrine system and other systems are summarized in Figure 11-16●. This overview does not consider all of the hormones associated with the digestive system and the control of digestive functions. These hormones will be detailed in Chapter 17.

✓ Insulin lowers the level of glucose in the blood and glucagon causes glucose levels to rise. What is this type of hormonal interaction called?

✓ The lack of which hormones would inhibit skeletal formation?

Chapter Review

KEY TERMS

adrenal cortex, p. 293
adrenal medulla, p. 293
endocrine cell, p. 282
glucagon, p. 297
hormone, p. 282
hypophyseal portal system, p. 286
hypophysis, p. 286
insulin, p. 297
pancreas, p. 296
peptide hormone, p. 283
pituitary gland, p. 286
steroid hormone, p. 283

SUMMARY OUTLINE

INTRODUCTION p. 282

1. In general, the nervous system performs short-term "crisis management," while the endocrine system regulates longer-term, ongoing metabolic processes. Endocrine cells release chemicals called **hormones** that alter the metabolic activities of many different tissues and organs simultaneously. (*Figure 11-1*)

AN OVERVIEW OF THE ENDOCRINE SYSTEM p. 282

Structure of Hormones p. 283

1. Hormones can be divided into three groups based on chemical structure: amino acid derivatives, peptide hormones, and lipid derivatives.

2. Amino acid derivatives are structurally similar to amino acids; they include *epinephrine, norepinephrine, thyroid hormones,* and *melatonin.*

3. **Peptide hormones** are chains of amino acids.

4. There are two classes of lipid derivatives. **Prostaglandins** are fatty acid–based, and **steroid hormones** are lipids structurally similar to cholesterol.

Mechanisms of Hormonal Action p. 283

5. Hormones exert their effects by modifying the activities of **target cells** (peripheral cells that are sensitive to that particular hormone). (*Figure 11-2*)

6. Receptors for amino acid–derived and peptide hormones are located on the cell membranes of target cells; in this case the hormone acts as a **first messenger** that causes a **second messenger** to appear in the cytoplasm. Thyroid and steroid hormones cross the cell membrane and bind to receptors in the cytoplasm or nucleus. (*Figure 11-3*)

Control of Endocrine Activity p. 285

7. The simplest patterns of endocrine control involve the direct negative feedback of changes in the extracellular fluid on the endocrine cells.

8. The most complex endocrine responses involve the hypothalamus. The hypothalamus regulates the activities of the nervous and endocrine systems via three mechanisms: (1) Its autonomic centers exert direct neural control over the endocrine cells of the adrenal medullae; (2) it acts as an endocrine organ itself by releasing hormones into the circulation; (3) it secretes **regulatory hormones** that control the activities of endocrine cells in the pituitary gland. (*Figure 11-4*)

THE PITUITARY GLAND p. 286

1. The pituitary gland releases nine important peptide hormones; all bind to membrane receptors and use cyclic-AMP as a second messenger. (*Figure 11-5*)

The Anterior Pituitary p. 286

2. Hypothalamic neurons release regulatory factors into the surrounding interstitial fluids. Their secretions then enter highly permeable capillaries.

INTEGUMENTARY SYSTEM

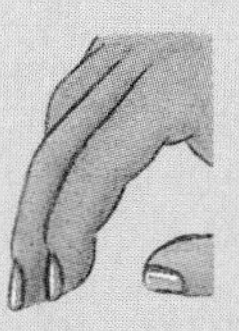

Protects superficial endocrine organs; epidermis synthesizes vitamin D_3

Sexual hormones stimulate sebaceous gland activity, influence hair growth, fat distribution, and apocrine sweat gland activity; PRL stimulates development of mammary glands; adrenal hormones alter dermal blood flow, stimulate release of lipids from adipocytes; MSH stimulates melanocyte activity

THE ENDOCRINE SYSTEM

FOR ALL SYSTEMS

Adjusts metabolic rates and substrate utilization; regulates growth and development

SKELETAL SYSTEM

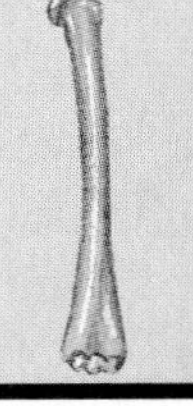

Protects endocrine organs, especially in brain, chest, and pelvic cavity

Skeletal growth regulated by several hormones; calcium mobilization regulated by parathyroid hormone and calcitonin; sex hormones speed growth and closure of epiphyseal plates at puberty, and help maintain bone mass in adults

MUSCULAR SYSTEM

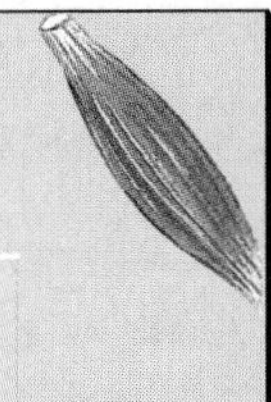

Skeletal muscles provide protection for some endocrine organs

Hormones adjust muscle metabolism, energy production, and growth; regulate Ca^{2+} and PO_4^{3-} levels in body fluids; speed skeletal muscle growth

NERVOUS SYSTEM

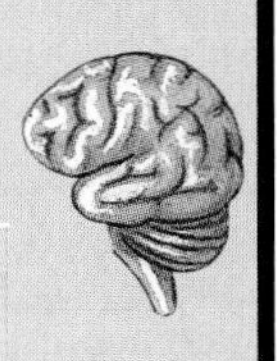

Hypothalamic hormones directly control pituitary and indirectly control secretions of other endocrine organs; controls adrenal medullae; secretes ADH and oxytocin

Several hormones affect neural metabolism; hormones help regulate fluid and electrolyte balance; reproductive hormones influence CNS development and behaviors

CARDIOVASCULAR SYSTEM

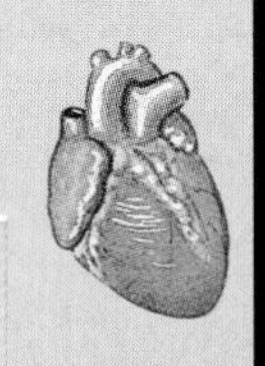

Circulatory system distributes hormones throughout the body; heart secretes ANP

Erythropoietin regulates production of RBCs; several hormones elevate blood pressure; epinephrine elevates heart rate and contractile force

LYMPHATIC SYSTEM

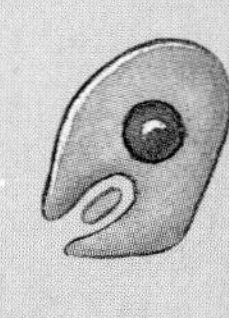

Lymphocytes provide defense against infection and, with other WBCs, assist in repair after injury

Glucocorticoids have anti-inflammatory effects; thymosins stimulate development of lymphocytes; many hormones affect immune function

RESPIRATORY SYSTEM

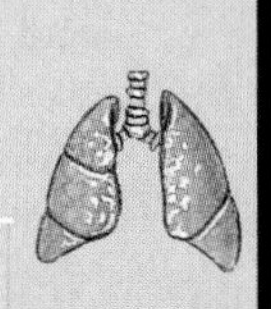

Provides O_2 and eliminates CO_2 generated by endocrine cells

Epinephrine and norepinephrine stimulate respiratory activity and dilate respiratory passageways

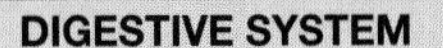

DIGESTIVE SYSTEM

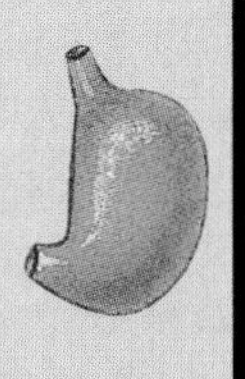

Provides nutrients and substrates to endocrine cells; endocrine cells of pancreas secrete insulin and glucagon

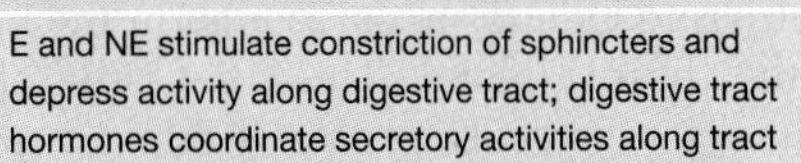

E and NE stimulate constriction of sphincters and depress activity along digestive tract; digestive tract hormones coordinate secretory activities along tract

URINARY SYSTEM

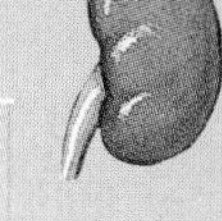

Kidney cells (1) release renin and erythropoietin when local blood pressure declines; (2) produce calcitriol

Aldosterone, ADH, and ANP adjust rates of fluid and electrolyte reabsorption in kidneys

REPRODUCTIVE SYSTEM

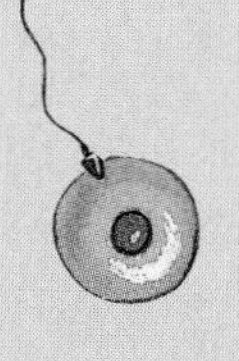

Steroid sex hormones and inhibin suppress secretory activities in hypothalamus and pituitary

Hypothalamic factors and pituitary hormones regulate sexual development and function; oxytocin stimulates uterine and mammary gland smooth muscle contractions

•Figure 11-16
Functional Relationships between the Endocrine System and Other Systems

3. The **hypophyseal portal system** ensures that all of the blood entering the *portal vessels* will reach target cells in the anterior pituitary before returning to the general circulation. (*Figure 11-6*)

4. The rate of regulatory hormone secretion by the hypothalamus is regulated through negative feedback mechanisms. (*Figure 11-7*)

5. The seven hormones of the **anterior pituitary** are: (1) **thyroid-stimulating hormone (TSH),** which triggers the release of thyroid hormones; (2) **adrenocorticotropic hormone (ACTH),** which stimulates the release of **glucocorticoids** by the adrenal gland; (3) **follicle-stimulating hormone (FSH),** which stimulates *estrogen* secretion and egg development in women and sperm production in men; (4) **luteinizing hormone (LH),** which causes ovulation and progestin production in women and androgen production in men; (5) **prolactin (PRL),** which stimulates the development of the mammary glands and the production of milk; (6) **growth hormone (GH),** which stimulates cell growth and replication by triggering the release of **somatomedins** from liver cells; and (7) **melanocyte-stimulating hormone (MSH),** which stimulates melanocytes to produce melanin in other species but is not normally secreted by the nonpregnant human adult.

The Posterior Pituitary p. 289

6. The **posterior pituitary** contains the axons of hypothalamic neurons that manufacture **antidiuretic hormone (ADH)** and **oxytocin.** ADH decreases the amount of water lost at the kidneys. In women, oxytocin stimulates smooth muscle cells in the uterus and contractile cells in the mammary glands. In men, it stimulates prostatic smooth muscle contractions. (*Figure 11-8; Table 11-1*)

THE THYROID GLAND p. 289

1. The thyroid gland lies near the **thyroid cartilage** of the larynx and consists of two lobes. (*Figure 11-9*)

Thyroid Follicles and Thyroid Hormones p. 289

2. The thyroid gland contains numerous **thyroid follicles.** Thyroid follicles release several hormones, including **thyroxine (TX** or **T_4)** and **triiodothyronine (T_3).**

3. Thyroid hormones exert a **calorigenic effect** which enables us to adapt to cold temperatures.

The C Cells of the Thyroid Gland: Calcitonin p. 292

4. The **C cells** of the follicles produce **calcitonin (CT),** which helps regulate calcium ion concentrations in body fluids.

THE PARATHYROID GLANDS p. 293

1. Four parathyroid glands are embedded in the posterior surface of the thyroid gland. The **chief cells** of the parathyroid produce **parathyroid hormone (PTH)** in response to lower than normal concentrations of calcium ions. These and the C cells of the thyroid gland maintain calcium ion levels within relatively narrow limits. (*Figures 11-10, 11-11*)

THE THYMUS p. 293

1. The thymus produces several hormones, called **thymosins,** which play a role in developing and maintaining normal immunological defenses.

THE ADRENAL GLANDS p. 293

1. A single **adrenal gland** lies along the superior border of each kidney. Each gland, surrounded by a fibrous capsule, can be subdivided into the superficial **adrenal cortex** and the inner **adrenal medulla.** (*Figure 11-12*)

The Adrenal Cortex, p. 293

2. The adrenal cortex manufactures steroid hormones called *adrenocortical steroids* (**corticosteroids**). The cortex produces (1) **glucocorticoids (GCs)**, notably **cortisol, corticosterone,** and **cortisone,** in response to ACTH—these hormones affect glucose metabolism; (2) **mineralocorticoids (MCs),** principally **aldosterone,** in response to *angiotensin II*—this hormone restricts sodium and water losses at the kidneys, sweat glands, digestive tract, and salivary glands; and (3) androgens of uncertain significance. (*Figure 11-12; Table 11-2*)

The Adrenal Medulla p. 295

3. The adrenal medulla produces epinephrine and norepinephrine. (*Figure 11-12; Table 11-2*)

THE KIDNEYS p. 295

1. Endocrine cells in the kidneys produce hormones important for calcium metabolism, blood volume, and blood pressure.

2. **Calcitriol** stimulates calcium and phosphate ion absorption along the digestive tract.

3. **Erythropoietin (EPO)** stimulates red blood cell production by the bone marrow.

4. **Renin** release leads to the formation of **angiotensin II,** the hormone that stimulates the adrenal production of aldosterone.

THE HEART p. 296

1. Specialized muscle cells in the heart produce *atrial natriuretic peptide (ANP)* when blood pressure and/or blood volume becomes excessive.

ENDOCRINE TISSUES OF THE DIGESTIVE SYSTEM p. 296

1. The linings of the digestive tract, the liver, and the pancreas produce exocrine secretions that are essential to the normal breakdown and absorption of food. These organs also produce a variety of hormones that coordinate digestive and metabolic activities.

The Pancreas p. 296

2. The **pancreas** contains both exocrine and endocrine cells. The **exocrine pancreas** secretes an enzyme-rich fluid that travels to the digestive tract. Cells of the **endocrine pancreas** form clusters called **pancreatic islets** (islets of Langerhans), containing **alpha cells** (which produce the hormone **glucagon**) and **beta cells** (which secrete **insulin**). (*Figure 11-13*)

3. Insulin lowers blood glucose by increasing the rate of glucose uptake and utilization; glucagon raises blood glucose by increasing the rates of glycogen breakdown and glucose synthesis in the liver. (*Figure 11-14*)

ENDOCRINE TISSUES OF THE REPRODUCTIVE SYSTEM p. 298

The Testes p. 298

1. The **interstitial cells** of the male testis produce androgens and **inhibin.** The androgen **testosterone** is the most important sexual hormone in the male. (*Table 11-3*)

The Ovaries p. 298

2. In women, ova (eggs) develop in **follicles;** follicle cells surrounding the eggs produce **estrogens** and inhibin. After ovulation, the cells reorganize into a **corpus luteum** that releases a mixture of estrogens and progestins, espe-

cially **progesterone.** If pregnancy occurs, the placenta functions as an endocrine organ. (*Table 11-3*)

THE PINEAL GLAND p. 298

1. The pineal gland synthesizes **melatonin.** Melatonin appears to (1) slow the maturation of sperm, eggs, and reproductive organs, (2) protect neural tissue from *free radicals*, and (3) establish daily circadian rhythms.

PATTERNS OF HORMONAL INTERACTION p. 299

1. The endocrine system functions as an integrated unit, and hormones often interact. These interactions may have (1) **antagonistic** (opposing) effects, (2) **synergistic** (additive) effects, (3) **permissive** effects, or (4) **integrative** effects, in which hormones produce different but complementary results.

Hormones and Growth p. 300

2. Normal growth requires the cooperation of several endocrine organs. Five hormones are especially important: growth hormone, thyroid hormones, insulin, parathyroid hormone, and gonadal hormones.

Hormones and Behavior p. 300

3. Many hormones affect the functional state of the nervous system, producing changes in mood, emotional states, and various behaviors.

Hormones and Aging p. 300

4. The endocrine system shows relatively few functional changes with advanced age. The most dramatic endocrine change is the decline in the concentration of reproductive hormones.

INTERACTIONS WITH OTHER SYSTEMS p. 302

1. The endocrine system affects all systems by adjusting metabolic rates and regulating growth and development. (*Figure 11-16*)

CHAPTER QUESTIONS

LEVEL 1 Reviewing Facts and Terms

Match each item in column A with the most closely related item in column B. Use letters for answers in the spaces provided.

Column A	Column B
___ 1. thyroid gland	a. islets of Langerhans
___ 2. pineal gland	b. atrophies by adulthood
___ 3. polyuria	c. atrial natriuretic peptide
___ 4. parathyroid gland	d. cell growth
___ 5. thymus gland	e. melatonin
___ 6. adrenal cortex	f. hypophysis
___ 7. heart	g. excessive urine production
___ 8. endocrine pancreas	h. calcitonin
___ 9. gonadotropins	i. secretes regulatory hormones
___ 10. hypothalamus	j. FSH and LH
___ 11. pituitary gland	k. secretes androgens, mineralocorticoids, and glucocorticoids
___ 12. growth hormone	l. stimulated by low calcium levels

13. Adrenocorticotropic hormone (ACTH) stimulates the release of:
(a) thyroid hormones by the hypothalamus
(b) gonadotropins by the adrenal glands
(c) somatotropins by the hypothalamus
(d) steroid hormones by the adrenal glands

14. FSH production in males supports:
(a) maturation of sperm by stimulating sustentacular cells
(b) development of muscles and strength
(c) production of male sex hormones
(d) increased desire for sexual activity

15. The hormone that induces ovulation in women and promotes the ovarian secretion of progesterone is:
(a) interstitial cell–stimulating hormone
(b) estradiol
(c) luteinizing hormone
(d) prolactin

16. The two hormones released by the posterior pituitary are:
(a) somatotropin and gonadotropin
(b) estrogen and progesterone
(c) growth hormone and prolactin
(d) antidiuretic hormone and oxytocin

17. The primary function of antidiuretic hormone (ADH) is to:
(a) increase the amount of water lost at the kidneys
(b) decrease the amount of water lost at the kidneys
(c) dilate peripheral blood vessels to decrease blood pressure
(d) increase absorption along the digestive tract

18. The element required for normal thyroid function is:
(a) magnesium
(b) calcium
(c) potassium
(d) iodine

19. Reduced fluid losses in the urine due to retention of sodium ions and water is a result of the action of:
    (a) antidiuretic hormone
    (b) calcitonin
    (c) aldosterone
    (d) cortisone
20. The adrenal medullae produce the hormones:
    (a) cortisol and cortisone
    (b) epinephrine and norepinephrine
    (c) corticosterone and testosterone
    (d) androgens and progesterone
21. What seven hormones are released by the anterior pituitary gland?
22. What effects do calcitonin and parathyroid hormone have on blood calcium levels?
23. (a) What three phases of the general adaptation syndrome (GAS) constitute the body's response to stress? (b) What endocrine secretions play dominant roles in the alarm and resistance phases?

LEVEL 2 Reviewing Concepts

24. What is the primary difference in the way the nervous and endocrine systems communicate with their target cells?
25. How can a hormone modify the activities of its target cells?
26. What possible results occur when a cell receives instructions from two different hormones at the same time?
27. How would blocking the activity of phosphodiesterase affect a cell that responds to hormonal stimulation by the cAMP second messenger system?

LEVEL 3 Critical Thinking and Clinical Applications

28. Roger M. has been suffering from extreme thirst; he drinks numerous glasses of water every day and urinates a great deal. Name two disorders that could produce these symptoms. What test could a clinician perform to determine which disorder is present?
29. Julie is pregnant and is not receiving any prenatal care. She has a poor diet consisting mostly of fast food. She drinks no milk, preferring colas instead. How will this situation affect Julie's level of parathyroid hormone?

12 BLOOD

"Out of sight, out of mind." There is probably nothing this old saying applies to more fittingly than blood. As long as we don't see it, we don't think or worry much about it. But looking at vials or bags of it in a hospital—or worse, watching even a drop of our own oozing from a cut or scrape—makes most of us feel distinctly uneasy; some people even faint at the sight. This anxiety probably comes from being reminded of how dependent we are on our relatively small supply of the precious fluid. If you suddenly lose more than about 15 to 20 percent of your supply, you will need to get more in a hurry. Fortunately, blood transfusions are now fairly routine matters—but as we will learn in this chapter, not just anyone's blood will do.

Chapter Outline and Objectives

Vocabulary Development

embolos, plug; *embolus*
erythros, red; *erythrocytes*
haima, blood; *hemostasis*
hypo-, below; *hypoxia*
karyon, nucleus; *megakaryocyte*
leukos, white; *leukocyte*
megas, big; *megakaryocyte*
myelos, marrow; *myeloid*
ox-, presence of oxygen; *hypoxia*
penia, poverty; *leukopenia*
punctura, a piercing; *venipuncture*
stasis, halt; *hemostasis*
thrombos, clot; *thrombocytes*
vena, vein; *venipuncture*

The living body is in constant chemical communication with its external environment. Nutrients are absorbed through the lining of the digestive tract, gases move across the delicate epithelium of the lungs, and wastes are excreted in the feces and urine. These chemical exchanges occur at specialized sites or organs because all parts of the body are linked by a transport network, the *cardiovascular system.*

Small embryos don't need cardiovascular systems because diffusion across their exposed surfaces can exchange materials rapidly enough to meet their demands. By the time the embryo has reached a few millimeters in length, however, developing tissues will consume oxygen and nutrients and generate waste products faster than they can be provided or removed by simple diffusion. If growth is to continue, the cardiovascular system must provide a mechanism for rapidly transporting nutrients, waste products, and cells within the body. The heart begins beating by the end of the third week of embryonic life, when most other organ systems have barely started their development. Once blood begins circulating, the embryo can more efficiently use the nutrients obtained from the maternal bloodstream, and the embryo doubles its size in the next week.

The cardiovascular system can be compared to the cooling system of a car. The basic components include a circulating fluid (blood), a pump (the heart), and an assortment of conducting pipes (the circulatory system). Although the cardiovascular system is far more complicated and versatile, both mechanical and biological systems can suffer from fluid losses, pump failures, or damaged pipes. This chapter considers the nature of the circulating blood. Chapter 13 focuses on the structure and function of the heart, and Chapter 14 examines the organization of blood vessels and the integrated functioning of the cardiovascular system. Chapter 15 considers the *lymphatic system,* a defense system intimately connected to the cardiovascular system.

FUNCTIONS OF BLOOD

The circulating fluid of the body is blood, a specialized connective tissue introduced in Chapter 4. ∞ *[p. 84]* Blood provides essential homeostatic services to each of the roughly 75 trillion individual cells in the human body. The functions of blood include:

1. ***Transportation of dissolved gases, nutrients, hormones, and metabolic wastes.*** Oxygen is carried from the lungs to the tissues, and carbon dioxide from the tissues to the lungs. Nutrients absorbed at the digestive tract or released from storage in adipose tissue or the liver are distributed throughout the body. Hormones are carried from endocrine glands toward their target tissues, and the wastes produced by tissue cells are absorbed by the blood and carried to the kidneys for excretion.
2. ***Regulation of the pH and electrolyte composition of interstitial fluids throughout the body.*** The blood absorbs and neutralizes the acids generated by active tissues, such as the lactic acid produced by skeletal muscles.
3. ***Restriction of fluid losses through damaged vessels or at other injury sites.*** Blood contains enzymes and factors that respond to breaks in the vessel walls by initiating the process of *blood clotting.* The blood clot that develops acts as a temporary patch and prevents further changes in blood volume.
4. ***Defense against toxins and pathogens.*** Blood transports *white blood cells,* specialized cells that migrate into peripheral tissues to fight infections or remove debris. It also delivers *antibodies,* special proteins that attack invading organisms or foreign compounds.
5. ***Stabilization of body temperature.*** Blood absorbs the heat generated by active skeletal muscles and redistributes it to other tissues. When body temperature is already high, that heat will be lost across the surface of the skin. If body temperature is too low, the warm blood is directed to the brain and to other temperature-sensitive organs.

COMPOSITION OF BLOOD

Blood is normally confined to the circulatory system, and it has a characteristic and unique composition (Figure 12-1●). **Plasma** (PLAZ-mah), the ground substance of blood, has a density only slightly greater than that of water. It contains dissolved proteins rather than the network of insoluble fibers found in loose connective tissue or cartilage. **Formed elements** are blood cells and cell fragments that are suspended in the plas-

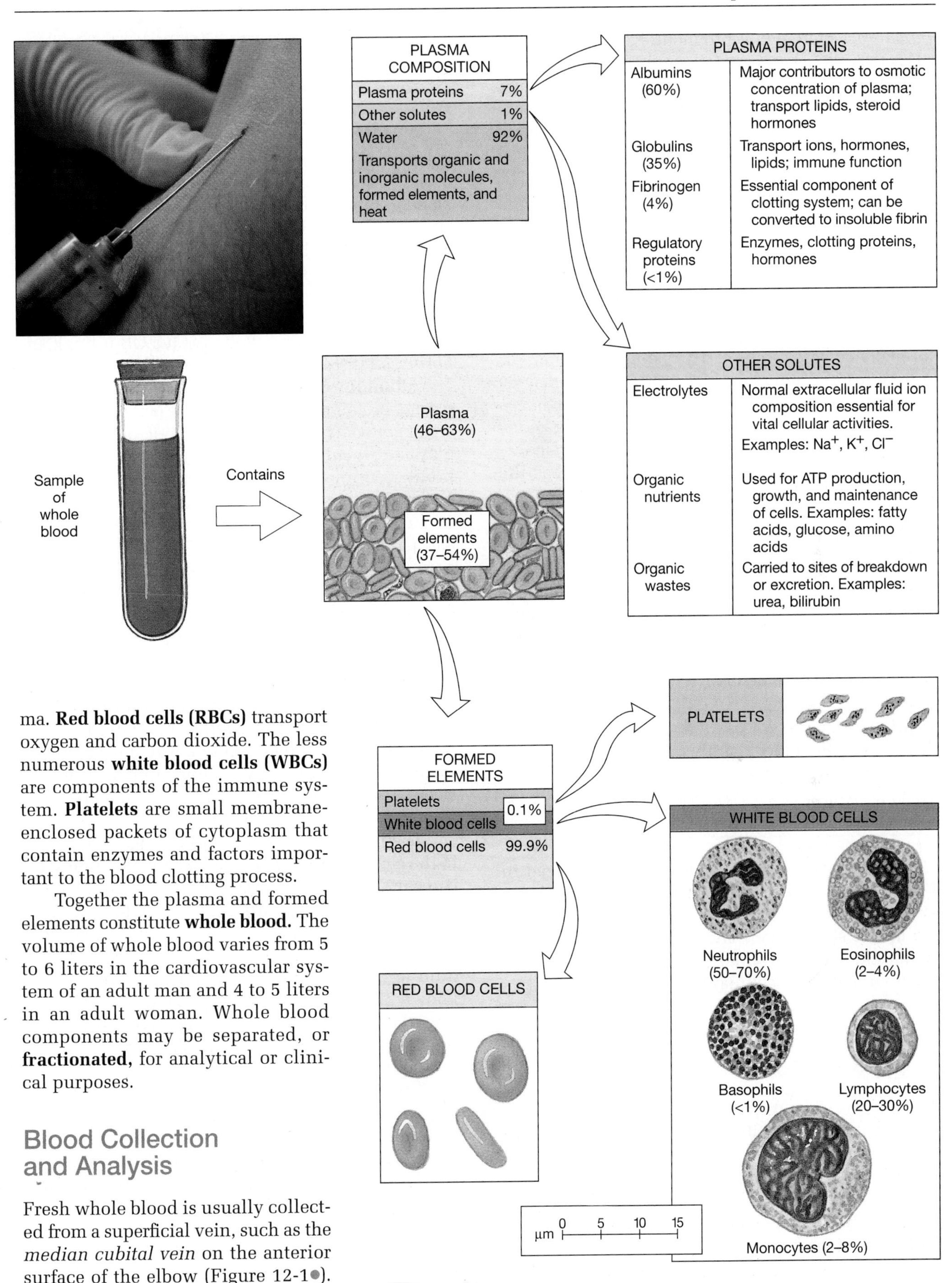

PLASMA COMPOSITION
Plasma proteins 7%
Other solutes 1%
Water 92%
Transports organic and inorganic molecules, formed elements, and heat
PLASMA PROTEINS
Albumins (60%)
Major contributors to osmotic concentration of plasma; transport lipids, steroid hormones
Globulins (35%)
Transport ions, hormones, lipids; immune function
Fibrinogen (4%)
Essential component of clotting system; can be converted to insoluble fibrin
Regulatory proteins (<1%)
Enzymes, clotting proteins, hormones
OTHER SOLUTES
Electrolytes
Normal extracellular fluid ion composition essential for vital cellular activities.
Examples: Na+, K+, Cl−
Organic nutrients
Used for ATP production, growth, and maintenance of cells. Examples: fatty acids, glucose, amino acids
Organic wastes
Carried to sites of breakdown or excretion. Examples: urea, bilirubin
Sample of whole blood
Contains
Plasma (46–63%)
Formed elements (37–54%)
FORMED ELEMENTS
Platelets
White blood cells
0.1%
Red blood cells 99.9%
PLATELETS
WHITE BLOOD CELLS
Neutrophils (50–70%)
Eosinophils (2–4%)
Basophils (<1%)
Lymphocytes (20–30%)
Monocytes (2–8%)
RED BLOOD CELLS
μm 0 5 10 15

Figure 12-1
The Composition of Whole Blood

ma. **Red blood cells (RBCs)** transport oxygen and carbon dioxide. The less numerous **white blood cells (WBCs)** are components of the immune system. **Platelets** are small membrane-enclosed packets of cytoplasm that contain enzymes and factors important to the blood clotting process.

Together the plasma and formed elements constitute **whole blood.** The volume of whole blood varies from 5 to 6 liters in the cardiovascular system of an adult man and 4 to 5 liters in an adult woman. Whole blood components may be separated, or **fractionated,** for analytical or clinical purposes.

Blood Collection and Analysis

Fresh whole blood is usually collected from a superficial vein, such as the *median cubital vein* on the anterior surface of the elbow (Figure 12-1). This procedure is called a **venipuncture** (VEN-e-punk-chur; *vena,* vein +

punctura, a piercing). Venipuncture is a popular sampling technique because (1) superficial veins are easy to locate; (2) the walls of veins are thinner than those of comparably sized arteries; and (3) blood pressure in the venous system is relatively low, and the puncture wound seals quickly. The most common clinical procedures examine venous blood.

Blood from peripheral capillaries can be obtained by puncturing the tip of a finger, the lobe of an ear, or (in infants) the sole of the great toe or heel of the foot. A small drop of capillary blood can be used to prepare a *blood smear,* a thin film of blood on the surface of a microscope slide. The blood smear can then be stained with special dyes so that the different types of formed elements can be easily distinguished.

An **arterial puncture,** or "arterial stick," may be required when checking the efficiency of gas exchange at the lungs. Samples are usually drawn from the *radial artery* at the wrist or the *brachial artery* at the elbow.

Whole blood from all of these sources has the same basic physical characteristics.

- **Temperature.** Blood temperature is roughly 38°C (100.4°F), slightly higher than normal body temperature.
- **Viscosity.** Blood viscosity is five times that of water, because interactions between dissolved proteins, formed elements, and the surrounding water molecules make plasma relatively sticky, cohesive, and resistant to flow.
- **pH.** Blood pH averages 7.4 and ranges from 7.35 to 7.45. It is therefore slightly alkaline.

PLASMA

The composition of whole blood is summarized in Figure 12-1●. Plasma contributes approximately 55 percent of the volume of whole blood, and water accounts for 92 percent of the plasma volume. Together, plasma and interstitial fluid account for most of the volume of extracellular fluid (ECF) in the body.

Differences Between Plasma and Interstitial Fluid

In many respects, the composition of the plasma resembles that of interstitial fluid (IF). In contrast to cytoplasm, for example, both have roughly the same concentrations of the major plasma ions (electrolytes). The chief differences between plasma and interstitial fluid involve the concentrations of dissolved proteins and respiratory gases (oxygen and carbon dioxide). The protein concentrations differ because plasma contains circulating plasma proteins that cannot cross the walls of blood vessels and thus cannot enter the interstitial fluids. The differences in the levels of respiratory gases will be discussed in Chapter 16.

Plasma Proteins

Plasma contains considerable quantities of dissolved proteins. As Figure 12-1● indicates, there are about 7 g of protein in each 100 ml of plasma; this amount is almost five times the concentration in interstitial fluid. The large size and globular shapes of most blood proteins prevent them from crossing capillary walls, and they remain trapped within the circulatory system. There are three primary classes of plasma proteins: *albumins* (al-BŪ-minz), *globulins* (GLOB-ū-linz), and *fibrinogen* (fī-BRIN-ō-jen) (see Figure 12-1●).

Albumins constitute roughly 60 percent of the plasma proteins. As the most abundant proteins, they are major contributors to the osmotic pressure of the plasma. **Globulins,** accounting for 35 percent of the protein population, include *immunoglobulins* and *transport proteins.* **Immunoglobulins** (i-mū-nō-GLOB-ū-linz), also called **antibodies,** attack foreign proteins and pathogens. **Transport proteins** bind small ions, hormones, or compounds that might otherwise be filtered out of the blood at the kidneys. One example is thyroid-binding globulin, which binds and transports thyroid hormones.

Both albumins and globulins can become attached to lipids, such as triglycerides, fatty acids, or cholesterol. These lipids are not water-soluble in themselves, but the protein-lipid combination readily dissolves in plasma, and in this way the circulatory system transports insoluble lipids to peripheral tissues. Globulins involved in lipid transport are called *lipoproteins* (lī-pō-PRŌ-tēnz).

The third type of protein, **fibrinogen,** functions in the clotting reaction. Under certain conditions, fibrinogen molecules interact to form large, insoluble strands of **fibrin** (FĪ-brin), the basic framework for a blood clot. If steps are not taken to prevent clotting in a plasma sample, fibrinogen will convert to fibrin. The fluid left after the clotting proteins are removed is known as **serum.** Another much less abundant, yet important plasma protein also involved in the clotting reaction is *prothrombin.* Clotting will be discussed in a later section.

The liver synthesizes and releases more than 90 percent of the plasma proteins, including all of the albumin, fibrinogen and prothrombin, and most of the globulins. The immunoglobulins, however, are synthesized elsewhere. Immunoglobulins (antibodies) are produced by *plasma cells,* cells of the immune system. Because the liver is the primary source of plasma proteins, liver disorders can alter the composition and functional properties of the blood. For example, some forms of liver disease can lead to uncontrolled bleeding, caused by inadequate synthesis of fibrinogen, prothrombin, and other plasma proteins involved in the clotting response.

FORMED ELEMENTS

The major cellular components of blood are red blood cells and white blood cells. In addition, blood contains the noncellular formed elements called *platelets,* small packets of cytoplasm that function in the clotting response (see Figure 12-1●).

Production of Formed Elements

Formed elements are produced through the process of **hemopoiesis** (hēm-ō-poi-Ē-sis), or *hematopoiesis.* Blood cells appear in the circulation during the third week of embryonic development. These cells divide repeatedly, increasing in number. The vessels of the embryonic yolk sac are the primary site of blood formation for the first 8 weeks of development. As other organ systems appear, some of the embryonic blood cells move out of the circulation and into the liver, spleen, thymus, and bone marrow. These embryonic cells differentiate into **stem cells,** which produce blood cells by their divisions. The liver and spleen are the primary sites of hemopoiesis from the second to fifth month of development. As the skeleton enlarges, the bone marrow becomes increasingly important, and it is the primary site after the fifth developmental month. In the adult, it is the only site of red blood cell production and the primary site of white blood cell formation.

Stem cells called **hemocytoblasts** produce all of the blood cells, but the process occurs in a series of steps (see Figure 12-7●, p. 321). Hemocytoblast divisions produce *myeloid* and *lymphoid* stem cells. These *progenitor cells* remain capable of division, but their daughter cells will differentiate only into specific types of blood cells. For example, one type of progenitor cell produces daughter cells that mature into red blood cells; another gives rise to certain types of white blood cells. We will consider the fates of the different progenitor cells as we discuss the formation of each type of formed element.

Red Blood Cells

Red blood cells (RBCs), or **erythrocytes** (e-RITH-rō-sītz; *erythros,* red), contain the pigment *hemoglobin,* which binds and transports oxygen and carbon dioxide. RBCs are the most abundant blood cells, accounting for 99.9 percent of the formed elements and giving whole blood its deep red color.

The number of erythrocytes in the blood of a normal individual staggers the imagination. A standard blood test checks the number per **microliter** (μl) of whole blood. One microliter, or cubic millimeter (mm^3), of whole blood from a man contains roughly 5.4 million erythrocytes; a microliter of blood from a woman contains about 4.8 million. There are approximately 260 million red blood cells in a single drop of whole blood, and 25 trillion in the blood of an average adult. Erythrocytes thus account for roughly one-third of the cells in the human body.

The **hematocrit** (hē-MA-tō-krit) is the percentage of whole blood occupied by cellular elements. In adult men, it averages 46 (range: 40–54); in adult women, 42 (range: 37–47). The difference in hematocrit between males and females reflects the fact that androgens stimulate red blood cell production, whereas estrogens have an inhibitory effect. Because whole blood contains roughly 1000 red blood cells for each white blood cell, the hematocrit closely approximates the volume of erythrocytes. For this reason, hematocrit values are often reported as the *volume of packed red cells (VPRC)* or simply the *packed cell volume (PCV).*

Many factors can alter the hematocrit. For example, dehydration increases the hematocrit by reducing plasma volume, and internal bleeding or problems with RBC formation may decrease it through the loss of red blood cells. As a result, the hematocrit alone does not provide specific diagnostic information. However, a change in hematocrit is an indication that other, more specific tests are needed. (Some of those tests will be considered later in the chapter.) AM *Polycythemia*

Structure of RBCs

Erythrocytes are cells specialized to transport oxygen and carbon dioxide within the bloodstream. As Figure 12-2● makes clear, each red blood cell has a thin central region and a thick outer margin. This unusual shape gives it a relatively large surface area that facilitates diffusion between its cytoplasm and surrounding plasma and allows it to squeeze through narrow capillaries.

Red blood cells lack several organelles found in most other cells. During their formation, red blood cells lose their mitochondria, ribosomes, and nuclei. Without a nucleus and ribosomes, our RBCs can neither undergo cell division nor synthesize proteins. Without mitochondria, they can obtain energy only through anaerobic metabolism, relying on glucose obtained from the surrounding plasma. This characteristic makes RBCs relatively inefficient from an energy standpoint, but it ensures that absorbed oxygen will be carried to peripheral tissues, not "stolen" by mitochondria in the cell.

Hemoglobin Structure and Function

A mature red blood cell consists of a cell membrane surrounding a compact mass of transport proteins. The cytoplasm contains water (66 percent) and pro-

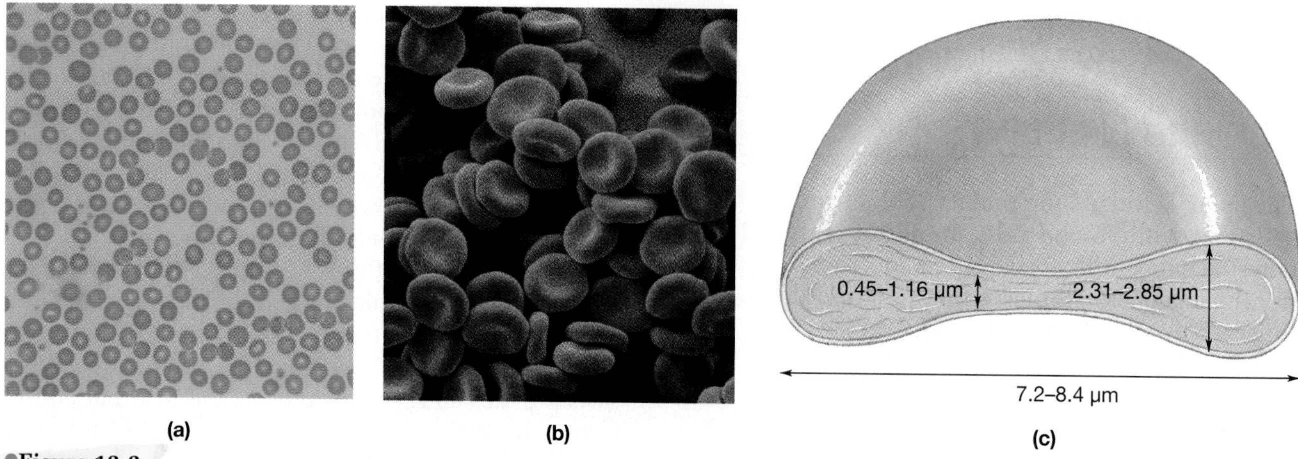

Figure 12-2
Anatomy of Red Blood Cells
(a) When viewed in a standard blood smear, red blood cells appear as two-dimensional objects because they are flattened against the surface of the slide. (LM × 320) **(b)** A scanning electron micrograph of red blood cells reveals their three-dimensional structure quite clearly. (SEM × 1195) **(c)** A sectional view of a mature red blood cell, showing average dimensions.

teins (about 33 percent). Molecules of **hemoglobin** (HĒ-mō-glō-bin) (**Hb**) account for over 95 percent of the erythrocyte's proteins and give the cell its red color. Hemoglobin is responsible for the cell's ability to transport oxygen and carbon dioxide.

Four globular protein subunits combine to form a single molecule of hemoglobin. Each of the subunits contains a single molecule of an organic pigment called **heme**. Each heme unit holds an iron ion in such a way that it can interact with an oxygen molecule. The iron-oxygen interaction is very weak, and the two can be easily separated without damage to either the hemoglobin or the oxygen molecule.

The actual amount of oxygen bound in each erythrocyte depends on the conditions in the surrounding plasma. When oxygen is abundant in the plasma, the hemoglobin molecules gain oxygen until all of the heme molecules are occupied. As the plasma oxygen levels decline, the hemoglobin molecules release their oxygen reserves. When plasma oxygen concentrations are falling, plasma carbon dioxide levels are usually rising. Under these conditions, the globin portion of each hemoglobin molecule begins to bind carbon dioxide molecules in a process that is just as reversible as the binding of oxygen to heme.

As red blood cells circulate, they are exposed to varying combinations of oxygen and carbon dioxide concentrations. At the lungs, diffusion brings oxygen into the plasma and removes carbon dioxide. The hemoglobin molecules in red blood cells respond by absorbing oxygen and releasing carbon dioxide. In the peripheral tissues, the situation is reversed because active cells are consuming oxygen and producing carbon dioxide. As blood flows through these areas, oxygen diffuses out of the plasma and carbon dioxide diffuses in. Under these conditions hemoglobin releases its stored oxygen and binds carbon dioxide.

Normal activity levels can be sustained only when tissue oxygen levels are kept within normal limits. If the hematocrit is low, or the hemoglobin content of the RBCs is reduced, the oxygen-carrying capacity of the blood is reduced and the condition of **anemia** exists. Anemia causes a variety of symptoms, including premature muscle fatigue, weakness, and a general lack of energy.

RBC Life Span and Circulation

An erythrocyte is exposed to severe physical stresses. A single round-trip of the circulatory system usually takes less than 30 seconds. In that time, a red blood cell is forced along vessels, where it bounces off the walls, collides with other red cells, and is squeezed through tiny capillaries. With all this mechanical battering and no repair mechanisms, a red blood cell has a relatively short life span of about 120 days. The continual elimination of red blood cells is usually unnoticed because new erythrocytes are entering the circulation at a comparable rate. About 1 percent of the circulating erythrocytes are replaced each day, and in the process approximately 3 million new erythrocytes enter the circulation *each second!*

Hemoglobin Conservation and Recycling As red blood cells age, they either rupture or are destroyed by phagocytic cells. If a damaged or aged erythrocyte ruptures, its hemoglobin breaks down into individual subunits small enough to pass through the filtration mechanism at the kidneys, where they may be lost in the urine. When abnormally large numbers of erythrocytes are breaking down in the circulation, the urine may develop a reddish or brown coloration. This condition, called *hemoglobinuria*, is usually prevented by specialized mechanisms that reclaim and

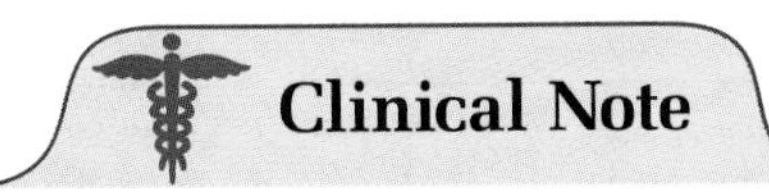

Abnormal Hemoglobin

Several inherited disorders are characterized by the production of abnormal hemoglobin. Two of the best known are *thalassemia* and *sickle cell anemia* (SCA).

The various forms of **thalassemia** (thal-ah-SĒ-mē-ah) result from an inability to produce adequate amounts of two of the four globular protein components of hemoglobin. This defect not only slows the rate of RBC production, but also results in mature RBCs that are fragile and short-lived. Individuals with severe thalassemia must undergo frequent transfusions. AM *Thalassemia*

Sickle cell anemia results from a mutation affecting the amino acid sequence of one of the globular proteins of the hemoglobin molecule. When the blood contains an abundance of oxygen, the hemoglobin molecules and the erythrocytes that carry them appear normal. But when the defective hemoglobin gives up enough of its stored oxygen, the adjacent hemoglobin molecules interact, and the cells change shape, becoming stiff and markedly curved (Figure 12-3●). This "sickling" does not affect the oxygen-carrying capabilities of the erythrocytes, but it causes them to become more fragile and easily damaged. Moreover, an RBC that has folded to squeeze into a narrow capillary may lose its oxygen, change shape, and become stuck. A circulatory blockage results, and nearby tissues become oxygen-starved. For more information on this condition and new treatment options, see the *Applications Manual.* AM *Sickle Cell Anemia*

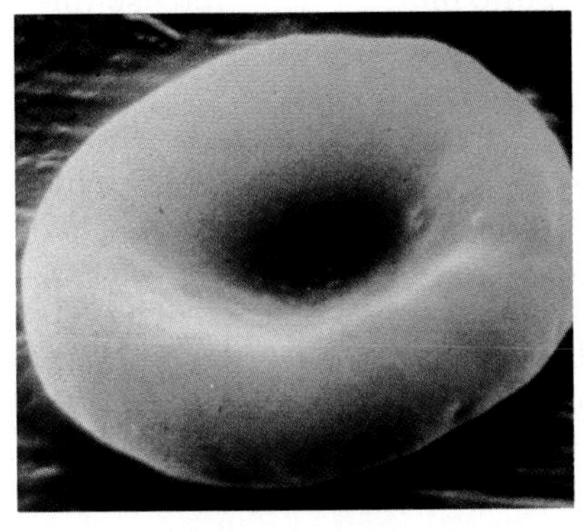

(a) Normal RBC

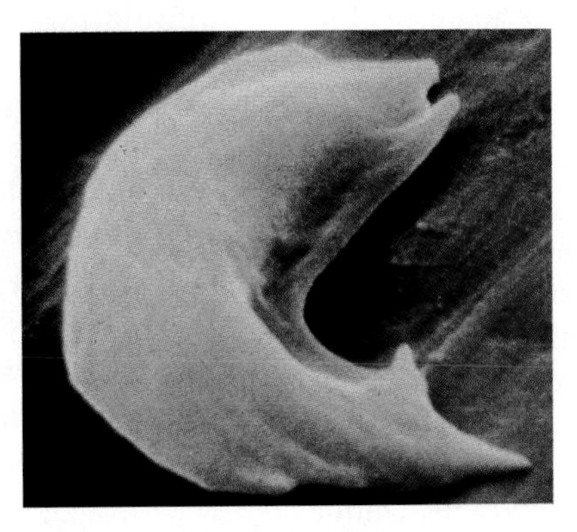

(b) Sickled RBC

●**Figure 12-3**
Sickling in Red Blood Cells
(a) When fully oxygenated, the cells of an individual with the sickling trait appear relatively normal. **(b)** At lower oxygen concentrations, the RBCs change shape, becoming relatively rigid and sharply curved. (SEM × 67,500)

recycle hemoglobin. Only about 10 percent of the red cells survive long enough to rupture, or **hemolyze** (HĒ-mō-līz), within the circulation. Phagocytic cells of the liver, spleen, and bone marrow monitor the condition of circulating erythrocytes, and they usually recognize and engulf erythrocytes *before* they hemolyze. These phagocytes also remove hemoglobin and red blood cell fragments from the circulation. (Phagocytosis and phagocytic cells were introduced in Chapter 3; additional details will be found in Chapter 15.) ∞ *[p. 56]*

Once a red blood cell has been engulfed and broken down by a phagocytic cell, the hemoglobin molecules begin to be recycled:

1. The globular proteins are disassembled into their component amino acids. These amino acids are either metabolized by the cell or released into the circulation for use by other cells.
2. Each heme is stripped of its iron and converted to **biliverdin** (bil-ē-VER-din), a substance with green coloration. (Bad bruises often appear greenish because biliverdin forms in the blood-filled tissues.) Biliverdin is then converted to **bilirubin** (bil-ē-ROO-bin) and released into circulation. Liver cells absorb the bilirubin and excrete it in the bile. If the bile ducts are blocked, bilirubin then diffuses into peripheral tissues, giving them a yellow color most apparent in the skin and eyes. This combination of yellow skin and eyes is called **jaundice** (JAWN-dis). AM *Bilirubin Tests and Jaundice*
3. Iron extracted from the heme molecules may be stored within the phagocytic cell or released into the bloodstream, where it binds to a plasma protein, **transferrin** (tranz-FER-in). Red blood cells developing in the bone marrow absorb the amino acids and transferrins from the circulation and use them to synthesize new hemoglobin molecules. Excess transferrins are removed and stored by the liver and bone marrow, and the iron is stored in special protein-iron complexes.

In summary, most of the components of an individual erythrocyte are recycled following hemolysis or phagocytosis. The entire system is remarkably efficient; although roughly 26 mg of iron are incorporated into hemoglobin molecules each day, a dietary supply of 1–2 mg will keep pace with the incidental losses that occur at the kidney and the digestive tract.

Abnormalities in iron uptake, metabolism, or excretion can cause serious clinical problems. Women are especially dependent on a normal dietary supply of iron, because their iron reserves are smaller than those of men. The body of a normal man contains around 3.5 g of iron in the ionic form Fe^{2+}. Of that amount, 2.5 g are bound to the hemoglobin of circulating red blood cells, and the rest is stored in the liver and bone marrow. In women, the total body iron content averages 2.4 g, with roughly 1.9 g incorporated into red blood cells. Thus a woman's iron reserves consist of only 0.5 g, half that of a typical man. If dietary supplies of iron are inadequate, hemoglobin production slows down, and symptoms of *iron deficiency anemia* appear. *Too much* iron can also cause problems because of excessive buildup in the liver and cardiac muscle tissue. Excessive iron deposition has recently been linked to heart disease. AM *Iron Deficiencies and Excesses*

Red Blood Cell Formation

Red blood cell formation, or **erythropoiesis** (e-rith-rō-poi-Ē-sis), occurs in the bone marrow, or **myeloid tissue** (MĪ-e-loyd; *myelos,* marrow), of the adult. **Red marrow,** where active blood cell production occurs, is found in portions of the vertebrae, sternum, ribs, skull, scapulae, pelvis, and proximal limb bones. Other marrow areas contain a fatty tissue known as **yellow marrow.** Under extreme stimulation, such as a severe and sustained blood loss, areas of yellow marrow can convert to red marrow, increasing the rate of red blood cell formation.

Stages in RBC Maturation During its maturation, a red blood cell passes through a series of stages (see Figure 12-7●, p. 321). **Hematologists** (hē-ma-TOL-o-jists), specialists in blood formation and function, have given specific names to key stages. **Erythroblasts** are very immature red blood cells that are actively synthesizing hemoglobin. After roughly 4 days of differentiation and hemoglobin production, the erythroblast sheds its nucleus and becomes a **reticulocyte** (re-TIK-ū-lō-sīt). After 2 days in the bone marrow, reticulocytes enter the circulation. At this time they can still be detected in a blood smear, and reticulocytes normally account for about 0.8 percent of the circulating erythrocytes. After 24 hours in circulation, the reticulocytes complete their maturation and become indistinguishable from other mature RBCs.

Regulation of Erythropoiesis For erythropoiesis to proceed normally, the myeloid tissues must receive adequate supplies of amino acids, iron, vitamin B_{12}, and other vitamins (B_6 and folic acid) required for protein synthesis. It is stimulated directly by *erythropoietin* and indirectly by several hormones, including thyroxine, androgens, and growth hormone. As noted above, estrogens have an inhibitory effect on erythropoiesis.

Erythropoietin, also called **EPO** or *erythropoiesis-stimulating hormome,* appears in the plasma when peripheral tissues, especially the kidneys, are exposed to low oxygen concentrations, a condition called **hypoxia** (hī-POKS-ē-a; *hypo-,* below + *ox-,* presence of oxygen). For example, EPO release occurs during anemia and when blood flow to the kidneys declines (Figure 12-4●). Once in the circulation, EPO travels to areas of red marrow, where it stimulates stem cells and developing erythrocytes.

Erythropoietin has two major effects: (1) It stimulates increased rates of mitotic divisions in erythroblasts and in the progenitor cells that produce erythroblasts; and (2) it speeds up the maturation of red blood cells, primarily by accelerating the rate of hemoglobin synthesis. Under maximum EPO stimulation, the bone marrow can increase the rate of red blood cell formation tenfold, to around 30 million per second.

This reserve is important when recovering from a severe blood loss. If EPO is administered to a normal individual, however, the hematocrit may rise to 65 or more, placing an intolerable strain on the heart. Comparable strains can occur in the practice of **blood doping,** in which athletes reinfuse packed red blood cells removed and stored at an earlier date. The goal is to

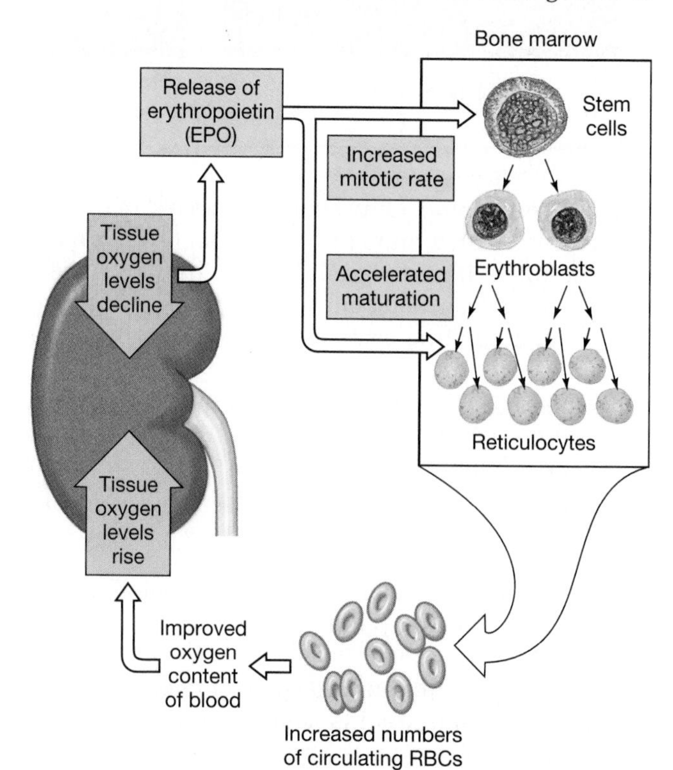

●**Figure 12-4**
The Control of Erythropoiesis

Tissues deprived of oxygen release EPO, which accelerates division of stem cells and the maturation of erythroblasts. More red blood cells then enter the circulation, improving the delivery of oxygen to peripheral tissues.

TABLE 12-1 RBC Tests and Related Terminology

Test	Determines	Terms Associated with Abnormal Values: Elevated	Depressed
Hematocrit (Hct)	Percentage of formed elements in whole blood Normal = 37–54%	Polycythemia (may result from erythrocytosis or leukocytosis)	Anemia
Reticulocyte count (Retic.)	Circulating percentage of reticulocytes Normal = 0.8%	Reticulocytosis	
Hemoglobin concentration (Hb)	Concentration of hemoglobin in blood Normal = 12–18 g/dl		Anemia
RBC count	Number of RBCs per μl of whole blood Normal = 4.4–6.0 million/μl	Erythrocytosis	Anemia
Mean corpuscular volume (MCV)	Average volume of single RBC Normal = 82–101 μm^3 (normocytic)	Macrocytic	Microcytic
Mean corpuscular hemoglobin concentration (MCHC)	Average amount of Hb in one RBC Normal = 27–34 pg/μl (normochromic)	Hyperchromic	Hypochromic

Note: For additional details see the Applications Manual.

increase hematocrit and improve performance, but the result can be sudden death from heart failure. AM *Erythrocytosis and Blood Doping*

BLOOD TESTS AND RBCS

Blood tests provide information about the general health of an individual, usually with a minimum of trouble and expense. Several common blood tests focus on red blood cells, the most abundant formed elements. These tests assess the number, size, shape, and maturity of circulating red blood cells. This information provides an indication of the erythropoietic activities under way and can also be useful in detecting problems, such as internal bleeding, that may not produce other obvious symptoms. Table 12-1 lists examples of important blood tests and related terms; for a detailed discussion, including sample calculations, see the *Applications Manual.* AM *Blood Tests and RBCs*

- ✓ What would be the effects of a decrease in the amount of plasma proteins?
- ✓ How would an individual's hematocrit change after a hemorrhage?
- ✓ How would the level of bilirubin in the blood be affected by diseases that damage the liver?
- ✓ How would an increase in the level of oxygen supplied to the kidneys affect the level of erythropoietin in the blood?

Blood Types

An individual's **blood type** is determined by the presence or absence of specific **surface antigens,** or *agglutinogens* (a-gloo-TIN-ō-jenz), in the erythrocyte cell membranes. The characteristics of RBC antigen molecules are genetically determined. There are at least 50 different kinds of antigens on the surfaces of RBCs. Three of particular importance have been designated antigens **A, B,** and **D (Rh).**

The red blood cells of a particular individual may have either, both, or neither antigen A or B on their surfaces. For example, **Type A** blood has antigen A only, **Type B** has antigen B only, **Type AB** has both, and **Type O** has neither. The average values for the U.S. population are Type O, 46 percent; Type A, 40 percent; Type B, 10 percent; and Type AB, 4 percent. There are variations in these values due to racial and ethnic differences (Table 12-2).

The presence or absence of the Rh antigen, sometimes called the *Rh factor,* is indicated by the terms **Rh-positive** (present) and **Rh-negative** (absent). In recording the complete blood type, clinicians usually omit the term *Rh* and report the data as O-negative, A-positive, and so forth.

Antibodies and Cross-Reactions Blood type is checked before an individual gives or receives blood. The surface antigens on a person's own red blood cells are ignored by its own immune system. However, plasma contains antibodies, or *agglutinins* (a-GLOO-ti-ninz), that will attack "foreign" surface antigens (Figure 12-5a●). For example, the plasma of individuals with Type A blood contains circulating *anti-B* antibodies (or simply, anti-B) that will attack Type B surface antigens. The plasma of Type B individuals contains antibodies (*anti-A*) that will attack Type A surface antigens. The red blood cells of an individual with Type O blood lack surface antigens A and B, so the plasma of such an in-

TABLE 12-2 Differences in Blood Group Distribution

Population	O	A	B	AB	Rh+
	Percentage with Each Blood Type				
U.S. (average)	46	40	10	4	85
Caucasian	45	40	11	4	85
African-American	49	27	20	4	95
Chinese	42	27	25	6	100
Japanese	31	39	21	10	100
Korean	32	28	30	10	100
Filipino	44	22	29	6	100
Hawaiian	46	46	5	3	100
Native North American	79	16	4	<1	100
Native South American	100	0	0	0	100
Australian Aborigines	44	56	0	0	100

dividual contains both anti-A and anti-B. At the other extreme, Type AB individuals lack antibodies sensitive to either A or B surface antigens.

When an antibody meets its specific antigen, a **cross-reaction** occurs. Initially the red blood cells clump together, a process called **agglutination** (a-gloo-ti-NĀ-shun), and they may also hemolyze (Figure 12-5b●). Clumps and fragments of red blood cells under attack form drifting masses that can plug small vessels in the kidneys, lungs, heart, or brain, damaging or destroying tissues. Such reactions can be avoided by ensuring that the blood types of the donor and the recipient are **compatible.** A donor must be chosen whose blood cells will not undergo cross-reaction with the plasma of the recipient.

TESTING FOR COMPATIBILITY

Testing for compatibility normally involves two steps: a determination of blood type and a cross-match test. Although some 50 antigens are known, the standard test for blood type categorizes a blood sample on the basis of the three most likely to produce dangerous cross-reactions. The test involves mixing drops of blood with solutions containing A, B, *and* Rh (D) antibodies and noting any cross-reactions. For example, if the red blood cells clump together when exposed to anti-A *and* anti-B, the individual has Type AB blood. If no reactions occur, the person must be Type O. The presence or absence of the Rh antigen is also noted, and the individual is classified as Rh-positive or Rh-negative. In the most common type, type O-positive (O+), the red blood cells do not have surface antigens A and B, but they do have antigen D.

Standard blood typing can be completed in a matter of minutes, and Type O blood can be safely administered in an emergency. For example, a patient with a severe gunshot wound may require 5 *liters* or more of blood before the damage can be repaired. Under these circumstances, it may not be possible to do more than collect blood, make certain that it is Type O, and administer it to the victim. Because their blood cells are unlikely to produce severe cross-reactions in a recipient, Type O individuals (especially O-negative) are sometimes called *universal donors*. It should be realized, however, that with at least 48 other possible agglutinogens on the cell surface, cross-reactions can occur, even to Type O blood. Whenever time and facilities permit, further testing is performed to ensure complete compatibility.

Cross-match testing involves exposing the donor's red blood cells to a sample of the recipient's plasma under controlled conditions. This procedure reveals the presence of significant cross-reactions involving other antigens and antibodies.

Because blood groups are inherited, blood tests are used as paternity tests and in crime detection. Results from such tests cannot prove that a particular individual *is* the father or criminal involved, but it can prove that he is *not* involved. For example, if the blood of an infant is Type O and the sample is Type AB, the two cannot be related.

HEMOLYTIC DISEASE OF THE NEWBORN

The plasma of an Rh-negative individual does not normally contain anti-Rh antibodies. These antibodies are present only if the individual has been **sensitized** by previous exposure to Rh-positive erythrocytes. Such exposure may occur accidentally, during a transfusion, but it may also accompany a normal pregnancy involving an Rh-negative mother and an Rh-positive fetus. The sensitization that causes this condition usually occurs at delivery, when bleeding occurs at the placenta and uterus, and the antibodies are not produced in significant amounts until after the birth of the infant. A sensitized mother will respond to a second Rh-positive fetus by producing massive amounts of anti-Rh antibodies. These attack and hemolyze the fetal red blood cells, producing a dangerous anemia. This increases the fetal demand for blood cells, and the resulting RBCs leave the bone marrow and enter the circulation before completing their development. These immature RBCs are called *erythroblasts,* and the condition is known as *erythroblastosis fetalis* (e-rith-rō-blas-TŌ-sis fē-TAL-is), also called *hemolytic disease of the newborn*, or *HDN*. Hemolytic Disease of the Newborn

✓ What blood types can be transfused into a person with Type AB blood?

✓ Why can't a person with type A blood receive blood from a person with type B blood?

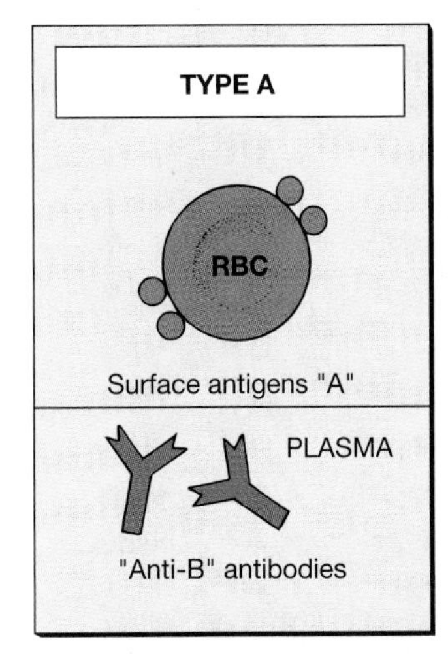

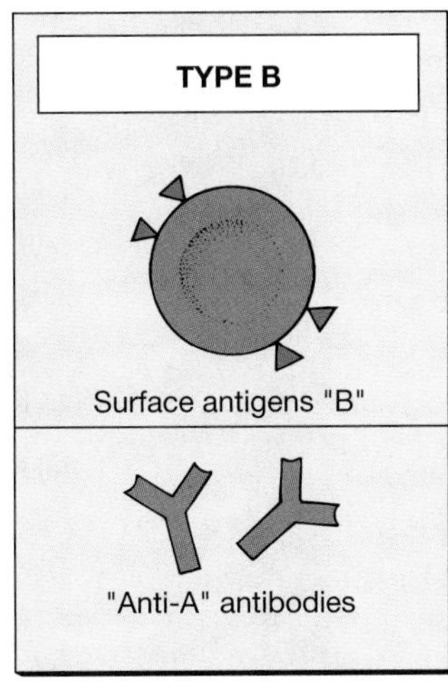

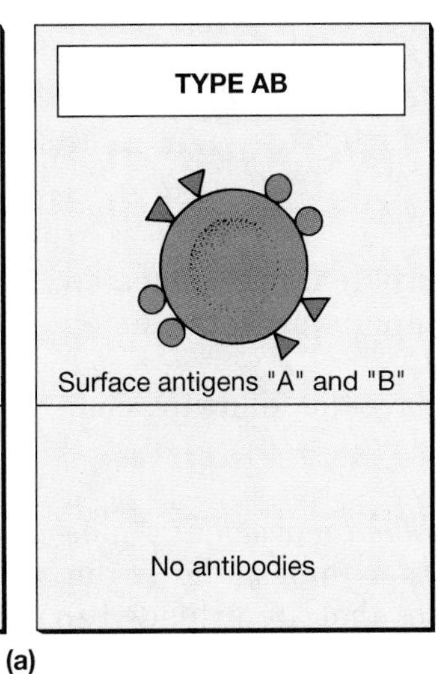

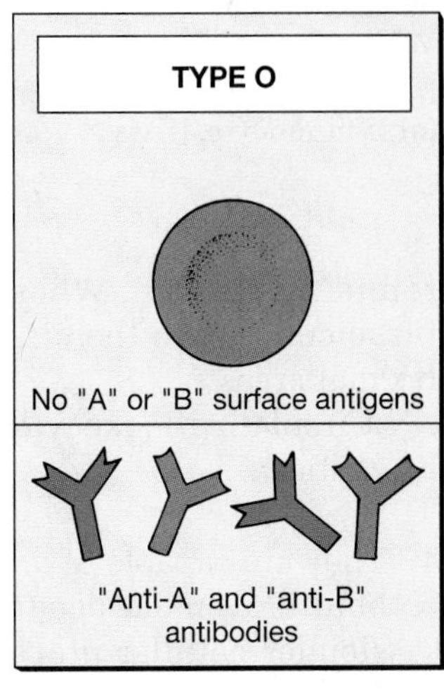

(a)

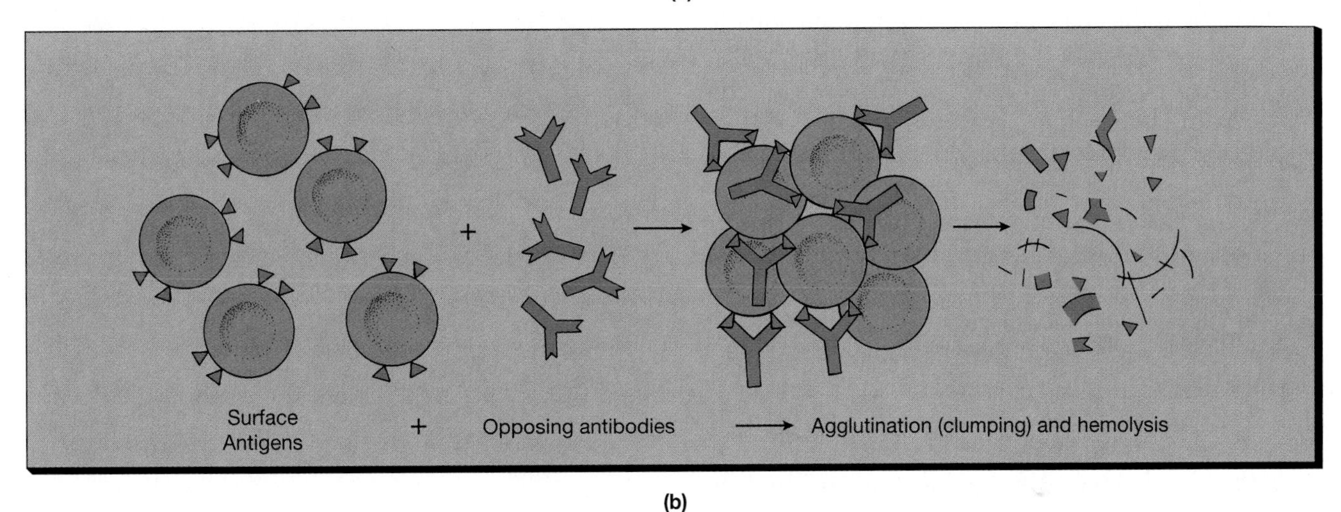

(b)

●**Figure 12-5**
Blood Typing and Cross-reactions

(a) The blood type depends on the presence of antigens on RBC surfaces. The plasma contains antibodies that will react with foreign antigens. **(b)** A cross-reaction occurs when antibodies encounter their target antigens. The result is extensive clumping of the affected RBCs, followed by hemolysis.

White Blood Cells

White blood cells, also known as WBCs or **leukocytes** (LOO-kō-sīts; *leukos,* white), can easily be distinguished from RBCs because each (1) has a nucleus and (2) lacks hemoglobin. White blood cells help defend the body against invasion by pathogens and remove toxins, wastes, and abnormal or damaged cells. Traditionally, leukocytes have been divided into two groups based on their appearance after staining: *granulocytes* (with abundant stained granules) and *agranulocytes* (with few if any stained granules). This categorization is convenient but somewhat misleading because the "granules" in granulocytes are actually secretory vesicles and lysosomes, and the "agranulocytes" contain lysosomes as well—they are just smaller and more difficult to see with the light microscope.

Typical leukocytes in the circulating blood are shown in Figure 12-6●. The granulocytes include neutrophils, eosinophils, and basophils; the two kinds of agranulocytes are monocytes and lymphocytes. A typical microliter of blood contains 6000–9000 leukocytes. Most of the white blood cells in the body, however, are found in peripheral tissues, and circulating leukocytes represent only a small fraction of the total population.

WBC Circulation and Movement

Unlike erythrocytes, leukocytes do not circulate for extended periods. The bloodstream provides rapid transportation to areas of invasion or injury, and as they travel along the miles of capillaries, leukocytes are sensitive to the chemical signs of damage to sur-

Figure 12-6
White Blood Cells

(a) Neutrophil. **(b)** Eosinophil. **(c)** Basophil. **(d)** Monocyte. **(e)** Lymphocyte. (LMs × 1500)

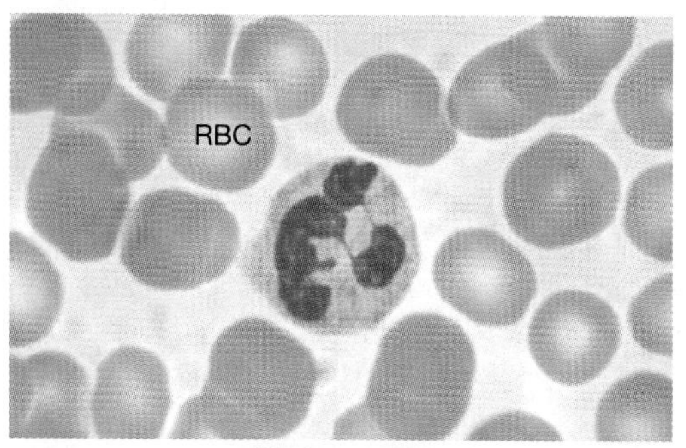

(a) Neutrophil

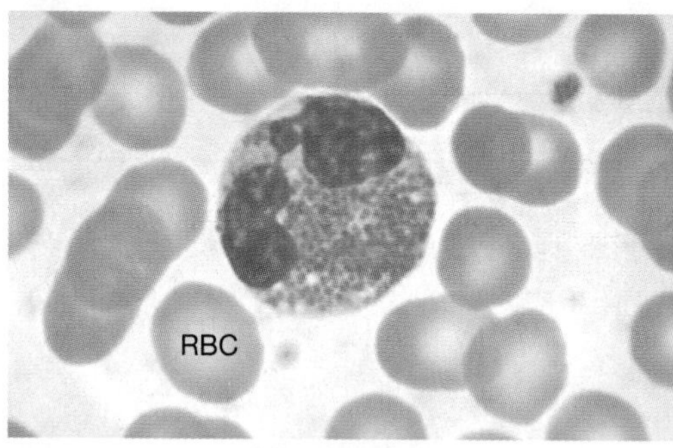

(b) Eosinophil

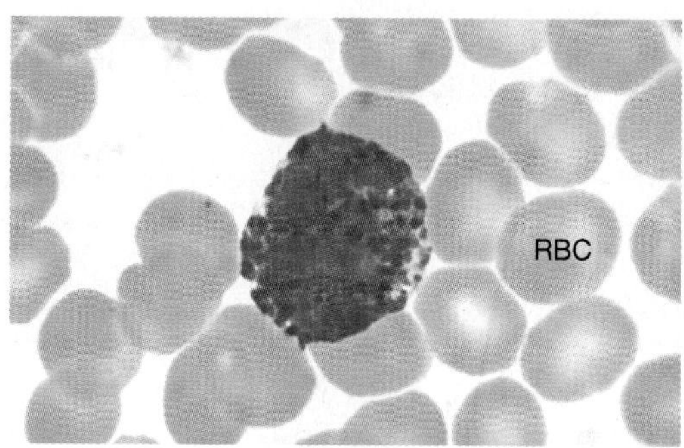

(c) Basophil

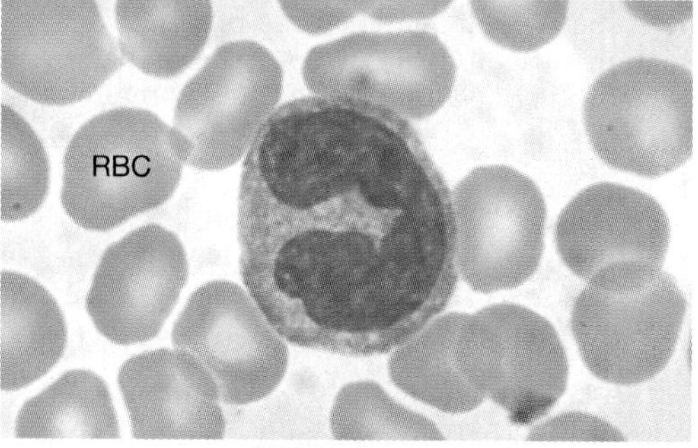

(d) Monocyte

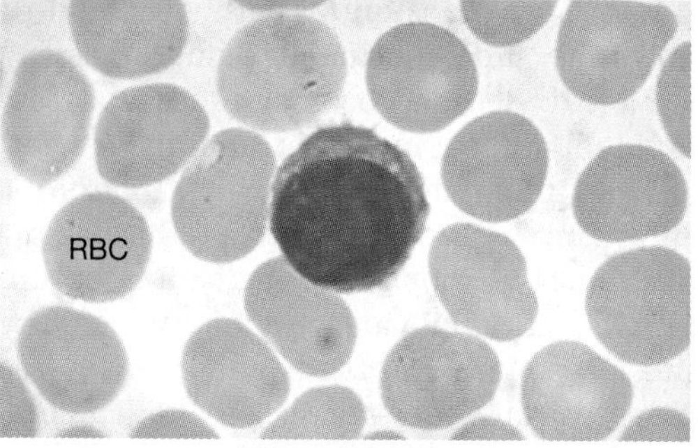

(e) Lymphocyte

rounding tissues. When problems are detected, these cells leave the circulation and enter the abnormal area.

Circulating leukocytes have the following characteristics:

1. They are capable of *amoeboid movement.* Amoeboid movement occurs as cytoplasm flows into slender cellular processes that are extended in front of the cell. This mobility allows them to move along the walls of blood vessels and, when outside of the bloodstream, through surrounding tissues.
2. They can migrate out of the bloodstream by squeezing between adjacent endothelial cells in the capillary wall. This process is known as **diapedesis** (dī-a-pe-DĒ-sis).
3. They are attracted to specific chemical stimuli. This characteristic, called **positive chemotaxis** (kē-mō-TAK-sis), guides them to invading pathogens, damaged tissues, and active white blood cells.
4. Some of the circulating WBCs, specifically neutrophils, eosinophils, and monocytes, are capable of *phagocytosis.* These cells may engulf pathogens, cell debris, or other materials in or out of the bloodstream. Neutrophils and eosinophils are sometimes called *microphages* to distinguish them from the larger phagocytes found in the blood and peripheral tissues, the monocytes and macrophages.

General Functions

Neutrophils, eosinophils, basophils, and monocytes contribute to the body's *nonspecific defenses.* These defenses respond to a variety of stimuli, but always in the same way—they do not discriminate between one threat and another. Lymphocytes, by contrast, are the cells responsible for *specific immunity:* the ability of the body to mount a counterattack against invading pathogens or foreign proteins *on an individual basis.* The interactions between white blood cells and the relationships between specific and nonspecific defenses will be discussed in Chapter 15.

Neutrophils

Fifty to 70 percent of the circulating white blood cells are **neutrophils** (NOO-trō-filz). This name was selected because their granules are chemically neutral and thus difficult to stain with either acid or basic dyes. A mature neutrophil (Figure 12-6a●) has a very dense, contorted nucleus that may be condensed into a series of lobes resembling beads on a chain.

Neutrophils are usually the first of the white blood cells to arrive at an injury site. They are very active phagocytes, specializing in attacking and digesting bacteria. Neutrophils usually have a short life span, surviving for about 12 hours. After ingesting 1-2 dozen bacteria, a neutrophil dies, but its breakdown releases chemicals that attract other neutrophils to the site.

Eosinophils

Eosinophils (ē-ō-SIN-ō-filz) were so named because their granules stain darkly with a red dye, eosin (Figure 12-6b●). They usually represent 2–4 percent of the circulating white blood cells. These cells are similar in size to neutrophils, but the combination of deep red granules and a bilobed (two-lobed) nucleus makes them easy to identify. Although they are phagocytic cells, eosinophils generally ignore bacteria and cellular debris. Instead they are attracted to foreign compounds that have reacted with circulating antibodies. Their numbers increase dramatically during an allergic reaction or a parasitic infection.

Basophils

Basophils (BĀ-sō-filz) have numerous granules that stain darkly with basic dyes, and in a standard blood smear the granules are a deep purple to blue color (Figure 12-6c●). They are somewhat smaller than the other granulocytes and are relatively rare, accounting for less than 1 percent of the leukocyte population. Basophils migrate to sites of injury and cross the capillary endothelium to accumulate within the damaged tissues, where they discharge their granules into the interstitial fluids. The granules contain histamine and *heparin,* a chemical that prevents blood clotting. The basophil release of histamine enhances the local inflammation initiated by the *mast cells* of damaged connective tissues. Other chemicals released by stimulated basophils attract eosinophils and other basophils to the area.

Monocytes

Monocytes (MON-ō-sīts) are nearly twice the size of a typical erythrocyte (Figure 12-6d●). The nucleus is large and often shaped like an oval or kidney bean. Monocytes normally account for 2–8 percent of the circulating leukocytes. Outside the bloodstream in peripheral tissues they are usually called *free macrophages,* to distinguish them from the immobile *fixed macrophages* found in many connective tissues. (Tissue macrophages were described in Chapter 4.) ∞ *[p. 81]* They are enthusiastic phagocytes that when active, release chemicals that attract and stimulate neutrophils, additional monocytes, and other phagocytic cells. Free macrophages also secrete substances that lure fibroblasts into the region. The fibroblasts then begin producing the scar tissue that will wall off the injured area.

Lymphocytes

Typical **lymphocytes** (LIM-fō-sīts) are roughly the same size as red blood cells and contain a relatively large nucleus surrounded by a thin halo of cytoplasm (Figure 12-6e●). Although lymphocytes account for 20–30 percent of the leukocyte population of the blood, this is only a minute segment of the entire *lymphocyte* population, because lymphocytes are the primary cells of the **lymphatic system,** a network of special vessels and organs distinct from those of the circulatory system.

The circulating blood contains three classes of lymphocytes, although they cannot be distinguished with a light microscope.

1. T cells. Lymphocytes called **T cells** attack foreign cells directly and stimulate or inhibit the activities of other lymphocytes.
2. B cells. Lymphocytes called **B cells** differentiate into **plasma cells,** tissue cells that secrete antibodies that can attack alien cells or proteins in distant portions of the body.
3. Natural Killer (NK) cells. Lymphocytes called **NK cells** (natural killer) are responsible for *immune surveillance,* the destruction of the body's own abnormal tissue cells.

The Differential Count and Changes in WBC Profiles

A variety of disorders, including pathogenic infection, inflammation, and allergic reactions, cause characteristic changes in circulating populations of WBCs. A **differential count** of the white blood cell population can be obtained by examining a stained blood smear. The values reported indicate the number of each type of cell encountered in a sample of 100 white blood cells.

The normal range for each cell type is indicated in Table 12-3. The term **leukopenia** (loo-kō-PĒ-nē-ah; *penia,* poverty) indicates inadequate numbers of white blood cells. *Leukocytosis* (loo-kō-sī-TŌ-sis) refers to excessive numbers. Leukocytosis with white blood cell counts of 100,000/µl or more usually indicates the presence of some form of **leukemia** (loo-KĒ-mē-ah), a cancer of blood forming tissues. Not all leukemias are characterized by leukocytosis; other indications are the presence of abnormal or immature WBCs. Unless treated, however, all leukemias are fatal; with treatment, many forms can be arrested or even cured. AM *The Leukemias*

TABLE 12-3 A Review of the Formed Elements of the Blood

Cell	*Abundance (Average per µl)*	*Functions*	*Remarks*
RED BLOOD CELLS	5.2 million (range: 4.4–6.0 million)	Transport oxygen from lungs to tissues and carbon dioxide from tissues to lungs	Remain in circulation; 120-day life expectancy; amino acids and iron recycled; produced in bone marrow
WHITE BLOOD CELLS	7000 (range: 6000–9000)		
Neutrophils	4150 (range: 1800–7300) Differential count: 50–70%	Phagocytic: Engulf pathogens or debris in tissues, release cytotoxic enzymes and chemicals	Move into tissues after several hours; may survive minutes to days, depending on tissue activity; produced in bone marrow
Eosinophils	165 (range: 0–700) Differential count: <2–4%	Attack antibody-labeled materials through release of cytotoxic enzymes, and/or phagocytosis	Move into tissues after several hours; survive minutes to days, depending on tissue activity; produced in bone marrow
Basophils	44 (range: 0–150) Differential count: <1%	Enter damaged tissues and release histamine and other chemicals that reduce inflammation	Survival time unknown; assist mast cells of tissues in producing inflammation; produced in bone marrow
Monocytes	456 (range: 200–950) Differential count: 2–8%	Enter tissues to engulf pathogens or debris	Move into tissues after 1–2 days; survive months or or longer; primarily produced in bone marrow
Lymphocytes	2185 (range: 1500–4000) Differential count: 20–30%	Cells of lymphatic system, providing defense against specific pathogens or toxins	Survive for months to decades; circulate from blood to tissues and back; produced in bone marrow and lymphatic tissues
PLATELETS	350,000 (range: 150,000–500,000)	Hemostasis: Clump together and stick to vessel wall (platelet phase); initiate coagulation cascade	Remain in circulation or in vascular organs; remain intact 7–12 days; produced by megakaryocytes in bone marrow

White Blood Cell Formation

Stem cells responsible for the production of white blood cells originate in the bone marrow. Neutrophils, eosinophils, and basophils complete their development in myeloid tissue; monocytes begin their differentiation in the bone marrow, enter the circulation, and complete their development when they become free macrophages in peripheral tissues. Each of these cell types goes through a characteristic series of maturational stages. Figure 12-7• summarizes the relationships between the various WBC populations and compares the formation of WBCs and RBCs.

Stem cells responsible for the production of lymphocytes, a process called **lymphopoiesis,** also originate in the bone marrow, but many of these subsequently migrate to peripheral **lymphoid tissues,** including the thymus, spleen, and lymph nodes. As a result, lymphocytes are produced in these organs as well as in the bone marrow.

Factors that regulate lymphocyte maturation are as yet incompletely understood. Prior to maturity, the "thymosins" produced by the thymus gland promote the differentiation and maintenance of different T cell populations. The importance of the thymus gland in adulthood, especially in aging, is uncertain. In the adult, production of B and T lymphocytes is primarily regulated by exposure to antigens (foreign proteins, cells, or toxins). When foreign antigens appear, lymphocyte production escalates.

Several hormones, called *colony-stimulating factors (CSFs),* are involved in the regulation of other white blood cell populations. Four CSFs have been identified, each targeting single stem cell lines or groups of stem cell lines.

Chemical communication between lymphocytes and other leukocytes assists in the coordination of the immune response. For example, active macrophages release chemicals that make lymphocytes more sensitive to antigens and accelerate the development of specific immunity. In turn, active lymphocytes release CSFs, reinforcing nonspecific defenses.

Platelets

Bone marrow contains enormous cells with large nuclei; these cells are called **megakaryocytes** (meg-a-KĀR-ē-ō-sits; *megas*, big + *karyon*, nucleus + *-cyte*, cell). As depicted in Figure 12-8●, megakaryocytes continually shed small membrane-enclosed packets of cytoplasm that enter the circulation. Now

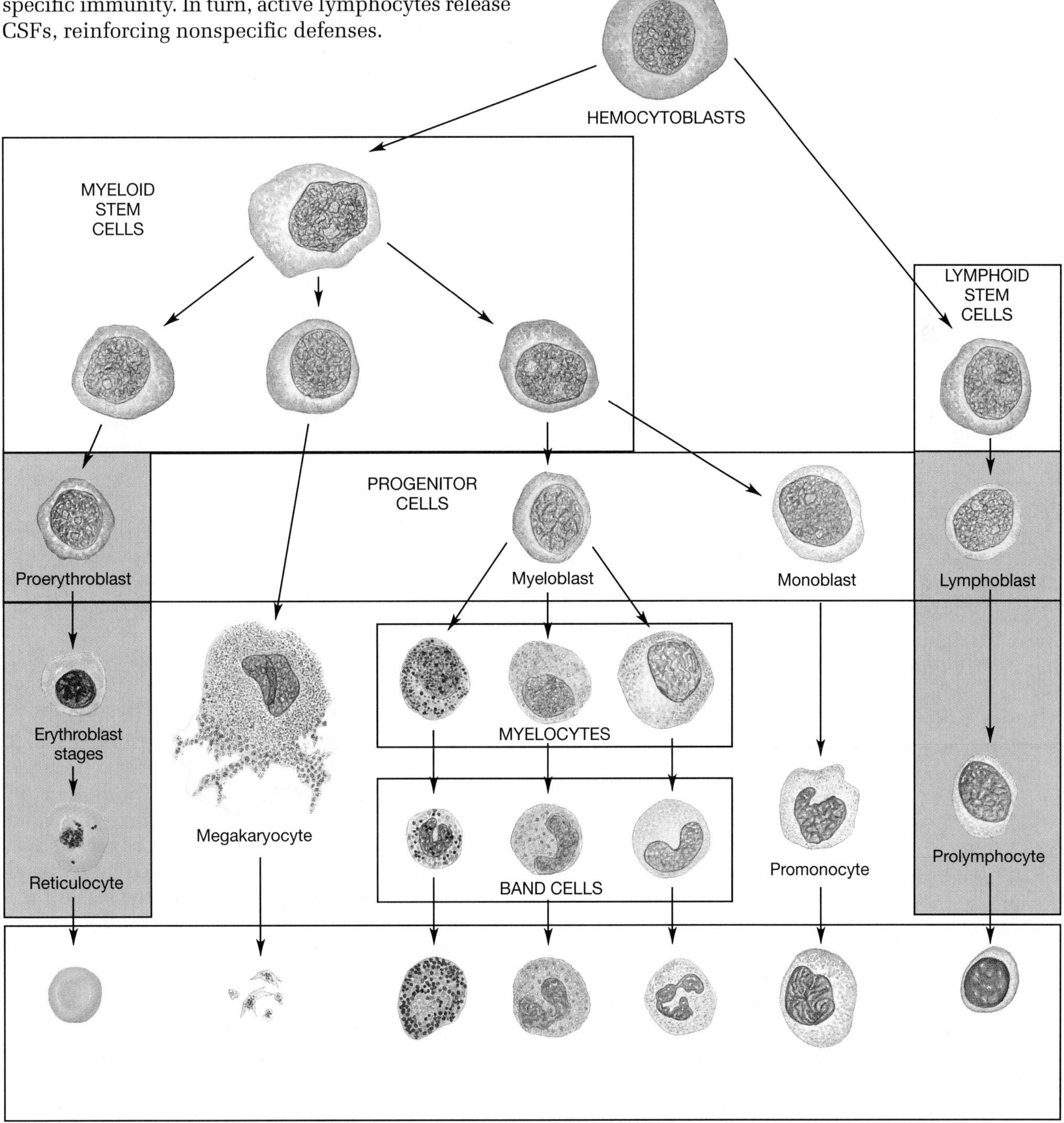

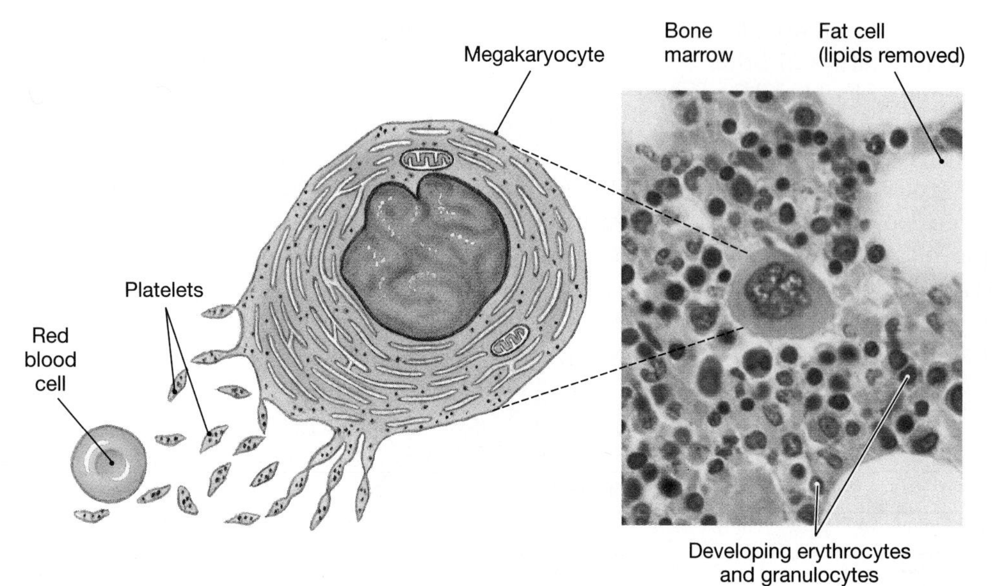

•Figure 12-8 Megakaryocytes and Platelet Formation

Megakaryocytes stand out in bone marrow sections because of their relatively enormous size and the unusual shape of their nuclei. These cells are continually shedding chunks of cytoplasm that enter the circulation as platelets. (LM × 673)

known as **platelets** (PLĀT-lets), they were once thought to be cells that had lost their nuclei, and histologists called them **thrombocytes** (THROM-bō-sīts; *thrombos,* clot). The term is still in use, although *platelet* is more suitable because these are cell fragments, not individual cells. Platelets initiate the clotting process and help close injured blood vessels. They are one participant in a vascular *clotting system,* discussed next, that also includes plasma proteins and the cells and tissues of the circulatory network.

Platelets are continually replaced, and an individual platelet circulates for 10–12 days before being removed by phagocytes. There are 150,000–500,000 platelets in each microliter of circulating blood; 350,000/μl represents the average concentration. An abnormally low platelet count (80,000/μl or less) is known as *thrombocytopenia* (throm-bō-sī-tō-PĒ-nē-ah) and usually indicates excessive platelet destruction or inadequate platelet production. Symptoms include bleeding along the digestive tract, within the skin, and occasionally inside the CNS.

Platelet counts in *thrombocytosis* (throm-bō-sī-TŌ-sis) may exceed 1,000,000/μl. Thrombocytosis usually results from accelerated platelet formation in response to infection, inflammation, or cancer.

✓ What type of white blood cell would you expect to find in the greatest numbers in an infected cut?

✓ What cell type would you expect to find in elevated numbers in a person producing large amounts of circulating antibodies to combat a virus?

✓ A sample of bone marrow has fewer than normal numbers of megakaryocytes. What body process would you expect to be impaired as a result?

HEMOSTASIS

The process of **hemostasis** (*haima,* blood + *stasis,* halt) prevents the loss of blood through the walls of damaged vessels. In doing so, it not only restricts blood loss but also establishes a framework for tissue repairs. The major steps in hemostasis include the vascular, platelet, and coagulation phases and clot retraction and removal.

Step 1: *The vascular phase.* Cutting the wall of a blood vessel triggers a contraction in the smooth muscle fibers in the vessel wall that decreases the vessel's diameter. This contraction produces a local **vascular spasm,** which can slow or even stop the loss of blood through the wall of a small vessel. This period of local vasoconstriction, called the **vascular phase** of hemostasis, lasts about 30 minutes.

Step 2: *The platelet phase.* During the vascular phase, the membranes of endothelial cells at the injury site become "sticky," and in small capillaries endothelial cells may stick together and block the opening completely. In larger vessels, platelets begin to attach to exposed endothelial surfaces. This attachment marks the start of the **platelet phase** of hemostasis. As more platelets arrive, they form a mass that may plug the break in the vascular lining.

Step 3: *The coagulation phase.* The vascular and platelet phases begin within a few seconds after the injury. The **coagulation** (kō-ag-ū-LĀ-shun) **phase** does not start until 30 seconds to several minutes later. Blood clotting, or **coagulation,** involves a complex sequence of steps leading to the conversion of circulating fibrinogen into the insoluble protein **fibrin.** As the fibrin network grows, blood cells and additional platelets are trapped within the fibrous

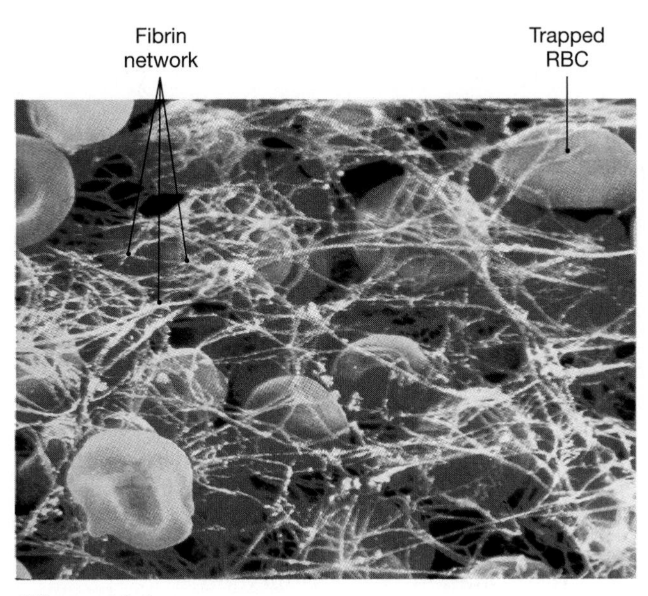

●**Figure 12-9**
Structure of a Blood Clot

A color-enhanced scanning electron micrograph showing the network of fibrin that forms the framework of a clot. (SEM × 3561) Red blood cells trapped in those fibers add to the mass of the blood clot and give it a red color. Platelets that stick to the fibrin strands gradually contract, shrinking the clot and tightly packing the RBCs.

tangle, forming a **blood clot** that effectively seals off the damaged portion of the vessel (Figure 12-9●).

The Clotting Process

Normal coagulation cannot occur unless the plasma contains the necessary **clotting factors,** which include calcium ions and 11 different plasma proteins. These proteins are converted to active enzymes that direct essential reactions in the clotting response. Most of the circulating clotting proteins are synthesized by the liver.

During the coagulation phase, the clotting proteins interact in sequence with the conversion of one protein into an enzyme that activates a second protein and so on, in a chain reaction, or *cascade.* Figure 12-10● provides an overview of the cascades involved in the *extrinsic, intrinsic,* and *common pathways.*

The Extrinsic, Intrinsic, and Common Pathways

When a blood vessel is damaged, both the extrinsic and intrinsic pathways respond. Clotting usually begins in 15 seconds and is initiated by the shorter and faster extrinsic pathway. The slower, intrinsic pathway reinforces the initial clot, making it larger and more effective.

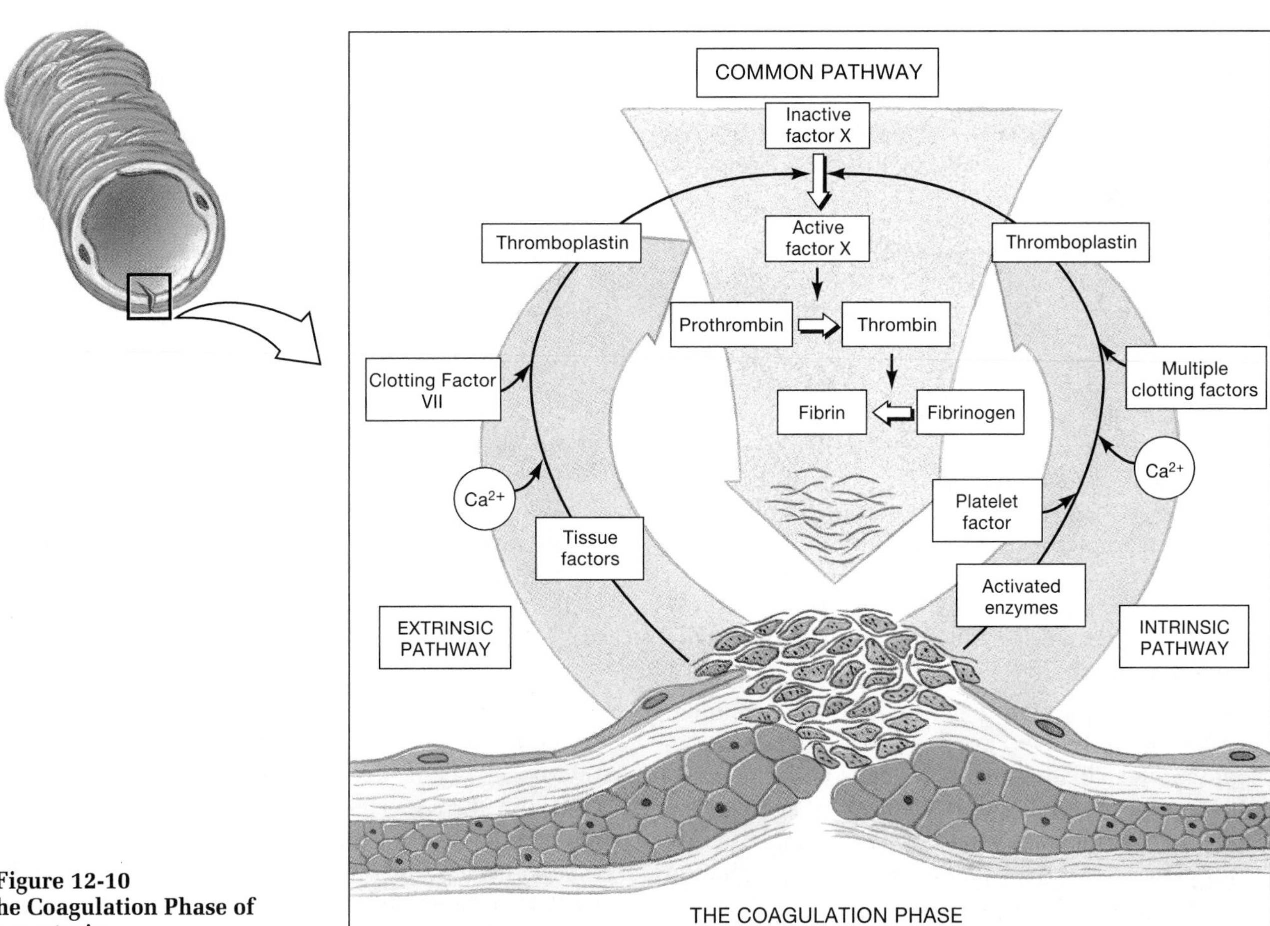

●**Figure 12-10**
The Coagulation Phase of Hemostasis

Clinical Note

Abnormal Hemostasis

The coagulation process involves a complex chain of events, and a disorder that affects any individual clotting factor may disrupt the entire process. As a result, there are many different clinical conditions involving the clotting system. AM *Testing the Clotting System*

Calcium ions and a single vitamin, **vitamin K**, have an effect on almost every aspect of the clotting process. For example, all three pathways (intrinsic, extrinsic, and common) require the presence of calcium ions. Adequate amounts of vitamin K must be present for the liver to be able to synthesize four of the clotting factors, including prothrombin. A vitamin K deficiency therefore leads to the breakdown of the common pathway, inactivating the clotting system. Roughly half of the vitamin K is absorbed directly from the diet, and the rest is synthesized by intestinal bacteria.

Excessive Coagulation

If the clotting system is inadequately controlled, clot formation will begin in the circulation, rather than at an injury site. These blood clots do not stick to the wall of the vessel but continue to drift around until plasmin digests them or they become stuck in a small blood vessel. A drifting blood clot is an example of an **embolus** (EM-bo-lus; *embolos*, plug). When an embolus becomes stuck in a blood vessel, it blocks circulation to the area downstream, killing the affected tissues. The blockage is called an **embolism.**

An embolus in the arterial system may get stuck in the capillaries in the brain, causing a stroke. If an embolus forms in the venous system it will probably become lodged in the capillaries in the lung, causing a condition known as a *pulmonary embolism*.

A *thrombus* begins to form when platelets stick to the wall of an intact blood vessel. Often the platelets are attracted to areas called **plaques,** where endothelial and smooth muscle cells contain large quantities of lipids. The blood clot gradually enlarges, projecting into the lumen of the vessel and reducing its diameter. Eventually the vessel may be completely blocked, or a large chunk of the clot may break off, creating an equally dangerous embolus.

Treatment of these conditions must be prompt to prevent irreparable damage to tissues whose vessels have been restricted or blocked by emboli or thrombi. The clots may be surgically removed or attacked by enzymes such as *streptokinase* (strep-tō-KĪ-nās) or *urokinase* (ū-rō-KĪ-nās) or by plasmin stimulated by administered t-PA (tissue plasminogen activator). Controversy exists over the relative benefits of t-PA versus streptokinase or urokinase when weighed against t-PA's relatively high cost.

Inadequate Coagulation

Hemophilia (hēm-ō-FĒL-ē-a) is one of many inherited disorders characterized by inadequate production of clotting factors. The incidence of this condition in the general population is about 1 in 10,000, with males accounting for 80–90 percent of those affected. In hemophilia, production of a single clotting factor (most often Factor VIII) is reduced; the severity of the condition depends on the degree of reduction. In severe cases extensive bleeding accompanies the slightest mechanical stresses, and hemorrhages occur spontaneously at joints and around muscles.

Transfusions of clotting factors can often reduce or control the symptoms of hemophilia, but plasma samples from many individuals must be pooled (combined) to obtain adequate amounts of clotting factors. Consequently, the procedure is very expensive and it also increases the risk of infection with blood-borne infections such as hepatitis or AIDS. Gene-splicing techniques have been used to manufacture Factor VIII, an essential component of the intrinsic clotting pathway. Although supplies are now limited, this procedure should eventually provide a safer and cheaper method of treatment.

- The **extrinsic pathway** begins with the release of **tissue factors** by damaged endothelial cells or peripheral tissues. The greater the damage, the more tissue factor is released and the faster clotting occurs. Tissue factor then combines with calcium ions and one of the clotting proteins (*Factor VII*) to form an enzyme called **tissue thromboplastin.**
- The **intrinsic pathway** begins with the activation of a clotting protein exposed to collagen fibers at the injury site. This pathway proceeds with the assistance of a platelet factor released by aggregating platelets. After a series of linked reactions involving various clotting proteins, the enzyme **platelet thromboplastin** is formed.
- The **common pathway** begins after thromboplastin from either the extrinsic or intrinsic pathways appears in the plasma. The first step involves the activation of a clotting protein (*Factor X*) responsible for the conversion of the clotting protein **prothrombin** into the enzyme **thrombin** (THROM-bin). Thrombin then completes the coagulation process by converting fibrinogen to fibrin.

Clot Retraction and Removal

Once the fibrin meshwork has appeared, platelets and red blood cells stick to the fibrin strands. The platelets then contract, and the entire clot begins to undergo **clot retraction.** This process reduces the size of the damaged area, making it easier for the fibroblasts, smooth muscle cells, and endothelial cells in the area to carry out the necessary repairs.

As the repairs proceed, the clot gradually dissolves. This process, called **fibrinolysis** (fī-brin-OL-i-sis), begins with the activation of a plasma protein, **plasminogen** (plaz-MIN-ō-jen), by **tissue plasminogen activator**, or **t-PA**, released by damaged tissues. Activation of plasminogen produces the enzyme **plasmin** (PLAZ-min), which begins digesting the fibrin strands and eroding the foundation of the clot.

✓ About half of our vitamin K is produced by bacteria in the intestine. How would you expect extended use of broad-spectrum antibiotics to affect blood clotting?

Chapter Review

KEY TERMS

cardiovascular system, p. 308
embolus, p. 324
erythrocyte, p. 311
fibrin, p. 310
fibrinolysis, p. 325
hematocrit, p. 311
hemoglobin, p. 312
hemopoiesis, p. 311
hemostasis, p. 322
leukocyte, p. 317
plasma, p. 308
platelets, p. 309

SUMMARY OUTLINE

INTRODUCTION p. 308

1. The cardiovascular system provides a mechanism for the rapid transport of nutrients, waste products, and cells within the body.

FUNCTIONS OF BLOOD p. 308

1. Blood is a specialized connective tissue. Its functions include (1) transporting dissolved gases, nutrients, hormones, and metabolic wastes; (2) regulating the pH and electrolyte composition of the interstitial fluids; (3) restricting fluid losses through damaged vessels; (4) defending against pathogens and toxins; and (5) stabilizing body temperature through the absorption and redistribution of heat.

COMPOSITION OF BLOOD p. 308

Blood Collection and Analysis p. 309

1. Blood contains **plasma, red blood cells (RBCs), white blood cells (WBCs),** and **platelets.** The plasma and **formed elements** constitute **whole blood,** which can be *fractionated* for analytical or clinical purposes. (*Figure 12-1*)

PLASMA p. 310

1. Plasma accounts for about 55 percent of the volume of blood; roughly 92 percent of plasma is water. (*Figure 12-1*)

Differences Between Plasma and Interstitial Fluid p. 310

2. Plasma differs from interstitial fluid because it has a higher dissolved oxygen concentration and large numbers of dissolved proteins. There are three classes of plasma proteins: *albumins, globulins,* and *fibrinogen.*

Plasma Proteins p. 310

3. **Albumins** constitute about 60 percent of plasma proteins. **Globulins** constitute roughly 33 percent of plasma proteins; they include **immunoglobulins (antibodies),** which attack foreign proteins and pathogens, and **transport proteins,** which bind ions, hormones and other compounds. **Fibrinogen** molecules function in the clotting reaction by interacting to form **fibrin;** removing fibrinogen from plasma leaves a fluid called **serum.**

FORMED ELEMENTS p. 311

Production of Formed Elements p. 311

1. **Hemopoiesis** is the process of blood cell formation. **Stem cells** called **hemocytoblasts** divide to form all of the blood cells.

Red Blood Cells p. 311

2. Red blood cells (**erythrocytes**) account for slightly less than half of the blood volume and 99.9 percent of the formed elements. The **hematocrit** value indicates the percentage of whole blood occupied by cellular elements.

3. RBCs transport oxygen and carbon dioxide within the bloodstream. They are highly specialized cells with large surface-to-volume ratios. Because RBCs lack mitochondria, ribosomes, and nuclei, they are unable to perform normal maintenance operations, and as a result, they usually degenerate after about 120 days in the circulation. (*Figure 12-2*)

4. Molecules of **hemoglobin (Hb)** account for over 95 percent of RBC proteins. Hemoglobin is a globular protein formed from four subunits. Each subunit contains a single molecule of **heme** and can reversibly bind an oxygen molecule. Damaged or dead RBCs are recycled by phagocytes.

5. **Erythropoiesis,** the formation of erythrocytes, occurs mainly within the **myeloid tissue** (bone marrow) in adults. RBC formation increases under **erythropoietin, EPO** (erythropoiesis-stimulating hormone) stimulation, which occurs when peripheral tissues are exposed to low oxygen concentrations. Stages in RBC development involve **erythroblasts** and **reticulocytes.** (*Figures 12-4, 12-7*)

6. One's **blood type** is determined by the presence or absence of specific **surface antigens** (*agglutinogens*) in the RBC cell membranes: antigens **A, B,** and **D (Rh)**. Antibodies (*agglutinins*) within the plasma will react with RBCs bearing different agglutinogens. Anti-Rh antibodies are synthesized only after an Rh-negative individual becomes **sensitized** to the Rh surface antigen. (*Figure 12-5; Table 12-2*)

White Blood Cells p. 317

7. White blood cells (**leukocytes**) defend the body against pathogens and remove toxins, wastes, and abnormal or damaged cells.

8. Leukocytes show **positive chemotaxis** (attraction to specific chemicals) and **diapedesis** (the ability to move through vessel walls).

9. Granular leukocytes are often subdivided into **neutrophils, eosinophils,** and **basophils**. Fifty to 70 percent of circulating WBCs are neutrophils, which are highly mobile phagocytes. The much less common eosinophils are phagocytes that are attracted to foreign compounds that have reacted with circulating antibodies. The relatively rare basophils migrate to damaged tissues and release histamines, aiding the inflammation response. (*Figure 12-6*)

10. Agranular leukocytes are subdivided into monocytes and lymphocytes. **Monocytes** migrating into peripheral tissues become free macrophages. **Lymphocytes,** the primary cells of the **lymphatic system,** include **T cells** (which attack foreign cells directly), **B cells** (which produce antibodies), and **NK cells** (which destroy abnormal tissue cells). (*Figure 12-6*)

11. Granulocytes and monocytes are produced by stem cells in the bone marrow. Stem cells responsible for **lymphopoiesis** (production of lymphocytes) also originate in the bone marrow, but many migrate to peripheral **lymphoid tissues.** (*Figure 12-7*)

12. Factors that regulate lymphocyte maturation are not completely understood. Several *colony-stimulating factors (CSFs)* are involved in regulating other WBC populations.

Platelets p. 321

13. **Megakaryocytes** in the bone marrow release packets of cytoplasm (**platelets**) into the circulating blood. Platelets are essential to the clotting process. (*Figure 12-8*)

HEMOSTASIS p. 322

1. The process of **hemostasis** prevents the loss of blood through the walls of damaged vessels.

2. The initial step, the **vascular phase,** is a period of local vasoconstriction resulting from a **vascular spasm** at the injury site. The **platelet phase** follows as platelets stick to damaged surfaces.

The Clotting Process p. 323

3. The **coagulation phase** occurs as factors released by endothelial cells or peripheral tissues (**extrinsic pathway**) and platelets (**intrinsic pathway**) interact with **clotting factors** to form a **blood clot**. (*Figures 12-9, 12-10*)

Clot Retraction and Removal p. 324

4. During **clot retraction,** platelets contract and pull the torn edges closer together. During **fibrinolysis,** the clot gradually dissolves through the action of **plasmin,** the activated form of circulating **plasminogen.**

CHAPTER QUESTIONS

LEVEL 1 Reviewing Facts and Terms

Match each item in column A with the most closely related item in column B. Use letters for answers in the spaces provided.

Column A	Column B
___ 1. interstitial fluid	a. hemocytoblasts
___ 2. hemopoiesis	b. high number of WBCs
___ 3. stem cells	c. agglutinogens
___ 4. hypoxia	d. neutrophil
___ 5. surface antigens	e. WBC migration
___ 6. antibodies	f. extracellular fluid
___ 7. diapedesis	g. monocyte
___ 8. leukopenia	h. low number of WBCs
___ 9. agranulocyte	i. low platelet count
___ 10. leukocytosis	j. agglutinins
___ 11. granulocyte	k. blood cell formation
___ 12. thrombocytopenia	l. low oxygen concentration

13. The formed elements of the blood include:
(a) plasma, fibrin, serum
(b) albumins, globulins, fibrinogen
(c) WBCs, RBCs, platelets
(d) a, b, and c are correct

14. Blood temperature is approximately _______, and the blood pH averages _______.
(a) 98.6°F, 7.0 (b) 104°F, 7.8
(c) 100.4°F, 7.4 (d) 96.8°F, 7.0

15. Plasma contributes approximately _______ percent of the volume of whole blood, and water accounts for _______ percent of the plasma volume.
(a) 55, 92 (b) 25, 55
(c) 92, 55 (d) 35, 72

16. When the clotting proteins are removed from plasma, _______ remains.
(a) fibrinogen (b) fibrin
(c) serum (d) heme

17. In an adult the only site of red blood cell production, and the primary site of white blood cell formation, is the:
(a) liver (b) spleen
(c) thymus (d) red bone marrow

18. The most numerous WBCs found in a differential count of a "normal" individual are:
(a) neutrophils (b) basophils
(c) lymphocytes (d) monocytes

19. Stem cells responsible for the process of lymphopoiesis are located in the:
(a) thymus and spleen
(b) lymph nodes
(c) red bone marrow
(d) a, b, and c are correct

20. The first step in the process of hemostasis is:
(a) coagulation (b) the platelet phase
(c) fibrinolysis (d) vascular spasm

21. The complex sequence of steps leading to the conversion of fibrinogen to fibrin is called:
(a) fibrinolysis (b) coagulation
(c) retraction (d) the platelet phase

22. What five major functions are performed by the blood?

23. What three primary classes of plasma proteins are found in the blood? What is the major function of each?

24. What type of antibodies (agglutinins) does the plasma contain for each of the following blood types?
(a) Type A (b) Type B
(c) Type AB (d) Type O

25. What three processes facilitate the movement of WBCs to areas of invasion or injury?

26. What contribution from the intrinsic and extrinsic pathways is necessary for the common pathway to begin?

27. Distinguish between an embolus and a thrombus.

LEVEL 2 Reviewing Concepts

28. Dehydration would cause:
(a) an increase in the hematocrit
(b) a decrease in the hematocrit
(c) no effect in the hematocrit
(d) an increase in plasma volume

29. Erythropoietin directly stimulates RBC formation by:
(a) increasing rates of mitotic divisions in erythroblasts
(b) speeding up the maturation of red blood cells
(c) accelerating the rate of hemoglobin synthesis
(d) a, b, and c are correct

30. A person with Type A blood has:
(a) Anti-A in the plasma
(b) Type B antigens in the plasma
(c) Type A antigens on the red blood cells
(d) Anti-B on the red blood cells

31. Hemolytic disease of the newborn may result if:
(a) an Rh-positive woman marries an Rh-negative man
(b) both the man and the woman are Rh-negative
(c) both the man and the woman are Rh-positive
(d) an Rh-negative woman carries an Rh-positive fetus

32. How do red blood cells (RBCs) differ from typical cells found in the body?

LEVEL 3 Critical Thinking and Clinical Applications

33. Which of the formed elements would you expect to see increase after donating a pint of blood?

34. Why do patients suffering from advanced kidney disease frequently become anemic?

13 The Heart

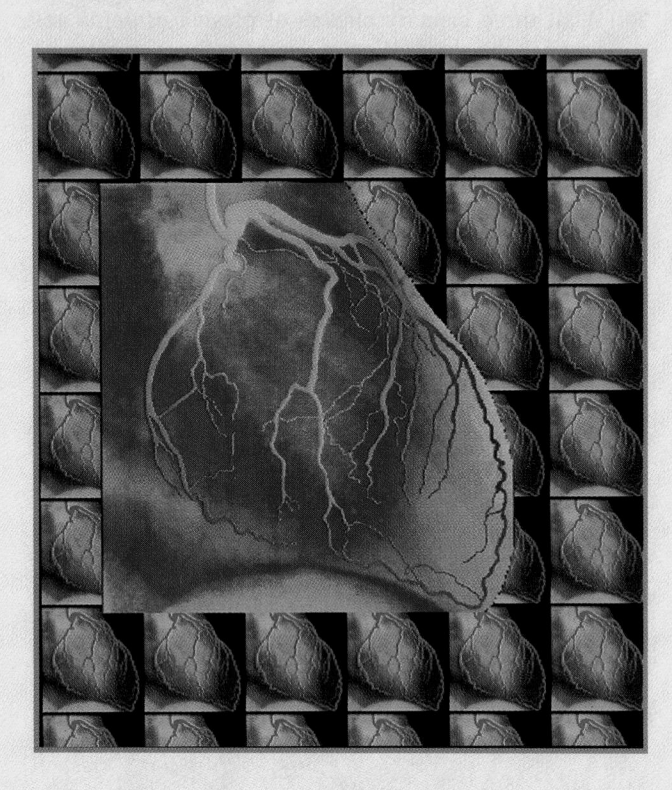

You know from previous chapters that hard-working muscles require a steady supply of blood to provide them with nutrients and oxygen. This is especially true for the hardest working muscle of all: your heart. Any substantial interruption in the flow of blood to this organ has grave consequences: what we commonly call a heart attack. Such attacks typically occur when there is an obstruction in one of the arteries that supply the heart muscle. Physicians can check the status of these vessels using scanning techniques that produce computer-enhanced images like those shown here.

Chapter Outline and Objectives

1. Describe the location and general features of the heart.
2. Trace the flow of blood through the heart, identifying the major blood vessels, chambers, and heart valves.
3. Identify the layers of the heart wall.

4. Describe the differences in the action potentials and twitch contractions of skeletal muscle fibers and cardiac muscle cells.
5. Describe the components and functions of the conducting system of the heart.
6. Explain the events of the cardiac cycle and relate the heart sounds to specific events in this cycle.

7. Define stroke volume and cardiac output and describe the factors that influence these values.

Vocabulary Development

anastomosis, outlet; *anastomoses*
atrion, hall; *atrium*
auris, ear; *auricle*
bi-, two; *bicuspid*
bradys, slow; *bradycardia*
cuspis, point; *bicuspid valve*
diastole, expansion; *diastole*
luna, moon; *semilunar valve*
mitre, a bishop's hat; *mitral valve*
semi-, half; *semilunar valve*
septum, wall; *interatrial septum*
systole, a drawing together; *systole*
tachys, swift; *tachycardia*
tri-, three; *tricuspid valve*
ventricle, little belly; *ventricle*

Every living cell relies on the surrounding interstitial fluid for oxygen, nutrients, and waste disposal. Conditions in the interstitial fluid are kept stable through continuous exchange between the peripheral tissues and circulating blood. If the blood remains stationary, its oxygen and nutrient supplies are quickly exhausted, its capacity to absorb wastes is soon saturated, and neither hormones nor white blood cells can reach their intended targets. All cardiovascular functions ultimately depend on the heart. This muscular organ beats approximately 100,000 times each day, pumping roughly 8000 liters of blood—enough to fill 40 55-gallon drums, or 8800 quart-sized milk cartons.

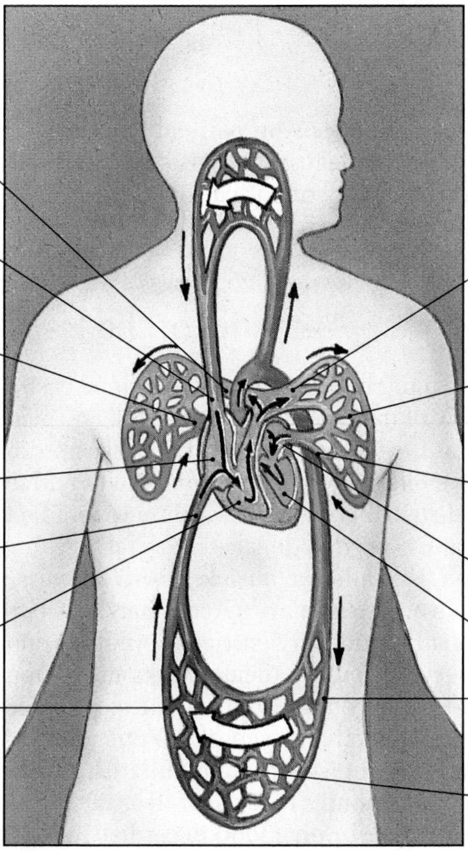

This chapter begins by examining the structural features that enable the heart to perform so reliably, even under widely varying physical demands. We will then consider the mechanisms that regulate heart activity to meet the body's ever-changing needs.

THE HEART AND THE CIRCULATORY SYSTEM

Blood flows through a network of blood vessels that extend between the heart and the peripheral tissues. The blood vessels can be subdivided into a **pulmonary circuit** that carries blood to and from the exchange surfaces of the lungs and a **systemic circuit** that transports blood to and from the rest of the body. Each circuit begins and ends at the heart. **Arteries,** or *efferent vessels,* carry blood away from the heart; **veins,** or *afferent vessels,* return blood to the heart. **Capillaries** are small, thin-walled vessels between the smallest arteries and veins.

As indicated in Figure 13-1●, blood travels through these circuits in sequence. For example, blood returning to the heart in the systemic veins must complete the pulmonary circuit before reentering the systemic arteries. The heart contains four muscular chambers, two associated with each circuit. The **right atrium** (Ā-trē-um; hall) receives blood from the systemic (body) circuit, and the **right ventricle** (VEN-tri-kl; "little belly") discharges it into the pulmonary (lungs) circuit. The **left atrium** collects blood from the pulmonary circuit, and the **left ventricle** ejects it into the systemic circuit. When the heart beats, the two ventricles contract at the same time and eject equal volumes of blood.

●**Figure 13-1**
An Overview of the Cardiovascular System
Blood flows through separate pulmonary and systemic circuits, driven by the pumping of the heart. Each circuit begins and ends at the heart and contains arteries, capillaries, and veins.

ANATOMY AND ORGANIZATION OF THE HEART

Despite its impressive workload, the heart is a small organ, roughly the size of a clenched fist. Lying near the center of the thoracic cavity, it is enclosed by the con-

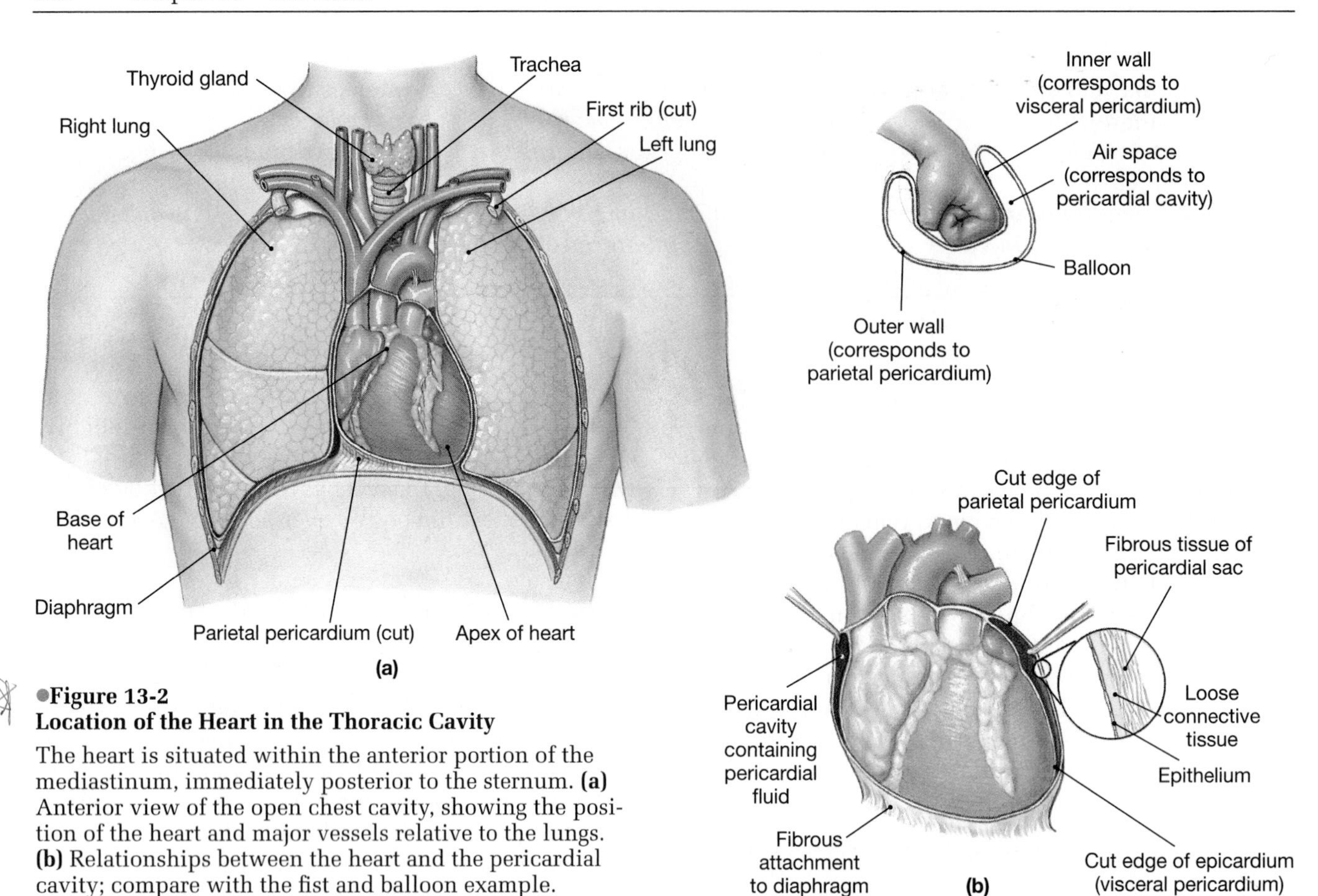

•Figure 13-2
Location of the Heart in the Thoracic Cavity
The heart is situated within the anterior portion of the mediastinum, immediately posterior to the sternum. **(a)** Anterior view of the open chest cavity, showing the position of the heart and major vessels relative to the lungs. **(b)** Relationships between the heart and the pericardial cavity; compare with the fist and balloon example.

nective tissues of the mediastinum, which also contains the thymus, esophagus, and trachea, and which divides the thoracic cavity into two pleural cavities. Figure 13-2a• shows an anterior view of the open chest cavity.

The heart lies near the anterior chest wall, directly behind the sternum and sits at an angle to the longitudinal axis of the body. It is surrounded by the **pericardial** (per-i-KAR-dē-al) **cavity,** one of the three ventral body cavities introduced in Chapter 1. ∞ *[p. 19]* This cavity is lined by a serous membrane called the **pericardium.** ∞ *[p. 87]* To visualize the relationship between the heart and the pericardial cavity, imagine pushing your fist toward the center of a large balloon (Figure 13-2b•). The balloon represents the pericardium, and your fist the heart. Your wrist, where the balloon folds back upon itself, corresponds to the *base* of the heart. The space inside the balloon is the pericardial cavity.

The pericardium can be subdivided into the *visceral pericardium* and the *parietal pericardium.* The **visceral pericardium,** or **epicardium,** covers the outer surface of the heart; the **parietal pericardium** lines the inner surface of the *pericardial sac* that surrounds the heart (Figure 13-2b•). The pericardial sac is reinforced by a dense network of collagen fibers that stabilizes the positions of the pericardium, heart, and associated vessels in the mediastinum. The slender gap between the opposing parietal and visceral surfaces is the pericardial cavity. This space normally contains a small quantity of pericardial fluid that acts as a lubricant, reducing friction as the heart beats.

Surface Anatomy of the Heart

The four chambers can be identified easily in a surface view of the heart (Figure 13-3a•). Several external features distinguish the atria from the ventricles. The atria have relatively thin muscular walls, and they are highly distensible. When the atrium is not filled with blood, its outer portion deflates into a rather lumpy and wrinkled flap called an **auricle** (AW-ri-kl; *auris,* ear). A deep groove, the **coronary sulcus,** marks the border between the atria and the ventricles. Another depression, the **interventricular sulcus,** marks the boundary line between the left and right ventricles (Figure 13-3a,b•). The connective tissue of the epicardium at the coronary and interventricular sulci usually contains substantial amounts of fat, as well as the major arteries and veins that supply blood to cardiac muscle tissue.

The great veins and arteries of the circulatory system are connected to the superior end of the heart at the **base.** The inferior, pointed tip of the heart is the **apex** (Ā-peks) (Figure 13-2a•). A typical heart measures approximately 12.5 cm (5 in.) from the attached base to the apex.

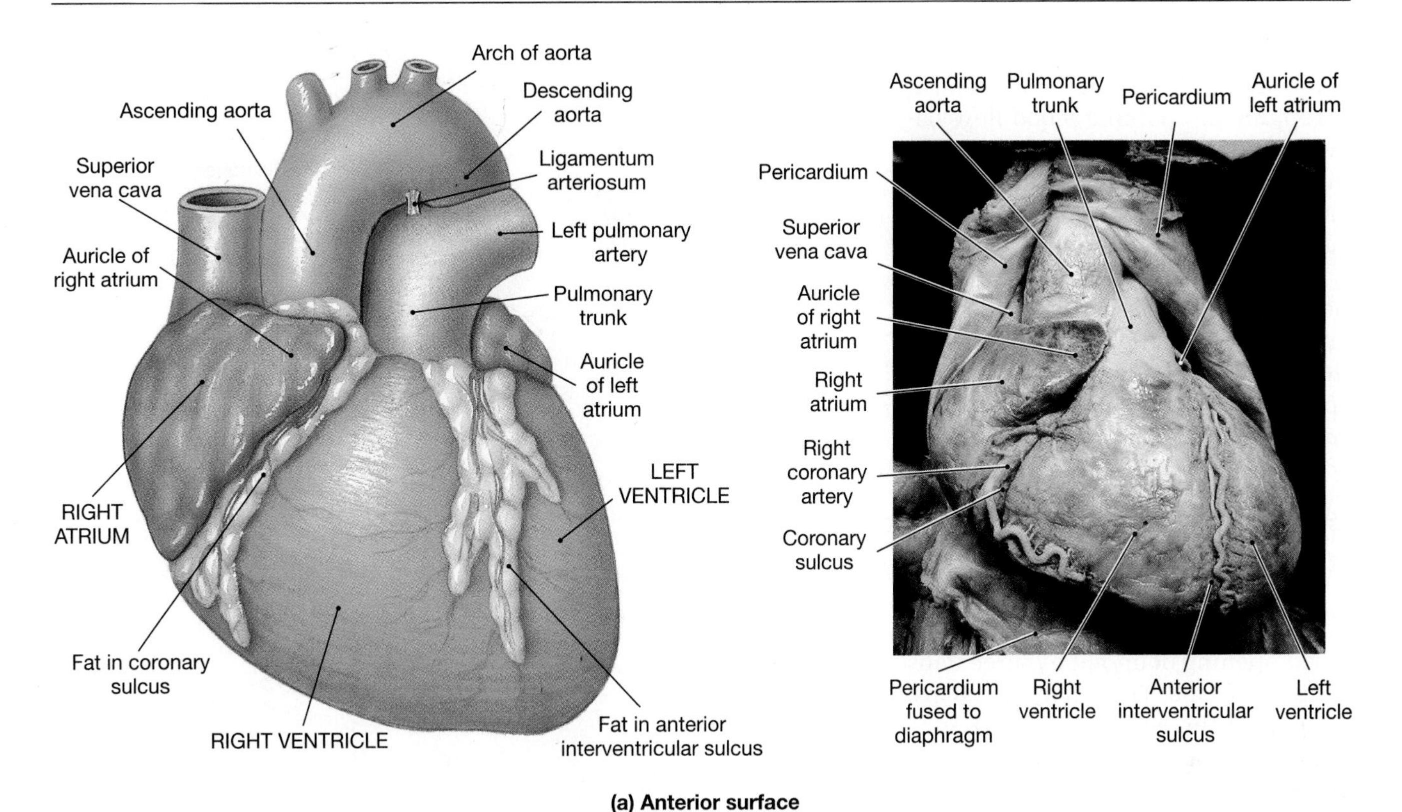

Figure 13-3
Surface Anatomy of the Heart
(a) Anterior view of the heart, showing major anatomical features. **(b)** The posterior surface of the heart. (Coronary arteries are shown in red, coronary veins in blue.)

The heart sits at an angle to the longitudinal axis of the body. It is also rotated slightly toward the left, so that the anterior surface (Figure 13-3a) primarily consists of the right atrium and right ventricle. The wall of the left ventricle forms much of the posterior surface (Figure 13-3b) between the base and the apex of the heart.

Internal Anatomy and Organization

Figure 13-4a illustrates the four internal chambers of the heart. The two atria are separated by the **interatrial septum** (*septum*, wall), not visible in the figure, and the two ventricles are divided by the **interventricular septum.** Each atrium communicates with the ventricle on the same side through folds of tissue, called **atrioventicular (AV) valves,** arranged so as to ensure a one-way flow of blood from the atria into the ventricles.

The right atrium receives blood from the systemic circuit via two large veins, the **superior vena cava** (VĒ-na CĀ-va) and **inferior vena cava.** The superior vena cava delivers blood from the head, neck, upper limbs, and chest. The inferior vena cava carries blood returning from the rest of the trunk, the viscera, and the lower limbs. The *coronary veins* of the heart return venous blood to the **coronary sinus,** which opens into the right atrium slightly below the connection with the inferior vena cava. From the fifth week of embryonic development until birth, an oval opening, the foramen ovale, penetrates the interatrial septum and permits blood flow between the two atria. The foramen ovale shunts blood away from the lungs while they are developing. At birth, the foramen ovale closes, and after 48 hours it is permanently sealed. A small depression, the *fossa ovalis,* persists at this site in the adult heart (Figure 13-4●). Occasionally, the foramen ovale remains open and the circulation to the lungs does not increase at birth. The tissues of the newborn infant soon become starved for oxygen, and the *cyanosis* that develops makes the infant appear to be a "blue baby."

Blood travels from the right atrium into the right ventricle through a broad opening bounded by three flaps of fibrous tissue. These flaps, or **cusps,** are part of the **right atrioventricular (AV) valve,** also known as the **tricuspid valve** (trī-KUS-pid; *tri-,* three + *cuspis,* point). Each cusp is braced by tendinous fibers, the **chordae tendineae** (KŌR-dē TEN-di-nē-ē; "tendinous cords"), that are connected to special **papillary** (PAP-i-ler-ē) **muscles** on the inner surface of the right ventricle (Figure 13-4●).

Blood leaving the right ventricle flows into the large **pulmonary trunk** that marks the start of the pulmonary circuit. The **pulmonary semilunar** (*semi-,* half + *luna,* moon; a crescent, or half-moon, shape) **valve** guards the entrance to this efferent trunk. Once within the pulmonary trunk, blood flows into the **left** and **right pulmonary arteries.** These vessels branch repeatedly within the lungs, supplying the capillaries where gas exchange occurs. From these respiratory capillaries blood collects into the **left** and **right pulmonary veins,** which return and empty into the left atrium.

Like the right atrium, the left atrium has an external auricle and a valve, the **left atrioventricular (AV) valve,** or **bicuspid** (bī-KUS-pid) **valve.** As the name *bicuspid* implies, the left AV valve contains a pair of cusps rather than a trio. Clinicians often use the term **mitral** (MĪ-tral; *mitre,* a bishop's hat) when referring to this valve.

The internal organization of the left ventricle resembles that of the right ventricle. A pair of papillary muscles brace the chordae tendineae that insert on the bicuspid valve. Blood leaving the left ventricle passes through the **aortic semilunar valve** and into the systemic circuit via the **ascending aorta.**

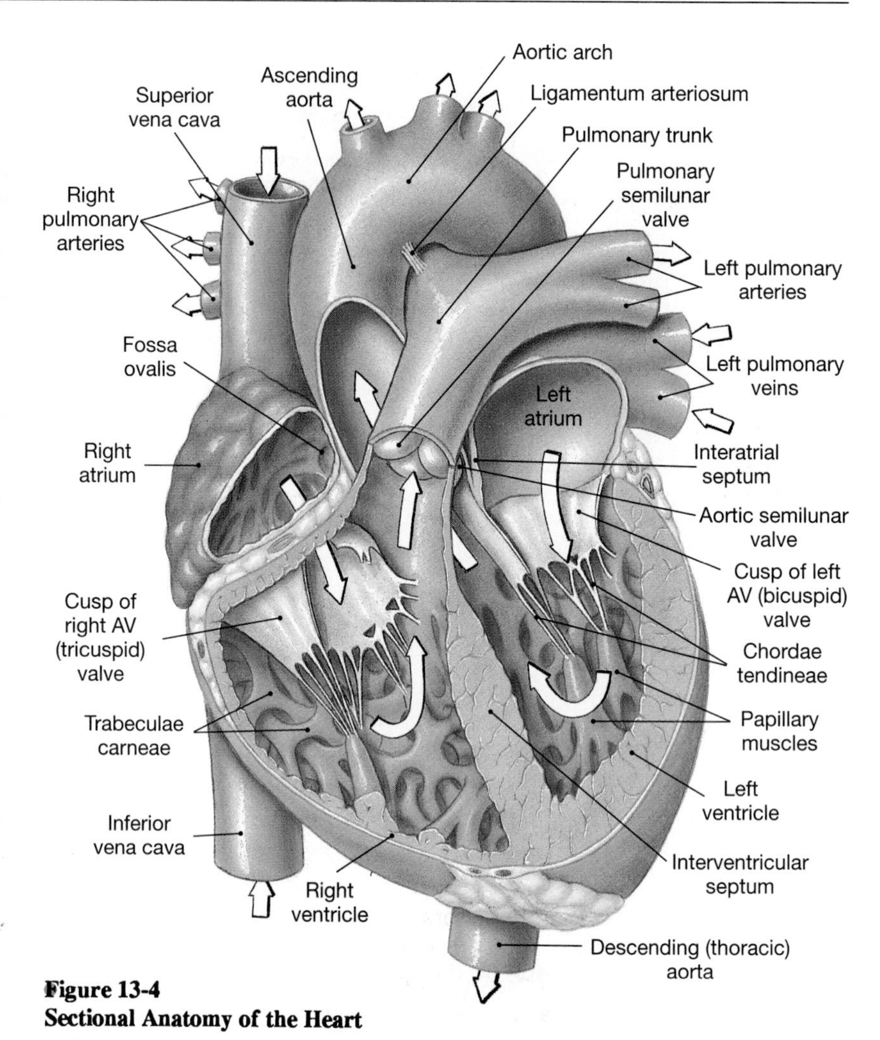

Figure 13-4
Sectional Anatomy of the Heart
A diagrammatic frontal section through the heart, showing major landmarks and the path of blood flow through the atria and ventricles.

Structural Differences between the Left and Right Ventricles

The function of an atrium is to collect blood returning to the heart and deliver it to the attached ventricle. The demands placed on the right and left atria are very similar, and the two chambers look almost identical. But the demands placed on the right and left ventricles are very different, and there are anatomical differences between the two.

The lungs are close to the heart, and the pulmonary arteries and veins are relatively short and wide. Thus the right ventricle normally does not need to push very hard to propel blood through the pulmonary circuit. The wall of the right ventricle is relatively thin, and in sectional view it resembles a pouch attached to the massive wall of the left ventricle. When the right ventricle contracts, it acts like a bellows pump, squeezing the blood against the mass of the left ventricle. This mechanism moves blood very efficiently with minimal effort, but it develops relatively low pressures.

A comparable pumping arrangement would not be suitable for the left ventricle, because six to seven times as much force must be exerted to push blood around the systemic circuit. The left ventricle has an extremely thick muscular wall, and it is round in cross-section. When this ventricle contracts, two things happen: The distance between the base and apex decreases, and the diameter of the ventricular chamber decreases. If you imagine the effects of simultaneously squeezing and rolling up the end of a toothpaste tube you will get the idea. As the powerful left ventricle contracts, it also bulges into the right ventricular cavity, helping to force blood out of the right ventricle. An individual whose right ventricular musculature has been severely damaged may survive because the contraction of the left ventricle helps push blood through the pulmonary circuit. AM *The Cardiomyopathies* and *Heart Transplants and Assist Devices*

The Heart Valves

Details of the structure and function of the heart valves are shown in Figure 13-5●.

The Atrioventricular Valves The atrioventricular valves prevent backflow of blood from the ventricles into the atria. The chordae tendineae and papillary muscles play an important role in the normal function of the AV valves. When a ventricle is filling with blood (a phase called *diastole*), the papillary muscles are relaxed and the AV valve offers no resistance to the flow of blood from atrium to ventricle (Figure 13-5a●). When the ventricle begins to contract, blood moving back toward the atrium swings the cusps together, closing the valve (Figure 13-5b●). During ventricular contraction (a phase called *systole*), tension in the papillary muscles and chordae tendineae keeps the cusps from swinging into the atrium. This action prevents the backflow, or **regurgitation,** of blood into the atrium each time the ventricle contracts. A small amount of regurgitation often occurs, even in normal individuals. The swirling action creates a soft but distinctive sound, called a *heart murmur.*

MITRAL VALVE PROLAPSE

Minor abnormalities in valve shape are relatively common. For example, an estimated 10 percent of normal individuals age 14–30 have some degree of **mitral valve prolapse.** In this condition, the mitral valve cusps do not close properly. The problem may involve abnormally long (or short) chordae tendineae or malfunctioning papillary muscles. Because the valve does not work perfectly, some regurgitation occurs during contraction of the left ventricle. Most individuals with mitral valve prolapse are completely asymptomatic, and they live normal healthy lives unaware of any circulatory malfunction.

The Semilunar Valves The pulmonary and aortic semilunar valves prevent backflow of blood from the pulmonary trunk and aorta into the right and left ventricles. The semilunar valves do not require muscular braces because the arterial walls do not contract, and the relative positions of the cusps are stable. When these valves close, the three symmetrical cusps support one another like the legs of a tripod (Figure 13-5a●).

VALVULAR HEART DISEASE

Serious valve problems are very dangerous, because they reduce pumping efficiency. If valve function deteriorates to the point that the heart cannot maintain adequate circulatory flow, symptoms of **valvular heart disease** appear. Congenital defects may be responsible, but often the condition develops after *carditis,* an inflammation of the heart. AM *Infection and Inflammation of the Heart*

One relatively common cause of carditis is *rheumatic* (roo-MA-tik) *fever,* an inflammation that may develop after infection by streptococcal bacteria. Valve problems may not appear until 10–20 years later, and the resulting disorder is known as *rheumatic heart disorder.* AM *RHD and Valvular Stenosis*

The Heart Wall

The wall of the heart contains three distinct layers: the *epicardium* (visceral pericardium), the *myocardium,* and the *endocardium* (Figure 13-6a●). The **epicardium,** which covers the outer surface of the heart, is a serous membrane that consists of an exposed epithelium and an underlying layer of loose connective tissue. The **myocardium,** or muscular wall of the heart, contains cardiac muscle tissue and associated connective tissues, blood vessels, and nerves. The inner surfaces of the heart, including the valves, are covered by a simple squamous epithelium, the **endocardium** (en-dō-KAR-dē-um), that is continuous with the epithelium of the attached blood vessels.

Cardiac Muscle Cells

Typical cardiac muscle cells, or *cardiocytes,* are detailed in Figure 13-6c,d●. These cells are relatively small and contain a single centrally located nucleus. Like skeletal muscle fibers, each cell contains myofibrils, and contraction involves the shortening of individual sarcomeres. Since cardiac muscle cells are almost totally dependent on aerobic metabolism to obtain the energy needed to continue contracting, they have many mitochondria and abundant reserves of myoglobin (to store oxygen). Energy reserves are maintained in the form of glycogen and lipids.

Each cardiac muscle cell contacts several others at specialized sites known as **intercalated** (in-TER-ka-lā-ted) **discs.** Tight junctions, desmosomes, and gap junctions at these sites give the cells mechanical stability and a means of communication through the

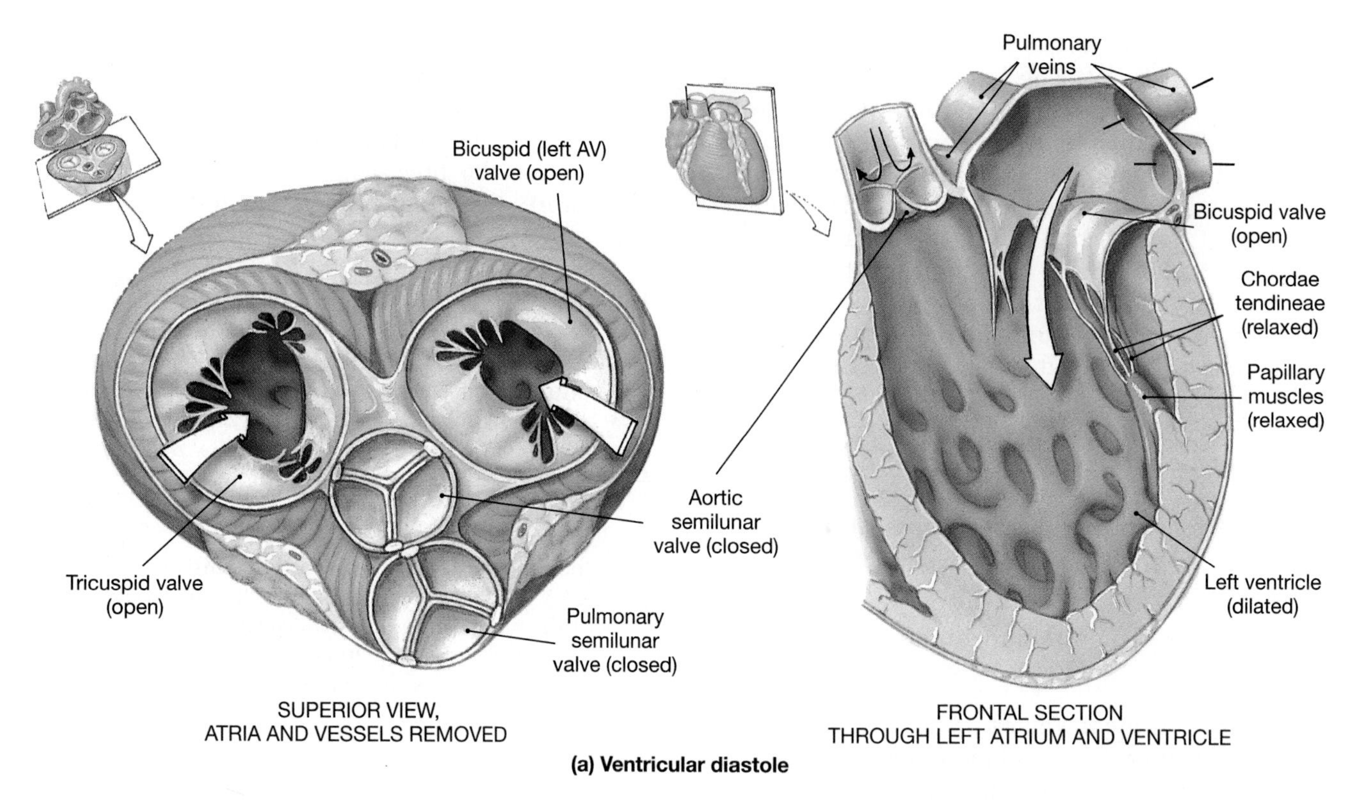

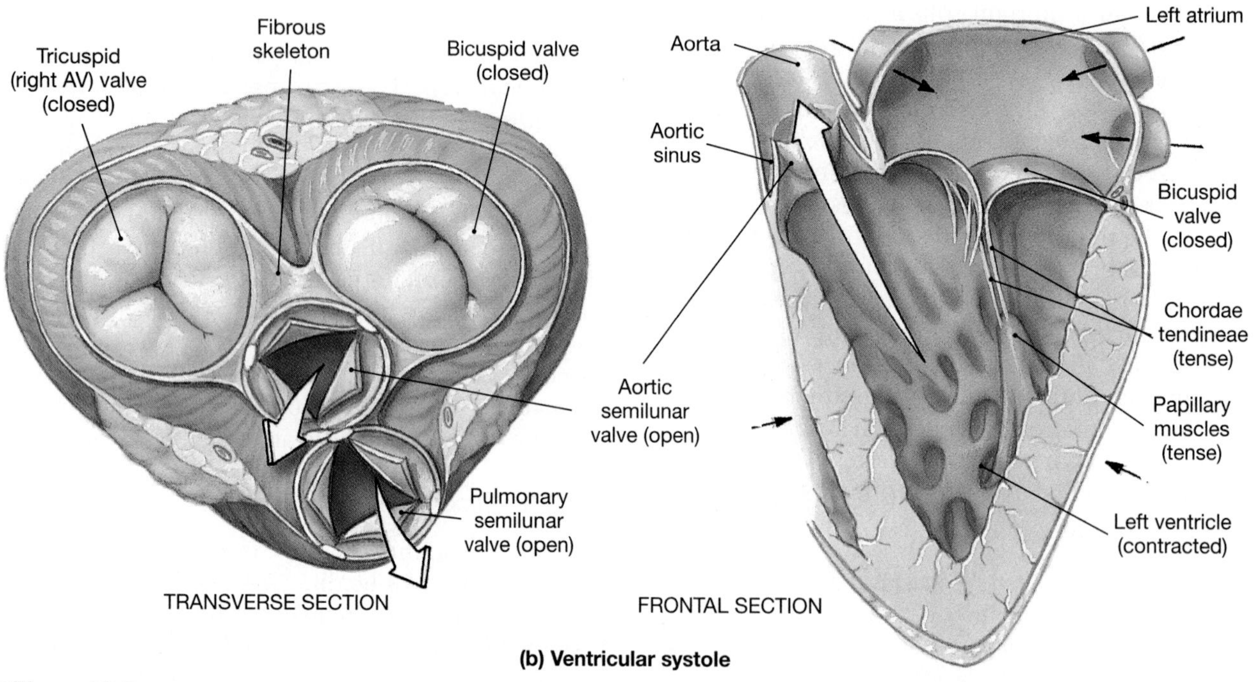

Figure 13-5
Valves of the Heart

(a) Valve position during ventricular relaxation (diastole), when the AV valves are open and the semilunar valves are closed. Note that the chordae tendineae are slack and the papillary muscles relaxed. **(b)** The appearance of the cardiac valves during ventricular contraction (systole), when the AV valves are closed and the semilunar valves are open. In the frontal section, note the role of the chordae tendineae and papillary muscles in preventing backflow through the bicuspid valve.

movement of ions and small molecules. ∞ *[p. 74]* As a result, action potentials travel from cell to cell without delay. In addition, myofibrils anchored to the discs of adjacent cells are locked together, increasing the efficiency of the cells as they "pull together." Because the cardiac muscle cells are mechanically, chemically, and electrically connected to one another in this way, the entire tissue resembles a single,

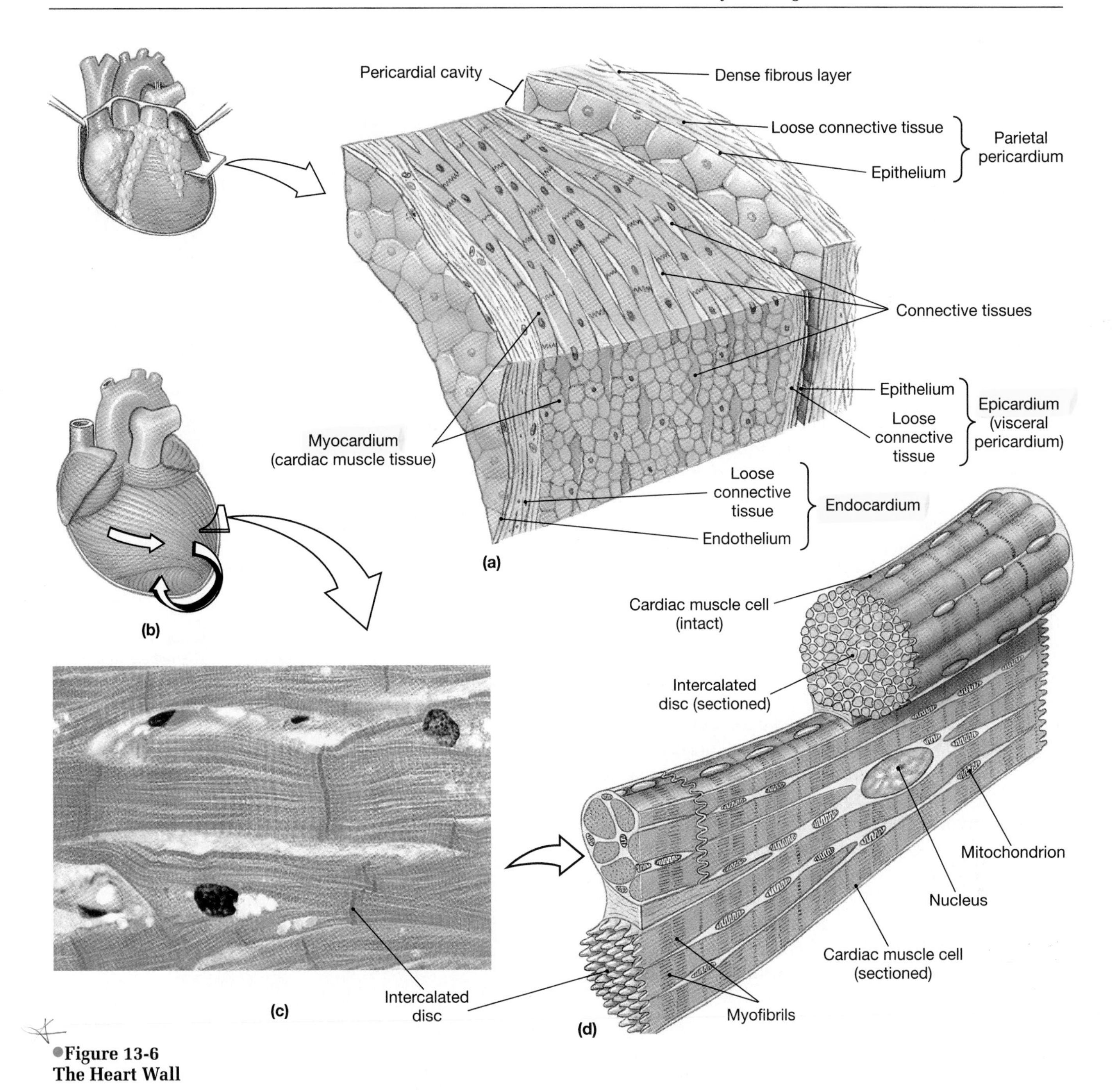

Figure 13-6
The Heart Wall

(a) A diagrammatic section through the heart wall showing the relative positions of the epicardium, myocardium, and endocardium. **(b)** Cardiac muscle tissue in the heart forms concentric layers that wrap around the atria and spiral within the walls of the ventricles. **(c,d)** Sectional and diagrammatic views of cardiac muscle tissue. Cardiac muscle cells are smaller than skeletal muscle fibers and also have a single, centrally placed nucleus, branching interconnections between cells, and intercalated discs. (LM × 575)

enormous muscle cell, and has been called a *functional syncytium* (sin-SIT-ē-um; a fused mass of cells).

The Fibrous Skeleton

The connective tissues of the heart include large numbers of collagen and elastic fibers that wrap around each cardiac muscle cell and also tie together adjacent cells. These fibers are in turn interwoven with more extensive sheets of fibrous tissue that separate concentric layers of cardiac muscle cells and encircle each of the heart valves. This internal connective tissue network is called the **fibrous skeleton** of the heart. The fibrous skeleton supports and stabilizes muscle cells and valves, and its elasticity limits overexpansion and helps the heart return to normal shape after contractions (see Figure 13-6). It also physically isolates the atrial muscles from those in the ventricles, a feature important to normal cardiac function.

The Blood Supply to the Heart

The heart works continuously, and cardiac muscle cells require reliable supplies of oxygen and nutrients. The **coronary circulation** supplies blood to the muscles of the heart. During maximum exertion, the oxygen demand rises considerably, and the blood flow to the heart may increase to nine times that of resting levels.

As Figure 13-7● illustrates, the coronary circulation involves an extensive network of vessels. The left and right **coronary arteries** originate at the base of the ascending aorta (Figure 13-7a,b●). Blood pressure here is the highest found anywhere in the systemic circuit, and this pressure ensures a continuous flow of blood to meet the demands of active cardiac muscle tissue. Each coronary artery gives rise to two branches (the *marginal* and *posterior interventricular [descending]* branches from the right, and the *circumflex* and *anterior interventricular [descending]* branches from the left). Small tributaries from these branches of the left and right coronary arteries form interconnections called **anastomoses** (a-nas-to-MŌ-sēz; *anastomosis,* outlet). Because the arteries are interconnected in this way, the blood supply to the cardiac muscle remains relatively constant, regardless of pressure fluctuations within the left and right coronary arteries. The **great** and **middle cardiac veins** carry blood away from the coronary capillaries. They drain into the **coronary sinus,** a large, thin-walled vein that lies in the posterior portion of the coronary sulcus. The coronary sinus communicates with the right atrium near the base of the inferior vena cava.

In a **myocardial** (mī-ō-KAR-dē-al) **infarction (MI),** or *heart attack,* the coronary circulation becomes blocked and the cardiac muscle cells die from lack of oxygen. The affected tissue then degenerates, creating a nonfunctional area known as an *infarct.* Heart attacks most often result from severe *coronary artery disease,* a condition characterized by the buildup of fatty deposits in the walls of the coronary arteries. AM *Coronary Artery Disease* and *Heart Attacks*

✓ Damage to the semilunar valves on the right side of the heart would interfere with blood flow to what vessel?

✓ What prevents the AV valves from opening back into the atria?

✓ Why are the left atrium and ventricle more muscular than the right atrium and ventricle?

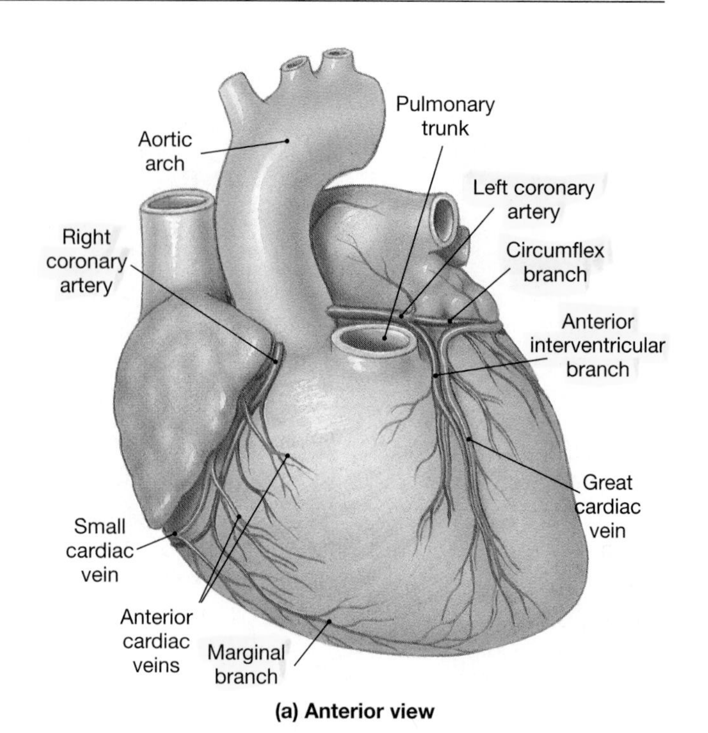

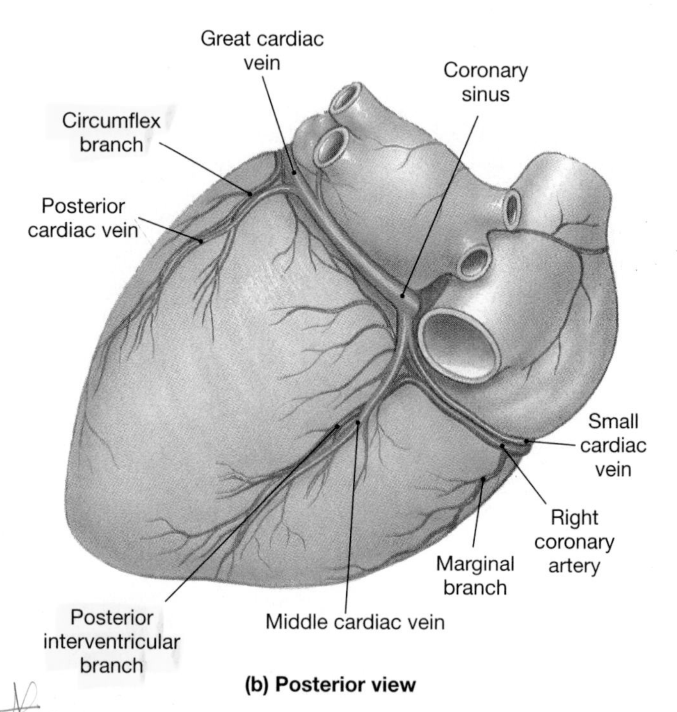

●**Figure 13-7**
Coronary Circulation
(a) Coronary vessels supplying the anterior surface of the heart. **(b)** Coronary vessels supplying the posterior surface of the heart.

THE HEARTBEAT

The heart's remarkably steady performance is a direct result of the unusual structural characteristics of its cardiac muscle tissue. Each time the heart beats, the contractions of individual cardiac muscle cells are coordinated and harnessed to ensure that blood flows in the right direction at the proper time. Two types of cardiac muscle cells are involved in a normal heartbeat. *Contractile cells* produce the powerful contractions that propel blood, while specialized muscle cells of the *conducting system* control and coordinate the activities of the contractile cells. (It would be useful to review the mechanics of skeletal muscle contraction detailed in Chapter 7.) [p. 159]

Contractile Cells

As in skeletal muscle, the first step in triggering a contraction in a cardiac muscle cell is the appearance of an action potential in the sarcolemma. In the 10-msec (millisecond) action potential of a skeletal muscle, a rapid depolarization is immediately followed by a rapid repolarization. In a cardiac muscle cell, the complete depolarization-repolarization process lasts 250–300 msec, some 30 times longer than the duration of an action potential in a skeletal muscle sarcolemma. Until the membrane repolarizes, it cannot respond to further stimulation, and the refractory period of a cardiac muscle cell membrane is relatively long. Thus a normal cardiac muscle cell is limited to a maximum rate of about 200 contractions per minute.

In skeletal muscle fibers, the refractory period ends before the muscle fiber develops peak tension and relaxes. As a result, twitches can summate, and tetanus can occur. In cardiac muscle cells, the refractory period continues until relaxation is under way. Summation is therefore not possible and tetanic contractions cannot occur in a normal cardiac muscle cell, regardless of the frequency and intensity of stimulation. This feature is absolutely vital, since a heart in tetany could not pump blood.

The Conducting System

In contrast to skeletal muscle, cardiac muscle tissue contracts on its own in the absence of neural or hormonal stimulation. This property, called *automaticity*, or *autorhythmicity*, also characterizes some types of smooth muscle tissue discussed in Chapter 7.

In the normal pattern of blood flow, each contraction follows a precise sequence: The atria contract first, followed by the ventricles. Cardiac contractions are coordinated by two types of specialized cardiac muscle cells that do not contract. **Nodal cells** are responsible for establishing the rate of cardiac contraction, and **conducting cells** distribute the contractile stimulus to the general myocardium.

Nodal Cells

Nodal cells are unusual because their cell membranes depolarize spontaneously and generate action potentials at regular intervals. Nodal cells are electrically coupled to one another, to conducting cells, and to normal cardiac muscle cells. As a result, when an action potential appears in a nodal cell, it sweeps through the conducting system, reaching all of the cardiac muscle tissue and causing a contraction. In this way, nodal cells determine the heart rate.

Not all nodal cells depolarize at the same rate, and the normal rate of contraction is established by the nodal cells that reach threshold first. These **pacemaker cells** are found in the **cardiac pacemaker,** or **sinoatrial** (sī-nō-Ā-trē-al) **node (SA node)**, which is embedded in the posterior wall of the right atrium near the entrance of the superior vena cava (Figure 13-8a●). Pacemaker cells depolarize rapidly and spontaneously, generating 70–80 action potentials per minute. This results in a heart rate of 70–80 beats per minute (bpm).

Conducting Cells

The stimulus for a contraction is usually generated at the SA node, but it must be distributed so that (1) the atria contract together, before the ventricles, and (2) the ventricles contract together, in a wave that begins at the apex and spreads toward the base. When the ventricles contract in this way, blood is pushed toward the base of the heart, into the aortic and pulmonary trunks.

The conducting network of the heart is illustrated in Figure 13-8a●. The SA nodal cells are electrically connected to those of the larger **atrioventricular** (ā-trē-ō-ven-TRIK-ū-lar) **node (AV node)** via conducting cells in the atrial walls. Although these nodal cells also depolarize spontaneously, they generate only 40–60 action potentials per minute. Under normal circumstances, before an AV cell depolarizes to threshold spontaneously, it is stimulated by an action potential generated by the SA node. However, if the AV node does not receive this action potential, it will then become the pacemaker of the heart and establish a heart rate of 40–60 beats per minute.

The AV node sits within the floor of the right atrium near the opening of the coronary sinus. From here the action potentials travel to the **AV bundle,** also known as the *bundle of His* (hiss). This bundle of conducting cells travels along the interventricular septum before dividing into **left** and **right bundle branches** that radiate across the inner surfaces of the left and right ventricles. At this point specialized **Purkinje** (pur-KIN-jē) **cells** (*Purkinje fibers*) convey the impulses to the contractile cells of the ventricular myocardium.

Pacemaker cells in the SA node usually generate 60–100 action potentials per minute. It takes roughly 50 msec for an action potential to travel from the SA node to the AV node over the conducting pathways (Figure 13-8b●). Along the way, the conducting cells pass the contractile stimulus to cardiac muscle cells of the right and left atria. The action potential then spreads across the atrial surfaces through cell-to-cell contact. The stimulus affects only the atria, because the fibrous skeleton electrically isolates the atria from the ventricles everywhere except at the AV bundle.

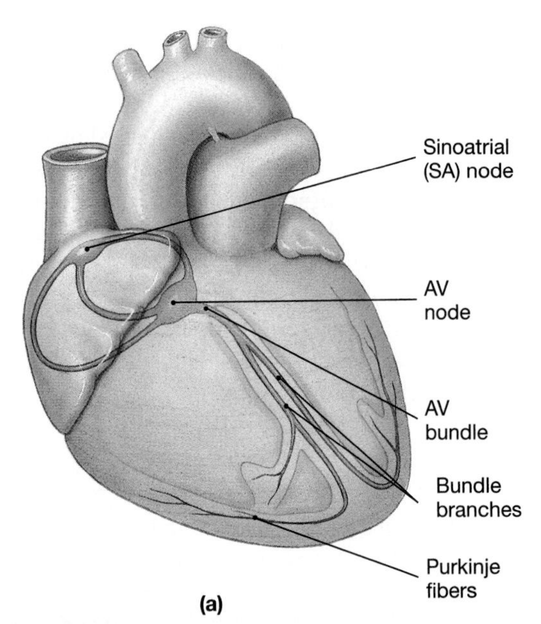

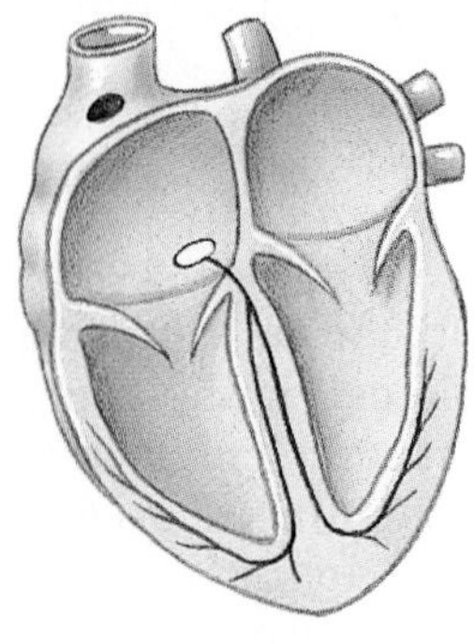

STEP 1:
SA node activity and atrial activation begin.
Time = 0

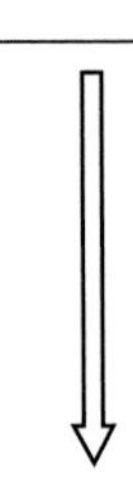

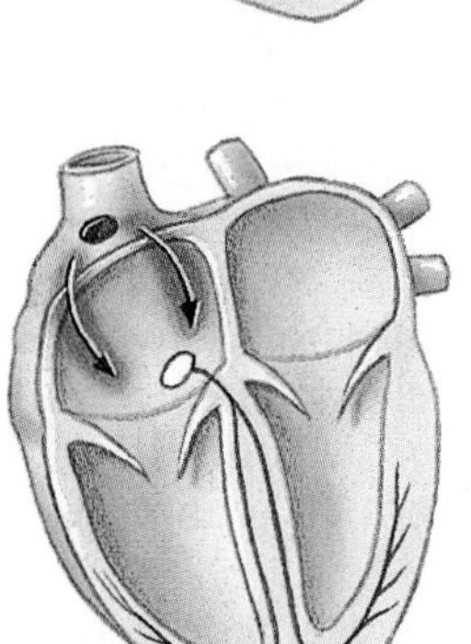

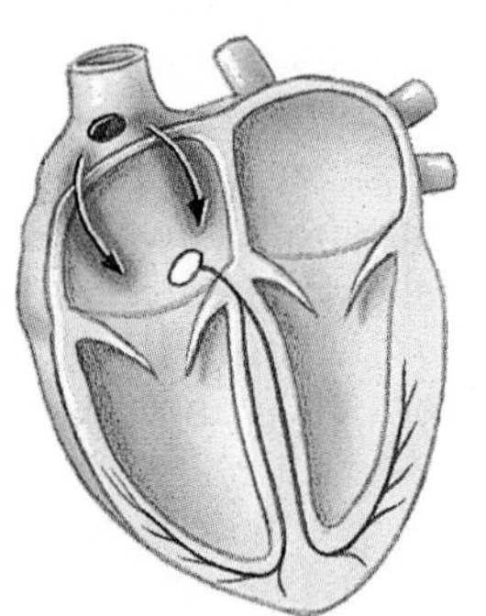

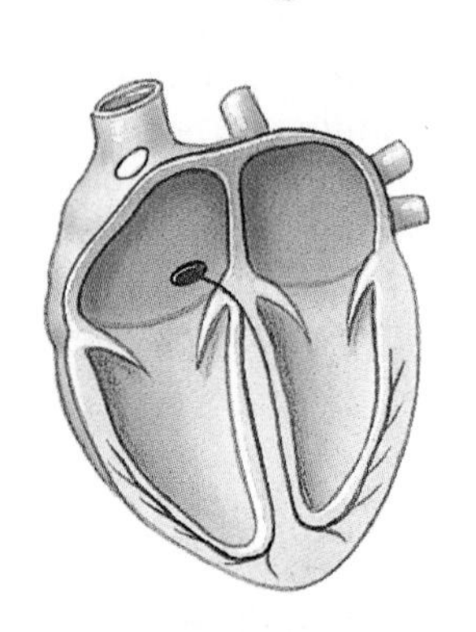

STEP 3:
There is a 100 msec delay at the AV node. Atrial contraction begins.
Elapsed time = 150 msec

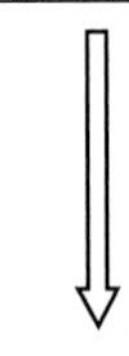

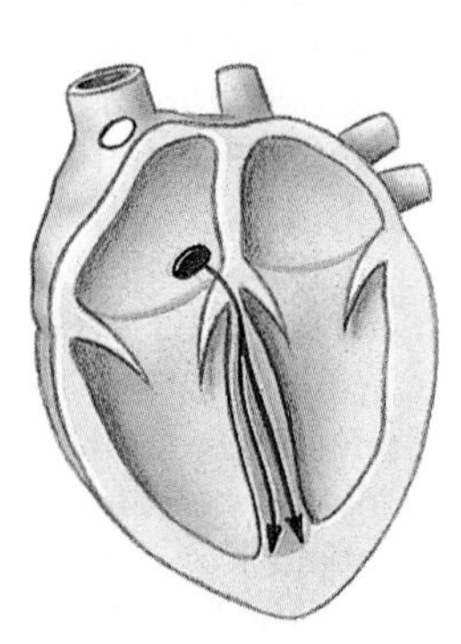

STEP 4:
The impulse travels along the interventricular septum, via the AV bundle and the bundle branches, to the Purkinje fibers.
Elapsed time = 175 msec

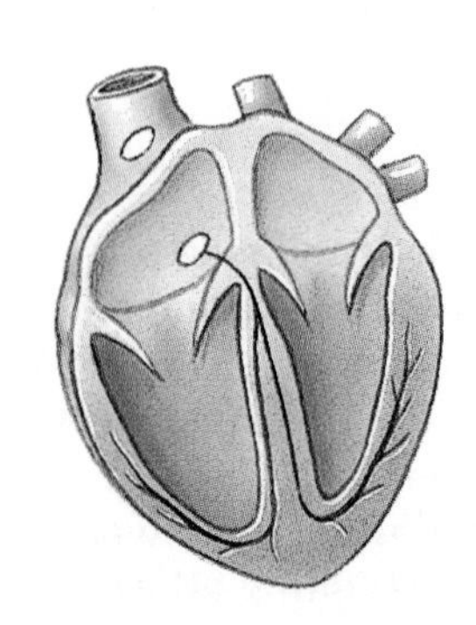

STEP 5:
The impulse is distributed by Purkinje fibers and relayed throughout the ventricular myocardium. Atrial contraction is completed, ventricular contraction begins.
Elapsed time = 225 msec
(b)

•Figure 13-8
The Conducting System of the Heart

(a) The stimulus for contraction is generated by pacemaker cells at the SA node. From there, impulses follow three different paths through the atrial walls to reach the AV node. After a brief delay, the impulses are conducted to the AV bundle (bundle of His), and then on to the left and right bundle branches, the Purkinje cells, and the ventricular myocardial cells. **(b)** The movement of the contractile stimulus through the heart.

At the AV node, the impulse slows down, and another 100 msec passes before it reaches the AV bundle. This delay is important, because the atria must be contracting and blood movement must be occurring before the ventricles are stimulated. Once the impulse enters the AV bundle it flashes down the septum, along the bundle branches, and into the ventricular myocardium via Purkinje cells. Within another 75 msec the stimulus to begin a contraction has reached all of the ventricular cardiac muscle cells.

Normal pacemaker activity results in an average heart rate of 70–80 beats per minute (bpm). A number of clinical problems are the result of abnormal pacemaker function. **Bradycardia** (brād-ē-KAR-dē-a; *bradys*, slow) is the term used to indicate a heart rate that is slower (less than 60 bpm) than normal, whereas **tachycardia** (tak-e-KAR-dē-a; *tachys*, swift) indicates a faster (100 or more bpm) than normal heart rate.

✓ Cardiac muscle does not undergo tetanus as skeletal muscle does. How does that feature affect the functioning of the heart?

✓ If the cells of the SA node were not functioning, what would be the effect on heart rate?

✓ Why is it important for the impulses from the atria to be delayed at the AV node before passing into the ventricles?

The Electrocardiogram

The electrical events occurring in the heart are powerful enough that they can be detected by electrodes on the body surface. A recording of these electrical activities constitutes an **electrocardiogram** (ē-lek-trō-KAR-dē-ō-gram), also called an **ECG** or **EKG.** Each time the heart beats, a wave of depolarization radiates through the atria, reaches the AV node, travels down the interventricular septum to the apex, turns, and spreads through the ventricular myocardium toward the base (Figure 13-9●).

By comparing the information obtained from electrodes placed at different locations, you can check the performance of specific nodal, conducting, and contractile components. For example, when a portion of the heart has been damaged, the affected muscle cells will no longer conduct action potentials, so an ECG will reveal an abnormal pattern of impulse conduction.

The appearance of the ECG tracing varies, depending on the placement of the monitoring electrodes, or *leads*. Figure 13-9● shows the important features of an electrocardiogram as analyzed with the leads in one of the standard configurations.

- The small **P wave** accompanies the depolarization of the atria. The atria begin contracting around 100 msec after the start of the P wave.
- The **QRS complex** appears as the ventricles depolarize. This is a relatively strong electrical signal because the mass of the ventricular muscle is much larger than that of the atria. The ventricles begin contracting shortly after the peak of the R wave.
- The smaller **T wave** indicates ventricular repolarization. You do not see a deflection corresponding to atrial repolarization because it occurs while the ventricles are depolarizing, and the electrical events are masked by the QRS complex.

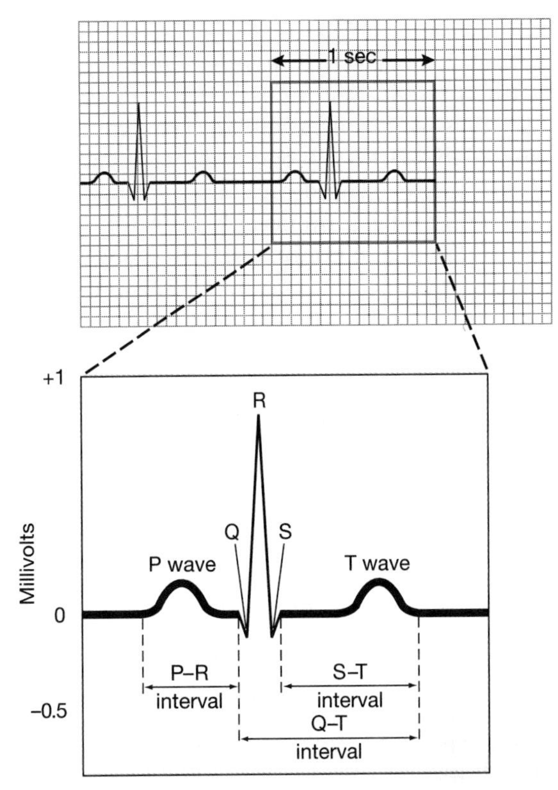

●**Figure 13-9**
An Electrocardiogram
An ECG printout is a strip of graph paper containing a record of the electrical events monitored by electrodes attached to the body surface. The placement of electrodes affects the size and shape of the waves recorded. This is an example of a normal ECG using three electrodes in standard position (left and right wrists, and left lower leg). The enlarged section indicates the major components of the ECG and the measurements most often taken during clinical analysis.

Analyzing an ECG involves measuring the size of the voltage changes and determining the durations and temporal relationships of the various components. Attention usually focuses on the amount of depolarization occurring during the P wave and the QRS complex. For example, a smaller than normal electrical signal may mean that the mass of the heart muscle has decreased, and excessively strong depolarizations may mean that the heart muscle has become enlarged.

ECG analysis is especially useful in detecting and diagnosing **cardiac arrhythmias** (a-RITH-mē-az), abnormal patterns of cardiac activity. Momentary arrhythmias are not inherently dangerous, and about 5 percent of the normal population experiences a few abnormal heartbeats each day. Clinical problems appear when the arrhythmias reduce the pumping efficiency of the heart. Serious arrhythmias may indicate damage to the myocardial musculature, injuries to the pacemakers or conduction pathways, exposure to drugs, or variations in the electrolyte composition of the extracellular fluids. AM *Interpreting Abnormal ECGs*

The Cardiac Cycle

The coordinated contractions of cardiac muscle tissue underlies the approximately 100,000 beats per day of the heart. The period between the start of one heartbeat and the beginning of the next is a single **cardiac cycle.** This cardiac cycle therefore includes alternate periods of contraction and relaxation. For any one chamber in the heart, the cardiac cycle can be divided into two phases. During contraction, or **systole** (SIS-to-lē), the chamber pushes blood into an adjacent chamber or into an arterial trunk. Systole is followed by the second phase, one of relaxation, or **diastole** (dī-AS-to-lē), during which the chamber fills with blood and prepares for the start of the next cardiac cycle.

Fluids will move from an area of high pressure to one of relatively lower pressure. During the cardiac cycle, the pressure within each chamber rises in systole and falls in diastole. An increase in pressure in one chamber will cause the blood to flow to another chamber or vessel where the pressure is lower. The atrioventricular and semilunar valves ensure that blood flows in one direction only during the cardiac cycle.

The correct pressure relationships depend on the careful timing of contractions. The elaborate pacemaking and conduction systems normally provide the required spacing between atrial systole and ventricular systole; if the atria and ventricles were to contract at the same moment, blood could not leave the atria because the AV valves would be closed. In the normal heart, atrial systole and atrial diastole are slightly out of phase with ventricular systole and diastole. Figure 13-10● shows the duration and timing of systole and diastole for a heart rate of 75 bpm (beats per minute).

The cardiac cycle begins with atrial systole. At the start of atrial systole, the ventricles are already filled to around 70 percent of capacity, and atrial systole essentially tops them off by providing the additional 30 percent. As atrial systole ends, ventricular systole begins. As ventricular pressures rise above those in the atrium, the AV valves swing shut. But blood can not begin moving into the arterial trunks until ventricular pressures exceed the arterial pressures. At this point, the blood pushes open the semilunar valves and flows into the aortic and pulmonary trunks. This blood flow continues for the duration of ventricular systole.

When ventricular diastole begins, ventricular pressures decline rapidly, and as they fall below those of the arterial trunks, the semilunar valves close. Ventricular pressures continue to drop, and as they fall below atrial pressures the AV valves open, and blood flows from the atria into the ventricles. Both the atria and the ventricles are now in diastole, blood now flows from the major veins through the relaxed atria and into the ventricles. By the time atrial systole marks the start of another cardiac cycle, the ventricles are roughly 70 percent filled. The relatively minor contribution atrial systole makes to ventricular volume explains why individuals can survive quite normally when their atria have been so severely damaged that they can no longer function. By contrast, damage to one or both ventricles can leave the heart un-

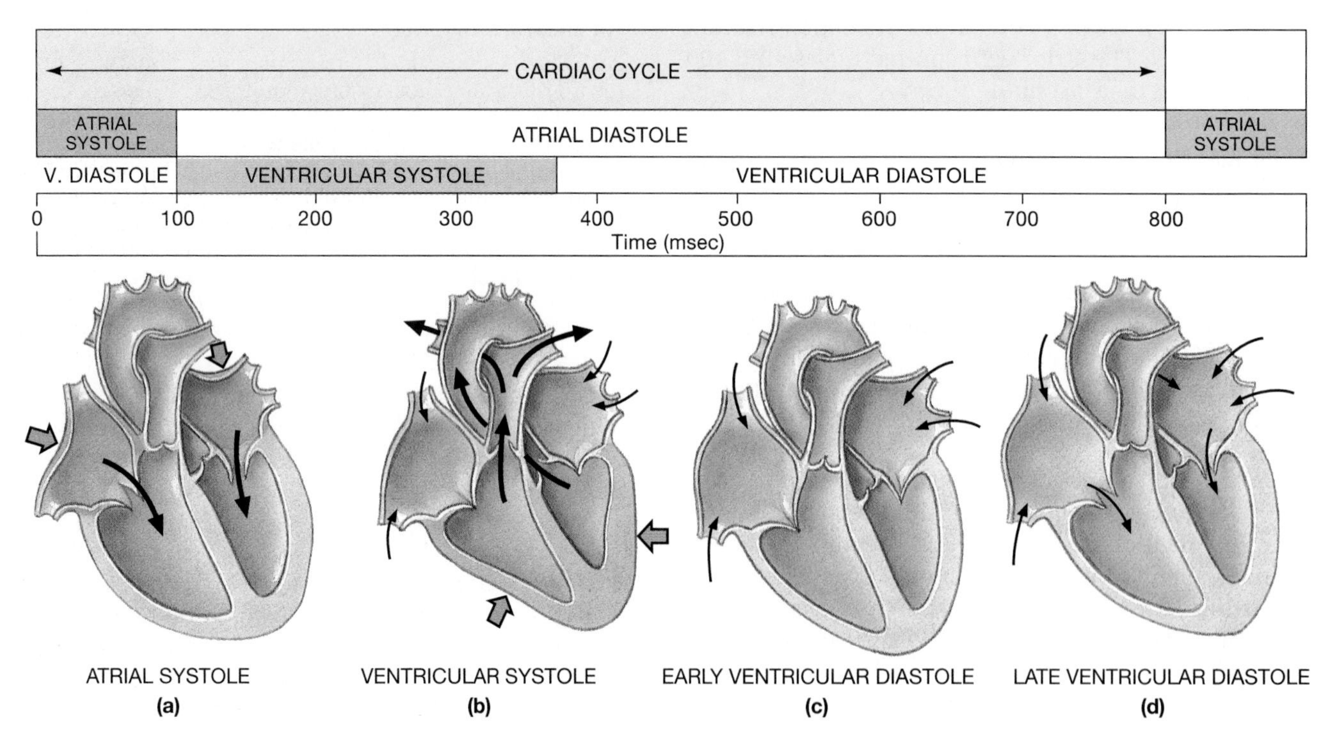

●**Figure 13-10**
The Cardiac Cycle

The atria and ventricles go through repeated cycles of systole and diastole. The timing of systole and diastole differs between the atria and ventricles. A cardiac cycle consists of one period of systole and diastole; we will consider a cardiac cycle as determined by the state of the atria. **(a)** Atrial systole: During this period the atria contract and the ventricles become filled with blood. **(b)** Ventricular systole: Blood is ejected into the pulmonary and aortic trunks. **(c)** Ventricular diastole: Once the AV valves open, passive filling of the ventricles occurs through the period of atrial systole in the next cardiac cycle. **(d)** Condition of the heart at the end of a cardiac cycle, with both the atria and ventricles in diastole.

able to maintain adequate cardiac output. A condition of **heart failure** then exists. AM *Heart Failure*

Heart Sounds

When you listen to your own heart with a stethoscope, you hear the familiar "lubb-dupp" that accompanies each heartbeat. These sounds accompany the action of the heart valves. The first heart sound ("lubb") lasts a little longer than the second. It marks the start of ventricular systole, and the sound is produced as the AV valves close and the semilunars open. The second heart sound, "dupp," occurs at the beginning of ventricular diastole, when the semilunar valves close.

Third and *fourth heart sounds* may be audible as well, but they are usually very faint and are seldom detectable in healthy adults. These sounds are associated with atrial contraction and blood flowing into the ventricles, rather than valve action.

✓ When pressure in the left ventricle is rising is the heart pumping blood? Explain.

✓ Why is it a problem if the heart beats too fast?

HEART DYNAMICS

Heart dynamics refers to the movements and forces generated during cardiac contractions. Each time the heart beats, the two ventricles eject equal amounts of blood. The amount ejected by a ventricle during a single beat is the **stroke volume (SV)**. Since the stroke volume may vary from beat to beat, physicians are often more interested in the **cardiac output (CO)**, or the amount of blood pumped by each ventricle in 1 minute.

Cardiac output can be calculated by multiplying the average stroke volume by the heart rate (HR):

CO	=	SV	×	HR
cardiac output (ml/min)		stroke volume (ml)		heart rate (bpm)

For example, if the average stroke volume is 80 ml and the heart rate is 70 beats per minute (bpm), the cardiac output will be:

$$CO = 80 \text{ ml} \times 70/\text{min} = 5600 \text{ ml/min (5.6 liters per minute)}$$

This represents a cardiac output equivalent to the total volume of blood of an average adult every minute. Cardiac output is highly variable, however, and a normal heart can increase both its rate of contraction and its stroke volume. When both the heart rate and stroke volume increase together, the cardiac output can increase by 600 to 700 percent, or up to 30 liters per minute.

Factors Controlling Cardiac Output

Cardiac output is precisely regulated so that peripheral tissues receive an adequate circulatory supply under a variety of conditions. The major factors that regulate cardiac output often affect both heart rate and stroke volume simultaneously. These primary factors include *blood volume reflexes, autonomic innervation,* and *hormones.* Secondary factors include the concentration of ions in the extracellular fluid and body temperature. Factors affecting cardiac output are diagrammed in Figure 13-11●.

Blood Volume Reflexes

Cardiac muscle contraction is an active process, but relaxation is entirely passive. The force necessary to return cardiac muscle to its precontracted length is provided by the blood pouring into the heart, aided by the elasticity of the fibrous skeleton. As a result, there is a direct relationship between the amount of blood entering the heart and the amount of blood ejected during the next contraction.

Two heart reflexes respond to changes in blood volume. One of these occurs in the right atrium and affects heart rate. The other is a ventricular reflex that affects stroke volume.

The **atrial reflex** (*Bainbridge reflex*) involves adjustments in heart rate that are triggered by an increase in the **venous return,** the flow of venous blood to the heart. When the walls of the right atrium are stretched, the cells of the SA node depolarize faster, and the heart rate increases. This effect is due to (1) the response of the nodal cells to stretching and (2) an increase in sympathetic activity in response to the stimulation of stretch receptors in the atrial walls.

The amount of blood pumped out of the ventricle each beat depends on the venous return and the *filling time.* **Filling time** is the duration of ventricular diastole, the period during which blood can flow into the ventricles. Filling time depends primarily on the heart rate: the faster the heart rate, the shorter the available filling time. Venous return changes in response to alterations in cardiac output, peripheral circulation, or other factors that affect the rate of blood flow through the venae cavae.

Over the range of normal activities, the greater the volume of blood entering the ventricles, the more powerful the contraction will be. The greater the degree of stretching in the walls of the heart, the more force will be developed when a contraction occurs. In a resting individual, the venous return is relatively low, the walls are not stretched significantly, and the ventricles develop little power. If the venous return suddenly increases, more blood flows into the heart, the myocardium stretches further, and the ventricles produce greater force on contraction. This general rule of "more in = more out" is often called the

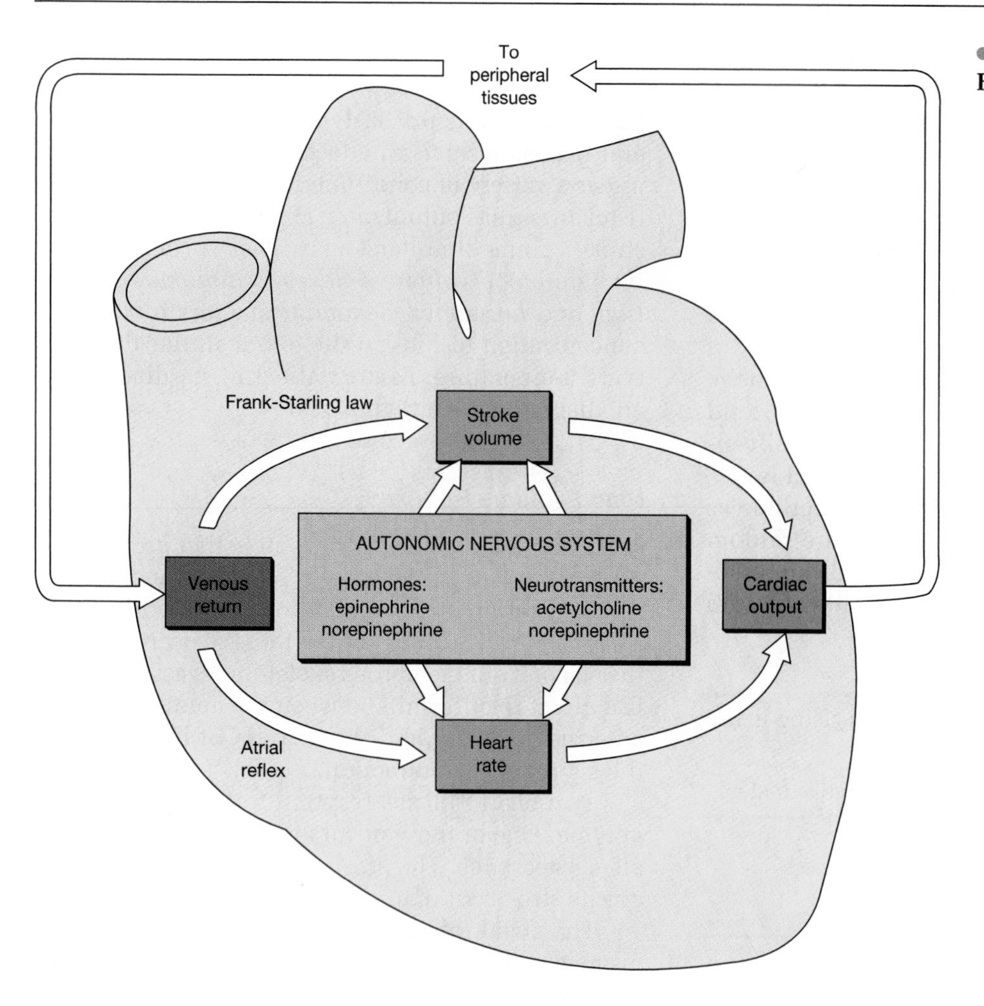

●**Figure 13-11**
Factors Affecting Cardiac Output

Frank-Starling law of the heart, in honor of the physiologists who first demonstrated the relationship.

Autonomic Innervation

The basic heart rate is established by the pacemaker cells of the SA node, but this rate can be modified by the autonomic nervous system (ANS). As shown in Figure 13-12●, both the sympathetic and parasympathetic divisions of the ANS innervate the heart. Postganglionic sympathetic neurons are found in the cervical and upper thoracic ganglia. The vagus nerve carries parasympathetic preganglionic fibers to small ganglia in the cardiac plexus. Both ANS divisions innervate the SA and AV nodes as well the atrial and ventricular cardiac muscle cells.

Autonomic Effects on Heart Rate Autonomic effects on heart rate primarily reflect the responses of the SA node to acetylcholine (ACh) and norepinephrine (NE). ACh released by parasympathetic motor neurons slows the rate of spontaneous depolarization, lowering the heart rate. NE released by sympathetic neurons accelerates the depolarization rate, increasing the heart rate. A more sustained rise in heart rate follows epinephrine (E) and norepinephrine release by the adrenal medullae during sympathetic activation.

Autonomic Effects on Stroke Volume Through the release of norepinephrine (NE), epinephrine (E), and acetylcholine (ACh), the ANS also affects stroke volume by altering the force of myocardial contractions.

- **Effects of NE and E.** Sympathetic release of NE at synapses in the myocardium, and the release of NE and E by the adrenal medullae stimulate cardiac muscle cell metabolism and increase the force and degree of contraction. The result is an increase in stroke volume.
- **Effects of ACh.** The primary effect of parasympathetic ACh release is inhibition, resulting in a decrease in the force of cardiac contractions. Because parasympathetic innervation of the ventricles is relatively limited, the atria show the greatest reduction in contractile force.

Both autonomic divisions are normally active at a steady background level, releasing ACh and NE both at the nodes and into the myocardium. Thus, cutting the vagus nerves increases the heart rate, and sympathetic blocking agents slow the heart rate. Through dual innervation and adjustments in autonomic tone, the ANS can make very delicate adjustments in cardiovascular function.

Coordination of Autonomic Activity The cardiac centers of the medulla oblongata contain the autonomic headquarters for cardiac control. (These centers were introduced in Chapter 8.) ∞ *[p. 224]* Stimulation of the **cardioacceleratory center** activates the necessary sympathetic motor neurons; the nearby **cardioinhibitory center** governs the activities of the parasympathetic motor neurons (Figure 13-12●). Information concerning the status of the cardiovascular system arrives at the cardiac centers over sensory fibers in the vagus and with the sympathetic nerves of the cardiac plexus.

The cardiac centers respond to changes in blood pressure and in the arterial concentrations of dissolved oxygen and carbon dioxide. These properties are monitored by baroreceptors and chemoreceptors innervated by the glossopharyngeal and vagus nerves (N IX and X). In functional terms, a decline in blood pressure or oxygen concentrations or an increase in carbon dioxide levels usually indicates that demands of peripheral tissues have increased. The cardiac centers then call for an increase in the cardiac output, and the heart works harder.

In addition to making automatic adjustments in response to sensory information, the cardiac centers can be influenced by higher centers, especially centers in the hypothalamus. As a result, changes in emotional state (rage, fear, arousal) have an immediate effect on heart rate.

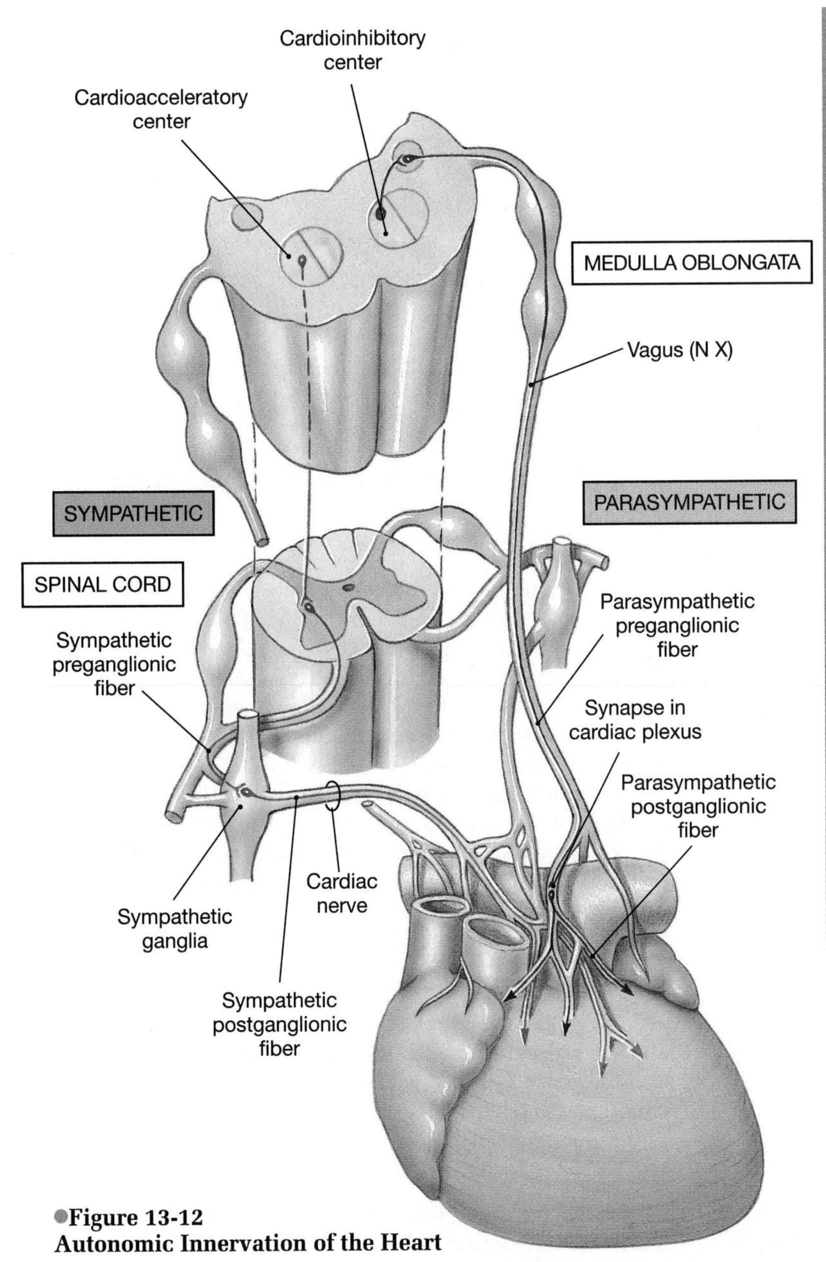

●**Figure 13-12**
Autonomic Innervation of the Heart

EXTRACELLULAR IONS, TEMPERATURE, AND THE HEART

Changes in extracellular calcium ion concentrations primarily affect the strength and duration of cardiac contractions and, thus, stroke volume. If such calcium concentrations are elevated *(hypercalcemia),* cardiac muscle cells become extremely excitable and their contractions become powerful and prolonged. In extreme cases, the heart goes into an extended state of contraction that is usually fatal. When calcium levels are abnormally low *(hypocalcemia),* the contractions become very weak and may cease altogether.

Abnormal extracellular potassium ion concentrations alter the resting potential at the SA node and primarily change the heart rate. When potassium concentrations are high (*hyperkalemia*), cardiac contractions become weak and irregular. When the extracellular concentration of potassium is abnormally low (*hypokalemia*), there is a reduction in heart rate.

Temperature changes affect metabolic operations throughout the body. For example, a lower temperature slows the rate of depolarization at the SA node, lowers the heart rate, and reduces the strength of cardiac contractions. An elevated body temperature accelerates the heart rate and the contractile force, one reason why your heart seems to be racing and pounding whenever you have a fever.

✓ What effect would stimulating the acetylcholine receptors of the heart have on cardiac output?

✓ What effect would a decreased venous return have on the stroke volume?

✓ How would increased sympathetic stimulation of the heart affect stroke volume?

Chapter Review

KEY TERMS

atrium, p. 329
atrioventricular valve, p. 331
cardiac cycle, p. 339
cardiac output, p. 341
diastole, p. 339
electrocardiogram (ECG, EKG), p. 339
endocardium, p. 333
epicardium, p. 333
myocardium, p. 333
pericardium, p. 330
systole, p. 339
ventricle, p. 329

SUMMARY OUTLINE

THE HEART AND THE CIRCULATORY SYSTEM p. 329

1. The circulatory system can be subdivided into the **pulmonary circuit** (which carries blood to and from the lungs) and the **systemic circuit** (which transports blood to and from the rest of the body). **Arteries** carry blood away from the heart; **veins** return blood to the heart. **Capillaries** are tiny vessels between the smallest arteries and veins. (*Figure 13-1*)

2. The heart has four chambers: the **right atrium** and **right ventricle** and the **left atrium** and **left ventricle.**

ANATOMY AND ORGANIZATION OF THE HEART p. 329

1. The heart is surrounded by the **pericardial cavity** (lined by the **pericardium**); the **visceral pericardium (epicardium)** covers the heart's outer surface, and the **parietal pericardium** lines the inner surface of the **pericardial sac** that surrounds the heart. (*Figure 13-2*)

Surface Anatomy of the Heart p. 330

2. A deep groove, the **coronary sulcus,** marks the boundary between the atria and ventricles. (*Figure 13-3*)

Internal Anatomy and Organization p. 331

3. The atria are separated by the **interatrial septum,** and the ventricles are divided by the **interventricular septum.** The right atrium receives blood from the systemic circuit via two large veins, the **superior vena cava** and **inferior vena cava.** (*Figure 13-4*)

4. Blood flows from the right atrium into the right ventricle via the **right atrioventricular (AV) valve (tricuspid valve).** This opening is bounded by three **cusps** of fibrous tissue braced by the tendinous **chordae tendineae** that are connected to **papillary muscles.**

5. Blood leaving the right ventricle enters the **pulmonary trunk** after passing through the **pulmonary semilunar valve.** The pulmonary trunk divides to form the **left** and **right pulmonary arteries.** The **left** and **right pulmonary veins** return blood to the left atrium. Blood leaving the left atrium flows into the left ventricle via the **left atrioventricular (AV) valve (bicuspid valve** or **mitral valve**). Blood leaving the left ventricle passes through the **aortic semilunar valve** and into the systemic circuit via the **ascending aorta.** (*Figure 13-5*)

6. Anatomical differences between the ventricles reflect the functional demands placed on them. The wall of the right ventricle is relatively thin, while the left ventricle has a massive muscular wall.

7. Valves normally permit blood flow in only one direction, preventing the **regurgitation** (backflow) of blood.

The Heart Wall p. 333

8. The bulk of the heart consists of the muscular **myocardium.** The **endocardium** lines the inner surfaces of the heart. The **fibrous skeleton** supports the heart's contractile cells and valves. (*Figures 13-5, 13-6a*)

9. **Cardiac muscle cells** are interconnected by **intercalated discs** that convey the force of contraction from cell to cell and conduct action potentials. (*Figure 13-6c,d*)

The Blood Supply to the Heart p. 336

10. The **coronary circulation** meets the high oxygen and nutrient demands of cardiac muscle cells. The **coronary arteries** originate at the base of the ascending aorta. Interconnections between arteries called **anastomoses** ensure a constant blood supply. The **great** and **middle cardiac veins** carry blood from the coronary capillaries to the **coronary sinus.** (*Figure 13-7*)

THE HEARTBEAT p. 336

1. Two general classes of cardiac cells are involved in the normal heartbeat: *contractile cells* and cells of the conducting system. The conducting system includes **nodal cells,** or **pacemaker cells,** and **conducting cells.**

Contractile Cells p. 337

2. Cardiac muscle cells have a long refractory period, so rapid stimulation produces isolated rather than tetanic contractions.

The Conducting System p. 337

3. The conducting system, composed of nodal cells and conducting cells, initiates and distributes electrical impulses within the heart. Nodal cells establish the rate of cardiac contraction, and conducting cells distribute the contractile stimulus to the general myocardium.

4. Unlike skeletal muscle, cardiac muscle contracts without neural or hormonal stimulation. Pacemaker cells found in the **cardiac pacemaker (sinoatrial [SA] node)** normally establish the rate of contraction. From the SA node the stimulus travels to the **atrioventricular (AV) node,** then to the **AV bundle,** which divides into **bundle branches.** From here **Purkinje cells** convey the impulses to the ventricular myocardium. (*Figure 13-8*)

The Electrocardiogram p. 339

5. A recording of electrical activities in the heart is an **electrocardiogram (ECG** or **EKG)**. Important landmarks of an ECG include the **P wave** (atrial depolarization), **QRS complex** (ventricular depolarization), and **T wave** (ventricular repolarization). (*Figure 13-9*)

The Cardiac Cycle p. 339

6. The **cardiac cycle** consists of **systole** (contraction), followed by **diastole** (relaxation). Both ventricles contract

at the same time and they eject equal volumes of blood. (*Figure 13-10*)

7. The closing of valves and rushing of blood through the heart cause characteristic heart sounds.

HEART DYNAMICS p. 341

1. Heart dynamics refers to the movements and forces generated during contractions. The amount of blood ejected by a ventricle during a single beat is the **stroke volume (SV)**; the amount of blood pumped each minute is the **cardiac output (CO)**.

Factors Controlling Cardiac Output p. 341

2. The major factors that affect cardiac output are blood volume reflexes, autonomic innervation and hormones. (*Figure 13-11*)

3. Blood volume reflexes are stimulated by changes in **venous return,** the amount of blood entering the heart. The **atrial reflex** accelerates the heart rate when the walls of the right atrium are stretched. Ventricular contractions become more powerful and increase stroke volume when the ventricular walls are stretched (the **Frank-Starling law of the heart**).

4. The basic heart rate is established by the pacemaker cells, but it can be modified by the ANS. (*Figure 13-12*)

5. ACh released by parasympathetic motor neurons lowers the heart rate and stroke volume. NE released by sympathetic neurons increases the heart rate and stroke volume.

6. Epinephrine and norepinephrine, hormones released by the adrenal medullae during sympathetic activation, increase both heart rate and stroke volume.

7. The **cardioacceleratory center** in the medulla oblongata activates sympathetic neurons; the **cardioinhibitory center** governs the activities of the parasympathetic neurons. The cardiac centers receive inputs from higher centers and from receptors monitoring blood pressure and the levels of dissolved gases.

CHAPTER QUESTIONS

LEVEL 1 Reviewing Facts and Terms

Match each item in column A with the most closely related item in column B. Use letters for answers in the spaces provided.

Column A	Column B
___ 1. epicardium	a. heart attack
___ 2. right AV valve	b. cardiac pacemaker
___ 3. left AV valve	c. tachycardia
___ 4. anastomoses	d. SV × HR
___ 5. myocardial infarction	e. tricuspid valve
___ 6. SA node	f. bradycardia
___ 7. systole	g. mitral valve
___ 8. diastole	h. interconnections between arteries
___ 9. cardiac output	i. visceral pericardium
___ 10. HR slower than usual	j. increased venous return
___ 11. HR faster than normal	k. contractions of heart chambers
___ 12. atrial reflex	l. relaxation of heart chambers

13. Blood supply to the muscles of the heart is provided by the:
(a) systemic circulation
(b) pulmonary circulation
(c) coronary circulation
(d) coronary portal system

14. The autonomic centers for cardiac function are located in the:
(a) myocardial tissue of the heart
(b) cardiac centers of the medulla oblongata
(c) cerebral cortex
(d) a, b, and c are correct

15. The simple squamous epithelium covering the valves of the heart constitutes the:
(a) epicardium (b) endocardium
(c) myocardium (d) fibrous skeleton

16. The structure that permits blood flow from the right atrium to the left atrium while the lungs are developing is the:
(a) foramen ovale (b) interatrial septum
(c) coronary sinus (d) fossa ovalis

17. Blood leaves the left ventricle by passing through the:
(a) aortic semilunar valve
(b) pulmonary semilunar valve
(c) mitral valve
(d) tricuspid valve

18. The QRS complex of the ECG appears as the:
(a) atria depolarize (b) ventricles depolarize
(c) ventricles repolarize (d) atria repolarize

19. In the cardiac cycle, during diastole the chambers of the heart:
(a) relax and fill with blood
(b) contract and push blood into an adjacent chamber
(c) experience a sharp increase in pressure
(d) reach a pressure of approximately 120 mm Hg

20. What role do the chordae tendineae and papillary muscles have in the normal function of the AV valves?

21. What are the principal valves found in the heart and what is the function of each?

22. Trace the normal pathway of an electrical impulse through the conducting system of the heart.

23. (a) What is the cardiac cycle? (b) What phases and events are necessary to complete the cardiac cycle?

LEVEL 2 Reviewing Concepts

24. Tetanic muscle contractions cannot occur in a normal cardiac muscle cell because:
    (a) cardiac muscle tissue contracts on its own
    (b) there is no neural or hormonal stimulation
    (c) the refractory period lasts until the muscle cell relaxes
    (d) the refractory period ends before the muscle cell reaches peak tension

25. The amount of blood forced out of the heart depends on:
    (a) the degree of stretching at the end of ventricular diastole
    (b) the contractility of the ventricle
    (c) the amount of pressure required to eject blood
    (d) a, b, and c are correct

26. The cardiac output cannot increase indefinitely because:
    (a) available filling time becomes shorter as the heart rate increases
    (b) cardiovascular centers adjust the heart rate
    (c) the rate of spontaneous depolarization decreases
    (d) the ion concentrations of pacemaker cell membranes decrease

27. Describe the association of the four muscular chambers of the heart with the pulmonary and systemic circuits.

28. What are the source and significance of the heart sounds?

29. (a) What effect does sympathetic stimulation have on the heart? (b) What effect does parasympathetic stimulation have on the heart?

LEVEL 3 Critical Thinking and Clinical Applications

30. A patient's ECG tracing shows a consistent pattern of two P waves followed by a normal QRS complex and T wave. What is the cause of this abnormal wave pattern?

31. Karen is taking the medication *verapamil,* a drug that blocks the calcium channels in cardiac muscle cells. What effect would you expect this medication to have on Karen's stroke volume?

14 Blood Vessels and Circulation

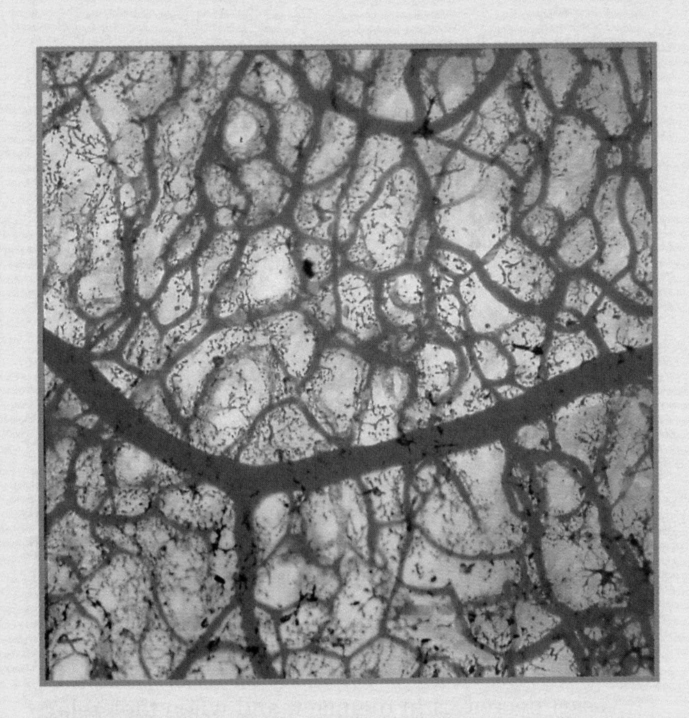

When we think of the cardiovascular system, we probably think first of the heart or of the great blood vessels that leave it and return to it. But the real work of the cardiovascular system is done here, in microscopic vessels that permeate most tissues. This is a network of capillaries, the delicate vessels that permit diffusion between the blood and the interstitial fluid.

Chapter Outline and Objectives

1 Distinguish among the types of blood vessels on the basis of their structure and function.

2 Explain the mechanisms that regulate blood flow through arteries, capillaries, and veins.

3 Discuss the mechanisms and various pressures involved in the movement of fluids between capillaries and interstitial spaces.

4 Describe the factors that influence blood pressure and how blood pressure is regulated.

5 Describe how central and local control mechanisms interact to regulate blood flow and pressure in tissues.

6 Explain how the activities of the cardiac, vasomotor, and respiratory centers are coordinated to control blood flow through the tissues.

7 Explain how the circulatory system responds to the demands of exercise and hemorrhaging.

8 Identify the major arteries and veins and the areas they serve.

9 Describe the age-related changes that occur in the cardiovascular system.

10 Discuss the structural and functional interactions among the cardiovascular system and other body systems.

Vocabulary Development

alveolus, sac; *alveoli*
baro-, pressure; *baroreceptors*
manometer, a device for measuring pressure; *sphygmomanometer*
porta, a gate; *portal vein*
pulmo-, lung; *pulmonary*
pulsus, stroke; *pulse*
saphenes, prominent; *saphenous vein*
skleros, hard; *arteriosclerosis*
sphygmos, pulse; *sphygmomanometer*
vaso-, vessel; *vasoconstriction*

The last two chapters examined the composition of blood and the structure and function of the heart, whose pumping action keeps blood in motion. We will now consider the vessels that carry blood to peripheral tissues and the nature of the exchange that occurs between the blood and interstitial fluids of the body.

Blood leaves the heart in the pulmonary and aortic trunks, each with a diameter of around 2.5 cm (1 in.). These vessels branch repeatedly, forming the major arteries that distribute blood to body organs. Within these organs further branching occurs, creating several hundred million tiny arteries that provide blood to more than 10 billion **capillaries** barely the diameter of a single red blood cell. These capillaries form extensive, branching networks: If all of the capillaries in the body were placed end to end they would circle the globe, with a combined length of 25,000 miles.

The vital functions of the cardiovascular system depend entirely on events at the capillary level: *All chemical and gaseous exchange between the blood and interstitial fluid takes place across capillary walls.* Tissue cells rely on capillary diffusion to obtain nutrients and oxygen and to remove metabolic wastes, such as carbon dioxide and urea.

This chapter considers the structural organization of the arteries, veins, and capillaries. We will then consider their functions and basic principles of cardiovascular regulation. The final section of the chapter examines the distribution of major blood vessels of the body.

THE ANATOMY OF BLOOD VESSELS

Blood flows to and from the lungs and other body organs through tubelike arteries and veins, with the heart providing the necessary propulsion. The large-diameter **arteries** that carry blood away from the heart branch repeatedly and gradually decrease in size until they become **arterioles** (ar-TĒ-rē-ōlz), the smallest vessels of the arterial system. From the arterioles, blood enters the capillary networks that service local tissues.

Blood flowing out of the capillary complex first enters the **venules** (VEN-ūlz), the smallest vessels of the venous system. These slender vessels subsequently merge with their neighbors to form small **veins**. Blood then passes through medium-sized and large veins before reaching the venae cavae (in the systemic circuit) or the pulmonary veins (in the pulmonary circuit) (see Figure 13-1●, p. 329).

Structure of Vessel Walls

The walls of arteries and veins contain three distinct layers (Figure 14-1●):

1. The **tunica interna** (in-TER-na), or *tunica intima*, is the innermost layer of a blood vessel. It includes the endothelial lining of the vessel and an underlying layer of connective tissue dominated by elastic fibers.
2. The **tunica media**, the middle layer, contains smooth muscle tissue in a framework of collagen and elastic fibers. When these smooth muscles contract, the vessel decreases in diameter, and when they relax, the diameter increases.
3. The outer **tunica externa** (eks-TER-na), or *tunica adventitia* (ad-ven-TISH-ē-a), forms a sheath of connective tissue around the vessel. Its collagen fibers may intertwine with those of adjacent tissues, stabilizing and anchoring the blood vessel.

The multiple layers in their walls give arteries and veins considerable strength, and the muscular and elastic components permit controlled alterations in diameter as blood pressure or blood volume changes. Because the blood is under pressure, a weakness in the wall can lead to the rupture of the vessel, like a blowout in an old tire. AM *Aneurysms*

Arteries and veins often lie side by side in a narrow band of connective tissue, as in Figure 14-1●. The figure clearly shows the greater wall thickness characteristic of arteries. The thicker tunica media of an artery contains more smooth muscle and elastic fibers than does that of a vein. These contractile and elastic components resist the pressure generated by the heart as it forces blood into the arterial network.

Arteries

In traveling from the heart to the capillaries, blood passes through *elastic arteries*, *muscular arteries*, and *arterioles*. **Elastic arteries** are large, extremely resilient vessels with diameters of up to 2.5 cm (1 in.). Some examples are the pulmonary and aortic trunks and their major arterial branches. Their relatively thin walls contain a tunica media dominated by elastic fibers rather than smooth muscle cells. As a result, elastic arteries are able to absorb the pressure shock generated in systole, when the ventricles contract and blood leaves the heart. Over this period, when pressures are rising quickly, elastic arteries stretch rather than break. During ventricular diastole, arterial blood pressure declines, and the elastic fibers recoil to their original dimensions. The net result is that arterial expansion cushions the rise in pressure during ventricular systole, and arterial recoil slows the decline in pressure during ventricular diastole. If the arteries were solid pipes, rather than elastic tubes, pressures would rise much higher during systole, and they would fall much lower during diastole.

Muscular arteries, also known as *medium-sized arteries* or *distribution arteries*, distribute blood to peripheral organs. A typical muscular artery has a diameter of approximately 0.4 cm (0.15 in.). The carotid artery of the neck is one example. The thick tunica media in a muscular artery contains more smooth muscle and fewer elastic fibers than an elastic artery.

Arterioles are much smaller than muscular arteries, with an average diameter of about 30 µm. The tunica media of an arteriole of this size consists of one to three layers of smooth muscle fibers. The smooth muscle fibers within the walls of muscular arteries and arterioles enable these vessels to change their diameter, thereby altering the blood pressure and the rate of flow through the dependent tissues.

Capillaries

Capillaries are the only blood vessels whose walls permit exchange between the blood and the surrounding interstitial fluid. Because the walls are relatively thin, the diffusion distances are small and exchange can occur quickly. In addition, blood flows through capillaries relatively slowly, allowing sufficient time for diffusion or active transport of materials across the capillary walls.

A typical capillary consists of a single layer of endothelial cells inside a delicate basement membrane. The average diameter of a capillary is a mere 8 µm, very close to that of a single red blood cell. In most regions the endothelium forms a complete lining, and most substances enter or leave the capillary by diffusing through gaps between adjacent endothelial cells. In some areas, notably the choroid plexus of the brain, the hypothalamus, and filtration sites at the kidneys, small pores in the endothelial cells permit the passage of relatively large molecules, including proteins.

Capillary Beds

Capillaries function as part of an interconnected network called a **capillary bed**, simplified in Figure 14-2a●. Upon reaching its target area a single arteriole usually gives rise to dozens of capillaries that will in turn collect into several *venules*, the smallest vessels of the venous system. The entrance to each capillary is guarded by a band of smooth muscle, the **precapillary sphincter**. Contraction of the smooth muscle fibers narrows the diameter of the capillary entrance and reduces the flow of

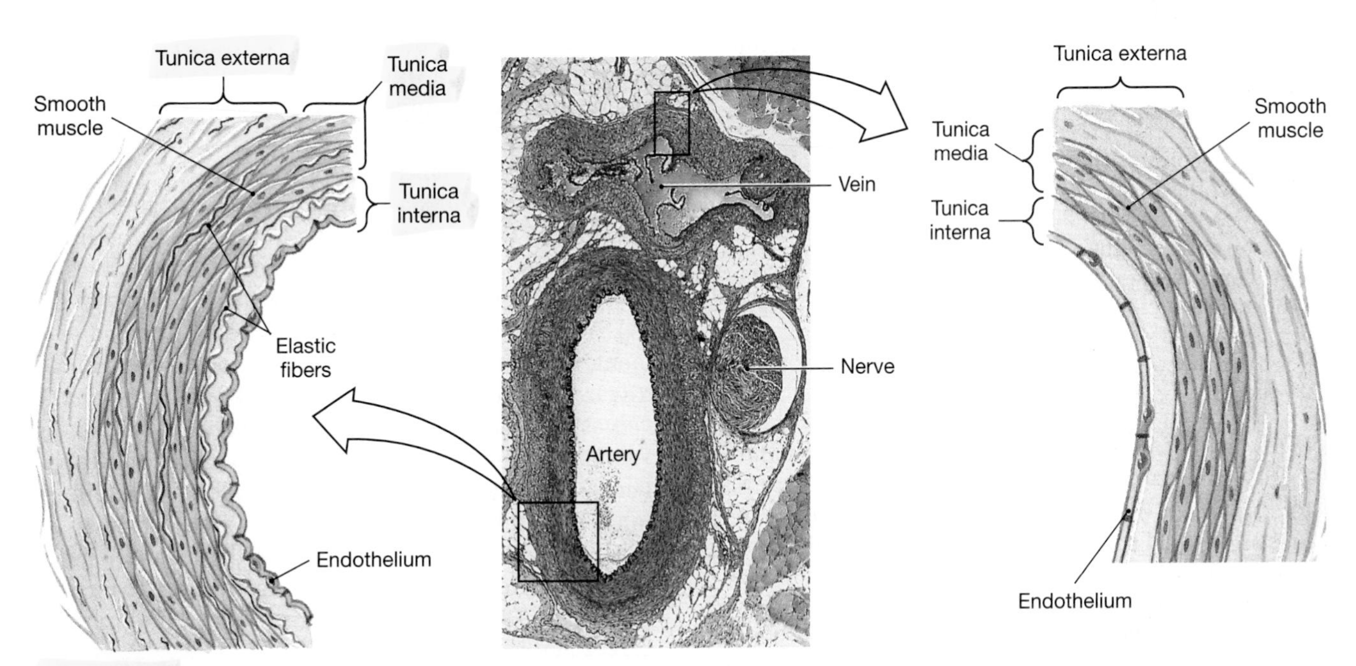

●**Figure 14-1**
A Comparison of a Typical Artery and a Typical Vein
(LM × 74)

Arteriosclerosis

Arteriosclerosis (ar-tē-rē-ō-skle-RŌ-sis; *skleros*, hard) is a thickening and toughening of arterial walls. Although this may not sound life-threatening, complications related to arteriosclerosis account for roughly one-half of all deaths in the United States. There are many different forms of arteriosclerosis; for example, arteriosclerosis of coronary vessels is responsible for *coronary artery disease* (CAD), and arteriosclerosis of arteries supplying the brain can lead to strokes. ∞ [p. 336]

Atherosclerosis (ath-er-ō-skle-RŌ-sis; *athero-*, a pasty deposit) is a type of arteriosclerosis characterized by changes in the endothelial lining. Many factors may be involved in the development of atherosclerosis. One major factor is lipid levels in the blood. Atherosclerosis tends to develop in persons whose blood contains elevated levels of plasma lipids, specifically cholesterol. Circulating cholesterol is transported to peripheral tissues in lipoproteins, protein-lipid complexes. (The various types of lipoproteins and their interrelationships are discussed in Chapter 18.) When cholesterol-rich lipoproteins remain in circulation for an extended period, circulating monocytes then begin removing them from the bloodstream. Eventually the monocytes become filled with lipid droplets, and they attach themselves to the endothelial walls of blood vessels. These cells then release growth factors that stimulate the divisions of smooth muscle fibers near the tunica interna, and the vessel wall thickens and stiffens.

Other monocytes then invade the area, migrating between the endothelial cells. As these changes occur, the monocytes, smooth muscle fibers, and endothelial cells begin phagocytizing lipids as well. The result is a **plaque**, a fatty mass of tissue that projects into the lumen of the vessel. At this point, the plaque has a relatively simple structure, and there is evidence that the process can be reversed, if appropriate dietary adjustments are made.

If the conditions persist, the endothelial cells become swollen with lipids, and gaps appear in the endothelial lining. Platelets now begin sticking to the exposed collagen fibers, and the combination of platelet adhesion and aggregation leads to the formation of a localized blood clot that will further restrict blood flow through the artery. The structure of the plaque is now relatively complex; plaque growth may be halted, but the structural changes are permanent.

Elderly individuals, especially elderly men, are most likely to develop atherosclerotic plaques. In addition to advanced age and male sex, other important risk factors include high blood cholesterol levels, high blood pressure, and cigarette smoking. Other factors that may promote development of atherosclerosis in both men and women include diabetes mellitus, obesity, and stress. For more information on the causes and treatment of arteriosclerosis, see the *Applications Manual*. *Arteriosclerosis*

blood. Relaxation of the sphincter dilates the opening, allowing blood to enter the capillary more rapidly.

Although blood usually flows from the arterioles to the venules at a constant rate, the blood flow within a single capillary can be quite variable. Each precapillary sphincter goes through cycles of activity, alternately contracting and relaxing perhaps a dozen times each minute. As a result of this cyclical change, called **vasomotion**, the blood flow within any one capillary occurs in a series of pulses rather than as a steady and constant stream. The net effect is that blood may reach the venules by one route now and by a quite different route later. This process is regulated at the tissue level, as smooth muscle fibers respond to local changes in the composition of the interstitial fluid. This regulation at the tissue level is called *autoregulation*.

Under certain conditions, the blood will completely bypass the capillary bed through an *arteriovenous anastomosis* (a-nas-tō-MŌ-sis; anastomosis, outlet), a vessel that connects an arteriole to a venule (Figure 14-2a●).

Veins

Veins collect blood from all tissues and organs and return it to the heart. Veins are classified on the basis of their internal diameters. The smallest, the venules, resemble expanded capillaries, and venules with diameters smaller than 50 μm lack a tunica media altogether. **Medium-sized veins** range from 2 to 9 mm in diameter. In these veins, the tunica media contains several smooth muscle layers, and the relatively thick tunica externa has longitudinal bundles of elastic and collagen fibers. **Large veins** include the two venae cavae and their tributaries within the abdominopelvic and thoracic cavities. In these vessels, the thin tunica media is surrounded by a thick tunica externa composed of elastic and collagenous fibers.

Veins have relatively thin walls because they do not have to withstand much pressure. In venules and medium-sized veins, the pressure is so low that it cannot oppose the force of gravity. In the limbs, medium-sized veins contain **valves** that act like the valves in the heart, preventing the backflow of blood (Figure 14-3●). As long as the valves function normally, any movement that distorts or compresses a vein will push blood toward the heart. If the walls of the veins near the valves weaken, or become stretched and distorted, the valves may not work properly. Blood then pools in the veins, and the vessels become distended. The effects range from mild discomfort and a cosmetic problem, as in superficial *varicose veins* in the thighs and legs, to painful distortion of adjacent tissues, as in *hemorrhoids*. *Problems with Venous Valve Function*

✓ Several small, thin-walled vessels have very little smooth muscle tissue in the tunica media. What type of vessels are these? *(cont.)*

✓ How would relaxation of the precapillary sphincters affect the blood flow through a tissue?

✓ Why are valves found in veins but not in arteries?

CIRCULATORY PHYSIOLOGY

The components of the cardiovascular system (the blood, heart, and blood vessels) are functionally integrated to maintain an adequate blood flow through peripheral tissues and organs. Under normal circumstances, blood flow is equal to cardiac output. When cardiac output goes up, so does capillary blood flow; when cardiac output declines, blood flow is reduced. Two factors, *pressure* and *resistance*, affect the flow rates of blood through the capillaries.

Pressure

Liquids, including blood, will flow from an area of higher pressure toward an area of relatively lower pressure. The flow rate is directly proportional to the pressure difference; the greater the difference in pressure, the faster the flow. In the systemic circuit, the overall pressure difference is measured between the base of the ascending aorta (as blood leaves the left ventricle) and the entrance to the right atrium (as it returns). This pressure difference, called the *circulatory pressure*, averages around 100 mm Hg (mm Hg = millimeters of mercury, a standardized unit of pressure). This relatively high circulatory pressure is needed primarily to force blood through the arterioles and into the capillaries.

Circulatory pressure is often divided into three components: (1) *arterial pressure*, (2) *capillary pressure*, and (3) *venous pressure*. The term **blood pressure** will be used when referring to arterial pressure, rather than to the total circulatory pressure.

Resistance

A *resistance* is a force that opposes movement. For circulation to occur, the circulatory pressure must be greater than the *total peripheral resistance*, the resistance of the entire circulatory system. Because the resistance of the venous system is very low, attention focuses on the **peripheral resistance**, the resistance of the arterial system.

Neural and hormonal control mechanisms regulate blood pressure, keeping it relatively stable. Adjustments in the peripheral resistance of vessels supplying specific organs allow the rate of blood flow to be precisely controlled. For example, Chapter 7 dis-

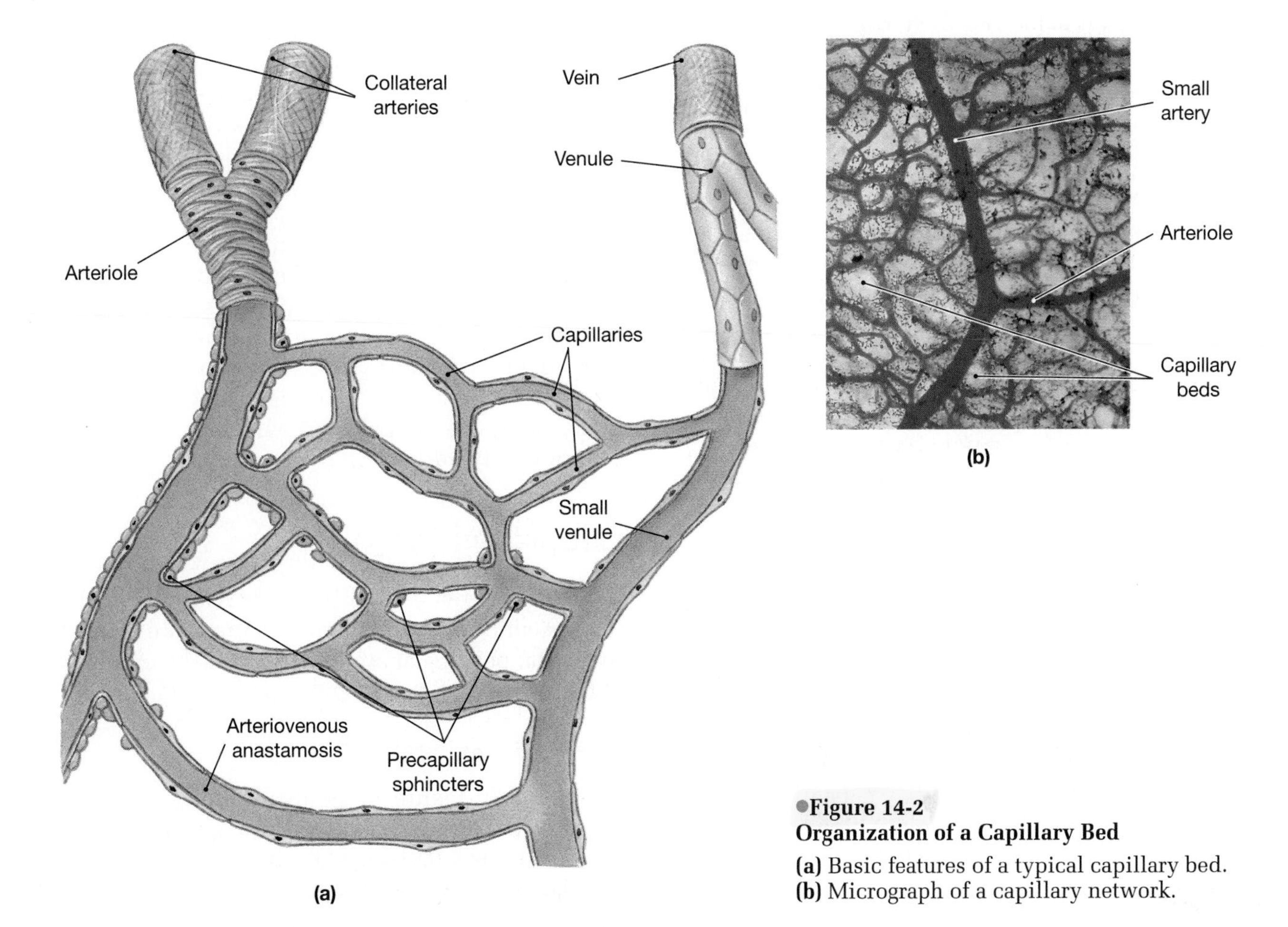

•Figure 14-2
Organization of a Capillary Bed
(a) Basic features of a typical capillary bed.
(b) Micrograph of a capillary network.

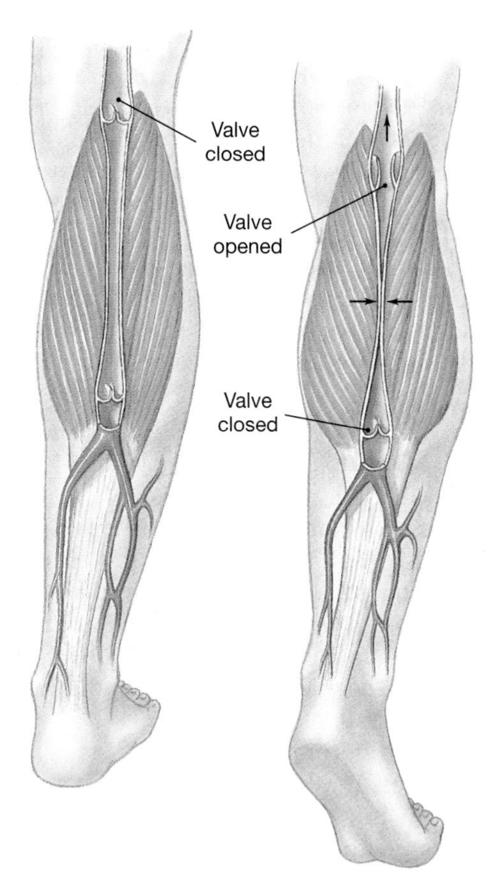

Figure 14-3
Function of Valves in the Venous System
Valves in the walls of medium-sized veins prevent the backflow of blood. Venous compression caused by the contraction of adjacent skeletal muscles helps maintain venous blood flow.

cussed the increase in blood flow to skeletal muscles during exercise. That increase occurs because there is a drop in the peripheral resistance of the arteries supplying active muscles.

Sources of peripheral resistance include *vascular resistance*, *viscosity*, and *turbulence*. Of these, only vascular resistance can be adjusted by the nervous or endocrine system to regulate blood flow. Viscosity and turbulence, which contribute to peripheral resistance, are normally constant.

Vascular Resistance

Vascular resistance is the resistance of the blood vessels, and it is the largest component of peripheral resistance. *The most important factor in vascular resistance is friction between the blood and the vessel walls.* The amount of friction depends on the diameter of the vessel and its length. Vessel length is constant, and vascular resistance is controlled by changing the diameter of blood vessels by contracting or relaxing smooth muscle in the vessel walls.

Large arteries such as the *aorta*, *brachiocephalic artery*, or *carotid artery* contribute little to the peripheral resistance, and most of the resistance occurs in the arterioles and capillaries. As noted earlier in the chapter, arterioles are extremely muscular; an arteriole 30 μm in diameter is wrapped in a 20 μm thick layer of smooth muscle. Local, neural, and hormonal stimuli that stimulate or inhibit the arteriolar smooth muscle tissue can adjust the diameters of these vessels, and a small change in diameter can produce a very large change in resistance.

Viscosity

Viscosity is resistance to flow. It is caused by interactions among molecules and suspended materials in a liquid. Liquids of low viscosity, such as water, flow at low pressures, whereas thick, syrupy liquids such as molasses flow only under relatively high pressures. Whole blood has a viscosity about five times that of water, because of the presence of plasma proteins and suspended blood cells. Under normal conditions, the viscosity of the blood remains stable, but disorders that affect the hematocrit or the plasma protein content can change blood viscosity and increase or decrease peripheral resistance.

Turbulence

Blood flow through a vessel is usually smooth, with the slowest flow near the walls and the greatest speed at the center of the vessel. High flow rates, irregular surfaces, or sudden changes in vessel diameter upset this smooth flow, creating eddies and swirls. This phenomenon, called *turbulence*, slows the rate of flow and increases resistance.

Turbulence normally occurs when blood flows between the chambers of the heart and from the heart into the aortic and pulmonary trunks. In addition to increasing resistance, this turbulence generates the *third* and *fourth heart sounds* that can often be heard through a stethoscope. Turbulent blood flow across damaged or misaligned heart valves is resonsible for the sound of *heart murmurs.* ∞ [p. 333]

Circulatory Pressure

Figure 14-4● details the blood pressure throughout the systemic circuit. Systemic pressures are highest in the aorta, peaking at around 120 mm Hg, and lowest at the venae cavae, averaging about 2 mm Hg.

Arterial Blood Pressure

The graph in Figure 14-4● also indicates that blood pressure in large and small arteries rises and falls, rising during ventricular systole and falling during ventricular diastole. **Systolic pressure** is the peak blood

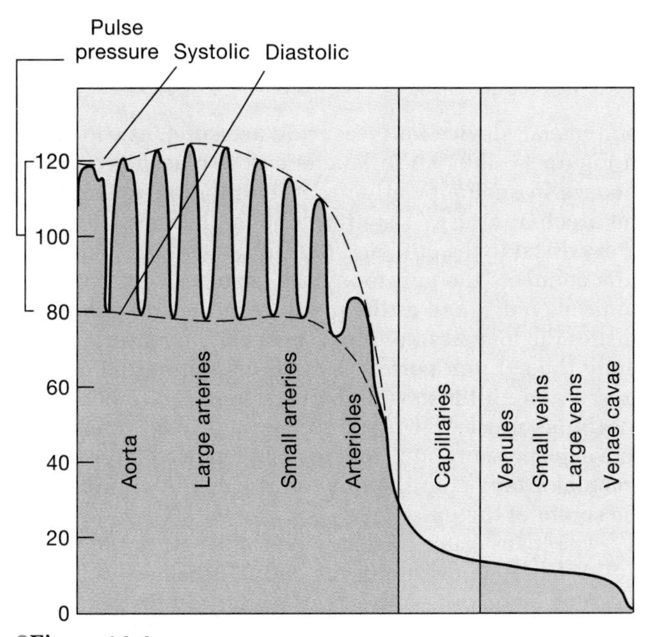

●**Figure 14-4**
Pressures within the Circulatory System

Notice the general reduction in circulatory pressures within the systemic circuit and the elimination of the pulse pressure within the arterioles.

pressure measured during ventricular systole, and **diastolic pressure** is the minimum blood pressure at the end of ventricular diastole. The difference between the systolic and diastolic pressures is the **pulse pressure** (*pulsus*, stroke).

The pulse pressure becomes smaller as the distance from the heart increases. As already discussed, the average pressure declines because of friction between the blood and the vessel walls. The pulse pressure fades because arteries are elastic tubes rather than solid pipes. Much like a puff of air expands a partially inflated balloon, the elasticity of the arteries allows them to expand with blood during systole. Because the aortic semilunar valve prevents the return of blood to the heart during diastole, this blood is pushed along its way by the recoil of the arteries. The magnitude of this phenomenon, called **elastic rebound**, is greatest near the heart and fades in succeeding arterial sections. By the time blood reaches a precapillary sphincter, there are no pressure oscillations, and the blood pressure remains steady at about 35 mm Hg. Along the length of a typical capillary, blood pressure gradually falls from about 35 mm Hg to roughly 18 mm Hg, the pressure at the start of the venous system. *Hypertension and Hypotension*

Capillary Pressures and Capillary Dynamics

The blood pressure within a capillary, or **capillary pressure**, pushes against the capillary walls, just as it does in the arteries. But unlike other portions of the circulatory system, capillary walls are quite permeable to small ions, nutrients, organic wastes, dissolved gases, and water.

Capillary Exchange Since Chapter 3 described the forces that move water and solutes across membranes, only a brief overview is provided here. ∞ *[p p. 50–55]* Solute molecules tend to diffuse across the capillary lining, driven by their individual concentration gradients. Water-soluble materials, including ions and small organic molecules such as glucose, amino acids, or urea, diffuse through small spaces between adjacent endothelial cells. Larger water-soluble molecules, such as plasma proteins, cannot normally leave the bloodstream. Lipid-soluble materials, including steroids, fatty acids, and dissolved gases, diffuse across the endothelial lining, passing through the membrane lipids.

Water molecules will move when driven by either *hydrostatic pressure* or *osmotic pressure*. Hydrostatic pressure is a physical force that pushes water molecules from an area of high pressure to an area of lower pressure. At a capillary, the hydrostatic pressure, or capillary blood pressure (*BP*), is greatest at the arteriolar end and least where it empties into a venule. The tendency for water and solutes to move out of the blood is therefore greatest at the start of a capillary, where the capillary pressure is highest, and declines along the length of the capillary as capillary pressure falls.

Osmosis, by contrast, is the movement of water across a semipermeable membrane separating two solutions of different solute concentrations. Water will move into the solution with the higher solute concentration, and the force of this water movement is called *osmotic pressure*. ∞ *[p. 52]* Because blood contains more dissolved proteins than the interstitial fluid ∞ *[p. 310]*, water tends to move from the interstitial fluid into the blood (Figure 14-6●). Thus, capillary blood pressure tends to push water *out of* the capillary, while capillary osmotic pressure (*COP*) forces tends to pull it back *in*.

Normally, hydrostatic and osmotic forces are not in balance along the length of a capillary, and the arterioles deliver more fluid to the capillaries than the venules carry away (Figure 14-6●). The water and dissolved materials that leave the plasma flow through the tissues and eventually enter a network of **lymphatic vessels** that drain into the venous system. (This topic will be discussed further in Chapter 15.)

CAPILLARY DYNAMICS AND BLOOD VOLUME

Any condition that affects hydrostatic or osmotic pressures in the blood or tissues will shift the direction of fluid movement. For example, consider what happens at the capillary level to an accident victim who has lost blood through hemorrhaging or a person lost in

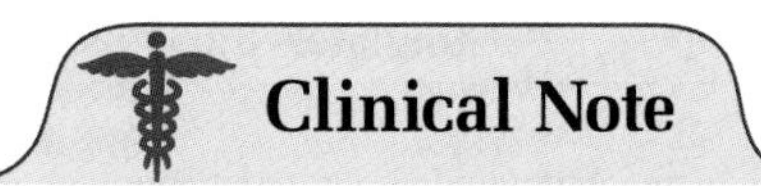

Clinical Note

Checking the Pulse and Blood Pressure

The pulse can be felt within any of the large or medium-sized arteries. The usual procedure involves squeezing an artery with the fingertips against a relatively solid mass, preferably a bone. When the vessel is compressed, the pulse is felt as a pressure against the fingertips.

Figure 14-5a● indicates the locations used to check the pulse. The inside of the wrist is often used because the *radial artery* can easily be pressed against the distal portion of the radius. Other accessible arteries include the *temporal, facial, carotid, brachial, femoral, popliteal,* and *posterior tibial* arteries. Firm pressure exerted at one of these arteries near the base of a limb can reduce or eliminate arterial bleeding distal to the site; the locations are called *pressure points.*

Blood pressure is determined with a *sphygmomanometer* (sfig-mō-ma-NOM-e-ter; *sphygmos*, pulse + *manometer*, device for measuring pressure), as shown in Figure 14-5b●. An inflatable cuff is placed around the arm in such a position that its inflation squeezes the brachial artery. A stethoscope is placed over the artery distal to the cuff, and the cuff is then inflated. A tube connects the cuff to a glass chamber containing liquid mercury, and as the pressure in the cuff rises it pushes the mercury up into a vertical column. A scale along the column permits one to determine the cuff pressure in millimeters of mercury (mm Hg). Inflation continues until cuff pressure is roughly 30 mm Hg above the pressure sufficient to completely collapse the brachial artery, stop the flow of blood, and eliminate the sound of the pulse.

The investigator then slowly lets the air out of the cuff. When the pressure in the cuff falls below systolic pressure, blood can again enter the artery. At first, blood enters only at peak systolic pressures, and the stethoscope picks up the sound of blood pulsing through the artery. As the pressure falls further, the sound changes because the vessel is remaining open for longer and longer periods. When the cuff pressure falls below diastolic pressure, blood flow becomes continuous and the sound of the pulse becomes muffled or disappears completely. Thus the pressure at which the pulse appears corresponds to the peak systolic pressure; when the pulse fades, the pressure has reached diastolic levels. The distinctive sounds heard during this test are called **sounds of Korotkoff** (sometimes spelled *Korotkov* or *Korotkow*). When the blood pressure is recorded, systolic and diastolic pressures are usually separated by a slashmark, as in "120/80" ("one twenty over eighty") or "110/75." A reading of 120/80 would give a pulse pressure of 40 (mm Hg).

Temporal artery
Facial artery
Brachial artery
Radial artery
Femoral artery
Popliteal artery
Posterior tibial artery
Dorsalis pedis artery

(a)

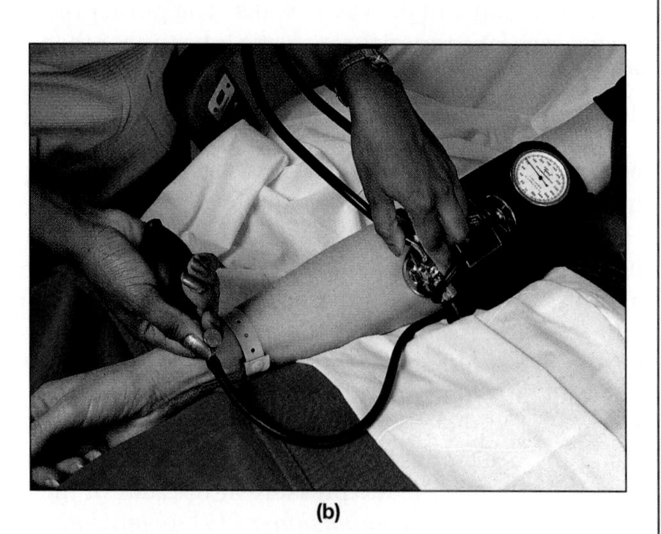

(b)

●**Figure 14-5**
Checking the Pulse and Blood Pressure

(a) Pressure points used to check the presence and strength of the pulse. **(b)** Use of a sphygmomanometer to check arterial blood pressure.

the desert, feeling the effects of dehydration. In both cases, the decrease in blood volume causes a drop in blood pressure. In the second case, however, the loss in blood volume is also accompanied by a rise in the blood osmotic pressure, because as water is lost the blood becomes more concentrated. In either case, a net movement of water into the bloodstream will occur, and the blood volume will increase at the expense of interstitial fluids.

On the other hand, *edema* (e-DĒ-ma) is an abnormal accumulation of interstitial fluid in the tissues. Although it has many different causes, the underlying problem is a disturbance in the normal balance between hydrostatic and osmotic forces at the capillary level. For example, a localized edema often occurs around a bruise. The fluid shift occurs because damaged capillaries at the injury site allow plasma proteins into the interstitial fluid. This leakage decreases the osmotic pressure of the blood and elevates that of the tissues. More water then moves into the tissue, resulting in edema. In the U.S. population, serious cases of edema most often result from an increase in the blood pressure in the arterial system, the venous system, or both. This condition often occurs during *congestive heart failure.* AM *Heart Failure*

Venous Pressure

Although blood pressure at the start of the venous system is only about one-tenth that at the start of the arterial system, the blood must still travel through a vascular network as complex as the arterial system before returning to the heart. However, venous pressures are low, and the veins offer little resistance. As a result once blood enters the venous system, pressure declines very slowly. As blood travels through the venous system toward the heart, the veins become larger, resistance drops further, and the flow rate increases. Pressures at the entrance to the right atrium fluctuate, but they average around 2 mm Hg. This means that the driving force pushing blood through the venous system is a mere 16 mm Hg (18 mm Hg in the venules –2 mm Hg in the venae cavae = 16 mm Hg) as compared to the 85 mm Hg pressure acting along the arterial system (120 mm Hg at the aorta –35 mm Hg at the capillaries).

When you are lying down, a 16 mm Hg pressure gradient is sufficient to maintain venous flow, but when you are standing the venous blood from the body below the heart must overcome gravity as it ascends within the inferior vena cava. Two factors help to overcome gravity and propel venous blood toward the heart:

1. **Muscular compression.** The contractions of skeletal muscles near a vein compress it, helping to push blood toward the heart. The valves in medium-sized veins ensure that blood flow occurs in one direction only (see Figure 14-3●, p. 352).
2. **The respiratory pump.** During inspiration, decreased pressure in the thoracic cavity draws air into the lungs. This drop in pressure also pulls blood into the venae cavae and atria, increasing venous return. On exhalation, the increased pressure that forces air out of the lungs compresses the venae cavae, pushing blood into the right atrium.

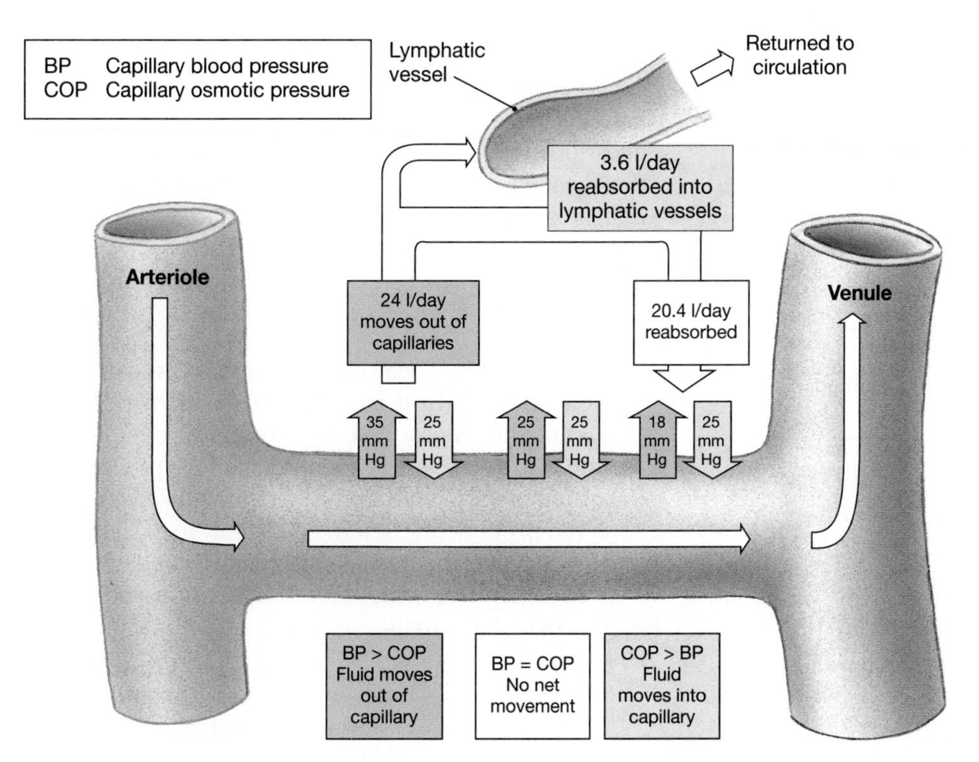

●**Figure 14-6**
Forces Acting across Capillary Walls

At the arterial end of the capillary, blood hydrostatic pressure (BP) is stronger than capillary osmotic pressure (COP), and fluid moves out of the capillary. Near the venule, BP is lower than COP, and fluid moves into the capillary.

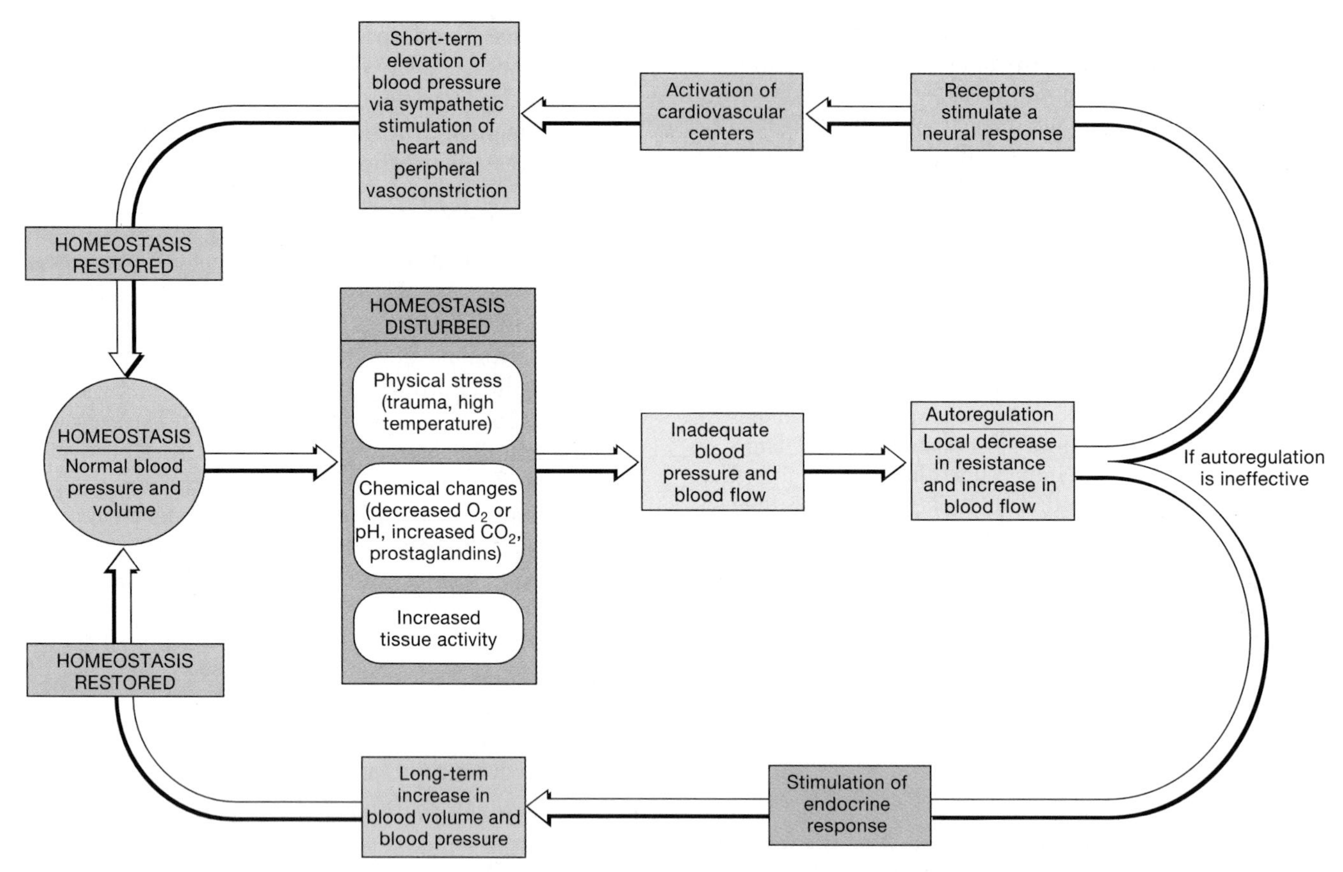

•Figure 14-7
Local, Neural, and Endocrine Adjustments That Maintain Blood Pressure and Blood Flow

During exercise, both factors cooperate to elevate venous return and push cardiac output to maximal levels. However, when an individual stands at attention, with knees locked and leg muscles immobile, these mechanisms are impaired and the reduction in venous return leads to a fall in cardiac output. This in turn reduces the blood supply to the brain, sometimes enough to cause *fainting*, a temporary loss of consciousness. The person then collapses, and in the horizontal position both venous return and cardiac output return to normal.

✓ In a normal individual, where would you expect the blood pressure to be greater, in the aorta or the inferior vena cava? Explain.

✓ While standing in the hot sun, Sally begins to feel light-headed and faints. Explain.

CARDIOVASCULAR REGULATION

Homeostatic mechanisms regulate cardiovascular activity to ensure that tissue blood flow, also called *tissue perfusion*, meets the demand for oxygen and nutrients. The three variable factors that influence tissue blood flow are cardiac output, peripheral resistance, and blood pressure. Cardiac output was discussed in Chapter 13, and peripheral resistance and blood pressure were considered earlier in this chapter (p. 351).

Most cells are relatively close to capillaries (within 135 µm). When a group of cells becomes active, the circulation to that region must increase to deliver the necessary oxygen and nutrients and to carry away the waste products and carbon dioxide that they generate. The goal of cardiovascular regulation is to ensure that these blood flow changes occur (1) at an appropriate time, (2) in the right area, and (3) without drastically altering blood pressure and blood flow to any vital organs.

Factors involved in the regulation of cardiovascular function (Figure 14-7•) include:

- ***Local factors.*** Local factors change the pattern of blood flow within capillary beds in response to chemical changes in the interstitial fluids. For example, at any given moment some of the precapillary sphincters are open and others are closed. As long as the proportion remains constant, the total blood flow through the tissues will remain constant. This is an example of *autoregulation* at the tissue level.
- ***Neural mechanisms.*** Neural control mechanisms respond to changes in arterial pressure or blood gas levels at specific sites. When those changes occur, the cardiovascular centers (cardiac and va-

somotor) of the autonomic nervous system adjust cardiac output and peripheral resistance to maintain adequate blood flow.

- ***Endocrine factors.*** The endocrine system releases hormones that enhance short-term adjustments and direct long-term changes in cardiovascular performance.

Short-term responses adjust cardiac output and peripheral resistance to stabilize blood pressure and tissue blood flow. Long-term adjustments involve alterations in blood volume that affect cardiac output and the transport of oxygen and carbon dioxide to and from active tissues.

Autoregulation of Blood Flow

Under normal resting conditions, cardiac output remains stable, and peripheral resistance within individual tissues is adjusted to control local blood flow. When a precapillary sphincter constricts, blood flow decreases; when it relaxes, blood flow increases. Precapillary sphincters can respond automatically to alterations in the local environment. For example, when oxygen is abundant, the smooth muscles contract and slow down the flow of blood. Oxygen levels decline in active tissues as the cells absorb O_2 for use in aerobic respiration. As tissue oxygen supplies dwindle, carbon dioxide levels rise, and the pH falls. This combination causes the smooth muscle cells in the precapillary sphincter to relax and blood flow to increase. The appearance of specific chemicals in the interstitial fluids can produce the same effect. For example, in the inflammation response, vasodilation occurs at an injury site because histamine, bacterial toxins, and prostaglandins cause a relaxation of precapillary sphincters.

Factors that promote dilation of precapillary sphincters are called *vasodilators* and those that stimulate constriction of precapillary sphincters are called *vasoconstrictors*. Together such factors control blood flow within a single capillary bed. When present in high concentrations, these factors also affect arterioles, increasing or decreasing blood flow to all of the capillary beds in a given region. Such an event will often trigger a neural response, because significant changes in blood flow to one region of the body will have an immediate effect on circulation to other regions.

Neural Control of Blood Pressure and Blood Flow

The nervous system is responsible for adjusting cardiac output and peripheral resistance to maintain adequate blood flow to vital tissues and organs. The cardiovascular centers responsible for these regulatory activities include the *cardiac centers* and *vasomotor centers* of the medulla oblongata. As noted in Chapter 13, the cardiac centers include a *cardioacceleratory center* that increases cardiac output through sympathetic innervation and a *cardioinhibitory center* that reduces cardiac output through parasympathetic innervation. ∞ [p. 343]

The vasomotor center of the medulla oblongata primarily controls the diameters of the arterioles. Inhibition of the vasomotor center leads to **vasodilation** (*vaso-*, vessel), a dilation of arterioles that reduces peripheral resistance. Stimulation of the vasomotor center causes **vasoconstriction** (the constriction of peripheral arterioles) and, with very strong stimulation, *venoconstriction* (constriction of peripheral veins), both of which increase peripheral resistance.

The cardiovascular centers detect changes in tissue demand by monitoring arterial blood, especially (1) blood pressure and (2) pH and dissolved gas concentrations. The *baroreceptor reflexes* respond to changes in blood pressure, and the *chemoreceptor reflexes* respond to changes in chemical composition.

Baroreceptor Reflexes

Baroreceptor reflexes (*baro-*, pressure) are autonomic reflexes that adjust cardiac output and peripheral resistance to maintain normal arterial pressures. When blood pressure climbs, the increased output from the baroreceptors travels to the medulla oblongata, where it inhibits the cardioacceleratory center, stimulates the cardioinhibitory center, and inhibits the vasomotor center (Figure 14-8●).

Under the command of the cardioinhibitory center, the vagus nerves release ACh, which reduces the rate and strength of the cardiac contractions, lowering cardiac output. The inhibition of the vasomotor center leads to the dilation of peripheral arterioles throughout the body. This combination of reduced cardiac output and decreased peripheral resistance then reduces the blood pressure.

This pattern is reversed if blood pressure becomes abnormally low. In that case, a corresponding reduction in baroreceptor output stimulates the cardioacceleratory center, inhibits the cardioinhibitory center, and stimulates the vasomotor center. The cardioacceleratory center stimulates sympathetic neurons innervating the SA node, AV node, and general myocardium. This stimulation increases heart rate and stroke volume, leading to an immediate increase in cardiac output. Vasomotor activity, also carried by sympathetic motor neurons, produces a rapid vasoconstriction, increasing peripheral resistance. These adjustments, increased cardiac output and increased peripheral resistance, work together to elevate blood pressure.

Baroreceptors monitor the degree of stretch in the walls of distensible organs. The three major baroreceptor populations respond to alterations in blood pressure at key locations within the cardiovascular system.

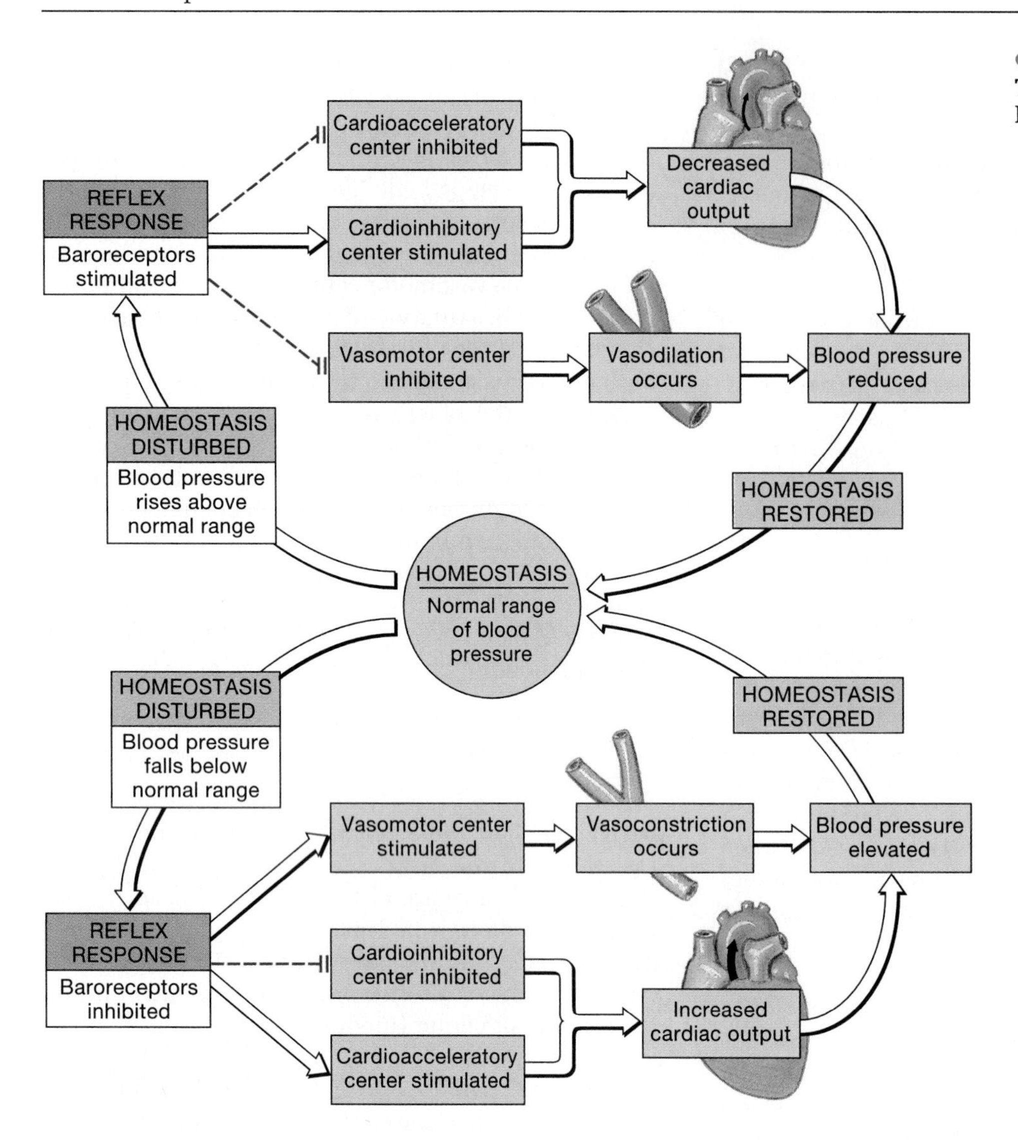

Figure 14-8
The Carotid and Aortic Sinus Baroreceptor Reflexes

1. *Aortic baroreceptors* are located within the **aortic sinuses**, pockets in the walls of the ascending aorta adjacent to the heart (Figure 13-5, p. 334).
2. *Carotid sinus baroreceptors* are located in the walls of the **carotid sinuses**, expanded chambers near the bases of the internal carotid arteries of the neck (Figure 14-15, p. 368). Because pressure changes at this location affect the blood flow to the brain, the carotid sinus reflex is both extremely sensitive and quite important.
3. *Atrial baroreceptors*, in the wall of the right atrium, monitor the blood pressure there and at the venae cavae, the end of the systemic circuit. The responses produced by the atrial reflex differ from those of the aortic and carotid reflexes. Under normal circumstances, the heart pumps blood into the aorta at the same rate that it is arriving at the right atrium. When blood pressure rises in the atrium, it means that a circulatory traffic jam exists, with blood arriving at the heart faster than it is being pumped out. The atrial baroreceptors solve the problem by stimulating the cardioacceleratory center, increasing cardiac output until the backlog of venous blood is removed and atrial pressure returns to normal.

Chemoreceptor Reflexes

The **chemoreceptor reflexes** (Figure 14-9) respond to changes in the carbon dioxide levels, oxygen levels, or pH in the blood and cerebrospinal fluid. The chemoreceptors involved are sensory neurons found in the **carotid bodies**, located in the neck near the carotid sinus, and the **aortic bodies**, situated near the arch of the aorta. These receptors monitor the composition of the arterial blood. Additional chemoreceptors, found on the ventrolateral surfaces of the medulla oblongata, monitor the composition of the cerebrospinal fluid.

Activation of chemoreceptors occurs through a drop in pH, a rise in CO_2 levels, or a fall in plasma O_2. Any of these changes leads to a stimulation of the cardioacceleratory and vasomotor centers. This elevates arterial pressure and increases blood flow through peripheral tissues. Chemoreceptor output also affects the respiratory centers in the medulla oblongata. As a result, a rise in blood flow and blood pressure is associated with an elevated respiratory rate. Coordination of cardiovascular and respiratory activity is vital, because accelerating tissue blood flow is useful only if the circulating blood contains adequate oxygen. In addition, a rise in the respiratory rate accelerates venous return through the action of the respiratory pump.

Influence of the ANS and Higher Brain Centers

The cardiac and vasomotor centers may also be influenced by the activities of other areas of the brain. Stimulation by sympathetic neurons of the ANS, assisted by the release of epinephrine and norepinephrine by the adrenal medullae, acts on these centers to increase cardiac output and cause vasoconstriction. In contrast, parasympathetic stimulation affects the cardioinhibitory center, reducing cardiac output. It does not directly affect the vasomotor center, but vasodilation occurs as sympathetic activity declines.

The activities of higher brain centers can also affect blood pressure. Our thought processes or emotional states can produce significant changes in blood pressure by influencing cardiac output. For example, strong emotions of anxiety, fear, or rage are accompanied by an elevation in blood pressure, caused by cardiac stimulation and vasoconstriction.

Hormones and Cardiovascular Regulation

The endocrine system provides both short-term and long-term regulation of cardiovascular performance. Epinephrine (E) and norepinephrine (NE) from the adrenal medullae stimulate cardiac output and peripheral vasoconstriction. Other regulatory hormones include antidiuretic hormone (ADH), angiotensin II, erythropoietin (EPO), and atrial natriuretic peptide (ANP), introduced in Chapter 11. ∞ [pp. 289, 295–296] Although ADH and angiotensin II affect blood pressure, all four hormones are concerned primarily with the long-term regulation of blood volume, as diagrammed in Figure 14-10 ●.

Angiotensin II

Angiotensin II appears in the blood following the release of renin by specialized kidney cells stimulated by a fall in blood pressure. Renin starts a chain reaction that ultimately converts an inactive protein, *angiotensinogen*, to the hormone angiotensin II. Angiotensin II causes an extremely powerful vasoconstriction that elevates blood pressure almost at once. It also stimulates the secretion of ADH by the pituitary and aldosterone by the adrenal cortex, and the two complement one another. ADH stimulates water conservation at the kidneys, and aldosterone stimulates the reabsorption of sodium ions and water

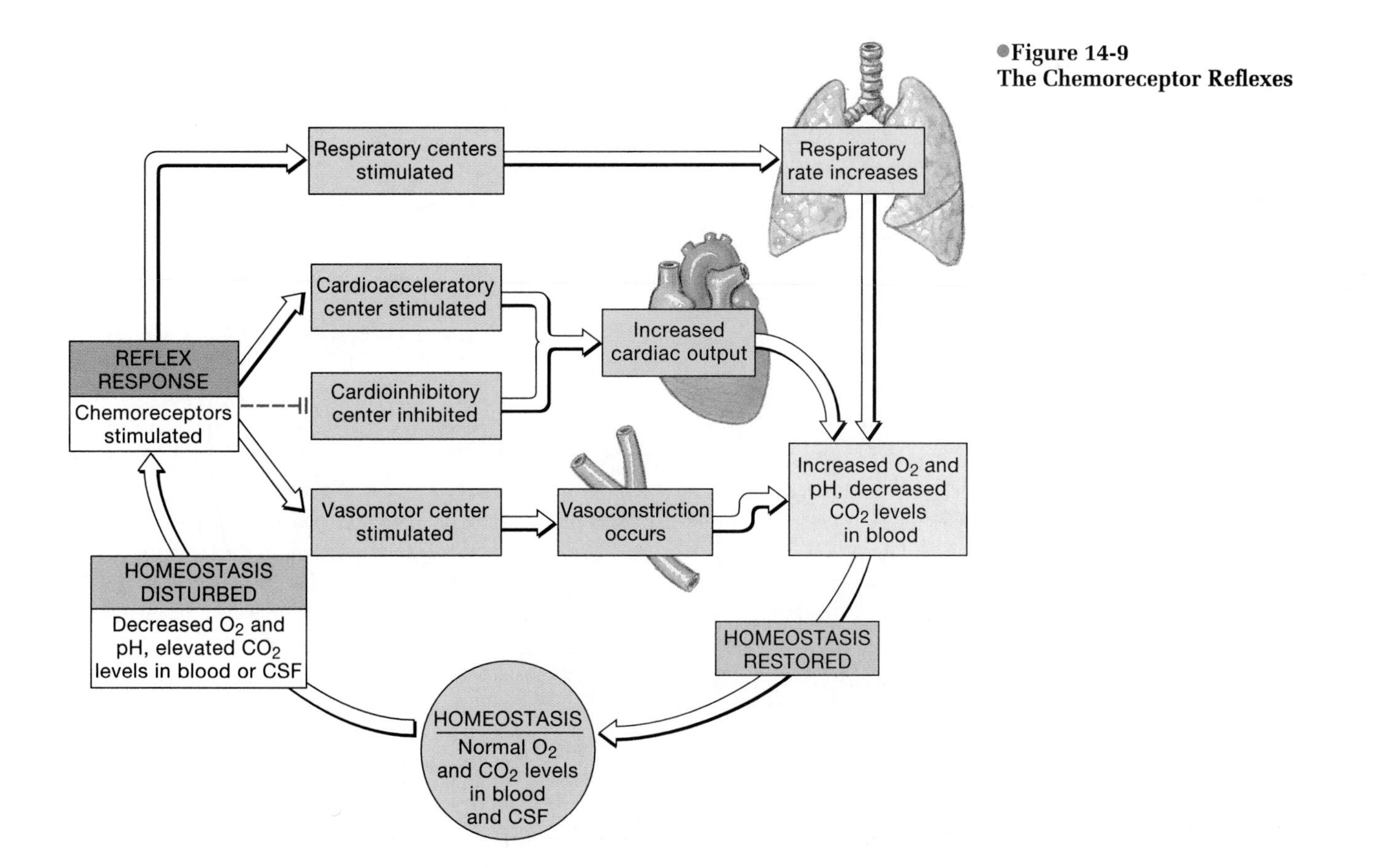

●**Figure 14-9**
The Chemoreceptor Reflexes

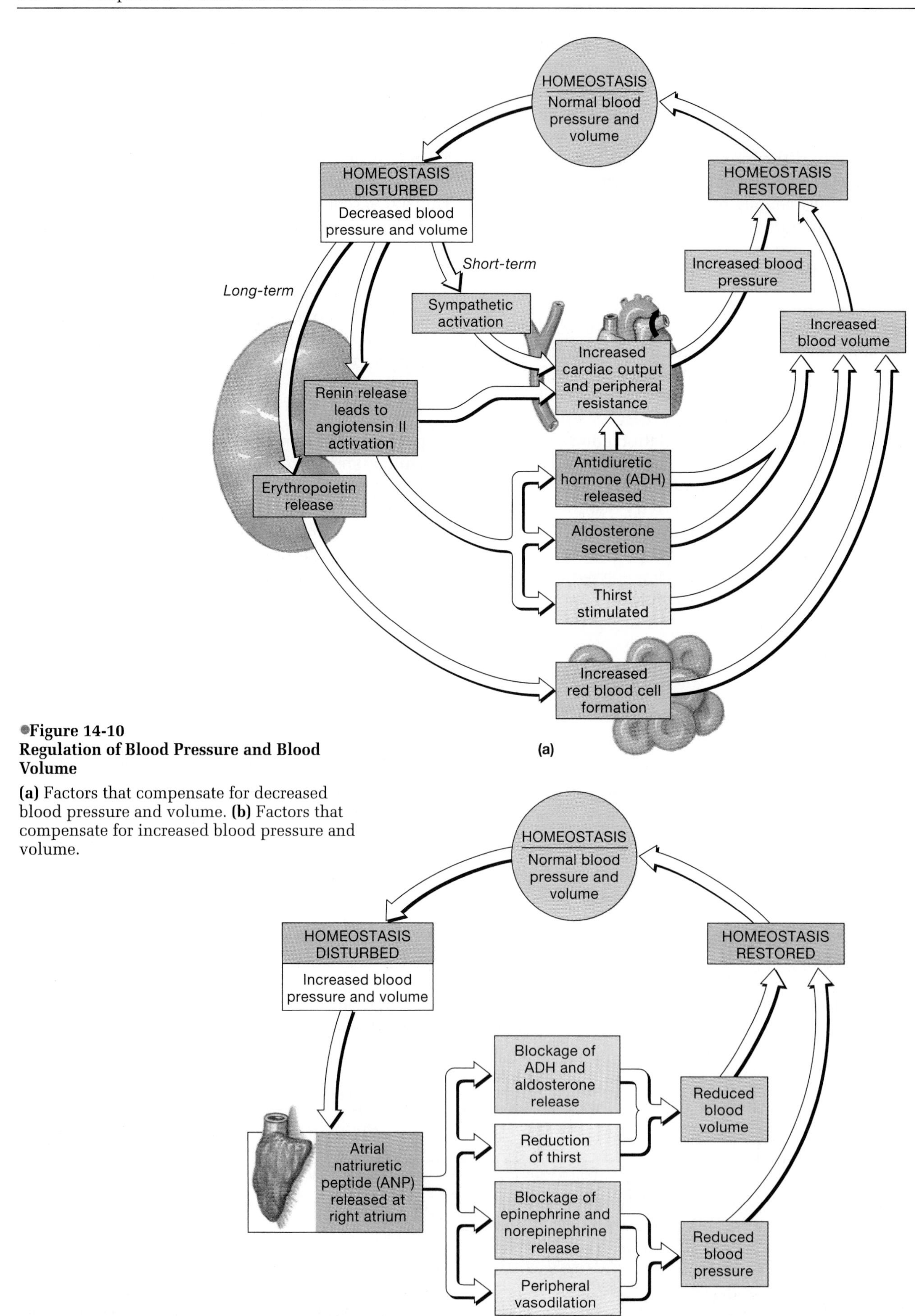

•Figure 14-10
Regulation of Blood Pressure and Blood Volume

(a) Factors that compensate for decreased blood pressure and volume. **(b)** Factors that compensate for increased blood pressure and volume.

from the urine. In addition, angiotensin II stimulates thirst, and the presence of ADH and aldosterone ensures that the additional water consumed will be retained, elevating the plasma volume.

Antidiuretic Hormone

In addition to its water-conserving effect on the kidneys, ADH also responds to an increase in the osmotic concentration of the plasma. The immediate result is a peripheral vasoconstriction that elevates blood pressure.

Erythropoietin

Erythropoietin (EPO) is released by the kidneys if the blood pressure declines or if the oxygen content of the blood becomes abnormally low. This hormone stimulates red blood cell production, elevating the blood volume and improving the oxygen-carrying capacity of the blood.

Atrial Natriuretic Peptide

In contrast to the three hormones just described, ANP release is stimulated by *increased* blood pressure (Figure 14-10b•). This hormone is produced by specialized cardiac muscle cells in the atrial walls when they are stretched by excessive venous return. ANP reduces blood volume and blood pressure by (1) promoting the loss of sodium ions and water at the kidneys, (2) increasing water losses at the kidneys by blocking the release of ADH and aldosterone, (3) reducing thirst, (4) blocking the release of E and NE, and (5) stimulating peripheral vasodilation. As blood volume and blood pressure decline, the stress on the atrial walls is removed, and ANP production ceases.

PATTERNS OF CARDIOVASCULAR RESPONSE

In our day-to-day lives the cardiovascular system operates as an integrated complex of quite diverse components. Two common stresses, exercise and blood loss, illustrate the adaptability of this system and its ability to maintain homeostasis. We will also consider the physiological mechanisms involved in shock and heart failure, two important cardiovascular disorders.

Exercise and the Cardiovascular System

At rest the cardiac output averages around 5.6 liters per minute. During exercise, both cardiac output and the pattern of blood distribution change markedly. As exercise begins, a number of interrelated changes occur.

- *Extensive vasodilation occurs* as the rate of skeletal muscle oxygen consumption increases. Peripheral resistance drops, blood flow through the capillaries increases, and blood enters the venous system at an accelerated rate.
- *The venous return increases* as skeletal muscle contractions squeeze blood along the peripheral veins and an increased breathing rate pulls blood into the venae cavae (the respiratory pump).
- *Cardiac output rises* as a result of the increased venous return. This increase occurs in direct response to ventricular stretching (the Frank-Starling law) and in a reflexive response to atrial stretching (the atrial reflex). The increased cardiac output keeps pace with the elevated demand, and arterial pressures are maintained despite the drop in peripheral resistance.

This regulation by venous feedback gradually increases cardiac output to about double resting levels. Over this range, typical of light exercise, the pattern of blood distribution remains relatively unchanged. At higher levels of exertion, other physiological adjustments occur as the sympathetic nervous system stimulates the cardiac and vasomotor centers. Cardiac output increases, and major changes in the peripheral distribution of blood increase the flow to active skeletal muscles. When one is exercising at maximal levels, the blood essentially races between the skeletal muscles, the lungs, and the heart. Only the blood supply to the brain remains unaffected.

EXERCISE, CARDIOVASCULAR FITNESS, AND HEALTH

Cardiovascular performance improves significantly with training. Trained athletes have larger hearts than nonathletes and greater average stroke volume. These are important functional changes. Cardiac output is equal to the stroke volume times the heart rate, so for the same cardiac output, an individual with a larger stroke volume will have a slower heart rate. A professional athlete at rest can maintain normal blood flow to peripheral tissues at a heart rate as low as 50 bpm (beats per minute), compared with about 80 bpm for a nonathlete.

Regular exercise has several beneficial effects. Even a moderate exercise routine (jogging 5 miles per week, for example) can lower total blood cholesterol levels. High cholesterol is one of the major risk factors for atherosclerosis, which leads to cardiovascular disease and strokes. In addition, a healthy lifestyle with regular exercise, a balanced diet, weight control, and no smoking reduces stress, lowers blood pressure, and slows plaque formation. Large-scale statistical studies indicate that regular moderate exercise may cut the incidence of heart attacks almost in half.

Cardiovascular Response to Hemorrhaging

In Chapter 12 we considered the local circulatory reaction to a break in the wall of a blood vessel. ∞ [p. 322] When the clotting response fails to prevent a significant blood loss and arterial blood pressure falls, the entire cardiovascular system begins making adjustments. The immediate goal is to maintain adequate blood pressure and peripheral blood flow. The long-term goal is to restore normal blood volume. Short-term and long-term responses are diagrammed in Figure 14-10a●.

Elevation of Blood Pressure

Short-term responses appear almost as soon as the pressures start to decline, when the carotid and aortic reflexes increase cardiac output and cause peripheral vasoconstriction. When you donate blood at a blood bank, the amount collected is usually 500 ml, roughly 10 percent of the total blood volume. Such a loss initially causes a drop in cardiac output, but the vasomotor center quickly improves venous return and restores cardiac output to normal levels by a combination of vasoconstriction and mobilization of the *venous reserve* (large reservoirs of slowly moving venous blood in the liver, bone marrow, and skin). This venous compensation can restore normal arterial pressures and peripheral blood flow after losses that amount to 15–20 percent of the total blood volume. With a more substantial blood loss, cardiac output is maintained by increasing the heart rate, often to 180 to 200 bpm. Sympathetic activation assists by constricting the muscular arteries and arterioles and elevating blood pressure.

Sympathetic activation also causes the secretion of epinephrine and norepinephrine by the adrenal medullae. At the same time, ADH is released by the posterior pituitary, and at the kidneys the fall in blood pressure causes the release of renin, initiating the activation of angiotensin II. The epinephrine and norepinephrine increase cardiac output, and in combination with ADH and angiotensin II they cause a powerful vasoconstriction that elevates blood pressure and improves peripheral blood flow.

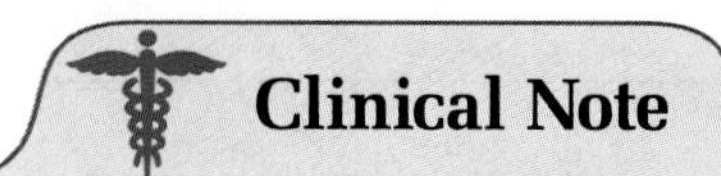

Shock

Shock is an acute circulatory crisis marked by low blood pressure (hypotension) and inadequate peripheral blood flow. Severe and potentially fatal symptoms develop as vital tissues become starved for oxygen and nutrients. Common causes of shock are (1) a fall in cardiac output after hemorrhaging or other fluid losses, (2) damage to the heart, (3) external pressure on the heart, or (4) extensive peripheral vasodilation. AM *Heart Failure*

All forms of shock share the same six basic symptoms:

1. Hypotension, with systolic pressures below 90 mm Hg.
2. Pale, cool, and moist ("clammy") skin. The skin is pale and cool because of peripheral vasoconstriction; the moisture reflects sympathetic activation of the sweat glands.
3. Confusion and disorientation, caused by a fall in blood pressure at the brain.
4. Rapid, weak pulse.
5. Cessation of urination, because the reduced blood to the kidneys slows or stops urine production.
6. Acidosis, caused by lactic acid generation in oxygen-deprived tissues.

There are several types of shock (*septic shock*, *anaphylactic shock*, and *toxic shock syndrome*, for example). One important form of shock, called **circulatory shock**, is caused by a severe reduction in blood volume. Through mechanisms already noted, the cardiovascular system can cope with fluid losses of up to about 30 percent of the total blood volume before symptoms of circulatory shock appear.

When blood volume declines by more than 35 percent, homeostatic mechanisms are unable to cope with the situation, and a vicious cycle begins: Low blood pressure and low venous return lead to decreased cardiac output and myocardial damage, further reducing cardiac output. When the mean arterial blood pressure falls to about 50 mm Hg, carotid sinus baroreceptors trigger a massive activation of sympathetic vasoconstrictors. Prevention of fatal consequences requires immediate treatment that concentrates on (1) preventing further fluid losses and (2) providing fluids through transfusions. Additional information on the causes and treatment of shock will be found in the *Applications Manual*. AM *Shock*

Restoration of Blood Volume

After a serious hemorrhage, several days may pass before the blood volume returns to normal. When the short-term responses are unable to maintain normal cardiac output and blood pressure, the decline in capillary blood pressure promotes the reabsorption of fluid from the interstitial spaces. Over this period, ADH and aldosterone promote fluid retention and reabsorption at the kidneys, preventing further reductions in blood volume. Thirst increases, and additional water is absorbed across the digestive tract. This intake of fluid elevates plasma volume and ultimately replaces the interstitial fluids "borrowed" at the capillaries. Erythropoietin targets the bone marrow, stimulating the maturation of red blood cells, which increase the blood volume and improve oxygen delivery to peripheral tissues.

✓ How would applying a small pressure to the common carotid artery affect your heart rate?

✓ Why does blood pressure increase during exercise?

✓ What effect would vasoconstriction of the renal artery have on blood pressure and blood volume?

THE BLOOD VESSELS

The circulatory system is divided into the *pulmonary circuit* and the *systemic circuit*. The pulmonary circuit is composed of arteries and veins that transport blood between the heart and the lungs. This circuit begins at the right ventricle and ends at the left atrium. From the left ventricle, the arteries of the systemic circuit transport oxygenated blood and nutrients to all organs and tissues, ultimately returning deoxygenated blood to the right atrium. Figure 14-11● summarizes the primary circulatory routes within the pulmonary and systemic circuits.

In the following descriptions of the various blood vessels, note that arteries and veins on the left and right sides are usually identical except near the heart, where large vessels connect to the atria or ventricles. For example, the distribution of the *left* and *right subclavian, axillary, brachial,* and *radial arteries* parallels the *left* and *right subclavian, axillary, brachial,* and *radial veins.* In addition, vessels often change names as they pass specific anatomical boundaries. For example, the *external iliac artery* becomes the *femoral artery* as it leaves the trunk and enters the thigh. Lastly, note that arteries and veins often make anastomotic, or direct, connections. These connections reduce the impact of a temporary or permanent blockage (*occlusion*) of a single vessel. For example, the *radial* and *ulnar arteries* form an anastomosis from which the *digital arteries* originate.

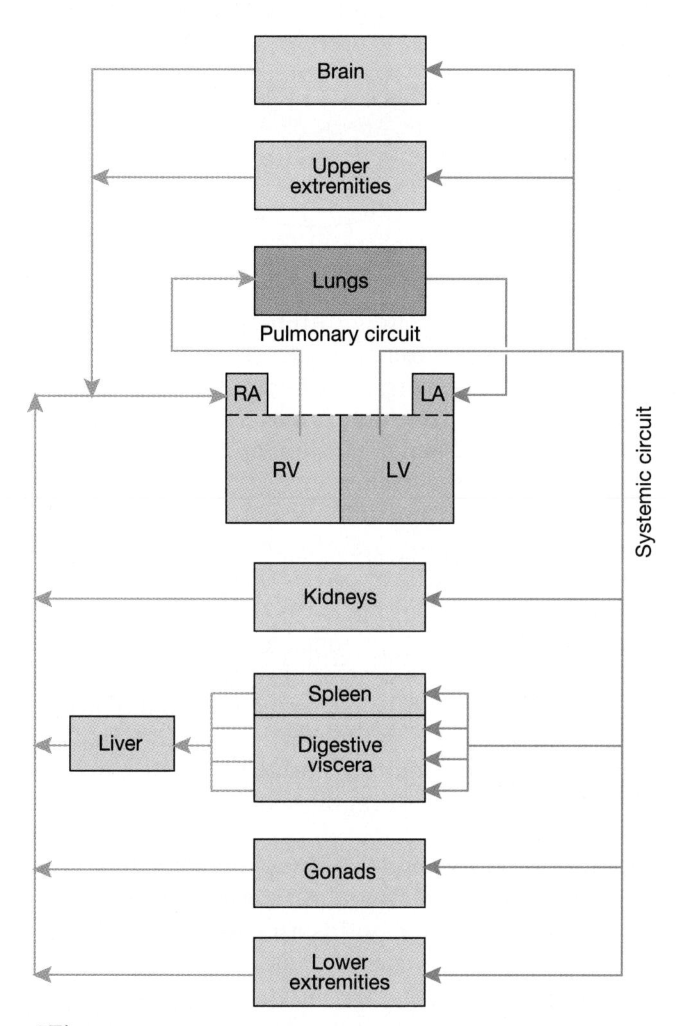

●Figure 14-11
An Overview of the Pattern of Circulation

The Pulmonary Circulation

Blood entering the right atrium has just returned from a trip to peripheral capillary beds, where oxygen was released and carbon dioxide absorbed. After traveling through the right atrium and ventricle, blood enters the **pulmonary trunk** (*pulmo-*, lung), the start of the pulmonary circuit. In this circuit, oxygen stores will be replenished, carbon dioxide excreted, and the "renewed" blood returned to the heart for distribution in the systemic circuit.

Figure 14-12● details the anatomy of the pulmonary circuit. The arteries of the pulmonary circuit differ from those of the systemic circuit in that they carry deoxygenated blood. (For this reason, color-coded diagrams usually show the pulmonary arteries in blue, the same color as systemic veins.) As the pulmonary trunk curves over the superior border of the heart it gives rise to the **left** and **right pulmonary arteries**. These large arteries enter the lungs before branching repeatedly, giving rise to smaller and smaller arteries. The smallest branches, the pulmonary arterioles, provide blood to capillary networks that surround small air pockets, or **alveoli** (al-VĒ-ōl-ī; *alveolus*, sac). The walls of alveoli are thin enough for gas exchange to occur between the capillary blood and inspired air, a process detailed in Chapter 16. As it leaves the alveolar capillaries, oxygenated blood enters venules that in turn unite to form larger vessels carrying blood to the **pulmonary veins**. These four veins, two from each lung, empty into the left atrium, completing the pulmonary circuit.

The Systemic Circulation

The systemic circulation supplies the capillary beds in all other parts of the body. This circuit, which at any given moment contains about 84 percent of the total blood volume, begins at the left ventricle and ends at the right atrium.

Systemic Arteries

Figure 14-13● indicates the relative locations of major systemic arteries. The detailed distribution of these vessels and their branches will be found in Figures 14-14 to 14-16●.

The **ascending aorta** begins at the aortic semilunar valve of the left ventricle, and the *left* and *right*

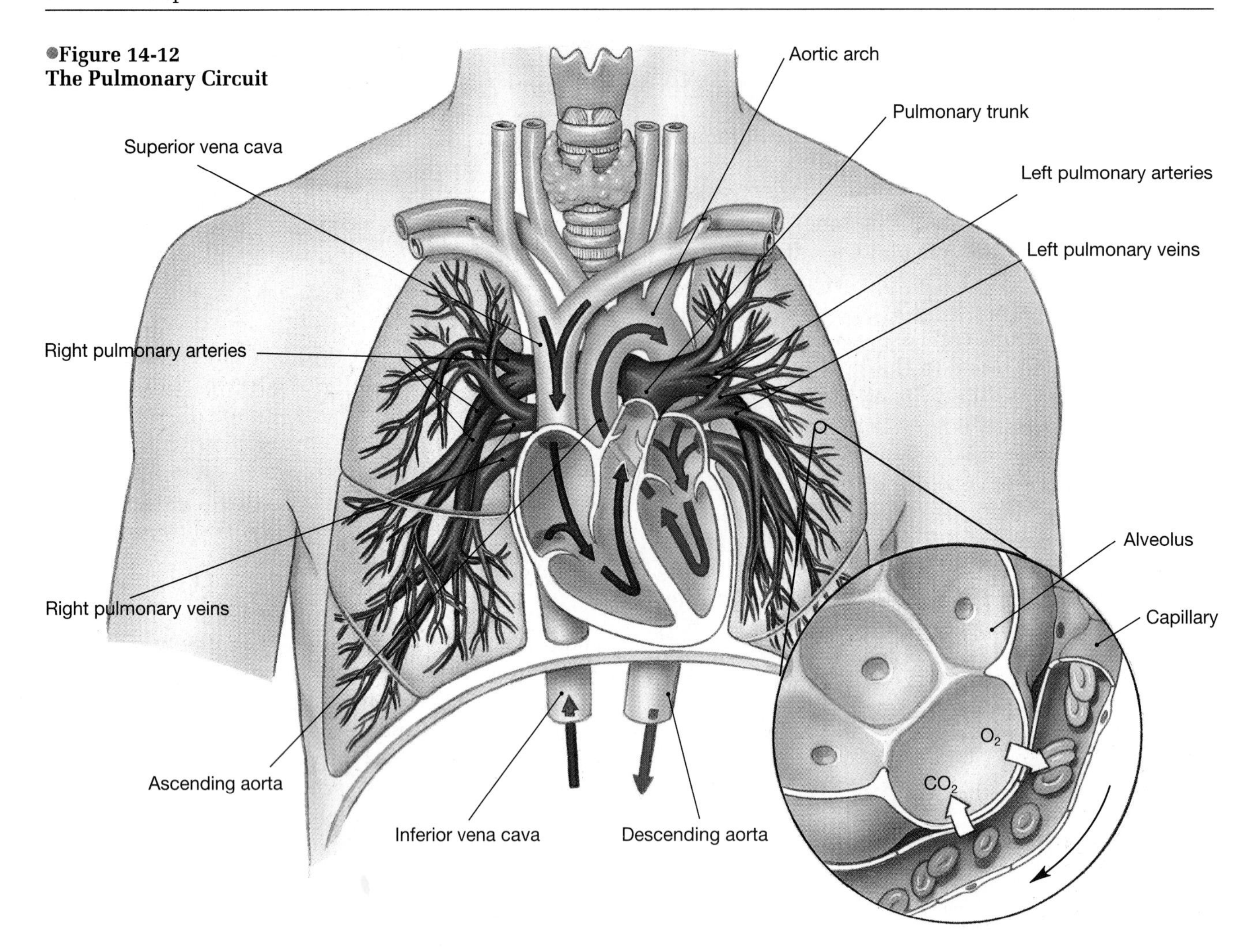

•Figure 14-12 The Pulmonary Circuit

coronary arteries originate near its base, as was illustrated in Figure 13-7•, p. 336. The **aortic arch** curves across the superior surface of the heart, connecting the ascending aorta with the caudally directed **descending aorta**.

Arteries of the Aortic Arch Three elastic arteries originate along the aortic arch (Figure 14-14•). These arteries, the **brachiocephalic** (brāk-ē-ō-se-FAL-ik), the **left common carotid**, and the **left subclavian** (sub-CLĀ-vē-an), deliver blood to the head, neck, shoulders, and upper limbs. The brachiocephalic artery ascends for a short distance before branching to form the **right common carotid artery** and the **right subclavian artery**. Thus, as the figure makes clear, there is only one brachiocephalic artery, and the left common carotid and left subclavian arteries arise separately from the aortic arch. In terms of their peripheral distribution, however, the vessels on the left side are mirror images of those on the right side. Because the following descriptions focus on major branches found on both sides of the body, the terms *right* and *left* will not be used. Figures 14-13• and 14-14a• illustrate the major branches of these arteries, and additional details are included in the flow chart in Figure 14-14b•.

The Subclavian Arteries The subclavian arteries supply blood to the upper limbs, chest wall, shoulders, back, and central nervous system. Before a subclavian artery leaves the thoracic cavity (Figure 14-14•) it gives rise to an *internal thoracic artery*, which supplies the pericardium and anterior wall of the chest, and a **vertebral artery**, which provides blood to the brain and spinal cord.

After passing the first rib, the subclavian gets a new name, the **axillary artery** (Figure 14-14•). The axillary artery crosses the axilla (armpit) to enter the arm, where its name changes again, becoming the **brachial artery**. The brachial artery provides blood to the arm before branching to create the **radial artery** and **ulnar artery** of the forearm. These arteries are connected by anastomoses at the palm, and the *digital arteries* originate from these arterial connections.

The Carotid Artery and the Blood Supply to the Brain Figure 14-15a • follows the arterial supply to the brain. The common carotids ascend deep in the tissues of the neck. A carotid artery can usually be located by pressing gently along either side of the trachea until a strong pulse is felt. Each common carotid artery divides into an **external carotid** and an

Figure 14-13
An Overview of the Arterial System

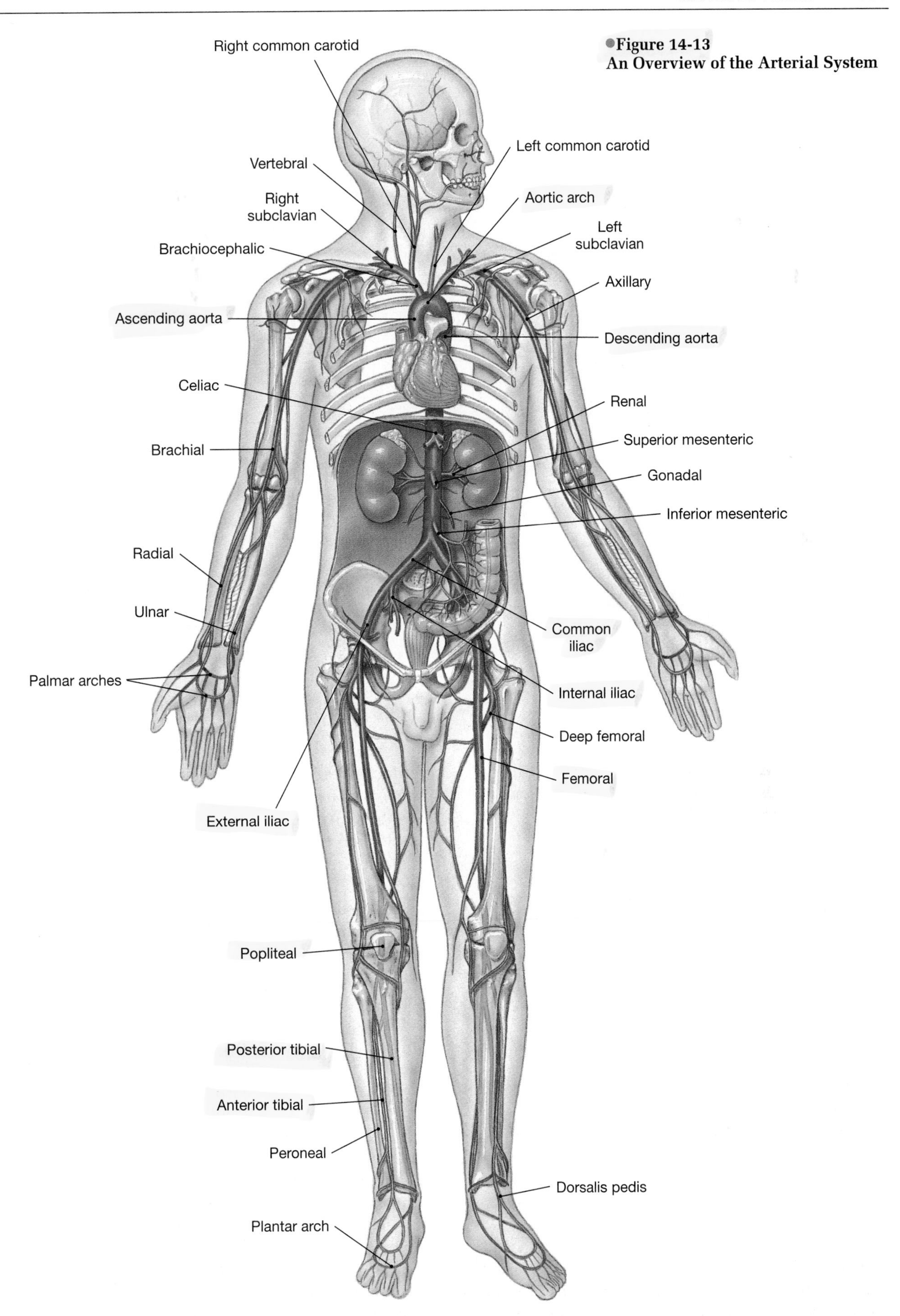

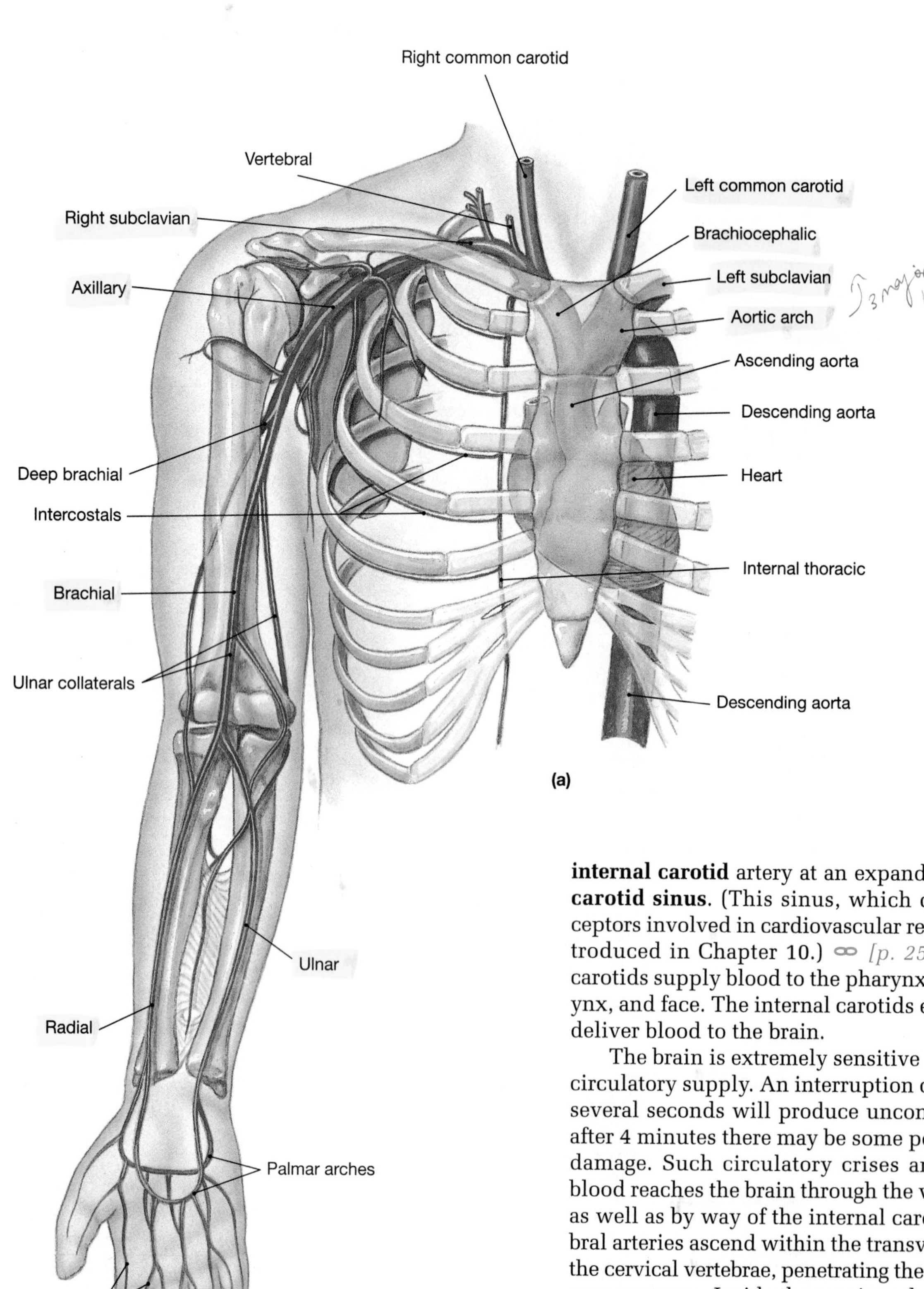

●**Figure 14-14**
Arteries of the Chest and Upper Limb
(a) Diagrammatic view. **(b)** Flow chart.

internal carotid artery at an expanded chamber, the **carotid sinus**. (This sinus, which contains baroreceptors involved in cardiovascular regulation, was introduced in Chapter 10.) ∞ *[p. 252]* The external carotids supply blood to the pharynx, esophogus, larynx, and face. The internal carotids enter the skull to deliver blood to the brain.

The brain is extremely sensitive to changes in its circulatory supply. An interruption of circulation for several seconds will produce unconsciousness, and after 4 minutes there may be some permanent neural damage. Such circulatory crises are rare, because blood reaches the brain through the vertebral arteries as well as by way of the internal carotids. The vertebral arteries ascend within the transverse foramina of the cervical vertebrae, penetrating the skull at the foramen magnum. Inside the cranium they fuse to form a large **basilar artery** that continues along the ventral surface of the brain. This gives rise to the vessels indicated in Figure 14-15b●.

Normally, the internal carotids supply the arteries of the anterior half of the cerebrum, and the rest of the brain receives blood from the vertebral arteries. But this circulatory pattern can easily change, because the internal carotids and the basilar artery are inter-

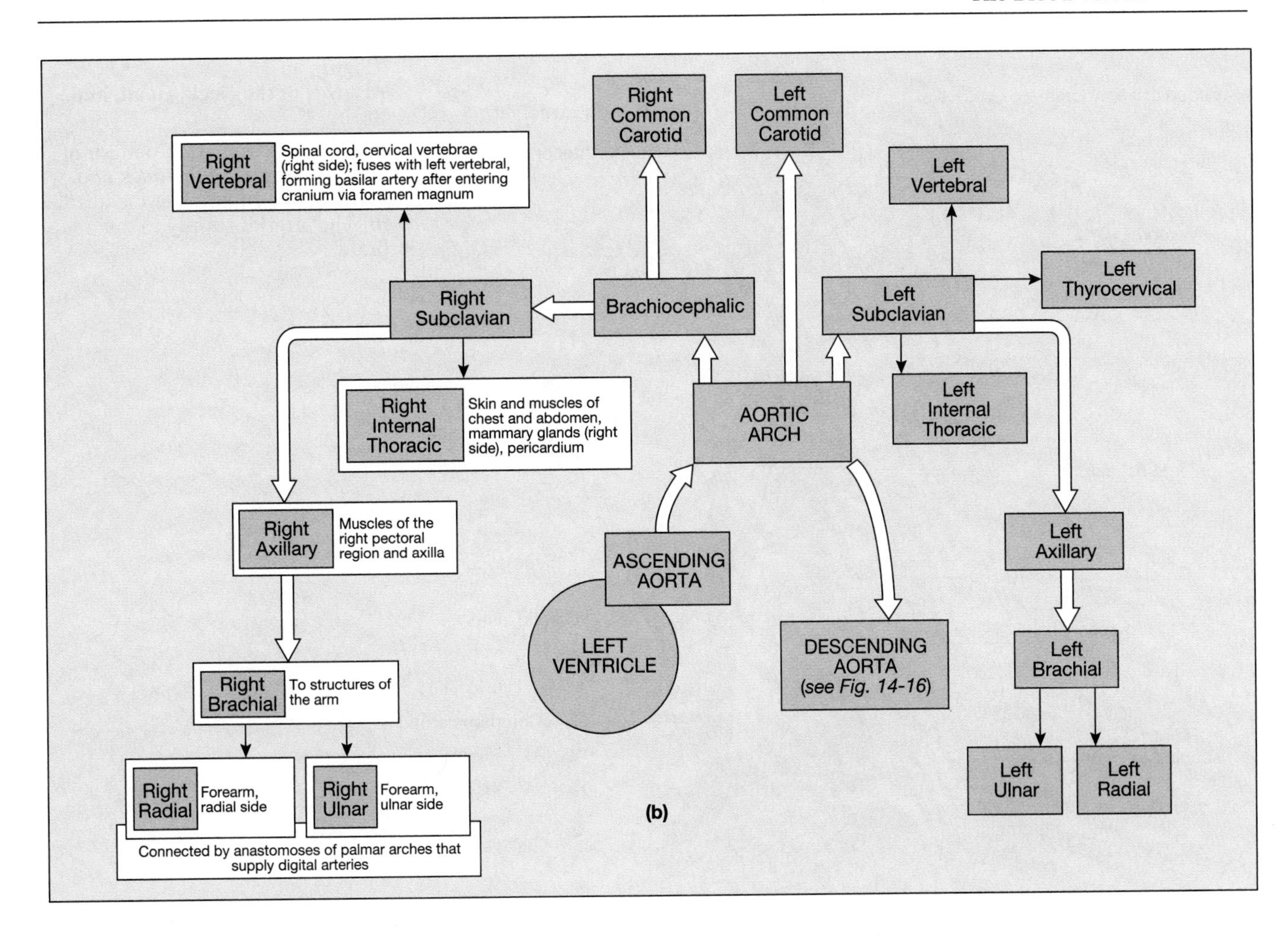

connected in a ring-shaped anastomosis, the **cerebral arterial circle**, or *circle of Willis*, that encircles the infundibulum, or stalk, of the pituitary gland. With this arrangement, the brain can receive blood from either the carotids or the vertebrals, and the chances for a serious interruption of circulation are reduced. However, a plaque, blood clot, or rupture in one of the smaller arteries supplying the brain will injure or kill the dependent tissues. Symptoms of a *stroke*, or *cerebrovascular accident (CVA)*, then appear. The *Applications Manual* contains a discussion of current views on the causes and treatment of strokes. AM *The Treatment of Cerebrovascular Disease*

The Descending Aorta The diaphragm divides the descending aorta into a superior **thoracic aorta** and an inferior **abdominal aorta** (Figure 14-16●). The thoracic aorta travels within the mediastinum, providing blood to the intercostal arteries that carry blood to the spinal cord and the body wall. This artery also gives rise to small arteries that end in capillary beds in the esophagus, pericardium, and other mediastinal structures. Near the diaphragm, the **phrenic** (FREN-ik) **arteries** deliver blood to the muscular diaphragm that separates the thoracic and abdominopelvic cavities. The branches of the thoracic aorta are detailed in Figure 14-16a,b●.

The abdominal aorta (Figure 14-16●) delivers blood to elastic arteries that distribute blood to visceral organs of the digestive system and to the kidneys and adrenal glands. The **celiac** (SĒL-ē-ak) **artery**, **superior mesenteric** (mez-en-TER-ik) **artery**, and **inferior mesenteric artery** arise on the anterior surface of the abdominal aorta and branch in the connective tissues of the mesenteries. These three vessels provide blood to all of the digestive organs in the abdominopelvic cavity. The celiac divides into three branches that deliver blood to the liver (the *common hepatic artery*), spleen (the *splenic artery*) and stomach (the *left gastric*, common hepatic, and splenic arteries). The superior mesenteric artery supplies the pancreas, small intestine, and most of the large intestine; and the inferior mesenteric delivers blood to the last portion of the large intestine and rectum.

Paired **gonadal** (gō-NAD-al) **arteries** originate between the superior and inferior mesenteric arteries; in males they are called *testicular arteries*; in females, *ovarian arteries*.

The **suprarenal arteries** and **renal arteries** arise along the lateral surface of the abdominal aorta and travel behind the peritoneal lining to reach the adrenal glands and kidneys. Small **lumbar arteries** begin on the posterior surface of the aorta and supply the spinal cord and the abdominal wall.

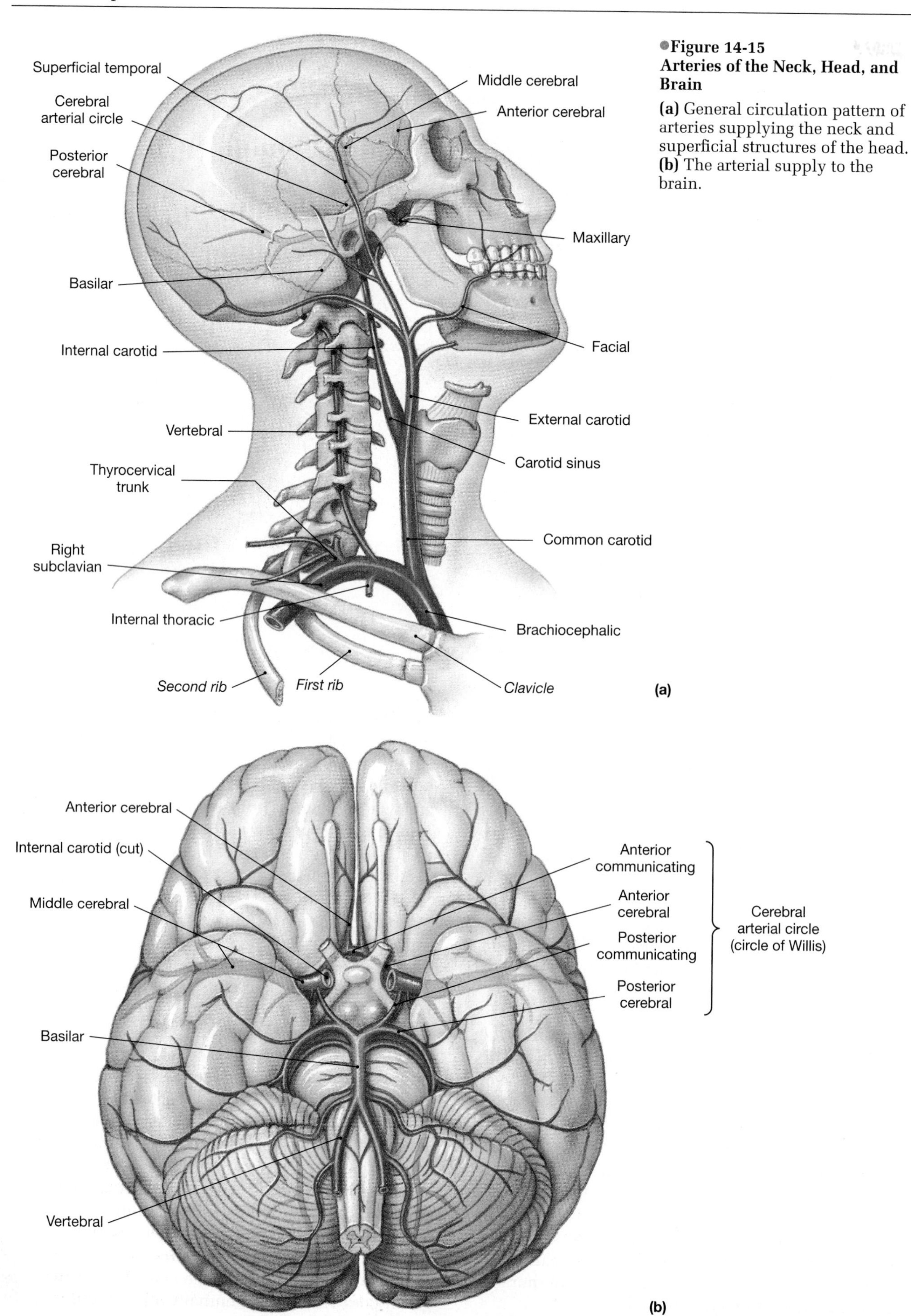

Figure 14-15
Arteries of the Neck, Head, and Brain

(a) General circulation pattern of arteries supplying the neck and superficial structures of the head. **(b)** The arterial supply to the brain.

Figure 14-16
Major Arteries of the Trunk
(a) Diagrammatic view. **(b)** Flow chart.

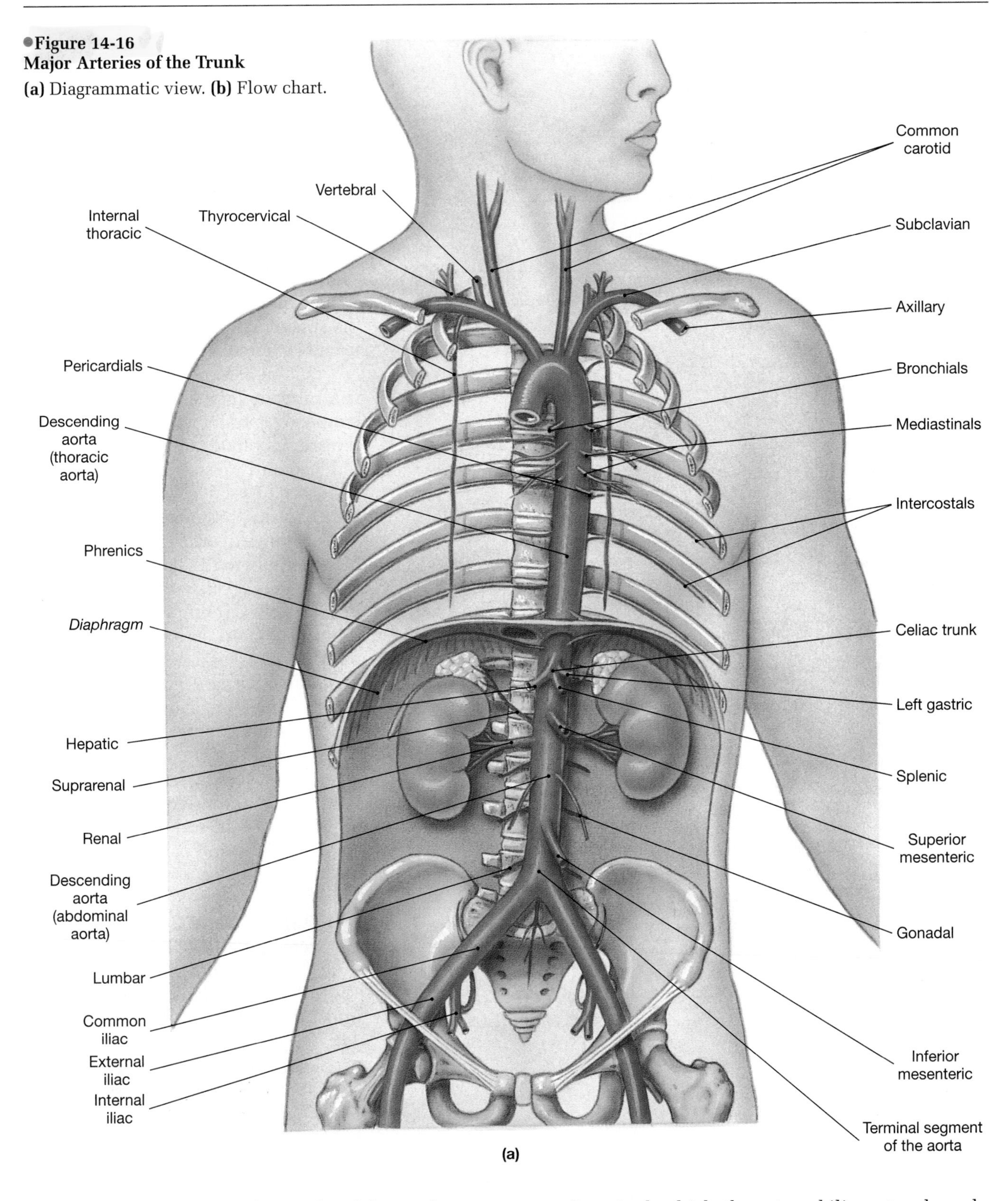

Near the level of vertebra L_4 the abdominal aorta divides to form a pair of muscular arteries. These **common iliac** (IL-ē-ak) **arteries** carry blood to the pelvis and lower limbs (see Figure 14-13 , p. 365). As it travels along the inner surface of the ilium, each common iliac divides to form an **internal iliac artery** that supplies smaller arteries of the pelvis and an **external iliac artery** that enters the lower limb.

Once in the thigh, the external iliac artery branches, forming the **femoral artery** and the **deep femoral artery**. When it reaches the leg, the femoral artery becomes the **popliteal artery**, which almost immediately branches to form the **anterior tibial**, **posterior tibial**, and **peroneal arteries**. These arteries are connected by two anastomoses, one on the top of the foot (the *dorsalis pedis*) and one on the bottom (the *plantar arch*).

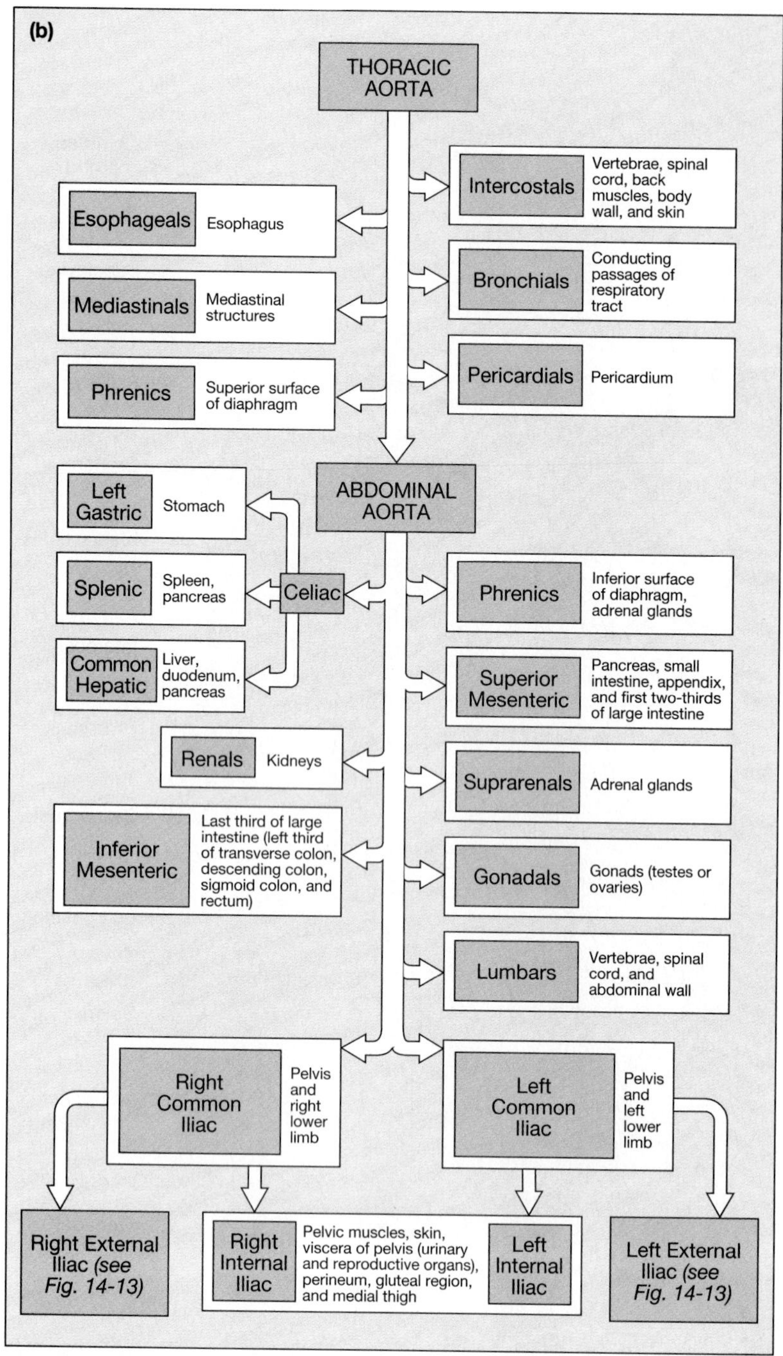

•Figure 14-16 *(continued)*
Major Arteries of the Trunk
(b) Flow chart.

Systemic Veins

Figure 14-17• illustrates the major vessels of the venous system. Complementary arteries and veins often run side by side, and in many cases they have comparable names. For example, the axillary arteries run alongside the axillary veins. In addition, arteries and veins often travel in the company of peripheral nerves that have the same names and innervate the same structures.

One significant difference between the arterial and venous systems concerns the distribution of major veins in the neck and limbs. Arteries in these areas are not found at the body surface; instead, they are deep beneath the skin, protected by bones and surrounding soft tissues. In contrast, there are usually two sets of peripheral veins, one superficial and the other deep. The superficial veins are so close to the surface that they can be seen quite easily. This location makes them easy targets for obtaining blood samples, and most blood tests are performed on venous blood collected from the superficial veins of the upper limb, usually where they cross the elbow.

This dual-venous drainage helps control body temperature. When body temperature becomes abnormally low, the arterial blood supply to the skin is reduced and the superficial veins are bypassed. Blood entering the limbs then returns to the trunk in the deep veins. When overheating occurs, the blood supply to the skin increases and the superficial veins dilate. This is one reason why superficial veins in the limbs become prominent during periods of heavy exercise, or when sitting in a sauna, hot tub, or steam bath.

The branching pattern of peripheral veins is much more variable than that of arteries. Arterial pathways are usually direct, because developing arteries grow toward active tissues. By the time blood reaches the venous system, pressures are low, and routing variations make little functional difference.

The Superior Vena Cava The **superior vena cava** (SVC) receives blood from the head and neck (Figure 14-18•), and the chest, shoulders, and upper limbs (Figure 14-19•).

Venous return from the head and neck. Small veins in the neural tissue of the brain empty into a network of thin-walled channels, the **dural sinuses.** The largest and most important sinus, the **superior sagittal sinus**, is found within the fold of dura mater lying between the cerebral hemispheres. Most of the blood leaving the brain passes through one of the dural sinuses and enters one of the **internal jugular veins** that penetrate the jugular foramen and descend in the deep tissues of the neck. The more superficial **external jugular veins** collect blood from the overlying structures of the head and neck. These veins travel just beneath the skin, and a *jugular ve-*

●Figure 14-17
An Overview of the Venous System

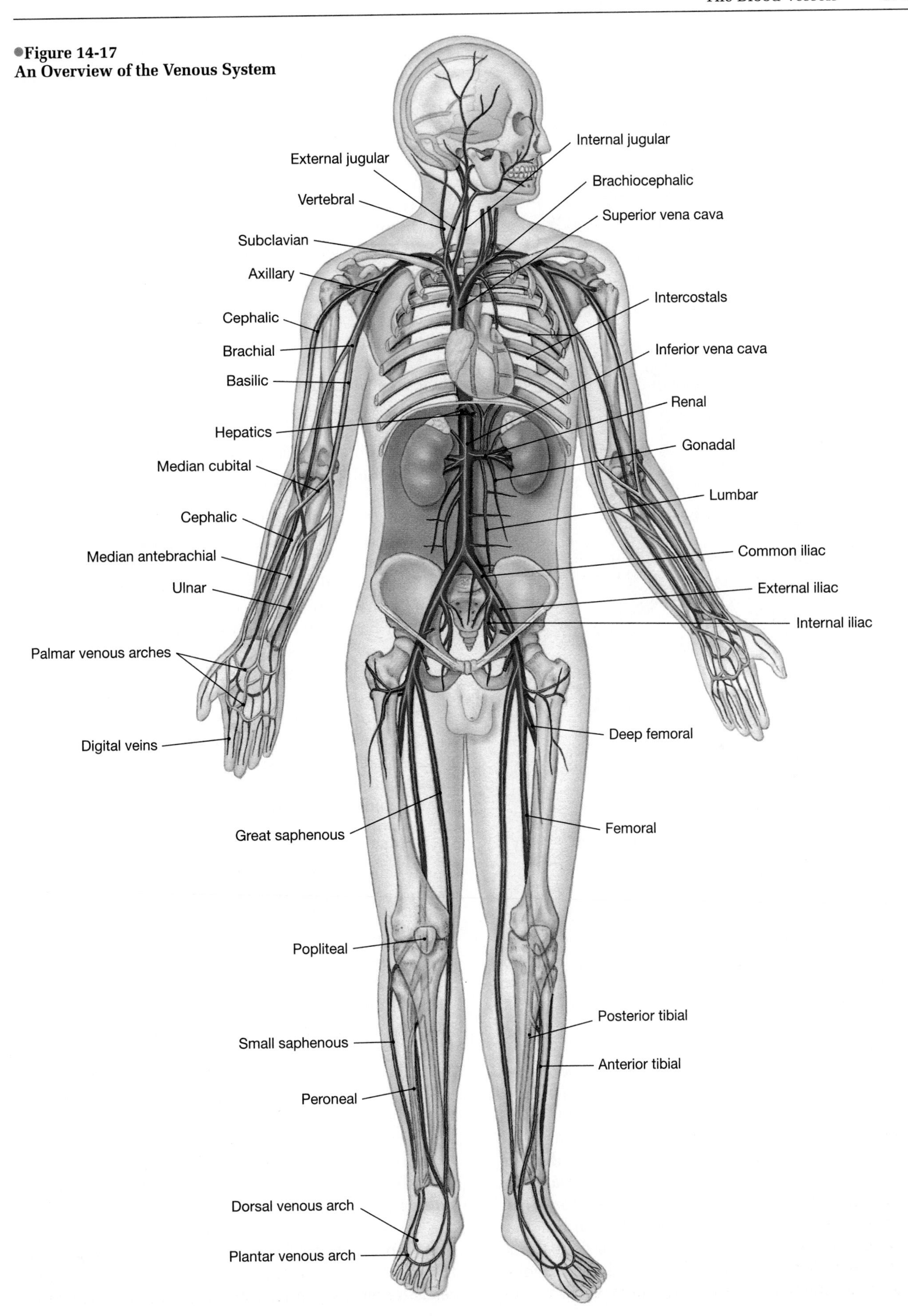
Internal jugular
External jugular
Brachiocephalic
Vertebral
Superior vena cava
Subclavian
Axillary
Intercostals
Cephalic
Brachial
Inferior vena cava
Basilic
Renal
Hepatics
Gonadal
Median cubital
Lumbar
Cephalic
Common iliac
Median antebrachial
External iliac
Ulnar
Internal iliac
Palmar venous arches
Deep femoral
Digital veins
Great saphenous
Femoral
Popliteal
Posterior tibial
Small saphenous
Anterior tibial
Peroneal
Dorsal venous arch
Plantar venous arch

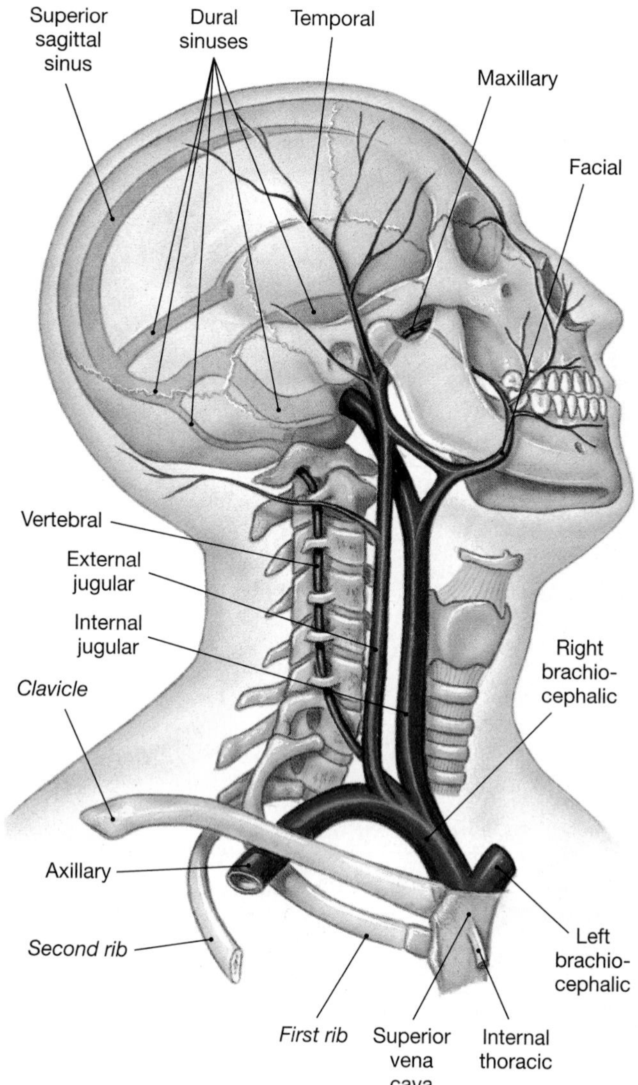

•**Figure 14-18**
Major Veins of the Head and Neck
Veins draining superficial and deep portions of the head and neck.

nous pulse (JVP) can sometimes be seen at the base of the neck. **Vertebral veins** drain the cervical spinal cord and the posterior surface of the skull, descending within the transverse foramina of the cervical vertebrae in company with the vertebral arteries.

Venous Return from the Limbs and Chest. The major veins of the upper body are indicated in Figure 14-19•, and information concerning the venous tributaries of the superior vena cava can be found in Figure 14-20a•. A venous network in the palms collects blood from the digital veins. These vessels drain into the **cephalic vein** and the **basilic vein**. The deeper veins of the forearm consist of a **radial vein** and an **ulnar vein**. After crossing the elbow, these veins fuse to form the **brachial vein**. As the brachial vein continues toward the trunk, it receives blood from the cephalic and basilic veins before entering the axilla as the **axillary vein**.

The axillary vein then continues into the trunk, and at the level of the first rib it becomes the **subclavian vein**. After traveling a short distance inside the thoracic cavity, the subclavian meets and merges with the external and internal jugular veins of that side. This fusion creates the large **brachiocephalic vein**, also known as the *innominate vein*. Near the heart, the two brachiocephalic veins (one from each side of the body) combine to create the superior vena cava, or SVC. The superior vena cava receives blood from the thoracic body wall via the **azygos** (AZ-i-gos) **vein** before arriving at the right atrium.

The Inferior Vena Cava The **inferior vena cava** (IVC) collects most of the venous blood from organs below the diaphragm (a small amount reaches the superior vena cava via the azygos vein). A flow chart of the tributaries of the IVC is shown in Figure 14-20b•, and the veins of the abdomen are illustrated in Figure 14-19•. Refer to Figure 14-17• for the veins of the lower limbs.

Blood leaving the capillaries in the sole of each foot collects into a network of **plantar veins**. The plantar network provides blood to the **anterior tibial vein**, the **posterior tibial vein**, and the **peroneal vein**, the deep veins of the leg. A *dorsal venous arch* drains blood from capillaries on the superior surface of the foot. This arch is drained by two superficial veins, the **great saphenous vein** (sa-FĒ-nus; *saphenes*, prominent) and the **small saphenous vein**. (Surgeons use segments of the great saphenous vein, the largest superficial vein, as a bypass vessel during *coronary bypass surgery*.) There are extensive interconnections between the plantar arch and the dorsal arch, and the path of blood flow can easily shift from superficial to deep veins.

At the knee, the small saphenous, tibial, and peroneal veins unite to form the **popliteal vein**. When it reaches the femur, it becomes the **femoral vein**. Immediately before penetrating the abdominal wall, the femoral, great saphenous, and **deep femoral** veins unite. The large vein that results penetrates the body wall as the **external iliac vein**. As it travels across the inner surface of the ilium, the external iliac fuses with the **internal iliac vein**, which drains the pelvic organs. The resulting **common iliac vein** then meets its counterpart from the opposite side to form the inferior vena cava (IVC).

Like the aorta, the inferior vena cava lies posterior to the abdominopelvic cavity. As it ascends to the heart (Figure 14-19•), it collects blood from several **lumbar veins**. In addition, the IVC receives blood from the *gonadal*, *renal*, *suprarenal*, *phrenic*, and *hepatic veins* before reaching the right atrium.

The Hepatic Portal System You may have noticed that the list of veins did not include any names that refer to digestive organs other than the liver. Instead

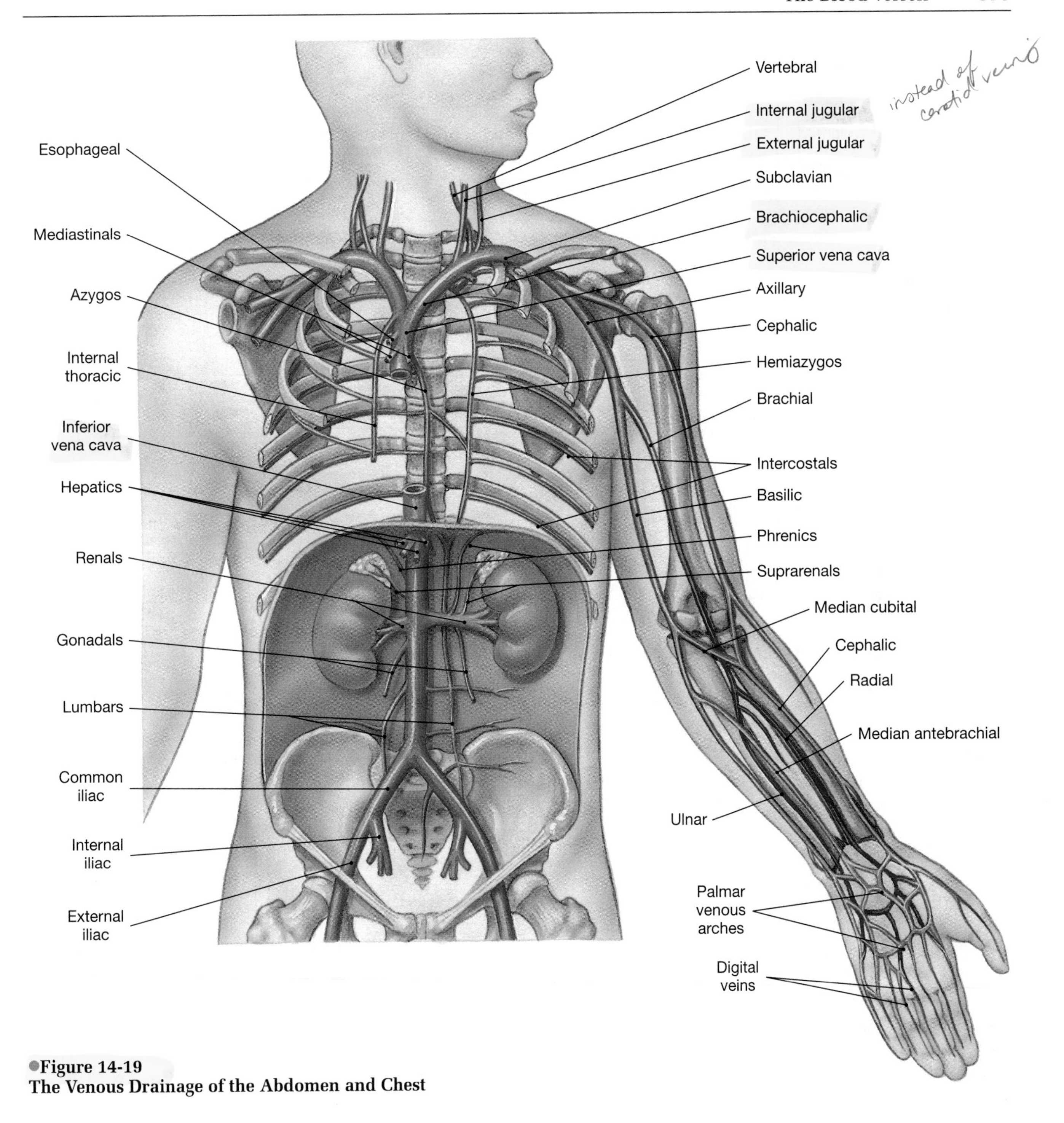

Figure 14-19
The Venous Drainage of the Abdomen and Chest

of traveling directly to the inferior vena cava, blood leaving the capillaries supplied by the celiac, superior, and inferior mesenteric arteries flows to the liver via the **hepatic portal system** (*porta*, a gate). Blood in this system is quite different from that in other systemic veins, because the hepatic portal vessels contain substances absorbed by the digestive tract. For example, levels of blood glucose, amino acids, fatty acids, and vitamins in the hepatic portal vein often exceed those found anywhere else in the circulatory system.

A portal system carries blood from one capillary bed to another and, in the process, prevents its contents from mixing with the entire bloodstream. The liver regulates the concentrations of nutrients, such as glucose or amino acids, in the circulating blood. When digestion is under way, the digestive tract absorbs high concentrations of nutrients, along with various wastes and an occasional toxin. The hepatic portal system delivers these compounds directly to the liver, where liver cells absorb them for storage, metabolic conversion, or excretion. After passing through the liver capillaries, blood collects into the hepatic veins that empty into the inferior vena cava. Because blood goes to the liver first, the composition of the blood in the

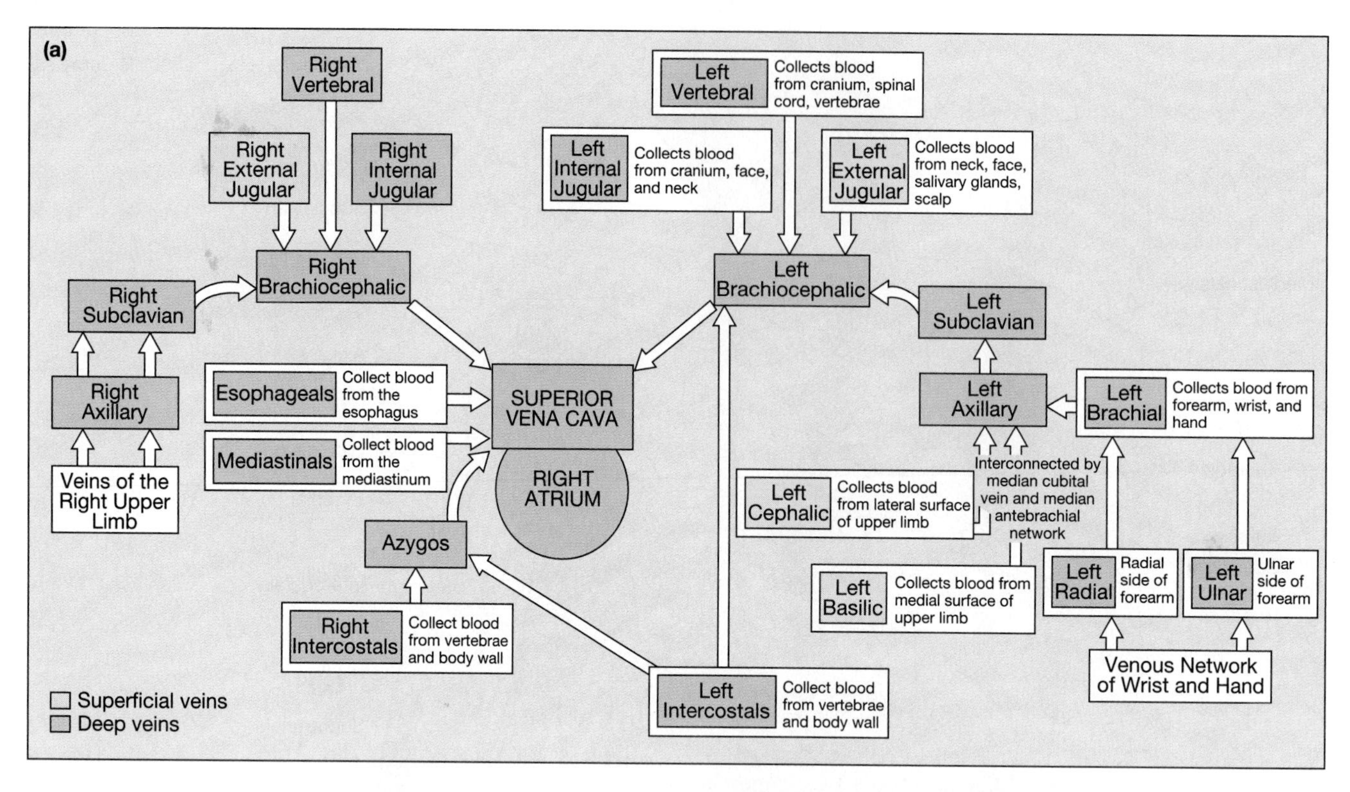

Figure 14-20
Flow Chart of Circulation to the Superior and Inferior Venae Cavae
(a) Tributaries of the SVC. **(b)** Tributaries of the IVC.

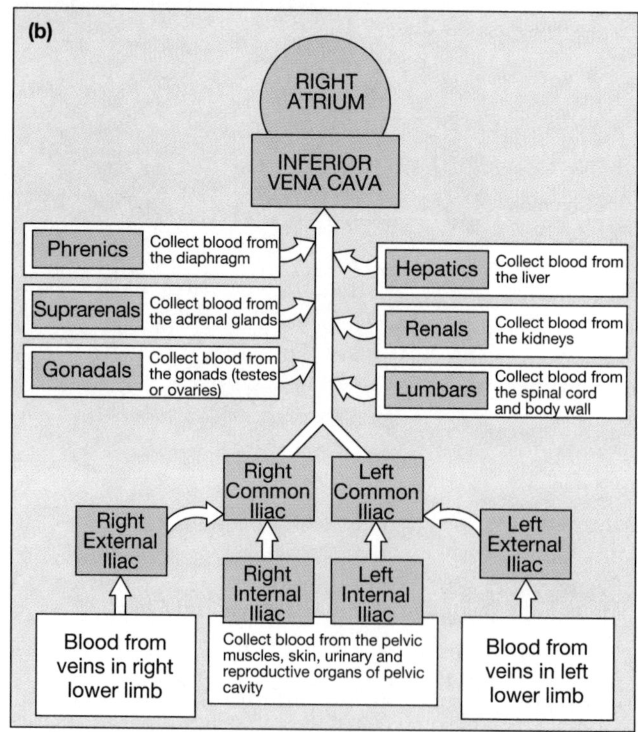

general circulation remains relatively stable, regardless of the digestive activities under way.

Blood delivered to the liver by the hepatic portal system often contains high concentrations of absorbed nutrients, but it is venous blood that contains relatively little oxygen. The liver cells obtain oxygen from arterial blood from the hepatic artery, a branch of the celiac trunk. Liver cells are thus exposed to a mixture of portal blood and arterial blood.

Figure 14-21 details the anatomy of the hepatic portal system. The system ends with the **hepatic portal vein**, which empties into the liver capillaries. It begins in the capillaries of the digestive organs. Blood from capillaries along the lower portion of the large intestine enters the **inferior mesenteric vein**. As it nears the liver, veins from the spleen, the lateral border of the stomach, and the pancreas fuse with the inferior mesenteric, forming the **splenic vein**. The **superior mesenteric vein** also drains the lateral border of the stomach, through an anastomosis with one of the branches of the splenic vein. In addition, the superior mesenteric collects blood from the entire small intestine and two-thirds of the large intestine. The hepatic portal vein forms through the fusion of the superior mesenteric and splenic veins. Of the two, the superior mesenteric normally contributes the greater volume of blood and most of the nutrients.

✓ Blockage of which branch of the aortic arch would interfere with the blood flow to the left arm?

✓ Why would compression of the common carotid artery cause a person to lose consciousness?

✓ Grace is in an automobile accident and ruptures her celiac artery. What organs would be affected most directly by this injury?

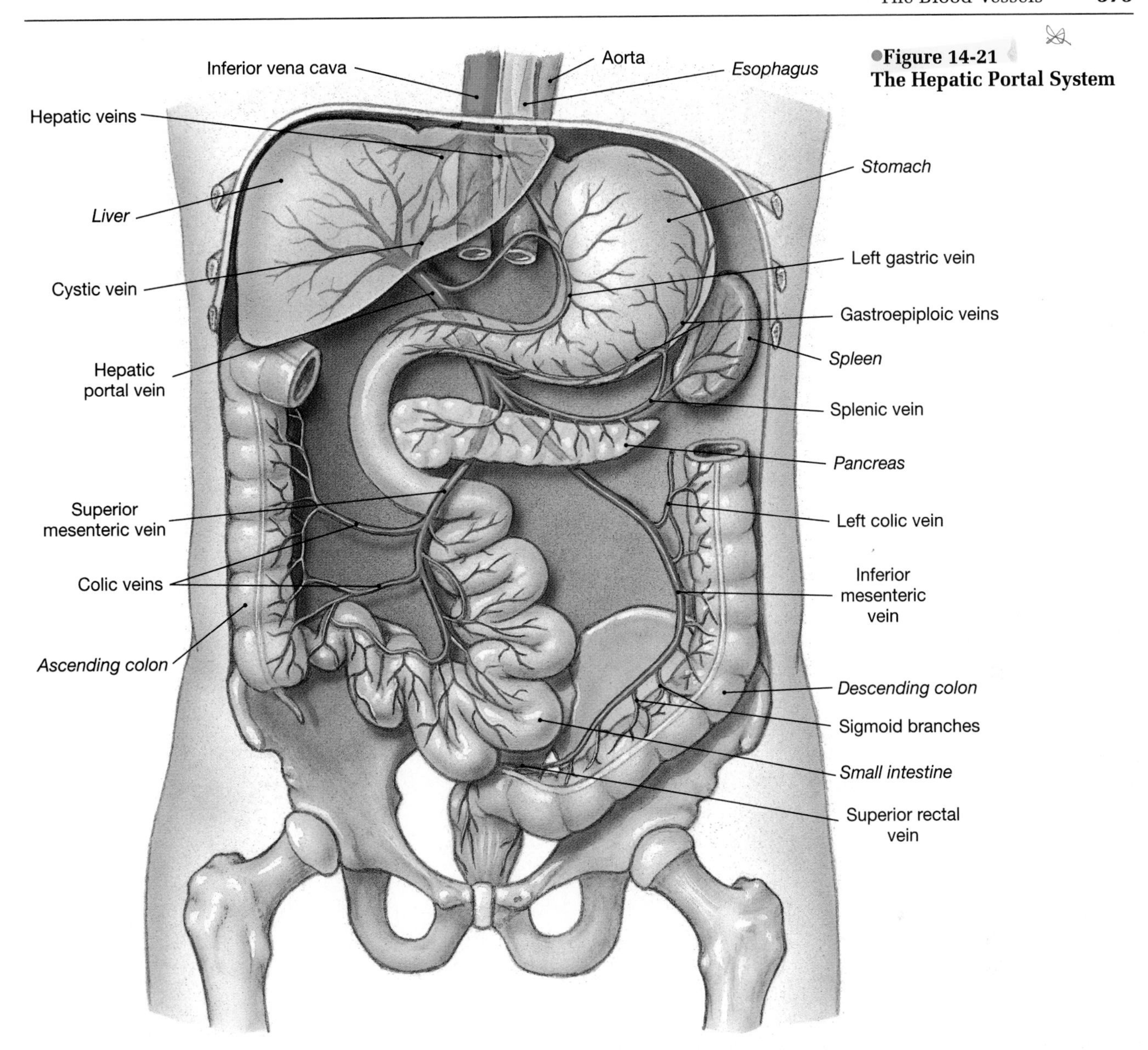

Figure 14-21 The Hepatic Portal System

Fetal Circulation

There are significant differences between the fetal and adult circulatory systems that reflect differing sources of respiratory and nutritional support. The embryonic lungs are collapsed and nonfunctional, and the digestive tract has nothing to digest. All of the embryonic nutritional and respiratory needs are provided by diffusion across the placenta.

Placental Blood Supply

Fetal circulation is diagrammed in Figure 14-22. Blood flow to the placenta is provided by a pair of **umbilical arteries** that arise from the internal iliac arteries and enter the umbilical cord. Blood returns from the placenta in the **umbilical vein**, bringing oxygen and nutrients to the developing fetus. The umbilical vein delivers blood to capillaries within the developing liver and to the inferior vena cava by the *ductus venosus*. When the placental connection is broken at birth, blood flow ceases along the umbilical vessels, and they soon degenerate.

Circulation in the Heart and Great Vessels

One of the most interesting aspects of circulatory development reflects the differences between the life of an embryo or fetus and that of an infant. Throughout embryonic and fetal life, the lungs are collapsed; yet following delivery, the newborn infant must be able to extract oxygen from inspired air rather than across the placenta.

Although the interatrial and interventricular septa of the heart develop early in fetal life, the interatrial partition remains functionally incomplete up to the time of birth. The interatrial opening, or *foramen ovale*, is associated with an elongate flap that acts as a valve. Blood can flow freely from the right atrium to the left atrium, but any backflow will close the valve and iso-

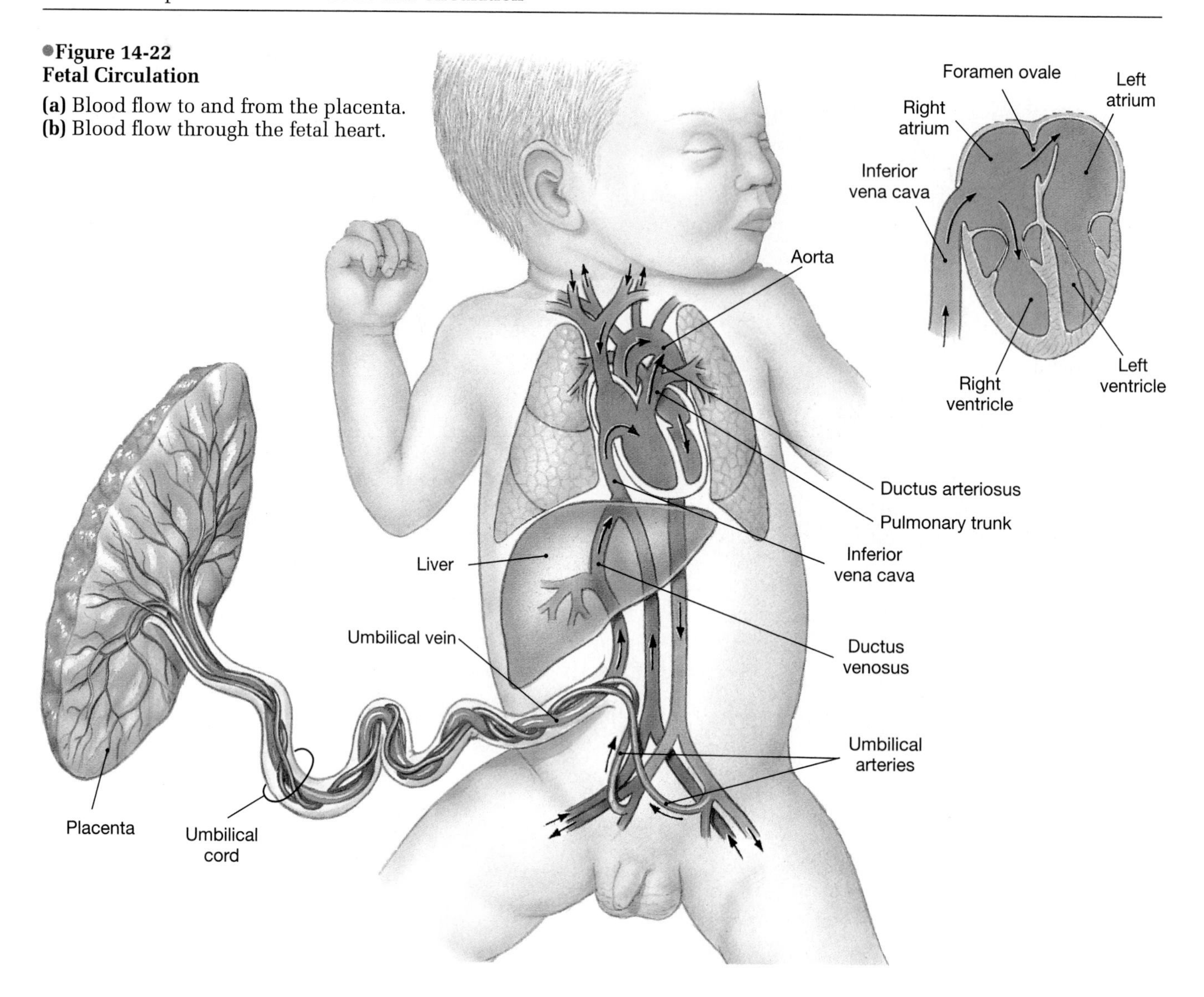

●Figure 14-22
Fetal Circulation
(a) Blood flow to and from the placenta.
(b) Blood flow through the fetal heart.

late the two chambers. Thus blood can enter the heart at the right atrium and bypass the pulmonary circuit altogether. A second short-circuit exists between the pulmonary and aortic trunks. This connection, the *ductus arteriosus*, consists of a short, muscular vessel.

With the lungs collapsed, the capillaries are compressed and little blood flows through the lungs. During diastole, blood enters the right atrium and flows into the right ventricle, but it also passes into the left atrium via the foramen ovale. About 25 percent of the blood arriving at the right atrium bypasses the pulmonary circuit in this way. In addition, over 90 percent of the blood leaving the right ventricle passes through the ductus arteriosus and enters the systemic circuit, rather than continuing to the lungs.

Circulatory Changes at Birth

At birth, dramatic changes occur in circulatory patterns. When the infant takes its first breath, the lungs expand, and so do the pulmonary vessels. Within a few seconds, the smooth muscles in the ductus arteriosus contract, isolating the pulmonary and aortic trunks, and blood begins flowing through the pulmonary circuit. As pressures rise in the left atrium, the valvular flap closes the foramen ovale and completes the circulatory remodeling. In the adult, the interatrial septum bears a shallow depression, the *fossa ovalis*, that marks the site of the foramen ovale (see Figure 13-4●, p. 332) The remnants of the ductus arteriosus persist as a fibrous cord, the *ligamentum arteriosum*.

If the proper circulatory changes do not occur at birth or shortly thereafter, problems will eventually develop because the blood flow to the lungs will not be sufficient to provide adequate amounts of oxygen. Treatment may involve surgical closure of the foramen ovale, the ductus arteriosus or both. Other forms of congenital heart defects result from abnormal cardiac development or inappropriate connections between the heart and major arteries and veins. *Congenital Circulatory Problems*

AGING AND THE CARDIOVASCULAR SYSTEM

The capabilities of the cardiovascular system gradually decline with age. Major changes affect all parts of the cardiovascular system: blood, heart, and vessels.

In the blood age-related changes may include (1) decreased hematocrit; (2) constriction or blockage of peripheral veins by formation of a *thrombus* (stationary blood clot); the thrombus can become detached, pass through the heart, and become wedged in a small artery, most often in the lungs, causing a *pulmonary embolism*; (3) pooling of blood in the veins of the legs because valves are not working effectively.

In the heart age-related changes include (1) a reduction in the maximum cardiac output; (2) changes in the activities of the nodal and conducting cells; (3) a reduction in the elasticity of the fibrous skeleton; (4) a progressive atherosclerosis that can restrict coronary circulation; and (5) replacement of damaged cardiac muscle cells by scar tissue.

In blood vessels age-related changes are often related to arteriosclerosis, a thickening and toughening of the arterial wall. For example, (1) the inelastic walls of arteries become less tolerant of sudden pressure increases, which may lead to a localized dilation, or *aneurysm*, whose subsequent rupture may cause a stroke, infarct, or massive blood loss, depending on the vessel involved; (2) calcium salts can be deposited on weakened vascular walls, increasing the risk of a stroke or infarct; (3) thrombi can form at atherosclerotic plaques.

INTEGRATION WITH OTHER SYSTEMS

The cardiovascular system is both anatomically and functionally linked to all other sysems, as vessel distribution makes clear. Figure 14-23● summarizes the physiological relationships between the cardiovascular system and other organ systems. The most extensive communication occurs between the cardiovascular and lymphatic systems. Not only are the two systems physically interconnected, but cell populations of the lymphatic system use the cardiovascular system as a highway to move from one part of the body to another. The next chapter examines the lymphatic system in detail.

Chapter Review

KEY TERMS

anastomosis, p. 350
arteriole, p. 348
blood pressure, p. 351
capillary, p. 348
capillary exchange, p. 353
hepatic portal system, p. 372
peripheral resistance, p. 351
pulmonary circuit, p. 363
respiratory pump, p. 355
systemic circuit, p. 363
vasoconstriction, p. 357
vasodilation, p. 357
venule, p. 348

SUMMARY OUTLINE

INTRODUCTION p. 348

1. Blood flows through a network of arteries, veins, and capillaries. All chemical and gaseous exchange between the blood and interstitial fluid takes place across capillary walls.

THE ANATOMY OF BLOOD VESSELS p. 348

1. Arteries and veins form an internal distribution system, propelled by the heart. **Arteries** branch repeatedly, decreasing in size until they become **arterioles**; from the arterioles blood enters the capillary networks. Blood flowing from the capillaries enters small **venules** before entering larger **veins**.

Structure of Vessel Walls p. 348

2. The walls of arteries and veins contain three layers: the **tunica interna**, **tunica media**, and outermost **tunica externa**. *(Figure 14-1)*

Arteries p. 348

3. In general, the walls of arteries are thicker than those of veins. The arterial system includes the large **elastic arteries**, medium-sized **muscular arteries**, and smaller arterioles. As we proceed toward the capillaries, the number of vessels increases, but the diameter of the individual vessels decreases and the walls become thinner.

Capillaries p. 349

4. Capillaries are the only blood vessels whose walls permit exchange between blood and interstitial fluid.

5. Capillaries form interconnected networks called **capillary beds**. A **precapillary sphincter** (a band of smooth muscle) adjusts the blood flow into each capillary. Blood flow within a capillary changes as **vasomotion** occurs. *(Figure 14-2)*

Veins p. 350

6. Venules collect blood from the capillaries and merge into **medium-sized veins** and then **large veins**. The arterial system is a high-pressure system; blood pressure in veins is much lower. Valves in these vessels prevent the backflow of blood. *(Figure 14-3)*

INTEGUMENTARY SYSTEM

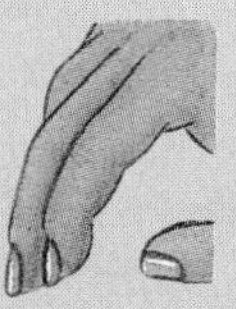

Mast cell stimulation produces localized changes in blood flow and capillary permeability

Delivers immune system cells to injury sites; clotting response seals breaks in skin surface; carries away toxins from sites of infection; provides heat

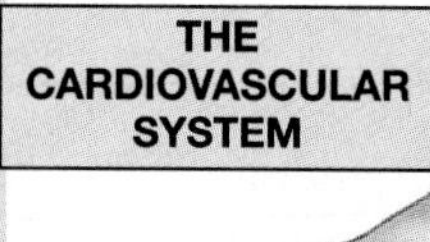

THE CARDIOVASCULAR SYSTEM

FOR ALL SYSTEMS

Delivers oxygen, hormones, nutrients, and white blood cells; removes carbon dioxide and metabolic wastes; transfers heat

SKELETAL SYSTEM

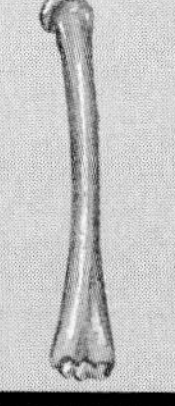

Stores calcium needed for normal cardiac muscle contraction; protects blood cells developing in bone marrow

Provides Ca^{2+} and PO_4^{3-} ions for bone deposition; delivers EPO to bone marrow, parathyroid hormone and calcitonin to osteoblasts and osteoclasts

MUSCULAR SYSTEM

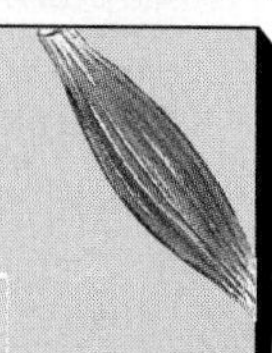

Skeletal muscle contractions assist in moving blood through veins; protects superficial blood vessels, especially in neck and limbs

Delivers oxygen and nutrients, removes carbon dioxide, lactic acid, and heat during skeletal muscle activity

NERVOUS SYSTEM

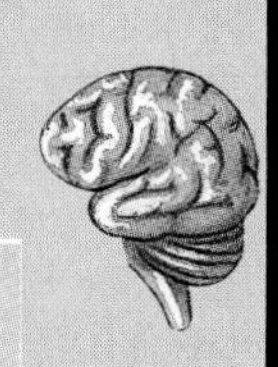

Controls patterns of circulation in peripheral tissues; modifies heart rate and regulates blood pressure; releases ADH

Endothelial cells maintain blood-brain barrier, help generate cerebrospinal fluid

ENDOCRINE SYSTEM

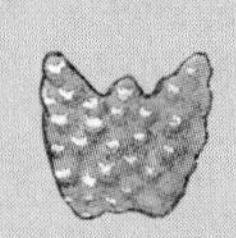

Erythropoietin regulates production of RBCs; several hormones elevate blood pressure; epinephrine stimulates cardiac muscle, elevating heart rate and contractile force

Distributes hormones throughout the body; atria secrete ANP

LYMPHATIC SYSTEM

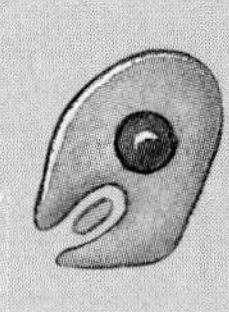

Defends against pathogens or toxins in blood; fights infections of cardiovascular organs; returns tissue fluid to circulation

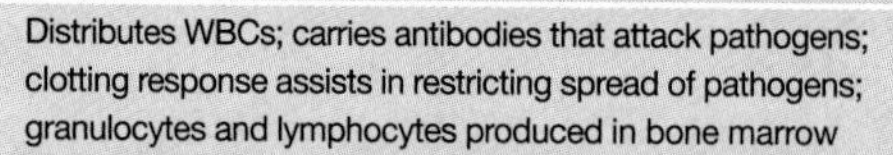

Distributes WBCs; carries antibodies that attack pathogens; clotting response assists in restricting spread of pathogens; granulocytes and lymphocytes produced in bone marrow

RESPIRATORY SYSTEM

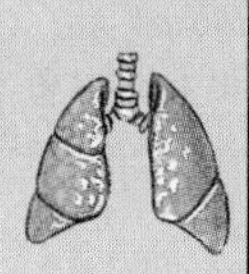

Provides oxygen to cardiovascular organs and removes carbon dioxide

RBCs transport oxygen and carbon dioxide between lungs and peripheral tissues

DIGESTIVE SYSTEM

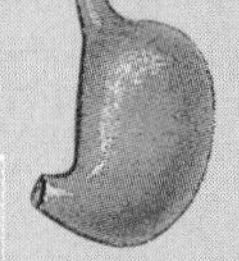

Provides nutrients to cardiovascular organs; absorbs water and ions essential to maintenance of normal blood volume

Distributes digestive tract hormones; carries nutrients, water, and ions away from sites of absorption; delivers nutrients and toxins to liver

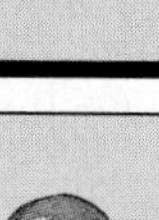

URINARY SYSTEM

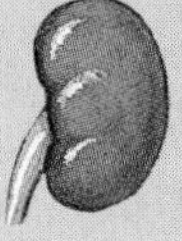

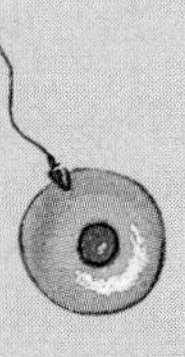

Estrogens may maintain healthy vessels and slow development of atherosclerosis with age

Distributes reproductive hormones; provides nutrients, oxygen, and waste removal for developing fetus; local blood pressure changes responsible for physical changes during sexual arousal

CIRCULATORY PHYSIOLOGY p. 351

Pressure p. 351

1. Flow is proportional to the difference in pressure; blood will flow from an area of higher pressure to one of relatively lower pressure.

Resistance p. 351

2. For circulation to occur, the *circulatory pressure* must be greater than the *total peripheral resistance* (the resistance of the entire circulatory system). For blood to flow into peripheral capillaries, **blood pressure** (arterial pressure) must be greater than the **peripheral resistance** (the resistance of the arterial system). Neural and hormonal control mechanisms regulate blood pressure.

3. The most important determinant of peripheral resistance is the diameter of arterioles.

Circulatory Pressure p. 352

4. The high arterial pressures overcome peripheral resistance and maintain blood flow through peripheral tissues. Capillary pressures are normally low; small changes in capillary pressure determine the rate of fluid movement into or out of the bloodstream. Venous pressure, normally low, determines venous return and affects cardiac output and peripheral blood flow.

5. Arterial pressure rises during ventricular systole and falls during ventricular diastole. The difference between these two pressures is **pulse pressure**. (*Figures 14-4, 14-5*)

6. At the capillaries, solute molecules diffuse across the capillary lining, and water-soluble materials diffuse through small spaces between endothelial cells. Water will move when driven by either hydrostatic or osmotic pressure. The direction of water movement is determined by the balance between these two opposing pressures. (*Figure 14-6*)

7. **Valves, muscular compression**, and the **respiratory pump** help the relatively low venous pressures propel blood toward the heart. (*Figure 14-3*)

CARDIOVASCULAR REGULATION p. 356

1. Homeostatic mechanisms ensure that tissue blood flow (perfusion) delivers adequate oxygen and nutrients.

2. Blood flow varies directly with cardiac output, peripheral resistance, and blood pressure.

3. Local, autonomic, and endocrine factors influence the coordinated regulation of cardiovascular function. Local factors change the pattern of blood flow within capillary beds in response to chemical changes in interstitial fluids. Central nervous system mechanisms respond to changes in arterial pressure or blood gas levels. Hormones can assist in short-term adjustments (changes in cardiac output and peripheral resistance) and long-term adjustments (changes in blood volume that affect cardiac output and gas transport). (*Figure 14-7*)

Autoregulation of Blood Flow p. 357

4. Peripheral resistance is adjusted at the tissues by local factors that result in the dilation or constriction of precapillary sphincters.

Neural Control of Blood Pressure and Blood Flow p. 357

5. **Baroreceptor reflexes** are autonomic reflexes that adjust cardiac output and peripheral resistance to maintain normal arterial pressures. Baroreceptor populations include the aortic and carotid sinuses and atrial baroreceptors. (*Figure 14-8*)

6. **Chemoreceptor reflexes** respond to changes in the oxygen or carbon dioxide levels in the blood and cerebrospinal fluid. Sympathetic activation leads to stimulation of the cardioacceleratory and vasomotor centers; parasympathetic activation stimulates the cardioinhibitory center. Epinephrine and norepinephrine stimulate cardiac output and peripheral vasoconstriction. (*Figure 14-9*)

Hormones and Cardiovascular Regulation p. 359

7. The endocrine system provides short-term regulation of cardiac output and peripheral resistance with epinephrine and norepinephrine from the adrenal medullae. Hormones involved in long-term regulation of blood pressure and volume are antidiuretic hormone (ADH), angiotensin II, erythropoietin (EPO), and atrial natriuretic peptide (ANP). (*Figure 14-10*)

8. ADH and angiotensin II also promote peripheral vasoconstriction in addition to their other functions. ADH and aldosterone promote water and electrolyte retention and stimulate thirst. EPO stimulates red blood cell production. ANP encourages sodium loss, fluid loss, reduces blood pressure, inhibits thirst, and lowers peripheral resistance.

PATTERNS OF CARDIOVASCULAR RESPONSE p. 361

Exercise and the Cardiovascular System p. 361

1. During exercise, blood flow to skeletal muscles increases at the expense of circulation to nonessential organs, and cardiac output rises. Cardiovascular performance improves with training. Athletes have larger stroke volumes, slower resting heart rates, and greater cardiac reserves than nonathletes.

Cardiovascular Response to Hemorrhaging p. 361

2. Blood loss causes an increase in cardiac output, mobilization of venous reserves, peripheral vasoconstriction, and the liberation of hormones that promote fluid retention and the manufacture of erythrocytes.

THE BLOOD VESSELS p. 363

1. The peripheral distributions of arteries and veins are usually identical on both sides of the body, except near the heart. (*Figure 14-11*)

The Pulmonary Circulation p. 363

2. The pulmonary circuit includes the **pulmonary trunk**, the **left** and **right pulmonary arteries**, and the **pulmonary veins** that empty into the left atrium. (*Figure 14-12*)

The Systemic Circulation, p. 363

3. The **ascending aorta** gives rise to the coronary circulation. The **aortic arch** communicates with the **descending aorta**. (*Figures 14-13 to 14-16*)

4. Arteries in the neck and limbs are deep beneath the skin; in contrast, there are usually two sets of peripheral veins, one superficial and one deep. This dual-venous drainage is important for controlling body temperature.

5. The **superior vena cava** receives blood from the head, neck, chest, shoulders, and arms. (*Figures 14-17 to 14-20*)

6. The **inferior vena cava** collects most of the venous blood from organs below the diaphragm. (*Figure 14-20*)

7. The **hepatic portal vein** collects blood from visceral organs in the abdominopelvic cavity. The hepatic portal vein delivers blood to capillary networks in the liver. (*Figure 14-21*)

Fetal Circulation p. 375

8. The placenta receives blood from the two **umbilical arteries**. Blood returns to the fetus via the **umbilical vein** that delivers blood to the *ductus venosus* within the liver. (*Figure 14-22a*)

9. Prior to delivery, blood bypasses the pulmonary circuit by flowing (1) from the right atrium into the left atrium through the *foramen ovale*, and (2) from the pulmonary trunk into the aortic arch via the *ductus arteriosus*. (*Figure 14-22*)

AGING AND THE CARDIOVASCULAR SYSTEM p. 377

1. Age-related changes in the blood can include (1) decreased hematocrit; (2) constriction or blockage of peripheral veins by a *thrombus* (stationary blood clot); (3) pooling of blood in the veins of the legs because the valves are not working effectively.

2. Age-related changes in the heart include (1) a reduction in the maximum cardiac output, (2) changes in the activities of the nodal and conducting cells, (3) a reduction in the elasticity of the fibrous skeleton, (4) a progressive atherosclerosis that can restrict coronary circulation, and (5) replacement of damaged cardiac muscle cells by scar tissue.

3. Age-related changes in blood vessels, often related to arteriosclerosis, include (1) inelastic walls of arteries are less tolerant of sudden pressure increases, which may lead to an *aneurysm*; (2) calcium salts can deposit on weakened vascular walls, increasing the risk of a stroke or infarct; (3) thrombi can form at atherosclerotic plaques.

INTEGRATION WITH OTHER SYSTEMS p. 377

1. The cardiovascular system delivers oxygen, nutrients, and hormones to all the body systems. (*Figure 14-23*)

CHAPTER QUESTIONS

LEVEL 1 Reviewing Facts and Terms

Match each item in column A with the most closely related item in column B. Use letters for answers in the spaces provided.

Column A	Column B
___ 1. diastolic pressure	a. drains the liver
___ 2. arterioles	b. largest superficial vein in body
___ 3. hepatic vein	c. carotid sinus
___ 4. renal vein	d. minimum blood pressure
___ 5. aorta	e. blood supply to leg
___ 6. precapillary sphincter	f. blood supply to pelvis
___ 7. medulla oblongata	g. peak blood pressure
___ 8. internal iliac artery	h. vasomotion
___ 9. external iliac artery	i. largest artery in body
___ 10. baroreceptors	j. drains the kidney
___ 11. systolic pressure	k. vasomotor center
___ 12. saphenous vein	l. smallest arterial vessels

13. Blood vessels that carry blood away from the heart are called:
(a) veins (b) arterioles
(c) venules (d) arteries

14. The layer of the arteriole wall that provides the properties of contractility and elasticity is the:
(a) tunica adventitia (b) tunica media
(c) tunica interna (d) tunica externa

15. The two-way exchange of substances between blood and body cells occurs only through:
(a) arterioles (b) capillaries
(c) venules (d) a, b, and c are correct

16. The blood vessels that collect blood from all tissues and organs and return it to the heart are the:
(a) veins (b) arteries
(c) capillaries (d) arterioles

17. Blood is compartmentalized within the veins because of the presence of:
(a) venous reservoirs (b) muscular walls
(c) clots (d) valves

18. The most important factor in vascular resistance is:
(a) the viscosity of the blood
(b) friction between the blood and the vessel walls
(c) turbulence due to irregular surfaces of blood vessels
(d) the length of the blood vessels

19. In a blood pressure reading of 120/80, the 120 represents ______ and the 80 represents ______.
(a) diastolic pressure; systolic pressure
(b) pulse pressure; mean arterial pressure
(c) systolic pressure; diastolic pressure
(d) mean arterial pressure; pulse pressure

20. Hydrostatic pressure forces water ________ a solution; osmotic pressure forces water _______ a solution.
(a) into, out of (b) out of, into
(c) out of, out of (d) a, b, and c are incorrect

21. The two factors that assist the relatively low venous pressures in propelling blood toward the heart are:
(a) ventricular systole and valvular closure
(b) gravity and vasomotion
(c) muscular compression and the respiratory pump
(d) atrial and ventricular contractions

22. The arteries of the pulmonary circuit differ from those of the systemic circuit in that they carry;
(a) oxygen and nutrients
(b) deoxygenated blood
(c) oxygenated blood
(d) oxygen, carbon dioxide, and nutrients
23. The two arteries formed by the division of the brachiocephalic artery are the:
(a) aorta and internal carotid
(b) axillary and brachial
(c) external and internal carotid
(d) common carotid and subclavian
24. The unpaired arteries supplying blood to the visceral organs include the:
(a) suprarenal, renal, lumbar
(b) iliac, gonadal, femoral
(c) celiac, superior and inferior mesenterics
(d) a, b, and c are correct
25. The artery generally used to feel the pulse at the wrist is the:
(a) ulnar (b) radial
(c) peroneal (d) dorsalis
26. The vein that drains the dural sinuses of the brain is the:
(a) cephalic (b) great saphenous
(c) internal jugular (d) superior vena cava
27. The vein that collects most of the venous blood from below the diaphragm is the:
(a) superior vena cava (b) great saphenous
(c) inferior vena cava (d) azygos
28. (a) What are the primary forces that cause fluid to move out of a capillary and into the interstitial fluid at its arterial end? (b) What are the primary forces that cause fluid to move into a capillary from the interstitial fluid at its venous end?
29. What two effects occur when the baroreceptor response to elevated blood pressure is triggered?
30. What factors affect the activity of chemoreceptors in the carotid and aortic bodies?
31. What circulatory changes occur at birth?
32. What age-related changes take place in the blood, heart, and blood vessels?

LEVEL 2 Reviewing Concepts

33. When dehydration occurs there is:
(a) accelerated reabsorption of water at the kidneys
(b) reabsorption of fluids from the interstitial fluid
(c) an increase in the blood osmotic pressure
(d) a, b, and c are correct
34. Increased CO_2 levels in tissues would promote:
(a) constriction of precapillary sphincters
(b) an increase in the pH of the blood
(c) dilation of precapillary sphincters
(d) decreased blood flow to tissues
35. Elevated levels of the hormones ADH and angiotensin II will produce:
(a) increased peripheral vasodilation
(b) increased peripheral vasoconstriction
(c) increased peripheral blood flow
(d) increased venous return
36. Relate the anatomical differences between arteries and veins to their functions.
37. Why do capillaries permit the diffusion of materials whereas arteries and veins do not?
38. Why is blood flow to the brain relatively continuous and constant?
39. An accident victim displays the following symptoms: hypotension; pale, cool, moist skin; confusion and disorientation. Identify her condition and explain why these symptoms occur. If you took her pulse, what would you find?

LEVEL 3 Critical Thinking and Clinical Applications

40. Bob is sitting outside on a warm day and is sweating profusely. His friend Mary wants to practice taking blood pressures, and he agrees to play patient. Mary finds that Bob's blood pressure is elevated, even though he is resting and has lost fluid from sweating (she reasons that fluid loss should lower blood volume and thus blood pressure). Mary asks you why Bob's blood pressure is high instead of low. What will you tell her?
41. People who suffer from allergies frequently take antihistamines and decongestants to relieve their symptoms. The medication's box warns that the medication should not be taken by individuals being treated for high blood pressure. Why?
42. Jill awakens suddenly to the sound of her alarm clock. Realizing she is late for class, she jumps to her feet, feels light-headed, and falls back on her bed. What probably caused this to happen? Why doesn't this happen all the time?

15 The Lymphatic System and Immunity

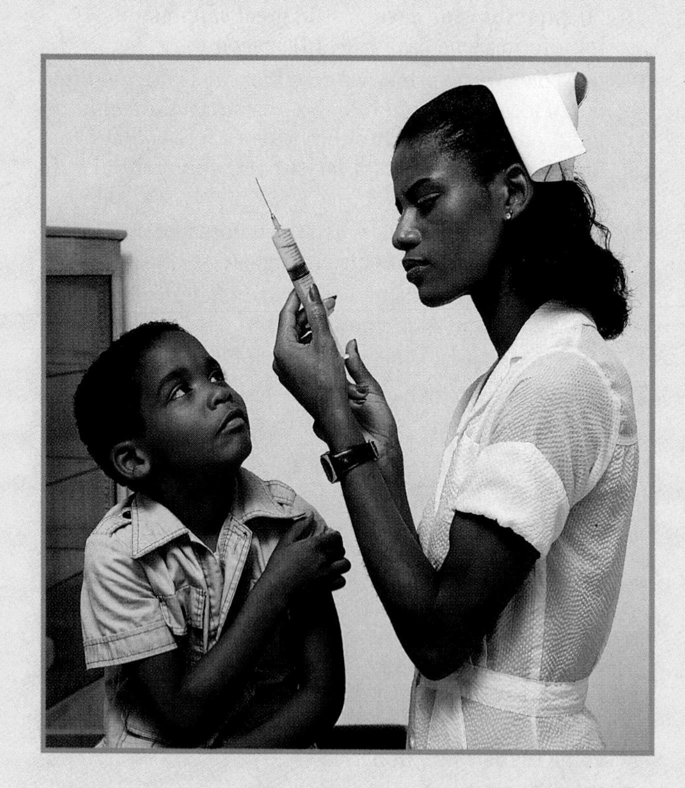

As children, all we notice is the sting of a vaccination. As adults, we realize that the shot hurts only for an instant (if at all), whereas the disease that it prevents can cripple or kill. In this chapter we will examine the defense mechanisms of the body.

Chapter Outline and Objectives

1. Identify the major components of the lymphatic system and explain their functions.
2. Discuss the importance of lymphocytes and describe where they are found in the body.

3. List the body's nonspecific defenses and explain how each functions.

4. Define specific resistance and identify the forms and properties of immunity.
5. Distinguish between cell-mediated immunity and antibody-mediated (humoral) immunity.
6. Discuss the different types of T cells and the role played by each in the immune response.
7. Describe the structure of antibody molecules and explain how they function.
8. Describe the primary and secondary immune responses to antigen exposure.

9. Relate allergic reactions and autoimmune disorders to immune mechanisms.
10. Describe the changes in the immune system that occur with aging.

11. Discuss the structural and functional interactions among the lymphatic system and other body systems.

Vocabulary Development

anamnesis, a memory; *anamnestic response*
apo-, away; *apoptosis*
chemo-, chemistry; *chemotaxis*
dia-, through; *diapedesis*
-gen, to produce; *pyrogen*
humor, a liquid; *humoral immunity*
lympha, water; *lymph*
pathos, disease; *pathogen*
pedesis, a leaping; *diapedesis*
ptosis, a falling; *apoptosis*
pyr, fire; *pyrogen*
taxis, arrangement; *chemotaxis*

The world is not always kind to the human body. Accidental bumps, cuts, and scrapes, chemical and thermal burns, extreme cold, and ultraviolet radiation are just a few of the hazards in the physical environment. Making matters worse, an assortment of viruses, bacteria, fungi, and parasites thrive in the environment. Many of these organisms are perfectly capable of not only surviving, but thriving inside our bodies—and potentially causing us great harm. These microorganisms, called **pathogens** (*pathos*, disease + *-gen*, to produce), are responsible for many human diseases. Each has a different mode of life and attacks the body in a characteristic way. For example, viruses spend most of their time hiding within cells, many bacteria multiply in the interstitial fluids, and the largest parasites burrow through internal organs. AM *The Nature of Pathogens*

Many different organs and systems work together in an effort to keep us alive and healthy. In this ongoing struggle, the **lymphatic system** plays a central role.

Lymphocytes, the dominant cells of the lymphatic system, were introduced in Chapter 12. ∞ [p. 319] These cells are vital to our ability to resist or overcome infection and disease. They respond to the presence of (1) invading pathogens, such as bacteria or viruses; (2) abnormal body cells, such as virus-infected cells or cancer cells; and (3) foreign proteins, such as the toxins released by some bacteria. Lymphocytes attempt to eliminate these threats or render them harmless by a combination of physical and chemical attack.

Lymphocytes respond to specific threats, such as a bacterial invasion of a tissue, by organizing a defense against that specific, or particular, type of bacterium. Such a *specific defense* of the body is known as an **immune response**. **Immunity** is the ability to resist infection and disease through the activation of specific defenses.

This chapter begins by examining the organization of the lymphatic system. We will then consider how the lymphatic system interacts with cells and tissues of other systems to defend the body against infection and disease.

Organization of the Lymphatic System

One of the least familiar organ systems, the lymphatic system includes the following three components:

1. **Vessels.** A network of **lymphatic vessels** that begins in peripheral tissues and ends at connections to the venous system.
2. **Fluid.** A fluid, called **lymph**, flows through the lymphatic vessels. Lymph resembles plasma but contains a much lower concentration of suspended proteins.
3. **Lymphoid organs.** *Lymphoid organs* are connected to the lymphatic vessels and contain large numbers of **lymphocytes**. Examples include the lymph nodes, spleen, and thymus.

Figure 15-1● provides the names and general locations of the vessels and organs of this system.

Functions of the Lymphatic System

The primary functions of the lymphatic system are:

- ***The production, maintenance, and distribution of lymphocytes.*** The lymphocytes are produced and stored within lymphatic organs, such as the spleen, thymus, and bone marrow.
- ***The return of fluid and solutes from peripheral tissues to the blood.*** The return of tissue fluids through the lymphatic system maintains normal blood volume and eliminates local variations in the composition of the interstitial fluid. The volume of flow is considerable—roughly 3.6 liters per day—and a break in a major lymphatic vessel can cause a rapid and potentially fatal decline in blood volume.
- ***The distribution of hormones, nutrients, and waste products from their tissues of origin to the general circulation.*** Substances unable to enter the bloodstream directly may do so via the lymphatic vessels. For example, lipids absorbed by the digestive tract often fail to enter the circulation at the capillary level. However, they still reach the bloodstream by passage along lymphatic vessels (a process explored further in Chapter 17).

Lymphatic Vessels

Lymphatic vessels, often called *lymphatics*, carry lymph from the peripheral tissues to the venous system in all parts of the body except the central nervous

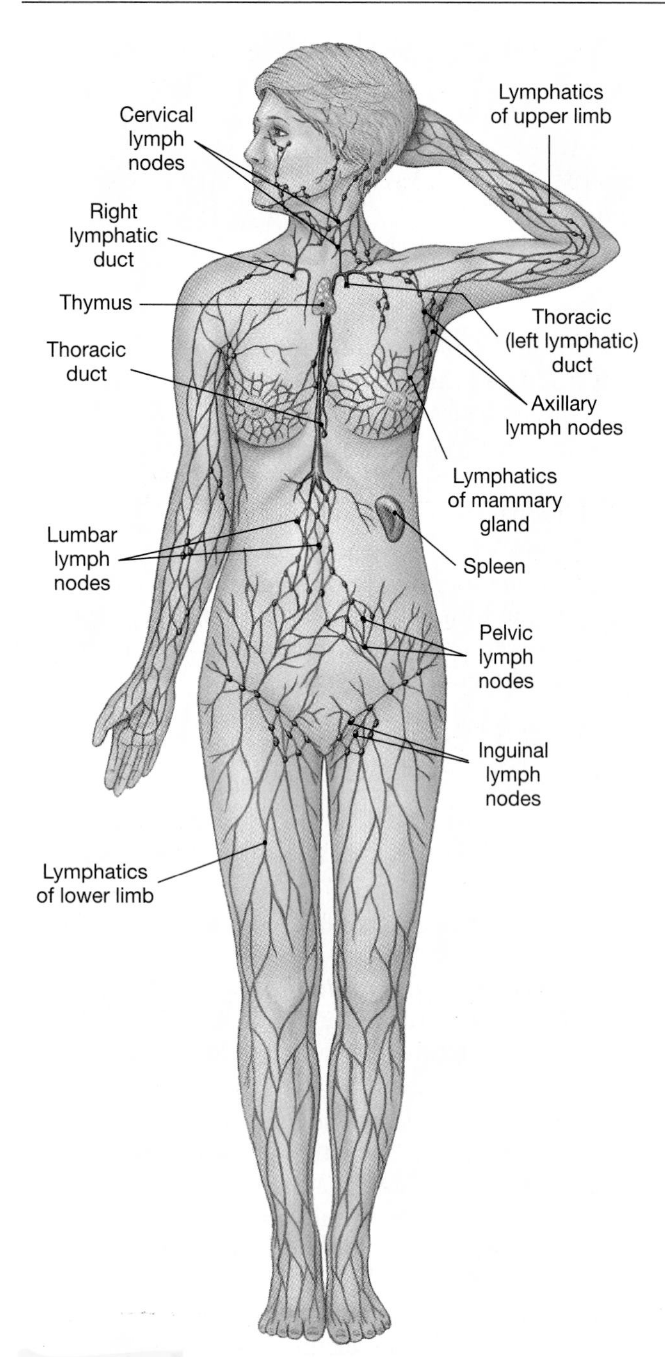

Figure 15-1
Components of the Lymphatic System

system. The smallest vessels begin as blind pockets and are called **lymphatic capillaries** (Figure 15-2a). The arrangement of the endothelial cells in a lymphatic capillary permits fluid entry into the lymphatic but prevents its return to the intercellular spaces.

From the lymphatic capillaries, lymph flows into larger lymphatic vessels that lead toward the trunk. The walls of these lymphatics contain layers comparable to those of veins, and, like veins, the larger lymphatics contain valves (Figure 15-2b). The valves in lymphatic vessels are closer together than those of veins. Pressures within the lymphatic system are extremely low, and the valves are essential to maintaining normal lymph flow.

The lymphatic vessels ultimately empty into two large collecting ducts (Figure 15-3). The **thoracic duct** collects lymph from the lower abdomen, pelvis, and lower limbs and from the left half of the head, neck, and chest. It empties its collected lymph into the venous system near the junction between the left internal jugular vein and the left subclavian vein. The smaller **right lymphatic duct**, which ends at the comparable location on the right side, delivers lymph from the right side of the body above the diaphragm.

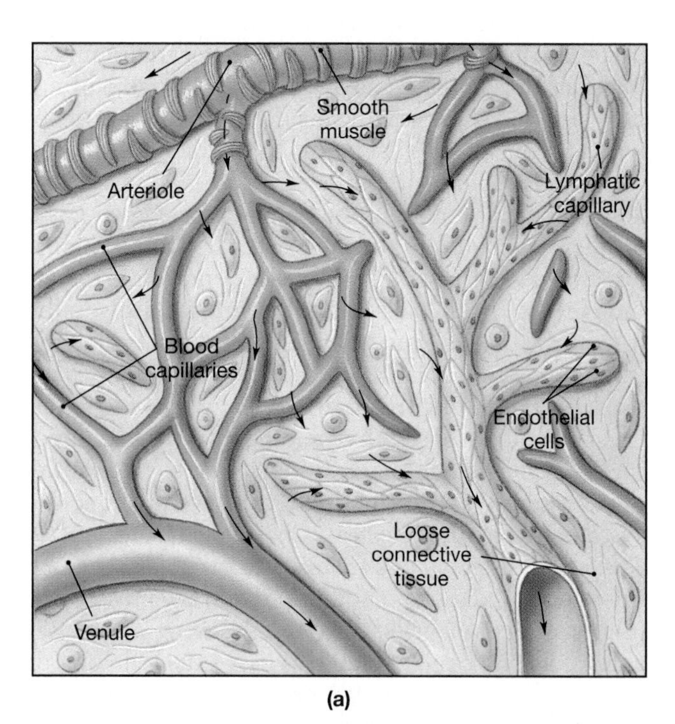

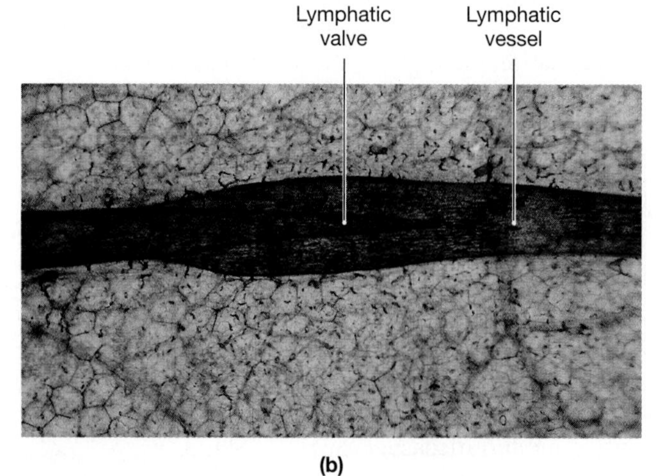

Figure 15-2
Lymphatic Capillaries

(a) A three dimensional view of the association of blood capillaries, tissue, interstitial fluid, and lymphatic capillaries. Arrows show the direction of interstitial fluid and lymph movement. **(b)** A valve within a small lymphatic vessel. (LM × 43)

The blockage of lymphatic drainage in a region can cause peripheral swelling due to the accumulation of interstitial fluid. This condition is called *lymphedema*. AM *Lymphedema*

Lymphocytes

Lymphocytes were introduced in Chapter 12 because they account for roughly 25 percent of the circulating white blood cell population. [p. 319] But circulating lymphocytes are only a small fraction of the total lymphocyte population. The body contains around 10^{12} lymphocytes, with a combined weight of over a kilogram. At any given moment most of the lymphocytes are found within lymphoid organs or other tissues. The bloodstream provides a rapid transport system for lymphocytes moving from one tissue to another.

Types of Lymphocytes

There are three classes of lymphocytes in the blood: **T cells** (**t**hymus-dependent), **B cells** (**b**one marrow-derived), and **NK cells** (**n**atural **k**illers). Each lymphocyte class has distinctive functions.

T Cells Approximately 80 percent of circulating lymphocytes are T cells. *Cytotoxic T cells* directly attack foreign cells or body cells infected by viruses. These lymphocytes are the primary cells that provide *cell-mediated immunity*, or *cellular immunity*. *Helper T cells* stimulate the activities of both T cells and B cells and *suppressor T cells* inhibit both T cells and B cells. Helper and suppressor T cells are also called **regulatory T cells**.

B Cells B cells constitute 10–15 percent of circulating lymphocytes. B cells can differentiate into **plasma cells**, cells responsible for the production and secretion of **antibodies**. Antibodies are globular proteins that are often called **immunoglobulins**. [p. 310] Antibodies react with specific chemical targets, called **antigens**. Antigens are usually pathogens, parts or products of pathogens, or other foreign compounds. When an antigen-antibody complex forms, it starts a chain of events leading to the destruction of the target compound or organism. Because the blood is the primary distribution route for antibodies, B cells are said to be responsible for *antibody-mediated immunity*, or *humoral* ("liquid") *immunity*.

NK Cells The remaining 5–10 percent of circulating lymphocytes are NK cells. These lymphocytes will attack foreign cells, normal cells infected with viruses, and cancer cells that appear in normal tissues.

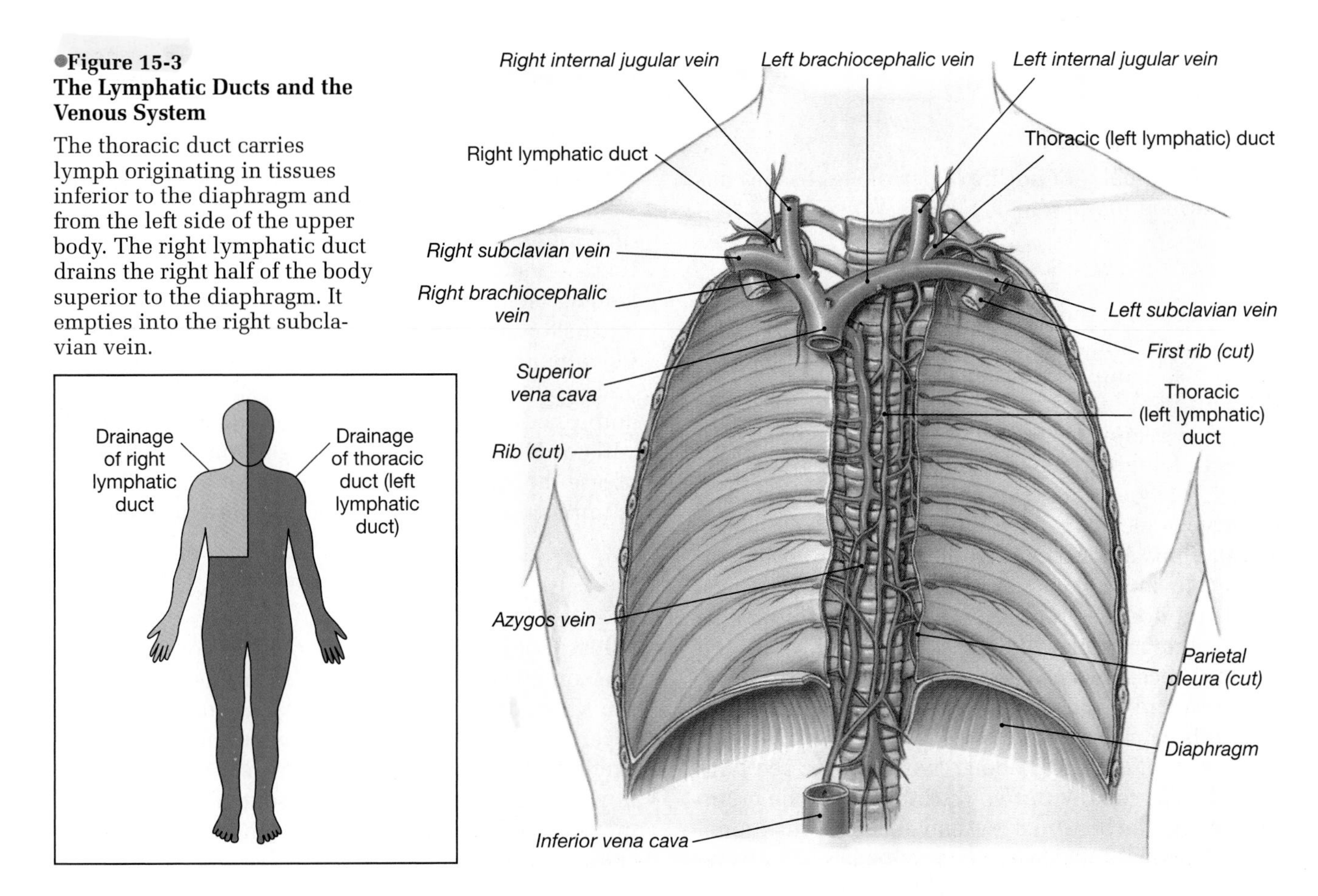

Figure 15-3
The Lymphatic Ducts and the Venous System
The thoracic duct carries lymph originating in tissues inferior to the diaphragm and from the left side of the upper body. The right lymphatic duct drains the right half of the body superior to the diaphragm. It empties into the right subclavian vein.

Figure 15-4
Derivation and Distribution of Lymphocytes
Hemocytoblast divisions produce lymphoid stem cells with two different fates. One group remains in the bone marrow, producing daughter cells that mature into B cells and NK cells. The second group migrates to the thymus, where subsequent divisions produce daughter cells that mature into T cells. All three lymphocyte types circulate throughout the body in the bloodstream, leaving the circulation to take temporary residence in peripheral tissues.

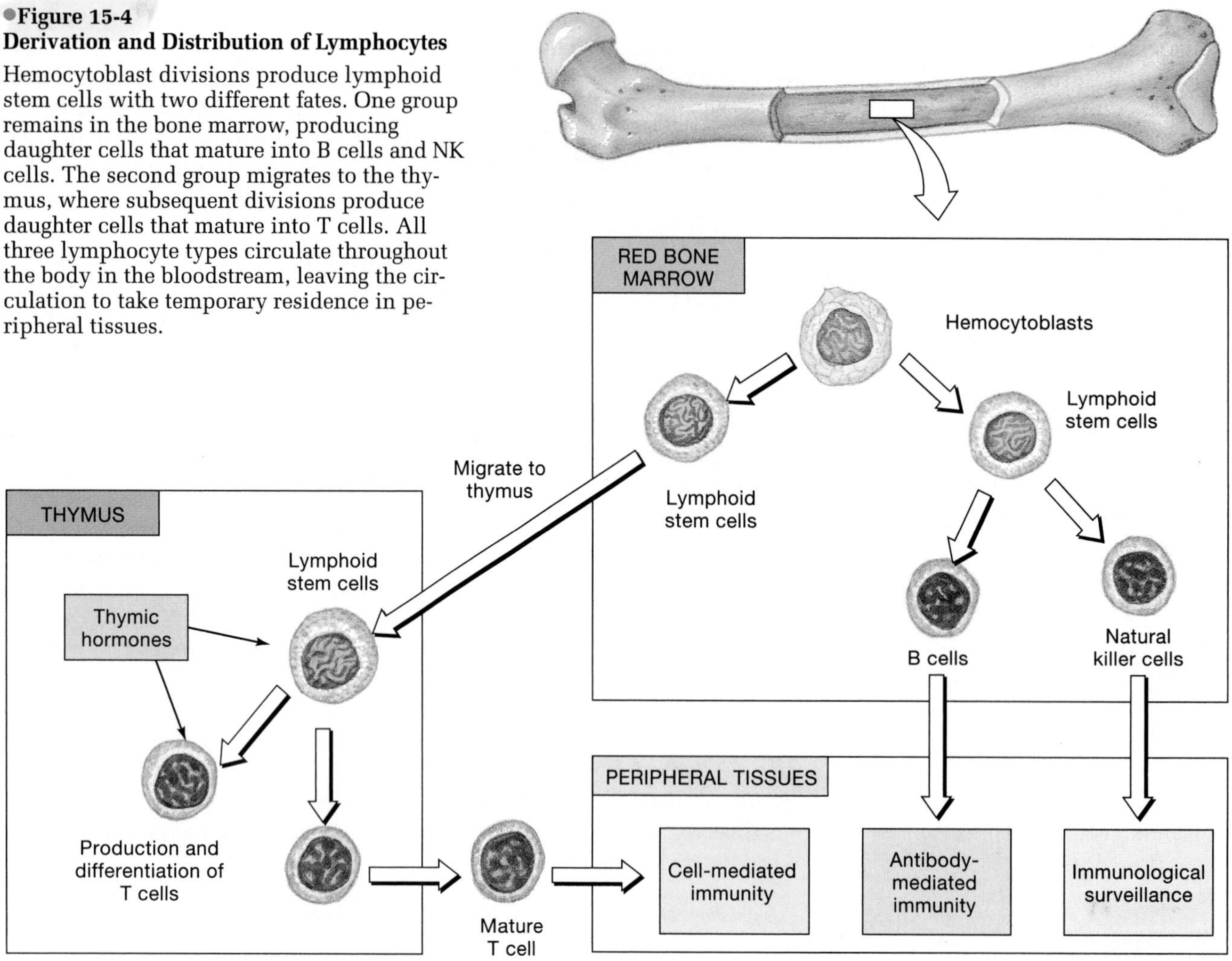

Their continual monitoring of peripheral tissues has been termed *immunological surveillance*.

Origin and Circulation of Lymphocytes

Lymphocytes in the blood, bone marrow, spleen, thymus, and peripheral lymphoid tissues are visitors, not residents. Lymphocytes move throughout the body; they wander through a tissue and then enter a blood vessel or lymphatic for transport to another site. In general, lymphocytes have relatively long life spans. Roughly 80 percent survive for 4 years, and some last 20 years or more. Throughout life, normal lymphocyte populations are maintained through the divisions of stem cells in the bone marrow and lymphoid tissues.

Lymphocyte production, or *lymphopoiesis*, involves the bone marrow and thymus (Figure 15-4●). In this process, each B cell and T cell gains the ability to respond to the presence of a specific antigen, and NK cells gain the ability to recognize abnormal cells. *Hemocytoblasts* in the bone marrow produce lymphoid stem cells with two distinct fates. One group remains in the bone marrow and generates functional NK cells and B cells that enter the circulation and travel throughout the body. The second group of lymphoid stem cells migrates to the thymus. Under the influence of hormones collectively known as the *thymosins*, these cells divide repeatedly, producing large numbers of T cells that reenter the circulation.

As these lymphocyte populations migrate through peripheral tissues, they retain the ability to divide and produce daughter cells of the same type. For example, a dividing B cell produces other B cells, not T cells or NK cells. As we shall see, the ability to increase the number of lymphocytes of a specific type is important to the success of the immune response.

Lymphoid Nodules

A **lymphoid nodule** is an area of loose connective tissue containing densely packed lymphocytes. A lymphoid nodule does not have a stable internal organization, and there is no surrounding capsule. Their size can increase or decrease, depending on how many lymphocytes are present at any given moment. In large lymphoid nodules, there is often a pale central region, called a *germinal center*, where lymphocytes are actively dividing. Lymphoid nodules are found beneath the epithelia lining the respiratory, digestive, and urinary tracts. Large nodules in the walls of the pharynx are called **tonsils**. There are usually five tonsils: a single *pharyngeal tonsil*, or *adenoids*, a pair of *palatine tonsils*, and a pair of *lingual tonsils*.

The lymphocytes in a lymphoid nodule are not always able to destroy bacterial or viral invaders, and if pathogens become established in a lymphoid nodule, an infection develops. Two examples are probably familiar to you: *tonsillitis*, an infection of one of the tonsils (usually the pharyngeal tonsil), and *appendicitis*, an infection of lymphoid nodules in the *appendix*, an organ of the digestive tract. AM *Infected Lymphoid Nodules*

Lymphoid Organs

Lymphoid organs have a stable internal structure and are separated from surrounding tissues by a fibrous capsule. Important lymphoid organs include the *lymph nodes*, the *thymus*, and the *spleen*.

Lymph Nodes

Lymph nodes are small, oval lymphoid organs ranging in diameter from 1 to 25 mm. They are covered by a dense, fibrous capsule (Figure 15-5●). One set of lymphatics delivers lymph to a lymph node, and another carries the lymph onward, toward the venous system. The lymph node functions like a kitchen water filter: It filters and purifies the lymph before it reaches the venous system. As lymph flows through a lymph node, at least 99 percent of the antigens present in the arriving lymph will be removed. As the antigens are detected and removed, T cells and B cells are stimulated, and an immune response initiated. Lymph nodes are located in regions where they can detect and eliminate harmful "intruders" before they reach vital organs of the body (Figure 15-1●, p. 384).

SWOLLEN GLANDS

Lymph nodes are often called *lymph glands*, and "swollen glands" usually accompany tissue inflammation or infection. Chronic or excessive enlargement of lymph nodes, a sign called *lymphadenopathy* (lim-fad-e-NOP-a-thē), may occur in response to bacterial or viral infections, endocrine disorders, or cancer.

Since the lymphatic capillaries offer little resistance to the passage of cancer cells, cancer cells often spread along the lymphatics and become trapped in the lymph nodes. Thus an analysis of swollen lymph nodes can provide information on the distribution and nature of the cancer cells, aiding in the selection of appropriate therapies. *Lymphomas*, an important group of lymphatic system cancers, are discussed in the *Applications Manual*. AM *Lymphomas*

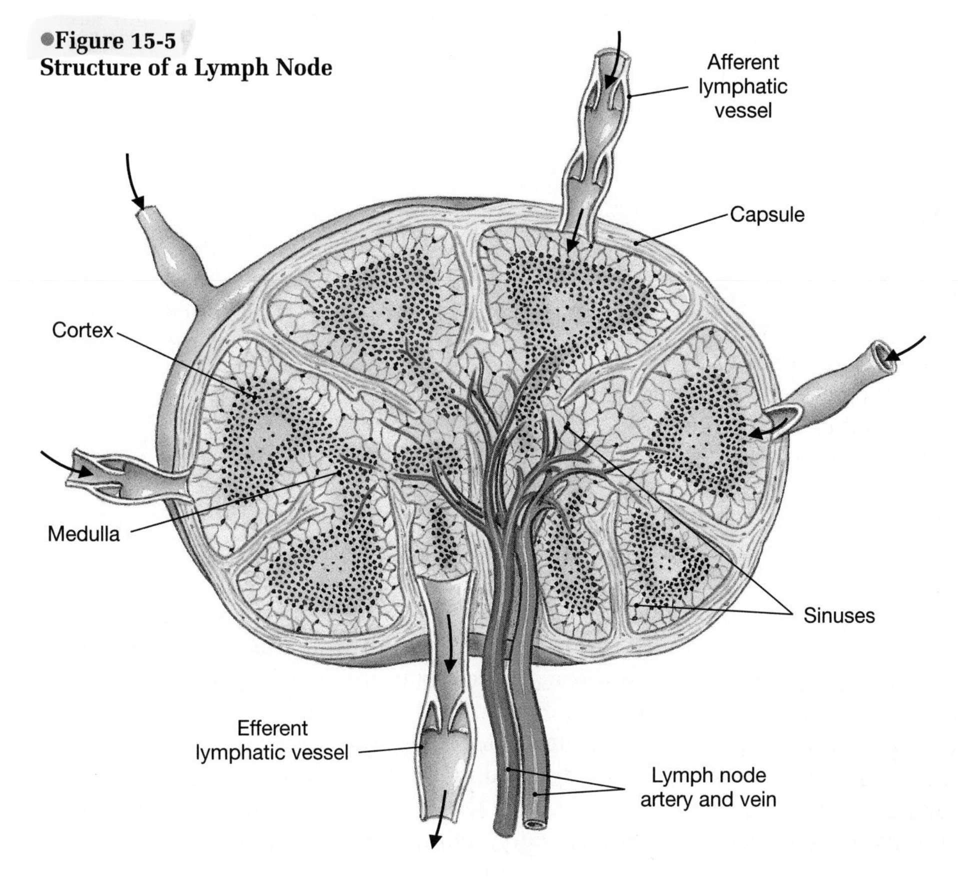

●**Figure 15-5**
Structure of a Lymph Node

The Thymus

The **thymus**, site of T cell maturation, lies behind the sternum (Figure 15-6●). The thymus reaches its greatest size (relative to body size) in the first year or two after birth and its maximum absolute size during puberty, when it weighs between 30 and 40 g (1.06 to 1.41 oz.). Thereafter the thymus gradually decreases in size.

The thymus has two lobes, each divided into *lobules* by fibrous partitions, or *septae* (*septum*, a wall). Each lobule consists of a densely packed outer *cortex* and a paler central *medulla*. Lymphocytes in the cortex are dividing, and as the T cells mature they migrate into the medulla, eventually entering one of the blood vessels in that region. Other cells within the lobules produce the thymic hormones collectively known as *thymosins*.

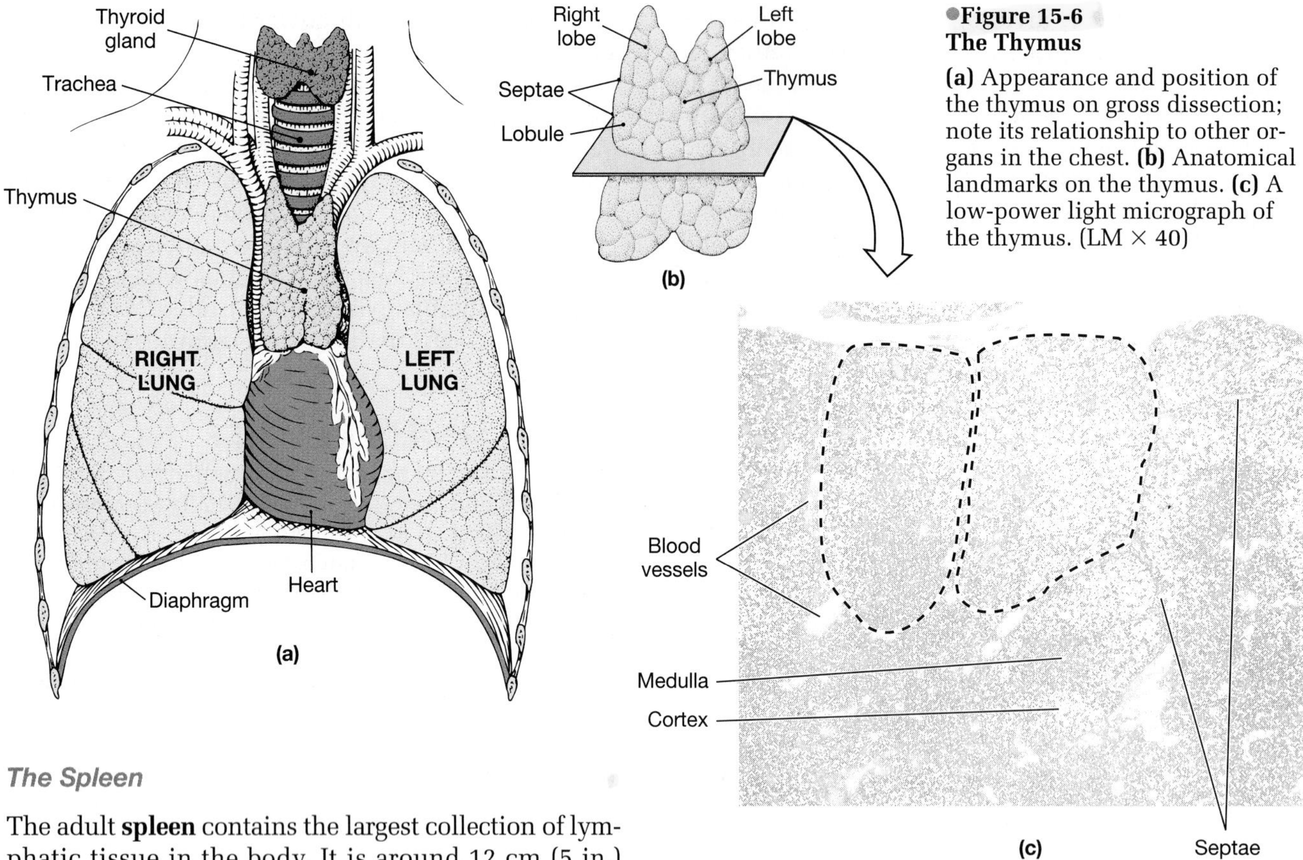

Figure 15-6 The Thymus

(a) Appearance and position of the thymus on gross dissection; note its relationship to other organs in the chest. **(b)** Anatomical landmarks on the thymus. **(c)** A low-power light micrograph of the thymus. (LM × 40)

The Spleen

The adult **spleen** contains the largest collection of lymphatic tissue in the body. It is around 12 cm (5 in.) long and can weigh about 160 g (5.6 oz.). As indicated in Figure 15-7a, it sits wedged between the stomach, the left kidney, and the muscular diaphragm. The spleen normally has a deep red color because of the blood it contains. The cellular components constitute the **pulp** of the spleen. Areas of *red pulp* contain large quantities of blood, whereas areas of *white pulp* resemble lymphoid nodules. As blood flows through the spleen, macrophages identify and engulf any damaged or infected cells. The presence of lymphocytes nearby ensures that any microorganisms or other abnormal antigens will stimulate an immune response.

The spleen performs similar functions for the blood that the lymph nodes perform for lymph: (1) removing abnormal blood cells and components, (2) and initiating immune responses by B cells and T cells in response to antigens in the circulating blood. In addition, the spleen stores iron from recycled red blood cells.

INJURY TO THE SPLEEN

An impact to the left side of the abdomen can distort or damage the spleen. Such injuries are known risks of contact sports, such as football or hockey, and more solitary athletic activities, such as skiing or sledding. The spleen tears so easily, however, that even a seemingly minor blow to the side may rupture the capsule. The result is serious internal bleeding and eventual circulatory shock.

Because the spleen is relatively fragile, it is very difficult to repair surgically. (Sutures usually tear out before they have been tensed enough to stop the bleeding.) Treatment for a severely ruptured spleen involves its complete removal, a process called a *splenectomy* (sple-NEK-to-mē).

The spleen may also be damaged through infection, inflammation, or invasion by cancer cells. These conditions and related symptoms are considered in the *Applications Manual.* AM *Disorders of the Spleen*

✓ How would blockage of the thoracic duct affect the circulation of lymph?

✓ If the thymus gland failed to produce thymosin, what particular population of lymphocytes would be affected?

✓ Why do lymph nodes enlarge during some infections?

The Lymphatic System and Body Defenses

The human body has multiple defense mechanisms, but these can be sorted into two general categories:

1. **Nonspecific defenses** do not discriminate between one threat and another. These defenses, which are present at birth, include *physical barriers, phagocytic cells, immunological surveillance, interfer-*

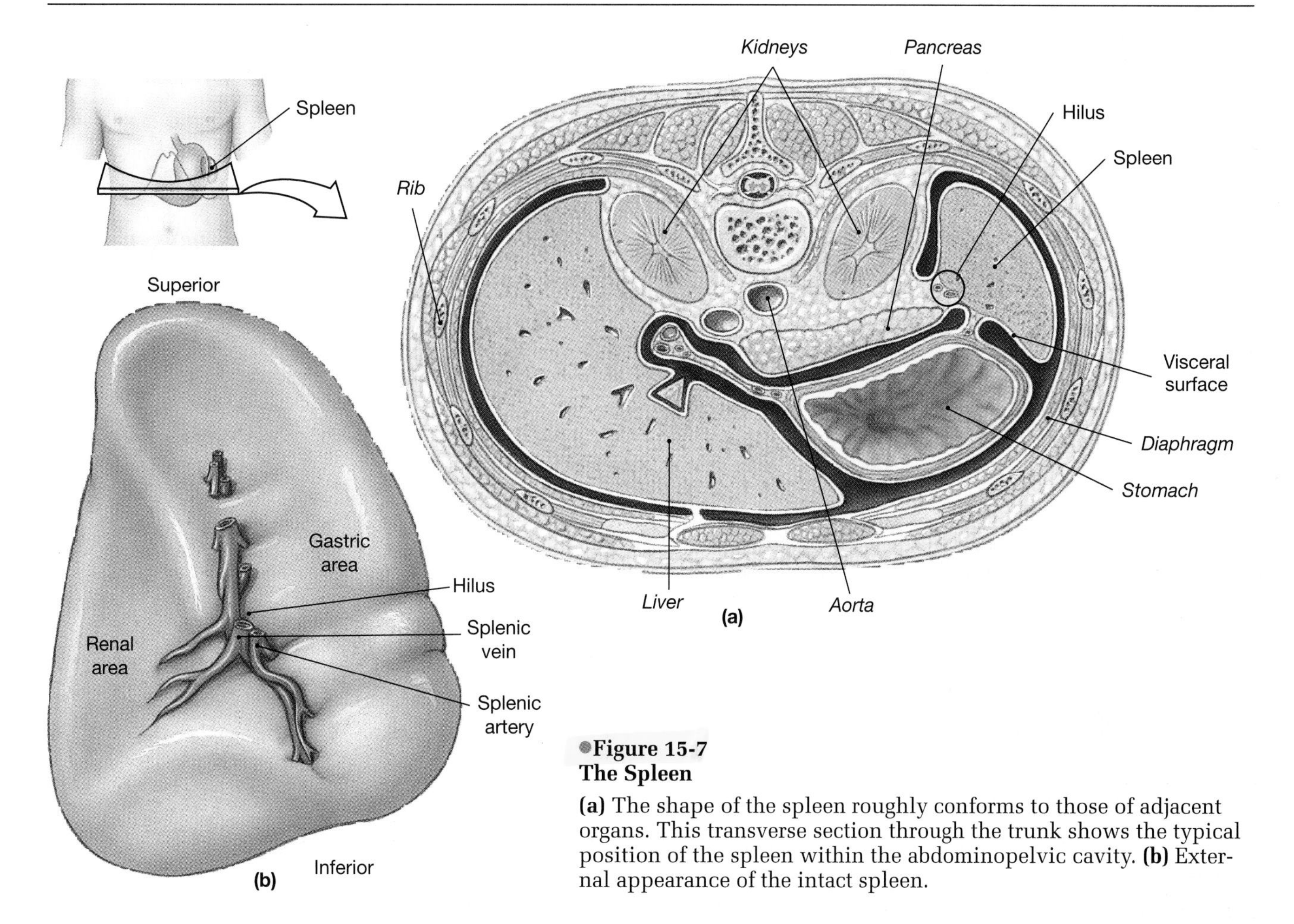

●**Figure 15-7**
The Spleen

(a) The shape of the spleen roughly conforms to those of adjacent organs. This transverse section through the trunk shows the typical position of the spleen within the abdominopelvic cavity. **(b)** External appearance of the intact spleen.

ons, complement, inflammation, and *fever.* They provide the body with a defensive capability known as *nonspecific resistance.*

2. **Specific defenses** provide protection against threats on an individual basis. For example, a specific defense may protect against infection by one type of bacteria but ignore other bacteria and viruses. Specific defenses are dependent upon the activities of lymphocytes. Together, they produce a state of protection known as **specific resistance**, or immunity.

Nonspecific and specific resistance do not function in complete isolation from each other; both are necessary to provide adequate resistance to infection and disease.

Nonspecific Defenses

Nonspecific defenses, summarized in Figure 15-8●, deny entrance, or limit the spread of microorganisms or other environmental hazards.

Physical Barriers

To cause trouble, a foreign (antigenic) compound or pathogen must enter the body tissues, and that means crossing an epithelium. The epithelial covering of the skin, described in Chapter 5, has multiple layers, a keratin coating, and a network of desmosomes that lock adjacent cells together. ∞ [p. 97] These create a very effective barrier that protects underlying tissues.

The exterior surface of the body has several layers of defense. The hairs found in most areas provide some protection against mechanical abrasion (especially on the scalp), and they often prevent hazardous materials or insects from contacting the skin's surface. The epidermal surface also receives the secretions of sebaceous and sweat glands. These secretions flush the surface, washing away microorganisms and chemical agents. The secretions also contain bactericidal chemicals, destructive enzymes (*lysozymes*), and antibodies.

The epithelia lining the digestive, respiratory, urinary, and reproductive tracts are more delicate, but they are equally well defended. Mucus bathes most surfaces of the digestive tract, and the stomach contains a powerful acid that can destroy many potential pathogens. Mucus moves across the lining of the respiratory tract, urine flushes the urinary passageways, and glandular secretions do the same for the reproductive tract. Special enzymes, antibodies, and an acidic pH add to the effectiveness of these secretions.

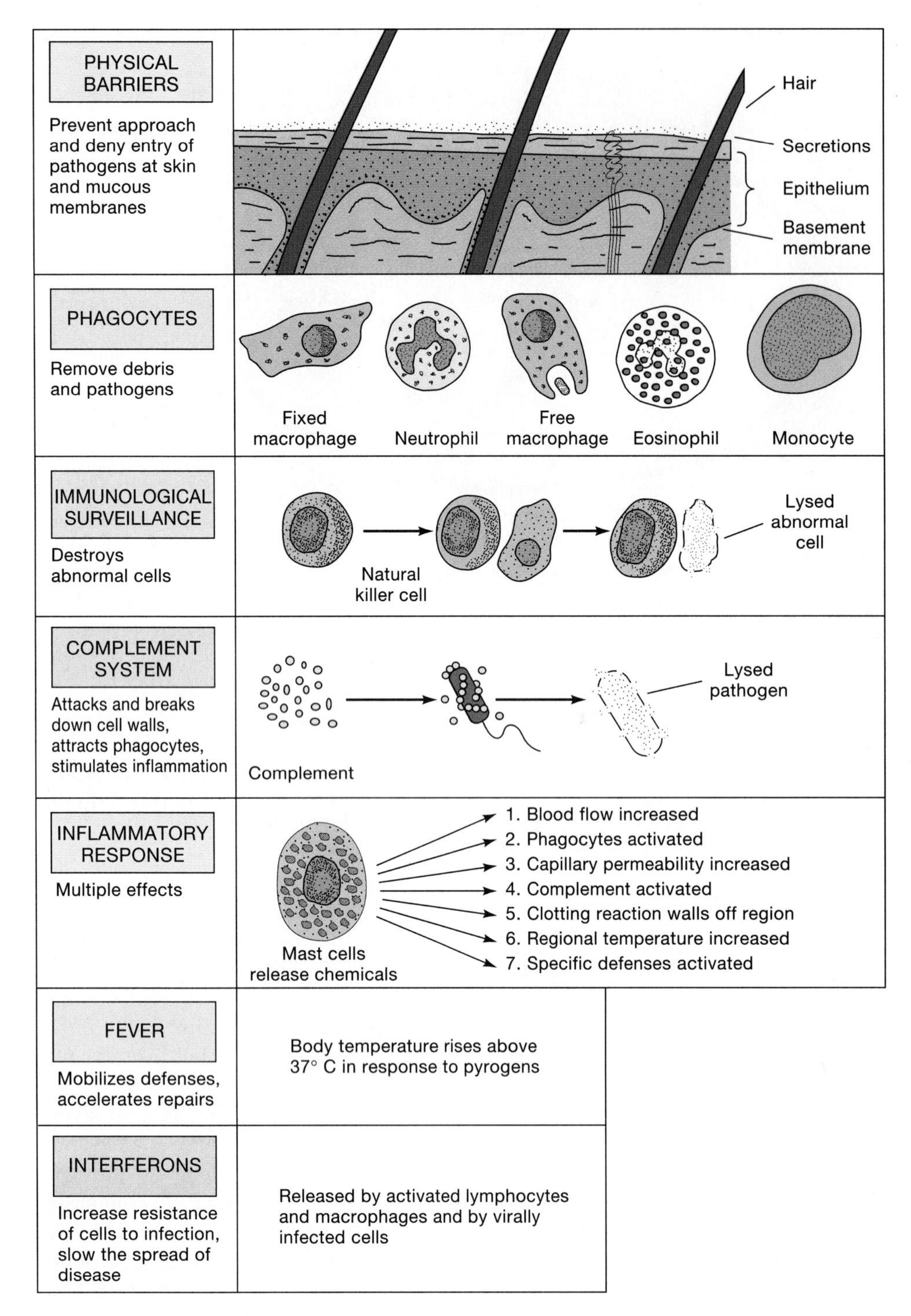

•Figure 15-8 Nonspecific Defenses

Phagocytes

Phagocytes in peripheral tissues remove cellular debris and respond to invasion by foreign compounds or pathogenic organisms. These cells represent the "first line" of cellular defense, often attacking and removing the microorganisms before lymphocytes become aware of the incident. Two general classes of phagocytic cells are found in the human body: *microphages* and *macrophages*.

Microphages are the neutrophils and eosinophils normally found in the circulating blood. These phagocytic cells leave the bloodstream and enter peripheral tissues subjected to injury or infection. As noted in Chapter 12, neutrophils are abundant, mobile, and quick to phagocytize cellular debris or invading bacteria. ∞ *[pp. 318–319]* The less abundant eosinophils target foreign compounds or pathogens that have been coated with antibodies.

The body also contains several different types of **macrophages**: large, actively phagocytic cells derived from the monocytes of the blood. Almost every tissue in the body shelters resident or visiting macrophages. In some organs, the macrophages have special names: For example, in the CNS they are called *microglia*. The relatively diffuse collection of phagocytic cells throughout the body is called the *monocyte-macrophage system*.

All phagocytic cells function in much the same way, although the items selected for phagocytosis may differ from one cell type to another. Mobile macrophages and microphages also share a number of other functional characteristics in addition to phagocytosis. They can all move through capillary walls by squeezing between adjacent endothelial cells, a process known as *diapedesis* (*dia*, through + *pedesis*, a leaping). They may also be attracted or repelled by chemicals in the surrounding fluids, a phenomenon called *chemotaxis* (*chemo-*, chemistry + *taxis*, arrangement). They are particularly sensitive to chemicals released by other body cells or by pathogens.

Immunological Surveillance

Our immune defenses generally ignore normal cells in the body's tissues, but abnormal cells are attacked and destroyed. The constant monitoring of normal tissues has been called **immunological surveillance**. This surveillance primarily involves the lymphocytes known as NK (natural killer) cells. These cells are sensitive to the presence of antigens characteristic of abnormal cell membranes. When they encounter these antigens on a cancer cell or a cell infected with viruses, NK cells secrete proteins that kill the abnormal cell by destroying its cell membrane. Killing the abnormal cells can slow the spread of a viral infection and may eliminate cancer cells before they spread to other tissues. Unfortunately, some cancer cells avoid detection, a process called *immunological escape*. Once immunological escape has occurred, cancer cells can multiply and spread without interference by NK cells.

Interferons

Interferons are small proteins released by activated lymphocytes, macrophages, and tissue cells infected with viruses. Cells exposed to these molecules respond by producing proteins that interfere with viral replication. In addition, interferons stimulate the activities of macrophages and NK cells. Interferons are examples of *cytokines* (SĪ-tō-kīnz), chemical messengers released by tissue cells to coordinate local activities. In effect, cytokines are the hormones of the immune system; they are released to alter the activities of cells and tissues throughout the body. Their role in the regulation of specific defenses will be discussed below.

Complement

The plasma contains 11 special *complement proteins* that form the **complement system**. The term *complement* refers to the fact that this system "complements," or supplements, the action of antibodies. These proteins interact with one another in chain reactions comparable to those of the clotting system. The reaction is begun when a complement binds to an antibody molecule or to bacterial cell walls. The bound complement protein then interacts with other complement proteins. Complement activation is known to (1) attract phagocytes, (2) enhance phagocytosis, (3) destroy cell membranes, and (4) promote inflammation.

Inflammation

Inflammation, a localized tissue response to injury introduced in Chapter 4, produces local sensations of swelling, redness, heat, and pain. ∞ *[p. 91]* Inflammation can be produced by any stimulus that kills cells or damages loose connective tissue. *Mast cells* within the affected tissue play a pivotal role in this process. ∞ *[p. 83]* When stimulated by mechanical stress or chemical changes in the local environment, mast cells release chemicals, including *histamine* and *heparin*, into the interstitial fluid. These chemicals initiate the process of inflammation.

Figure 15-9• summarizes the events that occur during inflammation of the skin. The goals of inflammation are:

- To perform a temporary repair at the injury site and prevent the access of additional pathogens.
- To slow the spread of pathogens away from the injury site.
- To mobilize a wide range of defenses that can overcome the pathogens and facilitate permanent repairs. The repair process is called *regeneration*.

Fever

A **fever** is the maintenance of a body temperature greater than 37.2°C (99°F). The hypothalamus, which contains nuclei that regulate body temperature, acts as the body's thermostat. ∞ *[p. 12]* Circulating proteins called *pyrogens* (PĪ-rō-jenz; *pyr*, fire + *-gen*, to produce) can reset the thermostat and cause a rise in body temperature. Pathogens, bacterial toxins, and antigen-antibody complexes may act as pyrogens or stimulate the release of pyrogens by macrophages.

Within limits, a fever may be beneficial. For example, high body temperatures may accelerate the activities of the immune system. However, high fevers (over 40°C, or 104°F) can damage many different physiological systems. For example, a high fever can cause CNS problems, such as nausea, disorientation, hallucinations, or convulsions.

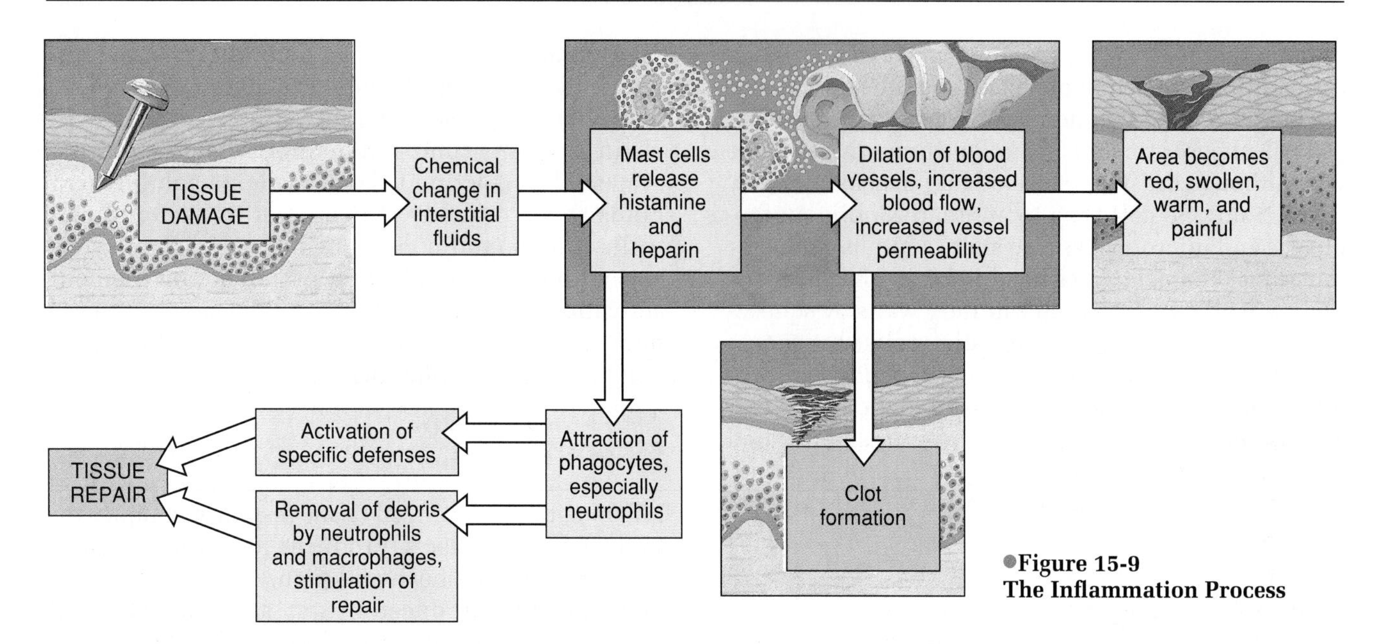

•**Figure 15-9**
The Inflammation Process

✓ What types of cells would be affected by a decrease in the monocyte-forming cells in the bone marrow?

✓ A rise in the level of interferon in the body would suggest what kind of infection?

✓ What effects do pyrogens have in the body?

SPECIFIC DEFENSES: THE IMMUNE RESPONSE

The body's specific defenses that produce specific resistance, or immunity, are provided by the coordinated activities of T cells and B cells, which respond to the presence of *specific* antigens. T cells provide a defense against abnormal cells and pathogens inside living cells; this process is called *cell-mediated immunity.* B cells provide a defense against antigens and pathogenic organisms in body fluids. This process is called *antibody-mediated immunity.*

Forms of Immunity

Immunity may be either *innate* or *acquired* (Figure 15-10•). **Innate immunity** is genetically determined; it is present at birth and has no relation to previous exposure to the antigen involved. For example, people are not subject to the same diseases as goldfish.

Acquired immunity may be either *active* or *passive.* **Active immunity** appears following exposure to an antigen, as a consequence of the immune response. The immune system has the *capability* of defending against an enormous number of antigens. However, the appropriate defenses are mobilized only after an individual's lymphocytes encounter a particular antigen. Active immunity may develop as a result of natural exposure to an antigen in the environment (*naturally acquired immunity*) or from deliberate exposure to an antigen (*induced active immunity*). Naturally acquired immunity normally begins to develop after birth, and it is continually enhanced as the individual encounters "new" pathogens or other antigens. You might compare this process to vocabulary development—a child begins with a few basic common words, and learns new ones on an as-needed basis. The purpose of induced active immunity, or *artificially acquired immunity,* is to stimulate antibody production under controlled conditions so that the individual will be able to overcome natural exposure to the pathogen at some time in the future. This is the basic principle behind immunization to prevent disease. AM *Immunization*

Passive immunity is produced by the transfer of antibodies from another individual. *Natural passive immunity* results when antibodies produced by the mother cross the placental barrier to provide protection against embryonic or fetal infections. In *induced passive immunity,* antibodies are administered to fight infection or prevent disease.

Properties of Immunity

Four general properties of immunity are recognized:

1. **Specificity.** A specific defense is activated by an antigen, and the response targets only that particular antigen in a process known as *antigen recognition.* Specificity occurs because the cell membrane of each T cell and B cell has receptors that will bind only one specific antigen, ignoring all others. Either lymphocyte will destroy or inactivate that antigen without affecting other antigens or normal tissues.

2. **Versatility.** In the course of a normal lifetime, an individual encounters tens of thousands of antigens. The immune system cannot anticipate which antigens it will encounter, so it must be ready to confront *any* antigen at *any* time. It does this by producing millions of different lymphocyte populations, each with different antigen receptors. Thus, the immune system can produce appropriate and specific responses to each antigen.
3. **Memory.** The immune system "remembers" antigens that it encounters. During the initial response to an antigen, lymphocytes sensitive to its presence undergo repeated cell divisions. Two kinds of cells are produced: some that attack the invader and others that remain inactive unless they are exposed to the same antigen at a later date. These *memory cells* enable the immune system to "remember" previously encountered antigens and launch a faster, stronger counterattack if one of them ever appears again.
4. **Tolerance. Tolerance** is said to exist when the immune system does not respond to a particular antigen. For example, the immune response targets foreign cells and compounds (*non-self* antigens), but it usually ignores normal tissues (*self* antigens).

An Overview of the Immune Response

The goal of the **immune response** is to destroy or inactivate pathogens, abnormal cells, and foreign molecules such as toxins. The process begins with the appearance of an antigen. Lymphocytes sensitive to a particular antigen have receptors in their membranes that can bind that particular antigen, a process known as **antigen recognition**. As Figure 15-11● shows, the precise nature of the resulting immune response varies depending on whether the activated lymphocyte is a T cell or a B cell.

T Cells and Cellular Immunity

Before an immune response can begin, T cells must be activated by exposure to an antigen. However, this activation seldom happens by direct lymphocyte-antigen interaction, and foreign compounds or pathogens entering a tissue often fail to stimulate an immediate response.

T cells recognize antigens when they are bound to receptors on the membranes of other cells. The receptors involved are called *major histocompatibility complex (MHC) proteins*. MHC proteins are found on the surfaces of all of our cells. To trigger an immune response, foreign cells (including bacteria) and foreign proteins must first be engulfed by macrophages. The macrophages break down the foreign antigens, creating antigenic fragments that are displayed on their cell surfaces, bound to MHC proteins. T cells that contact this macrophage membrane become activated, initiating an immune response. In the case of viruses, T cells can be activated by contact with an infected cell. An infected cell displays viral antigens on its surface, also bound to MHC proteins. The activation of T cells by these antigens can rally a defense against the viral infection.

●**Figure 15-10**
Types of Immunity

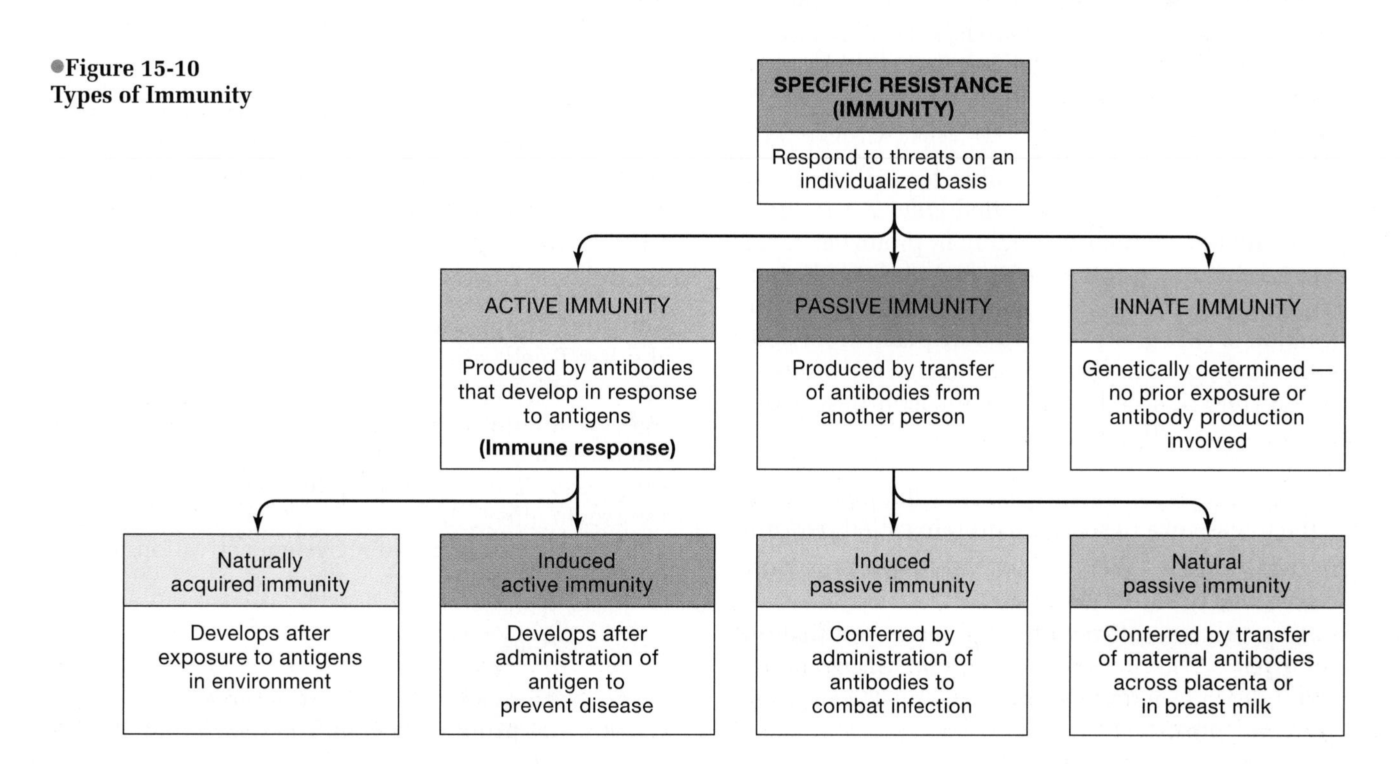

•Figure 15-11
An Overview of the Immune Response

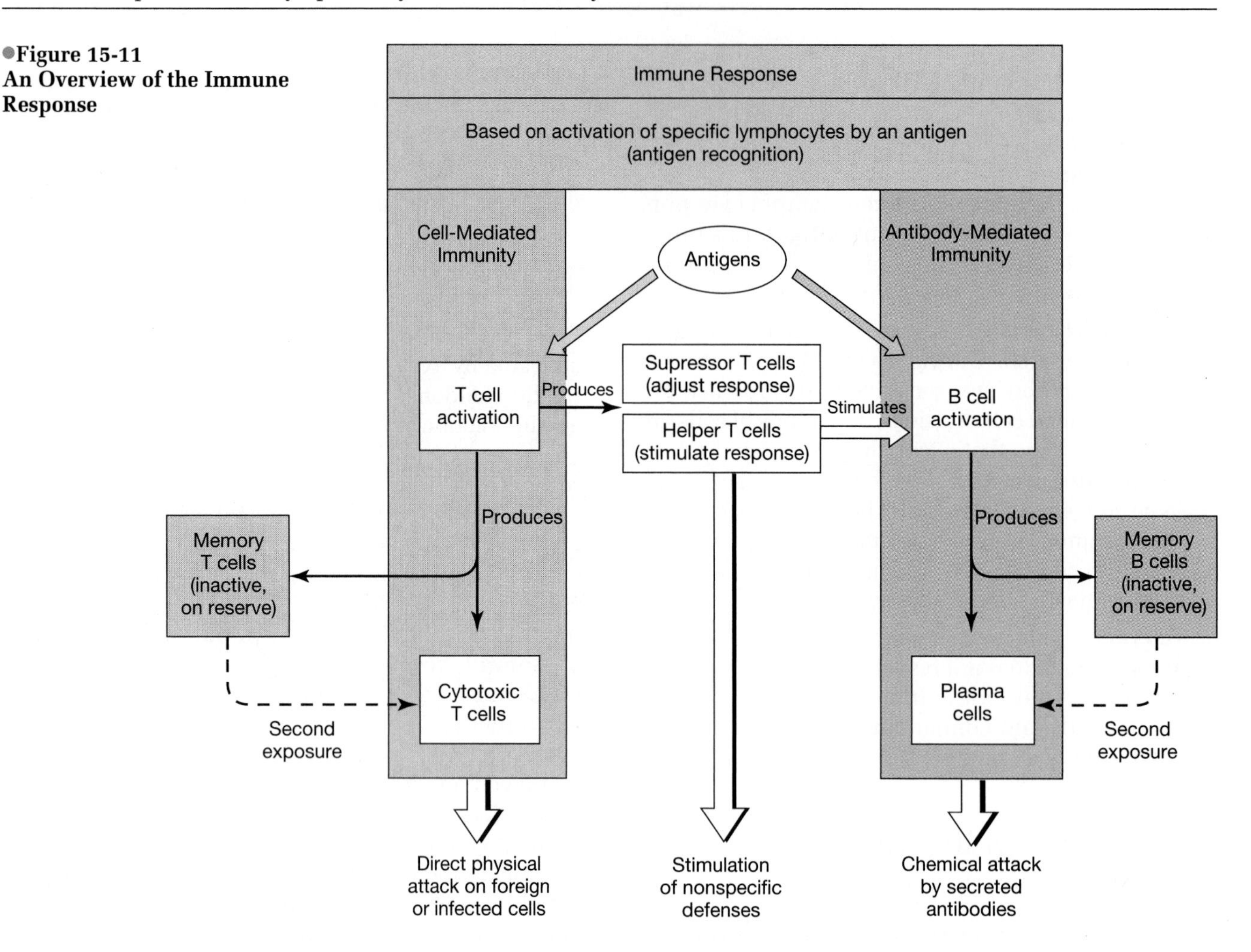

On activation, T cells divide and differentiate into cells with specific functions in the immune response. The major cell types are *cytotoxic T cells*, *memory T cells*, *suppressor T cells*, and *helper T cells*.

Cytotoxic T Cells

Cytotoxic T cells are responsible for cell-mediated immunity. These cells, also called *killer T cells*, track down and attack the bacteria, fungi, protozoa, or foreign tissues that contain the target antigen. For example, cytotoxic T cells are responsible for the rejection of skin grafts or organ transplants from other individuals. AM *Transplants and Graft Rejection*

A cytotoxic T cell may accomplish its destruction in several ways (Figure 15-12•):

- By rupturing the antigenic cell membrane through the release of a destructive protein called *perforin*.
- By killing the target cell by secreting a poisonous *lymphotoxin* (lim-fō-TOK-sin).
- By activating genes within the target cell that tell it to die. The process of genetically programmed cell death is called *apoptosis* (ap-op-TŌ-sis; *apo-*, away + *ptosis*, a falling).

Memory T Cells

During the cell divisions that follow T cell activation, some of the cells develop into cytotoxic T cells and others develop into **memory T cells**. Memory T cells remain "in reserve." If the same antigen appears a second time, these cells will immediately differentiate into cytotoxic T cells, producing a more rapid and effective cellular response.

Suppressor T Cells

Activated **suppressor T cells** depress the responses of other T cells and B cells by secreting *suppression factors*. This suppression does not occur immediately, because suppressor T cells take much longer to become activated than other types of T cells. As a result, suppressor T cells act *after* the initial immune response. In effect, these cells "put on the brakes" and limit the degree of immune system activation from a single stimulus.

Helper T Cells

On activation, some T cells undergo a series of divisions that produce **helper T cells**. Helper T cells release a variety of cytokines that (1) coordinate specific and nonspecific defenses and (2) stimulate cell-mediated immunity and antibody-mediated (*humoral*; *humor*, a liquid) immunity. The mechanism involved will be discussed in the next section.

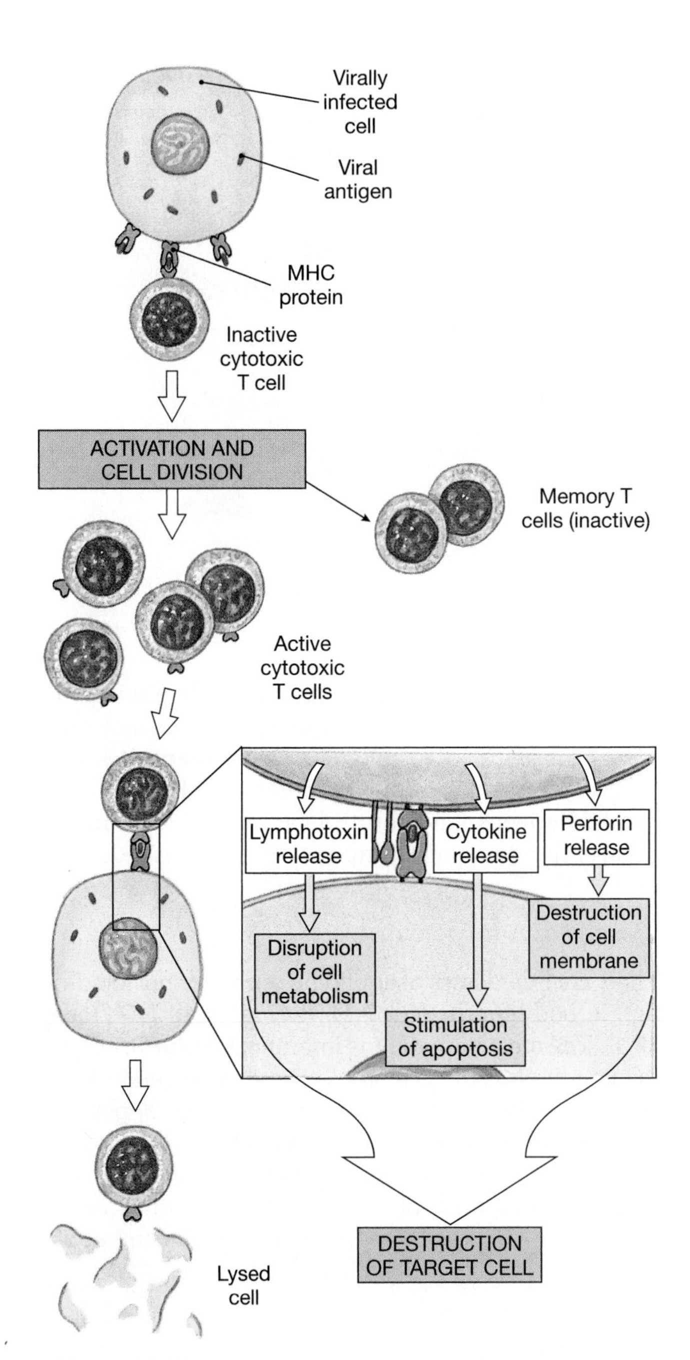

•Figure 15-12
The Activation of Cytotoxic T Cells
An inactive cytotoxic T cell must encounter an appropriate antigen bound to MHC proteins for activation. Once activated, it undergoes divisions that produce memory T cells and active cytotoxic T cells. When one of these active cells encounters a membrane displaying the target antigen, the cytotoxic T cell will use one of several methods to destroy the cell.

B Cells and Antibody-Mediated Immunity

B cells are responsible for launching a chemical attack on antigens by producing appropriate *antibodies*. B cell activation proceeds in a series of steps diagrammed in Figure 15-13•.

B Cell Activation

B cell activation primarily occurs through the activities of helper T cells. Activated helper T cells bind to inactive B cells and begin secreting cytokines that (1) promote B cell activation, (2) stimulate B cell division, (3) accelerate plasma cell production, and (4) enhance antibody production.

As Figure 15-13• shows, the activated B cell divides several times, producing daughter cells that differentiate into plasma cells and **memory B cells**. Plasma cells begin synthesizing and secreting large numbers of antibodies that have the same target as the antibodies on the surface of the sensitized B cell. Memory B cells perform the same role for antibody-

•Figure 15-13
The Activation of B Cells and Antibody Production
Activated helper T cells encountering a B cell sensitive to a specific antigen release cytokines that trigger the activation of the B cell. The activated B cell then goes through cycles of divisions producing memory B cells and B cells that differentiate into plasma cells. The plasma cells secrete antibodies.

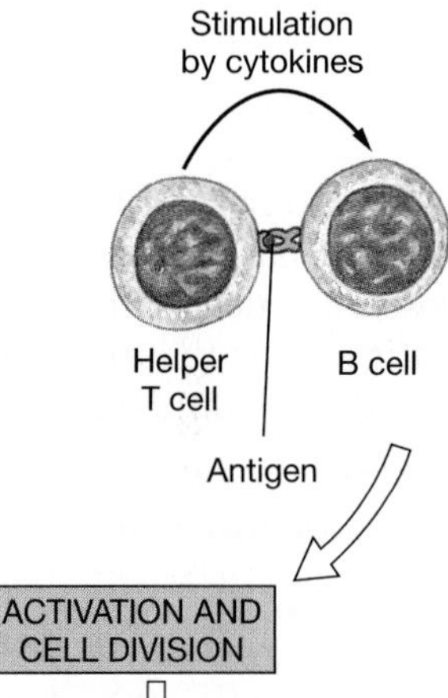

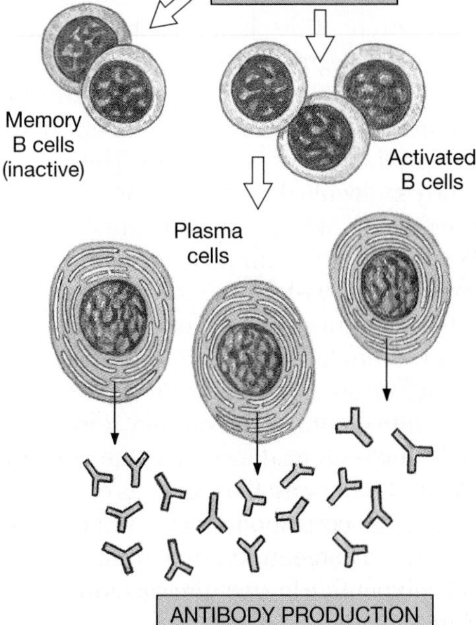

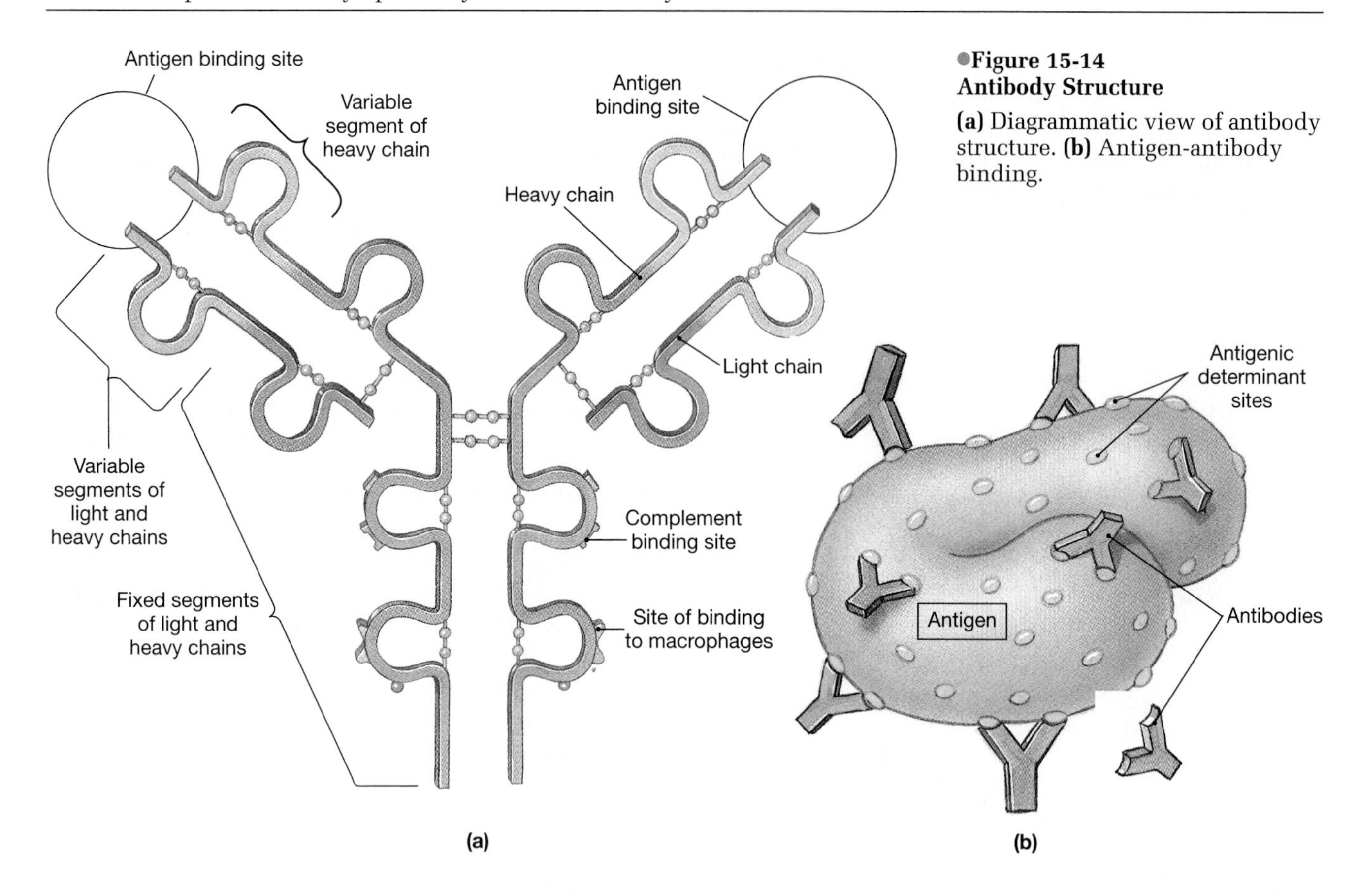

●Figure 15-14 Antibody Structure
(a) Diagrammatic view of antibody structure. **(b)** Antigen-antibody binding.

mediated immunity that the memory T cells perform for cellular-mediated immunity. They will remain in reserve to deal with subsequent exposure to the same antigens. At that time, they will respond and differentiate into antibody-secreting plasma cells.

Antibody Structure

Figure 15-14● presents different views of a single antibody molecule. The molecule consists of two parallel pairs of polypeptide chains, each with *fixed segments* and *variable segments*. The fixed segments resemble those of every other antibody molecule of that particular class (the various classes of antibodies are described in the following section). The specificity of the antibody molecule depends on the structure of the variable segments, which can be made from an enormous number of different amino acid sequences. Because of this molecular variability, it has been estimated that the 10 trillion or so B cells of a normal adult can produce an estimated 100 million different antibodies.

When an antibody molecule binds to its proper antigen, an **antigen-antibody complex** is formed. The specificity of that binding depends on the three-dimensional "fit" between the variable segments of the antibody molecule and the corresponding sites of the antigen.

Antigens come in a variety of sizes, ranging from a single polypeptide to an entire bacterium. Antibodies do not target the antigen as a whole, but instead focus on certain portions of its exposed surface called *antigenic determinant sites*. To be a complete antigen, a molecule must have at least two antigenic determinant sites, one for each arm of the antibody molecule. Most environmental antigens have multiple antigenic determinant sites; entire microorganisms may have thousands.

Classes of Antibodies

There are five classes of antibodies, or immunoglobulins (Ig), in body fluids: *IgG*, *IgM*, *IgA*, *IgE*, and *IgD* (Table 15-1). The most important is **immunoglobulin G**, or **IgG**, the largest class of antibodies. There are several different types of IgG, which together account for 80 percent of all immunoglobulins. The IgG antibodies are responsible for resistance against many viruses, bacteria, and bacterial toxins. Circulating *IgM* antibodies are responsible for the cross-reactions between incompatible blood types described in Chapter 12. ∞ *[p. 315]*

Antibody Function

The function of antibodies is to destroy antigens. The formation of an antigen-antibody complex may cause their elimination in several different ways.

1. **Neutralization**. Antibodies can bind to toxins or viruses, making them incapable of attaching to a cell. This mechanism is called *neutralization*.

AIDS

Acquired immune deficiency syndrome (AIDS), or *late-stage HIV disease,* is caused by a virus known as **human immunodeficiency virus (HIV)**. HIV is an example of a *retrovirus*, a virus that carries its genetic information in RNA rather than DNA. The virus enters the cell through receptor-mediated endocytosis. In the human body, the virus binds to CD4, a membrane protein characteristic of helper T cells. Several types of antigen-presenting cells, including those of the monocyte-macrophage line, are also infected by HIV, but it is the infection of helper T cells that leads to clinical problems.

Cells infected with HIV are ultimately killed. The gradual destruction of helper T cells impairs the immune response, because these cells play a central role in coordinating cellular and humoral responses to antigens. To make matters worse, suppressor T cells are relatively unaffected by the virus, and over time the excess of suppressing factors "turns off " the normal immune response. Circulating antibody levels decline, cellular immunity is reduced, and the body is left without defenses against a wide variety of microbial invaders.

With immune function so reduced, ordinarily harmless pathogens can initiate lethal infections, known as *opportunistic infections.* AIDS patients are especially prone to lung infections and pneumonia, often caused by *Pneumocystis carinii* or other fungi, and to a wide variety of bacterial, viral, and protozoan diseases. Because immunological surveillance is also depressed, the risk of cancer increases. One of the most common cancers seen in AIDS patients, though very rare in normal individuals, is **Kaposi's sarcoma**, characterized by rapid cell divisions in endothelial cells of cutaneous blood vessels.

Infection with HIV occurs through intimate contact with the body fluids of infected individuals. Although all body fluids carry the virus, the major routes of transmission involve contact with blood, semen, or vaginal secretions. Most AIDS patients become infected through sexual contact with an HIV-infected person (who may *not necessarily* be suffering from the clinical symptoms of AIDS). The next largest group of patients consists of intravenous drug users who shared contaminated needles. A relatively small number of individuals have become infected with the virus after receiving a transfusion of contaminated blood or blood products. Finally, an increasing number of infants are born with the disease, having acquired it in the womb from infected mothers.

AIDS is a public health problem of massive proportions. As of the end of 1995 an estimated 320,000 people have died from AIDS in the United States alone, and over 195,000 are living with AIDS. Estimates of the number of individuals infected with HIV in the United States range from 1 to 2 million. The virus has spread throughout the population. For example, a study performed in 1990 indicated that the incidence of infection at U.S. colleges and universities was 1 student in 500. The numbers worldwide are even more frightening: The World Health Organization estimates that over 20 million people are already infected, and the number is increasing rapidly in Africa, Asia, and South America. At the current rate of increase, there may be 40 million to 110 million infected individuals by the year 2000. *Every 15 seconds another person becomes infected with the HIV virus.*

The best defense against AIDS is to avoid sexual contact with infected individuals. *All forms of sexual intercourse carry the potential risk of viral transmission.* The use of *synthetic* condoms greatly reduces the chance of infection (although it does not completely eliminate it). Condoms that are not made of synthetic materials are effective in preventing pregnancy but do not block the passage of viruses.

Clinical symptoms of AIDS may not appear for 5–10 years or more following infection, and when they do appear they are often mild, consisting of lymphadenopathy and chronic but nonfatal infections. After a variable period of time, full-blown AIDS develops. Although AIDS is usually considered to be a fatal disease, and the vast majority of those who carry the virus will eventually die of the disease, a handful of individuals appear to be able to tolerate the virus for 15 years or more without developing symptoms of HIV disease. It appears that these individuals have been infected by less virulent viral strain.

Despite intensive efforts, a vaccine has yet to be developed that will provide immunity from HIV infection. However, three organizations (World Health Organization, the U.S. Army, and the National Institute for Allergy and Infectious Diseases) have announced that they either have chosen or are choosing sites for vaccine trials.

While efforts continue to prevent the spread of HIV, the survival rate for AIDS patients has been steadily increasing because new drugs are available that slow the progress of the disease, and improved antibiotic therapies help combat secondary infections. This combination is extending the life span of patients while the search for more effective treatment continues. For more information on the distribution of HIV infection, current and future drug therapies, and additional details on AIDS, consult the *Applications Manual.* AM *AIDS*

2. **Agglutination and precipitation**. When large numbers of antibodies bind to antigens, they can create large complexes. Antigenic cells may clump together, a process called **agglutination**. The clumping of red blood cells that occurs when incompatible blood types are mixed (Chapter 12) is an example of agglutination. *[p. 317]* Smaller antigens may form insoluble masses that settle out of body fluids. This process is called **precipitation**.
3. **Activation of complement.** Upon binding to an antigen, portions of the antibody molecule change shape, exposing areas that bind complement proteins. The bound complement molecules then activate the complement system, destroying the antigen.

TABLE 15-1 Classes of Antibodies

Class	Function	Remarks
IgG	Responsible for defense against many viruses, bacteria, and bacterial toxins	Largest class of antibodies, with several subtypes
IgM	Anti-A, anti-B, anti-D forms responsible for cross-reactions between incompatible blood types; other forms attack bacteria insensitive to IgG	First antibody type secreted following arrival of antigen; levels decline as IgG production accelerates
IgA	Attack pathogens before they enter the body tissues	Found in glandular secretions (tears, mucus, and saliva)
IgE	Accelerate inflammation on exposure to antigen	Bound to surfaces of mast cells and basophils; important in allergic response
IgD	Bind antigens to B cells	May play a role in B cell activation

4. **Attraction of phagocytes**. Antigens covered with antibodies attract eosinophils, neutrophils, and macrophages, which can phagocytize pathogens and destroy foreign or abnormal cell membranes.
5. **Enhancement of phagocytosis**. A coating of antibodies and complement proteins increases the effectiveness of phagocytosis.
6. **Stimulation of inflammation**. Antibodies may promote inflammation by stimulating basophils and mast cells. This can help mobilize nonspecific defenses and slow the spread of the infection to other tissues.

Primary and Secondary Responses to Antigen Exposure

The initial response to antigen exposure is called the **primary response**. When an antigen appears a second time, it triggers a more extensive and prolonged **secondary response**. The secondary response reflects the presence of large numbers of memory cells that are already "primed" for the arrival of the antigen.

Because the antigen must activate the appropriate B cells and the B cells must then respond by differentiating into plasma cells, the primary response does not appear immediately (Figure 15-15●). Instead, there is a gradual, sustained rise in the concentration of circulating antibodies, and the antibody activity, or *antibody titer*, in the blood does not peak until several weeks after the initial exposure. Thereafter the antibody concentration declines, assuming that the individual is no longer exposed to the antigen.

Memory B cells do not differentiate into plasma cells unless they are exposed to the same antigen a second time. If and when that exposure occurs, these cells respond immediately, dividing and differentiating into plasma cells that secrete antibodies in massive quantities. This represents the secondary response, or *anamnestic response* (an-am-NES-tik; *anamnesis*, a memory), to antigen exposure.

The secondary response produces an immediate rise in IgG concentrations to levels many times higher than those of the primary response. This response is so much faster and stronger than the primary response because the numerous memory cells constitute a kind of cellular rapid deployment force. They are primed to fight one particular enemy and can be mobilized and prepared for action on short notice. The secondary response appears even if the second exposure occurs years after the first, for memory cells are long-lived, potentially surviving for 20 years or more.

Hormones of the Immune System

The specific and nonspecific defenses of the body are coordinated by physical interaction and by the release of chemical messengers. Examples of physical interaction include antigen presentation by activated macrophages and helper T cells. Examples of chemical messengers include the release of cytokines by many cell types. Cytokines of the immune response are sometimes classified according to their sources: *lymphokines* secreted by lymphocytes and the *monokines* released by active macrophages and other antigen-presenting cells. The general term cytokine is preferable, however, since lymphocytes, macrophages, and cells involved with nonspecific defenses and tissue repair may secrete the same chemical messenger.

Table 15-2 contains examples of some of the cytokines identified to date. **Interleukins (Il)** are probably the most diverse and important chemical messengers in the immune system. These proteins have widespread effects that include increasing T cell and B cell sensitivities and enhancing nonspecific defenses, such

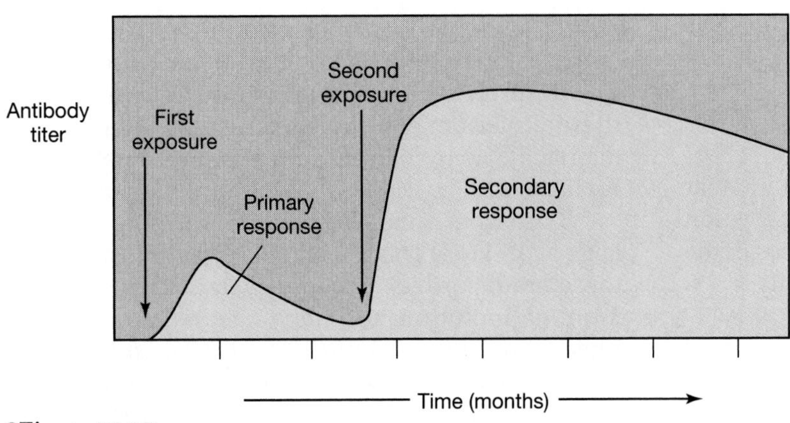

Figure 15-15
The Primary and Secondary Immune Responses

The primary response takes several weeks to develop peak antibody titers, and antibody concentrations do not remain elevated. The secondary response is characterized by a very rapid increase in antibody titer, to levels much higher than those of the primary response. Antibody activity remains elevated for an extended period following the second exposure to the antigen.

as inflammation or fever. Massive production of interleukins can cause problems at least as severe as those of the primary infection. For example, in *Lyme disease* the release of Il-1 by activated macrophages produces symptoms of fever, pain, skin rash, and arthritis that affect the entire body in response to a localized bacterial infection. AM *Lyme Disease*

Interferons (in-ter-FĒR-ons) make the synthesizing cell and its neighbors resistant to viral infection, thereby slowing the spread of the virus. These compounds may have other beneficial effects in addition to their antiviral activity.

Tumor necrosis factors (TNFs) slow tumor growth and kill sensitive tumor cells. In addition to their effects on tumor cells, tumor necrosis factors (1) stimulate production of neutrophils, eosinophils, and basophils, (2) promote eosinophil activity, (3) cause fever, and (4) increase T cell sensitivity to interleukins.

Phagocytic regulators include several cytokines that coordinate the specific and nonspecific defenses by adjusting the activities of phagocytic cells. These cytokines include factors that attract free macrophages and microphages to the area and prevent their premature departure.

Colony-stimulating factors (CSFs) are produced by a wide variety of cells including active T cells, cells of the monocyte-macrophage group, and fibroblasts. CSFs stimulate the production of blood cells in the bone marrow and lymphocytes in lymphoid tissues and organs.

TABLE 15-2 Examples of Chemical Mediators of the Immune Response

Compound	*Functions*
INTERLEUKINS	
Il-1	Stimulates T cells, promotes inflammation, causes fever
Il-2, -12	Stimulates T cells and NK cells
Il-3	Stimulates production of blood cells
Il-4, -5, -6, -7, -10, -11	Promote differentiation and growth of B cells, and stimulate plasma cell formation and antibody production
INTERFERONS	Activate other cells to prevent viral entry and replication, stimulate NK cells and macrophages
TUMOR NECROSIS FACTORS	Kill tumor cells, stimulate activities of T cells and eosinophils, inhibit parasites and viruses
PHAGOCYTIC REGULATORS	
Monocyte-chemotactic factor (MCF)	Attracts monocytes, activates them to macrophages
Macrophage-inhibitory factor (MIF)	Prevents macrophage migration from the area
COLONY-STIMULATING FACTORS (CSFs)	
M-CSF	Stimulates activity in the monocyte-macrophage line
GM-CSF	Stimulates production of both microphages and monocytes

✓ A decrease in the number of cytotoxic T cells would affect what type of immunity?

✓ How would a lack of helper T cells affect the humoral immune response?

✓ A sample of lymph contains an elevated number of plasma cells. On the basis of this observation would you expect the amount of antibodies in the blood to be increasing or decreasing? Why?

PATTERNS OF IMMUNE RESPONSE

The basic chemical and cellular interactions that follow the appearance of a foreign antigen have now been detailed. Figure 15-16● presents a broader, integrated view of the immune response and its relationship to nonspecific defenses.

Immune Disorders

Because the immune response is so complex, there are many opportunities for things to go wrong. A great variety of clinical conditions may result from disorders of immune function. **Autoimmune disorders** develop when the immune response mistakenly targets normal body cells and tissues. In an **immunodeficiency disease** either the immune system fails to develop normally or the immune response is blocked in some way. Autoimmune disorders and immunodeficiency diseases are relatively rare conditions—clear evidence of the effectiveness of the immune system's control mechanisms. A far more common, and usually far less dangerous, class of immune disorders are the **allergies**. We will consider autoimmune disorders and allergies below. AIDS, an important example of an immunodeficiency disease, was considered on p. 397. For more extended discussions of these conditions, including possible therapies, see related sections in the *Applications Manual.* AM *Immune Competence*

Autoimmune Disorders

Autoimmune disorders develop when the immune response mistakenly targets normal body cells and tissues. The immune system usually recognizes and ignores the antigens normally found in the body. The recognition system can malfunction, however, and when it does the activated B cells begin to manufacture antibodies against other cells and tissues. The symptoms produced depend on the identity of the antigen attacked by these misguided antibodies, called *autoantibodies*. For example, *rheumatoid arthritis* occurs when autoantibodies attack joint surfaces.

Many autoimmune disorders appear to be cases of mistaken identity. For example, proteins associated with the measles, influenza, and other viruses contain amino acid sequences that are similar to those of myelin proteins. As a result, antibodies that target these viruses may also attack myelin sheaths, producing the neurological complications sometimes associated with a vaccination or viral infection.

Similarly, unusual types of MHC proteins have been linked to at least 50 clinical conditions. Many of these are disorders considered in the *Applications Manual*, including *psoriasis, rheumatoid arthritis, myasthenia gravis, multiple sclerosis, narcolepsy, Type I diabetes, Graves' disease, Addison's disease, pernicious anemia, systemic lupus erythematosus,* and *chronic hepatitis.*

Allergies

Allergies are inappropriate or excessive immune responses to antigens. The sudden increase in cellular activity or antibody titers can have a number of unpleasant side effects. For example, neutrophils or cytotoxic T cells may destroy normal cells while attacking the antigen, or the antigen-antibody complex may trigger a massive inflammatory response. Antigens that trigger allergic reactions are often called **allergens.**

Four types of allergies are recognized: *immediate hypersensitivity* (Type I), *cytotoxic reactions* (Type II), *immune complex disorders* (Type III), and *delayed hypersensitivity* (Type IV). Immediate hypersensitivity is probably the most common type, and it includes "hay fever" and environmental allergies that may affect 15 percent of the U.S. population. The cross-reactions that occur following the transfusion of an incompatible blood type (Chapter 12) is an example of Type II hypersensitivity. ∞ *[p. 316]* For a discussion of other allergy types, consult the *Applications Manual.* AM *Immediate Hypersensitivity, Immune Complex Disorders,* and *Delayed Hypersensitivity*

Immediate hypersensitivity begins when B cells and helper T cells are first exposed and *sensitized* to an allergen. Sensitization leads to the production of large quantities of IgE antibodies. The tendency to produce IgE antibodies in response to an allergen may be genetically determined. The first exposure to an allergen does not produce allergic symptoms. However, the IgE antibodies that are produced at this time become attached to the cell membranes of basophils and mast cells throughout the body. When exposed to the same allergen at a later date, these cells are stimulated to release histamine, heparin, several cytokines, prostaglandins, and other chemicals into the surrounding tissues. The result is a sudden inflammation of the affected tissues.

The severity of the allergic reaction depends on the sensitivity of the individual and the location in-

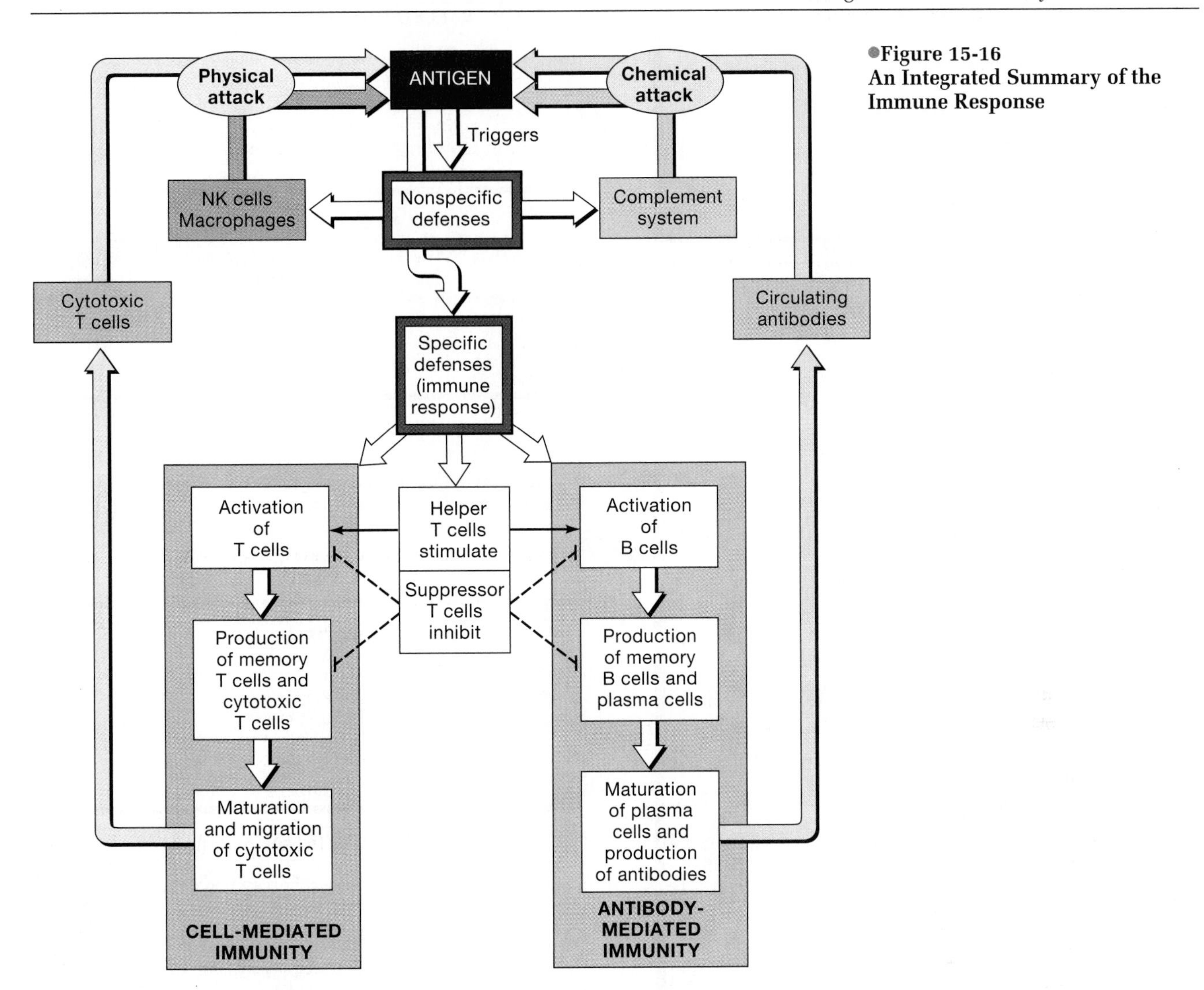

•Figure 15-16
An Integrated Summary of the Immune Response

volved. If the allergen exposure occurs at the body surface, the response is usually restricted to that area. If the allergen enters the circulation, the response may be more dramatic and occasionally lethal.

STRESS AND THE IMMUNE RESPONSE

Interleukin-1 is one of the first cytokines produced as part of the immune response. In addition to promoting inflammation, it also stimulates ACTH production by the anterior pituitary. This in turn leads to the secretion of glucocorticoids by the adrenal cortex. (The functions of ACTH and glucocorticoids were described in Chapter 11. ∞ *[p. 294]*) The anti-inflammatory effects of the glucocorticoids may help regulate the intensity of the immune response, but the long-term secretion of glucocorticoids, as in chronic stress, can suppress the immune response and lower resistance to disease. Glucocorticoids depress inflammation, inhibit phagocytes, and reduce interleukin production. The mechanisms involved are still under investigation, but it is clear that immune system depression due to chronic stress represents a serious threat to health.

Age and the Immune Response

With advancing age the immune system becomes less effective at combating disease. T cells become less responsive to antigens, so fewer cytotoxic T cells respond to an infection. Because the number of helper T cells is also reduced, B cells are less responsive, and antibody levels do not rise as quickly after antigen exposure. The net result is an increased susceptibility to viral and bacterial infection. For this reason, vaccinations for acute viral diseases, such as the flu (influenza), are strongly recommended for elderly individuals. The increased incidence of cancer in the elderly reflects the fact that immune surveillance declines, and tumor cells are not eliminated as effectively.

INTEGRATION WITH OTHER SYSTEMS

Figure 15-17• summarizes the interactions between the lymphatic system and other physiological sys-

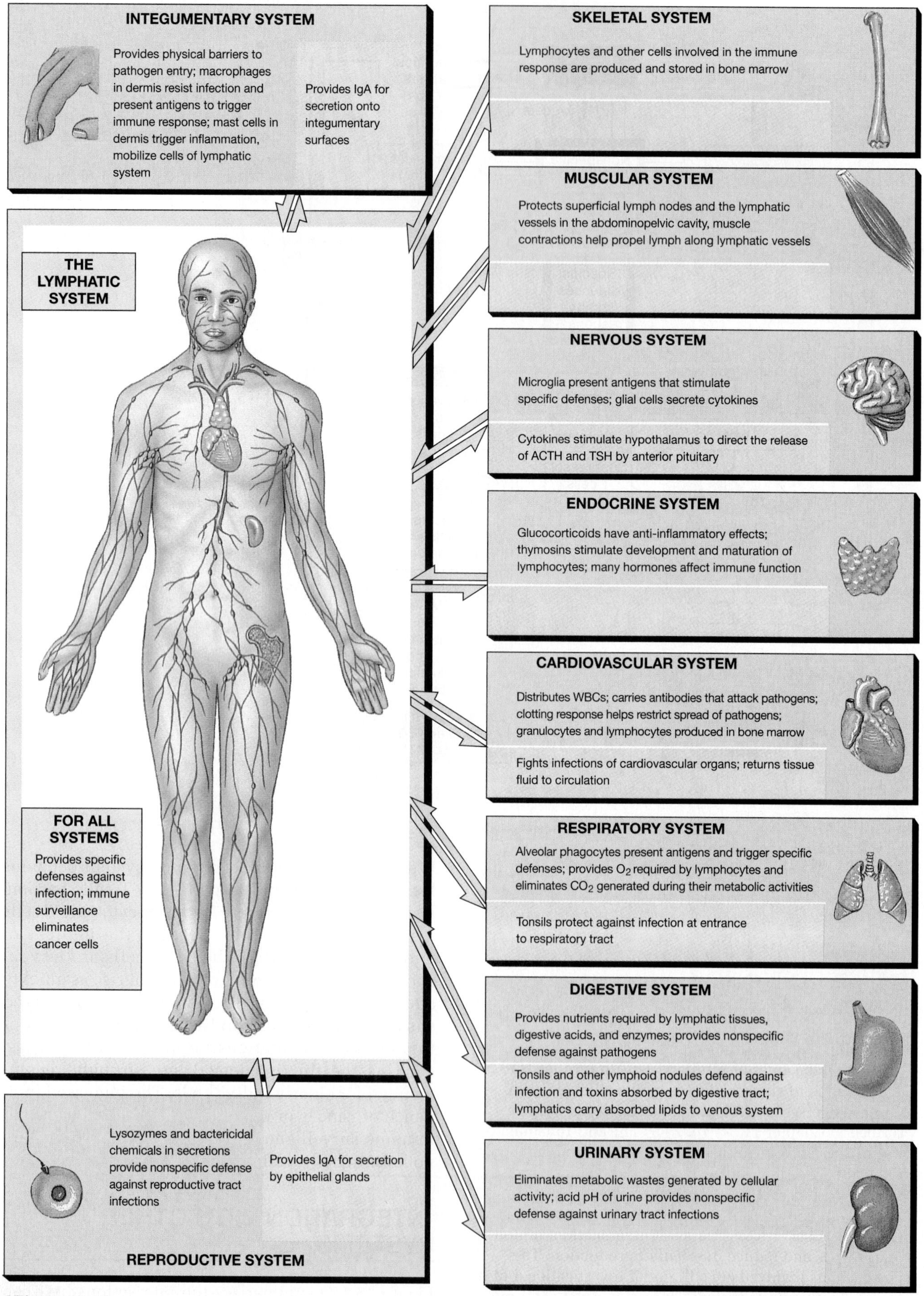

INTEGUMENTARY SYSTEM

Provides physical barriers to pathogen entry; macrophages in dermis resist infection and present antigens to trigger immune response; mast cells in dermis trigger inflammation, mobilize cells of lymphatic system

Provides IgA for secretion onto integumentary surfaces

SKELETAL SYSTEM

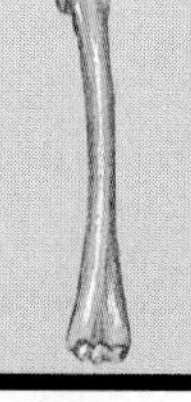

Lymphocytes and other cells involved in the immune response are produced and stored in bone marrow

MUSCULAR SYSTEM

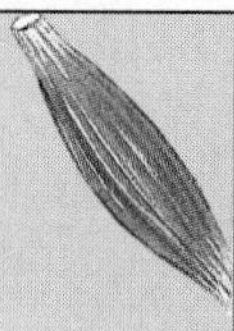

Protects superficial lymph nodes and the lymphatic vessels in the abdominopelvic cavity, muscle contractions help propel lymph along lymphatic vessels

THE LYMPHATIC SYSTEM

NERVOUS SYSTEM

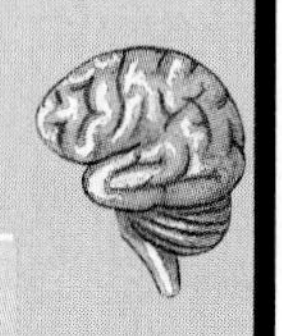

Microglia present antigens that stimulate specific defenses; glial cells secrete cytokines

Cytokines stimulate hypothalamus to direct the release of ACTH and TSH by anterior pituitary

ENDOCRINE SYSTEM

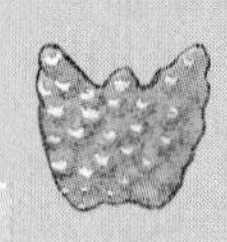

Glucocorticoids have anti-inflammatory effects; thymosins stimulate development and maturation of lymphocytes; many hormones affect immune function

CARDIOVASCULAR SYSTEM

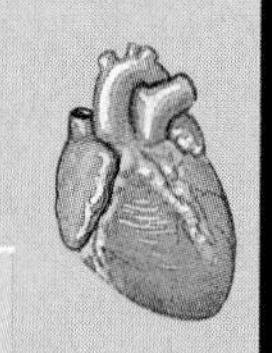

Distributes WBCs; carries antibodies that attack pathogens; clotting response helps restrict spread of pathogens; granulocytes and lymphocytes produced in bone marrow

Fights infections of cardiovascular organs; returns tissue fluid to circulation

FOR ALL SYSTEMS

Provides specific defenses against infection; immune surveillance eliminates cancer cells

RESPIRATORY SYSTEM

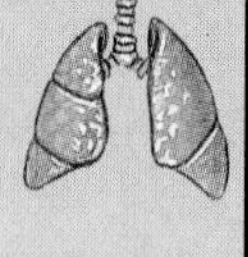

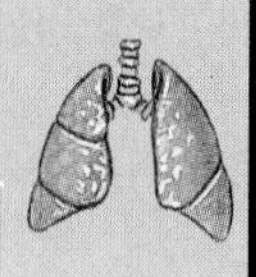

Alveolar phagocytes present antigens and trigger specific defenses; provides O_2 required by lymphocytes and eliminates CO_2 generated during their metabolic activities

Tonsils protect against infection at entrance to respiratory tract

DIGESTIVE SYSTEM

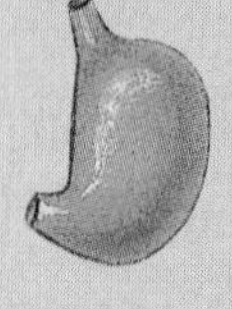

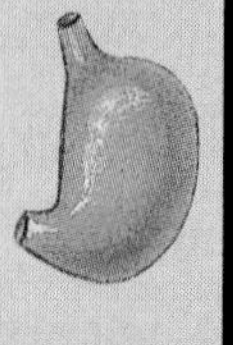

Provides nutrients required by lymphatic tissues, digestive acids, and enzymes; provides nonspecific defense against pathogens

Tonsils and other lymphoid nodules defend against infection and toxins absorbed by digestive tract; lymphatics carry absorbed lipids to venous system

REPRODUCTIVE SYSTEM

Lysozymes and bactericidal chemicals in secretions provide nonspecific defense against reproductive tract infections

Provides IgA for secretion by epithelial glands

URINARY SYSTEM

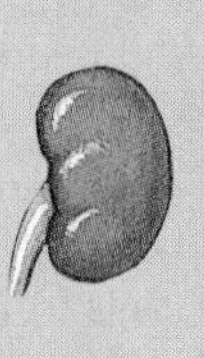

Eliminates metabolic wastes generated by cellular activity; acid pH of urine provides nonspecific defense against urinary tract infections

Figure 15-17
Functional Relationships between the Lymphatic System and Other Systems

tems. The relationships between the cells and tissues involved with the immune response and the nervous and endocrine systems are now the focus of intense research.

✓ Would the primary response or the secondary response be more affected by a lack of memory B cells for a particular antigen?

✓ What kind of immunity protects the developing fetus and how is it produced?

TECHNOLOGY, IMMUNITY, AND DISEASE

Our understanding of disease mechanisms has been profoundly influenced by recent advances in genetic engineering and protein analysis. The information and technical capabilities developed during the 1980s have already started to affect clinical procedures. This trend is sure to continue, and several lines of research are particularly exciting because of their broad potential application. These projects involve a mixture of genetic engineering, computer analysis, and protein biochemistry. Commercial products such as *monoclonal antibodies* are already in widespread clinical use. For a discussion of genetic engineering protocols and prospects, see the *Applications Manual.* AM Technology, Immunity and Disease

Chapter Review

KEY TERMS

allergen, p. 400
antibody, p. 396
antigen, p. 385
antigen-antibody complex, p. 395
antigen recognition, p. 393
B cells, p. 385
chemotaxis, p. 391
cytokines, p. 391
immunity, p. 392
immunoglobulin, p. 385
immunological surveillance, p. 391
lymphocyte, p. 385
monocyte-macrophage system, p. 391
NK cells, p. 385
pyrogen, p. 391
T cells, p. 385

SUMMARY OUTLINE

INTRODUCTION p. 383

1. The cells, tissues, and organs of the **lymphatic system** play a central role in the body's defenses against a variety of **pathogens**, or disease-causing organisms.

2. Lymphocytes, the primary cells of the lymphatic system, provide an **immune response** to specific threats to the body. **Immunity** is the ability to resist infection and disease through the activation of specific defenses.

ORGANIZATION OF THE LYMPHATIC SYSTEM p. 383

1. The lymphatic system includes a network of **lymphatic vessels** that carry **lymph** (a fluid similar to plasma but with a lower concentration of proteins). A series of **lymphoid organs** are connected to the lymphatic vessels. (*Figure 15-1*)

Functions of the Lymphatic System p. 383

2. The lymphatic system produces, maintains, and distributes lymphocytes (cells that attack invading organisms, abnormal cells, and foreign proteins). The system also helps maintain blood volume and eliminate local variations in the composition of the interstitial fluid.

Lymphatic Vessels p. 383

3. Lymph flows along a network of *lymphatics* that originate in the **lymphatic capillaries**. The lymphatic vessels empty into the **thoracic duct** and the **right lymphatic duct**. (*Figures 15-1 to 15-3*)

Lymphocytes p. 385

4. There are three classes of lymphocytes: **T cells** (thymus-dependent), **B cells** (bone marrow–derived), and **NK cells** (natural killer).

5. *Cytotoxic T cells* attack foreign cells or body cells infected by viruses; they provide *cellular immunity.* **Regulatory T cells** (*helper* and *suppressor T cells*) regulate and coordinate the immune response.

6. B cells can differentiate into **plasma cells**, which produce and secrete antibodies that react with specific chemical targets, or **antigens**. Antibodies in body fluids are also called **immunoglobulins**. B cells are responsible for *humoral immunity.*

7. NK cells attack foreign cells, normal cells infected with viruses, and cancer cells. They provide *immunological surveillance.*

8. Lymphocytes continually migrate in and out of the blood through the lymphoid tissues and organs. *Lymphopoiesis* (lymphocyte production) involves the bone marrow, thymus, and peripheral lymphoid tissues. (*Figure 15-4*)

Lymphoid Nodules p. 386

9. A **lymphoid nodule** consists of loose connective tissue containing densely packed lymphocytes.

Lymphoid Organs p. 387

10. Important lymphoid organs include the *lymph nodes*, the *thymus*, and the *spleen*. Lymphoid tissues and organs are distributed in areas especially vulnerable to injury or invasion.

11. **Lymph nodes** are encapsulated masses of lymphoid tissue containing lymphocytes. Lymph nodes monitor the

lymph before it drains into the venous system, removing antigens and initiating appropriate immune responses. (*Figure 15-5*)

12. The **thymus** lies behind the sternum. T cells become mature in the thymus. (*Figure 15-6*)

13. The adult **spleen** contains the largest mass of lymphoid tissue in the body. The cellular components form the **pulp** of the spleen. *Red pulp* contains large numbers of red blood cells, and areas of *white pulp* resemble lymphoid nodules. The spleen removes antigens and damaged blood cells from the circulation, initiates appropriate immune responses, and stores iron obtained from recycled red blood cells. (*Figure 15-7*)

The Lymphatic System and Body Defenses p. 388

14. The lymphatic system is a major component of the body's defenses. These fall into two categories: (1) **nonspecific defenses** that do not discriminate between one threat and another; and (2) **specific defenses** that protect against threats on an individual basis.

NONSPECIFIC DEFENSES p. 389

1. Nonspecific defenses prevent the approach, deny the entrance, or limit the spread of living or nonliving hazards. (*Figure 15-8*)

Physical Barriers p. 389

2. Physical barriers include hair, epithelia, and various secretions of the integumentary and digestive systems.

Phagocytes p. 390

3. There are two types of phagocytic cells: **microphages** (neutrophils and eosinophils) and **macrophages** (cells of the monocyte-macrophage system).

4. Phagocytes move between cells through *diapedesis*, and they show *chemotaxis* (sensitivity and orientation to chemical stimuli).

Immunological Surveillance p. 391

5. **Immunological surveillance** involves constant monitoring of normal tissues by NK cells sensitive to abnormal antigens on the surfaces of otherwise normal cells. Cancer cells with tumor-specific antigens on their surfaces are killed.

Interferons p. 391

6. **Interferons**, small proteins released by cells infected with viruses, trigger the production of antiviral proteins that interfere with viral replication inside the cell. Interferons are *cytokines*, chemical messengers released by tissue cells to coordinate local activities.

Complement p. 391

7. At least 11 complement proteins make up the **complement system**. They interact with each other in chain reactions to destroy target cell membranes, stimulate inflammation, attract phagocytes, and enhance phagocytosis.

Inflammation p. 391

8. Inflammation represents a coordinated nonspecific response to tissue injury. (*Figure 15-9*)

Fever p. 391

9. A **fever** (body temperature greater than 37.2°C or 99°F) can inhibit pathogens and accelerate metabolic processes.

SPECIFIC DEFENSES: THE IMMUNE RESPONSE p. 392

1. Specific defenses are provided by T cells and B cells. T cells provide **cell-mediated immunity**; B cells provide **antibody-mediated immunity**.

Forms of Immunity p. 392

2. Specific immunity may involve **innate immunity** (genetically determined and present at birth) or **active immunity** (which appears following exposure to an antigen). Active immunity may be *naturally acquired* or *induced*. **Passive immunity** (the transfer of antibodies from another person) may occur *naturally* or be *induced*. (*Figure 15-10*)

Properties of Immunity p. 392

3. Lymphocytes provide specific immunity, which has four general characteristics: specificity, versatility, memory, and tolerance. *Memory cells* enable the immune system to "remember" previous target antigens. **Tolerance** refers to the ability of the immune system to ignore some antigens, such as those of body cells.

An Overview of the Immune Response p. 393

4. The goal of the **immune response** is to destroy or inactivate pathogens, abnormal cells, and foreign molecules. It is based on the activation of lymphocytes by specific antigens through a process known as **antigen recognition**. (*Figure 15-11*)

T Cells and Cellular Immunity p. 393

5. Foreign antigens must usually be processed by macrophages and incorporated into their cell membranes bound to *MHC proteins* before they can activate T cells. T cells can also be activated by viral antigens displayed on the surfaces of virus-infected cells.

6. Activated T cells may differentiate into *cytotoxic T cells, memory T cells, suppressor T cells,* and *helper T cells.*

7. Cell-mediated immunity results from the activation of **cytotoxic**, or *killer*, **T cells**. Activated **memory T cells** remain on reserve to guard against future such attacks. (*Figure 15-12*)

8. **Suppressor T cells** depress the responses of other T and B cells. (*Figure 15-16*)

9. **Helper T cells** secrete cytokines that help coordinate specific and nonspecific defenses and regulate cellular and humoral immunity. (*Figure 15-13*)

B Cells and Antibody-Mediated Immunity p. 395

10. B cells, responsible for antibody-mediated immunity, normally become activated by helper T cells sensitive to the same antigen.

11. An activated B cell divides and produces **plasma cells** and **memory B cells**. Antibodies are produced by the plasma cells. (*Figure 15-13*)

12. An antibody molecule consists of two parallel pairs of polypeptide chains containing fixed and variable segments. (*Figure 15-14*)

13. When an antibody molecule binds to an antigen, they form an **antigen-antibody complex**. Antibodies focus on specific antigenic determinant sites.

14. There are five classes of antibodies in body fluids: (1) **immunoglobulin G (IgG)**, responsible for resistance against many viruses, bacteria, and bacterial toxins; (2) *IgM*, the first antibody type secreted after an antigen arrives; (3) *IgA*, found in glandular secretions; (4) *IgE*, which releases

chemicals that accelerate local inflammation; and (5) *IgD*, found on the surfaces of B cells. (*Table 15-1*)

15. Antibodies may destroy antigens through **neutralization**, **precipitation** and **agglutination**, activation of complement, attraction of phagocytes, enhancement of phagocytosis, and stimulation of inflammation.

16. The antibodies produced by plasma cells on first exposure to an antigen are the agents of the **primary response**. The maximum *antibody titer* appears during the **secondary** *(anamnestic)* **response** which follows subsequent exposure to the same antigen. (*Figure 15-15*)

Hormones of the Immune System p. 398

17. **Interleukins** increase T cell sensitivity to antigens exposed on macrophage membranes; stimulate B cell activity, plasma cell formation, and antibody production; and enhance nonspecific defenses.

18. **Interferons** slow the spread of a virus by making the synthesizing cell and its neighbors resistant to viral infections.

19. **Tumor necrosis factors** slow tumor growth and kill tumor cells.

20. Several **phagocytic regulators** adjust the activities of phagocytic cells in order to coordinate specific and nonspecific defenses. (*Table 15-2*)

PATTERNS OF IMMUNE RESPONSE p. 400

1. Foreign antigens may undergo physical or chemical attack by specific and nonspecific defenses. (*Figure 15-16*)

Immune Disorders p. 400

2. In an **immunodeficiency disease** either the immune system does not develop normally or the immune response is somehow blocked.

3. **Autoimmune disorders** develop when the immune response mistakenly targets normal body cells and tissues.

4. **Allergies** are inappropriate or excessive immune responses to **allergens** (antigens that trigger allergic reactions). There are four types of allergies: *immediate hypersensitivity* (Type I), *cytotoxic reactions* (Type II), *immune complex disorders* (Type III), and *delayed hypersensitivity* (Type IV).

Age and the Immune Response p. 401

5. With aging, the immune system becomes less effective at combating disease.

INTEGRATION WITH OTHER SYSTEMS p. 401

1. The lymphatic system has extensive interactions with the nervous and endocrine systems. (*Figure 15-17*)

CHAPTER QUESTIONS

LEVEL 1 Reviewing Facts and Terms

Match each item in column A with the most closely related item in column B. Use letters for answers in the spaces provided.

Column A	Column B
___ 1. humoral immunity	a. induce fever
___ 2. lymphoma	b. system of circulating proteins
___ 3. complement	c. CNS macrophages
___ 4. microphages	d. monocytes
___ 5. macrophages	e. genetically programmed cell death
___ 6. microglia	f. transfers of antibodies
___ 7. interferon	g. neutrophils, eosinophils
___ 8. pyrogens	h. secretion of antibodies
___ 9. innate immunity	i. present at birth
___ 10. active immunity	j. cytokine
___ 11. passive immunity	k. lymphatic system cancer
___ 12. apoptosis	l. exposure to antigen

13. Lymph from the lower abdomen, pelvis, and lower limbs is received by the:
(a) right lymphatic duct (b) inguinal duct
(c) thoracic duct (d) aorta

14. Lymphocytes responsible for providing cell-mediated immunity are called:
(a) macrophages (b) B cells
(c) plasma cells (d) cytotoxic T cells

15. B cells are responsible for:
(a) cellular immunity
(b) immunological surveillance
(c) antibody-mediated immunity
(d) a, b, and c are correct

16. Lymphoid stem cells that can form all types of lymphocytes occur in the:
(a) bloodstream (b) thymus
(c) bone marrow (d) spleen

17. Lymphatics are found in all portions of the body except the:
(a) lower limbs (b) central nervous system
(c) head and neck region (d) hands and feet

18. The largest collection of lymphoid tissue in the body is contained in the:
(a) adult spleen (b) adult thymus
(c) bone marrow (d) tonsils

19. Red blood cells that are damaged or defective are removed from the circulation by the:
(a) thymus (b) lymph nodes
(c) spleen (d) tonsils

20. Phagocytes move through capillary walls by squeezing between adjacent endothelial cells, a process known as:
(a) diapedesis (b) chemotaxis
(c) adhesion (d) perforation

21. Perforins are proteins associated with the activity of:
    (a) T cells (b) B cells
    (c) macrophages (d) plasma cells
22. Complement activation:
    (a) stimulates inflammation
    (b) attracts phagocytes
    (c) enhances phagocytosis
    (d) a, b, and c are correct
23. Inflammation:
    (a) aids in temporary repair at an injury site
    (b) slows the spread of pathogens
    (c) facilitates permanent repair
    (d) a, b, and c are correct
24. Memory B cells:
    (a) respond to a threat on first exposure
    (b) secrete large numbers of antibodies into the interstitial fluid
    (c) deal with subsequent injuries or infections that involve the same antigens
    (d) contain binding sites that can activate the complement system
25. What two large collecting vessels are responsible for returning lymph to the veins of the circulatory system? What areas of the body does each serve?
26. Give a function for each of the following:
    (a) cytotoxic T cells
    (b) helper T cells
    (c) suppressor (regulatory) T cells
    (d) plasma cells
    (e) NK cells
    (f) interferons
    (g) T cells
    (h) B cells
    (i) interleukins
27. What seven defenses, present at birth, provide the body with the defensive capability known as nonspecific resistance?

LEVEL 2 Reviewing Concepts

28. Compared with nonspecific defenses, *specific* defenses:
    (a) do not discriminate between one threat and another
    (b) are always present at birth
    (c) provide protection against threats on an individual basis
    (d) deny entrance of pathogens to the body
29. T cells and B cells can be activated only by:
    (a) pathogenic microorganisms
    (b) interleukins, interferons, and colony-stimulating factors
    (c) cells infected with viruses, bacterial cells, or cancer cells
    (d) exposure to a specific antigen at a specific site on a cell membrane
30. List and explain the four general properties of immunity.
31. How does the formation of an antibody-antigen complex cause elimination of an antigen?
32. What effects follow activation of the complement system?

LEVEL 3 Critical Thinking and Clinical Applications

33. An investigator at a crime scene discovers some body fluid on the victim's clothing. He carefully takes a sample and sends it to the crime lab for analysis. On the basis of analysis of immunoglobulins, could the crime lab determine whether the sample was blood plasma or semen? Explain.
34. Ted finds out that he has been exposed to the measles. He is concerned that he might have contracted the disease, so he goes to see his physician. The doctor takes a blood sample and sends it to a lab for antibody titers. The results show an elevated level of IgM antibodies to rubella (measles) virus, but very few IgG antibodies to the virus. Has Ted contracted the disease or not?

16 The Respiratory System

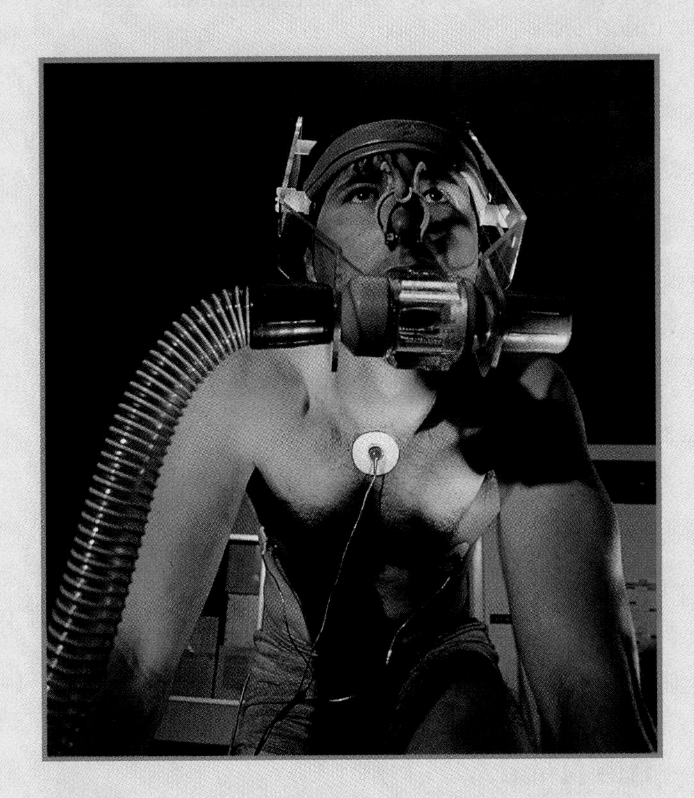

This is one way to go nowhere fast—pedaling on a stationary bicycle. All the effort, however, is hardly wasted. The rather awkward-looking array of hoses and wires attached to this rider is designed to monitor respiratory and cardiovascular performance during exercise. Similar equipment can be used in a clinical setting to assess respiratory function in resting or active individuals.

Chapter Outline and Objectives

Vocabulary Development

alveolus, a hollow cavity; *alveolus, alveolar duct*
ateles, imperfect; *atelectasis*
cricoid, ring-shaped; *cricoid cartilage*
ektasis, expansion; *atelectasis*
-ia, condition; *pneumonia*
kentesis, puncture; *thoracentesis*
oris, mouth; *oropharynx*
pneuma, air; *pneumothorax*
pneumon, lung; *pneumonia*
stoma, mouth; *tracheostomy*
thorac-, chest; *thoracentesis*
thyroid, shield-shaped; *thyroid cartilage*

Living cells need energy for maintenance, growth, defense, and replication. Our cells obtain that energy through aerobic respiration, a process that requires oxygen and produces carbon dioxide. ∞ *[p. 61]* Therefore, the cells in the body must have a way to obtain oxygen and eliminate carbon dioxide. Our respiratory exchange surfaces are inside the lungs, where diffusion occurs between the air and the blood. The exchange surfaces are relatively delicate because they must be very thin to encourage rapid diffusion. The cardiovascular system provides the link between the interstitial fluids of the body and the exchange surfaces of the lungs. The circulating blood carries oxygen from the lungs to peripheral tissues; it also accepts and transports the carbon dioxide generated by those tissues, delivering it to the lungs.

Our discussion of the respiratory system begins by following the air as it travels from the exterior toward the alveoli of the lungs. We will then consider the mechanics of breathing, or *pulmonary ventilation* (the physical movement of air into and out of the lungs), and the physiology of respiration, which includes the processes of breathing and gas transport and exchange between the air, blood, and tissues.

Functions of the Respiratory System

The respiratory system performs the following range of functions. It (1) moves air to and from the gas-exchange surfaces where diffusion can occur between air and circulating blood, (2) provides nonspecific defenses against pathogenic invasion, (3) permits vocal communication, and (4) helps control body fluid pH.

Organization of the Respiratory System

The **respiratory system** consists of the nose, nasal cavity, and sinuses; the *pharynx* (throat); the *larynx* (voice box); the *trachea* (windpipe); and the *bronchi* and *bronchioles* (conducting passageways) and *alveoli* (exchange surfaces) of the lungs (Figure 16-1●).

The Respiratory Tract

The **respiratory tract** consists of the airways that carry air to and from the exchange surfaces of the lungs. The respiratory tract can be divided into a *conducting portion* and a *respiratory portion.* The conducting portion begins at the entrance to the nasal cavity and continues through the pharynx, larynx, trachea, bronchi, and the larger bronchioles. The respiratory portion includes the smallest and most delicate bronchioles and the alveoli that are the site of gas exchange.

In addition to delivering air to the lungs, the conducting passageways filter, warm, and humidify the air, thereby protecting the alveoli from debris, pathogens, and environmental extremes. By the time the air reaches the alveoli, most foreign particles and pathogens have been removed, and the humidity and temperature are within acceptable limits.

The Nose

Air normally enters the respiratory system via the paired **external nares** (nostrils) that communicate with the **nasal cavity.** The **vestibule** (VES-ti-būl) is the portion of the nasal cavity contained within the flexible tissues of the external nose. Here coarse hairs guard the nasal cavity from large airborne particles such as sand, dust, and insects.

The maxillary, nasal, frontal, ethmoid, and sphenoid bones form the lateral and superior walls of the nasal cavity. The *nasal septum* divides the nasal cavity into left and right sides. The bony posterior septum includes portions of the vomer and the ethmoid bone. A bony **hard palate,** formed by the palatine and maxillary bones, separates the oral and nasal cavities. A fleshy **soft palate** extends behind the hard palate, marking the boundary line between the superior **nasopharynx** (nā-zō-FĀR-inks) and the rest of the pharynx. The nasal cavity opens into the nasopharynx at the **internal nares** (NĀ-rēz).

Superior, middle, and *inferior nasal conchae* project toward the nasal septum from the lateral walls of the nasal cavity (Figure 16-2●). To pass from the vestibule to the internal nares, air tends to flow between adjacent conchae. As the air eddies and swirls, like water flowing over rapids, small airborne particles come in contact with the mucus that coats the lining of the nasal cavity. In addition to promoting fil-

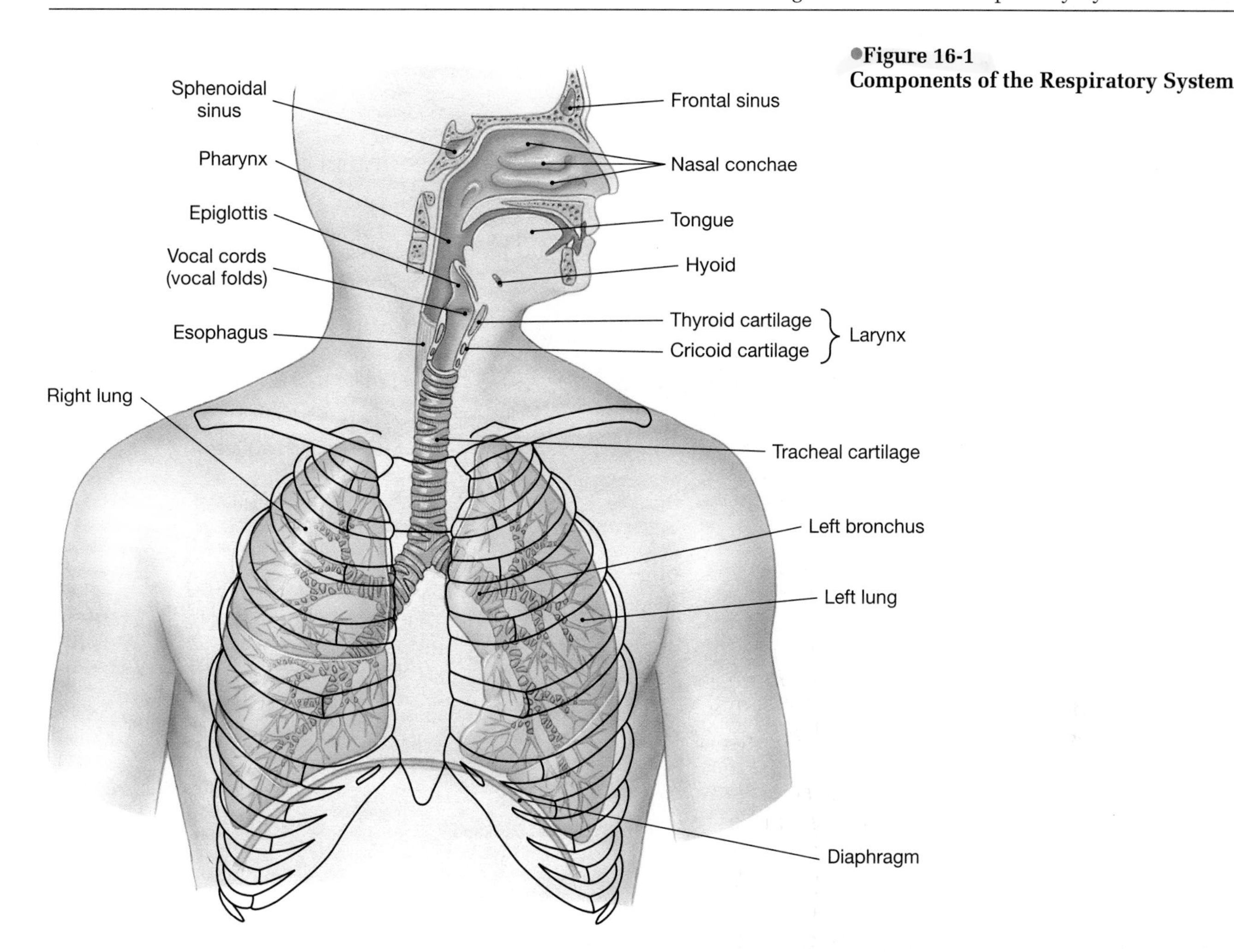

●Figure 16-1
Components of the Respiratory System

tration, the turbulence allows extra time for warming and humidifying the incoming air.

The ciliated lining of the nasal cavity, shown in Figure 16-3●, contains numerous *goblet cells.* The goblet cells and mucous glands beneath the epithelium produce mucus that bathes the exposed surfaces of the nasal cavity. Cilia sweep that mucus and any trapped debris or microorganisms toward the pharynx, where it can be swallowed and exposed to the acids and enzymes of the stomach. The respiratory surfaces are also flushed by mucus produced in the *paranasal sinuses* (the *frontal, sphenoid, ethmoid,* and *maxillary sinuses,* detailed in Figure 6-9●, p. 123), and by tears flowing through the nasolacrimal duct. Exposure to noxious vapors, large quantities of dust and debris, allergens, or pathogens usually causes a rapid increase in the rate of mucus production, and a "runny nose" develops.

CYSTIC FIBROSIS

Cystic fibrosis (CF) is the most common lethal inherited disease affecting Caucasians of Northern European descent, occurring at a frequency of 1 birth in 2,500. It occurs, with less frequency, in those with Southern European ancestry, in the Ashkenazi Jewish population, and in African-Americans. The condition results from a defective gene located on chromosome 7. Individuals with CF seldom survive past age 30; death is usually the result of a massive bacterial infection of the lungs and associated heart failure.

The most serious symptoms appear because the respiratory mucosa in these individuals produces a dense, viscous mucus that cannot be transported by the cilia of the respiratory tract. Mucus transport stops, and mucous plugs block the smaller respiratory passageways. These blockages make breathing difficult, and the inactivation of the normal respiratory defenses leads to frequent bacterial infections. For information on the genetic basis of this disorder and current strategies for treatment, refer to the *Applications Manual.* AM *Cystic Fibrosis*

The Pharynx

The **pharynx** is a chamber shared by the digestive and respiratory systems that extends between the internal nares and the entrances to the larynx and esophagus. Its three subdivisions are shown in Figure 16-2●. The soft palate separates the nasopharynx from the **oropharynx** (*oris,* mouth). The nasopharynx, lined by a typical respiratory epithelium, contains the entrances to the *auditory tubes* and the *pharyngeal tonsil.* The oropharynx extends between the soft palate and the base of the

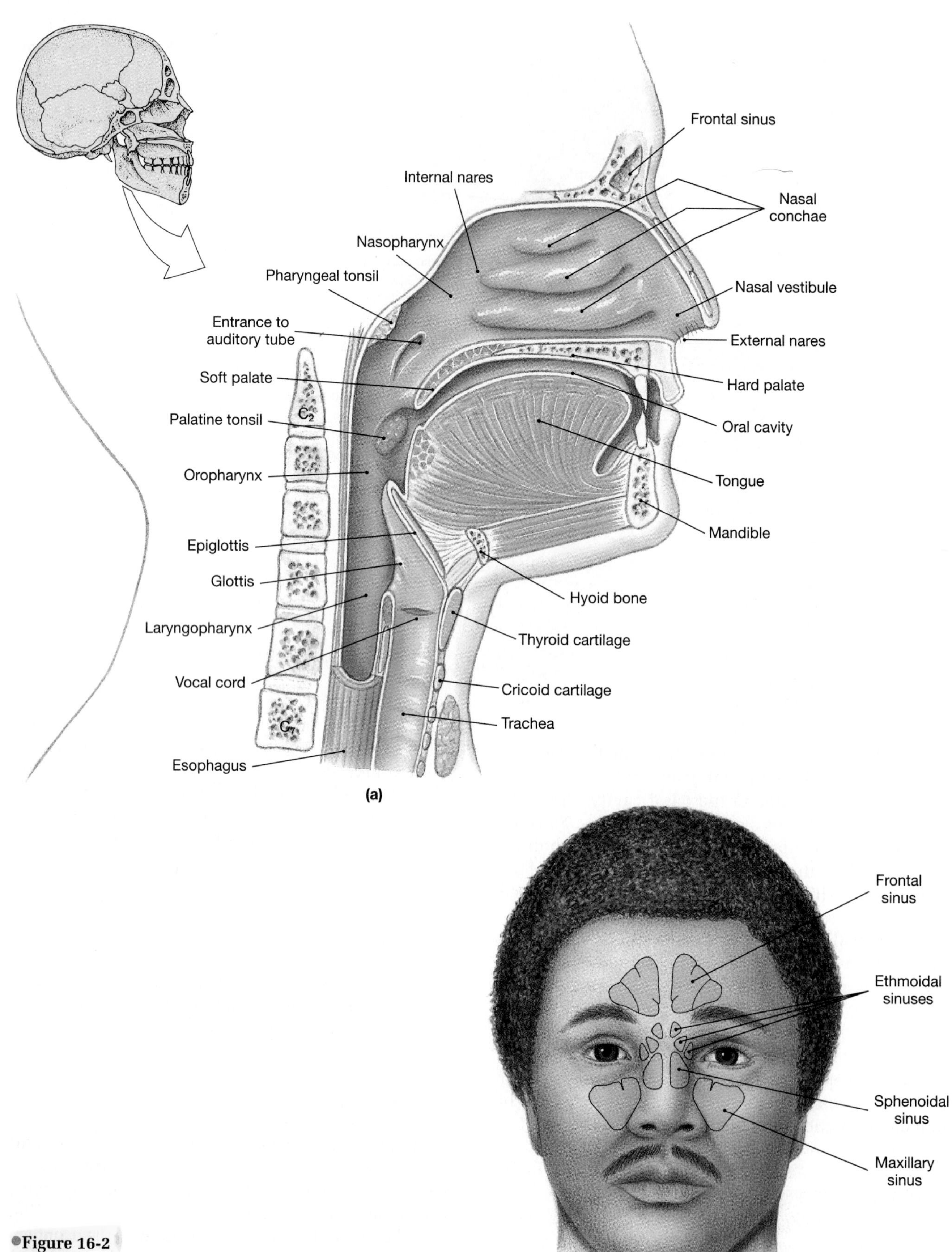

Figure 16-2
The Nose, Nasal Cavity, and Pharynx

(a) The nasal cavity and pharynx as seen in sagittal section, with the nasal septum removed. **(b)** The locations of the paranasal sinuses.

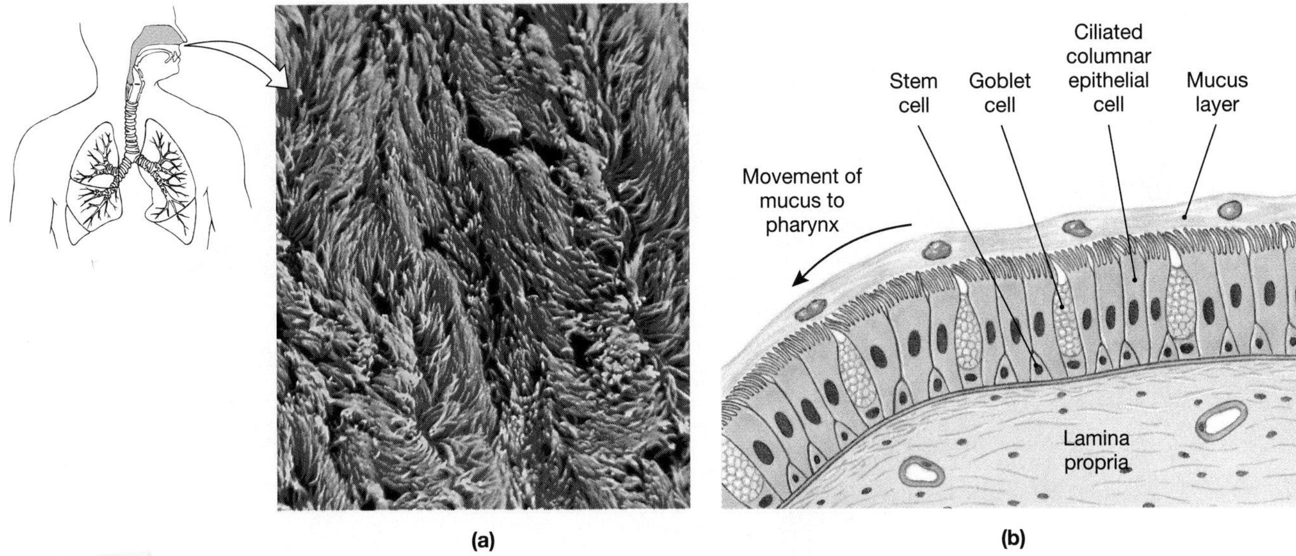

Figure 16-3
The Respiratory Epithelium

(a) A surface view of the epithelium, as seen with the scanning electron microscope. The cilia of the epithelial cells form a dense layer that resembles a shag carpet. The movement of these cilia propels mucus across the epithelial surface. (SEM × 1614) **(b)** Sketch showing the sectional appearance of the respiratory epithelium and its role in mucus transport.

tongue at the level of the hyoid bone. The palatine tonsils lie in the lateral walls of the oropharynx. The narrow **laryngopharynx** (lā-rin-gō-FĀR-inks) includes that portion of the pharynx lying between the hyoid bone and entrance to the esophagus. Materials entering the digestive tract pass through both the oropharynx and laryngopharynx. These regions are lined by a stratified squamous epithelium that can resist mechanical abrasion, chemical attack, and pathogenic invasion.

The Larynx

Incoming air leaves the pharynx by passing through a narrow opening, the **glottis** (GLOT-is), which is surrounded and protected by the **larynx** (LAR-inks). The larynx contains nine cartilages that are stabilized by ligaments or skeletal muscles or both (Figure 16-4). There are three large cartilages: The *epiglottis, thyroid cartilage,* and *cricoid cartilage.*

The elastic **epiglottis** (ep-i-GLOT-is) projects above the glottis. During swallowing the epiglottis folds back over the glottis, preventing the entry of liquids or solid food into the respiratory passageways. The curving **thyroid cartilage** (*thyroid;* shield-shaped) forms much of the anterior and lateral surfaces of the larynx. A prominent ridge on the anterior surface of this cartilage forms the "Adam's apple." The thyroid sits atop the **cricoid cartilage** (KRĪ-koyd; ring-shaped), which provides posterior support to the larynx. The thyroid and cricoid cartilages protect the glottis and the entrance to the trachea, and their broad surfaces provide sites for the attachment of important laryngeal muscles and ligaments.

There are also three pairs of smaller cartilages. The *arytenoid, corniculate,* and *cuneiform cartilages* are supported by the cricoid. Two pairs of ligaments, enclosed by folds of epithelium, extend across the larynx, between the thyroid cartilage and these cartilages, considerably reducing the size of the glottis. The upper pair, known as the **false vocal cords,** are relatively inelastic. They help to prevent foreign objects from entering the glottis, and they protect a more delicate pair of folds. These lower folds, the **true vocal cords,** contain elastic ligaments that extend between the thyroid cartilage and the arytenoids. Small muscles that insert on these cartilages change their position and alter the tension in these ligaments.

The Vocal Cords and Sound Production

The vocal cords vibrate when air passes through the glottis. These vibrations generate sound waves. As on a guitar or violin, short, thin strings vibrate rapidly, producing a high-pitched sound, and large, long strings vibrate more slowly, producing a low-pitched tone. Because children of both sexes have slender, short vocal cords, their voices tend to be high-pitched. At puberty, the larynx of a male enlarges more than that of a female. The true vocal cords of an adult male are thicker and longer and they produce lower tones than those of an adult female.

The pitch of the voice is regulated by the amount of tension in the vocal cords, which is controlled by skeletal muscles that change the position of the arytenoids. The volume depends on the force of the air movement. Further amplification and resonance occur

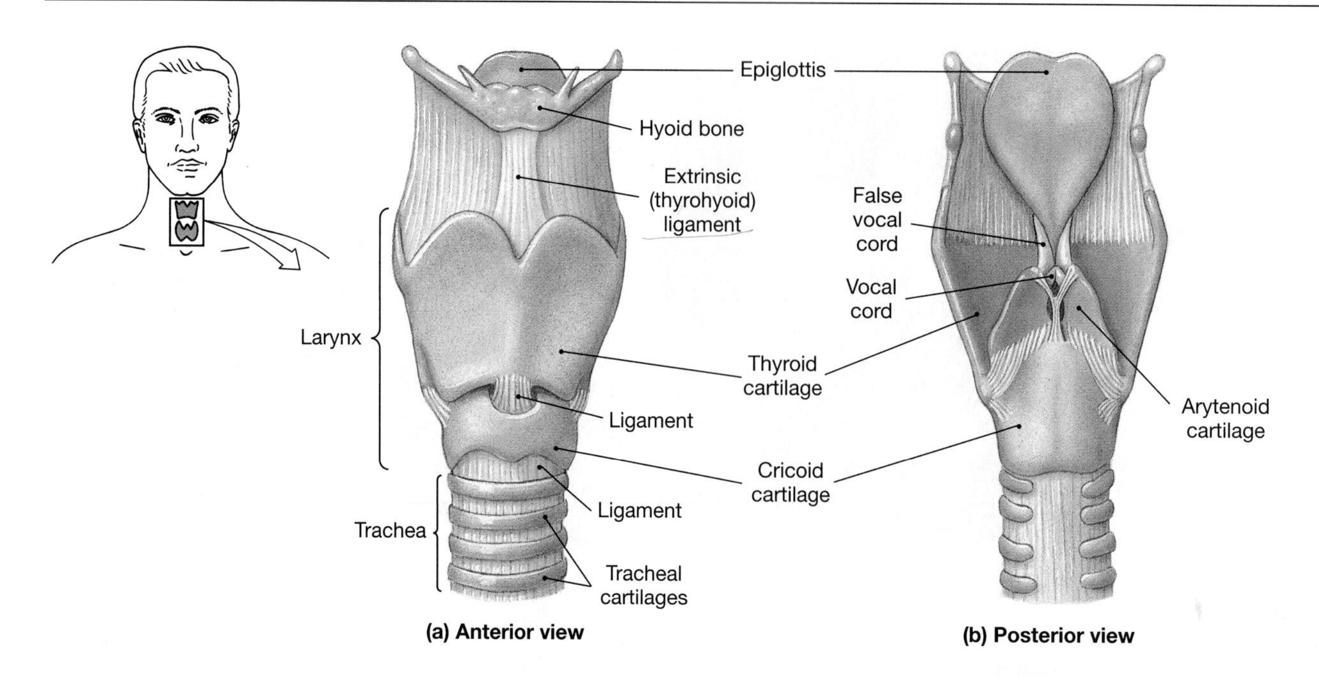

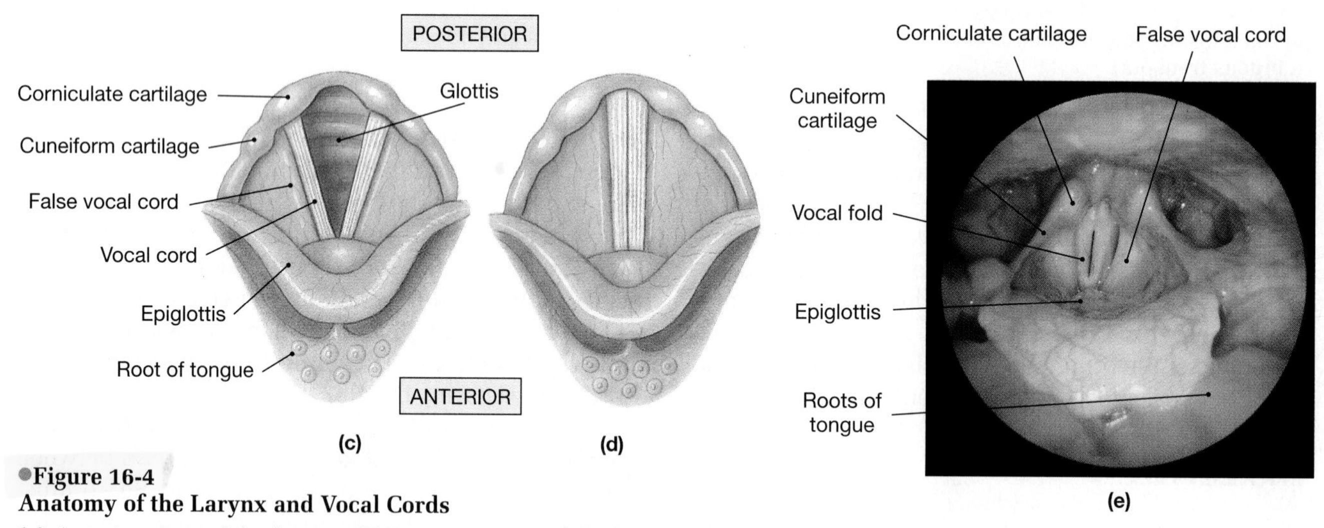

Figure 16-4
Anatomy of the Larynx and Vocal Cords
(a) Anterior view of the larynx. **(b)** Posterior view of the larynx. **(c)** Superior view of the larynx with the glottis open. **(d)** Superior view of the larynx with the glottis closed. **(e)** A fiber-optic view of the larynx with the glottis closed.

within the pharynx, the oral cavity, the nasal cavity, and the paranasal sinuses. The final production of distinct words further depends on voluntary movements of the tongue, lips, and cheeks.

The Trachea

The **trachea** (TRĀ-kē-a), or "windpipe,"is a tough, flexible tube with a diameter of around 2.5 cm (1 in.) and a length of approximately 11 cm (4.25 in.). It extends between the level of the sixth cervical vertebra, where it attaches to the cricoid cartilage of the larynx, and that of the fifth thoracic vertebra, where it branches to form a pair of *primary bronchi* (Figure 16-5●).

The walls of the trachea are supported by about 20 **tracheal cartilages.** These C-shaped cartilages stiffen the tracheal walls and protect the airway. They also prevent its collapse or overexpansion as pressures change in the respiratory system. The open portions of the C-shaped tracheal cartilages face posteriorly, toward the esophagus. Because the cartilages do not continue around the trachea, the posterior tracheal wall can easily distort, allowing large masses of food to pass along the esophagus. The diameter of the trachea is adjusted by the contractions of muscles under autonomic control. Sympathetic stimulation increases the diameter of the trachea and makes it easier to move large volumes of air along the respiratory passageways.

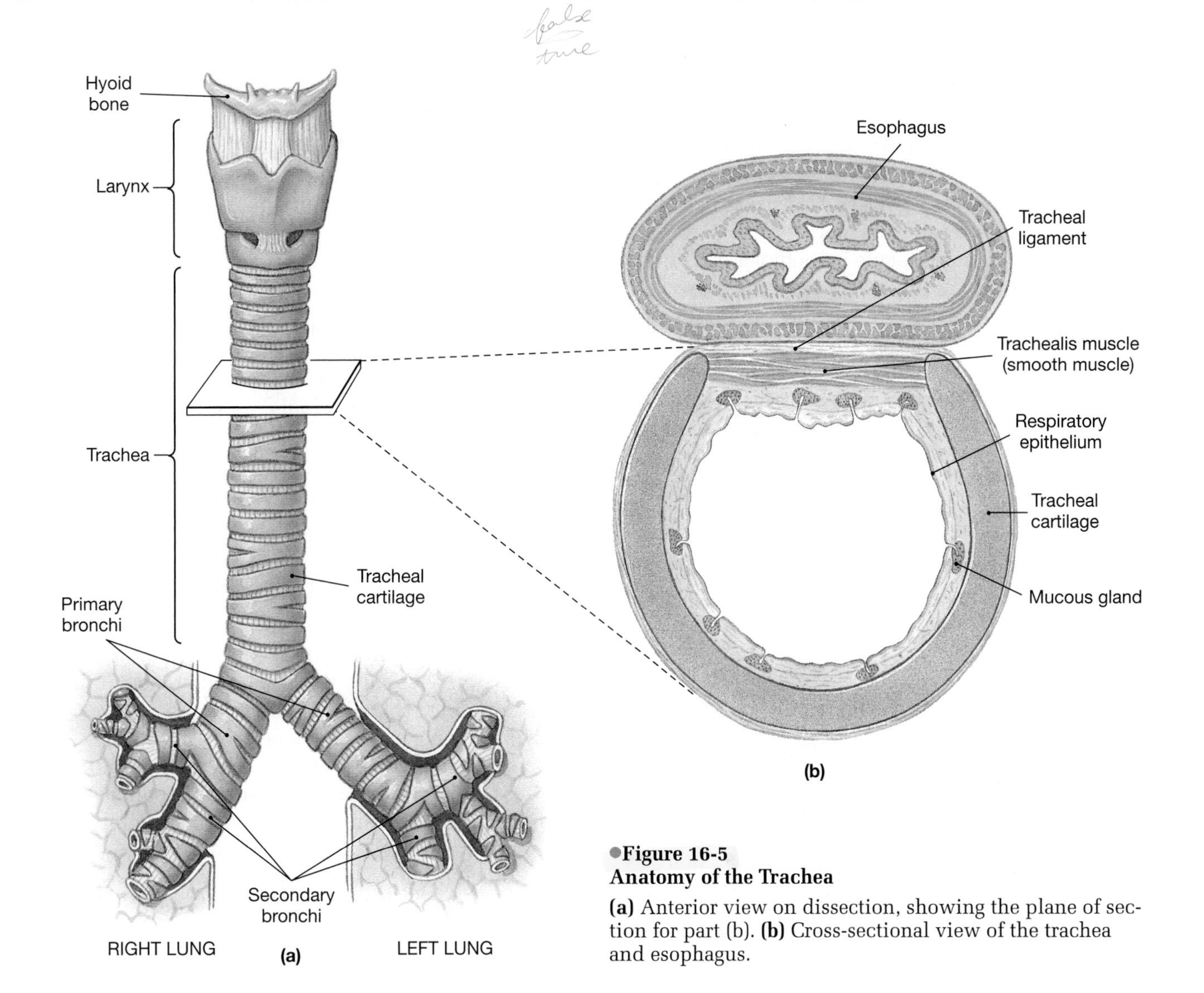

•Figure 16-5
Anatomy of the Trachea

(a) Anterior view on dissection, showing the plane of section for part (b). **(b)** Cross-sectional view of the trachea and esophagus.

TRACHEAL BLOCKAGE

Foreign objects that become lodged in the larynx or trachea are usually expelled by coughing. If the individual can speak or make a sound, the airway is still open, and no emergency measures should be taken. If the victim can neither breathe nor speak, an immediate threat to life exists. Unfortunately many victims become acutely embarrassed by this situation, and rather than seek assistance, they run to the nearest restroom and quietly expire.

In the *Heimlich* (HĪM-lik) *maneuver,* or *abdominal thrust,* a rescuer applies compression to the abdomen just beneath the diaphragm. This action elevates the diaphragm forcefully and may generate enough pressure to remove the blockage. This maneuver must be performed properly to avoid damaging to internal organs. Throughout the year, organizations such as the American Red Cross, the local fire department, or other charitable groups usually hold brief training sessions in the proper performance of the Heimlich maneuver.

If a tracheal blockage remains, professionally qualified rescuers may perform a *tracheostomy* (trā-kē-OS-to-mē; *stoma,* mouth). In this procedure, an incision is made through the anterior tracheal wall, and a tube is inserted. The tube bypasses the larynx and permits air to flow directly into the trachea.

The Bronchi

The trachea branches within the mediastinum, giving rise to the **right** and **left primary bronchi** (BRONG-kī) (Figure 16-5•). The histological organization of the primary bronchi resembles that of the trachea, complete with cartilaginous C-shaped rings. The primary bronchi and their branches form the *bronchial tree.* As it enters the lung, each primary bronchus gives rise to **secondary bronchi** that enter the lobes of that lung. The secondary bronchi divide to form 9–10 tertiary bronchi in each lung. Each tertiary bronchus branches repeatedly.

The cartilages of the secondary bronchi are quite massive, but as we proceed farther along the branches of the bronchial tree they become smaller and smaller. When the diameter of the passageway has narrowed to around 1 mm, cartilages disappear completely. This narrow passage represents a **bronchiole.**

Bronchioles

Bronchioles are to the respiratory system what arterioles are to the circulatory system. Varying the diameter of the bronchioles provides control over the amount of resistance to airflow and the distribution of air in the lungs.

Sympathetic activation leads to a relaxation of smooth muscles in the walls of bronchioles, leading to dilation of the respiratory passageways. Constriction of these smooth muscles can almost completely block the passageways. For example, acute bronchiolar constriction can occur during an asthma attack or an allergic reaction, due to inflammation of the bronchioles. AM *Asthma*

After further branching, we reach the level of the *terminal bronchioles,* with diameters of 0.3–0.5 mm. Each terminal bronchiole supplies air to a *lobule* of the lung (Figure 16-6●). A **lobule** is a segment of lung tissue that is bounded by connective tissue partitions and supplied by a single bronchiole, accompanied by branches of the pulmonary arteries and pulmonary veins. Within a lobule, a terminal bronchiole divides to form several *respiratory bronchioles.* These passages, the finest branches of the bronchial tree, deliver air to the respiratory surfaces of the lungs.

Alveolar Ducts and Alveoli

Figure 16-6a● shows the basic structure of a lobule. Respiratory bronchioles open into expansive chambers, called **alveolar ducts.** These passageways end at **alveolar sacs,** common chambers connected to individual **alveoli**. Each lung contains approximately 150 million alveoli, and their abundance gives the lung an open, spongy appearance (Figure 16-6b●).

In order to meet our metabolic requirements, the alveolar exchange surfaces of the lungs must be very large, equal to around 140 square meters—roughly the size of a tennis court. The alveolar epithelium primarily consists of an unusually thin and delicate simple squamous epithelium (Figure 16-6c●). Roaming **alveolar macrophages** (*dust cells*) patrol the epithelium, phagocytizing dust or debris that has reached the alveolar surfaces. **Surfactant** (sur-FAK-tant) **cells** produce an oily secretion, or **surfactant,** that forms a superficial coating over the alveolar epithelium. Surfactant is important because it reduces *surface tension* within the alveolus. Surface tension results from the attraction between water molecules at an air-water boundary. The alveolar walls are so delicate that without surfactant, the surface tension would be strong enough to collapse them. If surfactant levels are inadequate, because of injury or genetic abnormalities, each inhalation must be forceful enough to pop open the alveoli. An individual with this condition, called *respiratory distress syndrome,* is soon exhausted by the effort required to keep inflating and deflating the lungs. AM *Respiratory Distress Syndrome*

The Respiratory Membrane

Gas exchange occurs across the **respiratory membrane** of the alveoli. The respiratory membrane (Figure 16-6d●) consists of three components:

1. The squamous epithelial cells lining the alveolus.
2. The endothelial cells lining an adjacent capillary.
3. The fused basement membranes that lie between the alveolar and endothelial cells.

At the respiratory membrane, the total distance separating the alveolar air and the blood can be as little as 0.1 μm. Diffusion across the respiratory membrane proceeds very rapidly, because (1) the distance is small and (2) both oxygen and carbon dioxide are lipid-soluble. The membranes of the epithelial and endothelial cells thus do not pose a barrier to the movement of oxygen and carbon dioxide between the blood and alveolar air spaces.

Circulation to the Respiratory Membrane

The respiratory exchange surfaces receive blood from arteries of the *pulmonary circuit.* ∞ [p.329] The pulmonary arteries enter the lungs and branch, following the bronchi to the lobules. Each lobule receives an arteriole and a venule, and a network of capillaries surrounds each alveolus directly beneath the epithelium. After passing through the pulmonary venules, venous blood enters the pulmonary veins that deliver it to the left atrium.

Blood pressure in the pulmonary circuit is usually relatively low, with systemic pressures of 30 mm Hg or less. With pressures that low, pulmonary vessels can easily become blocked by small blood clots, fat masses, or air bubbles in the pulmonary arteries. Because the lungs receive the entire cardiac output, any drifting masses in the blood are likely to cause problems almost at once. Blockage of a branch of a pulmonary artery will stop blood flow to a group of lobules or alveoli. This condition is called a *pulmonary embolism.*

PNEUMONIA

Pneumonia (nū-MŌ-nē-a; *pneumon,* lung + *-ia,* condition) develops from a pathogenic infection of the lobules of the lung. As inflammation occurs within the lobules, respiratory function deteriorates as a result of fluid leakage into the alveoli or swelling and constriction of the respiratory bronchioles. When bacteria are involved, they are usually species normally found in the mouth and pharynx that have somehow managed to evade the respiratory defenses. As a result, pneumonia becomes more likely when the respiratory defenses have been compromised by other factors, such as epithelial damage from smoking or the breakdown of the immune system in AIDS. The most common pneumonia that develops in AIDS patients results from infection by the fungus *Pneumocystis carinii.* These organisms are normally found in the alveoli, but in healthy individuals the respiratory defenses are able to prevent infection and tissue damage.

Figure 16-6
Alveolar Organization

(a) Basic structure of a lobule. A network of capillaries surrounds each alveolus. **(b)** An SEM of the lung. **(c)** Diagrammatic view of alveolar structure. **(d)** The respiratory membrane.

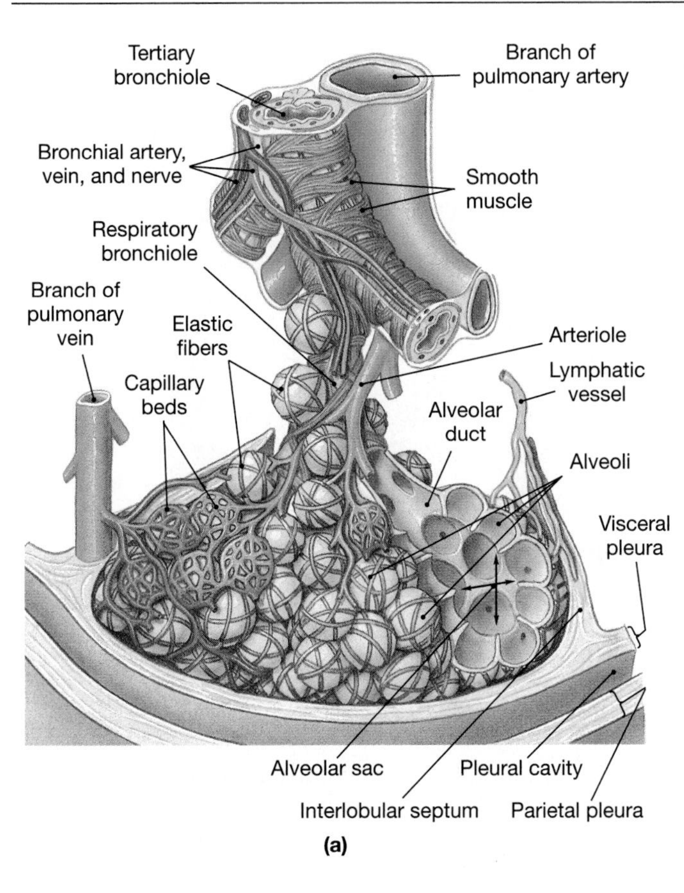

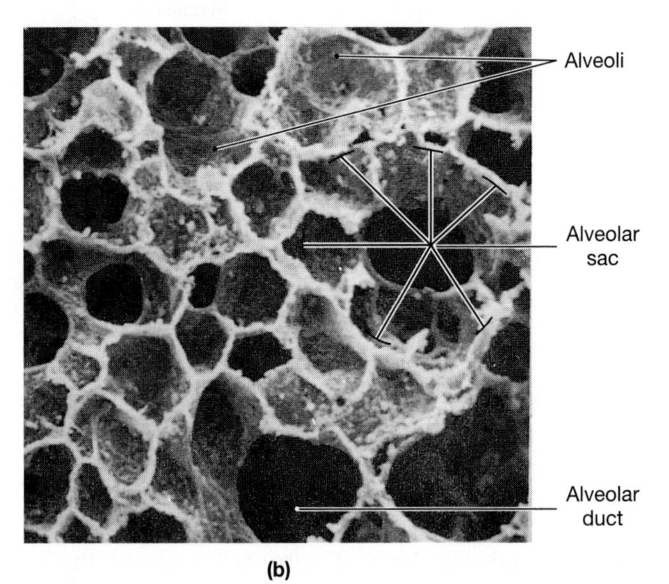

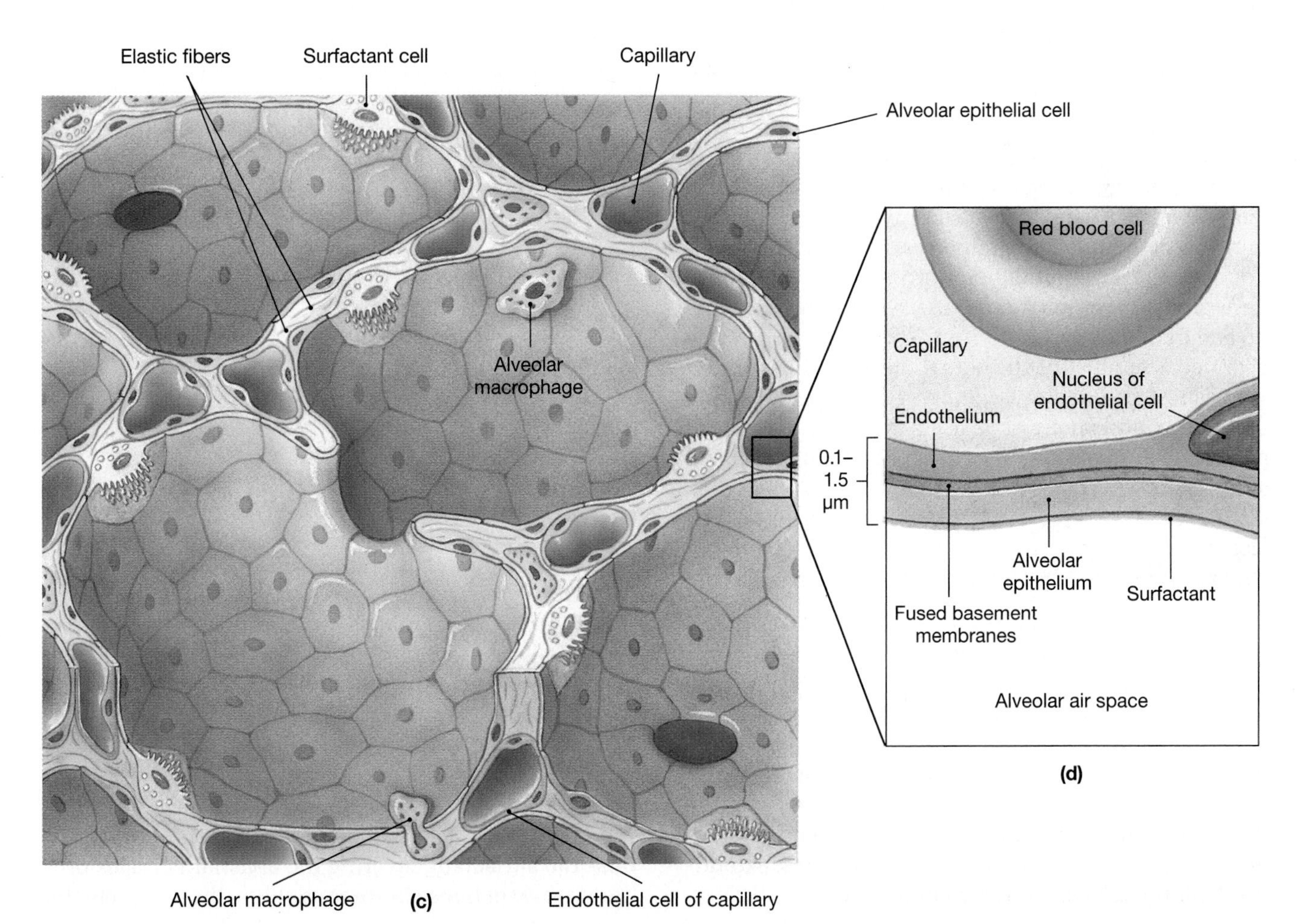

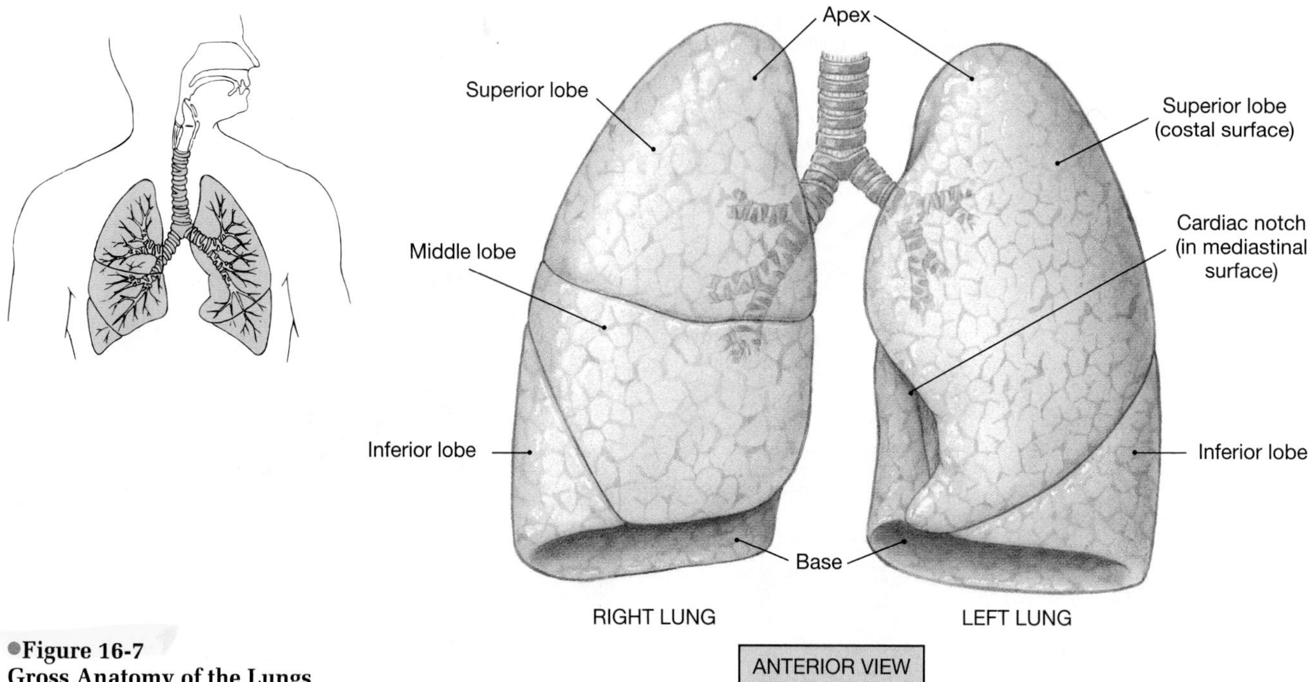

Figure 16-7
Gross Anatomy of the Lungs

The Lungs

The left and right lungs (Figure 16-7) occupy the left and right pleural cavities. Each lung has distinct **lobes** separated by deep fissures. The right lung has three lobes (*superior, middle,* and *inferior*) and the left lung has two (*superior* and *inferior*). The bluntly rounded *apex* of each lung extends into the base of the neck above the first rib, and the concave *base* rests on the superior surface of the diaphragm. The curving *costal surface* follows the inner contours of the rib cage. The *mediastinal surface* has a more irregular shape. The mediastinal surface of the left lung bears an indentation, the *cardiac notch,* that conforms to the shape of the pericardium, and both lungs bear grooves that mark the passage of vessels traveling to and from the heart.

Because most of the actual volume of each lung consists of air-filled passageways and alveoli, the lung has a light and spongy consistency. An abundance of elastic fibers gives the lungs the ability to tolerate large changes in volume.

The Pleural Cavities

The thoracic cavity has the shape of a broad cone. Its walls are the rib cage, and its floor is the muscular diaphragm. The mediastinum divides the thoracic cavity into two pleural cavities (Figure 16-8). Each lung occupies a single pleural cavity, lined by a serous membrane, or **pleura** (PLOO-ra). The *parietal pleura* covers the inner surface of the body wall and extends over the diaphragm and mediastinum. The *visceral pleura* covers the outer surfaces of the lungs, extending into the fissures between the lobes.

The pleural cavity actually represents a potential space rather than an open chamber, for the parietal and visceral layers are usually in close contact. A thin layer of fluid covering these surfaces provides lubrication that reduces friction and irritation of the pleura. Samples of pleural fluid are sometimes obtained for diagnostic purposes, using a long needle inserted between the ribs. This sampling procedure is called *thoracentesis* (thor-a-sen-TĒ-sis; *thorac-*, chest + *kentesis*, puncture). The fluid extracted is then examined for the presence of bacteria, blood cells, or other abnormal components.

Respiratory Changes at Birth

There are several important differences between the respiratory systems of a fetus and a newborn infant. Before delivery, the pulmonary vessels are collapsed, so pulmonary arterial resistance is high. The rib cage is compressed, and the lungs and conducting passageways contain only small amounts of fluid and no air. At birth, the newborn infant takes a truly heroic first breath through powerful contractions of the diaphragmatic and external intercostal muscles. The inspired air enters the passageways with enough force to push the contained fluids out of the way and inflate the entire bronchial tree and most of the alveolar complexes. The same drop in pressure that pulls air into the lungs pulls blood into the pulmonary circulation.

The exhalation that follows fails to completely empty the lungs, for the rib cage does not return to its former, fully compressed state. Cartilages and connective tissues keep the conducting passageways open, and the surfactant covering the alveolar surfaces prevents their collapse. Subsequent breaths complete the

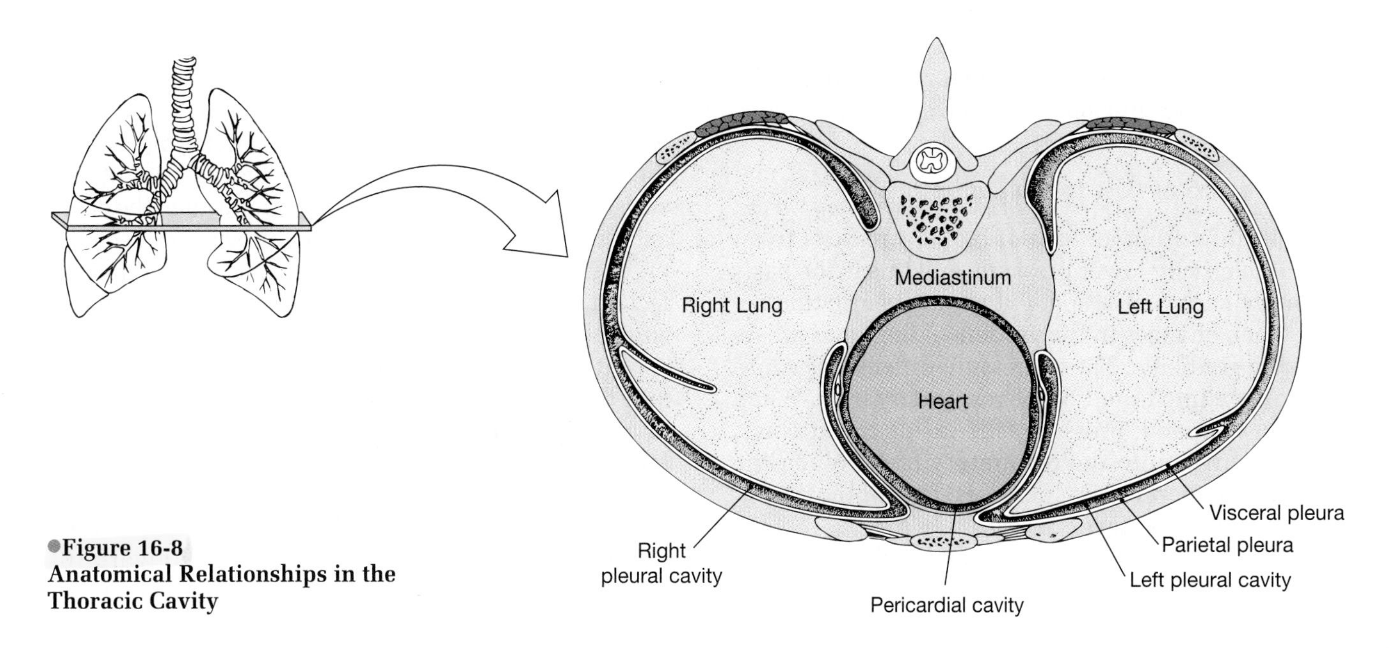

Figure 16-8
Anatomical Relationships in the Thoracic Cavity

inflation of the alveoli. These physical changes are sometimes used by pathologists to determine whether a newborn infant died before or shortly after delivery. Before the first breath, the lungs are completely filled with fluid, and they will sink if placed in water. After the first breath, even the collapsed lungs contain enough air to keep them afloat.

✓ When the tension in the vocal cords increases, what happens to the pitch of the voice?

✓ Why are the cartilages that reinforce the trachea C-shaped instead of complete circles?

✓ What would happen to the alveoli if surfactant were not produced?

RESPIRATORY PHYSIOLOGY

The process of respiration involves four integrated steps:

Step 1. *Pulmonary ventilation,* or breathing, which refers to the physical movement of air into and out of the lungs.

Step 2. *Gas diffusion across the respiratory membrane,* which separates the alveolar air from the blood within the alveolar capillaries.

Step 3. *The storage and transport of oxygen and carbon dioxide.* In this step, the blood stores and carries the respiratory gasses between the alveolar capillaries and the capillary beds in other tissues.

Step 4. *The exchange of oxygen and carbon dioxide between the blood and the interstitial fluids.* In this step, the blood delivers oxygen to body tissues and receives carbon dioxide for transport to the lungs.

Abnormalities affecting any single process will ultimately affect the gas concentrations of the interstitial fluids. If the oxygen content declines, the affected tissues will suffer from *hypoxia* (hī-POKS-ē-a), which places severe limits on the metabolic activities of peripheral tissues. If the supply of oxygen gets cut off completely, *anoxia* (a-NOKS-ē-a) results, and cells die very quickly. For example, much of the damage caused by strokes and heart attacks is the result of localized anoxia.

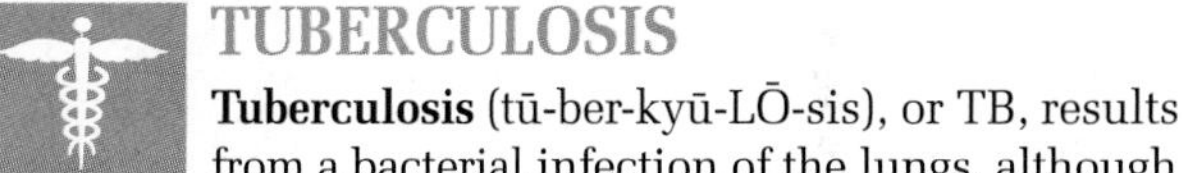

TUBERCULOSIS

Tuberculosis (tū-ber-kyū-LŌ-sis), or TB, results from a bacterial infection of the lungs, although other organs may be invaded as well. The bacteria, *Mycobacterium tuberculosis,* may colonize the respiratory passageways, the interstitial spaces, and/or the alveoli. Symptoms are variable, but usually include coughing and chest pain, with fever, night sweats, fatigue, and weight loss.

Tuberculosis is a major health problem throughout the world. It is the leading cause of death among infectious diseases, with roughly 3 million deaths each year from TB. An estimated 2 *billion* people are infected at this time, and there are 8 million cases diagnosed each year. Unlike other deadly diseases, such as AIDS, TB is transmitted through casual contact. Anyone alive and breathing is at risk for this disease; all it takes is exposure to the bacterium. For information on the symptoms and treatment of this disease, consult the *Applications Manual.* AM *Tuberculosis*

Pulmonary Ventilation

Pulmonary ventilation is the physical movement of air into and out of the respiratory tract. A single breath, or *respiratory cycle,* consists of an inhalation, or *inspira-*

tion, and an exhalation, or *expiration.* Breathing functions to maintain adequate **alveolar ventilation,** the movement of air into and out of the alveoli.

Pressure and Airflow to the Lungs

Air will flow from an area of higher pressure to an area of relatively lower pressure, and this is the basis for pulmonary ventilation. Pulmonary ventilation occurs when changes in the volume of the lungs create pressure gradients. These pressure differences cause air to move into or out of the respiratory tract. When the lungs expand, pressure falls inside the airways, and air moves into the respiratory tract. When the lungs contract, pressure increases, and air moves out of the respiratory tract.

The volume of the lungs depends on the volume of the pleural cavities. The parietal and pleural membranes are separated by only a thin film of pleural fluid, and, although the two membranes can slide across one another, they are held together by that fluid film. You encounter the same principle whenever you set a wet glass on a smooth surface. You can slide the glass quite easily, but when you try to lift it you encounter considerable resistance from this fluid bond. A comparable fluid bond exists between the parietal pleura and the visceral pleura covering the lungs. As a result, the surface of each lung sticks to the inner wall of the chest and the superior surface of the diaphragm. Movements of the chest wall or the diaphragm thus have a direct effect on the volume of the lungs. The basic principle is shown in Figure 16-9●.

At the start of a breath, pressures inside and outside the lungs are identical and there is no movement of air (Figure 16-9b●). When the thoracic cavity enlarges, the pressure inside the lungs decreases. Air now enters the respiratory passageways because the pressure inside the lungs (P_i) is lower than atmospheric pressure (pressure outside, or P_o) (Figure 16-9c●). During quiet breathing, enlargement of the thoracic cavity involves the contractions of the *diaphragm,* the muscular sheet that separates the thoracic and abdominopelvic cavities, aided by the *external intercostal muscles*:

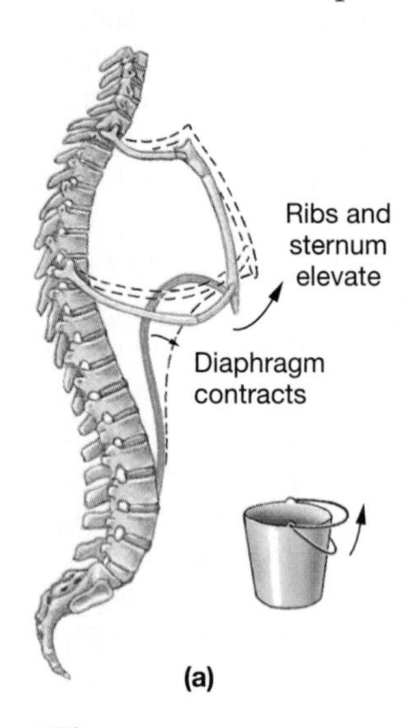

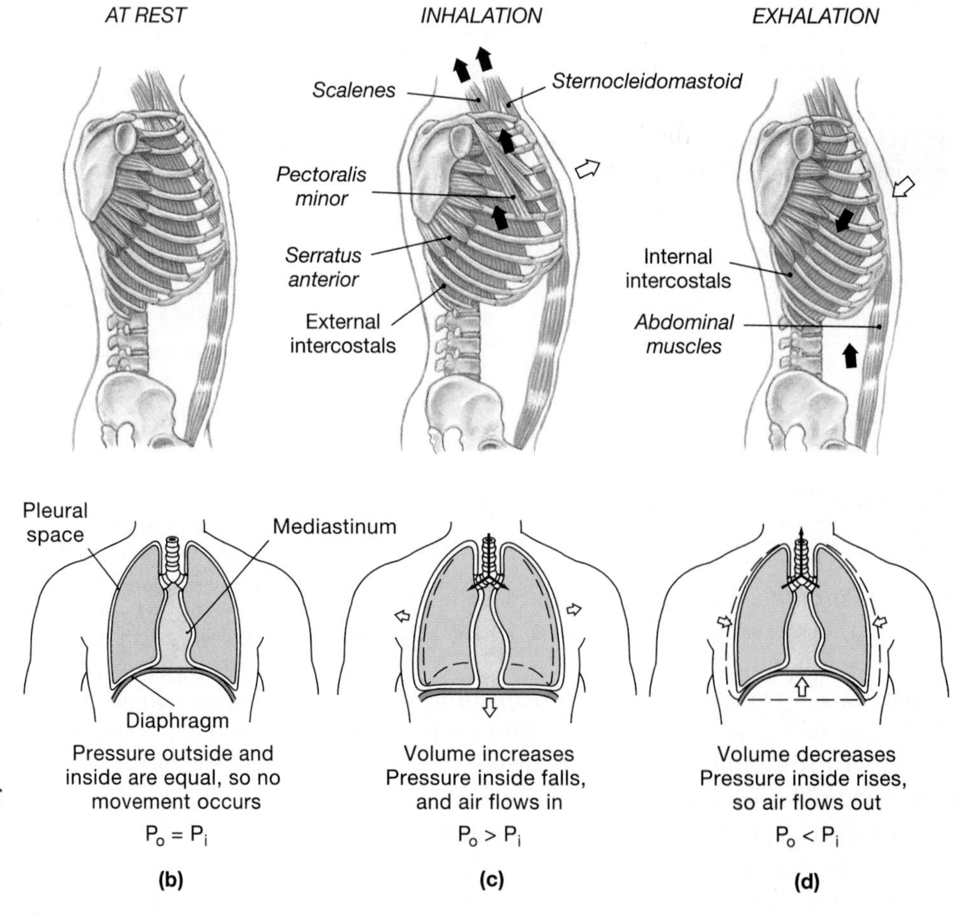

●Figure 16-9 Pressure and Volume Relationships in the Lungs

(a) As the ribs are elevated or the diaphragm depressed, the thoracic cavity increases in volume. **(b)** Anterior view at rest, with no air movement. **(c)** Inhalation: Elevation of the rib cage and depression of the diaphragm increase the size of the thoracic cavity. Pressure decreases, and air flows into the lungs. **(d)** Exhalation: When the rib cage returns to its original position, the volume of the thoracic cavity decreases. Pressure rises, and air moves out of the lungs. *Accessory muscles* may assist such movements of the rib cage to increase the depth and rate of respiration.

- The diaphragm forms the floor of the thoracic cavity. When relaxed, the diaphragm has the shape of a dome and projects upward into thoracic cavity, compressing the lungs. When the diaphragm contracts, it flattens and increases the volume of the thoracic cavity, expanding the lungs. Contraction of the diaphragm is controlled by the *phrenic nerves,* branches of the *cervical plexus.* ∞ *[p. 233]*
- The external intercostals elevate the rib cage. Because of the nature of the articulations between the ribs and the vertebrae, this movement increases the volume of the thoracic cavity. Contraction of the external intercostals is controlled by the *intercostal nerves,* branches of the thoracic spinal nerves. ∞ *[p. 322]* When a person is breathing heavily, accessory muscles, such as the sternocleidomastoid, can assist the external intercostals in elevating the ribs.

Downward movement of the rib cage and upward movement of the diaphragm reverse the process and reduce the size of the lungs. Pressure inside the lungs now exceeds atmospheric pressure, and air moves out of the lungs (Figure 16-9d●). During quiet breathing, expansion of the lungs stretches their elastic fibers. In addition, elevation of the rib cage stretches opposing skeletal muscles and the elastic fibers in the body wall. When the inspiratory muscles relax, these elastic components recoil, returning the diaphragm and rib cage to their original positions. When breathing heavily, contractions of the *internal intercostal muscles* and the abdominal muscles may assist in exhalation.

An injury to the chest wall that penetrates the parietal pleura or that damages the alveoli and the visceral pleura can allow air into the pleural cavity. This *pneumothorax* (noo-mō-THŌ-raks; *pneuma,* air) breaks the fluid bond between the pleurae and allows the elastic fibers to contract. The result is a collapsed lung, or *atelectasis* (at-e-LEK-ta-sis; *ateles,* imperfect + *ektasis,* expansion). Treatment for a collapsed lung involves removing as much of the air as possible before sealing the opening. This procedure restores the fluid bond and reinflates the lung. Lung volume can also be reduced by the accumulation of blood in the pleural cavity. This condition is called a *hemothorax.*

Modes of Breathing

Respiratory movements are usually classified as *quiet breathing* or *forced breathing,* depending on the pattern of muscle activity in the course of a single respiratory cycle. In *quiet breathing,* inhalation involves muscular contractions, but exhalation is passive. Inhalation results from the contraction of both the diaphragm and the external intercostals. Contraction of the diaphragm accounts for around 75 percent of the air movement in normal quiet breathing. That proportion can change, however, depending on the circumstances. For example, pregnant women increasingly rely on movements of the rib cage as expansion of the uterus forces abdominal organs against the diaphragm. In *forced breathing* both inhalation and exhalation are active.

ARTIFICIAL RESPIRATION

Artificial respiration is a technique to provide air to an individual whose respiratory muscles are no longer functioning. In *mouth-to-mouth resuscitation,* or *rescue breathing,* a rescuer provides ventilation by exhaling into the mouth or mouth and nose of the victim. After each breath, contact is broken to permit passive exhalation by the victim. Air provided in this way contains adequate oxygen to meet the needs of the victim. Trained rescuers may supply air through an *endotracheal tube* that is inserted into the trachea through the glottis. Mechanical ventilators, if available, can be attached to the endotracheal tube. If the victim's cardiovascular system is nonfunctional as well, a technique called *cardiopulmonary resuscitation (CPR)* is required to maintain adequate blood flow and tissue oxygenation.

Respiratory Volumes and Rates

As noted earlier, a **respiratory cycle** is a single cycle of inhalation and exhalation. The **tidal volume** is the amount of air moved into or out of the lungs during a single respiratory cycle. Only a small proportion of the air in the lungs is exchanged during a single quiet respiratory cycle; the tidal volume can be increased by inhaling more vigorously and exhaling more completely. The total volume of the lungs can be divided into *volumes* and *capacities* graphically shown in Figure 16-10●.

- *Expiratory reserve volume.* During a normal quiet respiratory cycle the tidal volume averages about 500 ml. The amount of air that could be voluntarily expelled at the end of a tidal cycle is about 1000 ml. This volume is called the **expiratory reserve volume** (ERV).
- *Inspiratory reserve volume.* The **inspiratory reserve volume** (IRV) is the amount of air that can be taken in over and above the tidal volume. Because lungs in males are larger than those of females, the IRV of males averages 3300 ml versus 1900 ml in females.
- *Vital capacity.* The sum of the inspiratory reserve volume, the expiratory reserve volume, and the tidal volume is the **vital capacity:** the maximum amount of air that can be moved into and out of the respiratory system in a single respiratory cycle.

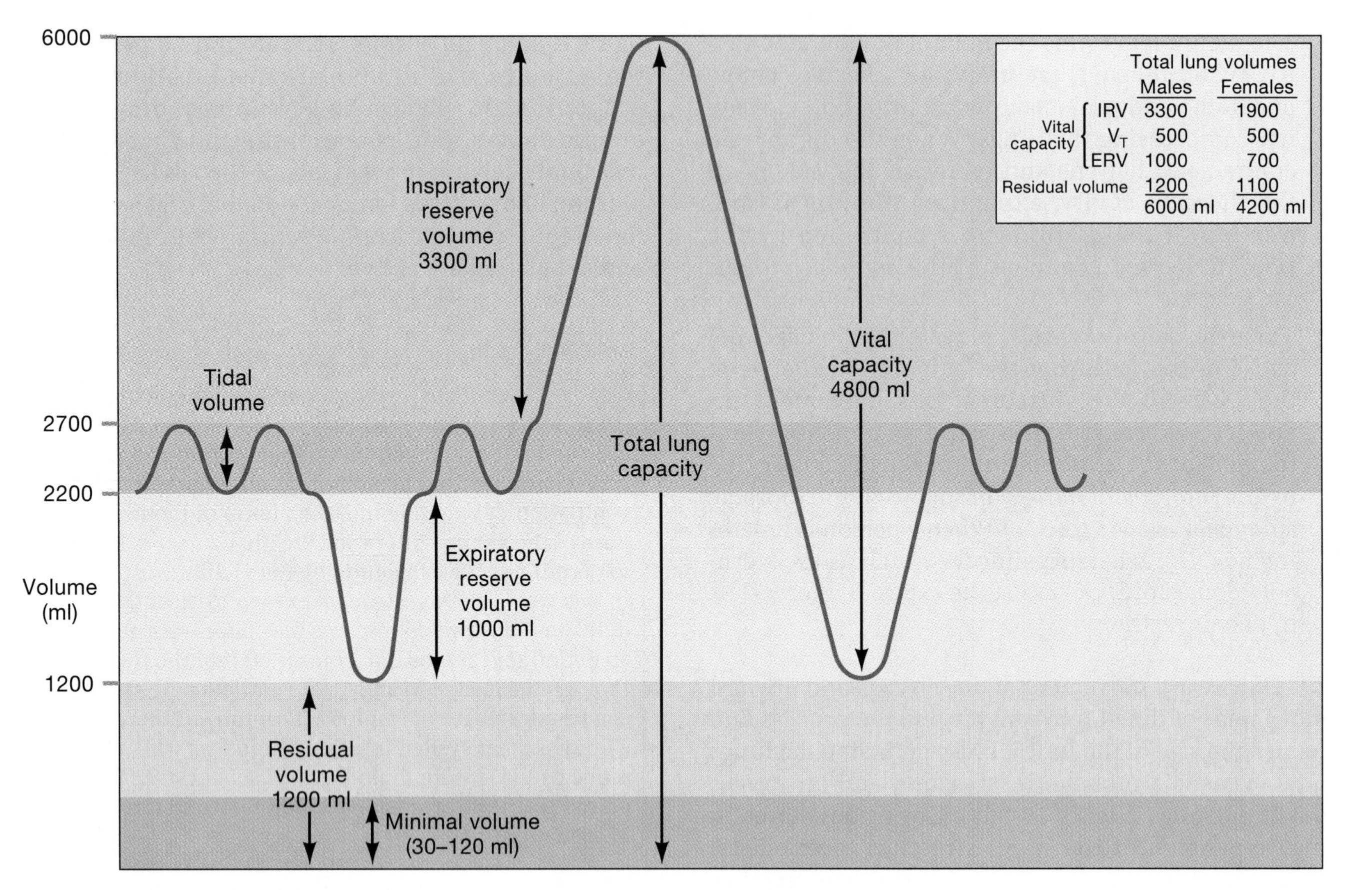

		Total lung volumes Males	Females
Vital capacity	IRV	3300	1900
	V_T	500	500
	ERV	1000	700
Residual volume		1200	1100
		6000 ml	4200 ml

Figure 16-10
Respiratory Volumes and Capacities
The graph diagrams the relationships between the respiratory volumes and capacities of an average male. The table compares the values for males and females.

- *Residual volume.* Roughly 1200 ml of air remains in the respiratory passageways and alveoli, even after exhausting the expiratory reserve volume. Most of this **residual volume** exists because the lungs are held against the thoracic wall, preventing their elastic fibers from contracting further.
- *Minimal volume.* When the chest cavity is opened, as in a pneumothorax, the lungs collapse and the amount of air in the respiratory system is reduced to the **minimal volume.** Some air remains in the lungs, even at minimal volume, because the surfactant coating the alveolar surfaces prevents their collapse.

Not all of the inspired air reaches the alveolar exchange surfaces within the lungs. A typical tidal inspiration pulls around 500 ml of air into the respiratory system. The first 350 ml travels along the conducting passageways and enters the alveolar spaces. The last 150 ml never gets farther than the conducting passageways and does not participate in gas exchange with the blood. The volume of air in the conducting passages is known as the **dead space** of the lungs.

PULMONARY FUNCTION TESTS

Pulmonary function tests monitor several aspects of respiratory function. A *spirometer* (spī-ROM-e-ter) measures parameters such as vital capacity, expiratory reserve, and inspiratory reserve. A *pneumotachometer* provides additional information by determining the rate of air movement; a *peak flow meter* records the maximum rate of forced expiration. Although these tests are relatively simple to perform, they have considerable diagnostic significance. For example, in *asthma* the constricted airways tend to close before an exhalation is completed. As a result, the functional residual capacity declines, and the narrow respiratory passageways reduce the flow rate as well. Air whistling through the constricted airways produces the characteristic "wheezing" that accompanies an asthmatic attack.

✓ Mark breaks a rib that punctures the chest wall on his left side. What would you expect to happen to his left lung as a result?

✓ In pneumonia, fluid accumulates in the alveoli of the lungs. How would vital capacity be affected?

Gas Exchange at the Respiratory Membrane

In pulmonary ventilation, the alveoli are supplied with oxygen, and the accumulated carbon dioxide is removed. The actual process of gas exchange occurs between the blood and alveolar air across the respiratory membrane. This process depends on (1) the *partial pressures* of the gases involved and (2) the diffusion of molecules from a gas into a liquid. (You may wish to review diffusion in Chapter 3 before proceeding.) ∞ [p. 50]

Mixed Gases and Partial Pressures

The air we breathe is not a single gas, but a mixture of gases. Nitrogen molecules (N_2) are the most abundant, accounting for about 78.6 percent of the atmospheric gas molecules. Oxygen molecules (O_2), the second most abundant, constitute roughly 20.8 percent of the atmospheric gas population. Most of the remaining 0.5 percent consists of water molecules, with carbon dioxide (CO_2) contributing a mere 0.04 percent to the total number of gas molecules in the atmosphere.

Atmospheric pressure at sea level is approximately 760 mm Hg. Each of the gases contributes to the total pressure in proportion to its relative abundance. The pressure contributed by a single gas is the **partial pressure** of that gas, abbreviated by the prefix *P*. All of the partial pressures added together equal the total pressure exerted by the gas mixture. In the case of the atmosphere, this relationship can be summarized as:

$$P_{N_2} + P_{O_2} + P_{H_2O} + P_{CO_2} = 760 \text{ mm Hg}$$

Because we know the individual percentages, the partial pressure for each gas can be calculated easily. For example, the partial pressure of oxygen, P_{O_2}, is 20.8 percent of 760 mm Hg, or roughly 159 mm Hg. The partial pressures for other atmospheric gases are given in Table 16-1. These values are important because the partial pressure of an individual gas determines its rate of diffusion between the alveolar air and the bloodstream. For example, the partial pressure of oxygen determines how much oxygen enters solution, but it has no effect on the rate of nitrogen or carbon dioxide diffusion.

DECOMPRESSION SICKNESS

Decompression sickness is a painful condition that develops when a person is suddenly exposed to a drop in atmospheric pressure. Nitrogen is the gas responsible for the problems experienced, because of its high partial pressure. When the pressure drops, nitrogen comes out of solution, forming bubbles like those in a shaken can of soda. The bubbles may form in joint cavities, in the bloodstream, and in the cerebrospinal fluid. The victims often curl up because of the pain in affected joints, and this reaction accounts for the common name for the condition, *the bends*. Decompression sickness most often affects SCUBA divers, who are breathing air under greater than normal pressures while submerged, when they return to the surface too quickly. It also may affect people in airplanes subject to sudden losses of cabin pressure. AM Decompression Sickness

Alveolar versus Atmospheric Air

As soon as air enters the respiratory tract its characteristics begin to change. For example, in passing through the nasal cavity, the air becomes warmer and the amount of water vapor increases. On reaching the alveoli, the incoming air mixes with air that remained in the alveoli after the previous respiratory cycle. The resulting alveolar gas mixture thus contains more carbon dioxide and less oxygen than does atmospheric air. As noted earlier, the last 150 ml of inspired air never gets farther than the conducting passageways and remains in the dead space of the lungs. During expiration, the departing alveolar air mixes with air in the dead space to produce yet another mixture that differs from both atmospheric and alveolar samples. The differences in composition between atmospheric (inspired) and alveolar air can be seen in Table 16-1●.

EMPHYSEMA

Emphysema (em-fi-SĒ-ma) is a chronic, progressive condition characterized by shortness of breath and an inability to tolerate physical exertion. The underlying problem is destruction of alveolar surfaces and inadequate surface area for oxygen and carbon dioxide exchange. In essence, respiratory bronchioles and alveoli are functionally eliminated. The alveoli gradually expand, their walls become incomplete, and adjacent alveoli merge to form larger, nonfunctional air

TABLE 16-1 Partial Pressures (mm Hg) and Normal Gas Concentrations in Air

Source of Sample	*Nitrogen (N_2)*	*Oxygen (O_2)*	*Carbon Dioxide (CO_2)*	*Water Vapor (H_2O)*
Inspired air (dry)	597 (78.6 percent)	159 (20.8 percent)	0.3 (0.04 percent)	3.7 (0.5 percent)
Alveolar air (saturated)	573 (75.4 percent)	100 (13.2 percent)	40 (5.2 percent)	47 (6.2 percent)
Expired air (saturated)	569 (74.8 percent)	116 (15.3 percent)	28 (3.7 percent)	47 (6.2 percent)

spaces. Emphysema has been linked to the inhalation of air containing fine particulate matter or toxic vapors, such as those found in cigarette smoke. There are also genetic factors that predispose individuals to this condition. Some degree of emphysema is a normal consequence of aging, and an estimated 66 percent of adult males and 25 percent of adult females have detectable areas of emphysema in their lungs. AM *Emphysema*

Partial Pressures within the Circulatory System

Figure 16-11● details the partial pressures of oxygen and carbon dioxide in the alveolar air and capillaries and in the arteries and veins of the pulmonary and systemic circuits. The blood delivered by the pulmonary arteries has a higher P_{CO_2} and a lower P_{O_2} than does alveolar air. Diffusion between the alveolar air and the pulmonary capillaries thus elevates the P_{O_2} of the blood while lowering its P_{CO_2}. By the time the blood enters the pulmonary venules, it has reached equilibrium with the alveolar air, so it departs the alveoli with a P_{O_2} of about 100 mm Hg and a P_{CO_2} of roughly 40 mm Hg.

Normal interstitial fluid has a P_{O_2} of 40 mm Hg and a P_{CO_2} of 45 mm Hg. As a result, oxygen diffuses out of the capillaries and carbon dioxide diffuses in until the capillary partial pressures are the same as those in the adjacent tissues. At a normal tissue P_{O_2}, the blood entering the venous system still contains

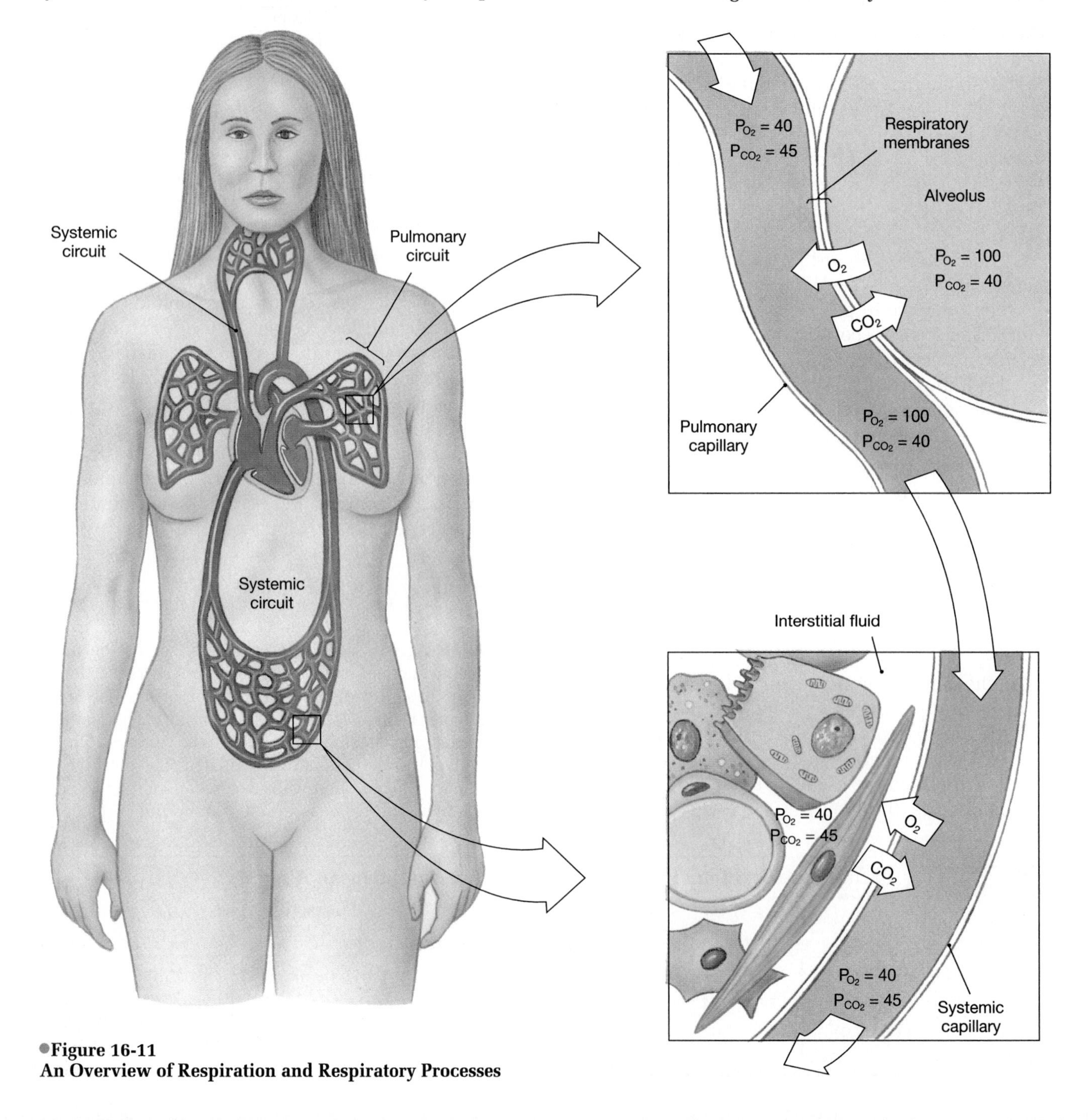

●**Figure 16-11**
An Overview of Respiration and Respiratory Processes

around 75 percent of its total oxygen content. The remaining oxygen represents a reserve that can be called upon when tissue activity increases or when the oxygen supply is temporarily reduced (as when holding your breath). The blood will replace the oxygen released in the tissues when it returns to the alveolar capillaries at the same time that the excess CO_2 is lost.

Gas Pickup and Delivery

Oxygen and carbon dioxide have limited solubilities in blood plasma. This limitation poses certain functional problems, for peripheral tissues need more oxygen and generate more carbon dioxide than the plasma can absorb and transport. The extra oxygen and carbon dioxide diffuse into the red blood cells, where the gas molecules are either tied up (in the case of oxygen) or used to manufacture soluble compounds (in the case of carbon dioxide). The important thing about these reactions is that they are *temporary* and *completely reversible.* When plasma oxygen or carbon dioxide concentrations are high, the excess molecules are removed by the red blood cells; when the plasma concentrations are falling, the red blood cells release their stored reserves.

Oxygen Transport

Only around 1.5 percent of the oxygen content of arterial blood consists of oxygen molecules in solution. All the rest is bound to hemoglobin (Hb) molecules, specifically to the iron atoms in the center of heme units. This reversible reaction can be summarized as:

$$Hb + O_2 \leftrightarrow HbO_2$$

The amount of oxygen retained by hemoglobin primarily depends on the P_{O_2} in its surroundings. As a result, the lower the oxygen content of a tissue, the more oxygen will be released by hemoglobin molecules as they circulate through the region. At a normal tissue P_{O_2} of 40 mm Hg, hemoglobin releases roughly 25 percent of its stored oxygen. Active tissues consume oxygen at an accelerated rate, and when the P_{O_2} declines it automatically increases the amount of oxygen released by hemoglobin molecules passing through local capillaries.

In addition to the effect of P_{O_2}, the amount of oxygen released by hemoglobin is also influenced by pH and temperature. Active tissues generate acids that lower the pH of the interstitial fluids. When the pH declines, the hemoglobin molecules release their bound oxygen molecules more readily. Hemoglobin also releases more oxygen when the temperature rises.

All three of these factors (P_{O_2}, pH, and temperature) are important during periods of maximal exertion. When a skeletal muscle works hard, its temperature rises, and the local pH and P_{O_2} decline. The combination makes the hemoglobin entering the area release much larger amounts of oxygen. Without this automatic adjustment, tissue P_{O_2} would fall to very low levels almost immediately, and the exertion would come to a premature halt.

CARBON MONOXIDE POISONING

Murder or suicide victims are sometimes found in their cars inside a locked garage; each winter, entire families are killed by leaky furnaces or space heaters. The cause of death is carbon monoxide poisoning. *Carbon monoxide* (CO) is found in the exhaust of automobiles, other petroleum-burning engines, oil lamps, and fuel-fired space heaters. Carbon monoxide competes with oxygen molecules for the binding sites on heme units. Unfortunately, the carbon monoxide usually wins, for it has a much stronger affinity for hemoglobin at very low partial pressures. The bond is extremely durable, and the attachment of a carbon monoxide molecule essentially inactivates that heme unit for respiratory purposes. If carbon monoxide molecules make up 0.1 percent of the components of inspired air, enough hemoglobin will be affected that survival will become impossible without medical assistance. Treatment may include (1) breathing pure oxygen, for under high partial pressures the oxygen molecules will "bump" the carbon monoxide from the hemoglobin, and, if necessary, (2) the transfusion of compatible red blood cells.

Carbon Dioxide Transport

Carbon dioxide is generated by aerobic metabolism in peripheral tissues. After entering the bloodstream, a carbon dioxide molecule may be (1) dissolved in the plasma, (2) bound to the hemoglobin of red blood cells, or (3) converted to a molecule of carbonic acid (Figure 16-12●). All of these are completely reversible reactions.

Plasma Transport Roughly 7 percent of the carbon dioxide absorbed by peripheral capillaries is transported in the form of dissolved gas molecules. The rest diffuses into the red blood cells.

Hemoglobin Binding Once within the red blood cells, some of the carbon dioxide molecules are bound to globin portions of hemoglobin molecules, forming **carbaminohemoglobin** (kar-bam-ē-nō-hē-mō-GLŌ-bin). Normally, about 23 percent of the carbon dioxide entering the blood in peripheral tissues will be transported as carbaminohemoglobin. Upon arrival at the pulmonary capillaries, the plasma P_{CO_2} declines, and the bound carbon dioxide is released.

Carbonic Acid Formation The rest of the carbon dioxide molecules, roughly 70 percent of the total, are converted to carbonic acid through the activity of the enzyme carbonic anhydrase. The carbonic acid molecules do not remain intact, however; almost immediately each of these molecules breaks down into

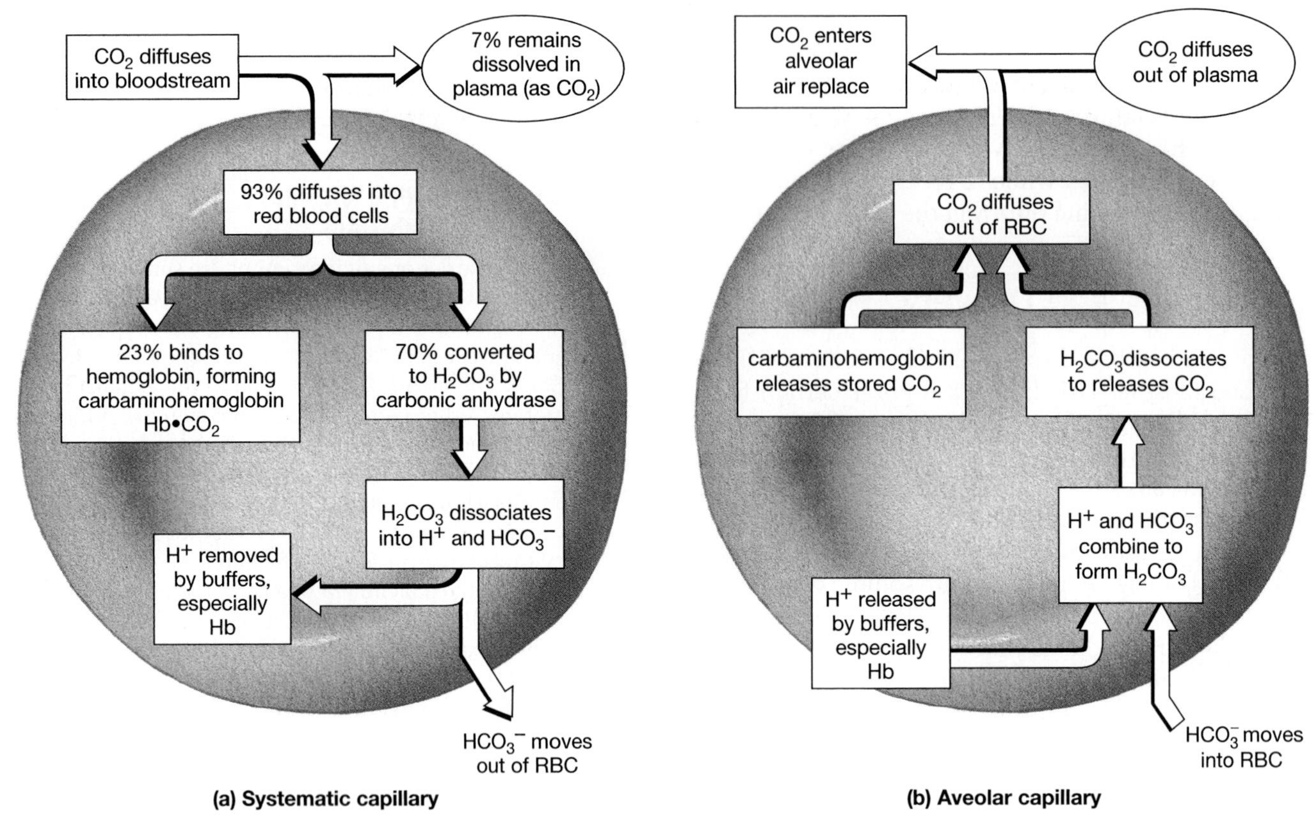

•Figure 16-12
Carbon Dioxide Transport in the Blood
(a) Carbon dioxide uptake in peripheral capillaries. **(b)** Carbon dioxide release at alveolar capillaries.

a hydrogen ion and a bicarbonate ion. The entire sequence can be summarized as:

$$CO_2 + H_2O \xleftrightarrow{\text{carbonic anhydrase}} H_2CO_3 \leftrightarrow H^+ + HCO_3^-$$

The reactions occur very rapidly and are completely reversible. Because most of the carbonic acid formed immediately dissociates into bicarbonate and hydrogen ions, we can ignore the intermediary step and summarize the reaction as:

$$CO_2 + H_2O \xleftrightarrow{\text{carbonic anhydrase}} H^+ + HCO_3^-$$

In peripheral capillaries this reaction proceeds vigorously, tying up large numbers of carbon dioxide molecules. The reaction is driven from left to right because carbon dioxide continues to arrive, diffusing out of the interstitial fluids, and the hydrogen ions and bicarbonate ions are continually being removed. Most of the hydrogen ions get tied up by hemoglobin molecules. Bicarbonate ions diffuse into the surrounding plasma, where they associate with sodium ions to form **sodium bicarbonate** ($NaHCO_3$).

When venous blood reaches the alveoli, carbon dioxide diffuses out of the plasma and the P_{CO_2} declines. Because all of the carbon dioxide transport mechanisms are reversible, as carbon dioxide diffuses out of the red blood cells the reactions shown in Figure 16-12 • proceed in the opposite direction. Hydrogen ions leave the hemoglobin molecules, and bicarbonate ions diffuse into the cytoplasm of the red blood cells, to be converted to water and CO_2.

✓ Why does it take more energy to breathe on a hot, humid day than on a cool, dry day?

✓ During exercise, hemoglobin releases more oxygen to the active skeletal muscles than it does when the muscles are at rest. Why?

✓ How would an obstruction of the airways affect the body's pH?

CONTROL OF RESPIRATION

Living cells are continually absorbing oxygen from the interstitial fluids and generating carbon dioxide. Under normal conditions, the cellular rates of absorption and generation are matched by the capillary rates of delivery and removal, and these rates are identical to those of oxygen absorption and carbon dioxide excretion at the lungs. If these rates become seriously unbalanced, the activities of the cardiovascular and respiratory systems must be adjusted.

The Respiratory Centers of the Brain

The **respiratory centers** integrating large-scale responses include three pairs of loosely organized nuclei in the reticular formation of the pons and medulla. These nuclei regulate the activities of the respiratory muscles and control the *respiratory rate* and the depth of breathing. The **respiratory rate** is the number of breaths per minute. The normal adult respiratory rate at rest ranges from 12 to 18 breaths per minute. Children breathe more rapidly, at around 18 to 20 breaths per minute.

The **respiratory rhythmicity center** of the medulla oblongata sets the pace for respiration. It can be subdivided into a *dorsal respiratory group (DRG),* or *inspiratory center,* and a *ventral respiratory group (VRG),* or *expiratory center.* Other areas of the brain, notably the respiratory centers of the pons, can adjust the output of the rhythmicity center. For example, these centers may adjust the respiratory rate and the depth of respiration in response to sensory stimuli, emotional states, or speech patterns.

Activities of the Respiratory Rhythmicity Center

The inspiratory center functions in every respiratory cycle, whether quiet or forced. During quiet respiration, the neurons of the inspiratory center gradually increase stimulation of the inspiratory muscles for 2 seconds and then the inspiratory center becomes silent for the next 3 seconds. During its inactivity, the respiratory muscles relax, and passive exhalation occurs. The inspiratory center will maintain this basic rhythm even in the absence of sensory or regulatory stimuli. The expiratory center remains inactive during quiet respiration and functions only during forced breathing, when it activates the accessory muscles involved in inhalation and exhalation. The relationships between the inspiratory and expiratory centers during quiet and forced ventilation are diagrammed in Figure 16-13●.

The performance of these respiratory centers can be affected by any factor that alters the metabolic or chemical activities of neural tissues. For example, elevated body temperatures or central nervous system stimulants, such as amphetamines or even caffeine, increase the respiratory rate. Decreased body temperature or CNS depressants, such as barbiturates or opiates, reduce the respiratory rate. Respiratory activities are also strongly influenced by reflexes triggered by mechanical or chemical stimuli.

Reflex Control of Respiration

Normal breathing occurs automatically, without conscious control. Two types of reflexes are involved: *mechanoreceptor reflexes* and *chemoreceptor reflexes.*

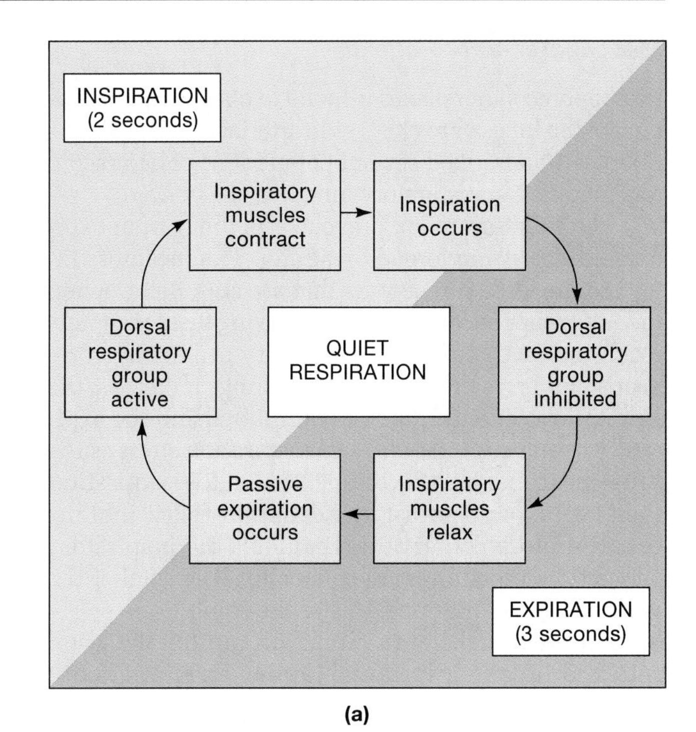

(a)

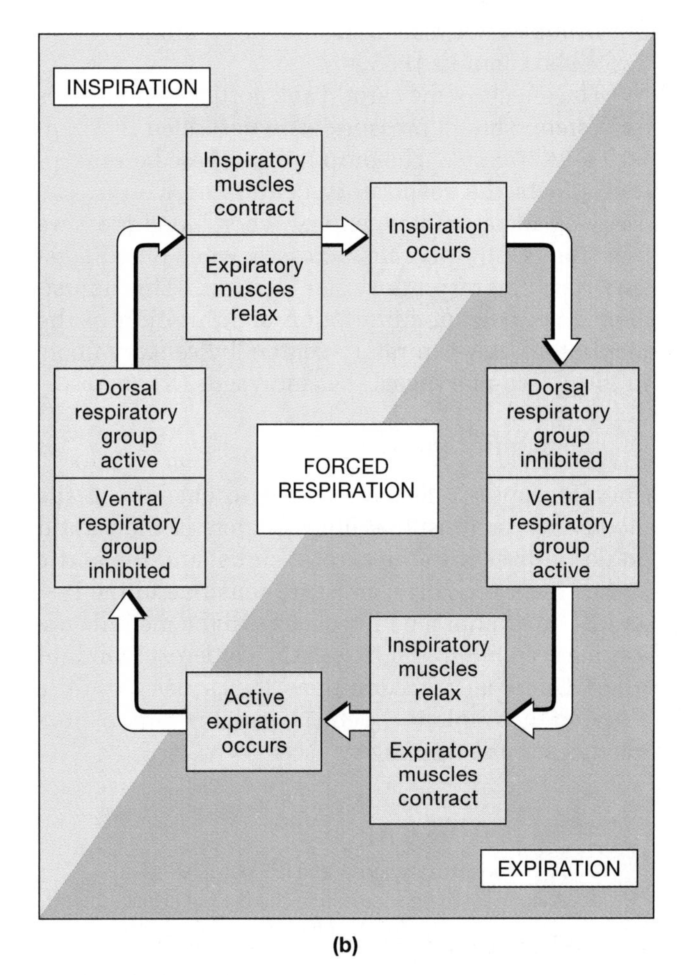

(b)

●**Figure 16-13**
Basic Regulatory Patterns
(a) Quiet respiration. **(b)** Forced respiration.

Mechanoreceptor Reflexes

Mechanoreceptor reflexes respond to changes in the volume of the lungs or to changes in arterial blood pressure. Chapter 10 described several populations of baroreceptors involved in respiratory function. [p. 253]

The **inflation reflex** prevents the lungs from overexpanding during forced breathing. The receptors involved are stretch receptors that are stimulated when the lungs expand. Sensory fibers leaving these receptors reach the inspiratory and expiratory centers over the vagus nerve. As the volume of the lungs increases, the inspiratory center is gradually inhibited and the expiratory center stimulated. Thus inspiration stops as the lungs near maximum volume, and active expiration then begins. In contrast, the **deflation reflex** inhibits the expiratory center and stimulates the inspiratory center when the lungs are collapsing. The smaller the volume of the lungs, the greater the inhibition.

Although neither the inflation nor the deflation reflex is involved in normal quiet breathing, both are important in regulating the forced ventilations that accompany strenuous exercise. Together the inflation and deflation reflexes are known as the *Hering-Breuer reflexes,* after the physiologists who described them in 1865.

The effects of the carotid and aortic baroreceptors on systemic blood pressure were described in Chapter 14. [p. 357] The output from these baroreceptors affects the respiratory centers as well as the cardiac and vasomotor centers. When blood pressure falls, the respiratory rate increases; when blood pressure rises the respiratory rate declines. This adjustment results from stimulation or inhibition of the inspiratory and expiratory centers by sensory fibers in the glossopharyngeal (IX) and vagus (X) nerves.

Chemoreceptor Reflexes

Chemoreceptor reflexes respond to changes in the blood and cerebrospinal fluid. Centers in the carotid bodies (adjacent to the carotid sinus) and the aortic bodies (near the aortic arch) are sensitive to the P_{CO_2} and P_{O_2} in arterial blood; receptors in the medulla oblongata respond to the P_{CO_2} in the cerebrospinal fluid (CSF). Under normal conditions, the P_{O_2} has very little effect on the respiratory centers, and it is carbon dioxide that sets the respiratory pace.

HYPERCAPNIA

The term *hypercapnia* (hī-per-KAP-nē-a) refers to an increase in the P_{CO_2} of arterial blood. Such a change stimulates the chemoreceptors in the carotid and aortic bodies and, since carbon dioxide crosses the blood-brain barrier quite rapidly, also the chemoreceptive neurons of the medulla oblongata. These receptors stimulate the respiratory center to produce **hyperventilation** (hī-per-ven-ti-LĀ-shun), an increase in the rate and depth of respiration. Breathing becomes more rapid, and a greater amount of air moves in and out of the lungs with each breath. Because more air moves in and out of the alveoli each minute, alveolar concentrations of carbon dioxide decline, accelerating the diffusion of carbon dioxide from the alveolar capillaries. If the arterial P_{CO_2} declines below normal levels, chemoreceptor activity declines and the respiratory rate falls. This **hypoventilation** continues until the carbon dioxide partial pressure returns to normal.

Respiratory Drive and P_{CO_2} Under normal conditions, carbon dioxide levels have a much more powerful effect on respiratory activity than does oxygen because arterial P_{O_2} does not usually decline enough to activate the oxygen receptors. But when arterial P_{O_2} does fall, the two receptor populations cooperate. Carbon dioxide is generated during oxygen consumption, so when oxygen concentrations are falling rapidly carbon dioxide levels are usually increasing.

Chemoreceptor reflexes are extremely powerful respiratory stimulators, and they cannot be consciously suppressed. For example, you can hold your breath before diving into a swimming pool and thereby prevent the inhalation of water. But you cannot actually commit suicide by holding your breath "till you turn blue." Once the P_{CO_2} rises to critical levels, you will be forced to take a breath. People will sometimes take deep, full breaths to extend their breath-holding times. Most believe that this helps by providing "extra" oxygen. But the real reason is that they are driving down levels of carbon dioxide. If the P_{CO_2} is reduced low enough, breath-holding ability may increase to the point that an individual becomes unconscious from oxygen starvation in the brain without ever feeling the urge to breathe.

Control by Higher Centers

Higher centers influence respiration through their effects on the respiratory centers of the pons and by direct control of respiratory muscles. For example, the contractions of respiratory muscles can be voluntarily suppressed or exaggerated; this control is necessary while talking or singing. The depth and pace of respiration also change following the activation of centers involved with rage, eating, or sexual arousal. These changes, directed by the limbic system, occur at an involuntary level. Figure 16-14• summarizes factors involved in the regulation of respiration.

SIDS

Sudden infant death syndrome (SIDS), also known as "crib death," kills an estimated 10,000 infants each year in the United States alone. Most crib deaths involve infants 2–4 months old, usually between midnight and 9:00 A.M., in the late fall or winter months. Eyewitness accounts indicate that the sleeping infant suddenly stops breathing, turns blue, and relaxes. There appear to be genetic factors involved, but controversy remains as to the relative importance of other factors, such as laryngeal spasms, cardiac arrhythmias, upper respiratory tract infections, viral infections, or CNS malfunctions. Position and environment may be involved; the problem has recently been linked to sleeping on the stomach on a soft surface. The age at the time of death corresponds with a period when the respiratory centers are establishing connections with other portions of the brain. It has recently been proposed that SIDS results from a problem in the interconnection process that disrupts the reflexive respiratory pattern.

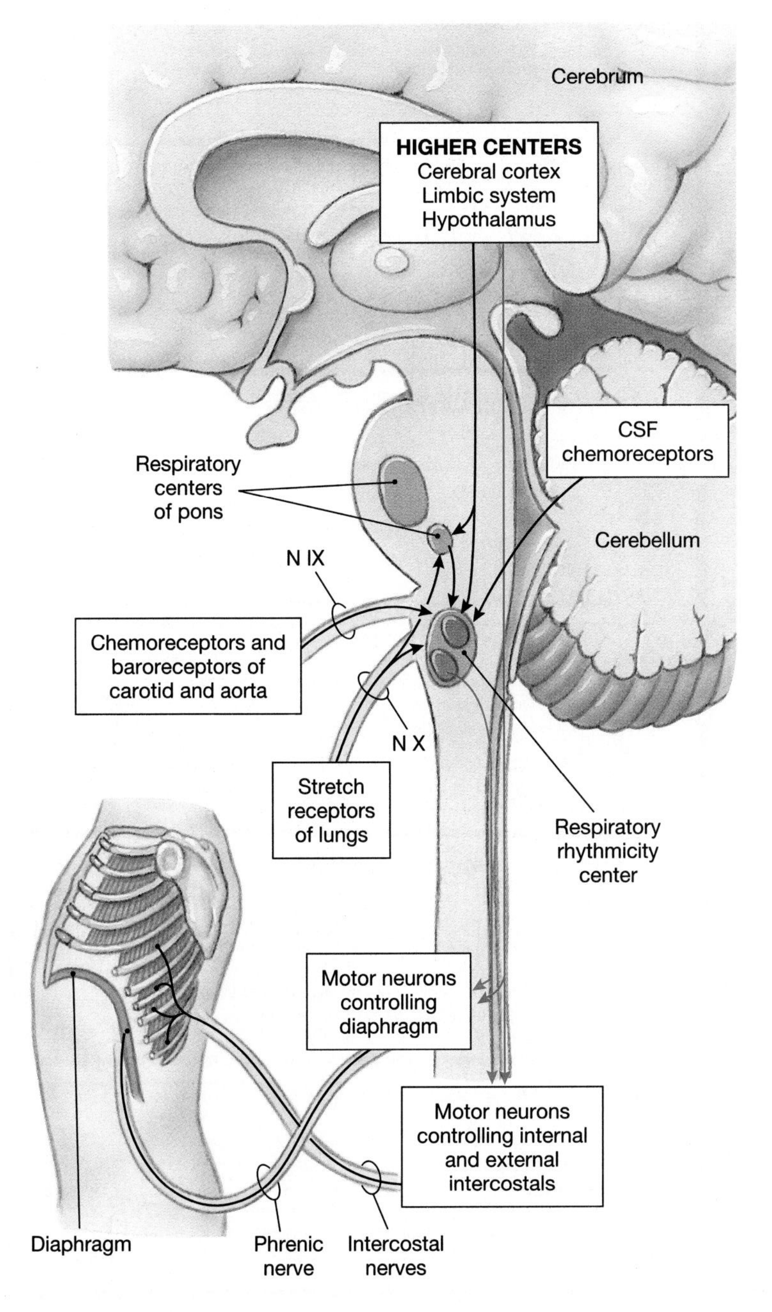

Figure 16-14
The Control of Respiration

AGING AND THE RESPIRATORY SYSTEM

Many factors interact to reduce the efficiency of the respiratory system in elderly individuals. Three examples are particularly noteworthy:

1. With increasing age, elastic tissue deteriorates throughout the body. This deterioration reduces the resilience of the lungs and lowers the vital capacity.
2. Movements of the chest cage are restricted by arthritic changes in the rib articulations and by decreased flexibility at the costal cartilages. In combination with the changes noted in (1), the stiffening and reduction in chest movement limit pulmonary ventilation. This restriction contributes to the reduction in exercise performance and capabilities with increasing age.
3. Some degree of emphysema is normally found in individuals age 50–70, but the extent varies widely depending on the lifetime exposure to cigarette smoke and other respiratory irritants. Comparative studies of nonsmokers and those who have smoked for varying lengths of time clearly show the negative effect of smoking on respiratory performance.

✓ Are peripheral chemoreceptors as sensitive to the levels of carbon dioxide as they are to the levels of oxygen?

✓ Strenuous exercise would stimulate what set of respiratory reflexes?

✓ Johnny is mad at his mother, so he tells her that he will just hold his breath until he turns blue and dies. Should Johnny's mother worry?

INTERACTIONS WITH OTHER SYSTEMS

The respiratory system has extensive anatomical connections to the cardiovascular system. It is functionally linked to all other systems, and Figure 16-15 details these interrelationships.

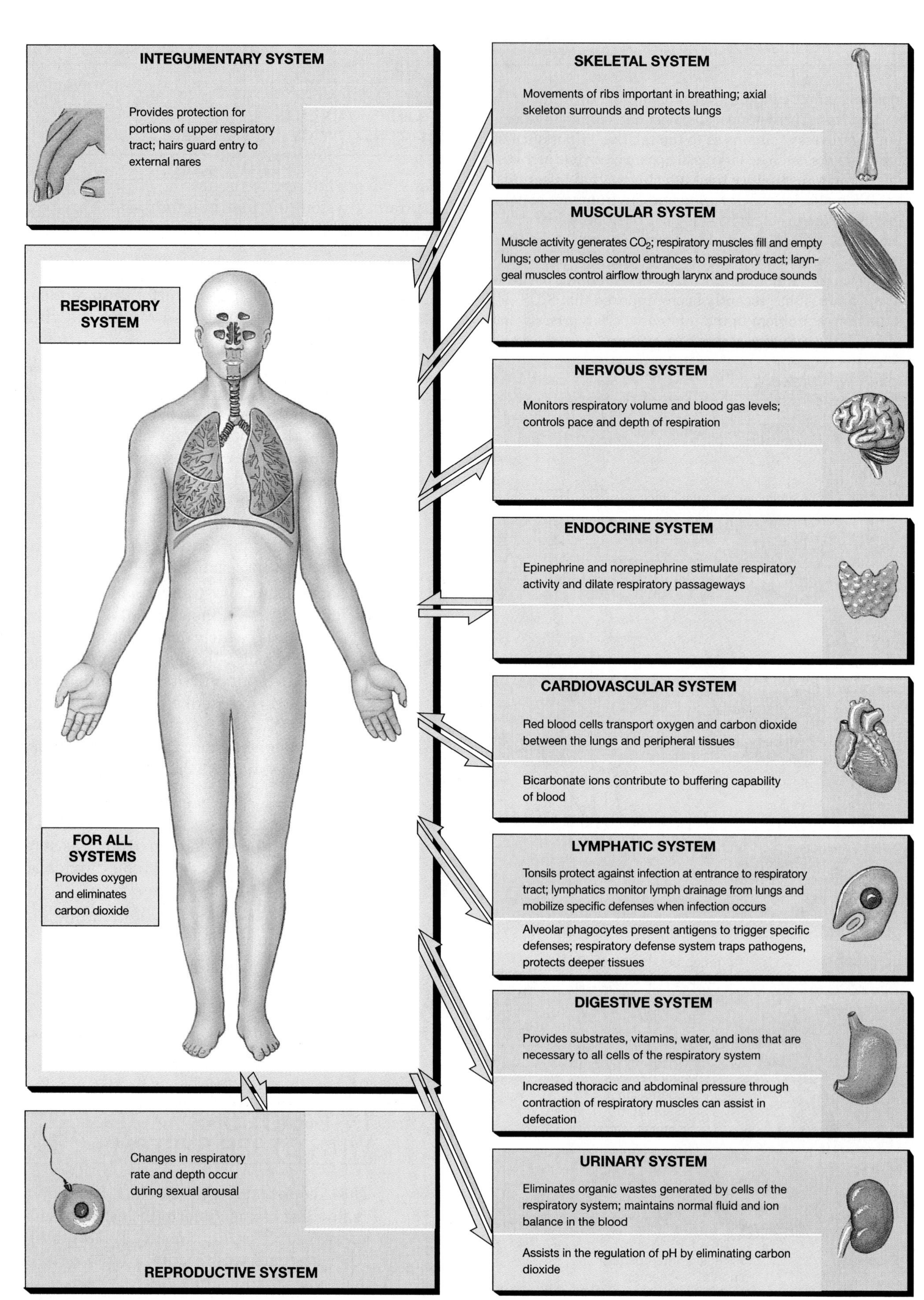

INTEGUMENTARY SYSTEM
Provides protection for portions of upper respiratory tract; hairs guard entry to external nares
SKELETAL SYSTEM
Movements of ribs important in breathing; axial skeleton surrounds and protects lungs
MUSCULAR SYSTEM
Muscle activity generates CO_2; respiratory muscles fill and empty lungs; other muscles control entrances to respiratory tract; laryngeal muscles control airflow through larynx and produce sounds
RESPIRATORY SYSTEM
NERVOUS SYSTEM
Monitors respiratory volume and blood gas levels; controls pace and depth of respiration
ENDOCRINE SYSTEM
Epinephrine and norepinephrine stimulate respiratory activity and dilate respiratory passageways
CARDIOVASCULAR SYSTEM
Red blood cells transport oxygen and carbon dioxide between the lungs and peripheral tissues
Bicarbonate ions contribute to buffering capability of blood
FOR ALL SYSTEMS
Provides oxygen and eliminates carbon dioxide
LYMPHATIC SYSTEM
Tonsils protect against infection at entrance to respiratory tract; lymphatics monitor lymph drainage from lungs and mobilize specific defenses when infection occurs
Alveolar phagocytes present antigens to trigger specific defenses; respiratory defense system traps pathogens, protects deeper tissues
DIGESTIVE SYSTEM
Provides substrates, vitamins, water, and ions that are necessary to all cells of the respiratory system
Increased thoracic and abdominal pressure through contraction of respiratory muscles can assist in defecation
Changes in respiratory rate and depth occur during sexual arousal
REPRODUCTIVE SYSTEM
URINARY SYSTEM
Eliminates organic wastes generated by cells of the respiratory system; maintains normal fluid and ion balance in the blood
Assists in the regulation of pH by eliminating carbon dioxide

•Figure 16-15
Functional Relationships between the Respiratory System and Other Systems

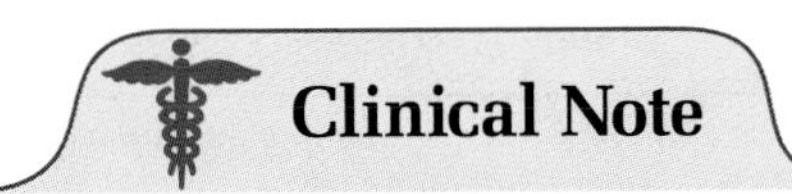

Lung Cancer, Smoking, and Diet

Lung cancer, or *pleuropulmonary neoplasm,* is an aggressive class of malignancies originating in the bronchial passageways or alveoli. These cancers affect the epithelial cells lining conducting passageways, mucous glands, or alveoli. Symptoms usually do not appear until the condition has progressed to the point that the tumor masses are restricting airflow or compressing adjacent mediastinal structures. Chest pain, shortness of breath, a cough or wheeze, and weight loss are common symptoms. Treatment programs vary depending on the cellular organization of the tumor and whether metastasis (cancer cell migration) has occurred, but surgery, radiation exposure, or chemotherapy may be involved. Deaths from lung cancer were rare at the turn of the century, but there were 29,000 in 1956, 105,000 in 1978, and 158,700 in 1996. These figures continue to rise, with the number of diagnosed cases doubling every 15 years. In recent years, between one in every five to eight new cancer cases detected are lung cancer, and, in 1996, 98,900 men and 78,100 women were diagnosed with this condition. The rate is decreasing for men but increasing markedly among women; the lung cancer rates in 1989 were 101,000 men and 54,000 women.

Lung cancers now account for 35 percent of all cancer deaths, making this condition the primary cause of cancer death in the U.S. population. Despite advances in the treatment of other forms of cancer, the survival statistics for lung cancer have not changed significantly. Even with early detection, the 5-year survival rates are only 30 percent (men) to 50 percent (women), and most lung cancer patients die within a year of diagnosis. Detailed statistical and experimental evidence has shown that *85–90 percent of all lung cancers are the direct result of cigarette smoking.* Claims to the contrary are simply unjustified and insupportable. The data are far too extensive to detail here, but the incidence of lung cancer for nonsmokers is 3.4 per 100,000 population, while the incidence for smokers ranges from 59.3 per 100,000 for those burning between a half-pack and a pack a day, to 217.3 per 100,000 for those smoking one to two packs a day. Before around 1970, this disease primarily affected middle-aged men, but as the number of women smokers has increased so has the number of women dying from lung cancer.

Smoking changes the quality of the inspired air, making it drier and contaminated with several carcinogenic compounds and particulate matter. The combination overloads the respiratory defenses and damages the epithelial cells throughout the respiratory system. Whether lung cancer develops appears to be related to the total cumulative exposure to the carcinogenic stimuli. The more cigarettes smoked, the greater the risk, whether those cigarettes are smoked over a period of weeks or years. The histological changes induced by smoking are reversible, and a normal epithelium will return if the stimulus is removed. At the same time, the statistical risks decline to significantly lower levels. Ten years after quitting, a former smoker stands only a 10 percent greater chance of developing lung cancer than a nonsmoker.

The fact that cigarette smoking often causes cancer is not surprising in view of the toxic chemicals contained in the smoke. What is surprising is that more smokers do not develop lung cancer. There is evidence that some smokers have a genetic predisposition for developing one form of lung cancer. Dietary factors may also play a role in preventing lung cancer, although the details are controversial. In terms of their influence on the risk of lung cancer, there is general agreement that (1) vitamin A has no effect, (2) beta-carotene and other vegetable components reduce the risk, and (3) a high-cholesterol, high-fat diet increases the risk.

Chapter Review

KEY TERMS

alveolus/alveoli, p. 414
bronchial tree, p. 413
larynx, p. 411
lungs, p. 416
nasal cavity, p. 408
partial pressure, p. 421
pharynx, p. 409
respiratory membrane, p. 414
respiratory rhythmicity center, p. 425
surfactant, p. 414
trachea, p. 412
vital capacity, p. 419

SUMMARY OUTLINE

INTRODUCTION p. 408

1. To continue functioning, body cells must obtain oxygen and eliminate carbon dioxide.

FUNCTIONS OF THE RESPIRATORY SYSTEM p. 408

1. The functions of the respiratory system include: (1) moving air to and from exchange surfaces where diffusion can occur between air and circulating blood; (2) defending the respiratory system and other tissues from pathogens; (3) permitting vocal communication; and (4) helping to control body fluid pH.

ORGANIZATION OF THE RESPIRATORY SYSTEM p. 408

1. The **respiratory system** includes the nose, nasal cavity, and sinuses; pharynx, larynx, trachea, and conducting passageways leading to the surfaces of the lungs. (*Figure 16-1*)

The Respiratory Tract p. 408

2. The **respiratory tract** consists of the conducting passageways that carry air to and from the alveoli.

The Nose p. 408

3. Air normally enters the respiratory system via the **external nares** that open into the **nasal cavity.** The **vestibule** (entrance) is guarded by hairs that screen out large particles. (*Figure 16-2*)

4. The **hard palate** separates the oral and nasal cavities. The **soft palate** separates the superior **nasopharynx** from the rest of the pharynx. The connections between the nasal cavity and nasopharynx represent the **internal nares.** (*Figure 16-2*)

5. Much of the respiratory epithelium is ciliated and produces mucus that traps incoming particles. (*Figure 16-3*)

The Pharynx p. 409

6. The **pharynx** is a chamber shared by the digestive and respiratory systems.

The Larynx p. 411

7. Inspired air passes through the **glottis** en route to the lungs; the **larynx** surrounds and protects the glottis. The **epiglottis** projects into the pharynx. (*Figure 16-4a,b*)

8. Air passing through the glottis vibrates the **true vocal cords** and produces sound. (*Figure 16-4c–d*)

The Trachea p. 412

9. The wall of the **trachea** ("windpipe") contains C-shaped **tracheal cartilages** that protect the airway. The posterior tracheal wall can distort to permit large masses of food to pass. (*Figure 16-5*)

The Bronchi p. 413

10. The trachea branches within the mediastinum to form the **right** and **left primary bronchi.** (*Figure 16-5*)

11. The primary bronchi, **secondary bronchi**, and their branches form the *bronchial tree.* As the **tertiary bronchi** branch within the lung the amount of cartilage in their walls decreases and the amount of smooth muscle increases. (*Figures 16-5, 16-6*)

Bronchioles p. 413

12. Each terminal **bronchiole** delivers air to a single pulmonary **lobule.** Within the lobule, the terminal bronchiole branches into respiratory bronchioles. (*Figure 16-6*)

Alveolar Ducts and Alveoli p. 414

13. The respiratory bronchioles open into **alveolar ducts** that end at **alveolar sacs**. Many alveoli are interconnected at each alveolar sac. (*Figure 16-6*)

The Respiratory Membrane p. 414

14. The **respiratory membrane** consists of (1) a simple squamous alveolar epithelium, (2) a capillary endothelium, and (3) their fused basement membranes. **Surfactant cells** scattered among the alveolar epithelial cells produce an oily secretion that keep the alveoli from collapsing. **Alveolar macrophages** patrol the epithelium and engulf foreign particles. (*Figure 16-6c,d*)

The Lungs p. 416

15. The lungs are made up of five **lobes**; the right lung has three and the left lung has two. (*Figure 16-7*)

The Pleural Cavities p. 416

16. Each lung occupies a single pleural cavity lined by a **pleura** (serous membrane). (*Figure 16-8*)

Respiratory Changes at Birth p. 416

17. Before delivery, the fetal lungs are fluid-filled and collapsed. After the first breath, the alveoli normally remain inflated for the life of the individual.

RESPIRATORY PHYSIOLOGY p. 417

1. Respiratory physiology focuses on a series of integrated processes: pulmonary ventilation, or breathing (movement of air into and out of the lungs); gas diffusion between the alveoli and circulating blood; and gas storage, transport, and exchange between the blood and interstitial fluids. If oxygen content declines, the affected tissues will suffer from *hypoxia;* if the oxygen supply is completely shut off, *anoxia* and tissue death result.

Pulmonary Ventilation p. 417

2. A single breath, or **respiratory cycle,** consists of an inhalation (inspiration) and an exhalation (expiration).

3. The relationship between the pressure inside the respiratory tract and atmospheric pressure determines the direction of airflow. (*Figure 16-9*)

4. The diaphragm and and the external intercostal muscles are involved in *quiet breathing,* when exhalation is passive. Accessory muscles become active during the active inspiratory and expiratory movements of *forced breathing,* when exhalation is active. (*Figure 16-9*)

5. The **vital capacity** includes the **tidal volume** plus the **expiratory reserve volume** and the **inspiratory reserve volume**. The air left in the lungs at the end of maximum expiration is the **residual volume.** (*Figure 16-10*)

Gas Exchange at the Respiratory Membrane p. 421

6. **Alveolar ventilation** is the amount of air reaching the alveoli each minute. Alveolar and atmospheric air differ in their composition. (*Figure 16-11; Table 16-1*)

Gas Pickup and Delivery p. 423

7. Blood entering peripheral capillaries delivers oxygen and absorbs carbon dioxide. The transport of oxygen and carbon dioxide in the blood involves reactions that are completely reversible.

8. Over the range of oxygen pressures normally present in the body, a small change in plasma P_{O_2} will mean a large change in the amount of oxygen bound or released.

9. Aerobic metabolism in peripheral tissues generates carbon dioxide. Roughly 7 percent of the CO_2 transported in the blood is dissolved in the plasma; another 23 percent is bound as **carbaminohemoglobin;** the rest is converted to carbonic acid, which dissociates into a hydrogen ion and a bicarbonate ion. (*Figure 16-12*)

CONTROL OF RESPIRATION p. 424

1. Large-scale changes in oxygen demand require the integration of cardiovascular and respiratory responses.

The Respiratory Centers of the Brain p. 425

2. The **respiratory centers** include three pairs of nuclei in the reticular formation of the pons and medulla oblongata. These nuclei regulate the respiratory muscles and control the **respiratory rate** and the depth of breathing. The **respiratory rhythmicity center** sets the basic pace for respiration. (*Figure 16-13*)

Reflex Control of Respiration p. 425

3. The **inflation reflex** prevents overexpansion of the lungs during forced breathing; the **deflation reflex** stimulates inspiration when the lungs are collapsing. Chemore-

ceptor reflexes respond to changes in the P_{O_2} and P_{CO_2} of the blood and cerebrospinal fluid.

Control by Higher Centers p. 426

4. Conscious and unconscious thought processes can affect respiration by affecting the respiratory centers or the motor neurons controlling respiratory muscles. (*Figure 16-14*)

AGING AND THE RESPIRATORY SYSTEM p. 427

1. The respiratory system is generally less efficient in the elderly because: (1) elastic tissue deteriorates, lowering the vital capacity of the lungs; (2) movements of the chest cage are restricted by arthritic changes and decreased flexibility of costal cartilages; and (3) some degree of emphysema is normal in the elderly.

INTERACTIONS WITH OTHER SYSTEMS p. 427

1. The respiratory system has extensive anatomical connections to the cardiovascular system. (*Figure 16-15*)

CHAPTER QUESTIONS

LEVEL 1 Reviewing Facts and Terms

Match each item in column A with the most closely related item in column B. Use letters for answers in the spaces provided.

Column A	Column B
___ 1. nasopharynx	a. no O_2 supply to tissues
___ 2. laryngopharynx	b. alveolar macrophages
___ 3. thyroid cartilage	c. produce oily secretion
___ 4. surfactant cells	d. covers inner surface of thoracic wall
___ 5. dust cells	e. low O_2 content in tissue fluids
___ 6. parietal pleura	f. inferior portion of pharynx
___ 7. visceral pleura	g. inspiration
___ 8. hypoxia	h. superior portion of pharynx
___ 9. anoxia	i. Adam's apple
___ 10. collapsed lung	j. covers outer surface of lungs
___ 11. inhalation	k. expiration
___ 12. exhalation	l. atelectasis

13. The structure that prevents the entry of liquids or solid food into the respiratory passageways during swallowing is the:
(a) glottis (b) arytenoid cartilage
(c) epiglottis (d) thyroid cartilage

14. The amount of air moved into or out of the lungs during a single respiratory cycle is the:
(a) respiratory minute volume
(b) tidal volume
(c) residual volume
(d) inspiratory capacity

LEVEL 2 Reviewing Concepts

15. When the diaphragm contracts, it tenses and moves inferiorly, causing:
(a) an increase in the volume of the thoracic cavity
(b) a decrease in the volume of the thoracic cavity
(c) decreased pressure on the contents of the abdominopelvic cavity
(d) increased pressure in the thoracic cavity

16. Gas exchange at the respiratory membrane is efficient because:
(a) the differences in partial pressure are substantial
(b) the gases are lipid-soluble
(c) the total surface area is large
(d) a, b, and c are correct

17. What is the functional significance of the decrease in the amount of cartilage and increase in the amount of smooth muscle in the lower respiratory passageways?

18. Why is breathing through the nasal cavity more desirable than breathing through the mouth?

19. How would you justify the statement, "the bronchioles are to the respiratory system what the arterioles are to the cardiovascular system"?

20. How are surfactant cells involved with keeping the alveoli from collapsing?

LEVEL 3 Critical Thinking and Clinical Applications

21. A decrease in blood pressure will trigger a baroreceptor reflex that leads to increased ventilation. What is the possible advantage of this reflex?

22. You spend the night at a friend's house during the winter. Your friend's home is quite old, and the hot-air furnace lacks a humidifier. When you wake up in the morning you have a fair amount of nasal congestion and you decide you might be coming down with a cold. After a steamy shower and some juice for breakfast, the nasal congestion disappears. Explain.

17 THE DIGESTIVE SYSTEM

They say you are what you eat. Perhaps that is what Giuseppe Arcimboddo had in mind when he painted this whimsical picture in the sixteenth century.

Chapter Outline and Objectives

Vocabulary Development

chylos,, juice; *chylomicron*
deciduus, falling off; *deciduous*
enteron, intestine; *myenteric plexus*
frenulum, a small bridle; *lingual frenulum*
hiatus, a gap or opening; *esophageal hiatus*
lacteus, milky; *lacteal*
odonto-, tooth; *periodontal ligament*
omentum, fat skin; *greater omentum*
rugae, wrinkles; *rugae*
sigmoides, Greek letter *S*; *sigmoid colon*
stalsis, constriction; *peristalsis*
vermis, a worm; *vermiform appendix*
villus, shaggy hair; *intestinal villus*

Few people give any serious thought to the digestive system unless it malfunctions. Yet we spend hours of conscious effort filling and emptying it. References to this system are part of our everyday language. We "have a gut feeling," "want to chew on" something, or find someone's opinions "hard to swallow." When something does go wrong with the digestive system, even something minor, most people seek treatment immediately. For this reason, television advertisements promote toothpaste and mouthwash, diet supplements, antacids, and laxatives on an hourly basis.

Chapter 16 discussed the respiratory system, which delivers the oxygen needed to "burn" metabolic fuels. The digestive system provides the fuel and performs many other chemical exchanges. It consists of a muscular tube, the **digestive tract** and various **accessory organs,** including the salivary glands, gallbladder, liver, and pancreas.

Digestive functions involve a series of interconnected steps:

Step 1: *Ingestion* occurs when foods enter the digestive tract through the mouth.

Step 2: *Mechanical processing* is the physical manipulation of solid foods, first by the tongue and the teeth and then by swirling and mixing motions of the digestive tract.

Step 3: *Digestion* refers to the chemical breakdown of food into small organic fragments that can be absorbed by the digestive epithelium.

Step 4: *Secretion* aids digestion through the release of water, acids, enzymes, and buffers by the digestive tract and accessory organs.

Step 5: *Absorption* is the movement of small organic molecules, electrolytes, vitamins, and water across the digestive epithelium and into the interstitial fluid of the digestive tract.

Step 6: *Excretion* is the elimination of waste products from the body. Within the digestive tract, these waste products are compacted and discharged through the process of *defecation* (def-e-KĀ-shun).

The lining of the digestive tract also plays a defensive role by protecting surrounding tissues against the (1) corrosive effects of digestive acids and enzymes and (2) pathogens that are either swallowed with food or residing inside the digestive tract. The digestive epithelium and its secretions provide a nonspecific defense against these bacteria, and bacteria reaching the underlying tissues are attacked by macrophages and other cells of the immune system.

AN OVERVIEW OF THE DIGESTIVE TRACT

Major anatomical subdivisions of the digestive tract are introduced in Figure 17-1●. The digestive tract begins with the oral cavity and continues through the pharynx, esophagus, stomach, small intestine, and large intestine before ending at the rectum and anus. Although these subdivisions have overlapping functions, each region has certain areas of specialization and shows distinctive histological features that reflect those specializations.

Histological Organization

The digestive tract has four major layers (Figure 17-2●): the *mucosa*, the *submucosa*, the *muscularis externa*, and the *serosa*.

1. The **mucosa**, or inner lining of the digestive tract is an example of a *mucous membrane.* It consists of an epithelial surface moistened by glandular secretions and an underlying layer of loose connective tissue, the *lamina propria.* Along most of the length of the digestive tract the mucosa is thrown into folds that (1) increase the surface area available for absorption and (2) permit expansion after a large meal. In the small intestine, the mucosa forms fingerlike projections, called *villi* (*villus*, shaggy hair), that further increase the area for absorption. Most of the digestive tract is lined by a simple columnar epithelium, often containing various types of secretory cells. Ducts opening onto the epithelial surfaces carry the secretions of glands located in the lamina propria, in the

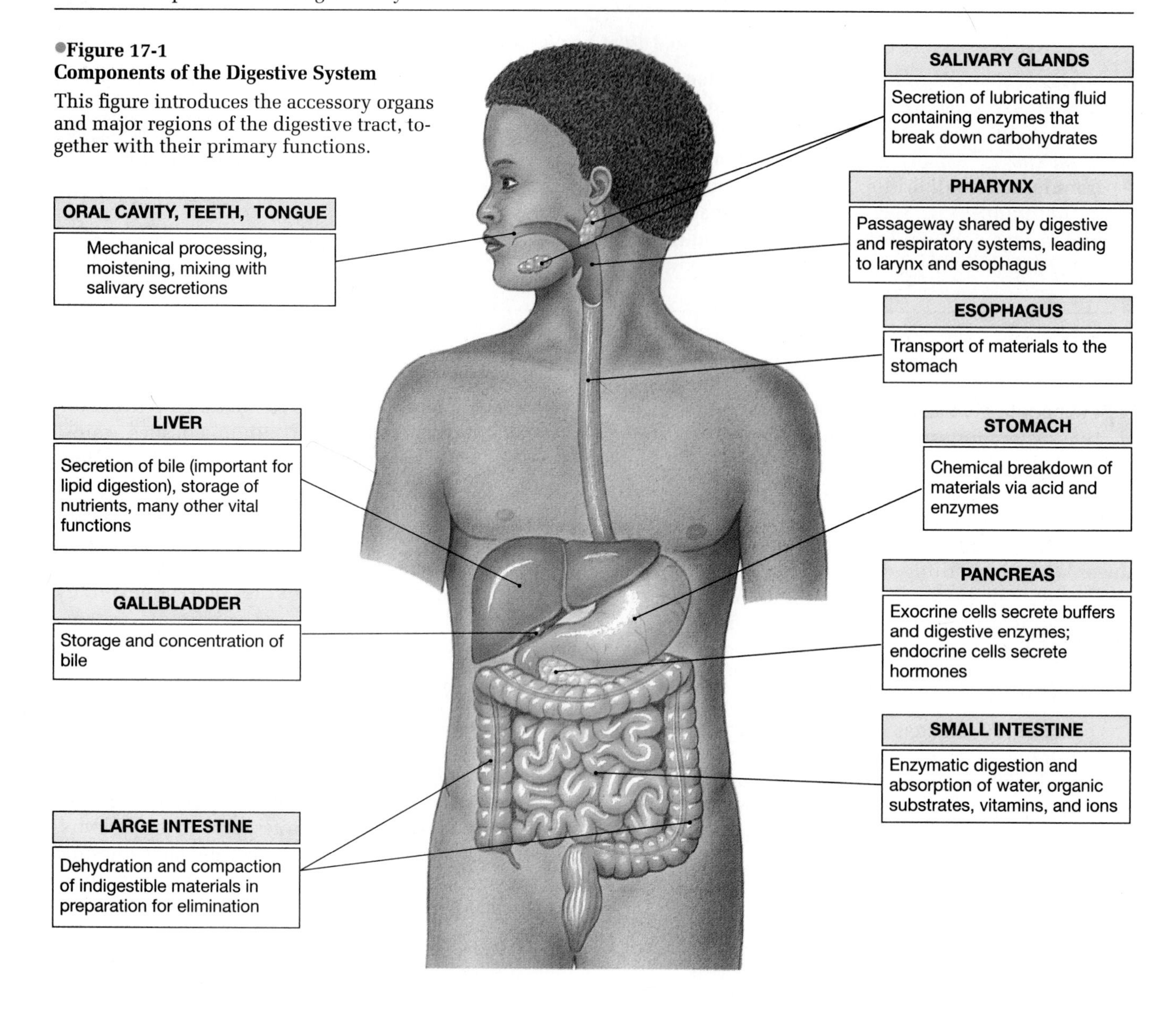
SALIVARY GLANDS
Secretion of lubricating fluid containing enzymes that break down carbohydrates
PHARYNX
Passageway shared by digestive and respiratory systems, leading to larynx and esophagus
ESOPHAGUS
Transport of materials to the stomach
STOMACH
Chemical breakdown of materials via acid and enzymes
PANCREAS
Exocrine cells secrete buffers and digestive enzymes; endocrine cells secrete hormones
SMALL INTESTINE
Enzymatic digestion and absorption of water, organic substrates, vitamins, and ions
ORAL CAVITY, TEETH, TONGUE
Mechanical processing, moistening, mixing with salivary secretions
LIVER
Secretion of bile (important for lipid digestion), storage of nutrients, many other vital functions
GALLBLADDER
Storage and concentration of bile
LARGE INTESTINE
Dehydration and compaction of indigestible materials in preparation for elimination

•Figure 17-1
Components of the Digestive System
This figure introduces the accessory organs and major regions of the digestive tract, together with their primary functions.

surrounding submucosa, or within accessory glandular organs. In most regions of the digestive tract, the outer portion of the mucosa contains a narrow band of smooth muscle and elastic fibers. Contractions of this layer, the *muscularis* (mus-kū-LAR-is) *mucosae,* move the mucosal folds and villi.

2. The **submucosa** is a second layer of loose connective tissue that surrounds the muscularis mucosae. It contains large blood vessels and lymphatics as well as a network of nerve fibers, sensory neurons, and parasympathetic motor neurons. This neural tissue, the *submucosal plexus,* helps control and coordinate the contractions of smooth muscle layers and also helps regulate the secretion of digestive glands.
3. The **muscularis externa** is a collection of smooth muscle cells arranged in an inner circular layer and an outer longitudinal layer. Contractions of these layers in various combinations agitate or propel materials along the digestive tract. These are autonomic reflex movements controlled primarily by a network of nerves, the *myenteric plexus* (*mys,* muscle + *enteron,* intestine), sandwiched between the inner and outer smooth muscle layers. Parasympathetic stimulation increases muscular tone and activity, and sympathetic stimulation promotes muscular inhibition and relaxation.
4. The **serosa,** a *serous membrane,* covers the muscularis externa along most portions of the digestive tract inside the peritoneal cavity (Figure 17-2•). This *visceral peritoneum* is continuous with the *parietal peritoneum* that lines the inner surfaces of the body wall. (These layers were introduced in Chapter 4.) ∞ [p. 87] In some areas, the parietal and visceral peritoneum are connected by double sheets of serous membrane called **mesenteries** (MEZ-en-ter-ēz), shown in Figure 17-2•. The loose connective tissue sandwiched between the epithelia provides an access route for the passage of the blood vessels, nerves, and lymphatics servicing the digestive tract.

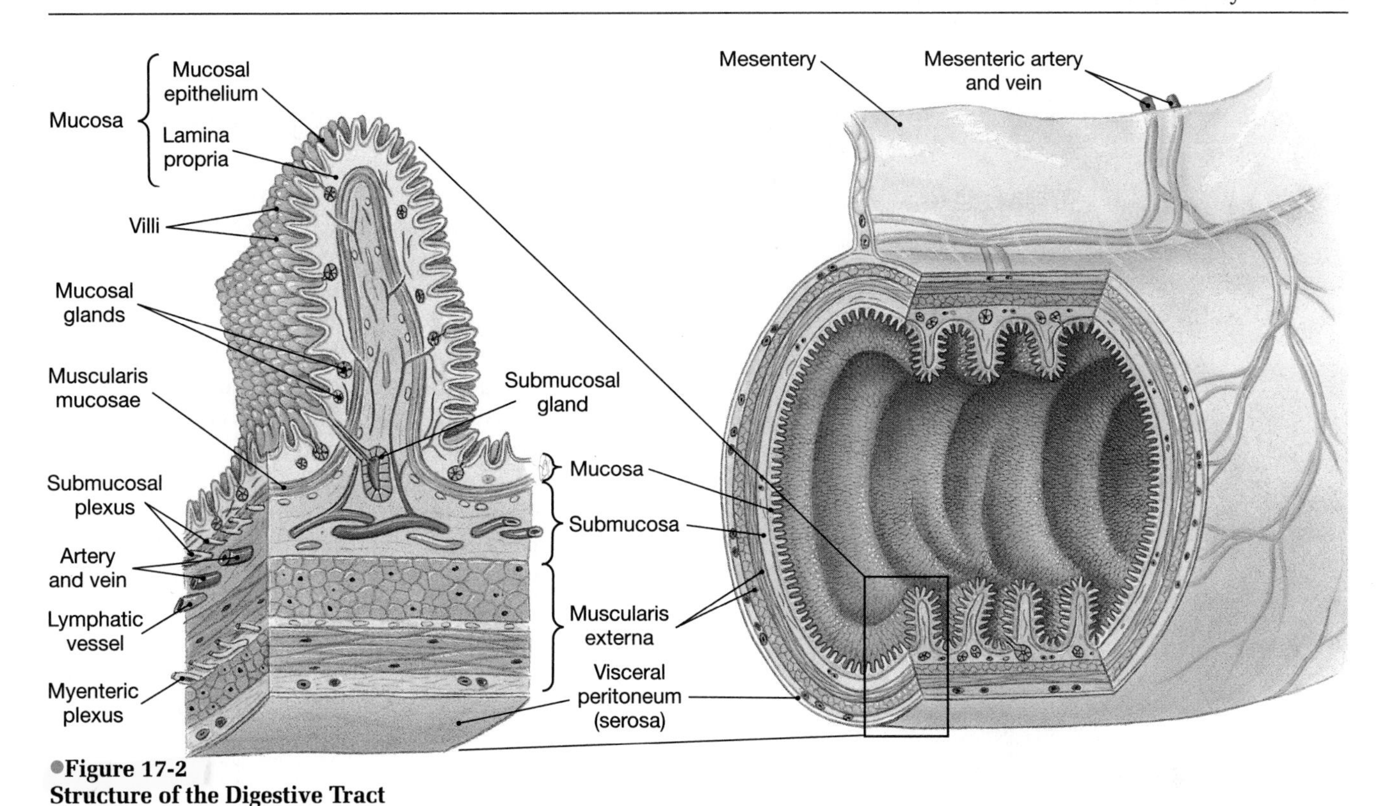

●Figure 17-2
Structure of the Digestive Tract
A diagrammatic view of a representative portion of the digestive tract. The features illustrated are those of the small intestine.

There is no serosa covering the muscularis externa of the oral cavity, pharynx, esophagus, and rectum. Instead, the muscularis externa is surrounded by a dense network of collagen fibers that firmly attaches these regions of the digestive tract to adjacent structures. This fibrous wrapping is called an *adventitia* (ad-ven-TISH-a).

ASCITES

The peritoneal lining continually produces peritoneal fluid that lubricates the opposing parietal and visceral surfaces. About 7 liters of fluid are secreted and reabsorbed each day, although the volume within the cavity at any one moment is very small. Under unusual conditions, the volume can increase markedly, causing the abdomen to swell, reducing blood volume and distorting visceral organs. This condition is called *ascites* (a-SĪ-tēz).

The Movement of Digestive Materials

Smooth muscle tissue is found within almost every organ, forming sheets, bundles, or sheaths around other tissues. The smooth muscle of the digestive tract shows rhythmic cycles of activity due to the presence of *pacesetter cells* that trigger waves of contraction. The coordinated contractions of the smooth muscle in the walls of the digestive tract play a vital role in moving materials along the tract, through *peristalsis* (*peri-*, around + *stalsis* constriction), and in mechanical processing, through *segmentation.*

Peristalsis and Segmentation

The muscularis externa propels materials from one part of the digestive tract to another through **peristalsis** (per-i-STAL-sis), waves of muscular contractions that move along the length of the digestive tract. During a peristaltic movement, the circular muscles first contract behind the digestive contents. Then longitudinal muscles contract, shortening adjacent segments. A wave of contraction in the circular muscles then forces the materials in the desired direction (Figure 17-3a●).

Regions of the small intestine also undergo **segmentation** movements that churn and fragment digestive materials. Over time, this action results in a thorough mixing of the contents with intestinal secretions. Because they do not follow a set pattern, segmentation movements do not propel materials in a particular direction (Figure 17-3b●).

THE ORAL CAVITY

The mouth opens into the oral cavity, the part of the digestive tract that receives food. The oral cavity (1) analyzes material before swallowing; (2) mechanically processes material through the actions of the teeth,

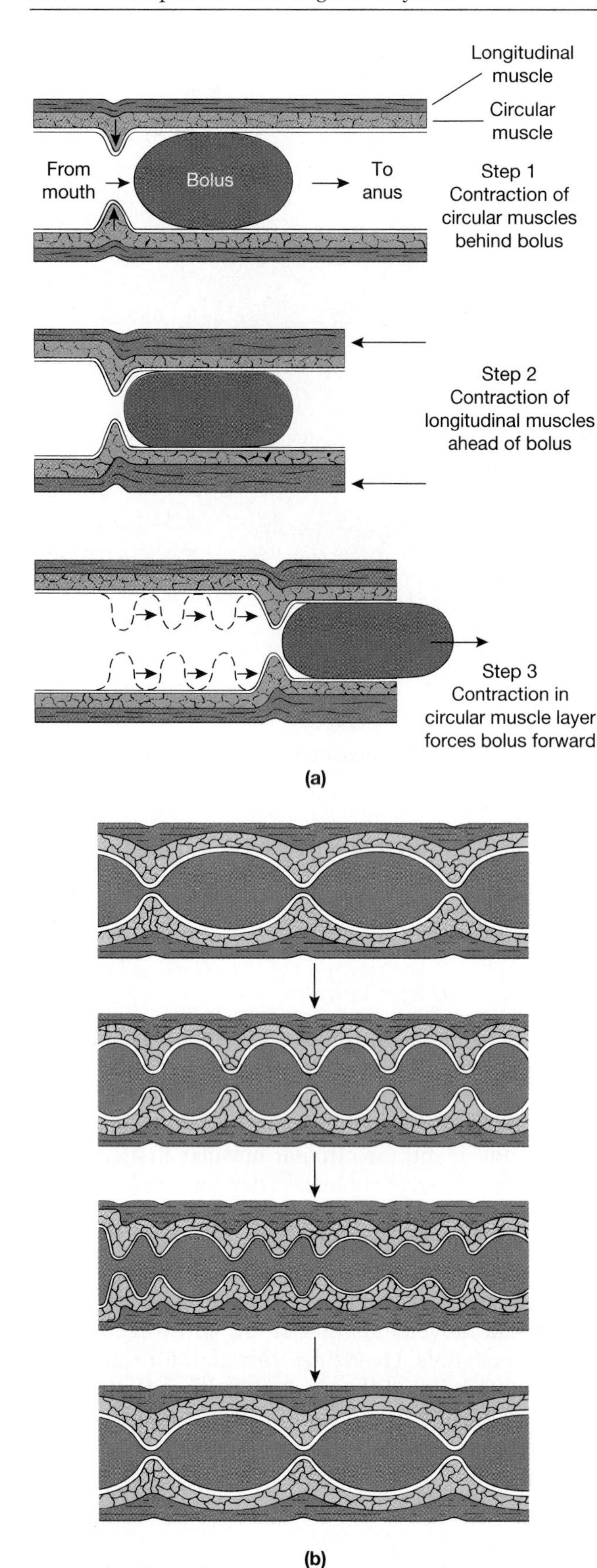

●Figure 17-3
Peristalsis and Segmentation

(a) Peristalsis propels materials along the length of the digestive tract. **(b)** Segmentation promotes mixing but does not produce net movement in any direction.

tongue, and palatal surfaces; (3) lubricates material by mixing it with mucus and salivary secretions; and (4) begins digestion of carbohydrates with the help of salivary enzymes.

Figure 17-4● shows the boundaries of the oral cavity, also known as the **buccal** (BUK-al) **cavity.** The **cheeks** form the lateral walls of this chamber; anteriorly they are continuous with the lips, or **labia** (LĀ-bē-a; singular, *labium*). The **vestibule,** a subdivision of the oral cavity, includes the space between the cheeks or lips and the teeth. A pink ridge, the gums, or **gingivae** (JIN-ji-vē), surrounds the bases of the teeth. The gums cover the tooth-bearing surfaces of the upper and lower jaws.

The **hard** and **soft palates** provide a roof for the oral cavity, while the tongue dominates its floor. The free anterior portion of the tongue is connected to the underlying epithelium by a thin fold of mucous membrane, the **lingual frenulum** (FREN-ū-lum; *frenulum,* a small bridle). The dividing line between the oral cavity and pharynx extends between the base of the tongue and the dangling **uvula** (Ū-vū-la).

The Tongue

A muscular tongue manipulates materials inside the mouth and may occasionally be used to bring things (such as ice cream) into the oral cavity. The primary functions of the tongue are (1) mechanical processing by compression, abrasion, and distortion; (2) manipulation to assist in chewing and prepare the material for swallowing; and (3) sensory analysis by touch, temperature, and taste receptors. Anatomical landmarks on the tongue are included in Figure 17-4●. Most of the tongue lies within the oral cavity, but the base of the tongue extends into the pharynx. A pair of prominent lateral swellings at the base mark the locations of the *lingual tonsils,* lymphoid nodules that help resist infections.

Salivary Glands

Figure 17-5● details the locations and relative sizes of the three pairs of salivary glands (compare Figure 17-4●). On each side, the large **parotid gland** lies below the zygomatic arch under the skin of the face. The **parotid duct** empties into the vestibule at the level of the second upper molar. The **sublingual glands** are located beneath the mucous membrane of the floor of the mouth, and numerous sublingual ducts open along either side of the lingual frenulum. The **submandibular glands** are in the floor of the mouth along the inner surfaces of the mandible; their ducts open into the mouth behind the teeth on either side of the lingual frenulum.

These salivary glands produce 1.0–1.5 liters of saliva each day, with a composition of 99.4 percent water, plus an assortment of ions, buffers, waste prod-

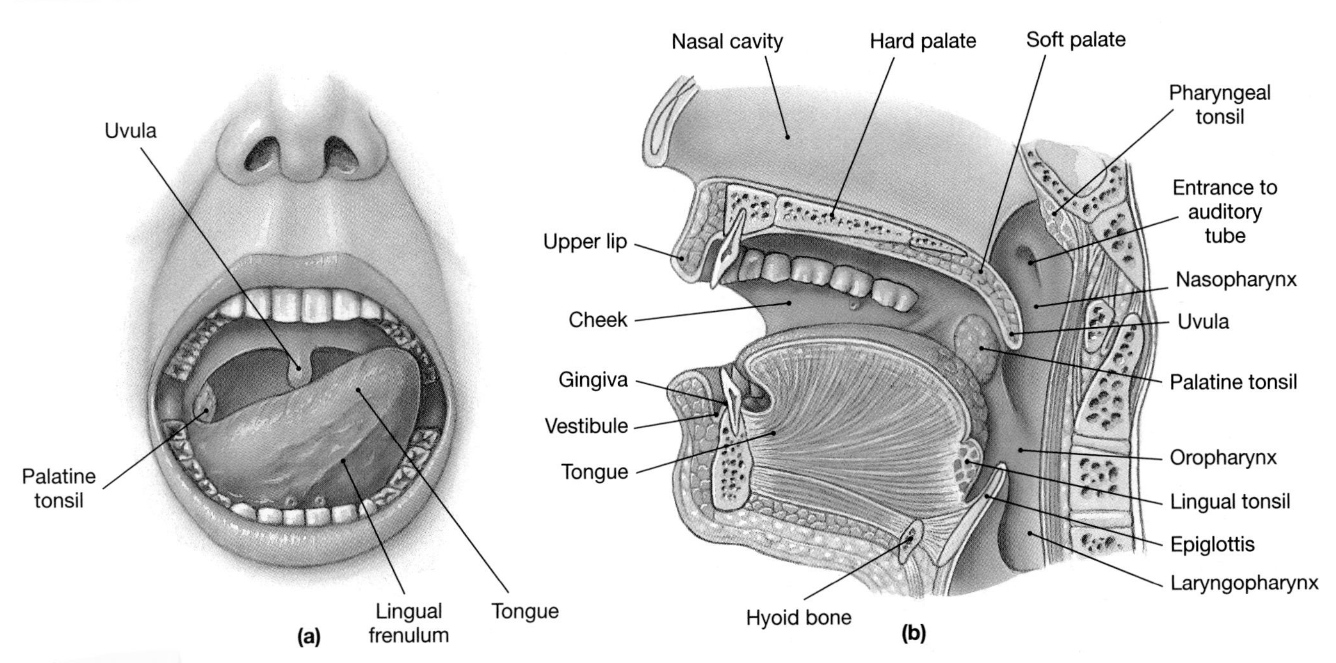

Figure 17-4
The Oral Cavity
(a) An anterior view of the oral cavity, as seen through the open mouth. **(b)** The oral cavity as seen in sagittal section.

ucts, metabolites, and enzymes. At mealtimes, large quantities of saliva lubricate the mouth and dissolve chemicals that stimulate the taste buds. Coating the food with slippery mucus reduces friction and makes swallowing possible. A continual background level of secretion flushes the oral surfaces, and salivary immunoglobulins (IgA) and lysozymes help to control populations of oral bacteria. When salivary secretions are reduced or eliminated, such as by radiation exposure, emotional distress, or other factors, the bacterial population in the oral cavity explodes. This condition soon leads to recurring infections and the progressive erosion of the teeth and gums.

MUMPS

The mumps virus most often targets the salivary glands, especially the parotid salivary glands, although other organs may also become infected. Infection typically occurs at 5–9 years of age. The first exposure stimulates antibody production and usually confers permanent immunity; active immunity can be conferred by immunization. In postadolescent males, the mumps virus may also infect the testes and cause sterility. Infection of the pancreas by the mumps virus may produce temporary or permanent diabetes; other organ systems, including the CNS, may be affected in severe cases. An effective mumps vaccine is available; widespread distribution has reduced the incidence of this disease in the United States.

Salivary Secretions

Each of the salivary glands produces a slightly different kind of saliva. The parotid glands produce a secretion rich in **salivary amylase,** an enzyme that breaks down complex carbohydrates, such as starches or glycogen, into smaller molecules that can be absorbed by the digestive tract. Saliva originating in the submandibular and sublingual salivary glands contains fewer enzymes but more buffers and mucus. At mealtimes, all three salivary glands increase their rates of secretion, and salivary production may reach

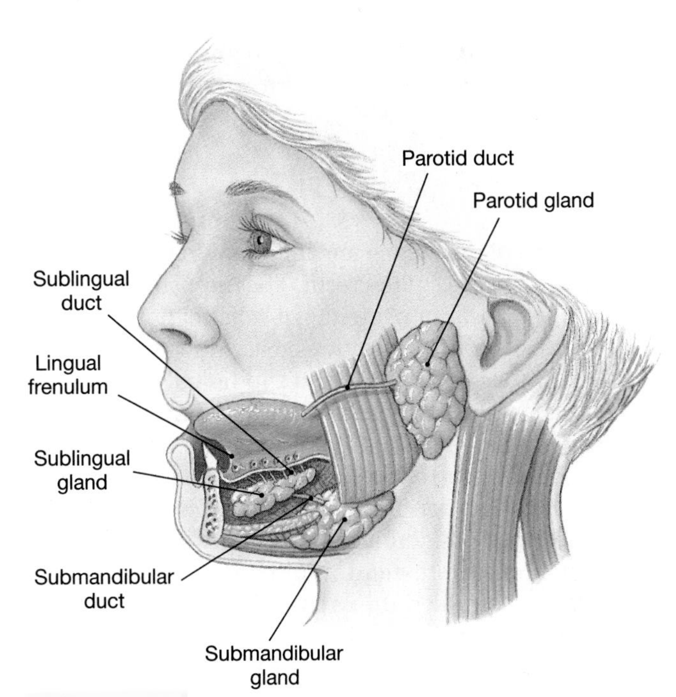

Figure 17-5
The Salivary Glands
Lateral view, showing the relative positions of the salivary glands and ducts on the left side of the head.

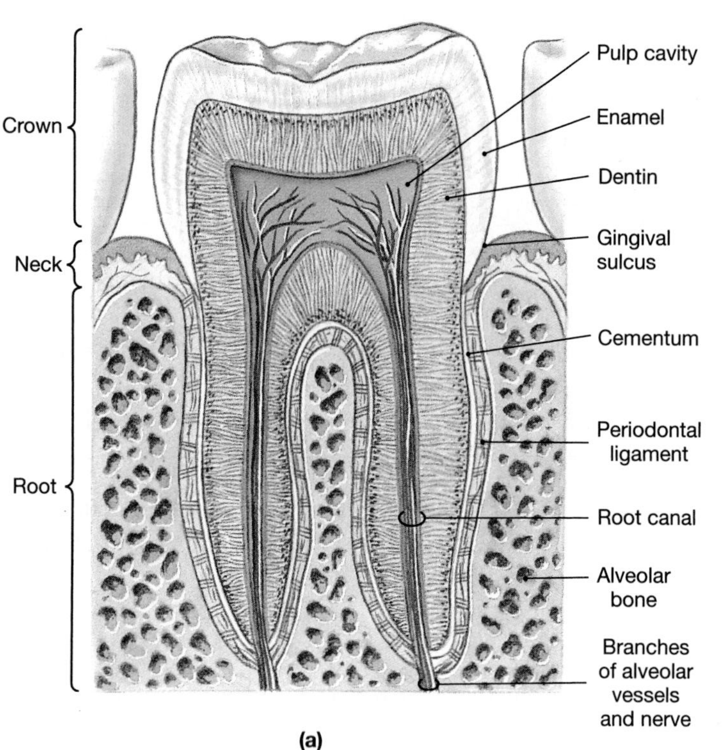

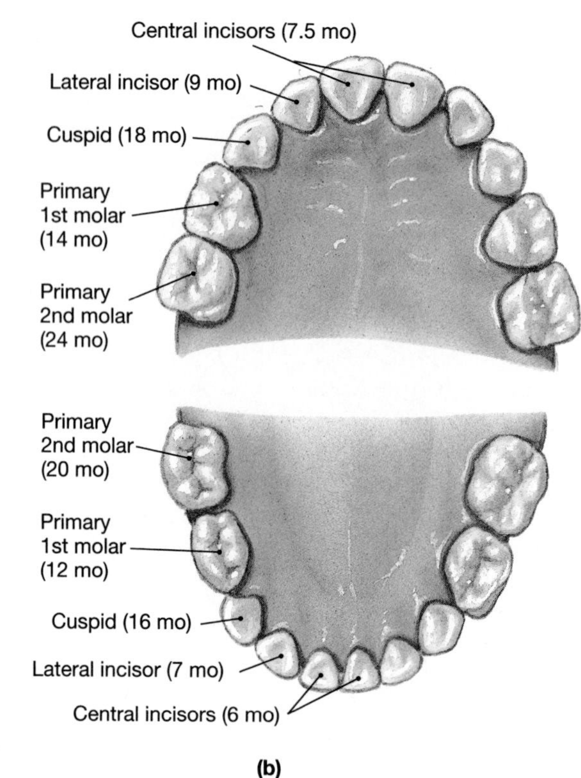

Figure 17-6
Teeth

(a) Diagrammatic section through a typical adult tooth. **(b)** The primary teeth of a child with the age of eruption given in months. **(c)** The normal orientation of adult teeth, with the age at eruption given in years.

7 ml per minute, with about half of that volume provided by the parotid glands. The pH also rises, shifting from slightly acidic (pH 6.7) to slightly basic (pH 7.5). These secretory activities are controlled by the autonomic nervous system.

Teeth

Movements of the tongue are important in passing food across the surfaces of the teeth. The opposing surfaces of the teeth perform chewing, or **mastication** (mas-ti-KĀ-shun), of food. Mastication breaks down tough connective tissues and plant fibers and helps saturate the materials with salivary lubricants and enzymes.

The bulk of each tooth consists of a mineralized matrix similar to that of bone (Figure 17-6a). This material, called **dentin** (DEN-tin), differs from bone in that it does not contain living cells. Instead, cytoplasmic processes extend into the dentin from cells within the central **pulp cavity.** The pulp cavity receives blood vessels and nerves via a narrow **root canal** at the base, or **root,** of the tooth. The root sits within a bony socket, or *alveolus* (a hollow cavity). Collagen fibers of the **periodontal ligament** (*peri-*, around + *odonto-*, tooth) extend from the dentin of the root to the surrounding bone. A layer of **cementum**

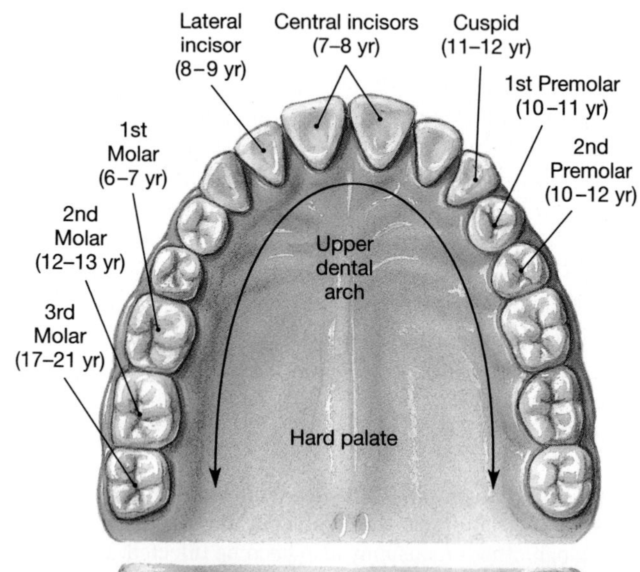

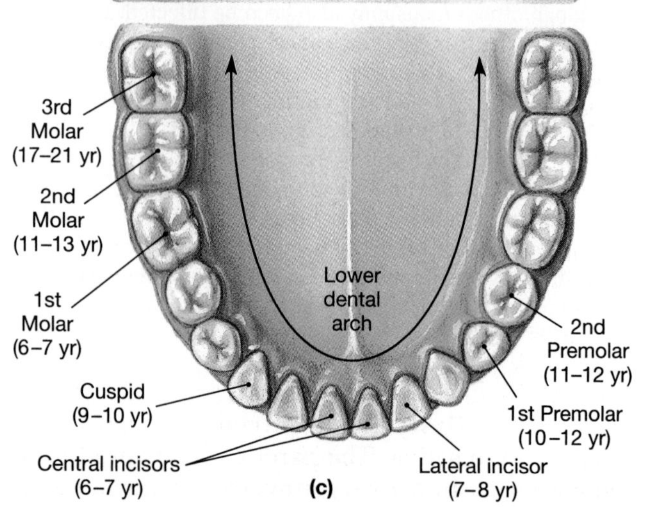

(se-MEN-tum) covers the dentin of the root, providing protection and firmly anchoring the periodontal ligament. Cementum also resembles bone, but it is softer, and remodeling does not occur following its deposition. Where the tooth penetrates the gum surface, epithelial cells form tight attachments to the tooth and prevent bacterial access to the easily eroded cementum of the root.

The **neck** of the tooth marks the boundary between the root and the **crown.** The dentin of the crown is covered by a layer of **enamel,** which contains calcium phosphate in a crystalline form, the hardest biologically manufactured substance. Adequate amounts of calcium, phosphates, and vitamin D_3 during childhood are essential if the enamel coating is to be complete and resistant to decay. Fluoride treatments or fluoridation of the water over the same period also helps, probably by increasing the density and hardness of the enamel layer.

Adult teeth are shown in Figure 17-6c●. There are four different types of teeth with specific functions. **Incisors** (in-SĪ-zerz), blade-shaped teeth found at the front of the mouth, are useful for clipping or cutting, as when nipping off the tip of a carrot stick. The **cuspids** (KUS-pidz), or *canines,* are conical with a sharp ridgeline and a pointed tip. They are used for tearing or slashing. A tough piece of celery might be weakened by the clipping action of the incisors, but then moved to one side to take advantage of the shearing action provided by the cuspids. **Bicuspids** (bī-KUS-pidz), or *premolars,* and **molars** have flattened crowns with prominent ridges. They are used for crushing, mashing, and grinding. A tough nut or sparerib will usually be shifted to the premolars and molars.

Dental Succession

During development, two sets of teeth begin to form. The first to appear are the **deciduous teeth** (de-SID-ū-us; *deciduus,* falling off), also known as *primary teeth, milk teeth,* or *baby teeth.* There are usually 20 deciduous teeth (Figure 17-6b●). These teeth will later be replaced by the adult **secondary dentition,** or *permanent dentition* (Figure 17-6c●). Three additional teeth appear on each side of the upper and lower jaws as the person ages, extending the length of the tooth rows posteriorly and bringing the permanent tooth count to 32. AM *Dental Problems and Dental Implants*

THE PHARYNX

The pharynx serves as a common passageway for solid food, liquids, and air. The three major subdivisions of the pharynx were discussed in Chapter 16. ∞ *[p. 409]* Food normally passes through the oropharynx and laryngopharynx on its way to the esophagus. The pharyngeal muscles cooperate with muscles of the oral cavity and esophagus to initiate the process of swallowing. The muscular contractions during swallowing force the food mass along the esophagus and into the stomach.

THE ESOPHAGUS

The esophagus is a muscular tube that begins at the pharynx and ends at the stomach. It has a length of approximately 25 cm (1 ft) and a diameter of about 2 cm (0.75 in.). The esophagus lies posterior to the trachea in the neck, passes through the mediastinum in the thoracic cavity, and enters the peritoneal cavity through an opening in the diaphragm, the *esophageal hiatus* (hī-Ā-tus; a gap or opening), before emptying into the stomach.

The esophagus is lined by a stratified squamous epithelium that can resist abrasion, hot or cold temperatures, and chemical attack. The secretions of mucous glands lubricate this surface and prevent materials from sticking to the sides of the esophagus as swallowing occurs.

Swallowing

Swallowing is a complex reflex that transports food from the pharynx to the stomach (Figure 17-7●). Before swallowing can occur, the food must have the proper texture and consistency. Once the material has been shredded or torn by the teeth, moistened with salivary secretions, and approved by the taste receptors, the tongue begins compacting the debris into a small mass, or **bolus.**

Swallowing begins with the compression of the bolus against the hard palate. The tongue then retracts, forcing the bolus into the pharynx and helping to elevate the soft palate, thus preventing the bolus from entering the nasopharynx (Figure 17-7a,b●). The bolus now comes in contact with the posterior pharyngeal wall (Figure 17-7c,d●). The larynx elevates, and the epiglottis folds to direct the bolus past the closed glottis. In less than a second the contraction of pharyngeal muscles forces the bolus through the entrance to the esophagus, which is guarded by the *upper esophageal sphincter.* Once within the esophagus, the bolus is pushed toward the stomach by a peristaltic wave. The approach of the bolus triggers the opening of the *lower esophageal sphincter* and the bolus enters the stomach (Figure 17-7e–h●).

For a typical bolus, the entire trip takes about 9 seconds to complete. Fluids may make the journey in a few seconds, arriving ahead of the peristaltic contractions; a relatively dry or bulky bolus travels much more slowly, and repeated peristaltic waves may be required to drive it into the stomach. A completely dry bolus cannot be swallowed at all, for friction with the walls of the esophagus will make peristalsis ineffective.

ESOPHAGITIS AND HIATAL HERNIAS

A weakened or permanently relaxed sphincter can cause inflammation of the esophagus, or *esophagitis* (ē-sof-a-JĪ-tis), as powerful gastric acids enter the lower esophagus. The esophageal epithelium has few defenses from acid and enzyme attack, and inflammation, epithelial erosion, and intense discomfort are the result. Occasional incidents of reflux, or backflow, from the stomach are responsible for the symptoms of "heartburn." This relatively common problem supports a multimillion dollar industry devoted to producing and promoting antacids.

The esophagus and major blood vessels pass from the thoracic cavity to the abdominopelvic cavity through an opening in the diaphragm called the esophageal hiatus. In a *hiatal hernia,* abdominal organs slide into the thoracic cavity through the opening used by the esophagus. The severity of the condition will depend on the location and size of the herniated organ(s). Hiatal hernias are actually very common, and most go unnoticed. When clinical problems develop, they usually occur because abdominal organs that have pushed into the thoracic cavity are exerting pressure on structures or organs there.

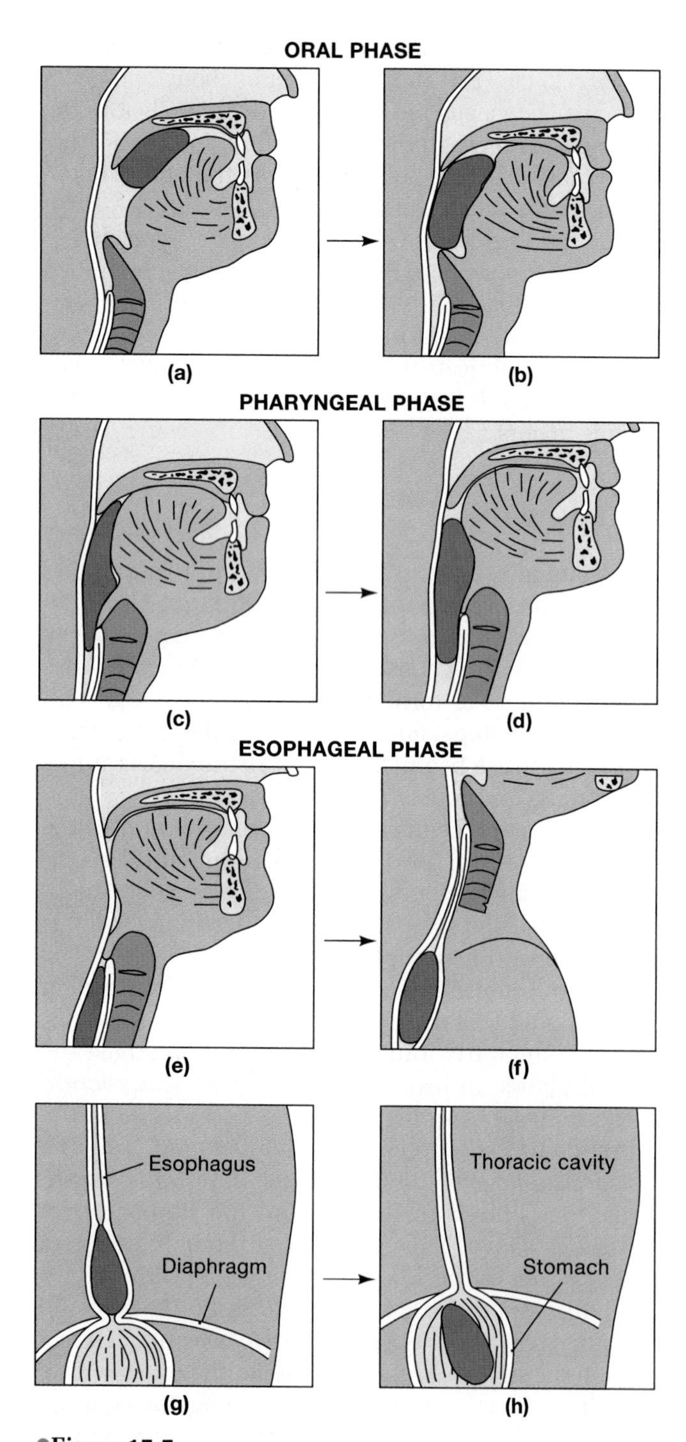

•Figure 17-7
The Swallowing Process

This sequence, based on a series of X-rays, shows the stages of swallowing and the movement of materials from the oral cavity to the stomach. (See also *Figure 16-4*)

✓ Would peristalsis or segmentation be more efficient in propelling intestinal contents from one place to another?

✓ What effect would a drug that blocks parasympathetic stimulation of the digestive tract have on peristalsis?

✓ What is occurring when the soft palate and larynx elevate and the glottis closes?

✓ What prevents the backflow of materials from the stomach into the esophagus?

THE STOMACH

The stomach, located within the left upper quadrant of the abdominopelvic cavity, receives the food from the esophagus. The stomach has four primary functions: (1) the temporary storage of ingested food, (2) the mechanical breakdown of resistant materials, (3) the beginning of digestion by breaking chemical bonds through the action of acids and enzymes, and (4) the production of *intrinsic factor,* a compound necessary for absorption of vitamin B_{12}. The agitation of ingested materials with the gastric juices secreted by the glands of the stomach produces a viscous, soupy mixture called **chyme** (kīm).

Figure 17-8a• indicates the principal anatomical landmarks of the stomach, a muscular organ with the shape of an expanded J. The esophagus connects to the stomach at the **cardia** (KAR-dē-a). The bulge of the stomach superior to the cardia is the **fundus** (FUN-dus) of the stomach, and the large area between the fundus and the curve of the J is the gastric **body.** The curve of the J, the **pylorus** (pī-LŌR-us), connects the stomach with the small intestine. A muscular **pyloric sphincter** regulates the flow of chyme between the stomach and small intestine.

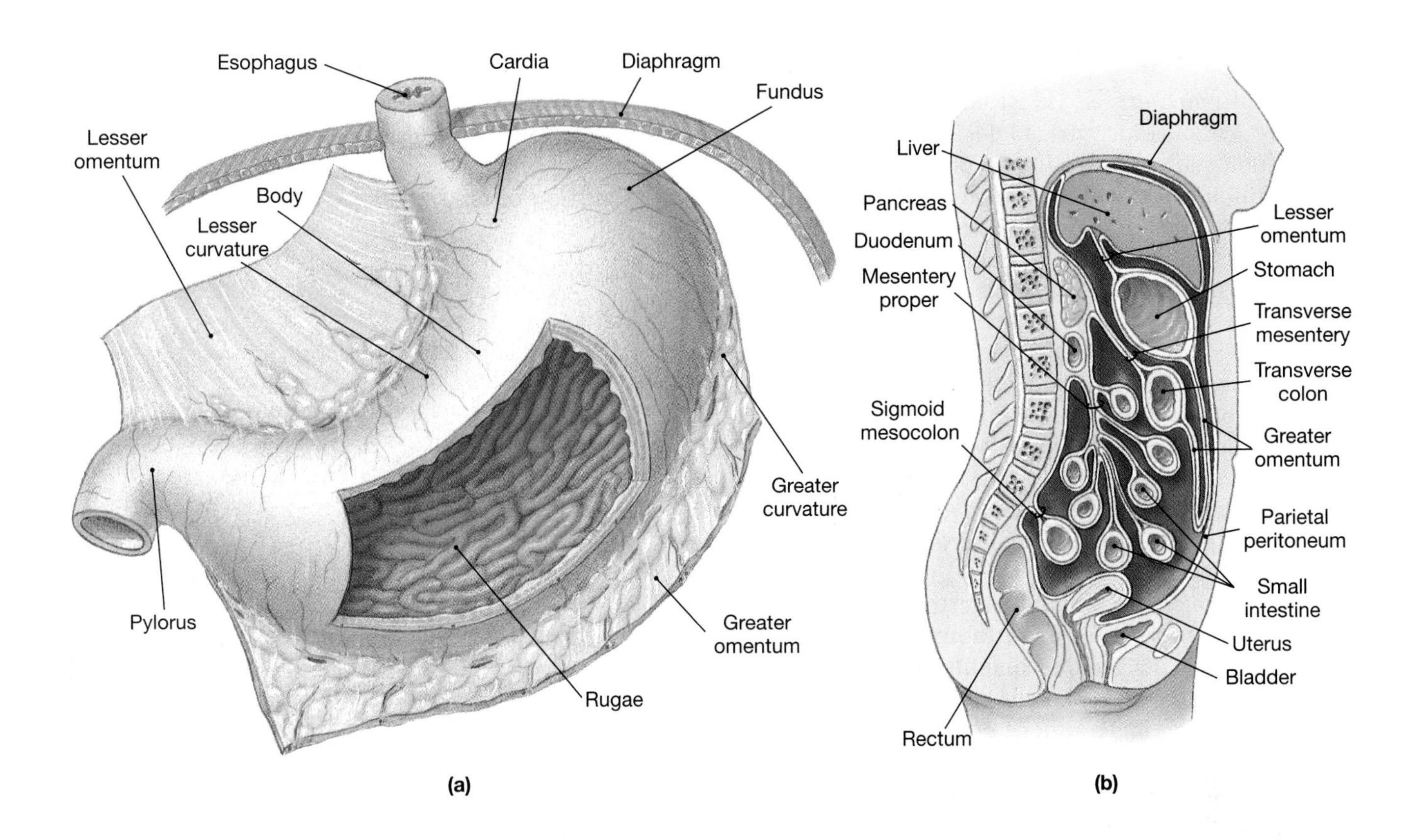

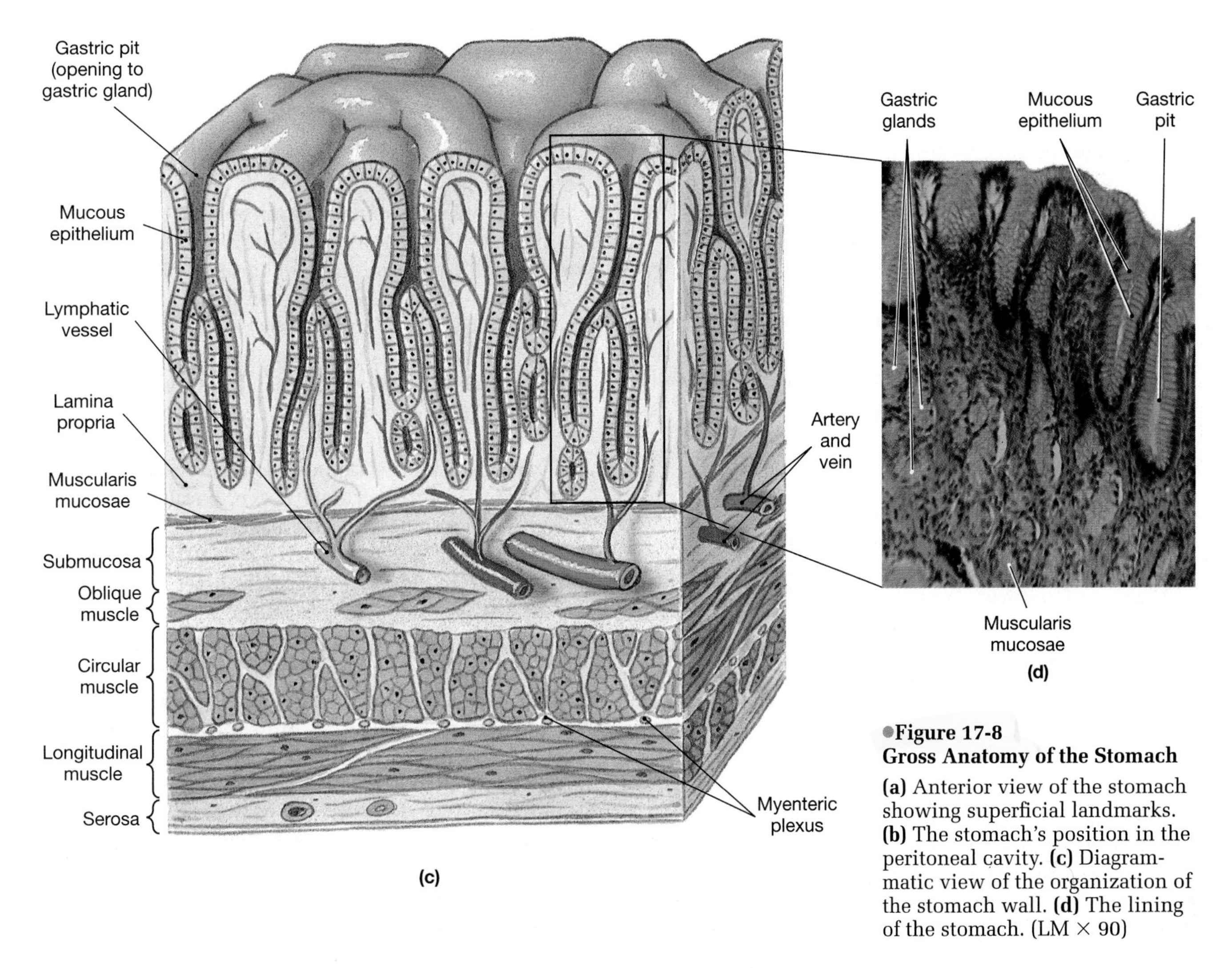

Figure 17-8
Gross Anatomy of the Stomach
(a) Anterior view of the stomach showing superficial landmarks. **(b)** The stomach's position in the peritoneal cavity. **(c)** Diagrammatic view of the organization of the stomach wall. **(d)** The lining of the stomach. (LM × 90)

The dimensions of the stomach are extremely variable. When empty, the stomach resembles a muscular tube with a narrow and constricted lumen. When full, it can expand to contain 1–1.5 liters. This degree of expansion is possible because the stomach wall contains thick layers of smooth muscle, and the mucosa of the relaxed stomach contains a number of prominent ridges and folds, called **rugae** (ROO-gē; wrinkles). As the stomach expands, the smooth muscle stretches and the rugae gradually disappear.

Unlike the two-layered muscularis externa of other portions of the digestive tract, that of the stomach contains a longitudinal layer, a circular layer, and an inner oblique layer. This extra layer adds strength and assists in the mixing and churning activities essential to forming chyme.

The visceral peritoneum covering the outer surface of the stomach is continuous with a pair of mesenteries. The **greater omentum** (ō-MEN-tum; *omentum*, a fatty skin) extends below the greater curvature and forms an enormous pouch that hangs over and protects the abdominal viscera (Figure 17-8b●). The much smaller **lesser omentum** extends from the lesser curvature to the liver.

The Gastric Wall

The stomach is lined by an epithelium dominated by mucous cells. The secreted mucus produced helps protect the stomach lining from the acids, enzymes, and abrasive materials it contains. Shallow depressions, called **gastric pits,** open onto the gastric surface (17-8c●). Each gastric pit communicates with **gastric glands** that extend deep into the underlying lamina propria (Figure 17-8d●). These glands are dominated by two types of secretory cells: **parietal cells** and **chief cells.** Together these cells secrete about 1500 ml of **gastric juice** each day.

Parietal Cells

Parietal cells secrete *intrinsic factor* and *hydrochloric acid* (HCl). **Intrinsic factor** facilitates the absorption of *vitamin* B_{12} across the intestinal lining. Hydrochloric acid lowers the pH of the gastric juice, kills microorganisms, breaks down cell walls and connective tissues in food, and activates the enzymatic secretions of the chief cells.

Chief Cells

Chief cells secrete **pepsinogen** (pep-SIN-ō-jen), an inactive form of the enzyme **pepsin.** The hydrochloric acid released by the parietal cells converts pepsinogen to pepsin. Pepsin is an example of a proteolytic, or protein splitting, enzyme that breaks down proteins. The stomachs of newborn infants also produce additional enzymes important for the digestion of milk, *rennin* and *gastric lipase.* Rennin coagulates milk proteins and gastric lipase initiates the digestion of milk fats.

GASTRITIS AND PEPTIC ULCERS

Inflammation of the gastric mucosa is called *gastritis* (gas-TRĪ-tis). This condition may develop after swallowing drugs, including alcohol and aspirin. Gastritis may also appear after severe emotional or physical stress, bacterial infection of the gastric wall, or the ingestion of strongly acid or alkaline chemicals.

A *peptic ulcer* develops when the digestive acids and enzymes manage to erode their way through the defenses of the stomach lining or proximal portions of the small intestine. The locations may be indicated by using the terms *gastric ulcer* (stomach) or *duodenal ulcer* (duodenum of the small intestine). Peptic ulcers result from the excessive production of acid or the inadequate production of the alkaline mucus that poses an epithelial defense. For the last decade, drugs such as *cimetidine (Tagamet)* have been used to inhibit acid production by parietal cells. Evidence now demonstrates that infections involving the bacterium *Helicobacter pylori* are responsible for a majority of peptic ulcers, and treatment for gastric ulcers today often involves the administration of antibiotic drugs. AM Gastric Ulcers

Regulation of Gastric Activity

The production of acid and enzymes by the stomach can be controlled by the central nervous system as well as by local hormonal mechanisms. Three stages can be identified, although considerable overlap exists between them (Figure 17-9●).

1. **The cephalic phase.** The sight, smell, taste, or thought of food initiates the *cephalic phase* of gastric secretion. This stage, which is directed by the CNS, prepares the stomach to receive food. Under the control of the vagus nerve, parasympathetic fibers innervate parietal cells, chief cells, and mucous cells of the stomach. In response to stimulation, the production of gastric juice accelerates, reaching rates of around 500 ml per hour. This phase usually lasts for a relatively brief period before the *gastric phase* commences.
2. **The gastric phase.** The *gastric phase* begins with the arrival of food in the stomach. Stimulation of stretch receptors in the stomach wall and chemoreceptors in the mucosa triggers the release of a hormone, **gastrin,** into the circulatory system. Proteins, alcohol in small doses, and caffeine are potent stimulators of gastric secretion because they excite the mucosal chemoreceptors. Both parietal and chief cells respond to the presence of gastrin

by accelerating their secretory activities. The effect on the parietal cells is the most pronounced, and the pH of the gastric contents drops sharply. This phase may continue for several hours while the ingested materials are processed by the acids and enzymes. Over this period, stomach contractions begin to swirl and churn the gastric contents, mixing the ingested materials with the gastric secretions to form chyme. As digestion proceeds, the contractions begin sweeping down the length of the stomach, and each time the pylorus contracts a small quantity of chyme squirts through the pyloric sphincter.

3. **The intestinal phase.** The *intestinal phase* begins when chyme starts to enter the small intestine. The purpose of this phase is to control the rate of gastric emptying and ensure that the secretory, digestive, and absorptive functions of the small intestine can proceed efficiently. Most of the regulatory controls are inhibitory, providing a brake for gastric activities. Both endocrine and neural mechanisms are involved. Intestinal hormones, such as *secretin*, *cholecystokinin* (CCK), and *gastric inhibitory peptide* (GIP), are released when chyme enters the small intestine. These hormones reduce gastric activity and give the small intestine time to deal with the arriving acids. Inhibitory reflexes that depress gastric activity are stimulated when the proximal portion of the small intestine becomes too full, too acidic, unduly irritated by the chyme, or filled with partially digested proteins, carbohydrates, or fats. For example, as the small intestine distends, inhibitory feedback slows the contractions in the stomach walls, inhibits the parasympathetic nervous system, and stimulates sympathetic innervation. The combination significantly reduces gastric activity.

Digestion in the Stomach

The stomach performs preliminary digestion of proteins by pepsin and, for a variable period, permits the digestion of carbohydrates by salivary amylase. This enzyme remains active until the pH throughout the material in the stomach falls below 4.5, usually within 1–2 hours after a meal.

As the stomach contents become more fluid and the pH approaches 2.0, pepsin activity increases and protein disassembly begins. Protein digestion is not completed in the stomach, but there is usually enough time for pepsin to break down complex proteins into smaller peptide and polypeptide chains before the chyme enters the small intestine.

Although digestion begins in the stomach, there is little if any nutrient absorption because (1) the epithelial cells are covered by a blanket of alkaline mucus and are not directly exposed to the chyme; (2) the epithelial cells lack the specialized transport mechanisms found in cells lining the small intestine; (3) the gastric lining is impermeable to water; and (4) digestion has not proceeded to completion by the time chyme leaves the stomach.

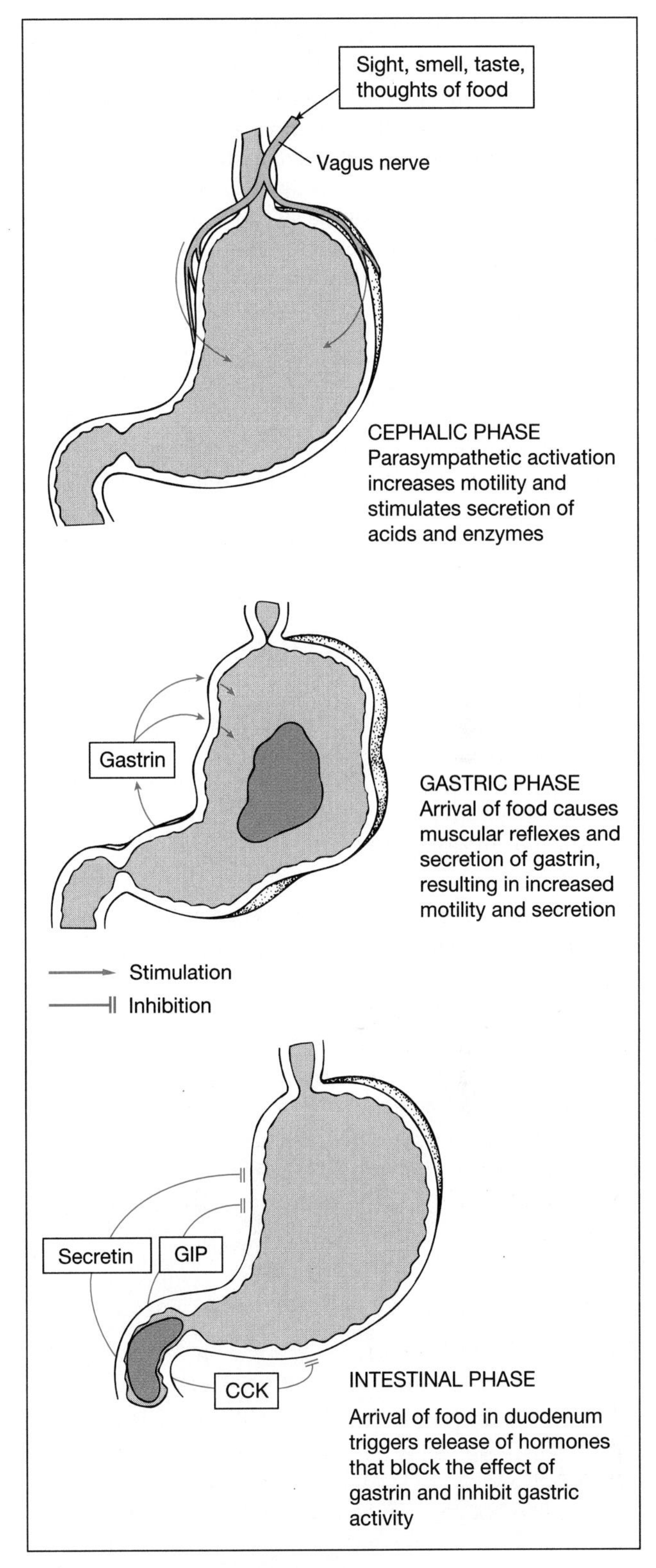

●**Figure 17-9**
The Phases of Gastric Secretion

STOMACH CANCER

Stomach cancer is one of the most common lethal cancers, responsible for roughly 15,000 deaths in the United States each year. Because the symptoms may resemble those of gastric ulcers, the condition may not be reported in its early stages. Diagnosis usually involves X-rays of the stomach at various degrees of distension. The mucosa can also be visually inspected using a flexible instrument called a *gastroscope.* Attachments permit the collection of tissue samples for histological analysis. Treatment of gastric cancer involves the surgical removal of part or all of the stomach. Even a total *gastrectomy* (gas-TREK-to-mē) can be survived, because the only absolutely vital function of the stomach is the secretion of intrinsic factor. Protein breakdown can still be performed by the small intestine, although at reduced efficiency, and the loss of gastric functions such as food storage and acid production is not life-threatening.

THE SMALL INTESTINE

The small intestine is about 6 meters (20 ft) long and has a diameter ranging from 4 cm at the stomach to about 2.5 cm at the junction with the large intestine. It has three subdivisions: the *duodenum,* the *jejunum,* and the *ileum* (Figure 17-10●).

- The **duodenum** (doo-A-dē-num) is the 25 cm (1 ft) closest to the stomach. This portion receives chyme from the stomach and exocrine secretions from the pancreas and liver.
- A rather abrupt bend marks the boundary between the duodenum and the **jejunum** (je-JOO-num). The jejunum, which is supported by a sheet of mesentery, is about 2.5 meters (8 ft) in length. The bulk of chemical digestion and nutrient absorption occurs in the jejunum. One rather drastic approach to weight control involves the surgical removal of a significant portion of the jejunum. AM: *Drastic Weight Loss Techniques*
- The jejunum leads us to the third segment, the **ileum** (IL-ē-um). The ileum ends at a sphincter, the *ileocecal valve,* which controls the flow of chyme from the ileum into the *cecum* of the large intestine.

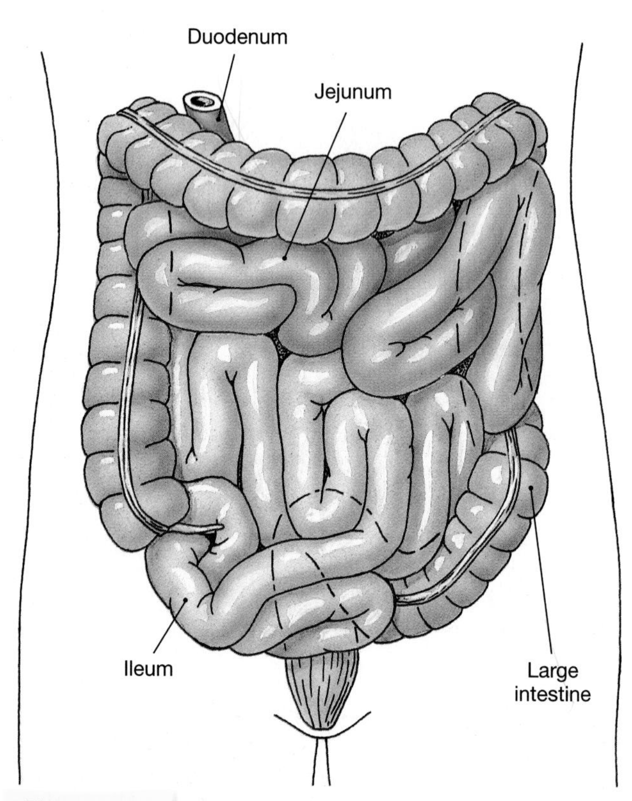

●Figure 17-10
Location of the Small Intestine

The small intestine fits in the relatively small peritoneal cavity because it is well packed, and the position of each of the segments is stabilized by mesenteries attached to the dorsal body wall (Figure 17-8b●, p. 441).

The Intestinal Wall

The intestinal lining bears a series of transverse folds called *plicae* (PLĪ-sē) (Figure 17-11a●). The lining of the intestine is also thrown into a series of fingerlike projections, the **villi** (Figure 17-11b●). These villi are covered by a simple columnar epithelium carpeted with microvilli. If the small intestine were a simple tube with smooth walls, it would have a total absorptive area of around 3300 square centimeters, or roughly 3.6 square feet. Instead, the epithelium contains folds, each fold supports a forest of villi, and each villus is covered by epithelial cells blanketed in microvilli. This arrangement increases the total area for absorption to approximately 2 million square centimeters, or more than 2200 square feet!

Each villus contains a network of capillaries (Figure 17-11c●) that transports respiratory gases and carries absorbed nutrients to the hepatic portal circulation. In addition to capillaries and nerve endings, each villus contains a terminal lymphatic called a **lacteal** (LAK-tē-al; *lacteus,* milky). This name refers to the pale, cloudy appearance of the lymph in these channels. Lacteals transport materials that are unable to cross the walls of local capillaries. For example, absorbed fatty acids are assembled into protein-lipid packages that are too large to diffuse into the bloodstream. These packets, called *chylomicrons* (*chylos;* juice), reach the circulation by passage through the lymphatic system.

At the bases of the villi are found the entrances to intestinal glands that secrete a watery *intestinal juice.* In the duodenum, large intestinal glands secrete an alkaline mucus that helps buffer the acids in chyme. Intestinal glands also contain endocrine cells responsible for the production of intestinal hormones considered in a later section.

Intestinal Movements

Once the chyme is within the small intestine, segmentation contractions must mix it with mucous secretions and enzymes before absorption can occur. As absorption occurs, weak peristaltic contractions slowly move the remaining materials along the length of the small intestine. These contractions are local reflexes not under CNS control, and the effects are limited to within a few centimeters of the site of the original stimulus. More elaborate reflexes coordinate activities along the entire length of the small intestine. Two examples are the *gastroenteric reflex* and the *gastroileal reflex.*

Distension of the stomach initiates the *gastroenteric* (gas-trō-en-TER-ik) *reflex,* which immediately accelerates glandular secretion and peristaltic activity in all segments. The increased peristalsis distributes materials along the length of the small intestine and empties the duodenum.

The *gastroileal* (gas-trō-IL-ē-al) *reflex* is a response to circulating levels of the hormone *gastrin.* The entry of food into the stomach triggers the release of gastrin, which relaxes the ileocecal valve at the entrance to the large intestine. Because the valve is relaxed,

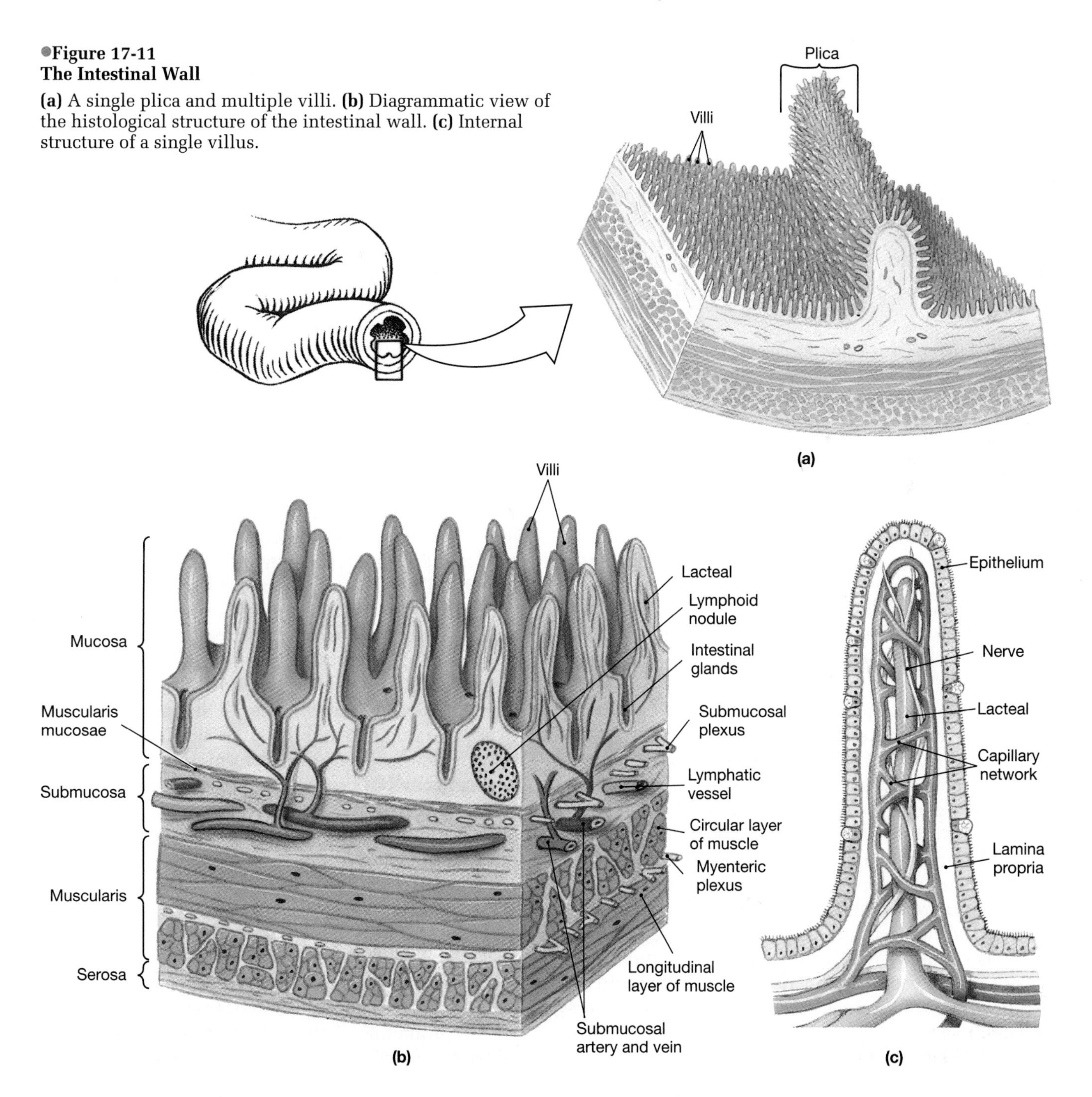

●Figure 17-11
The Intestinal Wall

(a) A single plica and multiple villi. **(b)** Diagrammatic view of the histological structure of the intestinal wall. **(c)** Internal structure of a single villus.

the increased peristalsis pushes materials from the ileum into the large intestine. On average, it takes about 5 hours for ingested food to pass from the duodenum to the end of the ileum, so the first of the materials to enter the duodenum after breakfast may leave the small intestine at lunch.

VOMITING

The responses of the digestive tract to chemical or mechanical irritation are rather predictable. Gastric secretion accelerates all along the digestive tract, and the intestinal contents are eliminated as quickly as possible. The *vomiting reflex* occurs in response to irritation of the soft palate, pharynx, esophagus, stomach, or proximal portions of the small intestine. During the preparatory phase, the pylorus relaxes, and the contents of the duodenum and proximal jejunum are discharged into the stomach by strong peristaltic waves that travel toward the stomach, rather than toward the ileum. Vomiting, or *emesis* (EM-e-sis), then occurs as the stomach regurgitates its contents through the esophagus and pharynx. As regurgitation occurs, the uvula and soft palate block the entrance to the nasopharynx. Increased salivary secretion assists in buffering the stomach acids and thereby prevents subsequent erosion of the teeth. In conditions marked by repeated vomiting, severe tooth damage can occur; this is one symptom of the eating disorder *bulimia,* discussed in the *Applications Manual.* Most of the force of vomiting comes from a powerful expiratory movement that elevates intra-abdominal pressures and presses the stomach against the tensed diaphragm. AM *Eating Disorders*

Intestinal Secretions

Roughly 1.8 liters of watery **intestinal juice** enter the intestinal lumen each day. Intestinal juice moistens the intestinal contents, assists in buffering acids, and dissolves both the digestive enzymes provided by the pancreas and the products of digestion. Much of this fluid arrives through osmosis, as water flows out of the mucosa, and the rest is provided by intestinal glands stimulated by activation of touch and stretch receptors in the intestinal walls.

Hormonal and CNS controls are important in regulating the secretions of the digestive tract and accessory organs. These regulatory mechanisms are focused on the duodenum, for it is there that the acids must be neutralized and the appropriate enzymes added. The submucosal glands protect the duodenal epithelium from gastric acids and enzymes. They increase their secretions in response to local reflexes and also to parasympathetic (vagal) stimulation. As a result of vagal activity, the duodenal glands begin secreting long before chyme reaches the pyloric sphincter. Sympathetic stimulation inhibits their activation, leaving the duodenal lining relatively unprepared for the arrival of the acid chyme. This is probably why duodenal ulcers can be caused by chronic stress or other factors that promote sympathetic activation.

Intestinal Hormones

Duodenal endocrine cells produce hormones that coordinate the secretory activities of the stomach, duodenum, pancreas, and liver.

Secretin (sē-KRĒ-tin) is released when the pH falls in the duodenum. This occurs when acid chyme arrives from the stomach. The primary effect of secretin is to increase the secretion of water and buffers by the pancreas and liver.

Cholecystokinin (kō-lē-sis-tō-KĪ-nin), or **CCK,** is also secreted when chyme arrives in the duodenum, especially when it contains lipids and partially digested proteins. This hormone also targets the pancreas and liver. In the pancreas, CCK accelerates the production and secretion of all types of digestive enzymes. At the liver, it causes the ejection of bile from the gallbladder into the duodenum. The presence of either secretin or CCK in high concentrations also reduces gastric motility and secretory rates.

Gastric inhibitory peptide, or **GIP,** is released when fats and glucose enter the small intestine. This peptide hormone inhibits gastric activity and causes the release of insulin from the pancreatic islets.

Functions of the major gastrointestinal hormones are summarized in Table 17-1, and their interactions are diagrammed in Figure 17-12●.

Digestion in the Small Intestine

In the stomach, food becomes saturated with gastric juices and exposed to the digestive effects of a strong acid and a proteolytic enzyme, pepsin. Most of the important digestive processes are completed in the small intestine, where the final products of digestion—simple sugars, fatty acids, and amino acids—are absorbed, along with most of the water content. Approximately 80 percent of all absorption takes place in the small intestine, with the rest divided between the stomach and the large intestine. However, the small intestine produces only a few of the enzymes needed to break down the complex materials found in the diet. Most of the enzymes and buffers are contributed by the liver and pancreas, discussed in the next section.

- ✓ What muscle regulates the flow of chyme from the stomach to the small intestine?
- ✓ When a person suffers from chronic ulcers in the stomach, treatment sometimes involves cutting the branches of the vagus nerve that serve the stomach. Why? *(cont.)*

TABLE 17-1 Important Gastrointestinal Hormones and Their Primary Effects

Hormone	*Stimulus*	*Origin*	*Target*	*Effects*
Gastrin	Vagal stimulation or arrival of food in the stomach	Stomach	Stomach	Stimulates production of acids and enzymes, increases motility
	Arrival of chyme containing large quantities of undigested proteins	Duodenum	Stomach	Stimulates gastric secretion and motion
Secretin	Arrival of acid chyme in the duodenum	Duodenum	Pancreas	Stimulates production of alkaline buffers
			Stomach	Inhibits gastric secretion and motility
Cholecystokinin (CCK)	Arrival of acid chyme containing lipids and partially digested proteins	Duodenum	Pancreas	Stimulates production of pancreatic enzymes
			Gallbladder	Stimulates contraction of gallbladder
			Duodenum	Causes relaxation of sphincter at base of bile duct
			Stomach	Inhibits gastric secretion and motion
Gastric-inhibitory peptide (GIP)	Arrival of chyme containing large quantities of glucose	Duodenum	Pancreas	Stimulates release of insulin by pancreatic islets

✓ How is the small intestine adapted for the absorption of nutrients?

✓ How would a meal that is high in fat affect the level of cholecystokinin (CCK) in the blood?

THE PANCREAS

The pancreas, shown in Figure 17-13a●, lies behind the stomach, extending laterally from the duodenum toward the spleen. It is an elongate, pinkish gray organ with a length of approximately 15 cm (6 in.) and a weight of around 80 g (3 oz). The surface of the pancreas has a knobbly texture, and its tissue is soft and easily torn.

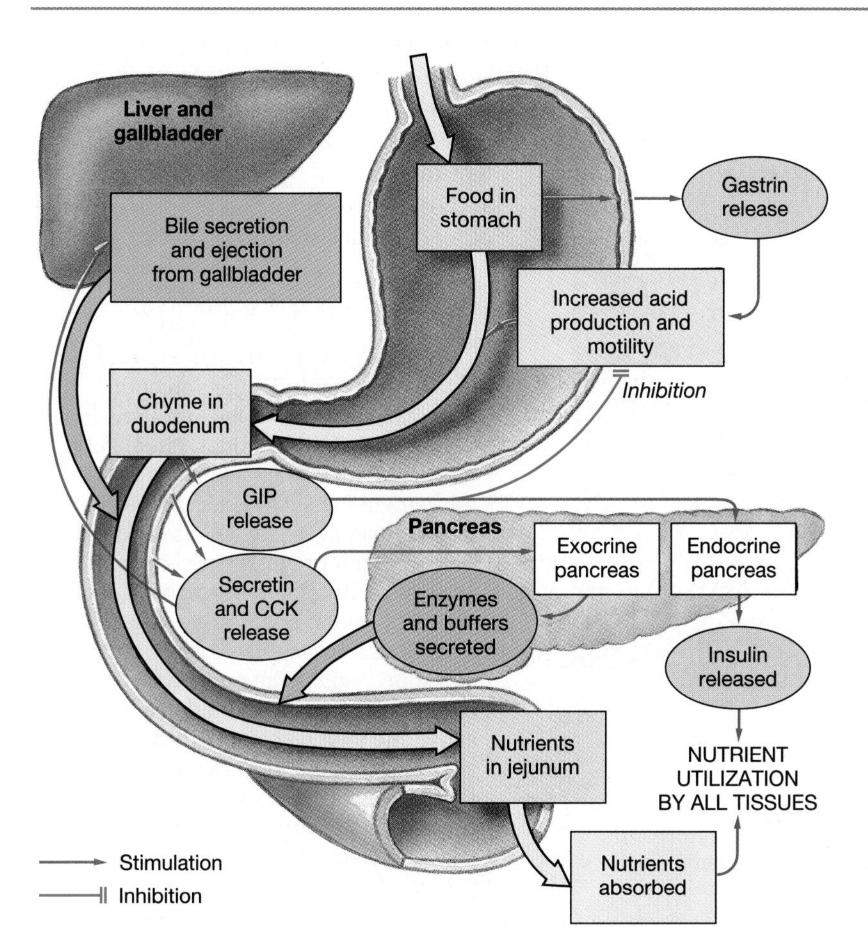

●**Figure 17-12**
The Activities of Major Digestive Tract Hormones
This diagram follows the primary actions of gastrin, GIP, secretin, and CCK.

Histological Organization

The pancreas is primarily an exocrine organ, producing digestive enzymes and buffers. Pancreatic islets, which secrete insulin and glucagon, account for only around 1 percent of the cellular population of the pancreas. The exocrine cells form the **pancreatic acini** (AS-i-nī) (Figure 17-13b●). The ducts draining the acini carry secreted enzymes and buffers into tributaries of the **pancreatic duct,** which carries these secretions to the duodenum. The pancreatic duct penetrates the duodenal wall with the *common bile duct* from the liver and gallbladder.

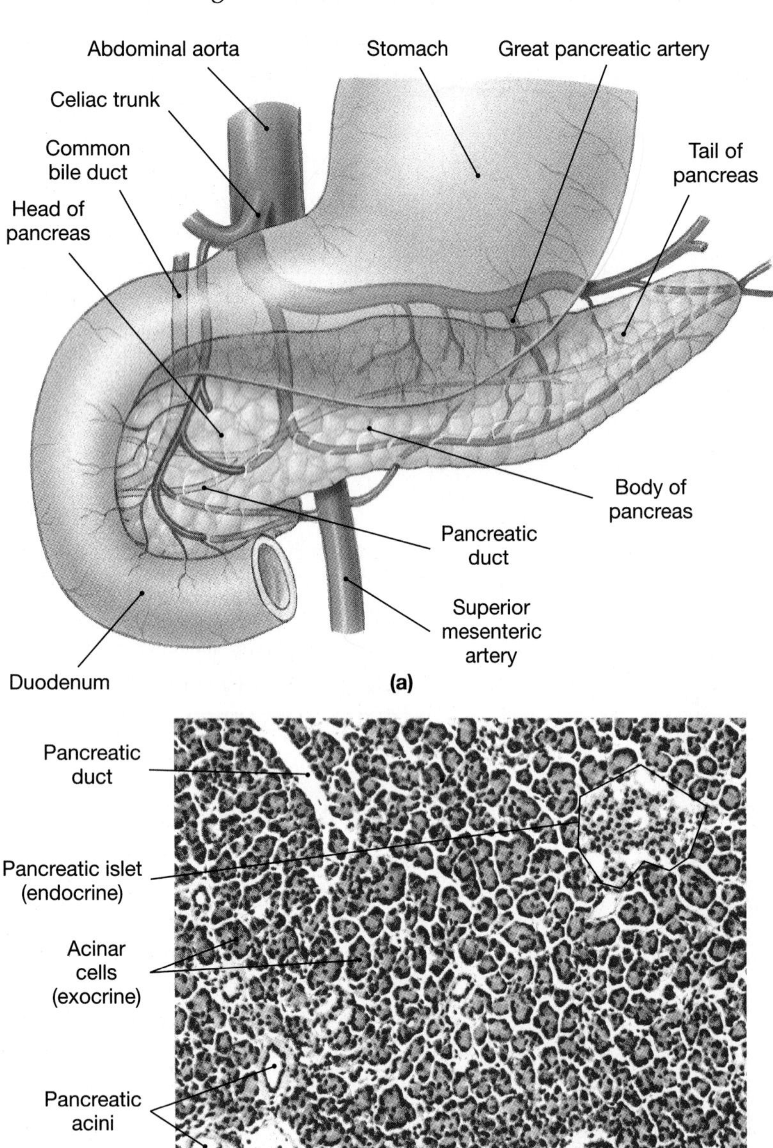

●**Figure 17-13**
The Pancreas

(a) Gross anatomy of the pancreas. The head of the pancreas is tucked into a curve of the duodenum that begins at the pylorus of the stomach. **(b)** A low-power light micrograph of the pancreas. This image shows the pancreatic duct and exocrine and endocrine tissues. (LM × 168)

Pancreatic enzymes are broadly classified according to their intended targets. **Lipases** (LĪ-pā-zez) attack lipids, **carbohydrases** (kar-bō-HĪ-drā-zez) digest sugars and starches, and **proteases** (prō-tē-ā-zez) (proteolytic enzymes) break proteins apart.

Control Of Pancreatic Secretion

The pancreatic acini produce a watery **pancreatic juice** in response to hormonal instructions from the duodenum. When acid chyme arrives in the small intestine, secretin is released, triggering the pancreatic production of an alkaline fluid with a pH of 7.5 to 8.8. Among its other components, this secretion contains buffers, primarily *sodium bicarbonate,* that help bring the pH of the chyme under control. A different intestinal hormone, cholecystokinin, controls the production and secretion of pancreatic enzymes. The specific enzymes involved are **pancreatic amylase,** similar to salivary amylase, **pancreatic lipase, nucleases** that break down nucleic acids, and several proteolytic enzymes.

Proteolytic enzymes account for around 70 percent of the total pancreatic enzyme production. The most abundant proteases are **trypsin** (TRIP-sin), **chymotrypsin** (kī-mō-TRIP-sin), and **carboxypeptidase** (kar-bok-sē-PEP-ti-dās). Together they shatter complex proteins into a mixture of short peptide chains and amino acids. The enzymes are quite powerful, and the pancreatic cells protect themselves by secreting them as inactive *proenzymes* that are activated by other enzymes within the intestinal tract.

PANCREATITIS

Pancreatitis (pan-krē-a-TĪ-tis) is an inflammation of the pancreas. Blockage of the excretory ducts, bacterial or viral infections, circulatory blockage, and drug reactions, especially those involving alcohol, are among the factors that may produce this condition. These stimuli provoke a crisis by injuring exocrine cells in at least a portion of the organ. Lysosomes within the damaged cells then activate the enzymes, and autodigestion begins. The proteolytic enzymes digest the surrounding, undamaged cells, activating their enzymes and starting a chain reaction. In most cases, only a portion of the pancreas will be affected, and the condition subsides in a few days. In 10–15 percent of pancreatitis cases, the process does not subside, and the enzymes may ultimately destroy the pancreas.

THE LIVER

The liver is the largest visceral organ, weighing about 1.5 kg (3.3 lb) and accounting for roughly 2.5 percent of the total body weight. This large, firm, reddish brown organ provides essential metabolic and synthetic services that fall into three general categories: *metabolic regulation, hematological regulation,* and *bile production.* These general functions will be detailed after a discussion of the anatomy of the liver.

Anatomy of the Liver

Figure 17-14● indicates the position and anatomical landmarks of the liver. The overall shape of the liver conforms to its surroundings, but it is divided into four lobes, two large (**left** and **right**) and two small (**caudate** and **quadrate**). A tough connective tissue fold, the *falciform ligament,* marks the division between the left and right lobes. Its thickened posterior margin is the *round ligament,* a fibrous band remnant of the fetal umbilical vein.

Lodged within a recess under the right lobe of the liver is the *gallbladder,* a muscular sac that stores and concentrates bile before it is excreted into the small intestine. The gallbladder and associated structures will be described in a later section.

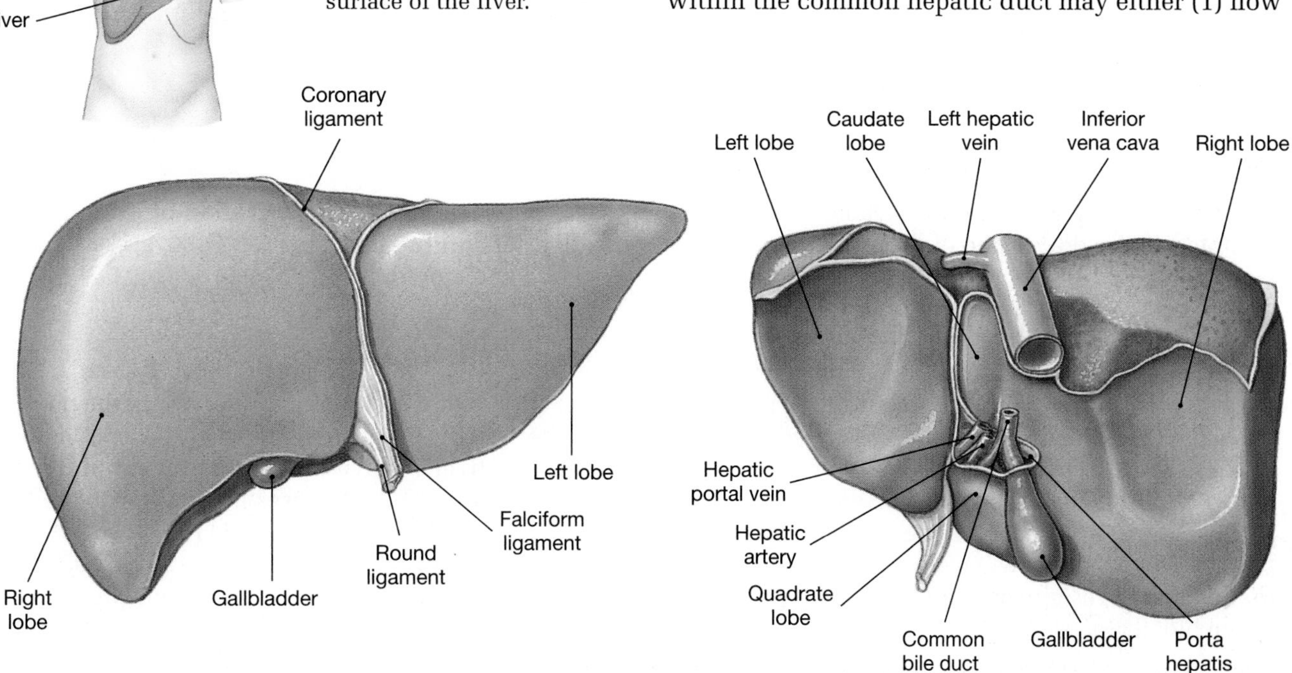

●**Figure 17-14 Anatomy of the Liver**

(a) Anatomical landmarks on the anterior surface of the liver. **(b)** The posterior surface of the liver.

Histological Organization

The basic functional unit of the liver is the **liver lobule.** There are approximately 100,000 lobules in the liver. Because their orientation usually varies, a single histological section seldom reveals all of their details. Important features are diagrammed in Figure 17-15●.

Liver cells, called *hepatocytes* (he-PAT-ō-sīts), within a lobule are arranged into a series of irregular plates like the spokes of a wheel. *Sinusoids,* specialized and highly permeable capillaries, form passageways between the adjacent plates that empty into the *central vein.* In addition to typical endothelial cells, the sinusoidal lining includes a large number of phagocytic *Kupffer* (KOOP-fer) *cells.* Part of the monocyte-macrophage system, these cells engulf pathogens, cell debris, and damaged blood cells.

Blood enters the sinusoids from branches of the hepatic portal vein and hepatic artery, and as blood flows past, the liver cells absorb and secrete materials into the bloodstream. Blood then leaves the sinusoids and enters the **central vein** of the lobule. The central veins of all of the lobules ultimately merge to form the *hepatic veins* that empty into the inferior vena cava. Liver diseases, such as the various forms of *hepatitis,* and conditions such as alcoholism can lead to degenerative changes in the liver tissue and constriction of the circulatory supply. AM *Hepatitis and Cirrhosis*

Bile is a secretory product released into a network of narrow channels called **bile canaliculi** between adjacent liver cells. These canaliculi carry bile away from the central vein, toward a network of ever-larger bile ducts within the liver until it eventually leaves the liver within the **common hepatic duct.** The bile within the common hepatic duct may either (1) flow

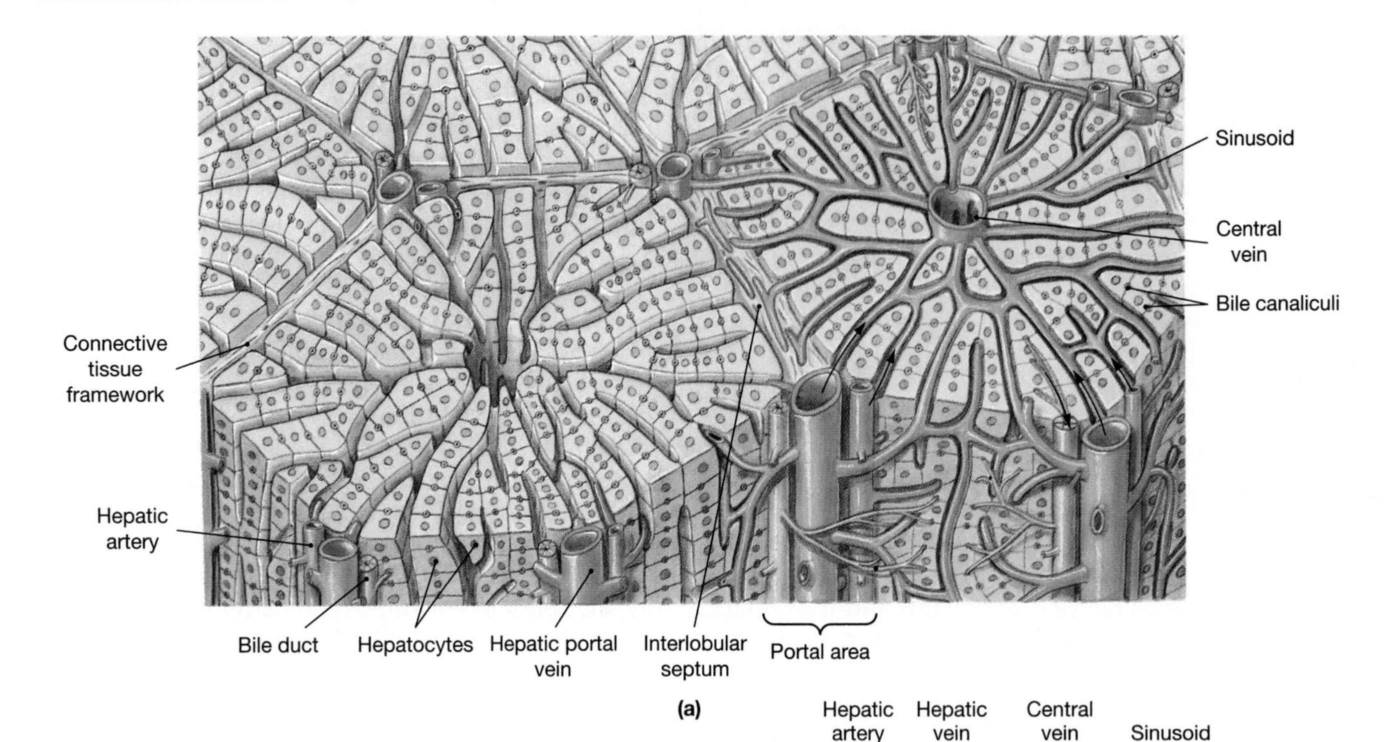

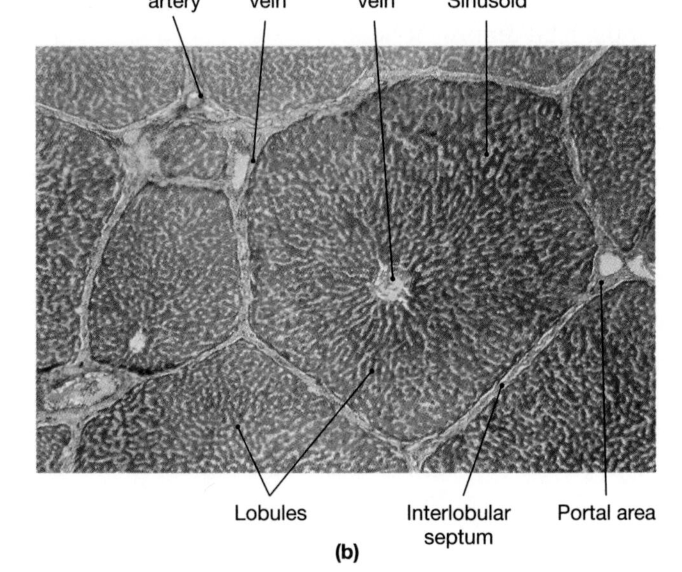

Figure 17-15
Liver Histology
(a) Diagrammatic view of lobular organization. **(b)** Light micrograph showing a typical liver lobule. (LM × 38)

into the **common bile duct** that empties into the duodenum or (2) enter the **cystic duct** that leads to the gallbladder. Liver cells produce roughly 1 liter of bile each day, but a sphincter at the intestinal end of the common bile duct stays opens only at mealtimes. When bile cannot flow along the common bile duct, however, it can enter the cystic duct for storage within the expandable gallbladder.

Liver Functions

Metabolic Regulation

The liver plays a vital role in metabolic regulation. As noted in the discussion of the hepatic portal system ∞ *[p. 372]*, all blood leaving the absorptive areas of the digestive tract flows through the liver before reaching the general circulation. This routing enables the liver cells (1) to extract absorbed nutrients or toxins from the blood before it reaches the rest of the body and (2) to monitor and adjust the circulating levels of organic nutrients. Excesses are removed and stored, and deficiencies are corrected by mobilizing stored reserves or synthesizing the necessary compounds. For example, when blood glucose levels rise, the liver removes glucose and synthesizes glycogen. When blood glucose levels fall, the liver breaks down glycogen and releases glucose into the circulation. Circulating toxins and metabolic waste products are also removed for later inactivation or excretion. Finally, fat-soluble vitamins (A, D, K, and E) are absorbed and stored.

Hematological Regulation

The liver is the largest blood reservoir in the body. In addition to the blood arriving over the hepatic portal vein, the liver receives about 25 percent of the cardiac output. As blood passes by, phagocytic cells in the liver remove aged or damaged red blood cells, debris, and pathogens from the circulation. Equally important, liver cells synthesize the plasma proteins that determine the osmotic concentration of the blood, transport nutrients, and establish the clotting and complement systems.

TABLE 17-2 Major Functions of the Liver

DIGESTIVE AND METABOLIC FUNCTIONS
Synthesis and secretion of bile
Storage of glycogen and lipid reserves
Maintenance of normal blood glucose, amino acid, and fatty acid concentrations
Synthesis and interconversion of nutrient types (e.g., transamination of amino acids or conversion of carbohydrates to lipids)
Synthesis and release of cholesterol bound to transport proteins
Inactivation of toxins
Storage of iron reserves
Storage of fat-soluble vitamins
OTHER MAJOR FUNCTIONS
Synthesis of plasma proteins
Synthesis of clotting factors
Synthesis of the inactive hormone angiotensinogen
Phagocytosis of damaged red blood cells (by Kupffer cells)
Blood storage (major contributor to venous reserve)
Absorption and breakdown of circulating hormones (insulin, epinephrine) and immunoglobulins
Absorption and inactivation of lipid-soluble drugs

Synthesis and Secretion of Bile

Bile is synthesized in the liver and excreted into the lumen of the duodenum. Bile consists mostly of water, ions, *bilirubin* (a pigment derived from hemoglobin), cholesterol, and an assortment of lipids collectively known as **bile salts.** The water and ions help to dilute and buffer acids in chyme as it enters the small intestine. Bile salts are synthesized from cholesterol in the liver and are required for the normal digestion and absorption of fats.

Other Liver Functions

To date, over 200 different functions have been assigned to the liver. (Table 17-2 contains a partial listing.) Therefore, any condition that severely damages the liver represents a serious threat to life. The liver has the ability to partially regenerate after injury, but often liver function does not fully recover. The extensive blood supply complicates the treatment of severe injuries to the liver, because bleeding can be difficult to control and the texture of the liver makes normal suturing ineffective. Thus liver transplants are becoming increasingly common since the development of immunosuppressive drugs, and there is now an 80 percent survival rate. Clinical trials are now under way testing an artificial liver known as *ELAD* (extracorporeal liver assist device) that may prove suitable for the long-term support of persons with chronic liver disease. AM *Liver Disease*

The Gallbladder

The gallbladder (Figure 17-16●) is a muscular organ shaped like a pear. It has two major functions, *bile storage* and *bile modification.* When filled to capacity, the gallbladder contains 40–70 ml of bile. As bile remains in the gallbladder its composition gradually changes. Water is absorbed, and the bile salts and

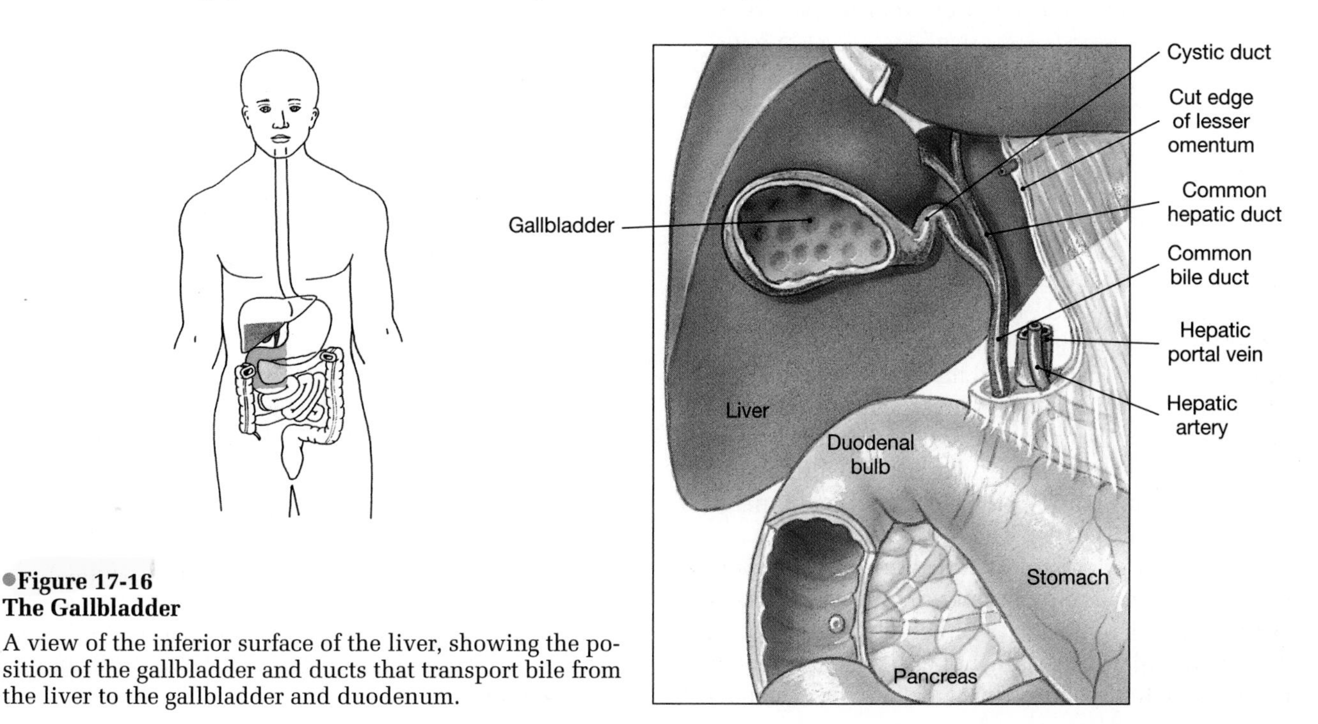

●**Figure 17-16**
The Gallbladder

A view of the inferior surface of the liver, showing the position of the gallbladder and ducts that transport bile from the liver to the gallbladder and duodenum.

other components of bile become increasingly concentrated. If they become too concentrated, the bile salts may precipitate, forming *gallstones* that can cause a variety of clinical problems. AM *Problems with Bile Storage and Secretion*

The Physiological Role of Bile

Most dietary lipids are not water-soluble. Mechanical processing along the digestive tract creates large drops containing various lipids that are much too massive to be attacked by digestive enzymes. Bile salts break the drops apart in a process called **emulsification** (ē-mul-si-fi-KĀ-shun), which creates tiny *emulsion droplets* with a superficial coating of bile salts. The formation of tiny droplets increases the surface area available for enzymatic attack. In addition, the layer of bile salts facilitates interaction between the lipids and lipid-digesting enzymes provided by the pancreas. (We will return to the mechanism of lipid digestion in a later section.)

The arrival of the intestinal hormone cholecystokinin, or CCK, in the circulating blood stimulates bile excretion. Cholecystokinin relaxes the biliary sphincter, and contractions of the walls of the gallbladder then push bile into the small intestine. CCK is released whenever chyme enters the intestine, but the amount secreted increases if the chyme contains large amounts of fat.

- ✓ A narrowing of the ileocecal valve would interfere with the movement of materials between what two organs?
- ✓ The digestion of which nutrient would be most impaired by damage to the exocrine pancreas?
- ✓ How would a decrease in the amount of bile salts in bile affect the digestion and absorption of fat?

THE LARGE INTESTINE

The horseshoe-shaped large intestine begins at the end of the ileum and ends at the anus. The large intestine lies below the stomach and liver and almost completely frames the small intestine (Figure 17-17●). The principal functions of the large intestine include (1) the reabsorption of water and compaction of feces, (2) the absorption of important vitamins liberated by bacterial action, and (3) the storing of fecal material prior to defecation.

The large intestine, often called the *large bowel,* has an average length of approximately 1.5 meters (5 ft) and a width of 7.5 cm (3 in.). It can be divided into three major regions: (1) the pouchlike *cecum,* the first portion of the large intestine; (2) the *colon,* the largest portion of the large intestine; and (3) the *rectum,* the last 15 cm (6 in.) of the large intestine and the end of the digestive tract.

The Cecum

Material arriving from the ileum first enters an expanded chamber, called a **cecum** (SĒ-kum). A muscular sphincter, the **ileocecal** (il-ē-ō-SĒ-kal) **valve,** guards the connection between the ileum and the cecum. The cecum usually has the shape of a rounded sac, and the slender **vermiform appendix** (*vermis,* a worm) attaches to the cecum along its posteromedial surface. The average appendix is almost 9 cm (3.5 in.) long, and its walls are dominated by lymphoid tissue. It is not firmly attached to the surrounding mesenteries, and it often wriggles and twists as its muscular walls contract. Inflammation of the appendix produces the symptoms of *appendicitis.* AM *Infected Lymphoid Nodules*

The Colon

The most striking external feature of the colon (Figure 17-17●) is the pouches, or **haustrae,** that permit considerable distension and elongation. Longitudinal bands of muscle, the **taenia coli** (TĒ-nē-a KŌ-lī), are visible on the outer surface of the colon just beneath the serosa. Muscle tone within these bands produces the haustrae.

The **ascending colon** begins at the ileocecal valve. It ascends along the right side of the peritoneal cavity until it reaches the inferior margin of the liver. It then turns horizontally, becoming the **transverse colon.** The transverse colon continues toward the left side, passing below the stomach and following the curve of the body wall. Near the spleen, it turns inferiorly to form the **descending colon.** The descending colon continues along the left side until it curves and recurves as the **sigmoid colon** (SIG-moyd; *sigmoides,* the Greek letter *S*). The sigmoid colon empties into the rectum.

COLON CANCER

Colon cancers are relatively common. There are approximately 149,000 cases diagnosed in the United States each year, and in 1995 there were an estimated 55,000 deaths from colon and rectal cancers. The mortality rate for these cancers remains high, and the best defense appears to be early detection and prompt treatment. The standard screening test involves checking the feces for blood. This is a simple procedure that can easily be performed on a stool (fecal) sample in the course of a routine physical. AM *Colon Inspection and Cancer*

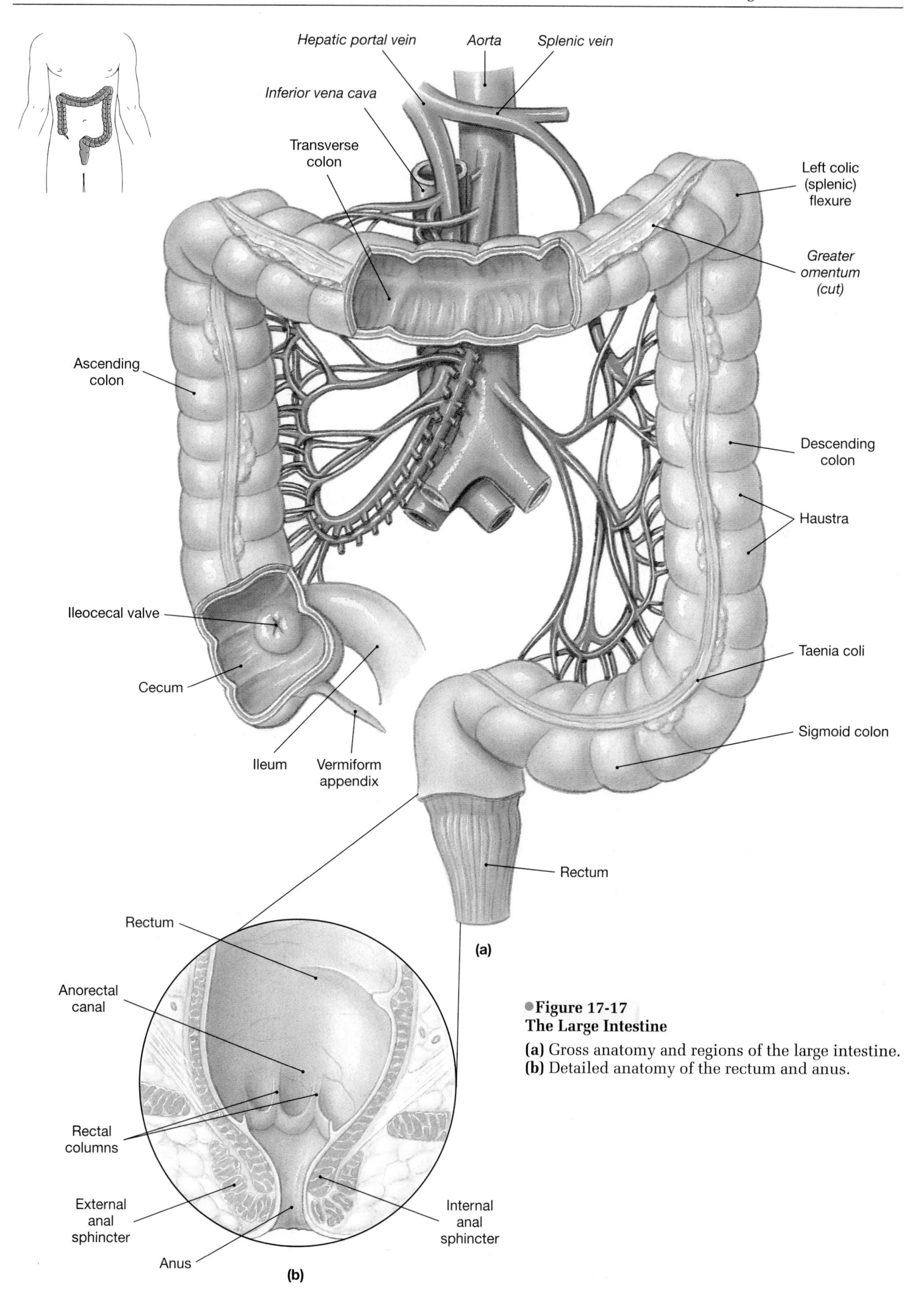

Figure 17-17
The Large Intestine

(a) Gross anatomy and regions of the large intestine. **(b)** Detailed anatomy of the rectum and anus.

Physiology of the Large Intestine

The major functions of the large intestine include absorption and preparation of the fecal material for elimination.

Absorption in the Large Intestine

The reabsorption of water is an important function of the large intestine. Although roughly 1500 ml of watery material arrives in the colon each day, some 1300 ml of water are recovered from it and only about 200 ml of feces are ejected. The remarkable efficiency of digestion can best be appreciated by considering the average composition of fecal wastes: 75 percent water, 5 percent bacteria, and the rest a mixture of indigestible materials, small quantities of inorganic matter, and the remains of epithelial cells.

In addition to reabsorbing water, the large intestine absorbs a variety of other substances from the chyme.

1. ***Vitamins.*** Bacteria within the colon generate three vitamins that supplement the dietary supply:
   - *Vitamin K,* a fat-soluble vitamin needed by the liver to synthesize four clotting factors.
   - *Biotin,* a water-soluble vitamin important in glucose metabolism.
   - *Vitamin B_5,* a water-soluble vitamin required in the manufacture of steroid hormones and some neurotransmitters.

   Deficiencies of biotin or vitamin B_5 are extremely rare after infancy because the intestinal bacteria produce enough to make up for any shortage in the diet. Vitamin K deficiencies, which interfere with blood clotting, may result from inadequate dietary fats, metabolic problems, or chronic diarrhea.
2. ***Bilirubin products.*** Chapter 12 discussed the breakdown of heme and its release as bilirubin in the bile. [p. 313] Inside the large intestine, bacteria convert the bilirubin into other products, some of which are absorbed and excreted in the urine. Others, on exposure to oxygen, are further modified into the pigments that give feces a brown color.
3. ***Bile salts.*** Most of the bile salts remaining in the material reaching the cecum will be reabsorbed and transported to the liver for secretion at a later date.
4. ***Toxins.*** Bacterial action breaks down peptides remaining in the feces into various compounds. Some of these will be reabsorbed and processed by the liver into relatively nontoxic compounds and eventually will be excreted at the kidneys. Other bacterial activities are responsible for the odor of feces or result in the generation of hydrogen sulfide (H_2S), a gas that produces a "rotten egg" odor.

Indigestible carbohydrates are not altered by intestinal enzymes and arrive in the colon intact. These molecules provide a nutrient source for colonic bacteria, whose metabolic activities are responsible for the small quantities of intestinal gas, or *flatus.* Beans often trigger gas because they contain a high concentration of indigestible polysaccharides.

Movements of the Large Intestine

The gastroileal reflex moves material into the cecum at mealtimes. Movement from the cecum to the transverse colon occurs very slowly, allowing hours for the reabsorption of water. Movement from the transverse colon through the rest of the large intestine results from powerful peristaltic contractions, called *mass movements,* that occur a few times each day. The normal stimulus is distension of the stomach and duodenum, and the commands are relayed over the intestinal nerve plexuses. The contractions force fecal materials into the rectum and produce the urge to defecate.

DIVERTICULOSIS

In *diverticulosis* (dī-ver-tik-ū-LŌ-sis), pockets (*diverticula*) form in the mucosa, usually in the sigmoid colon. These get forced outward, probably by the pressures generated during defecation. If the pockets push through weak points in the muscularis externa, they form semi-isolated chambers that are subject to recurrent infection and inflammation. The infections cause pain and occasional bleeding, and the condition is known as *diverticulitis* (dī-ver-tik-ū-LĪ-tis). Inflammation of other portions of the colon is called *colitis* (ko-LĪ-tis). AM *Inflammatory Bowel Disease*

The Rectum

The **rectum** (REK-tum) forms the end of the digestive tract (see Figure 17-17b). The last portion of the rectum, the **anorectal** (ā-nō-REK-tal) **canal,** contains small longitudinal folds joined by transverse folds that mark the boundary between the columnar epithelium of the rectum and a stratified squamous epithelium similar to that found in the oral cavity. Very close to the **anus,** the opening of the anorectal canal, the epidermis becomes keratinized and identical to that on the surface of the skin.

The circular muscle layer of the muscularis externa in this region forms the **internal anal sphincter.** The **external anal sphincter** guards the exit of the anorectal canal. This sphincter, which consists of skeletal muscle fibers, is under voluntary control.

Defecation

The rectal chamber is usually empty except when one of those powerful peristaltic contractions forces fecal materials out of the sigmoid colon. Distension of the rectal wall then triggers the *defecation reflex*. The defecation reflex involves two positive feedback loops:

1. Stretch receptors in the rectal walls order a series of peristaltic contractions in the colon and rectum, moving feces toward the anus.
2. The sacral parasympathetic system, also activated by the stretch receptors, stimulates peristalsis via motor commands distributed by the pelvic nerves.

Movement of feces through the anorectal canal requires relaxation of the internal anal sphincter, but when it relaxes the external sphincter automatically clamps shut. Thus the actual release of feces requires conscious effort to open the external sphincter voluntarily. If the commands do not arrive, the peristaltic contractions cease until additional rectal expansion triggers the defecation reflex a second time.

In addition to opening the external sphincter, consciously directed activities such as tensing the abdominal muscles or making expiratory movements while closing the glottis (called the *Valsalva maneuver*) elevate intra-abdominal pressures and help to force fecal materials out of the rectum. Such pressures also force blood into the network of veins in the lamina propria and submucosa of the anorectal canal causing them to stretch. Repeated incidents of straining to force defecation can cause them to be permanently distended, producing *hemorrhoids*.

DIARRHEA AND CONSTIPATION

Diarrhea (dī-a-RĒ-a) exists when an individual has frequent, watery bowel movements. Diarrhea results when the colonic mucosa becomes unable to maintain normal levels of absorption or if the rate of fluid entry into the colon exceeds its maximum reabsorptive capacity. Bacterial, viral, or protozoan infection of the colon or small intestine may cause acute bouts of diarrhea lasting several days. Severe diarrhea can be life-threatening because of cumulative fluid and ion losses. In *cholera* (KOL-e-ra), bacteria bound to the intestinal lining release toxins that stimulate a massive fluid secretion across the intestinal epithelium. Without treatment, the victim may die of acute dehydration in a matter of hours. AM *Diarrhea*

Constipation is infrequent defecation, usually involving dry and hard feces. Constipation occurs when fecal materials are moving through the colon so slowly that excessive water reabsorption occurs. The feces then become extremely compact, difficult to move, and highly abrasive. Inadequate dietary fiber and fluids, coupled with a lack of exercise, are the usual causes. Constipation can usually be treated by oral administration of stool softeners, such as *Colace*, laxatives, or *cathartics* (ka-THAR-tiks) that promote defecation. These compounds promote water movement into the feces, increase fecal mass, or irritate the lining of the colon to stimulate peristalsis. For example, indigestible fiber adds bulk to the feces, retaining moisture and stimulating stretch receptors that promote peristalsis. This is one of the benefits of "high-fiber" cereals. Active movement during exercise also assists in the movement of fecal materials through the colon.

DIGESTION AND ABSORPTION

A typical meal contains a mixture of carbohydrates, proteins, lipids, water, electrolytes, and vitamins. The digestive system handles each of these components differently. Large organic molecules must be broken down through digestion before absorption can occur. Water, electrolytes, and vitamins can be absorbed without preliminary processing, but special transport mechanisms are often involved.

The Processing and Absorption of Nutrients

Food contains large organic molecules, many of them insoluble. The digestive system first breaks down the physical structure of the ingested material and then disassembles the component molecules into smaller fragments. This disassembly produces small organic molecules that can be absorbed. Once absorbed, these will be used by the body to generate ATP and to synthesize complex carbohydrates, proteins, and lipids. This section will focus on the mechanics of digestion and absorption; the fate of the compounds in the body will be considered in Chapter 18.

Foods are usually complex chains of simpler molecules. In a typical dietary carbohydrate, the basic molecules are simple sugars. In a protein, the building blocks are amino acids, and in lipids they are usually fatty acids. Digestive enzymes break the bonds between the component molecules in a process called *hydrolysis*. (The hydrolysis of carbohydrates, lipids, and proteins was detailed in Chapter 2.) ∞ *[p. 34]*

Digestive enzymes differ in respect to their specific targets. Carbohydrases break the bonds between sugars, proteinases split the linkages between amino acids, and lipases separate the fatty acids from glycerides. Specific enzymes within each class may be even more selective, breaking bonds involving specific molecular participants. For example, a carbohydrase might ignore all bonds except those connecting two glucose molecules. Table 17-3 reviews the major digestive enzymes and their functions.

TABLE 17-3 Digestive Enzymes and Their Functions

Enzyme	*Source*	*Optimal pH*	*Target*	*Products*
Amylase	Salivary glands, pancreas	6.7–7.5	Bonds between carbohydrates	Disaccharides and trisaccharides
Pepsin	Chief cells of stomach	1.5–2.0	Bonds between amino acids in proteins	Short polypeptides
Trypsin, chymotrypsin, carboxypeptidase	Pancreas	7–8	Bonds between amino acids in proteins	Short peptide chains
Pancreatic lipase	Pancreas	7–8	Triglycerides	Fatty acids and monoglycerides
Nuclease	Pancreas	7–8	Nucleic acids	Nitrogenous bases and simple sugars

Carbohydrate Digestion and Absorption

Carbohydrate digestion begins in the mouth through the action of salivary amylase. Amylase breaks down complex carbohydrates into smaller fragments, producing a mixture primarily composed of disaccharides (two sugars) and trisaccharides (three sugars). Salivary amylase continues to digest the starches and glycogen in the meal for an hour or two before stomach acids render it inactive. In the duodenum, the remaining complex carbohydrates are broken down through the action of pancreatic amylase.

Epithelial Processing and Absorption Before they are absorbed, disaccharides and trisaccharides are fragmented into simple sugars by enzymes found on the surfaces of the intestinal microvilli. The intestinal epithelium then absorbs simple sugars through carrier-mediated transport mechanisms, such as facilitated diffusion. ∞ [p. 55] Simple sugars entering an intestinal cell diffuse through the cytoplasm and across the basement membrane to enter the interstitial fluid. They then enter the intestinal capillaries for delivery to the hepatic portal vein.

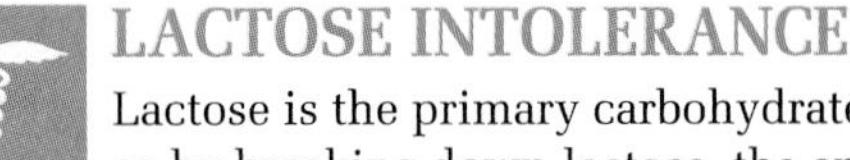

LACTOSE INTOLERANCE

Lactose is the primary carbohydrate in milk, so by breaking down lactose, the enzyme *lactase* provides essential services throughout infancy and early childhood. The intestinal mucosa often stops producing lactase by adolescence, in which case the individual becomes *lactose intolerant*. Individuals who are lactose intolerant can have a variety of unpleasant digestive problems after eating a meal containing milk and other dairy products. AM *Lactose Intolerance*

Lipid Digestion and Absorption

Chapter 2 introduced the structure of *triglycerides,* the most abundant dietary lipids (see Figure 2-15●). ∞ [p. 37] A triglyceride molecule consists of three fatty acids attached to a single molecule of glycerol. Triglycerides and other dietary fats are relatively unaffected by conditions in the stomach and enter the duodenum in the form of large lipid drops. As noted earlier in the chapter, bile salts emulsify these drops into tiny droplets that can be attacked by pancreatic lipase. This enzyme breaks the triglycerides apart, and the lipids released interact with bile salts in the surrounding chyme to form small complexes called **micelles** (mī-SELZ) (Figure 17-18●). When a micelle contacts the intestinal epithelium, the enclosed lipids diffuse across the cell membrane and enter the cytoplasm. The intestinal cells use the arriving lipids to manufacture new triglycerides that are then coated with proteins. This step creates a soluble complex known as a **chylomicron** (kī-lō-MĪ-kron). The chylomicrons are secreted into the interstitial fluids, where they enter the intestinal lacteals through the large gaps between adjacent endothelial cells. From the lacteals they proceed along the lymphatics, through the thoracic duct, and finally enter the circulation at the left subclavian vein.

Protein Digestion and Absorption

Proteins have very complex structures, and several different techniques are used to disassemble them. The mechanical processing in the oral cavity increases the surface area of food exposed to gastric juices following ingestion. Placing the bolus into a strongly acid environment kills most pathogenic microorganisms, breaks down cell walls, and provides the proper environment for the efforts of pepsin, the proteolytic enzyme secreted by chief cells of the stomach. Pepsin does not complete the process, but it does reduce the relatively huge proteins of the chyme into smaller polypeptide fragments.

After the acid bath has ended and the pH has risen in the duodenum, pancreatic enzymes come into play. Working together, each with its own specificities, trypsin, chymotrypsin, elastase, and carboxypeptidase

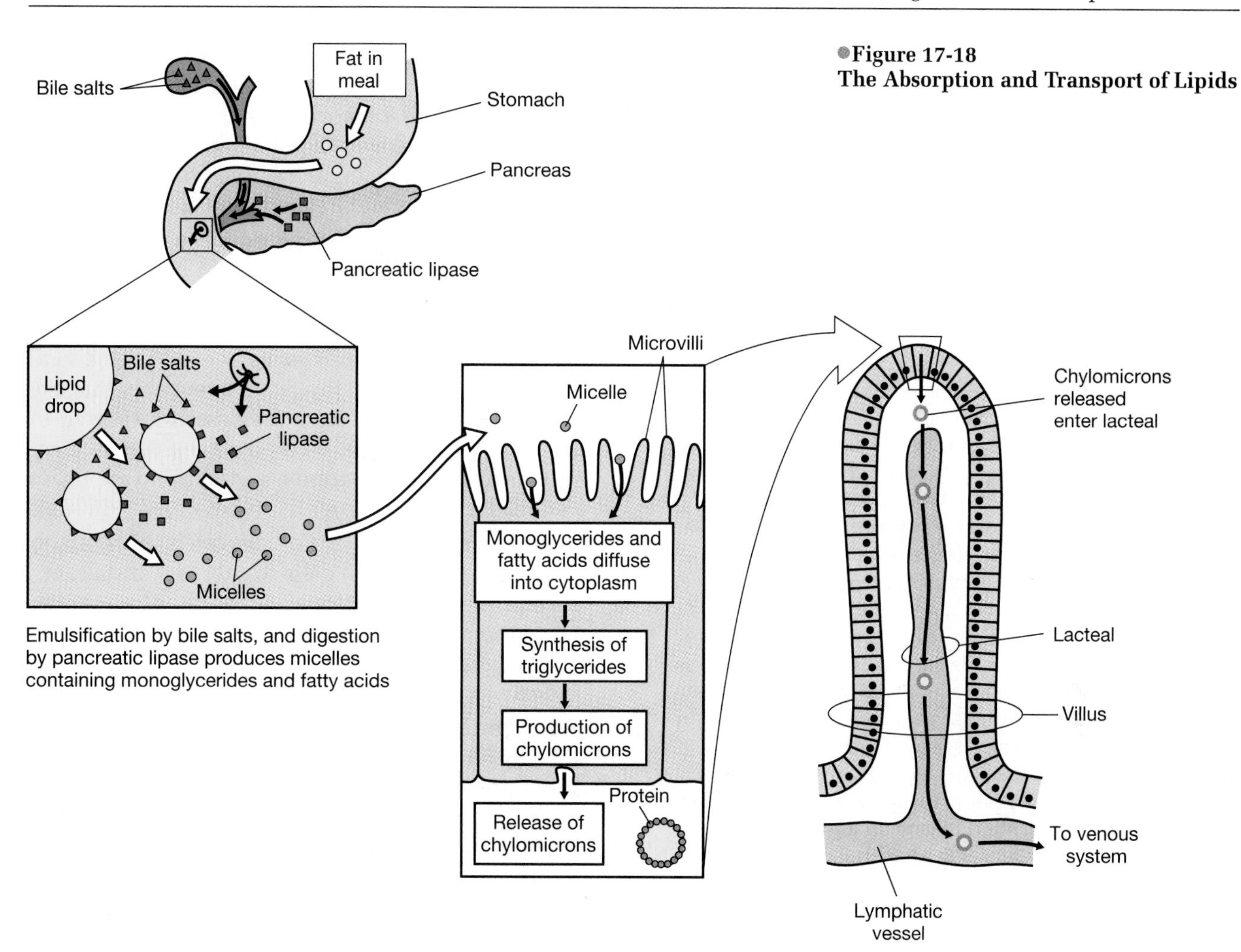

Figure 17-18
The Absorption and Transport of Lipids

complete the disassembly of the fragments into a mixture of short peptide chains and individual amino acids. Enzymes on the surfaces of the microvilli complete the process by breaking the peptide chains into their component amino acids, and the amino acids are absorbed through carrier-mediated transport. Carrier proteins at the inner surface of the cell then dump the absorbed amino acids into the interstitial fluid. Once within the interstitial fluids, most of the amino acids diffuse into intestinal capillaries.

Water and Electrolyte Absorption

Each day, 2–2.5 liters of water enter the digestive tract in the form of food or drink. The salivary, gastric, intestinal, and accessory gland secretions provide another 6–7 liters. Out of that total, only about 150 ml are lost in the fecal wastes. This water conservation occurs passively, following osmotic gradients; water always tends to flow into the solution containing the higher concentration of solutes.

The epithelial cells are continually absorbing dissolved nutrients and ions, and these activities gradually lower the solute concentration of the intestinal contents. As the solute concentration decreases, water moves into the surrounding tissues, "following" the solutes and maintaining osmotic equilibrium. The absorption of sodium and chloride ions is the most important factor promoting water movement. Other ions absorbed in smaller quantities are calcium, potassium, magnesium, iodine, bicarbonate, and iron. Calcium absorption occurs under hormonal control, requiring the presence of parathyroid hormone and calcitriol. Regulatory mechanisms governing the absorption or excretion of the other ions are poorly understood.

Absorption of Vitamins

Vitamins are organic compounds related to lipids and carbohydrates that are required in very small quantities. The nine **water-soluble vitamins** function primarily as participants in enzymatic reactions. All but one, **vitamin B_{12}**, are easily absorbed by the digestive epithelium. Vitamin B_{12} cannot be absorbed by the intestinal mucosa unless it has been bound to *intrinsic factor,* a protein secreted by the parietal cells of the stomach. The bacteria residing in the intestinal tract are an important source for several water-soluble vitamins.

Fat-soluble vitamins enter the duodenum in fat droplets, mixed with dietary lipids. They remain in association with those lipids when micelles form. The fat-soluble vitamins are then absorbed from the micelles with the products of lipid digestion. One fat-soluble vitamin, vitamin K, is also produced by bacterial action and absorbed in the colon. (This vitamin was introduced in Chapter 13 in the discussion of blood clotting.) ∞ [p. 323]

MALABSORPTION SYNDROMES

Difficulties in the absorption of all classes of compounds will result from damage to the accessory glands or the intestinal mucosa. If the accessory organs are functioning normally but their secretions cannot reach the duodenum, the condition is called *biliary obstruction* (bile duct blockage) or *pancreatic obstruction* (pancreatic duct blockage). Alternatively, the ducts may remain open but the glandular cells are damaged and unable to continue normal secretory activities. One example, *pancreatitis,* was noted earlier in the chapter.

Even with the normal enzymes present in the lumen, absorption will not occur if the mucosa cannot function properly. A genetic inability to manufacture specific enzymes will result in discrete patterns of malabsorption—*lactose intolerance* is a good example. Mucosal damage due to ischemia (an interruption of the blood supply), radiation exposure, toxic compounds, or infection will affect absorption in general and will deplete nutrient and fluid reserves as a result.

AGING AND THE DIGESTIVE SYSTEM

Essentially normal digestion and absorption occur in elderly individuals. But there are many changes in the digestive system that parallel age-related changes already described for other systems.

- ***The rate of epithelial stem cell division declines.*** The digestive epithelium becomes more susceptible to damage by abrasion, acids, or enzymes. Peptic ulcers therefore become more likely. In the mouth, esophagus, and anus the stratified epithelium becomes thinner and more fragile.
- ***Smooth muscle tone decreases.*** General motility decreases, and peristaltic contractions are weaker. This change slows the rate of chyme movement and promotes constipation. Sagging of the walls of haustrae in the colon can produce symptoms of *diverticulitis.* Straining to eliminate compacted fecal materials can stress the less-resilient walls of blood vessels, producing hemorrhoids. Problems are not restricted to the lower digestive tract. For example, weakening of muscular sphincters can lead to esophageal reflux and frequent bouts of "heartburn."
- ***The effects of cumulative damage become apparent.*** A familiar example would be the gradual loss of teeth due to *dental caries* ("cavities") or *gingivitis* (inflammation of the gums). Cumulative damage can involve internal organs as well. Toxins such as alcohol and other injurious chemicals that are absorbed by the digestive tract are transported to the liver for processing. The liver cells are not immune to these compounds, and chronic exposure can lead to *cirrhosis* or other types of liver disease.
- ***Cancer rates increase.*** Cancers are most common in organs where stem cells divide to maintain epithelial cell populations. Rates of colon cancer and stomach cancer rise in the elderly; oral and pharyngeal cancers are particularly common in elderly smokers.
- ***Changes in other systems have direct or indirect effects on the digestive system.*** For example, the reduction in bone mass and calcium content in the skeleton is associated with erosion of the tooth sockets and eventual tooth loss. The decline in olfactory and gustatory sensitivity with age can lead to dietary changes that affect the entire body.

✓ What component of a meal would increase the number of chylomicrons in the lacteals?

✓ The absorption of which vitamin would be impaired by removal of the stomach?

✓ Why is it that diarrhea is potentially life-threatening but constipation is not?

INTEGRATION WITH OTHER SYSTEMS

The digestive system is functionally linked to all other systems, and it has extensive anatomical connections to the nervous, cardiovascular, endocrine, and lymphatic systems. Figure 17-19● summarizes the physiological relationships between the digestive system and other organ systems.

INTEGUMENTARY SYSTEM

Provides vitamin D_3 needed for the absorption of calcium and phosphorus

Provides lipids for storage by adipocytes in subcutaneous layer

DIGESTIVE SYSTEM

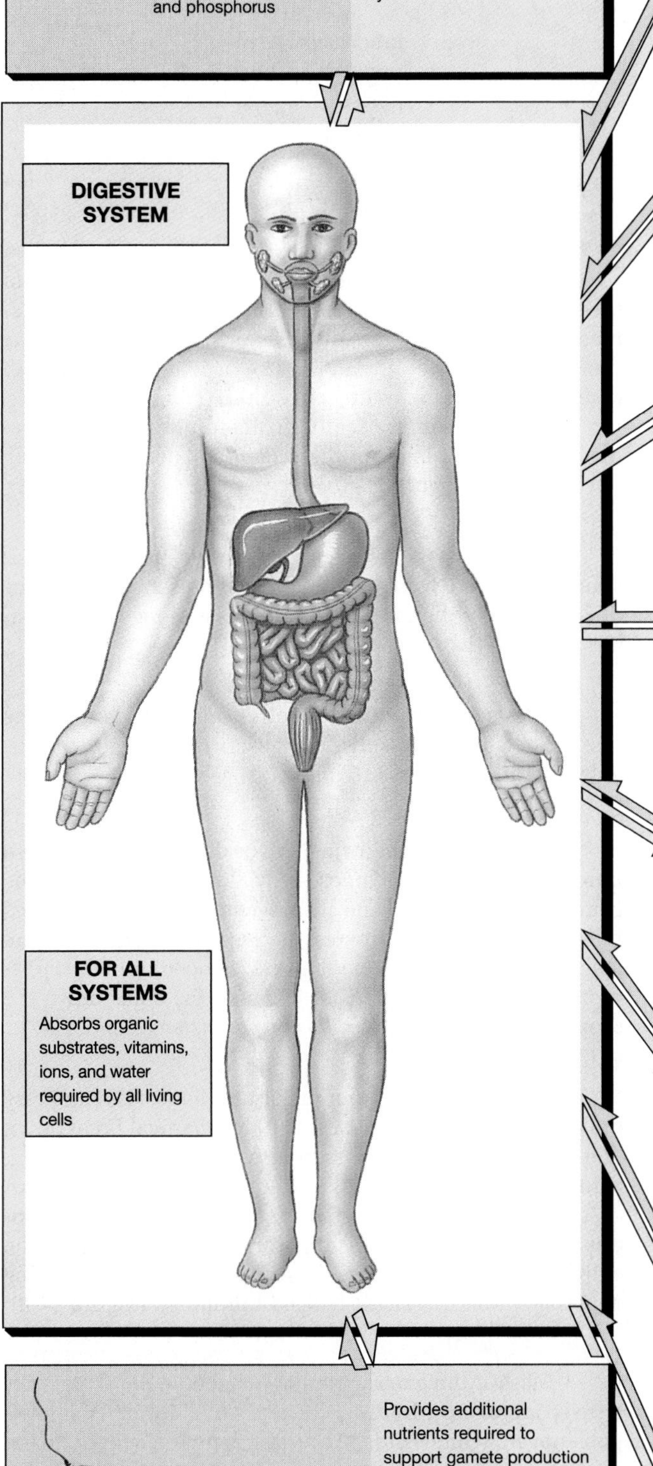

FOR ALL SYSTEMS

Absorbs organic substrates, vitamins, ions, and water required by all living cells

SKELETAL SYSTEM

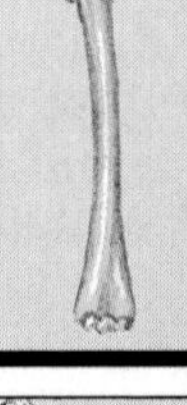

Skull, ribs, vertebrae, and pelvic girdle support and protect parts of digestive tract; teeth important in mechanical processing of food

Absorbs calcium and phosphate ions for incorporation into bone matrix; provides lipids for storage in yellow marrow

MUSCULAR SYSTEM

Protects and supports digestive organs in abdominal cavity; controls entrances and exits to digestive tract; liver metabolizes lactic acid from active muscles

Liver regulates blood glucose and fatty acid levels, removes lactic acid from circulation

NERVOUS SYSTEM

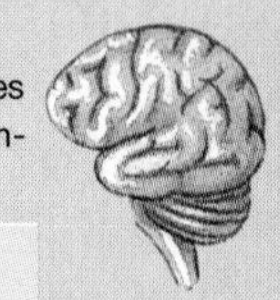

ANS regulates movement and secretion; reflexes coordinate passage of materials along tract; control over skeletal muscles regulates ingestion and defecation; hypothalamic centers control hunger, satiation, and feeding behaviors

Provides substrates essential for neurotransmitter synthesis

ENDOCRINE SYSTEM

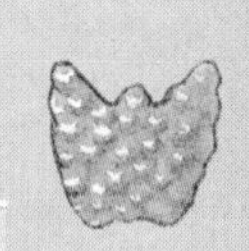

Epinephrine and norepinephrine stimulate constriction of sphincters and depress digestive activity; hormones coordinate activity along tract

Provides nutrients and substrates to endocrine cells; endocrine cells of pancreas secrete insulin and glucagon; liver produces angiotensinogen

CARDIOVASCULAR SYSTEM

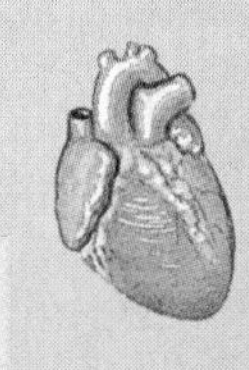

Distributes hormones of the digestive tract; carries nutrients, water, and ions from sites of absorption; delivers nutrients and toxins to liver

Absorbs fluid to maintain normal blood volume; absorbs vitamin K; liver excretes heme (as bilirubin), synthesizes coagulation proteins

LYMPHATIC SYSTEM

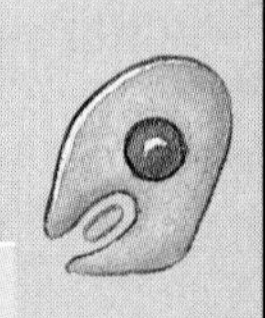

Tonsils and other lymphoid nodules along digestive tract defend against infection and provide defense against toxins absorbed from the tract; lymphatics carry absorbed lipids to venous system

Secretions of digestive tract (acids and enzymes) provide nonspecific defense against pathogens

RESPIRATORY SYSTEM

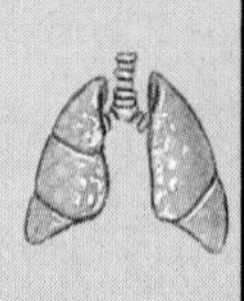

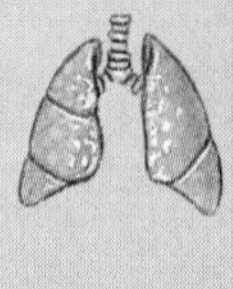

Increased thoracic and abdominal pressure through contraction of respiratory muscles can assist in defecation

Pressure of digestive organs against the diaphragm can assist exhalation and limit inhalation

URINARY SYSTEM

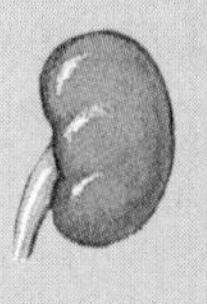

Excretes toxins absorbed by the digestive epithelium; excretes some bilirubin produced by liver

Absorbs water needed to excrete waste products at the kidneys; absorbs ions needed to maintain normal body fluid concentrations

REPRODUCTIVE SYSTEM

Provides additional nutrients required to support gamete production and (in pregnant women) embryonic and fetal development

Figure 17-19
Functional Relationships between the Digestive System and Other Systems

Chapter Review

KEY TERMS

bile, p. 451
chylomicrons, p. 456
chyme, p. 440
digestion, p. 433
duodenum, p. 444
gastric glands, p. 442
lacteal, p. 444
mesentery, p. 434
mucosa, p. 433
pancreatic juice, p. 448
peristalsis, p. 435
villus, p. 444

SUMMARY OUTLINE

INTRODUCTION p. 433

1. The digestive system consists of the muscular **digestive tract** and various **accessory organs.**

2. Digestive functions include ingestion, mechanical processing, digestion, secretion, absorption, compaction, and excretion.

AN OVERVIEW OF THE DIGESTIVE TRACT p. 433

1. The digestive tract includes the oral cavity, pharynx, esophagus, stomach, small intestine, large intestine, rectum, and anus. (*Figure 17-1*)

Histological Organization p. 433

2. The epithelium and underlying connective tissue, the *lamina propria,* form the **mucosa** (mucous membrane) of the digestive tract. Proceeding outward, one enters the **submucosa,** the **muscularis externa,** and a layer of loose connective tissue called the *adventitia.* Within the peritoneal cavity, the muscularis externa is covered by a serous membrane called the **serosa.** (*Figure 17-2*)

3. Double sheets of peritoneal membrane called **mesenteries** suspend the digestive tract.

The Movement of Digestive Materials p. 435

4. The neurons that innervate the smooth muscle of the muscularis externa are not under voluntary control.

5. The muscularis externa propels materials through the digestive tract through the contractions of **peristalsis. Segmentation** movements in areas of the small intestine churn digestive materials. (*Figure 17-3*)

THE ORAL CAVITY p. 435

1. The functions of the oral cavity are (1) analysis of potential foods; (2) mechanical processing using the teeth, tongue, and palatal surfaces; (3)lubrication by mixing with mucus and salivary secretions; and (4) digestion by salivary enzymes.

2. The oral cavity, or **buccal cavity,** is lined by oral mucosa. The **hard** and **soft palates** form its roof, and the tongue forms its floor. (*Figure 17-4*)

The Tongue p. 436

3. The primary functions of the tongue include (1) mechanical processing, (2) manipulation to assist in chewing and swallowing, and (3) sensory analysis.

Salivary Glands p. 436

4. The *parotid, sublingual,* and *submandibular glands* discharge their secretions into the oral cavity. Saliva lubricates the mouth, dissolves chemicals, flushes the oral surfaces, and helps control bacteria. Salivation is usually controlled by the ANS. (*Figure 17-5*)

Teeth p. 438

5. **Mastication** (chewing) occurs through the contact of the opposing surfaces of the teeth. The **periodontal ligament** anchors the tooth in an alveolar socket. **Dentin** forms the basic structure of a tooth. The **crown** is coated with **enamel,** and the **root** with **cementum.** (*Figure 17-6a*)

6. The 20 primary teeth, or **deciduous teeth,** are replaced by the 32 teeth of the **secondary dentition** during development. (*Figure 17-6b,c*)

THE PHARYNX p. 439

1. The pharynx serves as a common passageway for solid food, liquids, and air. Pharyngeal muscle contractions during swallowing propel the food mass along the esophagus and into the stomach.

THE ESOPHAGUS p. 439

1. The esophagus carries solids and liquids from the pharynx to the stomach through an opening in the diaphragm, the *esophageal hiatus.*

Swallowing p. 439

2. **Deglutition** (swallowing) can be divided into *oral, pharyngeal,* and *esophageal phases.* Swallowing begins with the compaction of a **bolus** and its movement into the pharynx, followed by the elevation of the larynx, reflection of the epiglottis, and closure of the glottis. After opening of the *upper esophageal sphincter,* peristalsis moves the bolus down the esophagus to the *lower esophageal sphincter.* (*Figure 17-7*)

THE STOMACH p. 440

1. The stomach has four major functions: (1) temporary bulk storage of ingested matter, (2) mechanical breakdown of resistant materials, (3) disruption of chemical bonds using acids and enzymes, and (4) production of intrinsic factor.

2. The four regions of the stomach are the **cardia, fundus, body,** and **pylorus.** The **pyloric sphincter** guards the exit from the stomach. In a relaxed state the stomach lining contains numerous **rugae** (ridges and folds). (*Figure 17-8*)

The Gastric Wall p. 442

3. Within the **gastric glands, parietal cells** secrete **intrinsic factor** and hydrochloric acid. **Chief cells** secrete **pepsinogen,** which acids in the gastric lumen convert to the enzyme **pepsin.**

Regulation of Gastric Activity p. 442

4. Gastric secretion includes: (1) the **cephalic phase,** which prepares the stomach to receive ingested materials; (2) the **gastric phase,** which begins with the arrival of food in the stomach; and (3) the **intestinal phase,** which controls the rate of gastric emptying. (*Figure 17-9*)

THE SMALL INTESTINE p. 444

1. The small intestine includes the **duodenum,** the **jejunum,** and the **ileum.** A sphincter, the ileocecal valve, marks the transition between the small and large intestines. (*Figure 17-10*)

The Intestinal Wall p. 444

2. The intestinal mucosa bears transverse folds called *plicae,* and small projections called intestinal **villi.** These increase the surface area for absorption. Each villus contains a terminal lymphatic called a **lacteal.** (*Figure 17-11*)

3. Some of the smooth muscle cells in the musularis externa of the small intestine contract periodically, without stimulation, to produce brief, localized peristaltic contractions that slowly move materials along the tract. More extensive peristaltic activities along the entire length of the small intestine are coordinated by the *gastroenteric* and the *gastroileal reflexes.*

4. On average, it takes about 5 hours for materials to pass from the duodenum to the end of the ileum. Along the way, absorptive effectiveness is enhanced by segmentation movements that stir and mix the intestinal contents.

Intestinal Secretions p. 446

5. Intestinal glands secrete **intestinal juice,** mucus, and hormones. Intestinal juice moistens the chyme, helps to buffer acids, and dissolves digestive enzymes and the products of digestion.

6. Intestinal hormones include **secretin, cholecystokinin (CCK), gastrin, gastric inhibitory peptide (GIP),** and others. (*Figure 17-12; Table 17-1*)

Digestion in the Small Intestine p. 446

7. Most of the important digestive and absorptive functions occur in the small intestine. Digestive enzymes and buffers are provided by the pancreas, liver, and gallbladder.

THE PANCREAS p. 447

1. The **pancreatic duct** penetrates the wall of the duodenum to deliver pancreatic secretions. (*Figure 17-13a*)

Histological Organization p. 448

2. Within each lobule, ducts branch repeatedly before ending in the **pancreatic acini** (blind pockets). (*Figure 17-13b*)

3. The pancreas has two functions: endocrine (secreting insulin and glucagon into the blood) and exocrine (secreting water, ions, and digestive enzymes into the small intestine). Pancreatic enzymes include **lipases, carbohydrases,** and **proteases.**

Control of Pancreatic Secretion p. 448

4. The pancreatic acini produce a watery **pancreatic juice** in response to hormonal instructions from the duodenum. When acid chyme arrives in the small intestine: (1) secretin triggers the pancreatic production of a fluid containing buffers, primarily sodium bicarbonate, that help bring the pH of the chyme under control; and (2) cholecystokinin is released.

5. CCK stimulates the pancreas to produce and secrete **pancreatic amylase, pancreatic lipase, nucleases,** and several proteolytic enzymes, notably **trypsin, chymotrypsin,** and **carboxypeptidase**

THE LIVER p. 449

1. The liver is the largest and most versatile visceral organ in the body. (*Figure 17-14*)

Anatomy of the Liver p. 449

2. The **liver lobule** is the organ's basic functional unit. Blood is supplied to the lobules by the hepatic artery and hepatic portal vein. Within the lobules, blood flows past **hepatocytes** through *sinusoids* to the **central vein. Bile canaliculi** carry bile away from the central vein and toward bile ducts. (*Figure 17-15*)

3. The bile ducts from each lobule unite to form the **common hepatic duct,** which meets the **cystic duct** to form the **common bile duct,** which empties into the duodenum.

Liver Functions p. 450

4. The liver performs metabolic and hematological regulation and produces **bile.** (*Table 17-2*)

The Gallbladder p. 451

5. The gallbladder stores and concentrates bile for release into the duodenum. During **emulsification,** bile salts break apart large drops of lipids and make them accessible to pancreatic lipases. (*Figure 17-16*)

THE LARGE INTESTINE p. 452

1. The main functions of the large intestine are to (1) reabsorb water and compact the feces, (2) absorb vitamins liberated by bacteria, and (3) store fecal material prior to defecation. (*Figure 17-17a*)

The Cecum p. 452

2. The **cecum** collects and stores material from the ileum and begins the process of compaction. The **vermiform appendix** is attached to the cecum.

The Colon p. 452

3. The colon has a larger diameter and a thinner wall than the small intestine. It bears **haustrae** (pouches) and the **taenia coli** (longitudinal bands of muscle).

Physiology of the Large Intestine p. 454

4. The large intestine reabsorbs water and other substances, such as vitamins, bilirubin products, bile salts, and toxins. Bacteria residing in the large intestine are responsible for intestinal gas, or *flatus.*

5. Distension of the stomach and duodenum stimulates peristalsis, or mass movements, of feces from the colon into the rectum.

The Rectum p. 454

6. The **rectum** terminates in the **anorectal canal** leading to the **anus.** (*Figure 17-17b*)

7. Muscular sphincters control the passage of fecal material to the anus. Distension of the rectal wall triggers the *defecation reflex.* Under normal circumstances the release of feces cannot occur unless the external anal sphincter is voluntarily relaxed.

DIGESTION AND ABSORPTION p. 455

The Processing and Absorption of Nutrients p. 455

1. The digestive system breaks down the physical structure of the ingested material and then disassembles the component molecules into smaller fragments through hydrolysis. (*Table 17-3*)

2. Amylase breaks down complex carbohydrates into disaccharides and trisaccharides. These are broken down into monosaccharides by enzymes at the epithelial surface and are absorbed by the intestinal epithelium through facilitated diffusion and cotransport.

3. Triglycerides are emulsified into large lipid drops. The resulting fatty acids and monoglycerides interact with bile salts to form **micelles,** from which they diffuse across the intestinal epithelium. The intestinal cells absorb fatty acids and monoglycerides and synthesize new triglycerides. These are packaged in chylomicrons that are released into the interstitial fluid and transported to the venous system by lymphatics. (*Figure 17-18*)

4. Protein digestion involves the gastric enzyme pepsin and the various pancreatic proteases. Peptidases liberate amino acids that are absorbed by the intestinal epithelium and released into the interstitial fluids.

Water and Electrolyte Absorption p. 457

5. About 2–2.5 liters of water are ingested each day, and digestive secretions provide 6–7 liters. Nearly all is reabsorbed by osmosis.

6. Various processes are responsible for the movement of cations (such as sodium and calcium) and anions (such as chloride and bicarbonate).

Absorption of Vitamins p. 457

7. The nine **water-soluble vitamins** are important as cofactors in enzymatic reactions. **Fat-soluble vitamins** are enclosed within fat droplets and are absorbed with the products of lipid digestion.

AGING AND THE DIGESTIVE SYSTEM, p. 458

1. Age-related changes include a thinner and more fragile epithelium due to a reduction in epithelial stem cell division and weaker peristaltic contractions as smooth muscle tone decreases.

INTEGRATION WITH OTHER SYSTEMS p. 458

1. The digestive system has extensive anatomical connections to the nervous, cardiovascular, endocrine, and lymphatic systems. (*Figure 17-19*)

CHAPTER QUESTIONS

LEVEL 1 Reviewing Facts and Terms

Match each item in column A with the most closely related item in column B. Use letters for answers in the spaces provided.

Column A	Column B
___ 1. pyloric sphincter	a. serous membrane sheet
___ 2. liver cells	b. moves materials along digestive tract
___ 3. mucosa	c. regulates flow of chyme
___ 4. mesentery	d. increases muscular activity of digestive tract
___ 5. chief cells	e. starch digestion
___ 6. palate	f. inhibits muscular activity of digestive tract
___ 7. parietal cells	g. inner lining of digestive tract
___ 8. parasympathetic stimulation	h. roof of oral cavity
___ 9. sympathetic stimulation	i. pepsinogen
___ 10. peristalsis	j. hydrochloric acid
___ 11. bile salts	k. hepatocytes
___ 12. salivary amylase	l. emulsification of fats

13. The enzymatic breakdown of large molecules into their basic building blocks is called:
(a) absorption (b) secretion
(c) mechanical digestion (d) chemical digestion

14. The activities of the digestive system are regulated by:
(a) hormonal mechanisms
(b) local mechanisms
(c) neural mechanisms
(d) a, b, and c are correct

15. The layer of the peritoneum that lines the inner surfaces of the body wall is the:
(a) visceral peritoneum (b) parietal peritoneum
(c) greater omentum (d) lesser omentum

16. Protein digestion in the stomach results primarily from secretions released by:
(a) hepatocytes (b) parietal cells
(c) chief cells (d) goblet cells

17. The part of the gastrointestinal tract that plays the primary role in the digestion and absorption of nutrients is the:
(a) large intestine (b) small intestine
(c) stomach (d) cecum and colon

18. The duodenal hormone that stimulates the production and secretion of pancreatic enzymes is:
(a) enterokinase (b) gastrin
(c) secretin (d) cholecystokinin

19. The essential metabolic and synthetic service(s) provided by the liver is (are):
(a) metabolic regulation
(b) hematological regulation
(c) bile production
(d) a, b, and c are correct

20. Bile release from the gallbladder into the duodenum occurs only under the stimulation of:
(a) cholecystokinin (b) secretin
(c) gastrin (d) enterokinase

21. The major function(s) of the large intestine is (are):
    (a) reabsorption of water and compaction of chyme into feces
    (b) absorption of vitamins liberated by bacterial action
    (c) storage of fecal material prior to defecation
    (d) a, b, and c are correct
22. The part of the colon that empties into the rectum is the:
    (a) ascending colon (b) descending colon
    (c) transverse colon (d) sigmoid colon
23. What are the primary digestive functions?
24. What is the purpose of the transverse or longitudinal folds in the mucosa of the digestive tract?
25. Name and describe the layers of the digestive tract, proceeding from the innermost to the outermost layer.
26. What are the four primary functions of the oral (buccal) cavity?
27. What specific function does each of the four types of teeth perform in the oral cavity?
28. What three subdivisions of the small intestine are involved in the digestion and absorption of food?
29. What are the primary functions of the pancreas, liver, and gallbladder in the digestive process?
30. What are the three major functions of the large intestine?
31. What five age-related changes occur in the digestive system?

LEVEL 2 Reviewing Concepts

32. If the lingual frenulum is too restrictive, an individual:
    (a) has difficulty tasting food
    (b) cannot swallow properly
    (c) cannot control movements of the tongue
    (d) cannot eat or speak normally
33. The gastric phase of secretion is initiated by:
    (a) distension of the stomach
    (b) an increase in the pH of the gastric contents
    (c) the presence of undigested materials in the stomach
    (d) a, b, and c are correct
34. A drop in pH to 4.0 in the duodenum stimulates secretion of:
    (a) secretin (b) cholecystokinin
    (c) gastrin (d) a, b, and c are correct
35. Differentiate between the action and outcome of peristalsis and segmentation.
36. How does the stomach promote and assist in the digestive process?
37. What changes in gastric function occur during the three phases of gastric secretion?

LEVEL 3 Critical Thinking and Clinical Applications

38. Sometimes patients suffering from gallstones will develop pancreatitis. How could this occur?
39. Barb suffers from Crohn's disease, a regional inflammation of the intestine that is thought to have some genetic basis, although the actual cause is as yet unknown. When the disease flares up she experiences abdominal pain, weight loss, and anemia. What part(s) of the intestine is (are) probably involved, and what is the cause of her symptoms?

18 NUTRITION AND METABOLISM

Human beings are remarkably adaptable animals—they can maintain homeostasis near the equator, where the year-round temperature averages 27°C and near the poles, where the temperature remains below 0°C most of the time. The key to survival under such varied environmental conditions is the preservation of normal body temperature. In part this can be accomplished by shedding or adding layers of clothing, but hormonal and metabolic adjustments are also crucially important. Thus, the metabolic rate rises in cold climates, generating additional heat that warms the body. This chapter examines several aspects of metabolism, including nutrition, energy use, heat production, and the regulation of body temperature under normal and abnormal conditions.

Chapter Outline and Objectives

1. Define metabolism and explain why cells need to synthesize new organic structures.
2. Describe the basic steps in glycolysis, the TCA cycle, and the electron transport system.
3. Describe the pathways involved in lipid metabolism.
4. Discuss protein metabolism and the use of proteins as an energy source.
5. Discuss nucleic acid metabolism.

6. Explain what constitutes a balanced diet and why it is important.
7. Discuss the functions of vitamins, minerals, and other important nutrients.

8. Describe the significance of the caloric value of foods.
9. Define metabolic rate and discuss the factors involved in determining an individual's metabolic rate.
10. Discuss the homeostatic mechanisms that maintain a constant body temperature.

Vocabulary Development

genesis, an origin; *thermogenesis*
glykus, sweet; *glycolysis*
lipos, fat; *lipogenesis*
lysis, breakdown; *glycolysis*
neo-, new; *gluconeogenesis*
therme, heat; *thermogenesis*
vita, life; *vitamin*

Metabolism refers to all the chemical reactions of the body. Living cells are chemical factories, using chemical reactions to break down organic molecules, and to obtain energy, usually in the form of ATP. To carry out these processes, cells in the human body must also obtain water, vitamins, ions, and oxygen. Oxygen is absorbed at the lungs, but all the other substances, usually called *nutrients*, are obtained by absorption at the digestive tract. The cardiovascular system distributes oxygen and nutrients to cells throughout the body.

The energy released inside a cell supports growth, cell division, contraction, secretion, and various other special functions that vary from cell to cell. Each tissue type contains different populations of cells, and the energy and nutrient requirements of any two tissues (such as loose connective tissue and cardiac muscle) are quite different. Furthermore, when cells, tissues, and organs change their patterns or levels of activity, they also change their nutrient requirements. Thus our nutrient requirements vary from moment to moment (resting versus active), hour to hour (asleep versus awake), and year to year (growing child versus adult).

When organic nutrients, such as carbohydrates or lipids, are abundant, energy reserves are built up. Different tissues and organs are specialized to store excess nutrients; the storage of lipids in adipose tissue is one familiar example. These reserves can then be called upon when the diet cannot provide the right quantity or quality of nutrients. The endocrine system, with the assistance of the nervous system, adjusts and coordinates the metabolic activities of the body's tissues and controls the storage and release of nutrient reserves.

The absorption of nutrients from food is called *nutrition*. The mechanisms involved in absorption through the lining of the digestive tract were detailed in Chapter 17. ∞ *[pp. 455–458]* This chapter considers what happens to nutrients after they are inside the body.

CELLULAR METABOLISM

Figure 18-1● reviews the ways cells use organic nutrients absorbed from the extracellular fluid. After the mechanical and chemical digestion of food, simple sugars, amino acids, and lipids are absorbed by the digestive tract and distributed by the bloodstream. After diffusing into the interstitial fluid, these nutrients cross the cell membrane to join other nutrients already in the cytoplasm. All of the cell's metabolic operations rely on the resulting *nutrient pool*.

Catabolism breaks down organic molecules, releasing energy that can be used to synthesize ATP or other high-energy compounds. Catabolism proceeds in a series of steps. In general, preliminary processing occurs in the cytosol, where enzymes break down large organic molecules into smaller fragments. For example, carbohydrates are broken down to short carbon chains, triglycerides are split into fatty acids and glycerol, and proteins are broken down to individual amino acids.

Relatively little ATP is formed during these initial steps. However, the simple molecules produced can be absorbed and processed by mitochondria, and the mitochondrial steps release significant amounts of energy. As mitochondrial enzymes break the covalent bonds that hold these molecules together, they capture roughly 40 percent of the energy released. The captured energy is used to convert ADP to ATP, and the rest escapes as heat that warms the interior of the cell and the surrounding tissues.

Anabolism, the synthesis of new organic molecules, involves the formation of new chemical bonds. The ATP produced by mitochondria provides energy to support anabolism, as well as other cell functions. Those additional functions, such as contraction, ciliary or cell movement, active transport, or cell division, vary from one cell to another. For example, muscle fibers need ATP to provide energy for contraction, and gland cells need ATP to synthesize and transport their secretions.

Cells synthesize new organic components for three basic reasons:

1. ***To perform structural maintenance or repairs.*** All cells must expend energy to perform ongoing maintenance and repairs, because most organic molecules and structures in the cell are temporary rather than permanent. Their removal and replacement are part of the process of *metabolic turnover*.
2. ***To support growth.*** Cells preparing for division enlarge and synthesize additional proteins and organelles.
3. ***To produce secretions.*** Secretory cells must synthesize their products and deliver them to the interstitial fluid.

The nutrient pool is the source for the organic molecules for both catabolism and anabolism. As you might expect, the cell tends to conserve the materials needed to build new compounds and breaks down

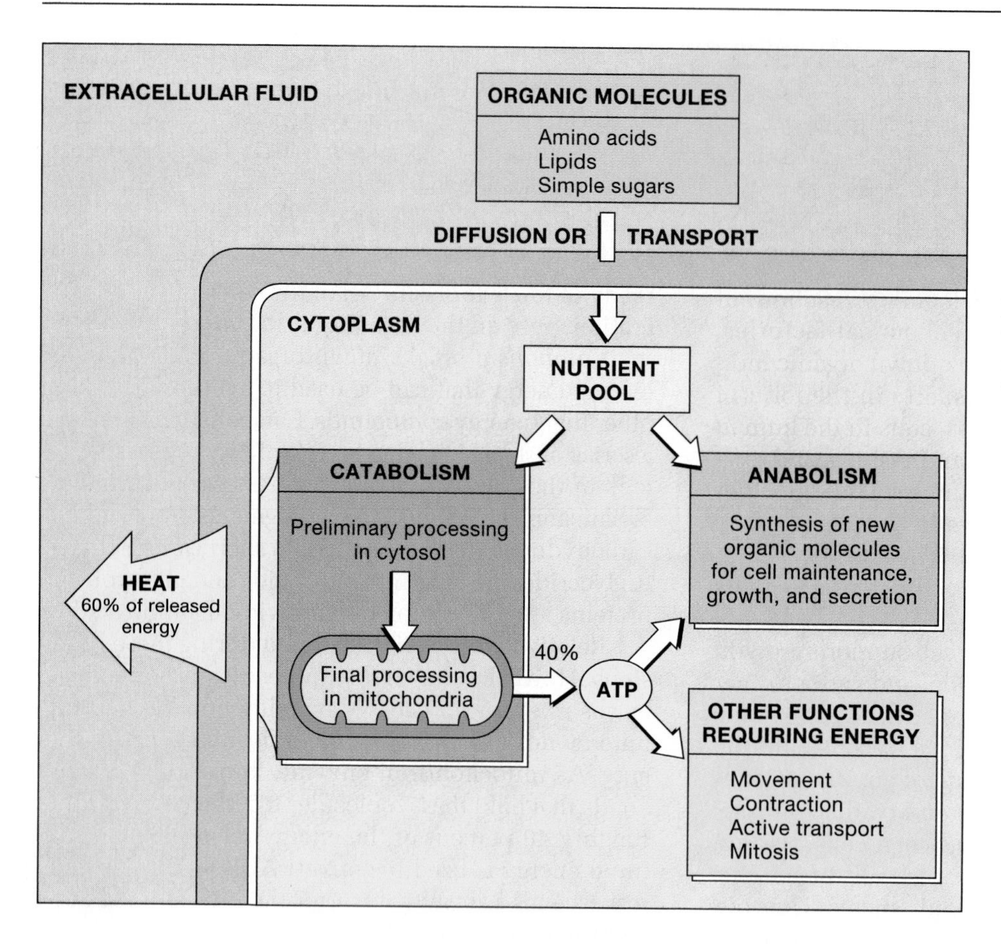

Figure 18-1
Cellular Metabolism

The cell obtains organic molecules from the extracellular fluid and breaks them down to obtain ATP. Only about 40 percent of the energy released through catabolism is captured in ATP; the rest is radiated as heat. The ATP generated through catabolism provides energy for all vital cellular activities, including anabolism.

the rest to provide ATP. The cell is continually replacing membranes, organelles, enzymes, and structural proteins. These anabolic activities require more amino acids than lipids, and relatively few carbohydrates. Catabolic activities, however, tend to process these organic molecules in the reverse order. In general, *a cell with excess carbohydrates, lipids, and amino acids will break down carbohydrates first.* Lipids are a second choice, and amino acids are seldom broken down.

The next sections will examine some of the major catabolic and anabolic reactions that occur within our cells.

Carbohydrate Metabolism

Carbohydrates, most familiar to us as sugars and starches, are important sources of energy. Most cells generate ATP and other high-energy compounds through the breakdown of carbohydrates, especially glucose. The complete reaction sequence can be summarized as:

$$\underset{\text{glucose}}{C_6H_{12}O_6} + \underset{\text{oxygen}}{6\,O_2} \rightarrow \underset{\text{carbon dioxide}}{6\,CO_2} + \underset{\text{water}}{6\,H_2O}$$

The breakdown occurs in a series of small steps, and several of the steps release sufficient energy to support the conversion of ADP to ATP. During the complete catabolism of glucose, a typical cell gains 36 ATP.

Although most of the actual energy production occurs inside mitochondria, the first steps take place in the cytosol. The steps involved were outlined in Chapter 7. [p. 166] In this reaction sequence, called *glycolysis,* six-carbon glucose molecules are broken down into two three-carbon molecules of *pyruvic acid.* The steps are said to be *anaerobic* because oxygen is not needed. The pyruvic acid can be absorbed by mitochondria and used for energy production. Mitochondrial activity, which requires oxygen, is called *cellular respiration,* or *aerobic metabolism.*

Glycolysis

Glycolysis (glī-KOL-i-sis; *glykus,* sweet + *lysis,* breakdown) is the breakdown of glucose to pyruvic acid. In this process, a series of enzymatic steps breaks the six-carbon glucose molecule into two three-carbon molecules of pyruvic acid (CH_3-CO-COOH).

Glycolysis requires (1) glucose molecules, (2) appropriate cytoplasmic enzymes, (3) ATP and ADP, and (4) **NAD** (**n**icotinamide **a**denine **d**inucleotide), a *coenzyme* that removes hydrogen atoms during a complex reaction. Coenzymes are organic molecules, usually derived from vitamins, that must be present for an enzymatic reaction to occur. If the cell lacks any of these four participants, glycolysis cannot take place.

The basic steps in glycolysis are summarized in Figure 18-2●. This reaction sequence provides a net gain of two ATP for each glucose molecule converted to two pyruvic acid molecules. A few highly specialized cells, such as red blood cells, lack mitochondria and derive all of their ATP through glycolysis. Skeletal muscle fibers rely on glycolysis for energy production during periods of active contraction, and most cells can survive brief periods of hypoxia using the ATP provided by glycolysis. When oxygen is readily available, however, mitochondrial activity provides most of the ATP required by our cells.

Mitochondrial Energy Production

Glycolysis yields an immediate net gain of two ATP for the cell, but a great deal of additional energy is still locked in the chemical bonds of pyruvic acid. The ability to capture that energy depends on the availability of oxygen. If oxygen supplies are adequate, mitochondria absorb the pyruvic acid molecules and break them down completely. The hydrogen atoms are removed by coenzymes, and they will ultimately be the source of most of the energy gain for the cell. The carbon and oxygen atoms are removed and released as carbon dioxide.

Once inside the mitochondrion, a pyruvic acid molecule first loses one carbon atom in a complicated reaction involving a molecule called *coenzyme A* (or *CoA*). This reaction yields one molecule of carbon dioxide and one molecule of **acetyl-CoA** (as-e-til-CŌ-ā). Acetyl-CoA consists of a two-carbon *acetyl group* ($CH_3C{=}O$) bound to coenzyme A. Next, the acetyl group is transferred from CoA to a four-carbon molecule, producing *citric acid*.

The TCA Cycle The formation of citric acid is the first step in a sequence of enzymatic reactions called the **tricarboxylic** (trī-kar-bok-SIL-ik) **acid (TCA) cycle**, also known as the *citric acid cycle*. The purpose of

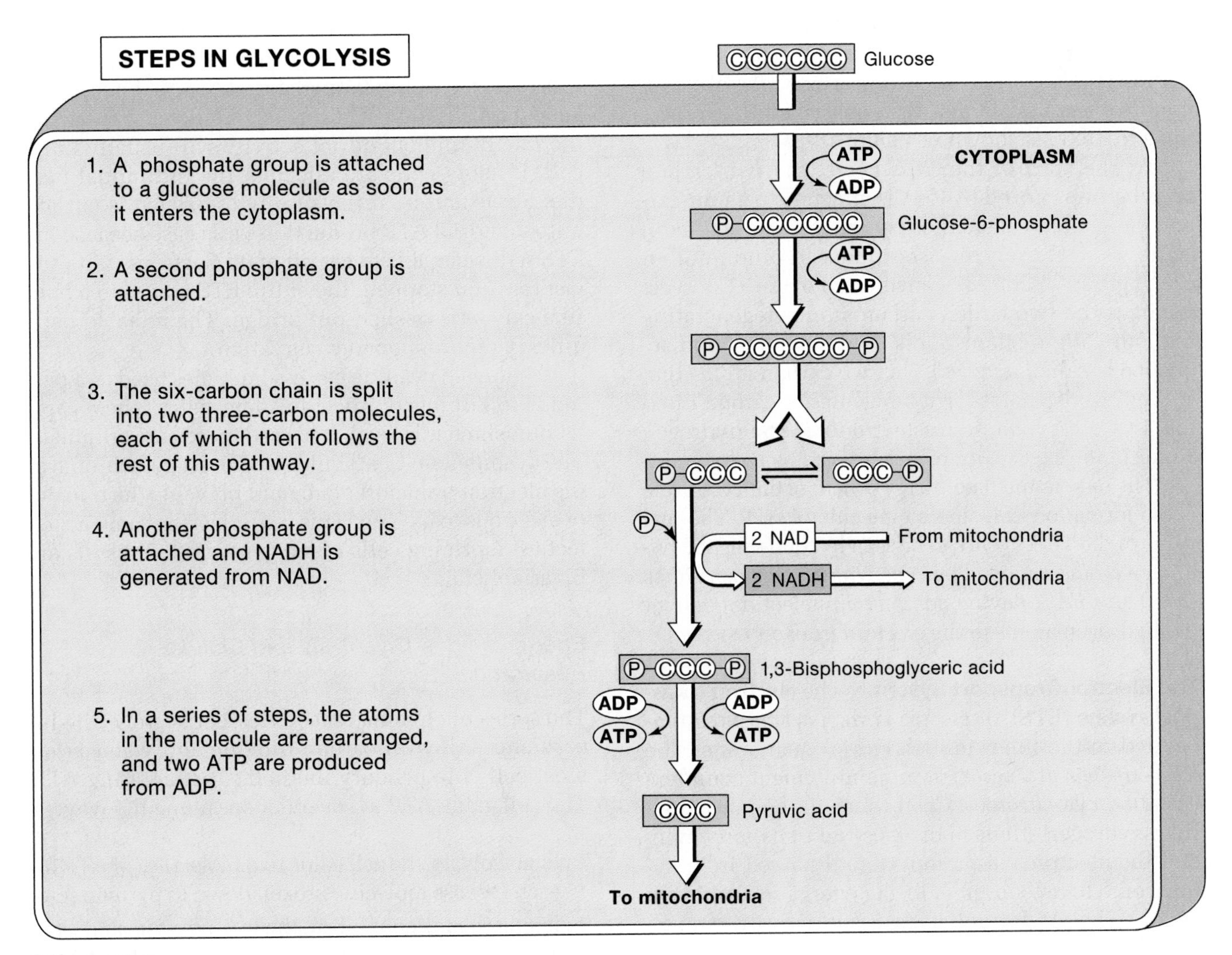

●**Figure 18-2**
Glycolysis

Glycolysis breaks down a six-carbon glucose molecule into two three-carbon molecules of pyruvic acid. This process involves a series of enzymatic steps. There is a net gain of two ATP for each glucose molecule converted to pyruvic acid.

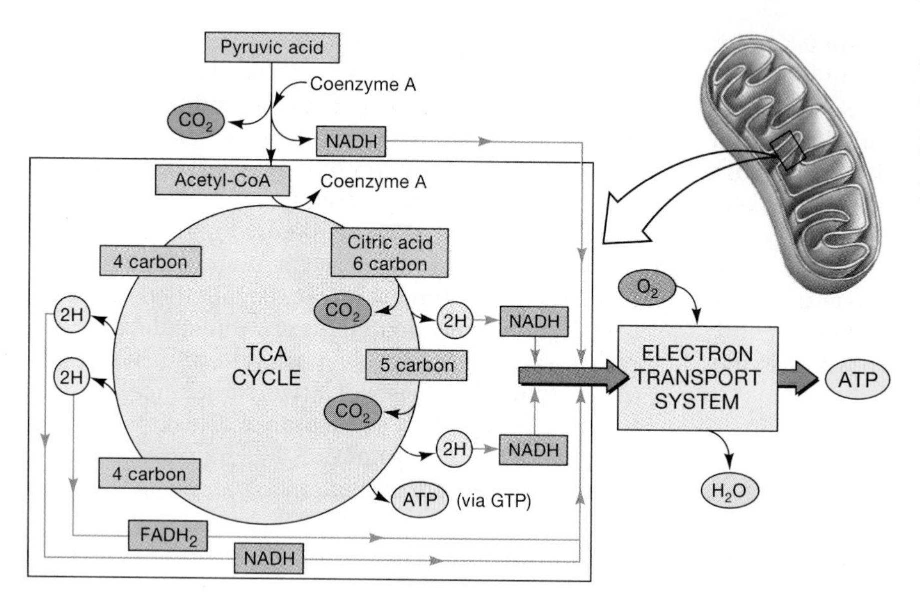

Figure 18-3
The TCA Cycle

The TCA cycle completes the breakdown of organic molecules begun by glycolysis and other catabolic pathways.

this cycle is to remove hydrogen atoms from organic molecules and transfer them to coenzymes. The electrons in these hydrogen atoms contain energy that can be used to generate ATP. The overall pattern of the TCA cycle, sometimes known as the *Krebs cycle*, in honor of the biochemist who described these reactions in 1937, is shown in Figure 18-3.

At the start of the TCA cycle, the two-carbon acetyl group carried by CoA is attached to a four-carbon molecule to make the six-carbon citric acid molecule. CoA is then released intact to bind another acetyl group. A complete revolution of the TCA cycle removes the two added carbon atoms, regenerating the four-carbon chain. (This is why the reaction sequence is called a *cycle*.) The two carbon atoms that are removed generate two molecules of carbon dioxide (CO_2), a metabolic waste product. The hydrogen atoms of the acetyl group are removed by coenzymes.

The only immediate energy benefit of the TCA cycle is the formation of a single molecule of ATP. The real value of the TCA cycle can be seen by following the hydrogen atoms removed by coenzymes. The coenzymes (NAD and *FAD*, *flavine adenine dinucleotide*) deliver the hydrogen atoms to the *electron transport system*.

The Electron Transport System The **electron transport system (ETS)**, or *electron transport chain*, is embedded in the inner mitochondrial membrane. The ETS consists of a series of protein-pigment complexes called *cytochromes*. Coenzymes in the matrix deliver hydrogen atoms to the electron transport chain, and the electrons are removed and passed from cytochrome to cytochrome, losing energy in a series of small steps. At several steps along the way, enough energy is released to attach a phosphate group to ADP, forming ATP. At the end of the electron transport system an oxygen atom accepts the electrons, creating an oxygen ion (O^{2-}). This ion is very reactive, and it quickly combines with hydrogen ions (H^+) to form a molecule of water.

The electron transport system is the single most important mechanism for the generation of ATP, and it provides roughly 95 percent of the ATP needed to keep our cells alive. Halting or significantly reducing the rate of mitochondrial activity will usually kill a cell. If many cells are affected, the individual may die. For example, if the supply of oxygen is cut off, mitochondrial ATP production will cease because the ETS will be unable to get rid of its electrons. With the last reaction stopped, the entire ETS comes to a halt, like cars at a washed-out bridge. The affected cells quickly die from energy starvation.

Hydrogen cyanide gas is sometimes used as a pesticide to kill rats or mice; in some states where capital punishment is legal, it is used to execute criminals. The cyanide ion (CN^-) binds to the last cytochrome in the electron transport chain and prevents the transfer of electrons to oxygen. This result has the same net effect as depriving cells of oxygen, and the results are invariably fatal.

Energy Yield of Glycolysis and Cellular Respiration

The series of chemical reactions that begin with glucose and end with carbon dioxide and water is for most cells the primary method for generating ATP. The cell gains ATP at several steps along the way:

- In glycolysis, the cell gains two molecules of ATP for each glucose molecule broken down to pyruvic acid.
- Inside the mitochondria, the two pyruvic acid molecules derived from each glucose molecule are fully broken down in the TCA cycle. Two revolutions of the TCA cycle, each yielding a molecule of ATP, provide a net gain of two additional molecules of ATP.

- For each molecule of glucose broken down, activity at the electron transport chain will provide 32 molecules of ATP.

Summing up, for each glucose molecule processed, a typical cell gains 36 molecules of ATP. *All but two of them are produced by the mitochondria.*

CARBOHYDRATE LOADING

Although other nutrients can be broken down to provide substrates for the TCA cycle, carbohydrates require the least processing and preparation. It is not surprising, therefore, that athletes have tried to devise ways of exploiting these compounds as ready sources of energy.

Eating carbohydrates *just before* exercising does not improve performance and may actually decrease endurance by slowing the mobilization of existing energy reserves. Runners or swimmers preparing for lengthy endurance contests, such as a marathon or 5-km swim, do not eat immediately before competing, and for 2 hours before the race their diet is limited to drinking water. However, these athletes often eat carbohydrate-rich meals for 3 days before the event. This *carbohydrate loading* increases the carbohydrate reserves of muscle tissue that will be called upon during the competition.

Maximum effects can be obtained by exercising to exhaustion for 3 days before starting the high-carbohydrate diet; this practice is called *carbohydrate depletion/loading.* There are a number of potentially unpleasant side effects to carbohydrate depletion/loading, including muscle and kidney damage, and sports physiologists recommend that athletes use this routine fewer than three times per year.

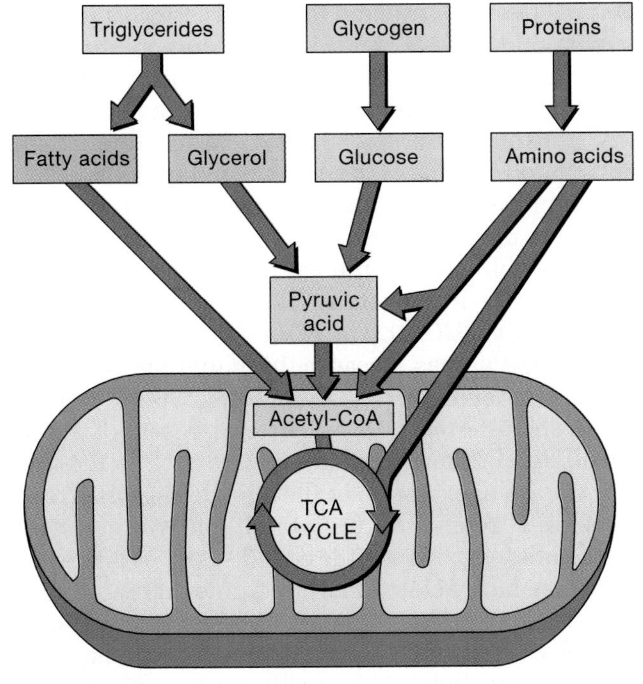

•Figure 18-4
Alternative Catabolic Pathways

Other Catabolic Pathways

Aerobic metabolism is relatively efficient and capable of generating large amounts of ATP. It is the cornerstone for normal cellular metabolism, but it has one obvious limitation: The cell must have adequate supplies of both oxygen and glucose. Cells can survive only briefly without oxygen, but low glucose concentrations have a much smaller effect on most cells because they can break down other nutrients to provide organic molecules for the TCA cycle, as shown in Figure 18-4•. Many cells can switch from one nutrient source to another as the need arises. For example, many cells can shift from glucose-based to lipid-based ATP production when necessary.

Cells break down proteins for energy only when lipids or carbohydrates are unavailable; this makes sense because proteins make up the enzymes and organelles that the cell needs to survive. Nucleic acids are present only in small amounts, and they are seldom catabolized for energy, even when the cell is dying of acute starvation. This restraint makes sense, as it is the DNA in the nucleus that determines all of the structural and functional characteristics of the cell. We will consider the catabolism of other compounds in later sections as we discuss lipid, protein, and nucleic acid metabolism.

Gluconeogenesis

The synthesis of glucose from nonglucose precursors is called **gluconeogenesis** (gloo-kō-nē-ō-JEN-e-sis; *glykus*, sweet + *neo-*, new + *genesis*, an origin). Because some of the steps in glycolysis are not reversible, carbohydrate synthesis involves a different set of regulatory enzymes, and carbohydrate breakdown and synthesis are independently regulated. Pyruvic acid or other three-carbon molecules can be used as starting materials. As indicated in Figure 18-5•, this reaction sequence enables a cell to create glucose molecules from other carbohydrates, glycerol, or some amino acids. But acetyl-CoA cannot be used to make glucose because the reaction that removes the carbon dioxide molecule (*adecarboxylation*) between pyruvic acid and acetyl-CoA cannot be reversed. Fatty acids and many amino acids cannot be converted to glucose because their catabolic pathways produce acetyl-CoA.

Glucose molecules created by gluconeogenesis can be used to manufacture other simple sugars, complex carbohydrates, or nucleic acids. In the liver and in skeletal muscle, glucose molecules are stored as glycogen. Glycogen is an important energy reserve that can be broken down when the cell cannot obtain enough glucose from the interstitial fluid. Although glycogen molecules are large, glycogen reserves take up very little space because they form compact, insoluble granules.

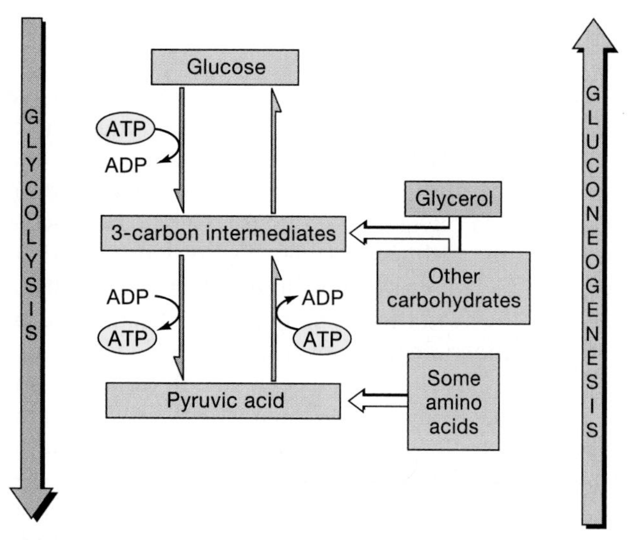

Figure 18-5
Carbohydrate Metabolism

This flow chart diagrams the major pathways for glycolysis and gluconeogenesis. Some amino acids, other carbohydrates, and glycerol can be converted to glucose.

Lipid Metabolism

Lipid molecules, like carbohydrates, contain carbon, hydrogen, and oxygen, but in different proportions. Because triglycerides are the most abundant lipid in the body, our discussion will focus on pathways for triglyceride breakdown and synthesis.

Lipid Catabolism

During lipid catabolism, or *lipolysis*, lipids are broken down into pieces that can be converted to pyruvic acid or channeled directly into the TCA cycle (Figure 18-4●). A triglyceride is first split into its component parts through hydrolysis. This step yields one molecule of glycerol and three of fatty acids. Glycerol enters the TCA cycle after cytoplasmic enzymes convert it to pyruvic acid. The catabolism of fatty acids, known as *beta-oxidation*, involves a completely different set of enzymes that break them down into two-carbon fragments. The fragments enter the TCA cycle or combine to form *ketone bodies*, short carbon chains discussed in a later section. Beta-oxidation occurs inside mitochondria, so the carbon chains can enter the TCA cycle immediately. The cell gains 144 ATP from the breakdown of an 18-carbon fatty acid molecule, almost 1.5 times the energy obtained from the breakdown of three glucose molecules.

Lipids and Energy Production

Lipids are important as an energy reserve because they can provide large amounts of ATP. Being insoluble, they can be stored in compact droplets in the cytosol. This feature also makes it difficult for water-soluble enzymes to reach them, and as a result, lipid reserves are harder to mobilize than carbohydrate reserves. In addition, most lipids are processed inside mitochondria, and mitochondrial activity is limited by the availability of oxygen. The net result is that lipids cannot provide large amounts of ATP in a short amount of time. However, cells with modest energy demands can shift over to lipid-based energy production when glucose supplies are limited. Skeletal muscle fibers normally cycle between lipid and carbohydrate metabolism. At rest, when energy demands are low, they break down fatty acids. When active, and energy demands are both large and immediate, skeletal muscle fibers shift over to metabolizing glucose.

Lipid Synthesis

The synthesis of lipids is known as **lipogenesis** (li-pō-JEN-e-sis; *lipos*, fat). It usually begins with acetyl-CoA, and almost any organic molecule can be used because lipids, amino acids, and carbohydrates can easily be converted to acetyl-CoA (Figure 18-6●). Our cells cannot *build* every fatty acid they can break down. *Linoleic acid*, an 18-carbon, unsaturated fatty acid, cannot be synthesized at all. Thus, a diet poor in linoleic acid slows growth and alters the appearance of the skin. *Arachidonic* and *linolenic acids* are other long-chain unsaturated fatty acids that the human body cannot synthesize. These three **essential fatty acids** must be included in the diet, because they are needed to synthesize prostaglandins and phospholipids for cell membranes.

Lipid Transport and Distribution

Lipids circulate through the bloodstream as *lipoproteins* and *free fatty acids*. **Lipoproteins** are lipid-protein complexes that contain large insoluble glycerides and cholesterol with a superficial coating dominated by phospholipids and proteins. The proteins and phospholipids make the entire complex soluble, and the proteins help regulate lipid absorption by cells.

Two major groups of lipoproteins that transport cholesterol between the liver and other tissues are **low-density lipoproteins (LDLs)** and **high-density lipoproteins (HDLs)**. The LDLs deliver cholesterol to peripheral tissues. Because this cholesterol may end up in arterial plaques, LDL cholesterol is often called "bad cholesterol." HDL cholesterol transports excess cholesterol from peripheral tissues back to the liver for storage or excretion in the bile. Because HDL cholesterol is returning cholesterol that will not cause circulatory problems, it is called "good cholesterol."

Free fatty acids (FFA) are lipids that can diffuse easily across cell membranes. Those circulating in the blood are usually bound to albumin, the most abundant plasma protein. Liver cells, cardiac muscle cells, skeletal muscle fibers, and many other body cells can

Clinical Note

Dietary Fats and Cholesterol

Elevated cholesterol levels are associated with the development of *atherosclerosis* (Chapter 14) and *coronary artery disease* (CAD) (Chapter 13). ∞ [pp. 350, 336] Current nutritional advice suggests reducing cholesterol intake to under 300 mg per day. This amount represents a 40 percent reduction for the average American adult. As a result of rising concerns about cholesterol, such phrases as "low in cholesterol," "contains no cholesterol," and "cholesterol-free" are now widely used in the advertising and packaging of foods. What do they really mean in terms of individual health and diet planning? Before answering that question we must consider some basic information about cholesterol and about lipid metabolism in general.

- ***Cholesterol has many vital functions in the human body***. It serves as a waterproofing for the epidermis, a lipid component of all cell membranes, a key constituent of bile, and the precursor of several steroid hormones and one vitamin (vitamin D). Because cholesterol is so important, dietary restrictions should have the goal of keeping cholesterol levels within acceptable limits. The goal is *not* the elimination of cholesterol in the diet or the circulating blood.
- ***The cholesterol content of the diet is not the only source for circulating cholesterol.*** The human body can manufacture cholesterol from the acetyl-CoA obtained through glycolysis or the breakdown of other lipids. If the diet contains an abundance of saturated fats, serum cholesterol levels will rise because excess lipids are broken down to acetyl-CoA that can be used to make cholesterol. This means that when trying to lower serum cholesterol by dietary control, one must restrict other lipids as well.
- ***Genetic factors affect each individual's cholesterol level.*** If the dietary supply of cholesterol is reduced, the body synthesizes more to maintain "acceptable" concentrations in the blood. The acceptable level depends on the genetic programming of the individual. Because individuals differ in genetic makeup, their cholesterol levels can vary even on similar diets. In virtually all instances, however, dietary restrictions can lower blood cholesterol substantially.
- ***Cholesterol levels vary with age and physical condition.*** At age 19, three out of four males have cholesterol levels below 170 mg/dl. Cholesterol levels in females of this age are slightly higher, typically at or below 175 mg/dl. With increasing age, the cholesterol values gradually climb, and over age 70 the values are 230 mg/dl (males) and 250 mg/dl (females). Cholesterol levels are considered unhealthy if they are higher than those of 90 percent of the population in that age group. For males, this value ranges from 185 mg/dl at age 19 to 250 mg/dl at age 70. For females, the comparable values are 190 mg/dl and 275 mg/dl.

To determine whether you need to do anything about your cholesterol level without performing calculations regarding age, weight, and sex, just remember three simple rules:

1. Individuals of any age with total cholesterol values below 200 mg/dl probably do not need to change their lifestyles unless they have a family history of coronary artery disease and atherosclerosis.
2. Those with cholesterol levels between 200 and 239 mg/dl should modify their diets, lose weight (if overweight), and have annual checkups.
3. Cholesterol levels over 240 mg/dl warrant drastic changes in dietary lipid consumption, perhaps coupled with drug treatment. Drug therapies are always recommended in cases where the serum cholesterol level exceeds 350 mg/dl. Examples of drugs used to lower cholesterol levels include *cholestyramine*, *colestipol*, and *lovastatin*.

When cholesterol levels are high, or when an individual has a family history of atherosclerosis or CAD, further tests may be performed to determine the relative amounts of cholesterol circulating in LDLs and HDLs. A high total cholesterol value linked to a high LDL spells trouble. In effect, an unusually large amount of cholesterol is being exported to peripheral tissues. Problems can also exist if the individual has high total cholesterol—or even normal total cholesterol—and low HDL levels (below 35 mg/dl). In this case, excess cholesterol delivered to the tissues cannot easily be returned to the liver for excretion. In either event, the amount of cholesterol in peripheral tissues, and especially in arterial walls, is likely to increase. A standard guideline is that the ratio of total cholesterol to HDL cholesterol should be less than 4.5 to 1. If it is more, the individual is at risk for developing atherosclerosis.

metabolize free fatty acids. They are an important energy source during periods of starvation, when glucose supplies are limited.

Protein Metabolism

There are roughly 100,000 different proteins in the human body, with varied forms, functions, and structures. All contain varying combinations of the same 20+ amino acids. Under normal conditions, there is a continual turnover of cellular proteins in the cytosol. Peptide bonds are broken, and the free amino acids are used to manufacture new proteins. If other energy sources are inadequate, mitochondria can break down amino acids in the TCA cycle to generate ATP. Not all amino acids enter the TCA cycle at the same point, so the ATP benefits vary. However, the average yield is comparable to that of carbohydrate catabolism.

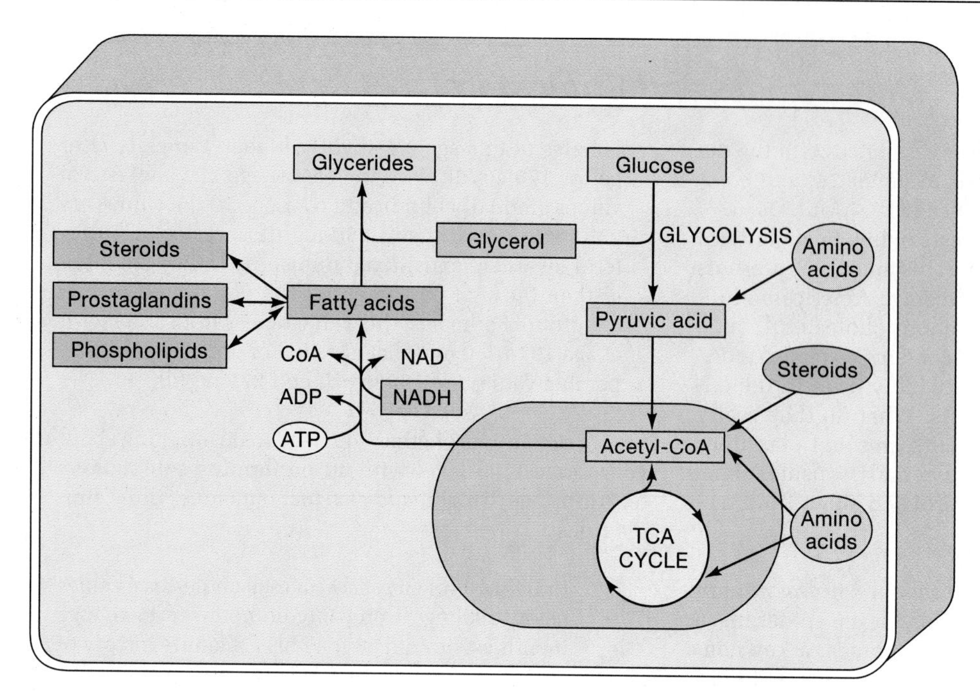

Figure 18-6
Lipid Synthesis

Pathways of lipid synthesis begin with acetyl-CoA. Molecules of acetyl-CoA can be strung together in the cytosol, yielding fatty acids. Those fatty acids can be used to synthesize glycerides or other lipid molecules. Lipids can also be synthesized from amino acids or carbohydrates.

Amino Acid Catabolism

The first step in amino acid catabolism is the removal of the amino group, which requires a coenzyme derivative of **vitamin B_6** (*pyridoxine*). The amino group may be removed by *transamination* (trans-am-i-NĀ-shun) or *deamination* (dē-am-i-NĀ-shun). Transamination attaches the amino group of an amino acid to another carbon chain, creating a "new" amino acid. Transaminations enable a cell to synthesize many of the amino acids needed for protein synthesis. Cells of the liver, skeletal muscles, heart, lung, kidney, and brain, which are particularly active in protein synthesis, perform large numbers of transaminations.

There are several inherited metabolic disorders that result from an inability to produce specific enzymes involved with amino acid metabolism. *Phenylketonuria* (fen-il-kē-tō-NOO-rē-a), or PKU, is an example. Individuals with PKU cannot convert phenylalanine to tyrosine, because of a defect in the enzyme *phenylalanine hydroxylase*. This reaction is an essential step in the synthesis of norepinephrine, epinephrine, and melanin. If PKU is not detected in infancy, CNS development is inhibited, and severe brain damage results. AM *Phenylketonuria*

Deamination is the removal of an amino group in a reaction that generates an ammonia molecule (NH_3). Ammonia is highly toxic, even in low concentrations. The liver, the primary site of deamination, has the enzymes needed to deal with the problem of ammonia generation. Liver cells take the ammonia and convert it to **urea**, a relatively harmless, water-soluble compound that is excreted in the urine. The fate of a carbon chain after deamination in the liver depends on its structure. The carbon chains of some amino acids can be converted to pyruvic acid and then used for gluconeogenesis. Others are converted to acetyl-CoA and broken down or are converted to **ketone bodies**. Ketone bodies are organic acids that are also produced when lipid catabolism is under way. *Acetone* is an example of a ketone body generated by the body. It is a small molecule that can diffuse into the alveoli of the lungs, giving the breath a distinctive odor.

Ketone bodies are not metabolized by the liver, and they diffuse into the general circulation. The ketone bodies are used by other cells, which reconvert them into acetyl-CoA for breakdown in the TCA cycle and the production of ATP. The increased production of ketone bodies that occurs during protein and lipid catabolism by the liver results in high ketone body concentrations in body fluids, a condition called **ketosis** (kē-TŌ-sis).

KETOACIDOSIS

A ketone body is also called a "keto acid" because it dissociates in solution, releasing a hydrogen ion. As a result, the appearance of ketone bodies represents a threat to the plasma pH. During prolonged starvation, ketone levels continue to rise. Eventually, the pH buffering capacities are exceeded, and a dangerous drop in pH occurs. This acidification of the blood is called *ketoacidosis* (kē-tō-as-i-DŌ-sis). In severe ketoacidosis, the pH may fall below 7.05, a value that can disrupt normal tissue activities and cause coma, cardiac arrhythymias, and death.

In *diabetes mellitus* most peripheral tissues cannot utilize glucose because of a lack of insulin. ∞ [p. 297] Under these circumstances, the cells survive by catabolizing lipids and proteins. The result is the production of large numbers of ketone bodies, and this condition leads to *diabetic ketoacidosis*, the most common and life-threatening form of ketoacidosis.

Proteins and Energy Production

Several factors make protein catabolism an impractical source of quick energy:

- Proteins are more difficult to break apart than are complex carbohydrates or lipids.
- One of the byproducts, ammonia, is a toxin that can damage cells.
- Proteins form the most important structural and functional components of any cell. Extensive protein catabolism therefore threatens homeostasis at the cellular and systems levels.

Amino Acids and Protein Synthesis

The basic mechanism for protein synthesis was detailed in Chapter 3 (Figures 3-17• and 3-18•). ∞ [pp. 63, 64–65] The human body can synthesize roughly half of the different amino acids needed to build proteins. There are ten **essential amino acids**. Eight of them (*isoleucine, leucine, lysine, threonine, tryptophan, phenylalanine, valine,* and *methionine*) cannot be synthesized at all; the other two (*arginine* and *histidine*) can be synthesized in amounts that are insufficient for growing children. The other amino acids, which can be synthesized on demand, are called the **nonessential amino acids**.

Protein deficiency diseases develop when an individual does not consume adequate amounts of all essential amino acids. All amino acids must be available if protein synthesis is to occur. Every tRNA must appear at the proper location bearing its individual amino acid; as soon as the amino acid called for by a particular codon is missing, the entire process comes to a halt. Regardless of the energy content of the diet, if it is deficient in essential amino acids the individual will be malnourished to some degree. Examples of protein deficiency diseases include *marasmus* and *kwashiorkor*. Although over 100 million children worldwide have symptoms of these disorders, neither condition is common in the United States today. AM *Protein Deficiency Diseases*

Nucleic Acid Metabolism

Living cells contain both DNA and RNA. But because the genetic information contained in the DNA of the nucleus is absolutely essential to the long-term survival of the cell, the DNA in the nucleus is never catabolized for energy, even if the cell is dying of starvation. RNA molecules are broken down and replaced on a regular basis, but most nucleotides are recycled rather than broken down further. When the nucleotides *are* broken down, only the sugars, cytosine, and uracil can enter the TCA cycle and be used to generate ATP. Adenine and guanine cannot be catabolized at all. Instead they are excreted as **uric acid**, another relatively nontoxic waste product, that is far less soluble than urea. Urea and uric acid are called *nitrogenous wastes* because they contain nitrogen atoms. AM *Gout*

Nucleic Acid Synthesis

All cells synthesize RNA, but DNA synthesis occurs only in cells that are preparing for mitosis (cell division) or meiosis (gamete production). The process of DNA replication was described in Chapter 3. ∞ [p. 66] Messenger RNA, transfer RNA, and ribosomal RNA are transcribed by different forms of RNA polymerase. Messenger RNA is manufactured as needed, when specific genes are activated. Although several ribosomes can be reading the same message at any given moment, a strand of mRNA has a life span measured in minutes or hours. Molecules of ribosomal and transfer RNA in the cytosol are broken down and replaced on a regular basis. Ribosomes are more durable than mRNA strands—the half-life of a ribosome is just over 5 days. Because each cell contains roughly 100,000 ribosomes, their replacement involves a considerable amount of synthetic activity.

✓ How would a diet that is deficient in vitamin B_6 affect protein metabolism?

✓ Elevated levels of uric acid in the blood could be an indicator of increased metabolism of what macromolecule?

✓ Why are high-density lipoproteins (HDLs) considered to be beneficial?

A Summary of Cellular Metabolism

Figure 18-7• summarizes the major pathways of cellular metabolism. Although this diagram follows the reactions in a "typical" cell, no one cell can perform all of the anabolic and catabolic operations and interconversions required by the body as a whole. As differentiation proceeds, each cell type develops its own complement of enzymes which determines its metabolic capabilities. In the presence of such cellular diversity, homeostasis can be preserved only when the metabolic activities of tissues, organs, and organ systems are coordinated.

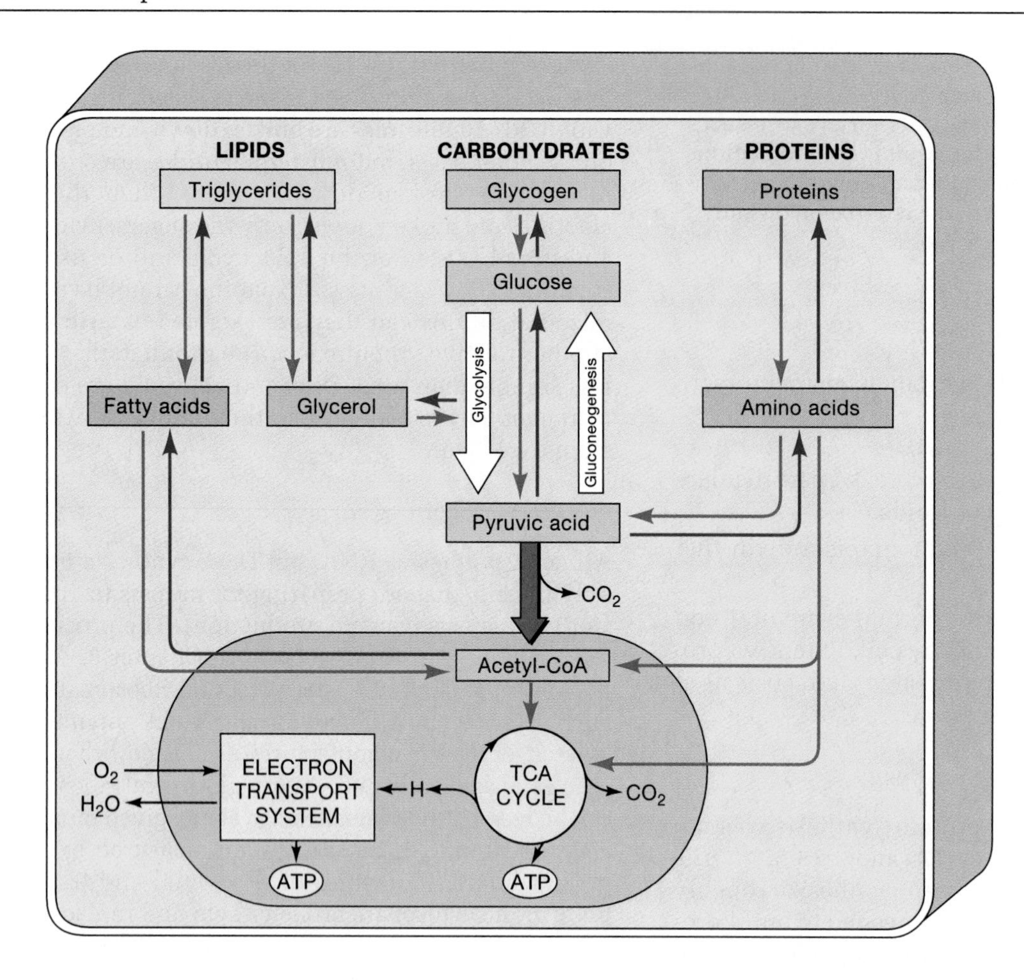

Figure 18-7 A Summary of Pathways of Catabolism and Anabolism
This diagram provides an overview of major catabolic (red) and anabolic (blue) pathways.

DIET AND NUTRITION

Homeostasis can be maintained indefinitely only if the digestive tract absorbs fluids, organic substrates, minerals, and vitamins at a rate that keeps pace with cellular demands. The absorption of nutrients from food is called **nutrition.**

The individual requirement for each nutrient varies from day to day and from person to person. *Nutritionists* attempt to analyze a diet in terms of its ability to meet the needs of a specific individual. A *balanced diet* contains all of the ingredients necessary to maintain homeostasis, including adequate substrates for energy generation, essential amino acids and fatty acids, minerals, and vitamins. In addition, the diet must include enough water to replace losses in urine, feces, and evaporation. A balanced diet prevents **malnutrition**, an unhealthy state resulting from inadequate or excessive intake of one or more nutrients. AM *Eating Disorders*

The Basic Food Groups

For several decades, the traditional American method of avoiding malnutrition was to include members of each of the four **basic food groups** in the diet. These were the *milk and dairy group*, the *meat group*, the *vegetable and fruit group*, and the *bread and cereal group*. Each group differs from the others in the typical balance of proteins, carbohydrates, and lipids contained, as well as in the amount and identity of vitamins and minerals.

Recently, the four groups have been increased to six by separating the fruit group and establishing a *fats, oils, and sweets group*. The six groups are now arranged in a *food pyramid*, with the bread and cereal group at the bottom (Figure 18-8). This arrangement serves to emphasize the need to restrict dietary fats, oils, and sugar and to increase consumption of breads and cereals, which are rich in complex carbohydrates (polysaccharides such as starch).

It should be realized that such groupings are artificial at best, and misleading at worst. What is important is to obtain nutrients in sufficient *quantity* (adequate to meet energy needs) and *quality* (including essential amino acids, fatty acids, vitamins, and minerals). There is nothing magical about the number six—since 1940 there have been 11, 7, 4, and 6 food groups advocated by the U.S. government at various times. The key is making intelligent choices about what you eat. The wrong selections can lead to malnutrition even if all six groups are represented.

For example, consider the case of the essential amino acids. Some members of the meat and milk groups, such as beef, fish, poultry, eggs, and milk, contain all of the essential amino acids in sufficient quantities. They are said to contain *complete proteins*. Many plants contain adequate *amounts* of protein, but they contain *incomplete proteins* that are deficient in one or more of the essential amino acids. True vegetarians, who restrict themselves to the fruit and vegetable groups (with or without the bread and cereal), must become adept at juggling the constituents of their meals to include a combination of ingredients that will meet all of their amino acid requirements. Even with a proper balance of amino acids, the vegetarian faces a significant problem, since vitamin B_{12} is obtained only from animal products.

Minerals, Vitamins, and Water

Minerals, vitamins, and water are essential components of the diet. The body cannot synthesize minerals, and our cells can generate only a small quantity of water and very few vitamins.

Minerals

Minerals are inorganic ions released through the dissociation of electrolytes, such as sodium chloride. Minerals are important for the following reasons:

1. Ions such as sodium and chloride contribute to the osmotic concentration of body fluids. Potassium is important in maintaining the osmotic concentration inside body cells.

Figure 18-8
The Food Pyramid

A Guide to Daily Food Choices

Nutrient Group	*Provides*	*Deficiencies*
Fats, oils, sweets	Calories	Usually deficient in most minerals and vitamins
Milk, yogurt, cheese	Complete proteins, fats, carbohydrates, calcium, potassium, magnesium, sodium, phosphorus, vitamins A, B_{12}, pantothenic acid, thiamine, riboflavin	Dietary fiber, vitamin C
Meat, poultry, fish, dry beans, eggs, nuts	Complete proteins, fats, potassium, phosphorus, iron, zinc, vitamins E, thiamine, B_6	Carbohydrates, dietary fiber, several vitamins
Vegetables and fruits	Carbohydrates, vitamins A, C, E, folacin, and dietary fiber, potassium	Often low in fats, calories, and protein
Bread, cereal, rice, pasta	Carbohydrates, vitamins E, thiamine, niacin, folacin, calcium, phosphorus, iron, sodium, dietary fiber	Fats

TABLE 18-1 Minerals and Mineral Reserves

Mineral	*Significance*	*Total Body Content*	*Primary Route of Excretion*	*Recommended Daily Intake*
BULK MINERALS				
Sodium	Major cation in body fluids; essential for normal membrane function	110 g, primarily in body fluids	Urine, sweat, feces	0.5–1.0 g
Potassium	Major cation in cytoplasm; essential for normal membrane function	140 g, primarily in cytoplasm	Urine	1.9–5.6 g
Chloride	Major anion in body fluids	89 g, primarily in body fluids	Urine, sweat	0.7–1.4 g
Calcium	Essential for normal muscle and neuron function, bone structure	1.36 kg, primarily in skeleton	Urine, feces	0.8–1.2 g
Phosphorus	As phosphate in high-energy compounds, nucleic acids, and bone matrix	744 g, primarily in skeleton	Urine, feces	0.8–1.2 g
Magnesium	Cofactor of enzymes, required for normal membrane functions	29 g (skeleton, 17 g; cytoplasm and body fluids, 12 g)	Urine	0.3–0.4 g
TRACE MINERALS				
Iron	Component of hemoglobin, myoglobin, cytochromes	3.9 g, 1.6 stored (ferritin or hemosiderin)	Urine (traces)	10–18 mg
Zinc	Cofactor of enzyme systems, notably carbonic anhydrase	2 g	Urine, hair (traces)	15 mg
Copper	Required as cofactor for hemoglobin systhesis	127 mg	Urine, feces (traces)	2–3 mg
Manganese	Cofactor for some enzymes	11 mg	Feces, urine (traces)	2.5–5 mg

2. Ions in various combinations play major roles in important physiological processes, including:
   - The maintenance of membrane potentials (Chapters 7 and 8).
   - Action potential generation (Chapter 8).
   - Neurotransmitter release (Chapters 7 and 8).
   - Muscle contraction (Chapters 7 and 13).
   - The construction and maintenance of the skeleton (Chapter 6).
   - The transport of respiratory gases (Chapter 16).
   - Buffer systems (Chapters 2 and 19).
   - Fluid absorption (Chapter 17).
   - Waste removal (Chapter 19).
3. Ions are essential to several important enzymatic reactions. For example, the enzyme that breaks down ATP in a contracting skeletal muscle requires the presence of calcium and magnesium ions, and an enzyme required for the conversion of glucose to pyruvic acid needs both potassium and magnesium ions.

The major minerals and a summary of their functional roles are presented in Table 18-1. Significant reserves of several important minerals in the body helps reduce the effects of dietary variations in supply. The reserves are often relatively small, however, and chronic dietary reductions can lead to various clinical problems. Alternatively, because storage capabilities are limited, a dietary excess of mineral ions can prove equally dangerous. AM *Iron Deficiencies and Excesses*

Vitamins

Vitamins (*vita*, life) are related to lipids and carbohydrates. They can be assigned to either of two groups, depending on their chemical structure and characteristics: fat-soluble vitamins and water-soluble vitamins.

Fat-Soluble Vitamins Vitamins A, E, and K are **fat-soluble vitamins**. These vitamins are absorbed primarily from the digestive tract along with the lipid contents of micelles. The term *vitamin D* is used to

TABLE 18-2 The Fat-Soluble Vitamins

Vitamin	Significance	Sources	Daily Requirement	Effects of Deficiency	Effects of Excess
A	Maintains epithelia; required for synthesis of visual pigments	Leafy green and yellow vegetables	1 mg	Retarded growth, night blindness, deterioration of epithelial membranes	Liver damage, skin peeling, CNS effects (nausea, anorexia)
D (steroids including D_3 or cholecalciferol	Required for normal bone growth, calcium and phosphorus absorption at gut and retention at kidneys	Synthesized in skin exposed to sunlight	None*	Rickets, skeletal deterioration	Calcium deposits in many tissues, disrupting functions
E (tocopherols)	Prevents breakdown of vitamin A and fatty acids	Meat, milk, vegetables	12 mg	Anemia, other problems suspected	None reported
K	Essential for liver synthesis of prothrombin and other clotting factors	Vegetables; production by intestinal bacteria	0.07–0.14 mg	Bleeding disorders	Liver dysfunction, jaundice

**Unless sunlight exposure is inadequate for extended periods and alternative sources (fortified milk products) are unavailable.*

refer to a group of steroids, including vitamin D_3, or *cholecalciferol*. Unlike the other fat-soluble vitamins, which must be obtained by absorption across the digestive tract, the skin can usually synthesize adequate amounts of vitamin D_3 when exposed to sunlight. Current information concerning the fat-soluble vitamins is summarized in Table 18-2.

Because they dissolve in lipids, fat-soluble vitamins normally diffuse into cell membranes and other lipids in the body, including the lipid inclusions in the liver and adipose tissue. The body therefore contains a significant reserve of these vitamins, and normal metabolic operations can continue for several months after dietary sources have been cut off. As Table 18-2 points out, *too much* of a vitamin may produce effects just as unpleasant as *too little*. *Hypervitaminosis* (hī-per-vī-ta-min-Ō-sis) occurs when the dietary intake exceeds the abilities to store, utilize, or excrete a particular vitamin. This condition most often involves one of the fat-soluble vitamins because the excess is retained and stored in body lipids.

Water-Soluble Vitamins Most of the **water-soluble vitamins** (Table 18-3) are components of coenzymes. Water-soluble vitamins are rapidly exchanged between the fluid compartments and the circulating blood, and excessive amounts are readily excreted in the urine. For this reason, hypervitaminosis involving water-soluble vitamins is relatively uncommon. All of the water-soluble vitamins except B_{12} can be easily absorbed by the intestinal epithelium. The B_{12} molecule is large, and as you will recall from Chapter 17 it must be bound to the *intrinsic factor* secreted by the gastric mucosa before absorption can occur. ∞ [p. 442]

Because these vitamins are not stored in large quantities, insufficient intake may lead to initial symptoms of vitamin deficiency within a period of days to weeks. The condition that results is termed a *deficiency disease*, or *avitaminosis* (ā-vī-ta-min-Ō-sis). Avitaminosis involving either fat-soluble or water-soluble vitamins can be caused by various factors other than dietary deficiencies. An inability to absorb a vitamin from the digestive tract, inadequate storage, or excessive demand may all produce the same result. The bacteria within our intestines help prevent deficiency diseases by producing five of the nine water-soluble vitamins, in addition to fat-soluble vitamin K.

Water

Daily water requirements average 2500 ml (10 cups), or roughly 40 ml/kg body weight. The specific requirement varies with environmental and metabolic activity. For example, exercise increases metabolic

TABLE 18-3 The Water-Soluble Vitamins

Vitamin	*Significance*	*Sources*	*Daily Requirement*	*Effects of Deficiency*	*Effects of Excess*
B_1 (thiamine)	Coenzyme in decarboxylation reactions	Milk, meat, bread	1.9 mg	Muscle weakness, CNS and cardiovascular problems including heart disease; called *beriberi*	Hypotension
B_2 (riboflavin)	Part of FAD	Milk, meat	1.5 mg	Epithelial and mucosal deterioration	Itching, tingling sensations
Niacin (nicotinic acid)	Part of NAD	Meat, bread, potatoes	14.6 mg	CNS, GI, epithelial, and mucosal deterioration; called *pellagra*	Itching, burning sensations, vasodilation, death after large dose
B_5 (pantothenic acid)	Part of acetyl-CoA	Milk, meat	4.7 mg	Retarded growth, CNS disturbances	None reported
B_6 (pyridoxine)	Coenzyme in amino acid and lipid metabolism	Meat	1.42 mg	Retarded growth, anemia, convulsions, epithelial changes	CNS alterations, perhaps fatal
Folacin (folic acid)	Coenzyme in amino acid and nucleic acid metabolism	Vegetables, cereal, bread	0.1 mg	Retarded growth, anemia, gastrointestinal disorders	Few noted except at massive doses
B_{12} (cobalamin)	Coenzyme in nucleic acid metabolism	Milk, meat	4.5 μg	Impaired RBC production causing *pernicious anemia*	Polycythemia
Biotin	Coenzyme in decarboxylation reactions	Eggs, meat, vegetables	0.1–0.2 mg	Fatigue, muscular pain, nausea, dermatitis	None reported
C (ascorbic acid)	Coenzyme; delivers hydrogen ions, antioxidant	Citrus fruits	60 mg	Epithelial and mucosal deterioration; called *scurvy*	Kidney stones

energy requirements, and it also accelerates water losses due to evaporation and perspiration. The temperature rise accompanying a fever has a similar effect, and for each degree the temperature rises above normal the daily water loss increases by 200 ml. Thus the advice "drink plenty of fluids" when one is sick has a solid physiological basis.

Most of the daily water ration is obtained by eating or drinking. The food consumed provides roughly 48 percent, and another 40 percent is obtained by drinking fluids. But a small amount of water—called *metabolic water*—is produced in the mitochondria during the operation of the electron transport system. The actual amount produced per day varies depending on the composition of the diet. A typical mixed diet in the United States contains 46 percent carbohydrates, 40 percent lipids, and 14 percent protein. This diet would produce roughly 300 ml of water per day (slightly more than 1 cup), about 12 percent of the average daily water requirement.

Diet and Disease

Diet has a profound influence on general health. We have already considered the effects of too many or too few nutrients, hypervitaminosis or avitaminosis, and above or below normal concentrations of minerals. More subtle, long-term problems may be encountered when the diet includes the wrong proportions or combinations of nutrients. The average American diet contains too many calories, and too great a proportion of those calories are provided by lipids. This diet increases the incidence of obesity, heart disease, atherosclerosis, hypertension, and diabetes in the U.S. population.

✓ In terms of servings per day, which of the six food groups is most important?

✓ What is the difference between foods described as complete proteins and incomplete proteins?

✓ How would a decrease in the amount of bile salts in the bile affect the amount of vitamin A in the body?

BIOENERGETICS

When chemical bonds are broken, energy is released. Inside cells, some of that energy may be captured as ATP, but much of it is lost to the environment as heat. The unit of energy measurement is the **calorie (c)** (KAL-o-rē), defined as the amount of energy required to raise the temperature of 1 g of water one degree centigrade. One gram of water is not a very practical measure when you are interested in the metabolic operations that keep a 70-kg human alive, so the **kilocalorie** (KIL-o-kal-o-rē) (kc), or simply **Calorie** (with a capital *C*), is used instead. Each Calorie represents the amount of energy needed to raise the temperature of 1 *kilo*gram of water one degree centigrade. When you turn to the back of a dieting guide to check the caloric value of various foods, the numbers indicate Calories, not calories. AM *Perspectives on Dieting*

Food and Energy

In living cells, organic molecules are oxidized to carbon dioxide and water. Oxidation also occurs when something burns, and this process can be experimentally observed and measured. A known amount of material (food) is placed in a chamber, called a *calorimeter* (kal-o-RIM-e-ter), that is filled with oxygen and surrounded by a known volume of water. Once the material is completely burned and only ash remains in the chamber, the number of Calories released can be determined by comparing the water temperatures before and after the test. The burning, or catabolism, of lipids releases a considerable amount of energy, roughly 9.46 Calories per gram (C/g). In contrast, the catabolism of carbohydrates releases 4.18 C/g and the catabolism of protein, 4.32 C/g. Most foods are mixtures of fats, proteins, and carbohydrates, and the values in a "Calorie counter" vary as a result.

Metabolic Rate

It is possible to examine the metabolic state of an individual and determine how many Calories are being utilized. The result can be expressed as Calories per hour, Calories per day, or Calories per unit of body weight per day, but what is actually measured is the sum total of all of the varied anabolic and catabolic processes occurring in the body. This value represents the **metabolic rate** of the individual at that time. This will change according to the activity under way—sprinting and sleeping measurements are quite different. In an attempt to reduce the variations, the testing conditions are standardized to determine the **basal metabolic rate (BMR)**. Ideally, the BMR would represent the minimum, resting energy expenditures of an awake, alert person. An average individual has a BMR of 70 C per hour or about 1680 C per day. Although the test conditions are standardized, there are many uncontrollable factors that can influence the BMR, including age, sex, physical condition, body weight, and genetic differences such as variations among ethnic groups.

The actual daily energy expenditure for each individual varies, depending on the activities undertaken. For example, a person leading a sedentary life may have near-basal energy demands, but a single hour of swimming can increase the daily caloric requirements by 500 C or more. If the daily energy intake exceeds the total energy demands, the excess will be stored, primarily as triglycerides in adipose tissue. If the daily caloric expenditures exceed the dietary supply, there will be a net reduction in the body's energy reserves and a corresponding loss in weight. This relationship accounts for the significance of calorie counting and exercise in a weight-control program.

Thermoregulation

The BMR (basal metabolic rate) is an estimate of the rate of energy use. Our cells capture only a part of that energy as ATP, and the rest is "lost" as heat. However, heat loss serves an important homeostatic pur-

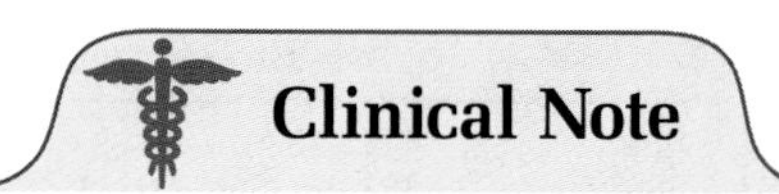

Alcohol: A Risky Diversion

Alcohol production and sales are big business throughout the Western world. Beer commercials on television, billboards advertising various brands of liquors, characters on screen enjoying a drink—all demonstrate the significance of alcohol in our society. Most people are unaware of the medical consequences of this cultural fondness for alcohol. Problems with alcohol are usually divided into those stemming from alcohol abuse and those involving alcoholism. The boundary between these conditions is rather hazy. *Alcohol abuse* is the general term for overuse and the resulting behavioral and physical effects of overindulgence. *Alcoholism* is chronic alcohol abuse with the physiological changes associated with addiction to other CNS-active drugs. Alcoholism has received the most attention in recent years, although alcohol abuse—especially when combined with driving an automobile—is now in the limelight.

Consider the following:

- Alcoholism affects more than 10 million people in the United States alone.
- Alcoholism is probably the most expensive health problem today, with an annual estimated direct cost of more than $110 billion. Indirect costs, in terms of damage to automobiles, property, and innocent pedestrians, are unknown.
- An estimated 25–40 percent of U.S. hospital patients are undergoing treatment related to alcohol consumption. There are approximately 200,000 deaths annually due to alcohol-related medical conditions. Some major clinical conditions are caused almost entirely by alcohol consumption. For example, alcohol is responsible for 60–90 percent of all liver disease in the United States.
- Alcohol affects all physiological systems. Major clinical symptoms of alcoholism include: (1) disorientation and confusion (nervous system); (2) ulcers, diarrhea, and cirrhosis (digestive system); (3) cardiac arrhythmias, cardiomyopathy, anemia (cardiovascular system); (4) depressed sexual drive and testosterone levels (reproductive system); and (5) itching and angiomas (integumentary system).
- The toll on newborn infants has risen steadily over the past 30 years as the number of women drinkers increased. Women consuming 1 ounce of alcohol per day during pregnancy have a higher rate of spontaneous abortion and produce children with lower birth weights than women who consume no alcohol. Heavier drinking causes *fetal alcohol syndrome (FAS)*. This condition is marked by characteristic facial abnormalities, a small head, slow growth, and mental retardation.
- Perhaps most disturbing of all, the problem of alcohol abuse is considerably more widespread than alcoholism. Although the medical effects are less well documented, they are certainly significant.

Several factors interact to produce alcoholism. The primary risk factors are gender (males are more likely to become alcoholics than females) and a family history of alcoholism. There does appear to be a genetic component, but the relative importance of genes versus social environment has been difficult to assess. It is likely that alcohol abuse and alcoholism can result from a variety of factors.

Treatment may consist of counseling and behavior modification. To be successful, treatment must involve a total avoidance of alcohol. Supporting groups, such as Alcoholics Anonymous, can be very helpful in providing a social framework for abstinence. Use of the drug *disulfiram (Antabuse)* has not proved to be as successful as originally anticipated. Antabuse sensitizes the individual to alcohol so that a drink produces intense nausea, and it was anticipated that this would be an effective deterrent. Clinical tests indicated that it could increase the time between drinks, but not prevent drinking altogether.

pose. Humans are subject to vast changes in environmental temperatures, but our complex biochemical systems have a major limitation: The enzyme systems will operate over only a relatively narrow range of temperatures. Therefore, our bodies have anatomical and physiological mechanisms that keep body temperatures within acceptable limits, regardless of the environmental conditions. This homeostatic process is called **thermoregulation** (*therme*, heat). Failure to control body temperature can result in a series of physiological changes, as indicated in Figure 18-9●.

Mechanisms of Heat Transfer

Heat exchange with the environment involves four basic processes: *radiation*, *conduction*, *convection*, and *evaporation* (Figure 18-10●).

1. ***Radiation.*** Warm objects lose heat as radiation. When we feel the heat from the sun, we are experiencing radiation. Our bodies lose heat the same way, but in smaller amounts. Over half of our heat loss occurs through radiation.
2. ***Conduction.*** *Conduction* refers to the direct transfer of energy through physical contact. When you sit down on a cold plastic chair in an air-conditioned room, you are immediately aware of this process. Conduction is usually not an effective mechanism for gaining or losing heat.
3. ***Convection.*** Convection is the result of conductive heat loss to the air that overlies the surface of the body. Warm air is lighter than cool air, and so it rises. As the body conducts heat to the air next to the skin, that air warms and rises; cooler air re-

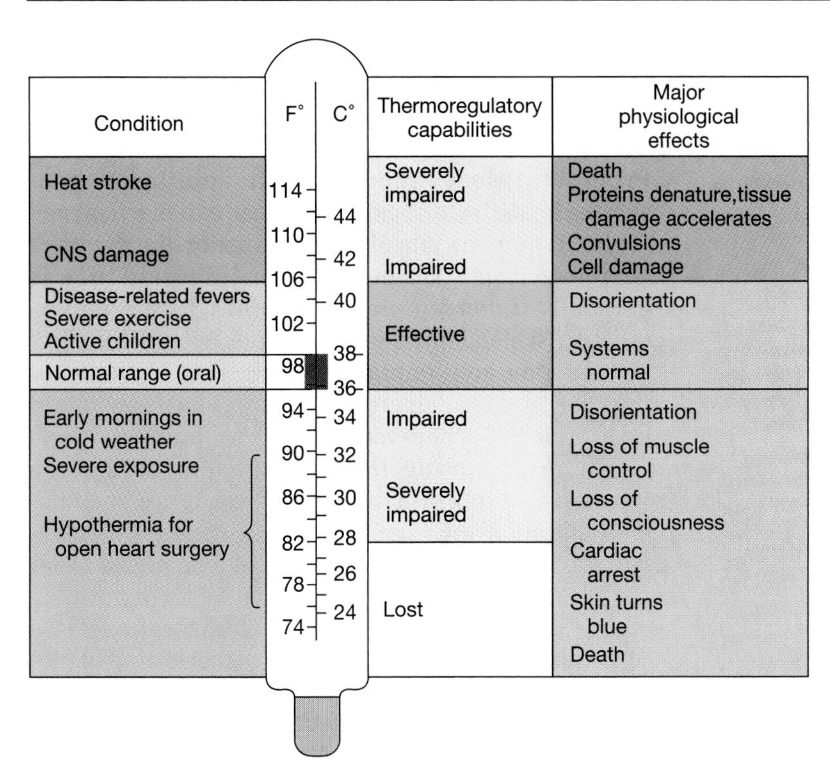

Condition
F°
C°
Thermoregulatory capabilities
Major physiological effects
Heat stroke
CNS damage
Disease-related fevers
Severe exercise
Active children
Normal range (oral)
Early mornings in cold weather
Severe exposure
Hypothermia for open heart surgery
114
110
106
102
98
94
90
86
82
78
74
44
42
40
38
36
34
32
30
28
26
24
Severely impaired
Impaired
Effective
Impaired
Severely impaired
Lost
Death
Proteins denature, tissue damage accelerates
Convulsions
Cell damage
Disorientation
Systems normal
Disorientation
Loss of muscle control
Loss of consciousness
Cardiac arrest
Skin turns blue
Death

Figure 18-9
Normal and Abnormal Variations in Body Temperature

places it, and as it in turn becomes warmed, the cycle repeats.

4. ***Evaporation.*** When water evaporates, it changes from a liquid to a vapor. This process absorbs energy, roughly 580 calories (0.58 C) per gram of water evaporated. Each hour, 20–25 ml of water crosses epithelia to be evaported from the alveolar surfaces of the lungs and the surface of the skin. This *insensible perspiration* remains relatively constant; at rest it accounts for roughly one-fifth of the average heat loss. The sweat glands responsible for *sensible perspiration* have a tremendous range of activity, ranging from virtual inactivity to secretory rates of 2–4 liters per hour. This is equivalent to an entire day's resting water loss in under an hour. A maximal secretion rate would, if it were completely evaporated, remove 2320 C per hour!

To maintain a constant body temperature, the individual must lose heat as fast as it is generated by metabolic operations. At rest over half of that loss occurs through radiation, 20 percent through evaporation, 15 percent through convection, and the rest through conduction. Altering these rates requires coordination of many different systems. These adjustments are made by the **heat-loss center** and **heat-gain center** of the hypothalamus. The heat-loss center adjusts activity in the parasympathetic division of the autonomic nervous system, and the heat-gain center directs its responses through the sympathetic division. The overall effect is to control temperature by influencing two events: the rate of heat production and the rate of heat loss to the environment. These may be further supported by behavioral changes or modifications, such as adding or removing articles of clothing.

Promoting Heat Loss When the temperature at the heat-loss center exceeds its thermostat setting, three major effects result:

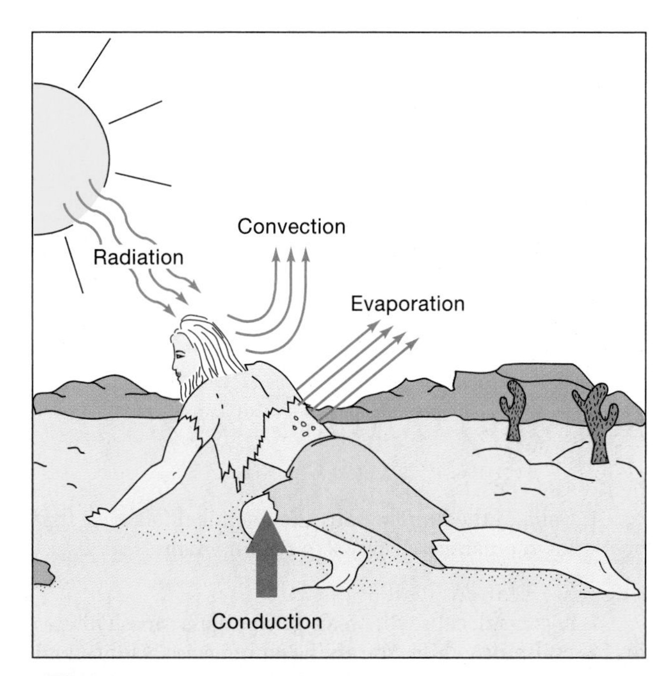

Figure 18-10
Routes of Heat Gain and Loss

1. Peripheral blood vessels dilate, and warm blood flows to the surface of the body. The skin takes on a reddish color and rises in temperature, and heat loss through radiation and convection increases.
2. Sweat glands are stimulated and, as perspiration flows across the body surface, evaporative heat losses accelerate.
3. The respiratory centers are stimulated and the depth of respiration increases. Often the individual begins respiring through the mouth, increasing evaporative losses through the lungs.

The efficiency of heat loss by evaporation varies with environmental conditions, especially the "relative humidity" of the air. If the air is saturated (100 percent humidity), it already holds as much water vapor as it will accept at that temperature. Under these conditions, evaporation is ineffective as a cooling mechanism. This is why humid, tropical conditions can be so uncomfortable—people perspire continually but remain warm and wet. AM *Heat Exhaustion and Heat Stroke*

Restricting Heat Loss The function of the heat-gain center of the brain is to prevent **hypothermia** (hī-pō-THER-mē-uh), or below-normal body temperature. When body temperature falls below acceptable levels, the heat-loss center is inhibited, and the heat-gain center is activated. Blood flow to the skin decreases, and with the circulation restricted it may take on a bluish or pale coloration. In addition, the pattern of blood flow changes. In warm weather, blood flows in a superficial venous network. In cold weather, blood is diverted to a network of deep veins that lie beneath an insulating layer of subcutaneous fat. AM *Hypothermia and Accidental Hypothermia*

Promoting Heat Production In addition to conserving heat, the heat-gain center has two mechanisms for increasing the rate of heat production. In *shivering thermogenesis* (ther-mō-JEN-e-sis), muscle tone is gradually increased until the point where brief, oscillatory skeletal muscle contractions occur. This shivering stimulates energy consumption by skeletal muscles, and it can increase the rate of heat generation by as much as 400 percent.

In *nonshivering thermogenesis* hormones are released that increase the metabolic activity of all tissues. Epinephrine from the adrenal gland immediately increases the rates of glycolysis in liver and skeletal muscle and the rate of aerobic metabolism in most tissues. The heat-gain center also stimulates the release of thyroxine by the thyroid gland, accelerating carbohydrate use and the breakdown of all other substrates. These effects develop gradually, over a period of days to weeks. AM *Thermoregulatory Problems of Infants*

✓ How would the BMR (basal metabolic rate) of a pregnant woman compare with her BMR in the nonpregnant state?

✓ Under what conditions would evaporative cooling of the body be ineffective?

✓ What effect would vasoconstriction of peripheral blood vessels have on body temperature on a hot day?

Chapter Review

KEY TERMS

aerobic metabolism, p. 466
anaerobic metabolism, p. 466
basal metabolic rate (BMR), p. 479
electron transport system, p. 468
glycogen, p. 469
glycolysis, p. 466
TCA (Krebs) cycle, p. 467
metabolism, p. 465
metabolic turnover, p. 465
nutrient, p. 465
thermogenesis, p. 482
vitamin, p. 476

SUMMARY OUTLINE

INTRODUCTION p. 465

1. Cells in the human body are chemical factories that break down organic substrates to obtain energy.

CELLULAR METABOLISM p. 465

1. In general, cells will break down excess carbohydrates first, then lipids, while conserving amino acids. Only about 40 percent of the energy released through catabolism is captured in ATP; the rest is released as heat. (*Figure 18-1*)

2. Cells synthesize new compounds (1) to perform structural maintenance or repair, (2) to support growth, and (3) to produce secretions.

Carbohydrate Metabolism p. 466

3. Most cells generate ATP and other high-energy compounds through the breakdown of carbohydrates.

4. Glycolysis and aerobic metabolism provide most of the ATP used by typical cells. In **glycolysis**, each molecule

of glucose yields two molecules of pyruvic acid, and two molecules of ATP. (*Figure 18-2*)

5. In the presence of oxygen, the pyruvic acid molecules enter the mitochondria, where they are broken down completely in the TCA cycle. The carbon and oxygen atoms are lost as carbon dioxide, and the hydrogen atoms are passed by coenzymes to the electron transport system. (*Figure 18-3*)

6. Cytochromes pass electrons along the electron transport chain of the **electron transport system** to eventually generate ATP and water.

7. For each glucose molecule completely broken down through aerobic pathways, a typical cell gains 36 molecules of ATP.

Other Catabolic Pathways p. 469

8. Cells can break down other nutrients to provide molecules for the TCA cycle if supplies of glucose are limited. (*Figure 18-4*)

9. **Gluconeogenesis**, the synthesis of glucose, enables a cell to create glucose molecules from other carbohydrates, glycerol, or some amino acids. Glycogen is an important energy reserve when the cell cannot obtain enough glucose from the extracellular fluid. (*Figure 18-5*)

Lipid Metabolism p. 470

10. During *lipolysis* (lipid catabolism), lipids are broken down into pieces that can be converted into pyruvic acid or channeled into the TCA cycle.

11. Triglycerides are the most abundant lipids in the body. Triglycerides are split into glycerol and fatty acids. The glycerol enters the glycolytic pathways, and the fatty acids enter the mitochondria for entry into the TCA cycle.

12. *Beta-oxidation* is the breakdown of fatty acid molecules into two-carbon fragments. The fragments may be used in the TCA cycle or converted to *ketone bodies.*

13. Lipids cannot provide large amounts of ATP in a short amount of time. However, cells can shift to lipid-based energy production when glucose reserves are limited.

14. In **lipogenesis**, the synthesis of lipids, almost any organic molecule can be used to form glycerol. **Essential fatty acids** cannot be synthesized and must be included in the diet. (*Figure 18-6*)

15. Lipids circulate as **lipoproteins** (lipid-protein complexes that contain large glycerides and cholesterol) and as **free fatty acids (FFA)** (water-soluble lipids that can diffuse easily across cell membranes).

Protein Metabolism p. 471

16. If other energy sources are inadequate, mitochondria can break down amino acids. In the mitochondria, the amino group may be removed by *transamination* or *deamination.* The resulting carbon skeleton may enter the TCA cycle to generate ATP or be converted to **ketone bodies**.

17. Protein catabolism is an impractical source for quick energy.

18. Roughly half of the amino acids needed to build proteins can be synthesized. There are ten **essential amino acids** that need to be acquired through the diet.

Nucleic Acid Metabolism p. 473

19. DNA in the nucleus is never catabolized for energy. RNA molecules are broken down and replaced regularly; usually they are recycled as new nucleic acids.

A Summary of Cellular Metabolism p. 473

20. No one cell can perform all of the anabolic and catabolic operations necessary to support life. Homeostasis can be preserved only when metabolic activities of different tissues are coordinated. (*Figure 18-7*)

DIET AND NUTRITION p. 474

1. **Nutrition** is the absorption of nutrients from food. A *balanced diet* contains all of the ingredients necessary to maintain homeostasis; it prevents **malnutrition**.

The Basic Food Groups p. 474

2. The six **basic food groups** are the milk; meat; vegetable; fruit; fats, oils, and sweets; and bread and cereal groups. These are arranged in a *food pyramid* with the bread and cereal group forming the base. (*Figure 18-8*)

Minerals, Vitamins, and Water p. 475

3. **Minerals** act as cofactors in various enzymatic reactions. They also contribute to the osmolarity of body fluids, and they play a role in membrane potentials, action potentials, neurotransmitter release, muscle contraction, construction and maintenance of the skeleton, transport of gases, buffer systems, fluid absorption, and waste removal. (*Table 18-1*)

4. Vitamins are needed in very small amounts. Vitamins A, D, E, and K are the **fat-soluble vitamins**; taken in excess, they can lead to *hypervitaminosis.* **Water-soluble vitamins** are not stored in the body; lack of adequate dietary supplies may lead to *deficiency disease (avitaminosis).* (*Tables 18-2, 18-3*)

5. Daily water requirements average about 40 ml/kg body weight. Water is obtained from food, drink, and metabolic generation.

Diet and Disease, p. 479

6. A balanced diet can improve general health. Most Americans consume too many calories, mostly lipids.

BIOENERGETICS p. 479

1. The energy content of food is usually expressed as **Calories** per gram (C/g). Less than half of the energy content of glucose or any other organic nutrient can be captured by our cells.

Food and Energy p. 479

2. The catabolism of lipids releases 9.46 C/g, about twice the amount as equivalent weights of carbohydrates and proteins.

Metabolic Rate p. 479

3. The total of all the anabolic and catabolic processes under way in the body represents the **metabolic rate** of a individual. The **basal metabolic rate (BMR)** is the rate of energy utilization at rest.

Thermoregulation p. 479

4. The homeostatic regulation of body temperature is **thermoregulation**. Heat exchange with the environment involves four processes: *radiation, conduction, convection,* and *evaporation.* (*Figures 18-9, 18-10*)

5. The hypothalamus acts as the body's thermostat, containing the **heat-loss center** and the **heat-gain center**.

6. Mechanisms for increasing heat loss include both physiological mechanisms (superficial blood vessel dilation, increased perspiration, and increased respiration) and behavioral adaptations.

7. Body heat may be conserved by decreased blood flow to the dermis. Heat may be generated by *shivering thermogenesis* and *nonshivering thermogenesis.*

CHAPTER QUESTIONS

LEVEL 1 Reviewing Facts and Terms

Match each item in column A with the most closely related item in column B. Use letters for answers in the spaces provided.

Column A	Column B
___ 1. glycogen formation	a. glycogenesis
___ 2. lipid catabolism	b. essential amino acid
___ 3. synthesis of lipids	c. below-normal body temperature
___ 4. linoleic acid	d. unit of energy
___ 5. deamination	e. fat-soluble vitamins
___ 6. phenylalanine	f. water-soluble vitamins
___ 7. ketoacidosis	g. lipolysis
___ 8. A, D, E, K	h. nitrogenous waste
___ 9. B complex and vitamin C	i. essential fatty acid
___ 10. calorie	j. removal of an amino group
___ 11. uric acid	k. decrease in pH
___ 12. hypothermia	l. lipogenesis

13. Cells synthesize new organic components to:
    (a) perform structural maintenance and repairs
    (b) support growth
    (c) produce secretions
    (d) a, b, and c are correct
14. During the complete catabolism of one molecule of glucose, a typical cell gains:
    (a) 4 ATP (b) 18 ATP
    (c) 36 ATP (d) 144 ATP
15. The breakdown of glucose to pyruvic acid is:
    (a) glycolysis
    (b) gluconeogenesis
    (c) cellular respiration
    (d) oxidative phosphorylation
16. Glycolysis yields an *immediate* net gain of _________ for the cell.
    (a) 1 ATP (b) 2 ATP
    (c) 4 ATP (d) 36 ATP
17. The electron transport chain and phosphorylation associated with it usually yield a total of _________ molecules of ATP in the complete catabolism of one glucose molecule.
    (a) 2 (b) 4
    (c) 32 (d) 36
18. The synthesis of glucose from simpler molecules is called:
    (a) glycolysis (b) glycogenesis
    (c) gluconeogenesis (d) beta-oxidation
19. The lipoproteins that transport excess cholesterol from peripheral tissues back to the liver for storage or excretion in the bile are the:
    (a) chylomicrons (b) VLDLs
    (c) LDLs (d) HDLs
20. The removal of an amino group in a reaction that generates an ammonia molecule is called:
    (a) ketoacidosis (b) transamination
    (c) deamination (d) denaturation
21. A complete protein contains:
    (a) the proper balance of amino acids
    (b) all the essential amino acids in sufficient quantities
    (c) a combination of nutrients selected from the food pyramid
    (d) N compounds produced by the body
22. All minerals and most vitamins:
    (a) are fat-soluble
    (b) cannot be stored by the body
    (c) cannot be synthesized by the body
    (d) must be synthesized by the body because they are not present in adequate amounts in the diet
23. The basal metabolic rate (BMR) represents the:
    (a) maximum energy expenditure when exercising
    (b) minimum, resting energy expenditure of an awake, alert person
    (c) minimum amount of energy expenditure during light exercise
    (d) muscular energy expenditure added to the resting energy expenditure
24. Over half of the heat loss from our bodies is attributable to:
    (a) radiation (b) conduction
    (c) convection (d) evaporation
25. Define the terms *metabolism*, *anabolism*, and *catabolism*.
26. What is a lipoprotein? What are the major groups of lipoproteins, and how do they differ?
27. Why are vitamins and minerals essential components of the diet?
28. What energy yields in Calories per gram are associated with the catabolism of carbohydrates, lipids, and proteins?
29. What is the basal metabolic rate (BMR)?
30. What four basic mechanisms are involved in the body's thermoregulation and the environment?

LEVEL 2 Reviewing Concepts

31. The function of the TCA cycle is to:
    (a) produce energy during periods of active muscle contraction
    (b) break six-carbon chains into three-carbon fragments
    (c) prepare the glucose molecule for further reactions
    (d) remove hydrogen atoms from organic molecules and transfer them to coenzymes
32. During periods of fasting or starvation, the presence of ketone bodies in the circulation causes:
    (a) an increase in the pH (b) a decrease in the pH
    (c) a neutral pH (d) diabetes insipidus
33. What happens during the process of glycolysis? What conditions are necessary for this process to take place?
34. Why is the TCA cycle called a cycle? What substance(s) enter(s) the cycle, and what substance(s) leave(s) it?
35. How are lipids catabolized in the body? How is beta-oxidation involved with lipid catabolism?
36. How can the food pyramid be used as a tool to obtain nutrients in sufficient quantity and quality? Why are the dietary fats, oils, and sugars at the top of the pyramid and the bread and cereal at the bottom?
37. How is the brain involved in body temperature regulation?
38. Articles in popular magazines sometimes refer to "good cholesterol" and "bad cholesterol." To what types and functions of cholesterol might these terms refer? Explain your answer.

LEVEL 3 Critical Thinking and Clinical Applications

39. Why is a starving person more susceptible to infectious disease than one who is well nourished?
40. The drug *colestipol* binds bile salts in the intestine, forming complexes that cannot be absorbed. How would this drug affect cholesterol levels in the blood?

19 The Urinary System

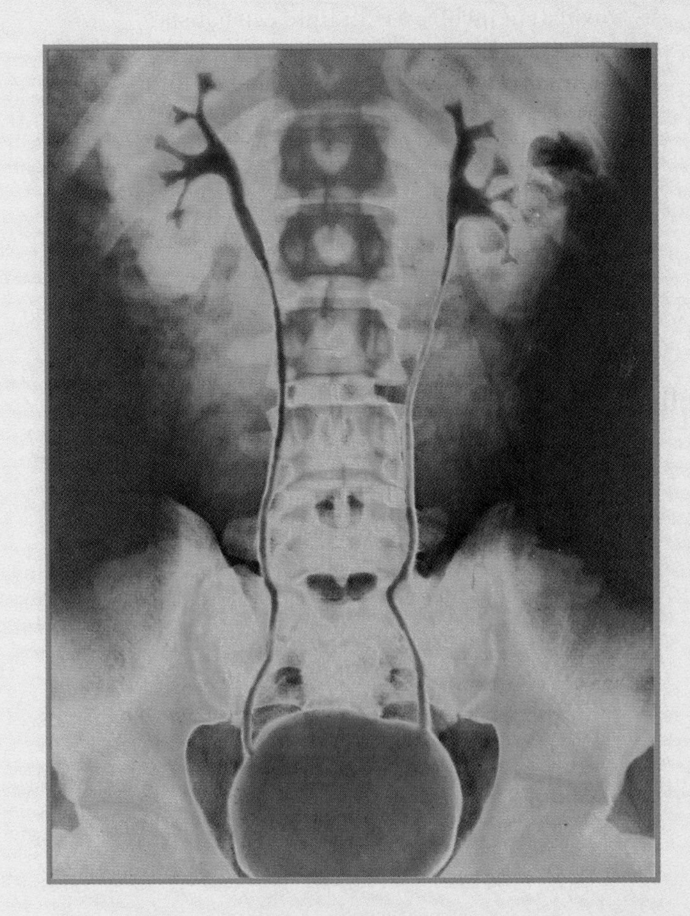

Urine formed in the kidneys takes a long journey through the ureters down to the urinary bladder—an excruciatingly painful journey if one is passing a "kidney stone." A radiographic view (such as this color-enhanced photo) can be used to detect the presence of kidney stones, a condition known as nephrolithiasis.

Chapter Outline and Objectives

Vocabulary Development

calyx, a cup of flowers; *minor calyx*
fenestra, a window; *fenestrated capillaries*
glomus, a ball; *glomerulus*
juxta, near; *juxtaglomerular apparatus*
podon, foot; *podocyte*
rectus, straight; *vasa recta*
retro-, behind; *retroperitoneal*
vasa, vessels; *vasa recta*

The human body contains trillions of cells bathed in extracellular fluid. Previous chapters compared these cells to factories that burn nutrients to obtain energy. One can imagine what would happen if real factories were built as close together as cells in the body. Each would generate piles of garbage, and the smoke they produced, together with the depletion of oxygen, would drastically reduce air quality. In short, there would be a serious pollution problem.

Comparable problems do not develop within the body as long as the activities of the digestive, cardiovascular, respiratory, and urinary systems are coordinated. The digestive tract absorbs nutrients from food and excretes solid wastes, and the liver adjusts the nutrient concentration of the circulating blood. The cardiovascular system delivers these nutrients and oxygen from the respiratory system to peripheral tissues. As blood leaves these tissues it carries the carbon dioxide and cellular waste products to sites of excretion. The carbon dioxide is eliminated at the lungs, as described in Chapter 16. Most of the organic waste products in the blood are removed and excreted in urine produced by the kidneys of the urinary system.

The urinary system (Figure 19-1●) performs vital excretory functions and eliminates the dissolved organic waste products generated by cells throughout the body. But it also has other essential functions that are often overlooked. A more complete list includes the following:

1. **Regulating blood volume and blood pressure** by (a) adjusting the volume of water lost in the urine and (b) releasing the hormones erythropoietin and renin. ∞ [p. 296]
2. **Regulating plasma concentrations of sodium, potassium, chloride, and other ions** by controlling the quantities lost in the urine and controlling calcium ion levels by the synthesis of calcitriol. ∞ [p. 295]
3. **Stabilizing blood pH** by controlling the loss of hydrogen ions (H^+) and bicarbonate ions (HCO_3^-) in the urine.
4. **Conserving valuable nutrients** (such as glucose and amino acids) by preventing their excretion in the urine, while eliminating organic waste products, especially the nitrogenous wastes *urea* and *uric acid.*

These activities must be carefully regulated to keep the composition of the blood within acceptable limits. A disruption of any one of these functions will have immediate and potentially fatal consequences. This

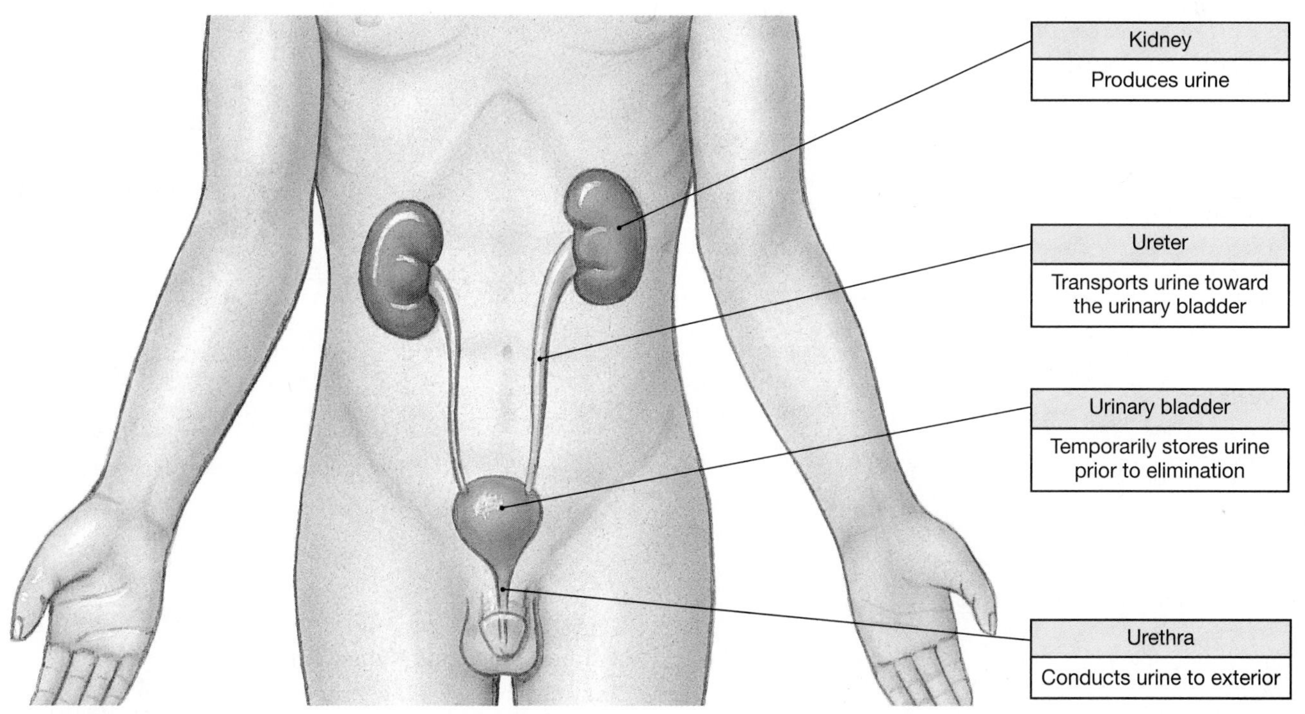

●**Figure 19-1**
Components of the Urinary System

chapter examines the functional organization of the urinary system and describes the major regulatory mechanisms that control urine production and concentration.

ORGANIZATION OF THE URINARY SYSTEM

The components of the urinary system are indicated in Figure 19-1●. The two **kidneys** produce **urine**, a liquid containing water, ions, and small soluble compounds. Urine leaving the kidneys travels along the paired **ureters** (ū-RĒ-terz) to the **urinary bladder** for temporary storage. When urination, or **micturition** (mik-tu-RI-shun), occurs, contraction of the muscular bladder forces the urine through the **urethra** and out of the body.

THE KIDNEYS

The kidneys are located on either side of the vertebral column between the last thoracic and third lumbar vertebrae. The right kidney often sits slightly lower than the left (see Figure 19-2a●), and both lie between the muscles of the dorsal body wall and the peritoneal lining (Figure 19-2b●). This position is called **retroperitoneal** (re-trō-per-i-tō-NĒ-al; *retro-*, behind) because the organs are behind the peritoneum.

The position of the kidneys is maintained by (1) the overlying peritoneum, (2) contact with adjacent organs, and (3) supporting connective tissues. In effect, each kidney is packed in a protecting, soft cushion of adipose tissue that prevents the jolts and shocks of day-to-day existence from disturbing normal kidney function. If the kidney is displaced, a condition called a *floating kidney*, the ureters or renal blood vessels may become twisted or kinked, an extremely dangerous event.

Superficial and Sectional Anatomy

As you might expect, each kidney is shaped like a kidney bean. An indentation, called the **hilus**, is the point of entry for the *renal artery* and exit for the *renal vein* and *ureter*. The surface of the organ is covered by a dense fibrous *renal capsule*. The inner portion of

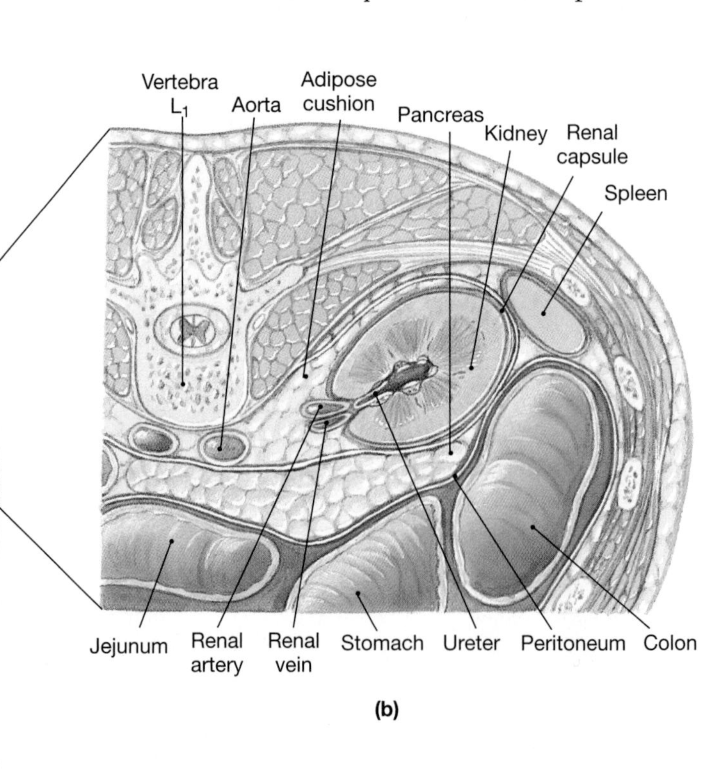

●**Figure 19-2**
An Introduction to Kidney Anatomy

(a) Anterior view of the trunk, showing the positions of the kidneys and other components of the urinary system. **(b)** Sectional view at the level indicated in part (a).

the capsule folds inward at the hilus and lines an internal cavity, the *renal sinus*.

Seen in section (Figure 19-3●), the kidney can be divided into an outer renal **cortex** and an inner renal **medulla**. The medulla contains 6 to 18 conical **renal pyramids**, whose tips, or **papillae**, project into the renal sinus. **Renal columns** extend from the cortex inward toward the renal sinus between adjacent renal pyramids.

Urine production begins in the renal pyramids and overlying areas of renal cortex. Ducts within each renal papilla discharge urine into a cup-shaped drain, called a **minor calyx** (KĀ-liks; *calyx*, a cup of flowers; plural *calyces*). Four or five minor calyces (KĀL-i-sēz) merge to form the **major calyces**, both of which combine to form a large, funnel-shaped chamber, the **renal pelvis**. The renal pelvis is connected to the ureter at the hilus of the kidney.

The Nephron

Urine production begins in the cortex at microscopic structures called **nephrons** (NEF-rons), the basic functional unit in the kidney. Each nephron consists of a **renal tubule** that is roughly 50 mm long. The tubule has two *convoluted* (coiled or twisted) segments separated by a simple U-shaped tube. The convoluted segments are in the cortex, and the tube extends partially or completely into the medulla. For clarity, the nephron shown in Figure 19-4● has been shortened and straightened out.

An Overview of the Nephron

The nephron begins at a *renal corpuscle* (KŌR-pus'l), that contains a capillary knot, or *glomerulus* (glo-MER-ū-lus; *glomus*, a ball) within an expanded chamber. Blood arrives at the glomerulus via the *afferent arteriole* and departs in the *efferent arteriole*. Filtration occurs in the renal corpuscle as blood pressure forces fluid and dissolved solutes out of the capillaries and into the surrounding chamber. This movement produces a protein-free solution known as a **filtrate**.

From the renal corpuscle, the filtrate enters the renal tubule, a long passageway that is subdivided into different regions. Major segments include the *proximal convoluted tubule*, the *loop of Henle* (HEN-lē), and the *distal convoluted tubule*. As the filtrate travels along the tubule, its composition is gradually changed and it is called *tubular fluid*. The changes that occur and the urine that results depend on the specialized activities under way in each segment of the nephron.

Each nephron empties into a *collecting duct*, the start of the **collecting system**. The collecting duct leaves the cortex and descends into the medulla, carrying tubular fluid from many nephrons toward a *papillary duct* that delivers the fluid, now called *urine*, into the renal pelvis.

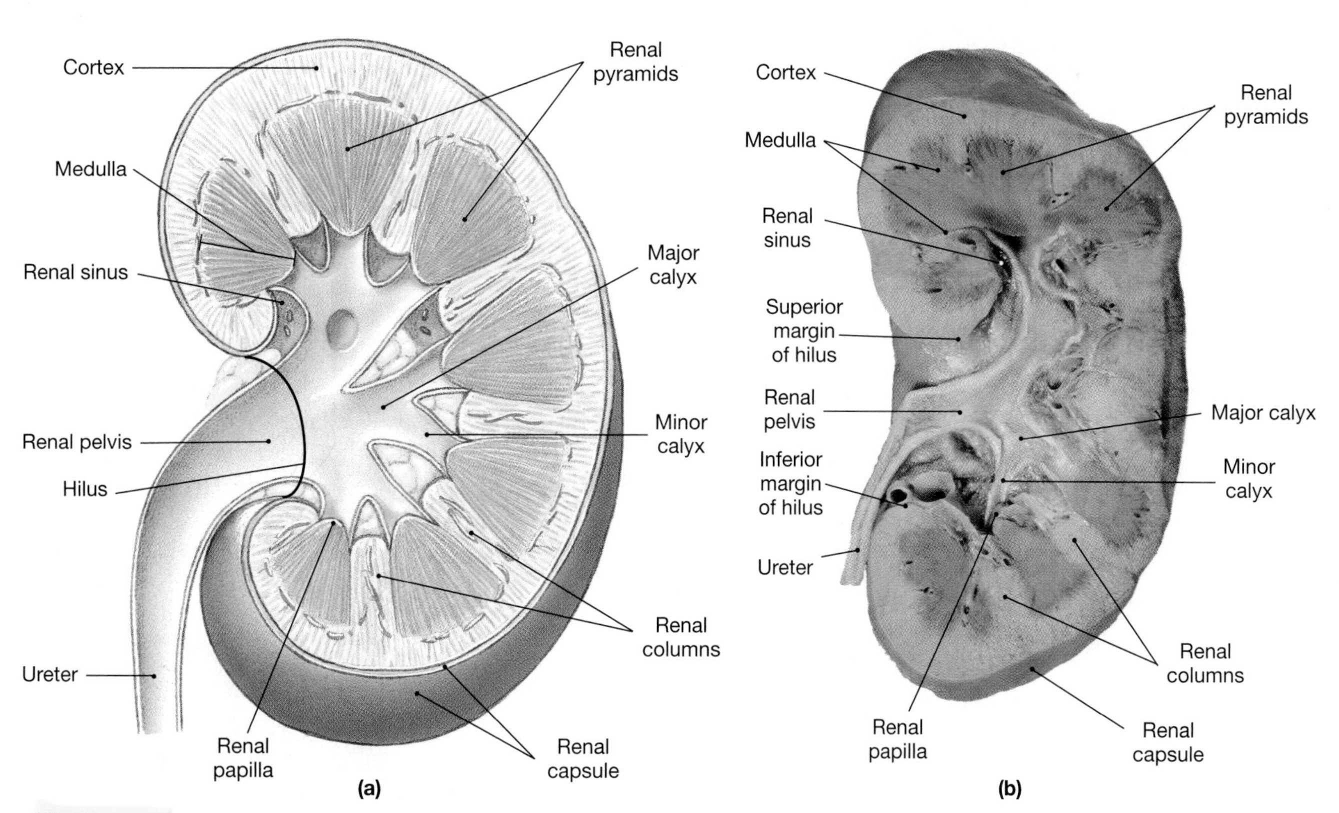

●**Figure 19-3**
Structure of the Kidney
Diagrammatic **(a)** and sectional **(b)** views of a frontal section through the left kidney.

Functions of the Nephron

Urine is very different from the filtrate produced at the renal corpuscle. The role of each segment of the nephron in the conversion of filtrate to urine is indicated in Figure 19-4●. The renal corpuscle is the site of filtration. During filtration, blood pressure forces water and small solutes across the glomerular walls. The functional advantage of this process is that it is passive and does not require an expenditure of energy. The disadvantage of filtration is that any filter with pores large enough to permit passage of organic waste products is unable to *prevent* the passage of water, ions, and nutrients such as glucose, fatty acids, and amino acids. These substances, along with most of the water, must be reclaimed and the waste products must be excreted in a relatively concentrated solution. The segments of the renal tubule are responsible for:

- Reabsorbing all of the useful organic molecules from the filtrate.
- Reabsorbing over 90 percent of the water in the filtrate.
- Secreting into the tubular fluid any waste products that were missed by the filtration process.

Additional water and salts will be removed in the collecting system before the urine is released into the

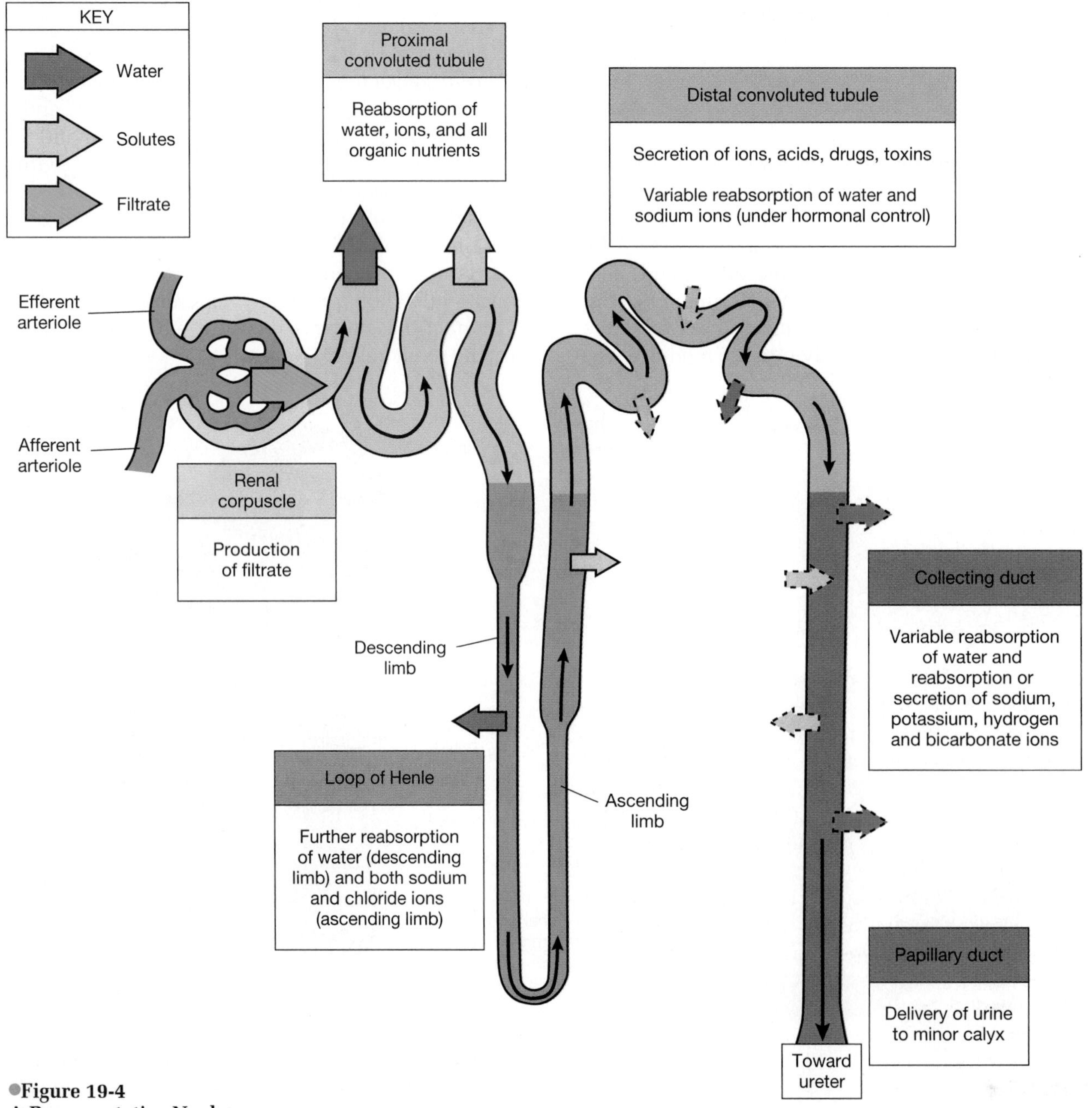

●**Figure 19-4**
A Representative Nephron
Diagrammatic view indicating major structures and functions of each segment of the nephron and collecting system.

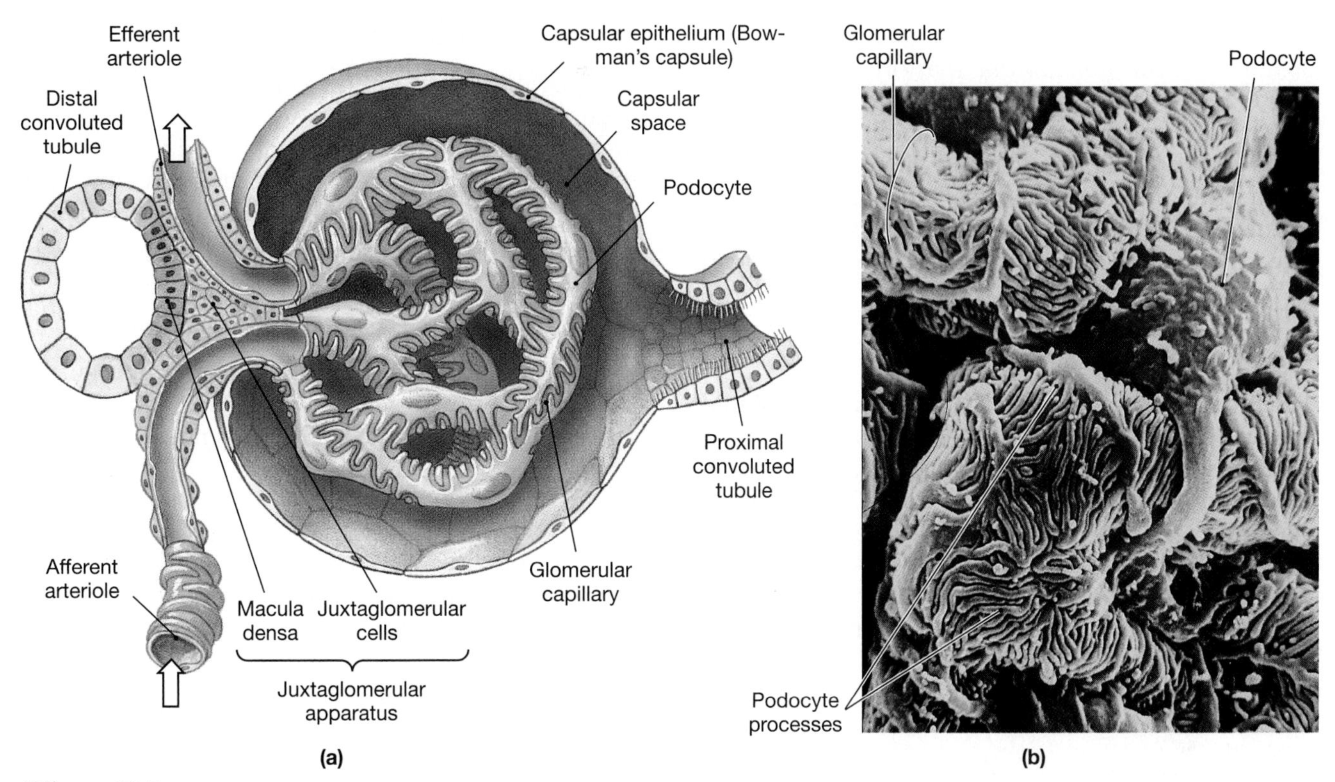

Figure 19-5
The Renal Corpuscle

(a) The renal corpuscle, showing important structural features. **(b)** Electron micrograph of the glomerular surface, showing individual podocytes and their processes. (SEM × 27,248) **(c)** A diagrammatic view of a section from a glomerulus, showing the composition of the filtration membrane.

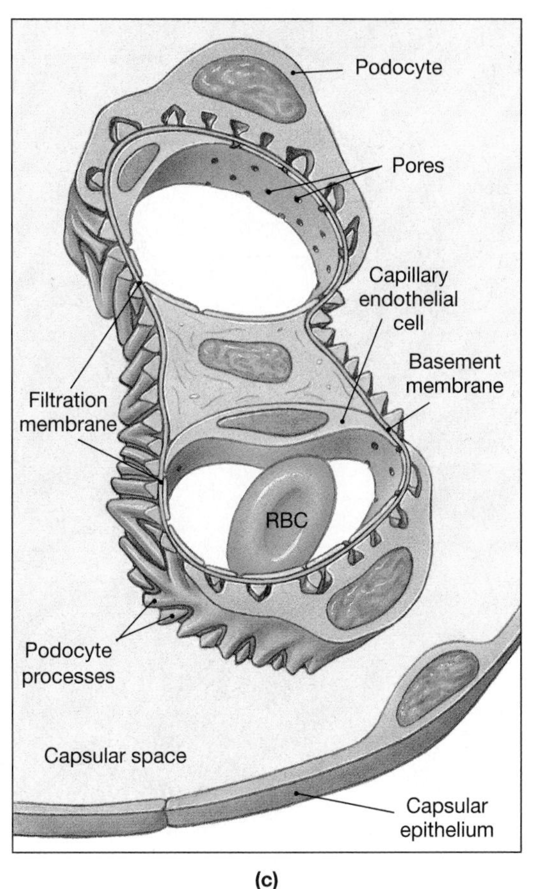

renal sinus. Table 19-2 (p. 495) summarizes the functions of the different regions of the nephron and collecting system.

The Renal Corpuscle

The **renal corpuscle** (Figure 19-5) consists of (1) the capillary knot of the **glomerulus** and (2) the expanded initial segment of the renal tubule, a region known as **Bowman's capsule**. The glomerulus projects into the Bowman's capsule much as the heart projects into the pericardial cavity (Figure 19-5a). A *capsular epithelium* lines the wall of the capsule and a *glomerular epithelium* covers the glomerular capillaries. The two are separated by the **capsular space**, which receives the filtrate and empties into the renal tubule. The glomerular epithelium consists of cells called **podocytes** (PŌ-do-sīts, *podon*, foot). Podocytes have long cellular processes that wrap around individual capillaries. A thick basement membrane separates the endothelial cells of the capillaries from the podocytes. The glomerular capillaries are called *fenestrated* (FEN-e-strā-ted; *fenestra*, a window) because their endothelial cells contain pores (Figure 19-5c). To enter the capsular space, a solute must be small enough to pass through (1) the pores of the endothelial cells, (2) the fibers of the basement membrane, and (3) the slits between the slender processes of the podocytes (Figure 19-5b,c). The fenestrated capillary, basement membrane, and podocyte processes create a *filtration membrane* that prevents the passage of blood cells and most plasma proteins but permits the movement of water, metabolic wastes, ions, glucose, fatty acids, amino acids, vitamins, and other solutes into the capsular space. Most of the valuable solutes will be reabsorbed by the proximal convoluted tubule. AM *Conditions Affecting Filtration* and *Glomerulonephritis*

The Proximal Convoluted Tubule

The filtrate next moves into the **proximal convoluted tubule (PCT)** (Figures 19-4, 19-5a●). The cells lining the PCT actively absorb organic nutrients, plasma proteins, and ions from the tubular fluid. These materials are then released into the interstitial fluid surrounding the renal tubule. As this transport occurs, the solute concentration of the interstitial fluid increases, while that of the tubular fluid decreases. Water then moves out of the tubular fluid by osmosis, reducing the volume of tubular fluid.

The Loop of Henle

The last portion of the proximal convoluted tubule bends sharply and connects to the **loop of Henle** (Figure 19-4●, p. 490). This loop can be divided into a *descending limb* that travels toward the renal pelvis and an *ascending limb* that returns to the cortex. The ascending limb, which is impermeable to water and solutes, actively transports sodium and chloride ions out of the tubular fluid. As a result, the interstitial fluid of the medulla contains an unusually high solute concentration. The descending limb is permeable to water, and as it descends into the medulla water moves out of the tubular fluid by osmosis.

The Distal Convoluted Tubule

The ascending limb of the loop of Henle ends where it bends and comes in close contact with the glomerulus and its vessels. At this point, the **distal convoluted tubule (DCT)** begins, and it passes between the afferent and efferent arterioles (Figure 19-5a●)

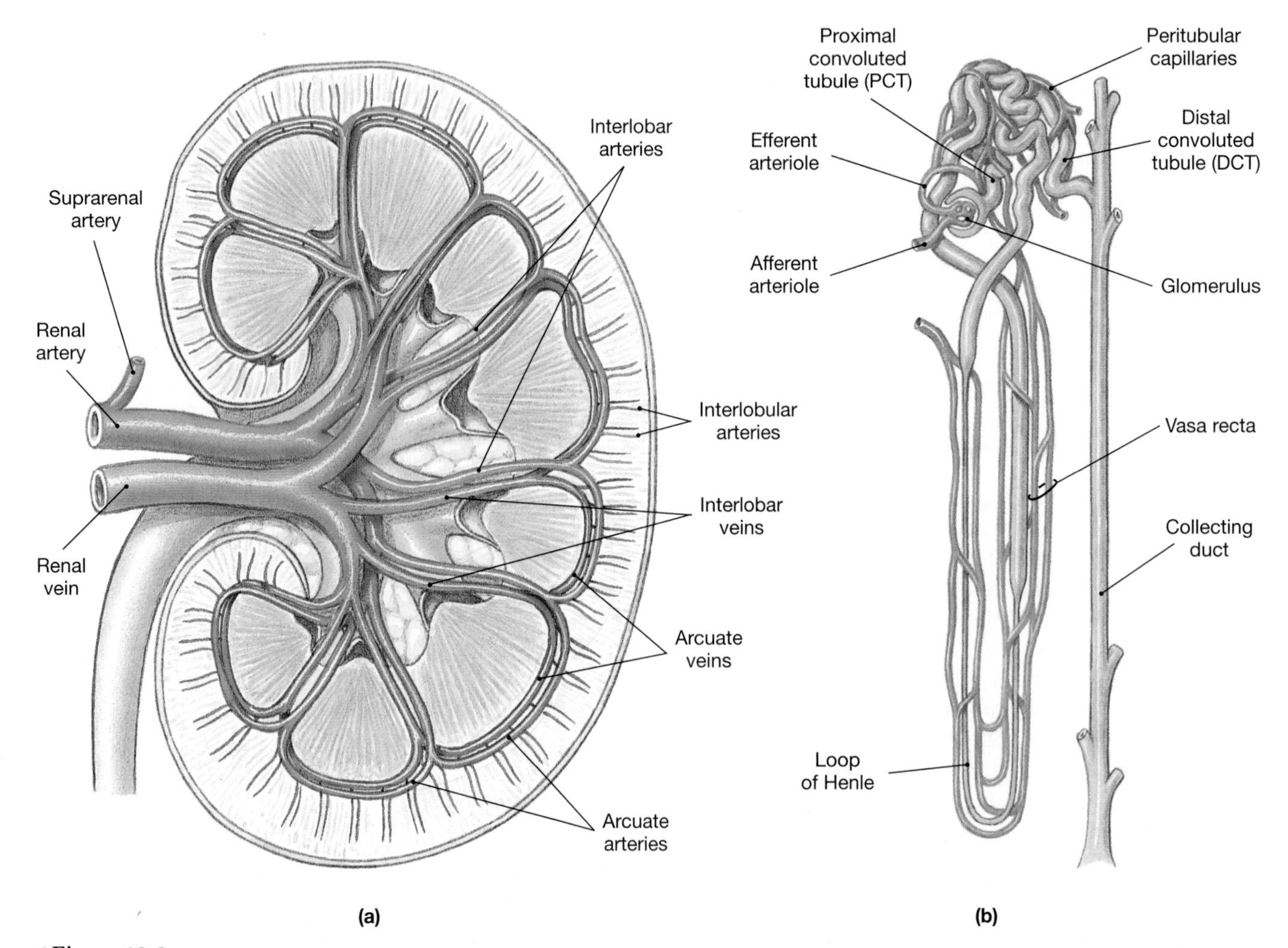

●**Figure 19-6**
Blood Supply to the Kidneys
(a) Sectional view, showing major arteries and veins; compare with *Figure 19-3*. **(b)** Circulation to an individual nephron.

The distal convoluted tubule is an important site for (1) The active secretion of ions, acids, and other materials; and (2) The selective reabsorption of sodium ions from the tubular fluid. In the final portions of the DCT an osmotic flow of water may assist in concentrating the tubular fluid.

The cells of the DCT closest to the glomerulus are unusually tall, and their nuclei are clustered together. This region, diagrammed in Figure 19-5a●, is called the *macula densa* (MAK-ū-la DEN-sa). The cells of the macula densa are closely associated with unusual smooth muscle fibers, the *juxtaglomerular cells* (*juxta*, near) in the wall of the afferent arteriole. Together the macula densa and **juxtaglomerular cells** form the **juxtaglomerular apparatus,** an endocrine structure that secretes two hormones, *renin* and *erythropoietin*, introduced in Chapter 11. ∞ *[p. 296]*

The Collecting System

The distal convoluted tubule, the last segment of the nephron, opens into the collecting system. The collecting system consists of *collecting ducts* and *papillary ducts* (Figure 19-4●, p. 490). Each **collecting duct** receives tubular fluid from many nephrons, and several collecting ducts merge to form a **papillary duct** that delivers urine to a minor calyx. In addition to transporting tubular fluid from the nephron to the renal pelvis, the collecting system can make final adjustments to the composition of the urine by reabsorbing water and reabsorbing or secreting sodium, potassium, hydrogen, and bicarbonate ions.

The Blood Supply to the Kidneys

In normal individuals, about 1200 ml of blood flows through the kidneys each minute, or some 20–25 percent of the cardiac output. This is a phenomenal amount of blood for organs with a combined weight of less than 300 g (10.5 oz)! Figure 19-6a● diagrams the path of blood flow to each kidney. Each kidney receives blood from a *renal artery* that originates from the abdominal aorta. As the renal artery enters the renal sinus, it divides into branches that supply a series of **interlobar arteries** that radiate outward between the lobes (Figure 19-6a●). They then turn, arching along the boundary lines between the cortex and medulla as the **arcuate** (AR-kū-āt) **arteries.** Each of the arcuates gives rise to a number of **interlobular arteries** supplying portions of the adjacent lobe.

Blood reaches each glomerulus through an **afferent arteriole** and leaves in an **efferent arteriole** (Figure 19-6b●). It then travels to the **peritubular capillaries** that supply the proximal and distal convoluted tubules. The peritubular capillaries provide a route for the pickup or delivery of substances that are reabsorbed or secreted by these portions of the nephron.

The efferent arterioles and peritubular capillaries are further connected to a series of long, slender capillaries that accompany the loops of Henle into the medulla. These capillaries, known as the **vasa recta** (*rectus*, straight), absorb and transport solutes and water reabsorbed by the loops of Henle or collecting ducts. Under normal conditions, the removal of solutes and water by the vasa recta balances the rates of solute and water reabsorption in the medulla.

Blood from the peritubular capillaries and vasa recta enters a network of venules and small veins that converge on the **interlobular veins.** In a mirror image of the arterial distribution, blood continues to converge and empty into the **arcuate**, **interlobar**, and **renal veins** (Figure 19-6a●).

✓ How is the position of the kidneys different from other organs in the abdominal region?

✓ Why don't plasma proteins pass into the capsular space under normal circumstances?

✓ Damage to what part of the nephron would interfere with the control of blood pressure?

BASIC PRINCIPLES OF URINE PRODUCTION

The primary goal in urine production is maintaining homeostasis by regulating the volume and composition of the blood. This process involves the excretion and elimination of dissolved solutes, specifically the following metabolic waste products:

1. **Urea.** This is the most abundant organic waste, and roughly 21 grams of urea are generated each day. Most of it is produced during the breakdown of amino acids.
2. **Creatinine.** Creatinine is generated in skeletal muscle tissue through the breakdown of *creatine phosphate*, a high-energy compound that plays an important role in muscle contraction. The body generates roughly 1.8 g of creatinine each day.
3. **Uric acid.** Approximately 480 mg of uric acid are produced each day during the breakdown and recycling of RNA.

Since these waste products must be excreted in solution, their elimination is accompanied by an unavoidable water loss. The kidneys can minimize this water loss by producing a urine that is four to five times more concentrated than normal body fluids. If the kidneys could not concentrate the filtrate produced by glomerular filtration, water losses would lead to fatal dehydration in hours. At the same time, the kidneys ensure that the urine excreted does not contain potentially useful organic substrates, such as sugars or amino acids, that are found in blood plasma.

To accomplish these goals, the kidneys rely on three distinct processes:

1. **Filtration.** In filtration, blood pressure forces water across a filtration membrane. Solute molecules small enough to pass through the membrane are carried along by the surrounding water molecules.
2. **Reabsorption.** Reabsorption is the removal of water and solute molecules from the filtrate after it enters the renal tubule. This is a selective process, whereas filtration occurs solely on the basis of size. Solute reabsorption may involve simple diffusion or the activity of carrier proteins in the tubular epithelium. Water reabsorption occurs passively, through osmosis. Reabsorbed water and solutes reenter the circulation at the peritubular capillaries and vasa recta.
3. **Secretion.** Secretion is the transport of solutes across the tubular epithelium and into the filtrate. Secretion is necessary because filtration does not force all of the dissolved materials out of the plasma, and blood entering the peritubular capillaries may still contain undesirable substances.

Together these processes create a fluid that is very different from other body fluids. Table 19-1 indicates the efficiency of the renal system by comparing the composition of urine and plasma. The kidneys can continue to work efficiently only as long as filtration, reabsorption, and secretion proceed in proper balance. Any disruption in this balance has immediate and potentially disastrous effects on the composition of the circulating blood. If both kidneys fail to perform their assigned roles, death will occur within a few days unless medical assistance is provided.

All segments of the nephron and and collecting system participate in the process of urine formation. Most regions perform a combination of reabsorption and secretion, but the balance between the two shifts from one region to another. As indicated in Table 19-2:

- Filtration occurs exclusively in the renal corpuscle, across the capillary walls of the glomerulus.
- Nutrient reabsorption occurs primarily at the proximal convoluted tubule.
- Active secretion occurs primarily at the distal convoluted tubule.
- The loop of Henle and the collecting system interact to regulate the amount of water and the number of sodium and potassium ions lost to the urine.

We will now take a closer look at events under way in each of the segments of the nephron and collecting system.

TABLE 19-1 Significant Differences between Urinary and Plasma Solute Concentrations

Component	*Urine*	*Plasma*
IONS (mEq/l)		
Sodium (Na^+)	147.5	138.4
Potassium (K^+)	47.5	4.4
Chloride (Cl^-)	153.3	106
Bicarbonate (HCO_3^-)	1.9	27
METABOLITES AND NUTRIENTS (mg/dl)		
Glucose	0.009	90
Lipids	0.002	600
Amino acids	0.188	4.2
Proteins	0.000	7.5 g/dl
NITROGENOUS WASTES (mg/dl)		
Urea	1800	305
Creatinine	150	8.6
Ammonia	60	0.2
Uric acid	40	3

Filtration at the Glomerulus

Filtration Pressure

Chapter 14 introduced the forces acting across capillary walls, and you should consider reviewing that discussion before proceeding. [p. 353] Blood pressure at the glomerulus tends to force water and solutes out of the bloodstream and into the capsular space. For filtration to occur, this outward force must exceed any opposing pressures, such as the osmotic pressure of the blood. The net force promoting filtration is called the *filtration pressure.* The filtration pressure is very low (around 10 mm Hg), and kidney filtration will stop if glomerular blood pressure falls significantly. Reflexive changes in the diameters of the afferent arterioles, the efferent arterioles, and the glomerular capillaries can compensate for minor variations in blood pressure. These changes can occur automatically or in response to sympathetic stimulation. More serious declines in systemic blood pressure can reduce or even stop glomerular filtration. As a result, hemorrhaging, shock, or dehydration can cause a dangerous or even fatal reduction in kidney function. Because the kidneys are more sensitive to blood pressure than are other organs, it is not surprising to find that they control many of the homeostatic mechanisms responsible for regulating blood pressure and blood volume. Examples such as the renin-angiotensin system are considered later in this chapter.

The Glomerular Filtration Rate

Glomerular filtration is the process of filtrate production at the glomerulus; it is driven by the filtration pressure. The **glomerular filtration rate (GFR)** is the amount of filtrate produced in the kidneys each minute. Each kidney contains around 6 square meters of filtration surface, and the GFR averages an astounding *125 ml per minute.* This means that almost 20 percent of the fluid delivered to the kidneys by the renal arteries leaves the bloodstream and enters the capsular spaces. In the course of a single day, the glomeruli generate about 180 liters (50 gal) of filtrate, roughly 70 times the total plasma volume. But as the filtrate passes through the renal tubules, over 99 percent of it is reabsorbed. Tubular reabsorption is obviously an extremely important process. An inability to reclaim the water entering the filtrate, as in *diabetes insipidus*, can quickly cause death by dehydration. (This condition, caused by inadequate ADH secretion, was discussed in Chapter 11.) [p. 289]

Glomerular filtration is the vital first step essential to all kidney functions. If filtration does not occur, waste products are not excreted, pH control is jeopardized, and an important mechanism for blood volume regulation is eliminated. Filtration depends on adequate circulation to the glomerulus and maintaining normal filtration pressures. If local adjustments fail to maintain acceptable filtration pressures and the GFR declines, hormonal adjustments are initiated by the kidney. The result is a rise in filtration pressures and a restoration of normal glomerular filtration rates.

Hormonal Regulation of the Glomerular Filtration Rate The *renin-angiotensin system* changes the glomerular filtration rate by its effects on blood pressure and volume. When glomerular blood pressure declines, so does the GFR. Under these conditions, the juxtaglomerular apparatus releases the enzyme *renin* into the blood. Renin starts a chain reaction that ultimately involves many different systems, bringing about a coordinated increase in blood volume and blood pressure. [p. 359] As glomerular blood pressure rises, so do the filtration pressure and the GFR.

TABLE 19-2 The Functions of the Nephron and Collecting System in the Kidney

Region	*Primary Function*
Renal corpuscle	Filtration of plasma to initiate urine formation
Proximal convoluted tubule (PCT)	Reabsorption of ions, organic molecules, vitamins, water
Loop of Henle	Descending limb: reabsorption of water from tubular fluid Ascending limb: reabsorption of ions; assists in creation of the medullary concentration gradient
Distal convoluted tubule (DCT)	Reabsorption of sodium ions; secretion of acids, ammonia, drugs
Collecting tubule	Reabsorption of water, sodium ions
Collecting duct	Reabsorption of water, sodium ions, bicarbonate
Papillary duct	Conduction of urine to minor calyx

Reabsorption and Secretion Along the Renal Tubule

Reabsorption and secretion at the kidney involve a combination of diffusion, osmosis, and carrier-mediated transport. In carrier-mediated transport, a specific substrate binds to a carrier protein that facilitates its movement across the cellular membrane. This movement may or may not require ATP. [p. 54]

The Proximal Convoluted Tubule

The cells lining the PCT actively absorb organic nutrients, plasma proteins, and ions from the filtrate and transport them into the interstitial fluid. As these materials are absorbed and transported, osmotic forces pull water across the wall of the PCT and into the surrounding interstitial fluid. The PCT usually reclaims 60–70 percent of the volume of filtrate produced at the glomerulus, along with virtually all of the glucose, amino acids, and other organic nutrients. The PCT also actively reabsorbs ions, including sodium, potassium, calcium, magnesium, bicarbonate, phosphate, and sulfate ions. The ion pumps involved are individually regulated and may be influenced by circulating ion or hormone levels. For example, the presence of parathyroid hormone stimulates calcium ion reabsorption. [p. 293]

Although reabsorption represents the primary function of the PCT, a few substances, such as hydrogen ions, can be actively secreted into the tubular fluid. Active secretion can play an important role in the regulation of blood pH, a topic considered in a later section. A few compounds in the tubular fluid, including urea and uric acid, are ignored by the PCT and by other segments of the renal tubule. As water and other nutrients are removed, the concentration of these waste products gradually rises in the tubular fluid.

The Loop of Henle

Roughly 60–70 percent of the volume of the filtrate produced at the glomerulus has been reabsorbed before the tubular fluid reaches the loop of Henle. In the process, the useful organic molecules, along with many mineral ions, have been reclaimed. The loop of Henle will reabsorb more than half of the remaining water, as well as two-thirds of the sodium and chloride ions remaining in the tubular fluid.

The descending and ascending limbs of the loop of Henle have different functions: The ascending limb reabsorbs sodium and chloride ions, and the descending limb reabsorbs water. The two work together (Figure 19-7●). The ascending limb, which is impermeable to water, pumps sodium and chloride ions out of the tubular fluid as it flows by and dumps them into the interstitial fluid. Over time, a *concentration gradient* is created in the medulla, with the highest concentration of solutes (roughly four times that of plasma) near the bend in the loop of Henle. The descending limb is permeable to water but not to solutes. As tubular fluid flows along the descending limb, it passes through this concentration gradient. The result is an osmotic flow of water out of the tubular fluid and into the interstitial fluid.

Roughly half the volume of filtrate that enters the loop of Henle is reabsorbed in the descending limb, and most of the sodium and chloride ions are removed in the ascending limb. With the loss of the sodium and chloride ions, the solute concentration of the filtrate declines to around one-third that of plasma. However, waste products such as urea now make up a significant percentage of the remaining solutes. In essence, most of the water and solutes have been removed, leaving the waste products behind.

The Distal Convoluted Tubule and the Collecting System

By the time the filtrate reaches the **distal convoluted tubule**, roughly 80 percent of the water and 85 percent of the solutes have already been reabsorbed. The distal convoluted tubule is connected to a collecting duct that drains into the renal pelvis (Figure 19-7●).

As filtrate passes through the DCT and collecting duct, final adjustments are made in its composition and concentration. Composition depends on the type of solutes present; concentration depends on the vol-

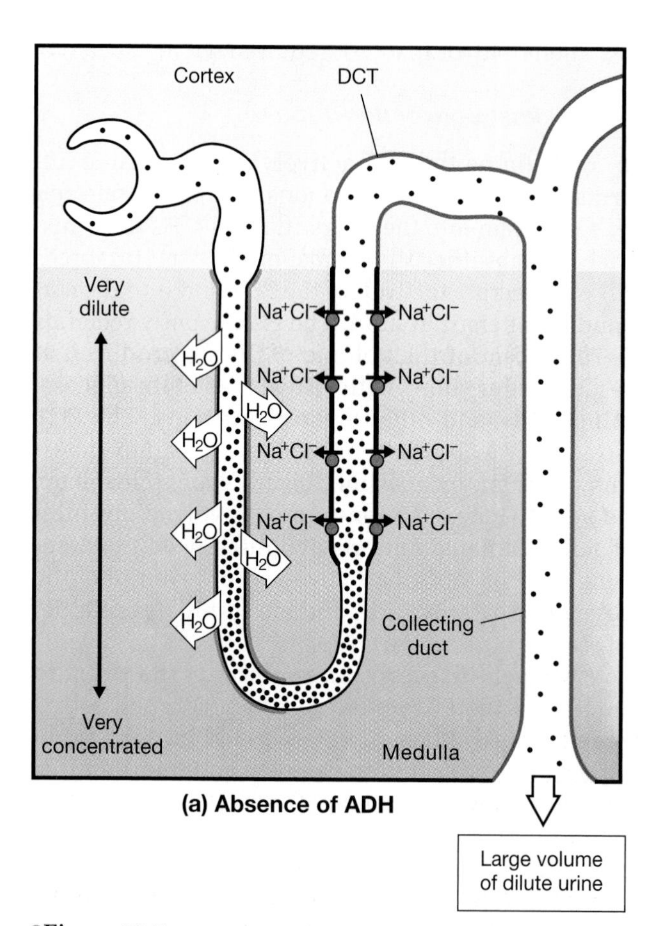

(a) Absence of ADH

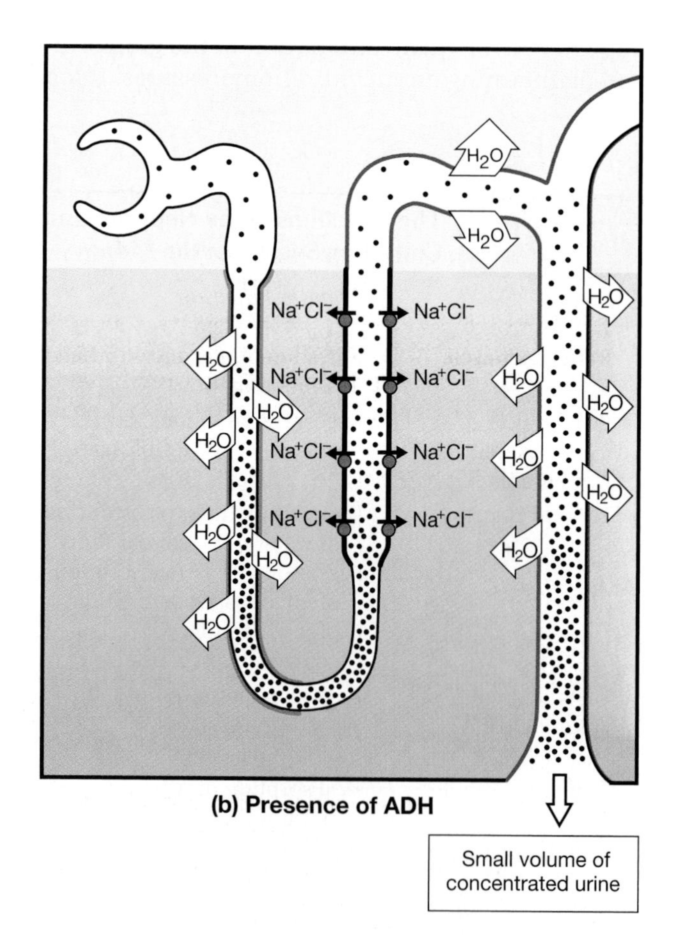

(b) Presence of ADH

●**Figure 19-7**
The Effects of ADH on the DCT and Collecting Duct
(a) Permeabilities and urine production without ADH. **(b)** Permeabilities and urine production with ADH.

ume of water in which they are dissolved. Because the DCT and collecting duct are impermeable to solutes, changes in filtrate composition can occur only through active reabsorption or secretion. The DCT is primarily concerned with active secretion.

Throughout most of the DCT, the tubular cells actively transport sodium ions out of the tubular fluid in exchange for potassium ions or hydrogen ions. The DCT and collecting ducts contain ion pumps that regulate the rates of sodium ion reabsorption and potassium ion secretion in response to the hormone *aldosterone*. Aldosterone secretion occurs (1) in response to circulating ACTH from the anterior pituitary and (2) in response to elevated potassium ion concentrations in the extracellular fluid. The higher the aldosterone levels, the more sodium ions are reclaimed, and the more potassium ions are lost.

The amount of water reabsorbed along the DCT and collecting duct is controlled by circulating levels of *antidiuretic hormone* (ADH). In the absence of ADH, the distal convoluted tubule and collecting duct are impermeable to water. The higher the level of circulating ADH, the greater the water permeability and the more concentrated the urine. Water moves out of the DCT and collecting duct because in each case the tubular fluid contains fewer solutes than the surrounding interstitial fluid. As noted above, the fluid arriving at the DCT has a solute concentration only around one-third that of the cortex, because the ascending limb of the loop of Henle has removed most of the sodium and chloride ions. The fluid passing along the collecting duct travels into the medulla, where it passes through the concentration gradient established by the loop of Henle.

If circulating ADH levels are low, little water reabsorption will occur, and virtually all of the water reaching the DCT will be lost in the urine (Figure 19-7a●). If circulating ADH levels are high, as in Figure 19-7b●, the DCT and collecting duct will be very permeable to water. In this case, the individual will produce a small quantity of urine with a solute concentration four to five times that of extracellular fluids. AM Diuretics

The Properties of Normal Urine

The general characteristics of normal urine are listed in Table 19-3, but the composition of the 1.2 liters of urine excreted each day depends on the metabolic and hormonal events under way. Because the composition and concentration of the urine vary independently, an individual can produce a small quantity of concentrated urine or a large quantity of dilute urine and still excrete the same amount of dissolved materials. For this reason, physicians often request a 24-hour urine collection rather than a single sample. This enables them to assess both quantity and composition accurately. AM Urinalysis

TABLE 19-3 General Characteristics of Normal Urine

pH	6.0 (range: 4.5–8)
Specific gravity	1.003–1.030
Osmolarity	855–1335 mOsm
Water content	93–97 percent
Volume	1200 ml/day
Color	clear yellow
Odor	varies depending on composition
Bacterial content	sterile

The Control of Kidney Function

Renal function is regulated in three ways: (1) by automatic adjustments in glomerular pressures, through changes in the diameters of the afferent and efferent arterioles; (2) through activities of the sympathetic division of the ANS; and (3) through the effects of hormones. The hormonal mechanisms make complex, long-term adjustments in blood pressure and blood volume that stabilize the GFR, in part by regulating transport mechanisms and water permeabilities of the DCT and collecting duct.

The Local Regulation of Kidney Function

Local, automatic changes in the diameters of the afferent arterioles, the efferent arterioles, and the glomerular capillaries can compensate for minor variations in blood pressure. For example, a reduction in blood flow and a decline in glomerular pressure trigger the dilation of the afferent arteriole and glomerular capillaries and the constriction of the efferent arteriole. This combination keeps glomerular blood pressure and blood flow within normal limits. As a result, filtration rates remain relatively constant despite a general decline in arterial blood pressures.

Sympathetic Activation and Kidney Function

Sympathetic activation has three different effects on kidney function. Over the short term, sympathetic activity primarily serves to shift blood away from the kidneys and lower the GFR during a sudden crisis or short-term emergency. Two different mechanisms are involved:

1. Sympathetic activation lowers the glomerular filtration rate by constricting the afferent arterioles and reducing blood flow to the glomerular capillaries. The sympathetic activation triggered by an

FOCUS A Summary of Urine Formation

Figure 19-8 summarizes the the major steps involved in the reabsorption of water and the production of concentrated urine.

Step 1. Glomerular filtration produces a filtrate resembling blood plasma but containing few plasma proteins. This filtrate has the same solute concentration as plasma or interstitial fluid.

Step 2. In the proximal convoluted tubule, 60–70 percent of the water and almost all of the dissolved nutrients are reabsorbed. The osmolarity of the tubular fluid remains unchanged.

Step 3. In the PCT and descending loop of Henle, water moves into the surrounding interstitial fluid, leaving a small fluid volume (roughly 20 percent of the original filtrate) of highly concentrated tubular fluid.

Step 4. The ascending limb is impermeable to water and solutes. The tubular cells actively pump sodium and chloride ions out of the tubular fluid. Because only sodium and chloride ions are removed, urea now accounts for a higher proportion of the solutes in the tubular fluid.

Step 5. The final composition and concentration of the tubular fluid will be determined by the events under way in the DCT and the collecting ducts. These segments are impermeable to solutes, but ions may be actively transported into or out of the filtrate under the control of hormones such as aldosterone.

Step 6. The concentration of urine is controlled by variations in the water permeabilities of the DCT and the collecting ducts. These segments are impermeable to water unless exposed to antidiuretic hormone (ADH). In the absence of ADH, no water reabsorption occurs, and the individual produces a large volume of dilute urine. At high concentrations of ADH, the collecting ducts become freely permeable to water, and the individual produces a small volume of highly concentrated urine.

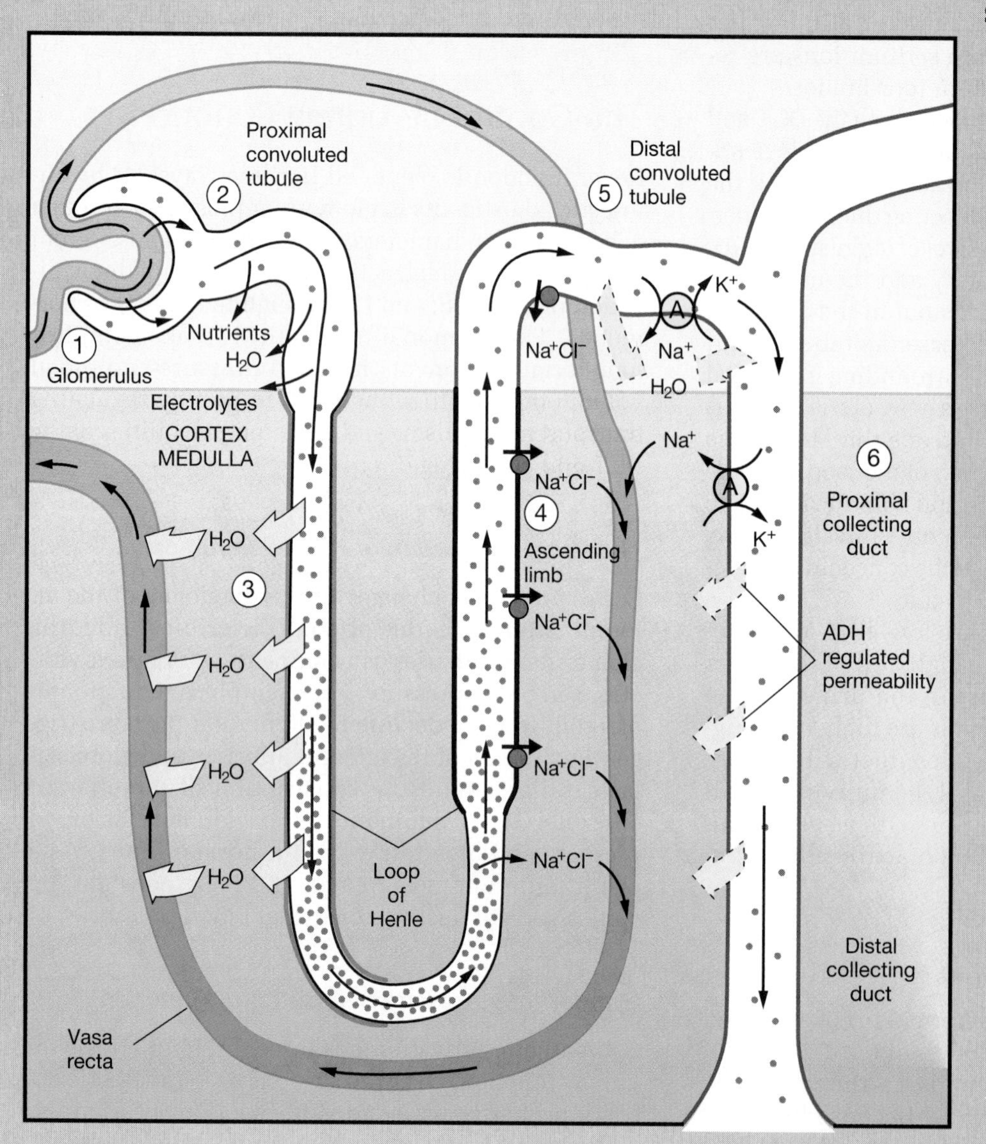

Figure 19-8 Major Steps in Urine Production

acute fall in blood pressure or a heart attack can override the hormonal regulatory mechanisms, and sustained sympathetic stimulation at high levels may cause kidney damage due to reduced blood flow and oxygen starvation.

2. Sympathetic activation of the vasomotor center changes the regional pattern of blood circulation, and this change can further reduce the GFR. For example, the dilation of superficial vessels in warm weather shunts blood away from the kid-

neys, and glomerular filtration declines temporarily. The effect becomes especially pronounced during periods of strenuous exercise. As the blood flow increases to the skin and skeletal muscles, it decreases to the kidneys. At maximal levels of exertion, renal blood flow may be less than one-quarter of normal resting levels. If you read the newspapers carefully after a marathon or ultramarathon, you will find that acute kidney disorders are relatively common; the symptoms usually (but not always) disappear within 48 hours.

Over the long term, sympathetic activation stimulates an increase in blood pressure and blood volume by triggering the release of renin by the juxtaglomerular apparatus. After a severe hemorrhage or dehydration, the hormonal response is vital to restoring normal filtration rates and kidney function.

The Hormonal Control of Kidney Function

The major hormones involved in regulating kidney function are angiotensin II, aldosterone, ADH, and ANP. These hormones have been discussed in earlier chapters, so only a brief overview will be provided here. ∞ *[pp. 289, 294, 359]* The secretion of angiotensin II, aldosterone, and ADH is integrated by the *renin-angiotensin system.*

The Renin-Angiotensin System If the glomerular pressures remain low because of a decrease in blood volume, a fall in systemic pressures, or a blockage in the renal artery or its tributaries, the juxtaglomerular apparatus releases *renin* into the circulation. Renin converts inactive *angiotensinogen* to *angiotensin I,* which a *converting enzyme* activates to *angiotensin II.* Figure 19-9● diagrams the primary effects of this potent hormone.

Angiotensin II has the following effects :

- *In peripheral capillary beds,* it causes a brief but powerful vasoconstriction, elevating blood pressure in the renal arteries.
- *At the nephron,* it triggers the contraction of the efferent arterioles, elevating glomerular pressures and filtration rates.
- *In the CNS,* it triggers the release of ADH, which in turn stimulates the reabsorption of water and sodium ions and causes the sensation of thirst.
- *At the adrenal gland,* it stimulates the secretion of aldosterone by the cortex and epinephrine and norepinephrine (NE) by the adrenal medullae. The result is a sudden, dramatic increase in systemic blood pressure. At the kidneys, aldosterone stimulates sodium reabsorption along the DCT and collecting system.

ADH Antidiuretic hormone (1) increases the water permeability of the DCT and collecting duct, stimulating the reabsorption of water from the tubular fluid, and (2) causes the sensation of thirst, leading to the consumption of additional water. ADH release occurs under angiotensin II stimulation; it also occurs independently, when hypothalamic neurons are stimulated by a fall in blood pressure or an elevation in the solute concentration of the circulating blood. The nature of the receptors involved has not been determined, but these specialized hypothalamic neurons have been named *osmoreceptors.*

Aldosterone Aldosterone secretion stimulates the reabsorption of sodium ions and the secretion of potassium ions along the DCT and collecting duct. Aldosterone secretion primarily occurs (1) under angiotensin II stimulation ∞ *[p. 359]*, and (2) in response to a rise in the potassium ion concentration of the blood.

Atrial Natriuretic Peptide The actions of atrial natriuretic peptide (ANP) oppose those of the renin-angiotensin system (Figure 14-10b●, p. 360). ANP is released by atrial cardiac muscle cells when blood volume and blood pressure are too high. The actions of ANP that affect the kidneys include (1) a decrease in the rate of sodium ion reabsorption in the DCT, leading to increased sodium ion loss in the urine, (2) dilation of the glomerular capillaries, which results in increased glomerular filtration and urinary water loss, and (3) inactivation of the renin-angiotensin system through inhibition of renin, aldosterone, and ADH secretion. The net result is an accelerated loss of sodium ions and an increase in the volume of urine produced. This combination lowers blood volume and blood pressure.

✓ How would a decrease in blood pressure affect the glomerular filtration rate (GFR)?

✓ If the nephrons lacked a loop of Henle, what would be the effect on the volume and solute (osmotic) concentration of the urine they produced?

THE TREATMENT OF RENAL FAILURE

Many different conditions can result in renal failure. Management of chronic renal failure typically involves restricting water and salt intake and reducing caloric intake to a minimum, with few dietary proteins. This combination reduces strain on the urinary system by (1) minimizing the volume of urine produced and (2) preventing the generation of large quantities of nitrogenous wastes. Acidosis, a common problem in patients with renal failure, can be countered with infusions of bicarbonate ions. If drugs, infusions, and dietary controls cannot stabilize the composition of the blood, more drastic measures are taken, such as *dialysis* or *kidney transplantation.*

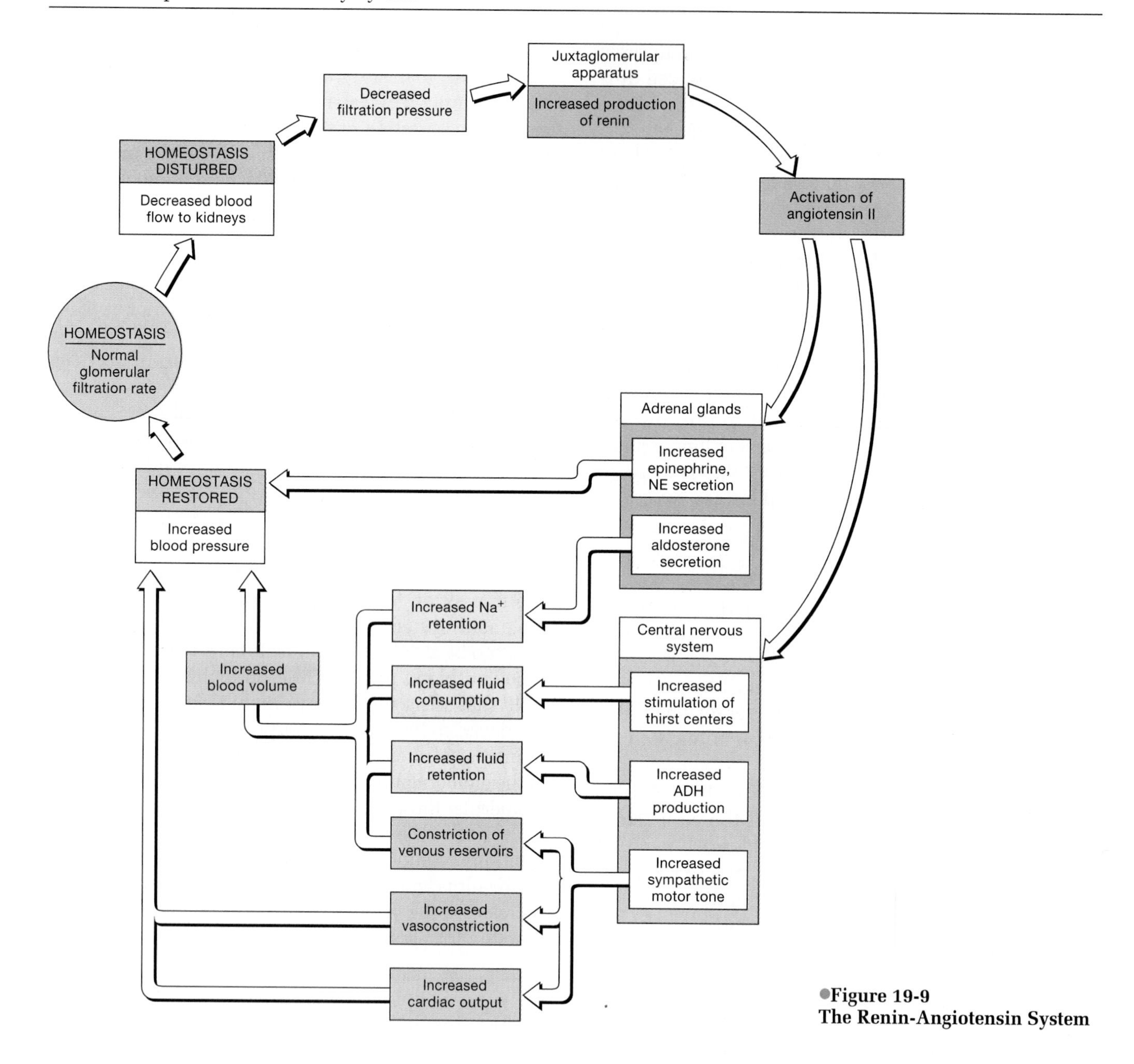

•Figure 19-9
The Renin-Angiotensin System

In dialysis, the functions of damaged kidneys are partially compensated for by diffusion between the patient's blood and a *dialysis fluid* whose composition is carefully regulated. The process, which takes several hours, must be repeated two or three times each week. In a kidney transplant, the kidney of a healthy donor is surgically inserted into the patient's body and connected to the circulatory system. If successful, the transplanted kidney(s) can take over all of the normal kidney functions. These procedures are considered in greater detail in the *Applications Manual.* AM *Advances in the Treatment of Renal Failure*

URINE TRANSPORT, STORAGE, AND ELIMINATION

Filtrate modification and urine production end when the fluid enters the renal pelvis. The rest of the urinary system is responsible for the transport, storage, and elimination of the urine.

The Ureters and Urinary Bladder

The ureters (Figures 19-1, p. 487, and 19-10a•) extend from the kidneys to the urinary bladder, a distance of about 30 cm (12 in.). Like the kidneys, the ureters are retroperitoneal, and they penetrate the posterior wall of the bladder without entering the peritoneal cavity. Under normal conditions, peristaltic contractions begin at the apex of the renal papillae and sweep along the minor and major calyces toward the renal pelvis. Similar contractions move urine out of the renal pelvis and along the ureter to the bladder.

In the male, the base of the urinary bladder lies between the rectum and the pubic symphysis (Figure

•Figure 19-10
Organs Responsible for the Conduction and Storage of Urine

(a) The ureter, urinary bladder, and urethra in the male. **(b)** The same organs in the female. **(c)** The urinary bladder in a male.

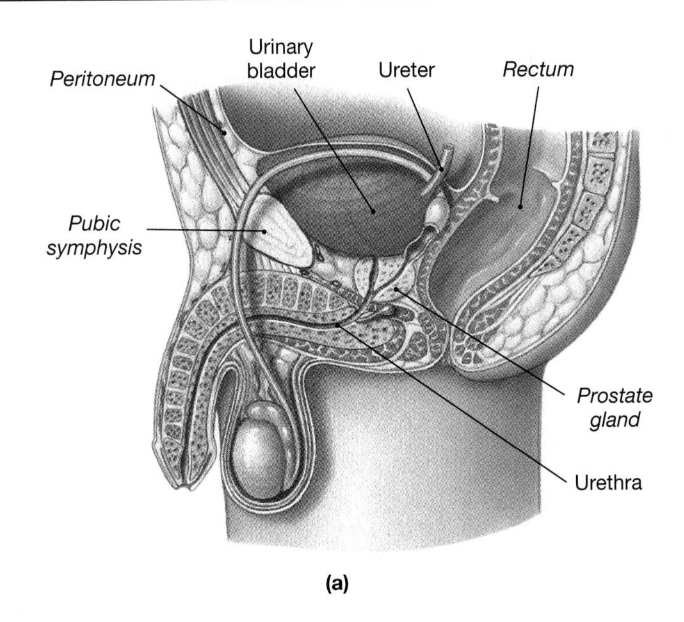

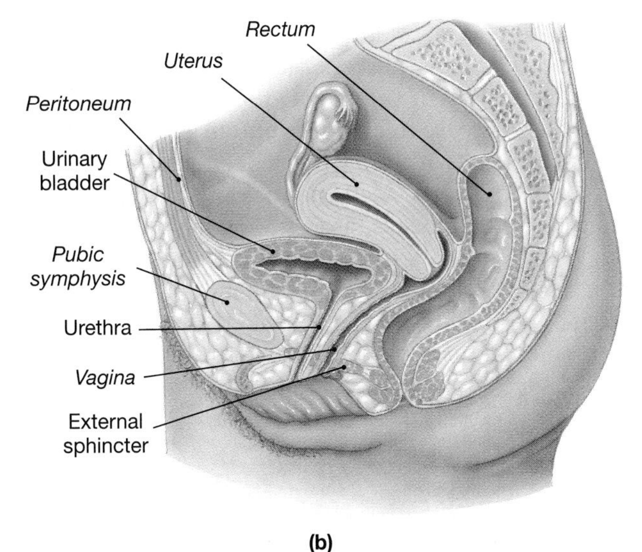

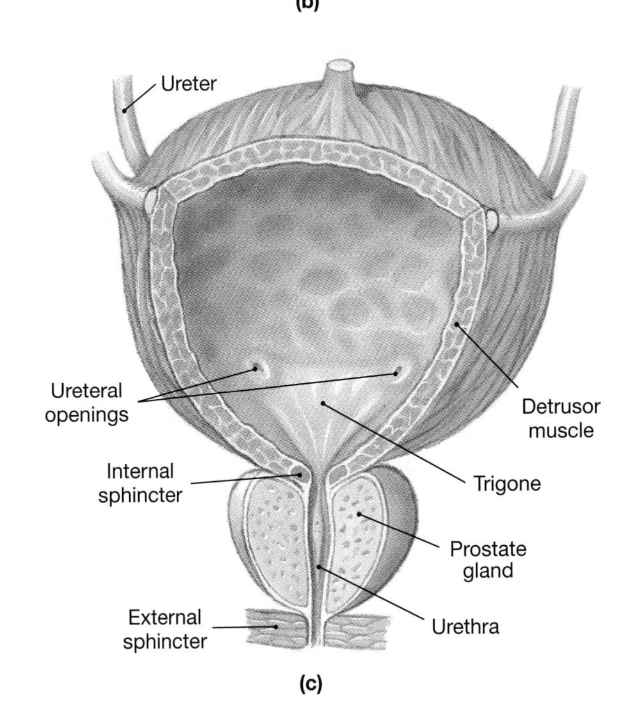

19-10a•). In the female (Figure 19-10b•), the urinary bladder sits inferior to the uterus and anterior to the vagina. Its dimensions vary, depending on the state of distension, but the full urinary bladder can contain up to a liter of urine.

In sectional view (Figure 19-10c•), the triangular area bounded by the openings of the ureters and the entrance to the urethra constitutes the *trigone* (TRĪ-gōn) of the bladder. The urethral entrance lies at the apex of this triangle, at the lowest point in the bladder. The area surrounding the urethral entrance, called the *neck* of the urinary bladder, contains a muscular sphincter that also extends along the proximal portions of the urethra. This **internal urethral sphincter** provides involuntary control over the discharge of urine from the bladder.

A *transitional epithelium* lines the renal pelvis, the ureter, and the urinary bladder. This stratified epithelium can tolerate a considerable amount of stretching, as indicated in Figure 4-5b•. ∞ *[p. 79].* The bladder wall contains both longitudinal and circular smooth muscle layers that form the powerful *detrusor* (de-TROO-sor) muscle of the bladder. Contraction of this muscle compresses the urinary bladder and expels its contents into the urethra.

The Urethra

In the female, the urethra is very short, extending 2.5–3.0 cm (about 1 in.) from the bladder to the vestibule. In the male, the urethra extends from the neck of the urinary bladder to the tip of the penis, about 18–20 cm (7–8 in.) in length. In both sexes, as the urethra passes through the urogenital diaphragm a circular band of skeletal muscle forms the **external urethral sphincter**. This sphincter consists of skeletal muscle fibers, and its contractions are under voluntary control. AM *Problems with the Conducting System* and *Urinary Tract Infections*

The Micturition Reflex and Urination

Urine reaches the urinary bladder by the peristaltic contractions of the ureters. The process of urination is coordinated by the **micturition reflex** (Figure 19-11•). Stretch receptors in the wall of the urinary bladder are stimulated as it fills with urine. Afferent sensory fibers in the pelvic nerves carry the resulting impulses to the sacral spinal cord. The increased level of activities by the sensory receptors and the sensory fibers (1) facilitates parasympathetic motor neurons in the sacral spinal cord and (2) stimulates interneurons that relay sensations to the cerebral cortex. As a result, we become consciously aware of the fluid pressure in the urinary bladder.

The urge to urinate usually appears when the bladder contains about 200 ml of urine. The micturition reflex begins to function when the stretch receptors have provided adequate stimulation to the parasympathetic motor neurons. At this time, the motor neurons stimulate the smooth muscle in the bladder wall. These commands travel over the pelvic nerves and produce a sustained contraction of the urinary bladder.

This contraction elevates fluid pressures inside the bladder, but urine ejection cannot occur unless both the internal and external sphincters are relaxed. We control the time and place of urination by voluntarily relaxing the external sphincter. When this sphincter relaxes, so does the internal sphincter. If the external sphincter does not relax, the internal sphincter remains closed, and the bladder gradually relaxes. A further increase in bladder volume begins the cycle again, usually within an hour. Each increase in urinary volume leads to an increase in stretch receptor stimulation that makes the sensation more acute. Once the volume of the urinary bladder exceeds 500 ml, the micturition reflex may generate enough pressure to force open the internal sphincter. This opening leads to a reflexive relaxation in the external sphincter, and urination occurs despite voluntary opposition or potential inconvenience. Normally, after micturition, less than 10 ml of urine remain in the bladder.

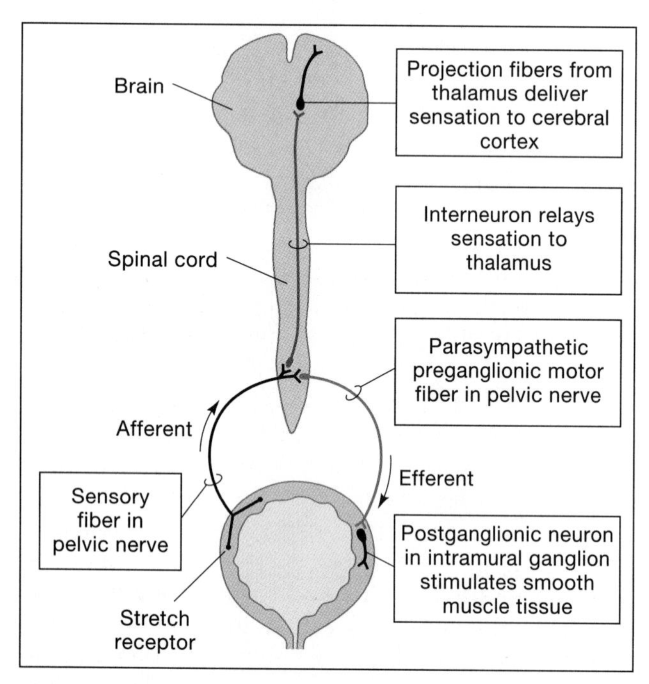

Figure 19-11
The Micturition Reflex

Basic components of the reflex arc involved in the micturition reflex.

INCONTINENCE

Incontinence (in-KON-ti-nens) is the inability to control urination voluntarily. Infants lack voluntary control over urination because the necessary corticospinal connections have yet to be established. Incontinence may develop in otherwise normal adults because of trauma to the internal or external sphincter. For example, childbirth can stretch and damage the sphincter muscles, and some women develop stress incontinence. In this condition, elevated intra-abdominal pressures, caused simply by a cough or sneeze, can overwhelm the sphincter muscles, causing urine to leak out. Incontinence may also develop in older individuals, because of a general loss of muscle tone.

Damage to the CNS, the spinal cord, or the nerve supply to the bladder or external sphincter may also produce incontinence. For example, incontinence often accompanies Alzheimer's disease, and it may also result from a stroke or spinal injury. In most cases, the affected individual develops an *automatic bladder*. The micturition reflex remains intact, but voluntary control of the external sphincter is lost, and the individual cannot prevent the reflexive emptying of the bladder. Damage to the pelvic nerves can eliminate the micturition reflex entirely, because these nerves carry both afferent and efferent fibers of this reflex arc. In this case, the bladder becomes greatly distended with urine. It remains filled to capacity, and the excess trickles into the urethra in an uncontrolled stream. Insertion of a catheter is often required to facilitate the discharge of urine.

INTEGRATION WITH OTHER SYSTEMS

Along with the urinary system, the integumentary, respiratory, and digestive systems are sometimes considered to form an anatomically diverse *excretory system*:

1. ***Integumentary system.*** Water and electrolyte losses in perspiration can affect plasma volume and composition. The effects are most apparent when losses are extreme, as in maximum sweat production. Small amounts of metabolic wastes are also excreted in perspiration.
2. ***Respiratory system.*** The lungs excrete the carbon dioxide generated by living cells. Small amounts of other compounds, such as acetone and water, evaporate into the alveoli and are eliminated during exhalation.
3. ***Digestive system.*** Small amounts of metabolic waste products are excreted in liver bile, and a variable amount of water is lost in feces.

These excretory activities affect the composition of body fluids. The respiratory system, for example, is the primary site of carbon dioxide excretion. But the excretory functions of these systems are not regulated as closely as are those of the kidneys, and under normal circumstances the effects of integumentary and digestive excretory activities are minor compared with those of the urinary system.

Figure 19-12● summarizes the functional relationships between the urinary system and other systems. Many of these relationships will be explored further in the next section, which considers fluid, pH, and electrolyte balance.

✓ What process is responsible for the movement of urine from the kidney to the urinary bladder?

✓ An obstruction of a ureter by a kidney stone would interfere with the flow of urine between what two points?

✓ The ability to control the micturition reflex depends on one's ability to control what muscle?

FLUID, ELECTROLYTE, AND ACID-BASE BALANCE

The next time you see a small pond, take a moment to think about the fish it contains. They live out their lives totally dependent on the quality of their isolated environment. If evaporation removes too much of the pond water, oxygen and food supplies run out, and the fish suffocate or starve. Most of the fish in a freshwater pond will die if the water becomes too salty; those in a saltwater pond will be killed if their environment becomes too dilute. The pH of the pond water, too, is a vital factor—that is why acid rain is such a problem.

The cells of our bodies live in a pond whose shores are the exposed surfaces of the skin. Most of the weight of the human body is water. Water accounts for up to 99 percent of the volume of extracellular fluid (ECF), and it is an essential ingredient of cytoplasm. All of a cell's operations rely on water as a diffusion medium for the distribution of gases, nutrients, and waste products. If the water content of the body changes, cellular activities are jeopardized. For example, when the water content of the body reaches very low levels, proteins denature, enzymes cease functioning, and cells die.

The ionic concentrations and pH of the body's water are as important as its absolute quantity. If concentrations of calcium or potassium ions in the ECF become too high, cardiac arrhythmias develop, and the individual's life is in jeopardy. A pH outside the normal range can lead to a variety of dangerous problems. Low pH is especially dangerous because hydrogen ions break chemical bonds, change the shapes of complex molecules, disrupt cell membranes, and impair tissue functions.

This section considers the dynamics of exchange between the various body fluids, such as blood plasma and interstitial fluid, and between the body and the external environment. Several different but interrelated types of homeostasis are involved:

- A person is in *fluid balance* when the amount of water gained each day is equal to the amount lost to the environment. Maintaining normal fluid balance involves regulating body water content and distribution.
- *Electrolyte balance* exists when there is neither a net gain nor a net loss of any ion in body fluids.
- *Acid-base balance* exists when the production of hydrogen ions precisely offsets their loss. While acid-base balance exists, the pH of body fluids remains within normal limits.

This section provides an overview that integrates earlier discussions of fluid, electrolyte, and acid-base balance in the body. Few other chapters have such wide-ranging clinical importance: *Treatment of any serious illness affecting the nervous, cardiovascular, respiratory, urinary, or digestive system must always include steps to restore normal fluid, electrolyte, and acid-base balance.*

Fluid and Electrolyte Balance

Figure 19-13● details the distribution of water in the body. Nearly two-thirds of the total body water content is found inside living cells, as the fluid medium of the **intracellular fluid (ICF)**, introduced in Chapter 3. ∞ *[p. 57]* The **extracellular fluid (ECF)** contains the rest of the body water. The largest subdivisions of the ECF are (1) the *tissue fluid*, or *interstitial fluid*, in peripheral tissues and (2) the *plasma* of the circulating blood. Minor components of the ECF include lymph, cerebrospinal fluid (CSF), synovial fluid, serous fluids (pleural, pericardial, and peritoneal fluids), aqueous humor, perilymph, and endolymph.

Exchange between these subdivisions of the ECF occurs primarily across the endothelial lining of capillaries. Fluid may also travel from the interstitial spaces to the plasma via the channels of the lymphatic system. Although there are regional variations in the identity and quantity of dissolved electrolytes, proteins, nutrients, and waste products within the ECF, these variations are relatively minor compared with the compositional differences *between* the ECF and the ICF.

In many respects, the ICF and ECF behave as distinct entities, and they are often called **fluid compartments**. As noted in earlier chapters, the principal ions

INTEGUMENTARY SYSTEM

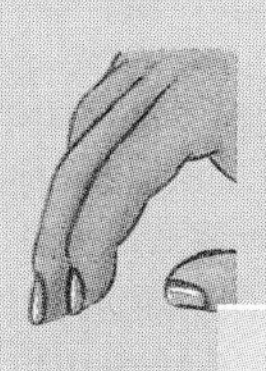

Sweat glands assist in elimination of water and solutes, especially sodium and chloride ions; keratinized epidermis prevents excessive fluid loss through skin surface; epidermis produces vitamin D_3, important for the renal production of calcitriol

SKELETAL SYSTEM

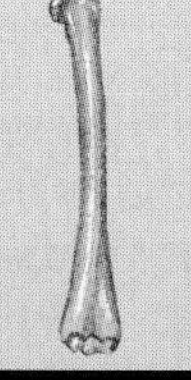

Axial skeleton provides some protection for kidneys and ureters; pelvis protects urinary bladder and proximal urethra

Conserves calcium and phosphate needed for bone growth

MUSCULAR SYSTEM

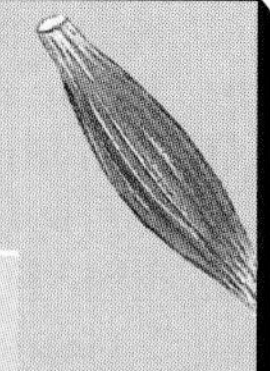

Sphincter controls urination by closing urethral opening; muscle layers of trunk provide some protection for urinary organs

Removes waste products of protein metabolism; assists in regulation of calcium and phosphate concentrations

URINARY SYSTEM

NERVOUS SYSTEM

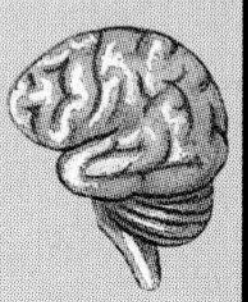
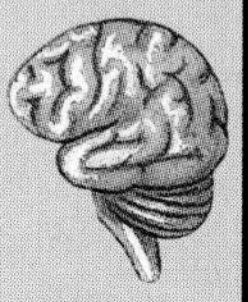

Adjusts renal blood pressure; monitors distension of bladder and controls urination

ENDOCRINE SYSTEM

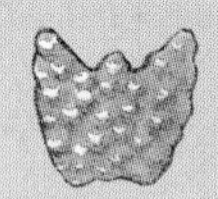

Aldosterone and ADH adjust rates of fluid and electrolyte reabsorption in kidneys

Kidney cells release renin when local blood pressure declines and erythropoietin (EPO) when renal O_2 levels decline

CARDIOVASCULAR SYSTEM

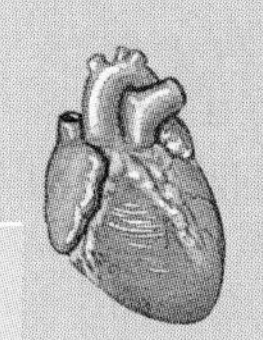

Delivers blood to capillaries when filtration occurs; accepts fluids and solutes reabsorbed during urine production

Releases renin to elevate blood pressure and erythropoietin to accelerate red blood cell production

FOR ALL SYSTEMS

Excretes waste products, maintains normal body fluid pH and ion composition

LYMPHATIC SYSTEM

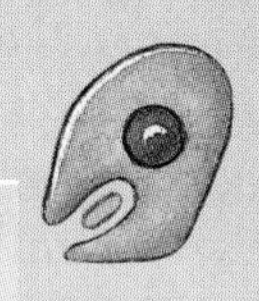

Provides specific defenses against urinary tract infections

Eliminates toxins and wastes generated by cellular activities; acid pH of urine provides nonspecific defense against urinary tract infections

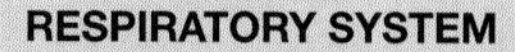

RESPIRATORY SYSTEM

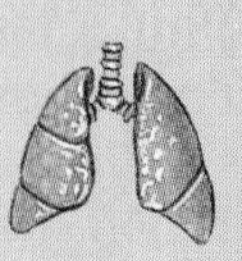
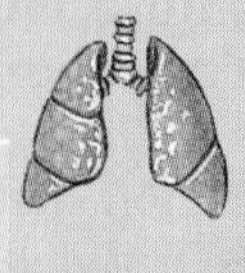

Assists in the regulation of pH by eliminating carbon dioxide

Assists in elimination of CO_2; provides bicarbonate buffers that assist in pH regulation

REPRODUCTIVE SYSTEM

Accessory organ secretions may have antibacterial action that helps prevent urethral infections in males

Urethra in males carries semen to the exterior

DIGESTIVE SYSTEM

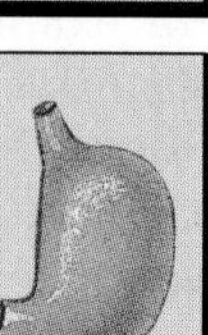

Absorbs water needed to excrete wastes at kidneys; absorbs ions needed to maintain normal body fluid concentrations; liver removes conjugated bilirubin

Excretes toxins absorbed by the digestive epithelium; excretes some bilirubin and nitrogenous wastes produced by the liver

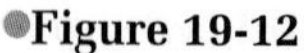

Figure 19-12
Functional Relationships between the Urinary System and Other Systems

in the ECF are sodium, chloride, and bicarbonate. The ICF contains an abundance of potassium, magnesium, and phosphate ions, plus large numbers of negatively charged proteins.

Despite these differences in the concentration of specific substances, the intracellular and extracellular osmolarities are identical. The cell membranes are freely permeable to water, and osmosis eliminates any minor concentration differences almost at once.

Fluid Balance

Water circulates freely within the extracellular fluid compartment. At capillary beds throughout the body, capillary blood pressure forces water out of the plasma and into the interstitial spaces. Some of that water is reabsorbed along the distal portion of the capillary bed, and the rest circulates into lymphatic vessels for transport to the venous circulation.

Water moves back and forth across the epithelial surfaces lining the peritoneal, pleural, and pericardial cavities and through the synovial membranes lining joint capsules. The flow rate is significant; for example, roughly 7 liters of peritoneal fluid are produced and reabsorbed each day. Water also moves between the blood and the cerebrospinal fluid, the aqueous and vitreous humors of the eye, and the perilymph and endolymph of the inner ear.

Roughly 2500 ml of water are lost each day through urine, feces, and perspiration. The losses due to perspiration vary, depending on the activities undertaken, but the additional deficits can be considerable, reaching well over 4 liters an hour. [p. 481] Water losses are normally balanced by the gain of fluids through eating (48 percent), drinking (40 percent), and metabolic generation (12 percent).

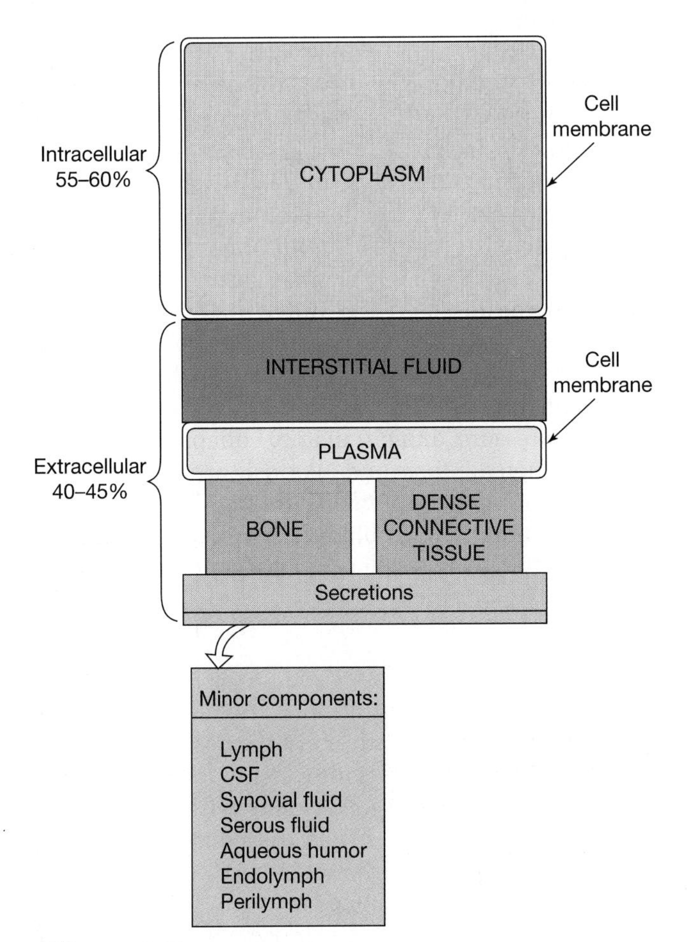

Figure 19-13
Body Fluid Compartments
The major body fluid compartments in a normal individual. For information concerning the chemical composition of body fluids, see Appendix III.

Fluid Shifts

Water movement between the ECF and ICF is called a *fluid shift*. Fluid shifts occur relatively rapidly, reaching equilibrium within a period of minutes to hours. These shifts occur in response to changes in the osmolarity of the extracellular fluid.

- If the ECF becomes more concentrated (hypertonic) with respect to the ICF, water will move from the cells into the ECF until osmotic equilibrium is restored.
- If the ECF becomes more dilute (hypotonic) with respect to the ICF, water will move from the ECF into the cells, and the volume of the ICF will increase accordingly.

In summary, if the osmolarity of the ECF changes, a fluid shift between the ICF and ECF will tend to oppose the change. Because the volume of the ICF is much greater than that of the ECF, the ICF acts as a "water reserve." In effect, instead of a large change in the composition of the ECF there are smaller changes in both the ECF and ICF.

Electrolyte Balance

An individual is in electrolyte balance when the rates of gain and loss are equal for each of the individual electrolytes in the body. Electrolyte balance is important for the following reasons:

- ***A gain or loss of electrolytes can cause a gain or loss in water.***
- ***The concentrations of individual electrolytes affect a variety of cell functions.*** Many examples of ion effects on cell function were encountered in earlier chapters. For example, the effects of high or low calcium and potassium ion concentrations on cardiac muscle tissue were noted in Chapter 13. [p. 343]

Two cations, sodium and potassium, merit special attention because (1) they are major contributors to the osmolarities of the ECF and ICF, and (2) they have direct effects on the normal functioning of living cells. Regulatory mechanisms involving a third important ion, calcium, were discussed in Chapters 6 and 11. ∞ [pp. 115, 292]

Sodium is the dominant cation within the extracellular fluid. Because more than 90 percent of the osmolarity of the ECF results from the presence of sodium salts, principally sodium chloride (NaCl) and sodium bicarbonate ($NaHCO_3$), alterations in the osmolarity of extracellular fluids usually reflect changes in the concentration of sodium ions. Potassium is the dominant cation in the intracellular fluid; extracellular potassium concentrations are normally low. In general:

- ***The most common problems with electrolyte balance are caused by an imbalance between sodium gains and losses.***
- ***Problems with potassium balance are less common but significantly more dangerous than those related to sodium balance.***

Sodium Balance The amount of sodium in the ECF represents a balance between sodium ion absorption at the digestive tract and sodium ion excretion at the kidneys and other sites. The rate of uptake varies directly with the amount included in the diet. The kidneys are responsible for regulating sodium ion losses, in response to circulating levels of aldosterone (decreased sodium loss) and ANP (increased sodium loss).

Whenever the rate of sodium intake or output changes, there is a corresponding gain or loss of water that tends to keep the sodium concentration constant. For example, eating a heavily salted meal will not raise the sodium ion concentration of body fluids because as sodium chloride crosses the digestive epithelium, osmosis brings additional water into the body. (This is why individuals with high blood pressure are told to restrict their salt intake; dietary salt will be absorbed, and because "water follows salt" the blood volume and blood pressure will increase.)

Potassium Balance Potassium ions are the primary cations of the intracellular fluid. *Roughly 98 percent of the potassium content of the body lies within the ICF. The potassium concentration of the ECF,* which is relatively low, *represents a balance* between (1) the rate of entry across the digestive epithelium and (2) the rate of loss into the urine. The rate of entry is proportional to the amount of potassium in the diet. Urinary potassium losses are controlled by adjusting the rate of active secretion *along the distal convoluted tubules of the kidneys.* The secondary rate is primarily determined by aldosterone levels. The ion pumps sensitive to this hormone reabsorb sodium ions from the filtrate in exchange for potassium ions from the interstitial fluid. When potassium levels rise in the ECF, aldosterone levels climb, and additional potassium ions are lost in the urine. When potassium levels fall in the ECF aldosterone levels fall, and potassium ions are conserved.

✓ How would eating a meal high in salt affect the amount of fluid in the intracellular fluid compartment?

✓ What effect would being lost in the desert for a day without water have on your blood osmolarity?

Acid-Base Balance

The pH of body fluids represents a balance between the acids, bases, and salts in solution. This pH normally remains within relatively narrow limits, usually from 7.35 to 7.45. Any deviation outside the normal range is extremely dangerous because changes in hydrogen ion concentrations disrupt the stability of cell membranes, alter protein structure, and change the activities of important enzymes.

When the pH falls below 7.35, a state of **acidosis** exists. **Alkalosis** exists if the pH increases above 7.45. These conditions affect virtually all systems, but the nervous system and cardiovascular system are particularly sensitive to pH fluctuations. For example, severe acidosis can be deadly because (1) CNS function deteriorates, and the individual becomes comatose; (2) cardiac contractions grow weak and irregular, and symptoms of heart failure develop; and (3) peripheral vasodilation produces a dramatic drop in blood pressure, and circulatory collapse may occur.

Although acidosis and alkalosis are both dangerous, in practice, problems with acidosis are much more common than problems with alkalosis because several different types of acids are generated by normal cellular activities.

Acids in the Body

Carbon dioxide concentration is the most important factor affecting the pH in body tissues. In solution, carbon dioxide interacts with water to form molecules of carbonic acid (H_2CO_3). As noted in Chapter 16, the carbonic acid molecules then dissociate to produce hydrogen ions and bicarbonate ions. ∞ [p. 424] The complete reaction sequence is:

$$CO_2 + H_2O \leftrightarrow H_2CO_3 \leftrightarrow H^+ + HCO_3^-$$

This reaction occurs spontaneously in body fluids, but it occurs very rapidly in the presence of *carbonic anhydrase,* an intracellular enzyme found in red blood cells, liver and kidney cells, parietal cells of the stomach, and in many other cell types.

Because most of the carbon dioxide in solution is converted to carbonic acid, and most of the carbonic acid dissociates, there is a direct relationship between the P_{CO_2} and the pH (Figure 19-14●). When carbon dioxide concentrations rise, additional hydrogen ions and bicarbonate ions are released, and the pH goes down. (Remember that the smaller the pH value, the greater the acidity.)

At the alveoli, carbon dioxide diffuses into the atmosphere, the number of hydrogen ions and bicarbonate ions declines, and the pH rises. This process, which effectively removes hydrogen ions from solution, will be considered in more detail later in the chapter.

Metabolic acids are generated during normal metabolism. Some are generated during the catabolism of amino acids, carbohydrates, or lipids. Examples include lactic acid produced during anaerobic metabolism and ketone bodies produced when breaking down fatty acids. Under normal conditions, metabolic acids are recycled or excreted rapidly, and significant accumulations do not occur.

Buffers and Buffer Systems

The acids discussed above, produced in the course of normal metabolic operations, must be controlled by the buffers and buffer systems in body fluids. *Buffers*, introduced in Chapter 2, are dissolved compounds that can provide or remove hydrogen ions and thereby stabilize the pH of a solution. ∞ *[p. 32]* Buffers include *weak acids* that can donate hydrogen ions and *weak bases* that can absorb them. A **buffer system** consists of a combination of a weak acid and its dissociation products: a hydrogen ion (H^+) and an anion. There are three major buffer systems, each with slightly different characteristics and distribution.

Protein buffer systems contribute to the regulation of pH in the ECF and ICF. Protein buffer systems, depend on the ability of amino acids to respond to alterations in pH by accepting or releasing hydrogen ions. If the pH climbs, the carboxyl group (–COOH) of the amino acid can dissociate, releasing a hydrogen ion. If the pH drops, the amino group ($-NH_2$) can accept an additional hydrogen ion, forming an $-NH_3^+$ group.

The plasma proteins and hemoglobin in red blood cells make substantial contributions to the buffering capabilities of the blood. In the interstitial fluids, extracellular protein fibers and dissolved amino acids also help regulate pH. In the intracellular fluids of active cells, structural and other proteins provide an extensive buffering capability that prevents destructive pH changes when organic acids are produced by cellular metabolism.

The **carbonic acid–bicarbonate buffer system** is an important buffer system in the ECF. Carbon dioxide generation occurs in all living tissues. As detailed above, most of it is converted to carbonic acid, which then dissociates into a hydrogen ion and a bicarbonate ion. The net effect is that $CO_2 + H_2O \leftrightarrow H^+ + HCO_3^-$. The carbonic acid and its dissociation products form the carbonic acid–bicarbonate buffer system. If hydrogen ions are removed, they will be replaced through the combination of water and carbon dioxide; if hydrogen ions are added, most will be removed through the formation of carbon dioxide and water.

The primary role of the carbonic acid–bicarbonate buffer system is to prevent pH changes caused by metabolic acids. The hydrogen ions released through the dissociation of these acids combine with bicar-

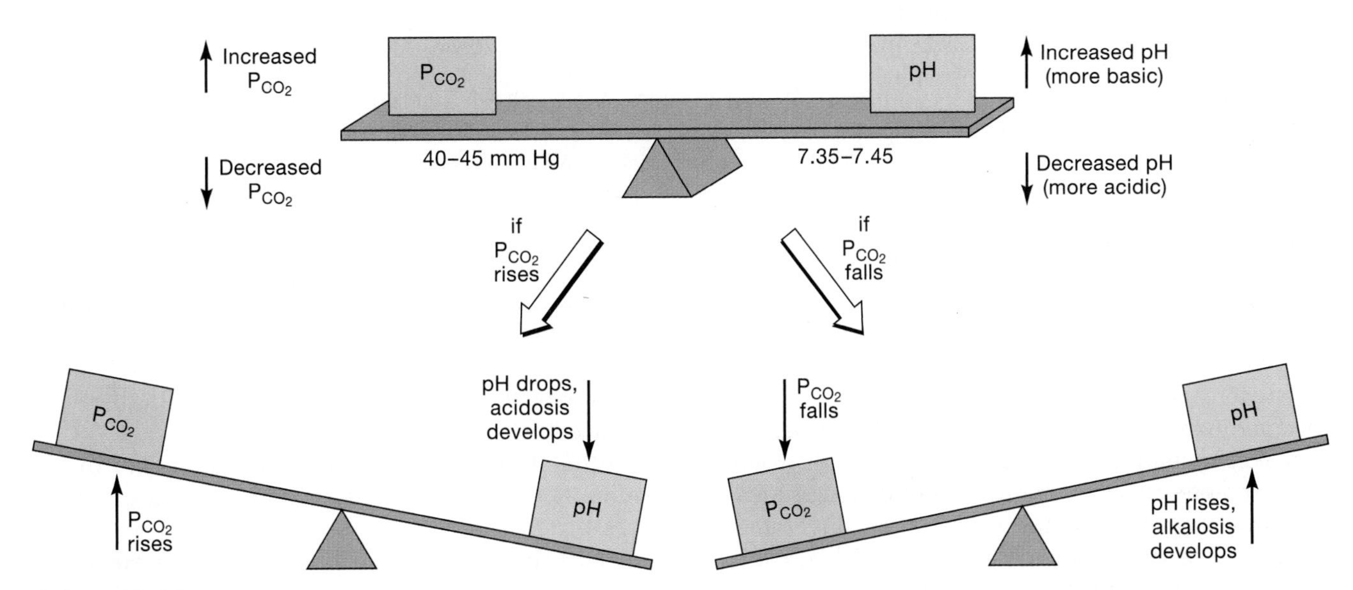

●**Figure 19-14**
Basic Relationships between Carbon Dioxide and Plasma pH

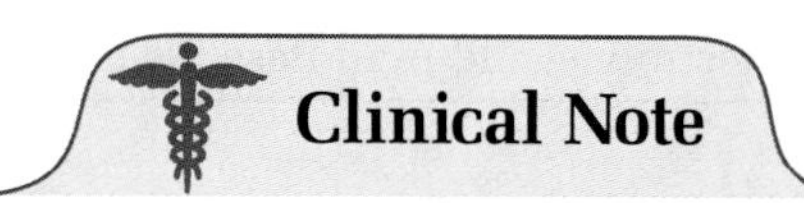

Disturbances of Acid-Base Balance

Together, buffer systems, respiratory compensation, and renal compensation maintain normal acid-base balance. These mechanisms are usually able to control pH very precisely, so that the pH of the extracellular fluids seldom varies more than 0.1 pH units, from 7.35 to 7.45. When buffering mechanisms are severely stressed, the pH wanders outside of these limits, producing symptoms of alkalosis or acidosis.

The primary source of the problem is usually indicated by the name given to the resulting condition. *Respiratory disorders* result from abnormal carbon dioxide levels in the ECF. These conditions are usually directly related to an imbalance between the rate of carbon dioxide removal at the lungs and its generation in other tissues. *Metabolic disorders* are caused by the generation of organic or fixed acids or by conditions affecting the concentration of bicarbonate ions in the extracellular fluids. The problems facing individuals with respiratory disorders are quite different from those facing people with metabolic disorders. Pulmonary compensation alone can often restore normal acid-base balance in individuals suffering from respiratory disorders. In contrast, compensation mechanisms for metabolic disorders may be able to stabilize pH, but other aspects of acid-base balance (buffer system function, bicarbonate levels, and P_{CO_2}) remain abnormal until the underlying metabolic problem is corrected.

Respiratory Acidosis

Respiratory acidosis develops when the respiratory system is unable to eliminate all of the CO_2 generated by peripheral tissues. The primary symptom is low plasma pH due to *hypercapnia*, an elevated plasma P_{CO_2}. As carbon dioxide levels climb, hydrogen and bicarbonate ion concentrations rise as well. Other buffer systems can tie up some of the hydrogen ions, but once the combined buffering capacity has been exceeded the pH begins to fall rapidly. *Respiratory acidosis represents the most frequent challenge to acid-base balance.* The usual cause is *hypoventilation*, an abnormally low respiratory rate. Our tissues generate carbon dioxide at a rapid rate, and even a few minutes of hypoventilation can cause acidosis, reducing the pH of the ECF to as low as 7.0. Under normal circumstances, the chemoreceptors monitoring the P_{CO_2} of the plasma and CSF will eliminate the problem by calling for an increase in the depth and rate of respiration.

Respiratory Alkalosis

Problems with **respiratory alkalosis** are relatively uncommon. This condition develops when respiratory activity lowers plasma P_{CO_2} to below normal levels, a condition called *hypocapnia*. A temporary hypocapnia can be produced by hyperventilation, when increased respiratory activity leads to a reduction in the arterial P_{CO_2}. Continued hyperventilation can elevate the pH to levels as high as 7.8–8. This condition usually corrects itself, for the reduction in P_{CO_2} removes the stimulation for the chemoreceptors, and the urge to breathe fades until carbon dioxide levels have returned to normal. Respiratory alkalosis caused by hyperventilation seldom persists long enough to cause a clinical emergency.

Metabolic Acidosis

Metabolic acidosis is the second most common type of acid-base imbalance. The most frequent cause is the production of a large number of metabolic acids such as lactic acid or ketone bodies. Metabolic acidosis can also be caused by an impaired ability to excrete hydrogen ions at the kidneys, and any condition accompanied by severe kidney damage can result in metabolic acidosis. Compensation for metabolic acidosis usually involves a combination of respiratory and renal mechanisms. Hydrogen ions interacting with bicarbonate ions form carbon dioxide molecules that are eliminated at the lungs, while the kidneys excrete additional hydrogen ions into the urine and generate bicarbonate ions that are released into the ECF.

Metabolic Alkalosis

Metabolic alkalosis occurs when bicarbonate ion concentrations become elevated. The bicarbonate ions then interact with hydrogen ions in solution, forming carbonic acid, and the reduction in H^+ concentrations then causes symptoms of alkalosis. The secretion of HCl by parietal cells of the stomach commonly results in alkalosis. Cases of severe metabolic alkalosis are relatively rare. Compensation involves a reduction in pulmonary ventilation, coupled with the increased loss of bicarbonates in the urine.

The alterations in blood chemistry characteristic of the major types of acid-base disorders are summarized in Table 19-4.

bonate ions, producing water and carbon dioxide. The carbon dioxide can then be excreted at the lungs. This buffering system can cope with large amounts of acid because body fluids contain an abundance of bicarbonate ions, primarily in the form of dissolved molecules of *sodium bicarbonate*, $NaHCO_3$. This readily available supply of bicarbonate ions is known as the *bicarbonate reserve*. When hydrogen ions enter the ECF, the bicarbonate ions that combine with them are replaced by the bicarbonate reserve.

The *phosphate buffer system* consists of an anion, $H_2PO_4^-$, that is a weak acid. In solution, it reversibly dissociates into a hydrogen ion and HPO_4^{2-}. In the ECF, the phosphate buffer system plays only a supporting role in the regulation of pH, primarily because the concentration of bicar-

TABLE 19-4 Acid-Base Disorders

Disorder	pH (normal = 7.35–7.45)	Remarks	Treatment
RESPIRATORY ACIDOSIS	Decreased (below 7.35)	Usually caused by hypoventilation and CO_2 buildup in tissues and blood	Improve ventilation, sometimes with bronchodilation and mechanical assistance
METABOLIC ACIDOSIS	Decreased (below 7.35)	Caused by organic or fixed acid buildup, impaired H^+ excretion at kidneys, or bicarbonate loss in urine or feces	Bicarbonate administration (gradual) with other steps as needed to correct primary cause
RESPIRATORY ALKALOSIS	Increased (above 7.45)	Usually caused by hyperventilation and reduction in plasma CO_2 levels	Reduce respiratory rate, allow rise in P_{CO_2}
METABOLIC ALKALOSIS	Increased (above 7.45)	Usually caused by prolonged vomiting and associated gastric acid loss	pH below 7.55, no treatment; pH above 7.55 may require ammonium chloride administration

bonate ions far exceeds that of phosphate ions. However, the phosphate buffer system is quite important in buffering the pH of the intracellular fluids, where the concentration of phosphate ions is relatively high.

Maintenance of Acid-Base Balance

The maintenance of acid-base balance involves controlling hydrogen ion losses and gains. In this process, the pulmonary and renal mechanisms support the buffer systems by (1) secreting or absorbing hydrogen ions, (2) controlling the excretion of acids and bases, and, when necessary, (3) generating additional buffers. It is the *combination* of buffer systems and these pulmonary and renal mechanisms that maintains pH within narrow limits.

Pulmonary Contributions to pH Regulation The lungs contribute to pH regulation by their effects on the carbonic acid–bicarbonate buffer system. Increasing or decreasing the rate of respiration can have a profound effect on the buffering capacity of body fluids by lowering or raising the P_{CO_2}. Changes in P_{CO_2} have a direct effect on the concentration of hydrogen ions in the plasma, since the additional carbon dioxide molecules will form carbonic acid which immediately dissociates.

Mechanisms responsible for the control of respiratory rate were discussed in Chapter 16 ∞ [p. 424], and only a brief summary will be presented here. A rise in P_{CO_2} stimulates chemoreceptors in the carotid and aortic bodies, and within the CNS; a fall in the P_{CO_2} inhibits them. Stimulation of the chemoreceptors leads to an increase in the respiratory rate. As the rate of respiration increases, more CO_2 is lost at the lungs, and the P_{CO_2} returns to normal levels. When the P_{CO_2} of the blood or CSF declines, respiratory activity becomes depressed, the breathing rate falls, and the P_{CO_2} in the extracellular fluids rises.

Changes in respiratory rate affect pH because when the P_{CO_2} rises, the pH declines, and when the P_{CO_2} decreases, the pH increases (Figure 19-14●). A change in the respiratory rate that helps stabilize pH is called **respiratory compensation**.

Renal Contributions to pH Regulation Glomerular filtration puts hydrogen ions, carbon dioxide, and the other components of the carbonic acid–bicarbonate and phosphate buffer systems into the filtrate. The kidney tubules then modify the pH of the filtrate by secreting hydrogen ions or reabsorbing bicarbonate ions. A change in the rates of hydrogen ion and bicarbonate ion secretion or absorption in response to changes in plasma pH is called **renal compensation**.

- ✓ What effect would a decrease in the pH of the body fluids have on the respiratory rate?
- ✓ How would a prolonged fast affect the body's pH?
- ✓ Why can prolonged vomiting produce alkalosis?

AGING AND THE URINARY SYSTEM

In general, aging is associated with an increased incidence of kidney problems. Age-related changes in the urinary system include:

1. ***A decline in the number of functional nephrons.*** The total number of kidney nephrons drops by 30–40 percent between ages 25 and 85.
2. ***A reduction in the GFR.*** This results from decreased numbers of glomeruli, cumulative damage to the filtration apparatus in the remaining glomeruli, and reductions in renal blood flow.
3. ***Reduced sensitivity to ADH.*** With age, the distal portions of the nephron and collecting system become less responsive to ADH. Less reabsorption of water and sodium ions occurs, and more potassium ions are lost in the urine.
4. ***Problems with the micturition reflex.*** Several factors are involved in such problems:
   - The sphincter muscles lose muscle tone and become less effective at voluntarily retaining urine. Loss of muscle tone leads to problems with incontinence, often involving a slow leakage of urine.
   - The ability to control micturition is often lost following a stroke, Alzheimer's disease, or other CNS problems affecting the cerebral cortex or hypothalamus.
   - In males, **urinary retention** may develop secondary to enlargement of the prostate gland. In this condition, swelling and distortion of surrounding prostatic tissues compress the urethra, restricting or preventing the flow of urine.

Chapter Review

KEY TERMS

acidosis, p. 506
aldosterone, p. 497
alkalosis, p. 506
angiotensin I, II, p. 499
buffer system, p. 507
glomerulus, p. 489
juxtaglomerular apparatus, p. 493
micturition, p. 488
nephron, p. 489
peritubular capillaries, p. 493
renin, p. 493

SUMMARY OUTLINE

INTRODUCTION p. 487

1. The functions of the urinary system include (1) eliminating organic waste products, (2) regulating plasma concentrations of ions, (3) regulating blood volume and pressure by adjusting the volume of water lost and releasing hormones, (4) helping to stabilize blood pH, and (5) conserving nutrients.

ORGANIZATION OF THE URINARY SYSTEM p. 488

1. The urinary system includes the **kidneys**, the **ureters**, the **urinary bladder**, and the **urethra**. The kidneys produce **urine** (a fluid containing water, ions, and soluble compounds); during **urination** urine is forced out of the body. (*Figure 19-1*)

THE KIDNEYS p. 488

1. The left kidney extends superiorly slightly more than the right kidney. Both lie in a retroperitoneal position. (*Figure 19-2*)

SUPERFICIAL AND SECTIONAL ANATOMY P. 488

2. The **hilus** provides entry for the renal artery and exit for the renal vein and ureter.

3. The ureter communicates with the **renal pelvis**. This chamber branches into two **major calyces**, each connected to four or five **minor calyces** that enclose the **renal papillae**. (*Figure 19-3*)

The Nephron p. 489

4. The **nephron** (the basic functional unit in the kidney) includes the *renal corpuscle* and a *renal tubule* that empties into the **collecting system** via a *collecting duct*. From the renal corpuscle the **filtrate** travels through the *proximal convoluted tubule*, the *loop of Henle*, and the *distal convoluted tubule*. (*Figure 19-4*)

5. Nephrons are responsible for (1) production of filtrate, (2) reabsorption of nutrients, and (3) reabsorption of water and ions.

6. The renal tubule begins at the **renal corpuscle**. It includes a knot of intertwined capillaries called the **glomerulus** surrounded by the **Bowman's capsule**. Blood arrives via the **afferent arteriole** and departs in the **efferent arteriole**. (*Figure 19-5*)

7. At the glomerulus, **podocytes** cover the basement membrane of the capillaries that project into the **capsular space**. The processes of the podocytes are separated by narrow slits. (*Figure 19-5*)

8. The **proximal convoluted tubule (PCT)** actively reabsorbs nutrients, plasma proteins, and electrolytes from the filtrate. They are then released into the surrounding interstitial fluid. (*Figure 19-4*)

9. The **loop of Henle** includes a descending limb and an ascending limb. The ascending limb is impermeable to water and solutes; the descending limb is permeable to water. (*Figure 19-4*)

10. The ascending limb delivers fluid to the **distal convoluted tubule (DCT)**, which actively secretes ions and reabsorbs sodium ions from the urine. The **juxtaglomerular apparatus**, which releases renin and erythropoietin, is located at the start of the DCT. (*Figure 19-5a*)

11. The collecting ducts receive urine from nephrons and merge into a **papillary duct** that delivers urine to a minor calyx. The collecting system makes final adjustments to the urine by reabsorbing water or reabsorbing or secreting various ions.

The Blood Supply to the Kidneys p. 493

12. The blood vessels of the kidneys include the **interlobar**, **arcuate**, and **interlobular arteries**, and the **interlobar**, **arcuate**, and **interlobular veins**. Blood travels from the efferent arteriole to the **peritubular capillaries** and the **vasa recta**. Diffusion occurs between the capillaries of the vasa recta and the tubular cells through the **interstitial fluid** that surrounds the nephron. (*Figure 19-6*)

BASIC PRINCIPLES OF URINE PRODUCTION p. 493

1. The primary goal in urine production is the excretion and elimination of dissolved solutes, principally metabolic waste products, such as **urea**, **creatinine**, and **uric acid**.

2. Urine formation involves filtration, reabsorption, and secretion. (*Tables 19-1, 19-2*)

Filtration at the Glomerulus p. 494

3. *Glomerular filtration* occurs as fluids move across the wall of the glomerulus into the capsular space, in response to blood pressure in the glomerular capillaries. The **glomerular filtration rate (GFR)** is the amount of filtrate produced in the kidneys each minute. Any factor that alters the filtration (blood) pressure will change the GFR and affect kidney function.

4. Dropping filtration pressures stimulate the juxtoglomerular apparatus to release renin. Renin release increases blood volume and blood pressure.

Reabsorption and Secretion Along the Renal Tubule p. 495

5. The cells of the PCT normally reabsorb 60–70 percent of the volume of the filtrate produced in the renal corpuscle. The PCT generally reabsorbs sodium and other ions, water, and almost all of the nutrients in the filtrate. It also secretes various substances.

6. Water and ions are reclaimed from the filtrate by the loop of Henle. The ascending limb reabsorbs sodium and chloride ions, and the descending limb reabsorbs water. A concentration gradient in the medulla encourages the osmotic flow of water out of the filtrate and into the interstitial fluid. As water is lost by osmosis and the filtrate volume decreases, the urea concentration rises. (*Figure 19-7*)

7. The DCT performs final adjustments by actively secreting or absorbing materials. Sodium ions are actively absorbed, in exchange for potassium and hydrogen ions discharged into the filtrate. Aldosterone secretion increases the rate of sodium reabsorption and potassium loss.

8. The amount of water in the urine of the collecting ducts is regulated by ADH secretions. In the absence of ADH, the DCT, collecting tubule, and collecting duct are impermeable to water. The higher the ADH level in circulation, the more water is absorbed, and the more concentrated the urine. (*Figure 19-7*)

9. More than 99 percent of the filtrate produced each day is reabsorbed before reaching the renal pelvis. (*Table 19-3*)

10. Each segment of the nephron and collecting system contributes to the production of hypertonic urine. (*Figure 19-8*)

The Control of Kidney Function p. 497

11. Renal function may be regulated by automatic adjustments in glomerular pressures through changes in the diameters of the afferent and efferent arterioles.

12. Sympathetic activation (1) produces a powerful vasoconstriction of the afferent arterioles, decreasing the GFR and slowing the production of filtrate; (2) alters the GFR by changing the regional pattern of blood circulation; and (3) stimulates the release of renin by the juxtaglomerular apparatus.

13. Hormones that regulate kidney function include angiotensin II, aldosterone, ADH, and ANP. (*Figure 19-9*)

URINE TRANSPORT, STORAGE, AND ELIMINATION p. 500

1. Filtrate modification and urine production end when the fluid enters the renal pelvis. The rest of the urinary system is responsible for transporting, storing, and eliminating the urine.

The Ureters and Urinary Bladder p. 500

2. The ureters extend from the renal pelvis to the urinary bladder. Peristaltic contractions by smooth muscles move the urine. (*Figure 19-1, 19-10*)

3. Internal features of the bladder include the **trigone**, the **neck**, and the **internal urethral sphincter**. Contraction of the **detrusor** muscle compresses the bladder and expels the urine into the urethra. (*Figure 19-10*)

The Urethra p. 501

4. In both sexes, as the urethra passes through the urogenital diaphragm a circular band of skeletal muscles forms the **external urethral sphincter**, which is under voluntary control. (*Figure 19-10*)

The Micturition Reflex and Urination p. 501

5. The process of urination is coordinated by the **micturition reflex**, which is initiated by stretch receptors in the bladder wall. Voluntary urination involves coupling this reflex with the voluntary relaxation of the external sphincter, which allows the opening of the internal sphincter. (*Figure 19-11*)

INTEGRATION WITH OTHER SYSTEMS p. 502

1. The urinary, integumentary, respiratory, and digestive systems are sometimes considered as an anatomically diverse *excretory system*. The system's components work together to perform all of the excretory functions that affect the composition of body fluids. (*Figure 19-12*)

FLUID, ELECTROLYTE, AND ACID-BASE BALANCE p. 503

1. All of our cells' operations depend on water as a diffusion medium for dissolved gases, nutrients, and waste products. Maintenance of normal volume and composition in the extracellular and intracellular fluids is vital to life. Three types of homeostasis are involved: *fluid balance*, *electrolyte balance*, and *acid-base balance*.

Fluid and Electrolyte Balance p. 503

2. The **intracellular fluid (ICF)** contains nearly two-thirds of the total body water; the **extracellular fluid (ECF)** contains the rest. Exchange occurs between the ICF and ECF, but the two **fluid compartments** retain their distinctive characteristics. (*Figure 19-13*)

3. Water circulates freely within the ECF compartment. At capillary beds, hydrostatic pressure forces water from the plasma into the interstitial spaces. Water moves back and forth across the epithelial lining of the peritoneal, pleural, and pericardial cavities; through synovial membranes lining joint capsules; and between the blood and cerebrospinal fluid, the aqueous and vitreous humors of the eye, and the perilymph and endolymph of the inner ear.

4. Water losses are normally balanced by gains through eating, drinking, and metabolic generation.

5. If the ECF becomes hypertonic relative to the ICF, water will move from the ICF into the ECF (a fluid shift) until osmotic equilibrium has been restored. If the ECF becomes hypotonic relative to the ICF, water will move from the ECF into the cells and the volume of the ICF will increase accordingly.

6. Electrolyte balance is important because total electrolyte concentrations affect water balance and because the levels of individual electrolytes can affect a variety of cell functions. Problems with electrolyte balance generally result from an imbalance between sodium gains and losses. Problems with potassium balance are less common but more dangerous.

7. The rate of sodium uptake across the digestive epithelium is directly proportional to the amount of sodium in the diet. Sodium losses occur mainly in the urine and through perspiration. The rate of sodium reabsorption along the DCT is regulated by aldosterone levels; aldosterone stimulates sodium ion reabsorption.

8. Potassium ion concentrations in the ECF are very low. Potassium excretion increases (1) when sodium ion concentrations decline and (2) as ECF potassium concentrations rise. The rate of potassium excretion is regulated by aldosterone; aldosterone stimulates potassium ion excretion.

Acid-Base Balance p. 506

9. The pH of normal body fluids ranges from 7.35 to 7.45; variations outside this relatively narrow range produce **acidosis** or **alkalosis**.

10. Carbonic acid is the most important factor affecting the pH of the ECF. In solution, CO_2 reacts with water to form carbonic acid; the dissociation of carbonic acid releases hydrogen ions. An inverse relationship exists between the concentration of CO_2 and pH. (*Figure 19-14*)

11. Organic acids include metabolic products such as lactic acid and ketone bodies.

12. A **buffer system** consists of a weak acid and its dissociation products. There are three major buffer systems: (1) *protein buffer systems* in the ECF and ICF; (2) the **carbonic acid–bicarbonate buffer system**, most important in the ECF; and (3) the *phosphate buffer system* in the intracellular fluids and urine.

13. In protein buffer systems, the component amino acids respond to changes in H^+ concentrations. Blood plasma proteins and hemoglobin in red blood cells help prevent drastic changes in pH.

14. The carbonic acid–bicarbonate buffer system prevents pH changes due to organic acids in the ECF.

15. In **respiratory compensation**, the lungs help regulate pH by affecting the carbonic acid–bicarbonate buffer system; changing the respiratory rate can raise or lower the P_{CO_2} of body fluids, affecting the buffering capacity.

16. In the process of **renal compensation**, the kidneys vary their rates of hydrogen ion secretion and bicarbonate ion resorption depending on the pH of extracellular fluids.

AGIPAGNG AND THE URINARY SYSTEM p. 510

1. Aging is usually associated with increased kidney problems. Age-related changes in the urinary system include (1) declining number of functional nephrons, (2) reduced GFR, (3) reduced sensitivity to ADH, and (4) problems with the micturition reflex (**urinary retention** may develop in men whose prostate glands are inflamed).

CHAPTER QUESTIONS

LEVEL 1 Reviewing Facts and Terms

Match each item in column A with the most closely related item in column B. Use letters for answers in the spaces provided.

Column A	Column B
___ 1. urination	a. site of urine production
___ 2. renal capsule	b. capillaries around loop of Henle
___ 3. hilus	c. causes sensation of thirst
___ 4. medulla	d. accelerated sodium reabsorption
___ 5. nephrons	e. voluntary control
___ 6. renal corpuscle	f. glomerular epithelium
___ 7. external sphincter	g. fibrous tunic
___ 8. internal sphincter	h. blood leaves glomerulus
___ 9. aldosterone	i. dominant cation in ICF
___ 10. podocytes	j. renal pyramids
___ 11. efferent arteriole	k. blood to glomerulus
___ 12. afferent arteriole	l. interstitial fluid
___ 13. vasa recta	m. contains glomerulus
___ 14. ADH	n. exit for ureter
___ 15. ECF	o. dominant cation in ECF
___ 16. sodium	p. involuntary control
___ 17. potassium	q. micturition

18. After the filtrate leaves the glomerulus it empties into the:
    (a) distal convoluted tubule
    (b) loop of Henle
    (c) proximal convoluted tubule
    (d) collecting duct
19. The distal convoluted tubule (DCT) is an important site for:
    (a) active secretion of ions
    (b) active secretion of acids and other materials
    (c) selective reabsorption of sodium ions from the tubular fluid
    (d) a, b, and c are correct
20. The endocrine structure that secretes renin and erythropoietin is the:
    (a) juxtaglomerular apparatus
    (b) vasa recta
    (c) Bowman's capsule
    (d) adrenal gland
21. The primary purpose of the collecting system is to:
    (a) transport urine from the bladder to the urethra
    (b) selectively reabsorb sodium ions from tubular fluid
    (c) transport urine from the renal pelvis to the ureters
    (d) make final adjustments to the osmotic concentration and volume of urine
22. A person is in fluid balance when:
    (a) the ECF and ICF are isotonic
    (b) there is no fluid movement between compartments
    (c) the amount of water gained each day is equal to the amount lost to the environment
    (d) a, b, and c are correct
23. The primary components of the extracellular fluid are:
    (a) lymph and cerebrospinal fluid
    (b) blood plasma and serous fluids
    (c) interstitial fluid and plasma
    (d) a, b, and c are correct
24. All the homeostatic mechanisms that monitor and adjust the composition of body fluids respond to changes:
    (a) in the ICF (b) in the ECF
    (c) inside the cell (d) a, b, and c are correct
25. The most common problems with electrolyte balance are caused by an imbalance between gains and losses of:
    (a) calcium ions (b) chloride ions
    (c) potassium ions (d) sodium ions
26. What is the *primary* function of the urinary system?
27. What structures are included as component parts of the urinary system?
28. What are fluid shifts? What is their function, and what factors can cause them?
29. What three major hormones mediate major physiological adjustments that affect fluid and electrolyte balance? What are the primary effects of each hormone?

LEVEL 2 Reviewing Concepts

30. The urinary system regulates blood volume and pressure by:
(a) adjusting the volume of water lost in the urine
(b) releasing erythropoietin
(c) releasing renin
(d) a, b, and c are correct
31. The balance of solute and water reabsorption in the renal medulla is maintained by the:
(a) segmental arterioles and veins
(b) lobar arteries and veins
(c) vasa recta
(d) arcuate arteries
32. The higher the plasma concentration of aldosterone, the more efficiently the kidney will:
(a) conserve sodium ions
(b) retain potassium ions
(c) stimulate urinary water loss
(d) secrete greater amounts of ADH
33. When pure water is consumed:
(a) the ECF becomes hypertonic with respect to the ICF
(b) the ECF becomes hypotonic with respect to the ICF
(c) the ICF becomes hypotonic with respect to the plasma
(d) water moves from the ICF into the ECF
34. Increasing or decreasing the rate of respiration alters pH by:
(a) lowering or raising the partial pressure of carbon dioxide
(b) lowering or raising the partial pressure of oxygen
(c) lowering or raising the partial pressure of nitrogen
(d) a, b, and c are correct
35. What interacting controls operate to stabilize the glomerular filtration rate (GFR)?
36. Describe the micturition reflex.
37. Differentiate among fluid balance, electrolyte balance, and acid-base balance, and explain why each is important to homeostasis.
38. Why should a person with a fever drink plenty of fluids?
39. Exercise physiologists recommend that adequate amounts of fluid be ingested before, during, and after exercise. Why is adequate fluid replacement during conditions of extensive sweating important?

LEVEL 3 Critical Thinking and Clinical Applications

40. Why do long-haul trailer truck drivers frequently experience kidney problems?
41. For the past week Susan has felt a burning sensation in the urethral area when she urinates. She checks her temperature and finds that she has a low-grade fever. What unusual substances are likely to be present in her urine?
42. Carlos suffers from advanced arteriosclerosis. An analysis of his blood indicates elevated levels of aldosterone and decreased levels of ADH. Explain.

20 THE REPRODUCTIVE SYSTEM

Of all our organ systems, only one isn't needed for survival. Indeed, it doesn't even seem to confer any direct benefit on us, at least not physically. In a sense, our reproductive systems don't exist for us at all—they exist for our kind, for the human race. In this chapter we'll study the system that has enabled human beings to exist on the earth for a very long time—and with luck, to stick around for at least a while longer.

Chapter Outline and Objectives

Vocabulary Development

crypto, hidden; *cryptorchidism*
diplo, double; *diploid*
gyne, woman; *gynecologist*
haplo, single; *haploid*
labium, lip; *labia minora*
lutea, yellow; *corpus luteum*
meioun, to make smaller; *meiosis*
men, month; *menopause*
metra, uterus; *endometrium*
myo-, muscle; *myometrium*
oon, an egg; *oocyte*
orchis, testis; *cryptorchidism*
pausis, cessation; *menopause*
pellucidus, translucent; *zona pellucida*
rete, a net; *rete testis*
tetras, four; *tetrad*

An individual life span can be measured in decades, but the human species has survived for hundreds of thousands of years through the activities of the reproductive system. The entire process of reproduction seems almost magical; many primitive societies even failed to discover the basic link between sexual activity and childbirth and assumed that cosmic forces were responsible for producing new individuals. Although our society has a much clearer view of the reproductive process, a sense of wonder remains. Sexually mature males and females produce individual reproductive cells that are brought together through sexual intercourse. The fusion of these reproductive cells starts a chain of events leading to the appearance of an infant that will mature as part of the next generation.

The next two chapters will consider the mechanics of this remarkable process. We will begin by examining the anatomy and physiology of the reproductive system, which produces, stores, nourishes, and transports functional male and female reproductive cells, or **gametes** (GAM-ēts). The next chapter begins with **fertilization,** the fusion of a **sperm** from the father and **ovum** (Ō-vum) from the mother. All the cells in the body are the mitotic descendants of a single **zygote** (ZĪ-gōt), the cell created by the fusion of a sperm and an ovum. The gradual transformation of that single cell into a functional adult occurs through the process of *development,* the topic of the next chapter.

AN OVERVIEW OF THE REPRODUCTIVE SYSTEM

Both male and female reproductive systems include:

- Reproductive organs, or **gonads** (GŌ-nadz), that produce gametes and hormones.
- Ducts that receive and transport the gametes.
- Accessory glands and organs that secrete fluids into these or other excretory ducts.
- Perineal structures associated with the reproductive system, collectively known as the **external genitalia** (jen-i-TĀ-lē-a).

The functional roles of the male and female reproductive systems are quite different. In the adult male, the gonads, or **testes** (TES-tēz; singular *testis*), secrete *androgens*, principally testosterone, and produce one-half billion sperm each day. After storage, mature sperm travel along a lengthy duct system, where they are mixed with the secretions of accessory glands, creating **semen** (SĒ-men). During **ejaculation** (e-jak-ū-LĀ-shun) the semen is expelled from the body.

The female gonads, or **ovaries,** usually release only one egg per month. This gamete travels along short **uterine tubes** (*Fallopian tubes,* or *oviducts*) that terminate in a muscular chamber, the **uterus** (Ū-ter-us). A short passageway, the **vagina** (va-JĪ-na), connects the uterus with the exterior. During intercourse, the male ejaculation introduces semen into the vagina, and the sperm cells ascend the female reproductive tract. If a single sperm fuses with an egg, *fertilization* occurs and the resulting cell divides repeatedly. The uterus will enclose and support the developing embryo as it grows into a fetus and prepares for eventual delivery.

THE REPRODUCTIVE SYSTEM OF THE MALE

The principal structures of the male reproductive system are shown in Figure 20-1●. Proceeding from the testes, the sperm cells, or **spermatozoa** (sper-ma-tō-ZŌ-a), travel along the **epididymis** (ep-i-DID-i-mus), the **ductus deferens** (DUK-tus DEF-e-renz), the **ejaculatory** (ē-JAK-ū-la-tō-rē) **duct,** and the **urethra** before leaving the body. Accessory organs, notably the **seminal** (SEM-i-nal) **vesicles,** the **prostate** (PROS-tāt) **gland,** and the **bulbourethral** (bul-bō-ū-RĒ-thral) **glands,** secrete into the ejaculatory ducts and urethra. Externally visible structures include the **scrotum** (SKRŌ-tum), which encloses the testes, and the **penis** (PĒ-nis), an erectile organ that surrounds the distal portion of the urethra. Together the scrotum and penis constitute the external genitalia of the male.

The Testes

The *primary sex organs* of the male system are the testes. The testes hang within the scrotum, a fleshy pouch suspended below the perineum anterior to the anus. The scrotum is subdivided into two chambers, each containing a testis. Each testis has the shape of a

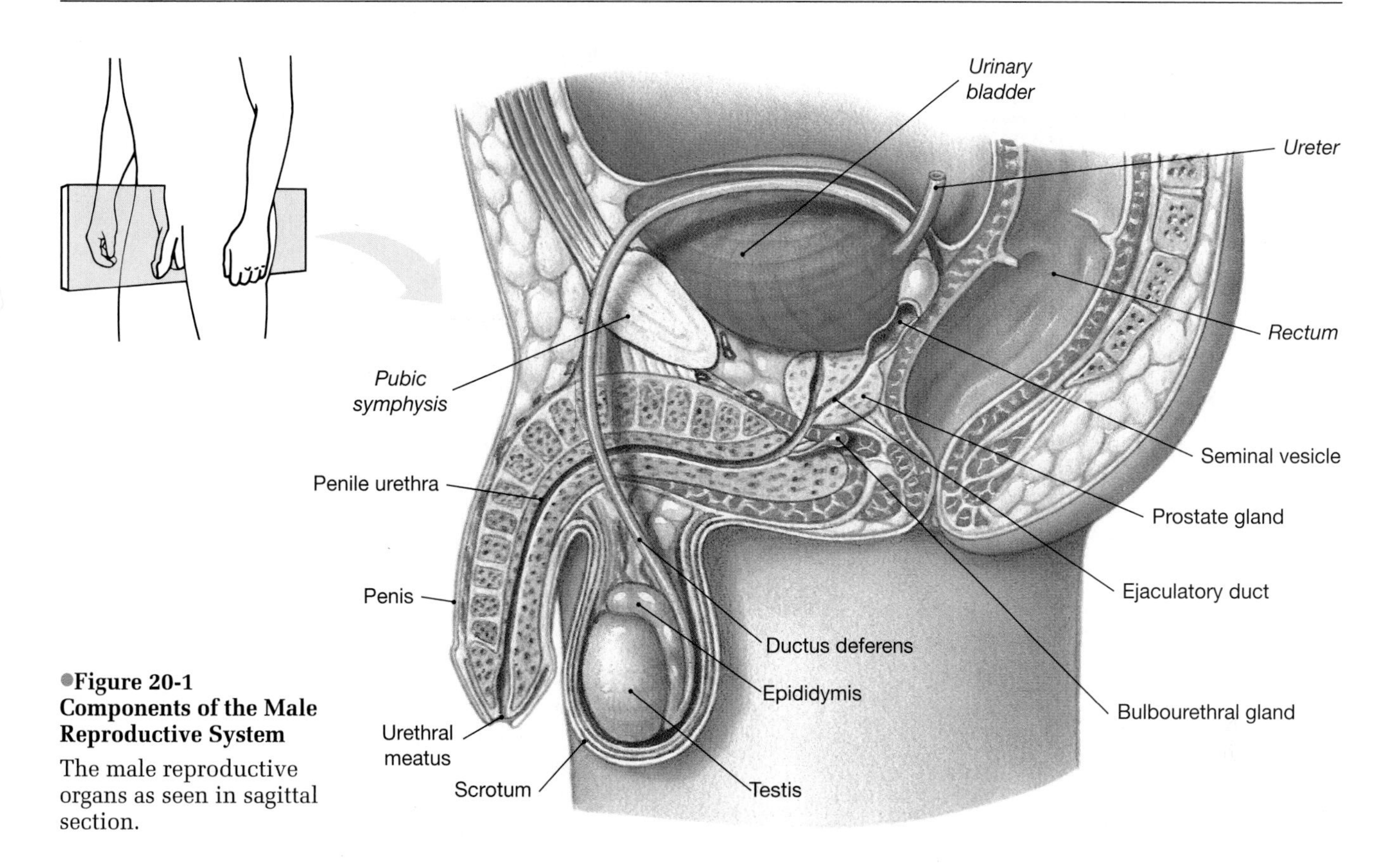

Figure 20-1 Components of the Male Reproductive System
The male reproductive organs as seen in sagittal section.

flattened oval roughly 5 cm (2 in.) in length and 2.5 cm (1 in.) in width. An epithelial lining on the inner surface of the scrotum and the outer surface of the testis prevents friction between the opposing surfaces.

The scrotum consists of a thin layer of skin, loose connective tissue, and smooth muscle (Figure 20-2a). Sustained contractions of the smooth muscle layer, the *dartos* (DAR-tōs), cause the characteristic wrinkling of the scrotal surface. Beneath the dermis is a layer of skeletal muscle, the **cremaster** (kre-MAS-ter) **muscle,** which can contract to pull the testes closer to the body. Normal sperm development requires temperatures around 1.1°C (2°F) below normal body temperature. When environmental temperatures rise, the cremaster relaxes, the testes move away from the body, and excess heat is lost across the surface of the scrotum. When the scrotum is cooled, as in an icy swimming pool, cremasteric contractions pull the testes closer to the body to keep them warm.

Each testis is wrapped in a dense fibrous capsule, the **tunica albuginea** (TŪ-ni-ka al-bū-JIN-ē-a). Collagen fibers from this wrapping extend into the testis, forming partitions, or *septa,* that subdivide the testis into roughly 250 *lobules.* Sperm production occurs in the approximately 800 slender, tightly coiled **seminiferous** (se-mi-NIF-e-rus) **tubules** (Figure 20-2b) that are distributed among the lobules. Each tubule averages around 80 cm (31 in.) in length, and a typical testis contains nearly one-half mile of seminiferous tubules. A maze of passageways known as the **rete** (RĒ-tē; *rete,* a net) **testis** provide passage for sperm cells from the seminiferous tubules to the epididymis.

The spaces between the tubules are filled with loose connective tissue, numerous blood vessels, and large **interstitial cells** that produce male sex hormones, or *androgens* (Figure 20-2c). The steroid *testosterone* is the most important androgen. (Testosterone and other sex hormones were introduced in Chapter 11.) [p. 298]

Sperm cells, or spermatozoa, are produced through **spermatogenesis** (sper-ma-tō-JEN-e-sis). This process begins with stem cells called *spermatogonia* at the outermost layer of cells in the seminiferous tubules (Figure 20-2c) and proceeds to the central tubular lumen.

Each seminiferous tubule also contains **sustentacular** (sus-ten-TAK-ū-lar) **cells** (*Sertoli cells*). The large sustentacular cells are attached to the tubular capsule and extend toward the lumen between the spermatocytes and spermatogonia (Figure 20-2c).

CRYPTORCHIDISM

In *cryptorchidism* (kript-ŌR-ki-dizm; *crypto,* hidden), one or both of the testis have not descended into the scrotum by the time of birth. Typically, the testes are lodged in the abdominal cavity or within the inguinal canal. This condition occurs in about 3 percent of full-term deliveries and in roughly 30 percent of premature births. In most instances, normal descent occurs a few weeks later, but the condition can be surgically corrected if it persists. Corrective measures are usually taken before puberty because cryptorchid (abdominal) testes will not

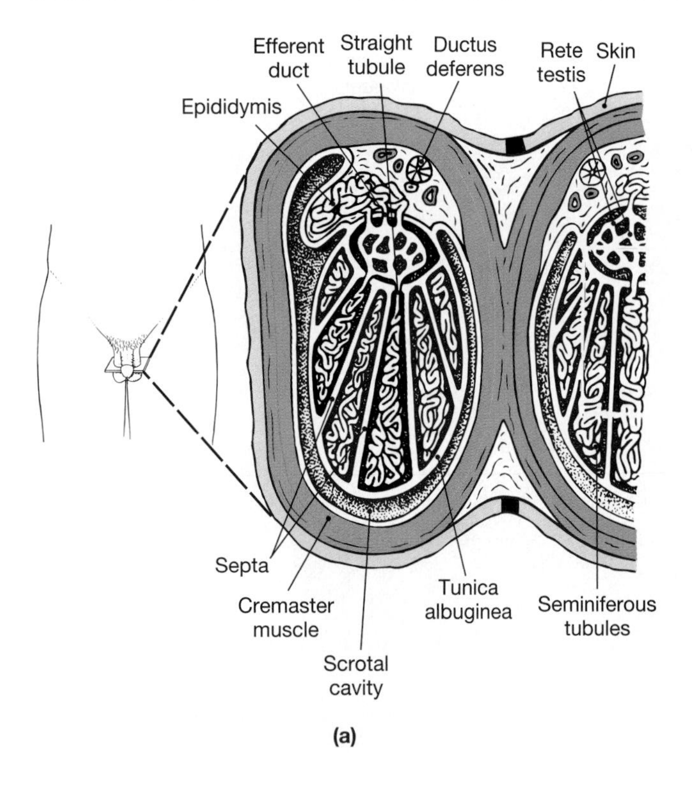

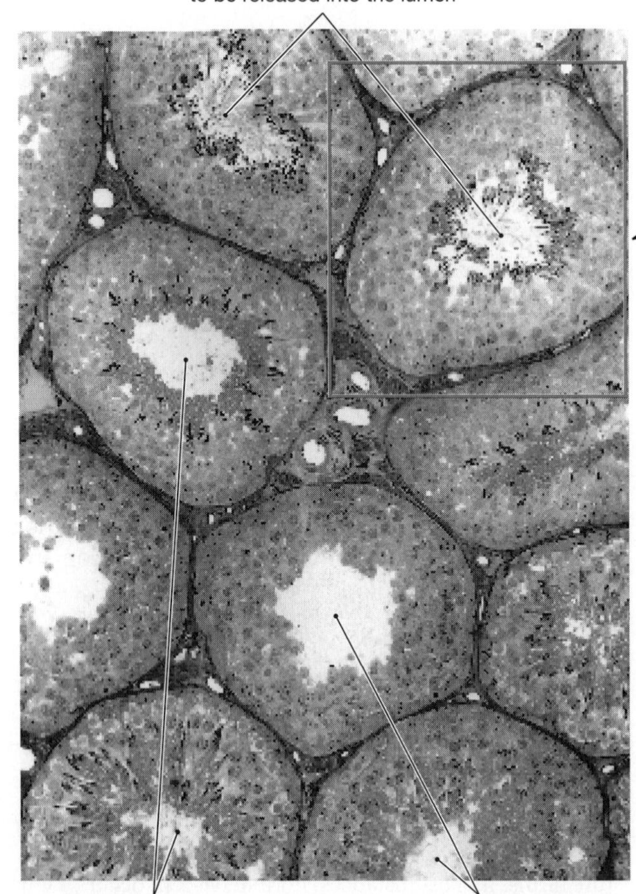

produce sperm, and the individual will be sterile. If the testes cannot be moved into the scrotum, they will usually be removed, because about 10 percent of those with uncorrected cryptorchid testes eventually develop testicular cancer. This surgical procedure is called a *bilateral orchiectomy* (ōr-kē-EK-to-mē; *orchis,* testis).

Spermatogenesis

Spermatogenesis involves three integrated processes:

- ***Mitosis.*** Stem cells called **spermatogonia** (sper-ma-to-GŌ-nē-a) undergo cell divisions throughout adult life. (See Chapter 3 for a review of mitosis.) ∞ [p. 66] The cell divisions produce daughter cells that are pushed to the lumen of the tube. These cells differentiate into *spermatocytes* (sper-MA-to-sīts) that prepare to undergo *meiosis.*
- ***Meiosis.*** **Meiosis** (mī-Ō-sis; *meioun,* to make smaller) is a special form of cell division involved in gamete production. Gametes contain half the number of chromosomes found in other cells. As a result, the fusion of a sperm and an egg yields a single cell with the normal number of chromosomes. In the seminiferous tubules, the meiotic divisions of spermatocytes produce immature gametes called *spermatids.*
- ***Spermiogenesis.*** Spermatids are small, relatively unspecialized cells. In the process of **spermiogenesis,** spermatids differentiate into physically mature spermatozoa.

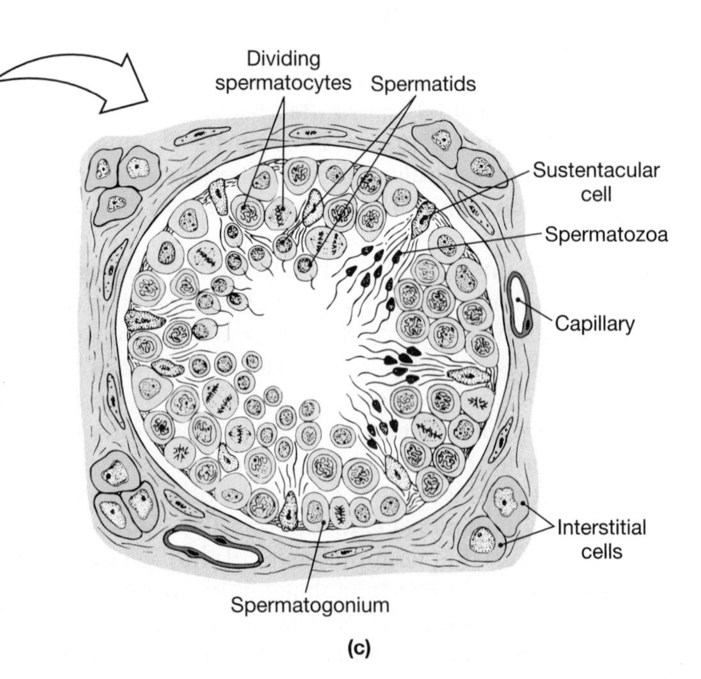

•Figure 20-2
Structure of the Testes

(a) Diagrammatic sketch and anatomical relationships of the testes. **(b)** A section through a coiled seminiferous tubule. **(c)** Cellular organization of a seminiferous tubule.

Mitosis and Meiosis Mitosis and meiosis differ significantly in terms of nuclear events and outcomes. Mitosis is part of the process of somatic cell division, which involves one division that produces two daughter cells, each containing 23 *pairs* of chromosomes. Each pair consists of one chromosome provided by the male parent and another by the female parent at the time of fertilization. Because they contain both members of each chromosome pair, they are called **diploid** (DIP-loyd; *diplo,* double) cells. Meiosis involves a pair of divisions and produces four cells, each of which contains 23 *individual chromosomes.* Because these cells contain only one member of each pair of chromosomes, they are called **haploid** (HAP-loyd; *haplo,* single) cells.

The mitotic divisions of the spermatogonia produce primary spermatocytes. As a primary spermatocyte prepares to begin meiosis, all of the chromosomes within the nucleus replicate themselves as if the cell were to undergo mitosis. As the prophase of the first meiotic division, **meiosis I,** arrives, the chromosomes condense and become visible (Figure 20-3●). As in mitosis, each chromosome consists of two duplicate *chromatids* (KRŌ-ma-tidz).

As meiosis begins, the primary spermatocyte contains 46 individual chromosomes, the same as any somatic cell in the body. The corresponding maternal and paternal chromosomes now come together. This event, known as **synapsis** (sin-AP-sis), produces 23 pairs of chromosomes, each member of the pair consisting of two chromatids. A matched set of four chromatids is called a **tetrad** (TET-rad; *tetras,* four). An exchange of genetic material, called *crossing-over,* can occur between the chromatids at this stage of meiosis. This exchange increases genetic variation among offspring.

During metaphase of meiosis I, the nuclear envelope disappears and the tetrads line up along the metaphase plate. As anaphase begins, the tetrads break up, and the maternal and paternal chromosomes separate. This is a major difference between mitosis and meiosis: In mitosis, each daughter cell receives one of the two copies of every chromosome, maternal and paternal, whereas in meiosis each daughter cell receives both copies of *either* the maternal chromosome *or* the paternal chromosome from each tetrad.

As anaphase proceeds, the maternal and paternal components are randomly distributed. For example, most of the maternal chromosomes may go to one daughter cell, and most of the paternal chromosomes to the other. As a result, telophase I ends with the formation of two daughter cells containing unique combinations of maternal and paternal chromosomes. In the testes, the daughter cells produced by the first meiotic division (meiosis I) are called **secondary spermatocytes.**

Each secondary spermatocyte contains 23 chromosomes. Each of these chromosomes still consists of two duplicate chromatids. These duplicates will be separated during **meiosis II.**

The interphase separating meiosis I and meiosis II is very brief, and the secondary spermatocyte soon enters prophase of meiosis II. The completion of metaphase II, anaphase II, and telophase II produces

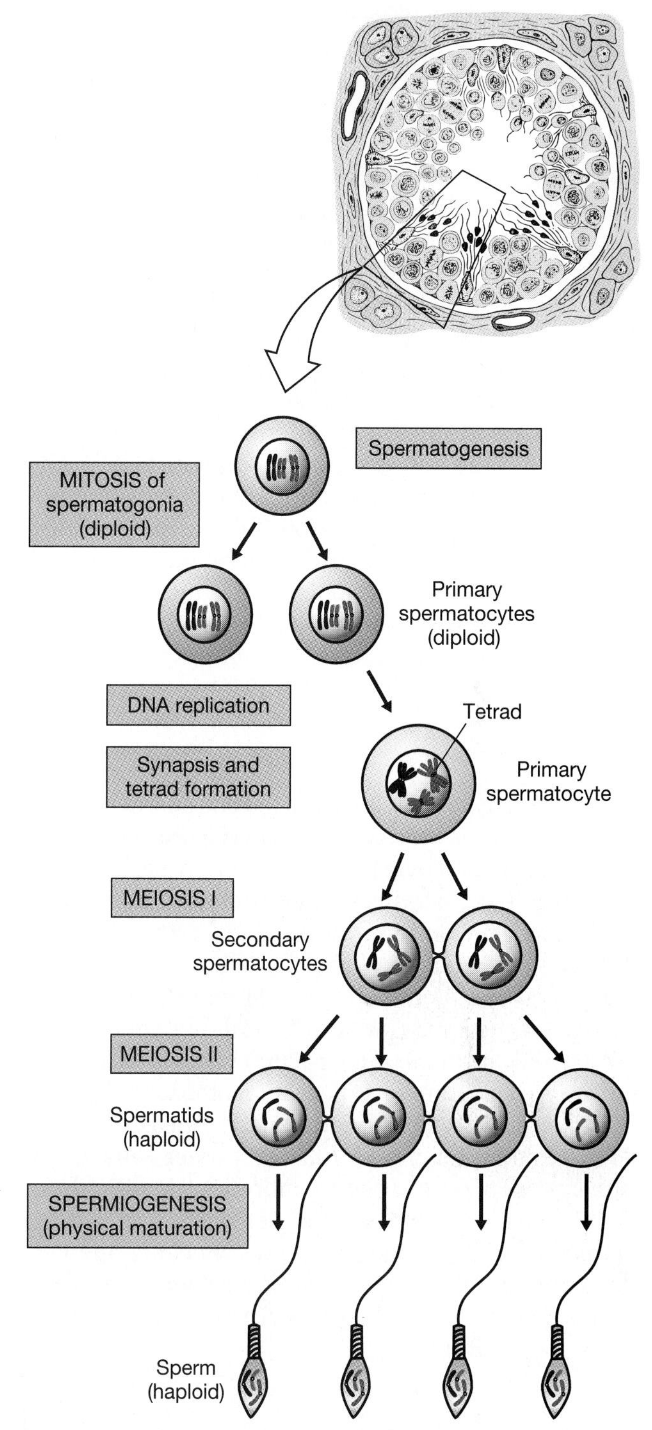

●**Figure 20-3**
Meiosis and the Formation of Spermatozoa
Each diploid primary spermatocyte that undergoes meiosis produces four haploid spermatids. Each spermatid then differentiates into a spermatozoon.

four **spermatids** (SPER-ma-tidz), each containing 23 chromosomes. In summary, for every primary spermatocyte that enters meiosis, four spermatids are produced (Figure 20-3●).

Spermiogenesis Each spermatid matures into a single **spermatozoon** (sper-ma-tō-ZŌ-on), or sperm cell, through the process of *spermiogenesis.* The entire process, from spermatogonial division to the release of a physically mature spermatozoon, takes approximately 9 weeks.

Sustentacular cells play a key role in spermiogenesis. Because there are no blood vessels inside the seminiferous tubules, all nutrients must enter by diffusion from the surrounding interstitial fluids. The large sustentacular cells control the chemical environment inside the seminiferous tubules and, because they surround the spermatids, provide nutrients and chemical stimuli that promote the production and differentiation of spermatozoa. They also help regulate spermatogenesis by producing *inhibin,* a hormone introduced in Chapter 11. [p. 298]

Anatomy of a Spermatozoon

There are three distinct regions to each sperm cell: the *head,* the *middle piece,* and the *tail* (Figure 20-4●). The **head** is a flattened oval filled with densely packed chromosomes. The tip forms the **acrosomal** (ak-rō-SŌ-mal) **cap,** which contains enzymes essential for fertilization. A very short **neck** attaches the head to the **middle piece,** which is dominated by the mitochondria providing the energy for moving the **tail.** The sperm cell's tail, the only example of a *flagellum* in the human body, moves the cell from one place to another. [p. 58]

The entire streamlined package measures only 60 μm in total length. Unlike most other cells, a mature spermatozoon lacks an endoplasmic reticulum, Golgi apparatus, lysosomes, peroxisomes, and inclusions, among other structures. Because the cell does not contain glycogen or other energy reserves, it must absorb nutrients (primarily fructose) from the surrounding fluid.

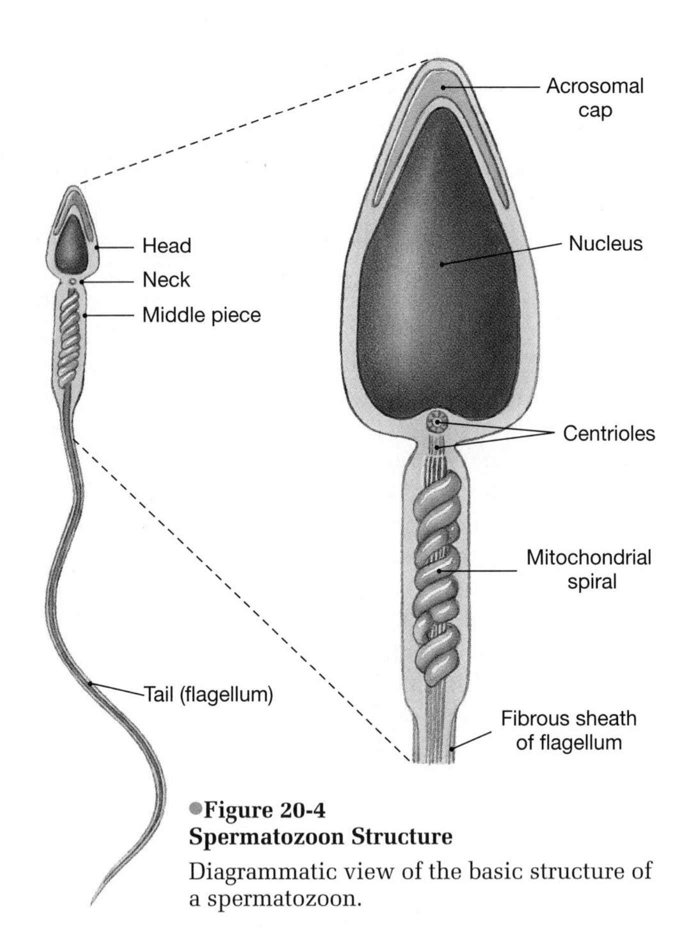

●**Figure 20-4**
Spermatozoon Structure
Diagrammatic view of the basic structure of a spermatozoon.

The Male Reproductive Tract

The testes produce physically mature spermatozoa that are, as yet, incapable of fertilizing an ovum. The other portions of the male reproductive system, sometimes called the *accessory structures,* are concerned with the functional maturation, nourishment, storage, and transport of spermatozoa.

The Epididymis

Late in their development, the spermatozoa become detached from the sustentacular cells and lie within the lumen of the seminiferous tubule. Although they have most of the physical characteristics of mature sperm cells, they are still functionally immature and incapable of coordinated locomotion. At this point, fluid currents transport them into the epididymis (see Figure 20-1●, p. 517). This elongate tubule, almost 7 meters (23 ft) long, is so twisted and coiled that it actually takes up very little space. During the 2 weeks that it takes for a spermatozoon to travel through the epididymis, it completes its physical maturation. Damaged or abnormal spermatozoa are recycled, and the epididymis absorbs cellular debris and organic nutrients. Mature spermatozoa then arrive at the ductus deferens.

Although the spermatozoa leaving the epididymus are physically mature, they remain immobile. To become active, motile, and fully functional they must undergo **capacitation,** and the epididymis secretes a substance that prevents premature capacitation. Capacitation occurs when the spermatozoa (1) mix with secretions of the seminal vesicles and (2) are exposed to conditions inside the female reproductive tract.

The Ductus Deferens

The ductus deferens, also known as the *vas deferens,* is 40–45 cm (16–18 in.) in length. It extends toward the abdominal cavity within a connective tissue sheath that also encloses the blood vessels, nerves, and lymphatics servicing the testis as well as part of the cremaster muscle. The entire complex is called the **spermatic cord.**

●Figure 20-5
The Ductus Deferens
(a) A posterior view of the prostate, showing subdivisions of the ductus deferens in relation to surrounding structures. **(b)** Light and scanning electron micrographs showing extensive layering with smooth muscle around the lumen of the ductus deferens. (LM × 34; SEM × 42)

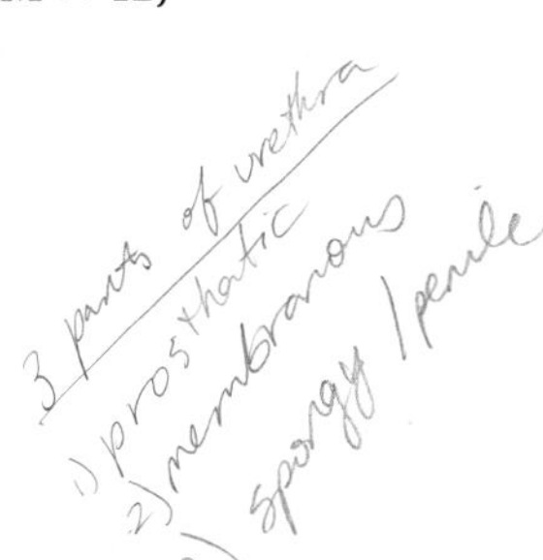

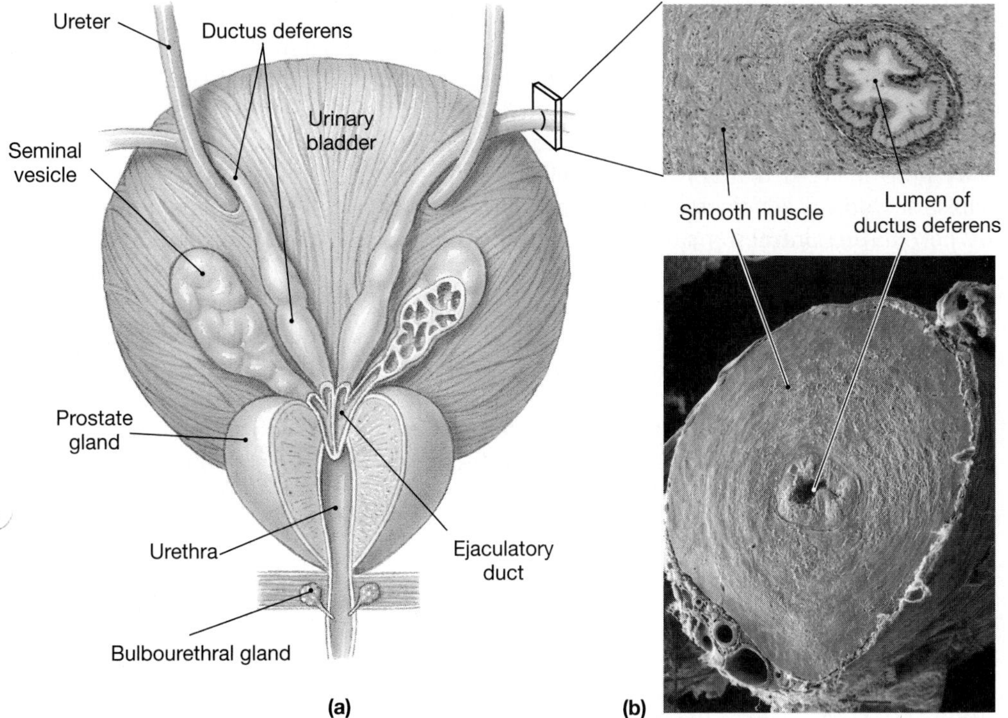

After passing through the *inguinal canal,* described in Chapter 7, the components of the spermatic cord go their separate ways. ∞ [p. 177] The ductus deferens curves downward alongside of the urinary bladder toward the prostate gland (Figures 20-1, p. 517, and 20-5●).

Peristaltic contractions in the muscular walls of the ductus deferens propel spermatozoa and fluid along the length of the duct. The ductus deferens can also store spermatozoa for up to several months. During this period the spermatozoa are in a state of suspended animation, remaining inactive with low metabolic rates.

The junction of the ductus deferens with the duct draining the seminal vesicle creates the *ejaculatory duct,* a relatively short (2 cm, or less than 1 in.) passageway that penetrates the muscular wall of the prostate and fuses with the ejaculatory duct from the other side before emptying into the urethra.

The Urethra

The urethra of the male extends from the urinary bladder to the tip of the penis, a distance of 15–20 cm (6–8 in.). The urethra in the male is a passageway used by both the urinary and reproductive systems.

The Accessory Glands

The fluids contributed by the seminiferous tubules and the epididymis account for only about 5 percent of the final volume of semen. The major fraction of seminal fluid is composed of secretions from the *seminal vesicles,* the *prostate gland,* and the *bulbourethral glands.* Major functions of these glandular organs include: (1) activating the spermatozoa, (2) providing the nutrients spermatozoa need for motility, (3) propelling spermatozoa and fluids along the reproductive tract through peristaltic contractions, and (4) producing buffers that counteract the acidity of the urethral and vaginal contents.

The Seminal Vesicles

Each seminal vesicle is a tubular gland with a total length of around 15 cm (6 in.). The body of the gland is coiled and folded into a compact, tapered mass roughly 5 cm by 2.5 cm (2 in. by 1 in.) (Figures 20-1, p. 516, and 20-5●).

The seminal vesicles contribute about 60 percent of the volume of semen. In particular, their secretions contain relatively high concentrations of fructose, a six-carbon sugar easily metabolized by spermatozoa. The secretions are also slightly alkaline, and this alkalinity helps neutralize acids in the prostatic secretions and within the vagina. When mixed with the secretions of the seminal vesicles, previously inactive but mature spermatozoa begin beating their flagella and become highly mobile.

The Prostate Gland

The *prostate gland* is a small, muscular, rounded organ with a diameter of about 4 cm (1.6 in.). As indicated in Figure 20-5●, the prostate gland surrounds the urethra as it leaves the urinary bladder. The prostatic wall produces a weakly acidic secretion that contributes about 30 percent of the volume of semen. In addition to several other compounds of uncertain significance, prostatic secretions contain **seminalplasmin** (sem-i-nal-PLAZ-min), an antibiotic that may help prevent urinary tract infections in males. These secretions are ejected into the prostatic urethra by peristaltic contractions of the muscular wall.

PROSTATITIS

Prostatic inflammation, or *prostatitis* (pros-ta-TĪ-tis), can occur at any age, but it most often afflicts older men. Prostatitis may result from bacterial infections, but the condition may also develop in the apparent absence of pathogens. Individuals with prostatitis complain of pain in the lower back, perineum, or rectum, sometimes accompanied by painful urination and the discharge of mucous secretions from the urethral meatus. Antibiotic therapy is usually effective in treating cases resulting from bacterial infection, but in other cases antibiotics may not provide relief. Prostatitis is taken seriously because the symptoms can resemble those of *prostate cancer.* Prostate cancer is the second most common cancer in men, and it is the number two cause of cancer deaths in males. AM *Prostatic Hypertrophy and Prostate Cancer*

The Bulbourethral Glands

The paired *bulbourethral glands,* or *Cowper's glands,* are round, with diameters approaching 10 mm (less than 0.5 in.) (Figure 20-5●). These glands secrete a thick, sticky, alkaline mucus that has lubricating properties.

Semen

A typical ejaculation releases 2–5 ml of semen. This volume of fluid, called an **ejaculate,** contains the following:

- ***Spermatozoa.*** A normal **sperm count** ranges from 20 million to 100 million spermatozoa per milliliter.
- ***Seminal fluid.*** **Seminal fluid,** the fluid component of semen, is a mixture of glandular secretions with a distinctive ionic and nutrient composition. In terms of total volume, the seminal fluid contains the combined secretions of the seminal vesicles (60 percent), the prostate (30 percent), the sustentacular cells and epididymis (5 percent), and the bulbourethral glands (less than 5 percent).
- ***Enzymes.*** Several important enzymes are present in the seminal fluid. For example, semen includes a protease that helps dissolve mucous secretions in the vagina, and *seminalplasmin,* an antibiotic enzyme from the prostate gland that kills a variety of bacteria.

Within a few minutes after ejaculation semen coagulates, liquefying again after a variable period. The function of this clotting is unknown.

The Penis

The penis is a tubular organ that surrounds the urethra (Figure 20-1●). It conducts urine to the exterior and introduces semen into the female vagina during sexual intercourse. The penis (Figure 20-6●) is divided into three regions: (1) the **root,** the fixed portion that attaches the penis to the body wall; (2) the **body (shaft),** the tubular portion that contains masses of erectile tissue; and, (3) the **glans,** the expanded distal end that surrounds the external urethral opening.

The skin overlying the penis resembles that of the scrotum. A fold of skin, the **prepuce** (PRĒ-pūs), or *foreskin,* surrounds the tip of the penis. The prepuce attaches to the relatively narrow **neck** of the penis and continues over the glans. *Preputial glands* in the skin of the neck and inner surface of the prepuce secrete a waxy material called *smegma* (SMEG-ma). Unfortunately, smegma can be an excellent nutrient source for bacteria. Mild inflammation and infections in this region are common, especially if the area is not washed frequently. One way of avoiding trouble is to perform a *circumcision* (ser-kum-SIZH-un) and surgically remove the prepuce. In Western societies this procedure is usually performed shortly after birth.

Most of the body, or shaft, of the penis consists of three columns of **erectile tissue** (Figure 20-6b●). Erectile tissue consists of a three-dimensional maze of vascular channels incompletely divided by sheets of elastic connective tissue and smooth muscle. On the anterior surface of the penis, there are two cylindrical **corpora cavernosa** (KŌR-po-ra ka-ver-NŌ-sa) that are bound to the pubis and ischium of the pelvis. The corpora cavernosa extend along as far as the neck of the penis. The relatively slender **corpus spongiosum** (spon-jē-Ō-sum) surrounds the urethra and extends all the way to the tip of the penis, where it forms the glans.

In the resting state, there is little blood flow into the erectile tissue because the arterial branches are constricted. During *arousal* parasympathetic stimulation dilates the walls of the arterial blood vessels to the erectile tissue, blood flow increases, the penis becomes engorged with blood, and **erection** occurs.

Hormones and Male Reproductive Function

Major reproductive hormones were introduced in Chapter 11, and the hormonal interactions in the male are diagrammed in Figure 20-7●. [p. 298] The anterior pituitary releases *follicle-stimulating hormone* **(FSH)** and a second peptide hormone, *luteinizing hormone* **(LH),** named after its effects in the female.

(This hormone was called *interstitial cell–stimulating hormone, ICSH,* in males before it was known to be identical to the hormone in females.) The pituitary release of these hormones occurs in the presence of *gonadotropin-releasing hormone* **(GnRH),** a hormone synthesized in the hypothalamus and carried to the anterior pituitary in the hypophyseal portal system.

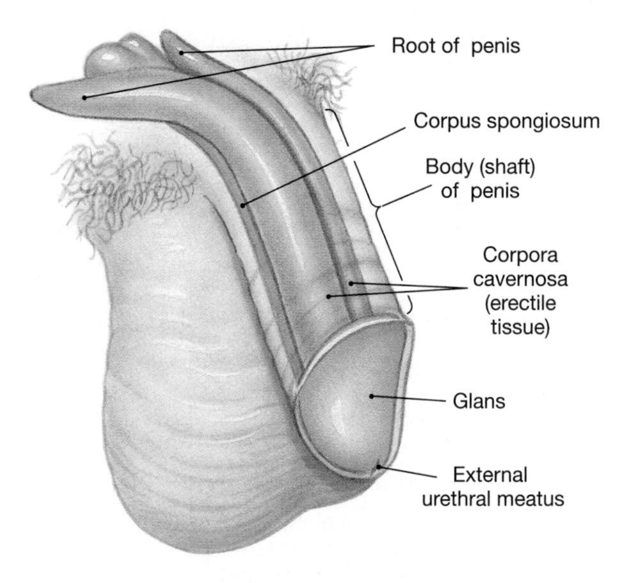

(a) Anterior and lateral view of penis

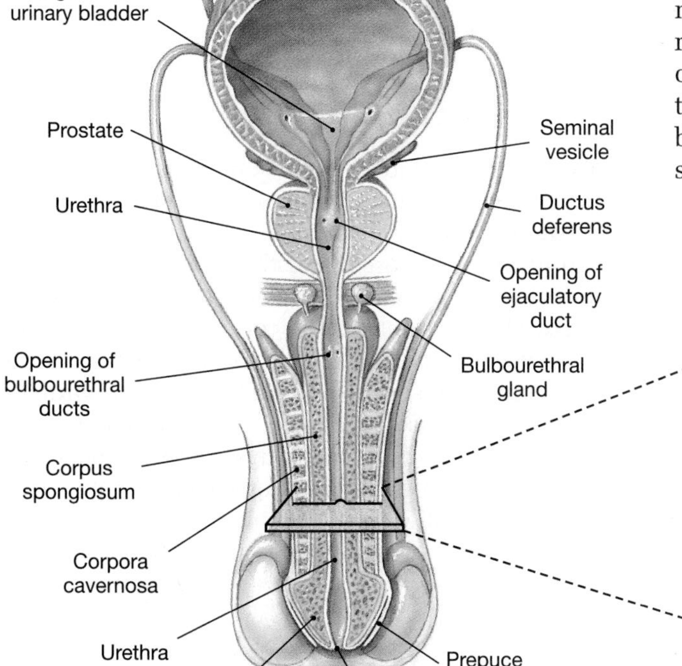

(b)

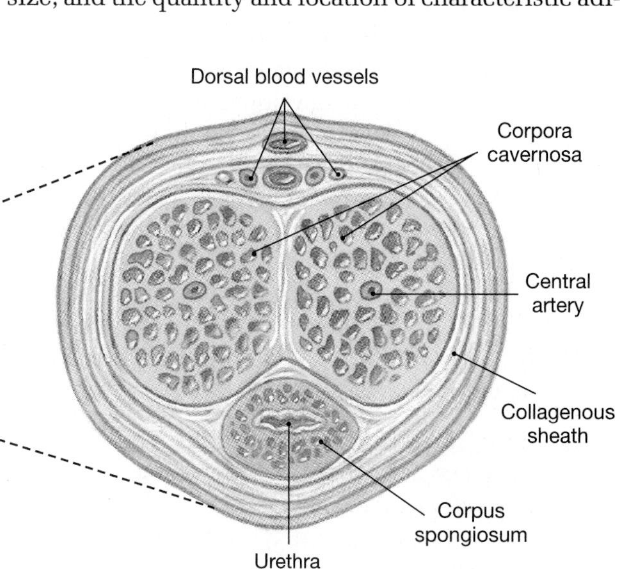

(c) Section through shaft of penis

•Figure 20-6
The Penis

(a) Anterior and lateral view of the penis, showing positions of the erectile tissues. **(b)** Frontal section through the penis and associated organs. **(c)** Sectional view through the penis.

FSH and Spermatogenesis

In the male, FSH targets primarily the sustentacular cells of the seminiferous tubules. Under FSH stimulation, and in the presence of testosterone from the interstitial cells, sustentacular cells promote spermatogenesis and spermiogenesis.

The rate of spermatogenesis is regulated by a negative feedback mechanism involving GnRH, FSH, and inhibin. Under GnRH stimulation, FSH promotes spermatogenesis along the seminiferous tubules. As spermatogenesis accelerates, however, so does the rate of inhibin secretion by the sustentacular cells of the testes. Inhibin inhibits FSH production in the anterior pituitary and may also suppress secretion of GnRH at the hypothalamus.

The net effect is that when FSH levels become elevated, inhibin production increases until the FSH levels return to normal. If FSH levels decline, inhibin production falls, and the rate of FSH production accelerates.

LH and Androgen Production

In the male, LH causes the secretion of testosterone and other androgens by the interstitial cells of the testes. Testosterone, the most important androgen, has numerous functions. It (1) promotes the functional maturation of spermatozoa; (2) maintains the accessory organs of the male reproductive tract; (3) determines the secondary sexual characteristics such as the distribution of facial hair, increased muscle mass and body size, and the quantity and location of characteristic adi-

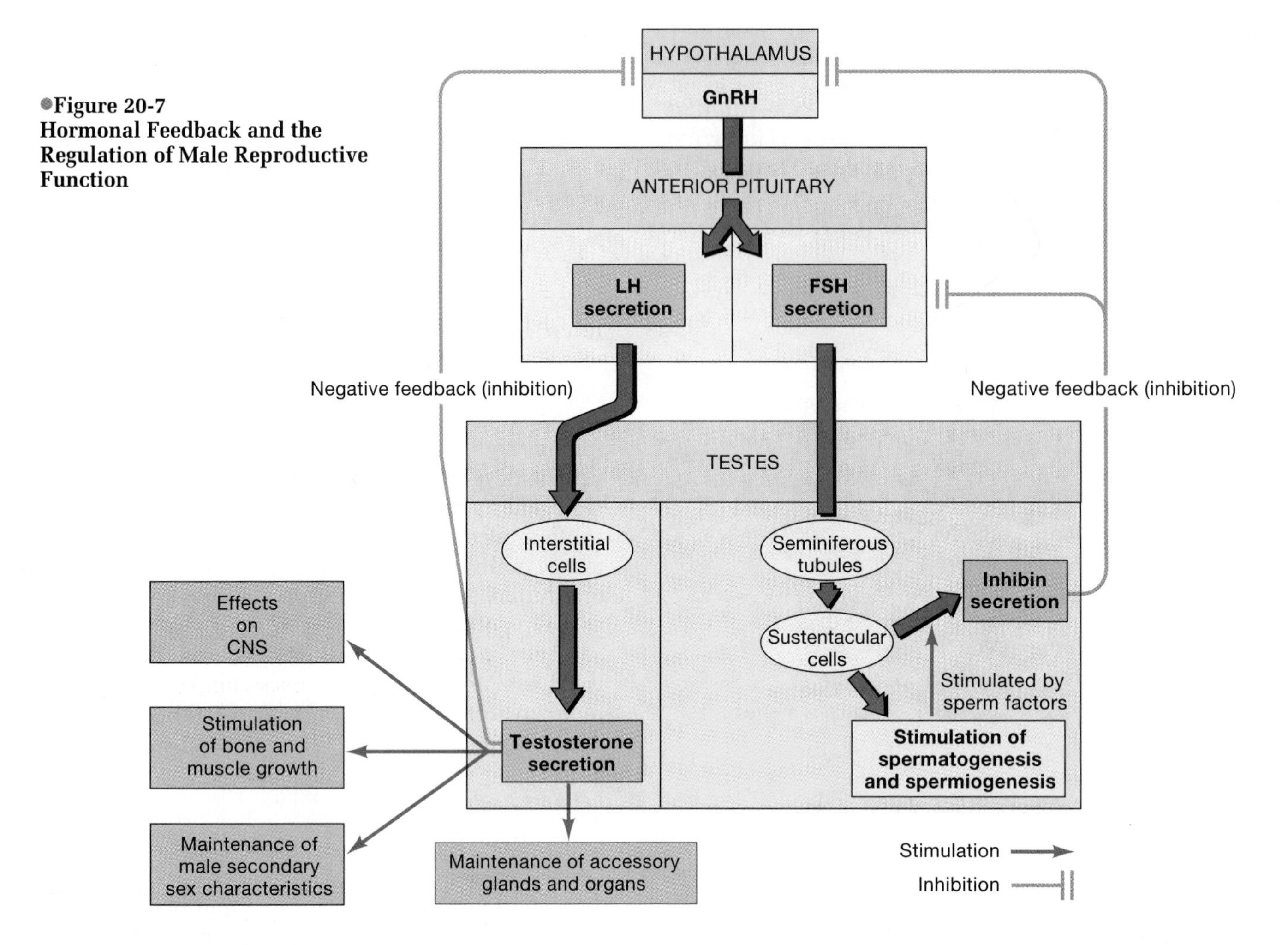

•Figure 20-7
Hormonal Feedback and the Regulation of Male Reproductive Function

pose tissue deposits; (4) stimulates metabolic operations throughout the body, especially those concerned with protein synthesis and muscle growth; and (5) influences brain development by stimulating sexual behaviors and sexual drive.

Testosterone production begins around the seventh week of embryonic development and reaches a peak after roughly 6 months of development. The early surge in testosterone levels stimulates the differentiation of the male duct system and accessory organs. Testosterone production accelerates markedly at puberty, initiating sexual maturation and the appearance of secondary sexual characteristics. In the adult, the level of testosterone is controlled by negative feedback. Above-normal testosterone levels inhibit the release of GnRH by the hypothalamus. This inhibition causes a reduction in LH secretion and lowers testosterone levels.

- ✓ On a warm day would the cremaster muscle be contracted or relaxed? Why?
- ✓ How would the lack of an acrosomal cap affect the ability of a spermatozoon to fertilize an ovum?
- ✓ What will occur if the arteries serving the penis dilate?
- ✓ What effect would low levels of FSH have on sperm production?

THE REPRODUCTIVE SYSTEM OF THE FEMALE

A woman's reproductive system must produce sex hormones and gametes and also protect and support a developing embryo and nourish the newborn infant. The primary sex organs of the female reproductive system are the *ovaries.* The internal and external accessory organs include the *uterine tubes* (*Fallopian tubes* or *oviducts*), the *uterus* (womb), the *vagina,* and the components of the external genitalia (Figure 20-8•). As in the male, various accessory glands secrete into the reproductive tract. Physicians specializing in the female reproductive system are called **gynecologists** (gī-ne-KOL-o-jists; *gyne*, woman).

The Ovaries

A typical ovary measures approximately 5 cm by 2.5 cm (2 in. by 1 in.). It has a pale white or yellowish coloration and a nodular consistency that resembles cottage cheese or lumpy oatmeal. The interior of the ovary can be divided into a superficial *cortex* and a deep *medulla.* The production of gametes occurs in the cortex, and the arteries, veins, lymphatics, and nerves within the relatively narrow medulla link the ovary with other body systems.

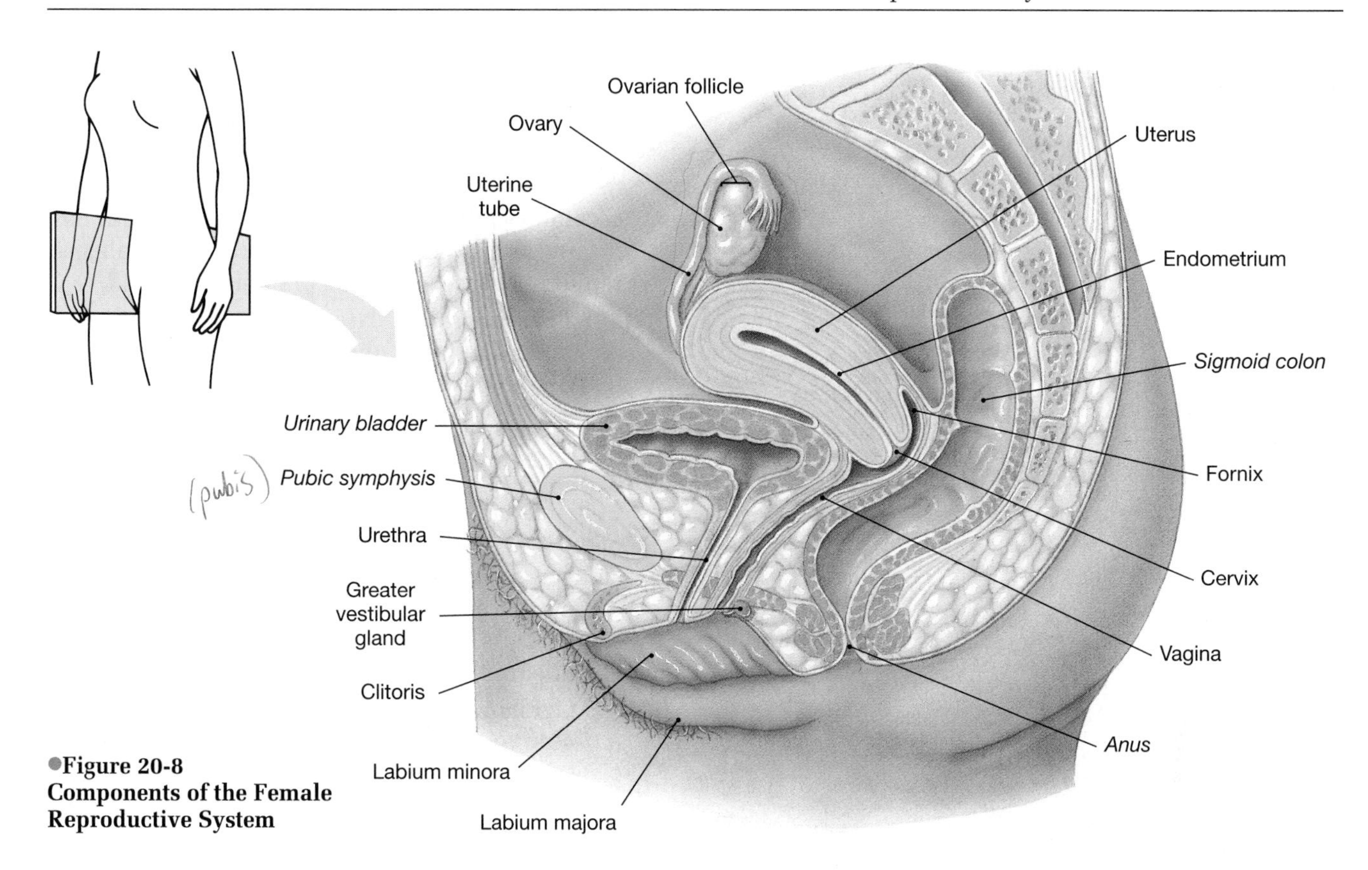

Figure 20-8 Components of the Female Reproductive System

The ovaries are responsible for (1) the production of female gametes, or **ova** (singular ovum), (2) the secretion of female sex hormones, including *estrogens* and *progestins,* and (3) the secretion of inhibin, involved in the feedback control of pituitary FSH production. AM *Ovarian Cancer*

Oogenesis

Ovum production, or **oogenesis** (ō-ō-JEN-e-sis; *oon,* egg), begins before birth, accelerates at puberty, and ends at *menopause* (*men,* month + *pausis,* cessation). Between puberty and menopause, oogenesis occurs on a monthly basis, as part of the *ovarian cycle.*

In the ovaries, stem cells, or **oogonia** (ō-o-GŌ-nē-a), complete their mitotic divisions before birth. Between the third and seventh months of fetal development, the daughter cells, or **primary oocytes** (Ō-o-sīts), prepare to undergo meiosis. They proceed as far as prophase of meiosis I, but at that time the process stops. The primary oocytes then remain in a state of suspended development until puberty, awaiting the hormonal signal to complete meiosis. Not all of the primary oocytes in the ovaries at birth survive until puberty. There are roughly 2 million in the ovaries at birth; by the time of puberty, about 400,000 remain. The rest of the primary oocytes degenerate, a process called *atresia* (a-TRĒ-zē-a).

Although the nuclear events under way during meiosis in the ovary are the same as those in the testis, the process differs in two important details.

1. The cytoplasm of the original oocyte is not evenly distributed during the meiotic divisions. Oogenesis produces one functional ovum, containing most of that cytoplasm, and three nonfunctional **polar bodies** that later disintegrate (Figure 20-9).
2. The ovary releases a *secondary oocyte,* rather than a mature ovum. The second meiotic division does not occur until *after* fertilization.

The Ovarian Cycle

Oogenesis occurs in the cortex within specialized structures called **ovarian follicles** (ō-VAR-ē-an FOL-i-klz). In the outer portions of the cortex, just beneath the capsule, there are scattered clusters of primary oocytes, each surrounded by a layer of follicle cells. The combination is known as a **primordial** (prī-MŌR-dē-al) **follicle.** At puberty, rising levels of FSH begin to activate a different group of primordial follicles each month. This monthly process is known as the **ovarian cycle.** Important steps in the ovarian cycle are shown in Figure 20-10.

Step 1: Formation of Primary Follicles. The cycle begins as the activated follicles develop into **primary follicles.** The follicle cells divide and form several concentric layers around the oocyte. As the wall of the follicle thickens further, a space opens up between the developing oocyte and the follicular cells. Within this space, called the **zona pellucida** (ZŌ-na pel-LŪ-si-da; *pellucidus,* translucent), microvilli

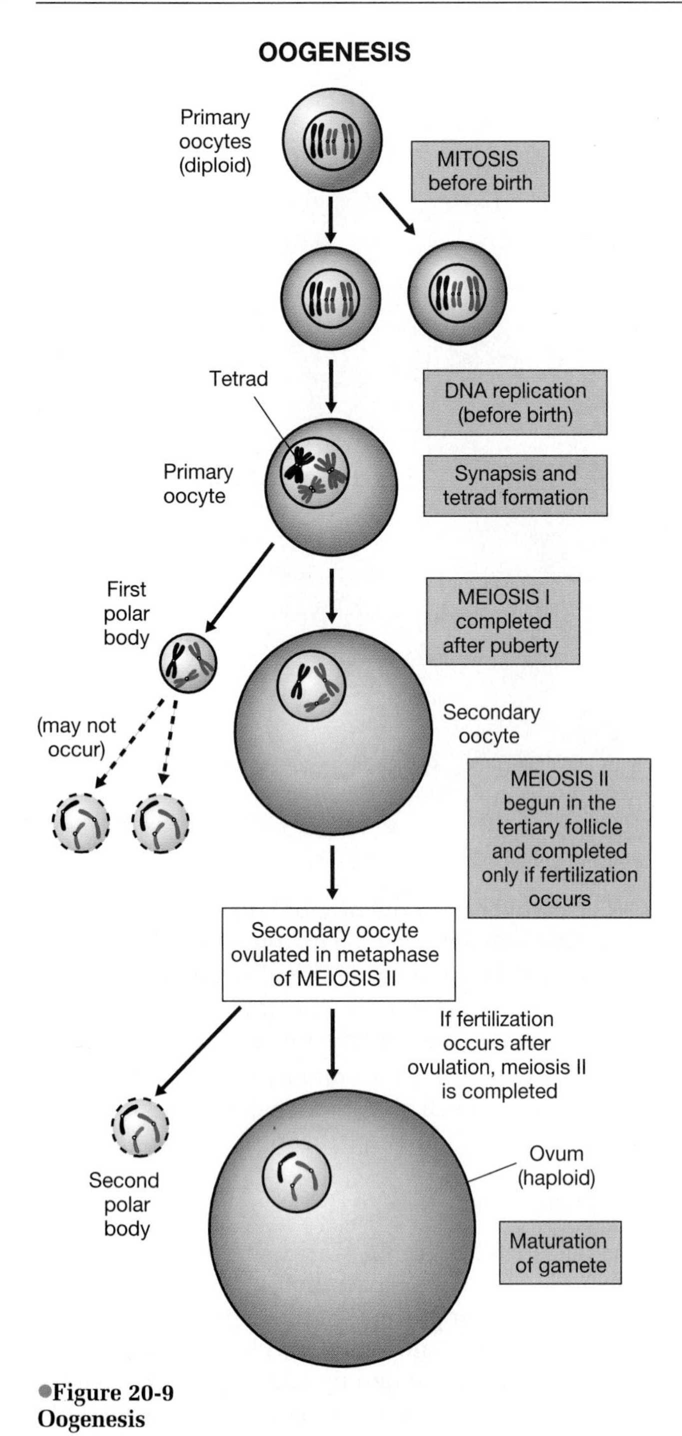

•Figure 20-9
Oogenesis

Diagrammatic view of meiosis and the production of an ovum.

originating at the surface of the oocyte contact those of the follicular cells. These microvilli increase the surface area available for absorption by roughly 35 times, and the follicular cells are continually providing the developing oocyte with nutrients.

Step 2: Formation of Secondary Follicles. Although many primordial follicles develop into primary follicles, usually only a few will take the next step. The transformation begins as the wall of the follicle thickens and the deeper follicular cells begin secreting small amounts of fluid. This *follicular fluid* accumulates in small pockets that gradually expand and separate the inner and outer layers of the follicle. At this stage, the complex is known as a **secondary follicle.** Although the oocyte continues to grow slowly, the follicle as a whole now enlarges rapidly because of this accumulation of fluid.

Step 3: Formation of Tertiary Follicles. Eight to 10 days after the start of the ovarian cycle, the ovaries usually contain only a single secondary follicle destined for further development. By days 10 to 14 of the cycle it has formed a mature **tertiary follicle,** or *Graafian* (GRAF-ē-an) *follicle,* roughly 15 mm in diameter. This complex spans the entire width of the ovarian cortex and stretches the ovarian capsule, creating a prominent bulge in the surface of the ovary. The oocyte, surrounded by a mass of follicular cells, projects into the expanded central chamber of the follicle, the **antrum** (AN-trum).

Step 4: Ovulation. As the time of egg release, or **ovulation** (ōv-ū-LĀ-shun), approaches, the follicular cells surrounding the oocyte lose contact with the follicular wall, and the oocyte floats within the central chamber. This event usually occurs at day 14 of a 28-day cycle. The follicular cells surrounding the oocyte are now known as the *corona radiata* (ko-RŌ-na rā-dē-A-ta). The distended follicular wall then ruptures, releasing the follicular contents, including the secondary oocyte, into the pelvic cavity. Because the corona radiata has a sticky surface, the oocyte usually attaches to the ovarian surface near the ruptured wall of the follicle. Contact with the entrance to the uterine tube or fluid currents established by its ciliated lining then sweeps the secondary oocyte into the uterine tube.

Step 5: Formation and Degeneration of the Corpus Luteum. The empty follicle collapses, and the remaining follicular cells invade the cavity and multiply to create an endocrine structure known as the **corpus luteum** (LŪ-tē-um; *lutea,* yellow). Unless pregnancy occurs, after about 12 days the corpus luteum begins to degenerate. The disintegration marks the end of the ovarian cycle, but almost immediately the activation of another set of primordial follicles begins the next ovarian cycle.

The Uterine Tubes

Each uterine tube (*Fallopian tube*) measures roughly 13 cm (5 in.) in length. The end closest to the ovary forms an expanded funnel, or **infundibulum** (in-fun-DIB-ū-lum; *infundibulum,* a funnel), with numerous fingerlike projections that extend into the pelvic cavity (see Figures 20-8 and 20-11•). The projections, called **fimbriae** (FIM-brē-ē), and the inner surfaces of the in-

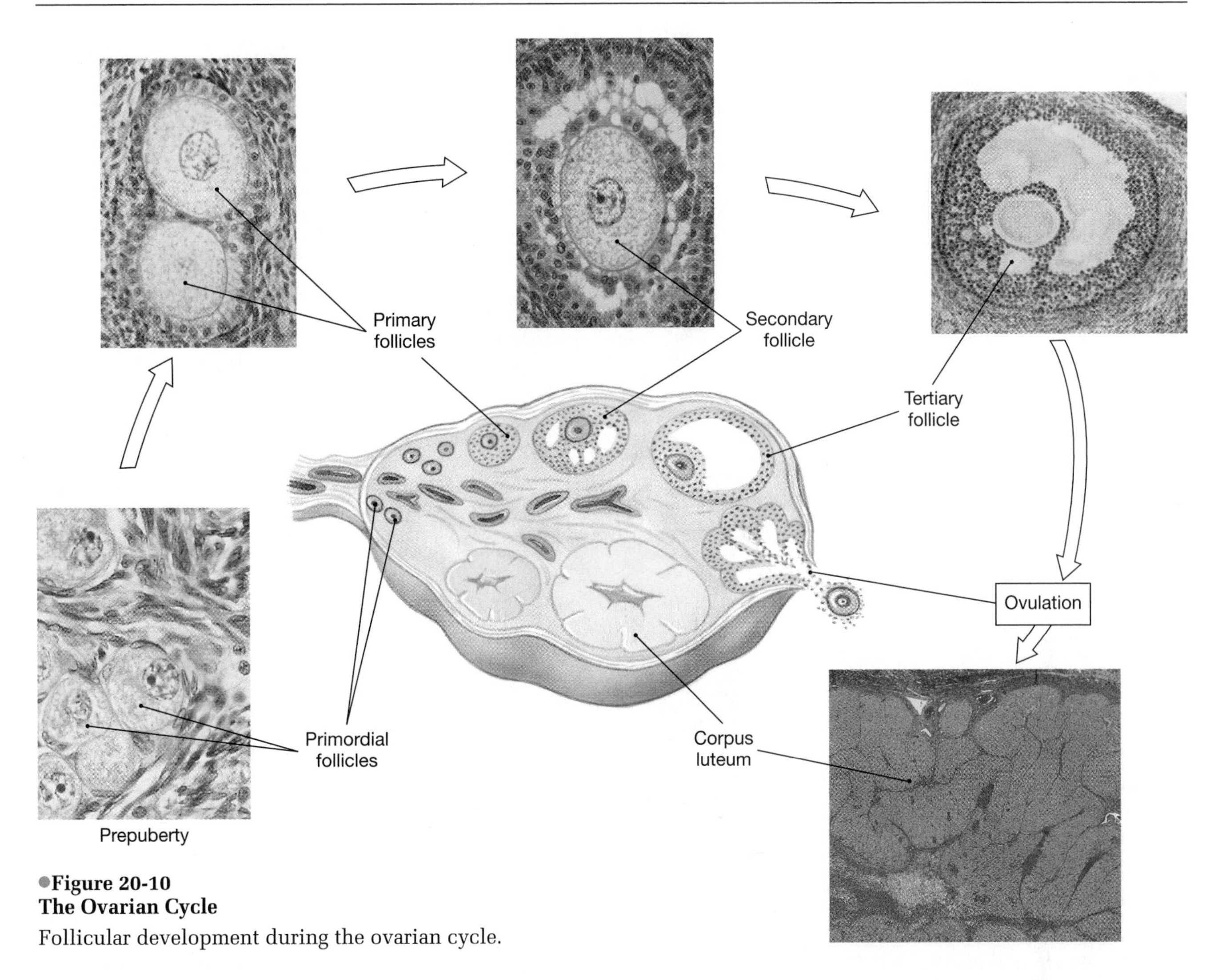

●**Figure 20-10**
The Ovarian Cycle
Follicular development during the ovarian cycle.

fundibulum are carpeted with cilia that beat toward the broad entrance to the uterine tube. Once inside the uterine tube, the ovum is probably transported by ciliary movement and peristaltic contractions. It normally takes 3–4 days for the secondary oocyte to travel from the infundibulum to the uterine chamber. *If fertilization is to occur, the secondary oocyte must encounter spermatozoa during the first 12–24 hours of its passage.* Unfertilized oocytes will degenerate, in the uterine tubes or uterus, without completing meiosis.

PELVIC INFLAMMATORY DISEASE (PID)

Pelvic inflammatory disease (PID) is a major cause of sterility in women. This condition, an infection of the uterine tubes, affects an estimated 850,000 women each year in the United States alone. Sexually transmitted pathogens are often involved, and as many as 50–80 percent of all first cases may be due to infection by *Neisseria gonorrhoeae,* the organism responsible for symptoms of *gonorrhea* (gon-ō-RĒ-a), a sexually transmitted disease discussed in a later section. PID may also result from invasion of the region by bacteria normally found within the vagina. Symptoms of pelvic inflammatory disease include fever, lower abdominal pain, and elevated white blood cell counts. In severe cases, the infection may spread to other visceral organs or produce a generalized peritonitis.

Sexually active women in the 15–24 age group have the highest incidence of PID. Although use of oral contraceptives decreases the risk of infection, the presence of an intrauterine device (IUD) may increase the risk by 1.4–7.3 times. Treatment with antibiotics may control the condition, but chronic abdominal pain may persist. In addition, damage and scarring of the uterine tubes may cause infertility by preventing the passage of a zygote to the uterus. Recently, another sexually transmitted bacterium, *Chlamydia,* has been identified as the probable cause of up to 50 percent of all cases of PID. Despite the fact that women with this infection may develop few if any symptoms, scarring of the uterine tubes may still produce infertility.

The Uterus

The uterus provides mechanical protection and nutritional support to the developing embryo and fetus (see Figures 20-8, p. 524, and 20-11●). The typical uterus is

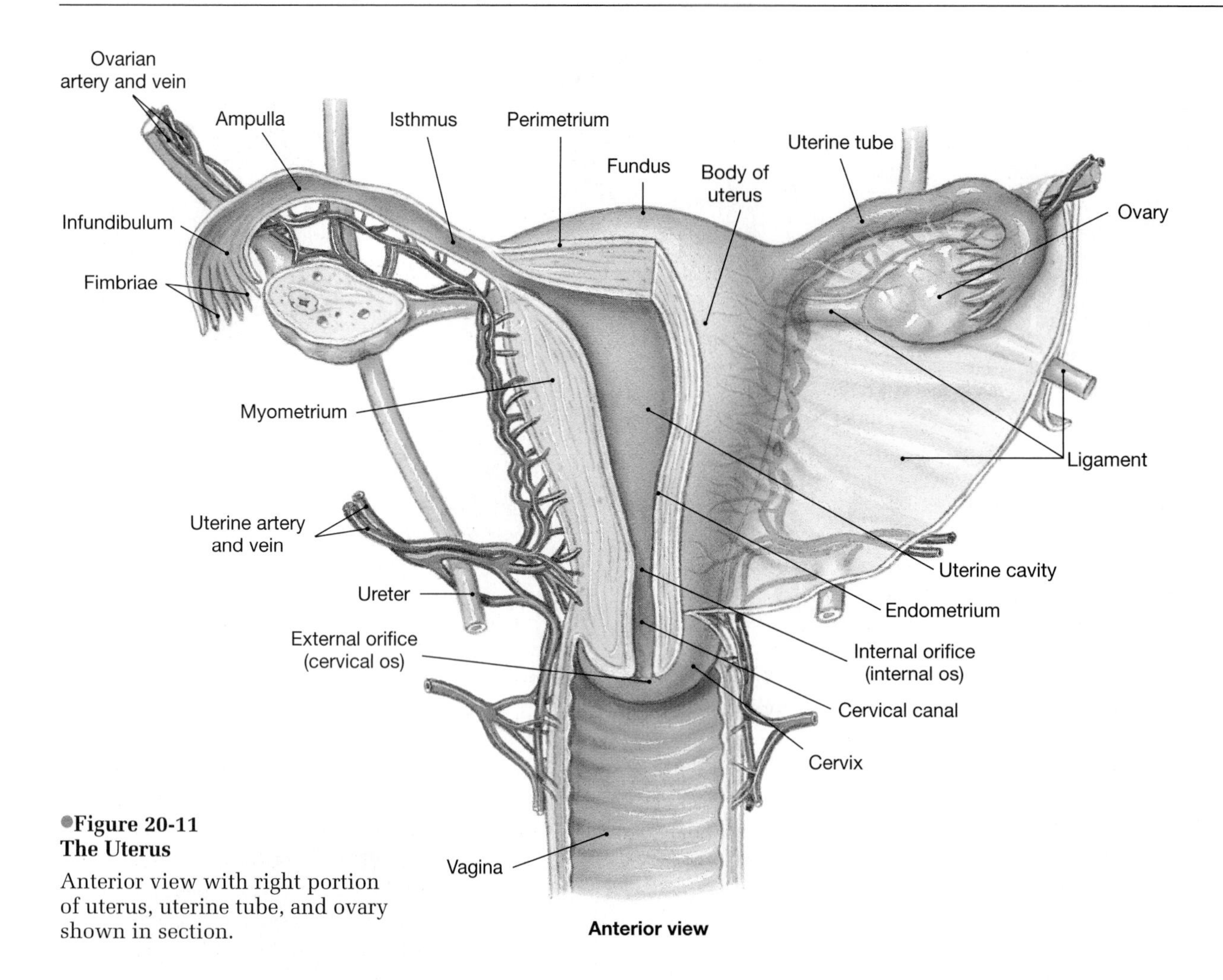

Figure 20-11
The Uterus
Anterior view with right portion of uterus, uterine tube, and ovary shown in section.

a small, pear-shaped organ about 7.5 cm (3 in.) in length with a maximum diameter of 5 cm (2 in.). It weighs 30–40 g (1–1.4 oz) and is stabilized by various ligaments.

The uterus can be divided into two regions: the *body* and the *cervix.* The **body** is the largest division of the uterus. The *fundus* is the rounded portion of the body superior to the attachment of the uterine tubes. The body ends at a constriction known as the **isthmus.** The **cervix** (SER-viks) is the inferior portion of the uterus. The cervix projects a short distance into the vagina, and the **uterine cavity** opens into the vagina at the **external orifice,** or *cervical os.*

In section, the thick uterine wall can be divided into an inner **endometrium** (en-dō-MĒ-trē-um), and a muscular **myometrium** (mī-ō-MĒ-trē-um; *myo-,* muscle + *metra,* uterus), covered by the *perimetrium,* a layer of visceral peritoneum (Figure 20-11). The endometrium of the uterus includes the epithelium lining the uterine chambers and the underlying connective tissues. Uterine glands opening onto the endometrial surface extend deep into the connective tissue layer almost all the way to the myometrium. The myometrium consists of a thick mass of interwoven smooth muscle cells.

The endometrium can be divided into a superficial *functional zone* and a deeper *basilar zone* that is adjacent to the myometrium. The structure of the basilar layer remains relatively constant over time, but that of the functional zone undergoes cyclical changes in response to sexual hormone levels. These alterations produce the characteristic features of the *uterine cycle.*

The Uterine Cycle

The **uterine cycle,** or *menstrual* (MEN-stru-al) *cycle,* is a repeating series of changes in the structure of the endometrium. This fascinating cycle of events begins with the **menarche** (me-NAR-kē), or first menstrual period at puberty, typically at age 11 to 12. The cycles continue until age 45 to 50, when **menopause** (MEN-ō-paws), the last menstrual cycle, occurs. Over the intervening three and a half to four decades the regular appearance of menstrual cycles will be interrupted only by unusual circumstances, such as illness, stress, starvation, or pregnancy. AM *Uterine Tumors and Cancer*

The uterine cycle averages 28 days in length, but it can range from 21 to 35 days in normal individuals. It can be divided into three stages: *menses,* the *proliferative phase,* and the *secretory phase.*

Menses The menstrual cycle begins with the onset of **menses** (MEN-sēz), a period marked by the wholesale destruction of the superficial layer, or *functional zone,* of the endometrium. The process is triggered by the decline in progesterone and estrogen concentrations as the corpus luteum disintegrates. The endometrial arteries constrict, reducing blood flow to this region, and the secretory glands, epithelial cells, and other tissues of the functional zone die of oxygen and nutrient deprivation. Eventually the weakened arterial walls rupture, and blood pours into the connective tissues of the functional zone. Blood cells and degenerating tissues break away and enter the uterine lumen, to be lost by passage into the vagina. This sloughing of tissue, which continues until the entire functional zone has been lost, is called **menstruation** (men-stru-Ā-shun). Menstruation usually lasts from 1 to 7 days, and over this period roughly 35 to 50 ml of blood are lost. Painful menstruation, or *dysmenorrhea,* may result from uterine inflammation and contraction or from conditions involving adjacent pelvic structures. AM *Premenstrual Syndrome*

The Proliferative Phase The **proliferative phase** begins in the days following the completion of menses as the surviving epithelial cells multiply and spread across the surface of the endometrium. This repair process is stimulated by the rising estrogen levels that accompany the growth of another set of ovarian follicles. By the time ovulation occurs, the functional zone is several millimeters thick and its new set of uterine, or endometrial, glands are manufacturing a mucus rich in glycogen. In addition, the entire functional zone is filled with small arteries that branch from larger trunks in the myometrium.

The Secretory Phase During the **secretory phase** of the cycle, the endometrial glands enlarge, steadily increasing their rates of secretion as the endometrium prepares for the arrival of a developing embryo. This activity is stimulated by the progestins and estrogens from the corpus luteum. This phase begins at the time of ovulation and persists as long as the corpus luteum remains intact. Secretory activities peak about 12 days after ovulation. Over the next day or two the glandular activity declines, and the menstrual cycle comes to a close. A new cycle then begins with the onset of menses and the disintegration of the functional zone. AM *Endometriosis*

AMENORRHEA

If menarche does not appear by age 16, or if the normal menstrual cycle of an adult becomes interrupted for 6 months or more, the condition of *amenorrhea* (ā-men-ō-RĒ-a) exists. *Primary amenorrhea* is the failure to initiate menses. This condition may indicate developmental abnormalities, such as nonfunctional ovaries or even the absence of a uterus, or an endocrine or genetic disorder. Transient *secondary amenorrhea* may be caused by severe physical or emotional stresses. In effect, the reproductive system gets "switched off" under these conditions. Examples of factors that can cause amenorrhea include drastic weight reduction programs, anorexia nervosa, and severe depression or grief. Amenorrhea has also been observed in marathon runners and others engaged in training programs that require sustained high levels of exertion and severely reduce body lipid reserves.

The Vagina

The vagina is a muscular tube extending between the uterus and the external genitalia (Figures 20-8, p. 525, and 20-12●). It has an average length of 7.5–9 cm (3–3.5 in.), but because the vagina is highly distensible its length and width are quite variable. The cervix of the uterus projects into the **vaginal canal.** The shallow recess surrounding the cervical protrusion is known as the **fornix** (FŌR-niks). The vagina lies parallel to the rectum, and the two are in close contact. After leaving the urinary bladder, the urethra turns and travels along the superior wall of the vagina.

The vaginal walls contain a network of blood vessels and layers of smooth muscle, and the lining is moistened by the secretions of the cervical glands and by the movement of water across the permeable epithelium. The vagina and vestibule are separated by an elastic epithelial fold, the **hymen** (HĪ-men), which may partially or completely block the entrance to the vagina. The two bulbocavernosus muscles pass on either side of the vaginal orifice, and their contractions constrict the entrance.

The vagina (1) serves as a passageway for the elimination of menstrual fluids; (2) receives the penis during coitus and holds spermatozoa prior to their passage into the uterus; and (3) in childbirth forms the lower portion of the birth canal through which the fetus passes on its way to an independent existence.

The vagina normally contains resident bacteria supported by the nutrients found in the cervical mucus. As a result of their metabolic activities, the normal pH of the vagina ranges between 3.5 and 4.5, and this acid environment restricts the growth of many pathogenic organisms. An infection of the vaginal canal, known as *vaginitis* (va-jin-Ī-tis), may be caused by fungal, bacterial, or parasitic organisms. In addition to any discomfort that may result, the condition may affect the survival of sperm and so reduce fertility. AM *Vaginitis*

The External Genitalia

The perineal region enclosing the female external genitalia is the **vulva** (VUL-va), or *pudendum* (Figure 20-12●). The vagina opens into the **vestibule,** a central

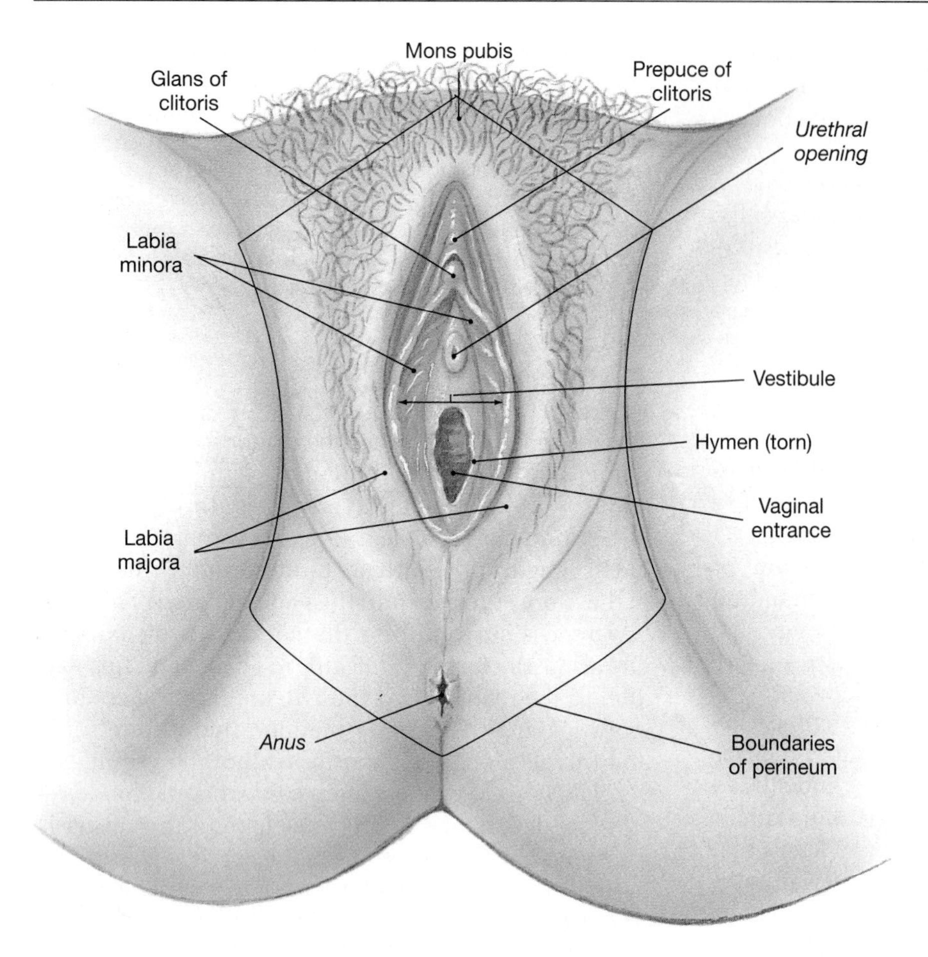

Figure 20-12
The Female External Genitalia
An external view of the female perineum.

space bounded by the **labia minora** (LĀ-bē-a mi-NOR-a; singular *labium*). The labia minora are covered with a smooth, hairless skin. The urethra opens into the vestibule just anterior to the vaginal entrance. Anterior to the urethral opening, the **clitoris** (KLI-tō-ris) projects into the vestibule. The clitoris is the female equivalent of the penis, derived from the same embryonic structures. Internally it contains erectile tissues that become engorged with blood during arousal. A small erectile *glans* sits atop the organ, and extensions of the labia minora encircle the body of the clitoris, forming the *prepuce.*

A variable number of small **lesser vestibular glands** discharge their secretions onto the exposed surface of the vestibule, keeping it moistened. During arousal, a pair of ducts discharges the secretions of the **greater vestibular glands** (Figure 20-8●) into the vestibule near the vaginal entrance. These mucous glands resemble the bulbourethral glands of the male.

The outer limits of the vulva are established by the *mons pubis* and the *labia majora.* The prominent bulge of the **mons pubis** is created by adipose tissue beneath the skin anterior to the pubic symphysis. Adipose tissue also accumulates within the fleshy **labia majora** that encircle and partially conceal the labia minora and vestibular structures.

The Mammary Glands

At birth, the newborn infant cannot fend for itself, and several key systems have yet to complete their development. While adjusting to an independent existence, the infant gains nourishment from the milk secreted by the maternal **mammary glands.** Milk production, or **lactation** (lak-TĀ-shun), occurs in the mammary glands of the breasts, specialized accessory organs of the female reproductive system (Figure 20-13●).

The mammary glands lie in the subcutaneous layer beneath the skin of the chest. Each breast bears a small conical projection, the **nipple,** where the ducts of underlying mammary glands open onto the body surface. The skin surrounding each nipple has a reddish brown coloration, and this region is known as the **areola** (a-RĒ-ō-la). Large sebaceous glands beneath the areolar surface give it a granular texture.

The glandular tissue of the breast consists of a number of separate lobes, each containing several secretory lobules. Within each lobe, the ducts leaving the lobules converge, giving rise to a single **lactiferous** (lak-TIF-e-rus) **duct**. Near the nipple, that lactiferous duct expands, forming an expanded chamber called a **lactiferous sinus.** There are usually 15–20 lactiferous sinuses opening onto the surface of each nipple.

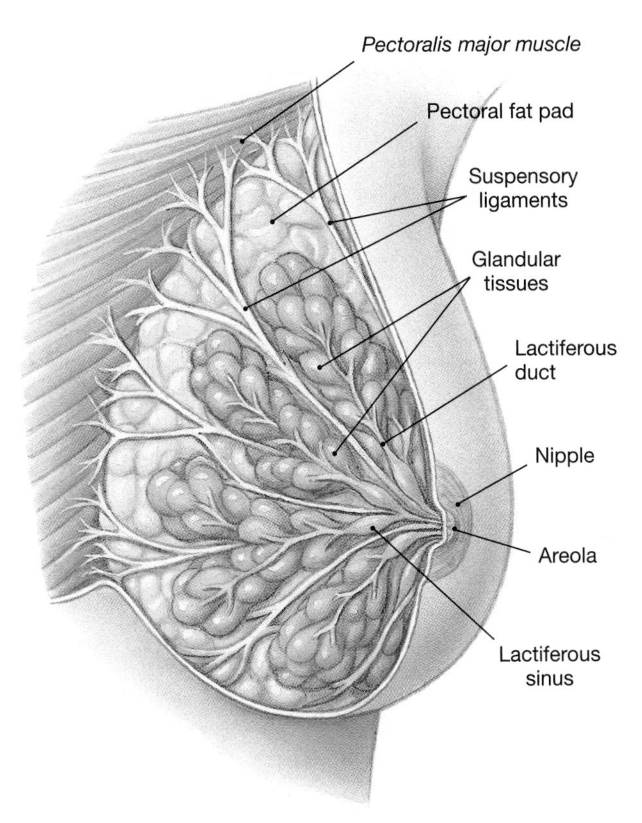

●**Figure 20-13**
The Mammary Glands of the Female Breast

Dense connective tissue surrounds the duct system and forms partitions that extend between the lobes and lobules. These bands of connective tissue, the *suspensory ligaments of the breast,* originate in the dermis of the overlying skin. A layer of loose connective tissue separates the mammary complex from the underlying muscles, and the two can move relatively independently.

✓ As the result of infections such as gonorrhea, scar tissue can block the lumen of each uterine tube. How would this blockage affect a woman's ability to conceive?

✓ What is the advantage of the normally acidic pH of the vagina?

✓ Which layer of the uterus is sloughed off during menstruation?

✓ Would blockage of a single lactiferous sinus interfere with delivery of milk to the nipple? Explain.

Hormones and the Female Reproductive Cycle

As in the male, the activity of the female reproductive tract is controlled by both pituitary and gonadal secretions. But the regulatory pattern is much more complicated, for a woman's reproductive system does not just produce functional gametes; it must also coordinate the ovarian and uterine cycles. Circulating hormones control the **female reproductive cycle** to ensure proper reproductive function. For example, it is obvious that a woman who fails to ovulate will be unable to conceive, even if her uterus is perfectly normal. A woman who ovulates normally, but whose uterus isn't ready to support an embryo, will be just as infertile. Because the processes are complex and difficult to study, many of the biochemical details still elude us, but the general patterns are reasonably clear.

Changes in circulating estrogen concentration are the primary mechanism for coordinating the female reproductive cycle, and the relationships are summarized in Figure 20-14●. The upper portion of this figure summarizes the ovarian cycle. Its cyclic pattern of hormonal regulation differs between the *preovulatory period* and *postovulatory period.*

BREAST CANCER

Breast cancer is the primary cause of death for women between the ages of 35 and 45, but it actually becomes even more common after age 50. There were approximately 46,000 deaths in the United States from breast cancer in 1995, and approximately 183,000 new cases reported. An estimated 12 percent of women in the United States will develop breast cancer at some point in their lifetimes, and the rate is steadily rising. The incidence is highest among Caucasians, somewhat lower in African Americans, and lowest in Asians and American Indians. Notable risk factors include (1) a family history of breast cancer, (2) a pregnancy after age 30, and (3) early menarche (first menstrual period) or late menopause (last menstrual period). Breast cancers in males are very rare; there are only around 300 deaths among males due to breast cancer each year in the U.S. The *Applications Manual* contains additional information concerning the incidence, detection, and treatment of breast cancer. AM *Breast Cancer*

Hormones and the Preovulatory Period

Follicular development begins under FSH stimulation, and each month some of the primordial follicles begin their development into primary follicles. As the follicular cells enlarge and multiply, they release steroid hormones collectively known as *estrogens,* the most important being **estradiol** (es-tra-DĪ-ol). Estrogens have multiple functions including (1) stimulating bone and muscle growth, (2) maintaining female secondary sex characteristics such as body hair distribution and the location of adipose tissue deposits, (3) affecting CNS activity, including sex-related behaviors and drives, (4) maintaining functional accessory reproductive glands and organs, and

●**Figure 20-14**
Hormonal Regulation of Ovarian Activity

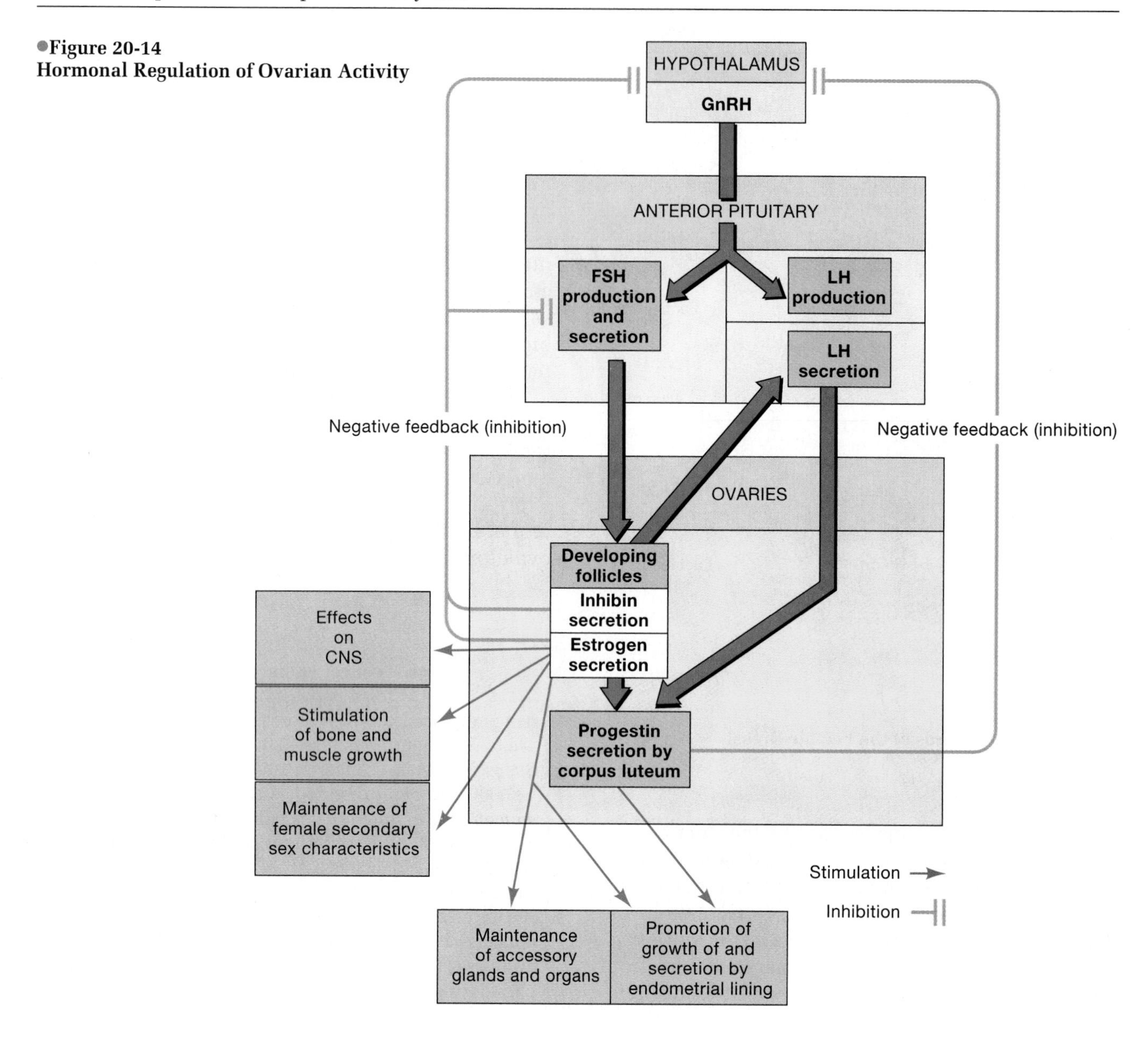

(5) initiating repair and growth of the endometrium (Figure 20-14●).

The upper portion of Figure 20-15● summarizes the hormonal events associated with the ovarian cycle. As follicular development proceeds, the concentration of circulating estrogens and inhibin rises, for the follicular cells are increasing in number and secretory activity. As estrogen and inhibin concentrations increase, they inhibit both the hypothalamic secretion of GnRH and the pituitary production and release of FSH. Estrogen also affects the rate of LH secretion. Although the synthesis of LH occurs under GnRH stimulation, the rate of release into the bloodstream depends on the circulating concentration of estrogens. Thus as the follicles develop and estrogen concentrations rise, the pituitary output of LH gradually increases. Despite a slow decline in FSH concentrations, the combination of estrogens, FSH, and LH continues to support follicular development and maturation.

Estrogen concentrations take a sharp upturn in the second week of the ovarian cycle, as this month's tertiary follicle enlarges in preparation for ovulation. At about day 14 estrogen levels peak, accompanying the maturation of that follicle. The high estrogen concentration then triggers a massive outpouring of LH from the anterior pituitary, which causes the rupture of the follicular wall and ovulation.

Hormones and the Postovulatory Period

After ovulation, LH stimulates the remaining follicular cells to form the corpus luteum, and the yellow color of this mass results from its lipid reserves. These compounds are used to manufacture steroid hormones known as **progestins** (prō-JES-tinz), predominantly the steroid **progesterone** (prō-JES-ter-ōn). Progesterone is the principal hormone of the postovulatory period. It prepares the uterus for pregnancy by stimulating the growth

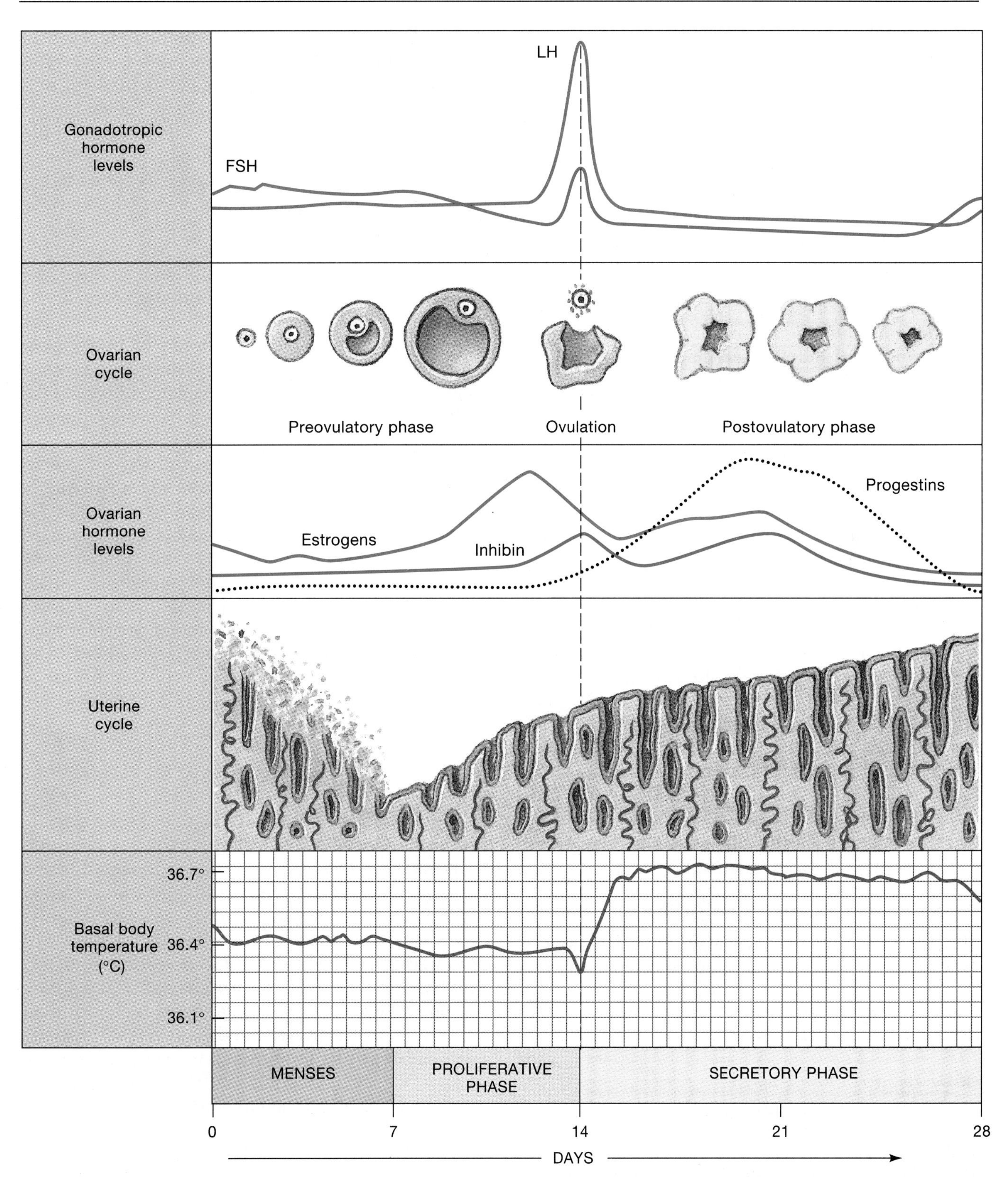

Figure 20-15
Hormonal Regulation of the Female Reproductive Cycle

and development of the blood supply and secretory glands of the endometrium, and it also stimulates metabolic activity and elevates basal body temperature.

Luteinizing hormone levels remain elevated for only two days, but that is long enough to stimulate the formation of the functional corpus luteum. Progesterone secretion continues at relatively high levels for the next week, but unless pregnancy occurs, the corpus luteum then begins to degenerate. Roughly 12 days after ovulation, the corpus luteum becomes nonfunctional, and progesterone and estrogen levels fall markedly. This decline stimulates the hypothalamic receptors, and GnRH production increases. This increase leads to an increase in FSH and LH production in the anterior pituitary, and the entire cycle begins again.

Hormones and the Uterine Cycle

The lower portion of Figure 20-15● follows changes in the endometrium during a single uterine cycle. The sudden declines in progesterone and estrogen levels that accompany the breakdown of the corpus luteum result in menses. The loss of endometrial tissue continues for several days, until rising estrogen levels stimulate the regeneration of the functional zone of the endometrium. The preovulatory phase continues until rising progesterone levels mark the arrival of the postovulatory phase. The combination of estrogen and progesterone then causes the enlargement of the endometrial glands and an increase in their secretions.

Hormones and Body Temperature

The hormonal fluctuations also cause physiological changes that affect the core body temperature. During the preovulatory period, when estrogen is the dominant hormone, the resting, or "basal," body temperature measured upon awakening in the morning is about 0.3°C (or 0.5°F) lower than it is during the postovulatory period, when progesterone dominates. At the time of ovulation, basal temperature declines sharply, making the temperature rise over the following day even more noticeable (Figure 20-15●). By keeping records of body temperature over a few menstrual cycles, a woman can often determine the precise day of ovulation. This information can be very important for those wishing to avoid or promote a pregnancy, for pregnancy can occur only if an ovum becomes fertilized within a day of its ovulation.

✓ What changes would you expect to observe in the ovulatory cycle if the LH surge did not occur?

✓ What effect would blockage of progesterone receptors in the uterus have on the endometrium?

✓ What event occurs in the menstrual cycle when the levels of estrogen and progesterone decline?

THE PHYSIOLOGY OF SEXUAL INTERCOURSE

Sexual intercourse, or **coitus** (KŌ-i-tus), introduces semen into the female reproductive tract. The following sections consider the process as it affects the reproductive systems of males and females.

Male Sexual Function

Male sexual function is coordinated by reflex pathways involving both divisions of the ANS. During **arousal,** erotic thoughts or the stimulation of sensory nerves in the genital region increase the parasympathetic outflow over the pelvic nerves, and erection occurs. Subsequent stimulation may initiate the secretion of the bulbourethral glands, lubricating the urethra and the surface of the glans.

During intercourse, the sensory receptors in the penis are rhythmically stimulated, eventually resulting in the coordinated processes of *emission* and *ejaculation*. **Emission** occurs under sympathetic stimulation. The process begins with peristaltic contractions of the ductus deferens, which push fluid and spermatozoa through the ejaculatory ducts and into the urethra. The seminal vesicles then contract, followed by waves of contraction in the prostate gland. While these contractions are proceeding, sympathetic commands close the sphincter at the entrance to the urinary bladder, preventing the passage of semen into the bladder.

Ejaculation occurs as powerful, rhythmic contractions appear in the *ischiocavernosus* and *bulbocavernosus* muscles, two superficial skeletal muscles of the pelvic floor. (The positions of these muscles can be seen in Figure 7-13●, p. 179.) Ejaculation is associated with intense pleasurable sensations, an experience known as male **orgasm** (ŌR-gazm). Among other changes, heart rate and blood pressure temporarily increase. After ejaculation, blood begins to leave the erectile tissue, and the erection begins to subside. This *detumescence* (de-tū-MES-ens) is mediated by the sympathetic nervous system.

Female Sexual Function

The phases of female sexual function are comparable to those of a male. During sexual arousal, parasympathetic activation leads to an engorgement of the erectile tissues of the clitoris and increased secretion of cervical mucous glands and the greater vestibular glands. Clitoral erection increases its sensitivity to stimulation, and the cervical and vestibular glands provide lubrication for the vaginal walls. A network of blood vessels in the vaginal walls becomes filled with blood at this time, and the vaginal surfaces are also moistened by fluid from underlying connective tissues. Parasympathetic stimulation also causes engorgement of blood vessels at the nipples, making them more sensitive to touch and pressure.

During intercourse, rhythmic contact with the clitoris and vaginal walls, reinforced by touch sensations from the breasts and other stimuli (visual, olfactory, and auditory) provides stimulation that can lead to orgasm. Female orgasm is accompanied by peristaltic contractions of the uterine and vaginal walls and, via impulses over the pudendal nerves, rhythmic contractions of the bulbocavernosus and ischiocavernosus muscles. The latter contractions give rise to the sensations of orgasm.

SEXUALLY TRANSMITTED DISEASES

Sexually transmitted diseases (STDs) are transferred from individual to individual, usually or exclusively by sexual intercourse. A variety of bacterial, viral, and fungal infections are included in this category. At least two dozen different STDs are currently recognized. All are unpleasant. *Chlamydia,* noted on p. 527, can cause PID and infertility. Other types of STDs are quite dangerous, and a few, including AIDS (p. 397) are deadly. The incidence of STDs has been increasing in the United States since 1984, primarily in urban centers and in urban minority populations. Acute poverty, coupled with drug use, prostitution, and the appearance of drug-resistant pathogens all contribute to the problem. The *Applications Manual* contains a detailed discussion of the most common forms of STDs, including: *gonorrhea, syphilis, herpes,* and *chancroid.* AM *Sexually Transmitted Diseases*

AGING AND THE REPRODUCTIVE SYSTEM

The aging process affects the reproductive systems of men and women. The most striking age-related changes in the female reproductive system occur at menopause, whereas changes in the male reproductive system occur more gradually and over a longer period of time.

Menopause

Menopause is usually defined as the time that ovulation and menstruation cease. It typically occurs at age 45–55, but in the years preceding it the ovarian and menstrual cycles become irregular. A shortage of primordial follicles is the underlying cause of these developments; by age 50, there are often no primordial follicles left to respond to FSH. In *premature menopause* this depletion occurs before age 40.

Menopause is accompanied by a sharp and sustained rise in the production of GnRH, FSH, and LH, while circulating concentrations of estrogen and progesterone decline. The decline in estrogen levels leads to reductions in the size of the uterus and breasts, accompanied by a thinning of the urethral and vaginal walls. The reduced estrogen concentrations have also been linked to the development of osteoporosis, presumably because bone deposition proceeds at a slower rate. A variety of neural effects are also reported, including "hot flashes," anxiety, and depression, but the hormonal mechanisms involved are not well understood. In addition, the risk of atherosclerosis and other forms of cardiovascular disease increase after menopause.

The symptoms accompanying and following menopause are sufficiently unpleasant that about 40 percent of menopausal women eventually seek medical assistance. Hormone replacement therapies involving a combination of estrogens and progestins can often prevent osteoporosis and the neural and vascular changes associated with menopause.

The Male Climacteric

Changes in the male reproductive system occur more gradually, over a period known as the *male climacteric.* Circulating testosterone levels begin to decline between ages 50 and 60, coupled with increases in circulating levels of FSH and LH. Although sperm production continues (men well into their eighties can father children), there is a gradual reduction in sexual activity in older men. This reduction may be linked to declining testosterone levels, and some clinicians are now tentatively suggesting the use of testosterone replacement therapy to enhance libido (sexual drive) in elderly men.

✓ Inability to contract the ischiocavernosus and bulbocavernosus muscles would interfere with what part of the male sex act?

✓ What changes occur in the female during sexual arousal as the result of increased parasympathetic stimulation?

✓ Why does the level of FSH rise and remain high during menopause?

INTEGRATION WITH OTHER SYSTEMS

Figure 20-16● summarizes the relationships between the reproductive system and other physiological systems. Normal human reproduction is a complex process that requires the participation of multiple systems. Hormones play a major role in coordinating these events, and Table 20-1 reviews the hormones discussed in this chapter. The reproductive process depends on various physical, physiological, and psychological factors, many of which require intersystem cooperation. For example, the male's sperm count must be adequate, the semen must have the correct pH and nutrients, and erection and ejaculation must occur in the proper sequence. For these steps to occur, the reproductive, digestive, endocrine, nervous, cardiovascular, and urinary systems must all be functioning normally.

Even when all else is normal, and fertilization occurs at the proper time and place, a normal infant will not result unless the zygote, a single cell the size of a pinhead, manages to develop into a full-term fetus weighing 3–4 kg. The last chapter in this text considers the process of development, focusing on the mechanisms that determine both the structure of the body and the distinctive characteristics of each individual.

INTEGUMENTARY SYSTEM

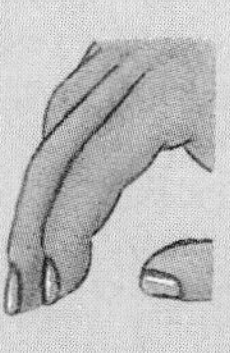

Covers external genitalia; provides sensations that stimulate sexual behaviors; mammary gland secretions provide nourishment for newborn

Reproductive hormones affect distribution of body hair and subcutaneous fat deposits

REPRODUCTIVE SYSTEM

FOR ALL SYSTEMS

Secretion of hormones with effects on growth and metabolism

SKELETAL SYSTEM

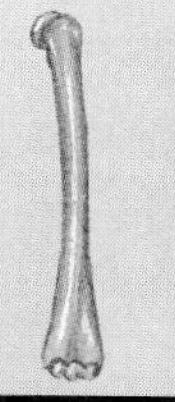

Pelvis protects reproductive organs of females, portion of ductus deferens and accessory glands in male

Sexual hormones stimulate growth and maintenance of bones; sex hormones at puberty accelerate growth and closure of epiphyseal plates

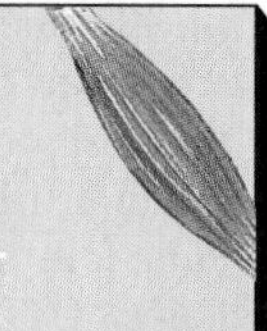

Contractions of skeletal muscles eject semen from male reproductive tract; muscle contractions during sexual act produce pleasurable sensations in both sexes

Reproductive hormones, especially testosterone, accelerate skeletal muscle growth

NERVOUS SYSTEM

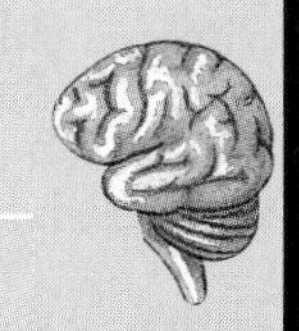

Controls sexual behaviors and sexual function

Sexual hormones affect CNS development and sexual behaviors

ENDOCRINE SYSTEM

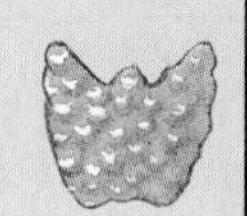

Hypothalamic regulatory factors and pituitary hormones regulate sexual development and function; oxytocin stimulates smooth muscle contractions in uterus and mammary glands

Steroid sex hormones and inhibin inhibit secretory activities of hypothalamus and pituitary

CARDIOVASCULAR SYSTEM

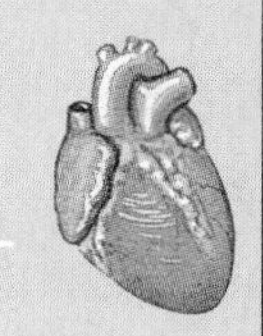

Distributes reproductive hormones; provides nutrients, oxygen, and waste removal for fetus; local blood pressure changes responsible for physical changes during sexual arousal

Estrogens may maintain healthy vessels and slow development of atherosclerosis

LYMPHATIC SYSTEM

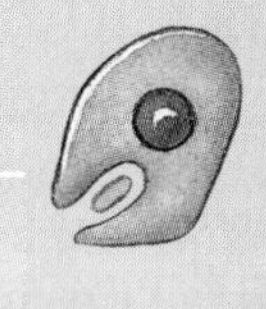

Provides IgA for secretion by epithelial glands; assists in repairs and defense against infection

Lysozymes and bactericidal chemicals in secretions provide nonspecific defense against reproductive tract infections

RESPIRATORY SYSTEM

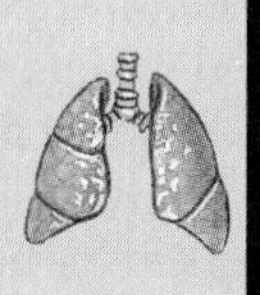

Provides oxygen and removes carbon dioxide generated by tissues of reproductive system

Changes in respiratory rate and depth occur during sexual arousal, under control of the nervous system

DIGESTIVE SYSTEM

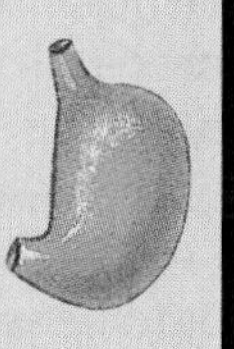

Provides additional nutrients required to support gamete production and (in pregnant women) embryonic and fetal development

URINARY SYSTEM

Urethra in males carries semen to exterior; kidneys remove wastes generated by reproductive tissues

Accessory organ secretions may have antibacterial action that helps prevent urethral infections in males

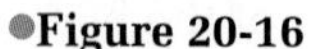

•Figure 20-16
Functional Relationships between the Reproductive System and Other Systems

TABLE 20-1 Hormones of the Reproductive System

Hormone	*Source*	*Regulation of Secretion*	*Primary Effects*
GONADOTROPIN-RELEASING HORMONE (GNRH)	Hypothalamus	*Male*: inhibited by testosterone *Female*: inhibited by estrogens and/or progestins	Stimulates FSH secretion, LH synthesis
FOLLICLE-STIMULATING HORMONE (FSH)	Anterior pituitary	*Male*: stimulated by GnRH, inhibited by inhibin *Female*: stimulated by GnRH, inhibited by estrogens and/or progestins	*Male*: stimulates spermatogenesis and spermiogenesis through effects on sustentacular cells *Female*: stimulates follicle development, estrogen production, and egg maturation
ESTROGENS (primarily estradiol)	Follicular and interstitial cells of ovaries	Stimulated by FSH	Stimulates LH secretion, maintains secondary sex characteristics and sexual behavior, stimulates repair of endometrium, inhibits secretion of GnRH
INHIBIN	Sustentacular cells of testes and follicle cells of ovaries	Stimulated by factors released by developing sperm (male) or developing follicles (female)	Inhibits secretion of FSH and possibly GnRH
LUTEINIZING HORMONE (LH)	Anterior pituitary	*Male*: stimulated by GnRH *Female*: production stimulated by GnRH, secretion by estrogens	*Male*: stimulates interstitial cells *Female*: stimulates follicular and interstitial cells
PROGESTINS (primarily progesterone)	Corpus luteum	Stimulated by LH	Stimulates endometrial growth and glandular secretion, inhibits GnRH secretion
ANDROGENS (primarily testosterone)	Interstitial cells of testes	Stimulated by LH	Maintains secondary sex characteristics and sexual behavior, promotes maturation of spermatozoa, inhibits GnRH secretion

Chapter Review

KEY TERMS

estrogens, p. 531
lactation, p. 530
meiosis, p. 518
menses, p. 528
oogenesis, p. 525
ovary, p. 516
ovulation, p. 526
progesterone, p. 532
seminiferous tubules, p. 517
spermatozoa, p. 516
testes, p. 516
testosterone, p. 517

FOCUS Birth Control Strategies

For physiological, logistical, financial, or emotional reasons most adults practice some form of conception control during their reproductive years. When the simplest and most obvious method, sexual abstinence, is unsatisfactory for some reason, another method of contraception must be used to avoid unwanted pregnancies. The selection process can be quite involved, for there are many methods available. Because each has specific strengths and weaknesses, the potential risks and benefits must be carefully analyzed on an individual basis.

Well over 50 percent of U.S. women age 15–44 are practicing some method of contraception; in 1995 an estimated 50 million American women were taking oral birth control pills. There are many different methods of contraception; only a few will be considered here.

Sterilization makes one unable to provide functional gametes for fertilization. Either sexual partner may be sterilized with the same net result. In a **vasectomy** (vaz-EK-to-mē), a segment of the ductus deferens is removed, making it impossible for spermatozoa to pass from the epididymis to the distal portions of the reproductive tract. The surgery can be performed in a physician's office in a matter of minutes. The spermatic cords are located as they ascend from the scrotum on either side, and after each cord is opened the ductus deferens is severed. After a 1-cm section is removed, the cut ends are usually tied shut. With the section removed, the cut ends do not reconnect; in time, scar tissue forms a permanent seal. A more recent vasectomy procedure often makes it possible to restore fertility at a later date. In this procedure, the cut ends of the ductus deferens are blocked with silicone plugs that can later be removed. After a vasectomy, the man experiences normal sexual function, for the epididymal and testicular secretions normally account for only around 5 percent of the volume of the semen. Spermatozoa continue to develop, but they remain within the epididymis until they degenerate. The failure rate for this procedure is 0.08 percent (a failure is defined as a resulting pregnancy).

In the female, the uterine tubes can be blocked through a surgical procedure known as a **tubal ligation.** Since the surgery involves entering the abdominopelvic cavity, complications are more likely than with vasectomy. As in a vasectomy, attempts may be made to restore fertility after a tubal ligation. The failure rate for this procedure is estimated at 0.45 percent.

Oral contraceptives manipulate the female hormonal cycle so that ovulation does not occur. The contraceptive pills produced in the 1950s contained relatively large amounts of progestins. These concentrations were adequate to suppress pituitary production of GnRH, so FSH was not released and ovulation did not occur. Unpleasant side effects included endometrial bleeding, and most of the oral contraceptive products developed subsequently added small amounts of estrogens. Current combination pills differ significantly from the earlier products because the hormonal doses are much lower, with only one-tenth the progestins and less than half the estrogens. The hormones are administered in a cyclic fashion, beginning five days after the start of menses and continuing for the next three weeks. Over the fourth week the woman takes placebo pills or no pills at all. Low-dosage combination pills are sometimes prescribed for women experiencing irregular menstrual cycles, for they create a 28-day cycle. There are now at least 20 different brands of combination oral contraceptives available, and over 200 million women are using them worldwide. In the United States, 25 percent of women under age 45 use the combination pill to prevent conception. The progestin-only "minipill" has proved to be less effective in preventing pregnancy. The failure rate for the combination oral contraceptives, when used as prescribed, is 0.24 percent over a 2-year period. Birth control pills are not without their risks, however. For example, women with severe hypertension, diabetes mellitus, epilepsy, gallbladder disease, heart trouble, or acne may find that their problems worsen when taking the combination pills. Women taking oral contraceptives are also at increased risk for venous thrombosis, strokes, pulmonary embolism, and (for women over 35) heart disease.

Two progesterone-only forms of birth control are now available. *Depoprovera* is injected every 3 months. The Silastic tubes of the Norplant system are saturated with progesterone and inserted under the skin. This method provides birth control for a period of approximately 5 years, but to date the relatively high cost has limited the use of this contraceptive method. Both Depoprovera and the Norplant system can cause irregular menstruation and temporary amenorrhea, but they are easy to use and extremely convenient.

The **condom,** also called a *prophylactic,* or "rubber," covers the body of the penis during intercourse and keeps spermatozoa from reaching the female reproductive tract. Synthetic condoms are also used to prevent transmission of sexually transmitted diseases, such as syphilis, gonorrhea, and AIDS. The condom failure rate has been estimated at over 6 percent. **Vaginal barriers** such as the *diaphragm, cervical cap,* and *vaginal sponge* rely on similar principles. A diaphragm, the most popular form of vaginal barrier in use at the moment, consists of a dome of latex rubber with a small metal hoop supporting the rim. Because vaginas vary in size, women choosing this method must be individually fitted. Before intercourse, the diaphragm is inserted so that it covers the external orifice, and it is usually coated with a small amount of spermicidal jelly or cream, adding to the effectiveness of the barrier. The failure rate for a properly fitted diaphragm is estimated at 5–6 percent. The cervical cap is smaller and lacks the metal rim. It, too, must be fitted carefully, but unlike the diaphragm it may be left in place for several days. The failure rate (8 percent) is higher than that for diaphragm use. The vaginal sponge consists of a small synthetic sponge saturated with a *spermicide* a sperm-killing foam or jelly. The failure rate for a contraceptive sponge is estimated at 6–10 percent.

An **intrauterine device (IUD)** consists of a small plastic loop or a T that can be inserted into the uterine chamber. The mechanism of action remains uncertain, but it is known that IUDs stimulate prostaglandin production in the uterus. The net result is an alteration in the chemical composition of uterine secretions, and the changes in the intrauterine environment lower the chances for fertilization and subsequent implantation. IUDs are in limited use today in the United States, but they remain popular in many other countries. The failure rate is estimated at 5–6 percent.

The **rhythm method** involves abstaining from sexual activity on the days ovulation might be occurring. The timing is estimated on the basis of previous patterns of menstruation and sometimes by following changes in basal body temperature. The failure rate for the rhythm method is very high, approaching 25 percent.

Sterilization, oral contraceptives, condoms, and vaginal barriers are the primary contraception methods for all age groups. But the relative proportion of the population using a particular method changes with age. Sterilization is most popular among older women, who may already have had children. Relative availability may also play a role. For example, a sexually active female under age 18 can buy a condom more easily than she can obtain a prescription for an oral contraceptive. But many of the observed changes occur because the relationship between risks and benefits varies for each age group.

When attempting to make a decision concerning the use and selection of contraceptives, many people simply examine the list of potential complications and make the "safest" choice. For example, media coverage of the potential risks associated with oral contraceptives made many women reconsider their use. But complex decisions should not be made on such a simplistic basis, and the risks associated with contraceptive use must be considered in light of their relative efficiencies. Pregnancy, although a natural phenomenon, has its risks, and the mortality rate for pregnant women in the United States averages around 8 deaths per 100,000 pregnancies. That average incorporates a broad range; the rate is 5.4 per 100,000 among women under 20, and 27 per 100,000 for women over 40. Although these risks are small, for pregnant women over age 35 the chances of dying from complications related to pregnancy are almost twice as great as the chances of being killed in an automobile accident and many times greater than the risks associated with the use of oral contraceptives. For women in third world countries, the comparison is even more striking. The mortality rate for pregnant women in parts of Africa is approximately 1 per 150 pregnancies. In addition to preventing pregnancy, combination birth control pills have also been shown to reduce the risks of ovarian and endometrial cancers and fibrocystic breast disease.

Before age 35, *any contraceptive method is safer than pregnancy.* In general, over this period the risks are proportional to the failure rates of each method. The notable exception involves individuals taking the pill who also smoke cigarettes. Younger women are more fertile, so despite a lower mortality rate for each pregnancy they are likely to have more pregnancies. As a result, birth control failures imply a higher risk in the younger age groups.

After age 35, the risk of complications associated with oral contraceptive use increases, while those of other methods remain relatively stable. Women over age 35 (smokers) or 40 (nonsmokers) are therefore often advised to seek other forms of contraception.

A number of experimental contraceptive methods are being investigated. For example, researchers are attempting to determine whether low doses of inhibin will suppress GnRH release and prevent ovulation. Another approach is to develop a method of blocking human chorionic gonadotropin (hCG) receptors at the corpus luteum. If the corpus luteum was unable to respond to hCG, normal menses would occur, despite implantation of a blastocyst.

Male contraceptives are also under development. *Gossypol,* a yellow pigment extracted from cottonseed oil, produces a dramatic decline in sperm count and sperm motility after 2 months. It can be administered topically, as it is readily absorbed through the skin. Fertility returns within a year after treatment is discontinued. Unfortunately, gossypol has not been approved as yet because of side effects such as a relatively high risk of permanent sterility (around 10 percent).

Weekly doses of testosterone suppress GnRH secretion over a period of 5 months. The result is a drastic reduction in the sperm count. The combination of a testosterone implant, comparable to that used in the Norplant system, with a GnRH antagonist, *cetrorelix,* effectively suppresses spermatogenesis. A new synthetic form of testosterone, *alpha-methyl-nortesosterone (MENT),* appears even more effective than testosterone in suppressing GnRH production.

A drug used to control blood pressure appears to cause a temporary, reversible sterility in males. This drug is now being evaluated to see if low dosages will affect fertility in normal males without affecting blood pressure. AM *Experimental Contraceptive Methods*

If contraceptive methods fail, options exist to either prevent implantation or terminate the pregnancy. The "morning-after pills" contain estrogens or progestins. They may be taken within 72 hours of intercourse, and they appear to act by altering the transport of the zygote or preventing its attachment to the uterine wall. The drug known as *RU-486 (Mifepristone)* blocks the action of progesterone at the endometrial lining. The result is a normal menses, and the degeneration of the endometrium whether or not a pregnancy has occurred.

Abortion refers to the termination of a pregnancy. Most clinicians discriminate between *spontaneous, therapeutic,* and *induced* abortions. *Spontaneous abortions,* or *miscarriages,* occur naturally, because of some developmental or physiological problem. *Therapeutic abortions* are performed when continuing the pregnancy represents a threat to the life and health of the mother. *Induced abortions* ("elective abortions") are performed at the request of the individual. Each year there are approximately 1.5 million induced abortions in the United States, roughly 1 abortion for every 3 births. Most involve unmarried or adolescent women. The ratio between abortions and deliveries for married women averages 1:10; for unmarried women and adolescents there are nearly twice as many abortions as deliveries. Induced abortions are currently legal during the first 3 months after conception, and many states permit abortions, sometimes with restrictions, until the fifth or sixth developmental month.

These operations are now the focus of considerable controversy, and opinions concerning the morality of abortion and current abortion laws must be left to the individual.

SUMMARY OUTLINE

INTRODUCTION p. 516

1. The human reproductive system produces, stores, nourishes, and transports functional **gametes** (reproductive cells). **Fertilization** is the fusion of a **sperm** from the father and an **ovum** from the mother to create a **zygote** (fertilized egg).

AN OVERVIEW OF THE REPRODUCTIVE SYSTEM p. 516

1. The reproductive system includes **gonads,** ducts, accessory glands and organs, and the **external genitalia.**

2. In the male, the **testes** produce sperm, which are expelled from the body in **semen** during **ejaculation.** The **ovaries** (gonads) of a sexually mature female produce an egg that travels along **uterine tubes** to reach the **uterus.** The **vagina** connects the uterus with the exterior.

THE REPRODUCTIVE SYSTEM OF THE MALE p. 516

1. The **spermatozoa** travel along the **epididymis,** the **ductus deferens,** the **ejaculatory duct,** and the **urethra** before leaving the body. Accessory organs (notably the **seminal vesicles, prostate gland,** and **bulbourethral glands**) secrete into the ejaculatory ducts and urethra. The **scrotum** encloses the testes, and the **penis** is an erectile organ. (*Figure 20-1*)

The Testes p. 516

2. The *dartos* muscle gives the scrotum a wrinkled appearance; the skeletal **cremaster muscle** pulls the testes closer to the body. The **tunica albuginea** surrounds each testis. Septa extend from the tunica albuginea to subdivide each testis into a series of lobules. **Seminiferous tubules** within each lobule are the sites of sperm production. Between the seminiferous tubules there are **interstitial cells** that secrete sex hormones. (*Figure 20-2*)

3. Seminiferous tubules contain **spermatogonia,** stem cells involved in **spermatogenesis,** and **sustentacular cells,** which sustain and promote the development of spermatozoa. (*Figure 20-3*)

4. Each spermatozoon has a **head, middle piece,** and **tail.** (*Figure 20-4*)

The Male Reproductive Tract p. 520

5. From the testis, the spermatozoa enter the epididymis, an elongate tubule. The epididymis monitors and adjusts the composition of the tubular fluid and serves as a recycling center for damaged spermatozoa. Spermatozoa leaving the epididymus are functionally mature, yet immobile.

6. The ductus deferens, or *vas deferens,* begins at the epididymis and passes through the inguinal canal as one component of the **spermatic cord.** The junction of the base of the seminal vesicle and the ductus deferens creates the ejaculatory duct, which empties into the urethra. (*Figure 20-5*)

7. The urethra extends from the urinary bladder to the tip of the penis and serves as a passageway used by both the urinary and reproductive systems.

The Accessory Glands p. 521

8. Each seminal vesicle is an active secretory gland that contributes about 60 percent of the volume of semen; its secretions contain fructose that is easily metabolized by spermatozoa. The **prostate gland** secretes fluid that make up about 30 percent of seminal fluid. Alkaline mucus secreted by the bulbourethral glands has lubricating properties. (*Figure 20-5*)

Semen p. 522

9. A typical ejaculation releases 2–5 ml of semen (an **ejaculate**), which contains 20–100 million sperm per milliliter.

The Penis p. 522

10. The skin overlying the penis resembles that of the scrotum. Most of the body of the penis consists of three masses of **erectile tissue.** Beneath the superficial layers there are two **corpora cavernosa** and a single **corpus spongiosum** that surrounds the urethra. Dilation of the erectile tissue with blood produces an **erection.** (*Figure 20-6*)

Hormones and Male Reproductive Function p. 522

11. Important regulatory hormones include **FSH** (follicle-stimulating hormone), **LH** (luteinizing hormone, identical to LH in females), and **GnRH** (gonadotropin-releasing hormone). Testosterone is the most important androgen. (*Figure 20-7*)

THE REPRODUCTIVE SYSTEM OF THE FEMALE p. 524

1. Principal organs of the female reproductive system include the ovaries, uterine tubes, uterus, vagina, and external genitalia. (*Figure 20-8*)

The Ovaries p. 524

2. The ovaries are the site of ovum production, or **oogenesis,** which occurs monthly in **ovarian follicles** as part of the **ovarian cycle.** As development proceeds one finds **primordial, primary, secondary,** and **tertiary follicles.** At **ovulation** an oocyte and the surrounding follicular walls of the **corona radiata** are released through the ruptured ovarian wall. (*Figures 20-9, 20-10*)

The Uterine Tubes p. 526

3. Each uterine tube has an **infundibulum** that opens into the uterine cavity. For fertilization to occur, the secondary oocyte must encounter spermatozoa during the first 12–24 hours of its passage from the infundibulum to the uterus. (*Figure 20-11*)

The Uterus p. 527

4. The uterus provides mechanical protection and nutritional support to the developing embryo. It is stabilized by various ligaments. Major anatomical landmarks of the uterus include the **body, cervix, external orifice**, and **uterine cavity**. The uterine wall can be divided into an inner **endometrium,** a muscular **myometrium,** and a superficial *perimetrium.* (*Figure 20-11*)

5. A typical 28-day **uterine cycle**, or *menstrual cycle,* begins with the onset of **menses** and the destruction of the functional zone of the endometrium. This process of **menstruation** continues from 1 to 7 days.

6. After menses, the **proliferative phase** begins, and the functional zone undergoes repair and thickens. During the **secretory phase** the endometrial glands are active and the uterus prepared for the arrival of an embryo. Menstrual activity begins at **menarche** and continues until **menopause.**

The Vagina p. 529

7. The vagina is a muscular tube extending between the uterus and external genitalia. A thin epithelial fold, the **hymen,** partially blocks the entrance to the vagina.

The External Genitalia p. 529

8. The components of the **vulva** include the **vestibule, labia minora, clitoris, labia majora,** and the **lesser** and **greater vestibular glands.** (*Figure 20-12*)

The Mammary Glands p. 530

9. A newborn infant gains nourishment from milk secreted by maternal **mammary glands.** (*Figure 20-13*)

Hormones and the Female Reproductive Cycle p. 531

10. Hormonal regulation of the **female reproductive cycle** involves coordinating the ovarian and uterine cycles.

11. **Estradiol,** one of the estrogens, is the dominant hormone of the preovulatory period. Ovulation occurs in response to peak levels of estrogen and LH. (*Figure 20-14*)

12. The hypothalamic secretion of GnRH triggers the pituitary secretion of FSH and the synthesis of LH. FSH initiates follicular development, and activated follicles and ovarian interstitial cells produce estrogens. **Progesterone,** one of the steroid hormones called **progestins,** is the principal hormone of the postovulatory period. Hormonal changes are responsible for the maintenance of the menstrual cycle. (*Figure 20-15*)

THE PHYSIOLOGY OF SEXUAL INTERCOURSE p. 534

Male Sexual Function p. 534

1. During **arousal** in the male, erotic thoughts or sensory stimulation or both lead to parasympathetic activity that produces erection. Stimuli accompanying **coitus** lead to **emission** and ejaculation. Strong muscle contractions are associated with **orgasm.**

Female Sexual Function p. 534

2. The phases of female sexual function resemble those of the male, with parasympathetic arousal and muscular contractions associated with orgasm.

AGING AND THE REPRODUCTIVE SYSTEM p. 535

Menopause p. 535

1. Menopause (the time that ovulation and menstruation cease in women) typically occurs around age 50. Production of GnRH, FSH, and LH rise, while circulating concentrations of estrogen and progesterone decline.

The Male Climacteric p. 535

2. During the *male climacteric,* between ages 50 and 60, circulating testosterone levels decline, while levels of FSH and LH rise.

INTEGRATION WITH OTHER SYSTEMS p. 535

1. Hormones play a major role in coordinating reproduction. (*Table 20-1*)

2. In addition to the endocrine and reproductive systems, reproduction requires the normal functioning of the digestive, nervous, cardiovascular, and urinary systems. (*Figure 20-16*)

CHAPTER QUESTIONS

LEVEL 1 Reviewing Facts and Terms

Match each item in column A with the most closely related item in column B. Use letters for answers in the spaces provided.

Column A	Column B
___ 1. gametes	a. production of androgens
___ 2. gonads	b. outer, muscular uterine wall
___ 3. interstitial cells	c. high concentration of fructose
___ 4. seminal vesicles	d. female erectile tissue
___ 5. prostate gland	e. secretes thick, sticky, alkaline mucus
___ 6. bulbourethral glands	f. painful menstruation
___ 7. prepuce	g. sexual intercourse
___ 8. corpus luteum	h. uterine lining
___ 9. endometrium	i. reproductive cells
___ 10. myometrium	j. female puberty
___ 11. dysmenorrhea	k. milk production
___ 12. menarche	l. secretes antibiotic
___ 13. clitoris	m. reproductive organs
___ 14. lactation	n. foreskin of penis
___ 15. coitus	o. endocrine structure

16. Perineal structures associated with the reproductive system are collectively known as:
(a) gonads
(b) sex gametes
(c) external genitalia
(d) accessory glands

17. The completion of the meiotic process in the male produces four spermatids, each containing:
(a) 23 chromosomes
(b) 23 pairs of chromosomes
(c) the diploid number of chromosomes
(d) 46 pairs of chromosomes

18. Erection of the penis occurs when:
    (a) sympathetic activation of penile arteries occurs
    (b) arterial branches are constricted, and muscular partitions are tense
    (c) the vascular channels become engorged with blood
    (d) a, b, and c are correct
19. In the male, the primary target of FSH is the:
    (a) sustentacular cells of the seminiferous tubules
    (b) interstitial cells of the seminiferous tubules
    (c) prostate gland
    (d) epididymis
20. The ovaries in the female are responsible for:
    (a) the production of female gametes
    (b) the secretion of female sex hormones
    (c) the secretion of inhibin
    (d) a, b, and c are correct
21. The process of ovum production, or oogenesis, begins:
    (a) before birth
    (b) after birth
    (c) at puberty
    (d) after puberty
22. In the female, the process of meiosis is not completed:
    (a) until birth
    (b) until puberty
    (c) unless and until fertilization occurs
    (d) until uterine implantation occurs
23. If fertilization is to occur, the ovum must encounter spermatozoa during the first ________ of its passage.
    (a) 1 to 5 hours
    (b) 6 to 11 hours
    (c) 12 to 24 hours
    (d) 25 to 36 hours
24. The part of the endometrium that undergoes cyclical changes in response to sexual hormonal levels is the:
    (a) serosa
    (b) basilar zone
    (c) muscular myometrium
    (d) functional zone
25. A sudden surge in LH concentration causes:
    (a) the onset of menses
    (b) the rupture of the follicular wall and ovulation
    (c) the beginning of the proliferative phase
    (d) the end of the uterine cycle
26. At the time of ovulation, the basal body temperature:
    (a) is not affected
    (b) increases noticeably
    (c) declines sharply
    (d) may increase or decrease a few degrees
27. Menopause is accompanied by:
    (a) sustained rises in GnRH, FSH, and LH
    (b) declines in circulating levels of estrogen and progesterone
    (c) thinning of the urethral and vaginal walls
    (d) a, b, and c are correct
28. What reproductive structures are common to both male and female?
29. What accessory organs and glands contribute to the composition of semen? What are the functions of each?
30. What are the primary functions of the epididymis in the male?
31. What are the primary functions of the ovaries in the female?
32. What are the three major functions of the vagina?

LEVEL 2 Reviewing Concepts

33. How does the human reproductive system differ functionally from all other systems in the body?
34. How is the process of meiosis involved in the development of the spermatozoon and the ovum?
35. Using an average uterine cycle of 28 days, describe each of the three phases of the menstrual cycle.
36. Describe the hormonal events associated with the uterine cycle.
37. How does the aging process affect the reproductive systems of men and women?

LEVEL 3 Critical Thinking and Clinical Applications

38. Diane has an inflammation of the peritoneum (peritonitis), which her doctor says resulted from a urinary tract infection. Why could this situation occur in females but not in males?
39. Rod suffers an injury to the sacral region of his spinal cord. Will he still be able to achieve an erection? Explain.
40. Women body builders and women suffering from eating disorders such as anorexia nervosa often cease having menstrual cycles, a condition known as amenorrhea. What does this relationship suggest about the role of body fat and menstruation? What benefit might there be in the discontinuance of menstruation under such circumstances?

21 Development and Inheritance

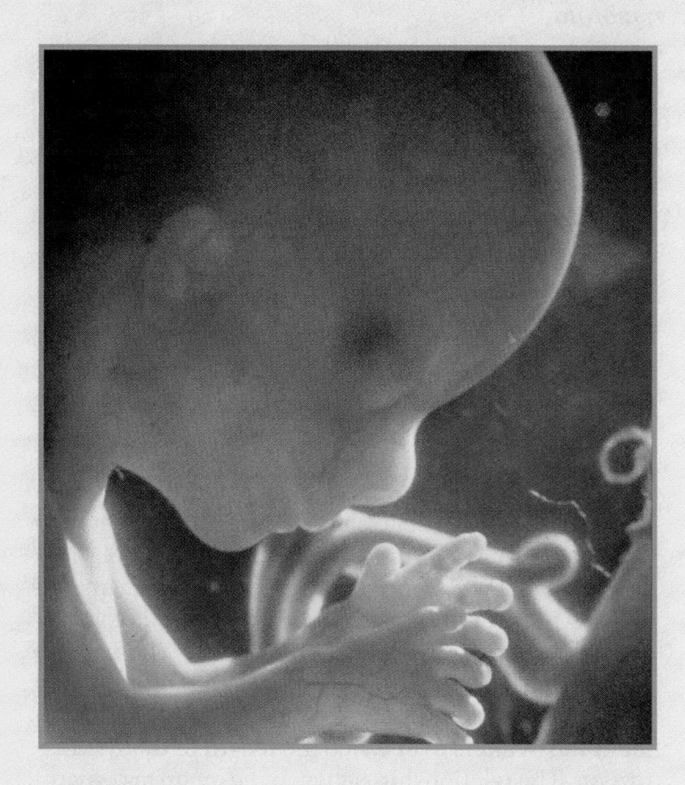

Many physiological processes last only a fraction of a second; others may take hours at most. But some important processes of life are measured in months, years, or decades. A human being develops in the womb for nine months, grows to maturity in 15 or 20 years, and may live the better part of a century. Birth, growth, maturation, aging, and death are all parts of a single, continuous process. And that process does not end with the individual, for human beings can pass at least some of their characteristics on to a new generation that will repeat the same cycle.

Chapter Outline and Objectives

Vocabulary Development

amphi, two-sided; *amphimixis*
blast, precursor; *trophoblast*
heteros, other; *heterozygous*
homos, same; *homozygous*
karyon, nucleus; *karyotyping*
koiloma, hollow; *blastocoele*
meros, part; *blastomere*
meso-, middle; *mesoderm*
mixis, mixing; *amphimixis*
morula, mulberry; *morula*
phainein, to display; *phenotype*
tropho-, food; *trophoblast*
typos, mark; *phenotype*
vitro, glass; *in vitro*

The process of **development** is the gradual modification of physical and physiological characteristics during the period from conception to physical maturity. The changes are truly remarkable—what begins as a single cell slightly larger than the period at the end of this sentence becomes an individual whose body contains trillions of cells organized into tissues, organs, and organ systems. The creation of different cell types in this process is called **differentiation.** Differentiation occurs through selective changes in genetic activity. As development proceeds, some genes are turned off and others turned on. The identities of these genes vary from one cell type to another.

A basic understanding of development provides important insights into anatomical structures. In addition, many of the mechanisms of development and growth (an increase in size) are similar to those responsible for the repair of injuries. This chapter will focus on major aspects of development and consider highlights of the developmental process. We will also consider the regulatory mechanisms and how developmental patterns can be modified—for good or ill.

AN OVERVIEW OF TOPICS IN DEVELOPMENT

Development involves (1) the division and differentiation of cells and (2) the changes that produce and modify anatomical structures. Development begins at fertilization, or **conception,** and can be divided into periods characterized by specific anatomical changes. **Embryology** (em-brē-OL-ō-jē) considers the developmental events that occur in the first 2 months after fertilization. Over this period the developing organism is called an **embryo.** After two months, the developing embryo becomes a **fetus**, and **fetal development** begins at the start of the ninth week and continues up to the time of birth. Embryological and fetal development are sometimes referred to collectively as **prenatal development,** the primary focus of this chapter. **Postnatal development** commences at birth and continues to maturity.

Although all human beings go through the same developmental stages, differences in genetic makeup produce distinctive individual characteristics. **Inheritance** refers to the transfer of genetically determined characteristics from generation to generation. **Genetics** is the study of the mechanisms responsible for inheritance. This chapter considers basic genetics as it applies to the appearance of inherited characteristics such as sex, hair color, and various diseases.

FERTILIZATION

Fertilization involves the fusion of two haploid gametes, producing a *zygote* containing the normal diploid number of chromosomes, 46. ∞ *[p. 519]* The functional roles of the spermatozoon and the ovum are very different. The spermatozoon simply delivers the paternal chromosomes to the site of fertilization. It is the ovum that must provide all of the nourishment and genetic programming to support embryonic development for nearly a week after conception. The volume of the ovum is therefore much greater than that of the spermatozoon. At the time of fertilization, the diameter of the ovum is over twice the length of the spermatozoon. The relationship between the ovum and sperm volumes is even more striking, on the order of 2000:1.

The sperm arriving in the vagina are already motile, but they cannot fertilize an egg until they undergo **capacitation** (ka-pas-i-TĀ-shun). This activation process, necessary for fertilization, is inhibited in the male reproductive tract until emission and ejaculation occur. Capacitation therefore normally occurs in the female reproductive tract.

Fertilization usually occurs in the upper one-third of the uterine tube within a day of ovulation. Contractions of the uterine musculature and ciliary currents in the uterine tubes aid the passage of sperm to the fertilization site. Of the 200 million spermatozoa introduced into the vagina in a typical ejaculate, only around 10,000 make it past the uterus, and fewer than 100 actually reach the secondary oocyte. A male with a sperm count below 20 million per milliliter will usually be sterile, because too few sperm survive to reach the oocyte. One or two spermatozoa cannot accomplish fertilization, because of the layer of cells surrounding the oocyte at ovulation.

The Oocyte at Ovulation

At ovulation, the secondary oocyte leaving the follicle is in metaphase of the second meiotic division. The cell's metabolic operations have been discontinued,

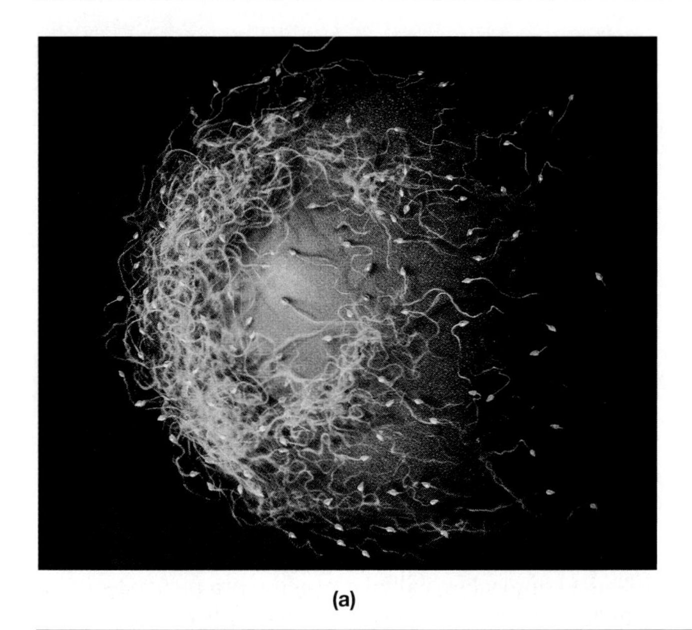
(a)

and the oocyte drifts in a sort of suspended animation, awaiting the stimulus for further development. If fertilization does not occur, the secondary oocyte disintegrates without completing meiosis.

Fertilization is complicated by the fact that when it leaves the ovary the secondary oocyte is surrounded by a layer of follicle cells, the *corona radiata* (Figure 21-1●). The corona radiata protects the oocyte as it passes through the ruptured follicular wall and into the infundibulum of the uterine tube. Although the physical process of fertilization requires only a single sperm in contact with the oocyte membrane, that spermatozoon must first penetrate the corona radiata. The acrosomal cap contains *hyaluronidase* (hī-al-u-RON-a-dās), an enzyme that breaks down the intercellular cement between adjacent cells of the corona radiata. Apparently, there must be at least a hundred or more sperm thrash-

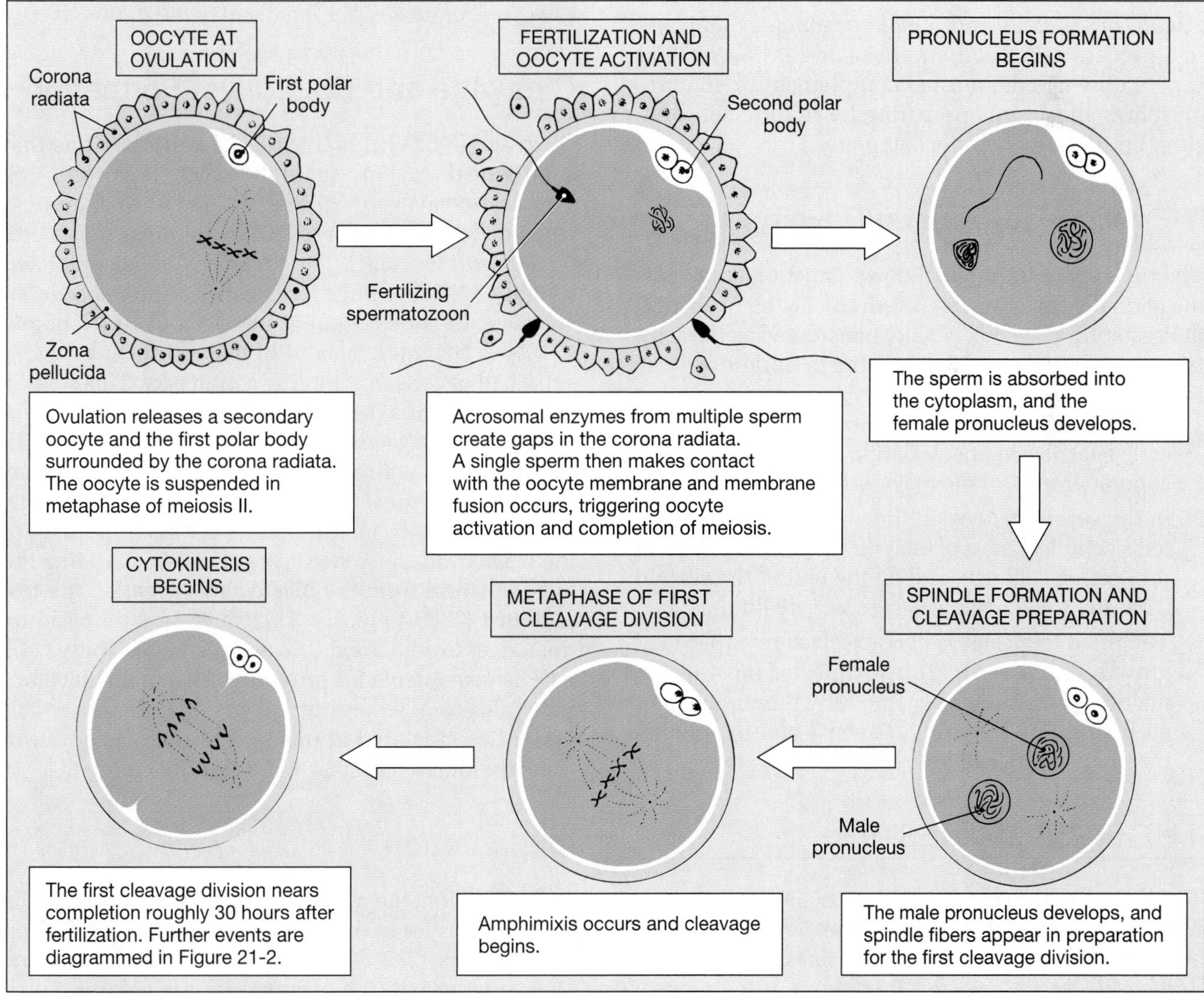

(b)

●**Figure 21-1**
Fertilization

(a) An oocyte at the time of fertilization. Note the difference in size between the gametes. **(b)** Fertilization and the preparations for cleavage.

ing around, bumping into the corona and releasing hyaluronidase, before the connections between the follicular cells begin to break down. No matter how many sperm slip through the gap, only a single spermatozoon will accomplish fertilization. When that sperm contacts the oocyte, their cell membranes fuse, and the sperm's nucleus enters the cytoplasm of the oocyte. This event initiates **oocyte activation.**

Activation of the oocyte involves a series of metabolic changes. For example, the metabolic rate of the oocyte increases rapidly and meiosis is completed, and vesicles within the oocyte fuse with the cell membrane and discharge their contents through exocytosis. Activation helps prevent fertilization by more than one sperm, a condition called *polyspermy.* If polyspermy does occur, the zygote will be incapable of normal development.

Once meiosis has been completed, the nuclear material of the sperm and ovum now reorganize into a *male pronucleus* and *female pronucleus* and fuse in a process called *amphimixis* (am-fi-MIK-sis). The zygote now has the normal complement of 46 chromosomes and begins preparing for mitotic cell division. Fertilization is now complete.

A Preview of Prenatal Development

The time spent in prenatal development is known as the period of **gestation** (jes-TĀ-shun). For convenience, the gestation period is usually considered as three integrated **trimesters,** each 3 months in duration:

1. The **first trimester** is the period of embryonic and early fetal development. During this period, the basic components of all the major organ systems appear.
2. In the **second trimester,** the organs and organ systems complete most of their development. The body proportions change, and by the end of the second trimester the fetus looks distinctively human.
3. The **third trimester** is characterized by rapid fetal growth. Early in the third trimester most of the major organ systems become fully functional, and an infant born 1 month or even 2 months prematurely has a reasonable chance of survival.

THE FIRST TRIMESTER

The events that occur in the first trimester are complex and vital to the survival of the embryo. Because accidents often happen, *the first trimester is the most dangerous period in life.* Only about 40 percent of conceptions produce embryos that survive the first trimester, and some surviving fetuses enter the second trimester already doomed or deformed by some developmental mistake. For this reason, pregnant women are usually warned to take great care to avoid drugs or other disruptive stresses during the first trimester. We will focus on four general processes that occur during this period: *cleavage, implantation, placentation,* and *embryogenesis.*

INFERTILITY

Infertility, or the inability to have children, has been the focus of media attention for the past 5 years. The reason is simple: Problems with infertility are relatively common. An estimated 10 to 15 percent of U.S. married couples are infertile, and another 10 percent are unable to have as many children as they desire. It is thus not surprising that reproductive physiology has become a popular field and the treatment of infertility has become a major medical industry. Recent advances in our understanding of reproductive physiology are providing new solutions to fertility problems. These approaches, called *assisted reproductive technologies (ART)* are detailed in the *Applications Manual.*
AM *Technology and the Treatment of Infertility*

Cleavage and Blastocyst Formation

Cleavage (KLĒV-ij) is a series of cell divisions that subdivide the cytoplasm of the zygote (Figure 21-2•). The first cleavage division produces two identical cells, called **blastomeres** (BLAS-tō-mērz; *blast,* precursor + *meros,* part). The first division is completed roughly 30 hours after fertilization, and subsequent cleavage divisions occur at intervals of 10–12 hours.

After several cycles of division, the embryo is a solid ball of cells, resembling a mulberry. This stage is called the **morula** (MOR-ū-la; morula, *mulberry*). After 5 days of cleavage, the blastomeres form a hollow ball, the **blastocyst,** with an inner cavity known as the *blastocoele* (BLAS-tō-sēl; *koiloma,* cavity). At this stage, you can begin to see differences between the cells of the blastocyst. The outer layer of cells, separating the outside world from the blastocoele, is called the **trophoblast** (TRŌ-fō-blast). The function is implied by the name: *tropho,* food + *blast,* precursor. These cells will be responsible for providing food to the developing embryo. A second group of cells, the **inner cell mass,** lies clustered at one end of the blastocyst. In time, the inner cell mass will form the embryo.

Implantation

At fertilization, the zygote is still 4 days away from the uterus. It arrives in the uterine cavity as a morula, and over the next 2–3 days blastocyst formation occurs. Over this period, the blastomeres are gaining nutrients from the fluid within the uterine cavity. This fluid, rich in glycogen, is secreted by the endometrial glands. When fully formed, the blastocyst contacts the endometrium, usually in the body or fundus of the uterus, and implantation occurs. Stages in the implantation process are diagrammed in Figure 21-3●; you may find it helpful to review the structure of the uterus (Figure 20-11●, p. 528) at this time.

Implantation begins as the surface of the blastocyst closest to the inner cell mass touches and adheres to the uterine lining (Day 7, Figure 21-3●). In this area, the superficial cells undergo rapid divisions, making the trophoblast several layers thick. Near the endometrial wall, the cell membranes separating the trophoblast cells disappear, creating a layer of cytoplasm containing multiple nuclei called the *syncytial trophoblast* (Day 8). The underlying cells remain intact and form a layer called the *cellular trophoblast.* The syncytial trophoblast erodes a path through the uterine epithelium. At first, this erosion creates a gap in the uterine lining, but the division and migration of epithelial cells soon repair the surface. With the conclusion of these repairs, the blastocyst loses contact with the uterine cavity, and further development occurs entirely within the functional zone of the endometrium.

Implantation usually occurs at the endometrial surface lining the uterine cavity. The precise location varies, although most often it is in the body of the uterus. This is not an ironclad rule, and in an *ectopic pregnancy* implantation occurs somewhere other than within the uterus. AM *Ectopic Pregnancies*

As implantation proceeds, the syncytial trophoblast continues to enlarge into the surrounding endometrium (Day 9). The digestion of uterine gland cells releases quantities of glycogen and other nutrients that are absorbed by the trophoblast and distributed by diffusion to the inner cell mass. These nutrients provide the energy needed to support the early stages of embryo formation. Trophoblastic extensions grow around endometrial capillaries, and as the capillary walls are destroyed, maternal blood begins to percolate through trophoblastic channels known as *lacunae.* Fingerlike *villi* extend away from the trophoblast into the surrounding endometrium, and these extensions gradually increase in size and complexity as development proceeds.

Formation of the Blastodisc

By the time of implantation, the inner cell mass is beginning to separate from the trophoblast. The separation gradually increases, creating a fluid-filled chamber called the **amniotic** (am-nē-OT-ik) **cavity.** The amniotic cavity can be seen in Day 9 of Figure 21-3●, and additional details from Days 10 to 12 are shown in Figure 21-4●. At this stage, the inner cell mass is organized into an oval, two-layered sheet of cells called a **blastodisc** (BLAS-tō-disk).

Gastrulation and Germ Layer Formation

A few days later, a third layer of cells begins forming through the process of **gastrulation** (gas-troo-LĀ-shun) (Day 12, Figure 21-4●). During gastrulation, cells on the blastodisc surface move toward the center line of

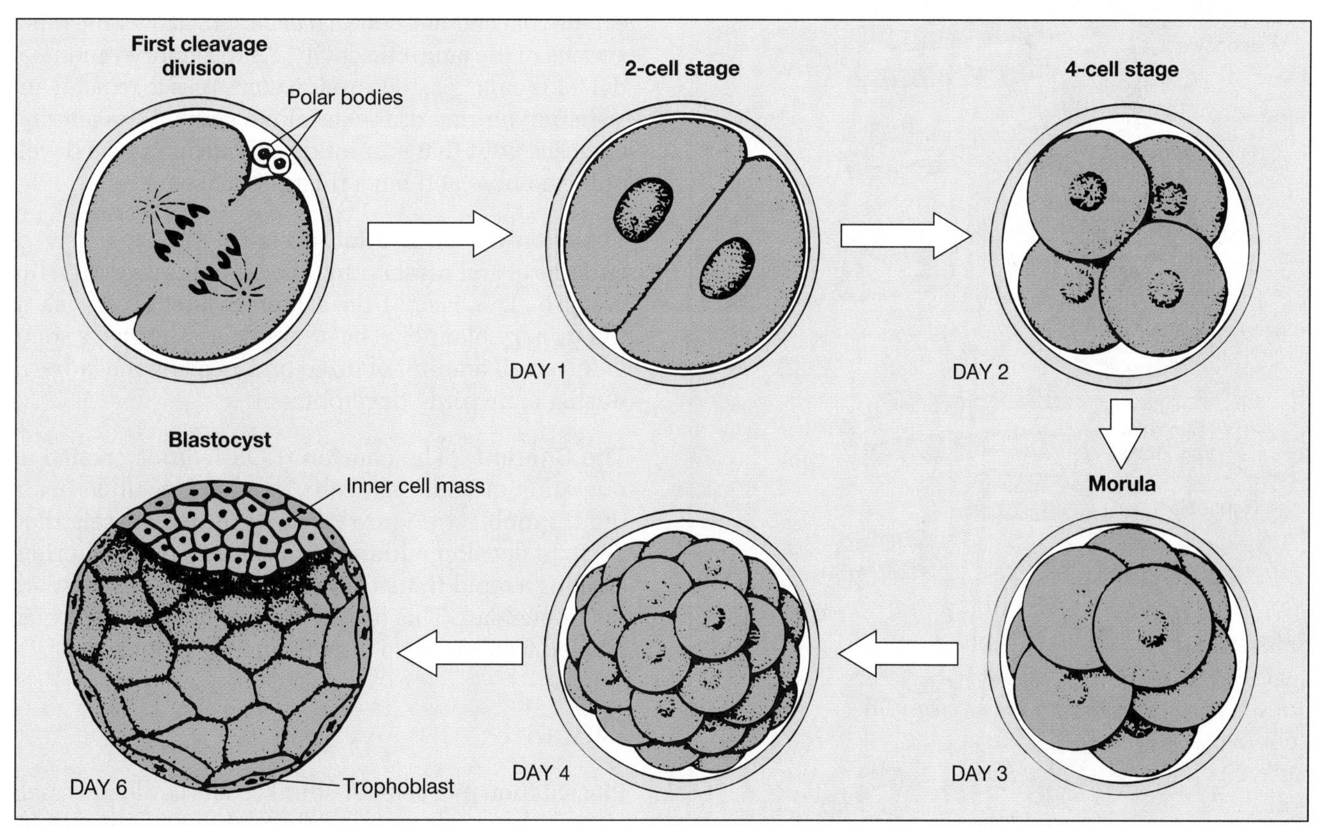

●**Figure 21-2**
Cleavage and Blastocyst Formation

●**Figure 21-3**
Stages in the Implantation Process

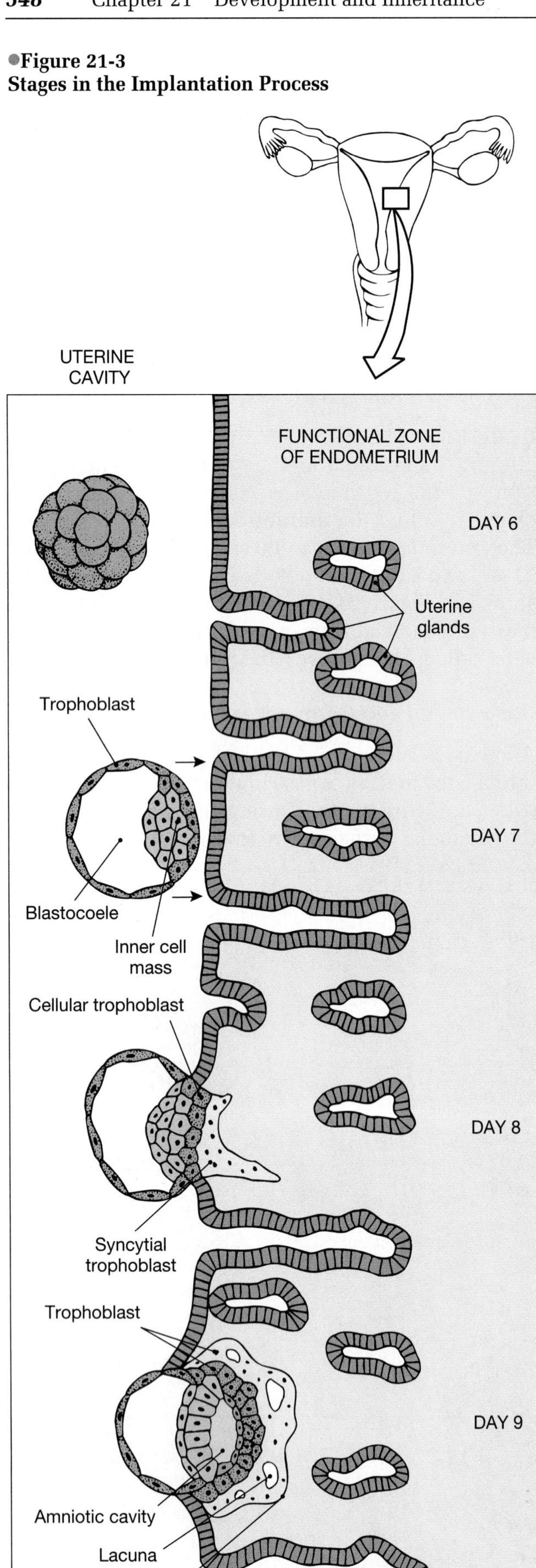

the blastodisc, to a region called the primitive streak, where they leave the surface to form a third, central layer. This movement creates three distinct embryonic layers with markedly different fates. The superficial layer in contact with the amniotic cavity is called the **ectoderm,** the layer facing the blastocoel is known as the **endoderm,** and the intervening, poorly organized layer is the **mesoderm** (*meso-*, middle). Table 21-1 lists the contributions each of these three **germ layers** makes to the body systems described in earlier chapters.

Formation of Extraembryonic Membranes

Germ layers also participate in the formation of four **extraembryonic membranes:** the *yolk sac,* the *amnion,* the *allantois,* and the *chorion.* Although these membranes support embryonic and fetal development, they leave few traces of their existence in adult systems. Figure 21-5● details stages in the development of the extraembryonic membranes.

Yolk sac The first of the extraembryonic membranes to appear is the **yolk sac,** which is already present 10 days after fertilization (Figure 21-4●). As gastrulation proceeds, mesodermal cells migrate around this pouch and complete the formation of the yolk sac (Figure 21-5a ●). Blood vessels soon appear within the mesoderm, and the yolk sac becomes an important site of blood cell formation.

Amnion The **amnion** (AM-nē-on) is completed when ectodermal and mesodermal cells spread over the inner surface of the amniotic cavity. As the embryo and later the fetus enlarges, the amnion continues to expand, increasing the size of the amniotic cavity. The amnion encloses fluid that surrounds and cushions the developing embryo and fetus (Figure 21-5c–e●).

The Allantois The **allantois** (a-LAN-tō-is) is a sac of endoderm and mesoderm that extends away from the embryo. The base of the allantois later gives rise to the urinary bladder. The allantois accumulates some of the small amount of urine produced by the kidneys during embryonic development.

The Chorion The **chorion** (KOR-ē-on) is created as migrating mesodermal cells form a layer underneath the trophoblast (Figure 21-5a ●). Blood vessels then begin to develop within the mesoderm of the chorion, creating a rapid-transit system linking the embryo with the trophoblast. This provides the nutrients and oxygen needed for continued growth and development.

Placentation

Placentation (pla-sen-TĀ-shun) occurs as blood vessels form in the chorion around the periphery of the blastocyst, and the **placenta** appears (see Figure 21-5a–e●).

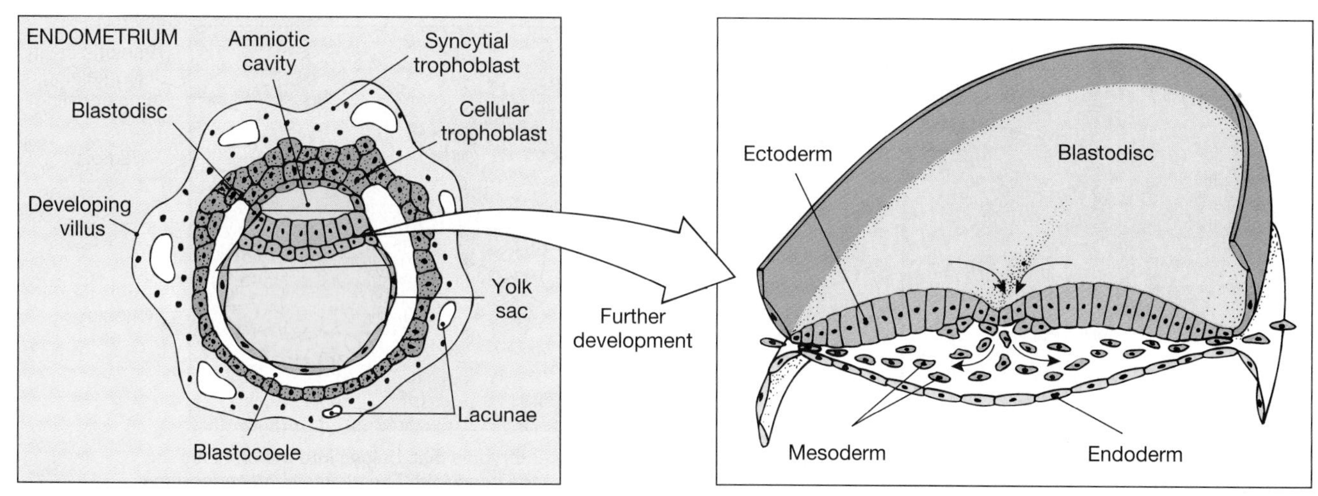

Day 10: The blastodisc begins as two layers. Migration of cells around the amniotic cavity is the first step in the formation of the amnion. Migration of cells facing the blastocoele creates a sac that hangs below the blastodisc. This is the first step in yolk sac formation.

Day 12: Migration of cells gives the blastodisc a third layer. From the time this process, called gastrulation, begins, the surface facing the amniotic cavity is called *ectoderm,* the layer facing the yolk sac is called the *endoderm,* and the migrating cells form the *mesoderm.*

Figure 21-4
Blastodisc Organization and Gastrulation

TABLE 21-1 The Fates of the Primary Germ Layers

Primary Germ Layer	*Developmental Contributions to the Body*
ECTODERMAL	*Integumentary system:* epidermis, hair follicles and hairs, nails, and glands communicating with the skin (apocrine and merocrine sweat glands, mammary glands, and sebaceous glands) *Skeletal system, muscular system:* pharyngeal cartilages and associated muscles *Nervous system:* all neural tissue, including brain and spinal cord *Endocrine system:* pituitary gland and the adrenal medullae *Respiratory system:* mucous epithelium of nasal passageways *Digestive system:* mucous epithelium of mouth and anus, salivary glands
MESODERMAL	*Skeletal system:* all components except some pharyngeal derivatives *Muscular system:* all components except some pharyngeal derivatives *Endocrine system:* adrenal cortex, endocrine tissues of heart, kidneys, and gonads *Cardiovascular system:* all components, including bone marrow *Lymphatic system:* all components *Urinary system:* the kidneys, including the nephrons and the initial portions of the collecting system *Reproductive system:* the gonads and the adjacent portions of the duct systems *Miscellaneous:* the lining of the body cavities (pleural, pericardial, peritoneal) and the connective tissues supporting all organ systems
ENDODERMAL	*Endocrine system:* thymus, thyroid, and pancreas *Respiratory system:* respiratory epithelium (except nasal passageways) and associated mucous glands *Digestive system:* mucous epithelium (except mouth and anus), exocrine glands (except salivary glands), liver, and pancreas *Urinary system:* urinary bladder and distal portions of the duct system *Reproductive system:* distal portions of the duct system, stem cells that produce gametes

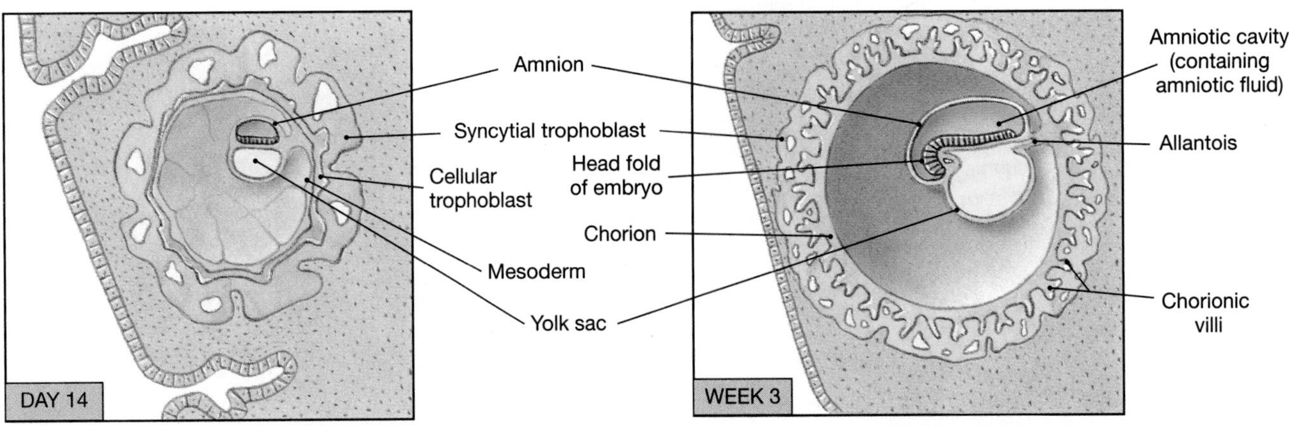

(a) Migration of mesoderm around the inner surface of the trophoblast creates the chorion. Mesodermal migration around the outside of the amniotic cavity, between the ectodermal cells and the trophoblast, creates the amnion. Mesodermal migration around the endodermal pouch below the blastodisc creates the definitive yolk sac.

(b) The embryonic disc bulges into the amniotic cavity at the head fold. The allantois, an endodermal extension surrounded by mesoderm, extends toward the trophoblast.

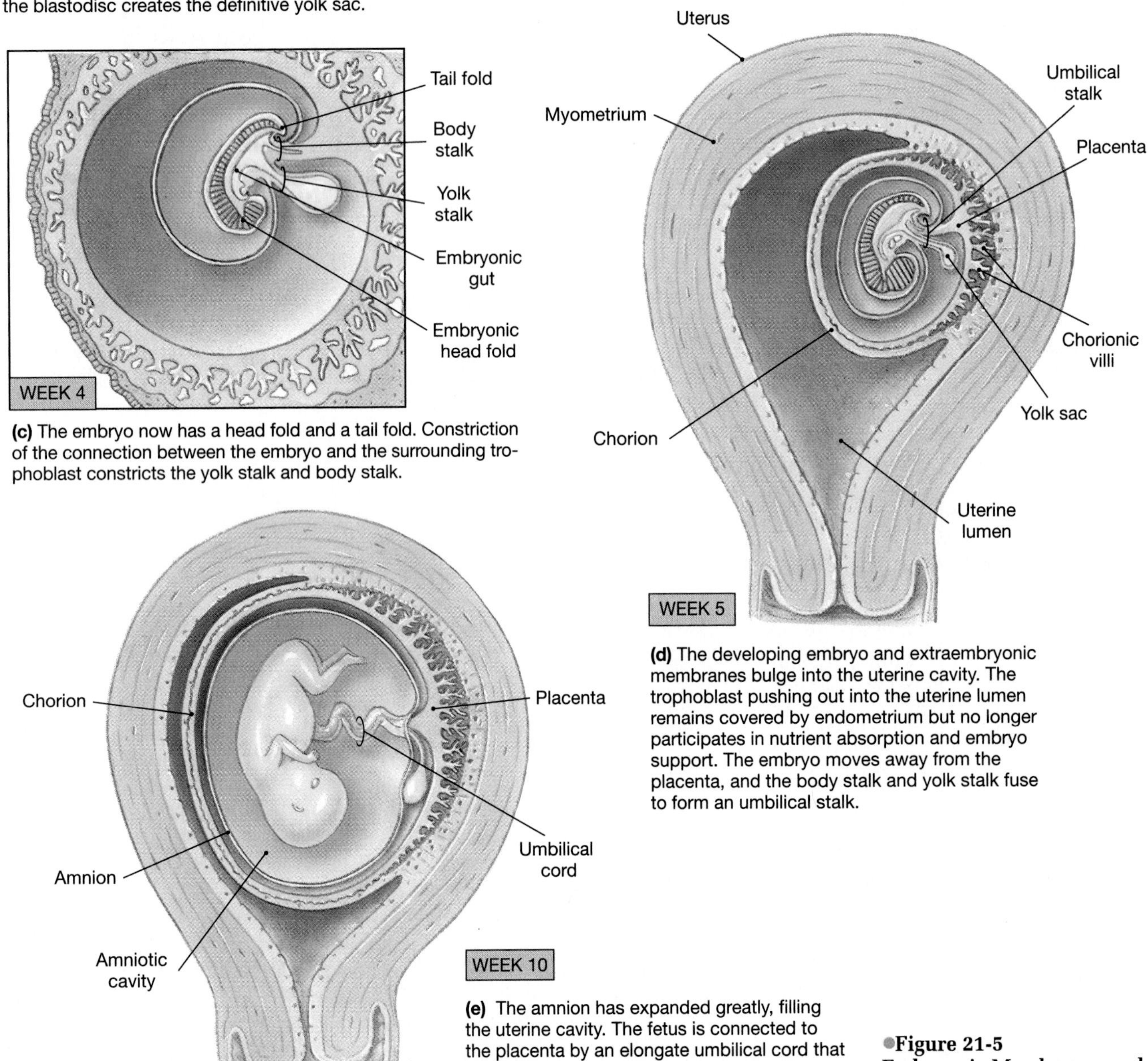

(c) The embryo now has a head fold and a tail fold. Constriction of the connection between the embryo and the surrounding trophoblast constricts the yolk stalk and body stalk.

(d) The developing embryo and extraembryonic membranes bulge into the uterine cavity. The trophoblast pushing out into the uterine lumen remains covered by endometrium but no longer participates in nutrient absorption and embryo support. The embryo moves away from the placenta, and the body stalk and yolk stalk fuse to form an umbilical stalk.

(e) The amnion has expanded greatly, filling the uterine cavity. The fetus is connected to the placenta by an elongate umbilical cord that contains a portion of the allantois, blood vessels, and the remnants of the yolk stalk.

Figure 21-5 Embryonic Membranes and Placenta Formation

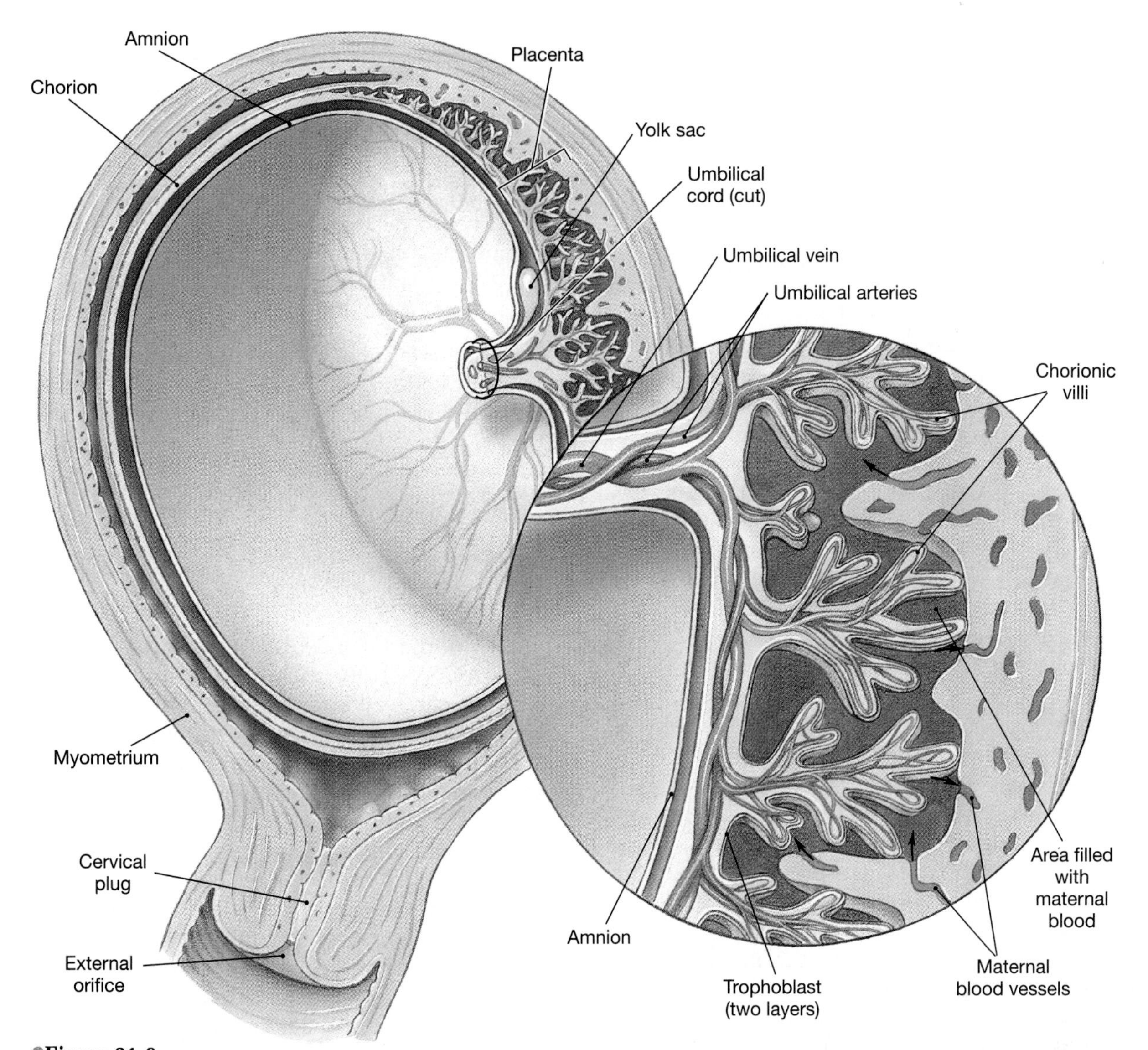

Figure 21-6
A Three-Dimensional View of Placental Structure
For clarity the uterus is shown after the embryo has been removed and the umbilical cord cut. Blood flows into the placenta through ruptured maternal blood arteries. It then flows around chorionic villi that contain fetal blood vessels. Fetal blood arrives over paired umbilical arteries and leaves over a single umbilical vein. Maternal blood reenters the venous system of the mother through the broken walls of small uterine veins. Note that no actual mixing of maternal and fetal blood occurs.

By the third week of development (Figure 21-5b), the mesoderm extends along each of the trophoblastic villi, forming *chorionic villi* in contact with maternal tissues. Embryonic blood vessels develop within each of the villi, and circulation through these chorionic blood vessels begins early in the third week of development, when the heart starts beating. These villi continue to enlarge and branch, forming an intricate network within the endometrium. Blood vessels continue to be eroded, and maternal blood flows slowly through the lacunae. Chorionic blood vessels pass close by, and exchange between the embryonic and maternal circulations occurs by diffusion across the trophoblast layers.

At first, the entire blastocyst is surrounded by chorionic villi. The chorion continues to enlarge, expanding like a balloon within the endometrium, and by the fourth week the embryo, amnion, and yolk sac are suspended within an expansive, fluid-filled chamber (Figure 21-5c). The only communication between the embryo and the chorion occurs at the site where mesoderm cells first came into contact with the inner wall of the trophoblast. As the end of the first trimester approaches, the fetus moves farther away from the placenta (Figure 21-5d,e). It remains connected by the **umbilical cord,** or *umbilical stalk,* that contains the allantois, blood vessels, and the yolk sac.

Placental Circulation

Figure 21-6 indicates the extent of the fetal circulation at the placenta near the end of the first trimester. Blood flows to the placenta through the paired **umbilical arteries** and returns in a single **umbilical vein.** The chorionic villi provide the surface area for active and passive exchange between the fetal and maternal bloodstreams. AM *Problems with Placentation*

Placental Hormones

In addition to its role in the nutrition of the fetus, the placenta acts as an endocrine organ. Hormones are synthesized by the syncytial trophoblast and released into the maternal circulation. The hormones produced include *human chorionic gonadotropin, human placental lactogen, placental prolactin, relaxin, progestins,* and *estrogens.*

Human chorionic (kō-rē-ON-ik) **gonadotropin (hCG)** appears in the maternal bloodstream soon after implantation has occurred. The presence of hCG in blood or urine samples provides a reliable indication of pregnancy, and kits sold for the early detection of pregnancy are sensitive for the presence of this hormone. Because of the hCG, the corpus luteum persists for 3–4 months and maintains its production of progesterone. As a result, the endometrial lining remains perfectly functional, and menses does not occur and terminate the pregnancy.

The decline in the corpus luteum does not trigger the return of menstrual periods, because by the end of the first trimester the placenta is secreting sufficient amounts of progestins to maintain the endometrial lining and the pregnancy. As the end of the third trimester approaches, estrogen production accelerates. The rising estrogen levels play a role in stimulating labor and delivery.

During the second trimester, the placenta also secretes **human placental lactogen (hPL)** and **placental prolactin,** which help prepare the mammary glands for milk production. Conversion of the mammary glands to active status requires the presence of placental hormones (hPl, placental prolactin, estrogen, and progestins) as well as several maternal hormones (growth hormone, prolactin, and thyroid hormones).

Relaxin is secreted by the placenta as well as the corpus luteum. Relaxin (1) increases the flexibility of the pubic symphysis, permitting expansion of the pelvis during delivery; (2) causes dilation of the cervix, making it easier for the fetus to enter the vaginal canal; and (3) suppresses the release of oxytocin by the hypothalamus, delaying the onset of labor contractions.

Embryogenesis

After gastrulation begins, folding and differential growth of the embryonic disc produce projecting bulges into the amniotic cavity called the *head fold* (Fig. 21-5b●) and *tail fold* (Fig. 21-5c●). The orientation of the embryo can now be seen, complete with dorsal and ventral surfaces and left and right sides. The changes in proportions and appearance that occur between the fourth developmental week and the end of the first trimester are summarized in Figure 21-7●.

Embryogenesis (em-brē-ō-JEN-e-sis) is the formation of a viable embryo. The first trimester is a critical period for development because events in the first 12 weeks establish the basis for organ formation, a process called **organogenesis.** Important developmental milestones are indicated in Table 21-2.

✓ What would happen if a spermatozoon did not become capacitated in the vagina?

✓ What is the fate of the inner cell mass of the blastocyst?

✓ Sue's doctor tells her that her pregnancy test indicates elevated levels of the hormone hCG (human chorionic gonadotropin). Is she pregnant or not?

✓ What are two important functions of the placenta?

THE SECOND AND THIRD TRIMESTERS

By the start of the second trimester (Figure 21-7c●), the rudiments of all the major organ systems have formed. Over the next 3 months, the fetus will grow to a weight of about 0.64 kg (1.4 lb). Figure 21-8●, shows a 4-month fetus as viewed with a fiber-optic endoscope and a 6-month fetus as seen in ultrasound. The changes in body form that occur during the first and second trimesters are shown in Figure 21-9●.

During the third trimester, the basic components of all the organ systems appear, and most become ready to fulfill their normal functions. The rate of growth begins to decrease, but in absolute terms this trimester sees the largest weight gain. In 3 months, the fetus puts on around 2.6 kg (5.7 lb), reaching a full-term weight of somewhere near 3.2 kg (7 lb). Important events in organ system development are highlighted in Table 21-2.

Pregnancy and Maternal Systems

The developing fetus is totally dependent on maternal organ systems for nourishment, respiration, and waste removal. These functions must be performed by maternal systems in addition to their normal operations. For example, the mother must absorb enough oxygen, nutrients, and vitamins for herself *and* her fetus, and she must eliminate all of the generated wastes. Although this is not a burden over the initial weeks of gestation, the demands become significant as the fetus grows larger. To survive under these conditions, the maternal systems must make major adjustments. In practical terms the mother must breathe, eat, and excrete for two. AM *Problems with the Maintenance of a Pregnancy*

- ***The respiratory rate goes up and the tidal volume increases.*** As a result, the lungs obtain the extra oxygen required and remove the excess carbon dioxide generated by the fetus.

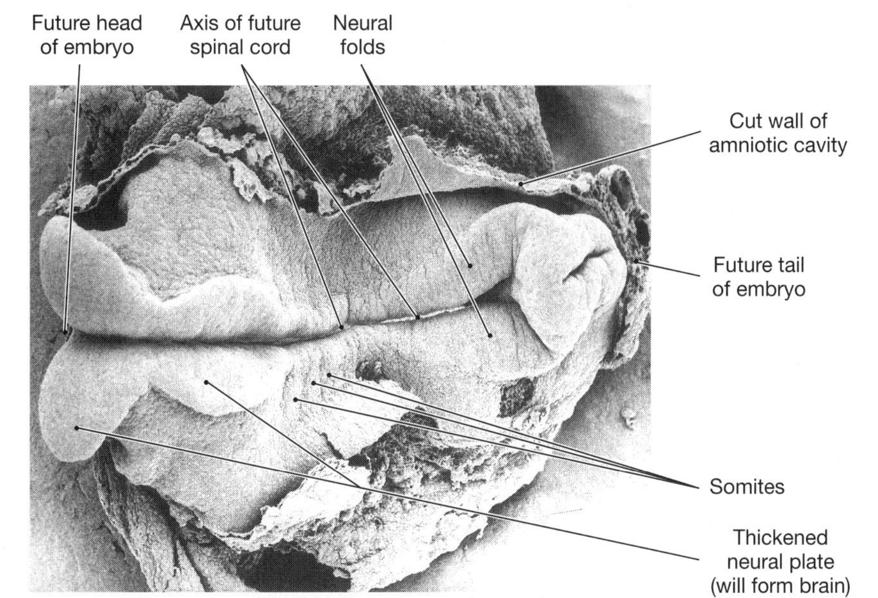

(a) 2 Weeks

Figure 21-7 The First Trimester

(a) Superior view of an SEM of an embryo at Week 2. **(b)** An SEM and a fiberoptic view of the lateral surfaces of embryos at Week 4-5. **(c)** Fiberoptic view of an embryo at 8 weeks. **(d)** Fiberoptic view of an embryo at 12 weeks.

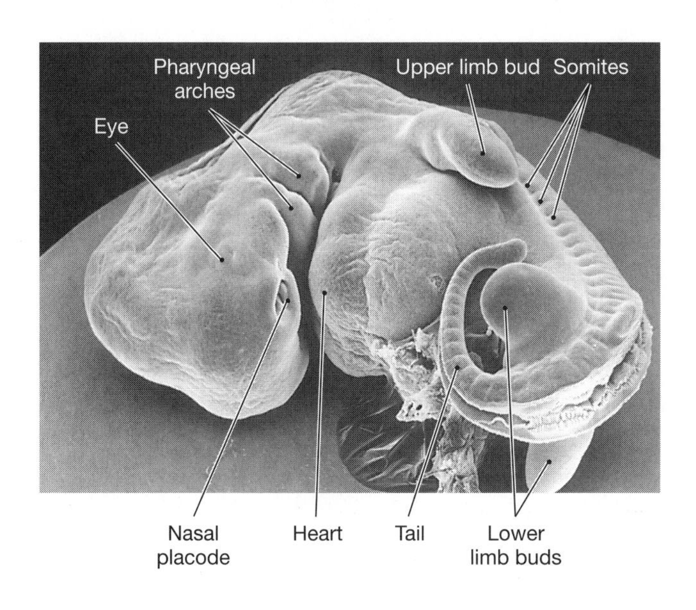

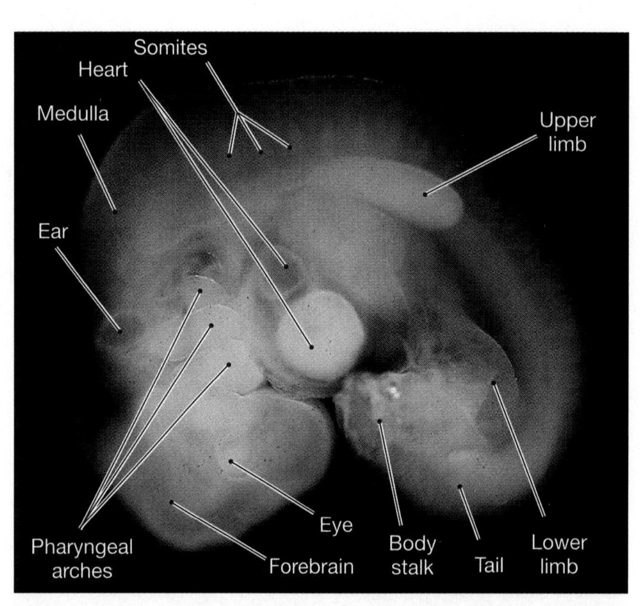

(b) 4 Weeks

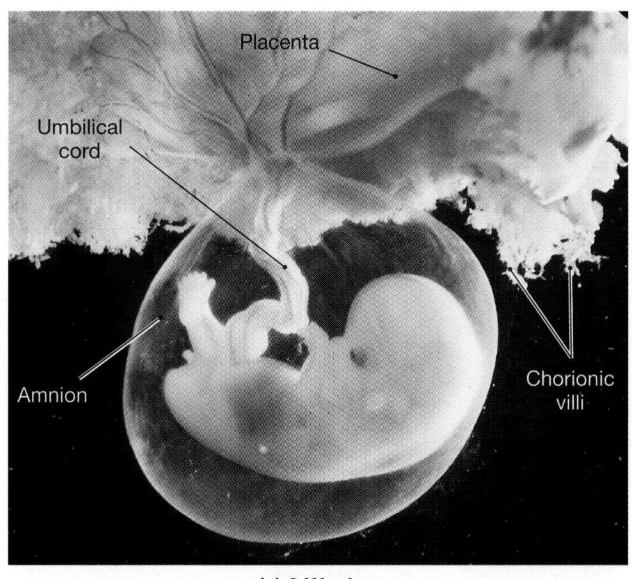

(c) 8 Weeks

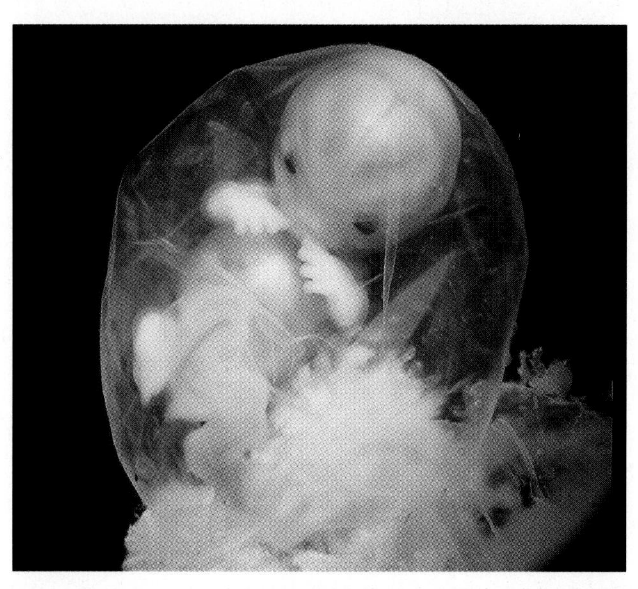

(d) 12 Weeks

TABLE 21-2 An Overview of Prenatal and Early Postnatal Development

Gestational Age (Months)	*Length and Weight*	*Integumentary System*	*Skeletal System*	*Muscular System*	*Nervous System*	*Special Sense Organs*
1	5 mm 0.02 g		(b) Somite formation	(b) Somite formation	(b) Neural tube	(b) Eye and ear formation
2	28 mm 2.7 g	(b) Nail beds, hair follicles, sweat glands	(b) Axial and appendicular cartilage formation	(c) Rudiments of axial musculature	(b) CNS, PNS organization, growth of cerebrum	(b) Taste buds, olfactory epithelium
3	78 mm 26 g	(b) Epidermal layers appear	(b) Ossification centers spreading	(c) Rudiments of appendicular musculature	(c) Basic spinal cord and brain structure	
4	133 mm 150 g	(b) Hair, sebaceous glands (c) Sweat glands	(b) Articulations (c) Facial and palatal organization	Fetus starts moving	(b) Rapid expansion of cerebrum	(c) Basic eye and ear structure (b) Peripheral receptor formation
5	185 mm 460 g	(b) Keratin production, nail production			(b) Myelination of spinal cord	
6	230 mm 823 g			(c) Perineal muscles	(b) CNS tract formation (c) Layering of cortex	
7	270 mm 1492 g	(b) Keratinization, nail formation, hair formation				(c) Eyelids open, retina sensitive to light
8	310 mm 2274 g		(b) Epiphyseal plate formation			(c) Taste receptors functional
9	346 mm 2912 g					
Postnatal development		Hair changes in consistency and distribution	Formation and growth of epiphyseal plates continue	Muscle mass and control increase	Myelination, layering, CNS tract formation continue	

Note: (b) = begin formation; (c) = complete formation.

Endocrine System	*Cardiovascular and Lymphatic Systems*	*Respiratory System*	*Digestive System*	*Urinary System*	*Reproductive System*
	(b) Heartbeat	(b) Trachea and lung formation	(b) Intestinal tract, liver, pancreas (c) yolk sac	(c) Allantois	
(b) Thymus, thyroid, pituitary, adrenal glands	(c) Basic heart structure, major blood vessels, lymph nodes and ducts (b) Blood formation in liver	(b) Extensive bronchial branching into mediastinum (c) Diaphragm	(b) Intestinal subdivisions, villi, salivary glands	(b) Kidney formation (adult form)	(b) Mammary glands
(c) Thymus, thyroid gland	(b) Tonsils, blood formation in bone marrow		(c) Gallbladder, pancreas		(b) Definitive gonads, ducts, genitalia
	(b) Migration of lymphocytes to lymphatic organs, blood formation in spleen			(b) Degeneration of embryonic kidneys	
	(c) Tonsils	(c) Nostrils open	(c) Intestinal subdivisions		
(c) Adrenal glands	(c) Spleen, liver, bone marrow	(b) Alveolar formation	(c) Epithelial organization, glands		
(c) Pituitary gland			(c) Intestinal plicae		(b) Testes descend
		Complete pulmonary branching and alveolar formation		Complete nephron formation at birth	Descent complete at or near time of delivery
	Cardiovascular changes at birth; immune system becomes operative thereafter	Alveoli inflate at birth			

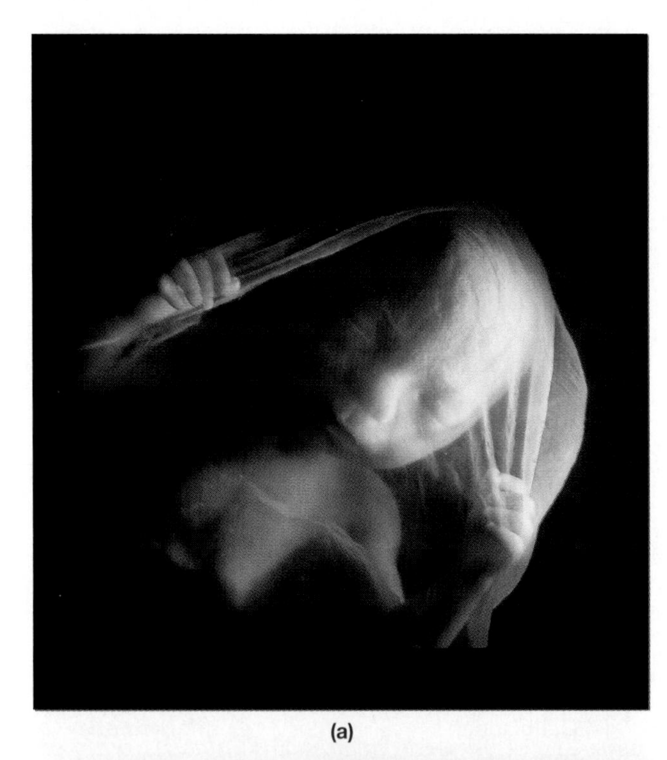

(a)

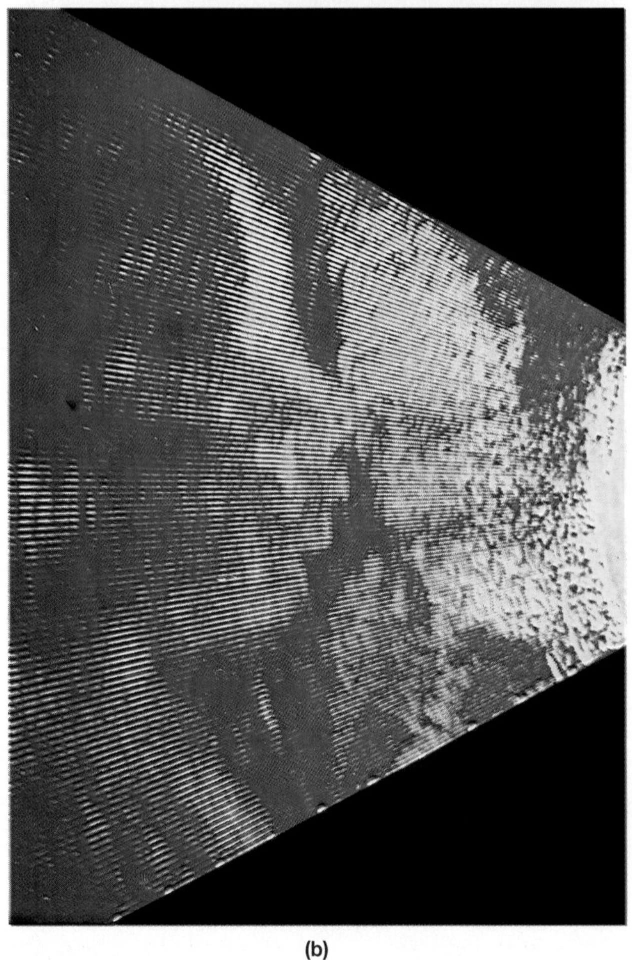

(b)

Figure 21-8
The Second and Third Trimesters
(a) A 4-month fetus seen through a fiber-optic endoscope.
(b) A 6-month fetus seen with ultrasound equipment.

- ***The maternal blood volume increases.*** This increase occurs because (1) blood flowing into the placenta reduces the volume in the rest of the systemic circuit, and (2) fetal activity lowers the blood P_{O_2} and elevates the P_{CO_2}. The combination stimulates the production of renin and erythropoietin (EPO), leading to an increase in maternal blood volume. By the end of gestation, the maternal blood volume has increased by almost 50 percent.
- ***The maternal requirements for nutrients and vitamins climb 10–30 percent.*** Pregnant women, who must "eat for two," are often hungry.
- ***The glomerular filtration rate increases by roughly 50 percent.*** This increase corresponds to the increase in blood volume, and it accelerates the excretion of metabolic wastes generated by the fetus. Because (1) the volume of urine produced increases and (2) the weight of the uterus presses down on the urinary bladder, pregnant women need to urinate frequently.
- ***The uterus undergoes a tremendous increase in size.*** Structural and functional changes in the expanding uterus are so important that we will discuss them in a separate section.
- ***The mammary glands increase in size and secretory activity begins.*** By the end of the sixth month of pregnancy, the mammary glands begin producing secretions that are stored in the duct system.

Structural and Functional Changes in the Uterus

At the end of gestation, a typical uterus will have grown from 7.5 cm (3 in.) in length and 60 g (2 oz) in weight to 30 cm (12 in.) in length and 1100 g (2.4 lb) in weight. It may then contain almost 5 liters of fluid, giving the organ with contents a total weight of roughly 10 kg (22 lb). This remarkable expansion occurs through the enlargement and elongation of existing cells (especially smooth muscle cells), rather than by an increase in the total number of cells in the uterus.

The tremendous stretching of the myometrium is associated with a gradual increase in the rates of spontaneous smooth muscle contractions. In the early stages of pregnancy, the contractions are weak, painless, and brief in duration. There are indications that the progesterone released by the placenta has an inhibitory effect on the uterine smooth muscle, preventing more extensive and powerful contractions.

Three major factors oppose the calming action of progesterone:

1. ***Rising estrogen levels.*** Estrogens, also produced by the placenta, increase the sensitivity of the uterine smooth muscles and make contractions

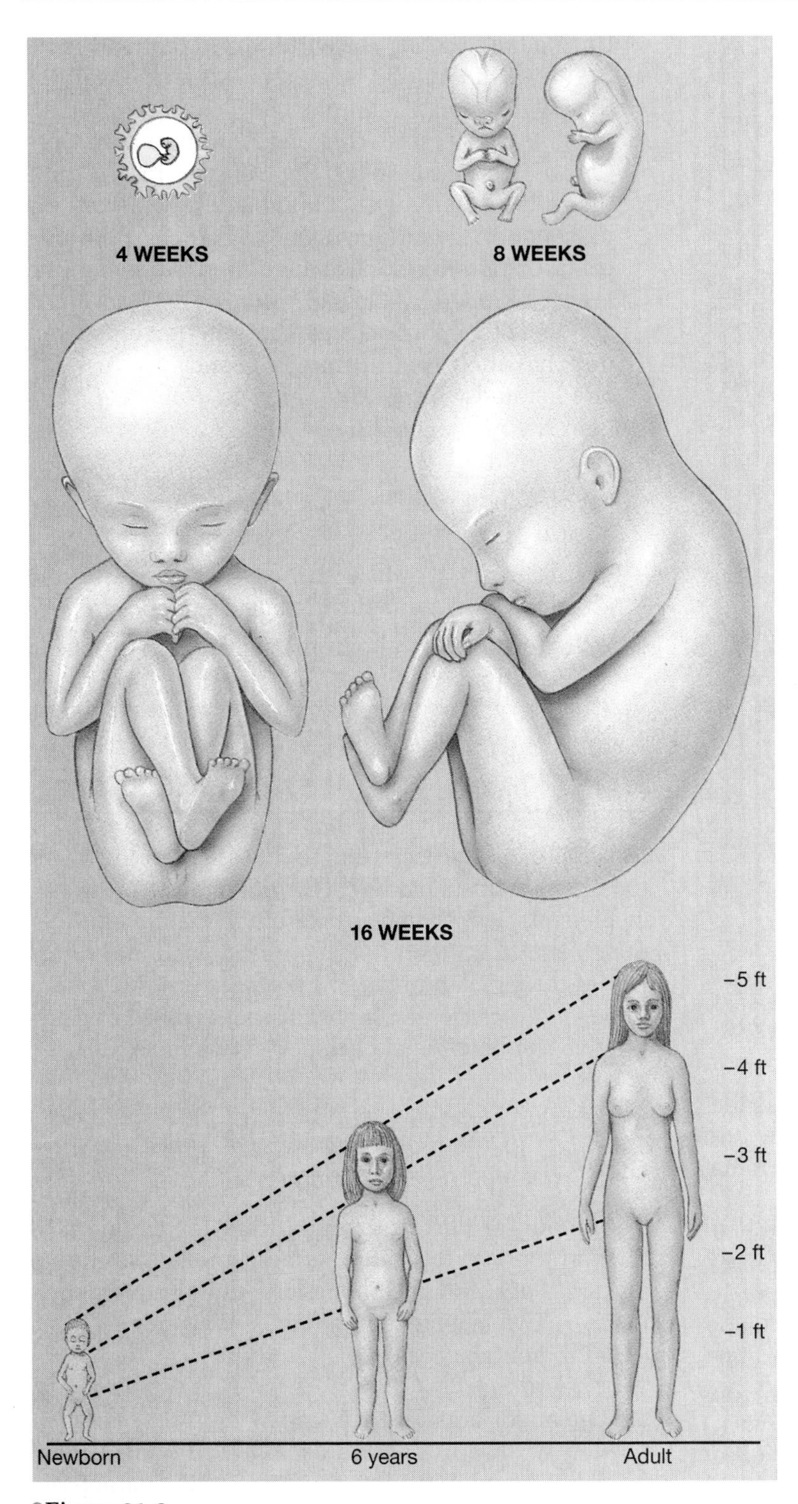

•Figure 21-9
Growth and Changes in Body Form

The views at 4, 8, and 16 weeks are presented at actual size. Notice the changes in body form and proportions as development proceeds. These changes do not stop at birth. For example, the head, which contains the brain and sense organs, is relatively large at birth.

more likely. Throughout pregnancy, progesterone exerts the dominant effect, but as the time of delivery approaches estrogen production accelerates, and the myometrium becomes more sensitive to stimulation.

2. ***Rising oxytocin levels.*** Rising oxytocin levels stimulate an increase in the force and frequency of uterine contractions. Oxytocin release is stimulated by high estrogen levels and by distortion of the uterine cervix.
3. ***Prostaglandin production.*** In addition to estrogens and oxytocin, uterine tissues late in pregnancy produce prostaglandins that stimulate smooth muscle contractions.

After 9 months of gestation, multiple factors interact to produce **labor contractions** in the myometrium of the uterine wall. Once begun, a positive feedback mechanism operates to ensure that the contractions continue until delivery has been completed.

Figure 21-10• diagrams important factors that stimulate and sustain labor. The actual trigger for the onset of labor may be events in the fetus rather than the mother. At the time labor begins, the fetal pituitary secretes oxytocin that is released into the maternal bloodstream at the placenta. The resulting increase in myometrial contractions and prostaglandin production, on top of maternal estrogens and oxytocin, may be the "last straw."

LABOR AND DELIVERY

The goal of labor is the forcible expulsion of the fetus, a process known as **parturition** (par-tū-RISH-un), or birth. During labor, each contraction begins near the top of the uterus and sweeps in a wave toward the cervix. These contractions are strong and occur at regular intervals. As parturition approaches, the contractions increase in force and frequency, changing the position of the fetus and moving it toward the cervical canal.

Stages of Labor

Labor has traditionally been divided into three stages (Figure 21-11•), the *dilation stage,* the *expulsion stage,* and the *placental stage.*

- The **dilation stage** (Figure 21-11a•) begins with the onset of labor, as the cervix dilates completely and the fetus begins to slide down the cervical canal. This stage may last 8 or more hours, but during this period the labor contractions occur at intervals of

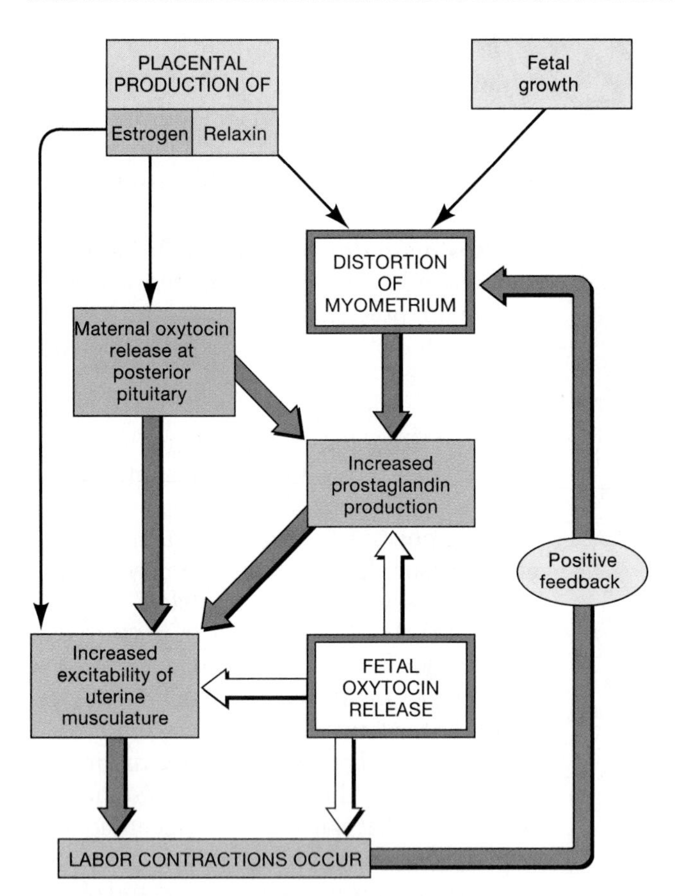

•**Figure 21-10**
Interacting Factors in Labor and Delivery

once every 10 to 30 minutes. Late in the process the amnion usually ruptures, an event sometimes referred to as "having the water break."

- The **expulsion stage** (Figure 21-11b•) begins as the cervix dilates completely, pushed open by the approaching fetus. Expulsion continues until the fetus has completed its emergence from the vagina, a period that usually lasts less than 2 hours. The arrival of the newborn infant into the outside world represents the birth, or **delivery**. If the vaginal entrance is too small to permit the passage of the fetus and there is acute danger of perineal tearing, the entryway may be temporarily enlarged by making an incision through the perineal musculature. After delivery, this **episiotomy** (e-pēz-ē-OT-o-mē) can be repaired with sutures, a much simpler procedure than dealing with a potentially extensive perineal tear. If unexpected complications arise during the dilation or expulsion stage, the infant may be removed by **cesarean section.** In such cases, an incision is made through the abdominal wall, and the uterus is opened just enough to allow passage of the infant's head. This procedure is performed during 15–25 percent of the deliveries in the United States—more often than necessary according to some studies. Efforts are now being made to reduce the frequency of both episiotomies and cesarean sections.
- During the third, or **placental,** stage of labor (Figure 21-11c•), the muscle tension builds in the walls of the partially empty uterus, and the organ gradually decreases in size. This uterine contraction tears the connections between the endometrium and the placenta. Usually within an hour after delivery the placental stage ends with the ejection of the placenta, or "afterbirth." The disruption of the placenta is accompanied by a loss of blood, perhaps as much as 500–600 ml, but because the maternal blood volume has increased during pregnancy the loss can be tolerated without difficulty. AM *Common Problems with Labor and Delivery*

✓ Why does a mother's total blood volume increase during pregnancy?

✓ What effect would a decrease in progesterone have on the uterus during late pregnancy?

✓ What role does relaxin play in inducing uterine contractions?

POSTNATAL DEVELOPMENT

Developmental processes do not cease at delivery, for the newborn infant has few of the anatomical, functional, or physiological characteristics of the mature adult. In the course of postnatal development, each individual passes through a number of **life stages** (*infancy, childhood, adolescence,* and *maturity*) each typified by a distinctive combination of characteristics and abilities.

The Neonatal Period, Infancy, and Childhood

The **neonatal period** extends from the moment of birth to 1 month thereafter. **Infancy** then continues to 2 years of age, and **childhood** lasts until puberty commences. Two major events are under way during these developmental stages:

1. The major organ systems other than those associated with reproduction become fully operational and gradually acquire the functional characteristics of adult structures.
2. The individual grows rapidly, and there are significant changes in body proportions.

Pediatrics is the medical specialty that focuses on this period of life, from birth through childhood, and also adolescence. Because infants and young children often cannot clearly describe the problems they are experiencing, pediatricians and parents must be skilled observers. Standardized testing procedures are used to assess an individual's developmental progress

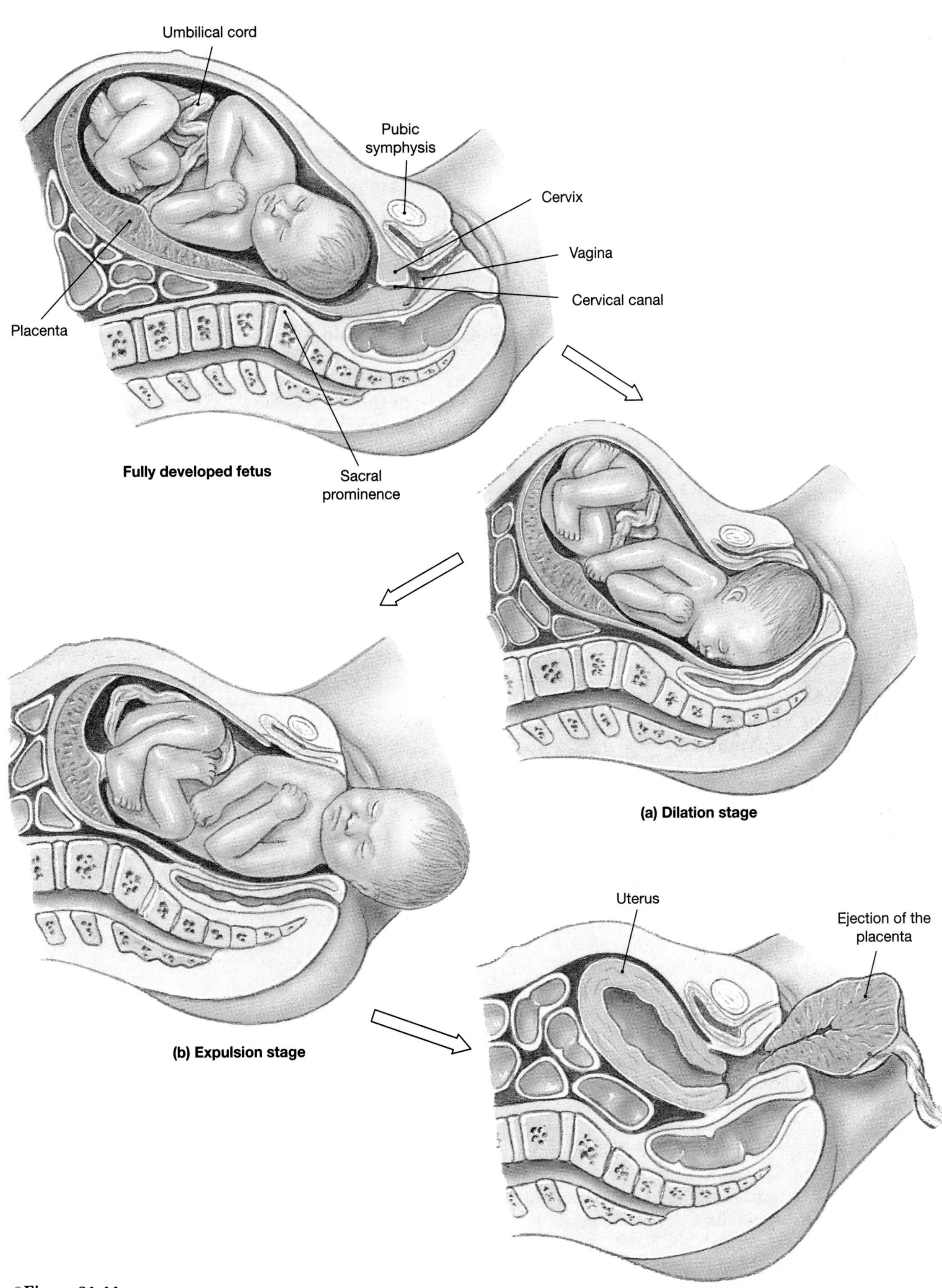

Figure 21-11 The Stages of Labor

relative to normal values. AM *Monitoring Postnatal Development*

The Neonatal Period

A variety of physiological and anatomical alterations occur as the fetus completes the transition to the status of a newborn infant, or **neonate.** Prior to delivery, transfer of dissolved gases, nutrients, waste products, hormones, and immunoglobulins occurred across the placenta. At birth, the newborn infant must become relatively self-sufficient, with the processes of respiration, digestion, and excretion performed by its own specialized organs and organ systems. The changes in the cardiovascular and respiratory systems at birth were detailed in Chapter 14. *[p. 375]* Typical heart rates of 120–140 beats per minute and respiratory rates of 30 breaths per minute in neonates are considerably higher than those of adults.

Before birth, the digestive system remains relatively inactive, although it does accumulate a mixture of bile secretions, mucus, and epithelial cells. This collection of debris is excreted in the first few days of life. Over that period the newborn infant begins to nurse.

As waste products build up in the arterial blood, they are filtered into the filtrate at the glomeruli of the kidneys. Glomerular filtration is normal, but the tubular fluid cannot be concentrated to any significant degree. As a result, urinary water losses are high, and neonatal fluid requirements are relatively greater than those of adults.

The neonate has little ability to control body temperature, particularly in the first few days after delivery. For this reason, newborn infants are usually kept bundled up in warm coverings. As the infant grows larger and increases the thickness of its insulating adipose "blanket," its metabolic rate also rises, and thermoregulatory abilities improve. Nevertheless, daily and even hourly alterations in body temperature continue throughout childhood.

Lactation and the Mammary Glands By the end of the sixth month of pregnancy the mammary glands are fully developed, and the gland cells begin producing a secretion known as **colostrum** (ko-LOS-trum). Colostrum, which is provided to the infant during the first 2 or 3 days of life, contains relatively more proteins and far less fat than milk. Many of the proteins are antibodies that help the infant ward off infections until its own immune system becomes fully functional. As colostrum production declines, milk production increases. Milk consists of a mixture of water, proteins, amino acids, lipids, sugars, and salts. It also contains large quantities of lysozymes, enzymes with antibiotic properties.

The actual secretion of the mammary glands is triggered when the infant begins to suck on the nipple. Stimulation of tactile receptors there leads to the release of oxytocin at the posterior pituitary. Oxytocin causes cells within the lactiferous ducts and sinuses to contract. This contraction results in the ejection of milk, or *milk let-down* (Figure 21-12●). The milk let-down reflex continues to function until weaning occurs, typically 1–2 years after birth.

Infancy and Childhood

The most rapid growth occurs during prenatal development, and after delivery the relative rate of growth continues to decline. Postnatal growth during infancy and childhood occurs under the direction of circulating hormones, notably growth hormone from the pituitary, adrenal steroids, and thyroid hormones. These hormones affect each tissue and organ in specific ways, depending on the sensitivities of the individual cells. As a result, growth does not occur uniformly, and the body proportions gradually change (see Figure 21-9●, p. 557).

Adolescence and Maturity

Adolescence begins at puberty, when three events interact to promote increased hormone production and sexual maturation:

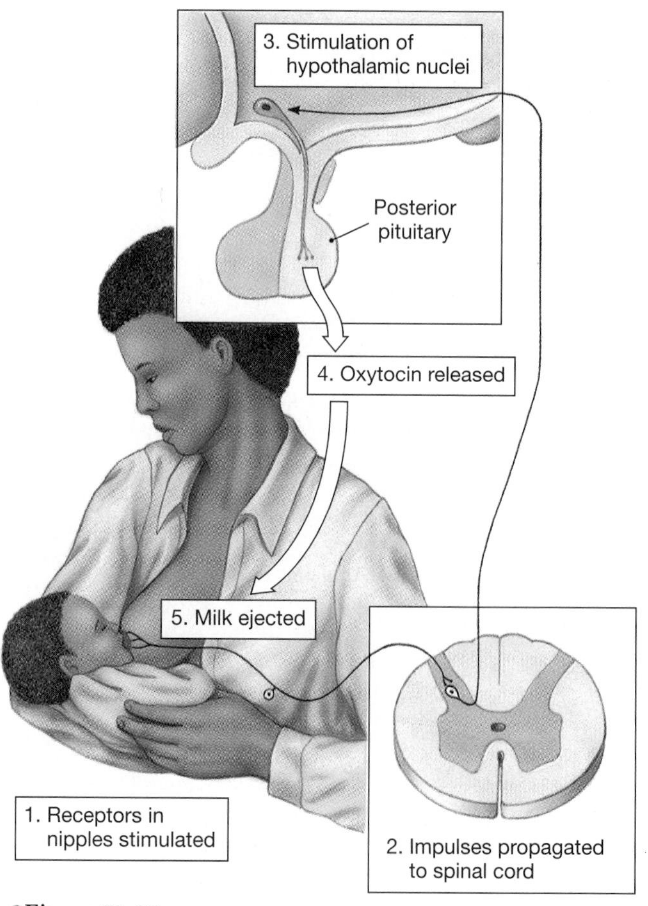

●**Figure 21-12**
The Milk Let-Down Reflex

1. The hypothalamus increases its production of gonadotropin-releasing hormone (GnRH).
2. The anterior pituitary becomes more sensitive to the presence of GnRH, and there is a rapid elevation in the circulating levels of FSH and LH.
3. Ovarian or testicular cells become more sensitive to FSH and LH. These changes initiate gametogenesis and the production of male or female sex hormones that stimulate the appearance of secondary sexual characteristics and behaviors.

In the years that follow, the continued background secretion of estrogens or androgens maintains these sexual characteristics. In addition, the combination of sex hormones and growth hormone, adrenal steroids, and thyroxine leads to a sudden acceleration in the growth rate. The timing of the increase in size varies between the sexes, corresponding to different ages at the onset of puberty. In girls, the growth rate is maximum between ages 10 and 13, whereas boys grow most rapidly between ages 12 and 15. Growth continues at a slower pace until ages 18 to 21, when most of the epiphyseal plates close. ∞ [p. 300]

The boundary between adolescence and maturity is very hazy, for it has physical, emotional, behavioral, and legal implications. **Maturity** is often associated with the end of growth in the late teens or early twenties. Although development ends at maturity, physiological changes continue. These changes are part of the process of aging, or **senescence.** Aging reduces the efficiency and capabilities of the individual, and even in the absence of other factors will ultimately lead to death. AM *Death and Dying*

GENETICS, DEVELOPMENT, AND INHERITANCE

Every somatic cell in the body carries copies of the original 46 chromosomes present in the fertilized egg or zygote. Those chromosomes and their component genes represent the individual's **genotype** (JĒN-ō-tīp). Through development and differentiation, the instructions contained within the genotype are expressed in many different ways. No single living cell or tissue makes use of all the information and instructions contained within the genotype. For example, in muscle fibers the genes important for excitable membrane formation and contractile proteins are active, while a different set of genes is operating in the cells of the pancreatic islets. But the instructions contained within the genotype determine the anatomical and physiological characteristics of each individual. Those visible characteristics are known as the individual's **phenotype** (FĒN-ō-tīp; *phainein,* to display + *typos*, mark).

Your genotype is derived from those of your parents, but not in a simple way. You are not an exact copy of either parent, nor are you an easily identifiable mixture of their characteristics. Our discussion will begin with the basic patterns and their implications and then will examine the mechanisms responsible for regulating the activities of the genotype during subsequent prenatal development.

Genes and Chromosomes

Chromosome structure and the functions of genes were introduced in Chapter 3. ∞ [pp. 62, 68] Chromosomes contain DNA, and genes are segments of DNA. Each gene carries the information needed to direct the synthesis of a specific polypeptide.

Every somatic cell contains 23 pairs of chromosomes. One member of each pair was contributed by the sperm, and the other by the ovum. The members of each pair are known as **homologous** (hō-MOL-o-gus) **chromosomes.** Twenty-two of those pairs are known as **autosomal** (aw-to-SŌ-mal) **chromosomes.** The chromosomes of the twenty-third pair are called the *sex chromosomes* because they differ in the two sexes.

Autosomal Chromosomes

The two chromosomes in an autosomal pair have the same structure and carry genes that affect the same traits. If one member of the pair contains three genes in a row, with number 1 determining hair color, number 2 eye color, and number 3 skin pigmentation, the other chromosome will carry genes affecting the same traits, and in the same sequence.

The various forms of any one gene are called **alleles** (a-LĒLS). If both chromosomes of a homologous pair carry the same allele of a particular gene, the individual is **homozygous** (hō-mō-ZĪ-gus; *homos*, same) for that trait. For example, if a zygote receives a gene for curly hair from the sperm and one for curly hair from the egg, the individual will be homozygous for curly hair. *If you are homozygous for a particular trait, your phenotype will have that characteristic.* Usually about 80 percent of an individual's total set of genes consists of homozygous alleles. Because the chromosomes of a homologous pair have different origins, one paternal and the other maternal, they do not *have* to carry the same alleles. When an individual has two different alleles carrying different instructions, the individual is **heterozygous** (het-er-ō-ZĪ-gus; *heteros*, other) for that trait. In that case, the phenotype will be determined by the interactions between the corresponding alleles.

- If an allele is **dominant,** it will be expressed in the phenotype *regardless of any conflicting instructions carried by the other allele.*

- If an allele is **recessive,** it will be expressed in the phenotype only if it is present on both chromosomes of a homologous pair. For example, the albino skin condition is characterized by an inability to synthesize a yellow-brown pigment, *melanin.* A single dominant allele determines normal skin coloration; two recessive alleles must be present to produce an albino individual.

Predicting Inheritance Not every allele can be neatly characterized as dominant or recessive. Some that can be are included in Table 21-3. If you restrict attention to these alleles, it is possible to predict the characteristics of individuals on the basis of those of their parents.

Dominant traits are traditionally indicated by capitalized abbreviations, and recessives are abbreviated in lower case. For a given trait, the possibilities are indicated by *AA* (homozygous dominant), *Aa* (heterozygous), or *aa* (homozygous recessive). The gametes involved in fertilization each contribute a single allele for a given trait. That allele must be one of the two contained by all other cells in the parental body. Consider, for example, the offspring of an albino mother and a normal father. Because albinism is a recessive trait, the maternal alleles can be abbreviated *aa*. No matter which of her ova gets fertilized, it will carry the recessive *a* gene. The father has normal coloration, and this is a dominant trait. He may therefore be homozygous *or* heterozygous for this trait, since *AA* or *Aa* will give rise to the same phenotype.

A simple box diagram known as a *Punnett square* lets us predict the probabilities that the children will have particular characteristics by showing us the possible combinations of parental alleles that they can inherit. In the square shown in Figure 21-13•, the maternal alleles are listed along the horizontal axis, and the paternal ones along the vertical axis. The possible combinations are indicated in the small boxes. Figure 21-13a• considers the possible offspring of an *aa* mother and an *AA* father. All of the children must have the genotype *Aa,* and they will all have normal skin coloration. Compare these results with those of Figure 21-13b•, for a heterozygous father (*Aa*). The heterozygous individual produces two types of gametes, *A* and *a,* and the egg may be fertilized by either one. As a result, there is a 50 percent probability that a child of such a father will inherit the genotype *Aa,* and so have normal skin color. The probability of inheriting the genotype *aa,* and thus having the albino phenotype, is also 50 percent. The Punnett square can also be used in reverse, to draw conclusions about the identity and genotype of a parent. For example, a man with the genotype *AA* cannot be the father of an albino child (*aa*).

TABLE 21-3 The Inheritance of Selected Phenotypic Characteristics

DOMINANT TRAITS
One allele determines phenotype, and the other is suppressed
normal skin coloration
brachydactyly (short fingers)
ability to taste phenylthiocarbamate (PTC)
free earlobes
curly hair
presence of Rh factor on red blood cell membranes
Both dominant alleles may be expressed (codominance)
presence of A or B antigens on red blood cell membranes
structure of serum proteins (albumins, transferrins)
structure of hemoglobin molecule
RECESSIVE TRAITS
albinism
blond hair
red hair (expressed only if individual is also homozygous for blond hair)
lack of A, B agglutinogens (Type O blood)
inability to roll the tongue into a U-shape
SEX-LINKED TRAITS
color blindness
POLYGENIC TRAITS
eye color
hair colors other than pure blond or red

Simple Inheritance In **simple inheritance,** phenotypic characters are determined by interactions between a single pair of alleles. The frequency of appearance of an inherited disorder resulting from simple inheritance can be predicted using a Punnett square. Although they are rare disorders in terms of overall numbers, more than 1200 different inherited conditions have been identified that reflect the presence of one or two abnormal alleles for a single gene. A partial listing is included in Table 21-4, along with the location where additional information can be obtained.

Polygenic Inheritance **Polygenic inheritance** involves interactions between alleles on several genes. Because multiple alleles are involved, the frequency of occurrence cannot easily be predicted using a simple Punnett square. Several important adult disorders, including hypertension and coronary artery disease, fall within this category. Many of the developmental disorders responsible for fetal mortalities and congenital malformations also result from multiple ge-

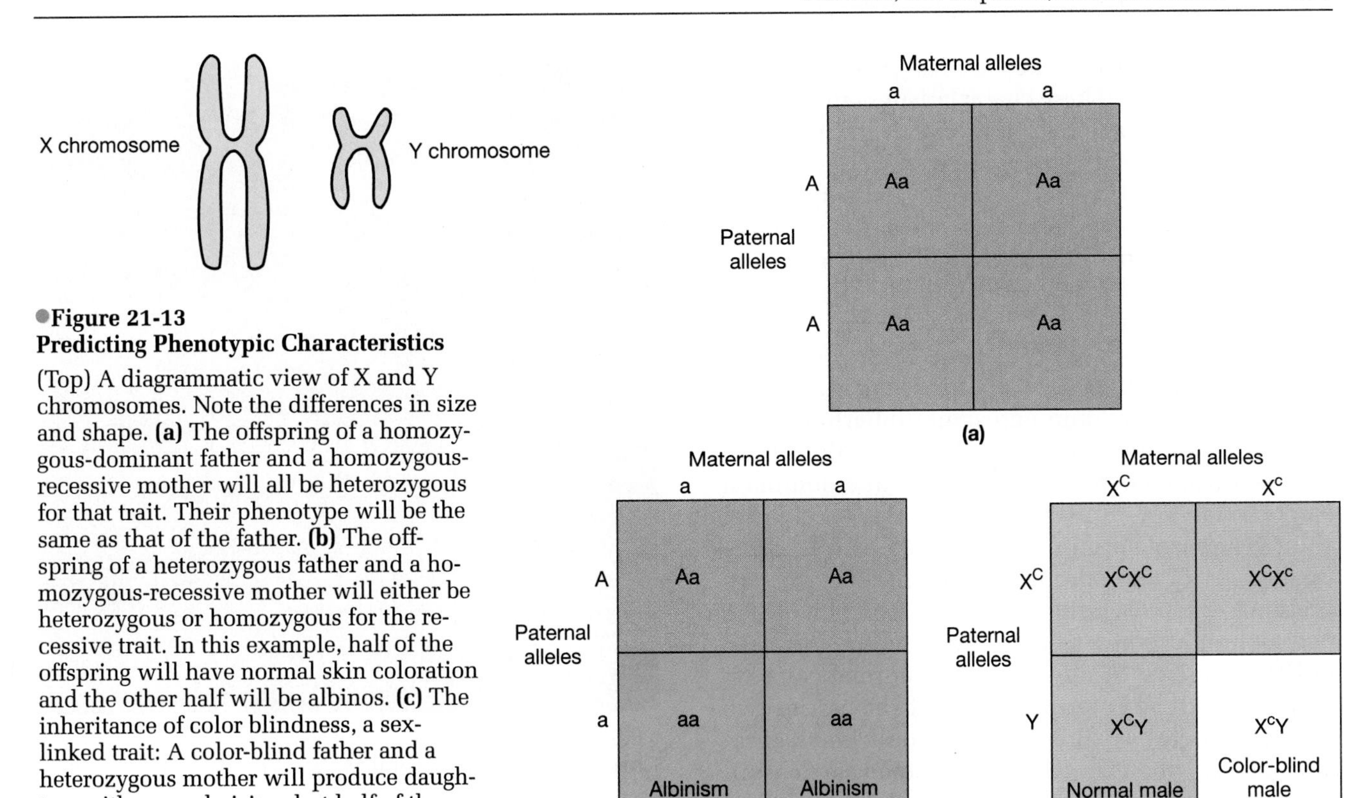

Figure 21-13
Predicting Phenotypic Characteristics

(Top) A diagrammatic view of X and Y chromosomes. Note the differences in size and shape. **(a)** The offspring of a homozygous-dominant father and a homozygous-recessive mother will all be heterozygous for that trait. Their phenotype will be the same as that of the father. **(b)** The offspring of a heterozygous father and a homozygous-recessive mother will either be heterozygous or homozygous for the recessive trait. In this example, half of the offspring will have normal skin coloration and the other half will be albinos. **(c)** The inheritance of color blindness, a sex-linked trait: A color-blind father and a heterozygous mother will produce daughters with normal vision, but half of the sons will be color-blind.

TABLE 21-4 Relatively Common Inherited Disorders

Disorder	*Page in Text or Applications Manual*
AUTOSOMAL DOMINANTS	
Adult polycystic kidney disease	*AM*
Marfan's syndrome	p. 83 and *AM*
Huntington's disease	p. 278 and *AM*
AUTOSOMAL RECESSIVES	
Deafness	p. 275
Albinism	p. 100
Sickle cell anemia	p. 314 and *AM*
Cystic fibrosis	p. 409 and *AM*
Phenylketonuria	p. 472 and *AM*
Tay-Sachs disease	*AM*
X-LINKED	
Duchenne's muscular dystrophy	p. 157 and *AM*
Myotonic muscular dystrophy	*AM*
Hemophilia (one form)	p. 324
Color blindness	p. 264

Note: For the diseases in blue, the genetic basis has been identified.

netic interactions. In these cases, the particular genetic composition of the individual does not by itself determine the onset of the disease. Instead, the conditions regulated by these genes establish a susceptibility to particular environmental influences. This means that not every individual with the genetic tendency for a particular condition will actually develop it. It is therefore difficult to track polygenic conditions through successive generations. However, because many inherited polygenic conditions are *likely* but not *guaranteed* to occur, steps can be taken to prevent a crisis. For example, hypertension may be prevented or reduced by controlling diet and fluid volume, and coronary artery disease may be prevented by lowering serum cholesterol concentrations.

Sex Chromosomes

The chromosomes of the twenty-third pair are called the **sex chromosomes** because they determine the genetic sex of the individual. Unlike other chromosomal pairs, the sex chromosomes are not necessarily identical in appearance and gene content. There are two different sex chromosomes, an **X chromosome** and a **Y chromosome.** The Y chromosome is considerably smaller than the X chromosome and contains fewer genes, but among those genes are dominant alleles that specify that an individual with that chromosome will be a male. The normal male chromosome pair is *XY*, and the female pair is *XX*. The ova produced by a woman will always carry *X*, and sperm may carry *X* or *Y*.

The X chromosome also carries genes that affect somatic structures. These characteristics are called **X-linked** because in most cases there are no corresponding alleles on the Y chromosome. They are also known as **sex-linked** traits because the responsible genes are located on the sex chromosomes. The best known X-linked or sex-linked characteristics are associated with noticeable diseases or defects that are caused by single alleles.

The inheritance of color blindness, a condition discussed in Chapter 10, exemplifies the differences between sex-linked and autosomal inheritance. [p. 264] A relatively common form of color blindness is associated with the presence of a dominant or recessive gene on the X chromosome. Normal color vision is determined by the presence of a dominant gene, *C*, and color blindness results from the presence of the recessive gene *c*. A woman, with her two X chromosomes, can be either homozygous, *CC*, *or* heterozygous, *Cc*, and still have normal color vision. She will be color-blind only if she carries two recessive alleles, *cc*. But a male has only one X chromosome, so whatever that chromosome carries will determine whether he has normal color vision or is color-blind. A Punnett square for an X-linked trait, as in Figure 21-13c●, reveals that the sons produced by a normal father and a heterozygous mother will have a 50 percent chance of being color-blind, while the daughters will all have normal color vision. A number of other clinical disorders are X-linked traits, including certain forms of *hemophilia, diabetes insipidus,* and *muscular dystrophy.* In several instances, advances in molecular genetic techniques have permitted us to locate specific genes on the X chromosome. This technique provides a relatively direct method of screening for the presence of a particular condition before the symptoms appear and even before birth.

The Human Genome Project

Few of the genes responsible for inherited disorders have been identified or even localized to a specific chromosome. However, that situation is changing rapidly, owing to the attention devoted to the **Human Genome Project (HGP).** This project, funded by the National Institutes of Health and the Department of Energy, is attempting to transcribe the entire human genome, chromosome by chromosome and gene by gene. The project began in October 1990 and was expected to take 10–15 years. Progress has been more rapid than expected, and it may actually take considerably less time.

The first step in understanding the human genome is to prepare a map of the individual chromosomes. **Karyotyping** (KAR-ē-ō-tī-ping; *karyon,* nucleus + *typos,* mark) is the determination of an individual's chromosome complement. Figure 21-14●, shows a set of normal human chromosomes. Each chromosome has characteristic banding patterns, and segments can be stained with special dyes. The banding patterns are useful as reference points when more-detailed genetic maps are prepared. The banding patterns themselves can be useful, as abnormal banding patterns are characteristic of some genetic disorders and several cancers, including a form of leukemia.

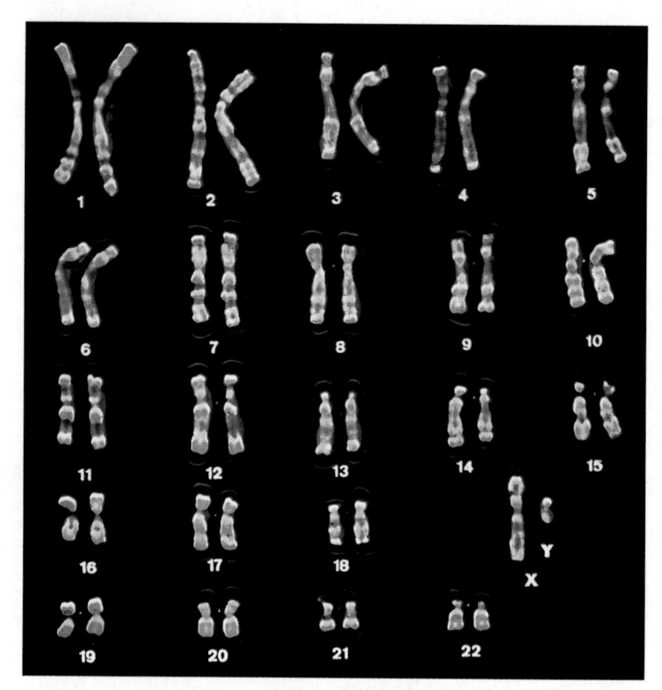

●**Figure 21-14**
Chromosomes of a Normal Male

As of 1996, HGP progress includes the following:

- Eight chromosomes—chromosomes 3, 11, 12, 16, 19, 21, 22, and the Y chromosome—have been mapped completely, and preliminary maps have been made for all other chromosomes.
- Over 3700 genes have been identified. Although a significant number, this is only a small fraction of the estimated 100,000 genes in the human genome.
- The specific genes responsible for more than 60 inherited disorders have been identified, including 5 of the disorders listed in Table 21-4. Genetic screening can now be done for many of these conditions.

The Human Genome Project is attempting to determine the normal genetic composition of a "typical" human being. Yet we are all variations on a basic theme. As we improve our abilities to manipulate our own genetic foundations, we will face troubling ethical and legal decisions. For example, few people object to the insertion of a "correct" gene into somatic cells to cure a specific disease. But what if we could insert that modified gene into a gamete and change not

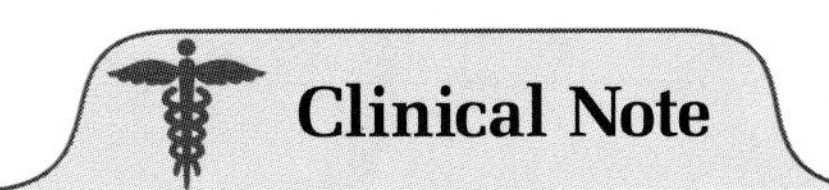

Chromosomal Abnormalities and Genetic Analysis

There are two types of abnormalities in autosomal chromosomes that do not invariably kill the individual before birth. These are *translocation defects* and *trisomy.*

- In a **translocation defect,** crossing-over occurs between different chromosome pairs. For example, a piece of chromosome 8 may become attached to chromosome 14. The genes moved to their new position may function abnormally, becoming inactive or overactive. A translocation between chromosomes 8 and 14 is responsible for *Burkitt's lymphoma,* a type of lymphatic system cancer.
- In **trisomy,** something goes wrong in meiosis II, and the gamete that contributes to the zygote carries an extra copy of one chromosome. Because there are three copies of this chromosome, rather than two, the condition is termed *trisomy.* The nature of the trisomy is indicated by the number of the chromosome involved. Zygotes with extra copies of chromosomes seldom survive. Individuals with trisomy 13 and trisomy 18 may survive until delivery, but rarely live longer than a year. The notable exception is trisomy 21.

Trisomy 21, or **Down syndrome,** is the most common viable chromosomal abnormality. Estimates of the frequency of appearance range from 1.5 to 1.9 per 1000 births for the U.S. population. The affected individual suffers from mental retardation and characteristic physical malformations, including a facial appearance that gave rise to the term *mongolism* once used to describe this condition. The degree of mental retardation ranges from moderate to severe, and few individuals with this condition lead independent lives. Anatomical problems affecting the cardiovascular system often prove fatal during childhood or early adulthood. Although some individuals survive to moderate old age, many develop Alzheimer's disease while still relatively young (before age 40).

For unknown reasons, there is a direct correlation between maternal age and the risk of having a child with trisomy 21. Below maternal age 25 the incidence of Down syndrome approaches 1 in 2000 births, or 0.05 percent. For maternal ages 30–34, the odds increase to 1 in 900, and over the next decade they go from 1 in 290 to 1 in 46, or more than 2 percent. These statistics are becoming increasingly significant, for many women have delayed childbearing until their mid-30s or later.

Abnormal numbers of sex chromosomes do not produce effects as severe as those induced by extra or missing autosomal chromosomes. In **Klinefelter syndrome,** the individual carries the sex chromosome pattern *XXY.* The phenotype is male, but the extra X chromosome causes reduced androgen production. As a result, the testes fail to mature, the individuals are sterile, and the breasts are slightly enlarged. The incidence of this condition among newborn males averages 1 in 750 births.

Individuals with **Turner syndrome** have only a single female sex chromosome; their sex chromosome complement is abbreviated *XO.* This kind of chromosomal deletion is known as **monosomy.** The incidence of this condition at delivery has been estimated as 1:10,000 live births. The condition may not be recognized at birth, for the phenotype is normal female, but maturational changes do not appear at puberty. The ovaries are nonfunctional, and estrogen production occurs at negligible levels.

Chromosomal Analysis

In **amniocentesis,** a sample of amniotic fluid is removed, and the fetal cells that it contains are analyzed. This procedure permits the identification of more than 20 congenital conditions. The needle inserted to obtain a fluid sample is guided into position using ultrasound. Unfortunately, amniocentesis has two major drawbacks:

1. Because the sampling procedure represents a potential threat to the health of the fetus and mother, amniocentesis is performed only when known risk factors are present. Examples of risk factors would include a family history of specific conditions, or in the case of Down syndrome, a maternal age over 35.
2. Sampling cannot safely be performed until the volume of amniotic fluid is large enough that the fetus will not be injured during the sampling process. The usual time for amniocentesis is at a gestational age of 14–15 weeks. It may take several weeks to obtain results once samples have been collected, and by the time the results are received, the option of therapeutic abortion may no longer be available.

An alternative procedure known as **chorionic villi sampling** analyzes cells collected from the villi during the first trimester. Although it can be performed at an earlier gestational age, this technique has largely been abandoned because of an associated increased risk of spontaneous abortion.

only that individual but all his or her descendants? And what if the gene did not correct or prevent a disorder, but "improved" the individual by increasing intelligence, height, vision, or altering some other phenotypic characteristic? These and other difficult questions will not go away, and in the years to come we will have to find answers with which we all can live.

✓ Curly hair is an autosomal dominant trait. What would be the phenotype of a person who is heterozygous for this trait?

✓ Joe has three daughters and complains that it's his wife's fault that he doesn't have any sons. What would you tell him?

Chapter Review

KEY TERMS

blastocyst, p. 546
embryo, p. 544
fetus, p. 544
genotype, p. 561
gestation, p. 546
heterozygous, p. 561
homozygous, p. 561
implantation, p. 546
parturition, p. 557
phenotype, p. 561
syncytium, p. 547
trophoblast, p. 546

SUMMARY OUTLINE

INTRODUCTION p. 544

1. **Development** is the gradual modification of physical and physiological characteristics from conception to maturity. The creation of different cell types is **differentiation.**

AN OVERVIEW OF TOPICS IN DEVELOPMENT p. 544

1. **Prenatal development** occurs before birth; **postnatal development** begins at birth and continues to maturity, when **senescence** (aging) begins. **Inheritance** refers to the transfer of genetically determined characteristics from generation to generation. **Genetics** is the study of the mechanisms of inheritance.

FERTILIZATION p. 544

1. Fertilization normally occurs in the uterine tube within a day after ovulation. Sperm cannot fertilize an egg until they have undergone **capacitation.**

The Oocyte at Ovulation p. 544

2. The acrosomal caps of the spermatozoa release hyaluronidase, an enzyme that separates cells of the corona radiata and exposes the oocyte membrane. When a single spermatozoon contacts that membrane, fertilization occurs and oocyte **activation** follows.

3. During activation, the secondary oocyte completes meiosis, and the penetration of additional sperm is prevented.

4. After activation, the *female pronucleus* and *male pronucleus* fuse in a process called *amphimixis.* (*Figure 21-1*)

A Preview of Prenatal Development p. 546

5. The 9-month **gestation** period can be divided into three **trimesters.**

THE FIRST TRIMESTER p. 546

1. **Cleavage** subdivides the cytoplasm of the zygote in a series of mitotic divisions; the zygote becomes a blastocyst. During **implantation,** the blastocyst burrows into the uterine endometrium. **Placentation** occurs as blood vessels form around the blastocyst and the **placenta** appears.

2. Infertility is relatively common, but recent advances are providing new solutions.

Cleavage and Blastocyst Formation p. 546

3. The **blastocyst** consists of an outer **trophoblast** and an **inner cell mass.** (*Figure 21-2*)

Implantation p. 546

4. Implantation occurs about 7 days after fertilization as the blastocyst adheres to the uterine lining. (*Figure 21-3*)

5. As the trophoblast enlarges and spreads, maternal blood flows through open *lacunae.* After **gastrulation** the **blastodisc** contains an embryo composed of **endoderm, ectoderm,** and an intervening **mesoderm.** It is from these three **germ layers** that the body systems differentiate. (*Figure 21-4; Table 21-1*)

6. The germ layers help form four **extraembryonic membranes:** the yolk sac, amnion, allantois, and chorion. (*Figure 21-5*)

7. The **yolk sac** is an important site of blood cell formation. The **amnion** encloses fluid that surrounds and cushions the developing embryo. The base of the **allantois** later gives rise to the urinary bladder. Circulation within the vessels of the **chorion** provides a rapid-transit system linking the embryo with the trophoblast.

Placentation p. 548

8. *Chorionic villi* extend outward into the maternal tissues, forming an intricate, branching network through which maternal blood flows. As development proceeds, the **umbilical cord** connects the fetus to the placenta. The placenta synthesizes hCG, estrogens, progestins, hPL, and relaxin. (*Figure 21-6*)

Embryogenesis p. 552

9. The first trimester is critical because events in the first 12 weeks establish the basis for **organogenesis** (organ formation). (*Figure 21-7; Table 21-2*)

THE SECOND AND THIRD TRIMESTERS p. 552

1. In the second trimester, vital organ systems are near functional completion. During the third trimester these organ systems become functional. (*Figures 21-8, 21-9*)

Pregnancy and Maternal Systems p. 552

2. The developing fetus is totally dependent on maternal organs for nourishment, respiration, and waste removal. Maternal adaptations include increased blood volume, respiratory rate, tidal volume, nutrient intake, and glomerular filtration.

Structural and Functional Changes in the Uterus p. 556

3. Progesterone produced by the placenta has an inhibitory effect on uterine muscles; its calming action is opposed by estrogens, oxytocin, and prostaglandins. At some point multiple factors interact to produce **labor contractions** in the uterine wall. (*Figure 21-10*)

LABOR AND DELIVERY p. 557

1. The goal of labor is **parturition,** the forcible expulsion of the fetus.

Stages of Labor p. 557

2. Labor can be divided into three stages: the **dilation stage, expulsion stage,** and **placental stage.** (*Figure 21-11*)

POSTNATAL DEVELOPMENT p. 558

1. Postnatal development involves a series of **life stages,** including **infancy, childhood, adolescence,** and **maturity. Senescence** begins at maturity and ends in the death of the individual.

The Neonatal Period, Infancy, and Childhood p. 558

2. The **neonatal period** extends from birth to 1 month of age. **Infancy** then continues to 2 years of age, and **childhood** lasts until puberty commences. During these stages major organ systems (other than reproductive) become fully operational and gradually acquire adult characteristics, and the individual grows rapidly.

3. In the transition from fetus to **neonate,** the respiratory, circulatory, digestive, and urinary systems begin functioning independently. The newborn must also begin thermoregulating.

4. Mammary glands produce protein-rich *colostrum* during the infant's first few days and then convert to milk production. These secretions are released as a result of the milk let-down reflex. (*Figure 21-12*)

Adolescence and Maturity p. 560

5. Adolescence begins at puberty when: (1) the hypothalamus increases its production of GnRH; (2) circulating levels of FSH and LH rise rapidly; and (3) ovarian or testicular cells become more sensitive to FSH and LH. These changes initiate gametogenesis, production of sex hormones, and a sudden acceleration in growth rate.

GENETICS, DEVELOPMENT, AND INHERITANCE p. 561

1. Every somatic cell carries copies of the original 46 chromosomes in the zygote; these represent the individual's **genotype.** The physical expression of the genotype is the **phenotype** of the individual.

Genes and Chromosomes p. 561

2. Every somatic human cell contains 23 pairs of chromosomes; each pair consists of **homologous chromosomes;** 22 pairs are **autosomal chromosomes,** and the twenty-third pair of chromosomes are called the sex chromosomes because they differ in the two sexes.

3. Chromosomes contain DNA, and genes are functional segments of DNA. The various forms of a gene are called **alleles.** If both homologous chromosomes carry the same allele of a particular gene, the individual is **homozygous;** if they carry different alleles, the individual is **heterozygous.**

4. Alleles are considered **dominant** or **recessive** depending on how their traits are expressed. (*Table 21-3*)

5. Combining maternal and paternal alleles in a *Punnett square* allows us to predict the probability of a particular phenotype among the offspring. (*Figure 21-13*)

6. In **simple inheritance,** phenotypic characters are determined by interactions between a single pair of alleles. **Polygenic inheritance** involves interactions among alleles on several chromosomes. (*Table 21-4*)

7. There are two different **sex chromosomes,** an **X chromosome** and a **Y chromosome.** The normal male sex chromosome complement is *XY;* that of females is *XX.* The X chromosome carries **X-linked** genes that affect somatic structures but have no corresponding alleles on the Y chromosome.

The Human Genome Project p. 564

8. The Human Genome Project has identified 3700 of our estimated 100,000 genes, including some of those responsible for inherited disorders. (*Figure 21-14; Table 21-4*)

CHAPTER QUESTIONS

LEVEL 1 Reviewing Facts and Terms

Match each item in column A with the most closely related item in column B. Use letters for answers in the spaces provided.

Column A

___ 1. gestation
___ 2. cleavage
___ 3. gastrulation
___ 4. chorion
___ 5. human chorionic gonadotropin
___ 6. birth
___ 7. episiotomy
___ 8. afterbirth
___ 9. senescence
___ 10. neonate
___ 11. phenotype
___ 12. homozygous recessive
___ 13. heterozygous
___ 14. male genotype
___ 15. female genotype
___ 16. trisomy 21

Column B

a. blastocyst formation
b. ejection of placenta
c. germ layer formation
d. indication of pregnancy
e. embryo-maternal circulatory exchange
f. visible characteristics
g. time of prenatal development
h. *aa*
i. Down syndrome
j. newborn infant
k. *Aa*
l. *XY*
m. *XX*
n. parturition
o. process of aging
p. perineal musculature incision

17. The gradual modification of anatomical structures during the period from conception to maturity is:
(a) development (b) differentiation
(c) embryogenesis (d) capacitation

18. Human fertilization involves the fusion of two haploid gametes, producing a zygote containing:
(a) 23 chromosomes
(b) 46 chromosomes
(c) the normal haploid number of chromosomes
(d) 46 pairs of chromosomes

19. The secondary oocyte leaving the follicle is in:
(a) interphase
(b) metaphase of the first meiotic division
(c) telophase of the second meiotic division
(d) metaphase of the second meiotic division
20. The process that establishes the foundation of all major organ systems is:
(a) cleavage (b) implantation
(c) placentation (d) embryogenesis
21. The zygote arrives in the uterine cavity as a:
(a) morula (b) trophoblast
(c) lacuna (d) blastomere
22. The surface that provides for active and passive exchange between the fetal and maternal bloodstreams is the:
(a) yolk stalk (b) chorionic villi
(c) umbilical veins (d) umbilical arteries
23. Milk let-down is associated with:
(a) events occurring in the uterus
(b) placental hormonal influences
(c) circadian rhythms
(d) reflex action triggered by suckling
24. If an allele must be present on both the maternal and paternal chromosomes in order to affect the phenotype, the allele is said to be:
(a) dominant (b) recessive
(c) complementary (d) heterozygous
25. Summarize the developmental changes that occur during the first, second, and third trimesters.
26. Identify the three stages of labor, and describe the events that characterize each stage.
27. Identify the three life stages that occur between birth and approximately age 10. Describe the characteristics of each stage and when it occurs.

LEVEL 2 Reviewing Concepts

28. Relaxin is a peptide hormone that:
(a) increases the flexibility of the symphysis pubis
(b) causes dilation of the cervix
(c) suppresses the release of oxytocin by the hypothalamus
(d) a, b, and c are correct
29. During adolescence, the events that interact to promote increased hormone production and sexual maturation result from activity of the:
(a) hypothalamus
(b) anterior pituitary
(c) ovaries and testicular cells
(d) a, b, and c are correct
30. In addition to its role in the nutrition of the fetus, what are the primary endocrine functions of the placenta?
31. Discuss the changes that occur in maternal systems during pregnancy. Why are these changes functionally significant?
32. During labor, what physiological mechanisms ensure that uterine contractions continue until delivery has been completed?
33. To what does the phrase "having the water break" refer during the process of labor?
34. What would you conclude about a trait in each of the following situations?
(a) Children who exhibit this trait have at least one parent who exhibits the same trait.
(b) Children exhibit this trait even though neither of the parents exhibits it.
(c) The trait is expressed more frequently in sons than in daughters.
(d) The trait is expressed equally in both daughters and sons.
35. Explain why more men than women are color-blind. What type of inheritance is involved?
36. Explain the goals and possible benefits of the Human Genome Project.

LEVEL 3 Critical Thinking and Clinical Applications

37. Hemophilia A, a condition in which the blood does not clot properly, is a recessive trait located on the X chromosome (X^h). A woman heterozygous for the trait marries a normal male. What is the probability that this couple will have hemophiliac daughters? What is the probability that this couple will have hemophiliac sons?
38. Explain why the normal heart and respiratory rates of neonates are so much higher than those of adults, even though adults are so much larger.
39. Sally gives birth to a baby with a congenital deformity of the stomach. She swears that it is the result of a viral infection that she suffered during the third trimester of pregnancy. Do you think this is a possibility? Explain.

Appendix 1

Answers to Concept Check Questions

CHAPTER 1

Page 3

1. *Metabolism* refers to all of the chemical operations under way in the body. Organisms rely on complex chemical reactions to provide the energy for responsiveness, growth, reproduction, and movement. 2. A *histologist* investigates the structure and properties of tissues. *Histology* is considered a form of microscopic anatomy because it requires the use of a microscope to reveal the cells that make up tissues.

Page 14

1. Physiological systems can function normally only under carefully controlled conditions. Homeostatic regulation serves to prevent potentially disruptive changes in the body's internal environment. 2. Positive feedback is useful in processes that must move quickly to completion once they have begun, such as blood clotting. It is harmful in situations where a stable condition must be maintained, because it will serve to increase any departure from the desired condition. For example, positive feedback in the regulation of body temperature would cause a slight fever to spiral out of control, with fatal results. For this reason physiological systems usually exhibit negative feedback, which tends to oppose any departure from the norm. 3. When homeostasis fails, organ systems function less efficiently or begin to malfunction. The result is the state that we call *disease*. If the situation is not corrected, death may result.

Page 21

1. The two eyes would be separated by a *midsagittal section*. 2. The body cavity inferior to the diaphragm is the *abdominopelvic* (or *peritoneal*) *cavity*.

CHAPTER 2

Page 29

1. Atoms combine with each other so as to gain a complete set of eight electrons in their outer energy levels. Oxygen atoms do not have a full outer energy level and so will readily react with many other elements to attain this stable arrangement. Neon already has a full outer energy level and thus has little tendency to combine with other elements. 2. Hydrogen can exist as three different isotopes: hydrogen-1, with a mass of 1; deuterium, with a mass of 2; and tritium, with a mass of 3. The heavier sample must contain a higher proportion of one or both of the heavier isotopes. 3. A water molecule is formed by polar covalent bonds. Water molecules are attracted to one another by hydrogen bonds.

Page 32, Set 1

1. Since this reaction involves a large molecule being broken down into two smaller ones, it is an example of a *decomposition* reaction. Because energy is released in the process, the reaction can also be classified as *exergonic*. 2. Removing the product of a reversible reaction would keep its concentration low compared with the concentration of the reactants. Thus the formation of product molecules would continue but the reverse reaction would slow down, resulting in a shift in the equilibrium toward the product.

Page 32, Set 2

1. An acid is a solute that releases H^+ ions in a solution and a base is a solute that removes H^+ from a solution. 2. The normal pH range for body fluids is 7.35 to 7.45. Fluctuations in pH outside of this range can break chemical bonds, alter the shape of molecules, and affect the functioning of cells, thereby causing harm to cells and tissues. 3. Stomach discomfort is often the result of excess stomach acidity ("acid indigestion"). Antacids contain a weak base that neutralizes the excessive acid.

Page 42

1. A C:H:O ratio of 1:2:1 would indicate that the molecule is a *carbohydrate*. The body uses carbohydrates chiefly as an energy source. 2. The heat of boiling will break bonds that maintain the protein's tertiary and/or quaternary structure. The resulting change in shape will affect the ability of the protein molecule to perform its normal biological functions. These alterations are known as *denaturation*. 3. DNA and RNA are nucleic acids. Both are composed of sequences of nucleotides. Each nucleotide consists of a five-carbon sugar, a phosphate group (PO_4^{3-}), and a nitrogenous base.

CHAPTER 3

Page 56

1. Active transport processes require the expenditure of cellular energy in the form of the high-energy bonds of ATP molecules. Passive transport processes (*diffusion, osmosis, filtration*, and *facilitated diffusion*) move ions or molecules across the cell membrane without any energy expenditure by the cell. 2. In order to transport H^+ ions against their concentration gradient—that is, from a region where they are less concentrated (the cells lining the stomach) to a region where they are more concentrated (the interior of the stomach)—energy must be expended. An *active transport* process must be involved. 3. This is an example of phagocytosis.

Page 59

1. The fingerlike projections on the surface of the intestinal cells are *microvilli*. They serve to increase the cells' surface area so that they can absorb nutrients more efficiently. 2. Cells that lack centrioles are unable to divide.

Page 61

1. SER functions in the synthesis of lipids such as steroids. Ovaries and testes would be expected to have a great deal of SER because they produce large amounts of steroid hormones. 2. The function of mitochondria is to produce energy for the cell in the form of ATP molecules. A large number of mitochondria in a cell would indicate a high demand for energy.

Page 65

1. The nucleus of a cell contains DNA that codes for production of all of the cell's polypeptides and proteins. Some of these proteins are structural proteins that are responsible for the shape and other physical characteristics of the cell. Other proteins are enzymes that govern cellular metabolism, direct the production of cell proteins, and control all of the cell's activities. 2. If a cell lacked the enzyme RNA-polymerase it would not be able to transcribe RNA from DNA. 3. The deletion of a base from a coding sequence of DNA during transcription would alter the entire mRNA base sequence after the deletion point. This would result in different codons on the messenger RNA that was transcribed from the affected region, and this, in turn, would result in the incorporation of a different series of amino acids into the polypeptide. Almost certainly the polypeptide product would not be functional.

Page 68

1. Cells that are preparing to undergo mitosis manufacture additional organelles and duplicate sets of their DNA. 2. The four stages of mitosis are prophase, metaphase, anaphase, and telophase. 3. If spindle fibers failed to form during mitosis, the cell would not be able to separate the chromosomes into two sets. If cytokinesis occurred, the result would be one cell with two sets of chromosomes and one cell with none.

CHAPTER 4

Page 80

1. No. A simple squamous epithelium does not provide enough protection against infection, abrasion, and dehydration and is not found in the skin surface. 2. The process described is *holocrine secretion*. 3. The presence of microvilli on the free surface of epithelial cells greatly increases the surface area for absorption. Cilia function to move materials over the surface of epithelial cells.

Page 85

1. Collagen fibers add strength to connective tissue. We would therefore expect vitamin C deficiency to result in the production of connective tissue that is weaker and more prone to damage. 2. The tissue is adipose (fat) tissue. 3. Cartilage lacks a direct blood supply, which is necessary for proper healing to occur. Materials that are needed to repair damaged cartilage must diffuse from the blood to the chondroblasts, a process that takes a long time and retards the healing process.

Page 88

1. Cell membranes are composed of lipid bilayers. Tissue membranes consist of a layer of epithelial tissue and a layer of connective tissue. 2. *Serous fluid* minimizes the friction between the serous membranes covering the surfaces of organs and the surrounding body cavity. 3. All of these regions are subject to mechanical trauma and abrasion—by food (pharynx and esophagus), feces (anus), and intercourse or childbirth (vagina).

Page 89

1. Since cardiac and skeletal muscle are both striated (banded), this must be *smooth muscle* tissue. 2. Only skeletal muscle tissue is voluntary. 3. Both skeletal muscle cells and neurons are called fibers because they are relatively long and slender.

CHAPTER 5

Page 100

1. Cells are constantly shed from the outer layers of the *stratum corneum*. 2. The splinter is lodged in the *stratum granulosum*. 3. When exposed to the ultraviolet radiation in sunlight or tanning lamps, melanocytes in the epidermis and dermis synthesize the pigment melanin, darkening the color of the skin.

Page 102

1. Contraction of the arrector pili muscles pulls the hair follicles erect, depressing the area at the base of the hair and making the surrounding skin appear higher. The result is known as "goose bumps" or "goose pimples." 2. Hair is a derivative of the epidermis, and if the epidermis is destroyed by the injury there will be no hair follicles to produce new hair. 3. The subcutaneous layer stabilizes the position of the

skin in relation to underlying tissues and organs; stores fat; and, because its lower region contains few capillaries and no vital organs, provides a useful site for the injection of drugs.

Page 105

1. If the duct of a sebaceous gland is blocked by infection, the result is a *furuncle* or *boil.* 2. Apocrine sweat glands produce a secretion containing several kinds of organic compounds. Some of these have an odor and others produce an odor when metabolized by skin bacteria. Deodorants are used to mask the odor of these secretions. 3. As a person ages, the blood supply to the dermis decreases and merocrine sweat glands become less active. These changes make it more difficult for the elderly to cool themselves in hot weather.

CHAPTER 6

Page 113

1. If the ratio of collagen to calcium in a bone increased, the bone would be more flexible and less strong. 2. Concentric layers of bone around a central canal is indicative of a Haversian system. Haversian systems make up compact bone. Since the ends (epiphyses) of long bones are primarily cancellous (spongy) bone, this sample most likely came from the shaft (diaphysis) of a long bone. 3. Since osteoclasts function in breaking down or demineralizing bone, the bone would have less mineral content and as a result it would be weaker.

Page 115

1. Long bones of the body, like the femur, have a plate of cartilage, called the epiphyseal plate, that separates the epiphysis from the diaphysis as long as the bone is still growing lengthwise. An X-ray would indicate whether the epiphyseal plate was still present. If it was, then growth was still occurring, and if not the bone had reached its adult length. 2. The increase in the male sex hormone, testosterone, that occurs at puberty, contributes to an increased rate of bone growth and the closure of the epiphyseal plates. Since the source of testosterone, the testes, is removed in castration, we would expect these boys to have a longer, though slower, growth period and be taller than they would have been if they had not been castrated. 3. Women who are pregnant need large amounts of calcium to support the needs of the developing fetus for bone growth. If the expectant mother does not include enough calcium in her diet, her body will mobilize the calcium reserves of her skeleton to provide for the needs of the fetus, resulting in weakened bones and an increased risk of fracture.

Page 118

1. The larger arm muscles of the weight lifter will apply more mechanical stress to the bones of the arms. In response to the stress, the bones will grow thicker. We would expect the jogger to have heavier thigh bones for similar reasons. 2. The sex hormones known as estrogens play an important role in moving calcium into bones. After menopause, the level of these hormones decreases dramatically and as a result, it is difficult to replace the calcium in bones that is being lost due to normal aging. Males do not show a decrease in sex hormone levels (androgens).

Page 122

1. The mastoid and styloid processes are projections found on the temporal bones of the skull. 2. The sella turcica contains the pituitary gland and is located in the sphenoid bone. 3. The occipital condyles of the occipital bone of the cranium articulate with the vertebral column.

Page 125

1. The bone that forms the superior portion of the orbit is the frontal bone, and the maxilla forms the inferior portion of the orbit. These would be the two bones fractured by the ball. 2. The paranasal sinuses function to make some of the heavier skull bones lighter and to produce mucus. 3. The most powerful muscles that are involved in closing the mouth attach to the mandible at the coronoid process. A fracture of the coronoid process would make it difficult for these muscles to function properly and close the mouth. 4. Since many muscles that move the tongue and the larynx are attached to the hyoid bone, you would expect a person with a fractured hyoid bone to have difficulty moving the tongue, breathing, and swallowing.

Page 129

1. The odontoid process is found on the second cervical vertebra, or axis, which is located in the neck. 2. Improper compression of the chest during CPR could and frequently does result in a fracture of the sternum or ribs. 3. In adults, the five sacral vertebrae fuse to form a single sacrum.

Page 132

1. The clavicle attaches the scapula to the sternum and thus restricts the scapula's range of movement. If the clavicle is broken, the scapula will have a greater range of movement and will be less stable. 2. The two rounded prominences on either side of the elbow are parts of the humerus (the lateral and medial epicondyles).

Page 135

1. The three bones that make up the coxa are the ilium, ischium, and the pubis. 2. Although the fibula is not part of the knee joint nor does it bear weight, it is an important point of attachment for many leg muscles. When the fibula is fractured, these muscles cannot function properly to move the leg and walking is difficult and painful. The fibula also helps stabilize the ankle joint. 3. Joey has most likely fractured the calcaneus (heel bone).

Page 141

1. Originally, the joint is a type of syndesmosis. 2. When the bones interlock, they form sutural joints. 3. (a) abduction; (b) supination; (c) flexion.

Page 143

1. Since the subscapular bursa is located in the shoulder joint, inflammation of this structure (bursitis) would be found in the tennis player. The condition is associated with repetitive motion that occurs at the shoulder, such as swinging a tennis racket. The jogger would be more at risk for injuries to the knee joint. 2. Mary has most likely fractured her ulna.

Page 147

1. There are seven major ligaments that stabilize the knee joint. As a result, a complete dislocation is rare. 2. Damage to the menisci in the knee joint would result in a decrease in the joint's stability. The individual would have a harder time locking the knee in place while standing and would have to use muscle contractions to stabilize the joint. When standing for long periods, the muscles would fatigue and the knee would "give out." We would also expect the individual to experience pain.

CHAPTER 7

Page 157

1. Since tendons attach muscles to bones, severing the tendon would disconnect the muscle from the bone so that when the muscle contracted nothing would happen. 2. Skeletal muscle appears striated when viewed under the microscope because it is composed of the myofilaments actin and myosin, which are arranged in such a way as to produce a banded appearance in the muscle. 3. You would expect to find the greatest concentration of calcium ions in the cisternae of the sarcoplasmic reticulum of the muscle.

Page 162

1. Since the ability of a muscle to contract depends on the formation of cross-bridges between the myosin and actin myofilaments, a drug that would interfere with cross-bridge formation would prevent the muscle from contracting. 2. Because the amount of cross-bridge formation is proportional to the amount of available calcium ions, increased permeability of the sarcolemma to calcium ions would lead to an increased intracellular concentration of calcium and a greater degree of contraction. In addition, since relaxation depends on decreasing the amount of calcium in the sarcoplasm, an increase in the permeability of the sarcolemma to calcium could result in a situation in which the muscle would not be able to relax completely. 3. Without acetylcholinesterase, the motor end plate would be continuously stimulated by the acetylcholine, and the muscle would be locked into contraction.

Page 164

1. The ability of the muscle to contract depends on the ability to form cross-bridges between the actin and myosin. If the myofilaments overlap very little, then very few cross-bridges are formed and the contraction is weak. If the myofilaments do not overlap at all, then no cross-bridges form and the muscle cannot contract. 2. A muscle unit with 1500 fibers is most likely from a large muscle involved in powerful, gross body movement. Muscles that control fine or precise movements, such as movement of the eye or the fingers, have only a few fibers per motor unit, whereas muscles of the legs, for instance, that are involved in powerful contractions have hundreds of fibers per motor unit. 3. There are two types of muscle contractions, isometric and isotonic. In an isotonic contraction, tension remains constant and the muscle shortens. In isometric contractions, however, the same events of contraction occur, but instead of the muscle's shortening, the tension in the muscle increases.

Page 168

1. The sprinter requires large amounts of energy for a relatively short burst of activity. To supply this demand for energy, the muscles switch to anaerobic metabolism. Anaerobic metabolism is not as efficient in producing energy as aerobic metabolism, and the process also produces acidic waste products. The combination of less energy and the waste products contributes to fatigue. Marathon runners, on the other hand, derive most of their energy from aerobic metabolism, which is more efficient and does not produce the level of waste products that anaerobic respiration does. 2. We would expect activities that require short periods of strenuous activity to produce a greater oxygen debt because this type of activity relies heavily on energy production by anaerobic respiration. Since lifting weights is more strenuous over the short term we would expect this type of exercise to produce a greater oxygen debt than swimming laps, which is an aerobic activity. 3. Individuals who are naturally better at endurance types of activities such as cycling or marathon running have a higher percentage of slow muscle fibers which are physiologically better adapted to this type of activity than fast fibers, which are less vascular and fatigue faster.

Page 169

1. The cell membranes of cardiac muscle cells are extensively interwoven and bound tightly to each other at intercalated discs, allowing them to efficiently "pull together." The intercalated discs also contain gap junctions, which allow ions and small molecules to flow directly from one cell to another. This results in the rapid passage of action potentials from cell to cell, resulting in their simultaneous contraction. 2. Cardiac muscle and smooth muscle are more affected by changes in the concentration of calcium ions in the extracellular fluid than skeletal muscle because in cardiac and smooth muscle, the majority of the calcium ions that trigger a contraction come from the extracellular fluid. In skeletal muscle, most of the cal-

cium ions come from the sarcoplasmic reticulum. 3. The actin and myosin filaments of smooth muscle are not as rigidly organized as they are in skeletal muscle. This allows smooth muscle to contract over a relatively large range of resting lengths.

Page 172

1. The opening between the stomach and the small intestine would be guarded by a circular muscle known as a sphincter muscle. The concentric circles of muscle fibers found in sphincter muscles are ideally suited for opening and closing openings or acting as valves in the body. 2. The *triceps brachii* extends the forearm and is an antagonist of the biceps brachii. 3. The name *flexor carpi radialis longus* tells you that this is a long muscle that lies next to the radius and functions to flex the hand.

Page 176

1. Contraction of the masseter muscle raises the mandible, while the mandible depresses when the muscle is relaxed. These movements are important in the process of chewing, or mastication. 2. You would expect the buccinator muscle, which forms the mouth for blowing, to be well-developed in a trumpet player.

Page 177

1. Damage to the external intercostal muscles would interfere with the process of breathing. 2. A blow to the rectus abdominis would cause the muscle to contract forcefully, resulting in flexion of the torso. In other words, you would "double up."

Page 186

1. When you shrug your shoulders you are contracting your levator scapulae muscles. 2. The rotator cuff muscles include the supraspinatus, infraspinatus, subscapularis, and teres minor. The tendons of these muscles help to enclose and stabilize the shoulder joint. 3. Injury to the flexor carpi ulnaris would impair the ability to flex and adduct the hand.

Page 188

1. The hamstring refers to a group of five muscles that collectively function in flexing the leg. These muscles are the biceps femoris, semimembranous, semitendinosus, gracilis, and sartorius. 2. The Achilles (calcaneal) tendon attaches the soleus and gastrocnemius muscles to the calcaneus (heel bone). When these muscles contract, they cause extension of the foot. A torn Achilles tendon would make extension of the foot difficult and the opposite action, flexion, would be more pronounced as a result of less antagonism from the soleus and gastrocnemius.

CHAPTER 8

Page 202

1. The afferent division of the nervous system is composed of nerves that carry sensory information to the brain and spinal cord. Damage to this division would interfere with a person's ability to experience a variety of sensory stimuli. 2. Sensory neurons of the peripheral nervous system are usually unipolar; thus this tissue is most likely associated with a sensory organ. 3. Microglial cells are small phagocytic cells that are found in increased number in damaged and diseased areas of the CNS.

Page 206

1. Depolarization of the neuron membrane involves the opening of the sodium channels and the rapid influx of sodium ions into the cell. If the sodium channels were blocked, a neuron would not be able to depolarize and conduct an action potential. 2. Action potentials are propagated along myelinated axons by saltatory propagation at speeds much higher than those seen along unmyelinated axons. An axon with a propagation speed of 50 msec must be myelinated.

Page 207

1. A neurotransmitter that opens the potassium channels but not the sodium channels would cause a hyperpolarization at the postsynaptic membrane. The transmembrane potential would be greater and it would be more difficult to bring the membrane to threshold.

Page 209

1. When an action potential reaches the presynaptic terminal of a cholinergic synapse, calcium channels are opened and the influx of calcium triggers the release of acetylcholine into the synapse to stimulate the next neuron. If the calcium channels were blocked, the acetylcholine would not be released and transmission across the synapse would cease. 2. The minimum number of neurons required for a reflex arc is two. One must be a sensory neuron to bring impulses to the central nervous system, and the other a motor neuron that can bring about a response to the sensory input.

Page 214

1. The ventral root of spinal nerves is composed of visceral and somatic motor fibers. Damage to this root would interfere with motor function. 2. Since the polio virus would be located in the somatomotor neurons, we would find it in the anterior gray horns of the spinal cord, where the cell bodies of these neurons are located. 3. All spinal nerves are classified as mixed nerves, because they contain both sensory and motor fibers. 4. The cerebrospinal fluid that surrounds the spinal cord is found in the subarachnoid space, which lies beneath the epithelium of the arachnoid layer and on top of the pia mater.

Page 216

1. The six regions in the adult brain and their major functions: (1) the *cerebrum*: conscious thought processes; (2) the *diencephalon*: the thalamic portion contains relay and processing centers for sensory information, and the hypothalamic portion contains centers involved with emotions, autonomic function, and hormone production; (3) the *midbrain*: processes visual and auditory information and generates involuntary motor responses; (4) the *pons*: contains tracts and relay centers that connect brain stem to cerebellum; (5) the *cerebellum*: adjusts voluntary and involuntary motor activities; and (6) the *medulla oblongata*: contains major centers concerned with the regulation of autonomic function, such as heart rate, blood pressure, respiration, and digestive activities. 2. The pituitary gland is attached to the floor of the diencephalon, or hypothalamus.

Page 222

1. Diffusion across the arachnoid granulations is the means by which cerebrospinal fluid reenters the bloodstream. If this process decreased, then excess fluid would start to accumulate in the ventricles and the volume of fluid in the ventricles would increase. 2. The primary motor cortex is located in the precentral gyrus of the frontal lobe of the cerebrum. 3. Damage to the temporal lobe of the cerebrum would interfere with the processing of olfactory (smell) and auditory (sound) impulses.

Page 224

1. All ascending sensory information, other than olfactory, passes through the thalamus before reaching our conscious awareness. 2. Changes in body temperature would stimulate the hypothalamus, a division of the diencephalon. 3. Even though the medulla oblongata is small, it contains many vital reflex centers including those that control breathing and regulate the heart and blood pressure. Damage to the medulla oblongata can result in a cessation of breathing, or changes in heart rate and blood pressure that are incompatible with life.

CHAPTER 9

Page 232

1. The glossopharyngeal nerve (cranial nerve IX) controls swallowing muscles and provides sensory information from the tongue. 2. Since the abducens nerve (cranial nerve VI) controls lateral movements of the eyes via the lateral rectus muscles, we would expect an individual with damage to this nerve to be unable to move his eyes laterally.

Page 236

1. Physicians use the sensitivity of stretch reflexes, such as the knee jerk, or *patellar reflex*, to test the general condition of the spinal cord, peripheral nerves, and muscles. 2. In a monosynaptic reflex, a sensory neuron synapses directly on a motor neuron and produces a rapid, stereotyped movement. More complicated responses occur with polysynaptic reflexes because the interneurons between the sensory and motor neurons may control several different muscle groups simultaneously. In addition, some interneurons may stimulate a muscle group, or groups, while others may inhibit other muscle groups. 3. A positive Babinski reflex is abnormal for an adult and indicates possible damage of descending tracts in the spinal cord.

Page 238

1. A tract within the posterior column of the spinal cord is responsible for carrying information about touch and pressure from the lower part of the body to the brain. 2. The anatomical basis for opposite-side motor control is that crossing-over occurs, and the pyramidal motor fibers innervate lower motor neurons on the opposite side of the body. 3. The superior portion of the motor cortex exercises control over the hand, arm, and upper portion of the leg. An injury to this area would affect the ability to control the muscles in those regions of the body.

Page 243

1. The sympathetic division of the autonomic nervous system is responsible for the physiological changes that occur in response to stress and increased activity. 2. The parasympathetic division is sometimes referred to as the anabolic system because parasympathetic stimulation leads to a general increase in the nutrient content of the blood. Cells throughout the body respond to the increase by absorbing the nutrients and using them to support growth and other anabolic activities. 3. Since most blood vessels receive sympathetic stimulation, a decrease in sympathetic stimulation would lead to a relaxation of the muscles in the walls of the vessels and vasodilation (the vessels would increase in diameter). This in turn would result in increased blood flow to the tissue. 4. A patient who is anxious about impending root canal would probably exhibit some or all of the following changes: a dry mouth, increased heart rate, increased blood pressure, increased rate of breathing, cold sweats, an urge to urinate or defecate, change in motility of the digestive tract (i.e., "butterflies in the stomach"), and dilated pupils. These changes would be the result of anxiety or stress causing an increase in sympathetic stimulation.

CHAPTER 10

Page 256

1. By the end of the lab period adaptation has occurred. In response to the constant level of stimulation, the receptor neurons have become less active, partially as the result of synaptic fatigue. 2. Since nociceptors are pain receptors, if they are stimulated, you would perceive a painful sensation in your affected hand. 3. Proprioceptors relay information about limb position and movement to the central nervous system, especially the cerebellum. Lack of this information would result in uncoordinated movements, and the individual probably would not be able to walk. 4. The taste receptors (taste buds) are only sensitive to molecules and ions that are in

solution. If you dry the surface of the tongue, there is no moisture for the sugar molecules or salt ions to dissolve in and they will not stimulate the taste receptors.

Page 267

1. The first layer of the eye to be affected by inadequate tear production would be the conjunctiva. Drying of this layer would produce an irritated, scratchy feeling. 2. When the lens is round you are looking at something closer to you. 3. Even with a congenital lack of cone cells in the eye you would still be able to see as long as you had functioning rod cells. Since cone cells function in color vision, you would see only black and white. 4. A deficiency or lack of vitamin A in the diet would affect the quantity of retinal that the body could produce and thus interfere with night vision.

Page 276

1. Without the movement of the round window, the perilymph would not be moved by the vibration of the stapes at the oval window, and there would be little or no perception of sound. 2. Loss of stereocilia (as a result of constant exposure to loud noises for instance) would reduce hearing sensitivity and could eventually result in deafness.

CHAPTER 11

Page 289

1. Epinephrine, norepinephrine, and peptide hormones represent the first messengers from the endocrine glands. Since these hormones cannot enter the target cells, intracellular cyclic-AMP acts as the second messenger from the endocrine glands. 2. Dehydration increases the osmotic pressure of the blood. The increase in blood osmotic pressure would stimulate the posterior pituitary to release more ADH. 3. Somatomedins are the mediators of growth hormone action. If the level of somatomedins is elevated, we would expect to see the level of growth hormone elevated as well. 4. Increased levels of cortisol would inhibit the cells that control ACTH release from the pituitary; therefore the level of ACTH would decrease. This is an example of a negative feedback mechanism.

Page 295

1. An individual who lacked iodine would not be able to form the hormone thyroxine. As a result we would expect to see the symptoms associated with thyroxins deficiency, such as decreased rate of metabolism, decreased body temperature, poor response to physiological stress, and an increase in the size of the thyroid gland (goiter). 2. Most of the thyroid hormone in the blood is bound to a protein called thyroid-binding globulins. This represents a large reservoir of thyroxine that guards against rapid fluctuations in the level of this important hormone. Because there is such a large amount stored in this way, it takes several days to deplete the supply of hormone, even after the thyroid gland has been removed. 3. Removal of the parathyroid glands would result in a decrease in the blood levels of calcium ion. This could be counteracted by increasing the amount of vitamin D and calcium in the diet. 4. One of the functions of cortisol is to decrease the cellular use of glucose while increasing the available glucose by promoting the breakdown of glycogen and the conversion of amino acids to carbohydrates. The net result is an elevation in the level of glucose in the blood.

Page 299

1. Insulin increases the rate of conversion of glucose to glycogen within skeletal muscle and liver cells. 2. Glucagon stimulates the conversion of glycogen to glucose in the liver. Increased amounts of glucagon would then lead to decreased amounts of liver glycogen. 3. The pineal gland receives neural input from the optic tracts and its secretion, melatonin, is influenced by light-dark cycles. Increased amounts of light inhibit the production and release of melatonin from the pineal gland.

Page 302

1. The type of hormonal interaction exemplified by the insulin and glucagon is antagonism. In this type of hormonal interaction, two hormones have opposite effects on their target tissues. 2. The hormones growth hormone, thyroid hormone, parathyroid hormone, and the gonadal hormones all play a role in formation and development of the skeletal system.

CHAPTER 12

Page 315

1. A decrease in the amount of plasma proteins in the blood may cause (1) a decrease in plasma osmotic pressure, (2) a decreased ability to fight infection, and (3) a decrease in the transport and binding of some ions, hormones, and other molecules. 2. The hematocrit measures the amount of formed elements (mostly red blood cells) as a percentage of the total blood. In hemorrhage the loss of blood, especially red blood cells, would cause the hematocrit to be less. 3. Diseases that damage the liver such as hepatitis or cirrhosis would impair the liver's ability to excrete bilirubin. As a result, the bilirubin would accumulate in the blood, producing a condition known as jaundice. 4. A decreased blood flow to the kidneys would trigger the release of erythropoietin. The elevated erythropoietin would lead to an increase in erythropoiesis (red blood cell formation).

Page 316

1. A person with Type AB blood can accept Type A, Type B, Type AB, or Type O blood. 2. If a person with Type A blood received a transfusion of Type B blood, the red cells would clump or agglutinate, potentially blocking blood flow to various organs and tissues.

Page 322

1. In an infected cut we would expect to find a large number of neutrophils. Neutrophils are phagocytic white cells that are usually the first to arrive at the site of an injury and that specialize in dealing with infectious bacteria. 2. The type of white blood cell that produces circulating antibodies is the B lymphocyte, and these would be found in increased numbers. 3. Megakaryocytes are the precursors of platelets, which play an important role in hemostasis and the clotting process. A decreased number of megakaryocytes would result in fewer platelets, which in turn would interfere with the ability to clot properly.

Page 325

1. The use of broad-spectrum antibiotics would lower the number of intestinal bacteria, and thus the amount of vitamin K produced. This decrease in vitamin K would lead to a decrease in the production of several clotting factors, most notably prothrombin. As a result, clotting time would increase.

CHAPTER 13

Page 336

1. The semilunar valves on the right side of the heart guard the opening to the pulmonary artery. Damage to these valves would interfere with the blood flow through this vessel. 2. When the ventricles begin to contract, they force the AV valves to close, which in turn pulls on the chordae tendineae, which then pull on the papillary muscles. The papillary muscles respond by contracting, counteracting the force that is pushing the valves upward. 3. The wall of the left ventricle is more muscular than that of the right ventricle because the left ventricle has to generate enough force to propel blood throughout all of the body's systems except the lungs. The right ventricle only has to generate enough force to propel the blood a few centimeters to the lungs. Since the left ventricle is so muscular it takes more force to push blood into the chamber against the normal tension of the muscle; this in turn requires that the left atrium be more muscular than the right atrium.

Page 338

1. Unlike the situation in skeletal muscle fibers, twitch summation in cardiac muscle cells does not occur and tetanus is not possible. The longer refractory period in cardiac muscle cells results in a relatively long relaxation period during which the heart's chambers may refill with blood. A heart in tetany could not fill with blood. 2. If these cells were not functioning, the heart would still continue to beat but at a slower rate. 3. If the impulses from the atria were not delayed at the AV node, they would be conducted through the ventricles so quickly by the bundle branches and Purkinje fibers that the ventricles would begin contracting immediately before the atria had finished their contraction. As a result the ventricles would not be as full of blood as they could be and the pumping of the heart would not be as efficient, especially during activity.

Page 341

1. When pressure in the left ventricle is rising, the heart is contracting but no blood is leaving the heart. During this initial phase of contraction, the AV valves and semilunar valves are both closed. The increase in pressure is the result of increased tension as the muscle contracts. When the pressure in the ventricle exceeds the pressure in the aorta, the aortic semilunar valves are forced open and the blood is rapidly ejected from the ventricle. 2. If the heart beats too quickly (tachycardia), there is not sufficient time for it to fill completely between the beats. Since the heart pumps blood proportionately to what enters, the less blood that enters, the less it will be able to pump. If it beats too fast, very little blood will enter circulation and tissues will suffer damage from lack of blood supply.

Page 343

1. Stimulating the acetylcholine receptors of the heart would cause the heart to slow down. Since the cardiac output is the product of stroke volume times the heart rate, if the heart rate decreases so will the cardiac output (assuming no change in the stroke volume). 2. The venous return fills the heart with blood, stretching the heart muscle. According to the Frank-Starling Law, the more the heart muscle is stretched, the more forcefully it will contract (to a point). The more forceful the contraction the more blood the heart will eject with each beat (stroke volume). Therefore, increased venous return will increase the stroke volume, assuming all other factors are constant. 3. Increased sympathetic stimulation of the heart will result in increased heart rate and increased force of contraction.

CHAPTER 14

Page 350

1. The blood vessels are veins. Arteries and arterioles have a relatively large amount of smooth muscle tissue in a thick, well-developed tunica media. 2. Relaxation of the precapillary sphincters would increase the blood flow to a tissue. 3. Blood pressure in the arterial system pushes blood into the capillaries. Blood pressure on the venous side is very low, and other forces help keep the blood moving. Valves prevent the blood from flowing backward whenever the venous pressure drops.

Page 356

1. In a normal individual, the pressure should be greatest in the aorta and least in the venae cavae. Blood, like other fluids, moves along a pressure gradient from high pressure to low pressure. If the pressure were higher in the inferior vena cava, the blood would flow backward. 2. When a person stands for periods of

time, blood tends to pool in the lower extremities. The venous return to the heart decreases, and in turn the cardiac output decreases, sending less blood to the brain, causing light-headedness and fainting. A hot day adds to the effect, because body water is lost through sweating.

Page 362

1. Pressure at this site would decrease blood pressure at the carotid sinus, where the carotid baroreceptors are located. This decrease causes a decreased frequency of action potentials along the glossopharyngeal nerve (IX) to the medulla, and more sympathetic impulses will be sent to the heart. The net result will be an increase in the heart rate. 2. In exercise (1) blood flow to muscles increases, (2) cardiac output increases, and (3) resistance in visceral tissues increases. 3. Vasoconstriction of the renal artery will decrease both blood flow and blood pressure at the kidney. In response, the kidney will increase the amount of renin that it releases, which in turn will lead to an increase in the level of angiotensin II. The angiotensin II will bring about increased blood pressure and increased blood volume.

Page 374

1. The left subclavian artery is the branch of the aorta that sends blood to the left shoulder and arm. 2. The common carotid arteries carry blood to the head. A compression of one of the common carotid arteries would result in decreased blood flow to the brain and loss of consciousness or even death. 3. Organs served by the celiac artery include the stomach, spleen, liver, and pancreas.

CHAPTER 15

Page 388

1. The thoracic duct drains lymph from the area beneath the diaphragm and the left side of the head and thorax. Most of the lymph enters the venous blood by way of this duct. A blockage of this duct would not only impair circulation of lymph through most of the body, it would also promote accumulation of fluid in the extremities (lymphedema). 2. The thymosins from the thymus play a role in the differentiation of stem lymphocytes into T lymphocytes. A lack of these hormones would result in an absence of T lymphocytes. 3. During an infection, the lymphocytes and phagocytes in the lymph nodes in the affected region undergo cell division to better deal with the infectious agent. This increase in the number of cells in the node causes the node to become enlarged or swollen.

Page 392

1. A decrease in the number of monocyte-forming cells in the bone marrow would result in a decreased number of macrophages in the body, since all of the different macrophages are derived from the monocytes. This would include the microglia of the CNS, the Kupffer's cells of the liver, and alveolar macrophages as well as others. 2. A rise in interferon would indicate a viral infection. Interferon is released from cells that are infected with viruses. It does not help the infected cell, but "interferes" with the virus's ability to infect other cells. 3. Pyrogens stimulate the temperature control area of the preoptic nucleus of the hypothalamus. The result is an increase in body temperature or fever.

Page 400

1. Cytotoxic T cells function in cell-mediated immunity. A decrease in the number of cytotoxic T cells would interfere with the ability to kill foreign cells and tissues as well as cells infected by viruses. 2. Helper T cells promote B cell division, the maturation of plasma cells, and the production of antibody by the plasma cells. Without the helper T cells the humoral immune response would take much longer to occur and would not be as efficient. 3. Since plasma cells produce and secrete antibodies, we would expect to see increased levels of circulating antibodies in the blood if the number of plasma cells were increased.

Page 403

1. The secondary response would be affected by the lack of memory B cells for a specific antigen. The ability to produce a secondary response depends upon the presence of memory B cells and T cells that are formed during the primary response to an antigen. These cells are not involved in the primary response, but are held in reserve against future contact with the same antigen. 2. The developing fetus is protected primarily by passive immunity, the product of IgG antibodies that cross the placenta from the mother's circulation. In addition, the fetus may show some degree of active cellular immunity by the third month of development.

CHAPTER 16

Page 417

1. Increased tension in the vocal cords will cause a higher pitch in the voice. 2. The tracheal cartilages are C-shaped to allow room for esophageal expansion when large portions of food or liquid are swallowed. 3. Without surfactant, surface tension in the thin layer of water that moistens their surfaces would cause the alveoli to collapse.

Page 420

1. On a hot, humid day, the air that we breathe contains more water vapor than on a cool, dry day, and this means that the partial pressure of oxygen in the air is less. Since gases expand when heated, on a hot day the same volume of gas would contain fewer molecules. Thus an individual must breathe deeper or faster (or both) to gain the same amount of oxygen as compared to a cool, dry day. 2. Since the rib penetrates the chest wall, atmospheric air will enter the thoracic cavity. This condition is called a pneumothorax. Pressure within the pleural cavity is normally lower than atmospheric pressure. When air enters the pleural cavity, the natural elasticity of the lung may cause it to collapse. The resulting condition is called atelectasis, or a collapsed lung. 3. Since the fluid produced in pneumonia takes up space that would normally be occupied by air, the vital capacity will be decreased.

Page 424

1. As skeletal muscles become more active they generate more heat and more acid waste products and so lower the pH of surrounding fluid. The combination of lower pH and higher temperature causes the hemoglobin to release more oxygen than it would under conditions of lower temperature and higher pH. 2. An obstruction of the airways would interfere with the body's ability to gain oxygen and eliminate carbon dioxide. Since most carbon dioxide is carried in the blood as bicarbonate ion that is formed from the dissociation of carbonic acid, an inability to eliminate carbon dioxide would result in an excess of hydrogen ions thus lowering the body's pH.

Page 427

1. Chemoreceptors are more sensitive to carbon dioxide. When this gas dissolves it produces hydrogen ions that lower pH and alter cell or tissue activity. 2. Strenuous exercise would stimulate the inflation and deflation reflexes, also known as the Hering-Breuer reflexes. In the inflation reflex, stimulation of stretch receptors within the lungs results in an inhibition of the inspiratory center and stimulation of the expiratory center. In contrast, collapse of the lungs initiates the deflation reflex. This reflex results in an inhibition of the expiratory center and stimulation of the inspiratory center. 3. Johnny's mother shouldn't worry. When Johnny holds his breath, the level of carbon dioxide in his blood will increase. This will lead to increased stimulation of the inspiratory center, forcing Johnny to breathe again.

CHAPTER 17

Page 440

1. Peristalsis would be more efficient in propelling intestinal contents. Segmentation is essentially a churning action that mixes intestinal contents with digestive fluids. 2. Parasympathetic stimulation increases muscle tone and motility in the digestive tract. A drug that blocks this activity would decrease the rate of peristalsis. 3. The process that is being described is swallowing. 4. The lower esophageal sphincter normally prevents the back flow of the stomach contents into the esophagus.

Page 446

1. The pyloric sphincter regulates the flow of chyme into the small intestine. 2. The vagus nerve contains parasympathetic motor fibers that can stimulate gastric secretions. This can occur even if food is not present in the stomach (cephalic phase of gastric digestion). Cutting the branches of the vagus that supply the stomach would prevent this type of secretion from occurring and decrease the chance of ulcer formation. 3. The small intestine has several adaptations that increase surface area to increase its absorptive capacity. First, walls of the small intestine are thrown into folds called the plicae. The tissue that covers the plicae forms fingerlike projections, the villi. The cells that cover the villi have an exposed surface that is covered by small fingerlike projections called the microvilli. In addition, the small intestine has a very rich blood and lymphatic supply to transport the nutrients that are absorbed. 4. It would increase.

Page 452

1. A narrowing of the ileocecal valve would interfere with the flow of chyme from the small intestine to the large intestine. 2. Damage to the exocrine pancreas would most affect the digestion of fats (lipids) because it is the primary source of lipases. 3. A decrease in the amount of bile salts would decrease the effectiveness of fat digestion and absorption.

Page 458

1. Chylomicrons are formed from the fats that are digested in a meal. A meal that is high in fat would increase the number of chylomicrons in the lacteals. 2. Removal of the upper portion of the stomach would interfere with the absorption of vitamin B_{12}. This vitamin requires intrinsic factor, a molecule produced by the parietal cells in the stomach. 3. Diarrhea is potentially life-threatening because a person could lose fluid and electrolytes faster than they can be replaced. This would result in dehydration and possibly death. Although it could be quite uncomfortable, constipation does not interfere with any major body process. The few toxic waste products that are normally eliminated by way of the digestive system can move into the blood and be eliminated by the kidneys..

CHAPTER 18

Page 473

1. Vitamin B_6 (pyridoxine) is an important coenzyme in the processes of deamination and transamination, the first step in processing amino acids in the cell. A deficiency in this vitamin would interfere with the ability to metabolize proteins. 2. Uric acid is the product of adenine and guanine nucleotide degradation in the body. The macromolecules that contain adenine and guanine are the nucleic acids. An increase in uric acid levels could indicate increased breakdown of nucleic acids. 3. HDLs are beneficial because they reduce the amount of fat (including cholesterol) in the bloodstream by transporting it back to the liver for storage or excretion in the bile.

Page 479

1. In terms of recommended servings, between 6 and 11 per day, the bread, cereal, rice, and pasta group is the most important. 2. Foods that contain all of the essential amino acids in nutritionally required amounts are said to contain complete proteins. Foods that are deficient in one or more of the essential amino acids contain incomplete proteins. 3. Bile salts are necessary for the digestion and absorption of fats and fat-soluble vitamins. Vitamin A is a fat-soluble vitamin. A decrease in the amount of bile salts in the bile would result in a decreased ability to absorb the vitamin A from food and result in a vitamin A deficiency.

Page 482

1. The BMR of a pregnant woman should be higher than the BMR of the woman in a nonpregnant state because of increased metabolism associated with support of the fetus as well as the added effect of fetal metabolism. 2. Evaporation is ineffective as a cooling mechanism under conditions of high relative humidity, when the air is holding large amounts of water vapor. 3. Vasoconstriction of peripheral vessels would decrease blood flow to the skin and decrease the amount of heat that the body can lose. As a result, the body temperature would increase.

CHAPTER 19

Page 493

1. Unlike most other organs, such as those of the digestive system, the kidneys lie beneath the peritoneal lining in a retroperitoneal position. 2. The slits created by the podocytes are so fine that they will allow only substances smaller than plasma proteins to pass into the capsular space. 3. Damage to the juxtaglomerular apparatus portion of the nephrons would interfere with the normal control of blood pressure.

Page 499

1. Decreases in blood pressure would reduce the blood hydrostatic pressure within the glomerulus and decrease the GFR. 2. If the nephrons lacked a loop of Henle, the kidneys would not be able to form a concentrated urine.

Page 503

1. Under normal conditions, peristaltic contractions move urine along the minor and major calyces toward the renal pelvis, out of the renal pelvis, and along the ureter to the bladder. 2. An obstruction of the ureters would interfere with the passage of urine from the renal pelvis to the urinary bladder. 3. In order to control the micturition reflex, one must be able to control the external urinary sphincter, a ring of skeletal muscle that acts as a valve.

Page 506

1. Consuming a meal high in salt would temporarily increase the osmolarity of the ECF. As a result some of the water in the ICF would shift to the ECF. 2. Fluid loss through perspiration, urine formation, and respiration would increase the osmolarity of body fluids.

Page 509

1. A decrease in the pH of body fluids would have a stimulating effect on the respiratory center in the medulla. The result would be an increase in the rate of breathing. This would lead to an elimination of more carbon dioxide, which would tend to cause the pH to increase. 2. In a prolonged fast, fatty acids are mobilized and large numbers of ketone bodies are formed. These molecules are acids that lower the body's pH. This would eventually lead to a condition known as ketoacidosis. 3. In vomiting, large amounts of stomach acid are lost from the body. This acid is formed by the parietal cells of the stomach by taking hydrogen ions from the blood. Excessive vomiting would lead to excessive removal of hydrogen ions from the blood to produce the acid, thus raising the body's pH and creating a condition called metabolic alkalosis.

CHAPTER 20

Page 524

1. The cremaster muscle as well as the dartos muscle would be relaxed on a warm day, so that the scrotal sac could descend away from the warmth of the body and cool the testes. 2. The acrosomal cap contains enzymes necessary for penetrating the cell layers around the ovum. Without these enzymes, fertilization would not occur. 3. Dilation of the arteries serving the penis will result in erection. 4. FSH is required for maintaining a high level of testosterone available to support spermatogenesis. Low levels of FSH would lead to low levels of testosterone in the seminiferous tubules and thus a lower rate of sperm production and a low sperm count.

Page 531

1. A blockage of the uterine tube would cause sterility. 2. The acidic pH of the vagina helps to prevent bacterial, fungal, and protozoal infections in this area. 3. The functional layer of the endometrium is sloughed off during menstruation. 4. Blockage of a single lactiferous sinus would not interfere with milk moving to the nipple because each breast usually has between 15 and 20 lactiferous sinuses.

Page 534

1. If the LH surge did not occur during an ovulatory cycle, ovulation and corpus luteum formation would not occur. 2. Progesterone is responsible for the functional maturation and secretion of the endometrial lining. Blocking progesterone receptors would inhibit endometrial development and make it unsuitable for implantation. 3. A sudden decline in the levels of estrogen and progesterone during the menstrual cycle signals the beginning of the menses.

Page 535

1. Inability to contract the ischiocavernosus and bulbocavernosus muscles would interefere with a male's ability to ejaculate and experience orgasm. 2. As the result of parasympathetic stimulation in females during sexual arousal, there is engorgement of the erectile tissues of the clitoris, increased secretion of cervical and vaginal glands, increased blood flow to the walls of the vagina, and engorgement of the blood vessels in the nipples. 3. At menopause, circulating estrogen levels begin to drop. Estrogen has an inhibitory effect on GnRH and FSH, and as the level of estrogen declines the levels of these two hormones rise and remain elevated.

CHAPTER 21

Page 552

1. A sperm cannot fertilize an ovum unless it has undergone capacitation in the female reproductive tract. 2. The inner cell mass of the blastocyst eventually develops into the embryo. 3. After fertilization, the developing trophoblasts and then later on the placenta produce and release the hormone hCG. She is pregnant. 4. Placental functions include (1) supplying the developing fetus with a route for gas exchange, nutrient transfer, and waste product elimination, and (2) producing hormones that affect maternal systems.

Page 558

1. Because blood flow through the placenta reduces the volume of blood in the systemic circuit, and this stimulates an increase in maternal blood volume. 2. Progesterone reduces uterine contractions. A decrease in progesterone at any time during the pregnancy can lead to uterine contractions and in late pregnancy, labor. 3. Relaxin has varied functions. In terms of uterine contractions, it suppresses the release of oxytocin by the hypothalamus and delays the onset of labor contractions.

Page 565

1. A person who is heterozygous for curly hair would have one dominant gene and one recessive gene. The person's phenotype would be "curly hair." 2. There are two different sex chromosomes, an X chromosome and a Y chromosome. The normal male chromosome pair is *XY*, and the female pair is *XX*. The ova produced during meiosis by Joe's wife will always carry *X*, and the sperm produced during meiosis by Joe may carry *X* or *Y*. As a result, the sex of Joe's children depends on which type of sperm cell fertilizes the ovum.

Appendix II

Weights and Measures

Accurate descriptions of physical objects would be impossible without a precise method of reporting the pertinent data. Dimensions such as length and width are reported in standardized units of measurement, such as inches or centimeters. These values can be used to calculate the volume of an object, a measurement of the amount of space it fills. **Mass** is another important physical property. The mass of an object is determined by the amount of matter it contains; on earth the mass of an object determines its weight.

Most U.S. readers describe length and width in terms of inches, feet, or yards; volumes in pints, quarts, or gallons; and weights in ounces, pounds, or tons. These are units of the **U.S. system** of measurement. Table 1 summarizes the familiar and unfamiliar terms used in the U.S. system. For reference purposes, this table also includes a definition of the "household units," popular in recipes and cookbooks. The U.S. system can be very difficult to work with, because there is no logical relationship between the various units. For example, there are 12 inches in a foot, 3 feet in a yard, and 1760 yards in a mile. Without a clear pattern of organization, converting feet to inches or miles to feet can be confusing and time-consuming. The relationships between ounces, pints, quarts, and gallons or ounces, pounds, and tons are no more logical.

In contrast, the **metric system** has a logical organization based on powers of 10, as indicated in Table 2. For example, a **meter (m)** represents the basic unit for the measurement of size. For measuring larger objects, data can be reported in terms of **dekameters** (*deka,* ten), **hectometers** (*hekaton,* hundred), or **kilometers (km**; *chilioi,* thousand); for smaller objects, data can be reported in **decimeters** (0.1 m; *decem,* ten), **centimeters (cm** = 0.01 m; *centum,* hundred), **millimeters (mm** = 0.001 m; *mille,* thousand), and so forth. Notice that the same prefixes are used to report weights, based on the **gram (g)**, and volumes, based on the **liter (l)**. This text reports data in metric units, usually with U.S. equivalents. You should use this opportunity to become familiar with the metric system, because most technical sources report data only in metric units, and most of the rest of the world uses the metric system exclusively. Conversion factors are included in Table 2.

The U.S. and metric systems also differ in their methods of reporting temperatures; in the United States, temperatures are usually reported in degrees Fahrenheit (°F), whereas scientific literature and individuals in most other countries report temperatures in degrees centigrade or Celsius (°C). The relationship between temperatures in degrees Fahrenheit and those in degrees Centigrade has been indicated at the bottom of Table 2.

TABLE 1 The U.S. System of Measurement

Physical Property	*Unit*	*Relationship to Other U.S. Units*	*Relationship to Household Units*
LENGTH	inch (in.)	1 in. = 0.083 ft	
	foot (ft)	1 ft = 12 in. = 0.33 yd	
	yard (yd)	1 yd = 36 in. = 3 ft	
	mile (mi)	1 mi = 5280 ft = 1760 yd	
VOLUME	fluidram (fl dr)	1 fl dr = 0.125 fl oz	
	fluid ounce (fl oz)	1 fl oz = 8 fl dr	= 6 teaspoons (tsp)
		= 0.0625 pt	= 2 tablespoons (tbsp)
	pint (pt)	1 pt = 128 fl dr = 16 fl oz = 0.5 qt	= 32 tbsp = 2 cups (c)
	quart (qt)	1 qt = 256 fl dr = 32 fl oz = 2 pt = 0.25 gal	= 4 c
	gallon (gal)	1 gal = 128 fl oz = 8 pt = 4 qt	
MASS	grain (gr)	1 gr = 0.002 oz	
	dram (dr)	1 dr = 27.3 gr = 0.063 oz	
	ounce (oz)	1 oz = 437.5 gr = 16 dr	
	pound (lb)	1 lb = 7000 gr = 256 dr = 16 oz	
	ton (t)	1 t = 2000 lb	

TABLE 2 The Metric System of Measurement

Physical Property	Unit	Relationship to Standard Metric Units	Conversion to U.S. Units	
LENGTH	nanometer (nm)	1 nm = 0.000000001 m (10^{-9})	= 4 × 10^{-8} in.	25,000,000 nm = 1 in.
	micrometer (μm)	1 μm = 0.000001 m (10^{-6})	= 4 × 10^{-5} in.	25,000 mm = 1 in.
	millimeter (mm)	1 mm = 0.001 m (10^{-3})	= 0.0394 in.	25.4 mm = 1 in.
	centimeter (cm)	1 cm = 0.01 m (10^{-2})	= 0.394 in.	2.54 cm = 1 in.
	decimeter (dm)	1 dm = 0.1 m (10^{-1})	= 3.94 in.	0.25 dm = 1 in.
	meter (m)	standard unit of length	= 39.4 in.	0.0254 m = 1 in.
			= 3.28 ft	0.3048 m = 1 ft
			= 1.09 yd	0.914 m = 1 yd
	dekameter (dam)	1 dam = 10 m		
	hectometer (hm)	1 hm = 100 m		
	kilometer (km)	1 km = 1000 m	= 3280 ft	
			= 1093 yd	
			= 0.62 mi	1.609 km = 1 mi
VOLUME	microliter (μl)	1 μl = 0.000001 l (10^{-6})		
		= 1 cubic millimeter (mm^3)		
	milliliter (ml)	1 ml = 0.001 l (10^{-3})	= 0.03 fl oz	5 ml = 1 tsp
		= 1 cubic centimeter (cm^3 or cc)		15 ml = 1 tbsp
				30 ml = 1 fl oz
	centiliter (cl)	1 cl = 0.01 l (10^{-2})	= 0.34 fl oz	3 cl = 1 fl oz
	deciliter (dl)	1 dl = 0.1 l (10^{-1})	= 3.38 fl oz	0.29 dl = 1 fl oz
	liter (l)	standard unit of volume	= 33.8 fl oz	0.0295 l = 1 fl oz
			= 2.11 pt	0.473 l = 1 pt
			= 1.06 qt	0.946 l = 1 qt
MASS	picogram (pg)	1 pg = 0.000000000001 g (10^{-12})		
	nanogram (ng)	1 ng = 0.000000001 g (10^{-9})		
	microgram (μg)	1 μg = 0.000001 g (10^{-6})	= 0.000015 gr	66,666 mg = 1 gr
	milligram (mg)	1 mg = 0.001 g (10^{-3})	= 0.015 gr	66.7 mg = 1 gr
	centigram (cg)	1 cg = 0.01 g (10^{-2})	= 0.15 gr	6.7 cg = 1 gr
	decigram (dg)	1 dg = 0.1 g (10^{-1})	= 1.5 gr	0.67 dg = 1 gr
	gram (g)	standard unit of mass	= 0.035 oz	28.35 g = 1 oz
			= 0.0022 lb	453.6 g = 1 lb
	dekagram (dag)	1 dag = 10 g		
	hectogram (hg)	1 hg = 100 g		
	kilogram (kg)	1 kg = 1000 g	= 2.2 lb	0.453 kg = 1 lb
	metric ton (mt)	1 mt = 1000 kg	= 1.1 t	
			= 2205 lb	0.907 mt = 1 t

Temperature	*Centigrade*	*Fahrenheit*
Freezing point of pure water	0°	32°
Normal body temperature	36.8°	98.6°
Boiling point of pure water	100°	212°
Conversion	°C → °F: °F = (1.8 × °C) + 32	°F → °C: °C = (°F – 32) × 0.56

The figure below spans the entire range of measurements that we will consider in this book. Gross anatomy traditionally deals with structural organization as seen with the naked eye or with a simple hand lens. A microscope can provide higher levels of magnification and reveal finer details. Before the 1950s, most information was provided by *light microscopy.* A photograph taken through a light microscope is called a **light micrograph (LM)**. Light microscopy can magnify cellular structures about 1000 times and show details as fine as 0.25 µm. The symbol µm stands for *micrometer*; 1 µm = 0.001 mm, or 0.00004 inch. With a light microscope one can identify cell types, such as muscle cells or neurons, and see large structures within the cell. Because individual cells are relatively transparent, thin sections taken through a cell are treated with dyes that stain specific structures, making them easier to see.

Although special staining techniques can show the general distribution of proteins, lipids, carbohydrates, and nucleic acids in the cell, many fine details of intracellular structure remained a mystery until investigators began using electron microscopy. This technique uses a focused beam of electrons, rather than a beam of light, to examine cell structure. In *transmission electron microscopy,* electrons pass through an ultrathin section to strike a photographic plate. The result is a **transmission electron micrograph (TEM)**. Transmission electron microscopy shows the fine structure of cell membranes and intracellular structures. In *scanning electron microscopy,* electrons bouncing off exposed surfaces create a **scanning electron micrograph (SEM)**. Although scanning microscopy cannot achieve as much magnification as transmission microscopy, it provides a three-dimensional perspective on cell structure.

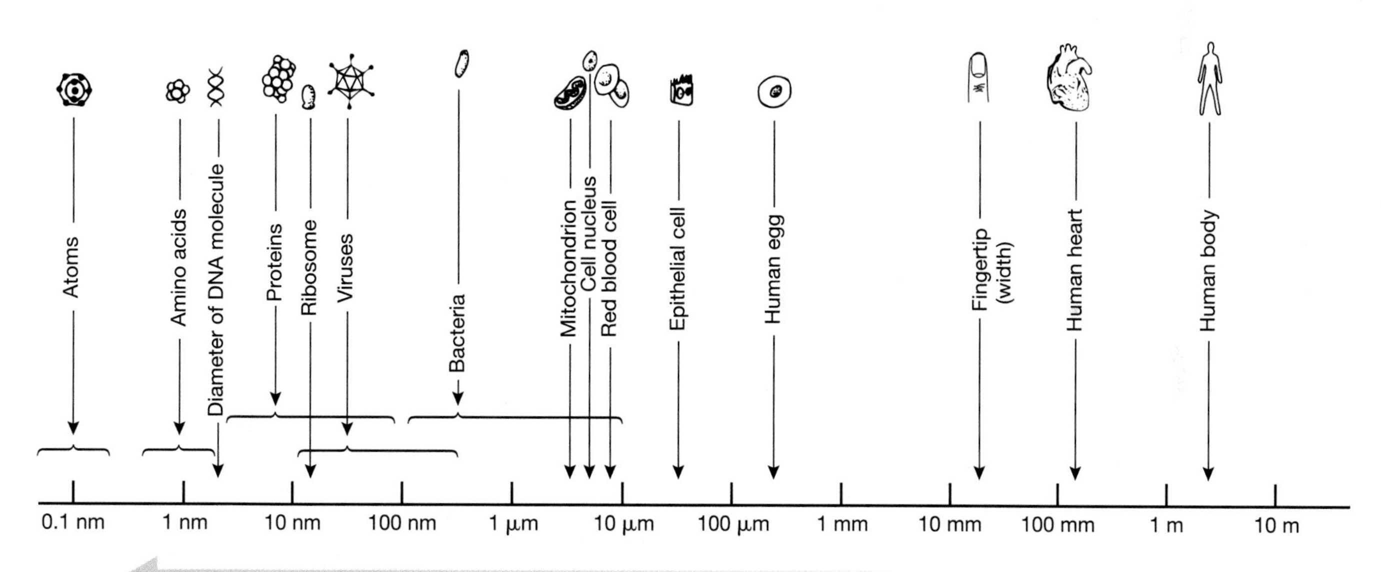

Appendix III

Normal Physiological Values

Tables 3 and 4 present normal averages or ranges for the chemical composition of body fluids. These should be considered approximations, rather that absolute values, as test results vary from laboratory to laboratory owing to differences in procedures, equipment, normal solutions, and so forth. Blanks in the tabular data appear where data were not available; sources used in the preparation of these tables are indicated below.

Sources

Ballenger, John Jacob. 1977. *Diseases of the Nose, Throat, and Ear.* Philadelphia: Lea and Febiger.

Braunwauld, Eugene, Kurt J. Isselbacher, Robert G. Petersdorf, Jean D. Wilson, Joseph B. Martin, and Anthony S. Fauci, eds. 1987. *Harrison's Principles of Internal Medicine,* 11th ed. New York: McGraw-Hill.

TABLE 3 The Chemistry of Blood, Cerebrospinal Fluid, and Urine

	Normal Ranges		
Test	*Blood*[a]	*CSF*	*Urine*
pH	S: 7.38—7.44	7.31—7.34	4.6—8.0
Osmolarity (mOsm/l)	S: 280—295	292—297	500—800
Electrolytes	(mEq/l unless noted)		(urinary loss per 24-hour period[b])
Bicarbonate	P: 21—28	20—24	
Calcium	S: 4.5—5.3	2.1—3.0	6.5—16.5 mEq
Chloride	S: 100—108	116—122	120—240 mEq
Iron	S: 50—150 μg/l	23—52 μg/l	40—150 μg
Magnesium	S: 1.5—2.5	2—2.5	4.9—16.5 mEq
Phosphorus	S: 1.8—2.6	1.2—2.0	0.8—2 g
Potassium	P: 3.8—5.0	2.7—3.9	35—80 mEq
Sodium	P: 136—142	137—145	120—220 mEq
Sulfate	S: 0.2—1.3		1.07—1.3 g
Metabolites	(mg/dl unless noted)		(urinary loss per 24-hour period[c])
Amino acids	P/S: 2.3—5.0	10.0—14.7	41—133 mg
Ammonia	P: 20—150 μg/dl	25—80 μg/dl	340—1200 mg
Bilirubin	S: 0.5—1.2	<0.2	0.02—1.9 mg
Creatinine	P/S: 0.6—1.2	0.5—1.9	1.01—2.5
Glucose	P/S: 70—110	40—70	16—132 mg
Ketone bodies	S: 0.3—2.0	1.3—1.6	10—100 mg
Lactic acid	WB: 5—20[d]	10—20	100—600 mg
Lipids (total)	S: 400—1000	0.8—1.7	0—31.8 mg
Cholesterol (total)	S: 150—300	0.2—0.8	1.2—3.8 mg
Triglycerides	S: 40—150	0—0.9	
Urea	P/S: 23—43	13.8—36.4	12.6—28.6
Uric Acid	S: 2.0—7.0	0.2—0.3	80—976 mg
Proteins	(g/dl)	(mg/dl)	(urinary loss per 24-hour period[c])
Total	S: 6.0—7.8	20—4.5	47—76.2 mg
Albumin	S: 3.2—4.5	10.6—32.4	10—100 mg
Globulins (total)	S: 2.3—3.5	2.8—15.5	7.3 mg (average)
Immunoglobulins	S: 1.0—2.2	1.1—1.7	3.1 mg (average)
Fibrinogen	P: 0.2—0.4	0.65 (average)	

[a] *S = serum, P = plasma, WB = whole blood.*

[b] *Because urinary output averages just over 1 liter per day, these electrolyte values are comparable to mEq/l.*

[c] *Because urinary metabolite and protein data approximate mg/l or g/l, they must be divided by 10 for comparison with CSF or blood concentrations.*

[d] *Venous blood sample*

Lentner, Cornelius, ed. 1981. *Geigy Scientific Tables,* 8th ed. Basel, Switzerland: Ciba-Geigy.
Davidsohn, Israel, and John Bernard Henry, eds. 1969. *Todd-Sanford Clinical Diagnosis by Laboratory Methods,* 14th ed. Philadelphia: W.B. Saunders.
Diem, K., and C. Lenter, eds. 1970. *Scientific Tables,* 7th ed. Basel, Switzerland: Ciba-Geigy.
Halsted, James A. 1976. *The Laboratory in Clinical Medicine: Interpretation and Application.* Philadelphia: W.B. Saunders.
Harper, Harold A. 1987. *Review of Physiological Chemistry.* Los Altos, Calif.: Lange Medical Publications.

TABLE 4 The Composition of Minor Body Fluids

	Normal Averages or Ranges					
Test	*Perilymph*	*Endolymph*	*Synovial Fluid*	*Sweat*	*Saliva*	*Semen*
pH			7.4	4—6.8	6.4[a]	7.19
Specific gravity			1.008–1.015	1.001–1.008	1.007	1.028
Electrolytes (mEq/l)						
Potassium	5.5–6.3	140–160	4.0	4.3–14.2	21	31.3
Sodium	143–150	12–16	136.1	0–104	14[a]	117
Calcium	1.3–1.6	0.05	2.3–4.7	0.2–6	3	12.4
Magnesium	1.7	0.02		0.03–4	0.6	11.5
Bicarbonate	17.8–18.6	20.4–21.4	19.3–30.6		6[a]	24
Chloride	121.5	107.1	107.1	34.3	17	42.8
Proteins (mg/dl)						
Total	200	150	1.72 g/dl	7.7	386[b]	4.5 g/dl
Metabolites (mg/dl)						
Amino acids				47.6	40	1.26 g/dl
Glucose	104		70–110	3.0	11	224 (fructose)
Urea				26–122	20	72
Lipids, total	12		20.9	[d]	25–500[c]	188

[a] *Increases under salivary stimulation.*
[b] *Primarily alpha-anylase, with some lysozomes.*
[c] *Cholesterol.*
[d] *Not present in eccrine secretions.*

Glossary

A

abdominopelvic cavity: Portion of the ventral body cavity that contains abdominal and pelvic subdivisions.
abduction: Movement away from the midline.
absorption: The active or passive uptake of gases, fluids, or solutes.
accommodation: Alteration in the curvature of the lens to focus an image on the retina; decrease in receptor sensitivity or perception following chronic stimulation.
acetabulum (a-se-TAB-ū-lum): Fossa on lateral aspect of pelvis that accommodates the head of the femur.
acetylcholine (ACh) (as-ē-til-KŌ-lēn): Chemical neurotransmitter in the brain and PNS; dominant neurotransmitter in the PNS, released at neuromuscular junctions and synapses of the parasympathetic division.
acetylcholinesterase (AChE): Enzyme found in the synaptic cleft, bound to the postsynaptic membrane, and in tissue fluids; breaks down and inactivates ACh molecules.
acetyl-CoA: An acetyl group bound to coenzyme A, a participant in the anabolic and catabolic pathways for carbohydrates, lipids, and many amino acids.
acetyl group: $CH_3C{=}O$.
acid: A compound whose dissociation in solution releases a hydrogen ion and an anion; an acid solution has a pH below 7.0 and contains an excess of hydrogen ions.
acidosis (a-sid-Ō-sis): An abnormal physiological state characterized by a plasma pH below 7.35.
acinus/acini (AS-i-nī): Histological term referring to a blind pocket, pouch, or sac.
actin: Protein component of microfilaments; form thin filaments in skeletal muscles and produce contractions of all muscles through interaction with thick (myosin) filaments; *see* **sliding filament theory**.
action potential: A conducted change in the membrane potential of excitable cells, initiated by a change in the membrane permeability to sodium ions; *see also* **nerve impulse**.
active transport: The ATP-dependent absorption or excretion of solutes across a cell membrane.
acute: Sudden in onset, severe in intensity, and brief in duration.
adduction: Movement toward the axis or midline of the body as viewed in the anatomical position.
adenine: One of the nitrogenous bases in the nucleic acids RNA and DNA.
adenoids: The pharyngeal tonsil.
adenosine: A nucleoside consisting of adenine and a 5-carbon sugar.
adenosine diphosphate (ADP): Adenosine with two phosphate groups attached.
adenosine phosphate (AMP): A nucleotide consisting of adenine plus a phosphate group (PO_4^{3-}); also known as adenosine monophosphate.
adenosine triphosphate (ATP): A high-energy compound consisting of adenosine with three phosphate groups attached; the third is attached by a high-energy bond.
adenylate cyclase: An enzyme bound to the inner surfaces of cell membranes that can convert ATP to cyclic-AMP. Also called *adenyl cyclase* and *adenylyl cyclase*.
adipocyte (AD-i-pō-sīt): A fat cell.
adipose tissue: Loose connective tissue dominated by adipocytes.
adrenal cortex: Superficial portion of adrenal gland that produces steroid hormones.
adrenal gland: Small endocrine gland secreting hormones, located superior to each kidney.
adrenal medulla: Core of the adrenal gland; a modified sympathetic ganglion that secretes hormones into the blood following sympathetic activation.
adrenergic (ad-ren-ER-jik): A synaptic terminal that releases norepinephrine when stimulated.
adrenocorticotropic hormone (ACTH): Hormone that stimulates the production and secretion of glucocorticoids by the adrenal cortex; released by the anterior pituitary.
adventitia (ad-ven-TISH-a): Superficial layer of connective tissue surrounding an internal organ; fibers are continuous with those of surrounding tissues, providing support and stabilization.
aerobic respiration: The complete breakdown of organic substrates into carbon dioxide and water, via pyruvic acid; a process that yields large amounts of ATP but requires mitochondria and oxygen.
afferent arteriole: An arteriole bringing blood to a glomerulus of the kidney.
afferent fiber: Axons carrying sensory information to the CNS.
agglutination (a-gloo-ti-NĀ-shun): Aggregation of red blood cells due to interactions between surface agglutinogens and plasma agglutinins.
agonist: A muscle responsible for a specific movement.
albinism: Absence of pigment in hair and skin caused by inability of melanocytes to produce melanin.
aldosterone: A mineralocorticoid produced by the adrenal cortex; stimulates sodium and water conservation at the kidneys; secreted in response to the presence of angiotensin II.
alkalosis (al-kah-LŌ-sis): Condition characterized by a plasma pH of greater than 7.45 and associated with relative deficiency of hydrogen ions or an excess of bicarbonate ions.
allantois (a-LAN-tō-is): One of the extraembryonic membranes; it provides vascularity to the chorion and is therefore essential to placenta formation; the proximal portion becomes the urinary bladder.
alleles (a-LĒLZ): Alternate forms of a particular gene.
allergen: An antigenic compound that produces a hypersensitivity response.
alpha cells: Cells in the pancreatic islets that secrete glucagon.
alveolar sac: An air-filled chamber that supplies air to several alveoli.
alveolus/alveoli (al-VĒ-o-lī): Blind pockets at the end of the respiratory tree, lined by a simple squamous epithelium and surrounded by a capillary network; gas exchange with the blood occurs here.
amino acids: Organic compounds whose chemical structure can be summarized as $R{-}CHNH_2COOH$.
amino group: NH_2.
amnesia: Temporary or permanent memory loss.
amnion (AM-nē-on): One of the extraembryonic membranes; surrounds the developing embryo/fetus.
amniotic fluid (am-nē-OT-ik): Fluid that fills the amniotic cavity; provides cushioning and support for the embryo/fetus.
amphiarthrosis (am-fē-ar-THRŌ-sis): An articulation that permits a small degree of independent movement.
amphimixis (am-fi-MIK-sis): The fusion of male and female pronuclei following fertilization.
ampulla/ampullae (am-PYŪL-la): A localized dilation in the lumen of a canal or passageway.
amylase: An enzyme that breaks down polysaccharides, produced by the salivary glands and pancreas.
anabolism (a-NAB-ō-lizm): The synthesis of complex organic compounds from simpler precursors.
anaerobic: Without oxygen.
anaphase (AN-a-fāz): Mitotic stage in which the paired chromatids separate and move toward opposite ends of the spindle apparatus.
anastomosis (a-nas-to-MŌ-sis): The joining of two tubes, usually referring to a connection between two peripheral vessels without an intervening capillary bed.
anatomical position: An anatomical reference position, the body viewed from the anterior surface with the palms facing forward; supine.
anatomy (a-NAT-ō-mē): The study of the structure of the body.
androgen (AN-drō-jen): A steroid sex hormone primarily produced by the interstitial cells of the testis and manufactured in small quantities by the adrenal cortex in either sex.
anemia (a-NĒ-mē-ah): Condition marked by a reduction in the hematocrit and/or hemoglobin content of the blood.
aneurysm (AN-ū-rizm): A weakening and localized dilation in the wall of a blood vessel.
angiogram (AN-jē-ō-gram): An X-ray image of circulatory pathways.
angiotensin I, II: Angiotensin II is a hormone that causes an elevation in systemic blood pressure, stimulates secretion of aldosterone, promotes thirst, and causes the release of ADH.
angiotensinogen: Blood protein produced by the liver that is converted to angiotensin I by the enzyme renin.
anion (AN-ī-on): An ion bearing a negative charge.
annulus (AN-ū-lus): A cartilage or bone shaped like a ring.
anoxia (a-NOKS-ē-a): Tissue oxygen deprivation.
antagonist: A muscle that opposes the movement of an agonist.
anterior: On or near the front or ventral surface of the body.
anterior pituitary: *See* **pituitary gland**.
antibiotic: Chemical agent that selectively kills pathogenic microorganisms.
antibody (AN-ti-bod-ē): A globular protein produced by plasma cells that will bind to

specific antigens and promote their destruction or removal from the body.
anticoagulant: Compound that slows or prevents clot formation by interfering with the clotting system.
anticodon: Triplet of nitrogenous bases on a tRNA molecule that interacts with an appropriate codon on a strand of mRNA.
antidiuretic hormone (ADH) (an-tī-dī-ū-RET-ik): Hormone synthesized in the hypothalamus and secreted at the posterior pituitary; causes water retention at the kidneys and an elevation of blood pressure.
antigen: A substance capable of inducing the production of antibodies.
antigen-antibody complex: The combination of an antigen and a specific antibody.
antigenic determinant site: A portion of an antigen that can interact with an antibody molecule.
antrum (AN-trum): A chamber or pocket.
anus: External opening of the anorectal canal.
aorta: Large, elastic artery that carries blood away from the left ventricle and into the systemic circuit.
aortic reflex: Baroreceptor reflex triggered by increased aortic pressures; leads to a reduction in cardiac output and a fall in systemic pressure.
aphasia: Inability to speak.
apocrine secretion: Mode of secretion in which the glandular cell sheds portions of its cytoplasm.
aponeurosis/aponeuroses (ap-ō-nū-RŌ-sēz): A broad tendinous sheet that may serve as the origin or insertion of a skeletal muscle.
appendix: A blind tube connected to the cecum of the large intestine.
appositional growth: Enlargement by the addition of cartilage or bony matrix to the outer surface.
aqueous humor: Fluid similar to perilymph or CSF that fills the anterior chamber of the eye.
arachnoid (a-RAK-noyd): The middle meninges that encloses CSF and protects the central nervous system.
arcuate (AR-kū-āt): Curving.
areola (a-RĒ-ō-la): Pigmented area that surrounds the nipple of a breast.
areolar: Containing minute spaces, as in areolar connective tissue.
arrector pili (a-REK-tōr PĪ-li): Smooth muscles whose contractions cause piloerection.
arrhythmias (a-RITH-mē-az): Abnormal patterns of cardiac contractions.
arteriole (ar-TĒ-rē-ol): A small arterial branch that delivers blood to a capillary network.
artery: A blood vessel that carries blood away from the heart and toward a peripheral capillary.
arthritis (ar-THRĪ-tis): Inflammation of a joint.
articular: Pertaining to a joint.
articular capsule: Dense collagen fiber sleeve that surrounds a joint and provides protection and stabilization.
articular cartilage: Cartilage pad that covers the surface of a bone inside a joint cavity.
ascending tract: A tract carrying information from the spinal cord to the brain.
association areas: Cortical areas of the cerebrum responsible for integration of sensory inputs and/or motor commands.
association neuron: *See* **interneuron**.
asthma (AZ-ma): Reversible constriction of smooth muscles around respiratory passageways, frequently caused by an allergic response.
astigmatism: Visual disturbance due to an irregularity in the shape of the cornea.
astrocyte (AS-trō-sīt): One of the glial cells in the CNS; responsible for the blood-brain barrier.
atherosclerosis (ath-er-ō-skle-RŌ-sis): Formation of fatty plaques in the walls of arteries, leading to circulatory impairment.
atom: The smallest stable unit of matter.
atomic number: The number of protons in the nucleus of an atom.
atomic weight: Roughly, the average total number of protons and neutrons in the atoms of a particular element.
atria: Thin-walled chambers of the heart that receive venous blood from the pulmonary or systemic circuit.
atrial natriuretic peptide (nā-trē-ū-RET-ik): Hormone released by specialized atrial cardiocytes when they are stretched by an abnormally large venous return; promotes fluid loss and reductions in blood pressure and venous return.
atrial reflex: Reflexive increase in heart rate following an increase in venous return; due to mechanical and neural factors; also called the *Bainbridge reflex.*
atrioventricular (AV) node (ā-trē-ō-ven-TRIK-ū-lar): Specialized cardiocytes that relay the contractile stimulus to the AV bundle, the bundle branches, the Purkinje fibers, and the ventricular myocardium; located at the boundary between the atria and ventricles.
atrioventricular (AV) valve: One of the valves that prevent backflow into the atria during ventricular systole.
atrophy (AT-rō-fē): Wasting away of tissues from lack of use, ischemia, or nutritional abnormalities.
auditory: Pertaining to the sense of hearing.
auditory ossicles: The bones of the middle ear: malleus, incus, and stapes.
autoimmunity: Immune system sensitivity to normal cells and tissues, resulting in the production of autoantibodies.
autolysis: Destruction of a cell due to the rupture of lysosomal membranes in its cytoplasm.
automaticity: Spontaneous depolarization to threshold, a characteristic of cardiac pacemaker cells.
autonomic ganglion: A collection of visceral motor neurons outside the CNS.
autonomic nerve: A peripheral nerve consisting of preganglionic or postganglionic autonomic fibers.
autonomic nervous system (ANS): Centers, nuclei, tracts, ganglia, and nerves involved in the unconscious regulation of visceral functions; includes components of the CNS and PNS.
autoregulation: Alterations in activity that maintain homeostasis in direct response to changes in the local environment; does not require neural or endocrine control.
autosomal (aw-to-SŌ-mal): Chromosomes other than the X or Y chromosomes.
avascular (ā-VAS-kū-lar): Without blood vessels.
axilla: The armpit.
axon: Elongate extension of a neuron that conducts an action potential.
axon hillock: In a multipolar neuron, the portion of the neural soma adjacent to the initial segment.
axoplasm (AK-so-plazm): Cytoplasm within an axon.

B

bacteria: Single-celled microorganisms, some pathogenic, that are common in the environment.
Bainbridge reflex: *See* **atrial reflex**.
baroreceptor reflex: A reflexive change in cardiac activity in response to changes in blood pressure.
baroreceptors (bar-ō-rē-SEP-torz): Receptors responsible for baroreception.
basal metabolic rate (BMR): The resting metabolic rate of a normal fasting subject under homeostatic conditions.
base: A compound whose dissociation releases a hydroxide ion (OH^-) or removes a hydrogen ion from the solution.
basement membrane: A layer of filaments and fibers that attach an epithelium to the underlying connective tissue.
basilar membrane: Membrane that supports the organ of Corti and separates the cochlear duct from the tympanic duct in the inner ear.
basophils (BĀ-sō-filz): Circulating granulocytes (WBCs) similar in size and function to tissue mast cells.
B cells: Lymphocytes capable of differentiating into the plasma cells that produce antibodies.
beta cells: Cells of the pancreatic islets that secrete insulin in response to elevated blood sugar concentrations.
beta oxidation: Fatty acid catabolism that produces molecules of acetyl-CoA.
bicarbonate ions: HCO_3^-; anion components of the carbonic acid-bicarbonate buffer system.
bicuspid (bī-KUS-pid): A sharp, conical tooth, also called a canine tooth.
bicuspid valve: The left AV valve, also known as the *mitral valve.*
bile: Exocrine secretion of the liver that is stored in the gallbladder and ejected into the duodenum.
bile salts: Steroid derivatives in the bile, responsible for the emulsification of ingested lipids.
bilirubin (bil-ē-ROO-bin): A pigment, byproduct of hemoglobin catabolism.
biopsy: The removal of a small sample of tissue for pathological analysis.
bladder: A muscular sac that distends as fluid is stored, and whose contraction ejects the fluid at an appropriate time; used alone, the term usually refers to the urinary bladder.
blastocyst (BLAS-tō-sist): Early stage in the developing embryo, consisting of an outer trophoblast and an inner cell mass.
blood-brain barrier: Isolation of the CNS from the general circulation; primarily the result of astrocyte regulation of capillary permeabilities.
blood clot: A network of fibrin fibers and trapped blood cells.
blood pressure: A force exerted against the vascular walls by the blood, as the result of the push exerted by cardiac contraction and the elasticity of the vessel walls. It is usually measured along one of the muscular arteries, with systolic pressure measured during

ventricular systole, and diastolic pressure during ventricular diastole.
blood-testis barrier: Isolation of the seminiferous tubules from the general circulation, due to the activities of the sustentacular (Sertoli) cells.
bolus: A compact mass; usually refers to compacted ingested material on its way to the stomach.
bone: *See* **osseous tissue**.
bowel: The intestinal tract.
brachial: Pertaining to the arm.
brachial plexus: Network formed by branches of spinal nerves C_5–T_1 en route to innervate the upper limb.
brachium: The arm.
brain stem: The brain minus the cerebrum, diencephalon, and cerebellum.
brevis: Short.
Broca's center: The speech center of the brain, usually found on the neural cortex of the left cerebral hemisphere.
bronchial tree: The trachea, bronchi, and bronchioles.
bronchitis (brong-KĪ-tis)**:** Inflammation of the bronchial passageways.
bronchodilation: Dilation of the bronchial passages; may be caused by sympathetic stimulation.
bronchus/bronchi: One of the branches of the bronchial tree between the trachea and bronchioles.
buccal (BUK-al)**:** Pertaining to the cheeks.
buffer: A compound that stabilizes the pH of a solution by removing or releasing hydrogen ions.
buffer system: Interacting compounds that prevent increases or decreases in the pH of body fluids; includes the carbonic acid–bicarbonate buffer system, the phosphate buffer system, and the protein buffer system.
bulbar: Pertaining to the brain stem.
bulbourethral glands (bul-bō-ū-RĒ-thral)**:** Mucous glands at the base of the penis that secrete into the penile urethra; also called Cowper's glands.
bundle branches: Specialized conducting cells in the ventricles that carry the contractile stimulus from the AV bundle to the Purkinje fibers.
bundle of His (hiss)**:** Specialized conducting cells in the interventricular septum that carry the contracting stimulus from the AV node to the bundle branches and thence to the Purkinje fibers. Also called AV bundle.
bursa: A small sac filled with synovial fluid that cushions adjacent structures and reduces friction.

C

calcaneal tendon: Large tendon that inserts on the calcaneus; tension on this tendon produces plantar flexion of the foot; also called the Achilles tendon.
calcaneus (kal-KĀ-nē-us)**:** The heelbone, the largest of the tarsal bones.
calcification: The deposition of calcium salts within a tissue.
calcitonin (kal-si-TŌ-nin)**:** Hormone secreted by C cells of the thyroid when calcium ion concentrations are abnormally high; restores homeostasis by increasing the rate of bone deposition and the renal rate of calcium loss.
calorie (c) (KAL-o-rē)**:** The amount of heat required to raise the temperature of one gram of water 1°C.
Calorie (C): The amount of heat required to raise the temperature of one kilogram of water 1°C.
calyx/calyces (KĀL-i-sēz)**:** A cup-shaped division of the renal pelvis.
canaliculi (kan-a-LIK-ū-lī)**:** Microscopic passageways between cells; bile canaliculi carry bile to bile ducts in the liver; in bone, canaliculi permit the diffusion of nutrients and wastes to and from osteocytes.
cancellous bone (KAN-sel-us)**:** Spongy bone, composed of a network of bony struts.
cancer: A malignant tumor that tends to undergo metastasis.
capillary: Small blood vessels, interposed between arterioles and venules, whose thin walls permit the diffusion of gases, nutrients, and wastes between the plasma and interstitial fluids.
capitulum (ka-PIT-ū-lum)**:** General term for a small, elevated articular process; used to refer to the rounded distal surface of the humerus that articulates with the radial head.
carbaminohemoglobin (kar-bam-ē-nō-hē-mō-GLŌ-bin)**:** Hemoglobin bound to carbon dioxide molecules.
carbohydrate (kar-bo-HĪ-drāt)**:** Organic compound containing carbon, hydrogen, and oxygen in a ratio that approximates 1:2:1.
carbon dioxide: CO2, a compound produced by the decarboxylation reactions of aerobic metabolism.
carbonic anhydrase: An enzyme that catalyzes the reaction $H_2O + CO_2 \rightleftharpoons H_2CO_3$; important in carbon dioxide transport, gastric acid secretion, and renal pH regulation.
carboxyl group (kar-BOKS-il)**:** –COOH, an acid group found in fatty acids, amino acids, etc.
cardia (KAR-dē-a)**:** The area of the stomach surrounding its connection with the esophagus.
cardiac cycle: One complete heartbeat, including atrial and ventricular systole and diastole.
cardiac output: The amount of blood ejected by the left ventricle each minute; normally about 5 liters.
cardiac reserve: The potential percentage increase in cardiac output above resting levels.
cardiocyte (KAR-dē-ō-sīt)**:** A cardiac muscle cell.
cardiopulmonary resuscitation (CPR): Method of artificially maintaining respiratory and circulatory function.
cardiovascular centers: Poorly localized centers in the reticular formation of the medulla of the brain; includes cardioacceleratory, cardioinhibitory, and vasomotor centers.
cardium: The heart.
carotid artery: The principal artery of the neck, servicing cervical and cranial structures; one branch, the internal carotid, represents a major blood supply for the brain.
carotid body: A group of receptors adjacent to the carotid sinus that are sensitive to changes in the carbon dioxide levels, pH, and oxygen concentrations of the arterial blood.
carotid sinus: A dilated segment at the base of the internal carotid artery whose walls contain baroreceptors sensitive to changes in blood pressure.
carotid sinus reflex: Reflexive changes in blood pressure that maintain homeostatic pressures at the carotid sinus, stabilizing blood flow to the brain.
carpus/carpal: The wrist.
cartilage: A connective tissue with a gelatinous matrix containing an abundance of fibers.
catabolism (ka-TAB-ō-lizm)**:** The breakdown of complex organic molecules into simpler components, accompanied by the release of energy.
catalyst (KAT-ah-list)**:** A substance that accelerates a specific chemical reaction but that is not altered by the reaction.
cation (KAT-ī-on)**:** An ion that bears a positive charge.
caudal/caudally: Closest to or toward the tail (coccyx).
cavernous tissue: Erectile tissue that can be engorged with blood; found in the penis and clitoris.
cecum (SĒ-kum)**:** An expanded pouch at the start of the large intestine.
cell: The smallest living unit in the human body.
cell-mediated immunity: Resistance to disease through the activities of sensitized T cells that destroy antigen-bearing cells by direct contact or through the release of lymphotoxins; also called cellular immunity.
central nervous system (CNS): The brain and spinal cord.
central sulcus: Groove in the surface of a cerebral hemisphere, between the primary sensory and primary motor areas of the cortex.
centriole: A cylindrical intracellular organelle composed of 9 groups of microtubules, 3 in each group; functions in mitosis or meiosis by organizing the microtubules of the spindle apparatus.
centromere (SEN-trō-mēr)**:** Localized region where two chromatids remain connected following chromosome replication; site of spindle fiber attachment.
cerebellum (ser-e-BEL-um)**:** Posterior portion of the metencephalon, containing the cerebellar hemispheres; includes the arbor vitae, cerebellar nuclei, and cerebellar cortex.
cerebral cortex: An extensive area of neural cortex covering the surfaces of the cerebral hemispheres.
cerebral hemispheres: Expanded portions of the cerebrum covered in neural cortex.
cerebral nuclei: Nuclei of the cerebrum that are important components of the extrapyramidal system.
cerebrospinal fluid (CSF): Fluid bathing the internal and external surfaces of the CNS; secreted by the choroid plexus.
cerebrum (SER-e-brum)**:** The largest portion of the brain, composed of the cerebral hemispheres; includes the cerebral cortex, the cerebral nuclei, and the white matter.
cervix: The lower part of the uterus.
chemoreception: Detection of alterations in the concentrations of dissolved compounds or gases.
chemotaxis (kē-mō-TAK-sis)**:** The attraction of phagocytic cells to the source of abnormal chemicals in tissue fluids.
cholecystokinin (CCK) (kō-lē-sis-tō-KĪ-nin)**:** Duodenal hormone that stimulates the contraction of the gallbladder and the secretion of enzymes by the exocrine pancreas.

cholesterol: A steroid component of cell membranes and a substrate for the synthesis of steroid hormones and bile salts.
cholinergic synapse (kō-lin-ER-jik): Synapse where the presynaptic membrane releases ACh on stimulation.
cholinesterase (kō-li-NES-te-rās): Enzyme that breaks down and inactivates ACh.
chondrocyte (KON-drō-sīt): Cartilage cell.
chordae tendineae (KOR-dē TEN-di-nē-ē): Fibrous cords that stabilize the position of the AV valves in the heart, preventing backflow during ventricular systole.
chorion/chorionic (KOR-ē-on) (ko-rē-ON-ik): An extraembryonic membrane, consisting of the trophoblast and underlying mesoderm, that forms the placenta.
choroid: Middle, vascular layer in the wall of the eye.
choroid plexus: The vascular complex in the roof of the third and fourth ventricles of the brain, responsible for CSF production.
chromatid (KRŌ-ma-tid): One complete copy of a single chromosome.
chromosomes: Dense structures, composed of tightly coiled DNA strands and associated histones, that become visible in the nucleus when a cell prepares to undergo mitosis or meiosis; normal human somatic cells contain 46 chromosomes apiece.
chylomicrons (kī-lō-MĪ-kronz): Relatively large droplets that may contain triglycerides, phospholipids, and cholesterol in association with proteins; synthesized and released by intestinal cells and transported to the venous blood via the lymphatic system.
chyme (kīm): A semifluid mixture of ingested food and digestive secretions that is found in the stomach and proximal small intestine as digestion proceeds.
ciliary body: A thickened region of the choroid that encircles the lens of the eye; it includes the ciliary muscle and the ciliary processes that support the suspensory ligaments of the lens.
cilium/cilia: A slender organelle that extends above the free surface of an epithelial cell, and usually undergoes cycles of movement; composed of a basal body and microtubules in a 9 × 2 array.
circulatory system: The network of blood vessels that are components of the cardiovascular system.
circumduction (sir-kum-DUK-shun): A movement at a synovial joint where the distal end of the bone describes a circle but the shaft does not rotate.
cisterna (sis-TUR-na): An expanded chamber.
cleavage (KLĒ-vij): Mitotic divisions that follow fertilization of the ovum and lead to the formation of a blastocyst.
clitoris (KLI-to-ris): A small erectile organ of the female that is the developmental equivalent of the male penis.
clot: A network of fibrin fibers and trapped blood cells; also called a thrombus if it occurs within the circulatory system.
clotting factors: Plasma proteins synthesized by the liver that are essential to the clotting response.
coccyx (KOK-siks): Terminal portion of the spinal column, consisting of relatively tiny, fused vertebrae.
cochlea (KOK-lē-a): Spiral portion of the bony labyrinth of the inner ear that surrounds the organ of hearing.
cochlear duct (KOK-lē-ar): Membranous tube within the cochlea that is filled with endolymph and contains the organ of Corti.
codon (KŌ-don): A sequence of three nitrogenous bases along an mRNA strand that will specify the location of a single amino acid in a peptide chain.
coelom (SĒ-lom): The ventral body cavity, lined by a serous membrane and subdivided during development into the pleural, pericardial, and abdominopelvic (peritoneal) cavities.
coenzymes (kō-EN-zīmz): Complex organic cofactors, usually structurally related to vitamins.
collagen: Strong, insoluble protein fiber common in connective tissues.
colliculus/colliculi (kol-IK-ū-lus): A little mound; in the brain, used to refer to one of the thickenings in the roof of the midbrain; the superior colliculus is associated with the visual system, and the inferior colliculi with the auditory system.
colon: The large intestine.
comminuted: Broken or crushed into small pieces.
commissure: A crossing over from one side to another.
common bile duct: Duct formed by the union of the cystic duct from the gallbladder and the bile ducts from the liver; terminates at the duodenal ampulla, where it meets the pancreatic duct.
common pathway: In the clotting response, the events that begin with the appearance of thromboplastin and end with the formation of a clot.
compact bone: Dense bone containing parallel osteons.
compensation curves: The cervical and lumbar curves that develop to center the body weight over the legs.
complement: Plasma proteins that interact in a chain reaction following exposure to activated antibodies or the surfaces of certain pathogens, and which promote cell lysis, phagocytosis, and other defense mechanisms.
compound: A molecule containing two or more elements in combination.
concentration: Amount (in grams) or number of atoms, ions, or molecules (in moles) per unit volume.
concentration gradient: Regional differences in the concentration of a particular substance.
concha/conchae (KONG-ka): Three pairs of thin, scroll-like bones that project into the nasal cavities; the superior and medial conchae are part of the ethmoid, and the inferior are separate bones.
condyle: A rounded articular projection on the surface of a bone.
cone: Retinal photoreceptor responsible for color vision.
congenital (kon-JEN-i-tal): Already present at the birth of an individual.
conjunctiva (kon-junk-TĪ-va): A layer of stratified squamous epithelium that covers the inner surfaces of the lids and the anterior surface of the eye to the edges of the cornea.
connective tissue: One of the four primary tissue types; provides a structural framework for the body that stabilizes the relative positions of the other tissue types; includes connective tissue proper, cartilage, bone, and blood; always has cell products, cells, and ground substance.
contractility: The ability to contract, possessed by skeletal, smooth, and cardiac muscle cells.
convergence: In the nervous system, the term indicates that the axons from several neurons innervate a single neuron; this is most common along motor pathways.
coracoid process (ko-RA-kōyd): A hook-shaped process of the scapula that projects above the anterior surface of the capsule of the shoulder joint.
cornea (KŌR-nē-a): Transparent portion of the fibrous tunic of the anterior surface of the eye.
cornification: The production of keratin by a stratified squamous epithelium; also called keratinization.
corona radiata (ko-RŌ-na rā-dē-A-ta): A layer of follicle cells surrounding a secondary oocyte at ovulation.
coronoid (kō-RŌ-noyd): Hooked or curved.
corpus/corpora: Body.
corpus callosum: Bundle of axons linking centers in the left and right cerebral hemispheres.
corpus luteum (LOO-tē-um): Progestin-secreting mass of follicle cells that develops in the ovary after ovulation.
cortex: Outer layer or portion of an organ.
Corti, organ of: Receptor complex in the cochlear duct that includes the inner and outer hair cells, supporting cells and structures, and the tectorial membrane; provides the sensation of hearing.
corticospinal tracts: Descending tracts that carry motor commands from the cerebral cortex to the anterior gray horns of the spinal cord.
corticosteroid: A steroid hormone produced by the adrenal cortex.
cortisol (KŌR-ti-sol): One of the corticosteroids secreted by the adrenal cortex; a glucocorticoid.
costa/costae: A rib.
cotransport: Membrane transport of a nutrient, such as glucose, in company with the movement of an ion, usually sodium; transport requires a carrier protein but does not involve direct ATP expenditure and can occur regardless of the concentration gradient for the nutrient.
covalent bond (kō-VĀ-lent): A chemical bond between atoms that involves the sharing of electrons.
coxa/coxae: The bones of the hip.
cranial nerves: Peripheral nerves originating at the brain.
cranium: The brainbox; the skull bones that surround the brain.
creatine phosphate: A high-energy compound present in muscle cells; during muscular activity the phosphate group is donated to ADP, regenerating ATP. Also called phosphocreatine and phosphorylcreatine.
creatinine: A breakdown product of creatine metabolism.
crenation: Cellular shrinkage due to an osmotic movement of water out of the cytoplasm.
cribriform plate: Portion of the ethmoid bone of the skull that contains the foramina used

by the axons of olfactory receptors en route to the olfactory bulbs of the cerebrum.
cricoid cartilage (KRĪ-koyd): Ring-shaped cartilage forming the inferior margin of the larynx.
crista/cristae: A ridge-shaped collection of hair cells in the ampulla of a semicircular canal; the crista and cupula form a receptor complex sensitive to movement along the plane of the canal.
cross-bridge: Myosin head that projects from the surface of a thick filament and that can bind to an active site of a thin filament in the presence of calcium ions.
cupula (KYŪ-pū-la): A gelatinous mass that sits in the ampulla of a semicircular canal in the inner ear, and whose movement stimulates the hair cells of the crista.
cutaneous membrane: The epidermis and papillary layer of the dermis.
cuticle: Layer of dead, cornified cells surrounding the shaft of a hair; for nails, *see* **eponychium**.
cyanosis: Bluish coloration of the skin due to the presence of deoxygenated blood in vessels near the body surface.
cystic duct: A duct that carries bile between the gallbladder and the common bile duct.
cytokinesis (sī-tō-ki-NĒ-sis): The cytoplasmic movement that separates two daughter cells at the completion of mitosis.
cytology (sī-TOL-ō-jē): The study of cells.
cytoplasm: The material between the cell membrane and the nuclear membrane.
cytosine: One of the nitrogenous base components of nucleic acids.
cytoskeleton: A network of microtubules and microfilaments in the cytoplasm.
cytosol: The fluid portion of the cytoplasm.
cytotoxic T cells: Lymphocytes of the cellular immune response that kill target cells by direct contact or through the secretion of lymphotoxins; also called killer T cells.

D

daughter cells: Genetically identical cells produced by somatic cell division.
decomposition reaction: A chemical reaction that breaks a molecule into smaller fragments.
defecation (def-e-KĀ-shun): The elimination of fecal wastes.
deglutition (deg-loo-TISH-un): Swallowing.
degradation: Breakdown, catabolism.
dehydration: A reduction in the water content of the body that threatens homeostasis.
dehydration synthesis: The joining of two molecules associated with the removal of a water molecule.
demyelination: The loss of the myelin sheath of an axon, usually due to chemical or physical damage to Schwann cells or oligodendrocytes.
denaturation: Irreversible alteration in the three-dimensional structure of a protein.
dendrite (DEN-drīt): A sensory process of a neuron.
dentin (DEN-tin): Bonelike material that forms the body of a tooth; it differs from bone in lacking osteocytes and osteons.
deoxyribonucleic acid (dē-ok-sē-rī-bo-nū-KLĀ-ik): **DNA strand:** a nucleic acid consisting of a chain of nucleotides containing the sugar deoxyribose and the nitrogen bases adenine, guanine, cytosine, and thymine; **DNA molecule:** two DNA strands wound in a double helix and held together by weak bonds between complementary nitrogen base pairs.
deoxyribose: A 5-carbon sugar resembling ribose but lacking an oxygen atom.
depolarization: A change in the membrane potential that moves it from a negative value toward 0 mV.
depression: Inferior (downward) movement of a body part.
dermis: The connective tissue layer beneath the epidermis of the skin.
development: Growth and the acquisition of increasing structural and functional complexity; includes the period from conception to maturity.
dialysis: Diffusion between two solutions of differing solute concentrations across a semipermeable membrane containing pores that permit the passage of some solutes and not others.
diapedesis (dī-a-pe-DĒ-sis): Movement of white blood cells through the walls of blood vessels by migration between adjacent endothelial cells.
diaphragm (DĪ-a-fram): Any muscular partition; often used to refer to the respiratory muscle that separates the thoracic from the abdominopelvic cavities.
diaphysis (dī-A-fi-sis): The shaft of a long bone.
diarthrosis (dī-ar-THRO-sis): A synovial joint.
diastolic pressure: Pressure measured in the walls of a muscular artery when the left ventricle is in diastole.
diencephalon (dī-en-SEF-a-lon): A division of the brain that includes the epithalamus, thalamus, and hypothalamus.
differential count: The determination of the relative abundance of each type of white blood cell, based on a random sampling of 100 WBCs.
differentiation: The gradual appearance of characteristic cellular specializations during development, as the result of gene activation or repression.
diffusion: Passive molecular movement from an area of relatively high concentration to an area of relatively low concentration.
digestion: The chemical breakdown of ingested materials into simple molecules that can be absorbed by the cells of the digestive tract.
digestive tract: An internal passageway that begins at the mouth and ends at the anus.
dilate: To increase in diameter; to enlarge or expand.
diploid (DIP-lōyd): Having a complete somatic complement of chromosomes (23 pairs in human cells).
disaccharide (di-SAK-ah-rīd): A compound formed by the joining of two simple sugars by dehydration synthesis.
distal: Movement away from the point of attachment or origin; for a limb, away from its attachment to the trunk.
distal convoluted tubule: Portion of the nephron closest to the collecting tubule and duct; an important site of active secretion.
divergence: In neural tissue, the spread of excitation from one neuron to many neurons; an organizational pattern common along sensory pathways of the CNS.
dizygotic twins (dī-zī-GOT-ik): Twins that result from the fertilization of two different ova.
dominant gene: A gene whose presence will determine the phenotype, regardless of the nature of its allelic companion.
dopamine (DŌ-pa-mēn): An important neurotransmitter in the CNS.
dorsal: Toward the back, posterior.
dorsal root ganglion: PNS ganglion containing the cell bodies of sensory neurons.
dorsiflexion: Elevation of the superior surface of the foot.
duct: A passageway that delivers exocrine secretions to an epithelial surface.
ductus arteriosus (ar-te-rē-Ō-sus): Vascular connection between the pulmonary trunk and the aorta that functions throughout fetal life; normally closes at birth or shortly thereafter and persists as the ligamentum arteriosum.
ductus deferens (DUK-tus DEF-e-renz): A passageway that carries sperm from the epididymis to the ejaculatory duct.
duodenal ampulla: Chamber that receives bile from the common bile duct and pancreatic secretions from the pancreatic duct.
duodenal papilla: Conical projection from the inner surface of the duodenum that contains the opening of the duodenal ampulla.
duodenum (doo-A-dē-num): The proximal 25 cm of the small intestine that contains short villi and submucosal glands.
dura mater (DŪ-ra MĀ-ter): Outermost component of the meninges that surround the brain and spinal cord.
dynamic equilibrium: Maintenance of normal body orientation as sudden changes in position (rotation, acceleration, etc.) occur.

E

eccrine glands (EK-rin): Sweat glands of the skin that produce a watery secretion.
ectoderm: One of the three primary germ layers; covers the surface of the embryo and gives rise to the nervous system, the epidermis and associated glands, and a variety of other structures.
ectopic (ek-TOP-ik): Outside of its normal location.
effector: A peripheral gland or muscle cell innervated by a motor neuron.
efferent arteriole: An arteriole carrying blood away from a glomerulus of the kidney.
efferent fiber: An axon that carries impulses away from the CNS.
ejaculation (e-jak-ū-LĀ-shun): The ejection of semen from the penis as the result of muscular contractions of the bulbocavernosus and ischiocavernosus muscles.
ejaculatory ducts (e-JAK-ū-la-to-rē): Short ducts that pass within the walls of the prostate and connect the ductus deferens with the prostatic urethra.
elastin: Connective tissue fibers that stretch and rebound, providing elasticity to connective tissues.
electrical coupling: A connection between adjacent cells that permits the movement of ions and the transfer of graded or conducted changes in the membrane potential from cell to cell.
electrocardiogram (ECG, EKG) (ē-lek-trō-KAR-dē-ō-gram): Graphic record of the electrical activities of the heart, as monitored at specific locations on the body surface.
electroencephalogram (EEG): Graphic record of the electrical activities of the brain.
electrolytes (ē-LEK-trō-līts): Soluble inorganic compounds whose ions will conduct an electrical current in solution.

electron: One of the three fundamental particles; a subatomic particle that bears a negative charge and normally orbits around the protons of the nucleus.
electron transport system: Cytochrome system responsible for most of the energy production in living cells; a complex bound to the inner mitochondrial membrane.
element: All of the atoms with the same atomic number.
elevation: Movement in a superior, or upward, direction.
embolism (EM-bō-lizm): Obstruction or closure of a vessel by an embolus.
embolus (EM-bo-lus): An air bubble, fat globule, or blood clot drifting in the circulation.
embryo (EM-brē-o): Developmental stage beginning at fertilization and ending at the start of the third developmental month.
embryology (em-brē-OL-o-jē): The study of embryonic development, focusing on the first 2 months after fertilization.
emmetropia: Normal vision.
emulsification (ē-mul-si-fi-KĀ-shun): The physical breakup of fats in the digestive tract, forming smaller droplets accessible to digestive enzymes; normally the result of mixing with bile salts.
enamel: Crystalline material similar in mineral composition to bone, but harder and without osteocytes, that covers the exposed surfaces of the teeth.
endocardium (en-dō-KAR-dē-um): The simple squamous epithelium that lines the heart and is continuous with the endothelium of the great vessels.
endochondral ossification (en-dō-KON-dral): The conversion of a cartilaginous model to bone; the characteristic mode of formation for skeletal elements other than the bones of the cranium, the clavicles, and sesamoid bones.
endocrine gland: A gland that secretes hormones into the blood.
endocytosis (EN-dō-sī-tō-sis): The movement of relatively large volumes of extracellular material into the cytoplasm via the formation of a membranous vesicle at the cell surface; includes pinocytosis and phagocytosis.
endoderm: One of the three primary germ layers; the layer on the undersurface of the embryonic disc that gives rise to the epithelia and glands of the digestive system, the respiratory system, and portions of the urinary system.
endogenous: Produced within the body.
endolymph (EN-dō-limf): Fluid contents of the membranous labyrinth (the saccule, utricle, semicircular canals, and cochlear duct) of the inner ear.
endometrium (en-dō-MĒ-trē-um): The mucous membrane lining the uterus.
endomysium (en-dō-MĪS-ē-um): A delicate network of connective tissue fibers that surrounds individual muscle cells.
endoneurium: A delicate network of connective tissue fibers that surrounds individual nerve fibers.
endoplasmic reticulum (en-dō-PLAZ-mik re-TIK-ū-lum): A network of membranous channels in the cytoplasm of a cell that function in intracellular transport, synthesis, storage, packaging, and secretion.
endothelium (en-dō-THĒ-lē-um): The simple squamous epithelium that lines blood and lymphatic vessels.
enzyme: A protein that catalyzes a specific biochemical reaction.
eosinophil (ē-ō-sin-ō-fil): A granulocyte (WBC) with a lobed nucleus and red-staining granules; participates in the immune response and is especially important during allergic reactions.
ependyma (ep-EN-di-mah): Layer of cells lining the ventricles and central canal of the CNS.
epicardium: Serous membrane covering the outer surface of the heart; also called the visceral pericardium.
epidermis: The epithelium covering the surface of the skin.
epididymis (ep-i-DID-i-mus): Coiled duct that connects the rete testis to the ductus deferens; site of functional maturation of spermatozoa.
epidural space: Space between the spinal dura mater and the walls of the vertebral foramen; contains blood vessels and adipose tissue; a frequent site of injection for regional anesthesia.
epiglottis (ep-i-GLOT-is): Blade-shaped flap of tissue, reinforced by cartilage, that is attached to the dorsal and superior surface of the thyroid cartilage; it folds over the entrance to the larynx during swallowing.
epimysium (ep-i-MĪS-ē-um): A dense investment of collagen fibers that surrounds a skeletal muscle and is continuous with the tendons/aponeuroses of the muscle and with the perimysium.
epineurium: A dense investment of collagen fibers that surrounds a peripheral nerve.
epiphyseal plate (e-pi-FI-sē-al): Cartilaginous region between the epiphysis and diaphysis of a growing bone.
epiphysis (e-PIF-i-sis): The head of a long bone.
epithelium (e-pi-THĒ-lē-um): One of the four primary tissue types; a layer of cells that forms a superficial covering or an internal lining of a body cavity or vessel.
equational division: The second meiotic division.
erythrocyte (e-RITH-rō-sīt): A red blood cell; an anucleate blood cell containing large quantities of hemoglobin.
erythropoietin (EPO)(e-rith-rō-poi-Ē-tin): Hormone released by tissues, especially the kidneys, exposed to low oxygen concentrations; stimulates erythropoiesis in bone marrow.
esophagus: A muscular tube that connects the pharynx to the stomach.
essential amino acids: Amino acids that cannot be synthesized in the body in adequate amounts, and must be obtained from the diet.
essential fatty acids: Fatty acids that cannot be synthesized in the body, and must be obtained from the diet.
estrogens (ES-trō-jenz): A class of steroid sex hormones that includes estradiol.
evaporation: Movement of molecules from the liquid to the gaseous state.
eversion (ē-VER-shun): A turning outward.
excitable membranes: Membranes that conduct action potentials, a characteristic of muscle and nerve cells.
excretion: Elimination from the body.
exocrine gland: A gland that secretes onto the body surface or into a passageway connected to the exterior.
exocytosis (EK-sō-sī-tō-sis): The ejection of cytoplasmic materials by fusion of a membranous vesicle with the cell membrane.
extension: An increase in the angle between two articulating bones; the opposite of flexion.
external auditory canal: Passageway in the temporal bone that leads to the tympanum of the inner ear.
external ear: The pinna, external auditory canal, and tympanum.
external nares: The nostrils; the external openings into the nasal cavity.
extracellular fluid: All body fluid other than that contained within cells; includes plasma and interstitial fluid.
extraembryonic membranes: The yolk sac, amnion, chorion, and allantois.
extrapyramidal system: Nuclei and tracts associated with the involuntary control of muscular activity.
extremities: The limbs.
extrinsic pathway: Clotting pathway that begins with damage to blood vessels or surrounding tissues and ends with the formation of tissue thromboplastin.

F

facilitated diffusion: Passive movement of a substance across a cell membrane via a protein carrier.
fascia (FASH-a): Connective tissue fibers, primarily collagenous, that form sheets or bands beneath the skin to attach, stabilize, enclose, and separate muscles and other internal organs.
fasciculus (fa-SIK-ū-lus): A small bundle, usually referring to a collection of nerve axons or muscle fibers.
fatty acids: Hydrocarbon chains ending in a carboxyl group.
feces: Waste products eliminated by the digestive tract at the anus; contains indigestible residue, bacteria, mucus, and epithelial cells.
fertilization: Fusion of egg and sperm to form a zygote.
fetus: Developmental stage lasting from the start of the third developmental month to delivery.
fibrillation (fi-bri-LĀ-shun): Uncoordinated contractions of individual muscle cells that impair or prevent normal function.
fibrin (FĪ-brin): Insoluble protein fibers that form the basic framework of a blood clot.
fibrinogen (fī-BRIN-o-jen): Plasma protein, soluble precursor of the fibrous protein fibrin.
fibrinolysis (fī-brin-OL-i-sis): The breakdown of the fibrin strands of a blood clot by a proteolytic enzyme.
fibroblasts (FĪ-brō-blasts): Cells of connective tissue proper that are responsible for the production of extracellular fibers and the secretion of the organic compounds of the extracellular matrix.
fibrocartilage: Cartilage containing an abundance of collagen fibers; found around the edges of joints, in the intervertebral discs, the menisci of the knee, etc.
fibula: The lateral, relatively small bone of the leg.
filtrate: Fluid produced by filtration at a glomerulus in the kidney.
filtration: Movement of a fluid across a membrane whose pores restrict the passage of solutes on the basis of size.

filtration pressure: Hydrostatic pressure responsible for the filtration process.
fimbriae (FIM-brē-ē): A fringe; used to describe the fingerlike processes that surround the entrance to the uterine tube.
fissure: An elongate groove or opening.
flagellum/flagella (fla-JEL-ah): An organelle structurally similar to a cilium, but used to propel a cell through a fluid.
flexion (FLEK-shun): A movement that reduces the angle between two articulating bones; the opposite of extension.
flexor reflex: A reflex contraction of the flexor muscles of a limb in response to an unpleasant stimulus.
flexure: A bending.
follicle (FOL-i-kl): A small secretory sac or gland.
follicle-stimulating hormone (FSH): A hormone secreted by the anterior pituitary; stimulates oogenesis (female) and spermatogenesis (male).
fontanel (fon-tah-NEL): A relatively soft, flexible, fibrous region between two flat bones in the developing skull.
foramen: An opening or passage through a bone.
forearm: Distal portion of the arm between the elbow and wrist.
forebrain: The cerebrum.
fossa: A shallow depression or furrow in the surface of a bone.
fourth ventricle: An elongate ventricle of the metencephalon (pons and cerebellum) and the myelencephalon (medulla) of the brain; the roof contains a region of choroid plexus.
fovea (FŌ-vē-a): Portion of the retina providing the sharpest vision, with the highest concentration of cones; also called the macula lutea.
fracture: A break or crack in a bone.
frontal plane: A sectional plane that divides the body into anterior and posterior portions; also called coronal plane.
fructose: A hexose (simple sugar containing 6 carbons) found in foods and in semen.
fundus (FUN-dus): The base of an organ.

G

gallbladder: Pear-shaped reservoir for the bile secreted by the liver.
gametes (GAM-ēts): Reproductive cells (sperm or eggs) that contain one-half of the normal chromosome complement.
gametogenesis (ga-mē-tō-JEN-e-sis): The formation of gametes.
gamma aminobutyric acid (GABA) (GAM-ma a-MĒ-nō-bū-TIR-ik): A neurotransmitter of the CNS whose effects are usually inhibitory.
ganglion/ganglia: A collection of nerve cell bodies outside of the CNS.
ganglionic neuron: An autonomic neuron whose cell body is in a peripheral ganglion and whose axon is a postganglionic fiber.
gap junctions: Connections between cells that permit electrical coupling.
gastric glands: Tubular glands of the stomach whose cells produce acid, enzymes, intrinsic factor, and hormones.
gastrin (GAS-trin): Hormone produced by enteroendocrine cells of the stomach, when exposed to mechanical stimuli or vagal stimulation, and the duodenum, when exposed to chyme containing undigested proteins.
gastrulation (gas-troo-LĀ-shun): The movement of cells of the inner cell mass that creates the three primary germ layers of the embryo.
gene: A portion of a DNA strand that functions as a hereditary unit and is found at a particular locus on a specific chromosome.
genetics: The study of mechanisms of heredity.
genotype (JĒN-ō-tīp): The genetic complement of a particular individual.
germinal centers: Pale regions in the interior of lymphoid tissues or nodules, where cell divisions are under way.
gestation (jes-TĀ-shun): The period of intrauterine development.
gingivae (JIN-ji-vē): The gums.
gland: Cell that produces exocrine or endocrine secretions.
glans penis: Expanded tip of the penis that surrounds the urethral opening; continuous with the corpus spongiosum.
glaucoma: Eye disorder characterized by rising intraocular pressures due to inadequate drainage of aqueous humor at the canal of Schlemm.
glenoid fossa: A rounded depression that forms the articular surface of the scapula at the shoulder joint.
glial cells (GLĒ-al): Supporting cells in the neural tissue of the CNS and PNS.
globular proteins: Proteins whose tertiary structure makes them rounded and compact.
glomerular filtration rate (GFR): The rate of filtrate formation at the glomerulus.
glomerulus (glo-MER-ū-lus): A ball or knot; in the kidneys, a knot of capillaries that projects into the enlarged, proximal end of a nephron; the site where filtration occurs, the first step in the production of urine.
glossopharyngeal nerve (glos-ō-fa-RIN-jē-al): Cranial nerve IX.
glottis (GLOT-is): The passage from the pharynx to the larynx.
glucagon (GLOO-ka-gon): Hormone secreted by the alpha cells of the pancreatic islets; elevates blood glucose concentrations.
glucocorticoids: Hormones secreted by the adrenal cortex to modify glucose metabolism; cortisol, cortisone, and corticosterone are important examples.
gluconeogenesis (gloo-kō-nē-ō-JEN-e-sis): The synthesis of glucose from protein or lipid precursors.
glucose: (GLOO-Kōs): A 6 carbon sugar, $C_6H_{12}O_6$, the preferred energy source for most cells and the only energy source for neurons under normal conditions.
glycerides: Lipids composed of glycerol bound to 1-3 fatty acids.
glycogen (GLĪ-kō-jen): A polysaccharide that represents an important energy reserve; a polymer consisting of a long chain of glucose molecules.
glycolysis (glī-KOL-i-sis): The cytoplasmic breakdown of glucose into lactic acid via pyruvic acid, with a net gain of 2 ATP.
goblet cell: A goblet-shaped, mucus-producing, unicellular gland found in certain epithelia of the digestive and respiratory tracts.
Golgi apparatus (GŌL-jē): Cellular organelle consisting of a series of membranous plates that give rise to lysosomes and secretory vesicles.
gonadotropic hormones: FSH and LH, hormones that stimulate gamete development and sex hormone secretion.
gonadotropin-releasing hormone (GnRH) (gō-nad-ō-TRŌ-pin): Hypothalamic releasing hormone that causes the secretion of FSH and LH by the anterior pituitary gland.
gonads (GŌ-nads): Organs that produce gametes and hormones.
granulocytes (GRAN-ū-lō-sīts): White blood cells containing granules visible with the light microscope; includes eosinophils, basophils, and neutrophils; also called granular leukocytes.
gray matter: Areas in the CNS dominated by neuron bodies, glial cells, and unmyelinated axons.
gray ramus: A bundle of postganglionic sympathetic nerve fibers that go to a spinal nerve for distribution to effectors in the body wall, skin, and extremities.
greater omentum: A large fold of the dorsal mesentery of the stomach that hangs in front of the intestines.
groin: The inguinal region.
gross anatomy: The study of the structural features of the human body without the aid of a microscope.
growth hormone (GH): Anterior pituitary hormone that stimulates tissue growth and anabolism when nutrients are abundant and restricts tissue glucose dependence when nutrients are in short supply.
guanine: One of the nitrogenous bases found in nucleic acids.
gustation (GUS-tā-shun): The sense of taste.
gyrus (JĪ-rus): A prominent fold or ridge of neural cortex on the surfaces of the cerebral hemispheres.

H

hair: A keratinous strand produced by epithelial cells of the hair follicle.
hair cells: Sensory cells of the inner ear.
hair follicle: An accessory structure of the integument; a tube lined by a stratified squamous epithelium that begins at the surface of the skin and ends at the hair papilla.
hair root: A thickened, conical structure consisting of a connective tissue papilla and the overlying matrix, a layer of epithelial cells that produces the hair shaft.
hallux: The big toe.
haploid (HAP-loyd): Possessing one-half of the normal number of chromosomes; a characteristic of gametes.
hard palate: The bony roof of the oral cavity, formed by the maxillary and palatine bones.
helper T cells: Lymphocytes whose secretions and other activities coordinate the cellular and humoral immune responses.
hematocrit (hē-MA-tō-krit): Percentage of the volume of whole blood contributed by cells; also called the packed cell volume (PCV) or the volume of packed red cells (VPRC).
heme (hēm): A special organic compound containing a central iron atom that can reversibly bind oxygen molecules; a component of the hemoglobin molecule.
hemocytoblasts: Stem cells whose divisions produce all of the various populations of blood cells.
hemoglobin (HĒ-mō-glō-bin): Protein composed of four globular subunits, each bound to a single molecule of heme; the protein found in red blood cells that gives them the ability to transport oxygen in the blood.
hemolysis: Breakdown (lysis) of red blood cells.

hemopoiesis (hēm-ō-poi-Ē-sis): Blood cell formation and differentiation.
hemorrhage: Blood loss.
hepatic duct: Duct carrying bile away from the liver lobes and toward the union with the cystic duct.
hepatic portal vein: Vessel that carries blood between the intestinal capillaries and the sinusoids of the liver.
hepatocyte (he-PAT-ō-sit): A liver cell.
heterozygous (het-er-ō-ZĪ-gus): Possessing two different alleles at corresponding loci on a chromosome pair; the individual's phenotype may be determined by one or both of the alleles.
hexose: A 6-carbon simple sugar.
high-density lipoprotein: A lipoprotein with a relatively small lipid content, thought to be responsible for the movement of cholesterol from peripheral tissues to the liver.
hilum or hilus (HĪ-lus): A localized region where blood vessels, lymphatics, nerves, and/or other anatomical structures are attached to an organ.
histamine (HIS-ta-min): Chemical released by stimulated mast cells or basophils to initiate or enhance an inflammatory response.
histology (his-TOL-ō-jē): The study of tissues.
holocrine (HŌ-lō-krin): Form of exocrine secretion where the secretory cell becomes swollen with vesicles and then ruptures.
homeostasis (hō-mē-ō-STĀ-sis): The maintenance of a relatively constant internal environment.
homologous chromosomes (hō-MOL-o-gus): The members of a chromosome pair, each containing the same gene loci.
homozygous (hō-mō-ZĪ-gus): Having the same allele for a particular character on two homologous chromosomes.
hormone: A compound secreted by one cell that travels through the circulatory system to affect the activities of cells in another portion of the body.
human chorionic gonadotropin (hCG): Placental hormone that maintains the corpus luteum for the first 3 months of pregnancy.
human placental lactogen (hPL): Placental hormone that stimulates the functional development of the mammary glands.
humoral immunity: Immunity resulting from the presence of circulating antibodies produced by plasma cells.
hydrogen bond: Weak interaction between the hydrogen atom on one molecule and a negatively charged portion of another molecule.
hydrolysis (hī-DROL-i-sis): The breakage of a chemical bond through the addition of a water molecule; the reverse of dehydration synthesis.
hydrostatic pressure: Fluid pressure.
hydroxyl group (hī-DROK-sil): OH^-.
hypertonic: A term used when comparing two solutions to refer to the solution with the higher osmolarity.
hypertrophy (HĪ-per-trō-fē): Increase in the size of tissue without cell division.
hyperventilation (hī-per-ven-ti-LĀ-shun): A rate of respiration sufficient to reduce the plasma P_{CO_2} to levels below normal.
hypodermic needle: A needle inserted through the skin to introduce drugs into the subcutaneous layer.
hypophyseal portal system (hī-pō-FI-sē-al): Network of vessels that carry blood from capillaries in the hypothalamus to capillaries in the anterior pituitary gland (hypophysis).
hypophysis (hī-POF-i-sis): The anterior pituitary gland.
hypothalamus: The floor of the diencephalon; region of the brain containing centers involved with the unconscious regulation of visceral functions, emotions, drives, and the coordination of neural and endocrine functions.
hypotonic: When comparing two solutions, used to refer to the one with the lower osmolarity.
hypoxia (hī-POKS-ē-a): Low tissue oxygen concentrations.

I

ileocecal valve (il-ē-ō-SĒ-kal): A fold of mucous membrane that guards the connection between the ileum and the cecum.
ileum (IL-ē-um): The last 2.5 m of the small intestine.
ilium (IL-ē-um): The largest of the three bones whose fusion creates a coxa.
immunity: Resistance to injuries and diseases caused by foreign compounds, toxins, and pathogens.
immunization: Developing immunity by the deliberate exposure to antigens under conditions that prevent the development of illness but stimulate the production of memory B cells.
immunodeficiency: An inability to produce normal numbers and types of antibodies and sensitized lymphocytes.
immunoglobulin (i-mū-nō-GLOB-ū-lin): A circulating antibody.
implantation (im-plan-TĀ-shun): The erosion of a blastocyst into the uterine wall.
inclusions: Aggregations of insoluble pigments, nutrients, or other materials in the cytoplasm.
incus (IN-kus): The central auditory ossicle, situated between the malleus and the stapes in the middle ear cavity.
infarct: An area of dead cells resulting from an interruption of circulation.
infection: Invasion and colonization of body tissues by pathogenic organisms.
inferior vena cava: The vein that carries blood from the parts of the body below the heart to the right auricle.
infertility: Inability to conceive.
inflammation: A nonspecific defense mechanism that operates at the tissue level, characterized by swelling, redness, warmth, pain, and some loss of function.
inflation reflex: A reflex mediated by the vagus nerve that prevents overexpansion of the lungs.
infundibulum (in-fun-DIB-ū-lum): A tapering, funnel-shaped structure; in the nervous system, refers to the connection between the pituitary gland and the hypothalamus; the infundibulum of the uterine tube is the entrance bounded by fimbriae that receives the ova at ovulation.
ingestion: The introduction of materials into the digestive tract via the mouth.
inguinal canal: A passage through the abdominal wall that marks the path of testicular descent and that contains the testicular arteries, veins, and ductus deferens.
inguinal region: The area near the junction of the trunk and the thighs that contains the external genitalia.
inhibin (in-HIB-in): A hormone produced by the sustentacular cells that inhibits the pituitary secretion of FSH.
initial segment: The proximal portion of the axon where an action potential first appears.
injection: Forcing of fluid into a body part or organ.
inner cell mass: Cells of the blastocyst that will form the body of the embryo.
innervation: The distribution of sensory and motor nerves to a specific region or organ.
insertion: Point of attachment of a muscle that is most movable.
inspiratory reserve volume: The maximum amount of air that can be drawn into the lungs over and above the normal tidal volume.
insoluble: Incapable of dissolving in solution.
insulin (IN-su-lin): Hormone secreted by the beta cells of the pancreatic islets; causes a reduction in plasma glucose concentrations.
integument (in-TEG-ū-ment): The skin.
intercalated discs (in-TER-ka-lā-ted): Regions where adjacent cardiocytes interlock and where gap junctions permit electrical coupling between the cells.
intercellular cement: Proteoglycans, especially hyaluronic acid, found between adjacent epithelial cells.
interferons (in-ter-FĒR-ons): Peptides released by virally infected cells, especially lymphocytes, that make other cells more resistant to viral infection and slow viral replication.
interleukins (in-ter-LOO-kins): Peptides released by activated monocytes and lymphocytes that assist in the coordination of the cellular and humoral immune responses.
internal ear: The membranous labyrinth that contains the organs of hearing and equilibrium.
internal nares: The entrance to the nasopharynx from the nasal cavity.
interneuron: An association neuron; neurons inside the CNS that are interposed between sensory and motor neurons.
interoceptors: Sensory receptors monitoring the functions and status of internal organs and systems.
interphase: Stage in the life of a cell during which the chromosomes are uncoiled and all normal cellular functions except mitosis are under way.
interstitial fluid (in-ter-STISH-al): Fluid in the tissues that fills the spaces between cells.
interstitial growth: Form of cartilage growth through the growth, mitosis, and secretion of chondrocytes inside the matrix.
intervertebral disc: Fibrocartilage pad between the bodies of successive vertebrae that acts as a shock absorber.
intracellular fluid: The cytosol.
intramembranous ossification (in-tra-MEM-bra-nus): The formation of bone within a connective tissue without the prior development of a cartilaginous model.
intrinsic factor: Glycoprotein secreted by the parietal cells of the stomach that facilitates the intestinal absorption of vitamin B_{12}.
intrinsic pathway: A pathway of the clotting system that begins with the activation of platelets and ends with the formation of platelet thromboplastin.
inversion: A turning inward.

ion: An atom or molecule bearing a positive or negative charge due to the acceptance or donation of an electron.

ionic bond (ī-ON-ik): Molecular bond created by the attraction between ions with opposite charges.

ionization (ī-on-i-ZĀ-shun): Dissociation; the breakdown of a molecule in solution to form ions.

iris: A contractile structure made up of smooth muscle that forms the colored portion of the eye.

ischemia (is-KĒ-mē-a): Inadequate blood supply to a region of the body.

ischium (IS-kē-um): One of the three bones whose fusion creates the coxa.

islets of Langerhans: *See* **pancreatic islets**.

isometric contraction: A muscular contraction characterized by rising tension production but no change in length.

isotonic: A solution having an osmolarity that does not result in water movement across cell membranes; of the same contractive strength.

isotonic contraction: A muscular contraction during which tension climbs and then remains stable as the muscle shortens.

isotope: Forms of an element whose atoms contain the same number of protons but different numbers of neutrons (and thus differ in atomic weight).

J

jejunum (je-JOO-num): The middle portion of the small intestine.

joint: An area where adjacent bones interact; an articulation.

juxtaglomerular apparatus: The macula densa and the juxtaglomerular cells; a complex responsible for the release of renin and erythropoietin.

K

karyotyping (KAR-ē-ō-tī-ping): The determination of the chromosomal characteristics of an individual or cell.

keratin (KER-a-tin): Tough, fibrous protein component of nails, hair, calluses, and the general integumentary surface.

keratinization (KER-a-tin-i-zā-shun): The production of keratin by epithelial cells.

ketone bodies: Keto acids produced during the catabolism of lipids and ketogenic amino acids; specifically acetone, acetoacetate, and beta-hydroxybutyrate.

kidney: A component of the urinary system; an organ functioning in the regulation of plasma composition, including the excretion of wastes and the maintenance of normal fluid and electrolyte balance.

killer T cells: *See* **cytotoxic T cells**.

kilocalorie (KIL-o-kal-o-rē): The amount of heat required to raise the temperature of a kilogram of water 1°C.

Kupffer cells (KOOP-fer): Stellate reticular cells of the liver; phagocytic cells of the liver sinusoids.

L

labia (LĀ-bē-a): Lips; labia majora and minora are components of the female external genitalia.

labrum: A lip or rim.

labyrinth: A maze of passageways; usually refers to the structures of the inner ear.

lacrimal gland (LAK-ri-mal): Tear gland on the dorsolateral surface of the eye.

lactation (lak-TĀ-shun): The production of milk by the mammary glands.

lacteal (LAK-tē-al): A terminal lymphatic within an intestinal villus.

lactic acid: Compound produced from pyruvic acid under anaerobic conditions.

lactiferous duct (lak-TIF-e-rus): Duct draining one lobe of the mammary gland.

lactiferous sinus: An expanded portion of a lactiferous duct adjacent to the nipple of a breast.

lacuna (la-KOO-na): A small pit or cavity.

lambdoidal suture (lam-DŌYD-al): Synarthrotic articulation between the parietal and occipital bones of the cranium.

lamellae (la-MEL-lē): Concentric layers of bone within an osteon.

lamina (LA-min-a): A thin sheet or layer.

lamina propria (LA-mi-na PRO-prē-a): Loose connective tissue that underlies a mucous epithelium and forms part of a mucous membrane.

Langerhans cells (LAN-ger-hanz): Cells in the epithelium of the skin and digestive tract that participate in the immune response by presenting antigens to T cells.

large intestine: The terminal portions of the intestinal tract, consisting of the colon, the rectum, and the anorectal canal.

laryngopharynx (lā-rin-gō-FAR-inks): Division of the pharynx inferior to the epiglottis and superior to the esophagus.

larynx (LAR-inks): A complex cartilaginous structure that surrounds and protects the glottis and vocal cords; the superior margin is bound to the hyoid bone and the inferior margin is bound to the trachea.

latent period: The time between the stimulation of a muscle and the start of the contraction phase.

lateral: Pertaining to the side.

lateral ventricle: Fluid-filled chamber within one of the cerebral hemispheres.

lens: The transparent body lying behind the iris and pupil and in front of the vitreous humor.

lesser omentum: A small pocket in the mesentery that connects the lesser curvature of the stomach to the liver.

leukemia (loo-KĒ-mē-ah): A malignant disease of the blood-forming tissues.

leukocyte (LOO-kō-sīt): A white blood cell.

ligament (LI-ga-ment): Dense band of connective tissue fibers that attach one bone to another.

ligamentum arteriosum: The fibrous strand found in the adult that represents the remains of the ductus arteriosus of the fetus.

limbic system (LIM-bik): Group of nuclei and centers in the cerebrum and diencephalon that are involved with emotional states, memories, and behavioral drives.

limbus (LIM-bus): The edge of the cornea, marked by the transition from the corneal epithelium to the ocular conjunctiva.

lingual frenulum: An epithelial fold that attaches the inferior surface of the tongue to the floor of the mouth.

lipase (LĪ-pās): A pancreatic enzyme that breaks down triglycerides.

lipid: An organic compound containing carbons, hydrogens, and oxygens in a ratio that does not approximate 1:2:1; includes fats, oils, and waxes.

lipolysis: The catabolism of lipids as a source of energy.

lipoprotein (lī-pō-PRŌ-tēn): A compound containing a relatively small lipid bound to a protein.

liver: An organ of the digestive system with varied and vital functions that include the production of plasma proteins, the excretion of bile, the storage of energy reserves, the detoxification of poisons, and the interconversion of nutrients.

lobule (LOB-ūl): The basic organizational unit of the liver at the histological level.

loose connective tissue: A loosely organized, easily distorted connective tissue containing several different fiber types, a varied population of cells, and a viscous ground substance.

lumbar: Pertaining to the lower back.

lumen: The central space within a duct or other internal passageway.

lungs: Paired organs of respiration, situated in the left and right pleural cavities.

luteinizing hormone (LH) (LOO-tē-in-ī-zing): Anterior pituitary hormone that in the female assists FSH in follicle stimulation, triggers ovulation, and promotes the maintenance and secretion of the endometrial glands; in the male, stimulates spermatogenesis; formerly known as interstitial cell-stimulating hormone in males.

lymph: Fluid contents of lymphatic vessels, similar in composition to interstitial fluid.

lymphatics: Vessels of the lymphatic system.

lymph nodes: Lymphatic organs that monitor the composition of lymph.

lymphocyte (LIM-fō-sīt): A cell of the lymphatic system that participates in the immune response.

lymphopoiesis: The production of lymphocytes.

lysis (LĪ-sis): The destruction of a cell through the rupture of its cell membrane.

lysosome (LĪ-so-sōm): Intracellular vesicle containing digestive enzymes.

lysozyme: An enzyme present in some exocrine secretions that has antibiotic properties.

M

macrophage: A phagocytic cell of the monocyte-macrophage system.

macula (MAK-ū-la): A receptor complex in the saccule or utricle that responds to linear acceleration or gravity.

macula densa (MAK-ū-la DEN-sa): A group of specialized secretory cells in a portion of the distal convoluted tubule adjacent to the glomerulus and the juxtaglomerular cells; a component of the juxtaglomerular apparatus.

macula lutea (LOO-tē-a): The fovea.

malleus (MAL-ē-us): The first auditory ossicle, bound to the tympanum and the incus.

malnutrition: An unhealthy state produced by inadequate dietary intake of nutrients, calories, and/or vitamins.

mamillary bodies (MAM-i-lar-ē): Nuclei in the hypothalamus concerned with feeding reflexes and behaviors; a component of the limbic system.

mammary glands: Milk-producing glands of the female breast.

marrow: A tissue that fills the internal cavities in a bone; may be dominated by hemopoietic cells (red marrow) or adipose tissue (yellow marrow).

mast cell: A connective tissue cell that when stimulated releases histamine, serotonin, and heparin, initiating the inflammatory response.
mastication (mas-ti-KĀ-shun): Chewing.
mastoid sinus: Air-filled spaces in the mastoid process of the temporal bone.
matrix: The ground substance of a connective tissue.
maxillary sinus (MAK-si-ler-ē): One of the paranasal sinuses; an air-filled chamber lined by a respiratory epithelium that is located in a maxillary bone and opens into the nasal cavity.
mechanoreception: Detection of mechanical stimuli, such as touch, pressure, or vibration.
medial: Toward the midline of the body.
mediastinum (mē-dē-as-TĪ-num): Central tissue mass that divides the thoracic cavity into two pleural cavities; includes the aorta and other great vessels, the esophagus, trachea, thymus, the pericardial cavity and heart, and a host of nerves, small vessels, and lymphatics; area of connective tissue attaching a testis to the epididymus, proximal portion of ductus deferens, and associated vessels.
medulla: Inner layer or core of an organ.
medulla oblongata: The most caudal of the brain regions.
medullary cavity: The space within a bone that contains the marrow.
megakaryocytes (meg-a-KĀR-ē-ō-sīts): Bone marrow cells responsible for the formation of platelets.
meiosis (mī-Ō-sis): Cell division that produces gametes with half of the normal somatic chromosome complement.
melanin (ME-la-nin): Yellow-brown pigment produced by the melanocytes of the skin.
melanocyte (me-LAN-ō-sīt): Specialized cell found in the deeper layers of the stratified squamous epithelium of the skin, responsible for the production of melanin.
melanocyte-stimulating hormone (MSH) (me-LAN-ō-sīt): Hormone of the anterior pituitary that stimulates melanin production.
melatonin (mel-a-TŌ-nin): Hormone secreted by the pineal gland; inhibits secretion of MSH and GnRH.
membrane potential: The potential difference, in millivolts, measured across the cell membrane; a potential difference that results from the uneven distribution of positive and negative ions across a cell membrane.
membranous labyrinth: Endolymph-filled tubes of the inner ear that enclose the receptors of the inner ear.
meninges (men-IN-jēz): Three membranes that surround the surfaces of the CNS; the dura mater, the pia mater, and the arachnoid.
meniscus (mēn-IS-kus): A fibrocartilage pad between opposing surfaces in a joint.
menses (MEN-sēz): The first menstrual period that normally occurs at puberty.
merocrine (MER-o-krin): A method of secretion where the cell ejects materials through exocytosis.
mesencephalic aqueduct: Passageway that connects the third ventricle (diencephalon) with the fourth ventricle (metencephalon).
mesenchyme: Embryonic/fetal connective tissue.
mesentery (MEZ-en-ter-ē): A double layer of serous membrane that supports and stabilizes the position of an organ in the abdominopelvic cavity and provides a route for the associated blood vessels, nerves, and lymphatics.
mesoderm: The middle germ layer that lies between the ectoderm and endoderm of the embryo.
messenger RNA (mRNA): RNA formed at transcription to direct protein synthesis in the cytoplasm.
metabolic turnover: The continual breakdown and replacement of organic materials within living cells.
metabolism (meh-TAB-ō-lizm): The sum of all of the biochemical processes under way within the human body at a given moment; includes anabolism and catabolism.
metabolites (me-TAB-ō-līts): Compounds produced in the body as the result of metabolic reactions.
metacarpals: The five bones of the palm of the hand.
metaphase (MET-a-faz): A stage of mitosis wherein the chromosomes line up along the equatorial plane of the cell.
metatarsal: One of the five bones of the foot that articulate with the tarsals (proximally) and the phalanges (distally).
micelle (mī-SEL): A spherical aggregation of bile salts, monoglycerides, and fatty acids in the lumen of the intestinal tract.
microglia (mī-KRŌG-lē-a): Phagocytic glial cells in the CNS, derived from the monocytes of the blood.
microphages: Neutrophils and eosinophils.
microtubules: Microscopic tubules that are part of the cytoskeleton, and are found in cilia, flagella, the centrioles, and spindle fibers.
microvilli: Small, fingerlike extensions of the exposed cell membrane of an epithelial cell.
micturition (mik-tu-RI-shun): Urination.
midbrain: The mesencephalon.
middle ear: Space between the external and internal ear that contains auditory ossicles.
midsagittal plane: A plane passing through the midline of the body that divides it into left and right halves.
mineralocorticoid: Corticosteroids produced by the adrenal cortex; steroids, such as aldosterone, that affect mineral metabolism.
mitochondrion (mī-tō-KON-drē-on): An intracellular organelle responsible for generating most of the ATP required for cellular operations.
mitosis (mī-TŌ-sis): The division of a single cell that produces two identical daughter cells; the primary mechanism of tissue growth.
mitral valve (MĪ-tral): The left AV, or bicuspid, valve of the heart.
molecular weight: The sum of the atomic weights of the atoms in a molecule.
molecule: A compound containing two or more atoms that are held together by chemical bonds.
monocytes (MON-ō-sīts): Phagocytic agranulocytes (white blood cells) in the circulating blood.
monoglyceride (mo-nō-GLI-se-rīd): A lipid consisting of a single fatty acid bound to a molecule of glycerol.
monosaccharide (mon-ō-SAK-ah-rīd): A simple sugar, such as glucose or ribose.
monosynaptic reflex: A reflex where the sensory afferent synapses directly on the motor efferent.
monozygotic twins: Twins produced through the splitting of a single fertilized egg (zygote).
morula (MOR-ū-la): A mulberry-shaped collection of cells produced through the mitotic divisions of a zygote.
motor unit: All of the muscle cells controlled by a single motor neuron.
mucosa (mū-KŌ-sa): A mucous membrane; the epithelium plus the lamina propria.
mucus: Lubricating secretion produced by unicellular and multicellular glands along the digestive, respiratory, urinary, and reproductive tracts.
multifactorial trait: A phenotypic character that reflects the interactions of many different genes.
multipolar neuron: A neuron with many dendrites and a single axon, the typical form of a motor neuron.
muscle: A contractile organ composed of muscle tissue, blood vessels, nerves, connective tissues, and lymphatics.
muscle tissue: A tissue characterized by the presence of cells capable of contraction; includes skeletal, cardiac, and smooth muscle tissue.
muscularis externa (mus-kū-LAR-is): Concentric layers of smooth muscle responsible for peristalsis.
muscularis mucosae: Layer of smooth muscle beneath the lamina propria; responsible for moving the mucosal surface.
myelin (MĪ-e-lin): Insulating sheath around an axon consisting of multiple layers of glial cell membrane; significantly increases conduction rate along the axon.
myelination: The formation of myelin.
myenteric plexus (mī-en-TER-ik): Parasympathetic motor neurons and sympathetic postganglionic fibers located between the circular and longitudinal layers of the muscularis externa.
myocardium: The cardiac muscle tissue of the heart.
myofibril: Organized collections of myofilaments in skeletal and cardiac muscle cells.
myoglobin (MĪ-ō-glō-bin): An oxygen-binding pigment especially common in slow skeletal and cardiac muscle fibers.
myogram: A recording of the tension produced by muscle fibers on stimulation.
myometrium (mī-ō-MĒ-trē-um): The thick layer of smooth muscle in the wall of the uterus.
myosin: Protein component of the thick myofilaments.

N

nail: Keratinous structure produced by epithelial cells of the nail root.
nares, external (NĀ-rēz): The entrance from the exterior to the nasal cavity.
nares, internal: The entrance from the nasal cavity to the nasopharynx.
nasal cavity: A chamber in the skull bounded by the internal and external nares.
nasolacrimal duct: Passageway that transports tears from the nasolacrimal sac to the nasal cavity.
nasolacrimal sac: Chamber that receives tears from the lacrimal ducts.

nasopharynx (nā-zō-FĀR-inks): Region posterior to the internal nares, superior to the soft palate, and ending at the oropharynx.
negative feedback: Corrective mechanism that opposes or negates a variation from normal limits.
neonate: A newborn infant, or baby.
neoplasm: A tumor, or mass of abnormal tissue.
nephron (NEF-ron): Basic functional unit of the kidney.
nerve impulse: An action potential in a neuron cell membrane.
neural cortex: An area where gray matter is found at the surface of the CNS.
neuroeffector junction: A synapse between a motor neuron and a peripheral effector, such as a muscle, gland cell, or fat cell.
neuroglia (noo-RŌG-lē-a): Cells of the CNS and PNS that support and protect the neurons.
neuromuscular junction: A specific type of neuroeffector junction.
neuron (NOO-ron): A cell in neural tissue specialized for intercellular communication via (1) changes in membrane potential and (2) synaptic connections.
neurotransmitter: Chemical compound released by one neuron to affect the membrane potential of another.
neurulation: The embryological process responsible for the formation of the CNS.
neutron: A fundamental particle that does not carry a positive or negative charge.
neutrophil (NOO-trō-fil): A phagocytic microphage that is very numerous and usually the first of the mobile phagocytic cells to arrive at an area of injury or infection.
nipple: An elevated epithelial projection on the surface of the breast, containing the openings of the lactiferous sinuses.
Nissl bodies: The ribosomes, Golgi, RER, and mitochondria of the perikaryon of a typical nerve cell.
nitrogenous wastes: Organic waste products of metabolism that contain nitrogen, such as urea, uric acid, and creatinine.
nociception (nō-sē-SEP-shun): Pain perception.
node of Ranvier: Area between adjacent glial cells where the myelin covering of an axon is incomplete.
noradrenaline: Hormone secreted by the adrenal medulla, released at most sympathetic neuroeffector junctions and at certain synapses inside the CNS; also called norepinephrine.
norepinephrine (nor-ep-i-NEF-rin): A neurotransmitter in the PNS and CNS and a hormone secreted by the adrenal medulla; also called noradrenaline.
nucleic acid (nū-KLĒ-ik): A polymer of nucleotides containing a pentose sugar, a phosphate group, and one of four nitrogenous bases that regulate the synthesis of proteins and make up the genetic material in cells.
nucleolus (noo-KLĒ-o-lus): Dense region in the nucleus that represents the site of RNA synthesis.
nucleoside: A nitrogenous base plus a simple sugar.
nucleotide: Compound consisting of a nitrogenous base, a simple sugar, and a phosphate group.
nucleus: Cellular organelle that contains DNA, RNA, and proteins; a mass of gray matter in the CNS.
nucleus pulposus (pul-PO-sus): Gelatinous central region of an intervertebral disc.
nutrient: An organic compound that can be broken down in the body to produce energy.

O

obesity: Body weight 10–20 percent above standard values as the result of body fat accumulation.
ocular: Pertaining to the eye.
oculomotor nerve (ok-ū-lō-MŌ-ter): Cranial nerve III, that controls the extrinsic oculomotor muscles other than the superior oblique and the lateral rectus.
olecranon: The proximal end of the ulna that forms the prominent point of the elbow.
olfaction: The sense of smell.
olfactory bulb (ol-FAK-tor-ē): Expanded ends of the olfactory tracts; the sites where the axons of NI synapse on CNS interneurons that lie beneath the frontal lobe of the cerebrum.
oligodendrocytes (o-li-gō-DEN-drō-sīts): CNS glial cells responsible for maintaining cellular organization in the gray matter and providing a myelin sheath in areas of white matter.
oncogene (ON-kō-jēn): A gene that can turn a normal cell into a cancer cell.
oncologists (on-KOL-ō-jists): Physicians specializing in the study and treatment of tumors.
oocyte (Ō-ō-sīt): A cell whose meiotic divisions will produce a single ovum and three polar bodies.
oogenesis (ō-ō-JEN-e-sis): Ovum production.
oogonia (ō-ō-GŌ-nē-a): Stem cells in the ovaries whose divisions give rise to oocytes.
ooplasm: The cytoplasm of the ovum.
opsin: A protein, one structural component of the visual pigment rhodopsin.
optic chiasm (OP-tik KĪ-asm): Crossing point of the optic nerves.
optic nerve: Nerve that carries signals from the eye to the optic chiasm.
optic tract: Tract over which nerve impulses from the retina are transmitted between the optic chiasm and the thalamus.
orbit: Bony cavity of the skull that contains the eyeball.
organelle (or-gan-EL): An intracellular structure that performs a specific function or group of functions.
organic compound: A compound containing carbon, hydrogen, and usually oxygen.
organogenesis: The formation of organs during embryological and fetal development.
organs: Combinations of tissues that perform complex functions.
origin: Point of attachment of a muscle that is least movable.
oropharynx: The middle portion of the pharynx, bounded superiorly by the nasopharynx, anteriorly by the oral cavity, and inferiorly by the laryngopharynx.
osmolarity (oz-mo-LAR-i-tē): The total concentration of dissolved materials in a solution, regardless of their specific identities, expressed in terms of moles.
osmoreceptor: A receptor sensitive to changes in the osmolarity of the plasma.
osmosis (oz-MŌ-sis): The movement of water across a semipermeable membrane toward a solution containing a relatively high solute concentration.
osmotic pressure: The force of osmotic water movement; the pressure that must be applied to prevent osmotic movement across a membrane.
osseous tissue: A strong connective tissue containing specialized cells and a mineralized matrix of crystalline calcium phosphate and calcium carbonate.
ossicles: Small bones.
ossification: The formation of bone.
osteoblast: A cell that produces the fibers and matrix of bone.
osteoclast (OS-tē-ō-klast): A cell that dissolves the fibers and matrix of bone.
osteocyte (OS-tē-ō-sīt): A bone cell responsible for the maintenance and turnover of the mineral content of the surrounding bone.
osteolysis (os-tē-ŌL-ī-sis): The breakdown of the mineral matrix of bone.
osteon (OS-tē-on): The basic histological unit of compact bone, consisting of osteocytes organized around a central canal and separated by concentric lamellae.
otic: Pertaining to the ear.
otoliths (otoconia) (ō-tō-KŌ-nē-a): Aggregations of calcium carbonate crystals in a gelatinous membrane that sits above one of the maculae of the vestibular apparatus.
oval window: Opening in the bony labyrinth where the stapes attaches to the membranous wall of the vestibular duct.
ovarian cycle (ō-VAR-ē-an): Monthly chain of events that leads to ovulation.
ovary: Female reproductive gland that produces gametes.
ovulation (ōv-ū-LĀ-shun): The release of a secondary oocyte, surrounded by cells of the corona radiata, following the rupture of the wall of a tertiary follicle.
ovum/ova (Ō-vum): The functional product of meiosis II, produced after fertilization of a secondary oocyte.
oxytocin (oks-i-TŌ-sin): Hormone produced by hypothalamic cells and secreted into capillaries at the posterior pituitary; stimulates smooth muscle contractions of the uterus or mammary glands in the female, but has no known function in males.

P

pacemaker cells: Cells of the SA node that set the pace of cardiac contraction.
Pacinian corpuscle (pa-SIN-ē-an): Receptor sensitive to vibration.
palate: Horizontal partition separating the oral cavity from the nasal cavity and nasopharynx; can be divided into an anterior bony (hard) palate and a posterior fleshy (soft) palate.
palpate: To examine by touch.
palpebrae (pal-PĒ-brē): Eyelids.
pancreas: Digestive organ containing exocrine and endocrine tissues; exocrine portion secretes pancreatic juice, endocrine portion secretes hormones, including insulin and glucagon.
pancreatic duct: A tubular duct that carries pancreatic juice from the pancreas to the duodenum.
pancreatic islets: Aggregations of endocrine cells in the pancreas.
pancreatic juice: A mixture of buffers and digestive enzymes that is discharged into the duodenum under the stimulation of the enzymes secretin and cholecystokinin.
papilla (pa-PIL-la): A small, conical projection.

paranasal sinuses: Bony chambers lined by respiratory epithelium that open into the nasal cavity; includes the frontal, ethmoidal, sphenoidal, and maxillary sinuses.
parasympathetic division: One of the two divisions of the autonomic nervous system; also known as the craniosacral division; generally responsible for activities that conserve energy and lower the metabolic rate.
parathyroid glands: Four small glands embedded in the posterior surface of the thyroid; responsible for parathyroid hormone secretion.
parathyroid hormone (PTH): Hormone secreted by the parathyroid gland when plasma calcium levels fall below the normal range; causes increased osteoclast activity, increased intestinal calcium uptake, and decreased calcium ion loss at the kidneys.
parietal: Referring to the body wall or outer layer.
parietal cell: Cells of the gastric glands that secrete HCl and intrinsic factor.
Parkinson's disease: Progressive motor disorder due to degeneration of the cerebral nuclei.
parotid glands (pa-ROT-id)**:** Large salivary glands that secrete a saliva containing high concentrations of salivary (alpha) amylase.
parturition (par-tū-RISH-un)**:** Childbirth, delivery.
patella (pa-TEL-ah)**:** The sesamoid bone of the kneecap.
pathogenic: Disease-causing.
pathologist (pa-THO-lo-jist)**:** A physician specializing in the identification of diseases on the basis of characteristic structural and functional changes in tissues and organs.
pedicel (PED-i-sel)**:** A slender process of a podocyte that forms part of the filtration apparatus of the kidney glomerulus.
pedicles (PE-di-kels)**:** Thick bony struts that connect the vertebral body with the articular and spinous processes.
pelvic cavity: Inferior subdivision of the abdominopelvic (peritoneal) cavity; encloses the urinary bladder, the sigmoid colon and rectum, and male or female reproductive organs.
pelvis: A bony complex created by the articulations between the coxae, the sacrum, and the coccyx.
penis (PĒ-nis)**:** Component of the male external genitalia; a copulatory organ that surrounds the urethra and that serves to introduce semen into the female vagina.
perforating canal: A passageway in compact bone that runs at right angles to the axes of the osteons, between the periosteum and endosteum.
perfusion: The blood flow through a tissue.
pericardial cavity (per-i-KAR-dē-al)**:** The space between the parietal pericardium and the epicardium (visceral pericardium) that covers the outer surface of the heart.
pericardium (per-i-KAR-dē-um)**:** The fibrous sac that surrounds the heart and whose inner, serous lining is continuous with the epicardium.
perichondrium (per-i-KON-drē-um)**:** Layer that surrounds a cartilage, consisting of an outer fibrous and an inner cellular region.
perilymph (PER-ē-limf)**:** A fluid similar in composition to cerebrospinal fluid; found in the spaces between the bony labyrinth and the membranous labyrinth of the inner ear.
perimysium (per-i-MĪS-ē-um)**:** Connective tissue partition that separates adjacent fasciculi in a skeletal muscle.
perineum (pe-ri-NĒ-um)**:** The pelvic floor and associated structures.
perineurium: Connective tissue partition that separates adjacent bundles of nerve fibers in a peripheral nerve.
periosteum (pe-rē-OS-tē-um)**:** Layer that surrounds a bone, consisting of an outer fibrous and inner cellular region.
peripheral nervous system (PNS): All neural tissue outside of the CNS.
peripheral resistance: The resistance to blood flow primarily caused by friction with the vascular walls.
peristalsis (per-i-STAL-sis)**:** A wave of smooth muscle contractions that propels materials along the axis of a tube such as the digestive tract, the ureters, or the ductus deferens.
peritoneal cavity: *See* **abdominopelvic cavity**.
peritoneum (per-i-tō-NĒ-um)**:** The serous membrane that lines the peritoneal (abdominopelvic) cavity.
peritubular capillaries: A network of capillaries that surrounds the proximal and distal convoluted tubules of the kidneys.
permeability: Ease with which dissolved materials can cross a membrane; if freely permeable, any molecule can cross the membrane; if impermeable, nothing can cross; most biological membranes are selectively permeable, some materials can cross.
pes: The foot.
petrous: Stony, usually used to refer to the thickened portion of the temporal bone that encloses the inner ear.
pH: The negative exponent of the hydrogen ion concentration.
phagocyte: A cell that performs phagocytosis.
phagocytosis (fa-gō-si-TŌ-sis)**:** The engulfing of extracellular materials or pathogens; movement of extracellular materials into the cytoplasm by enclosure in a membranous vesicle.
phalanx/phalanges (fa-LANKS)**:** A bone of the fingers or toes.
pharmacology: The study of drugs, their physiological effects, and their clinical uses.
pharyngotympanic tube: A passageway that connects the nasopharynx with the middle ear cavity; also called the Eustachian or auditory tube.
pharynx: The throat; a muscular passageway shared by the digestive and respiratory tracts.
phenotype (FĒN-ō-tīp)**:** Physical characteristics that are genetically determined.
phosphate group: PO_4^{3-}.
phospholipid (fos-fō-LIP-id)**:** An important membrane lipid whose structure includes hydrophilic and hydrophobic regions.
phosphorylation (fos-for-i-LĀ-shun)**:** The addition of a high-energy phosphate group to a molecule.
photoreception: Sensitivity to light.
physiology (fiz-ē-OL-o-jē)**:** Literally the study of function; considers the ways living organisms perform vital activities.
pia mater: The tough, outer meningeal layer that surrounds the CNS.
pigment: A compound with a characteristic color.
pineal gland: Neural tissue in the posterior portion of the roof of the diencephalon, responsible for the secretion of melatonin.
pinna: The expanded, projecting portion of the external ear that surrounds the external auditory canal.
pinocytosis (pi-nō-sī-TŌ-sis)**:** The introduction of fluids into the cytoplasm by enclosing them in membranous vesicles at the cell surface.
pituitary gland: Endocrine organ situated in the sella turcica of the sphenoid bone and connected to the hypothalamus by the infundibulum; includes the posterior pituitary and the anterior pituitary.
placenta (pla-SENT-a)**:** A complex structure in the uterine wall that permits diffusion between the fetal and maternal circulatory systems; also called the afterbirth.
plantar: Referring to the sole of the foot.
plasma (PLAZ-mah)**:** The fluid ground substance of whole blood; what remains after the cells have been removed from a sample of whole blood.
plasma cell: Activated B cells that secrete antibodies.
plasmalemma (plaz-ma-LEM-a)**:** Cell membrane.
platelets (PLĀT-lets)**:** Small packets of cytoplasm that contain enzymes important in the clotting response; manufactured in the bone marrow by cells called megakaryocytes.
pleura (PLOO-ra)**:** The serous membrane lining the pleural cavities.
pleural cavities: Subdivisions of the thoracic cavity that contain the lungs.
plexus (PLEK-sus)**:** A network or braid.
plica (PLĪ-ka)**:** A permanent transverse fold in the wall of the small intestine.
podocyte (PŌ-do-sīt)**:** A cell whose processes surround the glomerular capillaries and assist in the filtration process.
polar body: A nonfunctional packet of cytoplasm containing chromosomes eliminated from an oocyte during meiosis.
polar bond: A form of covalent bond in which there is an unequal sharing of electrons.
pollex: The thumb.
polymer: A large molecule consisting of a long chain of subunits.
polymorph: Polymorphonuclear leukocyte; a neutrophil.
polypeptide: A chain of amino acids strung together by peptide bonds; those containing over 100 peptides are called proteins.
polysaccharide (pol-ē-SAK-ah-rīd)**:** A complex sugar, such as glycogen or a starch.
polysynaptic reflex: A reflex with interneurons interposed between the sensory fiber and the motor neuron(s).
polyunsaturated fats: Fatty acids containing carbon atoms linked by double bonds.
pons: The portion of the brain anterior to the cerebellum.
popliteal (pop-LIT-ē-al)**:** Pertaining to the back of the knee.
positive feedback: Mechanism that increases a deviation from normal limits following an initial stimulus.
posterior: Toward the back; dorsal.
postganglionic fiber: The axon of a ganglionic neuron.
postovulatory phase: The secretory phase of the menstrual cycle.
potential difference: The separation of opposite charges; requires a barrier that prevents ion migration.

precentral gyrus: The primary motor cortex on a cerebral hemisphere, located rostral to the central sulcus.
preganglionic neuron: Visceral motor neuron inside the CNS whose output controls one or more ganglionic motor neurons in the PNS.
premolars: Bicuspids; teeth with flattened surfaces located anterior to the molar teeth.
preovulatory phase: A portion of the menstrual cycle; period of estrogen-induced repair of the functional zone of the endometrium through the growth and proliferation of epithelial cells in the glands not lost during menses.
prepuce (PRĒ-pūs)**:** Loose fold of skin that surrounds the glans penis (males) or the clitoris (females).
presynaptic membrane: The synaptic surface where neurotransmitter release occurs.
prime mover: A muscle that performs a specific action.
progesterone (prō-JES-ter-ōn)**:** The most important progestin secreted by the corpus luteum following ovulation.
progestins (prō-JES-tinz)**:** Steroid hormones structurally related to cholesterol.
prognosis: A prediction concerning the possibility or time course of recovery from a specific disease.
prolactin (prō-LAK-tin)**:** Hormone that stimulates functional development of the mammary gland in females; a secretion of the anterior pituitary gland.
pronation (prō-NĀ-shun)**:** Rotation of the forearm that makes the palm face posteriorly.
pronucleus: Enlarged egg or sperm nucleus that forms after fertilization but before amphimixis.
prophase (PRŌ-fāz)**:** The initial phase of mitosis, characterized by the appearance of chromosomes, breakdown of the nuclear membrane, and formation of the spindle apparatus.
proprioception (prō-prē-ō-SEP-shun)**:** Awareness of the positions of bones, joints, and muscles.
prostaglandin (pros-tah-GLAN-din)**:** Lipoid secreted by one cell that alters the metabolic activities or sensitivities of adjacent cells; sometimes called "local hormones."
prostate gland (PROS-tāt)**:** Accessory gland of the male reproductive tract, contributing roughly one-third of the volume of semen.
protein: A large polypeptide with a complex structure.
prothrombin: Circulating proenzyme of the common pathway of the clotting system; converted to thrombin by the enzyme thromboplastin.
proton: A fundamental particle bearing a positive charge.
protraction: To move anteriorly in the horizontal plane.
proximal: Toward the attached base of an organ or structure.
proximal convoluted tubule: The portion of the nephron between Bowman's capsule and the loop of Henle; the major site of active reabsorption from the filtrate.
pseudopodia (soo-dō-PŌ-dē-a)**:** Temporary cytoplasmic extensions typical of mobile or phagocytic cells.
pseudostratified epithelium: An epithelium containing several layers of nuclei but whose cells are all in contact with the underlying basement membrane.
puberty: Period of rapid growth, sexual maturation, and the appearance of secondary sexual characteristics; usually occurs at ages 10–15.
pubic symphysis: Fibrocartilaginous amphiarthrosis between the pubic bones of the coxae.
pubis (PŪ-bis)**:** The anterior, inferior component of the coxa.
pulmonary circuit: Blood vessels between the pulmonary semilunar valve of the right ventricle and the entrance to the left atrium; the blood circulation through the lungs.
pulmonary ventilation: Movement of air in and out of the lungs.
pulp cavity: Internal chamber in a tooth, containing blood vessels, lymphatics, nerves, and the cells that maintain the dentin.
pupil: The opening in the center of the iris through which light enters the eye.
purine: An N compound with a ring-shaped structure; examples include adenine and guanine, two nitrogen bases common in nucleic acids.
Purkinje cell (pur-KIN-jē)**:** Large, branching neuron of the cerebellar cortex.
Purkinje fibers: Specialized conducting cardiocytes in the ventricles.
pus: An accumulation of debris, fluid, dead and dying cells, and necrotic tissue.
P wave: Deflection of the ECG corresponding to atrial depolarization.
pyloric sphincter (pī-LOR-ic)**:** Sphincter of smooth muscle that regulates the passage of chyme from the stomach to the duodenum.
pylorus (pī-LOR-us)**:** Gastric region between the body of the stomach and the duodenum; includes the pyloric sphincter.
pyrimidine: An N compound with a ring-shaped structure; examples include cytosine, thymine, and uracil, nitrogenous bases common in nucleic acids.
pyruvic acid (pī-RŪ-vik)**:** 3-carbon compound produced by glycolysis.

Q

quaternary structure: Three-dimensional protein structure produced by interactions between individual protein subunits.

R

radiodensity: Relative resistance to the passage of X-rays.
radiopaque: Having a relatively high radiodensity.
ramus: A branch.
raphe (RĀ-fē)**:** A seam.
recessive gene: An allele that will affect the phenotype only when the individual is homozygous for that trait.
recombinant DNA: DNA created by splicing together a specific gene from one organism into the DNA strand of another organism.
rectum (REK-tum)**:** The last 15 cm (6 in.) of the digestive tract.
rectus: Straight.
red blood cell (RBC): *See* **erythrocyte**.
reductional division: The first meiotic division, which reduces the chromosome number from 46 to 23.
reflex: A rapid, automatic response to a stimulus.
reflex arc: The receptor, sensory neuron, motor neuron, and effector involved in a particular reflex; interneurons may or may not be present, depending on the reflex considered.
refraction: The bending of light rays as they pass from one medium to another.
refractory period: Period between the initiation of an action potential and the restoration of the normal resting potential; over this period the membrane will not respond normally to stimulation.
relaxation phase: The period following a contraction when the tension in the muscle fiber returns to resting levels.
relaxin: Hormone that loosens the pubic symphysis; a hormone secreted by the placenta.
renal: Pertaining to the kidneys.
renal corpuscle: The initial portion of the nephron, consisting of an expanded chamber that encloses the glomerulus.
renin: Enzyme released by the kidney cells when renal blood pressure or O_2 level declines; converts angiotensinogen to angiotensin I.
rennin: Gastric enzyme that breaks down milk proteins.
repolarization: Movement of the membrane potential away from + mV values and toward the resting potential.
residual volume: Amount of air remaining in the lungs after maximum forced expiration.
respiration: Exchange of gases between living cells and the environment; includes pulmonary ventilation, external respiration, internal respiration, and cellular respiration.
respiratory minute volume: The amount of air moved into and out of the respiratory system each minute.
resting potential: The membrane potential of a normal cell under homeostatic conditions.
rete (RĒ-tē)**:** An interwoven network of blood vessels or passageways.
reticular formation: Diffuse network of gray matter that extends the entire length of the brain stem.
reticulospinal tracts: Descending tracts that carry involuntary motor commands issued by neurons of the reticular formation.
retina: The innermost layer of the eye, lining the posterior cavity; also known as the neural tunic; contains the photoreceptors.
retinal (RET-in-al)**:** Visual pigment derived from vitamin A.
retraction: Movement posteriorly in the horizontal plane.
retroperitoneal (re-trō-per-i-tō-NĒ-al)**:** Situated behind or outside of the peritoneal cavity.
reverberation: Positive feedback along a chain of neurons so that they remain active once stimulated.
Rh factor: Surface antigen that may be present (Rh-positive) or absent (Rh-negative) from the surfaces of red blood cells.
rhodopsin (rō-DOP-sin)**:** The visual pigment found in the membrane discs of the distal segments of rods.
rhythmicity center: Medullary center responsible for the basic pace of respiration; includes inspiratory and expiratory centers.
ribonucleic acid (rī-bō-nū-KLĀ-ik)**:** A nucleic acid consisting of a chain of nucleotides that contain the sugar ribose and the nitrogen bases adenine, guanine, cytosine, and uracil.
ribose: A 5-carbon sugar that is a structural component of RNA.

ribosome: An organelle containing rRNA and proteins, that is essential to mRNA translation and protein synthesis.
right lymphatic duct: Lymphatic vessel delivering lymph from the right side of the head, neck, and chest to the venous system via the right subclavian vein.
rod: Photoreceptor responsible for vision under dimly lit conditions.
rough endoplasmic reticulum (RER): A membranous organelle that is a site of protein synthesis and storage.
round window: An opening in the bony labyrinth of the inner ear that exposes the membranous wall of the tympanic duct to the air of the middle ear cavity.
Ruffini corpuscles (rū-FĒ-nē): Receptors sensitive to tension and stretch in the dermis of the skin.
rugae (ROO-gē): Mucosal folds in the lining of the empty stomach that disappear as gastric distension occurs.

S

saccule (SAK-ūl): A portion of the vestibular apparatus of the inner ear, responsible for equilibrium.
sagittal plane: Sectional plane that divides the body into left and right portions.
salt: An inorganic compound consisting of a cation other than H^+ and an anion other than OH^-.
saltatory conduction: Relatively rapid conduction of an action potential between successive nodes of a myelinated axon.
sarcolemma: The cell membrane of a muscle cell.
sarcomere: The smallest contractile unit of a striated muscle cell.
sarcoplasm: The cytoplasm of a muscle cell.
Schlemm, canal of: Passageway that delivers aqueous humor from the anterior chamber of the eye to the venous circulation.
Schwann cells: Glial cells responsible for the neurilemma that surrounds axons in the PNS.
sciatic nerve (sī-A-tik): Nerve innervating the posteromedial portions of the thigh and leg.
sclera (SKLER-a): The fibrous outer layer of the eye, forming the white area of the anterior surface; a portion of the fibrous tunic of the eye.
scrotum (SKRŌ-tum): Loose-fitting, fleshy pouch that encloses the testes of the male.
sebaceous glands (se-BĀ-shus): Glands that secrete sebum, usually associated with hair follicles.
sebum (SĒ-bum): A waxy secretion that coats the surfaces of hairs.
secondary sex characteristics: Physical characteristics that appear at puberty in response to sex hormones but that are not involved in the production of gametes.
secretin (sē-KRĒ-tin): Duodenal hormone that stimulates pancreatic buffer secretion and inhibits gastric activity.
semen (SĒ-men): Fluid ejaculate containing spermatozoa and the secretions of accessory glands of the male reproductive tract.
semicircular canals: Tubular components of the vestibular apparatus responsible for dynamic equilibrium.
semilunar valve: A three-cusped valve guarding the exit from one of the cardiac ventricles; includes the pulmonary and aortic valves.
seminal vesicles (SEM-i-nal): Glands of the male reproductive tract that produce roughly 60 percent of the volume of semen.
seminiferous tubules (se-mi-NIF-e-rus): Coiled tubules where sperm production occurs in the testis.
senescence: Aging.
septae (SEP-tē): Partitions that subdivide an organ.
serotonin (ser-ō-TO-nin): A neurotransmitter in the CNS; a compound that enhances inflammation, released by activated mast cells and basophils.
serous cell/secretion: A cell that produces a watery secretion containing high concentrations of enzymes.
serous membrane: A squamous epithelium and the underlying loose connective tissue; the lining of the pericardial, pleural, and peritoneal cavities.
serum: Blood plasma from which clotting agents have been removed.
sesamoid bone: A bone that forms within a tendon.
sigmoid colon (SIG-mōyd): The S-shaped 18-cm portion of the colon between the descending colon and the rectum.
simple epithelium: An epithelium containing a single layer of cells above the basement membrane.
sinus: A chamber or hollow in a tissue; a large, dilated vein.
sinusoid (SĪ-nus-oyd): An extensive network of vessels found in the liver, adrenal cortex, spleen, and pancreas; similar in histological structure to capillaries.
skeletal muscle: A contractile organ of the muscular system.
skeletal muscle tissue: Contractile tissue dominated by skeletal muscle fibers; characterized as striated, voluntary muscle.
sliding filament theory: The concept that a sarcomere shortens as the thick and thin filaments slide past one another.
small intestine: The duodenum, jejunum, and ileum; the digestive tract between the stomach and large intestine.
smooth endoplasmic reticulum (SER): Membranous organelle where lipid and carbohydrate synthesis and storage occur.
smooth muscle tissue: Muscle tissue found in the walls of many visceral organs; characterized as nonstriated, involuntary muscle.
soft palate: Fleshy posterior extension of the hard palate, separating the nasopharynx from the oral cavity.
sole: The inferior surface of the foot.
solute: Materials dissolved in a solution.
solution: A fluid containing dissolved materials.
solvent: The fluid component of a solution.
soma (SŌ-ma): Body.
somatic (sō-MAT-ik): Pertaining to the body.
somatomedins: Compounds stimulating tissue growth, released by the liver following GH secretion.
spermatic cord: Spermatic vessels, nerves, lymphatics, and the ductus deferens, extending between the testes and the proximal end of the inguinal canal.
spermatids (SPER-ma-tidz): The product of meiosis in the male, cells that differentiate into spermatozoa.
spermatocyte (sper-MA-to-sīt): Cells of the seminiferous tubules that are engaged in meiosis.
spermatogenesis (sper-ma-to-JEN-e-sis): Sperm production.
spermatogonia (sper-ma-to-GŌ-nē-a): Stem cells whose mitotic divisions give rise to other stem cells and spermatocytes.
spermatozoa (sper-ma-to-ZŌ-a): Sperm cells; singular *spermatozoon*.
spermiogenesis: The process of spermatid differentiation that leads to the formation of physically mature spermatozoa.
sphincter (SFINK-ter): Muscular ring that contracts to close the entrance or exit of an internal passageway.
spinal nerve: One of 31 pairs of peripheral nerves that originate on the spinal cord from anterior and posterior roots.
spindle apparatus: A muscle spindle (intrafusal fibers) and its sensory and motor innervation.
spinous process: Prominent posterior projection of a vertebra, formed by the fusion of two laminae.
spleen: Lymphoid organ important for red blood cell phagocytosis, immune response, and lymphocyte production.
squamous (SKWĀ-mus): Flattened.
squamous epithelium: An epithelium whose superficial cells are flattened and platelike.
stapedius (stā-PĒ-dē-us): A muscle of the middle ear whose contraction tenses the auditory ossicles and reduces the forces transmitted to the oval window.
stapes (STĀ-pez): The auditory ossicle attached to the tympanum.
stereocilia: Elongate microvilli characteristic of the epithelium of the epididymis, portions of the ductus deferens, and the inner ear.
steroid: A ring-shaped lipid structurally related to cholesterol.
stimulus: An environmental alteration that produces a change in cellular activities; often used to refer to events that alter the membrane potentials of excitable cells.
stratified: Containing several layers.
stratum (STRA-tum): Layer.
stretch receptors: Sensory receptors that respond to stretching of the surrounding tissues.
subarachnoid space: Meningeal space containing CSF; the area between the arachnoid membrane and the pia mater.
subclavian (sub-CLĀ-vē-an): Pertaining to the region under the clavicle.
subcutaneous layer: The layer of loose connective tissue below the dermis; also called the hypodermis or superficial fascia.
sublingual glands (sub-LING-gwal): Mucus-secreting salivary glands situated under the tongue.
submandibular glands: Salivary glands nestled in depressions on the medial surfaces of the mandible; salivary glands that produce a mixture of mucins and enzymes (salivary amylase).
submucosa (sub-mū-KŌ-sa): Region between the muscularis mucosae and the muscularis externa.
submucosal glands: Mucous glands in the submucosa of the duodenum.
substrate: A participant (product or reactant) in an enzyme-catalyzed reaction.
sulcus (SUL-kus): A groove or furrow.
superior: Directional reference meaning *above*.

superior vena cava: The vein that carries blood from the parts of the body above the heart to the right atrium.
supination (su-pi-NĀ-shun): Rotation of the forearm so that the palm faces anteriorly.
supine (SU-pīn): Lying face up, with palms facing anteriorly.
suppressor T cells: Lymphocytes that inhibit B cell activation and plasma cell secretion of antibodies.
suprarenal gland (sū-pra-RĒ-nal): *See* **adrenal gland.**
surfactant (sur-FAK-tant): Lipid secretion that coats alveolar surfaces and prevents their collapse.
sustentacular cells (sus-ten-TAK-ū-lar): Supporting cells of the seminiferous tubules of the testis, responsible for the differentiation of spermatids, the maintenance of the blood-testis barrier, and the secretion of inhibin, ABP, and MIH.
sutural bones: Irregular bones that form in fibrous tissue between the flat bones of the developing cranium; also called Wormian bones.
suture: Fibrous joint between flat bones of the skull.
sympathetic division: Division of the autonomic nervous system responsible for "fight or flight" reactions; primarily concerned with the elevation of metabolic rate and increased alertness.
symphysis: A fibrous amphiarthrosis, such as those between adjacent vertebrae or between the pubic bones of the coxae.
symptom: Clinical term for an abnormality of function due to the presence of disease.
synapse (SIN-aps): Site of communication between a nerve cell and some other cell; if the other cell is not a neuron, the term *neuroeffector junction* is often used.
synaptic delay (sin-AP-tik): The period between the arrival of an impulse at the presynaptic membrane and the initiation of an action potential in the postsynaptic membrane.
syncytial trophoblast: Multinucleate cytoplasmic layer that covers the blastocyst; the layer responsible for uterine erosion and implantation.
syncytium: A multinucleate mass of cytoplasm, produced by the fusion of cells or repeated mitoses without cytokinesis.
syndrome: A discrete set of symptoms that occur together.
synergist (SIN-er-jist): A muscle that assists a prime mover in performing its primary action.
synovial cavity (si-NŌ-vē-ul): Fluid-filled chamber in a diarthrodial joint.
synovial fluid (si-NŌ-vē-ul): Substance secreted by synovial membranes that lubricates joints.
synovial membrane: An incomplete layer of fibroblasts confronting the synovial cavity, plus the underlying loose connective tissue.
synthesis (SIN-the-sis): Manufacture; anabolism.
system: An interacting group of organs that performs one or more specific functions.
systemic circuit: Vessels between the aortic semilunar valve and the entrance to the right atrium; the circulatory system other than vessels of the pulmonary circuit.
systole (SIS-to-lē): The period of cardiac contraction.
systolic pressure: Peak arterial pressures measured during ventricular systole.

T

tactile: Pertaining to the sense of touch.
taenia coli (TĒ-nē-a KŌ-lī): Three longitudinal bands of smooth muscle in the muscularis externa of the colon.
tarsus: The ankle.
TCA cycle: Aerobic reaction sequence occurring in the mitochondrial matrix; in the process organic molecules are broken down, carbon dioxide molecules are released, and hydrogen molecules are transferred to coenzymes that deliver them to the electron transport system.
T cells: Lymphocytes responsible for cellular immunity and for the coordination and regulation of the immune response; includes regulatory T cells (helpers and suppressors) and cytotoxic (killer) T cells.
tears: Fluid secretion of the lacrimal glands that bathes the anterior surfaces of the eyes.
tectorial membrane (tek-TŌR-ē-al): Gelatinous membrane suspended over the hair cells of the organ of Corti.
telophase (TEL-o-fāz): The final stage of mitosis, characterized by the disappearance of the spindle apparatus, the reappearance of the nuclear membrane and the disappearance of the chromosomes, and the completion of cytokinesis.
tendon: A collagenous band that connects a skeletal muscle to an element of the skeleton.
tertiary follicle: A mature ovarian follicle, containing a large, fluid-filled chamber.
testes (TES-tēz): The male gonads, sites of gamete production and hormone secretion.
testosterone (tes-TOS-ter-ōn): The principal androgen produced by the interstitial cells of the testes.
tetanic contraction: Sustained skeletal muscle contraction due to repeated stimulation at a frequency that prevents muscle relaxation.
tetanus: A tetanic contraction; also used to refer to a disease state resulting from the stimulation of muscle cells by bacterial toxins.
tetany: A tetanic contraction; also used to refer to abnormally prolonged contractions resulting from disturbances in electrolyte balance.
tetrad (TET-rad): Paired, duplicated chromosomes visible at the start of meiosis I.
tetraiodothyronine (tet-ra-ī-ō-dō-THĪ-rō-nēn): T_4, or thyroxine, a thyroid hormone.
thalamus: The walls of the diencephalon.
thermoreception: Sensitivity to temperature changes.
thermoregulation: Homeostatic maintenance of body temperature.
thick filament: A myosin filament in a skeletal or cardiac muscle cell.
thin filament: An actin filament in a skeletal or cardiac muscle cell.
thorax: The chest.
threshold: The membrane potential at which an action potential begins.
thrombin (THROM-bin): Enzyme that converts fibronogen to fibrin.
thrombocyte (THROM-bō-sīt): *See* **platelets.**
thrombus: A blood clot.
thymine: A compound found in DNA.
thymosins (thī-MO-sinz): Hormones of the thymus essential to the development and differentiation of T cells.
thymus: Lymphoid organ, site of T cell formation.
thyroid gland: Endocrine gland whose lobes sit lateral to the thyroid cartilage of the larynx.
thyroid hormones: Thyroxine (T_4) and triiodothyronine (T_3); hormones of the thyroid gland that stimulate tissue metabolism, energy utilization, and growth.
thyroid-stimulating hormone (TSH): Anterior pituitary hormone that triggers the secretion of thyroid hormones by the thyroid gland.
thyroxine (TX) (thī-ROKS-in): A thyroid hormone (T_4).
tidal volume: The volume of air moved in and out of the lungs during a normal quiet respiratory cycle.
tissue: A collection of specialized cells and cell products that perform a specific function.
titer: Plasma antibody concentration.
tonsil: A lymphoid nodule beneath the epithelium of the pharynx; includes the palatine, pharyngeal, and lingual tonsils.
topical: Applied to the body surface.
trabecula (tra-BEK-ū-la): A connective tissue partition that subdivides an organ.
trabeculae carneae (tra-BEK-ū-lē CAR-nē-ē): Muscular ridges projecting from the walls of the ventricles of the heart.
trachea (TRĀ-kē-a): The windpipe, an airway extending from the larynx to the primary bronchi.
tract: A bundle of axons inside the CNS.
transcription: The encoding of genetic instructions on a strand of mRNA.
transection: To sever or cut in the transverse plane.
translation: The process of peptide formation using the instructions carried by an mRNA strand.
triad (muscle cell): The combination of a T tubule and two cisternae of the sarcoplasmic reticulum.
tricarboxylic acid cycle (trī-kar-bok-SIL-ik): *See* **TCA cycle.**
tricuspid valve (trī-KUS-pid): The right atrioventricular valve that prevents backflow of blood into the right atrium during ventricular systole.
trigeminal nerve (trī-GEM-i-nal): Cranial nerve V, responsible for providing sensory information from the lower portions of the face, including the upper and lower jaws, and delivering motor commands to the muscles of mastication.
triglyceride (trī-GLIS-e-rīd): A lipid composed of a molecule of glycerol attached to three fatty acids.
trigone (TRĪ-gōn): Triangular region of the urinary bladder bounded by the exits of the ureters and the entrance to the urethra.
triiodothyronine: T_3, one of the thyroid hormones.
trochanter (trō-KAN-ter): Large processes near the head of the femur.
trochlea (TRŌK-le-a): A pulley.
trochlear nerve (TRŌK-lē-ar): Cranial nerve IV, controlling the superior oblique muscle of the eye.
trophoblast (TRŌ-fō-blast): Superficial layer of the blastocyst that will be involved with implantation, hormone production, and placenta formation.

troponin/tropomyosin (trō-PŌ-nin) (trō-pō-MĪ-ō-sin): Proteins on the thin filaments that mask the active sites in the absence of free calcium ions.
trunk: The thoracic and abdominopelvic regions.
T tubules: Transverse, tubular extensions of the sarcolemma that extend deep into the sarcoplasm to contact cisternae of the sarcoplasmic reticulum.
tuberculum (tū-BER-kū-lum): A small, localized elevation on a bony surface.
tuberosity: A large, roughened elevation on a bony surface.
tubulin: Protein subunit of microtubules.
tumor: A tissue mass formed by the abnormal growth and replication of cells.
tunica (TOO-ni-ka): A layer or covering; in blood vessels: t. externa, the outermost layer of connective tissue fibers that stabilizes the position of the vessel; t. interna, the innermost layer, consisting of the endothelium plus an underlying elastic membrane; t. media, a middle layer containing collagen and elastic and smooth muscle fibers in varying proportions.
T wave: Deflection of the ECG corresponding to ventricular repolarization.
twitch: A single contraction-relaxation cycle in a skeletal muscle.
tympanum (tim-PAN-um): Membrane that separates the external auditory canal from the middle ear; membrane whose vibrations are transferred to the auditory ossicles and ultimately to the oval window; the eardrum.

U

ulcer: An area of epithelial sloughing associated with damage to the underlying connective tissues and vasculature.
ultrasound: Diagnostic visualization procedure that uses high-frequency sound waves.
umbilical cord (um-BIL-i-kal): Connecting stalk between the fetus and the placenta; contains the allantois, the umbilical arteries, and the umbilical vein.
umbilicus: The navel.
unicellular gland: Goblet cells.
unipolar neuron: A sensory neuron whose soma lies in a dorsal root ganglion or a sensory ganglion of a cranial nerve.
unmyelinated axon: Axon whose neurilemma does not contain myelin and where continuous conduction occurs.
uracil: One of the compounds characteristic of RNA.
ureters (ū-RĒ-terz): Muscular tubes, lined by transitional epithelium, that carry urine from the renal pelvis to the urinary bladder.
urethra (ū-RĒ-thra): A muscular tube that carries urine from the urinary bladder to the exterior.
urinalysis: Analysis of the physical and chemical characteristics of the urine.
urinary bladder: Muscular, distensible sac that stores urine prior to micturition.
urination: The voiding of urine; micturition.
uterus (Ū-ter-us): Muscular organ of the female reproductive tract where implantation, placenta formation, and fetal development occur.
utricle (Ū-trē-kl): The largest chamber of the vestibular apparatus; contains a macula important for static equilibrium.
uvea: The vascular tunic of the eye.

V

vagina (va-JĪ-na): A muscular tube extending between the uterus and the vestibule.
vascular: Pertaining to blood vessels.
vasoconstriction: A reduction in the diameter of arterioles due to contraction of smooth muscles in the tunica media; an event that elevates peripheral resistance and that may occur in response to local factors, through the action of hormones, or from stimulation of the vasomotor center.
vasodilation (vaz-ō-dī-LĀ-shun): An increase in the diameter of arterioles due to the relaxation of smooth muscles in the tunica media; an event that reduces peripheral resistance and that may occur in response to local factors, through the action of hormones, or following decreased stimulation of the vasomotor center.
vasomotion: Alterations in the pattern of blood flow through a capillary bed in response to changes in the local environment.
vasomotor center: Medullary center whose stimulation produces vasoconstriction and an elevation in peripheral resistance.
vein: Blood vessel carrying blood from a capillary bed toward the heart.
venae cavae (VĒ-na CĀ-va): The major veins delivering systemic blood to the right atrium.
ventilation: Air movement in and out of the lungs.
ventral: Pertaining to the anterior surface.
ventricle (VEN-tri-kl): One of the large, muscular pumping chambers of the heart that discharges blood into the pulmonary or systemic circuit.
venule (VEN-ūl): Thin-walled veins that receive blood from capillaries.
vertebral canal: Passageway that encloses the spinal cord, a tunnel bounded by the neural arches of adjacent vertebrae.
vertebral column: The cervical, thoracic, and lumbar vertebrae, the sacrum, and the coccyx.
vesicle: A membranous sac in the cytoplasm of a cell.
vestibular membrane: The membrane that separates the cochlear duct from the vestibular duct of the inner ear.
vestibular nucleus: Processing center for sensations arriving from the vestibular apparatus; located near the border between the pons and medulla.
vestibule (VES-ti-būl): A chamber; in the inner ear, the term refers to the utricle, saccule, and semicircular canals.
villus: A slender projection of the mucous membrane of the small intestine.
virus: A pathogenic microorganism.
viscera: Organs in the ventral body cavity.
visceral: Pertaining to viscera or their outer coverings.
viscosity: The resistance to flow exhibited by a fluid due to molecular interactions within the fluid.
vital capacity: The maximum amount of air that can be moved into or out of the respiratory system; the sum of the inspiratory reserve, the expiratory reserve, and the tidal volume.
vitamin: An essential organic nutrient that functions as a coenzyme in vital enzymatic reactions.
vitreous humor: Gelatinous mass in the vitreous chamber of the eye.
voluntary: Controlled by conscious thought processes.
vulva (VUL-va): The female pudendum (external genitalia).

W

white blood cells (WBCs): Leukocytes; the granulocytes and agranulocytes of the blood.
white matter: Regions inside the CNS that are dominated by myelinated axons.
white ramus: A nerve bundle containing the myelinated preganglionic axons of sympathetic motor neurons en route to the sympathetic chain or a collateral ganglion.
Wormian bones: *See* **sutural bones.**

X

xiphoid process (ZĪ-fōid): Slender, inferior extension of the sternum.

Y

Y chromosome: The sex chromosome whose presence indicates that the individual is a genetic male.
yolk sac: One of the three extraembryonic membranes, composed of an inner layer of endoderm and an outer layer of mesoderm.

Z

zygote (ZĪ-gōt): The fertilized ovum, prior to the start of cleavage.

INDEX

E

F

Q

R

Illustration Credits

Photographs

CHAPTER 1 Chapter Opener Tony Stone Images **1-11a,b** Science Source/Photo Researchers **1-12a,c** CNRI/Photo Researchers **1-12b** Photo Researchers

CHAPTER 2 Chapter Opener David Spears/Science Photo Library **2-16c** PH Archives

CHAPTER 3 Chapter Opener P. Motta/Dept. of Anatomy/University La Sapienza/Photo Researchers **3-7a** Dana-Farber Cancer Institute **3-7b,c** David M. Phillips/Visuals Unlimited **3-13b** Dr. Birgit Satir, Yeshiva University **3-14** CNRI/ Science Source/Photo Researchers **3-15** Don Fawcett/ Harvard Medical School

CHAPTER 4 Chapter Opener David M. Phillips/Photo Researchers **4-3b** Custom Medical Stock Photo **4-4a** Wards Natural Science Establishment, Inc. **4-4b** PH Archives **4-4c** Frederic H. Martini **4-5a-d** Frederic H. Martini **4-8a** Science Source/Photo Researchers **4-8b** Frederic H. Martini **4-8c** John D. Cunningham/Visuals Unlimited **4-10a** Robert Brons/Biological Photo Service **4-10b** Science Source/Photo Researchers **4-10c** Ed Reschke/Peter Arnold, Inc. **4-11** Frederic H. Martini **4-13a** M. Abbey/Photo Researchers, Inc. **4-13a** Eric Grave/Phototake **4-13b** Phototake **4-13c** Frederic H. Martini **4-14** Frederic H. Martini

CHAPTER 5 Chapter Opener Paul Trimmer/The Image Bank **5-2** Biological Photo Service **5-3** PH Archives **5-4** Manfred Kage/Peter Arnold, Inc. **5-5** Frederic H. Martini

CHAPTER 6 Chapter Opener Superstock **6-3a** Frederic H. Martini **Focus Box:** *Pott's fracture:* SIU/Visuals Unlimited; *comminuted and epiphyseal fractures:* Visuals Unlimited; *transverse fracture:* Grace Moore/Medichrome; *spiral fracture:* Peter Arnold, Inc.; *Colles' fracture:* Scott Camazine/Photo Researchers; *greenstick fracture:* Patricia Barber, RBP/Custom Medical Stock Photo; *compression fracture:* L.V. Bergman & Associates, Inc.; *displaced fracture:* Custom Medical Stock Photo **6-6a,b** Ralph Hutchings **6-12a,b,c** Ralph Hutchings **6-14 thru 6-19** Ralph Hutchings **6-20** University of Toronto **6-21b thru 6-24** Ralph Hutchings

CHAPTER 7 Chapter Opener Anne Marie Weber **7-3a** Cormack, D. (ed.), *Ham's Histology,* 9/E. Philadelphia: J.B. Lippincott, 1987. By Permission. **7-8a** G.W. Willis/ Biological Photo Service **7-8b** Frederic H. Martini

CHAPTER 8 Chapter Opener Synapier/Science Photo Library/Photo Researchers **8-2** Wards Natural Science Establishment **8-5a** Biophoto Associates/Photo Researchers **8-5b** Photo Researchers **8-9** David Scott/Phototake **8-17a** Ralph Hutchings **8-17b** Ralph Hutchings **8-17c** Ralph Hutchings **8-22e** Larry Mulvehill/Photo Researchers

CHAPTER 9 Chapter Opener Will & Deni McIntyre/Photo Researchers **9-1** Ralph Hutchings

CHAPTER 10 Chapter Opener Bettman Archives **10-6** Frederic H. Martini **10-7a** Ralph Hutchings **10-10a** Ed Reschke/Peter Arnold, Inc. **10-10c** Custom Medical Stock Photo **10-20** Lennart Nilsson, *Behold Man,* Little, Brown & Co., 1973 **10-23b** Wards Natural Science Establishment, Inc.

CHAPTER 11 Chapter Opener Focus on Sports **11-5b** Manfred Kage/Peter Arnold, Inc. **11-9b** Frederick H. Martini **11-11b** Frederic H. Martini **11-12c** Frederic H. Martini **11-13b** Wards Natural Science Establishment, Inc.

CHAPTER 12 Chapter Opener John Riley/Tony Stone Images **12-1** Martin M. Rotker **12-2a** Frederic H. Martini **12-2b** Ed Reschke/Peter Arnold, Inc. **12-3a,b** Stanley Flegler/Visuals Unlimited **12-6a-e** Alfred Owczarzak/Biological Photo Service **12-8** Frederic H. Martini **12-9** Custom Medical Stock Photo

CHAPTER 13 Chapter Opener Lunagrafix, Inc./Photo Researchers **13-3** Ralph Hutchings **13-6** Peter Arnold, Inc.

CHAPTER 14 Chapter Opener Biphoto Associates/Photo Researchers **14-1** Michael J. Timmons **14-2b** Biophoto Associates/Photo Researchers **14-5** PH Archives

CHAPTER 15 Chapter Opener Jeffrey Reed/Medichrome **15-2** Frederic H. Martini **15-6c** Frederic H. Martini

CHAPTER 16 Chapter Opener D. Kirkland/Sygma **16-3a** Photo Researchers **16-4e** Phototake **16-6b** Adapted from Junqueira, Carneira, & Long, *Basic Histology,* 5/e. Norwalk, CT: Appleton-Century-Crofts, 1986.

CHAPTER 17 Chapter Opener Erich Lessing/ Art Resource **17-8d** Wards Natural Science Establishment, Inc. **17-13b** Carolina Biological Supply/Phototake NYC **17-15b** Wards Natural Science Establishment, Inc.

CHAPTER 18 Chapter Opener left Don King/The Image Bank **Chapter Opener right** Robert Semenlik/The Stock Market

CHAPTER 19 Chapter Opener Photo Researchers **19-3b** Ralph Hutchings **19-5b** David M. Phillips/Visuals Unlimited

CHAPTER 20 Chapter Opener Elyse Lewin/The Image Bank **20-2b** Wards Natural Science Establishment, Inc. **20-5b** Wards Natural Science Establishment, Inc. **20-10(1,2,3)** Frederick H. Martini **20-10(4)** Wards Natural Science, Inc. **20-10(5)** PH Archives

CHAPTER 21 Chapter Opener Pettit Format/Nestle/ Photo Researchers **21-1a** Francis Leroy, Biocosmos/Science Photo Library/Custom Medical Stock Photo **21-7a** Dr. Arnold Tamarin **21-7b** Dr. Arnold Tamarin **21-7c,d** Lennart Nilsson, *A Child is Born,* copyright 1977, Dell/Delacorte **21-8a** Lennart Nilsson, *Behold Man,* Little, Brown & Co., 1973 **21-8b** Photo Researchers **21-14** CNRI/Science Photo Library/Photo Researchers

Art

Illustrations by Ron Ervin: **3-12, 6-1, 6-7, 6-8, 6-9, 6-10, 6-11, 6-19b, 6-21c, 6-24b, 6-32, 6-33, 6-35, 7-10a,b, 7-13, 7-17a, 7-19**

Illustrations by Tina Sanders: **6-9, 8-19b, 8-21, 17-12**

Illustrations by Craig Luce: **1-12, 13-3, 13-4, 13-7, 13-8, 20-12**

Credits

Figures 2-6 and **15-14a** from Jacquelyn C. Black, *Microbiology: Principles and Applications*, 2/e, Englewood Cliffs, NJ: Prentice Hall, Inc., 1993 (**Figures 2-7, 18-7**).

Figure 21-13a,b Adapted from *Time*, January 17, 1994, pp. 50–51. Graphic by Nigel Holmes. Research by Leslie Dickstein. Source: Dr. Victor A. McKusick, Johns Hopkins University.

CLINICAL TOPICS IN
ESSENTIALS OF ANATOMY AND PHYSIOLOGY

The following topics appear throughout the text as Clinical Discussions (**CD**) *or Clinical Notes* (**CN**).